*The
Biological
World*

The Biological World

ALVIN NASON

Johns Hopkins University

ROBERT L. DEHAAN

Carnegie Institution of Washington

John Wiley & Sons, Inc.
New York
London
Sydney
Toronto

This book was set in Optima, with Palatino italic display, by York Graphic Services, Inc., and printed and bound by Halliday Lithographers, Inc. The designer was Jules Perlmutter. The drawings were designed and executed by John Balbalis. The depth editor was Arlen Sue Fox. Picture research was done by Stella Kupferberg. Joan Rosenberg supervised production. Endpaper photo by Walter Dawn.

Cover photo; from the book The Graphic Works of M. C. Escher by M. C. Escher. © Koninklijke Uitgerij Erven J. J. Tijl N.V., Zwolle, Holland, 1960. © American edition Meredith Press, New York. Reprinted by permission of Hawthorn Books, Inc., New York.

Library of Congress Cataloging in Publication Data

Nason, Alvin.
    The Biological World.

    Bibliography: p.
    1. Biology.   I. DeHaan, Robert Lawrence, 1930–   joint author.   II. Title.
QH308.2.N37   574   72-8573
ISBN 0-471-63045-4

Printed in the United States of America

10 9 8 7 6 5 4 3 2 1

We dedicate this book to our wives, Thelma and Genie

Preface

The much-discussed revolution in student values that has become so evident since the middle 1960s seems to be based fundamentally on the elevation of nature and the natural to the highest position in man's value system. The central ideas of campus philosophy, according to Daniel Yankelovich,* include a recognition of the interdependence of all things and species in nature, living (or wanting to live) close to nature, rejecting hypocrisy and social artifices, and stressing the importance of preserving the environment at the expense of technological growth. There are also noticeable anti-intellectual tendencies: a deemphasis of science in favor of the mystical and mysterious, a devaluation of detachment and objectivity as means of arriving at truth, and an emphasis on self-knowledge and introspection rather than systematic acquisition of information about measurable or tangible variables.

Although the biological world is very much a part of nature, why should today's student who holds some or all of these views want to study the subject as a scientific discipline.? The answer to that, it seems to us, is twofold. First, it is our belief that the world of living organisms is a fascinating one, intellectually stimulating in its own right, with sufficient wonders and marvels to stimulate the enthusiasm of even those wholly uninterested in nature. However, especially if one's desire is to comprehend and be close to nature, we believe it can be better appreciated in all its richness with a background

---

*Yankelovich, D., "The New Naturalism," *Saturday Review* (April 1, 1972), pp. 32–40.

of information that goes beyond the superficial and the obvious. One purpose of this book is to provide such a background of information. Second, as will become eminently clear in later pages of this volume, the biological world in which man lives is beset with problems of a magnitude never before seen. Many of these are basically biological problems whose solutions will require us to understand, better than we now do, how organisms function and how they relate to one another and to their environment. The only way to gain such understanding, it seems to us, is by asking questions of nature, that is, by the application of the methods of experimental science.

We have tried to organize this book in a way that recognizes the individual differences in the readers' level of background and interest and that allows each reader to get as much as possible out of the book at his or her own level. To this end we have used two main devices. First, each new term or concept is identified by boldface letters when it is introduced. A glance at any page, therefore, immediately calls attention to the ideas that are discussed there. Every boldface term is also defined in a sentence or two in the glossary-index at the end of the book. Second, many concepts are described and explained in straightforward and general terms and at a rather simple level in early chapters, and are introduced again in greater detail later in the book. Thus we permit some redundancy for the great pedagogical advantage of being able to approach the more complicated subjects at progressively graded levels. This is especially true of our treatment of topics such as cell structure, development, genetics, and cell physiology, all of which have a large component of molecular biology.

This book has the advantage of a substantial history. It arose out of Nason's *Essentials of Modern Biology,* published in 1968, and is therefore the grandchild of Nason's more extensive *Textbook of Modern Biology,* which appeared three years earlier. The current volume, however, is much more than merely a new edition of *Essentials;* it is a very different book. Roughly half of the chapters are new; the remainder have been extensively revised and updated from their original form. Most of the illustrations are new; all have been redone with an eye to greater clarity, more information, better aesthetic quality.

A more important difference, however, reflects a major change in the field of biology in recent years. Since the mid-1960s, when the original book was being written, biology has become permeated in profound ways with two concepts that were then only peripherally recognized. The environmental crisis is upon us. It is now commonly accepted that all organisms, including human beings, are totally dependent on one another and on their environment, earth. Thus ecology has become more than a separate field. It represents a new way of looking at biology: as a setting for all the rest of the discipline. Perhaps because of this recognition of the precariousness of our existence, we have also begun to appreciate more how we got here: the long slow process of evolution. A few years ago the study of evolution was largely the study of the fossil record and systematics, subjects that in the student's eye were notoriously dull. Today the concepts of molecular and organic evolution permeate all biological thinking. The problem of speciation and evolution is one of the most exciting in the field of biology as evidence is brought to bear from disciplines as wideranging as cytogenetics, immunochemistry, molecular biology, hematology, and radiation physics. Thus whenever it has seemed appropriate we have tried to approach each topic with its ecological and evolutionary significance in mind.

A book such as this is never the product solely of its authors. In addition to a large number of contributors of illustrative material, who are acknowledged individually in the credit list (p. 717), the book owes a great debt to others as well. Several teachers and investigators (and some friends) around the country read chapters or parts of the manuscript at various stages in its long gestation, and offered valuable advice and criticism: John Biggers, Harvard University; Marvin H. Cantor, San Fernando Valley State College; Richard E. Dickerson, California Institute of Technology; Douglas Fambrough, Carnegie Institution of Washington; Alan Gelperin, Princeton University; Leon J. Gorski, Central Connecticut State College; Helen Haberman, Goucher College; N. Scott McNutt, Harvard University; Douglas Pratt, University of Minnesota; Ronald Reeder, Carnegie Institution of Washington; Jay Savage, University of Southern California; Philip Siekevitz, Rockefeller University; William Sladen, Johns Hopkins University; and R. H. Whittaker, Cornell University.

Two persons cheerfully bore the brunt of the secretarial labor; Holly Meystre and Martha Sachs. We also thank Robert Rubin for his meticulous help with the glossary-index.

We owe an extra measure of thanks to our depth editor, Arlen Sue Fox, who has been indispensable in correcting, polishing, and unifying our often faulty and disparate styles of prose, to Stella Kupferberg, for searching out and selecting the numerous photographs, and, in particular, to John Balbalis, whose insightful and skilled illustrations have contributed immeasurably to both the aesthetic quality and utility of the book. To these and many other members of the staff of John Wiley & Sons we wish to express our heartfelt thanks for their patience, competence, and professionalism.

Finally, we must each acknowledge our debt of gratitude to our families—especially to Genie, Tracy, and Benjy—for their patience and tolerance during the long grumpy months of our preoccupation with "the book."

Baltimore, Md., November 1972

ALVIN NASON
ROBERT L. DEHAAN

Contents

xi

The
Biological
World

**part
one** *The Concepts of Biology*

chapter one | *Life and the World of Ideas*

"Life—human life included—is the outcome of an elaborate organization based on trivial ingredients and ordinary forces."

G. E. PALADE, *National Academy of Sciences, Proceedings,* 1964

Biology is the study of living things—of cats and caterpillars, of cactus and college students. It is the study of how organisms are made, how they function, and how they came to be. It is the study of how they are related to other living things and to nature. In short, biology is the science of life. But what do we mean by life? What are living things, and how do we distinguish them from nonliving things? The answer to this question is the underlying theme of this book.

Most of us have no trouble distinguishing living from nonliving things in our everyday lives. Even when we see a completely unfamiliar object we can usually tell whether it is alive or not. Living things have certain obvious properties that we associate with their being alive. We would all agree that most living things can:

1. **Grow**—increase in size and complexity during their lifetime.
2. **Reproduce**—make copies of themselves.
3. **Metabolize**—control the intake of materials from the environment; use them as fuel for energy and for synthesis of new structural elements; and excrete the waste by-products formed in the process.
4. **Respond** to stimuli from the environment.
5. **Move,** at least at some time during their life.
6. **Adapt** to changes of various kinds in their environment.

Anthyllis Vulnaria, a flowering plant indigenous to Europe.

3

Clearly, many nonliving objects have one or more of these characteristics. Salt crystals can "grow" in the proper environment. An automobile engine is designed to take in raw materials (fuel), "metabolize" it to yield energy for work, and eliminate the waste by-products (exhaust). Many modern machines and electronic appliances are able to "respond" to signals from their environment. Yet none of these objects is considered to be alive; thus no one of these characteristics alone, nor even all of them together, provides an adequate definition of life. They simply describe the major characteristics that most living organisms share.

Tremendous strides have been made during the last 20 years in analyzing the structure and function of living things in terms of chemistry and physics. This approach has produced two powerful scientific disciplines, **biochemistry** and **biophysics.** These are concerned primarily with analyzing the chemical and physical structure of cells at the **molecular level** (that is, in terms of the molecules that comprise cells), the chemical reactions that go on in living systems, and the relationships between molecular structure and function. From such studies on many different kinds of organism, certain basic patterns have become apparent that seem to be common to all living things.

We know, for example, that all organisms are made of similar chemical substances, and that all of the atoms and molecules that form these substances also occur in nonliving objects; we know that the chemical and physical reactions that take place in living material are governed by the same principles of chemistry and physics that are the basis for all reactions of matter and energy. That is, there are no special atoms in living matter, nor do its components follow unique laws. What produces the attributes of objects that we view as living is only the complexity of interactions exhibited by the peculiar combination of their atoms. The most important components of all living materials are **oxygen, hydrogen, carbon,** and **nitrogen,** combined into large complex molecules of **protein, fat (lipids), carbohydrate,** and **nucleic acid.** The

behavior of living matter is entirely comprehensible in terms of the properties of these varied large molecules. It is of no value to think of life as some indefinable or unique vital force that can never be investigated; the fact that life manifests greater complexity than most nonliving systems can be understood in terms of the peculiar chemistry of carbon and water and the fact that even the simplest organism has thousands of different kinds of molecule operating in coordinated fashion. On this basis, then, we can formulate a working definition of an organism or living system: **A living form is a highly organized, self-directing, complex system of chemically and physically defined structures capable of utilizing the matter and energy of its environment (by means of integrated and self-determined chains of physical and chemical reactions) for growth and reproduction.**

If living substance differs from nonliving material only in the degree of complexity of the reactions it can exhibit, then it becomes understandable that the complex molecules themselves could have developed from simpler components of inanimate matter. It is now believed by biologists that this gradual buildup, or **evolution,** of inanimate substance into a living system with the unique features of growth and self-duplication first took place about 4 billion years ago, and represents the origin of life on our planet (see Chapter 22).

The Hierarchy of Life

We have just said that the substances of organisms—lipids, carbohydrates, nucleic acids, proteins—are not themselves alive. It is only when they become elaborately organized and can interact in precise ways that the attributes appear that we call living.

This organization can be viewed as a hierarchy of structural patterns common to all living things. Starting with arrangements of atoms of carbon, oxygen, nitrogen, phosphorus, into relatively simple molecules such as amino acids, purines, and sugars, the hierarchy proceeds, step by step: first to **polymeric macromolecules** (Chapter 3), which are made of smaller repeating units, as are proteins, nucleic acids, carbohydrates, and lipids; then to **molecular** or **macromolecular aggregates** (Chapter 4), which may take the form of elementary structures such as fibrils, membranes, and particles (Chapter 5); to **aggregates** of such structures, which turn out to be cell organelles (see Chapter 5); to **whole cells** and **complexes of cells** forming tissues (Chapter 6), organs, and organisms; and beyond to the structure of **populations** of organisms (Chapter 20).

To understand any aspect of biology we generally need information from more than one (often several) of these levels; and all of the different levels of organization, from cells to societies, are the subject of biology. For example, the development of an embryo from egg to birth can be described purely in terms of the structure of the organism at different stages. However, an awareness of the behavior of the cells that comprise the tissues of the embryo, and of the movements of these cells as they form the various embryonic organs, can add great insights into the process. And an understanding of the molecular interrelationships among those cells leads to still deeper understanding. We have attempted to organize this book in a way that will lead the reader progressively through these levels.

Major Generalizations

Subdivisions of Biology

It is not sufficient to examine living things simply in terms of their component parts and mechanisms. The total organism, in view of its integrated complexity of structure and function, has an added dimension, its wholeness, that would be completely overlooked if we were to study it only in terms of its parts. This would be like examining the workings of an airplane by investigating in fine detail the structure and function of each of its individual parts (e.g., the engine, carburetor, wheels, rudder, wings) without giving attention to the over-all features, design, and coordinated behavior of the entire system. Any of the parts detached from the airplane is incapable of self-propelled, airborne flight; together, however, they form an entity possessing the unique features of a flying machine. Similarly, any analysis of a living thing must unveil fundamental molecular and progressively grosser mechanisms as well as study the intact and whole organism. The ultimate goal of biology is to understand the function and structure of living forms at all levels of organization.

To date on our planet several million different kinds of plants and animals have been recognized. They range in an almost infinite variety of sizes and shapes from the smallest microorganisms to the most complex higher forms such as man and the flowering plants (Fig. 1-1, pp. 6–7). Many organisms, as yet undiscovered and undescribed, undoubtedly exist, especially in the virtually unexplored seas that cover about 70 percent of the earth's surface. New species are continually evolving from existing organisms, and others are becoming extinct. Still others remain relatively stable, undergoing little or no evolutionary change during long periods of time.

The field of biology is almost indescribably large. Not only is there a large number of living forms, but each species, from the relatively simplest to the most intricate, is a highly complicated structural and functional entity. Of necessity, therefore, biology, like all other sciences, is divided into subdisciplines, which in turn are subdivided into still more highly specialized areas of study. The various disciplines can be regarded, for convenience, as falling into two principal groups that are inextricably linked with one another: the first includes the main areas determined by the **organisms** studied; the second covers areas delineated by the **approach** taken to the subject matter. Under the organismal division we include **zoology,** the study of animals; **botany,** which restricts itself to plants; and **micro-**

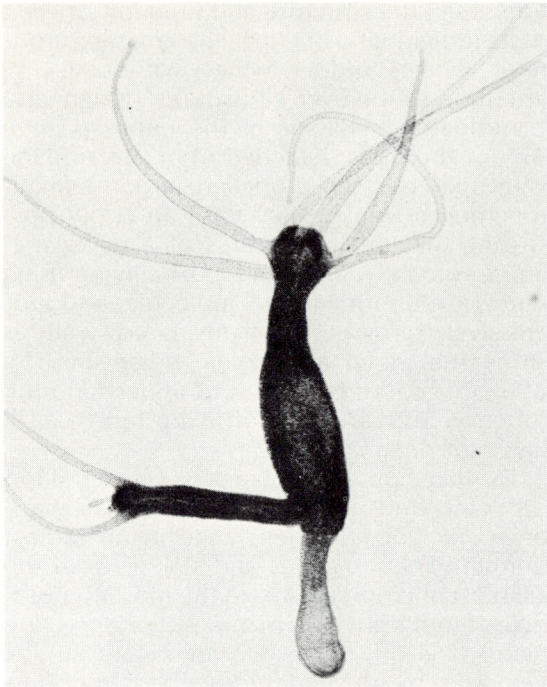

(a)

(b)

(c)

(d)

Figure 1-1 All organisms, like brothers, share
certain characteristics derived from their common
lineage. The tiny fresh-water hydra (a), garden
snail (b), and violin spider (c) look very different
from the desert lizard (d), and seem even more
unrelated to a Maguey cactus (e) or a flowering
plant such as clivia (f). However, the close
evolutionary relationships among all these forms can
be traced by fundamental similarities: the molecular
structure of the DNA that forms their chromosomes,
the cellular machinery they all possess for obtaining
energy from their environment, and a host of other
common characteristics.

(e)

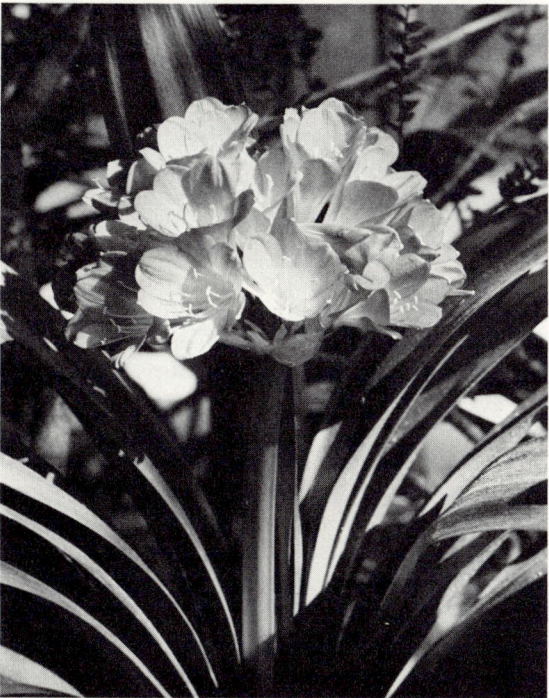

(f)

biology, the study of microorganisms. In the second major group we have **physiology,** the study of function; **morphology,** the study of form and structure of living things; and so on. Physiology, in turn, may be subdivided, for example, into **endocrinology,** the study of the function of the glands that secrete hormones; **neurophysiology,** the study of the function of nerves and the nervous system; and so on. These can also be further broken down according to the organism under investigation, say into plant or human physiology. The major disciplines of biology are summarized in Table 1-1 (p. 8).

*The Cell Is
the Basic Unit
of Life*

We spoke earlier of organizational levels in biology. Life can be understood in terms of atoms and molecules, cells and tissues, whole organisms and societies. We shall see in Chapters 3 to 5, however, that the minimum level of organization that we can define as alive is a cell; thus, by definition, we can say that the cell is the basic unit of life. Cells take many forms. There are free-living cells, or **protista;** within the tissues of the body there are muscle cells, nerve cells, blood cells, bone cells, and many others. Although different in many ways, most cells share certain characteristics. All cells are surrounded by a membrane, called a **plasma membrane,** that separates the cell from its external environment. Inside the membrane is a viscous, gel-like material called **cytoplasm,** containing a variety of tiny structures termed **organelles.** The largest single organelle of most cells is a **nucleus,** which contains **chromosomes.**

The cell concept is fundamentally a set of four generalizations:

1. **Virtually all organisms are made of cells.** This is true of plants, animals, and microorganisms; it is true of the animal man. Cells are the basic structural units of living things.
2. **Cells are the site of all life chemistry.** Organisms are alive because they can extract

table 1-1

Major Subdivisions of Biology According to Approach

Genetics: inheritance and variation
Physiology: function
Taxonomy: classification
Evolution: origin and changes
Morphology: form and structure
    Anatomy: gross structure
    Histology: tissue structure
    Cytology: cell structure
Biochemistry and Biophysics: structure and
    function at the molecular level
Embryology: embryo formation and development
Ecology: relationship of organisms to their living
    and nonliving environments
Paleontology: fossil organisms
Parasitology: parasites
Ethology: animal behavior

chemical energy from material around them. The chemical reactions whereby this takes place are referred to as **metabolism.** All metabolic reactions take place in cells.

3. **Cells arise only from preexisting cells.** Spontaneous generation of cells does not occur in the modern world. In a population of single-celled organisms such as **bacteria** or **protozoa,** each cell has arisen from the division (reproduction) of a parent cell. The cells of a multicellular animal or plant all come from the division of previous cells that come, ultimately, from the egg (or seed), which itself is a giant single cell formed by a special division in the body of the parent. The idea that all cells originate only from preexisting cells is absolutely fundamental to biology.

4. **Cells contain hereditary instructions that are passed from one generation to the next.** Every cell of every organism contains long threads of a special material called **deoxyribonucleic acid,** or **DNA;** DNA contains a code that determines the specific characteristics of the cell and how it will relate to other cells to form a multicellular

organism. Every cell makes a copy of its DNA strand and passes one of the copies into each daughter cell when it divides.

*The Gene
Is the Basic Unit
of Heredity
and Development*

An individual is the product of his heredity and his development. His heredity is the substance he receives from his parents. The essence of this substance is a set of instructions in the DNA molecule in the form of a sequence of serially aligned units, or **code-words,** called **genes.** The DNA is packed in an ovum, or egg cell, from the female parent and a sperm from the male; these unite to form a fertilized egg, or zygote, which contains instructions to make a new individual.

Shortly after its formation a zygote divides into two cells; each of these divides again to form four cells; and again to form eight cells. As division continues, the zygote—now called an embryo—undergoes a series of changes that lead, if its luck holds out, to the formation of an adult individual. These divisions and changes are called **development.** Development is controlled by the genes on DNA; the genes in the hereditary instructions determine whether a zygote develops into frog or whale, mulberry bush or man.

All genes are not active at the same time during development; rather, development takes place by a progression of activities of different genes at different times in various parts of the embryo. These separate lines of development lead various groups of embryonic cells to become different—to form heart cells, bone cells, or liver cells. This developmental process is known as **differentiation.**

Genes Mutate

The duplication of DNA in a cell is an amazingly perfect process. It is estimated that the DNA strand of an animal such as an insect may contain 20,000 genes, yet in most cases the DNA copy is an identical duplicate of the original. Occasionally, however, once or twice every million copies, a mistake is made somewhere in the sequence of genes. An incorrectly copied gene is called a **mutation.** Mutations generally show up as differences in the physical or chemical structure of an individual. Most mutations produce only very slight changes, often so slight that they cannot be detected, but occasionally a mutation causes a major change in an individual. Usually such major mutations produce harmful defects. Once a mutation occurs it is usually copied exactly, and is therefore passed on to subsequent generations. Because mutations are random events, different ones occur in different individuals. This means that every individual in a population carries a set of genes that is slightly different from that of every other individual.

All Organisms Evolve

Some combinations of genes are more advantageous than others to an organism in certain environmental conditions. If an advantage allows an organism with a given gene combination to survive and reproduce itself more often than others, individuals with that gene combination will gradually increase in the population over the course of many generations. This results in some heritable characteristics being "selected for," a process known as **natural selection.** Because new gene combinations are constantly introduced into populations by random mutation, every population experiences a continual, slow change in composition. The gradual change in populations that results from mutation and natural selection is called **evolution.**

There are about 2 million different species of organism alive today, and many millions that are known to have existed at earlier times but are now extinct. All of these, including man, have evolved, under the slow, inexorable force of natural selection, from one or a few primitive forms of life that arose some 3 billion years ago. Thus there is a basic unity among all living things; all organisms, like brothers, share certain characteristics derived from their common parental lineage.

All Living Systems Exhibit Homeostasis

One of the properties that distinguishes living things from inanimate matter is that organisms tend to adjust to changes in the external environment. Changes in temperature, acidity of the watery environment, availability of food or oxygen, all affect the lives of organisms, but most organisms have evolved mechanisms for buffering such changes, or for responding to them in such a way that the environment inside the organism experiences less change than that outside. This tendency to maintain the internal environment at a constant level is called **homeostasis,** from Greek words meaning "constant state."

An example of a homeostatic mechanism is the ability of animals to resist fluctuations in the availability of food. All animals take in food materials. Some of the products of digestion are built directly into an organism's own cells; some are burned to obtain energy; and some are excreted as waste products. But, in general, the amount of material used in these three processes equals the total amount consumed; a nice balance between intake and outflow is maintained. If food suddenly becomes unavailable, most organisms do not immediately stop functioning. Instead, they respond to the changed environmental situation by releasing stored food reserves, which—at least for a time—allow them to continue normal activities. That is, the internal functions are allowed to continue relatively unchanged despite a drastic change in the surroundings.

In this book we shall examine many examples of homeostatic mechanisms throughout the biological world and in our own bodies. In general, all exhibit a common pattern in which some external change produces a response designed automatically to compensate for that change; a mechanism known as **negative feedback.** A good example of a homeostatic feedback mechanism is the method by which we keep our body temperature at about 98.6°F (37.5°C), even in very cold or hot environments. Temperature-sensing elements in the brain, which function like the thermostat in a home heating system, respond to very small changes in body temperature. In a warm environment the body temperature starts to rise, but when it has gone up only a fraction of a degree the temperature-control region sends out signals that cause blood vessels near the surface of the skin to expand so that the body radiates heat away. Sweat glands are stimulated to secrete moisture, which absorbs heat from the body as it evaporates, and body temperature falls. In a cold environment, when body temperature begins to fall, surface vessels constrict, reducing heat radiation (and causing us to turn "blue"), and muscles are stimulated to twitch randomly; this shivering burns energy wastefully and produces heat that the rest of the body can absorb. In both cases information from a sensing element "feeds back" to a control center that responds with activities that balance the original change.

Organisms Depend on Their Ecological Relationships

No organism can long survive as an independent, isolated entity. All forms of life participate in a complex set of interrelationships with things in their environment—both living and nonliving—in satisfying their needs for food, shelter, the air they breathe, the water they drink, the mates they need to reproduce their species. All animals depend on plant life, both to utilize the sun's energy to convert inanimate chemicals into organic matter and to replenish the oxygen in the air (Fig. 1-2). These interactions with the environment on which all organisms depend are called **ecological relationships.** They are characterized not merely by the fact that one organism is dependent on another, but by the interdependencies of each on **all** the others and on the integrity of the **entire system.**

Figure 1-2   Red deer in their natural habitat.

All organisms respond to changes in the environment; the ability to respond is called **irritability.** Some organisms can initiate certain kinds of behavior as a result of spontaneous activity of their own cells. **Behavior** is the organized output of hundreds, millions, even billions of cells, interacting according to plans built into an organism by its genetic heritage and immediate history. Behavior obeys all of the biological laws. Behavior develops; it evolves. Behavioral traits can be inherited, and they can be influenced by environmental changes.

What
is Science?

The science of biology, like the rest of science, has grown to its present state as men have attempted to understand and control the world around them. Science is a way of organizing—and getting—information about the universe. Science generally progresses through three conceptual levels. (1) At the **descriptive** level the scientist asks merely what the system does—whether the system is a cell, an organism, or a society. What is its structure? What potentialities does it have? What are the relationships of its parts? How is it organized? (2) At the **experimental** level the scientist attempts to test **hypotheses,** or tentative explanations, for the phenomena he observes; to establish consistent rules or principles that are obeyed by the system. He attempts to establish from observation and experiments general laws that have **predictive** value. (3) At the **analytical** level the scientist attempts to understand the system and its functions in terms of simpler levels of organization. He examines societies in terms of individual behavior; individuals in terms of cellular interrelations; cells in terms of molecular interactions. That is, he attempts to establish **mechanisms** of action.

Method of Alternative Hypotheses

There are many things that are ordinarily associated with science: experiments, laboratories, numbers and equations, test tubes, instruments, computers. Science is often defined in beginning textbooks, like this one, in terms of "the experimental method"; that is, a person who does experiments is a scientist. There is, of course, some truth in these ideas. Test tubes and computers and mathematics are often used as tools in science, and most scientists often do perform experiments of one sort or another. But this misses the whole point. Science, in essence, is merely a way of asking questions. The object of science is to understand the universe: the stars in space; the objects we see around us, big and small, living and nonliving; our bodies; the earth on which we walk. A scientist is a person who asks questions about any aspect of the universe.

But, then, you may ask, isn't everyone a scientist? And the answer, of course, is yes—to some extent. Every child asks questions of the universe and, if he isn't squelched too soon by his elders, he will try to seek answers to those questions. But the difference in the way most of us ask questions and the way scientists do lies in the method of posing the question. Scientists use a special method that is called the **method of alternative hypotheses and strong inference.** In this method an investigator proceeds by postulating and testing hypotheses; a **hypothesis** is a tentative explanation put forth to account for an observed phenomenon. It is a tentative **model** of some aspect of that phenomenon. We use the word model in this sense not necessarily as a physical construction (like a stick model of a molecule). It may also be a well-thought-out idea or concept about a set of relationships, or it may be a series of chemical reactions or mathematical equations. It represents the scientist's estimate of one way that a phenomenon may be described or an explanation of how something works. The purpose of building models is as an aid to asking questions; it helps to see which model seems to fit the real world best.

Let us take a specific example. Most plants thrive in sunlight. Without sufficient sunshine their leaves turn pale; they wilt and die. Plants are an important source of food for animals and man because they contain many nutritional substances, including one called starch, a carbohydrate that animals are able to use as a fuel for energy. Where do plants get their starch? This is no idle question if we are, say, agricultural scientists attempting to increase the yield of food crops in a hungry world. There are at least two obvious possible answers. One is that they obtain starch through their roots from the soil around them, and that their need for sunlight is unrelated. The other is that they use the energy from the sun to manufacture starch in their own tissues. Here, then, we have two tentative explanations for an observed phenomenon, which we can pose formally as models or hypotheses:

1. Plants obtain starch by absorbing it through their roots.
2. Plants produce starch in their tissues with the aid of energy from absorbed sunlight.

What next? Clearly, we want an answer to the question, not merely two hypotheses, for our behavior depends on which hypothesis is valid. In the first case we ought to enrich the soil around plants with starch; farmers should use fertilizers containing carbohydrates. In the second case fertilizers should perhaps contain other things, but they should not contain starch or carbohydrates. Therefore we want a method to determine which, if either, is correct; we want to disprove one of the alternative hypotheses. The method, of course, is to design an experiment to do just that.

The primary purpose of a good experiment is to disprove a hypothesis. How does an experiment disprove a hypothesis? By demonstrating that predictions derived from the hypothesis are false. For example, from hypothesis 1 we would predict that a plant growing in soil deficient in starch or other

carbohydrates would die, no matter how much sunlight it received. We can easily set up an experiment to test this prediction. Artificial soil can be obtained in which the only nutrients present are those we, the experimenters, add, and we can carefully exclude all starch or other carbohydrates. We find then that plants thrive in such starchless soil when they are in sunlight; and even if they are depleted of their starch content by being kept in the dark for 24 hours, they rapidly regain it when they are returned to the light. Thus it cannot be true that plants obtain their starch by absorbing it through their roots; that is, we have disproved hypothesis 1, and we seem to move toward hypothesis 2.

Now, let us test hypothesis 2. If plants manufacture starch in their tissues, we should predict that protecting a part of a plant from sunlight will prevent starch production in that shielded area. Performing the simple experiment by covering part of a leaf with aluminum foil, we find that the result (shown for a leaf of a bean plant, Fig. 1-3), accords with the prediction; that is, it supports the hypothesis. Can we then say that the test has proved hypothesis 2 to be true? No! The reason we cannot lies in the nature of deductive logic.

A deductive inference is a logical statement in which a conclusion follows from a premise: **"If A . . . then B."** If plants obtain starch by absorbing it from the soil through their roots, **then** removing all starch from the soil should prevent a plant from increasing its content. Such statements can be proved false by an experimental test, but they cannot be proved true. Suppose the plants had died in the starchless soil, in accordance with the prediction of hypothesis 1. They might have died because we had failed to add some other essential element that we didn't know about to the artificial soil; or they might have died because of some accidental infection. Our second experiment could have disproved hypothesis 2 if the shielded area of the leaf formed just as much starch as the lighted regions. The fact that it didn't might mean that aluminum foil is poisonous to a leaf, or that it blocked out some essential element in

Figure 1-3 (a) A leaf from a bean plant that was kept in darkness for 24 hours to reduce the starch content; a portion of the leaf was then shielded from sunlight by aluminum foil, as shown, and the plant was kept in sunlight for 8 hours. (b) The same leaf has been stained with iodine, which turns blue-purple in contact with starch. Note that starch is still absent from the shielded area.

(a)   (b)

addition to sunlight. As long as there are other possible explanations for a result, however farfetched, a hypothesis can only be made **more probably** true. Despite commercials on television that state that "scientific tests prove" brand X is better than brand Y, an experiment can never prove such a hypothesis.

So now we have asked a question, tested a pair of alternative hypotheses, discarded one on the basis of that test, and are left with an answer to the question that is **probably** true. To increase the degree of probability of this answer, we must now set up a second pair of hypotheses to test. For example: a leaf is prevented from manufacturing starch either (1) by being shielded from light or (2) by the toxic effects of contact with aluminum. We must devise an experimental test to disprove these postulates (for example, we will shield the leaf with black paper or some other material instead of foil), again interpret the results, and again discard one or the other. If the results again support the idea that light is the essential ingredient, this adds further to the probability of truth of our original hypothesis 2. After many such experimental cycles, when a large number of alternative hypotheses have been discarded, the remain-

ing one, usually much modified, becomes generally accepted, and in day-to-day thinking may be treated as an established fact. Scientists are now confident that plants use energy from the sun to synthesize starch in the process of **photosynthesis** (Chapter 4).

We have spoken so far about plants and sunlight, although we have experimented on only one plant, a bean plant (Fig. 1-3). To make the **inductive generalization** (from a particular instance to the general), it is necessary at least to repeat these experiments on several different kinds of plant. Let us take one more example of experimentation from the history of biology to illustrate this point.

We now know that the whitish fluid called **semen,** produced by the males of most multicellular animals, contains **spermatozoa,** or sperm. Sperm are living cells with long, whiplike tails that can swim toward a female egg cell and unite with it to achieve fertilization; fertilization is what causes the egg to begin its development into a new individual (see Chapter 11). The Italian physician Lazarro Spallanzani lived during the eighteenth century, a time when scientists did not understand how semen was related to fertilization. Although sperm had been seen in semen almost two centuries earlier with the aid of

13 *Life and the World of Ideas*

the first primitive microscopes, their importance in fertilization was not recognized. Thus only two possibilities were considered:

1. Seminal fluid must make actual contact with an egg in order to fertilize it.
2. It is only necessary that a "vapor" from the semen make contact with the egg.

Physicians of the day, including Spallanzani, knew enough about the anatomy of the female reproductive tract to recognize that during intercourse semen would be deposited a considerable distance from the ovary, where (they knew) the eggs were housed. Because the role played by sperm was not recognized, the idea that they might be able to swim toward the egg (which is, of course, what actually happens) was so incredible that it was not even considered, and the vapor hypothesis was generally accepted. In 1785 Spallanzani published the report of his experimental test of the problem, which stands as a classic example of the scientific method. Some excerpts will illustrate why.[1]

"Is fertilization affected by the spermatic vapor? . . . It has been disputed for a long time and it is still being argued whether the visible and coarser parts of the semen serve in the fecundation [fertilization] of man and animals, or whether a very subtle part, a vapor which emanates therefrom and which is called aura spermatica, suffices for this function."

Thus he defines the problem, poses the question. Next he describes the controversy that has arisen from lack of any experimental evidence, and he states the alternative hypotheses:

"It cannot be denied that doctors and physiologists defend this last view, and are persuaded in this more by an apparent necessity than by reason or by experiments. . . . They reflect upon the orifice of the egg canals or the Fallopian tubes, so narrow that a very fine probe cannot enter there . . . , from which they conclude that the seminal liquid of the male, ejaculated into the organs of generation of the female, cannot arrive at the ovaries where the embryos are lodged; but that they must be fertilized by the part of the semen which evaporates. . . . Despite these reasons, many other authors hold the contrary opinion, and believe that fertilization is accomplished by means of the material part of the semen."

He suggests an experimental design and states the predictions arising from the two hypotheses, then goes on to describe details of his own work (see Fig. 1-4):

"Therefore, in order to decide the question, it is important to employ a convenient means to separate the vapor from the body

[1]Spallanzani, L., "Experiments upon the Generation of Animals and Plants." Translated in Gabriel, M. L., and Fogel, S. (eds.), *Great Experiments in Biology*, Prentice-Hall, Englewood Cliffs, N.J., 1955, pp. 189–193.

Figure 1-4 The "sperm vapor" experiment of Spallanzani. A batch of toads' eggs is suspended above a pool of seminal fluid in a cavity formed by two concave watch glasses.

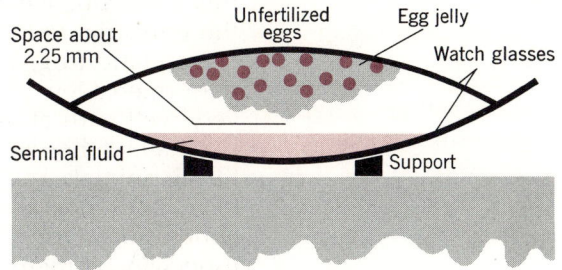

of the semen and to do this in such a way that the embryos are more or less enveloped by the vapor; for then if they are born, this would be an evident proof that the seminal vapor has been able to fertilize them; or on the other hand, they might not be born and then it will be equally sure that the spermatic vapor alone is insufficient and that the additional action of the material part of the sperm is necessary. . . ."

"In order to bathe [eggs] thoroughly with this spermatic vapor, I put into a watch glass a little less than 11 grains of seminal liquid from several toads. Into a similar glass, but a little smaller, I placed 26 [eggs] which, because of the viscosity of the jelly were tightly attached to the concave part of the glass. I placed the second glass on the first, and they remained united thus during five hours in my room where the temperature was 18° [C]. The drop of seminal liquid was placed precisely under the eggs, which must have been completely bathed by the spermatic vapor that arose; . . . for it could not have escaped outside of the watch crystals since they fitted together very closely."

It is hard to imagine a clearer or more explicit experimental description.

"But in spite of this, the eggs, subsequently placed in water, perished. . . . I touched another dozen eggs with the small remnant of semen which remained after evaporation . . . eleven of these tadpoles hatched successfully. . . ."

"The conjunction of these facts evidently proves that fertilization in the terrestrial toad is not produced by the spermatic vapor but rather by the material part of the semen."

Thus Spallanzani has disproved the spermatic vapor hypothesis; but note the qualification "in the terrestrial toad." He goes on to provide evidence for an inductive generalization; that is, to extend his results to other organisms:

"As might be supposed I did not do these experiments only on this toad, but I have repeated them in the manner described on the terrestrial toad with red eyes and dorsal tubercles, and also on the aquatic frog, and I have had the same results. I . . . even . . . performed a few of these experiments on the tree frog . . . they agree very well with all the others. . . ."

"Shall we however say that this is the universal process of nature for all animals and for man? The small number of facts which we have does not allow us, in good logic, to draw such a general conclusion: One can at most think that this is probably so, more especially as there is not a single fact to the contrary . . ."

This is an admirably cautious and accurate statement, indicating Spallanzani's awareness of the dangers of generalization.

". . . the question of the influence of the spermatic vapor in fertilization is at least definitely decided in the negative for several species of animals, and with a great probability for others."

Note that Spallanzani obviously recognized that his negative results conclusively disproved the spermatic vapor hypothesis, but nowhere did he claim more than probable verification of the opposite hypothesis.

Open Communication of Experimental Results

Any description of the science of biology must be based on the experiments and observations of thousands, or hundreds of thousands, of scientists. Often, in discussing a given topic, we attribute a piece of work to the men who actually did it, as we just did with Spallanzani. It would be confusing and terribly tiresome, however, to provide the name of every scientist who provided every fact with which this book is filled. More often it is convenient merely to state the observation, or to use an "if we . . ." sentence. "If we measure the volume of cells . . . , we will find . . . ," or "If we examine the bonds between hydrogen and oxygen atoms, we will see that. . . ."

These statements are not merely literary devices. They are ways of reminding you, the reader, that "we" (or you) could in fact measure the volume of these cells, or examine those chemical bonds between oxygen and hydrogen; and if we (or you) were to do so, using the same techniques and conditions the unnamed original scientist used, there is a very high probability that we would find what is stated. What such a sentence really says is, "Some time in the past (most likely in the last few years), Dr. X measured the volumes of certain cells under well-described and carefully controlled conditions, using techniques that we know to be sound; he found the results that are described; and we now confidently pass those results on to you."

How can we have such confidence? How do we know the methods Dr. X used were sound? How do we know that he interpreted those results correctly? The answers to these questions lie in another aspect of science, having to do with the open and free communication of results. As soon as Dr. X performed his set of experiments and obtained a series of interpretable results, he wrote a report and sent it as a manuscript to one of several hundred reputable scientific journals. In that report he described in detail the hypotheses he was testing, the methods he used, the measurements he made. He gave an explicit description of the results he obtained, and he discussed those results to show their relevance to work being done by other scientists in the same field. The editor of the journal who received the manuscript then sent copies to two or three reviewers, scientists in the same field who were familiar with the methods Dr. X used and who could be sure that Dr. X had taken all the necessary precautions, had made his measurements properly, had interpreted his results in a reasonable fashion. If the reviewers' opinions were favorable, the report would then be published in the journal, within a few months, for every interested member of the scientific community to read and criticize. Other scientists working on closely related problems—in this case having to do with the volumes of cells—might even repeat Dr. X's measurements on the same kinds of cell he used, or on different ones, to make sure his results were reliable, and these scientists would in turn publish their results.

It is on the basis of this elaborate process of self-correction and confirmation that science progresses and that textbooks can be written.

Why Science?

In this day, when we see all around us examples of man's misuse of his physical and chemical environment, it is easy to think of science and technological progress as the sources of most of our problems. We'll deal with this topic at length when we examine man's relationship to his environment (Chapters 20, 21). But what we have tried to show is that science is not technology; it is not test tubes, or biocides, or jet airplanes, or miracle drugs. Science is merely a way of looking at the world; a mode of thinking; a method of asking questions. It is—or should be—characterized by free inquiry, openness of mind, honesty, discipline. Perhaps what scientific inquiry is can best be illustrated by an example of what it is not. The following burlesque is attributed to Francis Bacon as a plea for experimental science in the seventeenth century.

"In the year of our Lord 1432 there arose a grievous quarrel among the brethren over the number of teeth in the mouth of a horse. For thirteen days the disputation raged without ceasing. All of the ancient books and chronicles were fetched out, and wondrous, ponderous erudition such as was never before heard of in this region was made manifest. At the beginning of the fourteenth day a youthful friar of goodly bearing asked his learned superiors for permission to add a word, and straightway, to the wonderment of the disputants, whose deep wisdom he sore vexed, he beseeched them to unbend in a manner coarse and unheard-of and to look in the open mouth of a horse and find the answer to their questionings. At this, their dignity being grievously hurt, they waxed exceeding wroth: and joining in a mightly uproar, they flew upon him and smote him, hip and thigh, and cast him out forthwith. For said they, surely Satan hath tempted his bold neophyte to declare unholy and unheard-of ways of finding the truth, contrary to all the teachings of the fathers."[2]

[2] From Welch, C., Evolution. In Grobman, A. B. (ed.), *Social Implications of Biological Education,* National Association of Biology Teachers, Washington, D.C., 1970, p. 120.

Man is now faced with enormous problems of worldwide scope. Many are self-inflicted; some result from the use or misuse of science and technology. Thus we are confronted with two alternatives: we can ignore our problems; or we can attempt to find solutions for them. If we select the latter alternative, we soon come to realize that finding solutions to problems requires asking questions. And to answer questions, we are again confronted with two alternatives. Like Bacon's disputants, we can seek answers from "authority"; or we can explore the real world and seek our answers from the universe around us, using the most effective tools at our disposal. This is the method of science—and therein, in our opinion, lies the hope of mankind.

Reading List

Bonner, J. T., *The Ideas of Biology*. Harper Torchbooks, New York, 1962.

Bronowski, J., *Science and Human Values* (revised ed.). Harper Torchbooks, New York, 1965.

Dubos, R., *Reason Awake: Science for Man*. Columbia University Press, New York, 1970.

Gabriel, M. L., and S. Fogel, *Great Experiments in Biology*. Prentice-Hall, Englewood Cliffs, N.J., 1955.

Platt, J. R. *The Excitement of Science,* Houghton Mifflin, Boston, 1962.

Platt, J. R. "Strong Inference." *Science,* **146:**347–353, October 16, 1964.

Russell, B. *On the Philosophy of Science,* Bobbs-Merril, Indianapolis, 1965.

"What drives life . . . is a little electric current, kept up by sunshine."

ALBERT SZENT-GYORGYI, 1960, *CHEMISTRY OF MUSCULAR CONTRACTION*

"But of the construction and growth and working of the body, as of all else that is of the earth earthy, physical science is, in my humble opinion, our only teacher and guide."

D'ARCY WENTWORTH THOMPSON, 1942, *ON GROWTH AND FORM*

part two | *The Molecular Basis of Life*

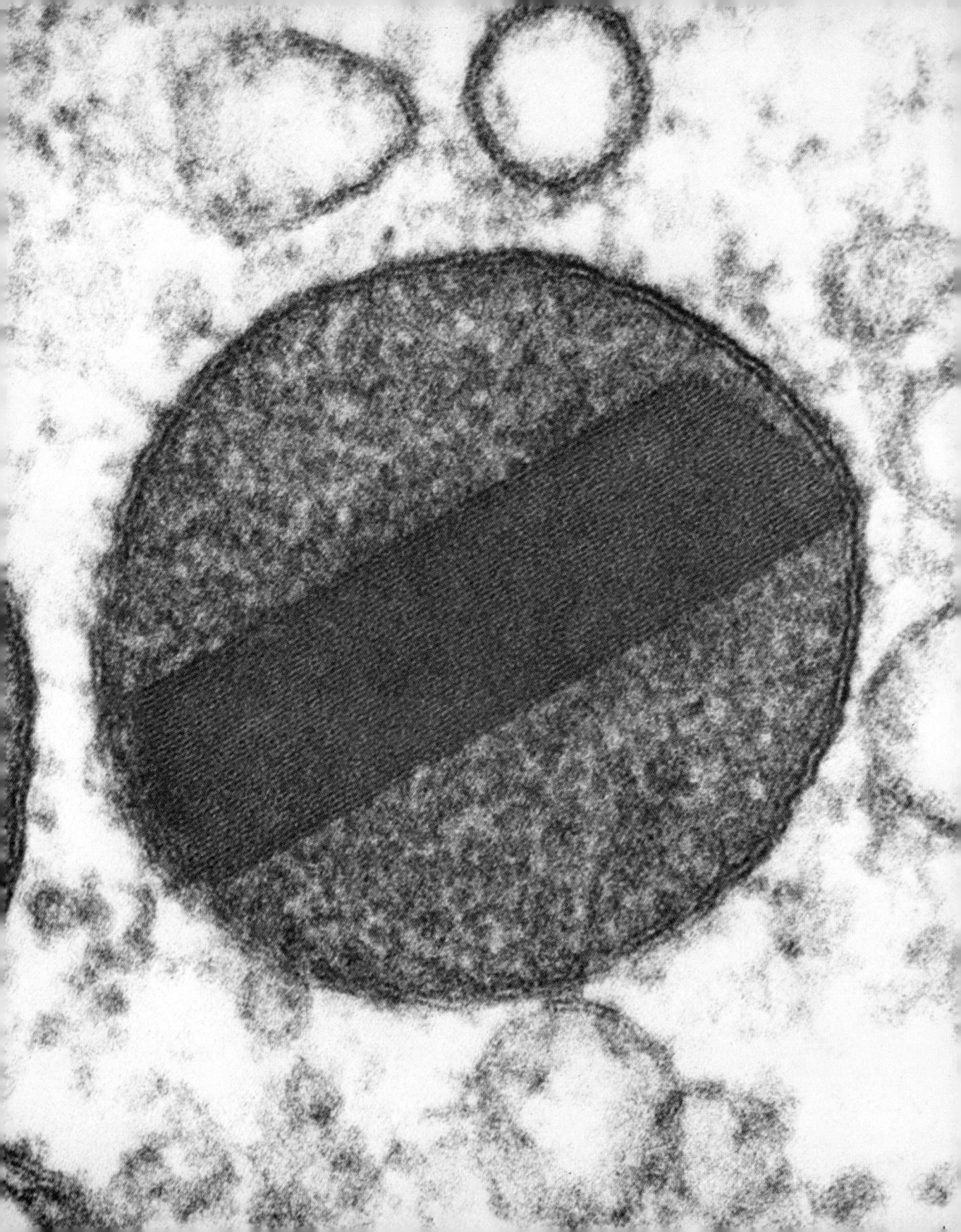

chapter
two

Chemical and Physical Principles

"Life is a relationship among molecules and not a property of any one molecule."

LINUS PAULING, 1960, *THE CHEMICAL BOND*

We saw in Chapter 1 that living systems are identified by their ability to do certain things: to convert raw materials into energy for their own use, to synthesize new protoplasmic components, to reproduce. Because these are essentially chemical processes, and because the structures involved are chemical structures, we need a certain minimum of chemical facts to understand them.

Students—especially nonscience students—are often turned off by introductory chemistry because of its bewildering complexity and the apparent unrelatedness of many facts and processes. The advantage biological chemistry has to offer is that every chemical process can be related to a biological process that is often familiar. For example, respiratory metabolism, as we shall see, involves a series of chemical reactions that result in the conversion of food materials into carbon dioxide and water, with the liberation of usuable energy. Somehow these reactions take on added significance and may be more readily understood when they are related to the biological phenomenon of respiration—breathing; their importance becomes clear when we understand how molecules of oxygen are extracted from the air, transported across the membranes of the lungs into the cells and used by the cells' metabolic machinery, and that this process takes place in special subcellular structures called **mitochondria.** When we recognize that an organism's respiratory machinery is not its lungs but the mitochondria in its cells,

Crystalline molecular array in a rat cell granule. Magnification 375,000X.

21

we can see respiration as a meaningful process whereby usable energy is obtained from the environment.

Because one of the primary characteristics of living systems is their ability to utilize energy—that is, to convert energy from one form to another—it may be wise to begin a discussion of the chemistry of life with the concept of mass-energy.

Mass-Energy

All the universe is energy; there is only energy. In some places some of that energy is condensed into electrons, protons, and neutrons, which are hooked together into atoms. And atoms in turn are bound into molecules that form substance we can touch and feel, which we call **mass-energy.**

But what do we mean by touch? Electrons not only are made of energy, they also exhibit a form of energy called **negative electric charge,** which has the property of attracting positive charges like a magnet and repelling other negative charges. When the atoms that make up our skin are brought close enough to the atoms of another surface, their electrons repel one another and we "feel" that surface. No real contact has been made; there are only forces of energy pushing on other forces of energy. But, you argue, we can see solid matter all around us. Well, what do we mean by see? If light (another form of energy) bounces off vibrating atoms (energy) into our eyes, it activates certain pigments; that is, it causes certain molecules to vibrate a little faster. Again, there has only been a transmission of energy—an interaction of energy on energy.

This is what Einstein's world-shaking equation is all about. It is usually written $E = mc^2$, which is simply a shorthand way of saying that energy and mass (i.e., matter) are interconvertible. Thus the universe is energy; matter is energy; there is only energy. But what do these thoughts about energy have to do with life or with biology?

Most living things (which we call **orga-**nisms) behave in certain ways that we have come to associate with their being alive. They may move, or breathe, or exhibit changes over time (like a flower blooming); they may respond in some way to being poked with a stick. These are indeed characteristics of living things, and each one of these processes requires the expenditure of energy. There is an even more important difference between living systems and nonliving matter: living systems tend to increase, to grow, to make more of themselves; inanimate matter tends inevitably to decrease, to dissipate itself. Observe a rock, a piece of metal, a chunk of dead wood. Given enough time it will rust or disintegrate. But, given time, something alive—an ant, a tree or a rabbit—will grow, will double itself 10 or 100 times over.

Now, what's the trick? What do organisms do differently than nonliving matter? As we shall see, both are obeying, in their own ways, the laws of physics. Inanimate matter passively loses energy; its organization becomes randomized; we say it decays. Living substance (which we call protoplasm), on the other hand, has evolved devices, amazing and wondrous mechanisms, for absorbing energy from its surroundings, which it puts to its own uses. Organisms obtain energy by absorbing matter (food) from their surroundings and breaking it down to liberate its energy; they also absorb energy directly from the sun. The major use they make of this energy is to create more protoplasm, more of themselves.

Given this picture of our universe as a sea of energy, we find that it seems to be constructed in such a way that it obeys a set of fundamental laws. These laws are called the **laws of thermodynamics;** thermodynamics is the study of energy relationships. According to the **first law of thermodynamics,** there is a fixed amount of energy in the universe. That is, energy can neither be created nor destroyed. It can be moved around; it can be transformed from one type to another; but when there is an increase in energy anywhere in the universe, that increase must be balanced by a decrease somewhere else.

The **second law of thermodynamics** is equally simple and just as profound. It deals with real systems in the real world. A **system** is defined as an assemblage of matter and energy. According to the second law, any real system in the universe, if left to itself, tends to lose energy and to become disorganized. The level of disorganization, or of **randomness,** is referred to as **entropy.** If I have a set of alphabet blocks that I pile into a single teetering column, carefully placing the blocks in order from A to Z, I have put the system (the pile of blocks) in a state of low entropy and high potential energy. The single vertical column and the serial order of the letters are a most improbable, nonrandom organization, so the level of randomness (entropy) is low. That the potential-energy level is high is easy to show if we nudge the pile slightly. The blocks come crashing down, to dissipate energy by striking the floor. Now the energy of the system (the blocks scattered at random on the floor) has decreased and its entropy has increased.

The point of the second law is that the entropy of the system now is higher than it was before I started piling up the blocks in the first place. If we could measure the total amount of energy lost from the system when the blocks fell and struck the floor, we would find that it wasn't quite so great as the energy we put into the system by lifting and organizing the blocks into a column. The difference is the increase in entropy. Let us put this idea in more general terms, and restate the second law: In any energy transfer, when a system increases in energy, that increase must be balanced by a decrease in energy somewhere else; moreover, there is an additional net loss of energy resulting from the transfer itself. That loss, which is energy that becomes unavailable for useful work, is measured by a gain in entropy.

The second law states that every system tends to lose energy. But it does not say that a system can never gain energy. Clearly, the blocks can be piled up again, but only by the input of additional energy. It will not happen spontaneously.

The Forms of Energy

Although the physicists' view of the universe is that it consists entirely of energy, in our daily lives it is convenient to deal with matter as we experience it, as what we can see and touch—as substance that has mass and occupies space. We can define energy in its simplest terms as the **capacity to produce motion in a body of matter.** A body of matter may be a stone, a house, a speck of dust; it may be the size of the sun or the size of an atom or an electron. The capacity to produce motion is called **potential energy;** the energy of the motion itself is **kinetic** energy.

Within this framework, there are really only a few fundamental forms that energy can take. For example, the universe is so constructed that all bodies of matter are attracted toward one another. That attraction we call **gravity;** it is an elemental form of energy. We know very little about what gravitational force is, but to demonstrate it we need only pick up a rock from the ground. In doing so we have pulled two bodies of matter (the rock and the earth) away from one another, and have had to supply enough energy to the rock-plus-earth system to overcome the force of that attraction. As long as we hold the rock up, that added energy is stored as potential energy. At any time we can let go of the rock and it immediately begins to move (i.e., to fall). The potential energy stored in the system is converted to kinetic energy.

Weight is a quantitative expression of this gravitational attraction between any object and a standard body, in our case the earth. The weight of a body is proportional to its mass; that is, the greater the mass, the greater the weight. On some other planet the weight of an object would be different from that on earth because of the different gravitational attraction, but the mass of the object would remain the same.

A second elemental form of energy is **electrostatic** or **electronic** energy. Ultimately, electronic energy (also called **electric** energy) derives from the structure of atoms. We'll discuss atomic structure in some detail on pp. 27–35; here we need only note that atoms are made of **electrons,** which are units of negative charge, **protons,** which carry positive charge, and usually **neutrons,** which are electrically neutral and therefore can be ignored for the moment. A body of matter that contains more electrons than protons carries a **net negative charge;** one that contains fewer electrons than protons is **positively charged.** Because the universe is so constructed that particles with the same charge repel one another but oppositely charged particles are strongly attracted, these two bodies, if given the opportunity, will move toward one another and thus, according to our definition, possess energy. There are many manifestations of electronic energy—electricity, magnetism, electrostatics. All, in one form or another, represent the tendency of negatively and positively charged particles to move together.

A third elemental form of energy is that displayed by any object in motion; we have called this **kinetic** energy. As any schoolboy knows, a marble in motion is able to cause a resting marble with which it collides to move. Thus some of the kinetic energy in the first marble is transferred to the second and causes motion in that body of matter. The amount of kinetic energy displayed by a moving object depends on its mass and velocity. When two objects with the same mass travel at different velocities, the faster moving body has the greater kinetic energy. When two different masses move at the same speed, the larger body has the greater kinetic energy.

One other form of energy with which we need be concerned for our discussion of biological energy relationships is **heat, or thermal energy.** What do we mean by heat? Subjectively, we know whether an object is hot or cold by feeling it. Physically, however, heat is the activity (i.e., motion) of the atoms and molecules that make up an object. Thus heat is really a form of kinetic energy. The rapid motion of the atoms of one object (e.g., a hot piece of metal) can increase the activity of the atoms of a second object with which it makes contact. This is merely another way of saying that heat flows from a hot object to a cold one. That this flow of heat represents the transfer of energy can be demonstrated by bringing a hot object in contact with a column of mercury in a thermometer. As heat flows into the mercury, we see the increasing agitation of its atoms as expansion, and the end of the column begins to move. This conforms to our definition of energy—the capacity to cause motion in a body of matter (the mercury).

If heat is merely molecular motion, then an object with no such motion would have no heat. That is precisely what is meant by **absolute zero,** the temperature at which molecular motion ceases ($-273.16°C$). For theoretical reasons, even at absolute zero a certain minimum of residual vibrational energy remains, known as **zero-point energy.** This cannot, however, be converted to thermal energy.

There are many manifestations of potential energy. The one with which we are most concerned is **chemical bond energy,** which is the electrostatic energy of electrons interacting. The ease with which chemical bond energy can be converted to kinetic energy—that is, can cause movement in a body of matter—can be shown in a few examples. The potential energy of a stretched slingshot, which derives from the forcible distortion of the bonds between the molecules of rubber, can be released and converted to the kinetic energy of a moving pellet. Potential energy stored in the chemical bonds of rocket fuel is transformed into both heat and the kinetic energy of a swiftly moving missile. Potential energy in the chemical bonds of food is converted, in a muscle cell, to contraction of muscle fibers and movement of a limb. All of these are examples of the fundamental fact that all forms of energy are interconvertible.[1]

The field of study concerned with how energy is stored in chemical bonds and how

[1]Some commonly used units of energy and energy conversions are given in Appendix 1, along with definitions of other scientific quantities.

it can be converted to other forms is called **chemical thermodynamics.** We will explore at least the basic concepts of chemical thermodynamics, (pp. 46–48), but first we must develop a certain background of information on the structure of matter and on how chemical reactions take place.

Physical Structure and Properties of Matter

Gases

Matter can exist in any of three physical states—gas, liquid, or solid. In the gaseous state it consists of rapidly moving molecules whose average distance from one another is so great that the volume occupied by the molecules themselves is negligible compared to the relatively vast empty spaces between them. A gas has neither a fixed volume nor a fixed shape, but distributes itself equally throughout any container or space in which it happens to be.

Gas molecules are only weakly attracted to one another, which accounts for their great freedom of movement and their completely disordered state. The molecules are constantly in rapid and random straight-line motion, frequently colliding with one another and the walls of the container. The collision of each molecule with a container wall produces a slight push or force, the sum of which is called **pressure.**

Because these molecules have mass and are in motion, they have kinetic energy. An increase in the temperature of a gas represents an increase in average velocity and therefore in the kinetic energy of the gas molecules themselves. The frequent collision of gas molecules with one another does not generally slow them down but results in a change in direction of the moving molecules. Virtually no net loss of kinetic energy occurs as long as the temperature remains unchanged, although energy transfers may take place between the colliding molecules. If kinetic energy were progressively lost in these frequent collisions, we would expect the molecules to lose their motion eventually and settle out. This is in fact what they do if temperature (i.e., kinetic energy) is not maintained, as when cooling steam turns to water.

Liquids

As a gas cools and compresses, its molecules, which were originally far apart, slow down and come closer together. As the molecules move more slowly, their attraction for one another increases and may become great enough to allow them to interact and settle out as a liquid. Although the molecules of a liquid still have movement, their freedom of motion is much less than in the gaseous state. They are also impeded by frequently colliding with each other because the molecules are so highly concentrated compared to the gas. Liquids, unlike gases, are practically incompressible because the amount of free space between their molecules is negligible. Liquids tend to maintain their volume but have no characteristic shape, assuming the form of their container.

We can calculate that the speed of movement of water molecules at room temperature is, on the average, something like 1000 mph. However, some molecules in liquids (as in gases) move faster and others slower than the average speed. Molecules near the surface that move upward and have enough kinetic energy to overcome the attractive forces of neighboring molecules escape into the atmosphere. This conversion from liquid to gas is called **evaporation.** Some of the vaporized molecules may also be knocked back randomly to the liquid in a reverse of evaporation, a process called **condensation.**

Solids

On further cooling, the molecules of a liquid assume an ordered arrangement whereby they are held together closely in definite spatial relationships to one another by strong attractive forces. This is the solid state. Unlike those of gases and liquids, the molecules of a solid almost entirely lack the freedom to move in any direction around each other from one position to another, although they do oscillate, or vibrate, within a limited space. For these reasons solids exhibit a characteristic incompressibility, hardness, and shape, but their molecular movement is still ample for kinetic energy to be measured as heat.

Diffusion

The spontaneous spreading, or migration, of molecules or particles (of gases, liquids, and, to a much lesser extent, solids) is termed **diffusion.** More precisely, diffusion is the spontaneous movement of molecules or other particles from a region of high concentration to one of lower concentration. The movement is entirely random, a result of the kinetic energy of the particles.

The phenomenon of diffusion is best illustrated by analogy to an imaginary model. Picture a large room containing a few hundred gravity-free basketballs (each representing a molecule) traveling through the air along random, straight-line pathways at different velocities (Fig. 2-1). They frequently collide with one another and with the ceiling and walls, only to bounce in a new direction from each collision. If we follow the course of any particular ball, it will show a haphazard, zigzag path. If a door is opened to

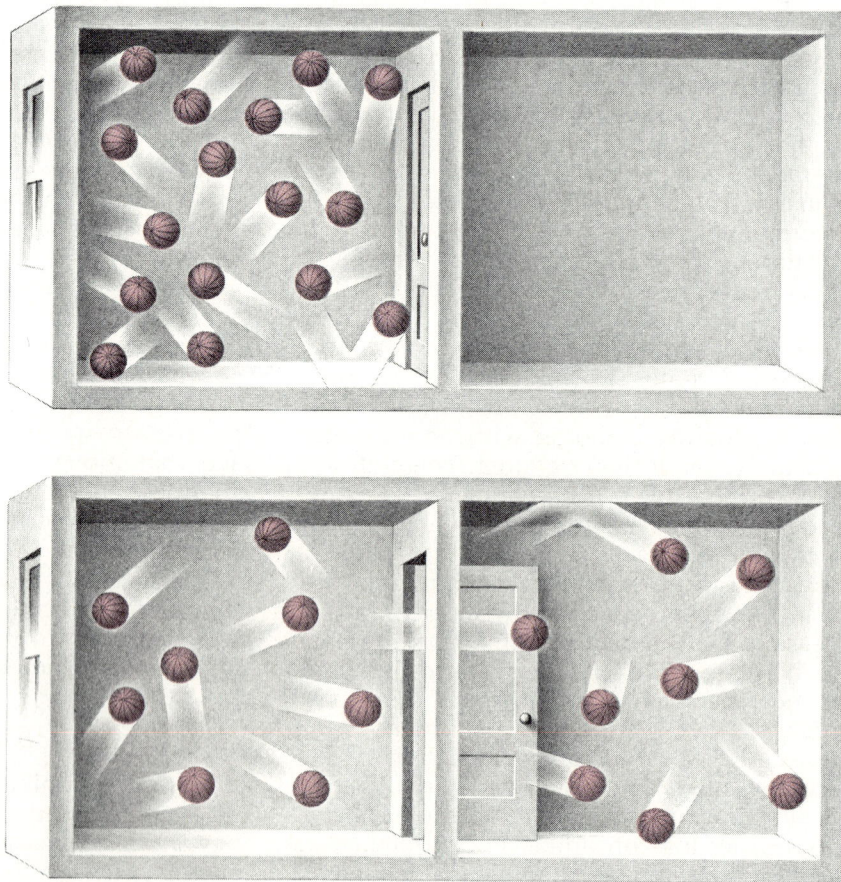

Figure 2-1   Schematic illustration of diffusion. In this imaginary model system basketballs represent molecules.

an adjoining room of equal size, some of the balls, by chance or probability alone, will bounce into the second room. Some balls that enter the adjoining room may also return to the original room by random movement. However, there is a greater chance at first that more balls will enter the adjoining room than will leave it because there are more balls in the original room. Thus diffusion, or a net movement of balls from a region of higher concentration (original room) to a region of lower concentration (adjoining room), will occur until the concentration is more or less the same in both rooms; when the number of balls is the same in both rooms, the frequency with which balls will bounce through the door in one direction will be the same, on the average, as that in the other direction.

We refer to this state, when the probability of particles moving in either direction is the same, as a **dynamic equilibrium,** or **steady state.** Note that the term does not refer to the equal concentration of balls in the two rooms, but to the equal probability or rate of movement in the two directions. If the walls of one of the rooms were made sticky, the number of balls in that room would have to be greater than in the other room (to make up for the balls that are trapped on the walls) before the frequency of movement through the door would be equal in both directions. Under these conditions the point of dynamic equilibrium would be attained with a higher concentration of balls in one room than the other. This concept will be very important when we discuss chemical reactions.

Different kinds of molecules in the same container diffuse essentially independently of one another at rates determined by the mass and speed (and therefore the kinetic energy) of each type. Diffusion of molecules as described for gases also applies to liquids and solids, although at a slower rate and to a more limited extent because of the more restricted freedom of motion of their molecules. In biological systems the movement of nutrients, oxygen, waste products, carbon dioxide, hormones, and most other substances occurs by diffusion within cells and into and out of cells.

Molecular and Atomic Structure
Elements and Compounds

The basic units or building blocks of molecules are called **atoms.** There are 92 naturally occurring kinds of atom in the universe. In addition, 11 new artificial atomic types have been produced by physicists in recent years. Atoms, needless to say, are very small, but they are not imaginary; they are not so small as to be immeasurable. A hydrogen atom, for example, is about 1×10^{-8} centimeters (cm) in diameter.[2] That is, it would take 100 million hydrogen atoms, lined up like tiny billiard balls, to fill out one centimeter (a little less than half an inch). In 1970, for the first time, A. V. Crewe and his co-workers were able to see single atoms of thorium (Fig. 2-2, next page) and uranium with a newly developed high-resolution scanning electron microscope with a resolving power of about 5 Å ($= 5 \times 10^{-8}$ cm).

The 92 naturally occurring atoms are found in a variety of chemical states. Depending on its particular properties, each kind of atom may occur singly, in chemical combination with atoms identical to itself, or in chemical combinations with one or more other kinds of atom, forming molecules that constitute all the matter of the universe. Any substance whose molecules are composed of only one kind of atom is an **element.** Although the molecules of an element may consist of one or more identical atoms, an element cannot be decomposed by any chemical treatment to a simpler material. An element, therefore, is chemically an irreducible **primary (or elementary) substance.** There are 92 naturally

---

[2] Throughout this book exponential notation (such as 1×10^{-2}) is used to deal with very large and very small numbers. For an explanation of exponents, see Appendix 1.

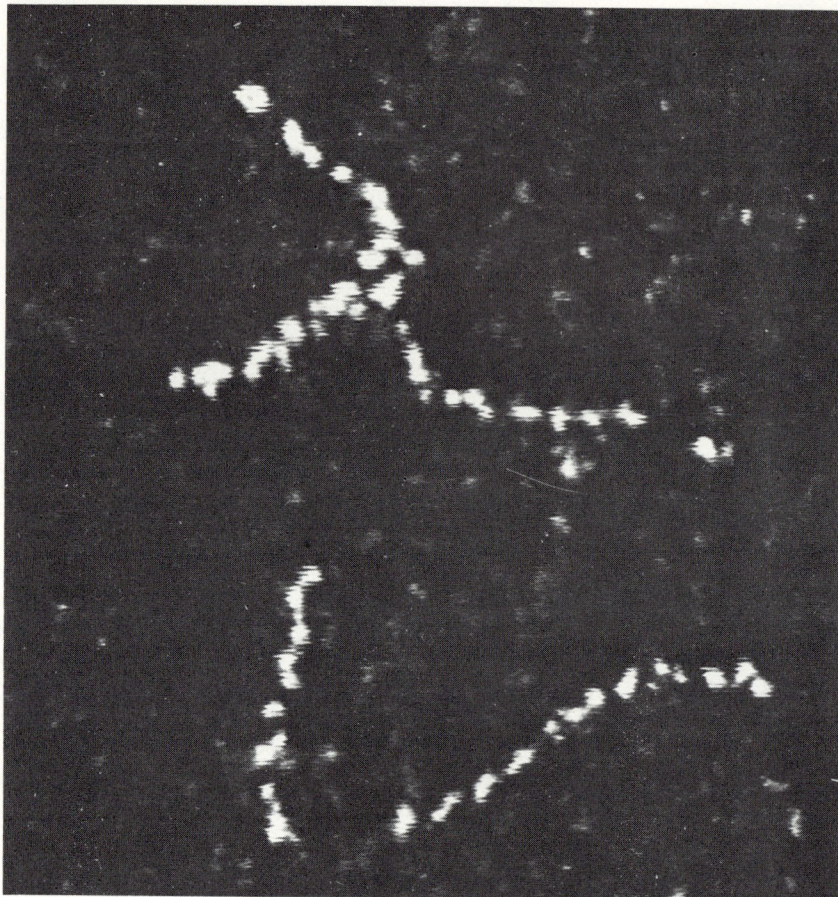

Figure 2-2 Chains of thorium atoms viewed with a high-resolution scanning electron microscope that has a resolving power of 5×10^{-8} cm (5 Å).

occurring elements, corresponding to the 92 atoms. By international scientific agreement, each element has been assigned a symbol to serve as a form of chemical shorthand for convenience in chemical formulas and equations. The abbreviations for the 20 elements known to be essential for life are shown in Appendix 1.

Substances whose molecules consist of two or more different kinds of atom in chemical combination with one another are called **compounds.** Their atoms are chemically linked in fixed and definite proportions and in a specific configuration or structural relationship. Unlike an element, a compound can be broken down by suitable chemical means to simpler substances, that is, to its constit-

uent atoms or elements. The molecules of a compound have their own individual properties, which are usually markedly different from those of their constituent atoms. Water (H_2O)[3], for example, which at room temperature is a liquid possessing its own characteristic properties (one of which is that it will not burn), can be broken down by appropriate chemical manipulation to yield hydrogen gas and oxygen gas, each possessing its own particular properties: hydrogen is highly flammable and oxygen supports combustion.

[3]Chemical notations for compounds are explained in Appendix 1.

Structure
of the Atom

How do atoms join in chemical combination to form molecules? What are the structural features of an atom that permit it to form chemical bonds with other atoms? And, especially, how is energy stored in the process?

Atoms consist of various combinations of three fundamental subatomic units: **electrons, protons,** and **neutrons.** An electron is an extremely light, cloudlike density of mass-energy, with a negative electric charge. A proton is much larger in mass (1850 times) than an electron and has a positive electric charge. Neutrons have a mass nearly equal to that of protons but, as their name implies, are electrically neutral. For each atom the number of electrons is equal to the number of protons in the nucleus. Because neutrons are neutral, and the number of positively charged protons equals that of the negatively charged electrons, each atom is also electrically neutral. Although other types of subatomic "particle" have been identified, the properties of atoms and their ability to react chemically with each other can be most simply explained on the basis of these units.

The core, or **nucleus,** of each of the 92 atoms consists of a characteristic number of protons and neutrons. (Hydrogen is an exception; it has one proton and one electron, but no neutrons.) The weight of an atom is primarily in the protons and neutrons packed into its dense core. Surrounding the nucleus is a space 10,000 times the diameter of the nucleus, in which the electrons move.

The simplest (and lightest) of all atoms is hydrogen, consisting of a positively charged nucleus of one proton, surrounded by a single electron cloud with an equal negative charge (Fig. 2-3). Its electron can be removed in a variety of ways, leaving the positively charged proton, which is called a **hydrogen ion** (designated as **H$^+$**). Hydrogen ions are responsible for the acidity of a solution and exert a marked effect on the activities of biological systems, as discussed in pp. 39–41.

The helium atom, the next lightest element,

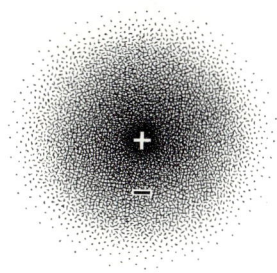

Figure 2-3   *Schematic model of a hydrogen atom, which consists of a positive proton surrounded by its negative electron cloud.*

has two protons that account for the two positive charges in its nucleus. These are balanced by two electrons surrounding the nucleus, producing a typical electrically neutral atom. The mass or weight of a helium nucleus, however, is four units instead of two, indicating that in addition to two protons two neutrons are also present. Uranium, one of the heaviest atoms known, possesses a nucleus of 92 protons and 146 neutrons, to give a weight of 238 (see p. 31 for a discussion of atomic weight).

When this concept of atomic structure was first adopted early in this century, largely through the work of the Danish physicist Niels Bohr, the electrons were viewed as small particles revolving around the nucleus like planets around the sun. This was in fact called the **planetary model** of the atom. The electrons revolving around the atomic nucleus were supposed to be held in orbit by a balance of forces: an electrostatic attraction between the negative electrons and the positive nucleus would draw the electrons in toward the center, and the rotational energy, like centrifugal force, represented a constant pull away from the center. The electron orbit was assumed to be the position at which these two forces just balanced one another (called **equilibrium**). Bohr found that the electrons were not distributed at random around the nucleus. Instead, their orbits seemed to be arranged in concentric shells at different distances from the center. This was explained by the finding that different electrons seemed to have different amounts of energy. If this were rotational energy, the electrons would fall into equilibrium at different distances from the center; the faster their rotational velocity (i.e., energy), the further from the center they would tend to orbit. Bohr calculated that the radius of the circular orbit of the hydrogen electron was .53 Å and that the velocity of the electron in orbit would be 2×10^8 cm/sec (4 million miles per hour).

We now know that this model is unacceptable for several reasons. The outward centrifugal force (called **angular momentum**) of a particle traveling at such enormous speeds in circular orbit should be readily measurable, but highly sensitive measurements have shown that the hydrogen electron has no angular momentum. Although some electrons of other atoms with more electron shells do have measurable angular momentum, the fact that some do not invalidates the Bohr model. Moreover, as we shall see in a moment, the Bohr planetary model also fails to predict the behavior of atoms when they combine to form molecules. Still, certain aspects of the model do conform to experimental facts. Electrons are arranged in concentric shells—now called **orbitals**—around the atomic nucleus. Instead of thinking of electrons as particles with discrete orbits, however, we have to think of them as vibrating clouds of mass-energy surrounding the nucleus. Atomic physicists now visualize a hydrogen electron as a hollow sphere of mass-energy vibrating in and out (like a guitar string), with a mean distance from the atomic center of .53 Å (Fig. 2-4).

The Chemical Properties of Atoms

Although the physical reasons are not obvious, atoms are constructed in such a way that there seems to be a maximum number of electrons that each orbital shell can contain at any one time. The innermost shell can contain only two electrons at most, and each

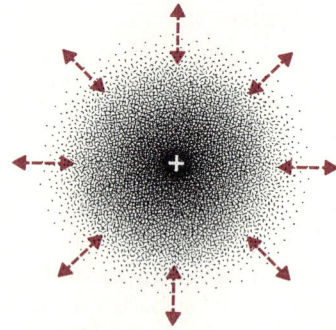

Figure 2-4   A hydrogen atom, showing the direction of vibration of the electron shell.

of the other shells is normally "full" when it contains eight electrons. The number of electron shells an atom has, therefore, depends on how many electrons it has. The number varies from one shell for hydrogen and helium to seven shells for the heavier atoms. The electron shell structure of some biologically important compounds is illustrated in Figure 2-5.

The chemical properties of atoms depend mainly on the number of electrons in the outermost energy level, or shell. If that shell is full, as in helium (which has two electrons) or neon (which has two in the inner shell and eight in the second) the element rarely undergoes chemical reactions because the electrons of the outermost shell are **stable.** This means that they rarely interact with the outer electrons of other atoms to form chemical bonds. Such atoms are designated **inert** elements (although it has recently been shown that even they can react chemically with other elements under appropriate conditions). On the other hand, an atom with only one electron in its outermost shell tends readily to give it up to another atom, leaving the next inner shell, in a stable configuration, as its outermost energy level. An atom with seven electrons in its outer shell has a strong tendency to accept one electron from another atom to produce the stable configuration of eight electrons. We speak of atoms with strong tendencies to donate or accept

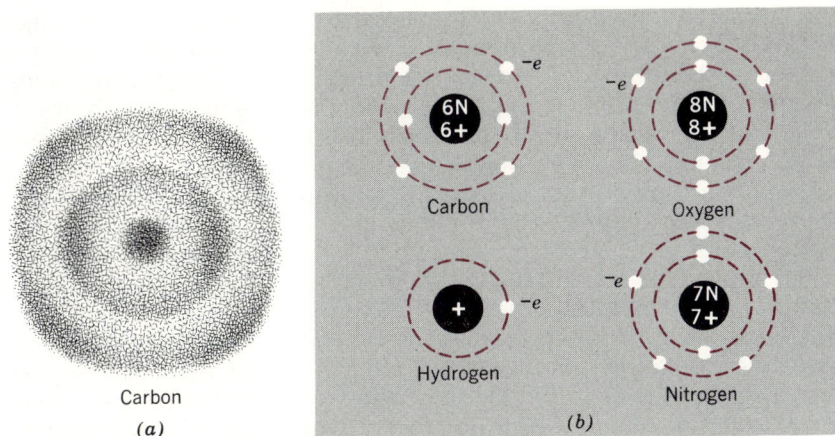

Figure 2-5 (a) Model of a carbon atom, showing the electron shells as clouds. (b) A schematic representation of four atoms important in biological systems. The inner disc represents the nucleus of the atom containing protons and neutrons. Here the electron shells are represented by circles containing e electrons. For example, carbon has six neutrons and six protons in its nucleus; its six electrons are arranged in two shells.

electrons as being highly **reactive;** it is these atoms that form **chemical bonds** (p. 32) easily and produce biologically important compounds. When the electrons of atoms interact to form chemical bonds, we say that a **chemical reaction** has occurred.

Because it is the outermost shell of electrons that mainly determines the chemical (i.e., reactive) properties of an atom, it is convenient to represent each atom by symbolizing the electrons in its outer shell as dots near the chemical symbol of the element. Fluorine and chlorine, each of which has seven electrons in its outer shell, are represented, by convention, like this:

$$:\ddot{F}\cdot \qquad :\ddot{C}l\cdot $$

Similarly, hydrogen with one electron in its outer shell, carbon with four, nitrogen with five, oxygen with six, and sodium with only one are shown as follows:

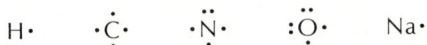

$$ H\cdot \qquad \cdot\dot{C}\cdot \qquad \cdot\ddot{N}\cdot \qquad :\ddot{O}\cdot \qquad Na\cdot $$

Atomic Weight and the Mole

The **atomic weight** of each element refers to the number of particles in an atom that contribute to its mass, that is, the total number of protons and neutrons (the electrons are so much lighter than the nuclear particles that their contribution to atomic mass is negligible). The atomic weight is written as a superscript immediately following the chemical symbol. For example, most oxygen atoms in the universe contain eight protons and eight neutrons. The atomic weight of oxygen is therefore 16, and the element is symbolized as O^{16}. Because the atomic weights of the elements are based on the relative weights of different elements to one another, one reference value or standard must be selected as the basis for comparison. Chemists and physicists have agreed to use the carbon atom (Fig. 2-5) as the arbitrary standard of reference. The atomic weight assigned to carbon is 12.000. It results in a value of about 1 (1.008) for the lightest atom, hydrogen, and approximately whole-number atomic weights for most of the other elements. Atomic weight is therefore defined as **the relative weight of any atom compared with the weight of the carbon atom, arbitrarily taken as 12.** Because carbon contains six protons and six neutrons, the weight of one proton or neutron is by definition $\frac{1}{12}$ the mass of the carbon atom.

The **molecular weight** of any element or compound is equal to the sum of the atomic weights of its constituent atoms. The unit of molecular weight is called a **dalton** after the early English physicist John Dalton, who was one of the pioneers of the atomic theory. For example, the molecular weight of methyl alcohol (wood alcohol), which is made up of one atom of carbon, four atoms of hydrogen, and one of oxygen, is the sum of the weight of these atoms: one carbon (12), one oxygen (16) and four hydrogen (one each), or 32 daltons. Although many simple compounds such as salts and sugars have molecular weights of only a few hundred, substances of great biological importance, such as proteins and nucleic acids, possess molecular weights that often range from a few thousand to as high as 10 million daltons or more.

The weight in grams of a compound equal to its molecular weight (e.g., 32 g of methyl alcohol) is called a **gram molecular** weight or **mole** of that substance. When a substance is dissolved in a liquid, its concentration is measured by its **molarity,** which is the number of moles of the compound dissolved in one liter (or 1000 milliliters) of solvent (e.g., 3.2 g of methyl alcohol, one-tenth of a gram-molecular weight, dissolved in one liter of water, would make a one-tenth molar solution, abbreviated .1M).

The
Chemical Bond

We have said that atoms are in particularly stable configuration when their outer electron shell is filled, and that there is a general tendency for atoms to form complete outer shells by reacting with other atoms (i.e., donating or accepting electrons).

When two or more atoms are bound to one another as a result of an electronic interaction, the force or attraction that holds them together is called a **chemical bond.** Each bond represents a certain amount of potential chemical energy. At one extreme the linkage may be so strong that very powerful forces are necessary to pry the atoms apart, whereas at the other the combination may be so weak that no effective bond can be said to exist. All degrees of bond strength exist between these two extremes, depending especially on how far the outermost electron shell is from the nucleus. Fluorine, for example, with seven electrons in its outer shell, binds an eighth electron more tightly (i.e., with greater energy) than does iodine, which also has seven electrons in its outer shell. This is because fluorine has only two electron shells; iodine has five. Thus the outer shell of fluorine is much closer to the nucleus than that of iodine (Fig. 2-6).

Such interaction of the outermost electrons of atoms to yield chemical bonds are, as we have said, called chemical reactions. They consist fundamentally of two types, those involving (1) loss or gain of electrons, in which electrons are transferred from one atom to another to form an **ionic bond,** and (2) a sharing of electrons between two atoms to form a **covalent bond.** Chemical reactions occur between atoms of elements or of compounds to yield new ionic or covalent bonds and therefore new molecules. The underlying principle is always that the tendency of atoms to form chemical bonds reflects the tendency of their electron clouds to rearrange themselves in the most stable patterns possible, those with an outermost shell configuration of eight electrons.

Ionic Bond

The transfer of electrons to form an ionic bond is illustrated by the reaction between sodium and chlorine. The element sodium (Na) is a highly reactive, silvery, soft metal. Its outermost energy shell has only a single

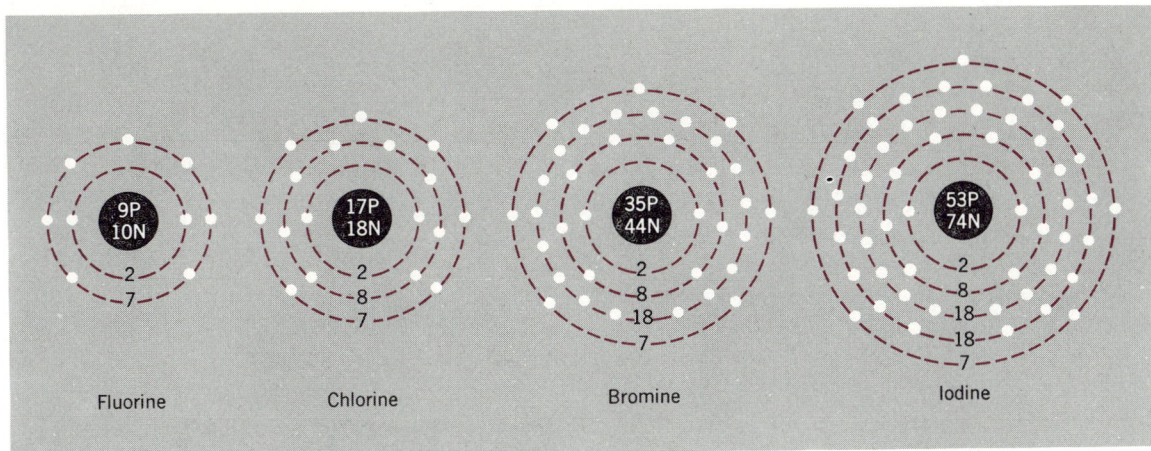

Figure 2-6   Atomic structure of fluorine, chlorine, bromine and iodine. Although these atoms differ in number of electron shells, they have similar chemical properties because all have seven electrons in their outermost shell. They are all strong electron acceptors, but fluorine forms the strongest bond because its outermost shell is closest to its nucleus.

electron. One way sodium might gain a complete outer shell would be to acquire seven more electrons from other atoms. But the sodium would then have an enormous excess of negative charge, and, since like charges repel each other, the electrons would tend to push each other away. In fact, as we said earlier, sodium tends readily to give up its single outer electron. That is, it is a good **electron donor.** Chlorine (Cl_2) is a greenish, poisonous gas, with seven electrons in its outermost shell. Much more energy would be required for another atom to withdraw an electron from the outer shell of chlorine than is needed for chlorine to gain an electron; it is therefore an **electron acceptor.** When sodium and chlorine react with one another, the single electron in the outermost shell of the sodium is transferred to the outer shell of the chlorine. By losing an electron the sodium is left with one more proton than it has electrons, and it therefore has a positive charge; but it has acquired a relatively stable configuration of eight electrons in what is now its outermost shell. The chlorine atom that gains the electron has now become negatively charged (18 electrons, 17 pro-

tons) and has acquired a similar stable configuration of outermost electrons (Fig. 2-7, next page).

Atoms that form ionic bonds are held together in much the same way as two magnets; the bonding strength is simply from the electrostatic attraction between the oppositely charged ions.

Atoms or molecules that acquire such an electric charge are called **ions,** and are symbolized by the appropriate chemical symbol followed by a superscript indicating the charge. In this case the sodium has been converted to a positive ion, Na^+, and the chlorine into a negative ion, Cl^-. Since these ions are of opposite electrical charge, they attract one another to form a substance called sodium chloride, better known as table salt. The reaction is conveniently represented as follows: $Na + Cl \longrightarrow Na^+ \ Cl^-$ (see Fig. 2-7). (The arrow should be read as "yields" or "produces.") This reaction states in words that two atoms of sodium brought together with a molecule of chlorine yields two molecules of NaCl (see Appendix 1).

The electrostatic bond holding the sodium and chloride ions together is an ionic bond.

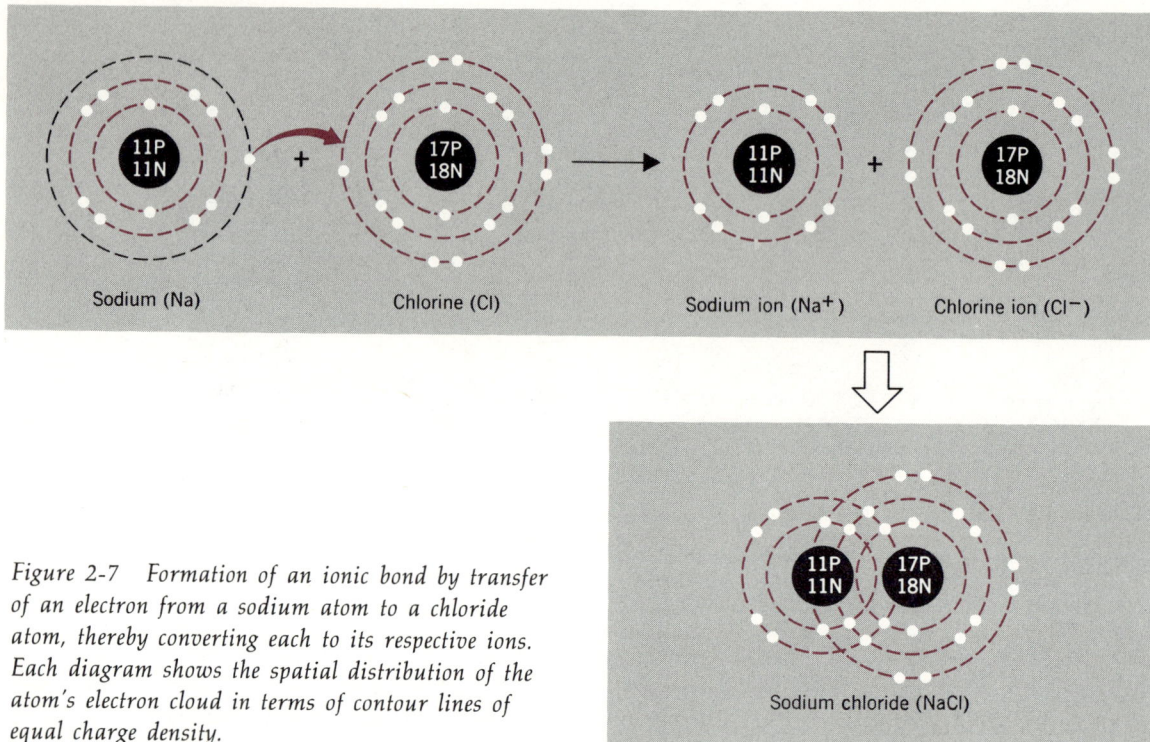

Figure 2-7   Formation of an ionic bond by transfer of an electron from a sodium atom to a chloride atom, thereby converting each to its respective ions. Each diagram shows the spatial distribution of the atom's electron cloud in terms of contour lines of equal charge density.

Again, ionic bonds involve the complete transference of an electron from one atom to another and are formed mainly between atoms that must gain or lose only one or two electrons in order to acquire a complete outer shell.

Covalent Bond

Most chemical bonds, especially those involved in biologically significant molecules, are not ionic; that is, electrons are not transferred completely from one atom to another to form ions. Instead, atoms are bound by sharing electrons between them to form pairs. Bonds based on shared electrons are called covalent bonds. For example, the two atoms of hydrogen in a molecule of hydrogen gas are covalently linked to one another by the mutual sharing of an electron to form a pair (Fig. 2-8). The electron shell of each hydrogen atom is therefore filled with its maximal (and most stable) electron number of two. The situation is similar for the oxygen molecule (O_2) except that the two atoms share two pairs of electrons to give a stable octet of electrons in the outermost shell around each of the two oxygen atoms. A water molecule (H_2O) consists of two atoms of hydrogen each covalently linked to an atom of oxygen (Fig. 2-8), resulting in a stable arrangement of outer electrons about each of the atoms.

The linkage of carbon to itself and to other atoms such as hydrogen, oxygen, and nitrogen (as in proteins, nucleic acids, carbohydrates, and fats) is of a covalent type. The carbon atom has a total of six electrons—two in its inner shell and four in its outer shell (Fig. 2-5). It most easily attains a stable octet of electrons by sharing electrons with other atoms. For example, methane gas (Fig. 2-8), which is believed to be one of the first carbon compounds formed in the early history of the earth, consists of molecules in which each carbon atom is covalently bound to four hydrogen atoms. Thus the carbon atom in each methane molecule has acquired a stable arrangement of eight electrons in its outer shell, and each hydrogen atom now shares its maximum of two electrons.

In a covalent linkage involving a sharing of a single pair of electrons between two atoms, the linkage is called a **single bond.** It is usually designated by a short, straight line or a pair of dots connecting the symbols of two atoms with one another. If two or three pairs of electrons are shared between two atoms,

then the linkage is called a **double bond** or **triple bond,** respectively, and is designated by double or triple straight lines:

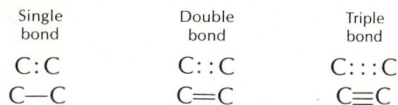

Single bond	Double bond	Triple bond
$C:C$	$C::C$	$C:::C$
$C{-}C$	$C{=}C$	$C{\equiv}C$

In oxygen gas, for example, each molecule consists of two atoms of oxygen joined by a double bond (they share two pairs of electrons) to form the outer stable electron octet. In carbon dioxide, abbreviated as CO_2 or $O{=}C{=}O$, which is an important compound in many biological reactions, an atom of carbon shares two pairs of electrons with each oxygen atom, to yield eight outer electrons for each atom.

Bond Energies

We noted earlier that chemical bonds hold atoms together with varying amounts of strength. What is the source of this bonding force? Where does the energy of the force reside? It should be apparent that all chemical bonds, both ionic and covalent, are the result of electrostatic forces, that is, of the attraction and repulsion between negatively charged electrons and positively charged atomic nuclei. The force required to drive the atoms apart is a measure of the strength of the bond, and is referred to as the **bond energy.** Covalent bond energies are somewhat more complex. Consider two hydrogen atoms covalently bonded to form a hydrogen molecule (H_2). On the basis of the Bohr planetary model, as the two atoms approach (Fig. 2-9*a*, next page), the repulsive forces between their electrons in orbit should increase and prevent them from joining. If, however, a situation—such as that drawn in Figure 2-9*b*—could exist in which the two electrons occupied positions simultaneously between the two positively charged nuclei, it would result instead in an attractive force between the nuclei. Each nucleus would be attracted to the double negative electron charge, and this attraction would outweigh the repulsion between the nuclei, which are further apart.

Figure 2-8   Covalent bonds between the two atoms of hydrogen in a hydrogen molecule, between the carbon and hydrogens of methane, and between the hydrogens and oxygen of water.

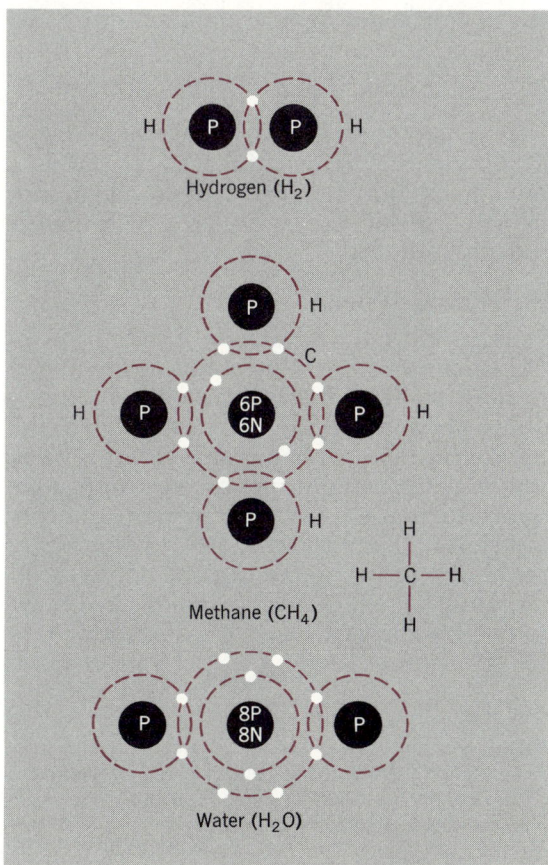

Hydrogen (H_2)

Methane (CH_4)

Water (H_2O)

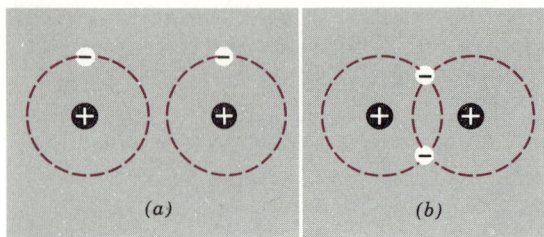

Figure 2-9   Hypothetical arrangement of pairs of hydrogen atoms.

As long as electrons are considered as negatively charged particles, the arrangement depicted in Figure 2-9*b* is energetically unstable. However, this difficulty disappears when we deal with electrons as wave phenomena. Recall the comparison of an electron with a vibrating guitar string. Each vibration of the string consists of a movement first to the right of center, then to left. A second string vibrating at the same frequency can be **in phase** with the first or **out of phase.** When the two orbitals are in phase the two electron clouds are capable of overlapping and producing a relatively high electron density and an attractive force between the two nuclei. When this happens the two nuclei are drawn into close proximity and both electron orbitals come to surround both nuclei. In the process of this fusion some of the orbital energy of the two electron clouds is lost. Another way of saying this is that the total energy of the molecule is less than the total energy of the two separate atoms (Fig. 2-10). Thus the molecule represents an energy "well." To separate the two atoms again, enough energy has to be supplied to lift the system out of that well; and the amount of energy that must be supplied represents the bond energy, just as with an ionic bond.

From the potential-energy graph at the bottom of Figure 2-10a, we can estimate that it would take 103,000 calories[4] of energy to split one mole of hydrogen molecules into their respective atoms. We can represent this fact by a simple thermochemical equation:

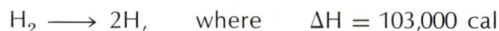

$$H_2 \longrightarrow 2H, \quad \text{where} \quad \Delta H = 103,000 \text{ cal}$$

ΔH (reads as "delta" H) stands for the difference (increase) in heat content and 103,000 cal measures the strength of the H—H bond.

To summarize, then, when a bond is formed, there is always a release of energy; and the stronger the bond, the greater the release of energy. To break a bond requires an input of the same amount of energy as was released in its formation. Thus any reaction in which weak bonds can be broken and strong ones formed in their place will behave as an energy source, since more energy will be released by the formation of the strong bonds than had to be supplied to break the weak bonds. This concept of the exchange of weak chemical bonds for stronger ones with the net liberation of usable energy is the basis for explaining how living systems obtain energy from fuel foods. We'll come back to this point when we deal with chemical thermodynamics and cellular metabolism.

Special Bonds

There are two special bonds of unusual importance in biological reactions: one is a low-energy bond called a **hydrogen bond;** the other is deceptively called a **high-energy phosphate bond.**

The Hydrogen Bond

Atoms that are covalently bonded into a molecular structure are generally incapable of forming any additional bonds. Hydrogen atoms in a molecule, however, have the unique property of being able to form weak bonds with certain highly electronegative atoms, such as O or N, that are parts of other molecules (Fig. 2-11, p. 38). The energy needed to break most covalent bonds is on the order of 50–100 kcal/mole;[5] hydrogen bonds, by contrast, require from about 1 to

---

[4]A calorie is a unit of heat equal to the amount of energy required to raise the temperature of water from 15.5 to 16.5°C. When a ΔH value is given in calories as in the example here (103,000 cal), it is always understood to be calories/mole (see Appendix 1).

[5]1 kcal = 1000 calories.

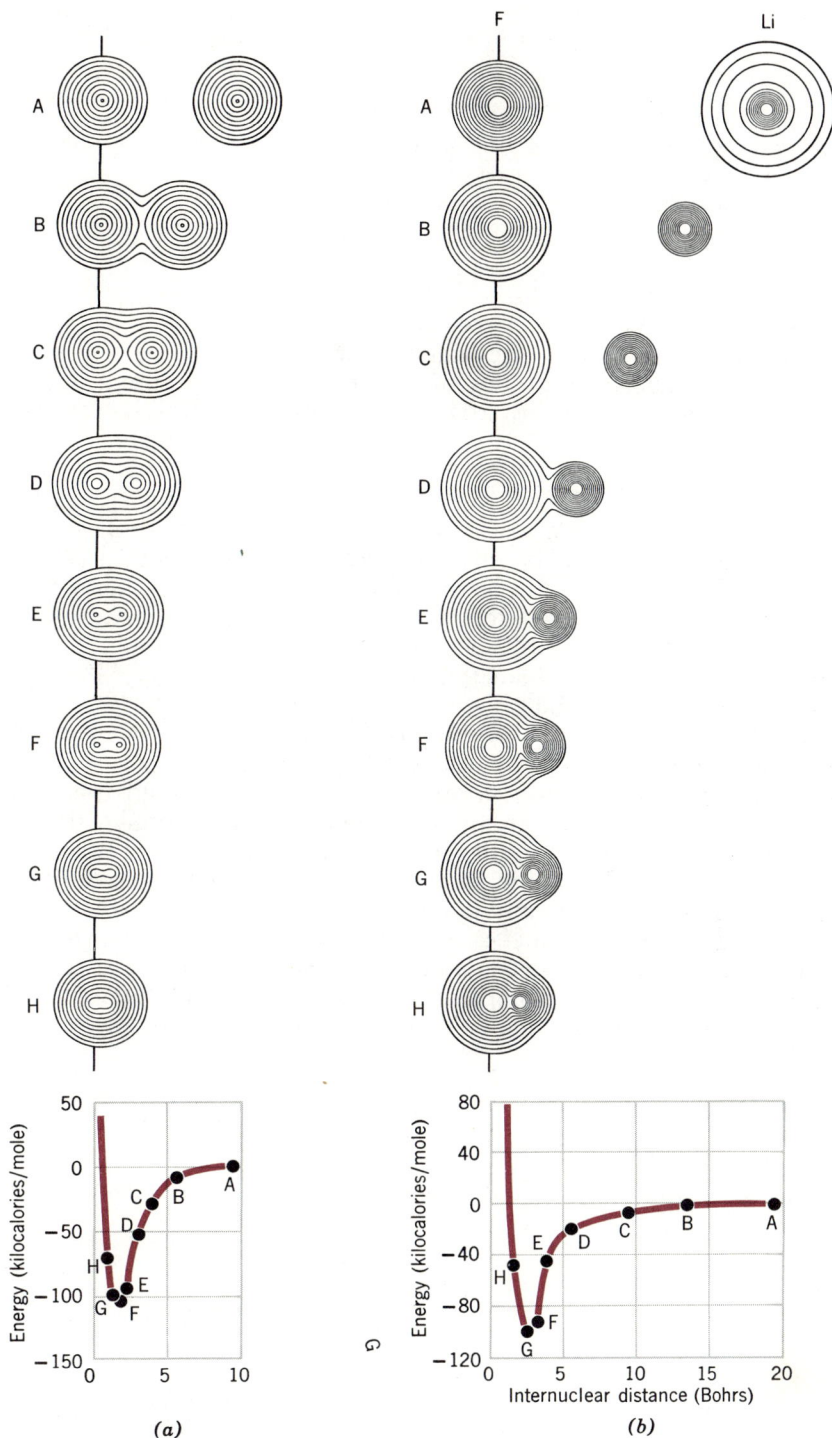

Figure 2-10 Formation of **diatomic** (*two-atom*) molecules can be visualized in **electron-density diagrams** of real atoms, plotted by high-speed computers. The two electron-density diagrams here show pairs of atoms at successively closer internuclear distances. The unit of internuclear distance is the Bohr (1 Bohr = .53 Å). (*a*) Two isolated hydrogen atoms come together to form the covalently bound H_2 molecule. (*b*) A neutral lithium atom and a neutral fluorine atom join to form the ionic bond of the lithium fluoride (LiF) molecule. Below each sequence is a corresponding graph giving the changes that take place in the total energy of the system during successive stages of molecular formation. The two atoms in each system reach a stable equilibrium at that point at which the system has minimum potential energy. This minimum-energy configuration is reached at stage G. In (*a*) this represents an internuclear distance of about 1.4 Bohrs (.74 Å); in (*b*) about 3.2 Bohrs. The loss of the outer rings from the lithium atom when it is still 13.9 Bohrs (7.5 Å) from fluorine (shown at B) represents the transfer of an electron which converts each atom to its ionic state.

$$R\text{---}N\text{---}H\bullet\bullet\bullet O\text{---}R$$
$$R\text{---}N\bullet\bullet\bullet H\text{---}O\text{---}R$$
$$R\text{---}O\bullet\bullet\bullet H\text{---}O\text{---}R$$

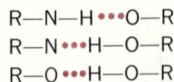

Figure 2-11 Diagrammatic illustration of a hydrogen bond formed between a nitrogen molecule and an oxygen molecule, or between two oxygens. R indicates the remainder of each molecule, and the dots represent hydrogen bonds.

10 kcal/mole. As we shall see, these weak bonds are of enormous importance in relation to the special properties of water and giant biological molecules such as proteins and nucleic acids.

The High-Energy Phosphate Bond

We said earlier that a reaction in which a weak bond can be broken and a strong one formed in its place is an energy source because more energy is released in the formation of the strong bond than is used to break the weak one. We can characterize any reaction involving the formation and breakage of several bonds as energy absorbing or energy producing merely by adding up the total energy liberated by the bonds formed and subtracting it from the total energy absorbed by the bonds broken. If the result is positive (i.e., more energy absorbed) the total reaction is energy absorbing; if the difference is negative the reaction has produced a net energy source. For example, an organic molecule called adenosine can combine with phosphoric acid to form adenosine phosphate

plus a molecule of water, according to the reaction:

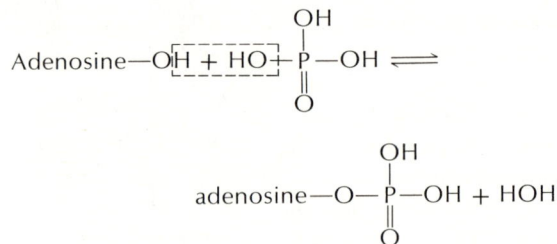

$$\text{Adenosine---OH} + \text{HO---}\underset{\underset{\displaystyle O}{\|}}{\overset{\overset{\displaystyle OH}{|}}{P}}\text{---OH} \rightleftharpoons$$

$$\text{adenosine---O---}\underset{\underset{\displaystyle O}{\|}}{\overset{\overset{\displaystyle OH}{|}}{P}}\text{---OH} + \text{HOH}$$

In the process the bonds formed are slightly weaker than those broken, and the total reaction requires an input of about 2 kcal to proceed. This is characteristic of a wide variety of similar phosphate reactions. The resulting compound, adenosine-phosphate can combine with a second molecule of phosphoric acid in a similar reaction, to form adenosine-diphosphate[6] (known as ADP):

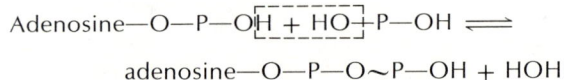

$$\text{Adenosine---O---P---OH} + \text{HO---P---OH} \rightleftharpoons$$

$$\text{adenosine---O---P---O}\sim\text{P---OH} + \text{HOH}$$

Although this reaction looks similar to the first, unlike the typical phosphate reaction, the new phosphate bond formed is substantially weaker than that broken, and the net reaction absorbs about 8 kcal. ADP can combine with yet another molecule of phosphoric acid to form adenosine-triphosphate (ATP), but again 8 kcal of energy are required for the reaction to proceed:

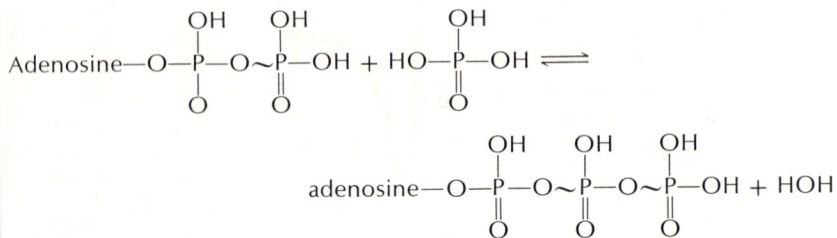

$$\text{Adenosine---O---}\underset{\underset{\displaystyle O}{|}}{\overset{\overset{\displaystyle OH}{|}}{P}}\text{---O}\sim\underset{\underset{\displaystyle O}{\|}}{\overset{\overset{\displaystyle OH}{|}}{P}}\text{---OH} + \text{HO---}\underset{\underset{\displaystyle O}{\|}}{\overset{\overset{\displaystyle OH}{|}}{P}}\text{---OH} \rightleftharpoons$$

$$\text{adenosine---O---}\underset{\underset{\displaystyle O}{\|}}{\overset{\overset{\displaystyle OH}{|}}{P}}\text{---O}\sim\underset{\underset{\displaystyle O}{\|}}{\overset{\overset{\displaystyle OH}{|}}{P}}\text{---O}\sim\underset{\underset{\displaystyle O}{\|}}{\overset{\overset{\displaystyle OH}{|}}{P}}\text{---OH} + \text{HOH}$$

[6]The prefix mono- means one, di- means two, and tri- means three molecules of the specified substance.

As indicated by the double arrows, all of these reactions can occur in both directions. ATP can split to form ADP plus a molecule of phosphoric acid, and ADP can be broken into adenosine-monophosphate (AMP) plus a molecule of phosphoric acid. In both cases the reaction proceeds with the liberation of 8 kcal of energy. As expected, however, when AMP breaks up into adenosine plus phosphoric acid, only 2 kcal are liberated. Because the splitting off of the end phosphate of either ATP or ADP liberates more energy than usual, these two molecules have come to be known as high-energy compounds. The bond between oxygen and phosphorus, as seen in the equations above, is drawn with a wavy line (O∼P) to indicate its special character, and is referred to as a **high-energy phosphate bond.** Although this is clearly a misnomer, the term is entrenched in the biochemical literature. We'll see in Chapter 4 that ATP is a crucial compound in cellular metabolism, and we refer to it many times. It is therefore useful to remember that the extra energy liberated from the splitting of this high-energy phosphate compound really results from the fact that the terminal O∼P bond is weaker than most and therefore absorbs less of the total reaction energy when it is broken.

Acids, Bases, and Ionization

A molecule is generally defined as an aggregate of atoms, bound together strongly enough to be considered as an entity, that is electrically neutral. Compounds like sodium chloride (NaCl), which are bound mainly by ionic bonds, have a strong tendency to dissociate into separate ions when dissolved in water. The result of this process, called **ionization,** is that sodium chloride does not exist in solution as molecules of NaCl, but instead forms two distinct entities, a Na^+ ion and a Cl^- ion. Even when water is removed and the Na^+ and Cl^- ions are bound by their electrostatic attractive forces, they do not reform into separate molecules of one sodium atom and one chlorine atom. Instead, many atoms of sodium and chlorine

bind together in a regular, three-dimensional arrangement to form a large salt crystal.

Substances that are ionized in water are especially important in biological systems. We shall see later (Chapters 15, 17) for example, that the signals carried by nerves and the stimuli that cause heart and muscle cells to contract result from electrical currents, and that these currents represent the movements of charged particles, ions, across the membranes of the nerve and muscle cells.

There are two classes of ionic compound to which we shall refer frequently, and that therefore deserve special mention. These are **acids** and **bases.** Acids are substances that yield protons or hydrogen ions, H^+, in solution. Bases, or **alkalis,** are substances that combine with H^+ or liberate OH^-, called **hydroxide** or **hydroxyl ions,** in solution. Acids have a typical sour taste and bases have a bitter taste.

Acids and bases are generally classified as **strong** or **weak.** Most of the molecules of a strong acid dissociate, producing relatively high concentrations of hydrogen ions and negative ions (other than OH^-); weak acids dissociate to a lesser extent. For example, hydrochloric acid, HCl, is a strong acid because when it is dissolved in water, a large proportion of its molecules are dissociated into H^+ and Cl^-; only a small fraction remain in the molecular form as HCl. Acetic acid, CH_3COOH, is a weak acid, which means that only a relatively small percentage of its molecules dissociate in aqueous solution into hydrogen ions and negative acetate ions ($CH_3COOH \rightleftharpoons CH_3COO^- + H^+$). Most of the molecules remain undissociated in the molecular form. Strong bases produce a relatively high concentration of hydroxyl ions (OH^-) and positive ions (other than H^+ ions); weak bases dissociate to a lesser extent. Sodium hydroxide, NaOH, is an example of a strong base (NaOH $\rightleftharpoons Na^+ + OH^-$), and ammonium hydroxide, NH_4OH, is a typically weak base ($NH_4OH \rightleftharpoons NH_4^+ + OH^-$).

When acidic and basic solutions are mixed together, they neutralize one another, forming water and ionic compounds known as **salts.** H^+ and OH^- have a great affinity for one another, combining very rapidly to produce water ($H^+ + OH^- \rightleftharpoons H_2O$). Salts, the other principal product of the reaction between an acid and a base, consist of the negative ion of the acid and the positive ion of the base. For example, in the reaction between the strong acid HCl and the strong base NaOH, the hydrogen ions and hydroxide ions combine to form water, and the Na^+ and Cl^- remain. If the water molecules are subsequently removed by evaporation, the remaining ions will aggregate to form a salt, sodium chloride.

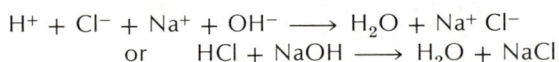

$$H^+ + Cl^- + Na^+ + OH^- \longrightarrow H_2O + Na^+ Cl^-$$
$$\text{or} \quad HCl + NaOH \longrightarrow H_2O + NaCl$$

Salts therefore represent the neutralization products of the reaction of a base with an acid.

What is pH?

Pure water consists almost entirely of undissociated H_2O molecules; only a tiny fraction of the water molecules are dissociated into H^+ and OH^- ions. In a liter of water, for example, only .0000001 mole of H^+ can be measured. We see in Appendix 1 that such numbers are conveniently expressed as exponents, or powers, of 10. For fractions the exponent is negative and is equal to the number of places 1 is to the right of the decimal point; thus the amount of H^+ in pure water is 10^{-7} moles.

Some years ago the Danish biochemist Sören Sörensen devised a **pH scale** for defining degrees of acidity (Fig. 2-12). He used the letter p to stand for power and H for the hydrogen ion. On the pH scale, pure water, with 10^{-7} moles of H^+ is said to be pH7; at pH7 the concentration of H^+ and OH^- are exactly the same, because in the dissociation of a molecule of H_2O one ion of each is produced. Pure water or any other solution with a pH value of 7 is said to be **neutral**. When substances that bind OH^- but liberate H^+ are added to water, the concentration of H^+ increases. Such substances are called acids, and the solution is referred to as **acidic**. Acidic solutions have pH values below 7, meaning a greater concentration of H^+. For example, pH3 means 10^{-3} moles of H^+ per liter of water, or .001 moles, which is of course a much greater concentration than .0000001 moles. Solutions with pH values above 7 have a greater concentration of OH^- than H^+ and are said to be **alkaline,** or **basic**.

Buffers

The pH of water undergoes very large changes with additions of small quantities of acid or base. For instance, the addition of a single drop of strong acid to a liter of water may decrease its pH by as much as 3 or 4 pH units. Yet when certain substances known as **buffers,** which can combine with added H^+ ions or OH^- ions, are also present, the addition of relatively large amounts of strong acid or base will cause much smaller changes in pH. Buffered solutions therefore resist changes in H^+ ion concentration. Blood in

Figure 2-12   The pH scale and hydrogen ion concentration.

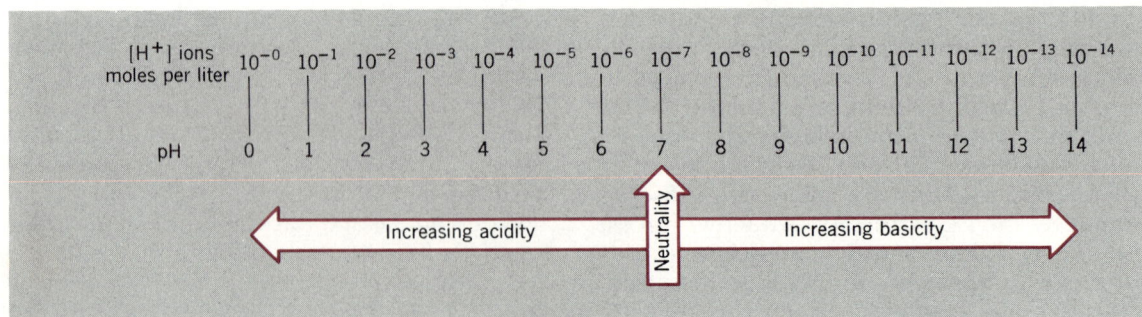

the human body, for instance, has a constant pH of 7.4, largely as a result of the buffering action of its constituent proteins. Fluctuations in blood pH of only a few tenths of a pH unit, as occurs in certain diseases, can be lethal.

Isotopes

Each element has a characteristic number of protons, indicated by its atomic number (Appendix 1). That is, every atom of hydrogen contains one proton, every atom of carbon has six, sodium 11, and so on. The number of neutrons in an atom, however, is not always the same, and therefore neither is the atomic weight. For example, most oxygen atoms, as we said earlier, contain eight protons and eight neutrons, and therefore have an atomic weight of 16. However, some oxygen atoms have nine neutrons and atomic weight of 17, and some have 10 neutrons, and therefore an atomic weight of 18. Atoms of an element that have different numbers of neutrons (and therefore have different atomic weights) are called **isotopes** of that element.

Thus O^{16}, O^{17}, and O^{18} are isotopes of oxygen and C^{12} and C^{13} are isotopes of carbon (Fig. 2-13). However, all the isotopes of each element behave the same chemically; that is, the number of neutrons in the atomic nucleus does not affect the chemical properties of an element. Any sample of oxygen we obtain from nature will contain all of the isotopes in proportion to their relative abundance on Earth: 99.76 percent of the atoms will be O^{16}; .039 percent will be O^{17}; and .20 percent will be O^{18}. The three other known isotopes of oxygen (O^{14}, O^{15}, and O^{19}) will amount to less than .001 percent.

Although the number of neutrons in an atom does not affect its chemical properties, it does influence the physical properties of the atomic nucleus. Neutrons in some unknown way bind the protons together and make the nucleus stable. With certain neutron-proton ratios, however, atoms are unstable and their nuclei disintegrate spontaneously, emitting neutrons, protons, or energy waves in the process, until they are transformed to another atom with a more stable configuration. Thus some isotopes are

Figure 2-13 Isotopes of carbon and of oxygen.

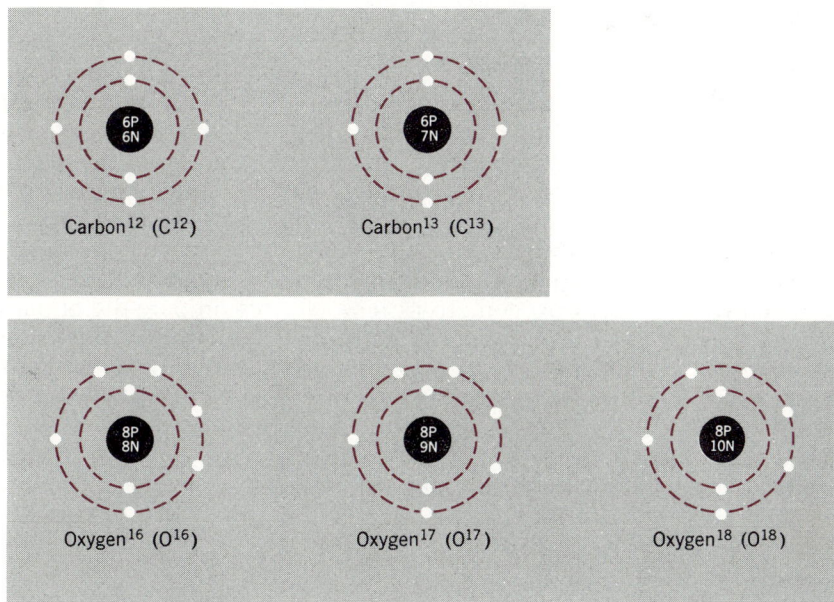

Carbon^{12} (C^{12})   Carbon^{13} (C^{13})

Oxygen^{16} (O^{16})   Oxygen^{17} (O^{17})   Oxygen^{18} (O^{18})

classified as **stable** isotopes; those that undergo spontaneous disintegration or decay are known as **radioactive** isotopes.

The process of radioactive decay occurs, at a rate unique for each isotope, by the emission of one or more characteristic kinds of energy wave or particle. These are known as **alpha, beta,** and **gamma** rays (or particles). We shall see later that these high-energy radiations can have profound effects on living material; they cause cancer and induce genetic mutations (Chapters 8–10). We shall also note, paradoxically, how such radioactive isotopes can be used by scientists as research tools of tremendous value (Chapter 7).

Chemical Reactions

Types of Reaction

Chemical reactions (i.e., the process of formation of chemical bonds) can be classified into two broad types: **oxidation-reduction** reactions, in which there is an electron transfer from one atom to another; and **recombination,** or **substitution,** reactions, in which there is no electron transfer.

Oxidation-Reduction Reactions

When a substance loses one or more electrons we refer to it as being **oxidized;** a molecule that gains electrons is **reduced.** The term oxidation was originally used years ago to designate reactions involving the formation of chemical bonds with oxygen; a substance is oxidized when it gains oxygen atoms and reduced when it loses oxygen atoms. For example, when charcoal (carbon) burns in air it combines chemically with oxygen (i.e., it is oxidized) to form the gas carbon dioxide. $C + O_2 \longrightarrow CO_2 + energy$. Since the $C=O$ bond is stronger than $O—O$, the amount of energy released in the formation of $O=C=O$ is much greater than that absorbed by breaking the $O=O$ (O_2), and the reaction is accompanied by a net liberation of energy in the form of light and heat (i.e., a flame).

It was later discovered that carbon and other molecules can combine with chlorine, sulfur, and certain other elements to form similar bonds with the release of similar amounts of energy. Clearly, it was not the combination of carbon with oxygen that was significant in the reaction, but something much more fundamental. Recall that the oxygen atom, with six electrons in its outer shell, has a strong tendency to pick up two more electrons; it is a strong electron acceptor. When it was realized that chlorine and sulfur share this property, the generalization could be deduced that any electron acceptor behaves as an **oxidizing agent** in that it causes other atoms to lose electrons while it itself is reduced. Any strong electron donor is a **reducing agent** because it causes other atoms to gain electrons, while it itself is oxidized.

Oxygen forms covalent bonds in almost all of its reactions; that is, it shares electrons with other atoms. Why are we now referring to oxygen as having gained electrons at the expense of other atoms? The answer is that although covalent bonds do involve shared electrons, the sharing need not be equal between the atoms involved. For example, hydrogen and oxygen react with one another to form water ($2H_2 + O_2 \rightleftharpoons 2H_2O$), in which each of the hydrogen atoms now shares a pair of electrons with oxygen. However, the pair of electrons is not distributed equally in the covalent bond between hydrogen and oxygen (in contrast to the equal sharing of the electron pair between the two hydrogen atoms of a hydrogen gas molecule, H_2). Instead the oxygen atom, because of strong electronegativity, tends to draw the shared pair of electrons more closely to itself. Compare the equal sharing of electrons (designated as dots) between the two nuclei of a hydrogen gas molecule, and the unequal sharing of electrons between the hydrogen and oxygen of water:

$$H : H \qquad H : O : H$$

By this unequal sharing of electrons in the covalent bond between the hydrogens and oxygen of the water molecule, the two hydrogen atoms have in effect lost their electrons to oxygen. The formation of H_2O from H_2 and O_2 can therefore be regarded as an oxidation-reduction reaction. The strong tendency of oxygen and certain other atoms such as chlorine and sulfur to monopolize the electron pairs in their covalent bonds with most other atoms accounts for the general application of the term "oxidation" to reactions dealing with the loss or unequal sharing of electrons, even though oxygen may not be involved.

Oxidation-reduction reactions may also involve ions as reactants and products. For example, iron occurs in two common ionic states, the **ferrous** ion (Fe^{2+}) and the **ferric** ion (Fe^{3+}). The Fe^{2+} ion differs from elemental iron (Fe) in having already lost two electrons, thus accounting for the net positive charge of two: Fe^{3+} is the result of a loss of three electrons. As part of a large molecule called heme (p. 72 and pp. 92–93), iron is bound to a particular protein to constitute **cytochrome.** In this important biological substance, the **reversible oxidation** of Fe^{2+} to Fe^{3+} plays an important role in the respiration and metabolism of living cells.

Recombination or Substitution Reactions

This second broad group of chemical reactions, in contrast to oxidation-reduction reactions, usually consists of a substitution, transfer, or recombination of atoms or groups of atoms between molecules, resulting in neither a net gain nor a net loss of electrons. This is illustrated by the reaction between the hypothetical molecules AB and CD:

$$A:B + C:D \longrightarrow A:C + B:D$$

In one sense the reaction is a splitting of molecule AB by the molecule CD, which is also split in the process. If CD represents a molecule of water, written as HOH, the reaction would be

$$AB + HOH \longrightarrow AOH + BH$$

Hydrolysis Reactions

When water is involved, as in the last equation, the reaction is spoken of as a **hydrolysis reaction**—the splitting of com-

pounds, by the introduction of water molecules, to form other compounds. For instance, the ionic compound **ferric chloride,** $FeCl_3$, is **hydrolyzed,** when dissolved in water, to form a brownish precipitate $Fe(OH)_3$, called **ferric hydroxide,** and HCl, or **hydrochloric acid.** In typical chemical shorthand the reaction is indicated as follows:

$$FeCl_3 + 3H_2O \longrightarrow Fe(OH)_3 + 3HCl$$

Hydrolysis reactions are also of great significance in biological processes. Digestion, for example, is essentially the result of hydrolysis—the splitting of large molecules such as proteins, starches, and fats into smaller molecules by water. We shall see in Chapter 4 that another transfer reaction, the transfer of phosphate groups, is one of the primary energy-storing events in cellular metabolism.

Collision Theory of Chemical Reactions

For chemical reactions to take place, the participating atoms or molecules must first collide with one another with sufficient force that their electrons can interact. Because in theory any atom or molecule can have some interaction with almost any other, the most important characteristic of chemical reactions is their **rate;** two molecules that react at a rate of one molecular interaction per year, or per hour, are for all practical purposes **nonreactive.** The factors that influence the rate of a chemical reaction include (1) the concentration of reacting substances; (2) the nature of the reacting substances and their suitability, from an energy viewpoint, to react with one another; (3) temperature; and (4) the presence of catalysts.

High concentrations mean that many molecules occupy a given volume, so that chances are good that they will collide with one another; the more collisions there are between molecules capable of reacting with each other, the higher the rate of chemical reactions will be. However, not every collision between molecules results in a chemical reaction. A second major requirement must also be fulfilled: the colliding molecules must be in an appropriate **activated** condition or energy state to undergo reaction.

We have said that two atoms bonded into a molecule have a lower total energy than when they are separate, so the bond energy can be viewed as an energy well (Fig. 2-10). Before a molecule can react with another, enough energy must be supplied to bring its atoms to the top of that energy well. This extra energy is called **activation energy,** the amount of energy over the average energy of a given quantity of molecules that is necessary for colliding molecules to undergo chemical reaction. Those with less than the necessary activation energy will simply bounce apart and go their separate ways.

To visualize the relationship of the chemical reaction to the activation energy of the reacting molecules, compare it to an energy barrier, as shown in Fig. 2-14a. For two colliding molecules, A and B, to react with one another they must have enough activation energy to reach the top of the barrier. At the moment of collision the pair is called an **activated complex** (AB), and, for an instant, it is at the very top of the energy barrier. It soon rearranges to yield the new molecules C and D. The situation is analogous to a sled on a hillside (Fig. 2-14b); if it is to slide down the opposite side of the hill, it must first receive a strong enough push (activation energy) to get it to the top of the hill (where it is equivalent to an activated complex). It can then readily slide down the opposite side. Similarly, many chemical reactions can be initiated by providing sufficient energy to get them over the barrier.

A rise in temperature, which also requires an input of energy, speeds up the rate of a chemical reaction in at least two ways. First, by increasing the velocity or kinetic energy of molecules, a higher temperature increases the frequency of collisions. Second, and most important, with increased velocity a greater proportion of the molecules have sufficiently high activation energy to react with one another when they collide.

Many chemical reactions can be remarkably accelerated by the addition of a substance that remains essentially unchanged at the end of the reaction. Such substances, known as

Figure 2-14   *Relationship of a chemical reaction to the activation energy of the reacting molecules A and B, formation of an intermediary activated complex (AB), and the energy differences between the reactants (A and B), the activated complex (AB), and the products (C and D).*

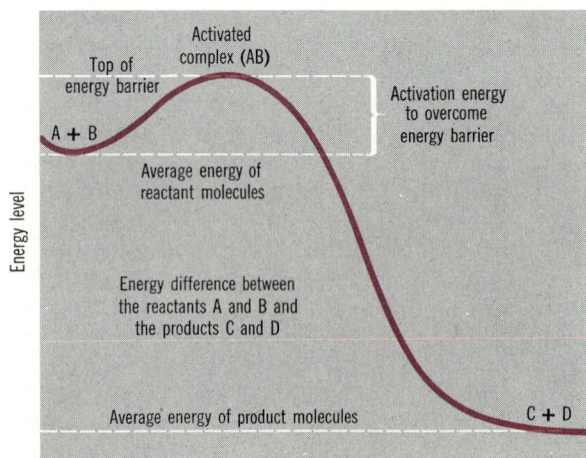

(a)

(b)

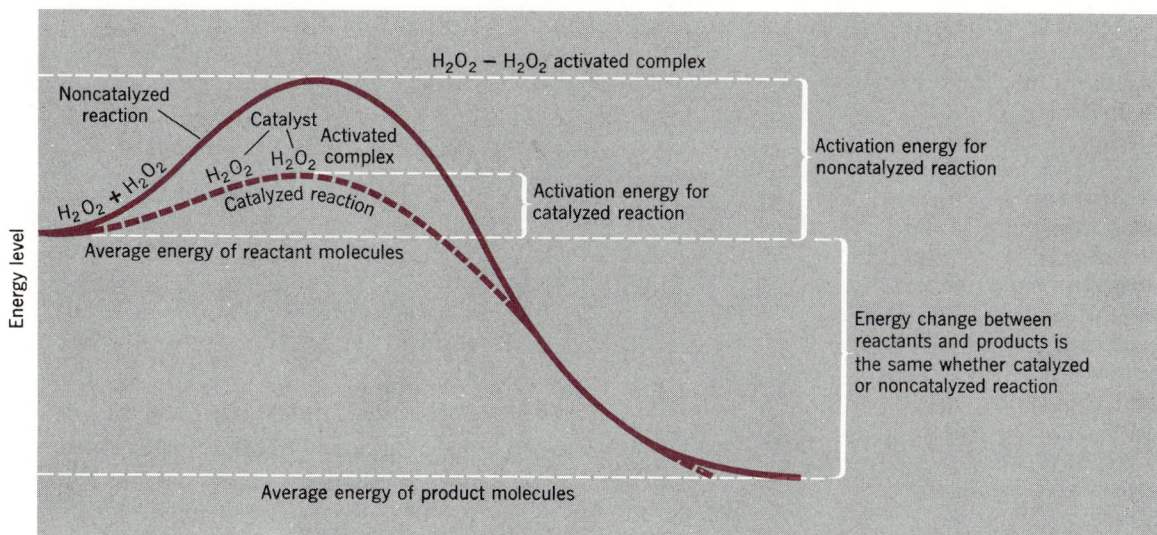

Figure 2-15 *Catalyzed and noncatalized reactions for hydrogen peroxide. In the noncatalyzed reaction (indicated by the solid line) two molecules of H_2O_2 possessing the required activation energy react with one another to form a presumed H_2O_2—H_2O_2 activated complex, which then rearranges to form the products H_2O and O_2. In the catalyzed reaction (dotted line) a different kind of activated compound is formed, which requires a lower activation energy, and the reaction rate is now faster.*

catalysts, are usually effective in minute (trace) amounts. For example, at room temperature hydrogen peroxide, H_2O_2 (a metabolic product formed in many cells), decomposes into water and oxygen at a very slow rate ($2H_2O_2 \longrightarrow 2H_2O + O_2$). If a small quantity of platinum or catalase, a specific biological catalyst, is added, the reaction is highly accelerated.

A catalyst increases the rate of a chemical reaction by participating in the formation of an activated complex in such a way as to reduce the activation energy that is required. This means that there is a lower energy barrier for the molecules to overcome. More molecules get over the lowered energy hump in a given period of time, resulting in a faster reaction rate. A catalyst is not itself used up in a reaction; it is used over and over again by the reacting molecules, which accounts for its effectiveness in small quantities. The energy relationships for a noncatalyzed and

catalyzed chemical reaction such as the decomposition of hydrogen peroxide are presented diagrammatically in Figure 2-15.

Most chemical reactions that take place in living cells would go at only a negligible rate if catalysts were not present. A special class of proteins made in cells act as biological catalysts. These are called **enzymes.** Enzymes (catalase is one) are crucial in all such biological processes as digestion, respiration, and the breakdown, buildup, and interconversion of carbohydrates, fats, proteins, and nucleic acids, as well as in the release and utilization of energy by living cells. Chapter 3 includes more detailed discussion of their properties, including mechanisms of action.

Chemical Equilibrium

Chemical reactions are rarely completed. That is, the reacting substances are not completely used up to form the products. In the course of a chemical reaction the concentrations of the reactants decrease progressively and those of the products correspondingly increase, until a point is finally reached when no net change occurs. This is known as an **equilibrium state;** it occurs because the reverse reaction is proceeding at the same rate as the forward reaction, resulting in no apparent net change in the concentration of reactants and products. It is exactly analogous to the dynamic equilibrium of diffusion described on p. 27.

When carbon dioxide (CO_2) and water (H_2O) react to form carbonic acid (H_2CO_3), the reverse reaction also occurs, as indicated in the following equation by arrows pointing in opposite directions:

$$CO_2 + H_2O \rightleftharpoons H_2CO_3$$

A chemical equilibrium therefore does not represent a static state or a system at rest. Instead, it is a dynamic state in which chemical reactions occur in both the forward and reverse directions at equal rates.

The effect of a catalyst, biological or otherwise, is to speed up the rate of a chemical reaction. In the case of reversible reactions it accelerates both the forward and the backward reactions equally, so that the equilibrium state is reached in a shorter time, but only if the catalyst is added at the beginning of a reaction. If it is added to a system already in the equilibrium state, it will have no effect.

Chemical Thermodynamics

When any chemical reaction takes place, certain rules regarding energy relationships always apply. The study of these energy relationships is the field of **chemical thermodynamics.** The two most fundamental properties of all chemical reactions are: energy is always liberated or absorbed in the process; and the total weight of the products is equal to the total weight of the reactants—that is, the total number of atoms is unchanged. When the bonds formed in the products of

a reaction are stronger than those of the reactants, the extra energy of formation is liberated (usually as heat). Such reactions are referred to as **exergonic,** and the energy released is called **reaction energy.** When the products of a reaction have weaker bonds than the reactants (i.e., more energy is required to break the initial bonds than is released in forming the final ones), energy must be supplied to the reaction mixture to make it go. Reactions that require energy are called **endergonic.**

Chemical Bond Energy

We have already said that chemical energy represents a form of potential energy and is present in substances as a result of the motion of electrons and arrangement of atoms in its molecules. As a good approximation, the chemical energy is viewed as being concentrated or stored in the bonds between the atoms that make up a molecule. Therefore any change in the chemical linkages between atoms (i.e., chemical reaction) results in either a net gain or net release of energy. Recall that energy must be provided to break a chemical bond; energy is liberated when bonds are formed. For example, in the reaction of hydrogen and oxygen gas to yield water, there is a net liberation of heat energy:

$$2H_2 + O_2 \rightleftharpoons 2H_2O + 57,000 \text{ cal/mole}$$

This means that the total energy released in forming the four O—H bonds (2H—O—H) was 57,000 calories per mole greater than that absorbed in breaking the two H—H and one O=O bonds of the reactants. The energy released is the difference in energy of the bond combinations before and after the reaction. To reverse the reaction, that is, to decompose water into hydrogen and oxygen gas, the same quantity of energy, at the very least, would have to be provided.

Free energy is the useful energy possessed by a chemical, and is the sum of a substance's heat content (as measured by the release of heat energy on complete oxidation of its chemical bonds) plus the energy it contains by virtue of its organization. As chemical reactions occur, free energy is transferred by making and breaking of bonds, but in every reaction some is also lost as an increase in entropy.

The liberation of free energy in a chemical reaction represents the difference in energy states between the reactants and the products. Free energy is called useful energy because it can perform useful work. A chemical reaction can proceed spontaneously if it is accompanied by a release of free energy. In other words, if the free-energy content of the reactants is greater than that of the products, the reaction can readily take place. In the reverse situation, when free energy is required, it must be provided to the system in order for the reaction to take place. A spontaneous chemical reaction can be compared to a waterfall. The forward reaction is analogous to the spontaneous descent of water over the falls; it moves from a higher to a lower energy value. The reverse reaction, from a lower to a higher free-energy level of the reacting chemical system, requires an input of energy from some outside source, just as the pumping of water to the top of the waterfall would require an energy input.

In all reactions, whether they liberate or consume energy, the activation-energy requirement must be fulfilled. Activation energy is an important factor in determining the *rate* of the chemical reactions, whereas the change in free energy indicates whether the reaction is *feasible* from an energy point of view.

It should be noted that the energy transformations that occur in chemical reactions represent neither creation nor destruction of energy, but are simply conversions of energy from one form to another. When energy is liberated it represents a transformation to another form of some of the energy stored in the chemical bonds of the reactants.

Energy Utilization

The net release of free energy during a chemical reaction means that it can be employed for the performance of useful work, provided a properly organized system is present to utilize the energy. If a system of suitable organization is not present, the useful energy released will be wasted, usually as heat. For instance, when gasoline is poured on the ground and ignited, it burns and emits energy that is dissipated in the form of heat and light. If it is instead burned in a highly organized system such as a gasoline engine, some of the liberated free energy can be transformed and utilized as mechanical energy, electricity, or possibly as some other form of energy, depending on the means available for transforming it. Similarly, in living cells the free or useful energy made available by respiration is captured and stored as chemical energy in the chemical bonds of certain unique, energy-transferring compounds. We have already been introduced to one of the most important of these compounds, called adenosine-triphosphate, or ATP. We shall return to a detailed discussion of its functions in Chapter 4, when we consider mechanisms of metabolism of fuel foods.

Energy Relationships in the Biological World

The free or useful energy made available during respiration in living cells is liberated principally from the transformation of compounds that contain much bond energy, like carbohydrates, to those with fewer or more stable bonds, which have available less total reaction energy. The higher-energy compounds originate almost entirely in the biological process of photosynthesis carried out by green plants (see Chapter 13). Here the radiant energy of the sun is used to convert the low-energy compound carbon dioxide to higher-energy carbon-oxygen compounds, such as carbohydrates, fats, and proteins. Organisms that are capable of manufacturing all the complex molecules they need are called **autotrophs.** Animals that ingest other organisms for nutrients and energy are termed **heterotrophs** (Chapter 21).

In all instances respiration results in the

liberation of energy, of which some is utilized and some is wasted. In the chain of biological events in which energy in the form of food is transformed within an organism and from one organism to another, less energy is passed on in each step than is received. Thus the energy originally obtained by plants from the sun is inexorably whittled down in the biological sequence that ultimately terminates in its complete conversion to a dissipated form—heat.

Energy transformation in the biological world therefore is not a cyclic process but a pathway running in only one direction—downhill toward zero useful energy, a one-way street. In this inevitable downhill conversion of sunlight to heat, a portion of the available energy serves as a power source for all activities characteristically associated with life.

Let us now turn to an examination of the structure of biological molecules, to see more clearly how the energy on which life processes run is trapped and stored in molecular bonds, and to learn more about the structural fabric of which living cells are made.

Reading List

Crewe, A. V., "A High-Resolution Scanning Electron Microscope," *Scientific American* (April 1971), pp. 26–35.

Companion, A. F., *Chemical Bonding.* McGraw-Hill, New York, 1969, Chapters 1–3.

Dickerson, R. E., *Molecular Thermodynamics.* W. A. Benjamin, New York, 1969, Chapters 1–3.

Haensel, V., and R. L. Burwell, Jr., "Catalysis," *Scientific American* (December 1971), pp. 46–58.

Kendall, H. W., and W. K. H. Panofsky, "The Structure of the Proton and the Neutron," *Scientific American* (June 1971), pp. 60–77.

Pauling, L., *The Chemical Bond.* Cornell University Press, Ithaca, N.Y., 1970, Chapters 1–4.

Thompson, D. T., *On Growth and Form.* Macmillan, New York, 1942, p. 13.

Wahl, A. C., "Chemistry by Computer," *Scientific American* (April 1970), pp. 54–70.

Wall, F. T., *Chemical Thermodynamics.* W. H. Freeman, San Francisco, 1958.

chapter three
Chemical Components of Living Systems

"Organisms and the cells that compose them are in all fundamental respects physical-chemical systems, and their activities occur in large part by the flow of chemical energy."

ERNEST J. DuPRAW, 1968, *CELL AND MOLECULAR BIOLOGY*

Living substance is a vastly complex mixture of ions, molecules, and particles organized into functional arrangements that together display the characteristics of a living system. The constituents, as we have said, are organized for the most part into a variety of units that in turn make up successively larger and more intricate units and systems (e.g., mitochondria, ribosomes, and so on), and, finally, the basic unit of life—the cell.

Chemical analysis of any cell or tissue shows that only about 20 of the 92 naturally occurring elements seem to be essential for their normal structure and function. These are listed in Appendix 1, and of course include carbon, hydrogen, oxygen, nitrogen, phosphorus, and sulfur.

The important biological substances range in size from relatively small molecules like amino acids or simple sugars to enormous ones such as carbohydrates, proteins, and nucleic acids. Amino acids have molecular weights in the range of 100 to 200; that of glucose is 180. Proteins and nucleic acids, in contrast, have molecular weights in the tens or hundreds of thousands, even into the millions. What, then, are the structure and properties of these molecules of which living systems are made?

49

Water

The major component of living substance is water. The adult human body is about 80 percent water; most animals and plants contain 65 to 90 percent water. If water did not exist, or if it had different chemical properties, it is doubtful whether anything we could define as "life" could have evolved.

The most important biological property of water is its **solvent power.** More different substances will dissolve in water, and in greater quantity, than in any other liquid. Substances dissolved in a liquid are called **solutes;** the mixture of a solute in a solvent is a **solution.** Solutions in which water, rather than some other liquid, is the solvent are called **aqueous** solutions. Cytoplasm is an aqueous solution (although usually with **gel** properties, as we shall see in a moment); most of the molecules and macromolecules we discuss are dissolved or suspended in aqueous solutions; in fact, essentially all chemical reactions of importance to biology take place in aqueous solutions.

The main reason that water is such an excellent solvent is because it has a **polar structure** (Fig. 3-1). That is, the shared hydrogen electrons are pulled close to the oxygen nu-

cleus with a geometry such that the hydrogen atoms are not symmetrical on each side of the oxygen, but form a **bond angle** of 105°. As a consequence, the molecule exhibits **polarity,** with the oxygen side being negative and the hydrogen side positive (like the poles of a magnet). This polar structure gives water a strong **ionizing,** or **dissociating,** capacity. Most compounds that are formed with ionic bonds become ionized to a greater or lesser extent when dissolved in water. Such compounds are termed **electrolytes,** and are classed as **strong** or **weak,** depending on how thoroughly they ionize. We'll see later that most of the giant molecules of living substance are weak electrolytes that owe their state of dissociation—hence many of their physical properties—to this ionizing capacity of the water in which they are dissolved.

Water has other unusual properties of great biological significance. Most liquids, for example, as they are cooled increase in density continuously until they freeze. In contrast, water reaches a maximum density at 4°C;[1] as it cools further its density **decreases.** That is, water at 4°C sinks to the bottom, while colder water, less dense, rises to the top; this accounts for the fact that bodies of water

[1] Recall that the freezing point of water is 0°C (32°F).

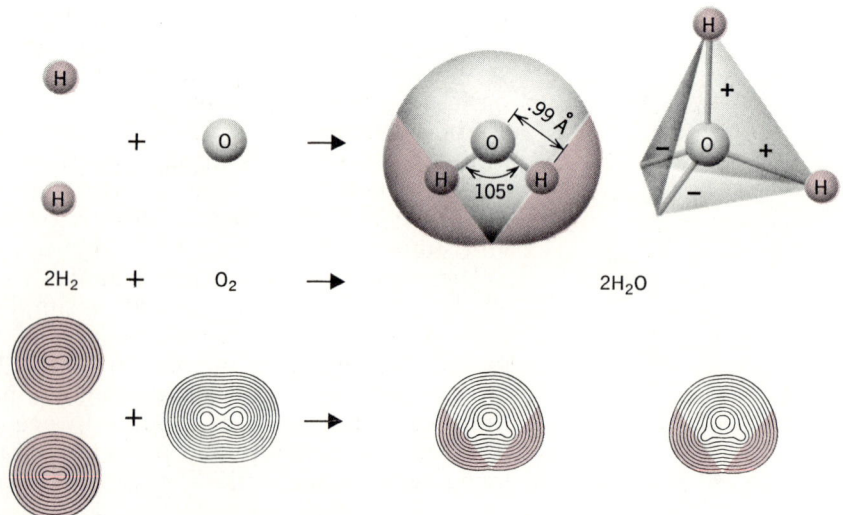

Figure 3-1   The structure of water. The powerful attractive force of the oxygen atom, with only six electrons in its outer shell, draws two hydrogen electrons in close. The H^+ protons are thus fixed in positions that form a bond angle of 105°.

$2H_2$ + O_2 → $2H_2O$

.99 Å

O

H 105° H

freeze from the top down. The layer of ice formed insulates the water below from heat loss, protecting organisms in the water from freezing.

Water therefore is not merely the primary biological solvent. It plays a major role in influencing the properties of the molecules it dissolves, and it participates in many of the most important biochemical reactions.

Colloids and the Gel State

When a solid dissolves in water it is dissociated into its individual molecules or ions. That is, particles of the solute of molecular dimensions are dispersed among the molecules of water. The molecular weight of most substances that form true solutions is generally, at most, a few hundred daltons (see p. 32), and therefore the size of the solute particles is usually less than about 10 Å. A molecule of NaCl, for example, is 2.8 Å; one of glucose is 7 Å. No matter how long a solution stands, the particles do not settle out, nor can they be filtered out with ordinary filters.

Substances with molecular weights in the thousands or more generally do not form true solutions. When dispersed in water, they exhibit particle sizes of 10 to 1000 Å; particles in this size range are termed **colloidal** particles. Light shining through such a mixture gives it a cloudy or opalescent appearance, as opposed to a true solution, which is optically clear. This is called the **Tyndall effect,** and results from the light-scattering properties of particles of these dimensions. It is the easiest way to recognize a **colloidal solution** (also called **colloidal sol,** or simply **sol**). A colloidal solution, then, is a system of particles of colloidal dimensions dispersed in a solvent medium.

Most substances of biological importance (proteins, nucleic acids, carbohydrates) have densities substantially greater than water. Colloidal suspensions of these materials would, in theory, settle out of solution, given enough time; but with only gravitational force acting on the particles it would take many years. Nevertheless, the fact that colloidal particles do settle out has been utilized in a procedure for separating high-molecular-weight substances that substitutes centrifugal force for that of gravity. This procedure is of great value in the laboratory for isolating and studying the components of cells. Instruments termed **ultracentrifuges** are now routinely available to biologists. They can attain rotor speeds of 60,000 revolutions per minute (rpm), giving centrifugal forces of the order of 500,000 times gravity. Proteins with molecular weights as low as 10,000 can be separated from suspension in a few hours by centrifugation; most cell organelles that are of colloidal dimensions (ribosomes, mitochondria, glycogen particles, etc.) can be separated in minutes by centrifugation at such speeds.

When colloidal particles are composed of giant, filamentous molecules, they can exist in colloidal solution either in the liquid (sol) state or in a semisolid condition known as a **gel.** Gelatin, a fibrous protein, is a good example of the gel state when it has solidified from a sol into the familiar moist, rubbery texture of Jell-O. Many long-chain proteins and high-molecular-weight nucleic acids are commonly found as gels, or can be converted easily to this state by modifications of the ionic content of the solvent. Presumably a gelled colloid forms an open matrix or semirigid, three dimensional lattice in which the solvent liquid is trapped. The best evidence suggests that the cytoplasm of a living cell is normally in the gel state (Chapter 5) and that it undergoes continual dynamic, localized sol-gel transformations. Apparently much of the behavior and chemistry of living systems is associated with the gel condition.

Principal Chemical Substances of the Cell

The chemical constituents of protoplasm are customarily classified in two broad categories: **inorganic** and **organic.** Inorganic substances are those that do not have carbon-hydrogen bonds. The most common inorganic compounds in cytoplasm are: water; dissolved gases, particularly oxygen and carbon dioxide; and salts and ions of elements such as iron, copper, zinc, manganese, phosphorus, calcium, magnesium, potassium, sodium, and chlorine. Some of the roles of inorganic substances in biological systems are considered in Chapters 4 and 14.

Organic substances do contain carbon-hydrogen bonds. The four principal constituents of protoplasm—carbohydrates, fats, proteins, and nucleic acids—are built either of chains of carbon atoms or of smaller molecules (amino acids, nucleotides), which are themselves made up primarily of chains of carbon atoms. Therefore let us begin with a look at the properties of carbon and the compounds that form from it.

Properties of Carbon and Its Compounds

An unusual property of the carbon atom is its ability to combine chemically with other carbon atoms to form chain-like molecules of varying lengths. Organic compounds are made up of a skeleton, or "backbone", of carbon atoms, ranging from one to many carbons, chemically linked to one another and to other atoms such as hydrogen, nitrogen, and sulfur.

The carbon atom has six electrons surrounding its nucleus—two in its inner shell and four in its outer shell (see Fig. 2-5). Like most other atoms, it has a tendency to acquire the stable configuration of eight in its outermost electron shell (p. 30), and can thus form four covalent bonds. The four bonds of carbon are equidistant from one another around the three-dimensional carbon atom and form four corners of an imaginary pyramid-like structure called a tetrahedron, with the carbon atom at the center. A simple organic compound like methane, CH_4, in which the carbon atom is covalently bonded to each of four hydrogen atoms, forms a three-dimensional molecule, as shown in Figure 3-2b and c. For convenience organic molecules are represented on paper as two-dimensional structures (Fig. 3-2a), but it is important to remember that they really are three-dimensional.

Recall from pp. 34–35 that the sharing of one pair of electrons between any two carbon atoms—called a single bond—is the most prevalent covalent bond in organic compounds. The carbons of most proteins,

Figure 3-2   Three representations of the geometry of the organic molecule methane (CH_4). (a) Two-dimensional **dash formula,** as commonly written. (b) Three-dimensional representation showing that the four bonds are directed toward the four corners of an imaginary tetrahedron. (c) A three-dimensional model showing the actual volumes and spatial relations of the atoms.

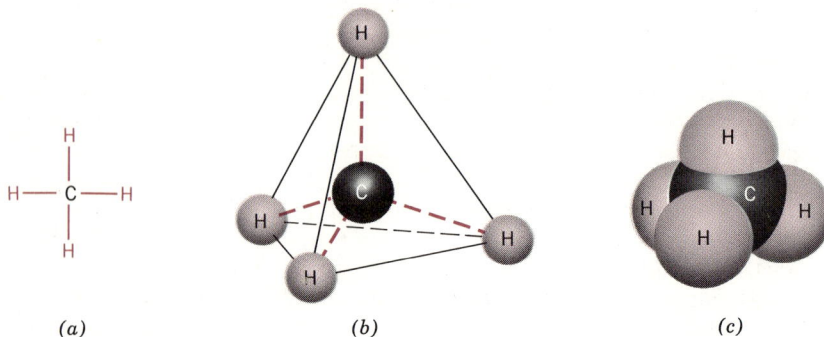

(a)   (b)   (c)

nucleic acids, carbohydrates, and fats are combined in this way. Two carbon atoms also may be chemically bound to one another by double and triple bonds. For example, the organic compound ethylene possesses a carbon-carbon double bond, and acetylene a carbon-carbon triple bond:

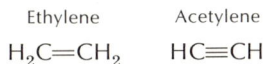

Ethylene	Acetylene
$H_2C{=}CH_2$	$HC{\equiv}CH$

Single bonds between carbon atoms are referred to as **saturated** bonds, whereas double or triple bonds are called **unsaturated** because the carbons can bond to additional atoms by chemical reaction.

In addition to forming bonds with itself and with the electropositive element hydrogen, carbon is able to react with electronegative elements like oxygen, nitrogen, phosphorus, sulfur, and chlorine, making it among the most reactive of all elements. Compounds of carbon, oxygen, nitrogen, and hydrogen alone account for about 99 percent of the dry weight of all living cells.

Aside from its versatility, the second outstanding property of carbon is that most of the compounds it forms are remarkably **inert** (i.e., do not react chemically with other compounds). Despite the fact that many organic reactions are **exergonic** (energy is released in the reaction), and these therefore tend to go far toward completion when once started (see p. 46), most carbon compounds still react spontaneously only at exceedingly low

rates with one another, with oxygen, or with water. This fact is of crucial biological importance; it accounts in part for the stability of cytoplasmic systems. However, the spontaneous nonreactivity of most carbon compounds can be readily converted to high reaction rates with the participation of biological catalysts, or enzymes.

Types of Organic Compound

Hydrocarbons

Organic compounds containing only carbon and hydrogen are called **hydrocarbons.** They may be **straight, branched,** or **cyclic** molecules. For example, normal (or **n-**) hexane (Fig. 3-3) is a saturated, straight-chain organic compound consisting of six carbon and 14 hydrogen atoms. **Iso**hexane also consists of six carbons and 14 hydrogens, but the carbon chain has what is called a branched configuration. Such compounds that have the same composition but different structures are said to be **isomers** of one another. This is analogous to the construction of two words such as *late* or *tale,* which are composed of the same four letters but are very different because of the arrangement of the letters. By virtue of this different arrangement of their atoms, isomers have very different physical

Figure 3-3 *Examples of three different six-carbon hydrocarbons. n-hexane and isohexane are isomers because they contain the same numbers of atoms. Cyclohexane is not isomeric because it has fewer hydrogen atoms.*

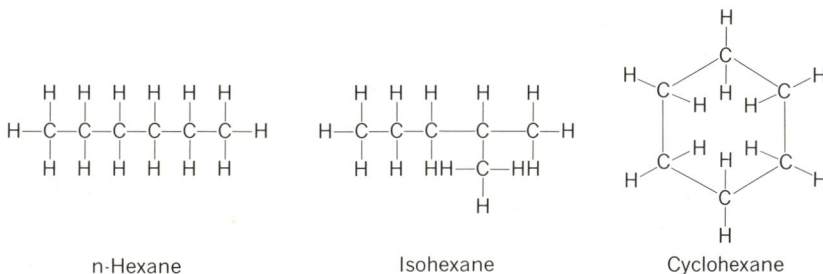

n-Hexane   Isohexane   Cyclohexane

and chemical properties. **Cyclo**hexane (Fig. 3-3) is another saturated hydrocarbon made of six carbon atoms, but these are attached to one another in a **ring** (a cyclic structure). Cyclohexane is not an isomer of hexane; count the hydrogens and you will see that it has 12 instead of 14. When the two ends of a carbon chain are joined in the circle, two hydrogens are eliminated to satisfy the four-bond requirement of each carbon atom.

Hydrocarbon Derivatives and Functional Groups

Nearly all organic compounds are derivatives of hydrocarbons; they have atoms of oxygen, nitrogen, phosphorus, sulfur, or occasionally certain metals linked to the carbon-carbon backbone. In most instances these additional atoms are chiefly responsible for whatever chemical reactivity the organic compound displays. For this reason they are called **functional groups** (the remainder of the hydrocarbon molecule may also undergo chemical reaction, but less readily).

In general, organic molecules that react similarly do so because they have the same functional groups. Therefore, despite the vast number of organic compounds, they can be classified into relatively few types; and the kinds of functional group a compound has in turn determine the kinds of chemical reaction it will undergo. Some of the common functional groups of organic compounds are listed in Table 3-1, which shows the structural formula, a more convenient shorthand form of writing the formula, and some of the compounds in which each of these functional groups is commonly found. In structural formulas of organic compounds the letter R and R′ are used to indicate different chains of carbon atoms of any length.

Alkanes

The alkanes are the simplest hydrocarbons. **Methane** (Fig. 3-2) is the single-carbon alkane. **Ethane** has two carbons, CH_3—CH_3, and in this case R stands for the CH_3 on the left. **Propane,** CH_3—CH_2—CH_3, is a three-carbon alkane; in the table R represents CH_3—CH_2—. The six-carbon alkane is **n-hexane** (Fig. 3-3).

Alcohols—The Hydroxyl Group

The —OH functional group is known as an **alcohol** or **hydroxyl** group. Substitution of the alcohol group (OH) for a hydrogen atom in the methane molecule CH_4 gives the compound CH_3OH, called **methyl alcohol,** or **methanol;** an OH in place of a hydrogen in ethane, CH_3CH_3, yields **ethyl alcohol,** or **ethanol,** CH_3CH_2OH,

$$
\begin{array}{ccc}
 & H & H \\
 & | & | \\
H\!-\!\!&C\!-\!\!&C\!-\!OH \\
 & | & | \\
 & H & H
\end{array}
$$

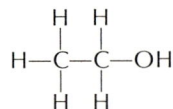

which is the alcohol of beer and wine. Some organic molecules contain several alcohol groups; **glycerol,** for example, a product of fat digestion, has three alcohol groups, one linked to each of its three carbons, and has the structure $CH_2OHCHOHCH_2OH$,

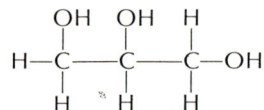

$$
\begin{array}{ccc}
OH & OH & H \\
| & | & | \\
H\!-\!C\!-\!\!&C\!-\!\!&C\!-\!OH \\
| & | & | \\
H & H & H
\end{array}
$$

Ethyl alcohol and derivatives of glycerol play important roles in cellular respiration and as constituents of fats. Alcohol groups also occur in a number of important hormones known as the **steroid** hormones, which are responsible for development of the sexual characteristics of many animals. Even at low concentrations in the blood, the alcohols have serious pharmacological effects. Methanol is a deadly poison. Ethanol acts as an intoxicant by interfering with the electrical activity of nerves, and depressing the function of certain brain centers. At high concentrations both alcohols denature (harden) protein irreversibly, and are therefore commonly used as components in biological fixatives (p. 114).

Aldehydes and Ketones—The Carbonyl Group

The carbon-oxygen double bond, $\geq\!C\!=\!O$,

table 3-1
Some Organic Functional Groups and Representative Compounds

GENERAL NAME	STRUCTURAL FORMULA	SHORTHAND FORM	COMPOUND	COMMON NAME
Alkanes	$R-\overset{\overset{H}{\mid}}{\underset{\underset{H}{\mid}}{C}}-H$	$R\ CH_3$	CH_4 C_2H_6 C_3H_8	Methane Ethane Propane
Alcohols	$R-\overset{\overset{H}{\mid}}{\underset{\underset{H}{\mid}}{C}}-OH$	$R\ CH_2OH$	CH_3OH C_2H_5OH C_3H_7OH	Methanol Ethanol Propanol
Carbonyls				
Aldehydes	$R-\overset{\overset{O}{\parallel}}{C}-H$	$R\ CHO$	$HCHO$ CH_3CHO	Formaldehyde Acetaldehyde
Ketones	$R-\overset{}{\underset{\underset{O}{\parallel}}{C}}-R'$	$R\ CO\ R'$	CH_3COCH_3	Acetone
Carboxyls	$R-\overset{\overset{O}{\parallel}}{C}-OH$	$R\ COOH$	$HCOOH$ CH_3COOH	Formic acid Acetic acid
Esters	$R-\overset{}{\underset{\underset{O}{\parallel}}{C}}-O-R'$	$R\ COO\ R'$	$CH_3COOC_2H_5$	Ethyl acetate
Sulfhydryls	$R-\overset{\overset{H}{\mid}}{\underset{\underset{H}{\mid}}{C}}-SH$	$R\ CH_2SH$	C_2H_5SH	Ethane thiol
Amines	$R-\overset{\overset{H}{\mid}}{\underset{\underset{H}{\mid}}{C}}-NH_2$	$R\ CH_2NH_2$	$C_2H_5NH_2$	Ethylamine
Organophosphates	$R-\overset{\overset{H}{\mid}}{\underset{\underset{H}{\mid}}{C}}-O-\overset{\overset{O}{\parallel}}{\underset{\underset{OH}{\mid}}{P}}-OH$	$R\ CH_2OPO(OH)_2$	$C_6H_{11}O_5PO_4$	Glucose-6-phosphate

is known as the **carbonyl** group. Organic molecules that contain a carbonyl group at the beginning or end of their carbon chains are called **aldehydes.** The simplest aldehyde, the compound **formaldehyde** ($HCHO$), is used as a biological preservative because of its tendency to cause cross-linking and hardening of proteins. **Acetaldehyde** (CH_3CHO) is a two-carbon aldehyde.

Formaldehyde   Acetaldehyde

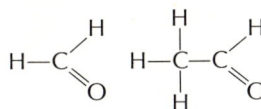

$$H-\overset{\overset{H}{\mid}}{\underset{\underset{O}{\parallel}}{C}} \qquad H-\overset{\overset{H}{\mid}}{\underset{\underset{H}{\mid}}{C}}-\overset{}{\underset{\underset{O}{\parallel}}{C}}$$

If the oxygen in a carbonyl group is double-bonded to a carbon anywhere but an end of a molecule, the molecule is called a **ketone**. The simplest ketone is **acetone** (CH_3COCH_3):

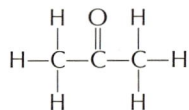

$$\begin{array}{ccccc} & H & O & H & \\ & | & \| & | & \\ H - & C - & C - & C & - H \\ & | & & | & \\ & H & & H & \end{array}$$

Aldehydes and ketones can readily be formed by the oxidation of corresponding alcohols; that is, by the removal of two hydrogen atoms. The oxidation of an alcohol group at the end of a carbon chain produces the related aldehyde, whereas the oxidation of an alcohol group in a nonterminal position forms the corresponding ketone. For example, the mild oxidation of ethyl alcohol, CH_3CH_2OH, produces acetaldehyde, CH_3CHO, whereas the oxidation of isopropyl alcohol, $CH_3CHOHCH_3$, yields acetone, CH_3COCH_3. These reactions are usually freely reversible, and several important oxidation-reduction reactions occurring in living cells involve the reversible transformation of carbonyl-containing compounds to their corresponding alcohols (e.g., the interconversion of acetaldehyde to ethyl alcohol in the fermentation of glucose by yeast, and pyruvic acid to lactic acid during muscle contractions).

Organic Acids—The Carboxyl Group

The $-C-C\begin{smallmatrix}OH\\ \diagup\\ \diagdown\\ O\end{smallmatrix}$ group, often written as —COOH, is known as the **carboxyl** group. This group can be formed by the oxidation of an aldehyde group. Compounds with a carboxyl are called **organic acids** because they ionize to produce H^+ ions ($-COOH \rightleftharpoons -COO^- + H^+$). They are weak acids, ionizing only to a relatively small extent in aqueous solutions.

Certain organic acids may have two and even three carboxyl groups attached to a single molecule and are designated as **dicarboxylic** and **tricarboxylic** acids, respectively. The three simplest carboxylic acids are the two-carbon form **acetic acid** (CH_3COOH) and two three-carbon acids, **pyruvic** and **lactic** acid (Fig. 3-4a). We shall refer frequently to these; they play important roles in cell respiration. Carboxylic acids composed of chains of four carbons or more tend to be insoluble in water and are called **fatty acids** (p. 62). **Stearic acid**, $C_{17}H_{35}COOH$ (Fig. 3-4b),

Figure 3-4 Organic acids. (a) The structure of acetic, lactic, and pyruvic acid; (b) two common fatty acids.

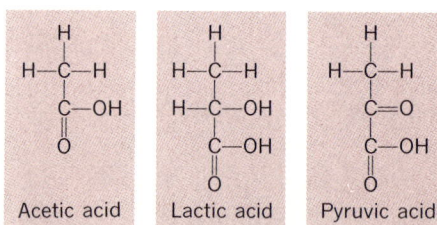

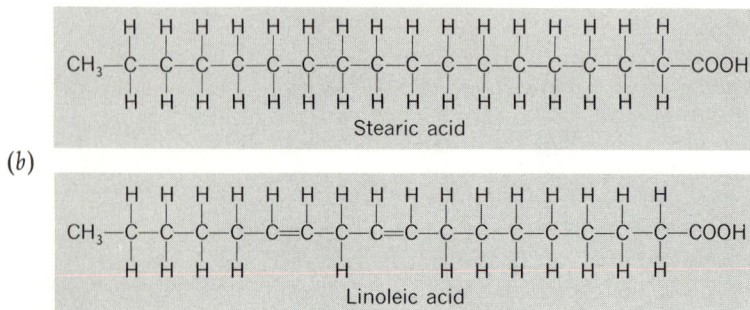

Acetic acid   Lactic acid   Pyruvic acid

(a)

(b)

Stearic acid

Linoleic acid

is a common saturated fatty acid (having only single bonds between the carbons) often found in the tissues of both plants and animals. **Linoleic** acid, $C_{17}H_{31}COOH$, is an example of a naturally occurring unsaturated fatty acid (with one or more double bonds).

Esters

An alcohol and an organic acid can react to form a compound known as an **ester.** The reaction involves the combination of the OH group of the alcohol with the H^+ from the carboxyl to produce a molecule of water and an ester that takes the generalized form R—O—C—R. For example, ethyl alcohol
$$\underset{O}{\overset{\parallel}{R-O-C-R}}$$
reacts with acetic acid to produce water and the ester **ethyl acetate.** (The H and OH atoms that form the water molecule are enclosed in the dashed box.)

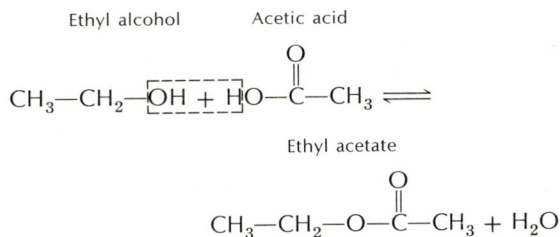

Ethyl alcohol   Acetic acid

$$CH_3-CH_2-\boxed{OH + H}O-\overset{\overset{\displaystyle O}{\parallel}}{C}-CH_3 \rightleftharpoons$$

Ethyl acetate

$$CH_3-CH_2-O-\overset{\overset{\displaystyle O}{\parallel}}{C}-CH_3 + H_2O$$

We shall see in a moment that lipids are esters of fatty acids and a three-carbon alcohol, glycerol.

Amino Acids—The Amine Group

The amine group, $-NH_2$, a derivative of ammonia (NH_3), is found in several different kinds of organic compound, the most important of which are **amino acids**—the main components of proteins—and **nucleotide bases** that make up the nucleic acids. Amino acids (Fig. 3-5) are compounds that contain both

Figure 3-5   *The structure of some representative amino acids. The basic structure NH_2—CH—COOH is common to all the amino acids (shown in color). They differ in the attached side group. Glycine is the simplest, the side group being a single H atom. Some amino acids have straight-chain side groups; others have rings.*

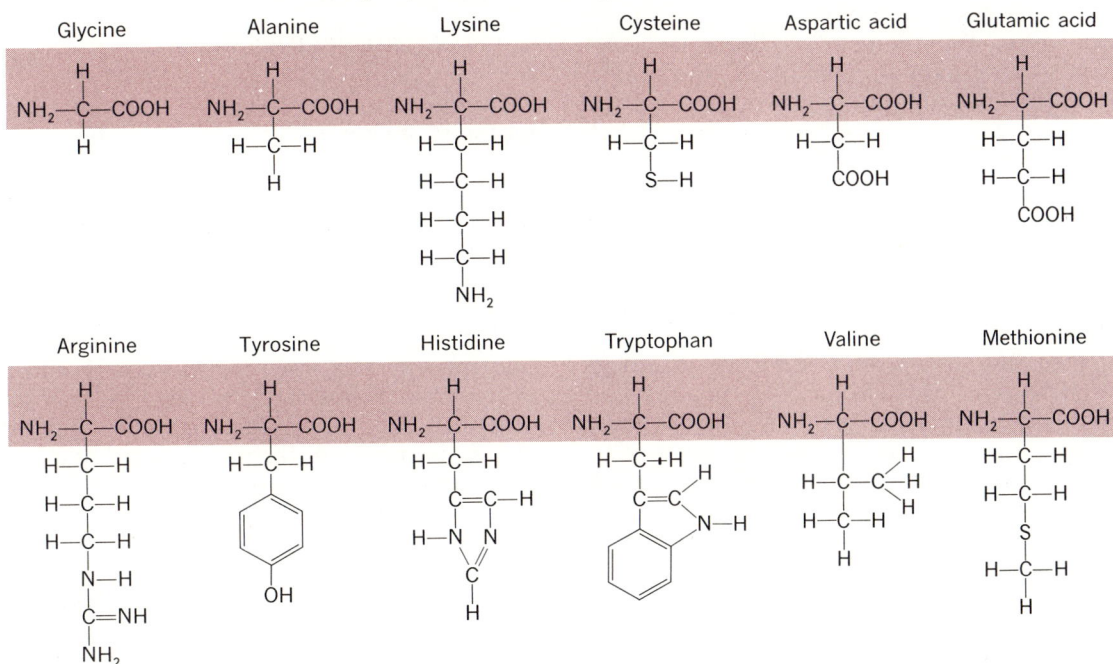

Figure 3-6 *Zwitterion forms of amino acids. At pH values near the isoelectric point (pH 6 to 7 in most amino acids) the negatively charged carboxyl group and positively charged amino group are both ionized. Lowering the pH (reaction shifted to the left) suppresses dissociation of the carboxyl group and causes the molecule to become positively charged. Raising the pH (reaction shifted to the right) causes dissociation of the bound H+ of the amino group and the molecule takes on a net negative charge. At the isoelectric point the net charge on the molecule is zero.*

an amino group and a carboxyl group, usually on the same carbon atom. Near neutral pH (see p. 40) both groups are ionized, and the molecule contains both a negative and a positive charge (Fig. 3-6). In this form the amino acid is termed **zwitterion** (from a German word meaning hybrid). A lower pH suppresses the dissociation of the carboxyl group and causes the molecule to become positively charged; a higher pH causes the bound H+ of the amino group to dissociate, and the molecule becomes negatively charged. The pH at which the positive and negative charges of an amino acid exactly balance each other is termed the **isoelectric point** usually about pH 6 to 7, and is characteristic for each amino acid. At that pH a molecule will not migrate in an electric field (the significance of this fact will become clear later). Of the 20 amino acids commonly found in proteins two have two carboxyl groups and are highly acidic; several have two amino groups and tend to be very basic. Amino acids are linked together by **peptide bonds** to form long chains that are proteins. We shall discuss peptide bonds and the structure of proteins shortly (pp. 64–68).

Organophosphates

Organophosphates—compounds of car-

bon, oxygen, and phosphorus—are of key importance in biochemical reactions. The general structure of organophosphates is

$$R-CH_2-O-\overset{\overset{\textstyle O^-}{|}}{\underset{\underset{\textstyle O}{|}}{P}}-O-R.$$ Phosphate esters,

formed by the combination of phosphoric acid (H_3PO_4) with an alcohol, are involved in several crucial steps in carbohydrate metabolism (Chapter 4). They form a special class of substances, the **phosphoglycerides,** that are components of most fats. Phosphate-ester bonds make up the backbone of the giant nucleic acid molecules in which all genetic information is coded. In addition, the phosphate-ester linkage of the high-energy phosphate compounds (see p. 89) release large amounts of energy when hydrolyzed. The energy-rich compound adenosine-triphosphate (ATP), present in every living organism, is universally employed as an energy-storage device.

Structure of the Cellular Macromolecules

Unless you intend to be a professional biologist or chemist it hardly seems necessary to learn by rote the name and structure of all 20 amino acids, or even of the several functional groups listed in the last section. We have said that the main difference between living and inanimate matter is in the size and complexity of the macromolecules and molecular aggregates that are found in protoplasm, and that carbohydrates, lipids, proteins, and nucleic acids are the primary units of which protoplasm is made. But just as atoms make molecules, amino acids are built into proteins. To put it more generally, these functional groups and small compounds are the subunits of the giant molecules that are the fabric of living cells and that give them the properties they exhibit. We feel that you will understand biology better—that you will read more intelligently a discussion of how cells or species or people behave—once you have formed an impression of the structural subunits.

Carbohydrates

The term carbohydrate means, literally, hydrate of carbon (a combination of carbon and water) and derives from the fact that for every carbon in a carbohydrate chain there are two hydrogen and one oxygen atom, or the equivalent of one molecule of water. Chemically, a carbohydrate is an organic compound that contains a carbonyl group (aldehyde or ketone) and two or more alcohol groups. Carbohydrates make up the bulk of living matter of plants; they are involved in both structure and food storage. They are the principal products of photosynthesis, the process whereby plants use the energy of sunlight to convert carbon dioxide and water into organic matter. Both plants and animals obtain energy for their life activities from the breakdown of carbohydrates through the process of **respiration.**

The carbohydrates are sometimes referred to as **saccharides,** a term that derives from the Greek word for sugar. Carbohydrates are composed of chains of carbon atoms, with a minimum of three carbons (called **trioses**) and more often five or six (**pentoses** and **hexoses**). These **short-chain** "simple sugars," such as glucose, are called **monosaccharides.** Two molecules of a monosaccharide may be linked together to form a **dissaccharide;** sucrose, table sugar, is such a compound. Carbohydrates made up of many monosaccharide molecules linked in long chains, such as starch, glycogen, or cellulose, are called **polysaccharides.** Polysaccharides also represent a class of compounds called **polymers,** which are large molecules made of repeating subunits, in this case monosaccharides.

Glucose ($C_6H_{12}O_6$), a six-carbon sugar, is the living world's most common hexose. It contains five hydroxyl groups and one aldehyde. In solution five of the carbons are connected into a six-membered ring across an oxygen atom. This ring, or **pyranose,** form is in equilibrium with a small amount of straight-chain form. These two structures are shown in Figure 3-7. In (c), the shorthand form of the ring structure, a carbon is assumed to be at each corner unless another atom (such as the oxygen) is shown. The hydrogen atoms are also not shown but are assumed to be at the empty end of each vertical line. The convention shown for numbering the carbon atoms allows each member of the chain to be identified individually. The glucose pyranose ring, for example, has carbons 1 and 5 linked by the oxygen atom.

Figure 3-7 Two forms of glucose: (a) Straight-chain form; (b) pyranose form. (c) This is the shorthand notation for the pyranose form, with carbon atoms numbered. Note that the ring structure is shown as a flat plane perpendicular to the page. The heavy part of the ring is nearest the reader. The orientation of the hydrogen (H) and hydroxyl (OH) groups is vertical, up or down from each carbon.

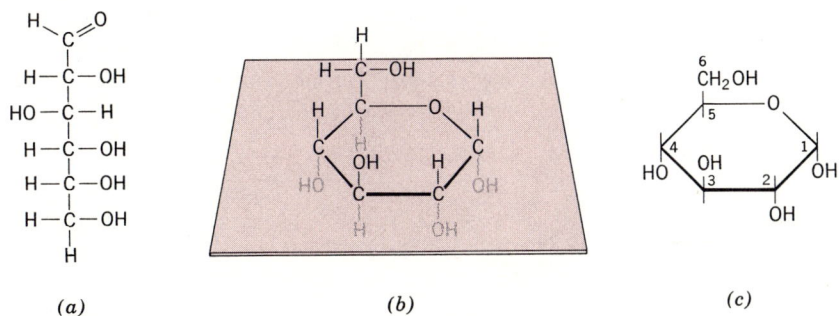

(a)   (b)   (c)

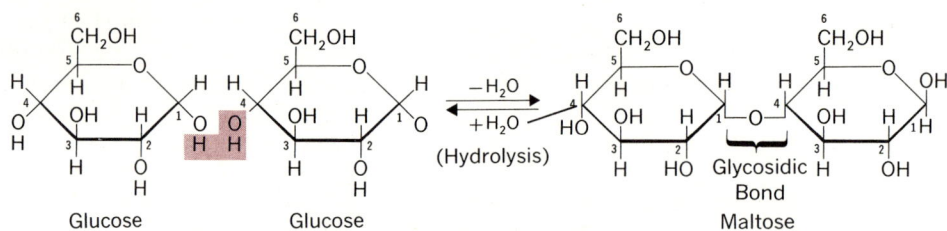

Figure 3-8 When two molecules of glucose are joined by a glycosidic bond to form maltose, a molecule of water is lost (read the reaction from left to right). The reverse reaction, in which maltose is split by the addition of a molecule of water, is hydrolysis.

Fructose, an isomer of glucose, has the same composition ($C_6H_{12}O_6$) but **fructopyranose** has carbons 2 and 5 linked by the oxygen atom.

Two molecules of glucose may be linked together by a **glycosidic** bond to form **maltose,** a sugar obtained from malt (steeped barley). This linkage is formed by removing a hydrogen atom from one molecule and a hydroxyl from the other, leaving an oxygen atom to join the two by a "1,4 linkage" (i.e., a linkage between carbon 1 of one monosaccharide and carbon 4 of the other). This amounts to removal of a molecule of water (Fig. 3-8). Thus the reverse of this reaction, breakdown of maltose into its component glucose residues, represents a hydrolysis. **Sucrose** is formed by a 1,4 glycosidic linkage between a glucose and a fructose. **Lactose,** the dissaccharide that occurs in milk, is formed from glucose and galactose.

The other biologically outstanding monosaccharides are the pentoses, **ribose** and **deoxyribose.** Ribose exists largely as a stable five-membered ring structure in which carbons 1 and 4 are linked by an oxygen. Deoxyribose (Fig. 3-9) is a constituent of DNA (see p. 78), and is distinguished from ribose by the fact that its carbon 2 lacks an oxygen atom—thus the name **de-oxy.** Deoxyribose also exists predominantly in the form of a five-membered ring.

Polysaccharides are chains of many monosaccharides linked by glycosidic bonds. The **cellulose** molecule (Fig. 3-10) is an unbranched chain of more than 10,000 glucose residues with a molecular weight of over 2 million. Cellulose is the major polysaccharide occurring in nature. It is probably the single most abundant organic compound found on our planet, comprising at least 50 percent of all the carbon in the plant world. Its main

Figure 3-9 The pentose (five-carbon) sugar deoxyribose, important both in energy metabolism and as structural components of DNA.

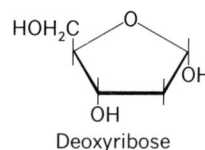

Deoxyribose

Figure 3-10 Portion of a cellulose molecule, an unbranched chain of glucose residues joined by glycosidic bonds.

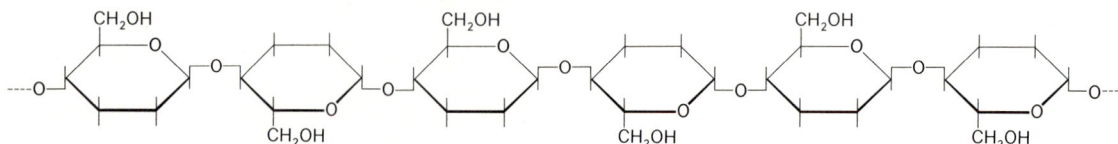

The Molecular Basis of Life

(a)

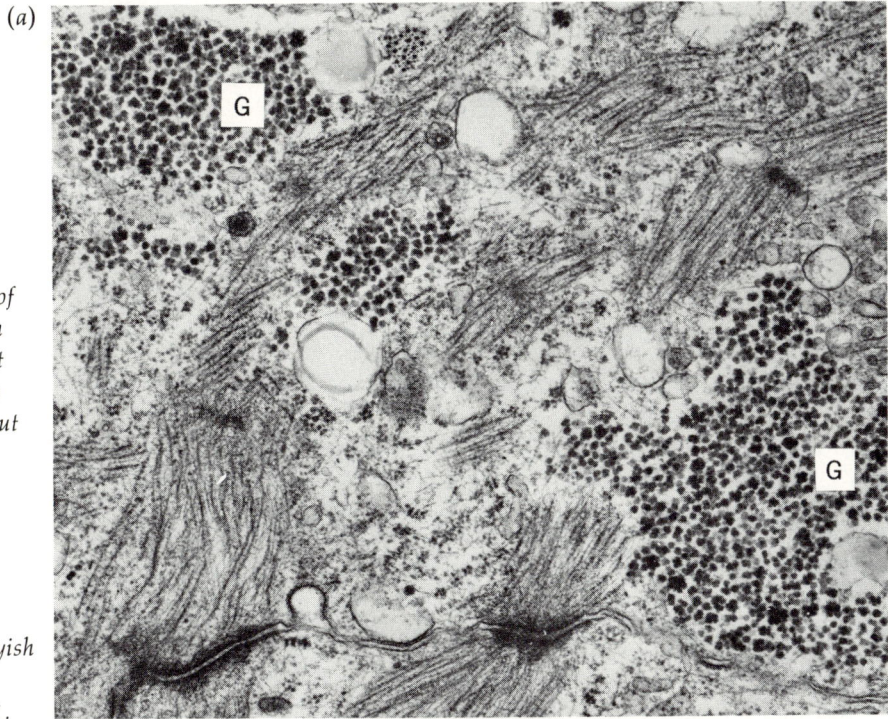

Figure 3-11 *(a) Pools of granular glycogen (G) in an early embryonic heart muscle cell. The glycogen particles range from about 100 to 200 Å in diameter. Magnification 50,000×. (b) Starch-storage bodies, called* **amyloplasts,** *in the cell of a bean seedling. The starch appears as a grayish granular material. Magnification 13,000×.*

(b)

function is structural; it occurs principally as the major component of the plant cell wall. Cotton, one of the purest forms of cellulose, is made up of at least 90 percent of the polysaccharide. Wood is essentially a collection of empty cell walls consisting of cellulose and a few other compounds such as lignin. Plant-cell walls in some forms have tensile strength exceeding even that of high-quality steel.

Glycogen (Fig. 3-11a) and **starch** (Fig. 3-11b) are polysaccharides that are the major food-storage products for animals and plants. Large molecules are useful for purposes of energy storage; a high molecular weight means that for a given amount of material the compound is present in a low molar concentration, and therefore does not contribute significantly to the osmotic pressure inside the cell (see Chapter 5). Starch is present in most plant cells as huge, microscopically visi-

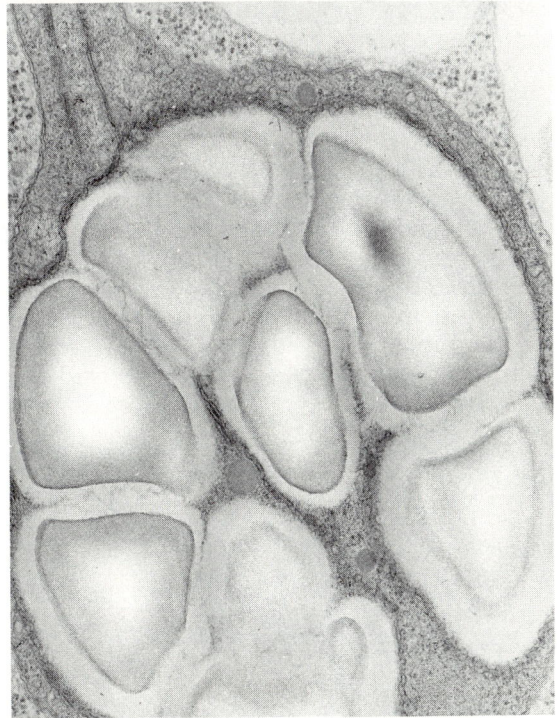

61   *Chemical Components of Living Systems*

ble granules. Like cellulose, it is a polysaccharide of glucose only, but it contains both branched and unbranched chains, each with several hundred monosaccharides. Glycogen is a polymer of about 30,000 glucose residues in the form of highly branched chains, with a molecular weight of about 5 million. In mammals glycogen granules occur mainly in the cells of liver, heart, and muscle tissue.

Lipids |

A second major group of biological macromolecules are the relatively small, water-insoluble compounds known as lipids. Like carbohydrates, lipids are important energy reserves. Excess foods may be converted in the body to fats, as many overweight people are aware. Lipids are insoluble in aqueous solution, and are present in cells primarily as components of membranes. The most commonly known lipids are **triglyceride fats.** Fats are made of two different types of subunit molecule, a three-carbon alcohol called glycerol and a series of long-chain compounds known as fatty acids. Fatty acids, as we said earlier, are hydrocarbons with a carboyxl group at one end (Fig. 3-4); the general formula is $CH_3(CH_2)_nC\diagup_{OH}^{O}$. Because in nature fatty acids are synthesized from two-carbon units (acetyl groups), all naturally occurring fatty acids have an even number of carbons, the most common being 16 or 18. Fatty acids can be either saturated, when their hydrocarbon chain contains only single bonds between the carbons (as in stearic acid), or

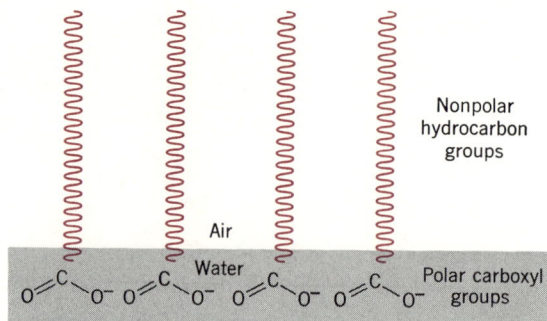

Figure 3-12   The arrangement of a monomolecular layer of fatty acids at the surface of water, highly schematized. The polar carboxyl groups extend into the water, and the nonpolar hydrocarbon ends wave above the surface. This configuration of molecules lowers the surface tension of water and increases its "wetting power."

unsaturated, when one or more double bonds are present (as in linoleic acid). The three-carbon acidic group of fatty acids makes the carboyxl end of the molecule extremely polar (i.e., electronegative) and water soluble **(hydrophylic);** the long-chain hydrocarbon is nonpolar and water insoluble **(hydrophobic).** Therefore they interact in a unique way with water: they form a **monomolecular layer** at the surface, with the carboxyl groups in the water and the hydrocarbon chains lined up like buoy flags above the surface (Fig. 3-12). We shall see that this configuration, only slightly modified, is incorporated into the structure of cellular membranes (Chapter 5).

Fats are esters of glycerol and fatty acids (Fig. 3-13). A molecule of fat may contain

Figure 3-13   A molecule of fat is an ester of glycerol and three fatty acids. $(CH_2)n$ means a chain of (CH_2), n carbons long.

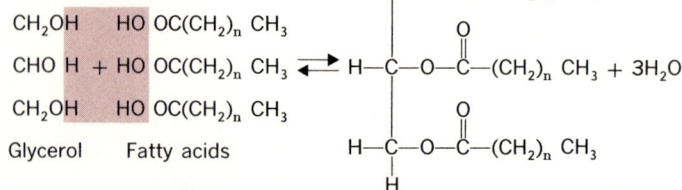

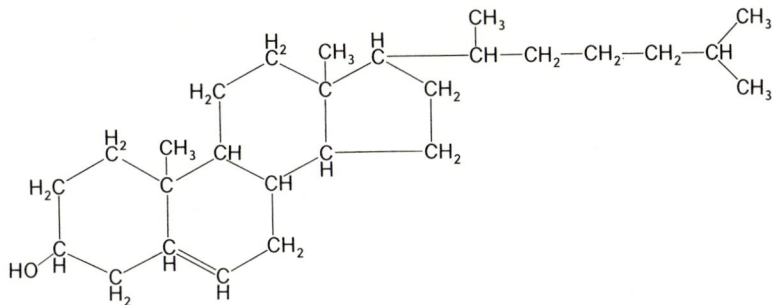

Figure 3-14   Cholesterol, a typical steroid.

Cholesterol

one, two, or three different kinds of fatty acid. Fats can be hydrolyzed to their component fatty acids and glycerol by heating in combination with strong alkali. This is a process known as **saponification,** which results in the formation of sodium or potassium salts of fatty acids, better known as soaps; this is how soaps were made from lye and animal fat throughout history.

Fats with a high level of unsaturation (that is, many double bonds) tend to have low melting points. Fats with sufficiently high unsaturation are "melted" (i.e., liquid) at room temperature and are termed **oils.** Recent evidence suggests that in some individuals, especially after a certain age, saturated fats are readily converted to a fatty coating, called **plaque,** that lines the walls of arteries and causes arteriosclerosis (hardening of the arteries). This is not true of unsaturated fats such as those of plant origin.

Other important lipids are **waxes,** which serve protective roles on plant leaves and the skin or fur of animals, and **phospholipids,** which are phosphate esters of glycerol. Phospholipids are of greatest importance as components of membranes. Like fatty acids, they are extremely **hydrophobic** at one end and **hydrophilic** at the other.

The last class of lipids we shall consider are **steroids,** a group of fat-soluble compounds of four fused rings of carbon atoms to which are usually attached a carbon chain of varying lengths. Steroids are typified by cholesterol (Fig. 3-14). The steroids usually have profound biological effects at very small concentrations and perform many different physiological activities in healthy animals. Not only do they regulate sexual development and function, but they also have important influences on cellular metabolism. Vitamin D, bile acids, adrenal hormones, and male and female sex hormones are all derivatives or modified versions of the basic steroid structure.

Proteins

Proteins are vastly more complex than either carbohydrates or lipids. Carbohydrates may be large, but they are composed of one or at most two subunits, repeated over and over in sequence. Variations do exist in the molecular structure of lipids, but they are minor when compared with those in proteins. Fats are relatively small. Although there is a wide variety of fatty acids, only three can combine with glycerol to form any particular fat. Proteins are made up of chains of 20 different amino acid subunits, which provides an opportunity for an almost infinite variety of different sequences.

A biochemist wishing to extract the proteins from animal or plant cells is likely to go through the following procedures. He first kills the cells with acid or another agent that permits them to be broken up into a dense suspension of opaque, whitish material. By centrifugation (p. 51) the solid material is sedimented into a compact mass or **pellet;** the clear solution remaining, called a **supernatant,** contains a host of inorganic salts and small organic molecules, including a slight amount of amino acids, nucleotides, short-chain carbohydrates, and other low-molecular-weight components. This **acid-soluble fraction** represents 2 to 3 percent of the dry weight of the original cell mass. The pellet, now containing only acid-insoluble material, is broken up in a warm mixture of alcohol and ether, which extracts from it all the lipid components. These are 10 to 15 percent of the original dry weight of the cells. The remaining mixture is centrifuged again, and the resulting pellet is suspended in a heated acid to dissolve out the nucleic acids. These total 10 to 20 percent of the original dry weight. The remaining insoluble material, representing 60 to 80 percent of the original dry weight of the cells, is almost pure protein (perhaps 100,000 different kinds), possibly 1000 of which are enzymes. This **protein precipitate** is then heated in hot, concentrated hydrochloric acid for several hours to hydrolyze it into its constituent subunits, resulting in a solution containing nothing but 20 different amino acids. Because all that is left of the proteins when they have been completely hydrolyzed is a solution of amino acids, these building-block molecules are often referred to in chemists' jargon as **residues.**

Not only are proteins the main structural components of protoplasm, they are also the major regulators and mediators of most of the activities carried out by cells. The biological catalysts or enzymes are all proteins; most animal hormones are proteins; contractile activity of cells is produced by proteins; the antibodies that protect us from disease-producing viruses and bacteria are proteins. All of the functions carried out by these substances depend on molecules with precise and reproducible structures; that is, molecules with a high degree of **structural specificity.** What is remarkable about proteins is that, despite their great size, enormous complexity, and bewildering variety, each protein molecule is built according to an exact plan with a precise sequence of subunits and a specific three-dimensional shape.

Chemistry of Proteins

Like glycosidic bonds, **peptide bonds** are formed between small building-block molecules (the amino acids) by removing water; in this process the peptide bond is established between the carboxyl carbon of one amino acid and the amino nitrogen of another (Fig. 3-15). With this structure, in the-

Figure 3-15 *Formation of a peptide bond. Two amino acids joined by a —C—NH linkage lose a molecule of water and form a dipeptide. The*

$$\underset{O}{\overset{\|}{\text{—C—NH}}}$$

splitting of such a bond with the addition of a molecule of water is another hydrolysis reaction.

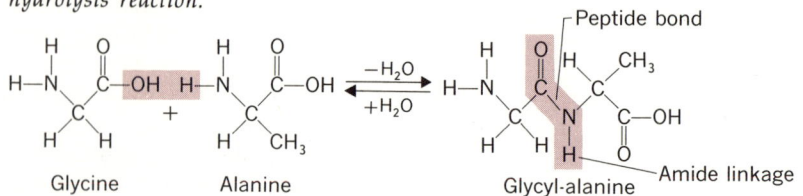

Figure 3-16 A hypothetical polypeptide: glycyl-aspartyl-lysyl-glutamyl-arginyl-histidyl-alanine. By convention, the segment of chain shown begins with its N-terminal group on the left and ends with its C-terminal group on the right.

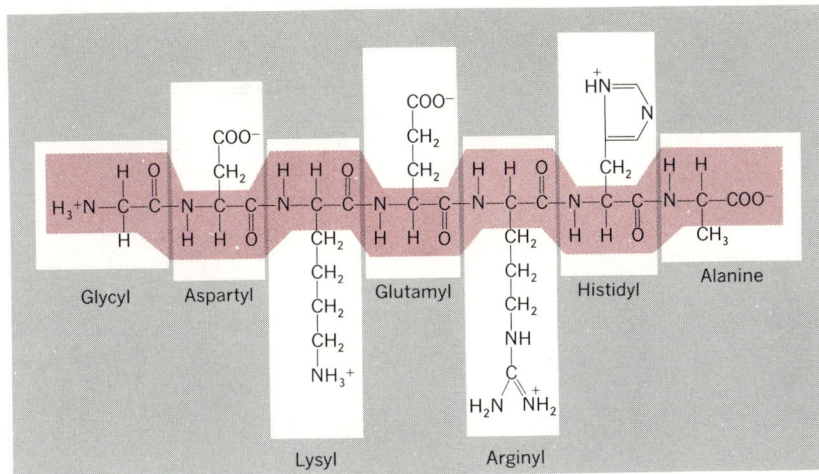

ory, any number of amino acids can be linked together into long chains called **polypeptides.** Figure 3-16 is a diagram of a hypothetical polypeptide composed of all the amino acids that have charged functional groups (p. 54) on their side chains. Note that the backbone of the chain is composed of a repeating sequence:

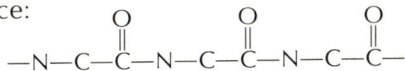

$$-N-C-C-N-C-C-N-C-C-$$

These atoms are not ionized, and are in fact incapable of taking on a charge unless the protein structure is broken. Thus the chemical and electrical properties of this (or any) polypeptide are determined solely by the extra amine or carboxyl groups that are not incorporated into the backbone chain; that is, those contributed by the dicarboxyl amino acids (aspartic and glutamic acid) and the basic amino acids lysine, arginine, and histidine. As we shall see, this fact is of critical importance to the way the chain folds and coils upon itself to take on a three-dimensional **(tertiary)** shape. Another fact to note about the polypeptide shown in Figure 3-16 is that its ends are different. The left-hand end has a free amino group; on the right the chain ends in a free carboxyl group. Because the carboxyl of one amino acid always links to the amine of another, no matter how many

more amino-acid residues are added on to the chain at either end it will still have an **N-terminal** group at one end and a **C-terminal** group at the other (N for nitrogen and C for carboxyl, of course).

Amino Acid Sequence—Primary Structure

If a biochemist has isolated the protein fraction from a culture of cells and hydrolyzed these proteins into their component amino acids, what further information can he obtain? He may, for example, wish to identify the amino acids that make up the proteins he has extracted; he can do this by any of several methods, perhaps the simplest being **paper chromatography** (this technique is described schematically in Fig. 3-17, p. 66). With such a complex mixture of proteins he will undoubtedly find all 20 amino acids and little more, which will only confirm for him that the fraction he has hydrolyzed is indeed protein. But suppose he is interested in only one protein, for example one with a specific enzyme activity. By dissolving the proteins before hydrolysis, and separating them according to size or charge differences, he can identify the one exhibiting the enzymatic properties that he wants. He may then hydrolyze it alone to determine its amino acid composition. Again, however, he will proba-

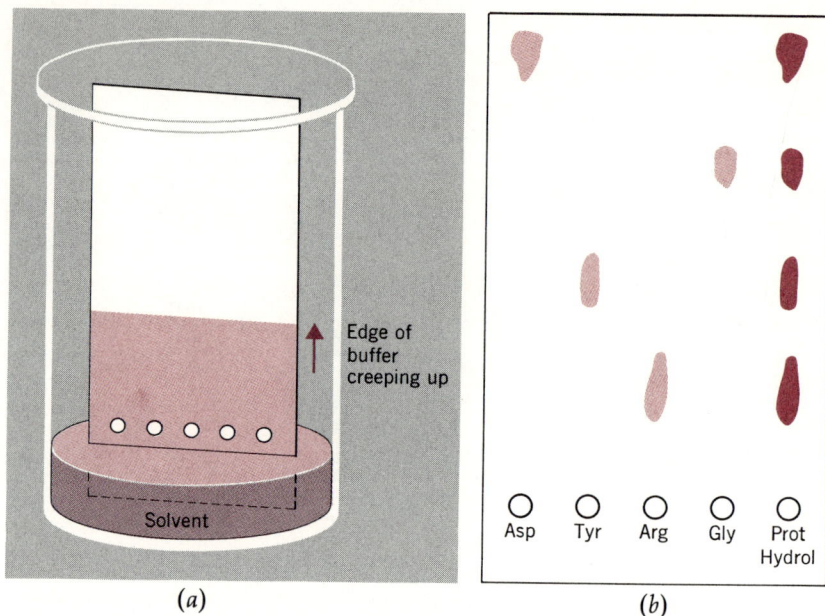

Figure 3-17 A simplified scheme for paper chromatography of amino acids.
A drop of an amino acid mixture resulting from acid hydrolysis of a protein
is applied to the corner of a rectangle of filter paper, and labeled. A drop of
a single purified amino acid is placed in each of the other labeled circles. The
paper is then dipped into a pool of appropriate solvent. (a) As the solvent
rises in the paper by capillary action, it carries the amino acids with it, but
each amino acid moves with the solvent at a different rate, depending on its
charge and molecular weight. After the solvent "front" has moved to the top,
the paper is dried and sprayed with a chemical that reacts with amino acids
to give a purple color. The unknown amino acids can then be identified by
comparing their positions with those of the known amino acids. In this scheme
only four comparisons are shown; in reality all 20 amino acids could be
identified. The filter paper with the separated amino acids on it (b) is called
a **chromatogram.**

bly be disappointed to find all 20 of the
amino acids. A few proteins are characterized
by a deficiency or excess of one or another
amino acids. (Insulin, for example, contains
no methionine but is very rich in glutamic
acid; and the relatively rare amino acid hy-
droxyproline is present in collagen but in few
other proteins. However, most have all 20 of
the common ones. In fact, if the biochemist
wishes to identify his single protein in a
unique way, he must know not only which
amino acids it contains but also the *order of*

their arrangement in the polypeptide chain.
That is, he must determine the **amino acid
sequence** of the protein.

Considering that even the smallest proteins
have molecular weights of several thousand,
and therefore may be made of dozens of
amino acid residues, it may seem like a for-
midable task for a chemist to try to determine
the linear sequence of these amino acids
along the protein chain. Indeed, although
proteins were known before the beginning of
this century, and their peptide structure was

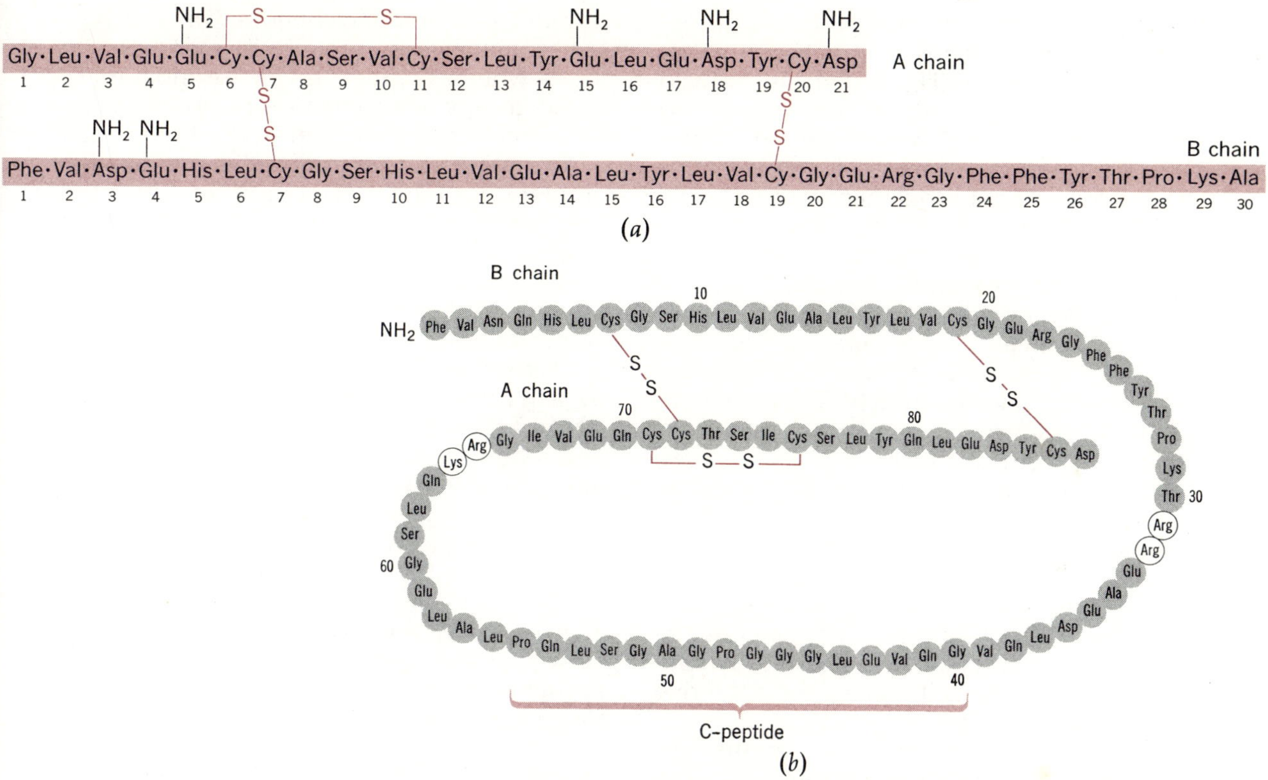

Figure 3-18 (a) *Amino acid composition and sequence of beef insulin showing the two distinct chains linked by disulfide bonds.* (b) *Structure of the original single long-chain molecule, synthesized in the cell, from which human insulin is derived by the action of a special protein-degrading enzyme. The part of the original precursor polypeptide that is digested away is labeled* **c peptide.** *The four residues drawn as open circles are the ones that differ from beef insulin. Note that the N-terminal and C-terminal ends of the two chains point in the same direction because of the way the chain is looped.*

worked out by the biochemist Emil Fischer in 1902, it has only been in the last 20 years that techniques have been available for "sequencing" proteins.

Frederick Sanger and his colleagues at Cambridge University in England spent more than 10 years working out the amino acid sequence of beef insulin (Fig. 3-18), the first protein to be so analyzed; for this work Sanger was awarded the Nobel Prize in 1958. Thus essentially everything we have learned about the actual structure of proteins has been learned in the last 15 to 20 years.

Insulin is a hormone secreted by special cells of the pancreas and is required for proper utilization of sugar by the body. When not enough is secreted, the disease **diabetes mellitus** results. This protein was carefully chosen for analysis by Sanger for three reasons: (1) It is easy to obtain in relatively pure form. (2) Because it lacks methionine, its purity can be determined. That is, any methionine in a hydrolysate must have come from "contaminants," traces of other proteins. (3) It is one of the smallest proteins known, with a molecular weight of less than 6000. Because the molecular weight of an average amino acid is about 110, Sanger could predict that he would be dealing with a sequence of approximately 50 amino acids. The actual number turned out to be 51 (Fig. 3-18a); he showed that the amino acid units are arranged in two distinct chains, one containing 30 amino acids units and the other 21. The two chains are parallel and held together by connecting bridges linking two sulfur atoms (S—S). Sanger also demonstrated that slight modifications in the structure of insulin, such as removal of one of its amino acids, could markedly decrease its biological activity as a hormone. Sanger and his group compared the sequences of insulin from different animal species (pig, sheep, horse, whale); they found them to be almost identical. Only in three positions (numbers 8, 9, and 10) in the A chain were variations found.

It has recently been shown that the A and B chains of insulin are synthesized in the cell as different parts of a single precursor molecule (Fig. 3-18b). Nonfunctional parts of the looped precursor chain are digested by special enzymes in the cell, leaving the definitive A and B chains intact.

The amino acid sequence, or **primary structure,** of a number of proteins has now been determined. Since 1960 sequencing has been completed for myoglobin (153 residues), human hemoglobin (two chains, 141 and 146 residues respectively), tobacco mosaic virus protein (158 residues), and several enzymes such as ribonuclease (124 residues), cytochrome C (104 residues), chymotrypsinogen (246 residues), and lysozyme (129 residues). Some of the most exciting results in this field currently are coming from sequence analysis of **immunoglobulins,** the proteins of which human antibodies are composed.

Why is it important to know the primary structure of a protein? There are at least two good reasons. First, it is the amino acid sequence—especially the order and spacing of residues with charged side groups—that determines the three-dimensional shape of the protein molecule; and, as will be seen when we examine the structures of enzymes and hormones, it is the three-dimensional configuration of these proteins that control their physiological roles. Depending on the protein, some residues seem to be merely space fillers, whereas others provide essential functional groups whose relative positions in the sequence remain constant over a broad span of evolution; for example, the mammalian pituitary hormone oxytocin differs from the frog hormone vasotocin by only one amino acid.

The second reason we wish to know the primary structure of proteins is that the amino acid sequence in each protein in a cell is determined in a precise and unique way by a particular nucleotide sequence in the DNA of the cell. If we are to understand how one sequence is translated into the other we must be able to determine both accurately. From these considerations we can begin to see emerging a concept of the control of cell function by information coded into the DNA, an idea we shall explore in detail in Chapters 4, 10, and 12. But first we must look more closely at the relation between the primary sequence and the three-dimensional structure of proteins.

Secondary and Tertiary Protein Structure

It is customary to classify proteins, according to their shape, as **fibrous** or **globular.** Fibrous proteins are generally arranged as long, linear molecules; most of the structural elements of the body fall into this group: collagen, the fibrous protein of tendon, ligaments, skin, etc.; myosin, one of the contractile muscle proteins; and keratin, the major protein of hair. Globular proteins are coiled and folded into nearly spherical shapes: the enzymes and protein hormones are all globular. The coiling of a linear chain represents its **secondary** configuration. Further twisting and folding of a coil is termed its **tertiary** structure. Information about the size and shape of protein molecules can be obtained by a variety of techniques: high-speed centrifugation, viscosity studies, X-ray scattering analysis. With the high resolution of modern electron microscopes, the shapes of large protein molecules can be seen directly (Fig. 3-19). Results from all of these methods suggest that across an enormous range of sizes (insulin, 6000 molecular weight; snail hemocyanin, 6.7 million) certain generalizations can be made: (1) Individual polypeptide chains are rarely larger than 50,000 to 100,000 molecular weight. Larger proteins tend to be built of multiple chains linked together. (2) Proteins are generally fairly rigid particles, having fixed and highly reproducible shapes. (3) Although the shapes of proteins vary from near-spherical to elongated fibers, many fibrous proteins are polymers with globular subunits.

A small protein of 10,000–15,000 molecular weight, composed of, say, 100 residues, would form a chain about 430 Å long, but only 2–3 Å thick. Such a molecule would not have the rigidity of typical proteins, nor would it give the appearance in the electron microscope of a fibrous protein such as collagen, which is about 15 Å thick (Fig. 3-19). Early protein chemists recognized that the linear chain is never fully extended, but must be folded or coiled in such a way as to produce the more compact, rigid structure found in nature; but the exact geometry of this folding was the subject of much speculation.

In 1951 Linus Pauling and Robert Corey provided the first clue from X-ray analyses of the bond angles and interatomic distances between the atoms of simple peptides. By making stick-and-ball scale models of the atoms involved, they showed that a peptide chain would rotate on itself in such a way as to form an extended coil, or **helix** (like the coil spring on a screen door, or a toy Slinky). Pauling noted that in this configuration the C=O group of every amino acid came into position next to an —NH group of another, at the right distance to form hydrogen bonds between them (Fig. 2-20, p. 70). We noted earlier that hydrogen bonds are among the weakest of chemical bonds. It was Pauling's insight, however, that if all or most of the atoms in a coil were linked to others by hydrogen bonds it would lend great stability to the structure. Subsequent work has confirmed that a wide variety of proteins are in fact coiled into this so-called **alpha helix** configuration.

Hydrogen bonds can be broken by many physical and chemical treatments. For example, raising the pH of a protein solution (i.e., making it more alkaline) tends to break hydrogen bonds, leading to the destabilization of a helix. Within limits this process is reversible; if the pH is lowered again, hydrogen bonds reform and the helical structure reappears. Heat also destroys hydrogen bonds,

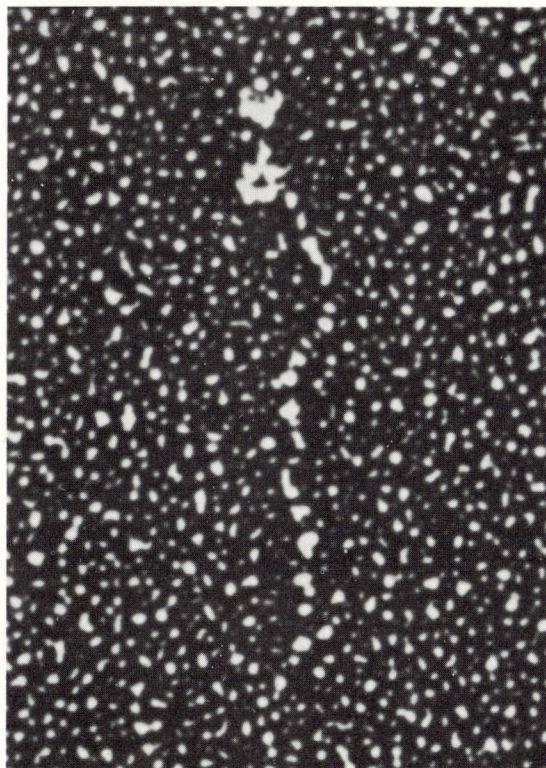

Figure 3-19 *Shadow-cast electron micrographs of mammalian myosin. Magnification 490,000×.* **Shadow-casting** *is done by evaporating a carbon or a metal such as platinum or gold or the preparation at a given angle. The metal will coat one side of the molecule and be absent from the other, thereby casting a "shadow" and producing a three-dimensional effect. From the length of the shadow and the known angle from which it was cast, we can calculate the height of the molecule. By measuring the other dimensions of the molecule, its volume can be calculated and therefore an estimate of its molecular weight can be obtained.*

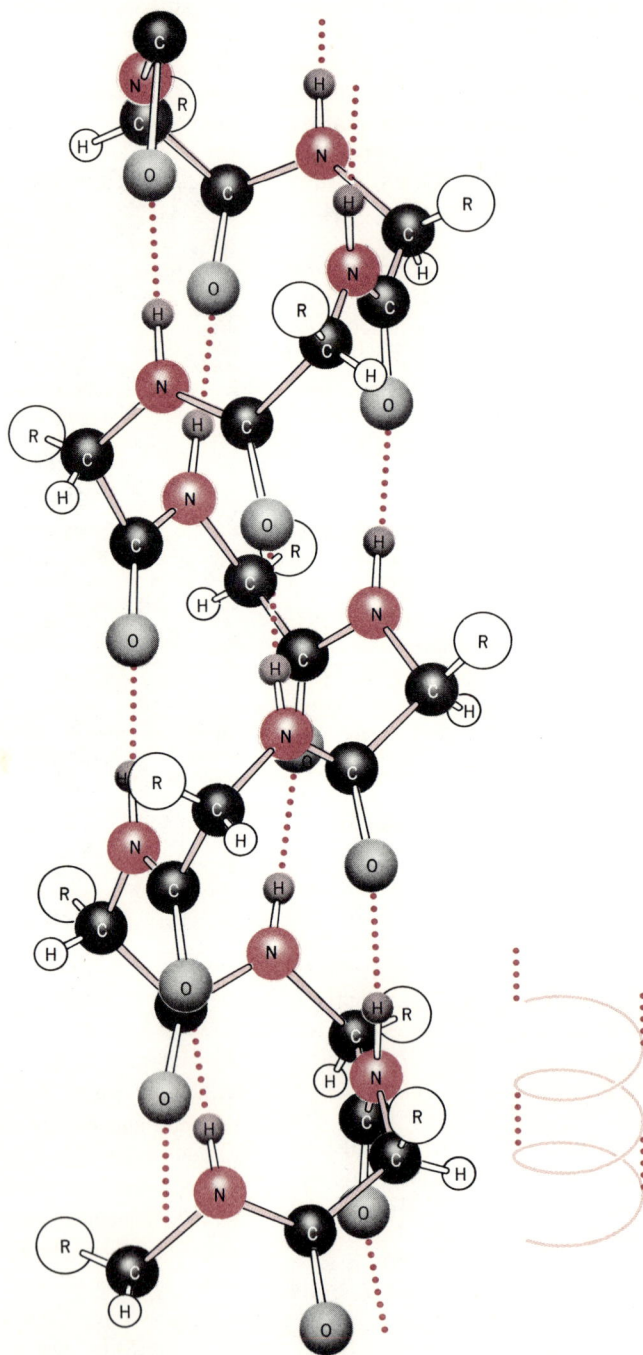

Figure 3-20 Model of a
right-handed alpha-helix.
Note that all C=O and
—NH groups form
hydrogen bonds (dotted
lines), and all R groups
(side chains) point away
from the center.

but beyond a temperature of about 60°C (140°F), for most proteins, the helical structure is irreversibly destroyed. The protein is then said to be **denatured.** The coagulation of egg whites on boiling is an example of denaturation; although boiling does not alter the primary structure (i.e., sequence) of the protein, it uncoils the peptide chains and releases the reactive groups. Denaturation deactivates all enzymes and protein hormones.

Look again at the illustration of the alpha helix (Fig. 3-20). Note that the hydrogen bonds extend between atoms of the backbone of the peptide chain. The side chains (R) of the amino acids are not involved, and in fact point away from the helix. During the last decade it has become clear that interactions between these chains also play an

important role in stabilizing the three-dimensional structure of proteins. For example, each cysteine group in the peptide has a sulfhydryl (—SH) group at the end of its side chain. Sulfhydryl groups are very reactive, especially with other sulfhydryls. Thus whenever two cysteine units are brought close together by folding of the peptide chain, no matter how far apart they may be spaced in the amino sequence, they form an S—S (sulfur-sulfur) linkage, called a **disulfide bond.** Such bonds are very strong, and hold the polypeptide in a folded configuration. S—S bonds bind the A and B chains of insulin together (Fig. 3-18). In a single-chain molecule such as ribonuclease (Fig. 3-21) disulfide bonds link parts of the chain in four different places, giving the long chain a relatively compact, nearly round over-all shape.

Figure 3-21 *Structure of the protein enzyme ribonuclease, showing disulfide bonds (—S—S—) cross-linking the peptide chain at four positions.*

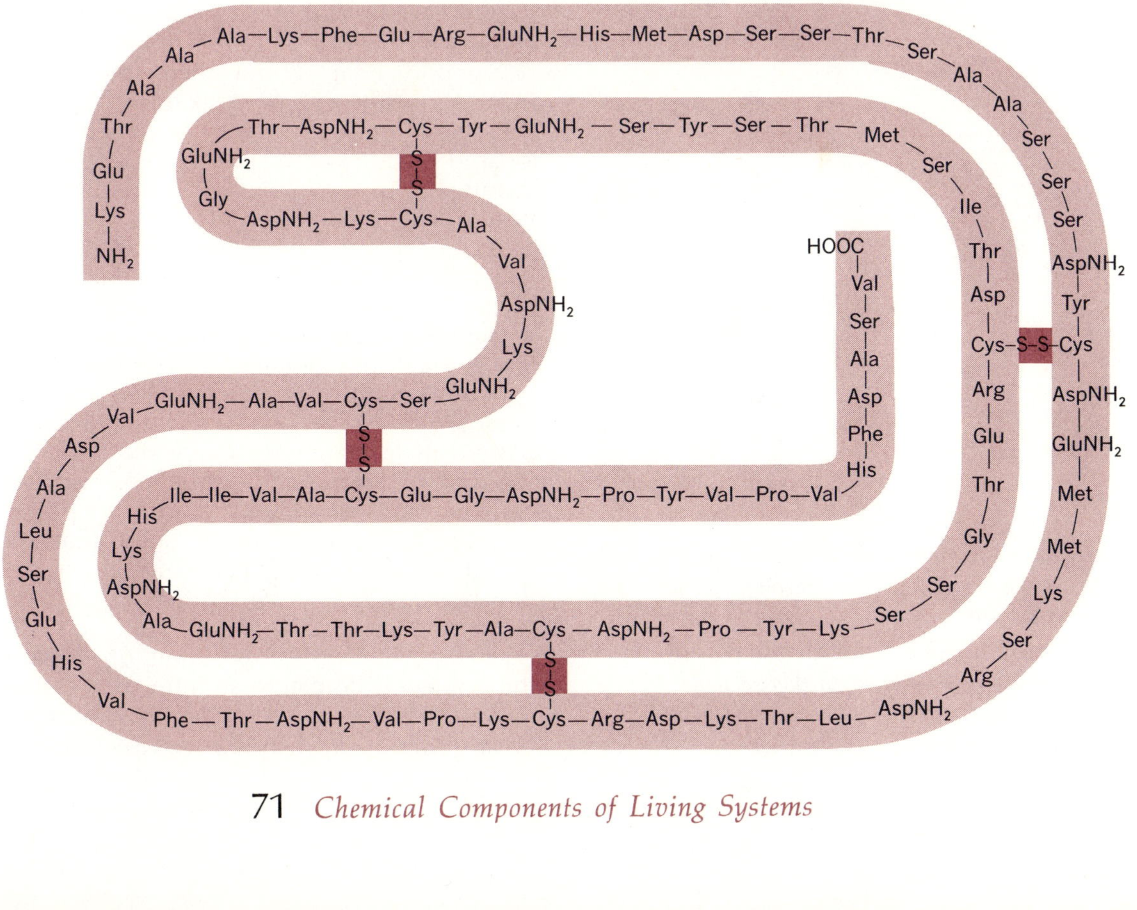

There are also other forces that lead to the folding of linear peptide chains into condensed globular structures. Some of the amino acids can form ionic bonds with one another across different regions of a peptide chain. The free carboxyl of glutamic acid, for example, can bind to the side-chain amino nitrogen of a nearby arginine residue to form an ionic linkage. Other amino acids have side chains made essentially of a hydrocarbon chain or other nonpolar group (valine, leucine, tryptophan). In a way comparable to fatty acids, these groups represent hydrophobic portions of the amino acids, which tend to orient away from an aqueous solvent (Fig. 3-12). The result is that proteins tend to fold into configurations in which these hydrophobic groups are together on the inside of a globular mass, with only the polar-end groups pointing out. Recent evidence shows that these **hydrophobic interactions** are of substantial importance in maintaining protein shapes.

The three-dimensional configuration of a protein can also be determined if it is bound to a nonprotein molecule. Such complexes are called **conjugated** proteins; the component that is not made of amino acid subunits is called a **prosthetic group.** The first protein whose complete three-dimensional structure was determined was myoglobin, which, joined to a prosthetic group called **heme,** is the oxygen-carrying pigment of muscle. This structural analysis was done with the aid of X-ray techniques. It had been known for some time that X-rays projected through any regular structure, such as a crystal, would produce diffraction patterns from which the three-dimensional characteristics of the structure could be analyzed down to dimensions of 1.5 to 2Å. After Max Perutz at Cambridge had demonstrated techniques for obtaining similar patterns from protein crystals, his colleague John Kendrew set about the laborious task of measuring thousands of such patterns from crystallized myoglobin and analyzing the data with the help of high-speed electronic computers. The picture that emerged was of a compact structure composed mainly of a tightly folded single polypeptide chain, about 70 percent of which is coiled into the alpha-helix form (Fig. 3-22).

Essentially all of the nonpolar side chains are located together in the interior of a molecule, forming a completely hydrophobic core. Within the myoglobin structure the heme prosthetic group contains an iron atom (p. 76), which is involved in the interaction of myoglobin with oxygen. Kendrew published a model of sperm whale myoglobin in 1960. Since that time detailed three-dimensional structures have been determined for human hemoglobin, which has a startlingly similar pattern of folds despite major differences in the primary amino acid sequence. Also, several enzymes—lysozyme, carboxypeptidase A, and ribonuclease—have been analyzed. Results for these have been especially important because they have led for the first time to an understanding, in terms of molecular geometry, of how enzymes function. Because this subject is crucial to an intelligent discussion of the rest of modern biology, we dwell for a moment on it in the next section.

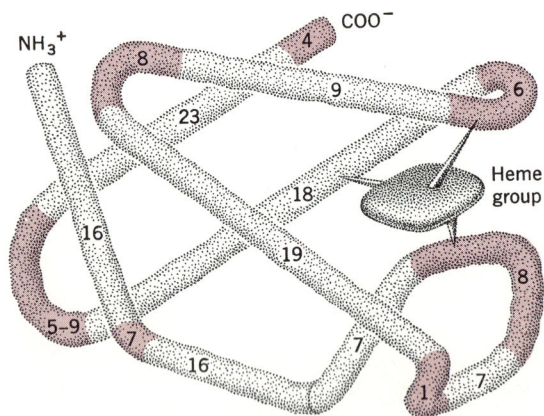

Figure 3-22 *Conformation (shape) of the myoglobin molecule. The shaded flat structure at the right is a heme group, which contains an atom of iron. The eight straight, uncolored portions are regions coiled into an alpha helix. The seven colored bands are not helical. The number of amino acids in each region is also shown.*

Enzymatic Role of Proteins

The most significant function of proteins in the living cell is their enzymatic role. Most enzymes are globular proteins ranging in molecular weight from approximately 10,000 up into the millions. The smallest enzymes are made of approximately 100 amino acid residues, the largest ones of many more. Enzymes are made by living cells and are responsible for the catalysis of virtually all biochemical reactions occurring in protoplasm.

The term enzyme is derived from Greek and means, literally "in yeast." It was originally coined in 1878, when it was generally but erroneously accepted that the yeast enzymes responsible for the fermentation of wine could function only in living cells. All attempts to extract these enzymes had failed. Louis Pasteur, the leading biochemical authority of that era, was the outstanding proponent of the idea that fermentation was an expression of life itself and could occur only in living cells. It was one of the few instances in which the brilliant Pasteur proved to be wrong, for in 1897 Eduard Buchner in Germany demonstrated for the first time that enzymes in cell-free extracts prepared by macerating yeast cells were capable of fermenting sugar to ethanol and carbon dioxide. Buchner's epic work stands as a landmark in the history of biology; it demolished an erroneous concept, and also ushered in a new and modern era of biology in which most of the functions of cells have been shown to occur—and can be studied—in the test tube. This is the era of biochemistry or molecular biology.

Mechanism of Enzyme Action. According to the modern collision theory of chemical reactions, a catalyst increases the rate of a chemical reaction by decreasing the required energy of activation for the reaction (see pp. 43–45). The enzyme as a catalyst accomplishes this by combining with the reacting molecules to form an intermediate stage called an **enzyme-substrate complex.** This highly unstable complex rapidly rearranges itself to yield the products of the reaction and the free, unaltered enzyme, thus making the latter available for recombination with more substrate. This process is repeated many times. One molecule of the enzyme catalase, for instance, can catalyze the breakdown of 5 million hydrogen peroxide molecules to water and oxygen in one minute. These relationships may be formulated as follows:

$$E + S \longrightarrow ES \longrightarrow E + \text{products}$$

E is the enzyme, S the substrate or reacting molecules, ES the enzyme-substrate complex.

Enzyme Specificity. Enzymes as biological catalysts act in fundamentally the same fashion as any other catalyst except for two important differences. First, each enzyme consists entirely or in part of proteins specific to that enzyme. Second, enzymes themselves are highly specific; each enzyme usually catalyzes only one kind of reaction (because each is a different protein). By contrast, a nonbiological catalyst like platinum catalyzes a wide variety of reactions.

In general, enzymes are extremely sensitive to high temperatures. Exposure to 100°C (212°F) for two minutes almost always results in a complete and irreversible loss of enzyme activity. Enzymes are also markedly influenced by pH or H^+ ion concentration. In most instances they exhibit their maximal catalytic activity within a characteristically narrow pH range. At a pH appreciably below or above this optimal level, enzymatic activity is considerably smaller or even absent. Different enzymes display different optimal pH levels, a phenomenon attributed largely to the specific nature of their proteins.

Enzymes display varying degrees of specificity with respect to the types of substrate and chemical reaction they catalyze. Some enzymes are so highly specific that they catalyze only one chemical reaction involving one reactant or substrate, whereas others may catalyze a number of related reactions in a wider range of reactants.

What accounts for this unique specificity of enzymes as compared to nonbiological catalysts? The combination that occurs between enzyme and substrate to form the enzyme-substrate complex depends on a particularly suitable spatial relationship between the substrate and an **active site** on the

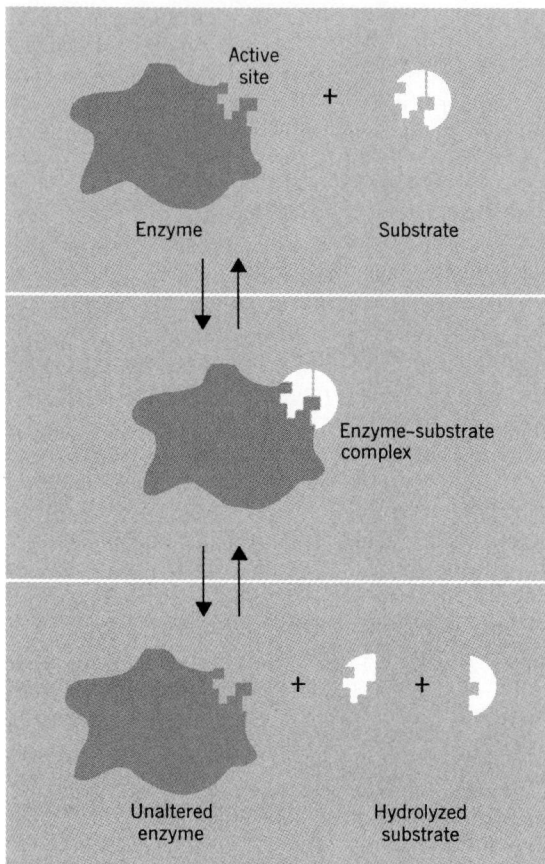

Figure 3-23 Schematic mechanisms of enzyme action, showing the molecular fit between the substrate and the active site on the enzyme protein molecule. Reading the reaction from top to bottom, a substrate is hydrolyzed. From bottom to top, two small molecules are synthesized into a larger product.

enzyme (Fig. 3-23). This has been compared to a lock and key; it implies that the configuration of atoms in the active site of the enzyme responsible for catalytic activity can only join with a particularly shaped substrate. The vast differences in size between most enzyme molecules and their relatively minute substrate molecules supports the notion that the enzyme-substrate complex must be limited to the specific active sites (usually one or at most a few) on the huge surface of the protein molecule. For example, catalase, a large molecule with a molecular weight of 250,000, has as its substrate hydrogen peroxide (H_2O_2), with a molecular weight of 34. For enzymes exhibiting a relatively broader specificity, the structural configuration of the active enzyme sites is pictured as being less specialized (like a skeleton key), allowing for union with a number of substrates. Recent evidence suggests that some enzymes may have flexible active sites, in the sense that the substrate induces a change in configuration or conformation of the enzyme molecule, like a hand closing around a handle.

Enzyme Classification. Enzymes can be conveniently classified in several major groups according to the chemical reactions they catalyze. These in turn can be subdivided into smaller groups, depending on the more specific aspects of their catalytic properties.

The present practice is to name an enzyme by adding the suffix **-ase** to the name of its substrate. For example, the enzyme responsible for the hydrolysis of the dissacharide sucrose to fructose and glucose is called **sucrase.** Not all enzymes, however, have been named according to this logical system. **Proteases** hydrolyze peptide bonds. **Oxidases** or **dehydrogenases** catalyze oxidation-reduction reactions, transferring hydrogen atoms (or electrons) from one molecule to another.

Hydrolases are enzymes that catalyze the splitting of molecules into smaller molecules by the introduction of water. They include enzymes that accelerate the hydrolysis of: disaccharides and polysaccharides into smaller units; peptide bonds of proteins, polypeptides, and dipeptides to yield amino acids; and fats to their component fatty acids and glycerol. Many hydrolases occur in living cells and are also secreted in the process of digestion in the alimentary tract. Digestion consists primarily of the hydrolysis of carbohydrates, proteins, and fats to their simpler structural units by specific hydrolases.

Transferases are enzymes that catalyze the transfer of chemical groups from one substrate to another. Typical of this group are **transaminases,** which mediate the transfer of the amino group of an amino acid to an acid containing a ketone group (see Chapter 4). Another subdivision of the transferases are **kinases,** which mediate the transfer of a phosphate group specifically from the biologically important high-energy compound adenosine-triphosphate (ATP) to particular substrates. For example, the metabolism of glucose by most cells involves first its phosphorylation by the transfer of a phosphate group from ATP, a reaction that cannot occur unless the enzyme **hexokinase** is present.

Apoenzymes, Coenzymes, Metal Components, and Prosthetic Groups

Coenzymes. Many enzymes, but not all, contain, in addition to their protein part (called an **apoenzyme**) a nonprotein **prosthetic group** of relatively small molecular weight (called a **coenzyme**), which is essential for enzymatic activity.

Dissociation of an enzyme into its apoenzyme and coenzyme causes a loss in enzymatic activity, which is restored in most instances when the components are again added together. The discovery and characterization of coenzymes has revealed the metabolic role of many vitamins, for we now know that coenzymes are made up in part of individual vitamins (Chapter 14). For example, vitamin B, or thiamine, is part of a coenzyme necessary for certain enzymatic reactions involving carbon dioxide (Chapter 4). The vitamin niacin, or nicotinic acid, is a component of two large coenzyme molecules called **nicotinamide adenine dinucleotide** and **nicotinamide adenine dinucleotide phosphate,** known as **NAD** and **NADP** respectively; these serve as coenzymes for several different enzymes catalyzing oxidation-reduction reactions (Chapter 4) in the cell. Other vitamins, such as riboflavin, pyridoxine, and pantothenic acid, are also components of coenzymes. Many enzymes also contain small quantities of a metal ion (e.g., copper, zinc, molybdenum, or iron) that may be necessary for enzymatic activity. Removal of the metal from these **metalloenzymes** causes a loss in their catalytic action.

Porphyrins. Porphyrins are found in a wide variety of conjugated proteins as parts of prosthetic groups, or coenzymes. They are present in many respiratory enzymes and several important nonenzymatic proteins, for example chlorophyll, the green pigment of plants (Chapter 13). They are widely distributed in nature, occurring in almost all living things in one form or another.

A porphyrin consists of four small ring units combined with one another to form a larger ring compound. The smaller rings, known as **pyrroles,** are made up of four carbons and one nitrogen:

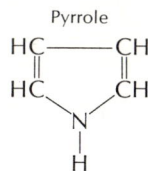

Pyrrole

and are linked by =C— groups to constitute the larger ring system, the porphyrin itself. In addition to their attachment to specific proteins, most naturally occurring porphyrins contain a metal component, thus they are called **metal porphyrins** (Fig. 3-24, p. 76).

The most biologically important types of metal porphyrin are **hemes,** which contain iron, and **chlorophylls,** which contain magnesium. In both groups the metal (designated as Me) is linked to each of the four nitrogen atoms of the pyrrole rings at the center of the porphyrin structure like the hub of a wheel. The fundamental structures of all porphyrins found in living things are essentially alike. They differ from one another largely in the type and arrangement of the side-chain groups attached to each of the constituent pyrrole rings.

Hemes. The principal hemes and their protein combinations in living systems include: **hemoglobin** (Chapter 15), the oxygen-transporting pigment of blood, which is a red-colored complex consisting of a heme and a specific globular protein; **cytochromes** (Chapter 4), which are red heme proteins important in the final stages of respiration or electron transfer; and two oxidizing enzymes, catalase and peroxidase, also heme proteins.

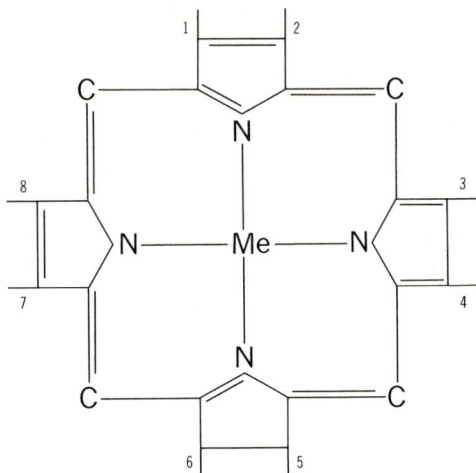

Figure 3-24   Diagrammatic chemical structure of metal porphyrin. For hemes the metal (Me) is iron, and for chlorophylls it is magnesium.

Figure 3-25   The structure of chlorophyll a. The four pyrrole rings of the porphyrin in which magnesium is bound are identified with roman numerals.

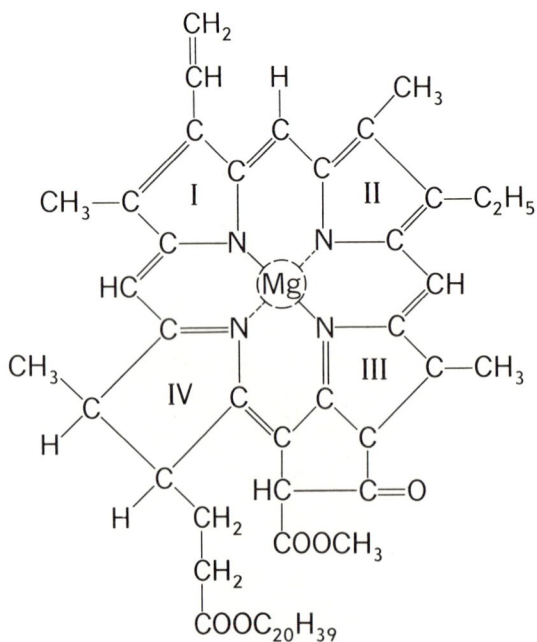

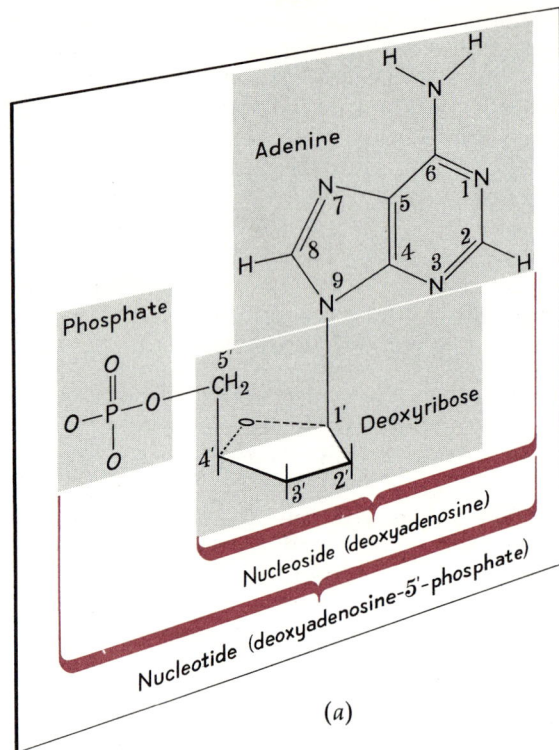

Figure 3-26 (a)

The heme iron of hemoglobin occurs exclusively in the **ferrous** (Fe^{2+}) form, a state that is absolutely essential for hemoglobin to function as the oxygen-carrying component of blood because oxygen combines reversibly with the Fe^{2+} (Chapter 4). In other heme proteins such as catalase and peroxidase the iron is in the **trivalent ferric** (Fe^{3+}) state. However, the heme iron of the cytochromes experiences a reversible change from the ferrous to the ferric state during its metabolic function as an electron carrier in respiration. A number of different cytochromes function in respiration involving molecular oxygen (**aerobic respiration,** Chapter 4).

Chlorophylls.   We shall discuss chlorophyll in detail in Chapter 13. Here let us say simply that four principal chlorophylls occur in the plant kingdom. The magnesium of the chlorophylls apparently does not undergo any change during photosynthesis. The most

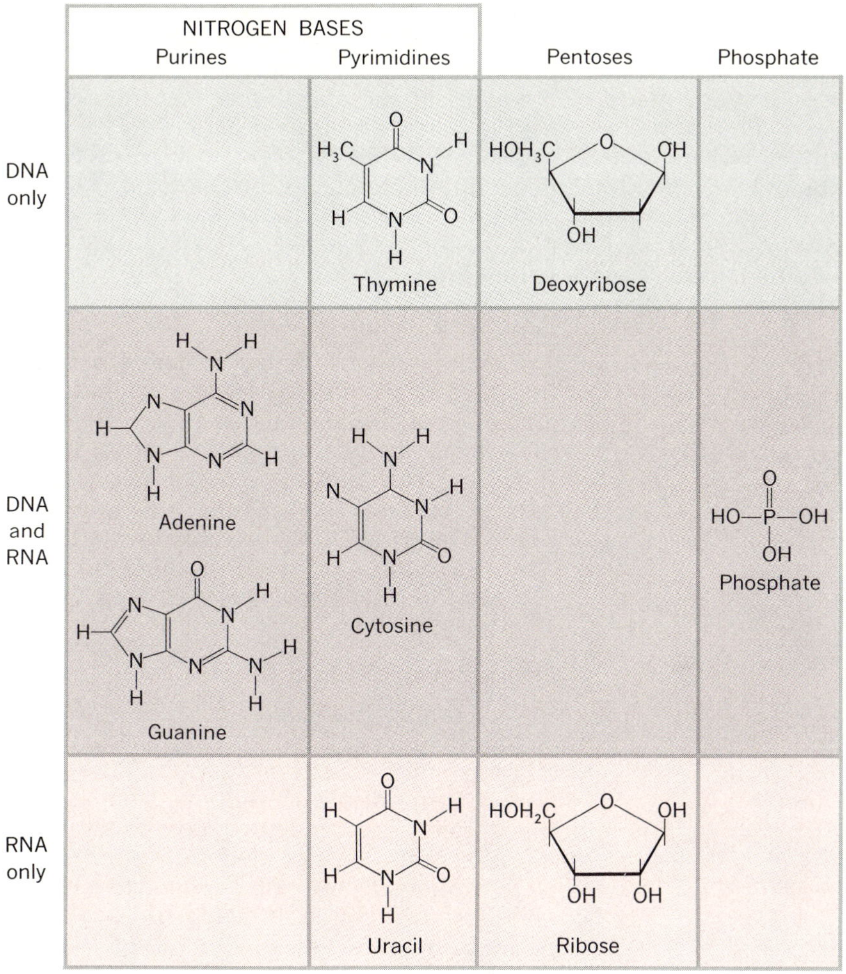

NITROGEN BASES		Pentoses	Phosphate
Purines	Pyrimidines		

DNA only — Thymine, Deoxyribose

DNA and RNA — Adenine, Guanine, Cytosine, Phosphate

RNA only — Uracil, Ribose

(b)

Figure 3-26   Components of DNA and RNA (these are discussed in some detail in the following pages). (a) AMP (adenosine monophosphate), a typical nucleotide. Note that adenine is a flat molecule in the plane of the page, whereas the ring of the deoxyribose forms a plane perpendicular to the page. (b) The building blocks of the nucleotides.

common of the chlorophylls is **chlorophyll *a*,** which is almost universally distributed in all green plants (Fig. 3-25). **Chlorophyll *b*** is the second most widely distributed; it is found in all higher plants and some algae. In the photosynthetic bacteria there are still other kinds of chlorophyll. All the chlorophylls, as magnesium-containing porphyrins, are essentially alike in their chemical structure. They differ mostly in the chemical make-up of their side chains connected to the pyrrole rings of the porphyrin structure.

Nucleic Acids

Nucleic acids are found in all living cells and are combined in nearly all instances with proteins. Chemically, nucleic acids, so called because they give an acid reaction in water, are huge, thread-like compounds of very great length with a molecular weight in the tens and hundreds of millions. They are chain-like substances built of hundreds and hundreds of similar but not identical smaller repeating units called **nucleotides.** Nucleotides, in turn, are made of **nucleosides**—which consist of an organic base and a molecule of ribose or deoxyribose sugar—and phosphate. A single nucleic acid strand made up of many nucleotides is called a **polynucleotide,** analogous to the polypeptide chain of a protein. Each nucleotide is made up of a phosphate group, a pentose sugar (always either ribose or deoxyribose), and an organic base called a **nucleotide base.** The sugar and phosphate groups can be regarded as the chemical backbone of nucleic acid; the bases can be viewed as important side branches. If the sugar is ribose, the nucleic acid is called **ribonucleic acid** or RNA; if the sugar is deoxyribose, the nucleic acid is called **deoxyribonucleic acid,** or **DNA.** Of the several kinds of organic base, there are only two that occur in nucleic acids: **purines** and **pyrimidines.**

Pyrimidines and Purines

The nucleic acid bases, so called because they give an alkaline (basic) reaction in water, are cyclic organic molecules of varying complexity that include nitrogen atoms as part of their ring structure. The major pyrimidines present in cells are **cytosine, thymine,** and **uracil.** The two purines are **adenine** and **guanine.** The structure of all of these building blocks of the nucleic acids is shown in Figure 3-26 (pp. 76 and 77).

Deoxyribonucleic Acid (DNA)

The deoxyribonucleic acids are found largely in the nuclei of cells as part of the chromosome structure. As such, they are always combined with proteins such as histones, and are therefore components of mixed compounds called **nucleoproteins.** DNA from nearly all sources characteristically contains four different kinds of base: two purines—adenine (A) and guanine (G)—and two pyrimidines—thymine (T) and cytosine (C). The total number of purines in any given DNA is always roughly equal to the total number of pyrimidines (A + G = T + C). Of even greater significance, as we shall see in Chapter 10, is the fact that the number of adenines is equal to the number of thymines (A = T) and the number of guanines is equal to the number of cytosines (G = C). The proportions of these bases (A + T/G + C) in DNA appear to be typical and constant for each particular species or organism; in some DNA molecules the content of A + T is greater than that of G + C, whereas in others the reverse is true. Both the proportions and the sequences of the nucleotides vary according to the given DNA, so that no two nucleic acids need be alike, just as no two

Figure 3-27 Portion of a single DNA polynucleotide chain. Nucleotides are linked together by bonds between their sugar (deoxyribose) and phosphate groups. The nucleosides do not form part of the backbone of the chain, but extend out to one side.

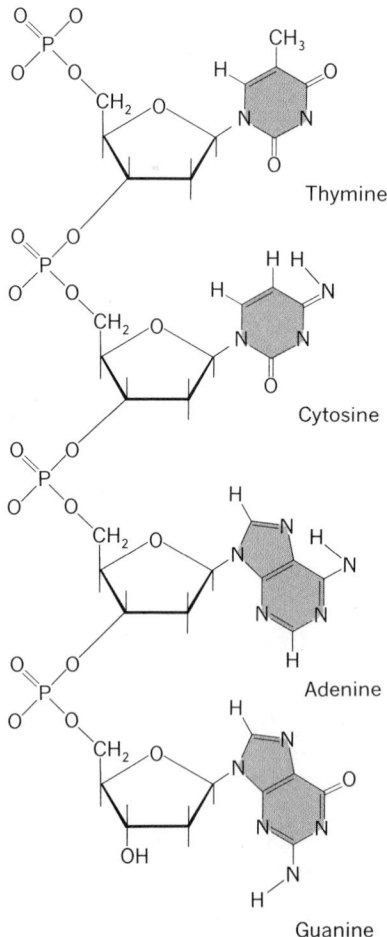

Thymine

Cytosine

Adenine

Guanine

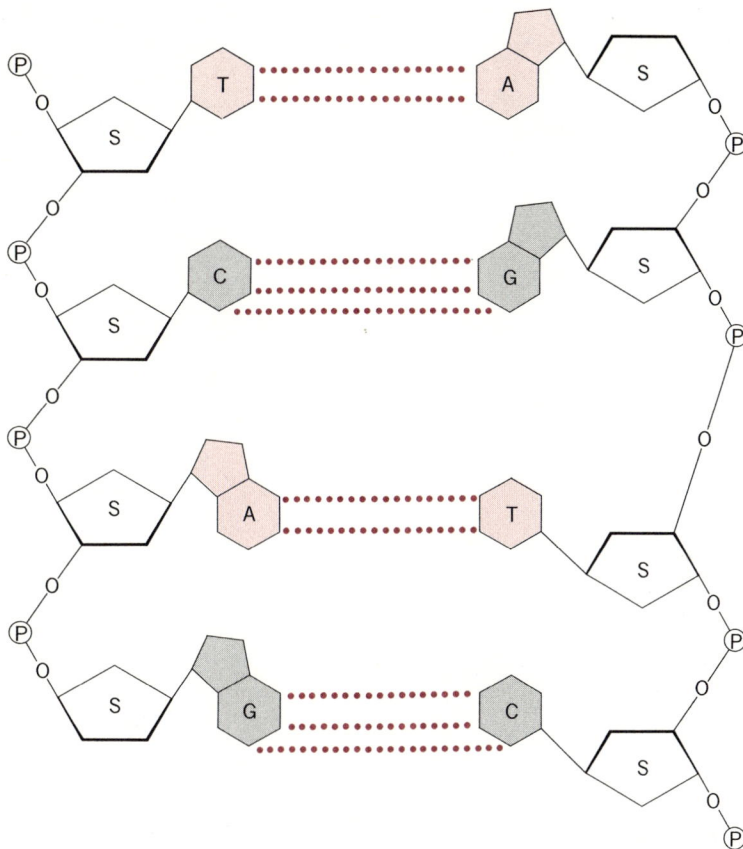

Figure 3-28 Segment of a DNA molecule uncoiled, showing that it consists of two polynucleotide chains, giving the molecule a ladderlike structure. The uprights of the ladder are the sugar-phosphate backbones; the cross-rungs are pairs of nitrogenous bases, hydrogen bonded. Note that each cross-rung has one purine (double shapes, A and G), and one pyrimidine (single shapes, C and T). When the purine is adenine (A), the paired pyrimidine must be thymine (T).

proteins are alike because they differ in the composition and order or arrangement of their amino acid units. Part of a DNA chain is illustrated in Figure 3-27.

This important information, together with certain physical data, led James Watson and Francis Crick in 1953 to propose a three-dimensional structure for DNA. Because A = T and G = C, they proposed that in a molecule of DNA every A must be bound to a T, and every G to a C (Fig. 3-28). They determined that the DNA molecule consists of two long, adjacently attached polynucleotide chains. For such chains to have a stable structure, Watson and Crick postulated that they would have to be aligned and coiled about one another to form a double helix, like two adjacently connected bannisters

of a spiral staircase (Fig. 3-29, p. 80). Because the purines (adenine and guanine) are larger molecules than the pyrimidines (thymine and cytosine), if they butted against one another they would hold the two strands so far apart that T could not bond with C. Watson and Crick therefore proposed that the two poly-nucleotide strands of the DNA double helix are reciprocally related as complementary copies of one another, because a base on one polynucleotide chain can only be paired with a specific base on the other (i.e., adenine with thymine and cytosine with guanine, Fig. 3-30, p. 80). Thus if the sequence of bases in one strand is TGTCA, that of the adjoining strand would be ACAGT (Fig. 3-29).

A vast quantity of data has since been obtained by research workers in genetics,

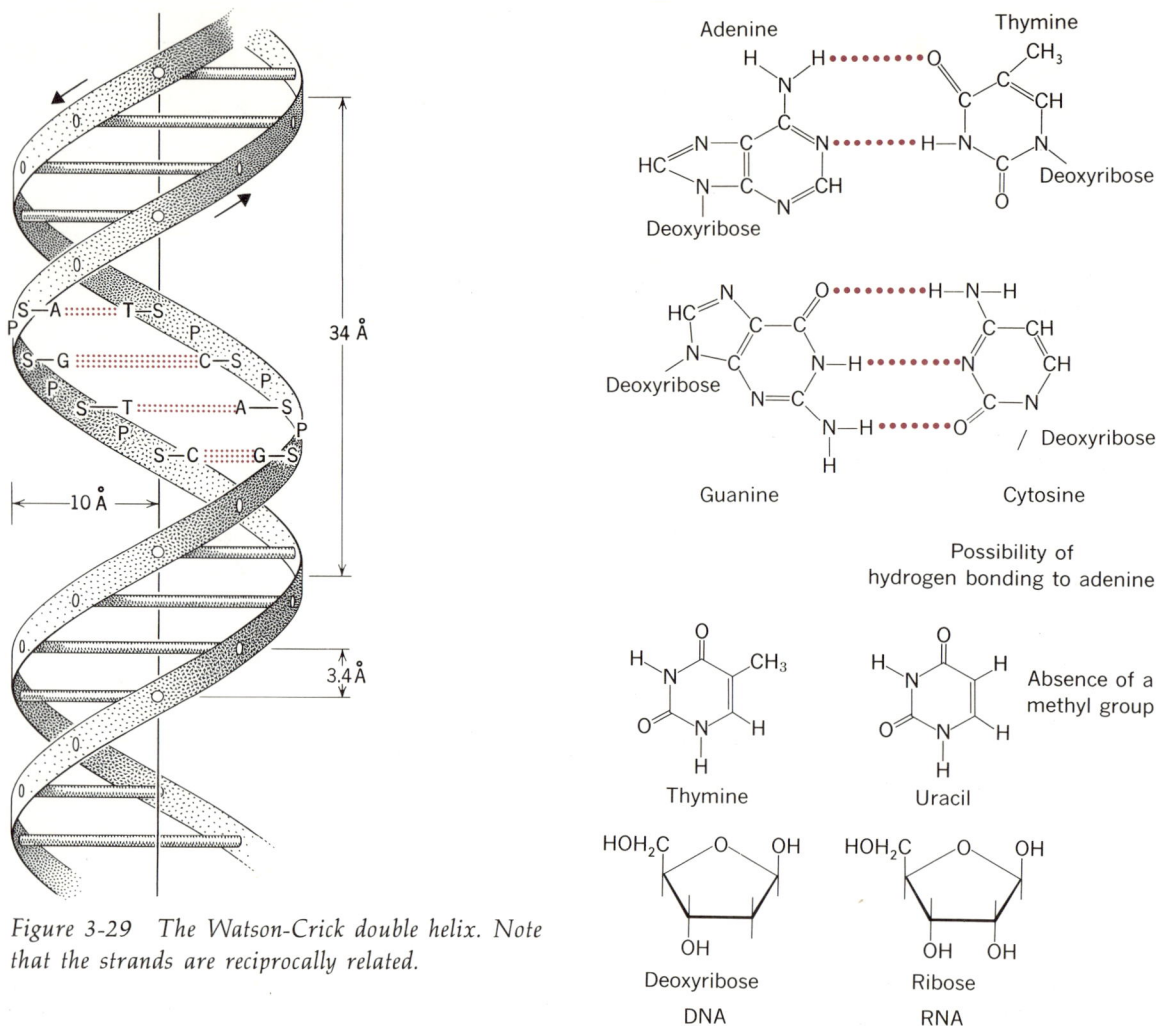

Figure 3-29 *The Watson-Crick double helix. Note that the strands are reciprocally related.*

biochemistry, and biophysics, confirming this molecular structure of DNA. For this epic contribution Watson and Crick were honored with the Nobel Prize in 1962 (see Chapter 10 for a more detailed discussion).

RNA

Our knowledge of the three-dimensional structure of RNA is considerably less than that of DNA. Ribonucleic acids are also universally present in living cells, in combination with proteins, as **ribonucleoproteins.** They occur largely in the cytoplasm of cells, and to a lesser extent in the nuclei. Three types of RNA are recognized in the cytoplasm; **ribosomal RNA** (about 80 percent of the total cellular RNA), **transfer RNA,** and **messenger RNA** (see Chapters 4 and 10). RNA strands are also polynucleotides similar to DNA, as already indicated, but with a number of important differences: RNA has ribose in place of deoxyribose; it has a slightly different base composition (thymine is replaced by the pyrimidine uracil); and the RNA molecule is generally believed to be a single-stranded structure, unlike DNA (Figs. 3-31, 3-32).

So far we have spoken mainly of the structure of the biological molecules. But living

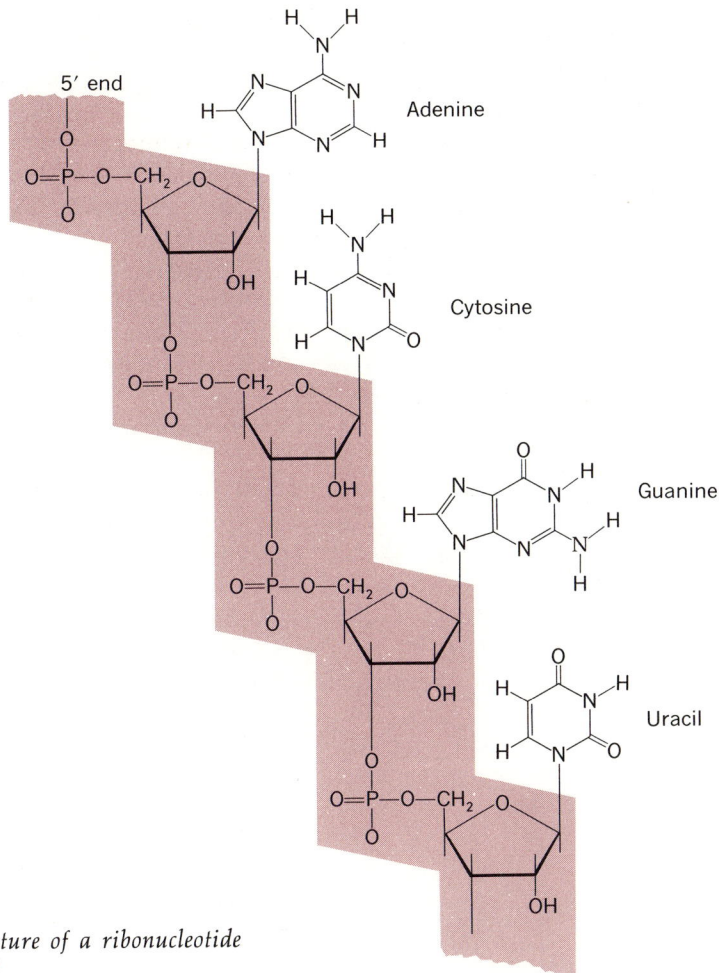

Figure 3-30
(left)   Hydrogen bonding
between purine and
pyrimidine bases.

Figure 3-31
(left)   Structures of
thymine and deoxyribose of
DNA, compared with
uracil and ribose of RNA.

Figure 3-32 (right)   Structure of a ribonucleotide
portion of an RNA strand.

systems are characterized mainly by the complexity of their chemical reactions (Chapter 1); that is, by their ability to use energy to build complex molecules out of simple subunits and then to extract the energy and subunits in a variety of biochemical reactions that we call **metabolism.** This is the topic we turn to in Chapter 4.

We have seen that the specificity of an enzyme depends on its shape; and its shape, in turn, is determined by its primary structure or amino acid sequence. We shall see in Chapter 4 that the sequence of every protein in a cell is determined in a precise and unique way by a nucleotide sequence in the DNA of the cell.

Reading List

Borek, E., *The Code of Life*. Columbia University Press, New York, 1969.

DuPraw, E. J., *Cell and Molecular Biology*. Academic Press, New York, 1968, p. 14.

Edelman, G. M., "The Structure and Function of Antibodies," *Scientific American* (August 1970), pp. 34–42.

Goldstein, L., *Cell Biology, A Book of Readings*. Wm. C. Brown, Dubuque, Iowa, 1966.

Haensel, V., and R. L. Burwell, Jr., "Catalysis," *Scientific American* (December 1971), pp. 46–61.

Haynes, R. H., and P. C. Hanawalt, *The Molecular Basis of Life*. W. H. Freeman, San Francisco, 1968. Readings from *Scientific American*.

Kamen, M. D., *A Tracer Experiment*. Holt, Rinehart & Winston, New York, 1964.

Kendrew, J. C., "The Three-Dimensional Structure of a Protein Molecule," *Scientific American* (December 1961), pp. 96–106.

Phillips, D. C., "The Three-Dimensional Structure of an Enzyme Molecule," *Scientific American* (November 1966), pp. 78–90.

Watson, J. D., *The Double Helix*. Signet Books, New York, 1968.

White, E. H., *Chemical Background for the Biological Sciences*. Prentice-Hall, Englewood Cliffs, N.J., 1970.

The Chemical Processes of Life

*"Respiration and photosynthesis are thus complementary elements in
the biological energy cycle, and they represent the most fundamental
and massive biochemical processes occurring on this planet."*

A. H. LEHNINGER, 1965, in J. A. MOORE (ed.), *IDEAS IN MODERN BIOLOGY*

*Respiration
and Metabolism*

The many chemical reactions and energy changes that occur in
living cells are collectively called **metabolism.** The process of metabolism may be broken into two broad subdivisions: **catabolism**
is the breakdown, or **degradation,** of large molecules into smaller
molecules and is usually accompanied by a release of energy;
anabolism, or **biosynthesis,** is the buildup, or **synthesis,** of large
molecules from smaller ones, frequently requiring an input of energy. The degradation, for example, of carbohydrates to carbon
dioxide and water by living cells is a catabolic process, whereas the
synthesis of proteins from amino acids is anabolic.

For many organisms, including ourselves, energy (stored in chemical bonds) and matter are provided by certain organic substances
in the diet such as carbohydrates, proteins, and fats. These undergo
various enzymatic reactions that liberate their stored energy and
make it available to the cells of the body. For green plants the source
of energy is sunlight, which is used to combine carbon dioxide and
water into energy-rich carbohydrates by the process of **photosynthesis.**

The term **respiration** is applied to one phase of metabolism
(p. 84). It is the integrated series of chemical reactions by which a
living cell obtains energy for its various life functions by oxidizing
foods or nutrients. In many organisms, including man, the oxidative
breakdown of carbohydrates is the principal energy source for most

activities of the cell. Although both respiration and **digestion** (another phase of metabolism) involve the degradation of large molecules to smaller ones, they are easily distinguished. In digestion large molecules are split into smaller molecules by enzymatic hydrolysis, and any energy released in the rupture of chemical bonds is liberated as heat. In respiration large molecules containing high-energy bonds are enzymatically broken down into smaller, lower-energy molecules in several ways, including oxidation-reduction reactions (p. 85). The most significant aspect of respiration is that an appreciable portion of the chemical energy released, instead of being given off as heat, is trapped in the form of other high-energy chemical bonds, to be utilized ultimately for various activities of the cell.

Although the vast welter of detailed biochemical reactions that constitute the various metabolic processes appears to be overwhelming, a number of basic patterns have become evident. First, a metabolic process generally consists of a sequence or series of successive enzymatic reactions, collectively called a **pathway** or **cycle,** whereby the product of one enzymatic reaction is a reactant or substrate of the next enzymatic step, and so on. For example, the metabolism of glucose to its ultimate products (carbon dioxide and water) in most organisms occurs mainly by way of about two dozen enzymatic steps called the **respiratory pathway** (see pp. 86–93).

Second, all cells capture useful energy by forming a unique high-energy-containing compound called **adenosine-triphosphate,** or **ATP** (see p. 89), which serves as the direct source of energy for all life activities. It is produced principally by the respiratory process and represents a rich source of energy stored in the form of chemical bonds.

Third, certain organic substances, especially coenzymes (p. 75), are widely distributed in biological systems and participate in numerous pathways of metabolism. The coenzymes called **NADP** and **NAD** (p. 88), for example, serve in a variety of oxidation-reduction reactions in the processes of respiration, fat metabolism (synthesis and degradation), and amino acid metabolism.

Fourth, in a number of biochemical reactions the substances or reactants must be in an activated state (p. 44). This is frequently achieved by combining with the coenzyme known as **coenzyme A (CoA)** in a reaction that requires an energy input, which is usually provided by ATP.

Fifth, the various pathways of metabolism in a living cell are in one or more ways interlinked. For convenience we discuss them separately, but always with the view that they are inevitably related to one another. For example, a derivative of coenzyme A, called **acetyl-CoA,** is a key intermediate in carbohydrate, fat, and protein metabolism.

Finally, the entire complement of enzymatic reactions making up the metabolism of the cell is under the basic direction and control of DNA in the nucleus. The vast quantity of information encoded in the molecular structure of DNA is ultimately expressed in the proteins (i.e., enzymes), the fundamental machinery of all living cells.

Over-All Features of Respiration

There are certain basic aspects of respiration common to all living systems:

1. **Respiration occurs in a series of many successive and coordinated reactions rather than in a single chemical step.** At certain of these steps small quantities of energy are released, with some of the energy being trapped by cells, rather than in one large, wasteful burst as might be expected from a single chemical reaction.
2. **Each of the many steps of respiration is catalyzed by a specific enzyme.**
3. **Among the key chemical steps in respiration are oxidation-reduction reactions** (Chapter 2). These consist of a transfer of electrons (or hydrogen atoms) from relatively high-energy bonds to other molecules to produce lower-energy bonds,

such as those of water and carbon dioxide, with a concomitant net release of energy. In respiration the hydrogens of glucose ($C_6H_{12}O_6$) are in effect transferred to molecular oxygen (O_2) by a series of enzymatic reactions to yield carbon dioxide (CO_2) and water (H_2O), plus large amounts of energy, as summarized in the equation $C_6H_{12}O_6 \longrightarrow 6CO_2 + 6H_2O + $ energy. Compare this to the nonbiological oxidation-reduction reaction between hydrogen gas (H_2) and O_2—i.e., when hydrogen burns—in which the energy-rich H—H bond of H_2 is broken and its hydrogens (or electrons) are transferred to O_2 to form water ($2H_2 + O_2 \longrightarrow 2H_2O + $ energy). The reaction is accompanied by an explosive release of energy as the electrons of the H atoms are pulled toward the oxygen.

4. **The physical and chemical organization of cells makes possible the relatively efficient capture and utilization of the useful energy released in respiration.** From one viewpoint respiration can be regarded as the flow of electrons from a high chemical energy level (i.e., from carbohydrates, with their energy-rich carbon-hydrogen bonds) to a lower energy level (e.g., carbon dioxide and water), like the flowing of a river downstream. Energy is released in the process and a portion of it is captured, depending on the organization of the surroundings; in cells the intricate subcellular chemical and physical structures such as the mitochondria (p. 133), and in the river, say, a paddle wheel with suitable attachments, trap and make available some of the energy that has been liberated in the downstream flow. Similarly, the energy liberated in a chemical reaction may be totally dissipated and wasted as heat unless it occurs in an organized environment suitable for its capture in a useful form. For example, the energy produced from gasoline burning on the ground is lost largely as heat. The combustion of the same quantity of gasoline, however, in a system engineered to trap that energy, such as an automobile engine, results in a conversion of a portion of the energy to a useful form—mechanical energy—that may be used to move the car. The cell, by virtue of its highly specialized and integrated chemical and physical organization, is able to use a significant portion of the energy liberated in respiration.

Principal Pathways of Metabolism

Carbohydrate Metabolism and the Respiratory Pathway

In most organisms carbohydrates (for example, glucose) are metabolized predominantly by the respiratory process. Respiration performs two important functions in the living cell. First, it carries out the all-important role of liberating and making available the energy stored in the carbon-hydrogen bonds of carbohydrate molecules. For this energy to be utilized successfully, it must be released in small parcels, a phenomenon that is attained in nature by a sequence of numerous enzymatic reactions instead of a single large energy-yielding reaction. A portion of the liberated energy is trapped by the forming of the "high-energy" phosphate bonds (p. 89) of the ATP molecule. It should be noted that ATP itself is present in relatively small concentrations in the cell at any given time; it is therefore not a means for storing energy but rather a way for making energy immediately available for cell function. It is the immediate and direct energy source for the entire array of activities manifested in all living systems. The major energy-storage substances in cells, in contrast, are fats and carbohydrates. The second important function of respiration is to convert carbohydrates to intermediate products that can be used as building blocks in the synthesis of other biologically important components of cells, such as lipids, proteins, nucleic acids, and other kinds of molecule (this is discussed on p. 96).

The Anaerobic and Aerobic Respiratory Pathways

In most organisms energy is obtained primarily from respiration of carbohydrates and fats. There are several pathways, but one well-established biochemical route at present appears to be the major respiratory pathway in plants, animals, and numerous microorganisms. This respiratory pathway has two phases: **anaerobic respiration** and **aerobic respiration.** The early sequence of steps in the respiration of most organisms, including man, is completely independent of oxygen, and for this reason is called anaerobic respiration.

Many living things, including man, possess in addition to anaerobic respiration a subsequent sequence of enzymatic reactions, collectively called aerobic respiration, which requires molecular oxygen. In aerobic respiration the principal products of anaerobic respiration (see p. 87) are further broken down to carbon dioxide and water, with the release of considerably more energy. Oxygen is directly involved in only the final step of the aerobic respiratory sequence, but is nevertheless essential. Aerobic respiration is therefore a continuation of anaerobic respiration, and is always preceded by that process; the interrelationship of these two processes is summarized in Figure 4-1. Together

Figure 4-1 Schematic diagram of the over-all respiratory pathway. In the absence of oxygen carbohydrates are broken down, by a stepwise series of reactions, to pyruvic acid. In the process a phosphate group (Pi) is added on to the end of each of two molecules of adenosine diphosphate (ADP), converting them to energy-rich molecules of adenosine triphosphate (ATP). In aerobic metabolism pyruvate is further oxidized through a cyclic sequence of reactions, called the **citric acid cycle,** to CO_2 and H_2O. The energy released in the oxidation is again trapped as bond energy in the formation of 36 more molecules of ATP.

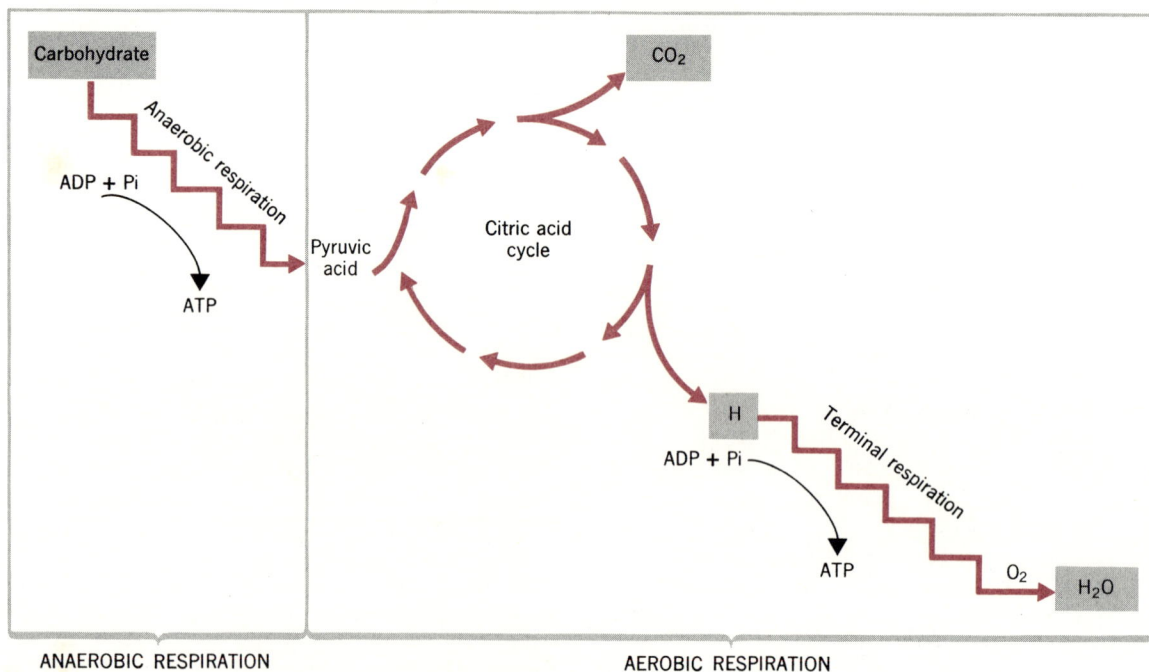

ANAEROBIC RESPIRATION   AEROBIC RESPIRATION

they total two dozen or so successive, highly integrated, specific enzymatic reactions.

The ability of certain microorganisms, such as yeast cells, to carry on a completely normal existence in the presence or absence of molecular oxygen helps to illustrate the relationship between anaerobic and aerobic respiration. When grown in the absence of oxygen, yeast derives its energy solely from the anaerobic respiration of a sugar such as glucose, degrading it to ethyl alcohol and carbon dioxide. If oxygen is provided, the alcohol is further metabolized by the enzymatic steps of aerobic respiration to carbon dioxide and water, releasing more useful energy. In man and most animals the products of anaerobic respiration are not alcohol and carbon dioxide but a substance called **pyruvic acid** (and under certain conditions a compound termed **lactic acid**). In contrast to yeast, oxygen is absolutely essential for life for most living forms, including man.

The fact that the anaerobic respiratory scheme is so widely distributed among living systems, from the most primitive to the most advanced, suggests that it is more ancient than aerobic respiration, from an evolutionary point of view. Similarly, the relatively limited distribution of aerobic respiration in what we consider to be primitive organisms (e.g., certain bacteria), and its widespread occurrence in the more advanced forms of life (e.g., higher plants and animals), point to aerobic respiration as a relatively recent evolutionary acquisition, established after the accumulation of appreciable molecular oxygen in the atmosphere as a result of photosynthesis. This viewpoint is also supported by the fact that the aerobic respiratory sequence utilizes the chemical products formed by the anaerobic pathway. We can speculate that, with the gradual depletion of complex organic energy and carbon sources from the primeval seas during the early history of the earth, organisms that evolved more efficient respiratory pathways for obtaining a complete liberation of energy from the remaining organic substances were probably more successful in the competition for survival.

Anaerobic Respiration

The term anaerobic respiration is often used interchangeably with the term **glycolysis.** The process, with minor modifications, is essentially the same in many widely divergent tissues and organisms. The steps in the pathway can be summarized in the following simplified outline and in Figure 4-2 (p. 90).

1. The six-carbon sugar glucose is split into two equal three-carbon molecules called **glyceraldehydes.**
2. The glyceraldehyde molecules are then oxidized by the removal of two hydrogen atoms to form two **glyceric acid** molecules. This takes place first by the addition of water to the aldehyde $\left(-C\big\langle{}^{O}_{H}\right)$ group of the glyceraldehyde molecule, followed by the oxidation removing two hydrogen atoms (which are transferred to the coenzyme NAD to form NADH).
3. Each of the two glyceric acid molecules then undergoes the removal of an H and OH, to form two pyruvic acid molecules and water.
4. In most higher animal tissues pyruvic acid is normally the major end product of anaerobic respiration. It is further metabolized by aerobic respiration to carbon dioxide and water. If the oxygen supply becomes limited, most of the pyruvic acid is converted to **lactic acid** in animal cells by the addition of two hydrogen atoms from NADH. In certain microorganisms, such as yeast, the pyruvic acid produced under anaerobic conditions experiences a removal of carbon dioxide to form a compound called **acetaldehyde.**
5. Acetaldehyde is then enzymatically reduced by accepting two hydrogens from NADH to yield NAD and **ethyl alcohol.** Thus ethyl alcohol and carbon dioxide instead of lactic acid are the end products of respiration in yeast cells grown in the absence of oxygen. For this reason anaerobic respiration in such organisms is also called **alcoholic fermentation.**

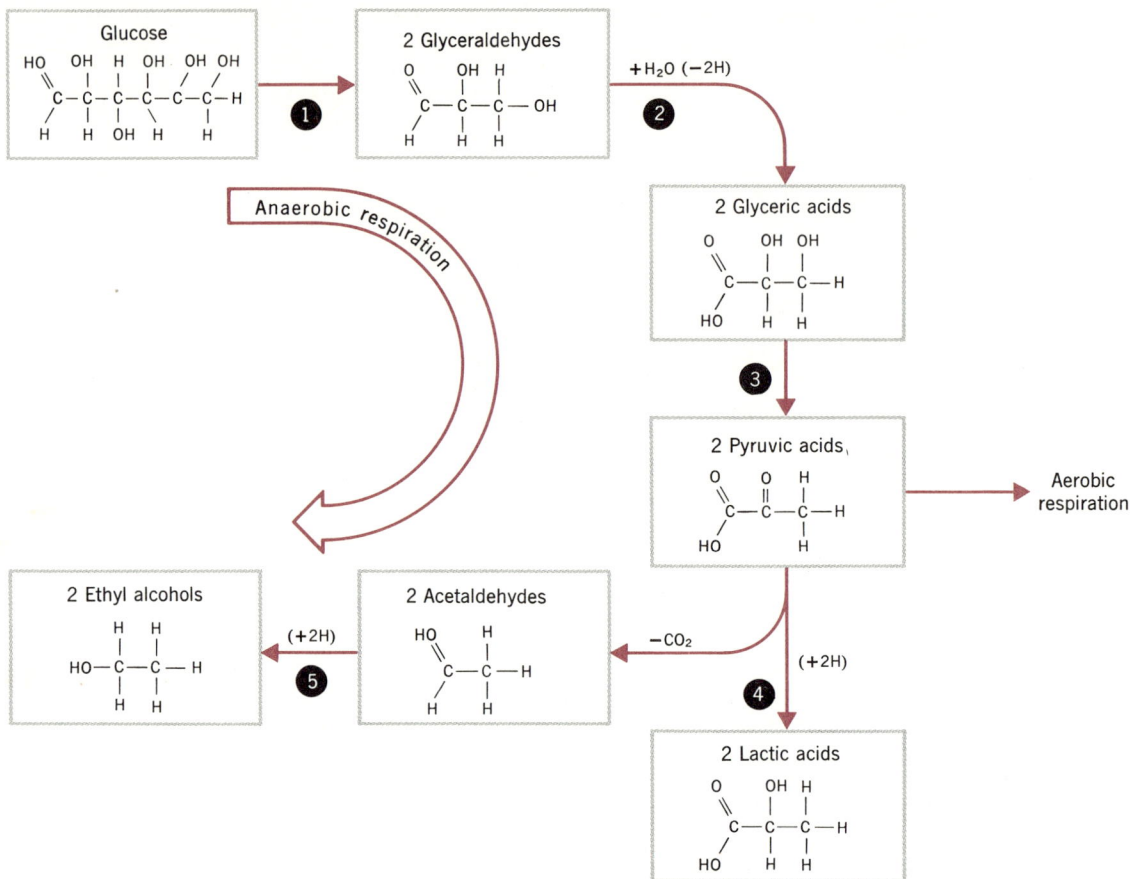

Figure 4-2   Simplified summary of the anaerobic respiratory pathway (called
glycolysis in animal and higher plant tissue and fermentation in
microorganisms).

Intensive research in cellular respiration by numerous biochemists over the years has revealed several common, underlying patterns. Not only does anaerobic respiration of sugar proceed by essentially the same pathway in nearly all organisms, but identical coenzymes and metal components are also involved. Some of the more important coenzymes are **nicotinamide adenine dinucleotide (NAD), nicotinamide adenine dinucleotide phosphate (NADP** which differs from **NAD** in its additional phosphate group), **adenosine diphosphate (ADP),** and **adenosine triphosphate (ATP).**

We now know that in most animal tissues the two hydrogens removed from glyceraldehyde in step 2 of the foregoing outline are enzymatically transferred to the coenzyme NAD, reducing it to what we conveniently designate **NADH.** The NADH subsequently donates the hydrogens for the enzymatic reduction of pyruvic acid to lactic acid (step 4) in animal tissues such as muscle under conditions of limited oxygen supply; and of acetaldehyde to ethyl alcohol (step 5) in yeast fermentation. In other cells and tissues NADP may serve in place of NAD, forming **NADPH.**

No molecular oxygen is consumed in anaerobic respiration, although two of the individual steps are oxidation-reduction reactions. NAD undergoes an alternate reduction (by glyceraldehyde) and oxidation (by pyruvate or acetaldehyde) during the course of anaerobic respiration. In this manner NAD accounts for the oxidation-reduction reactions of a large number of substrate molecules, even though it is present in cells in extremely small quantities.

The breakdown of glucose to pyruvic acid is far more detailed than indicated by this simplified version. Anaerobic respiration is in reality a process of breakdown not of free glucose as such, but of a phosphorylated derivative of glucose. The first step therefore involves the conversion of glucose to its appropriate phosphate derivative. This is accomplished by the enzymatic transfer of the terminal phosphate group from the important biological substance ATP (also designated adenosine-P~P~P[1]) to glucose to form glucose-phosphate and ADP (adenosine-P~P).

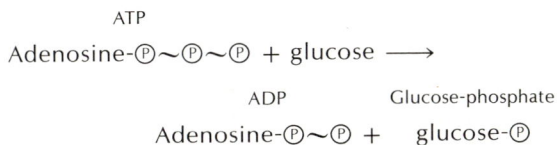

$$\text{ATP}$$
$$\text{Adenosine-}P\text{~}P\text{~}P + \text{glucose} \longrightarrow$$

$$\qquad\qquad \text{ADP} \qquad\quad \text{Glucose-phosphate}$$
$$\text{Adenosine-}P\text{~}P + \quad \text{glucose-}P$$

Glucose-phosphate is then enzymatically transformed to fructose-phosphate, which is enzymatically phosphorylated, again by ATP, to form fructose-diphosphate.

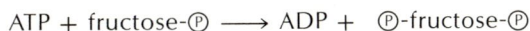

[1]The symbol P designates the phosphate group
$$\begin{array}{c} \text{O} \\ \| \\ -\text{P}-\text{O}. \\ | \\ \text{OH} \end{array}$$
Recall that the wavy line (~) represents a so-called high-energy bond, indicating that an unusually large net energy (8 kcal or more per gram-molecular weight at pH 7) is liberated when the phosphate group is split off. This energy is the result of the formation of products with considerably stronger or more stable bonds than those of the reactants. We refer repeatedly to "high-energy" bonds in this chapter; you should remember, however, that ~P is actually weaker than most; energy is released when the bond to phosphate is replaced by a more stable one (see pp. 36–39).

$$\qquad\qquad\qquad \text{Fructose-diphosphate}$$
$$\text{ATP} + \text{fructose-}P \longrightarrow \text{ADP} + \quad P\text{-fructose-}P$$

All subsequent steps in anaerobic respiration, until the formation of pyruvic acid, involve phosphorylated intermediates.

The ATP-ADP system also performs another major function: capturing useful energy released during respiration. ATP itself is exceptionally rich in chemical energy because the chemical bonds between two of its three phosphate groups (at the terminal and adjacent positions, as indicated by the wavy line) are relatively weak or unstable. Energy is released from the ATP molecule when these bonds are replaced by stronger ones. We are primarily interested in this terminal phosphate of ATP. It is formed from ADP, inorganic phosphate (Pi) from the mineral supply of the cell, and an adequate energy supply in the presence of a suitable enzyme system:

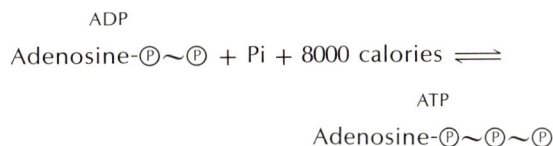

$$\qquad \text{ADP}$$
$$\text{Adenosine-}P\text{~}P + \text{Pi} + 8000 \text{ calories} \rightleftharpoons$$

$$\qquad\qquad\qquad\qquad\qquad \text{ATP}$$
$$\qquad\qquad\qquad \text{Adenosine-}P\text{~}P\text{~}P$$

Of the dozen or so reactions making up anaerobic respiration, only two provide useful energy trapped as ATP by the cell. ATP formation in Figure 4-2 and 4-3 occurs at step 2, following the oxidation of glyceraldehyde, and at step 3, in the removal of water to produce pyruvic acid. Considerably more ATP is generated during aerobic respiration, as described in pp. 93–94. The use of energy from ATP is analogous to the withdrawal of energy from a storage battery that is constantly being charged, in this case by the respiration of the cell. A more detailed schematic representation of anaerobic respiration is given in Figure 4-3 (next page).

The metabolic pathway of anaerobic respiration is comparable to a factory mass-production line. Each enzymatic reaction performs a specific operation on the substrate. In both the cell and the factory the sequence of events must be unbroken in order to complete the process; if one of the enzymes is poisoned or missing (or if one of

Glucose
— ATP
— ADP

Glucose—P

Fructose—P

ATP

ADP

Fructose–diphosphate

Two glyceraldehyde—P

— 2 NAD + Pi

2 NADH

Two diphosphoglyceric acids

— 2 ADP

2 ATP

Two phosphoglyceric acids

—2H$_2$O

Two phosphopyruvic acids

2 ADP

2 ATP

Two pyruvic acids

2 NADH

2 NAD

Two lactic acids

Two CO$_2$ +
Two acetaldehydes

2 NADH

2 NAD

Two ethyl alcohols

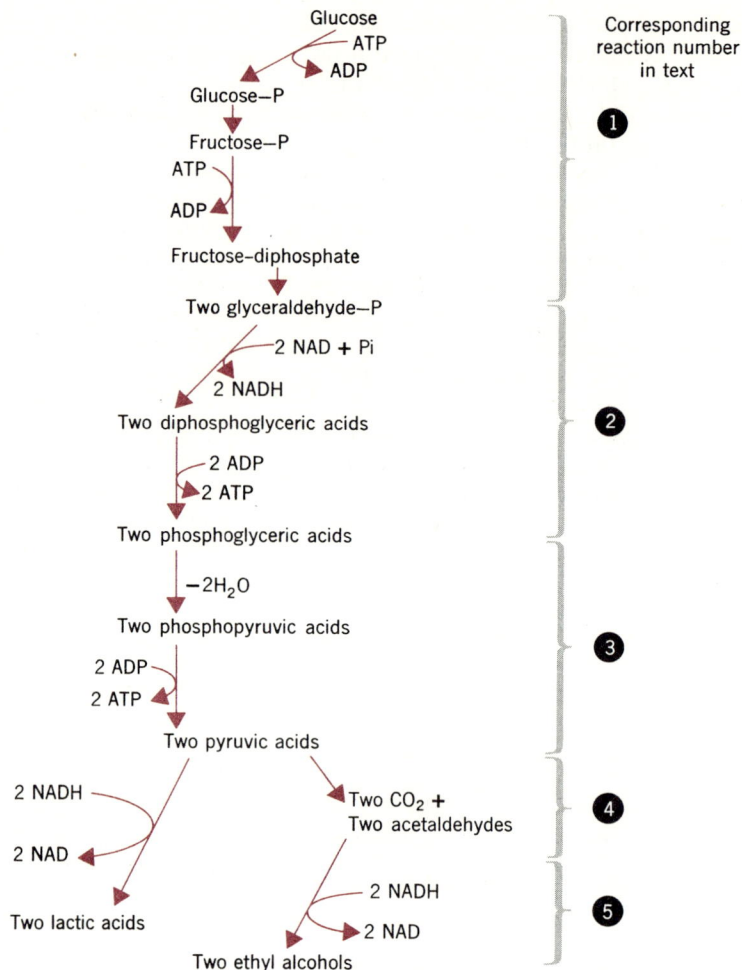

Figure 4-3 Summary of anaerobic respiration. The numbered brackets correspond to the steps in Figure 4-2 and text p. 87.

the machines in the factory breaks down), the entire production line may be brought to a standstill, with serious effects.

It should be noted that the product of any one of the enzymatic reactions is not as a rule used exclusively for the succeeding reaction only. Usually several different reactions compete for the same product, thereby constituting different branches and connecting links between various metabolic pathways.

Aerobic Respiration

For convenience, aerobic respiration can be divided into two main sequences of reactions known as the **citric acid,** or **Krebs, cycle;**

and the **terminal respiratory,** or **electron transport, pathway.** The first consists of a cyclic series of enzymatic reactions in which citric acid is one of several key intermediate products. The British biochemist and Nobel Prize winner Hans Krebs was responsible for a number of the major contributions to our knowledge of this process. The terminal respiratory pathway, in which several cytochromes (p. 75) participate, involves the stepwise transfer of hydrogens or electrons to oxygen from specific products of the citric acid cycle—NADH and succinic acid—to form water. The formation of carbon dioxide in aerobic respiration occurs during the citric acid cycle sequence of events, whereas the formation of most of the ATP produced in aerobic respiration takes place in the terminal respiratory sequence (Fig. 4-1).

Krebs Citric Acid Cycle. The fundamental biochemical changes occurring in the citric acid cycle is summarized in the following simplified version and in Figure 4-4. Pyruvic acid is the starting material.

1. **Oxidative decarboxylation of pyruvic acid.** This is actually a series of reactions involving the removal of two hydrogens (to NAD) and carbon dioxide from the pyruvic acid molecule. The resulting fragment of the pyruvic acid molecule then reacts with a coenzyme (CoA) to form an activated form of acetic acid called **acetyl-CoA** (p. 96). At least two enzymes and four different coenzymes, including NAD, participate in the reaction. The CoA molecule,

which includes in its structure the vitamin **pantothenic acid** (Chapter 14), is present in the general reserves of the cell.

2. **Condensation of acetyl-CoA and oxaloacetic acid.** The acetyl-CoA is now condensed by an enzymatic reaction with oxaloacetic acid, which is present in the general reserves of the cell, to form the six-carbon molecule **citric acid.** CoA is liberated in the reaction and returns to the general reserves of the cell. Citric acid is then rearranged by an enzymatic reaction to its isomer, **isocitric acid.**

3. **Conversion of isocitric acid to ketoglutaric acid.** The isocitric acid undergoes a transfer of two hydrogens to NADP, followed by the removal of carbon dioxide, to produce the five-carbon molecule ketoglutaric acid.

4. **Oxidative decarboxylation of ketoglutaric acid to succinic acid.** The conversion

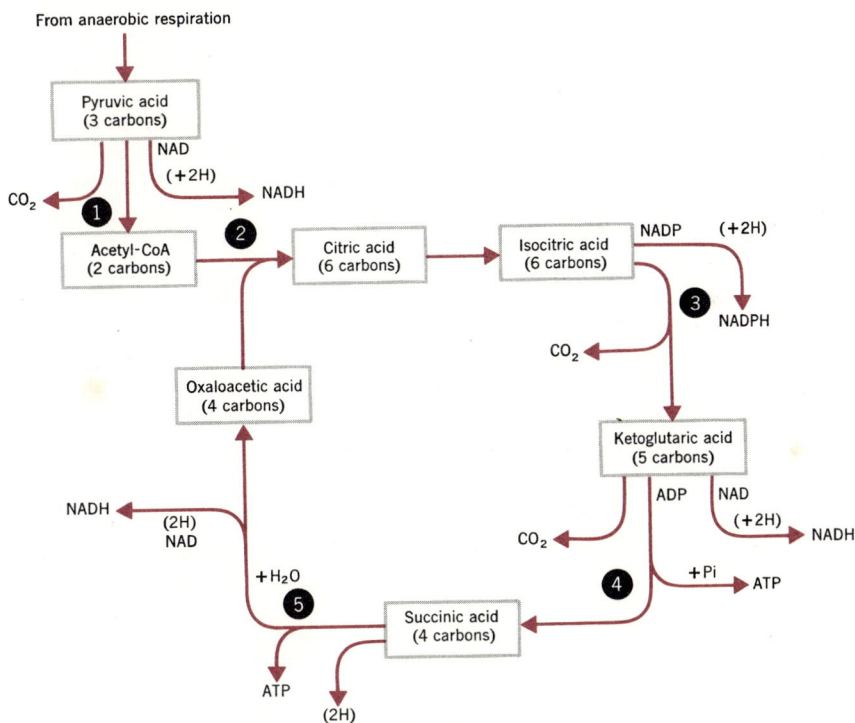

Figure 4-4   Summary of citric acid cycle.

of ketoglutaric acid to succinic acid first involves a sequence of enzymatic reactions that is analogous to the conversion of pyruvic acid to acetyl-CoA (above). Two hydrogens and a carbon dioxide are removed, and at least two enzymes and four different coenzymes, including NAD, participate in the reactions. The succinyl-CoA that is formed is then transformed by an enzymatic reaction, in the presence of inorganic phosphate and ADP, to succinic acid and ATP, with the liberation of CoA.

5. **Succinic acid to oxaloacetic acid.** In this final sequence of three enzymatic steps, the four-carbon molecule oxaloacetic acid is regenerated, thus completing the citric acid cycle. It involves a transfer from succinic acid of two hydrogens (ultimately to molecular oxygen via the terminal respiratory pathway), the addition of a molecule of water, and the transfer of two hydrogens to NAD.

The citric acid cycle operates continuously. Each cycle begins with the oxidative decarboxylation of pyruvic acid to an active acetate (acetyl-CoA) and its condensation with oxaloacetate, and ends with only the oxaloacetate remaining. The starting acetate molecule is thus broken down completely into carbon dioxide and hydrogens with each circuit of the cyclic pathway, whereas the oxaloacetate is regenerated to be used over and over again for the subsequent degradation of other acetate molecules. A small amount of oxaloacetic acid, therefore, functions in the respiration of large quantities of pyruvic acid.

The Terminal Respiratory (Electron Trans-
port) Pathway. The final stages of aerobic respiration are a sequence of enzymatic steps in which the electrons (hydrogens) produced in the previous stages of respiration are ultimately transferred to molecular oxygen; these are present as part of the structure of the NADH, NADPH, and succinic acid.

The union of hydrogen and oxygen to form water is an energy-liberating reaction. In fact, when the reaction occurs in a test tube it is explosive. In cells hydrogen is united with oxygen only after the hydrogens and electrons have first passed along a chain of complex molecules, the **electron transport pathway,** while small increments of energy are liberated. An appreciable portion of this energy is captured through the formation of ATP from ADP and Pi.

The passage of electrons from NADH and succinate (and indirectly from NADPH) to oxygen proceeds by several enzymatic reactions of the terminal respiratory pathway through the following regular chain of cofactors: two coenzymes, **FAD** and **Q**, and a series of cytochromes (p. 75) designated arbitrarily as cytochromes **b**, **c$_1$**, **c**, **a**, and **a$_3$**. The sequence is seen in Figure 4-5. The portion of the respiratory chain that extends from NADH or succinate to cytochrome c is called **cytochrome c reductase.** The remainder of the chain, extending from cytochrome c to molecular oxygen, is called **cytochrome oxidase.** During the passage of electrons through the terminal respiratory pathway, each of the cofactors experiences a reduction (i.e., a gain of electrons) followed by a reoxidation as the electrons (hydrogens) move on to the next component, finally combining

Figure 4-5   Sequence of components of the terminal respiratory chain extending from NADH or succinic acid to O_2.

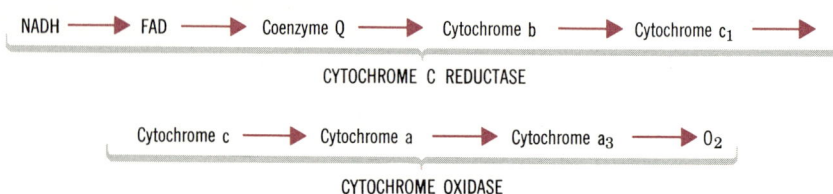

NADH ⟶ FAD ⟶ Coenzyme Q ⟶ Cytochrome b ⟶ Cytochrome c$_1$ ⟶

CYTOCHROME C REDUCTASE

Cytochrome c ⟶ Cytochrome a ⟶ Cytochrome a$_3$ ⟶ O$_2$

CYTOCHROME OXIDASE

with O_2 in the formation of water.

The terminal respiratory pathway is intimately associated with a system for making ATP from ADP and inorganic phosphate, and occurs exclusively on the inner membrane system of mitochondria (p. 133). The process by which ATP is formed using the energy liberated during the passage of electrons through the terminal respiratory chain is known as **oxidative phosphorylation.** It is the principal means for capturing an appreciable portion of energy liberated during respiration. For every molecule of NADH that is oxidized by oxygen via the terminal respiratory chain, three molecules of ATP are formed from the ADP and Pi present; the same is presumed to apply for the terminal respiration of NADPH. In the terminal oxidation of succinic acid, two molecules of ATP instead of three are formed per molecule of succinic acid oxidized.

Efficiency of Respiration

The fraction of the total free energy in a reaction that is actually made available for work is referred to as the **efficiency** of that reaction. The efficiency of anaerobic and aerobic respiration can be summarized as follows. The oxidation of glucose by anaerobic respiration results in the release of less than 10 percent of the total chemical energy stored in a sugar molecule; of the 686,000 calories per mole (180 g glucose), only 60,000 calories are liberated by the anerobic respiratory pathway. A portion of this energy is trapped through the formation of four moles of ATP. Because the production of each mole of ATP from ADP and inorganic phosphate represents the incorporation of about 8000 calories, a total of 32,000 of the 60,000 calories liberated by anaerobic respiration is captured in a form useful to the cell. The remainder is lost as heat. Actually, the **net yield of energy** in the absence of oxygen is only two moles of ATP instead of four because two ATPs are used to convert glucose to phosphorylated fructose in the early stages of anaerobic respiration (Fig. 4-3).

When pyruvic acid (or **pyruvate**) and NADH are further metabolized by the aerobic respiratory pathway, the NADH formed in the oxidation of phosphorylated glyceraldehyde (step 3, Fig. 4-3) in the anaerobic pathway passes on its hydrogens to oxygen by the terminal respiratory chain to yield three ATPs per mole. It is actually in the stepwise oxidation of the two moles of pyruvic acid and NADH (now containing a total of 626,000 calories) through the citric acid cycle and terminal respiratory chain that the release of the major portion of chemical energy originally present in the glucose occurs.

In the citric acid cycle itself ATP is formed directly only in step 4 (Fig. 4-4), whereas three moles of NADH, one mole of NADPH, and one mole of succinic acid are produced per mole of pyruvic acid metabolized. The subsequent transport of hydrogens from NADH, NADPH, and succinic acid via the terminal respiratory pathway to molecular oxygen, concomitant with the release of energy through phosphorylation, gives rise to ATP. For each mole of NADH or NADPH oxidized via the terminal respiratory chain, three moles of ATP are formed; the corresponding oxidation of succinic acid results in two moles of ATP.

Thus in the aerobic respiratory pathway, starting with the oxidation of pyruvic acid and proceeding successively through the citric acid cycle and the terminal respiratory chain, a total of 36 moles of ATP are produced (15 for each of two moles of pyruvic acid oxidized and three for each of the two moles of NADH arising in the anaerobic pathway). This represents about 288,000 calories (36×8000 calories per mole of ATP) of the 626,000 calories originally present in the two moles each of pyruvic acid and NADH that arose from the anaerobic breakdown of a single mole of glucose. The efficiency of energy captured in the aerobic respiratory scheme is therefore about 46 percent of the 626,000 calories. The over-all capture of useful energy in the total process of respiration (anaerobic and aerobic) is represented by the formation of 38 moles of ATP (two moles from the anaerobic respiratory pathway and 36 moles from the aerobic), corresponding to about 304,000 calories, or an over-all effi-

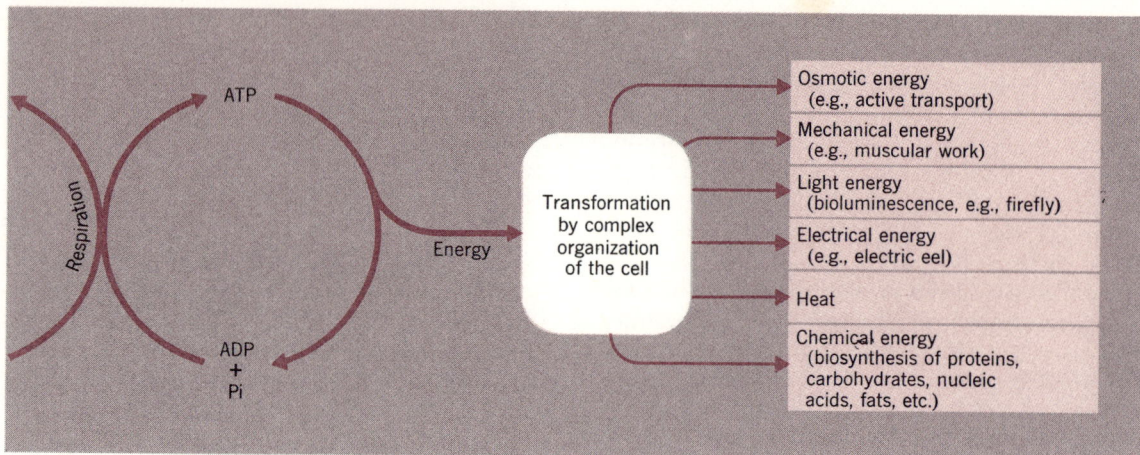

Figure 4-6 Schematic diagram of energy transformations in biological systems.

ciency of about 44 percent of the original 686,000 calories stored in a mole of sugar that the cell is able to capture in useful form (as ATP). The major uses of this energy are summarized in Figure 4-6.

It can be seen from the above values that the aerobic pathway yields about 20 times more available energy (38 moles of ATP per molecule of glucose) than the anaerobic pathway (two moles of ATP). A cell therefore would have to metabolize 20 times more glucose in the absence of oxygen to obtain the same amount of energy that is provided by respiration in the presence of air. Because aerobic organisms can extract considerably more useful energy from the same substrate for their life activities than anaerobic organisms can, they appear to have a distinct survival advantage.

Source of Metabolic Energy

What is the ultimate source of the energy stored in the complex molecules of carbohydrates, fats, and proteins from which all animal cells and most organisms derive their energy to perform their biological work? The answer is **sunlight.** Green plants, by possessing unique biochemical machinery in addition to the basic metabolic pathways con-

tained by most other living forms, are able to harness the energy of sunlight to convert carbon dioxide and water into carbohydrates (e.g., glucose) and certain other organic molecules. The process, called **photosynthesis,** consists of a series of biophysical and biochemical events that in green plants is ultimately accompanied by the release of molecular oxygen into the atmosphere (see Chapters 13 and 21).

Thus, by the process of photosynthesis, autotrophic green plants are able to push hydrogens or electrons in the opposite direction of the energy gradient, like pushing the water of a rapids upstream. Green plants trap the energy of sunlight and transform it into the stored chemical energy of carbon-hydrogen bonds of carbohydrates, starting with the lower-energy-containing substances CO_2 and H_2O. A glucose molecule, for example, in a sense possesses solar energy—in the form of chemical energy—in its structure. Photosynthesis by autotrophs, therefore ultimately provides the necessary high-energy carbon-hydrogen bonds for the metabolism of all other plants and animals.

In its over-all aspects the photosynthetic conversion of carbon dioxide, the energy of sunlight, and water to the level of carbohydrates is fundamentally the reverse of respiration, and is often written essentially as a reverse of the net process of aerobic respiration:

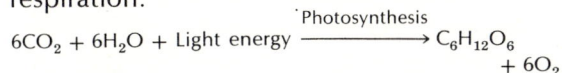

$$6CO_2 + 6H_2O + \text{Light energy} \xrightarrow{\text{Photosynthesis}} C_6H_{12}O_6 + 6O_2$$

The mechanisms and pathways by which this is attained, however, are not a simple reversal of the respiratory processes but are routes and reactions that are in part unique to photosynthesis.

From this over-all equation for photosynthesis it is evident that two basic requirements must be met. First, there must be a **source of energy.** This is provided by sunlight, which, as we shall soon see, is converted in part to ATP by a process termed **photophosphorylation,** which is somewhat similar to oxidative phosphorylation in mitochondria (p. 93). Second, there must be a source of hydrogens or electrons, which we speak of as **reducing power.** This is also provided by sunlight (and water molecules) to form NADPH, molecular oxygen being released as an end product. Photosynthesis therefore involves the net cleavage of water by the energy of sunlight and the stepwise utilization of its electrons or hydrogens (now raised to a higher energy level) for the transformation of carbon dioxide to organic compounds such as carbohydrates.

This takes place in two successive series of events, a **light phase** followed by a nonlight-requiring **dark phase.** The absorption of light by chlorophyll in the chloroplasts is considered to be the primary photochemical act of photosynthesis. It leads to the production of chemical energy in the form of ATP and reducing power as NADPH. These two stages are the light phase. ATP and NADPH are subsequently used by a series of enzymatic reactions that take place entirely in the dark to assimilate carbon dioxide to carbohydrates and its intermediates (and to fats, proteins, and other cellular constituents); this is the dark phase.

Actually, the assimilation of carbon dioxide (the dark phase) in both green plants and photosynthetic bacteria consists of enzymatic reactions that are not peculiar to photosynthesis. All the enzymes now known to participate in the conversion of carbon dioxide to carbohydrates have been found in a wide variety of organisms, many of which are not photosynthetic. In effect, the uniqueness of photosynthesis is its ability to convert sunlight into the energy of ATP and the reducing power of NADPH for the synthesis of cellular substances. The fundamental distinction between photosynthetic and nonphotosynthetic cells therefore involves the way in which they form ATP and reduced pyridine nucleotides. Photosynthetic cells can synthesize these compounds at the expense of light energy (and also by the usual metabolic reactions that occur in the dark, as in respiration), whereas nonphotosynthetic cells, including animal tissues and most microorganisms, cannot utilize light in this manner.

During the last two decades the demonstration of photosynthesis in a cell-free system of isolated plant chloroplasts (p. 142) has proved to be a major breakthrough in the study of photosynthetic processes. This, together with studies of some of the individual enzymatic steps in the process and the use of radioactive tracer techniques, has provided us with significant information. Further details of the process of photosynthesis are described in Chapter 13.

Lipid Metabolism

Our knowledge of lipid function and metabolism centers mostly about the fatty acids and the steroids (p. 63). Fatty acids are present in mammals largely in the form of triglycerides or fats (p. 62). The triglycerides themselves represent about 10 percent of the body weight and are distributed in different amounts in all tissues. In cells fats usually take the form of droplets in the cytoplasm.

The main role of fats is as a reservoir of chemical energy. They are in a considerably more reduced chemical state than either carbohydrates or proteins, possessing significantly more carbon-hydrogen bonds and therefore more stored chemical energy than either. Fats also function in higher animals as

$$\text{Fats} \begin{cases} \text{Glycerol} \longleftrightarrow \text{Glyceraldehyde} \\ \qquad\qquad\qquad\qquad \updownarrow \text{Pyruvic acid} \\ \text{Fatty acids} \longleftrightarrow \text{Acetyl-CoA} \longleftrightarrow \text{Citric acid cycle} \xleftarrow[\text{respiration}]{\text{Anaerobic}} \text{Glucose} \end{cases}$$

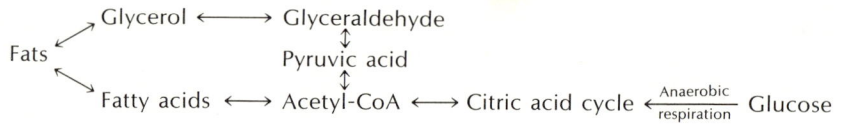

a structural component of living tissues, for example, in various membranes of cells. In addition, they insulate against excessive loss of heat to the environment, protect against mechanical injury, and apparently function in certain metabolic roles (e.g., as components of enzyme systems) that are only now beginning to be elucidated.

Fat Breakdown and Synthesis. Fats are enzymatically broken down, or hydrolyzed, to glycerol and fatty acids (p. 62). Glycerol is a three-carbon molecule and is transformed in the mitochondria into the three-carbon molecule of glyceraldehyde, which enters the glycolytic pathway. Fatty acids are degraded in mitochondria by a characteristic enzymatic pathway to two-carbon units combined with coenzyme A. This product is acetyl-CoA, which is identical to the acetyl-CoA obtained from carbohydrate respiration (p. 91). Thus acetyl-CoA is a link between carbohydrate and fat metabolism as outlined in the display above.

Because acetyl-CoA is also the building unit, or precursor material, of fatty acids and steroids, carbohydrate metabolism can provide precursors for the synthesis of necessary fats and steroids. The few fatty acids that cannot be synthesized in sufficient quantities in the body are termed **essential fatty acids,** and must be obtained from the diet.

Protein Breakdown and Amino Acid Metabolism

In the cell proteins may be enzymatically broken down or hydrolyzed into amino acids, and these may be enzymatically degraded still further in several ways, including removal of their amino group ($-NH_2$). The resulting organic products are eventually metabolized to pyruvic acid and acetyl-CoA. Thus acetyl-CoA as a product of amino acid breakdown is also a link joining protein metabolism, carbohydrate metabolism, and fat metabolism. This linkage is diagrammed in the display at the bottom of the page.

Man can synthesize only about 10 of the 20 or so naturally occurring amino acids. The 10 amino acids that cannot be synthesized in sufficient quantities and that are needed for the normal functioning of the animal are called **essential amino acids** and must be obtained from the diet (Table 4-1). Amino acids are the building blocks for the synthesis of proteins, an intricate process that is discussed in a later section of this chapter.

Central Metabolic Role of Acetyl-CoA

In summary, acetyl-CoA is an important intermediate in several fundamental metabolic processes. As a product of carbohydrate metabolism, fat metabolism, and amino acid metabolism, it can be utilized in a number of different ways. Acetyl-CoA may be completely broken down to carbon dioxide and water by the citric acid cycle and terminal respiratory chain, thus serving as an energy source. It is a building unit in fatty acid synthesis, steroid synthesis (and therefore in the

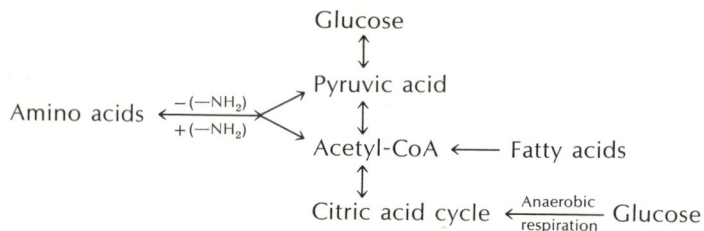

$$\begin{array}{c} \text{Glucose} \\ \updownarrow \\ \text{Amino acids} \underset{+(-NH_2)}{\overset{-(-NH_2)}{\rightleftarrows}} \quad \text{Pyruvic acid} \\ \updownarrow \\ \text{Acetyl-CoA} \longleftarrow \text{Fatty acids} \\ \updownarrow \\ \text{Citric acid cycle} \xleftarrow[\text{respiration}]{\text{Anaerobic}} \text{Glucose} \end{array}$$

table 4-1

Essential Amino Acids

Arginine	Methionine
Histidine	Phenylalanine
Isoleucine	Threonine
Leucine	Tryptophan
Lysine	Valine

they predetermine the composition, activity, and internal destiny of protoplasm.

Although the genetic and metabolic functions of nucleic acids in a sense are inseparable, it seems appropriate in a general chapter on cell metabolism to point out the salient features of the biosynthesis and breakdown of nucleic acids. The fundamental chemistry of these substances was presented in Chapter 3, and their central role in heredity is described in Chapters 8–10 on genetics. The nucleic acids DNA and RNA are intricately involved in protein synthesis. Briefly, it is the genetic makeup of the cell, in the form of information coded in its DNA, that uniquely determines the amino acid sequence, or primary structure, of proteins. The coded instructions for protein synthesis residing in the nucleotide sequence of DNA are transcribed by enzymatic synthesis of a so-called **messenger RNA (mRNA)** and conveyed to special RNA-containing structures in the cytoplasm called **ribosomes** (p. 130), where proteins are synthesized. It is on the ribosomes that the

formation of steroid hormones, p. 442), and carbohydrate synthesis. The important central role of acetyl-CoA in linking carbohydrate, lipid, and amino acid (and therefore protein) metabolism may be diagrammed as in Figure 4-7 below.

Nucleic Acid Metabolism

The nucleic acids play a central role in carrying genetic information and serving as the master control substances of the cell;

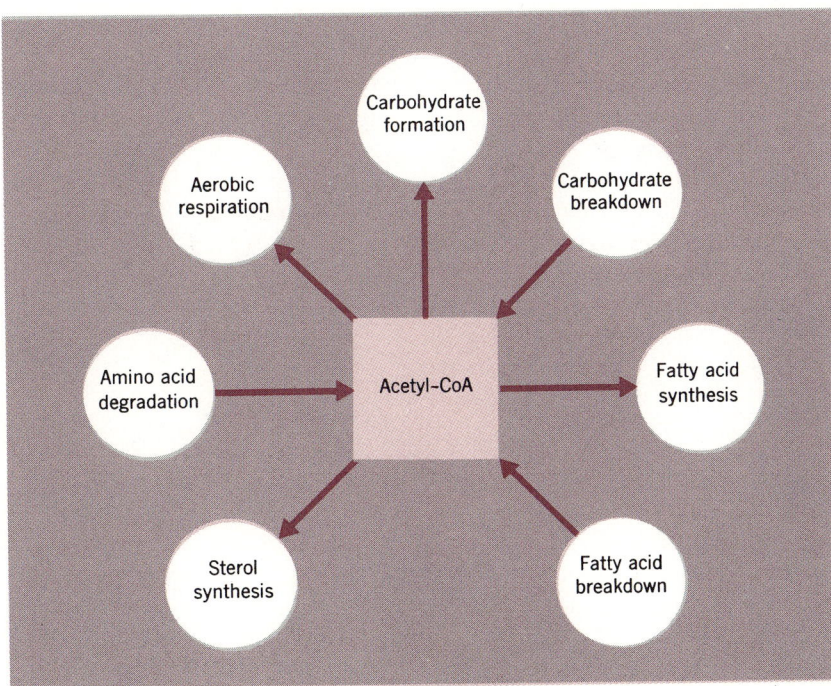

Figure 4-7 *Central role of acetyl-CoA in linking carbohydrate, lipid, and protein metabolism.*

nucleotide sequence of messenger RNA is translated into the amino acid sequence of the newly synthesized protein. Therefore it should be helpful to deal first with nucleic acid metabolism before going on to a description of protein synthesis.

Biosynthesis of DNA

A survey of the evidence that DNA is the genetic material of chromosomes is presented in Chapters 9 and 10. In Chapter 3 we described the broad features of DNA structure, indicating that its double-helical structure provides a molecular basis for the transmission of information from the parental DNA strands to the new, precisely synthesized complementary strands. The process includes the progressive unwinding of the parental DNA (as each of its two strands serves as a template for the enzymatic synthesis of the new strands) and the rewinding of the new (complementary) and old strands to give two new daughter double helices, identical to the original, with each containing one strand from the parental DNA (Fig. 4-8). We shall discuss replication in detail in Chapters 7 and 10, but let us introduce some concepts here.

The actual biosynthesis of each new DNA strand is catalyzed by the enzyme called **DNA polymerase,** originally discovered and purified from bacteria by Nobel Prize winner Arthur Kornberg and his colleagues. DNA polymerase catalyzes the formation of a new strand, or **polynucleotide,** of DNA from the deoxyribonucleoside[2] triphosphates of the four purine and pyrimidine bases—adenine, guanine, cytosine, and thymine (Fig. 4-9). Preformed DNA is required in this enzymatic synthesis, serving as a template for the production of a parallel strand of DNA. The preformed DNA also functions as a starting

[2]Recall (p. 78) that a nucleoside is composed of a purine or pyrimidine base plus ribose or deoxyribose. With three phosphates linked on, it becomes a nucleotide.

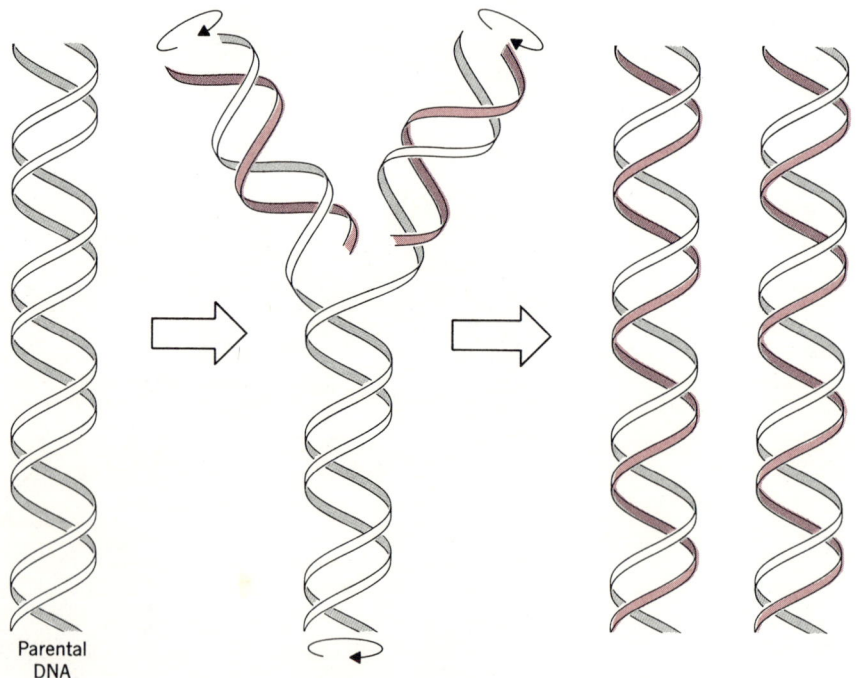

Figure 4-8 Replication of DNA. Each polynucleotide strand of the double-helix DNA structure unwinds and serves as a template for the formation of a new, complementary strand, with nucleotide units aligning in an order determined by specific base pairing.

Parental DNA

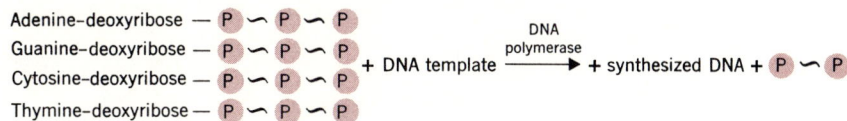

Figure 4-9 Enzymatic synthesis of DNA begins with the deoxyribonucleoside triphosphates of the four purine and pyrimidine bases; preformed DNA is an anchoring point and template for the new DNA.

point or primer to which mononucleotide units are enzymatically added (via DNA polymerase) in one direction, that is to a particular free end of a DNA chain.

The requirement for preformed DNA as a template is satisfied by a single strand of DNA. The native, intact double helix of DNA is entirely inactive in supporting DNA synthesis unless its two strands are either separated (presumably by unwinding), or broken **(nicked)** in one or more places on one or both strands. DNA polymerase binds to such nicks as well as to the appropriate free ends of DNA strands and proceeds to catalyze the synthesis of a single complementary DNA strand. Essentially the same process occurs in bacterial DNA, which exists as a closed, double-stranded circle. Replication of both DNA strands begins at a single point on the circle, with the parental DNA strands gradually unwinding as the site of DNA synthesis moves around the entire strand. In eucaryotic cells (p. 146) DNA polymerase occurs primarily in the nucleus and to a lesser extent in mitochondria. Procaryotic cells (p. 149), for example, bacteria, have no definite nucleus, and in such organisms the enzyme is apparently found largely attached to the cell membranes.

A second recently discovered enzyme, distinct from DNA polymerase, is called the **joining enzyme,** or **DNA ligase.** It has also been implicated in DNA synthesis. DNA ligase catalyzes the joining of the ends of two DNA segments or chains to form a longer chain, or the two ends of a single DNA chain to form a circular molecule. Thus it probably functions in the repair of breaks or nicks in DNA, in the formation of circular DNA found in bacteria, viruses, mitochondria, and chloro-

plasts (p. 142), in the recombination of genes during crossing-over of chromosomes (see p. 204), and in cooperation with DNA polymerase in the replication of DNA.

Most recently it has been discovered that when certain cancer-causing RNA viruses (viruses whose sole nucleic acids are RNA) infect a host cell, they are responsible for the appearance of an **RNA-dependent** DNA polymerase. This means the enzyme catalyzes the formation of DNA from the four deoxyribonucleoside triphosphates, but uses the preformed viral RNA instead of DNA as its template. The newly discovered polymerase thus provides a mechanism by which an RNA virus may insert stable genetic information (i.e., DNA) into a host cell.

Biosynthesis of RNA

In a broad sense protein synthesis is a reflection of the genetic messages encoded in DNA, for the amino acid composition and sequence of proteins (and therefore their structure) are determined by the genetic makeup of the cell (pp. 252–256). Proteins, primarily in the form of enzymes, are the fundamental regulators of all living cells, and are therefore the means by which genetic information is expressed and put to work. How is this genetic information conveyed from the DNA of the nucleus to the ribosomes in the cytoplasm where the proteins are actually made?

According to the latest experiments, the coded instructions for protein synthesis residing in the structure of DNA are **transcribed** (the process is called **transcription**) to a freshly synthesized form of RNA known as **messenger RNA (mRNA).** Messenger RNA is synthesized in the nucleus under the explicit

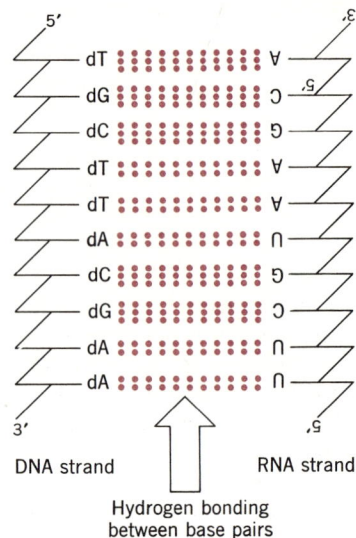

Figure 4-10 *Complementary bonding between segments of DNA chain and segments of RNA chain. Recall that A represents adenine; G, guanine; C, cytosine; U, uracil. The letters dA, dT, etc., refer to the deoxy form of nucleotide.*

direction of DNA. In brief, a portion of the DNA double helix molecule is temporarily dissociated into its two strands. Then, along the exposed sequence of bases of one of the two strands (the template) a complementary strand of mRNA is synthesized (Fig. 4-10). Thus information coded in DNA is transmitted into the ordered arrangement of bases of the mRNA. The mRNA moves out of the nucleus to the ribosomes, where, as we shall see (pp. 103–107), it serves as a template or mold for protein formation.

An enzyme called **RNA polymerase,** very similar in action to DNA polymerase, catalyzes the above synthesis of mRNA. All four ribonucleoside triphosphates adenosine triphosphate, or ATP; guanosine triphosphate, or GTP; cytosine triphosphate, or CTP; and uracil triphosphate, or UTP) are required (Fig. 4-11). Note that it is the ribonucleoside tri-

phosphates that are specifically utilized here (instead of the deoxyribonucleoside triphosphates in the DNA polymerase reaction), and that the uracil base is employed for RNA synthesis in place of the thymine base for DNA synthesis (both are complementary to adenine). It should also be noted that in addition to RNA polymerase a second protein, called the **sigma** or **initiation factor,** is necessary for RNA synthesis. The sigma factor is apparently important in the regulation of the transcription process, for it is somehow required to initiate the RNA chain. Once the chain is started, RNA polymerase catalyzes the covalent bonding of the ribonucleotides to one another in a linear sequence determined by the DNA template.

Messenger RNA is only a small fraction of the total cellular RNA and has a short half-life (i.e., the time it takes for half of it to be in-

Figure 4-11 *Enzymatic synthesis of mRNA starting with the ribonucleoside triphosphate of the four purine and pyrimidine bases and DNA as a template.*

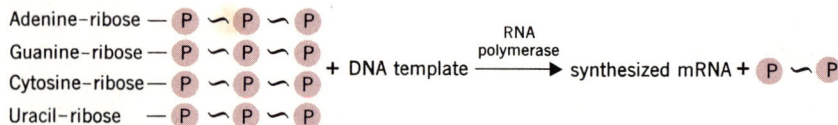

activated, or degraded) in bacterial cells (about two minutes). Therefore it is difficult to isolate from most cells, particularly in pure form, especially if each of the thousands of proteins in a cell requires its own specific RNA. Nevertheless, with cells that synthesize one or a few kinds of protein (e.g., red blood cells, which primarily synthesize hemoglobin, or bacterial cells infected with a DNA virus, which make only viral proteins) it is possible to isolate fractions rich in RNA.

One of the important techniques that has made possible the isolation of such fractions and the confirmation of their base sequences is the **hybridization** method. If messenger RNA has a base sequence along its length that is complementary to a DNA strand (as we would expect in the biosynthesis of mRNA) the two chains will tend to interact (by hydrogen bonding) through complementary base pairing in the same way that two strands interact in DNA synthesis to form the tightly coupled double helix. On this basis, any two given strands of DNA (or one of DNA and another of RNA) that have segments of complementary base sequences along their chains will tend to pair and rewind at their complementary regions to form "hybrid" double helices. The longer the segments of complementary bases, the greater the tendency of two strands to pair. In the laboratory it has been found that the double helical strands of DNA will unwind and separate when heated to about 40°C. Slow cooling permits association of the single strands to form a double helix, a process called **annealing.** By use of radioactive labeled strands it is possible to determine the extent to which hybrid double helices form. Thus by means of the hybridization technique (Fig. 4-12) it has also been possible to determine the similarities of DNA from two different organisms (and therefore speculate how closely they may be related in an evolutionary sense).

In addition to messenger RNA there are at least two other classes of RNA found in cells: **transfer RNA (tRNA)** and **ribosomal RNA.** Both are directly involved in protein synthesis, as we shall soon see. Transfer RNA is a low-molecular-weight form of RNA located in the cytoplasm. It acts essentially as a carrier of specific amino acids during protein synthesis on the ribosome. A specific tRNA exists for each amino acid. Some amino acids, in

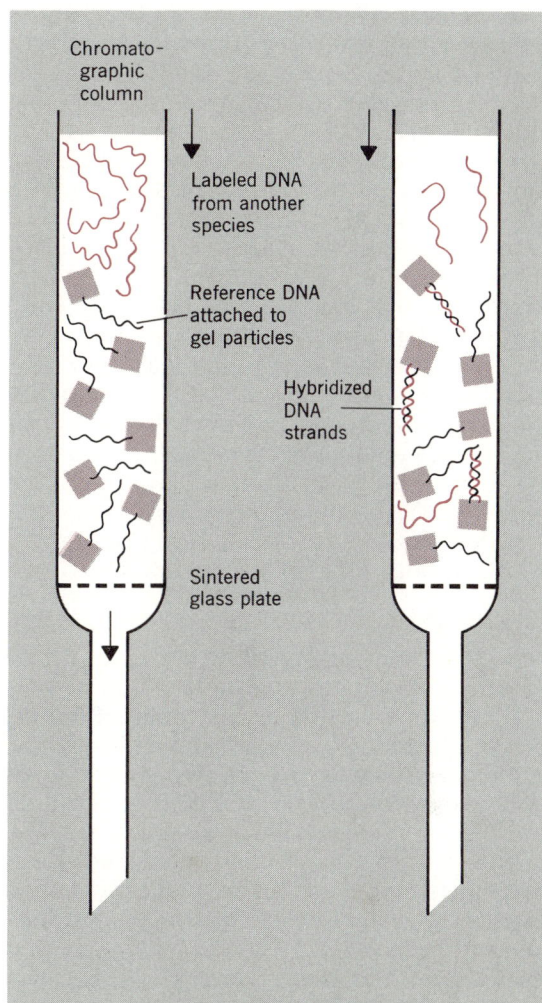

Figure 4-12   DNA-RNA hybridization. Single-stranded DNA is bound to nitrocellulose and packed as a porous gel in a funnel. Radioactive RNA (or DNA from another species) is then poured through the funnel to filter through the nitrocellulose gel. The number of strands of RNA that stick to the DNA is determined by the degree of complementarity between the two, and can be measured by the amount of radioactivity bound to the gel.

fact, have several distinctly different but nevertheless specific tRNAs. Ribosomal RNA constitutes up to 65 percent of the ribosomes (the remainder is protein), and is a large fraction of the total cellular RNA.

In higher organisms most of the RNA—messenger, transfer and ribosomal—is synthesized in the nucleus (a small quantity is also synthesized in the mitochondria and chloroplasts) by RNA polymerase, utilizing DNA as a template. It appears that in the biosynthesis of tRNA and ribosomal RNA only a very small segment of the DNA chain is used (e.g., less than 2 percent of the DNA for all ribosomal RNA, which accounts for more than half the total cellular RNA) compared to that used in mRNA formation.

Obviously DNA serves in at least two template roles, in the biosynthesis of complementary DNA strands and in the biosynthesis of complementary RNA strands. It is clear that DNA replication occurs at a particular stage in the division cycle of the nucleus—just before the onset of nuclear division (Chapter 7). RNA synthesis takes place at different times.

Breakdown of Nucleic Acids

Nucleic acids can be broken down to their component parts through the action of several enzymes known as **nucleases. Ribonuclease** specifically hydrolyzes ribonucleic acids, whereas deoxyribonuclease hydrolyzes only deoxyribonucleic acids. The liberated polynucleotides and nucleotides may then be further hydrolyzed to yield inorganic phosphate and the base-sugar residue called a **nucleoside.** The latter is probably subsequently hydrolyzed to form the free purine or pyrimidine base and sugar. The sugar may eventually be completely metabolized to carbon dioxide and water by way of the carbohydrate respiratory pathway.

The purine and pyrimidine bases undergo different fates. In higher animals we know that purines are only partially broken down before they are excreted, with small amounts appearing as urea or ammonia. Pyrimidines in higher animals are converted mainly to certain amino acids, which may be excreted as such or used in other metabolic pathways.

Amino acids, the building blocks of proteins, are joined in peptide linkages, or **polypeptide chains** (p. 65), to form the vast variety of highly specific enzymes of a cell. We know that certain genetic mechanisms must operate in protein synthesis to account for the specific order of arrangement of the amino acids in the polypeptide chains comprising the proteins, the binding of the polypeptide chains to one another, and their specific spatial relationships. We also know that peptide-bond formation is an energy-requiring process, a fact that implicated ATP participation long before we had any knowledge of the mechanism of protein synthesis.

One Gene, One Enzyme

The pioneering experiments of Nobel Prize winners G. W. Beadle and Edward L. Tatum in the 1940s were among the first to indicate that genes, as the units of inheritance, exercise their influence in metabolism by regulating chemical reactions in the cells, presumably by controlling the synthesis of enzyme proteins. The pink bread mold *Neurospora* can be grown in pure culture on a nutrient medium containing only essential inorganic salts, sugar, and the vitamin biotin (biotin cannot be synthesized by *Neurospora* and must therefore be provided). The mold is able to produce from these relatively simple nutrients the entire complement of its complex protoplasmic components, including the 20 amino acid building blocks of proteins, approximately a dozen water-soluble vitamins, and numerous other organic molecules of biological importance. It has two characteristics that are especially suited for genetic investigations: its life cycle is very short (approximately 10 days), consisting of an asexual stage and a sexual stage (see Fig. 14-10); and the spores can be isolated and grown separately in order to study the transmission of particular inherited characteristics.

Beadle and Tatum were able to demonstrate that many mutations (p. 190) in *Neurospora* abolished the ability of these organisms to synthesize an essential biological substance, for example, a vitamin or an amino acid, and were therefore unable to grow unless a specific organic compound was added to the minimal nutrient medium. Evidently the mutants had lost the ability to synthesize an essential substance because some part of the sequence of necessary enzymatic reactions had been blocked. That the mutants involved gene defects was amply established by the fact that the trait, the inability to synthesize a particular compound, behaved in its transmission from generation to generation as if controlled by a single gene or unit of inheritance.

The conclusions of Beadle and Tatum were formulated in their so-called **one gene, one enzyme** theory, proposing that each enzyme or other specific cellular protein is controlled by a specific gene. Subsequent progress has forced us to modify certain aspects of this idea, but the theory remains essentially sound. In fact, the structure of an enzyme in many instances is determined by more than one gene. We know that some proteins consist of more than one different polypeptide chain; each chain is determined by at least one separate gene. Therefore the Beadle-Tatum concept of one gene, one enzyme is presently more precisely stated as one gene, one polypeptide chain.

Mechanism of Enzyme (Protein)
Synthesis and its Genetic Control

The great breakthroughs in the study of protein synthesis began in the early 1950s. Living tissue, for example, rat liver, can be homogenized with mortar and pestle or food blender to yield a **cell-free homogenate,** consisting largely of membrane fragments and cytoplasmic components of broken cells. If such a homogenate is incubated with radioactive amino acids (radioactive for convenience in tracing), and if a respiratory substrate such as glucose and ATP are also added, the amino acids will be incorporated into newly formed proteins. When such homogenates were fractionated—separated into their various components by centrifugation (see p. 51) —it soon became evident that a soluble fraction of the homogenate was necessary, apparently as a source of essential cofactors, and that it was only the ribosome fraction (p. 252) that incorporated the amino acids into peptide linkages. In time the involvement and role of transfer RNA and messenger RNA were also elucidated. The widespread interest generated by these experiments quickly led to the findings by numerous workers of similar protein-synthesizing systems in many other types of cell. In particular, ribosomes from the bacterium *Escherichia coli* and immature red blood cells called **reticulocytes** (which synthesize the protein hemoglobin) proved to be considerably more active in amino acid incorporation than ribosomes from other cells.

The cumulative results of the vast amount of research on protein synthesis during the last 20 years has produced a remarkably detailed picture of the process, although there are still many unanswered questions. The numerous steps that collectively account for the biosynthesis of proteins (more precisely, of polypeptide chains) can be grouped into the following four stages.

1. **Activation of amino acids.** This first stage of protein synthesis occurs in the cytoplasm and is responsible for the activation of the different amino acids. Activation involves the formation of a bond between an amino acid and its corresponding tRNA to give **aminoacyl-tRNA,** also called **charged tRNA.** The reaction requires ATP and is catalyzed by a class of enzymes known accordingly as the **aminoacyl-tRNA synthetases,** each of which is highly specific for both the particular amino acid and its corresponding tRNA. The significance of this unusual specificity in contributing to the specific sequence of amino acid residues in the newly formed polypeptide chain will soon become evident. The activation reaction is summarized by the overall equation below.

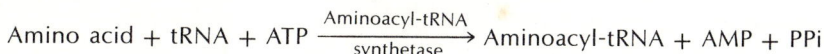

$$\text{Amino acid} + \text{tRNA} + \text{ATP} \xrightarrow[\text{synthetase}]{\text{Aminoacyl-tRNA}} \text{Aminoacyl-tRNA} + \text{AMP} + \text{PPi}$$

(Here PPi is a pyrophosphate formed by the cleavage of ATP to AMP). It takes place in two steps catalyzed by the same enzyme.

Transfer RNA, like messenger RNA and ribosomal RNA, consists of a single polynucleotide strand of characteristic molecular weight and base composition. Transfer RNA also shows considerable base pairing, presumably by looping the single-stranded RNA, to give a specific, three-dimensional conformation resembling a four-leaf clover (Fig. 4-13). One leaf or arm possesses the site to which the aminoacyl group is linked by the action of the aminoacyl synthetases described above. Another arm has a specific recognition site, or **anticodon,** that subsequently positions the aminoacyl-tRNA correctly on the ribosome (more precisely, on the mRNA attached to the ribosome) for transfer of the amino acid to the growing polypeptide chain. Each anticodon is apparently represented by a specific sequence of three nucleotides (i.e., a **base triplet**) that is different for each tRNA. The anticodon is believed to be complementary to a corresponding triplet called a **codon** (see Chapter 10) on mRNA. The base pairing of the anticodons of aminoacyl-tRNAs with the corresponding codons of mRNAs attached to the ribosomes thus accounts for the specific sequencing of amino acids in new proteins being synthesized. The role, if any, of the other two arms of tRNA is not clear.

2. **Initiation of the polypeptide chain.** The biosynthesis of all polypeptide chains (and therefore all proteins) is always initiated by a particular aminoacyl-tRNA, in which the activated amino acid is specifically

Figure 4-14   *The structure of a ribosome and its preparation for protein synthesis. The size of an intact inactive ribosome is estimated at about 200 Å or **70S (Svedberg units).** It is made of two unequal subunits, **30S** and **50S,** both composed of RNA and protein. To prepare for protein synthesis the ribosome dissociates into its subunits. Messenger RNA (mRNA) and a modified transfer RNA (fMet tRNA) both bind to the 30S subunit with the aid of specific initiation factors to form an initiation complex. The two subunits then reunite to form a functional 70S ribosome.*

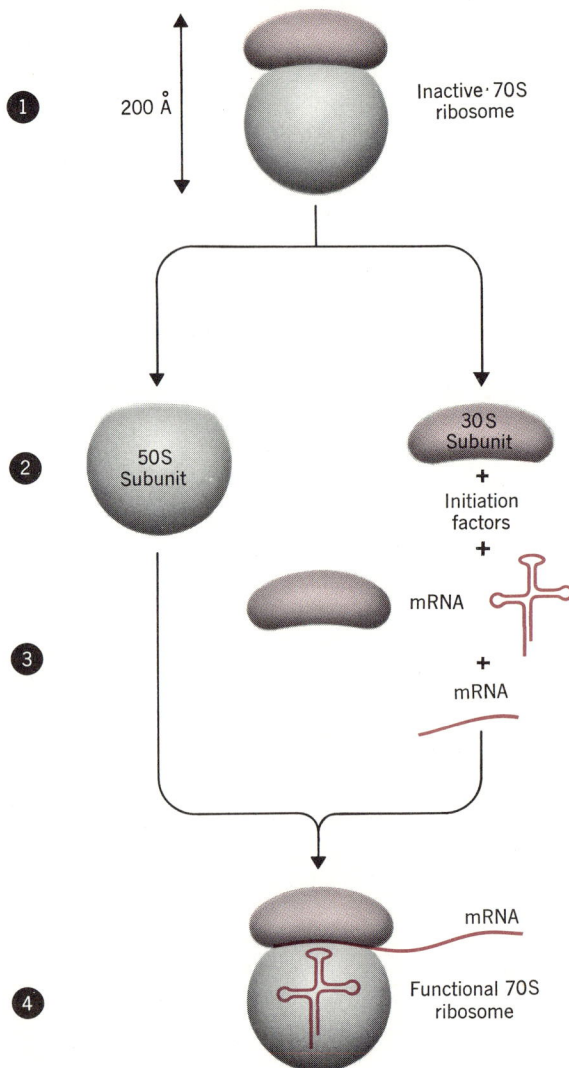

Figure 4-13   *Schematic representation of tRNA showing clover-like conformation formed by looping and base-pairing of the single-stranded polynucleotide to give the indicated four-armed structure of tRNA.*

N-formylmethionine (fMet, a derivative of the amino acid methionine). An inactive ribosome (Fig. 4-14, step 1) becomes functional, or capable of synthesizing polypeptide chains, by first dissociating into two unequal subunits (step 2) that bind to both N-formylmethionine-tRNA and its specific mRNA (step 3) and finally reassociating to give a functional ribosome (step 4). Thus the N-formylmethionine-tRNA is specifically bound by base pairing of its anticodon to the codon of messenger RNA on the functional ribosome to initiate the synthesis of the polypeptide chain. The mRNA is held in a groove between the two subunits of the functional ribosome; in effect, elongation of the newly initiated polypeptide chain by peptide bond formation occurs as the mRNA moves through the ribosomal groove (see stage 3 below). The conversion of an inactive ribosome to a functional one requires at least three specific **protein initiation factors,** which are released and used over and over again. Interestingly enough, although the biosynthesis of all proteins is always initiated by N-formylmethionine-tRNA, N-formylmethionine never appears in the completed polypeptide chain. The N-formyl group and the methionine are eventually removed by enzymatic action to give the finished polypeptide chain.

3. **Elongation.** Elongation of the newly initiated polypeptide chain occurs essentially in three successive steps (Fig. 4-15). First, the functional ribosome, which has moved along to the next codon of the attached

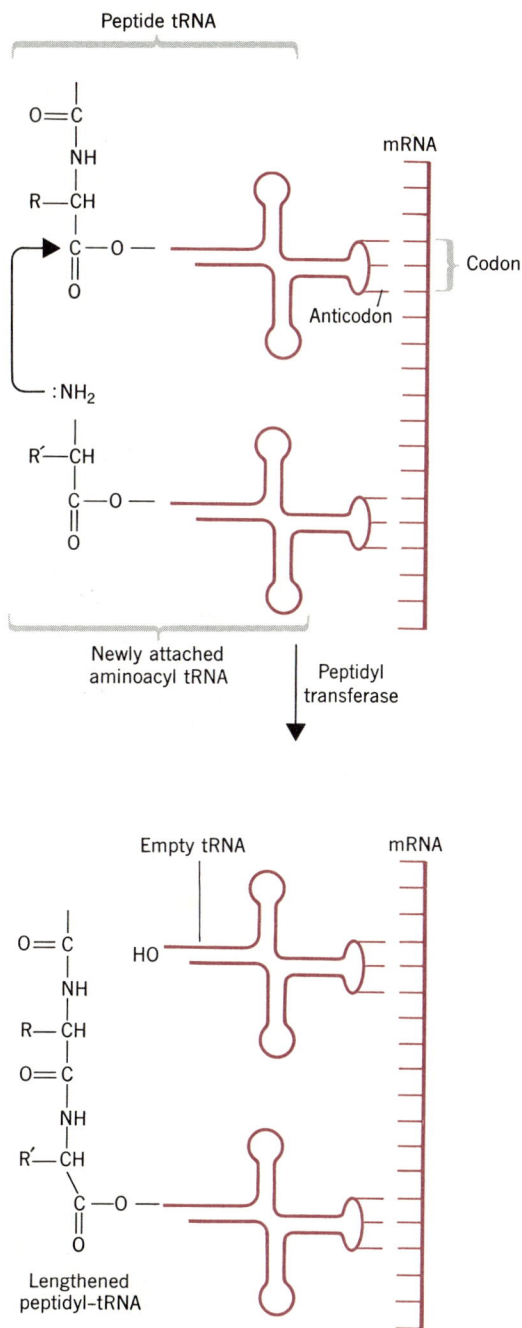

Figure 4-15 Chain elongation stage in protein synthesis. The newly bound aminoacyl-tRNA has formed a peptide bond with the end amino acid residue of the peptidyl-tRNA (the chemical process is indicated at the left of the diagram), thus releasing the latter tRNA. The mRNA now moves along a ribosomal groove to bring the next codon of mRNA into position for another addition of an amino acid residue to the lengthened peptidyl-tRNA.

mRNA chain, binds the corresponding amino acid as aminoacyl-tRNA, a process requiring GTP as a source of energy. Binding is specified by complementary pairing between the anticodon of the aminoacyl-tRNA and the mRNA codon at which the functional ribosome is positioned. Second, the peptide bond is now formed by an enzymatically catalyzed reaction to give a lengthened peptidyl-tRNA bound to the mRNA at its site of attachment to the ribosome. More precisely, the amino group of the newly bound aminoacyl-tRNA has formed a peptide bond with the C=O group of the end amino acid residue of the peptidyl-tRNA, thus releasing the latter tRNA. Finally, with the completion of each new peptide bond, both the newly elongated peptidyl-tRNA and the mRNA are moved along the ribosome (energy is again provided in the form of GTP) to bring the next codon of mRNA into position.

4. **Termination of the polypeptide chain.** Completion or termination of the polypeptide chain occurs when three special termination codons are reached in the mRNA (Fig. 4-16, step 1). The finished polypeptide is then detached from the

Figure 4-16   Termination steps and the release of the free polypeptide chain, free mRNA, free tRNA, and ribosome. The ribosome breaks into components, beginning the process again.

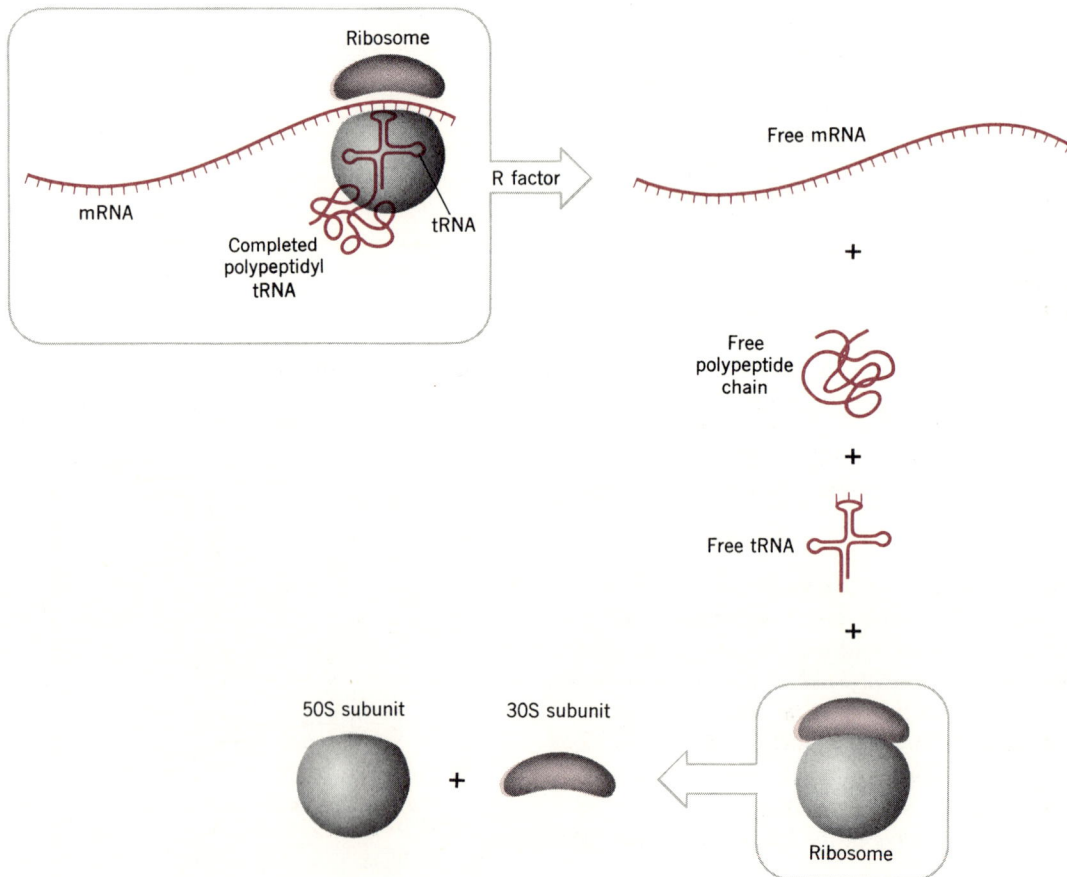

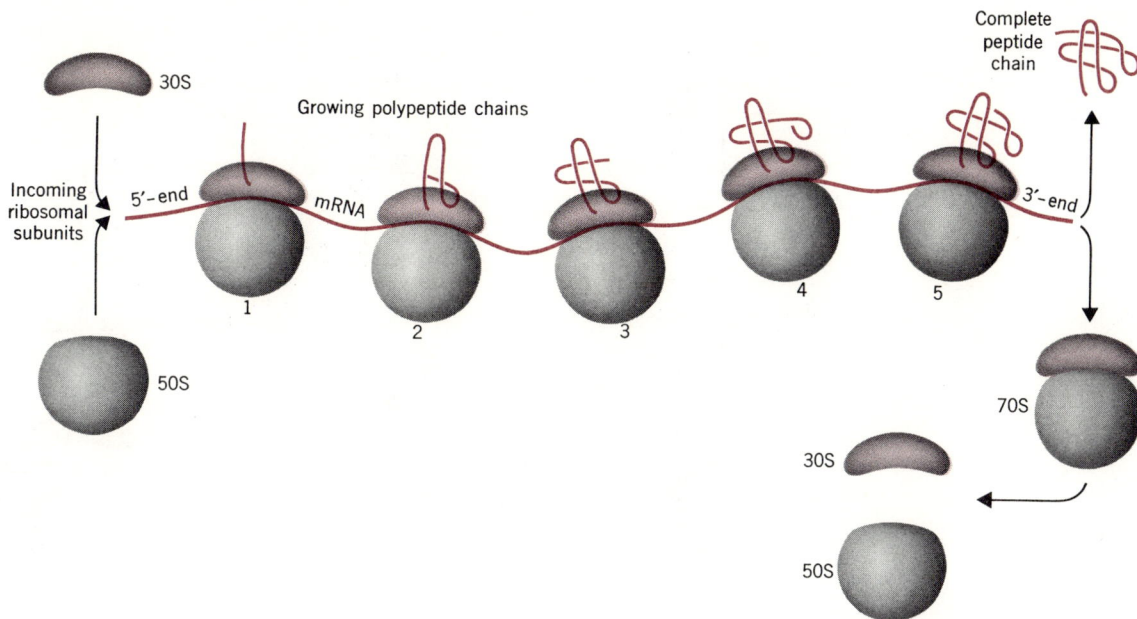

Figure 4-17 Simultaneous translation of mRNA by the several ribosomes of a polyribosome. The ribosomes function independently of one another, each forming a polypeptide chain as it moves along the mRNA molecule.

ribosome through the hydrolytic action of a **ribosomal protein release factor** (R factor) responsible for the cleavage of the tRNA of terminal aminoacyl residue on the newly completed polypeptide chain. The release of the polypeptide chain from the ribosome is also accompanied by the removal of the mRNA in a free form (step 2). Apparently two high-energy phosphate bonds are needed for the formation of each aminoacyl-tRNA and at least another (GTP) during the formation of each peptide bond, to make a total of three.

Ribosomes seem to work in groups, for when they are carefully isolated from cells they are obtained in clusters, called **polyribosomes,** of several to as many as 100. The ribosomes in a polyribosome appear to be held together in a linear order by a strand of mRNA, which they translate independently and almost simultaneously (Fig. 4-17) into identical copies of the polypeptide chain.

In brief, peptide bond formation or protein synthesis (summarized in Fig. 4-18, p. 108) occurs on the ribosomes and consists of these steps: (1) amino acids are activated in the cytoplasm by specific aminoacyl-tRNA synthetases to form aminoacyl-tRNAs; (2) the polypeptide chain is initiated by the formation of a functional ribosome to which both mRNA and N-formylmethionyl-tRNA are bound as a result of the reassociation of its two unequal ribosomal subunits. The N-formylmethionyl-tRNA and mRNA are held together by base pairing of the anticodon and corresponding codon; (3) the newly initiated peptide chain is elongated by sequential additions of new aminoacyl-tRNA to the next codon on mRNA, each followed by the enzymatic formation of a peptide bond to give a lengthened peptidyl-tRNA chain as the mRNA moves through the groove between the subunits of the functional ribosome; (4) the polypeptide chain terminates by special codons in the RNA, resulting in the release

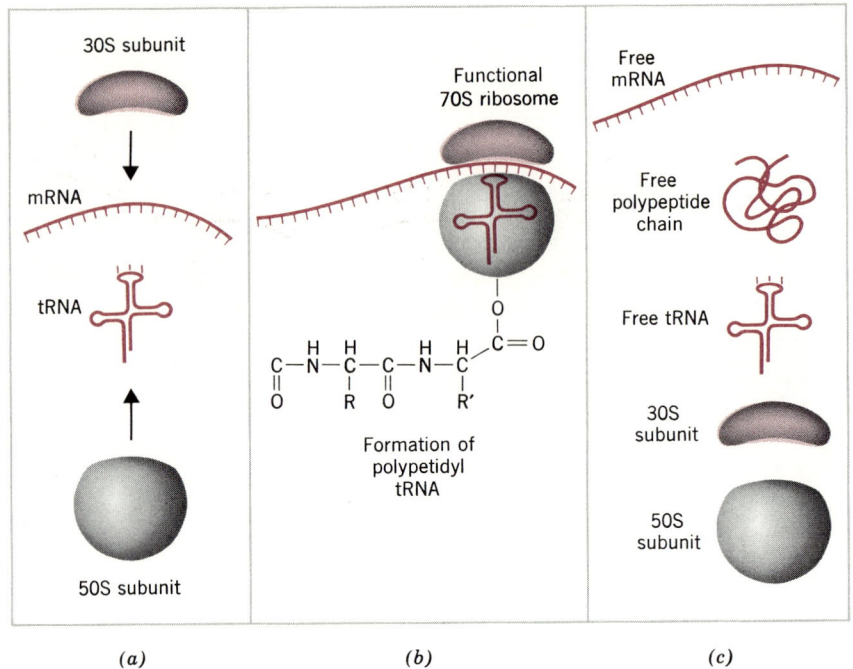

Figure 4-18 Schematic summary of the protein system.

(a)　　　　(b)　　　　(c)

of the completed polypeptide chain and free mRNA. At least three high-energy bonds are needed for the formation of a peptide bond.

If the protein is to be used for the most part within a cell, as in red blood cells, the ribosomes exist mostly in the cytoplasm as polyribosomes. If the protein is to be secreted outside the cell, as in the case of secretory cells, most of the ribosomes are on the endoplasmic reticulum.

Metabolic Pathways and Cell Structure

The elucidation of the various routes of metabolism inevitably has been correlated with the distribution and localization of these pathways among the subcellular components themselves (see Chapter 5). In many tissues and cells the enzymes of anaerobic respiration are located in the cytoplasmic matrix instead of in the mitochondria and endoplasmic reticulum. The aerobic respiratory route, however, appears to reside almost completely in the mitochondria. The enzymes of the citric acid cycle are relatively tightly bonded and organized with mitochondria. The terminal respiratory pathway is part of the membranous structures (cristae and membranes) of the mitochondria, and fatty acid synthesis occurs principally in the endoplasmic reticulum. Protein synthesis, as we understand it at present, takes place largely on the ribosomes. With respect to nucleic acid metabolism, DNA is synthesized mainly in the nucleus and RNA is synthesized in both the nucleus and the cytoplasm. Degradation probably occurs at similar localities.

Let us now examine some of these cellular structures, as we broaden our field of view from the purely molecular level, to include the organization and behavior of cells, and their interactions as tissues.

Reading List

Clark, B. F. C., and A. K. Marcker, "How Proteins Start," *Scientific American* (January 1968), pp. 36–42.

Conn, E. E., and P. K. Stumpf, *Outlines of Biochemistry* (2nd ed.). Wiley, New York, 1966.

Crick, F. H. C., "The Genetic Code III," *Scientific American* (October 1966), pp. 55–62.

Karlson, P., *Introduction to Modern Biochemistry.* Academic Press, New York, 1968.

Kornberg, A., "The Synthesis of DNA," *Scientific American* (October 1968), pp. 64–78.

Loewy, A. G., and P. Siekevitz, *Cell Structure and Function* (2nd ed.). Holt, Rinehart & Winston, New York, 1969.

McElroy, W. D., *Cellular Physiology and Biochemistry* (3rd ed.). Prentice-Hall, Englewood Cliffs, N.J., 1970.

Mosbach, K., "Enzymes Bound to Artificial Matrixes," *Scientific American* (March 1971), pp. 26–33.

Nomura, M., "Ribosomes," *Scientific American* (October 1969), pp. 28–35.

Porter, R. R., "The Structure of Antibodies," *Scientific American* (October 1967), pp. 81–90.

Racker, E., "The Membrane of the Mitochondrion," *Scientific American* (February 1968), pp. 32–39.

Ross, R., and P. Bornstein, "Elastic Fibers in the Body," *Scientific American* (June 1971), pp. 44–52.

Temin, H. M., "RNA-Directed DNA Synthesis," *Scientific American* (January 1972), pp. 24–43.

White, A., P. Handler, and E. L. Smith, *Principles of Biochemistry* (4th ed.). McGraw-Hill, New York, 1968.

White, E. H., *Chemical Background for the Biological Sciences.* Prentice-Hall, Englewood Cliffs, N.J., 1964.

Yanofsky, C., "Gene Structure and Protein Structure," *Scientific American* (May 1967), pp. 80–94.

"I took a good clear piece of Cork, and with a pen-knive sharpen'd as keen as a razor, I cut . . . an exceeding thin piece of it, and placing it on a black object plate, because it was itself a white body, and casting the light on it with a deep plano-convex glass, I could exceedingly plainly perceive it to be all perforated and porous, much like a honey-comb . . . in that these pores, or cells, were not very deep, but consisted of a great many little boxes, separated out of one continued long pore, by certain diaphragms. . . ."

ROBERT HOOKE, 1665, *MICROGRAPHIA*

part three | *The Cell as the Unit of Life*

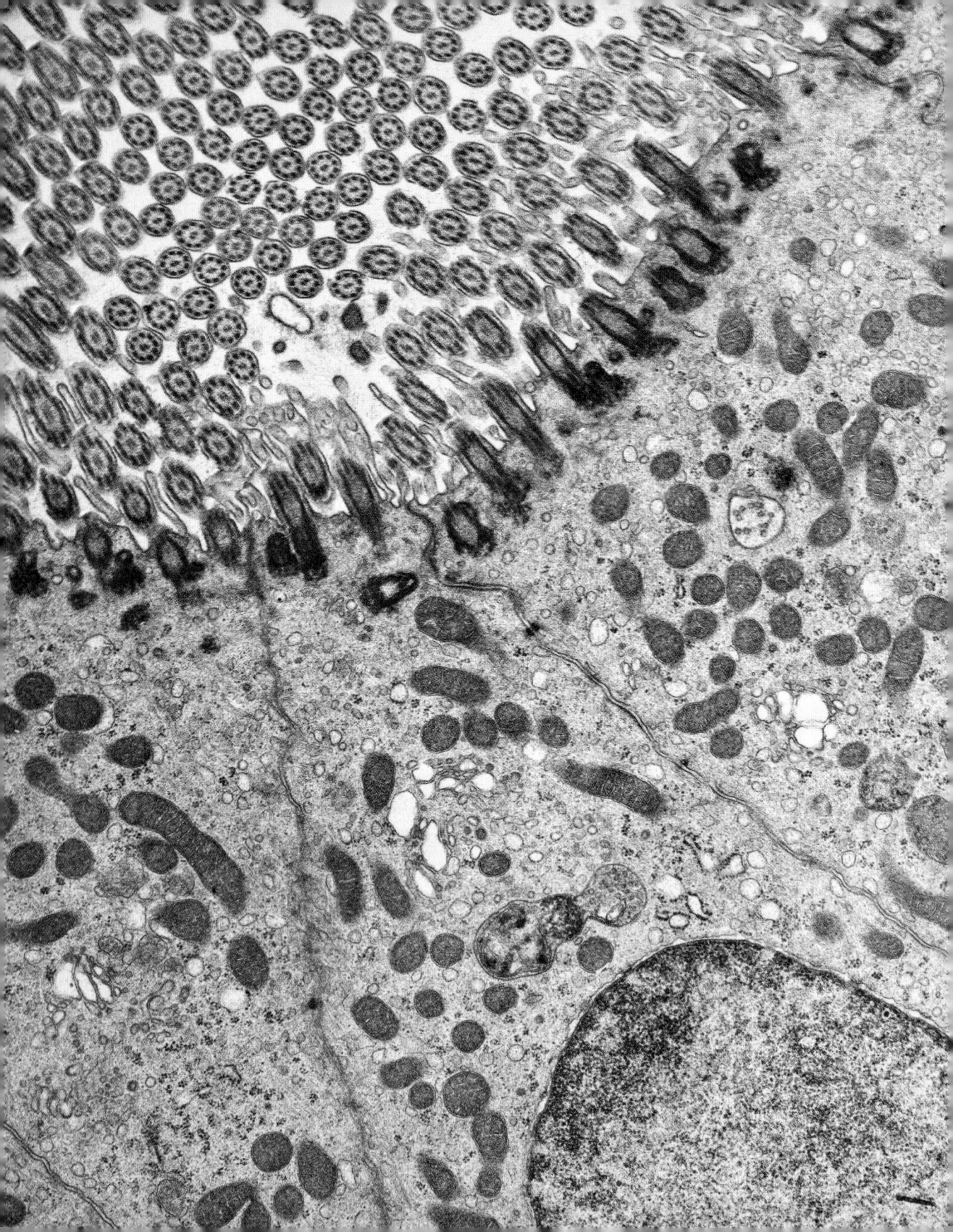

The Structure of the Cell

"The particles which constitute all animal organs . . . are little globules which may be distinguished under a microscope . . ."

C. F. WOLFF, 1759, *THEORIA GENERATIONIS*

In Chapter 1 we spoke of the hierarchy of biological levels common to all living things. Living systems, we noted, can be understood in terms of atoms and molecules; or of supramolecular aggregates and the particles, fibers, and membranes that make up cells; or, at higher levels, of the behavior of whole cells and cellular aggregates, of organs and tissues, or of organisms and populations of organisms.

Interestingly, within this hierarchy we begin to see life, with all the characteristics we have used to define it, only at the level of the cell. At lower levels of organization there are structures that exhibit one or another life process, but these occur only in the experimenter's test tube. Artificial lipid membranes, for example, can respond to an electrical stimulus as the membrane of a nerve does; molecular aggregates can perform typical metabolic steps. Viral particles can reproduce, but only in a parasitic relationship with a living cell. None of these can be self-sustaining. Therefore we have concluded that a cell is the minimum organization of matter that is alive.

Although life must have arisen and evolved in forms simpler than cells (Chapter 22) every form of life now existing, with minor exceptions, is either a cell or is made of cells. Thus, by definition, we say that the cell is the basic unit of life—the basic unit of structure and function in all existing organisms. In single-celled animals and plants the activities of the cell are the activities of the organism. In multi-

A cell from the retina of a rabbit. Magnification 20,000×.

113

cellular living things the coordinated activities of many cells account for the activities of the organ or organism. Spectacular evidence for this has been provided in recent years with development of techniques of cell culture. It has been found that almost any animal or plant tissue can be separated into its component cells, and that those isolated cells can be maintained alive, healthy, and active for weeks or months in a nutrient medium. In this chapter we discuss the structure of cells and the principal physiological roles associated with the main cellular constituents. In particular, we relate the biochemical activities discussed in Part 2 to cell structures.

Study of the Cell

Techniques of Investigation

Most cells are too small for us to see with the unaided eye; normally we can't detect an object smaller than .1 millimeter (1/250 of an inch). For these small dimensions, remember, we generally speak in terms of microns (μ): .1 mm = 100μ (Appendix 1). Although a few cells are quite large—the yolk of a chicken egg, which is a single cell, is usually more than an inch in diameter—most animal and plant cells range in size from 10 to 100μ. Thus it is understandable that **cytology,** the study of cell structure, has been based almost completely on observations with the microscope. The light microscope, the type most commonly used, extends our range of visibility about a thousandfold, so that we can see objects whose diameters are about 1/250,000 of an inch, or .1μ. This is the size of many bacteria, which are among the smallest cells known.

Techniques for preparing tissues for examination at high magnification, called the **histological** preparation of tissues (Fig. 5-1) are just as important as the microscope itself for exploration of cell structure. These techniques include **fixation, sectioning,** and **staining.** Fixation is the killing of cells or tis-

sues with chemicals that harden and preserve cell structure. Mixtures of such chemicals as alcohol, formaldehyde, and picric acid are called **fixatives.** Fixed tissues can be sliced into very thin sections on a machine called a **microtome.** A wide variety of dyes and stains can then be applied to color different parts of the cells, showing up structural details that could not be seen otherwise.

The light microscope (Fig. 5-2a, p. 116) made it possible to recognize and describe organisms in terms of their cellular nature; the electron microscope (Fig. 5-2b) permits us to explore living systems at still smaller levels of organization. Modern electron microscopes can magnify objects about 300,000 times, making it possible to examine structures only a few angstroms in size; the diameter of a DNA helix is about 20 Å and that of a globular hemoglobin molecule is about 55 Å. A small virus particle such as the one that causes infantile paralysis is about 300 Å.

With the electron microscope we have been able to describe the structure and organization of most of the components found in cells, often down to the level of macromolecular aggregates. We refer to the anatomy of cells at these levels as **cellular ultrastructure.** Our understanding of cell ultrastructure, coupled with insights into the functions of subcellular components derived from biochemical techniques, has permitted the growth of the field of **molecular biology,** which attempts to understand living systems at the level of interaction of molecules and macromolecular aggregates.

The Modern Cell Theory

The seemingly simple but all-important concept that the cell is the unit of life is the culmination of centuries of research by investigators in different parts of the world. This concept, the modern cell theory, explicitly states that all forms of life—plant, animal, and microbial—are made of cells (and their products) and arise only from preexisting cells. This is a fundamental cornerstone on which all the biological sciences are based.

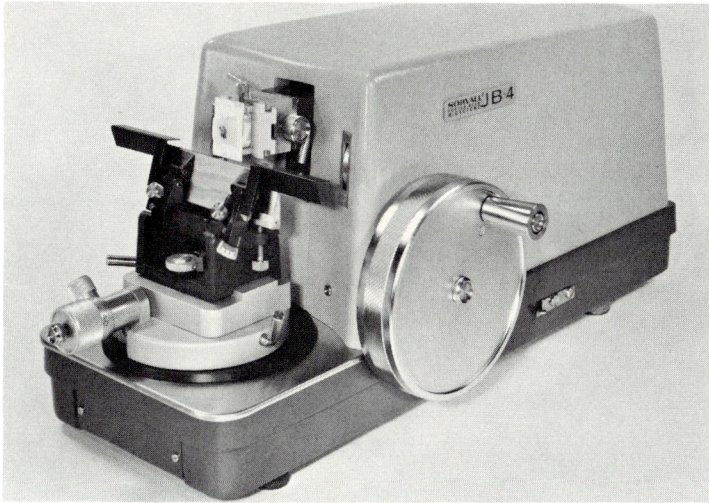

(a)

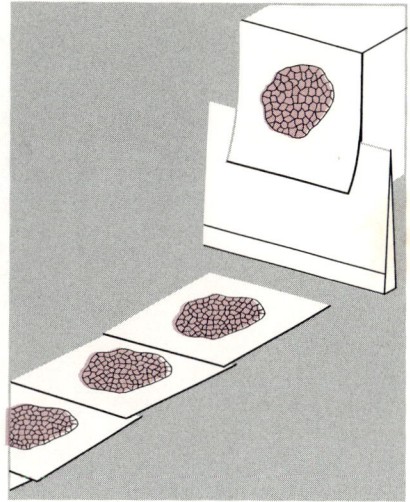

(b)

Figure 5-1 (a) Photograph of a modern microtome. (b) A tissue being sliced into sections with a microtome blade. Tissues are killed and hardened in a fixative, then impregnated with paraffin or plastic to hold them rigid. The block of tissue is passed across the blade of the microtome to slice off thin sections. Because paraffin softens as the blade cuts through it, slices tend to stick together to form ribbons of sections in series. Sections prepared for the light microscope are normally 5 to 10μ thick; those for the electron microscope are usually less than .5μ.

Historical Background

Probably the first observations of cells—really the remains of cells—magnified by a primitive microscope were reported by Robert Hooke, an English inventor, in 1665. Indeed, he introduced the term **cell** in describing the honeycomb-like structure of cork and other plant tissues (Fig. 5-3, p. 117). He was one of the first to give any thought to quantitative considerations of the cellular structure of tissues. A direct count of some 60 cells per $\frac{1}{8}$ of an inch of cork led him to deduce that there must be over 1 billion cells in a cubic inch.

Reports by numerous other microscopists with progressively improved lenses over the years led to the generalization, by the early nineteenth century, that all plants and animals are made of cells. The idea was finally widely accepted and firmly established as the cell theory through the independently published works of two German biologists, Matthias Schleiden and Theodor Schwann, in 1838 to 1839. Schleiden, a botanist, regarded the cell as the structural unit of plants, a theory that Schwann as a zoologist applied to animals.

Meanwhile there were other important findings on the nature of the cell. Robert Brown in England had reported the discovery of a round, distinct structure in cells, the **nucleus.** In 1839 Johannes Purkinje spoke of the living material of the cell as **protoplasm.**

A major contribution to the cell theory was the principle advanced by Rudolph Virchow in 1858 that new cells arise by division from preexisting cells—a conclusion that suggested an unbroken line of cell generations extending back in time to the very origins of life. The modern cell theory has inevitably directed our attention to the evolutionary

115 *The Structure of the Cell*

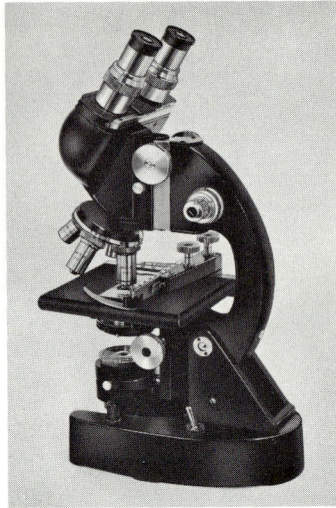

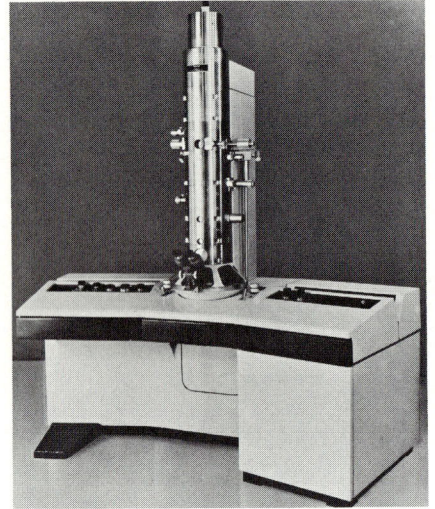

Figure 5-2 Comparison of (a) a compound light microscope, and (b) an electron microscope. The photographs above show the instruments themselves; the drawings below are schematic representations of the light or electron beams as they pass through the instruments. The maximal resolving power of a microscope is limited by the wavelength of the light it passes through a specimen; the shorter the wavelength, the smaller the detail that can be resolved. The wavelength of visible light ranges through the color spectrum between .4 and .7μ. The wavelength of a beam of electrons, which also behave as waves, is only .05 Å, or 1/10,000μ. This much shorter wavelength permits a corresponding increase in resolution in an electron microscope.

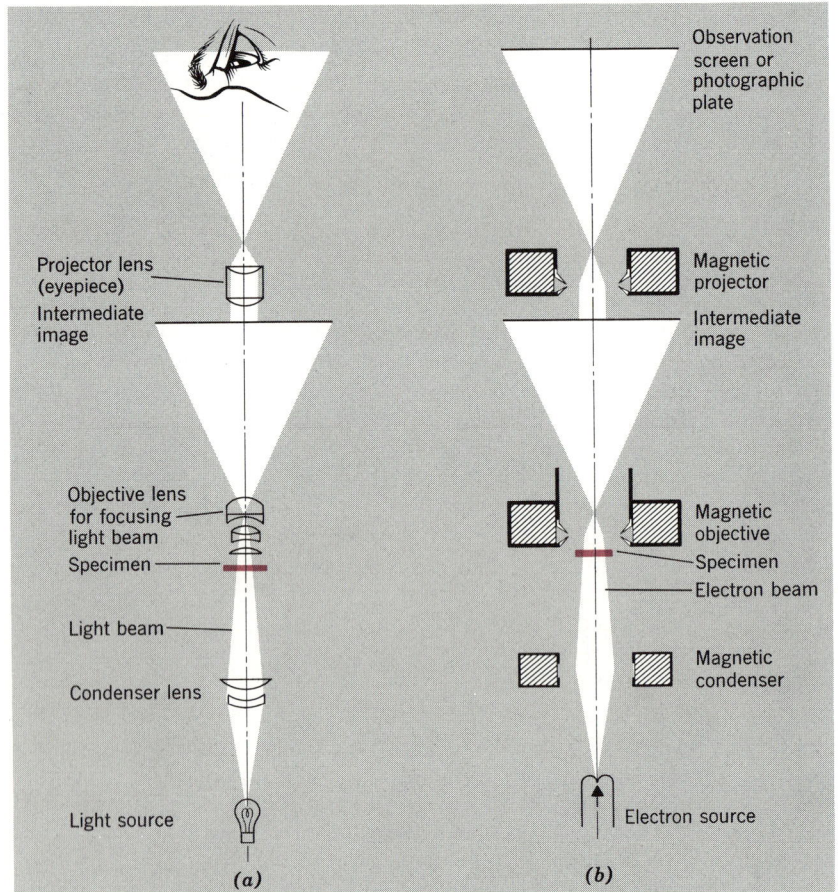

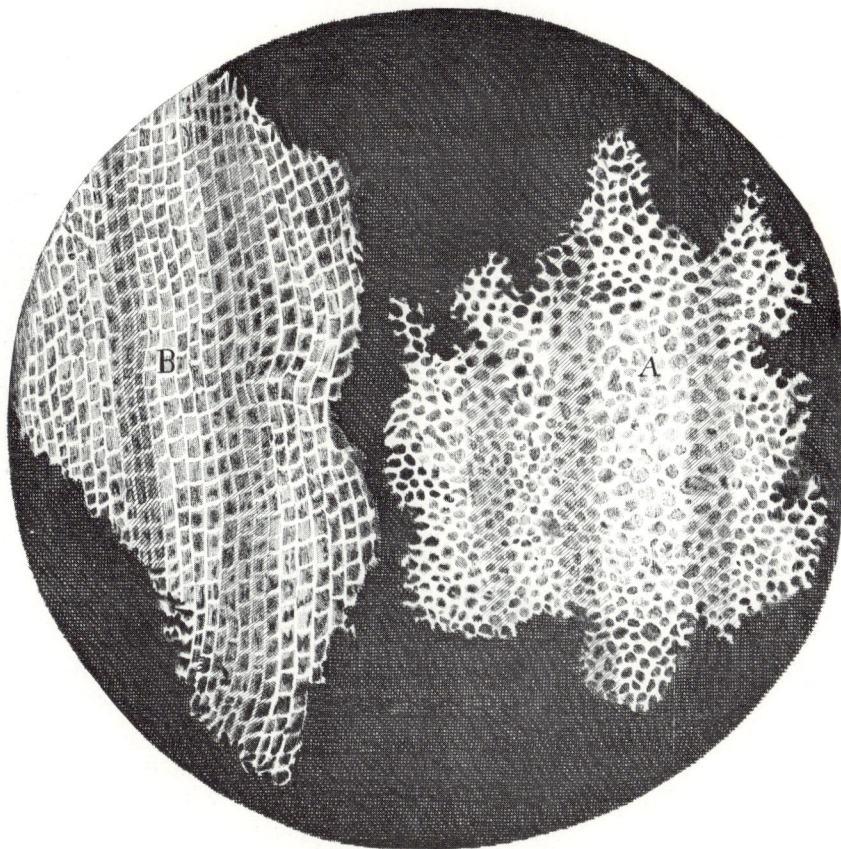

Figure 5-3 Cork tissue as Robert Hooke observed it under his primitive microscope. (From Hooke, Micrographia, 1665.)

implication that cells and therefore all organisms have a common ancestry.

The rapid development in the latter half of the nineteenth century of preservation (i.e., fixation) methods for cell structure and improved methods for sectioning and staining tissues resulted in a vast amount of descriptive cellular information. By the 1890s virtually every feature of the fixed and stained cell that could be resolved by the light microscope had been described. Basic patterns of cell change during growth and multiplication were becoming increasingly clear. The significance of the recently observed details of cell division and fertilization (the union of the nuclei of eggs and sperm) in terms of heredity and evolution were also beginning to be recognized.

Further confirmation for the idea that all tissues and organs are composed of or made by cells, and that the functions of these organs represent the aggregate behavior of the component cells, came in a dramatic way shortly after the beginning of the present century. Ross G. Harrison, then a young biologist at Johns Hopkins University, reported in 1907 that he had transplanted small fragments of spinal cord from frog embryos to a slip of glass and maintained them in a drop of lymph (a type of tissue fluid) from the lymph sacs of an adult frog (Fig. 5-4a, p. 118). The lymph clotted into a gel around the fragments, holding them against the glass. After observing these fragments for several days under the microscope, Harrison noted that individual nerve cells grew out of the main

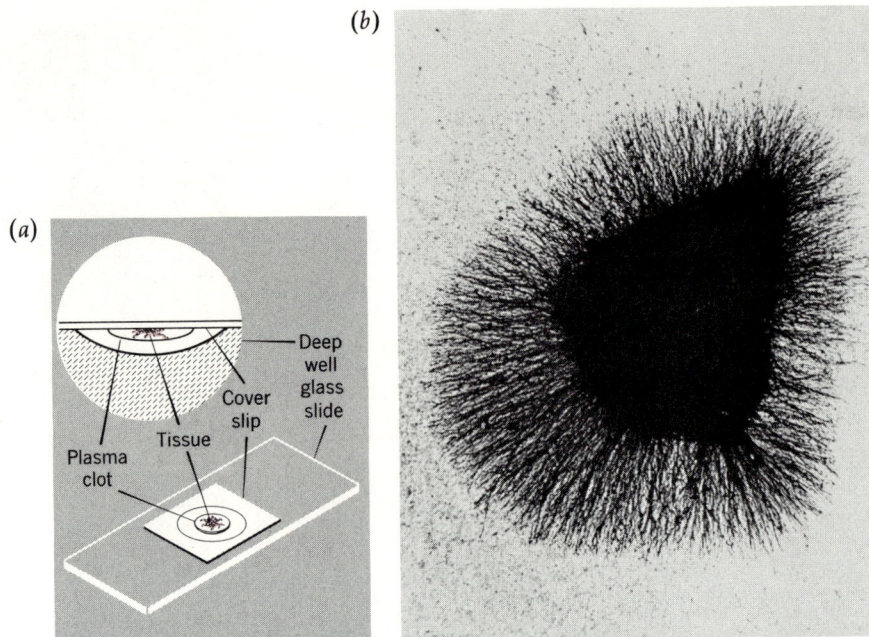

Figure 5-4 Culture of a fragment of nerve tissue. (a) The hanging drop method used by Ross Harrison. (b) The original explanted fragment is surrounded by a "halo" of nerve fibers that have grown out from it into the surrounding medium. Such a culture can be maintained in healthy condition for weeks. This culture has been treated with a purified extract called **NGF,** which increases the number of outgrowing fibers. Magnification 60×.

mass of tissue to form long fibers extending into the surrounding lymph clot (Fig. 5-4b). Not long after, one of Harrison's students applied these techniques to fragments of heart tissue explanted (removed) from chick embryos. He also noted that single cells emigrated from the fragment of the glass, and that these cells continued to beat rhythmically, even though they were completely isolated from any neighbors. He concluded that each heart-muscle cell must possess the capacity to contract and must have its own intrinsic rate of beat. In similar observations it was noted that a fragment of any given type of tissue will continue to perform the function of the tissue from which it was explanted.

Since that time, and especially in the last 20 years, relatively simple techniques have been devised for using enzymes to dissociate tissues into their component cells, and for growing those cells in nutrient culture media (Fig. 5-5). Live cells have been studied from virtually every tissue from a wide variety of animals, both embryonic and adult, including human, and recently from plant tissues as well. From such studies it has become increasingly clear that the properties, behavior, and metabolic characteristic of the tissue of origin derive from the component cells.

The opening of the twentieth century thus marked the beginning of the experimental approach and attempts to correlate cell structure with function and behavior. By the

1920s the study of cell components at the molecular level had been launched. These biochemical and biophysical approaches, together with the use of modern instruments, such as the electron microscope, and sophisticated chemical techniques, have led to a remarkable increase in our knowledge of cell biology.

Figure 5-5   (a) Organs may be removed from embryos of laboratory animals at appropriate stages; the heart from an embryonic chick is shown. The organ is cut into pieces and placed in a dissociating medium. After it has been stirred for 15 to 30 minutes the tissue comes apart into its component cells. The cells are suspended in a culture medium and distributed to culture plates. They adhere to the bottom, where they move about by ameboid activity, divide, and may (with heart cells) resume spontaneous rhythmic contractions.
(b) Photograph of freshly dissociated cells. Magnification 35×. (c) Cells growing in a sparse culture and (d) in a dense culture. Magnification 120×.
(e) A single heart cell stained to show its internal structure. Magnification 450×.

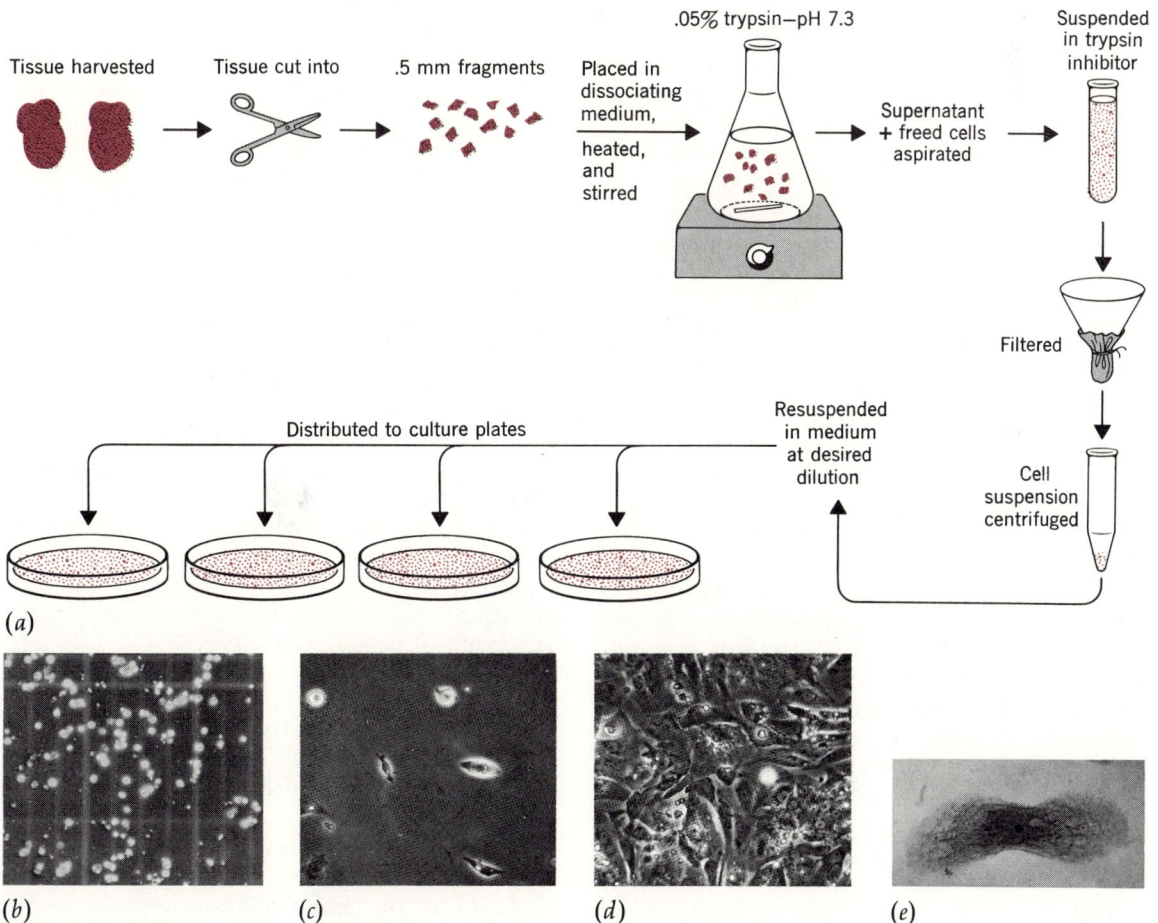

(a)

(b)   (c)   (d)   (e)

Cell Structure

Protoplasm

The concept of protoplasm that was once common implied that cells were constructed of a complex and fundamental living substance, essentially similar in all living things. In the last 25 years rapid progress in analyzing cellular ultrastructure in terms of modern biochemistry and biophysics has made the concept of protoplasm, as originally introduced, obsolete; the unit of life is the cell, and we can no longer regard it as simply a living substance, protoplasm. The term is still frequently employed, however; we shall continue to use it in this book to refer to the substance of the cell, but with the understanding that protoplasm is a highly integrated organization of diverse, complex subcellular structures of intricate chemical composition. These substances themselves are not alive, but together, by virtue of their organization and interrelationships, exhibit the characteristics of life. By analogy, an airplane is obviously not made of a fundamental aircraft substance; it is a complicated organization of specialized structures made of a variety of metals, alloys, rubber, plastics, and numerous other materials, none of which alone can fly, but when put together into complex specialized structures add up collectively to a typical flying machine.

The cell is a three-dimensional, gelatinous mass of molecular components. It is bounded by a thin specialized lipoprotein membrane, the **cell membrane** (sometimes called the **plasma membrane**), and contains within it a prominent **nucleus.** The correct term for the living contents of the cell—within the cell membrane but excluding the nucleus—is **cytoplasm.** The word **protoplast,** more or less synonymous with cell, emphasizes the living cell unit as distinct from secreted extracellular materials outside the cell membrane.

The Cell Membrane

The cell or plasma membrane, being at the cell periphery, is the structure with which the cell is in contact with the world around it. The cell adheres to neighbors or other surfaces by means of its membrane or membrane-associated materials. Substances that pass into the cell from the environment must traverse the membrane, and external stimuli are first "sensed" by the membrane.

Permeability studies on a wide variety of cell types indicate that lipids and many other substances soluble in lipids diffuse through natural membranes more rapidly than most water-soluble molecules, and that small molecules tend to penetrate faster than larger ones. These findings suggested that the cell surface itself must contain a large amount of lipid and that penetration is essentially by a **molecular sieve** mechanism, as though the surface contained holes of molecular dimensions through which dissolved molecules must pass in order to enter or leave the cell. When cells are placed in water or dilute solutions, they tend to swell by osmosis (Fig. 5-6). In a concentrated solution of sucrose, on the other hand, water tends to diffuse out of the cell, leading to shrinkage and crenation (wrinkling) of the cell surface. This kind of behavior is characteristic of a membrane with differential permeability, one that is more permeable to some molecules than it is to others.

Biologists have long believed that cells must be surrounded by an intact membrane that included lipids (fats), and explained many of the properties of cells in terms of characteristics they attributed to that membrane, despite the fact that no such membrane was visible even with the best light microscopes. In the 1930s studies of the permeability, surface tension, and electrical conductivity of cell surfaces led J. F. Danielli and Hugh Davson to propose a theoretical model of the composition of biological membranes. Their suggestion was that the cell membrane is organized like a sandwich, with a layer of lipid two molecules thick in the center, covered on both sides by a sheet or film of protein molecules. The lipids are arranged with the fatty acid portions pointing toward one another at the center of the membrane, and the water-soluble portion, called the **polar**

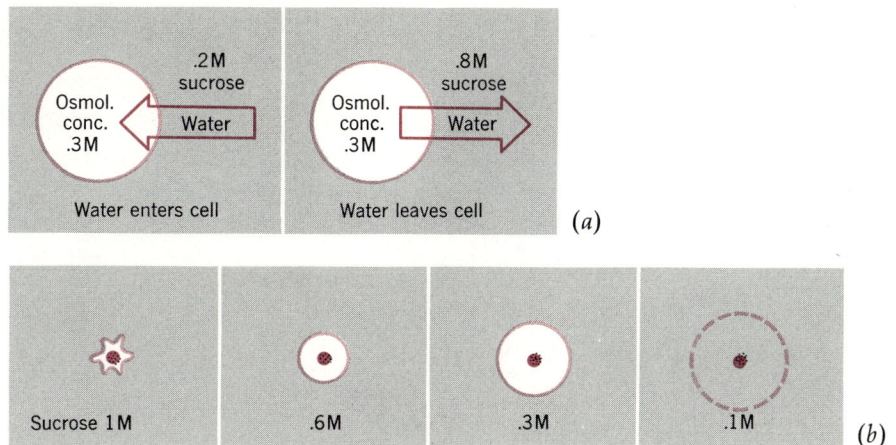

Figure 5-6 *Cells are sensitive to osmotic pressure.* (a) *When cells are immersed in a dilute solution of a substance, such as sucrose, that does not penetrate the cell membrane, the concentration of water is higher outside the cell than inside* (hypotonic); *water therefore diffuses into the cell. The opposite happens when a cell is placed in a sucrose solution of higher concentration than the cytoplasm* (hypertonic). (b) *Cells that are not surrounded by a rigid wall respond to this movement of water by shrinking in a highly concentrated (1M) solution; in solutions less concentrated (in sucrose) than the cytoplasm the entry of water causes the cell to swell (at about .3M) or even burst (at .1M). Some substances, such as amino acids, glucose, and many ions, can be pumped actively by the cell membrane into or out of cytoplasm; thus a cell can often compensate for osmotic pressure by using metabolic energy to run its membrane pumps.*

end (where the nitrogen and phosphate groups are attached), bound to the protein layers (Fig. 5-7, p. 122). On the basis of this **Davson-Danielli model,** and from the dimensions of typical lipid and protein molecules that were already known, it was predicted that the cell membrane should be only about 80 Å thick—60 Å for the bimolecular lipid layer plus about 10 Å for the protein coats on each side. No wonder the membrane hadn't been seen with the light microscope!

In recent years the structure of the plasma membrane of many cells has been studied at high magnification with the electron microscope. Electron micrographs made by J. David Robertson were among the first to show that the cell membrane manifests two parallel dark lines separated by a light space. His measurements of the total thickness of these surface structures were in the range of 75–100 Å; Robertson had the insight to see that these lines, two dark and one light, correspond with the Davson-Danielli model of the cell membrane: the two dark lines are the two protein sheets bound to the polar end of the phospholipid layer, with nonstained fatty acid chains forming a clear zone between them (Fig. 5-8, p. 123). Robertson termed this a **unit membrane.** The original micrographs of unit membranes were made on Schwann cells, which form the myelin sheath (Chapter 17) around nerves (Fig. 5-9, p. 124), and on membranes of red blood cells. A similar unit-membrane structure has been seen at

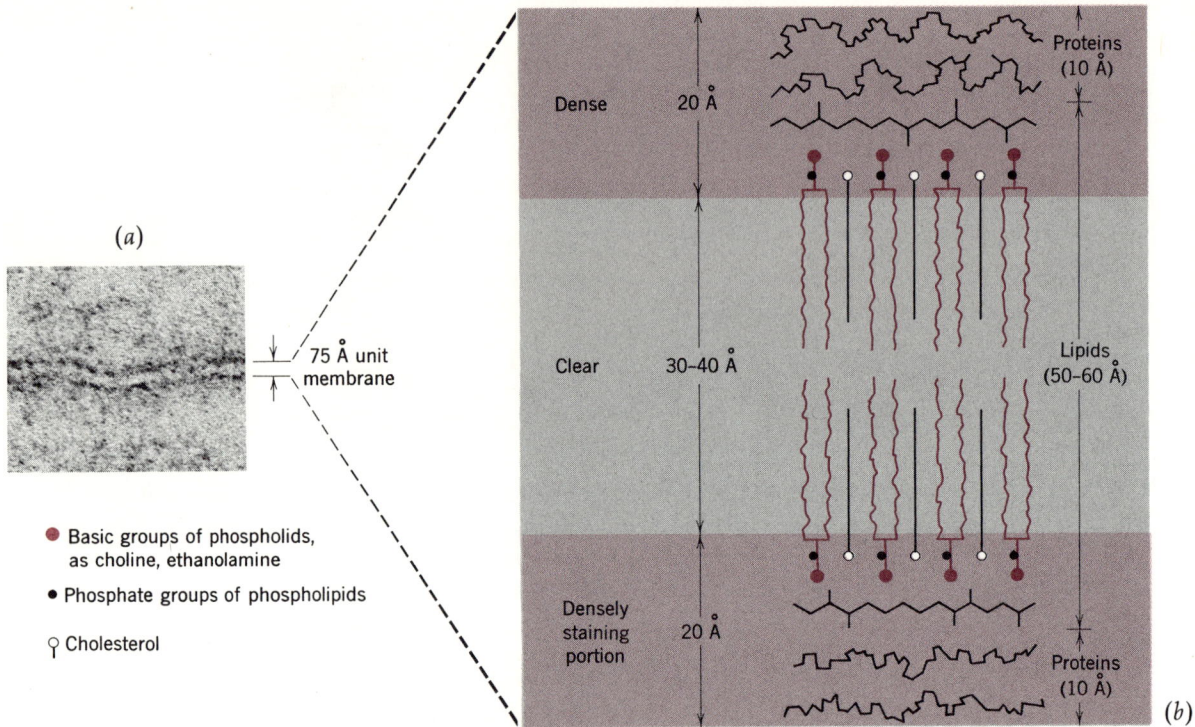

(a)

75 Å unit
membrane

● Basic groups of phospholids,
   as choline, ethanolamine

• Phosphate groups of phospholipids

♀ Cholesterol

Figure 5-7   (a) An electron micrograph of a tiny section of a unit membrane
(magnification 340,000 ×); (b) a model of its molecular structure according to
the unit membrane hypothesis. The two dense areas that stain with osmium
include the outer protein layer and the polar groups of the fatty acids. The
clear zone represents only the double area of the nonpolar fatty acid side
chains

the surface of a wide variety of plant and
animal cells under many conditions, and is
also typical of many membranous structures
found in cells. A method for examining the
surfaces of cells without fixing, by rapid
freezing (the **freeze-etch** method, Fig. 5-9c)
also suggests a unit-membrane-type structure.

Nonetheless, in the last 10 years evidence,
mainly from biochemical and physiological
experiments, has suggested that at least some
cell membranes—or perhaps some parts of
the surface of all cells—do not conform to
the unit-membrane model. For example, all
the lipids may be extracted from membranes
with ether. If the membranes are then prepared in the usual fashion for electron mi-

croscopy, they still appear as typical three-
layered, 80 Å unit membranes. But if it is the
lipid that normally keeps the two halves of
the sandwich apart, why—when that lipid is
removed—do the two protein layers not collapse together? A recent suggestion is that
there is protein or some other nonlipid material mixed with the lipid layer that keeps the
protein layers separated.

A different type of evidence that is hard to
reconcile with the unit-membrane model is
the finding that the enzyme phospholipase,
which hydrolyzes phospholipids (Chapter 3),
destroys many cell membranes. This indicates
that the lipid is not coated everywhere with
a protective layer of protein.

Finally, some membranes examined at very high magnification in the electron microscope simply do not share the typical unit-membrane appearance. The inner membrane of a mitochondrion, for example, or the surface membranes in certain specialized regions where cells touch, appear instead to be made of tiny globules linked together in a sheet.

These findings have recently led workers to suggest two alternatives to the unit-membrane model. One of these is the **subunit model** (Fig. 5-10a, p. 125), according to which globular aggregates of lipid and protein, with dimensions of 80-100 Å, are the subunits from which membranes are built, like a sheet of rounded bricks. The **liquid crystal model** visualizes masses of globular proteins "floating" in an oily layer of lipid (Fig. 5-10b). The evidence at present is too

Figure 5-8 (a) Electron micrograph of a human red blood cell unit plasma membrane. Magnification 145,000×. (b) A pair of unit membranes, prepared according to the freeze-etch method. Magnification 64,000×.

(a)

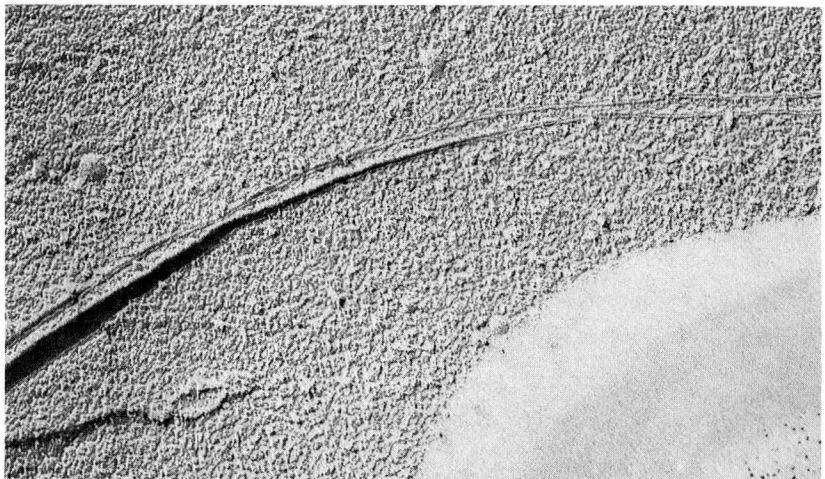

(b)

123 *The Structure of the Cell*

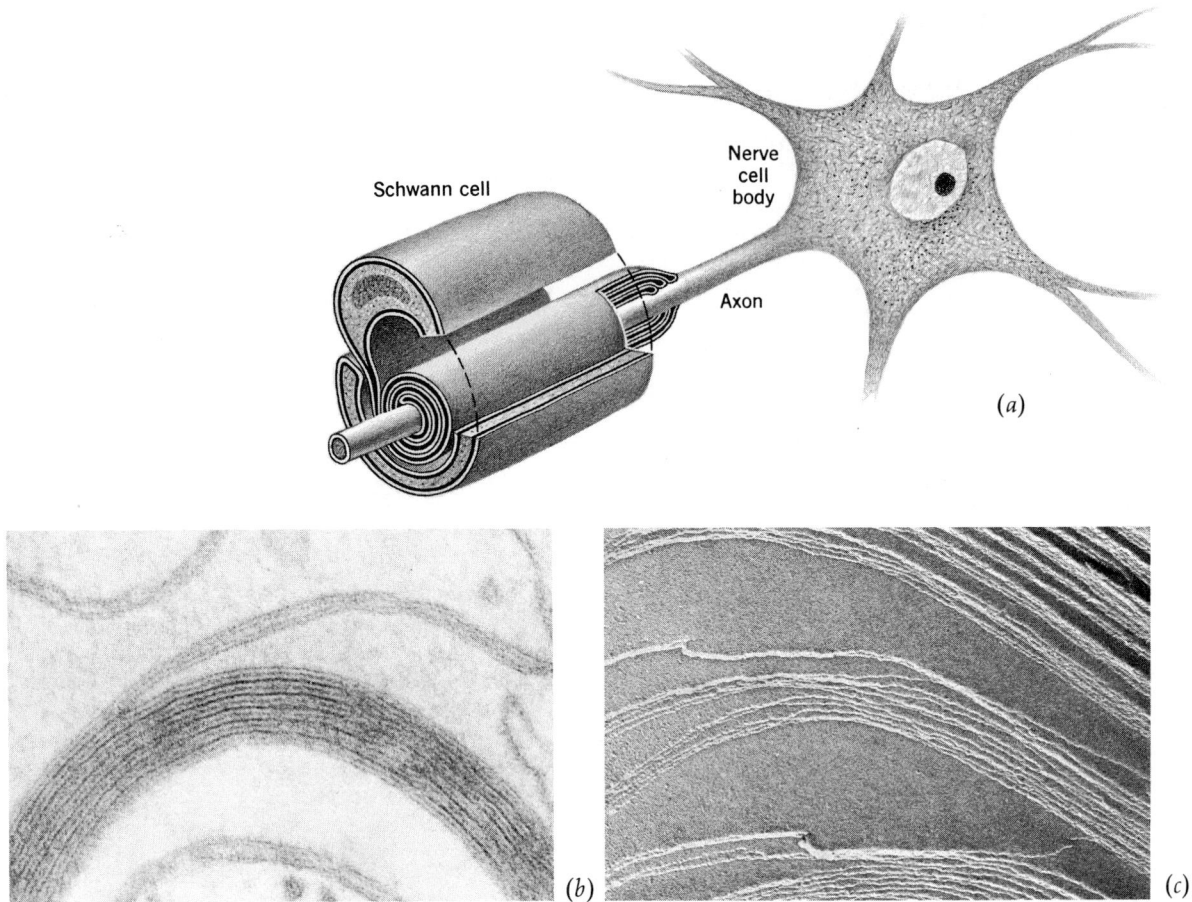

Figure 5-9 (a) *The structure of the myelin sheath. A Schwann cell wraps itself repeatedly around the long axon of a nerve cell, producing multiple layers of plasma membrane. (b) Electron micrograph of a section of such a myelin sheath. The stacked heavy dark lines correspond to the two fused inner layers of the Schwann cell unit membrane; the lighter lines represent the two fused outer layers. Magnification 170,000×. (c) Myelin sheath after* **freeze-etching.** *Magnification 64,000×. In this method no chemical fixatives are used. The specimen is rapidly frozen and "fractured" with a cold, chisel-like tool. The newly exposed surface is "shadowed" with carbon, and an electron micrograph is taken of the carbon "replica" surface.*

incomplete to decide whether each of these models is true for certain membranes or parts of membranes, or whether some intermediate model with properties of both is closer to the truth. Recent evidence suggests that lipoprotein subunits, or patches, serve as channels for passing materials across membranes.

The cell membrane is not merely a passive envelope around the cell, but rather an active, crucial component of the cell's total activities. Because everything that goes into

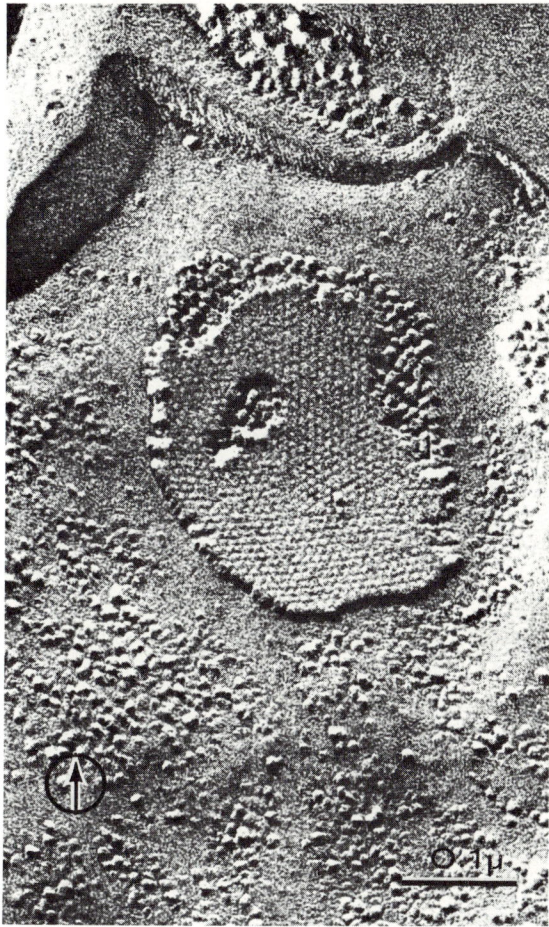

(a)

Figure 5-10 *Subunit membrane structure. (a) Many parts of membranes do not show a typical unit-membrane structure. Instead, they appear to contain globular subunits (arrow), which may be packed together to form a disc-like nexal region of cell contact, as shown in this freeze-etch preparation. Magnification 145,000×. (b) A model of a biological membrane based on recent evidence. The gray shapes represent globular protein of lipoprotein molecules lying on or inserted in the phospholipid bi-layer.*

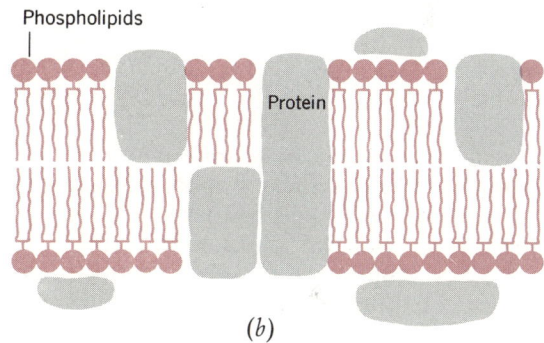

(b)

or out of a cell must pass through the membrane, one of its main roles is in regulating this traffic of materials. As we have mentioned, cell membranes are differentially permeable to dissolved materials (solutes) in the environment, although just how membranes maintain this selective permeability is not really understood.

Membranes are also able actively to transport materials from outside the cell to inside (and vice versa), even against a concentration gradient. After the introduction of radioactive tracer techniques in the 1940s it was discovered that a slow, constant flux of sodium, potassium, and chloride ions takes place across all typical plasma membranes. Yet this

process, **ion penetration,** does not normally result in ion accumulation and osmotic swelling because most cell membranes contain machinery for pumping ions out again, thereby maintaining a constant ionic concentration on both sides of osmotic pressure (Fig. 5-6). Such ion pumps require energy, and this is derived from the cell's metabolism. Cells exposed to metabolic inhibitors do swell, even in isotonic salt solutions. The pumping of materials across the cell membrane with the aid of metabolic energy is termed **active transport.**

Another method by which a cell may play an active role in determining what substances will move into it involves the active engulfing

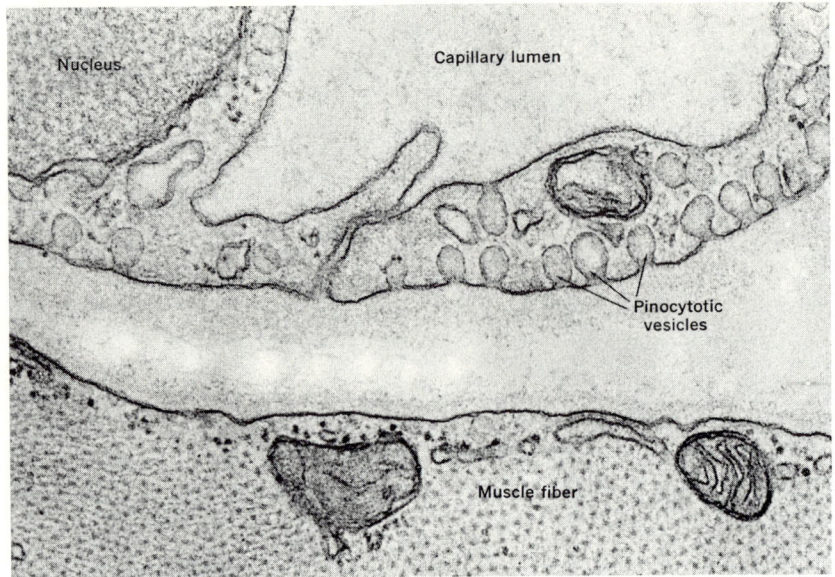

Figure 5-11 Electron micrograph of an endothelial cell in a capillary in muscle tissue. The small inpocketings are pinocytotic vesicles. Magnification 65,000×.

of material. This can happen when the plasma membrane forms small inpocketings, or **invaginations,** filled with external fluid, which may then be pinched off into the cytoplasm (Fig. 5-11) to form **vacuoles,** or **vesicles.** In turn, the membrane around the vesicles dissolves, releasing the fluid and other substances into the cytoplasm. This process is known as **pinocytosis.** When particles of solid matter are engulfed, the process is called **phagocytosis** (Fig. 5-14) (pp. 128, 129).

Isolated cells in culture adhere to the surface of the culture dish because their plasma membrane is sticky; cells in a tissue stick to their neighbors because of this surface adhesiveness. Where adjacent cells in a tissue meet, their membranes are often interlocked by fingerlike convolutions. With the electron microscope special adhesive sites on the convolutions called **desmosomes** (Fig. 5-12) can often be seen. Normally, the membranes of cells in contact are actually separated by a space of 200–300 Å filled with intercellular material, called **matrix.** Within the last few

years it has been shown that this space between cells may narrow to as little as 20 Å, or be obliterated altogether. These regions of close membrane contact are termed **gap,** or **nexal junctions** (Fig. 5-13). Evidence is beginning to accumulate that in certain tissues, especially in embryos, these specialized junctions may be regions through which certain ions and larger molecules can pass from cell to cell. The function of this form of communication from one cytoplasmic mass to another is one of the most fascinating current problems of cell biology.

Cytoplasm

We define cytoplasm to include all of the material inside the plasma membrane and outside the nucleus. One of the surprises revealed by electron microscopy was that cytoplasm is not merely a solution with particles floating in it. On the contrary, it is a highly organized matrix of molecular aggre-

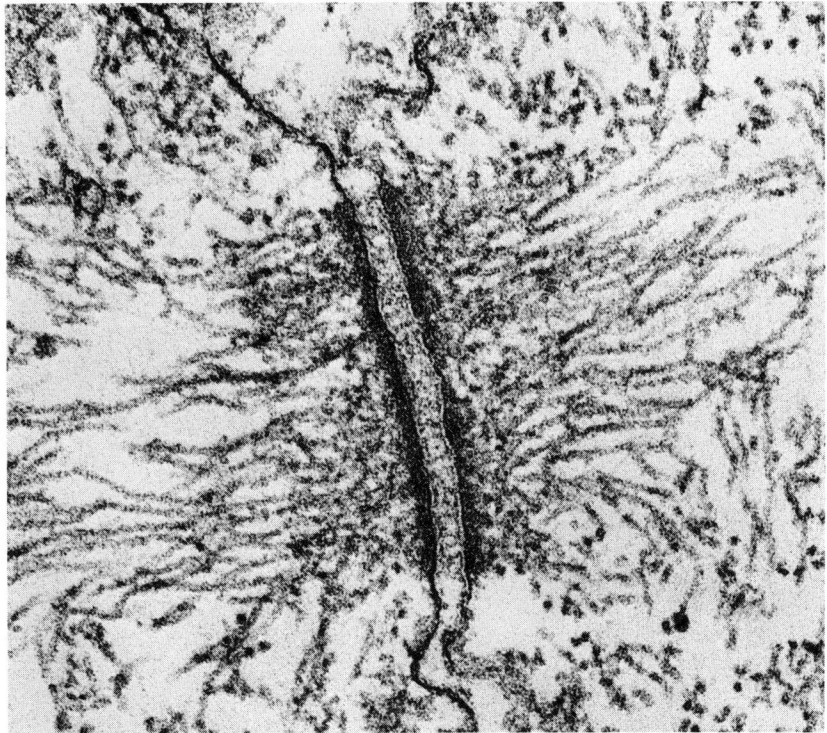

Figure 5-12   Electron micrograph of a desmosome from the skin of a larval newt. The clear space between the two halves of the desmosome is an intercellular gap of about 300 Å. Magnification 95,000×.

Figure 5-13   Nexal junctions, also called gap junctions, are regions where the intercellular gap is reduced from its usual dimension of 200–300 Å down to 20 Å or less. In this electron micrograph the nexal junction is filled with a black marker material. Magnification 200,000×.

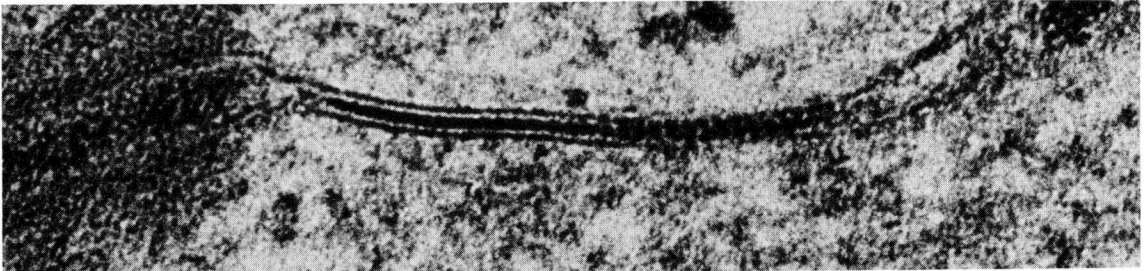

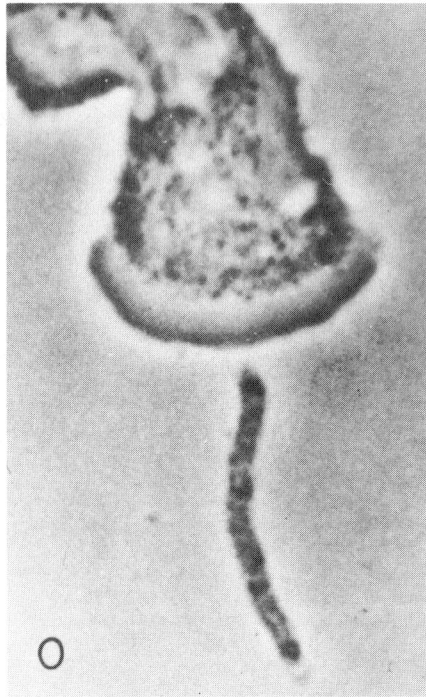

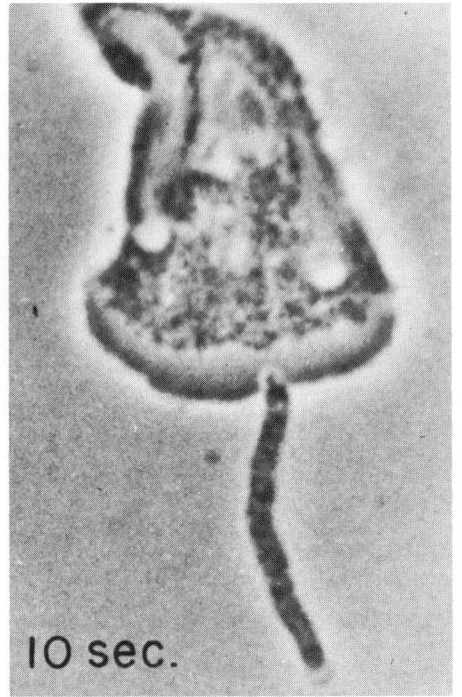

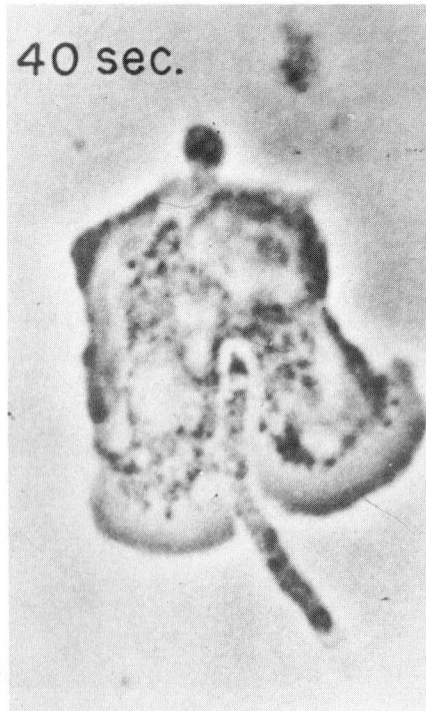

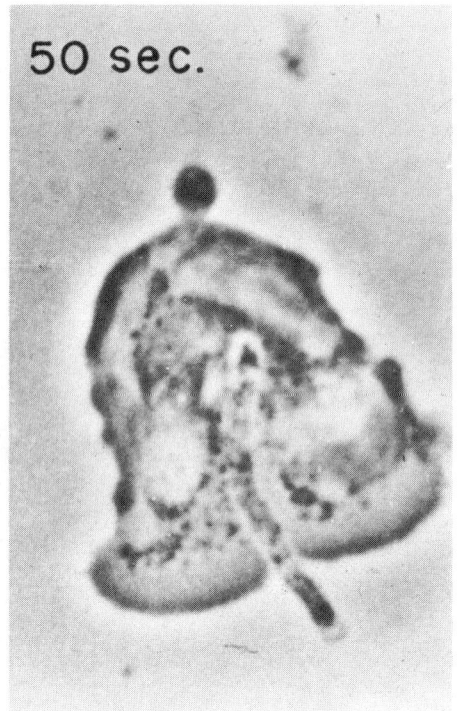

Figure 5-14 Frames
from a motion picture
of phagocytosis of a
bacillus by a human
white blood cell.
Magnification
1000×.

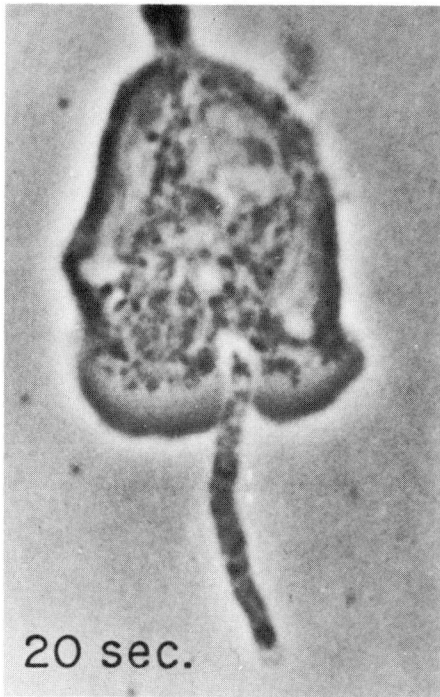

20 sec.

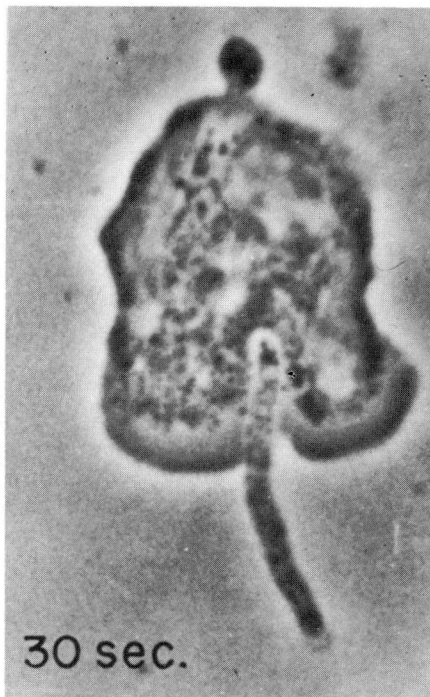

30 sec.

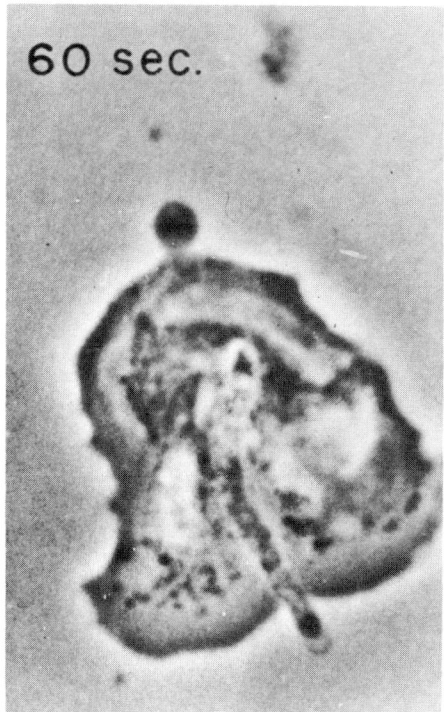

60 sec.

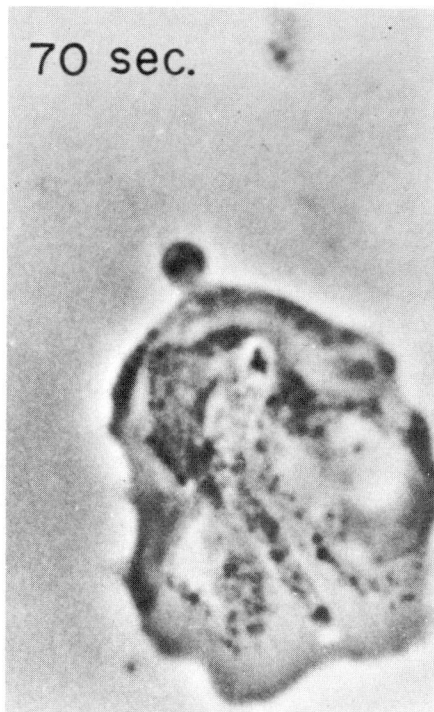

70 sec.

The Structure of the Cell

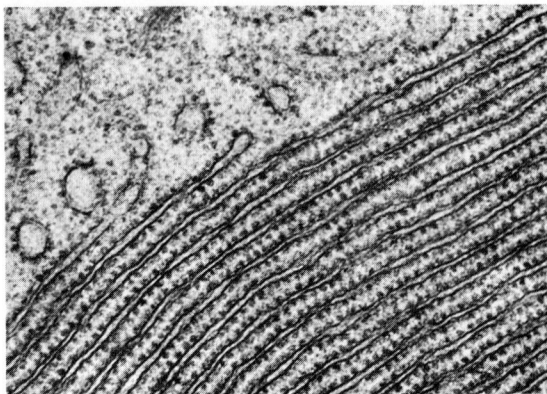

Figure 5-15 *The rough endoplasmic reticulum of a pancreatic cell, consisting of layers of broad, flat sacs. The outer surface of each sac is studded with ribosomes, which are 150 Å particles of ribonucleoprotein. Magnification 50,000×.*

gates or subcellular organelles, probably normally in a gelatinous state. Moreover, it is filled mainly with several different types of interconnected membranous structures showing the familiar unit-membrane pattern.

Endoplasmic Reticulum and Ribosomes

During the 1940s Keith Porter described a "lacelike reticulum"[1] in the cytoplasm. At high magnification cells revealed a system of tubules, swollen sacs (or **vacuoles**), and flattened spaces **(lamellae),** all composed of typical unit membranes. This system is the **endoplasmic reticulum** (often referred to as **ER**) and is present in all nucleate cells.

Some parts of the ER are lined on the outer surfaces with small (150–200 Å) particles of ribonucleoprotein called **ribosomes.** Membrane studded with these granules is called **rough ER** (Fig. 5-15); when no ribosomes line the membrane the ER is described as **smooth.** We have already seen that the ER and the ribosomes can be centrifuged out of the cytoplasm as the microsome fraction, (p. 103). and that it is on these ribosomes that protein synthesis takes place. In cells that are actively

[1] A reticulum is a net or mesh.

synthesizing proteins, such as secretory glandular cells and dividing cells, the rough ER is often very dense, essentially filling the cytoplasm. If such cells are fed amino acids labeled with radioactive tracers, the labeled amino acids can be traced in their progress as they are built into protein products. Such studies in the early 1960s showed that amino acids attach to the ribosomes, where they are incorporated into protein, and the newly formed protein is then transferred into the sacs and tubules of the ER. The new proteins may then be "packaged" in a membrane in the golgi apparatus (see next section), and then released into the cytoplasm or secreted by the cell. There is also evidence that the ER contains in its membranes enzymes capable of breaking down glycogen and a few other products. Apparently some of the energy used in protein synthesis and packaging is made available directly at the site by these ER membranes.

The Golgi Apparatus

The golgi apparatus, named for the Italian anatomist Camillo Golgi, who first described its appearance in the light microscope in 1898, consists of a series of flattened sacs and round vesicles of membranous material, apparently joined in some cells with the smooth ER (Fig. 5-16). Its relation to the endoplasmic reticulum suggests a role in cell synthesis (Fig. 5-17). The best current evidence is that the golgi system acts as a site of storage for newly synthesized materials within the cell and as a "packaging" system for secretions. For example, hormones, some enzymes, and other products that are later to be secreted are synthesized on the ribosomes, move into the channels of the ER, and through these to the golgi apparatus. There they may be enclosed in a membranous vesicle and finally transported to the plasma membrane for secretion.

It must be emphasized that these hypothesized functions of the golgi apparatus are not yet supported by strong evidence, and it may have other functions of greater importance to the cell. Only further research can provide answers to the many questions that arise about this intriguing cellular organelle.

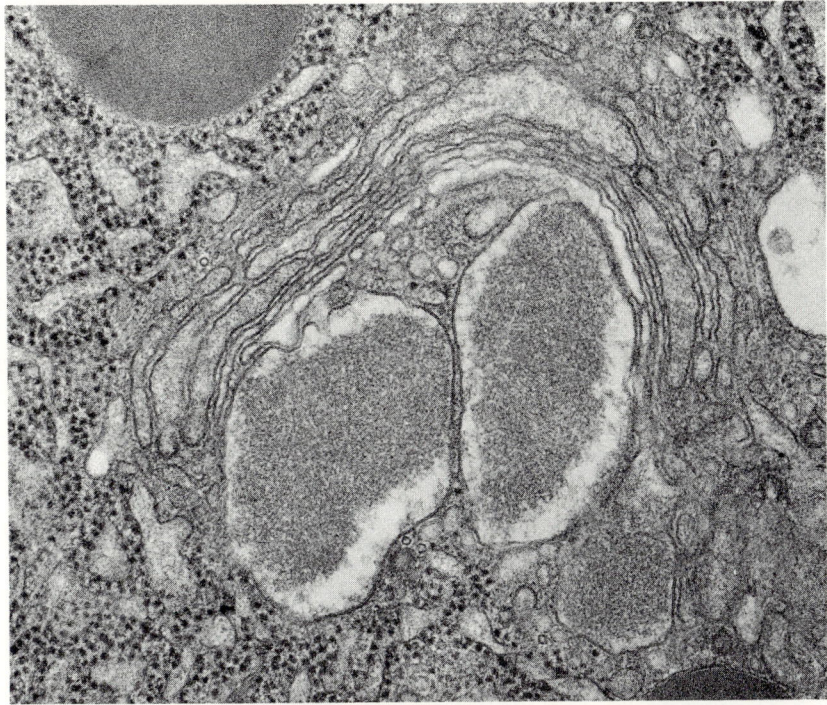

Figure 5-16 Electron micrograph of the golgi apparatus in a rat cell. It consists of layers (lamellae) of flattened sacs and round vesicles. Magnification 37,000×.

Figure 5-17 A model showing the close relationship between the ER and golgi apparatus.

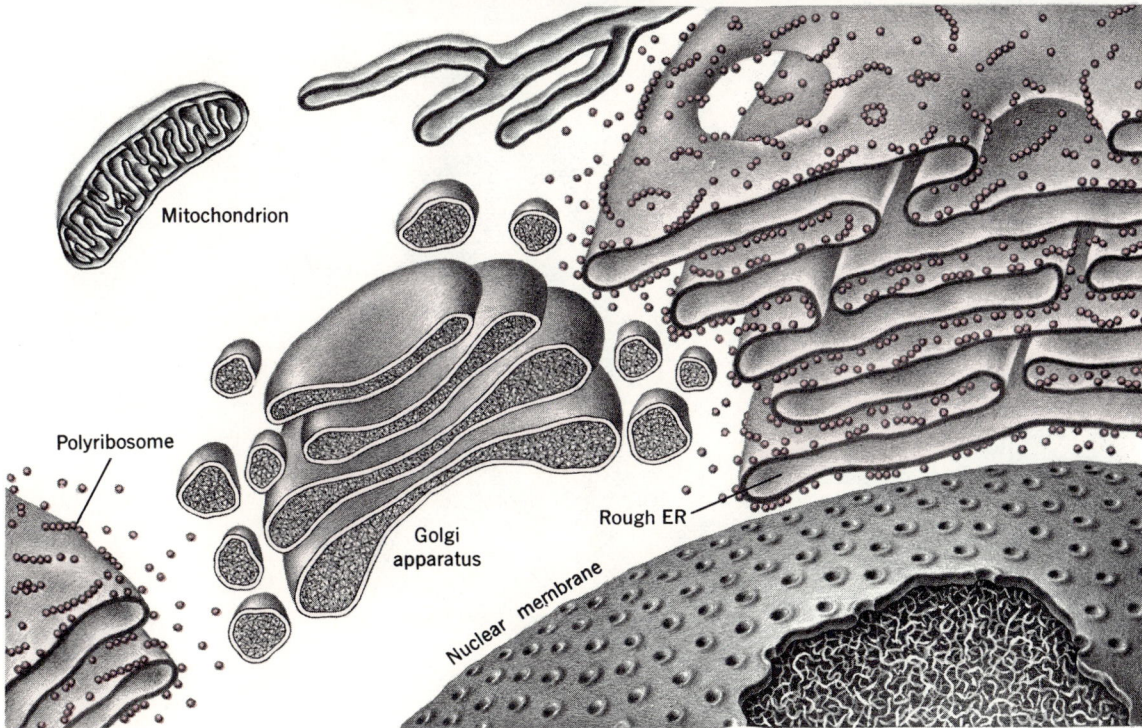

Mitochondrion

Polyribosome

Golgi apparatus

Rough ER

Nuclear membrane

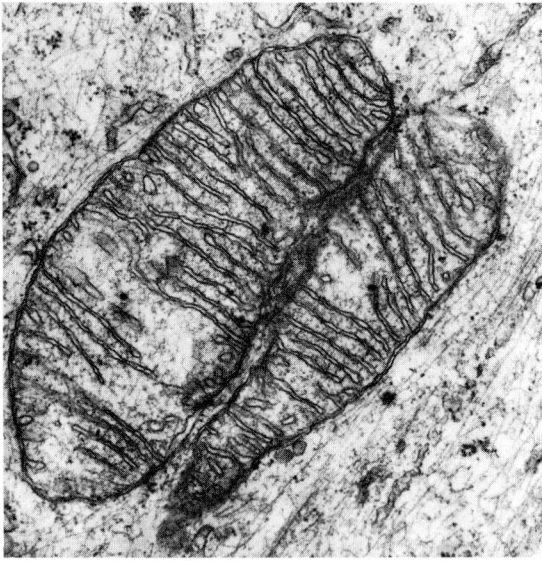

Figure 5-18   (a) Two mitochondria from a chick heart muscle cell showing the double mitochondrial membrane, cristae, and matrix. The fine fibers within the clear areas in the matrix represent mitochondrial DNA. Magnification 24,000×.
(b) A model of a single mitochondrion, on a scale of 150,000 to 1 (i.e., 100 Å = about 1.5 mm; or 1μ = 6 in.). This shows the smooth outer membrane and the inner membrane folded into the cristae. The entire surfaces of all the cristae are covered with a "fur" of tiny mushroom-like hairs or particles involved in oxidative phosphorylation. These are shown on only one small region of the drawing.

(a)

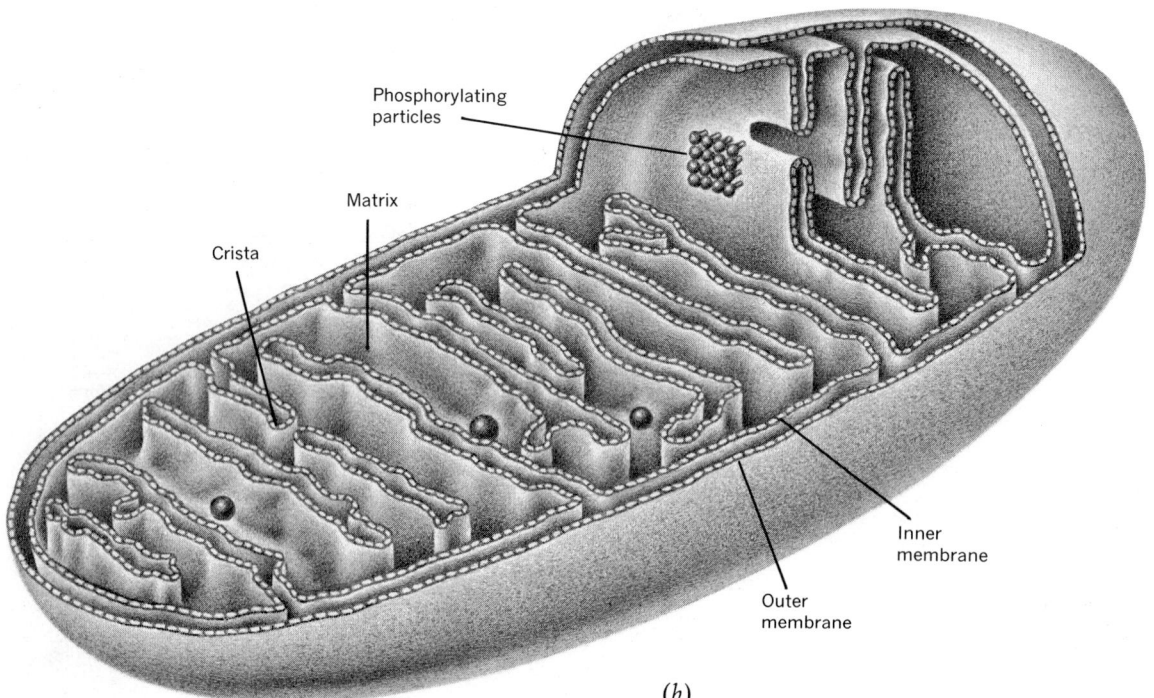

(b)

Mitochondria

The highly specialized and organized bodies called **mitochondria** (singular: **mitochondrion**) are another important component of the cytoplasm of living cells. Since the 1940s biologists have gained a vast insight into the architecture and workings of mitochondria, largely as the result of electron microscopy and extensive biochemical studies. These cytoplasmic structures carry out a variety of integrated chemical reactions, including those that supply most of the useful energy for a cell's vital activities. For this reason the mitochondria have been aptly called the "powerhouses" of the cell.

Mitochondria are barely visible under the light microscope, and in some cells may be so small that they can be seen only under the electron microscope (Fig. 5-18). They are generally spherical or rod-shaped and are much larger than ribosomes, ranging from .5 to 3μ in diameter by up to 10μ in length. They may change in size and shape, depending on the chemical and physical state of their surroundings, cell type, age, their own chemical activity, and the activity of the cell. There may be as many as 500 to 1000 mitochondria in the cytoplasm of such metabolically active cells as heart muscle and secretory cells.

A striking feature of mitochondria in the living cell is their constant motion. The phenomenon is attributed not to the movement of the mitochondria under their own power, but rather to the impact from the moving molecules and particles of the surrounding cytoplasm and to the streaming activity of cytoplasm that is often observable in metabolically active cells.

Electron microscopic examination of mitochondria (Fig. 5-18b) shows that each is bounded by a double membrane whose inner layer is usually folded into a characteristic system of ridges, or **cristae,** extending into the mitochondrial cavity filled with a matrix material. At very high magnification the outer membrane appears to have the typical unit-membrane structure. The inner membrane, in contrast, is apparently composed of an array of spherical units 40–45 Å in diameter. Recall

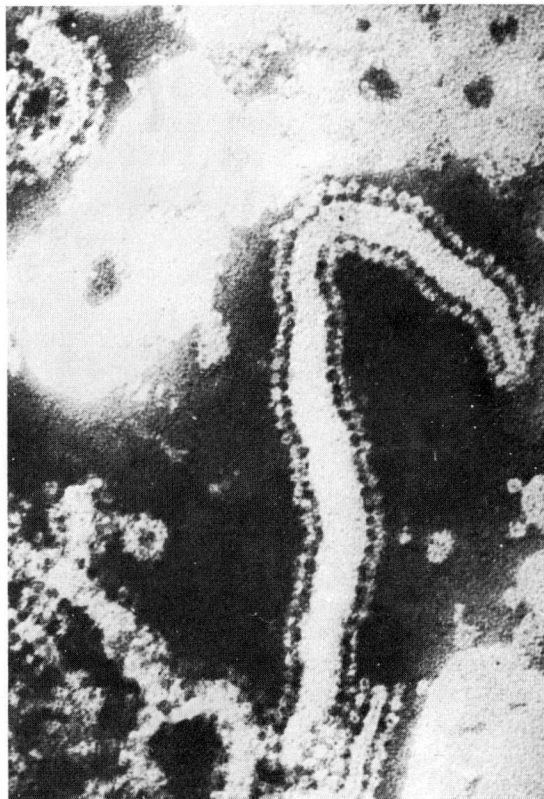

Figure 5-19   An electron-micrograph "negative image" of a cross-section of a mitochondrial crista, showing the "fur" of mushroom-like particles. Magnification 230,000 ×.

that we have described this as a typical sub-unit-membrane structure. Similar spherical subunits are found in chloroplast lamellae (see p. 142), and because both kinds of particle carry out phosphorylation the resemblance may be more than coincidental. With special staining, the crista membrane appears to be covered with a fine "fur" of tiny hairs, each with a ball 75–100 Å on the end of it (Fig. 5-19).

When mitochondria are isolated from cells and their inner and outer membranes are separated, it is found that the inner membrane (i.e., cristae) contains most of the enzymes of the citric acid cycle and entire elec-

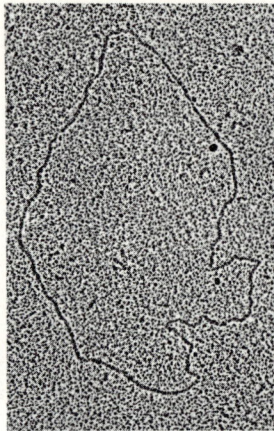

Figure 5-20 Circular DNA molecules isolated from mouse mitochondria. Each circle is about 5µ in circumference. Magnified about 30,000 times.

tron transport chain, and is able to carry out oxidative phosphorylation. That is, the subunits that make up the membrane are themselves made of protein with enzymic activity. It seems to be a general principle among all plant and animal cells that phosphorylation and electron flow is always associated with a lipoprotein membrane structure.

Mitochondria are apparently not synthesized by the cell in whose cytoplasm they lie. Early in the 1960s it was discovered that each mitochondrion contains a circular molecule of DNA (Fig. 5-20) that is distinguishable by its biochemical properties from the nuclear DNA of the cell and replicates at different times than does the cell chromatin. Thus it is almost as if each mitochondrion were a semi-independent, metabolically active microorganism living inside cells as a symbiont (i.e., to the mutual benefit of the mitochondrion and the host cell). This idea has been seriously proposed by several scientists, who suggest that all mitochondria (and chloroplasts) may be derived from bacteria-like microorganisms that, perhaps a billion years ago, fused with an ancestral cell (Chapter 22).

Lysosomes

The mid-1950s witnessed the discovery of a new group of membranous subcellular particles, called **lysosomes,** of about the same size as small mitochondria. Lysosomes are membrane-bounded structures or sacs (with no internal cristae) containing a host of hydrolytic enzymes (Chapter 2) that catalyze the digestion of most of the organic constituents of the living cell: proteins, nucleic acids, certain carbohydrates, and possibly fats. Disruption of the lysosomal membrane and subsequent release of the digestive enzymes leads to a rapid dissolution of the cell. Lysosomes are believed to function normally in digestion of food material stored in cells, in breakdown of foreign particles by white blood cells, in dissolution of the structures surrounding the egg cell in the course of fertilization by sperm, in bone-digesting activity of certain cells, and in the death and destruction of aged cells.

The Nucleus

Viewed with the light microscope, living, unstained cells always contain a large spherical or oval clear sac suspended in the surrounding cytoplasm, showing no obvious internal structures. With special optical systems, or when cells are fixed and stained, this central body, the **nucleus,** stands out as the most prominent structure in the cell. The nucleus is essentially a membrane-enclosed space containing the cell's DNA (deoxyribonucleic acid; see Chapter 3) and DNA-associated protein in a gel-like matrix of **nuclear sap.** (Fig. 5-21).

The nuclear membrane is a highly specialized structure, with paradoxical behavior. It exhibits semipermeable properties, indicating that it is an intact permeability barrier to some solutes. However, in the electron microscope large pores, called **annuli,** are apparent in the nuclear membrane (Fig. 5-22, p. 136). These are real holes; 85 Å particles of gold engulfed by amebas can pass through the pores into the nuclear sap. It may be that the pores open and close to let certain materials pass through.

In addition to acting as a permeability barrier between the cytoplasm and the nuclear

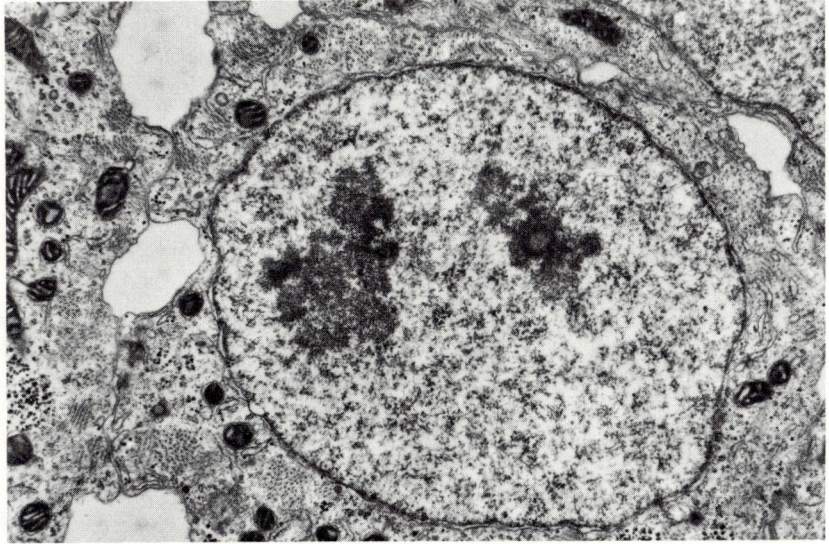

(a)

Figure 5-21 (a) Electron micrograph of a nucleus of a chick heart cell showing the typical structures. The dark masses inside the nucleus are the nucleoli, made of granular material (mainly RNA) condensed around an irregular network of loose strands. The patches of slightly dark material are heterochromatin. The nuclear membrane is perforated with many pore-like structures (annuli), through which the nuclear sap communicates with the cytoplasm. Magnification 11,000×. (b) A model of the nucleus of an embryonic cell of a honeybee. Chromatin fibers are shown in color. Microtubules anchor mitochondria and other cytoplasmic organelles to the nuclear membrane.

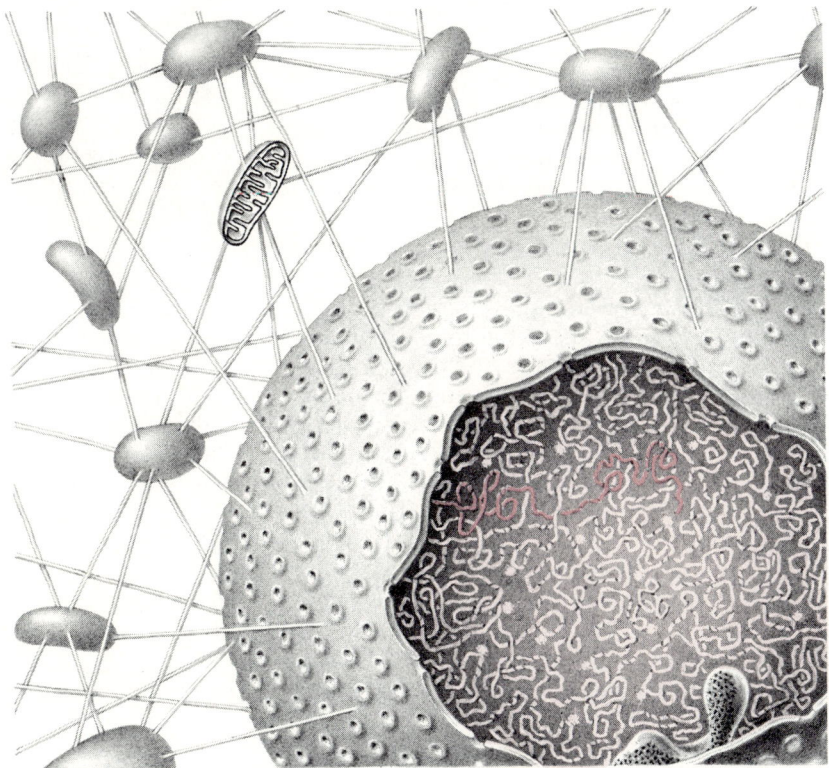

(b)

135 *The Structure of the Cell*

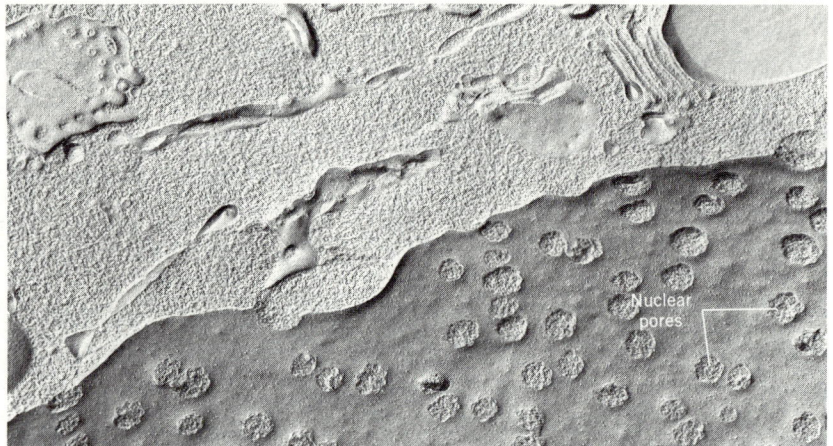

Figure 5-22 Nuclear pores. (a) Image of the annuli in a freeze-etch preparation of the nuclear membrane of an onion root tip cell. Magnification 25,000×. (b) Detailed view of the pores in a frog oöcyte nucleus. Magnification 155,000×.

(a)

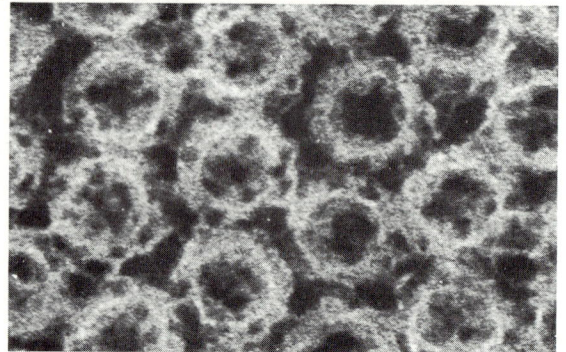

(b)

sap, the nuclear membrane seems to serve as an architectural center for the cell. Many of the structures in the cytoplasm are attached to the outside of this envelope, directly or by long fibers.

Early microscopists were disappointed when they examined sections of nondividing cells to find the nucleus itself was relatively structureless. Dyes known to stain DNA specifically would color the nuclear sap, whereas other stains could show up one or more clumps within the nucleus called **nucleoli** (Fig. 5-21a), which are rich in ribonucleic acids, or RNA (Chapter 3). But it was puzzling that the chromosomes (Chapter 7) that were known to be present in the nucleus could not be seen as distinct structures. Just before cell division began the chromosomes seemed to appear, as if by magic, forming out of the cell fluid.

It was not until the mid-1950s that Hans Ris, a biologist at the University of Chicago first took good electron micrographs of the nucleoprotein fibers in the nondividing nucleus. We call DNA in this form **chromatin** (Fig. 5-23). We now know that a strand of chromatin is a single molecule (i.e., a double helix) of DNA wrapped in protein. Its dimensions are 40–50 Å in diameter by up to 2 cm long. To visualize these dimensions better, calculate that if a chromatin thread were a

rope a half-inch in diameter, it would be 25 miles long! Most of the chromatin threads in the nondividing nucleus are coiled and wound into condensed masses. Just before cell division (termed **mitosis**) the amount of DNA in the chromatin doubles, and the coiling becomes tighter and tighter as the chromatin threads are condensed into discrete chromosomes. We shall examine this process in greater detail later, when we discuss cell division (Chapter 7); however, we can note at this point that it is this condensation through coiling that makes the chromosomes visible in the microscope just before mitosis. In the uncoiled state they are simply too thin to see.

The nucleus is the information center of the cell. The role of the nucleus, or rather of the DNA in the nucleus, is to provide instructions that guide all of the life processes of the cell. The nucleus plays a central role in cell division. It determines what **specialization** the cell will have, that is, how it will **differentiate** (Chapter 12). And the nucleus directs most of the metabolic activities of the cell. We now know that all of this information is coded into the chromatin in the linear sequence of nucleotides in the DNA helix (Chapters 3 and 4). Again, because of the importance of the DNA code and the **genes,** which are the units or code words of hereditary information, we reserve a detailed discussion of these matters for later sections on heredity (Chapter 10).

Figure 5-23   Chromatin fibers from a honeybee embryo cell, stretched and flattened onto a carbon film for electron microscopy. Each 230 Å fiber consists of a single DNA filament (double helix) packed into a lumpy protein sheath. Magnification about 100,000×.

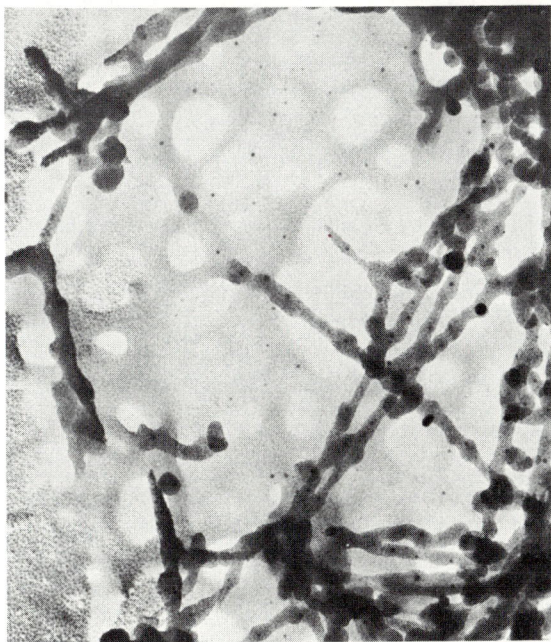

Characteristics of Animal Cells, Plant Cells, and Protists

We can easily distinguish most higher animals from plants. Animals are made up of certain characteristically organized systems (digestive, circulatory, nervous, etc.); are usually capable of movement under their own power; and are completely dependent on preformed organic substances for their energy and carbon supply (Chapter 4), to name only a few of their more obvious characteristics. Higher plants contain the green pigment chlorophyll, which gives them the unique ability to carry on photosynthesis; are usually immobile; and have their own characteristically organized systems.

An appreciable number of unicellular organisms, however, exhibit characteristics of both plants and animals. As a consequence, some biologists classify the very same organisms as plants that others place in the animal kingdom. Still others have been inclined to regard all unicellular organisms as neither plant nor animal but as a separate group called **protists** (see p. 145).

Evidence suggests that among many unicellular organisms a sharp differentiation between plants and animals has not yet occurred. The two separate lines of organisms that we call plants and animals may well have arisen in the course of evolution from a now-extinct ancient protist bearing a resemblance to certain present-day species. This view supports the basic concept of a common unity among all organisms and points to the different lines of evolution that life has taken in the course of time (Chapter 22).

Unique Features of the Animal Cell

Although all cells have a great many features in common (Fig. 5-24, p. 138), there are

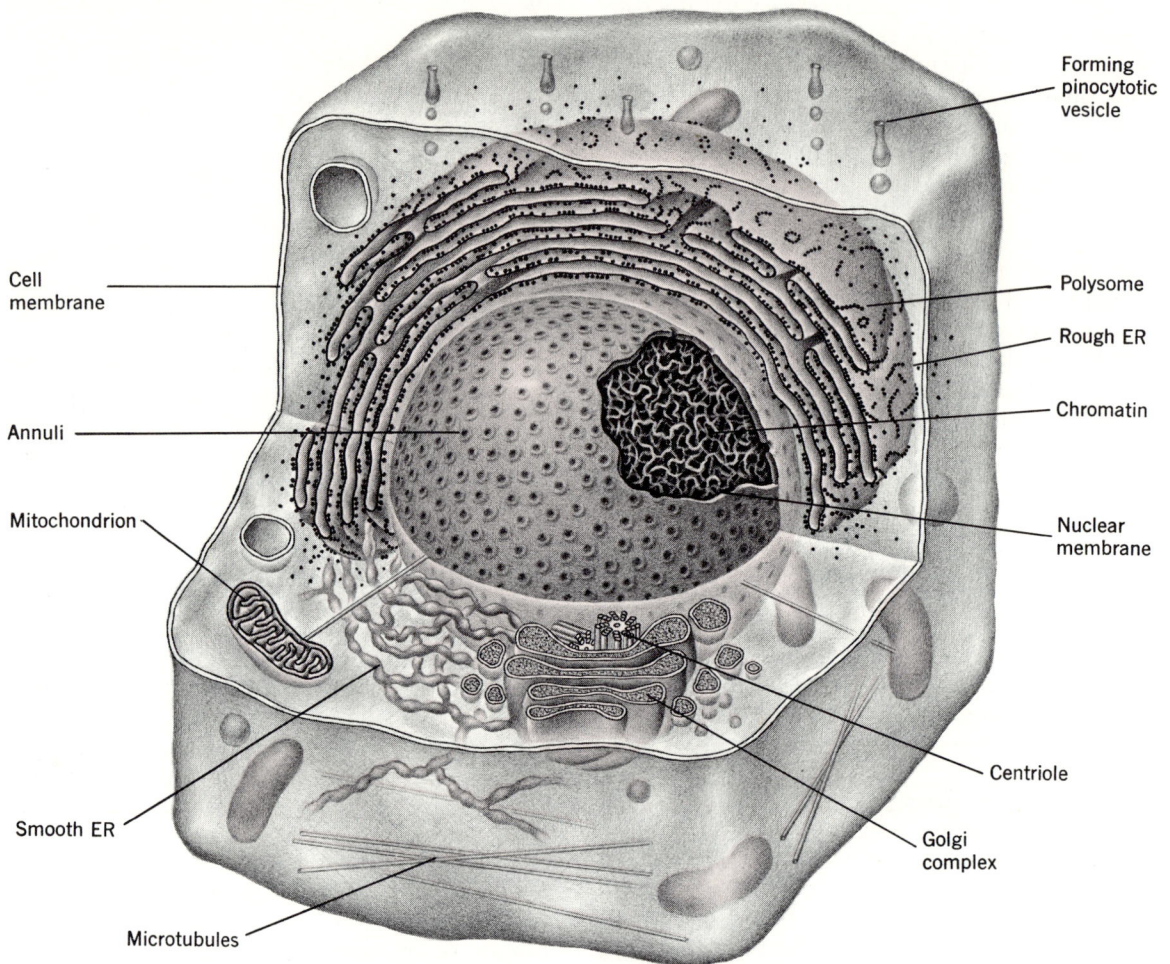

Figure 5-24 *The structure of a generalized eukaryotic cell. Chromatin fibers are surrounded by the nuclear membrane, which is penetrated by annuli, or pores. Rough ER is made of layers of membranous sacs studded with ribosomes; smooth ER is typically tubular. Mitochondria in the cytoplasm may be attached to the nucleus by microtubules. A golgi apparatus lies near the nucleus, partly surrounding a pair of centrioles. The plasma membrane is shown (top) forming pinocytotic vesicles.*

certain cellular features that differ between plants and animals.

Centrosome, Cilia, and Flagella

Virtually all animal cells and a number of lower or primitive plant cells contain a unique structure in the cytoplasm called a **centrosome.** It is usually present near the nucleus as a small, clear region with radiating, aster-like fibers and one or two deeply staining granules, called **centrioles,** at its center (Fig. 5-24). The cells of higher plants have no centrosome, but instead during cell division display two small clear areas called **polar caps**

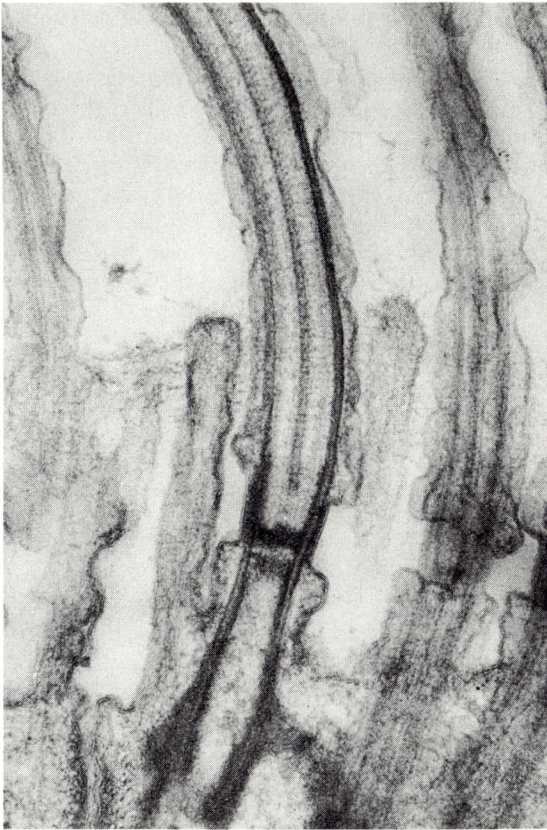

(a)

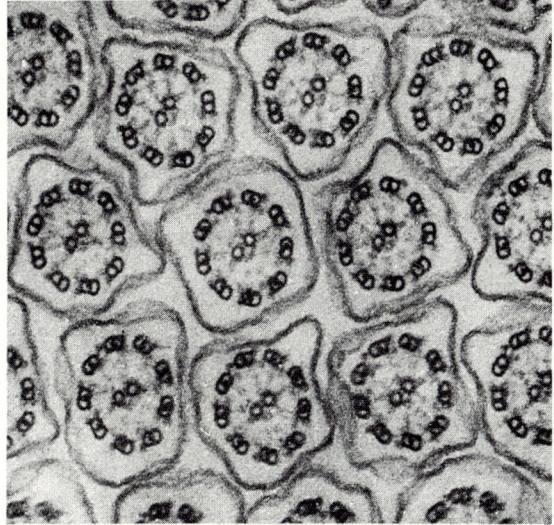

(b)

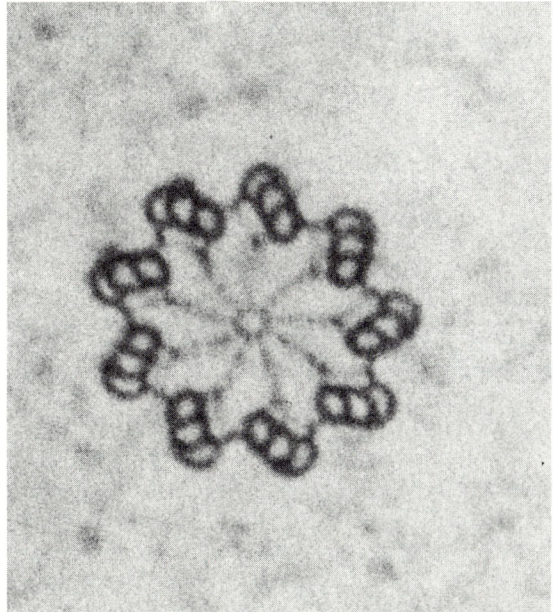

(c)

Figure 5-25 (a) Cilia on the outer surface of a cell in the gill of a fresh-water mussel. The shaft of the cilia extend outward from the basal bodies below. Magnification 34,000×. (b) Cross-section through a group of cilia like those shown in (a). Each cilium has a circle of nine double fibrils and a pair of single central tubules. Magnification 67,000×. (c) Cross-section through the basal body; note that the central microtubules do not extend this far into the cell. Magnification 174,000×.

(see p. 171), which apparently carry out the same function as the centrosome.

The centrosome has two major roles in the cell: (1) it organizes the construction of the mitotic apparatus before cell division; and (2) it controls the formation and activity of **cilia** and **flagella**, slender filamentous structures projecting from the external surface of the cell membrane of certain cells. Cilia and flagella always extend from a *basal body* in the cytoplasm (Fig. 5-25). This basal body is a centriole. Cilia are relatively short (about 10μ long) and usually present in large numbers, whereas flagella are considerably longer,

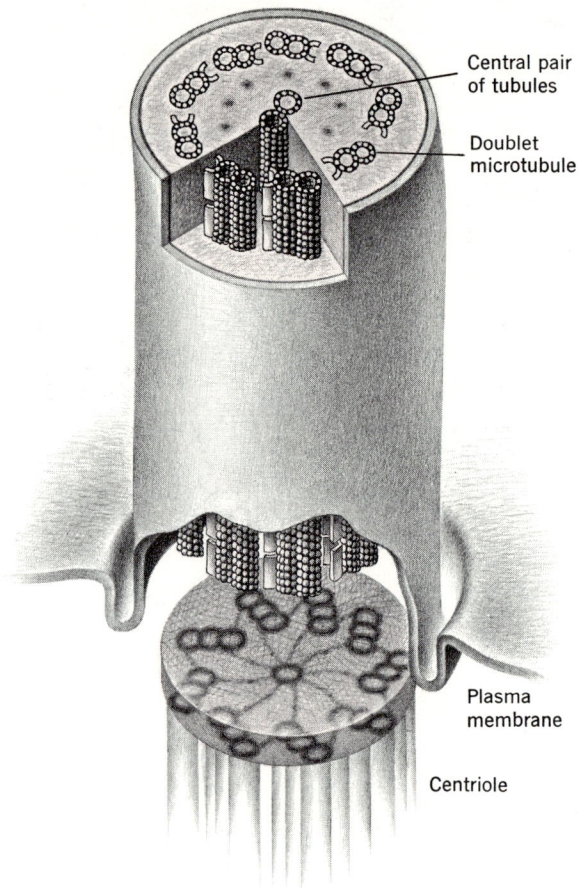

Figure 5-26 *A model of the structure of a cilium or flagellum, on a scale of 150,000:1.*

The extraordinary resolving power of the electron microscope has shown that cilia and flagella of all cells have a similar ultrastructure, among forms as varied as protozoa, plants, and people. Both cilia and flagella are enclosed by a semipermeable membrane, an extension of the plasma membrane of the cell. Within this membrane lies a bundle of 11 thin fibers, consisting always of a circle of nine double fibers around two central single **microtubules** (Fig. 5-26). This **9 + 2 pattern** has come to be recognized as one of the universal constants among the eukaryotes.

Unique Features of the Plant Cell

The Cell Wall

Structures that may be called **extraneous coats** are present on the outer surface of all cells, although in animal cells and protists they may be difficult to distinguish from the outer protein layer of the membrane. The higher plants secrete a well-developed, rigid covering of nonliving material, a **cell wall** (Fig. 5-27), outside the cell membrane. This is perhaps the most distinguishing characteristic of plant cells, surrounding every pro-

sometimes attaining a length of 150μ, and generally fewer in number. Cilia are found on certain unicellular organisms (e.g., paramecia, p. 151) and are responsible for the motility of these protozoa. Cells such as those lining the inner surface of the human trachea (windpipe) also have large quantities of active cilia; they beat in a coordinated movement, giving rise to currents of fluid at the outer cell surface, and thus propel out any minute foreign bodies that might gain entrance to the trachea. Flagella are also found on unicellular organisms and the great majority of sperm cells of both plants and animals. Their whiplike action accounts for the motility of these cells.

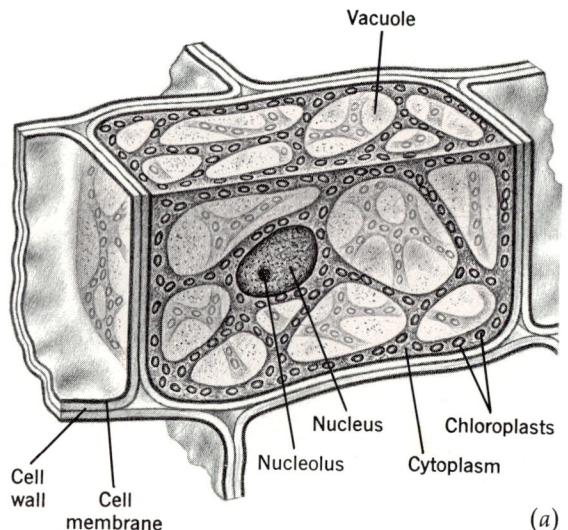

(a)

toplast. In most green plants it is composed primarily of a complex carbohydrate called **cellulose** (Chapter 3). Depending on the particular plant cell, the cell wall contains, in addition to cellulose, varying amounts of other substances including salts, lignin (a complex organic material responsible for the characteristic woody property of some plants), and certain waterproof, fat-like substances called **waxes** p. 63).

Unlike the cell membrane, the cell wall is permeable to most molecules and exercises no control in the passage of materials in and out of the cell. In effect it is a kind of plant skeleton, serving to protect, support, and maintain individual cells as well as a whole plant. Strands of cytoplasm penetrate the cell walls at regions called **plasmodesmata,** providing a continuous protoplasmic connecting system to adjoining cells and therefore to different parts of the plant. The presence in some bacteria and fungi of a definite cell-wall structure, although of apparently different composition from that of higher plant cells,

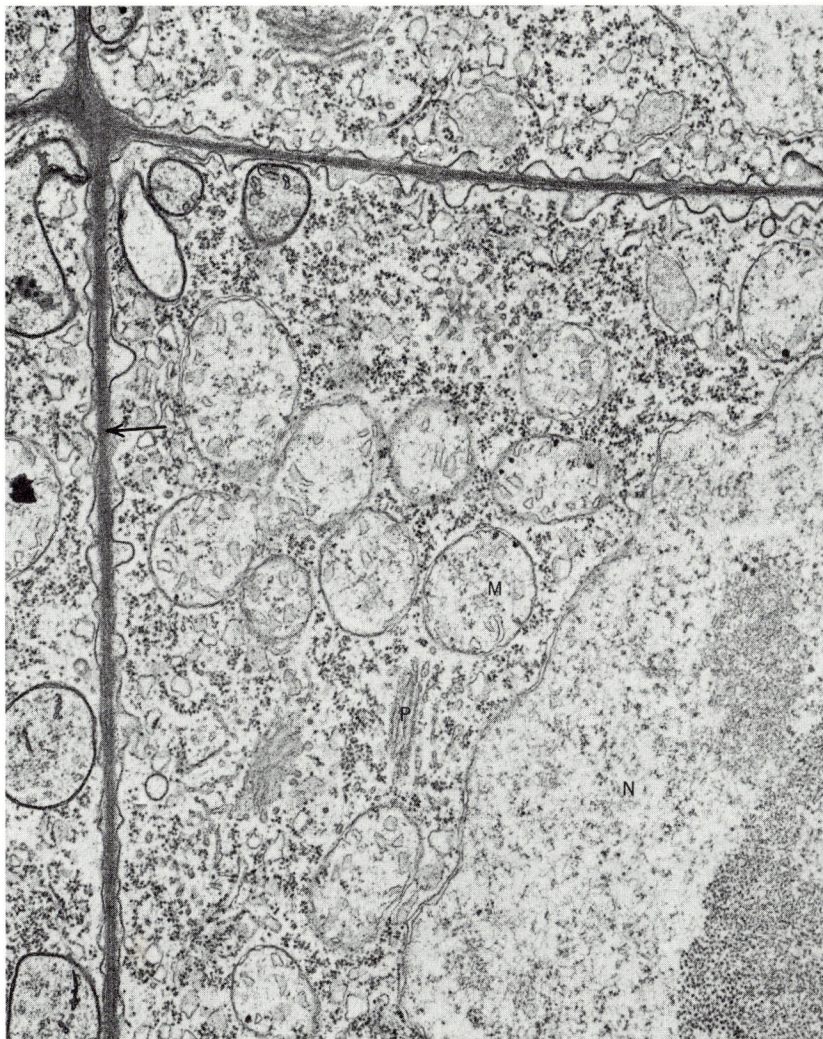

(b)

Figure 5-27   (a) A typical plant cell. (b) Electron micrograph of cells from an African violet flower, showing the thick cell wall (arrow), nucleus (N), mitochondria (M), and proplastids (P), which are stacks of membrane-forming chloroplasts. Magnification 22,000×.

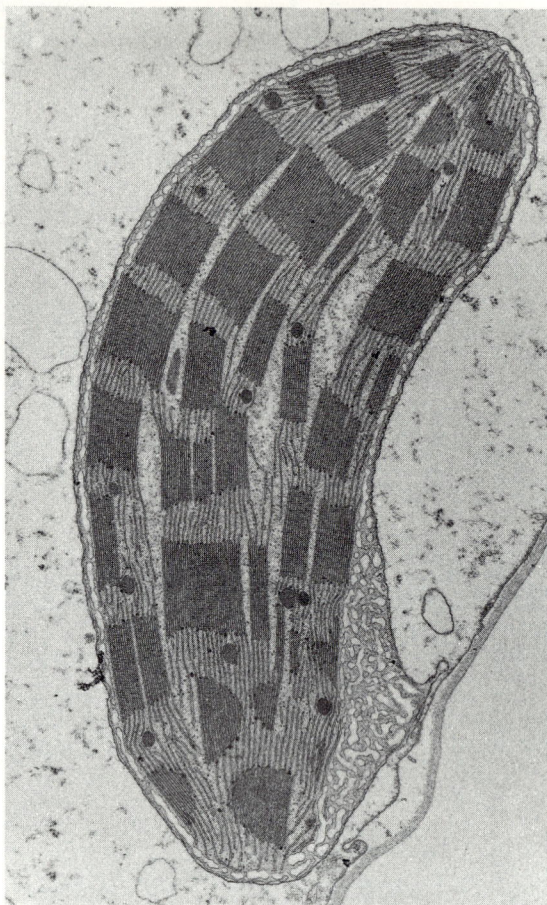

Figure 5-28   Electron micrograph of a chloroplast from a maize plant. The stacks of membrane discs form cylindrical units called grana. Magnification 17,000✕.

The pigmented plastids, the **chromoplasts,** are responsible for most of the varied colors of plants. The commonest of these, of course, is the **chloroplast** (Fig. 5-28), the site of photosynthesis, which contains the plant's green pigment **chlorophyll.**

Chloroplasts in higher plants range in size from 3 to 7μ. They often occupy relatively fixed positions along the cell membrane, but normally are able to reorient their positions relative to the direction of light, either by swinging from profile to full-face or by changing shape. In the electron microscope it can be seen that the chloroplast is enclosed by two concentric unit membranes that exhibit semipermeable properties. Internally, the chloroplast contains a large number of cylindrical units called **grana.** Each granum consists of a stack of disc-shaped compartments that contain the chlorophyll. The membranes that form the walls of these compartments appear to have a subunit structure (Fig. 5-29). Each subunit is thought to consist of functional arrangement of lipid, protein, and chlorophyll, and to be the basic unit of photophosphorylation.

Chloroplasts in many ways resemble mitochondria (Fig. 5-30): in their rich protein and fat composition; in their DNA content; and to a lesser extent in their enzyme constitution. Both chloroplasts and mitochondria are self-duplicating; that is, they are not dependent on division of the cell in which they reside for their own reproduction. However, chloroplasts, even more than mitochondria, have many characteristics of autonomous microorganisms or symbionts living inside the cells of the plants. They can divide and differentiate (become specialized); they are able to trap and utilize energy from their environment; and they contain distinctive types of DNA and RNA, the fundamental genetic materials (Chapter 3 and 10). With these facts in mind, Hans Ris has suggested that chloroplasts may be descended from a symbiotic microorganism that invaded primitive plant cells at some time in evolutionary history, and then lost the capacity to live independently, in a fashion similar to mitochondria (p. 134).

has been used by some biologists as a major criterion for designating these organisms as plants.

Plastids

Plastids are another kind of cytoplasmic organelle made of membranes. They occur in essentially all plants and algae, but never in animal cells. One group of colorless plastids, the **leucoplasts,** seem to function mainly in storage of starch granules and oil droplets in plant tissue that has not been exposed to light.

Figure 5-29   An isolated granum disc from a spinach chloroplast showing subunits on the surface. Magnification 110,000×.

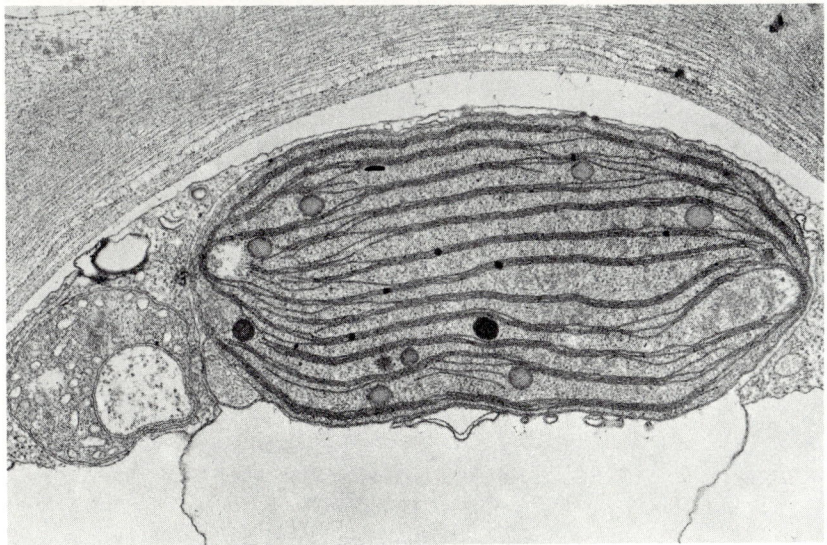

Figure 5-30   A chloroplast and a mitochondrion from an alga cell. The clear areas at each end of the chloroplast contain DNA fibrils; the chloroplast membranes are arranged in parallel bands, but in this primitive form, distinct grana are not present. Magnification 18,000×.

Summary

We have seen that the cell is the unit of structure and function in virtually all organisms, plant and animal. Especially since the advent of the electron microscope we have come to recognize that cells are highly organized units consisting of many specialized organelles; we have described the more important of these here, and have discussed briefly what roles these structures play.

Now let us turn to the questions of what different kinds of cell there are and how they are organized into functional groups to form tissues and organs.

Reading List

Cohen, S. S., "Are/Were Mitochondria and Chloroplasts Microorganisms?" *American Scientist* (1970), **58,** 281–289.

Fox, C. F., "The Structure of Cell Membranes," *Scientific American* (February 1972), pp. 31–38.

Goodenough, V. W., and R. P. Levine, "Genetic Activity of Mitochondria and Chloroplasts," *Scientific American* (November 1970), pp. 22–29.

Loewy, A. G., and P. Siekevitz, *Cell Structure and Function.* Holt, Rinehart & Winston, New York, 1969.

Luria, S. E., "The Recognition of DNA in Bacteria," *Scientific American* (January 1970), pp. 88–102.

McElroy, W. D., and C. P. Swanson, *Modern Cell Biology.* Prentice-Hall, Englewood Cliffs, N.J., 1968.

Neutra, M., and C. P. LeBlond, "The Golgi Apparatus," *Scientific American* (February 1969), pp. 100–107.

Nomura, M., "Ribosomes," *Scientific American* (October 1969), pp. 28–35.

Sharon, N., "The Bacterial Cell Wall," *Scientific American* (May 1969), pp. 92–98.

Solomon, A. K., "The State of Water in Red Cells," *Scientific American* (February 1971), pp. 89–96.

Swanson, C. P., *The Cell.* Prentice-Hall, Englewood Cliffs, N.J., 1969.

Wessels, N. K., "How Living Cells Change Shape," *Scientific American* (October 1971), pp. 76–82.

Types of Cells and Tissues

". . . but no one, so far as I know, has yet perceived that cellular tissue is the general matrix of all organization and that without this tissue no living body would be able to exist, nor could it have been formed."
                              J. B. De LARMACK, 1809, *PHILOSOPHIE ZOOLOGIQUE*

Nearly all living things are either **unicellular** (made of one cell) or **multicellular** (many cells). A third group, **noncellular** or **acellular** organisms, seems to contradict the thesis of the modern cell theory that all living things are made up of cells. In fact there is no contradiction, as we shall see shortly. Unicellular organisms, both plant and animal, are called **protista.** They vary in size, shape, and unique characteristics. Some types, such as bacteria, display a relatively simple organization that performs the basic activities common to all cells, whereas others, such as the protist *Paramecium* described later in this chapter, have additional specialized functions and structures.

In most multicellular organisms, termed **metazoa,** the cells tend to be highly specialized, to perform functions beyond the fundamental ones common to all cells. The liver cells of our bodies, for example, in addition to displaying the usual activities attributed to living systems (reproduction, growth, metabolism, and so on), also carry out the special function of secreting bile. Similarly, our muscle cells are highly specialized to contract and relax, thus performing mechanical work. We shall see later that multicellular organisms probably evolved from unicellular forms that began to aggregate into loose colonies. Several species of such **colonial** organisms still exist. Among the cells of a colony like the green alga *Gonium* (Fig. 6-1), for example, no differences can be noted; they are clearly unicellular

145

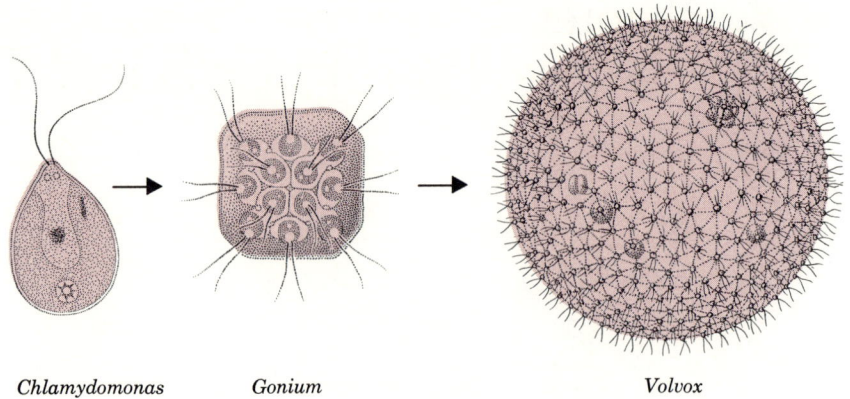

Figure 6-1 It is likely that multicellular organisms like Volvox *evolved from colonial organisms like* Gonium *and, ultimately, from a unicellular organism like* Chlamydomonas.

Chlamydomonas   *Gonium*   *Volvox*

organisms like *Chlamydomonas.* In the related species *Volvox,* however, cells of the colony are differentiated as either vegetative or reproductive, and each of the female reproductive cells can give rise to an entire new daughter colony. Although *Volvox* is considered a colonial protist, this characteristic of an aggregate of cells with different functions, in which some of the cells lose the capacity to reproduce the whole entity, is the main criterion for recognizing a true multicellular, or metazoan, organism.

Another distinction that must be made among organisms refers to the condition of their nucleus. All of the organisms classified as protista, animals, and plants have their chromosomes organized in a distinct nucleus, surrounded by a nuclear membrane, as described in Chapter 5. All such organisms, whether they are unicellular or metazoan, are referred to as **eukaryotic,** having a true nucleus. In two kinds of organism, the bacteria and the blue-green algae, the chromosomes are not surrounded by a nuclear membrane. These forms are not considered animals or plants; they are classified as a separate group, **prokaryotes** or **Monera,** those having a primitive nucleus.

Acellular Forms

What are acellular organisms? There are two forms of life that fall into this category: organisms that may be macroscopic in size but consist of a single mass of cytoplasm containing many nuclei, not divided into discrete cells by cell membranes; and organisms that lack one or more of the structures we associate with cellular organization—nucleus, cytoplasmic organelles, and plasma membrane.

Fungi

Numerous fungi, for example, are not made up of discrete cells. During most of their life cycle they contain a continuous mass of cytoplasm with many nuclei enclosed in a long, often branching, cylindrical cell wall (Fig. 6-2a). In some species the cytoplasm is compartmentalized by cross-walls, but even these compartments are often multinucleate (Fig. 6-2b). During certain stages of their life cycles other fungi, such as the slime mold *Physarum,* form shapeless masses of protoplasm surrounded by an outer membrane. These contain thousands of nuclei that are not compartmentalized by cell membranes. The multinucleate condition, referred to as a **syncytium,** is also found in higher organisms. In man, for example, skeletal muscle fibers contain many nuclei in the same cytoplasm. In all such cases the syncytium arises from the fusion of many single cells followed by the breakdown of their individual cell membranes. This process has been strikingly demonstrated in tissue culture in the fusion of isolated, individual embryonic muscle cells into syncytial fibers of contractile muscle (Chapter 18).

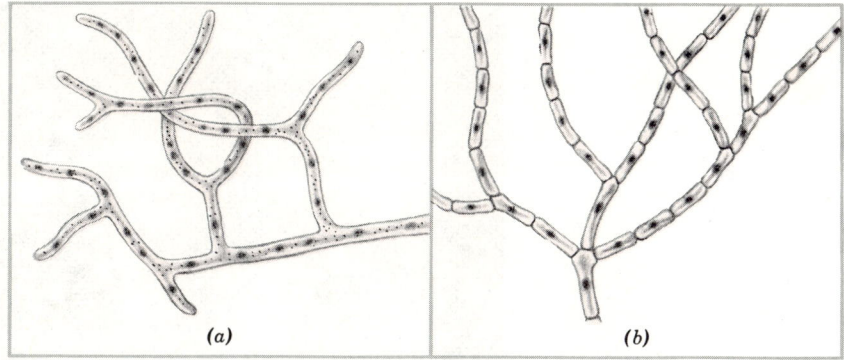

*Figure 6-2 (a) Many of the lower fungi are made of filaments, called **hyphae,** consisting of many nuclei in a cylindrical cell wall. (b) In related forms the hyphae are divided into separate compartments with one or a few nuclei in each. The dividing cell walls are called **septa,** and these forms have **septate hyphae.***

(a)   (b)

Viruses

Biologists who consider viruses living organisms must classify them as acellular; they are certainly the simplest acellular forms. Viruses are submicroscopic particles, 2000 Å or less (Fig. 6-3, pp. 148, 149). They are specific infective agents that invade plant, animal, and bacterial cells. Viruses cause such human diseases as smallpox, influenza, and poliomyelitis (infantile paralysis). There is no question that viruses are acellular, as they consists almost entirely of a coiled strand of DNA (or RNA) enveloped in a protein coat. They show no sign of a nuclear or plasma membrane, nor do they have cytoplasm or cytoplasmic organelles. The critical question, about which there is much disagreement, is whether they can be considered living organisms at all. They have never been shown to grow or reproduce other than within a living host cell. Moreover, they do not display typical cell metabolism.

The viruses that infect bacteria are called **bacteriophages** (or **phages;** pronounced to rhyme with **garage**). The greatest strides in the field of molecular genetics were taken through the study of these particles Chapter 10). It was early shown that a single phage particle (Fig. 6-4a, p. 150) invading a bacterial cell could transform and redirect the metabolism of that cell, within a few minutes, to the single purpose of reproducing phage particles like itself. In certain cases, within 30 minutes after infection, approximately 200 virus particles will have been formed by the host cell under the explicit direction of the DNA of the invading virus (Fig. 6-4b). The host cell shortly bursts or disintegrates and the virus particles, or **virions,** are released. In other instances a portion of the phage DNA may actually be incorporated into that of the host cell.

In most respects viruses fail to satisfy the structural or functional criteria we used in Chapter 1 to identify a living organism. We consider a unicellular organism alive because it is able to utilize materials from its environment to obtain energy, to reproduce, and to carry on its activities as an independent entity. Its subcellular organelles—mitochondria, chromosomes, ribosomes—are not themselves organisms, nor are they alive; they lack that independence. Because an infectious virus particle is just as dependent on a living cell for its reproduction as the cell's own genes are, the virus cannot be considered a living organism.

There is, however, another way of considering organisms that might cause us to modify this conclusion: a definition that emphasizes an organism's **evolutionary** independence rather than its functional integrity. According to this reasoning, to which we shall return in Chapter 23, **an organism is the unit element of a continuous lineage with an individual evolutionary history.**

In multicellular species, animals or plants, individual cell lines do not evolve independently, so their cells are not considered organisms. A virus, however, does have an independent evolutionary history for several rea-

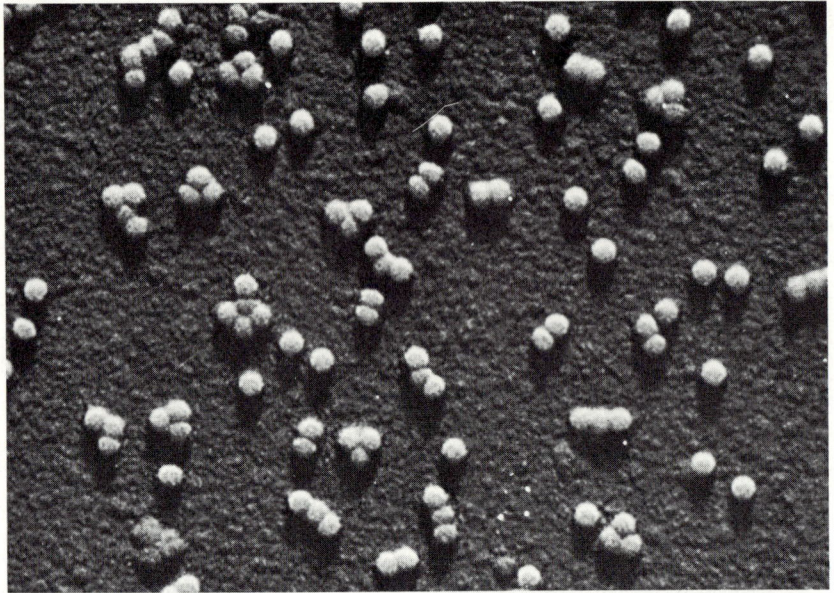

(a)

(b)

Figure 6-3 Electron micrographs of viruses. (a) Polio virus, magnification 90,000×; (b) crystals of tobacco necrosis virus, 158,000×; (c) tobacco mosaic virus, magnification 60,000×; (d) bacteriophage T4, magnification 82,000×.

sons. It is able to transfer from host to host. It can survive the death of the cell of which it is a parasite. It can mutate, and it can move between hosts of different species or even phyla. It can explore different evolutionary situations of the most unrelated kinds. Thus a virus definitely has more independence than any organelle in a cell; it is more an

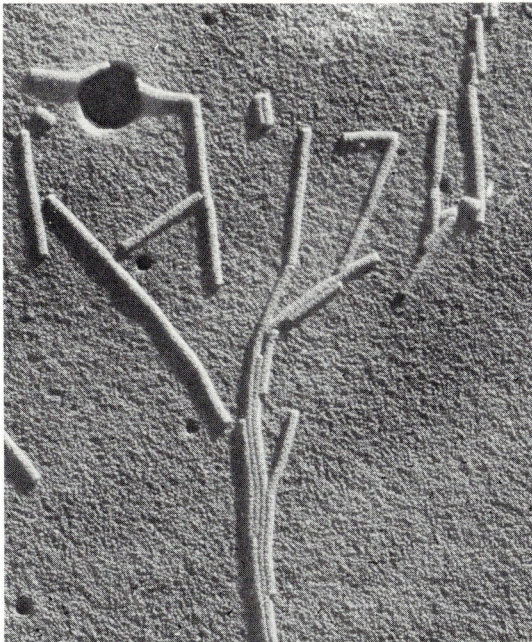

(c)

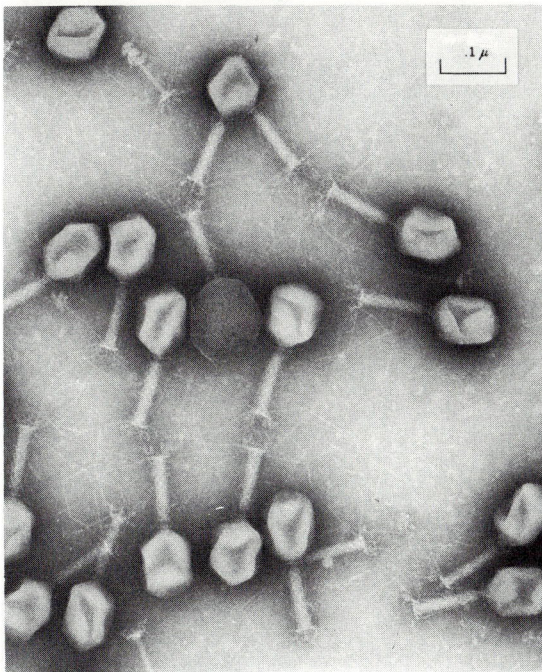

.1 μ

(d)

This apparent paradox is, of course, only a semantic one. It depends on our definition of an organism. However, it serves nicely to point up the intriguing properties of viral particles, and to emphasize the intellectual complexity of the concept of life.

Unicellular Organisms

The unicellular forms are presumably the most primitive in evolutionary terms. Both their fundamental similarity and the wide diversity of their features are illustrated by certain widely known organisms that typify the features they display. Bacteria are typical of the prokaryotes. Among the single-celled eukaryotes the major classes are distinguished mainly by their mode of locomotion: **ameboid cells,** typified by *Ameba;* the **flagellates,** exemplified by *Euglena;* and the **ciliated** forms such as *Paramecium.*

The first step in the evolution of the metazoa was the appearance of colonial organisms, formed by the aggregation of single-celled protists. Typical of those presently living are *Volvox,* a colonial flagellate, and the cellular slime molds such as *Dictyostelium* (Chapter 12), which is a colonial ameboid form. A few of the colonial organisms evolved the ability to fuse into a single multinuclear mass. This characteristic is illustrated in forms like the "true" slime mold *Physarum* or the alga *Acetabularia.*

Bacteria

Bacteria as a group are among the smallest single cells found in nature, ranging from about .1 to 1μ in width and depth, and .5 to 10μ in length. Bacteria used to be considered plants until they were recognized as prokaryotes. They have characteristic cell walls, which are not made of cellulose (as in cells of higher plants), but of other complex mucopolysaccharide substances. They can be classified generally, according to shape, into three general groups: rod-shaped (**bacillus,**

organism in that it is able to evolve independently. From this evolutionary point of view we would have to consider a virus an independent living organism.

Figure 6-4   (a) A model of a bacteriophage. The head, which is less than .1 mm long, contains a DNA double helix more than 50μ long. The contractile tail sheath consists of 144 protein globules in a coiled cylinder. Special tail fibers permit the virus to adhere to the bacterial surface. (b) A bacteriophage particle, called a **virion,** after its highly coiled DNA strand has been released by breaking open its head; the DNA is 500 times longer than the head. Magnification 38,000×.

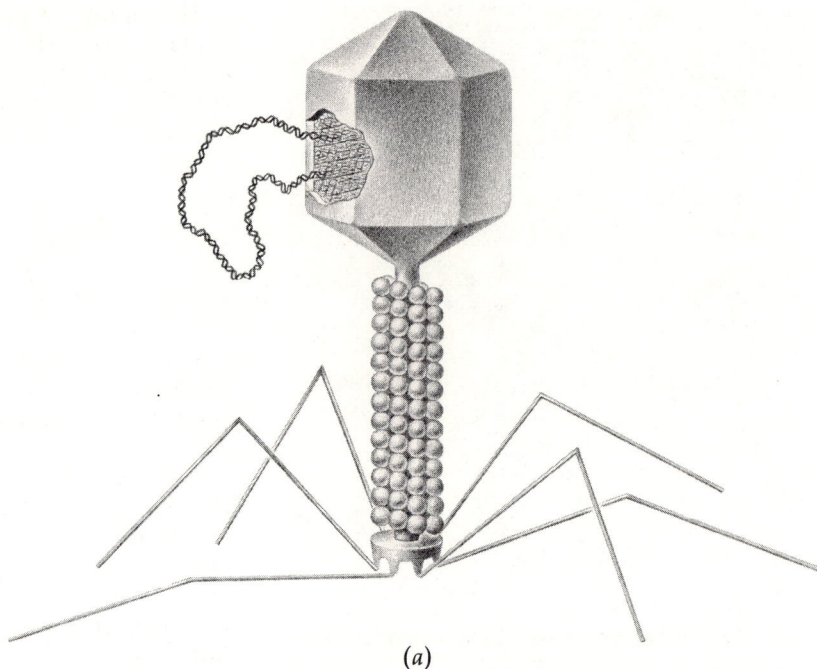

(a)

(b)

Fig. 6–5c, p. 153), spherical (**coccus,** Fig. 6-5a, p. 152), and spiral-shaped (**spirillum,** Fig. 6-5b, p. 152). They may occur singly, in groups of two, in long chains, or in regular clumps, depending on the species and the environmental conditions.

The intracellular organelles of prokaryotic bacteria (Fig. 6-6, p. 153) differ in important ways from those of other unicellular organisms. The most striking difference is that the bacterial chromatin lies in a nuclear region that is not surrounded by a nuclear membrane. This is also true of the blue-green algae. Some bacteria, and all blue-green algae, also contain chlorophyll in membrane-bound vesicles, but in these organisms the vesicles are not organized as distinct plastids (p. 142).

Prokaryotic cells contain ribosomes free in the cytoplasm, but they have no mitochondria, no endoplasmic reticulum, no golgi apparatus, and no lysosomes. The plasma membrane of these cells is folded into the cytoplasm at various points in structures called **mesosomes** and seems to carry out many of the membrane-associated enzymatic functions associated with the missing organ-

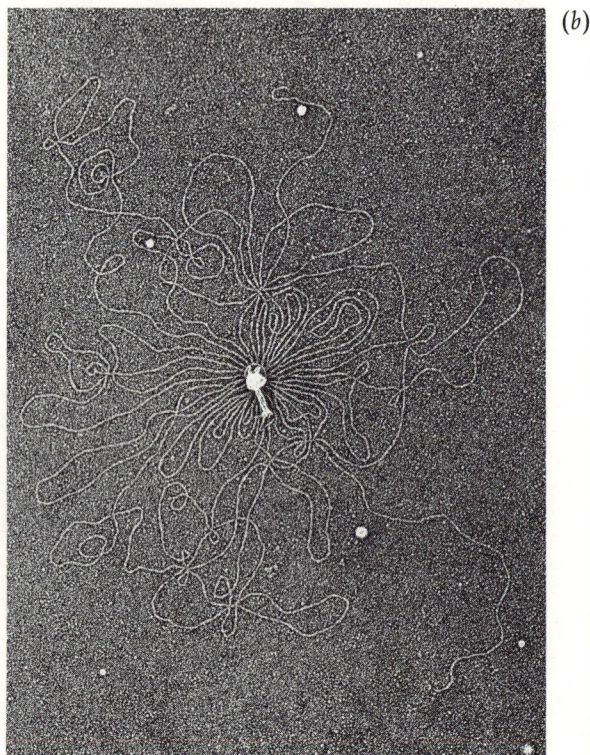

elles. Some bacteria have flagella, but they are simpler than the typical eukaryotic flagella (p. 140).

Ameba

The *Ameba* is typical of a free-living, fairly simple animal cell (Fig. 6-7a, p. 154), an irregular, gelatinous flowing mass of granular protoplasm of indefinite and constantly changing shape. It moves by extending fluid masses of cytoplasm called **pseudopods;** the type of movement is called **ameboid.** The cell contains a single nucleus and two special kinds of **vacuole,** a **food vacuole** and a **contractile vacuole.** The food vacuole is a small, clear, membranous sphere that forms around recently engulfed food particles, which are subsequently digested by enzymes secreted from the cytoplasm. The contractile vacuole forms as a clear spherical body that grows progressively larger as it fills with water from the surrounding cytoplasm, until it suddenly disappears by emptying its contents through the vacuolar and cell membranes. It therefore balances the intake of water by the organism. *Ameba* is called "simple" because it has no cellular organelles elaborately specialized for locomotion, digestion, or sensory functions, as do many other protista.

Euglena

Euglena (Fig. 6-7b) is a spindle-shaped organism. It is called **flagellate** because it is equipped at its front **(anterior)** end with a flagellum with which it propels itself through the water. It has a large, centrally located nucleus, numerous chloroplasts, a gullet-like mouth structure, and a light-sensitive, red-pigmented **eye spot.** These, it should be emphasized, are subcellular organelles that carry on some of the same functions as the organs of a multicellular form.

Because of its chlorophyll content, *Euglena* is sometimes classified as a plant. However, its flagellar structure, mobility, and lack of a cell wall are animal-like properties. This controversy can be ignored if we use the modern classification of protist.

Paramecium

The high degree of specialization achieved in a single-celled organism is well illustrated by *Paramecium* (Fig. 6-7c). This elongated animal cell has a rigid outer covering, called a **pellicle** through whose pores extend some 2500 cilia. By their coordinated and rhythmic beating, the cilia function to propel the animal in any direction. The organism contains at least two kinds of nuclei, one or more **micronuclei,** which function in sexual reproduction, and a **macronucleus,** which controls other cellular activities. It has an **oral groove** and fixed **mouth,** or **gullet,** for food ingestion, and forms food vacuoles that circulate within it while carrying on digestion. Waste products are discharged via an **anal pore;** water is removed by two contractile vacuoles, each with a set of radiating canals.

Volvox

Volvox (Figs. 6-1 and 6-8, p. 154) is another chlorophyll-containing flagellate with both plant-like and animal-like characteristics. It is considered by some biologists to be intermediate between the unicellular and multicellular forms; it is a colony of cells in which the first primitive signs of specialization have become evident. A *Volvox* colony consists of from several hundred to as many as 40,000 cells in a single layer, in the form of a hollow sphere held together by a gelatinous secretion. Each cell is connected to adjoining cells by strands of cytoplasm, and the hollow interior of the colony is filled with a watery fluid. Nearly all of the cells are equipped with two flagella and contractile vacuoles. The cytoplasm of each cell contains a centrally located nucleus and a single cup-shaped chloroplast. The colony moves through the water by waving its whip-like flagellae, rotating on an axis pointed in the direction of its movement. Some cells are equipped with large, light-sensitive eye spots, which are located near the anterior end of the axis, pointing forward. Smaller eye-spot cells are distributed in other parts of the sphere. Most of the cells divide by mitosis, but a few, located

Figure 6-5 (a) A vegetative cell of coccus Sarcina urea. Magnification 60,000×. (b) A spirochete, or spirillum bacterium, with an axial filament. 25,000×. (c) The bacillus Proteus mirabilis, a disease-producing bacterium from a patient with peritonitis; showing the rod shape and associated flagellae. 9,000×.

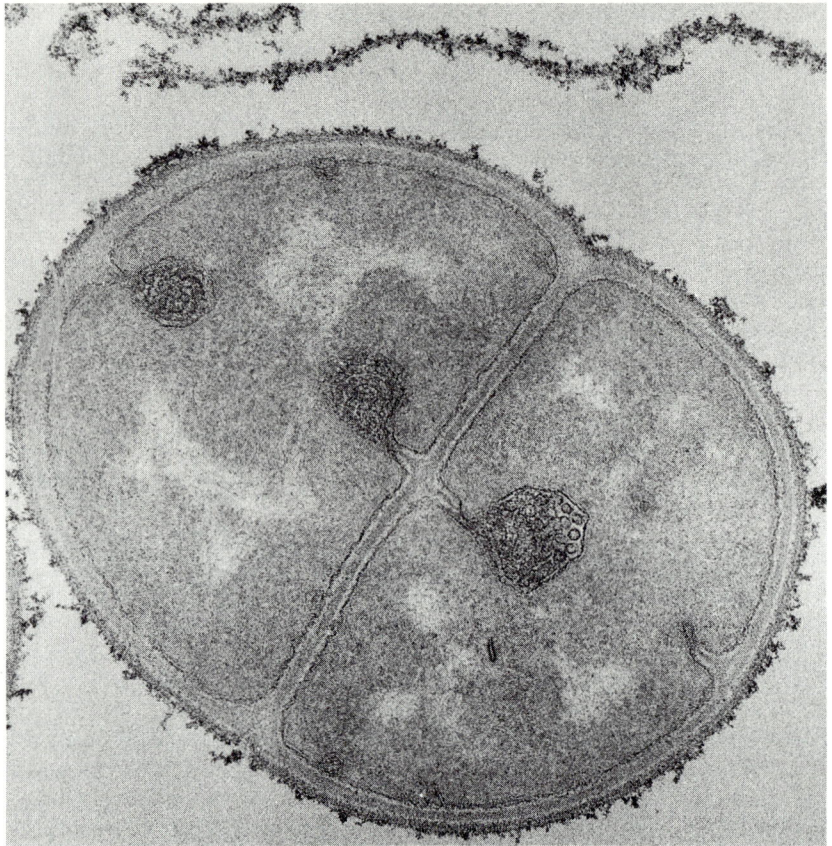

(a)

on the posterior (back) side of the colony, have the capacity to reproduce entire daughter colonies.

Acetabularia

Acetabularia is a single-celled green alga, found in tropical oceans, that is remarkable for its immense size; it may be several centimeters long. In the adult stage of its life cycle the organism, although a single cell, has a long stalk with an umbrella-like cap at one end and a set of root-like **rhizoids** at the other. The cell nucleus is located in one of the rhizoids. During its life cycle (Fig. 6-9, p. 155) which is approximately three years, the nucleus breaks down and the daughter nuclei spread through the stalk and cap. Several nuclei are soon surrounded by a membrane,

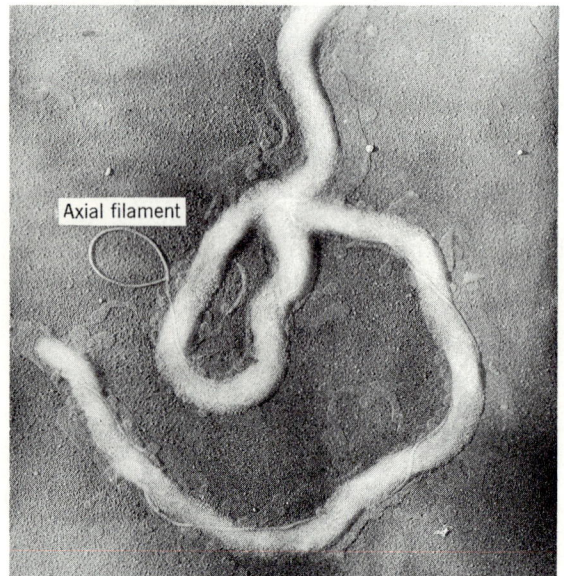

Axial filament

(b)

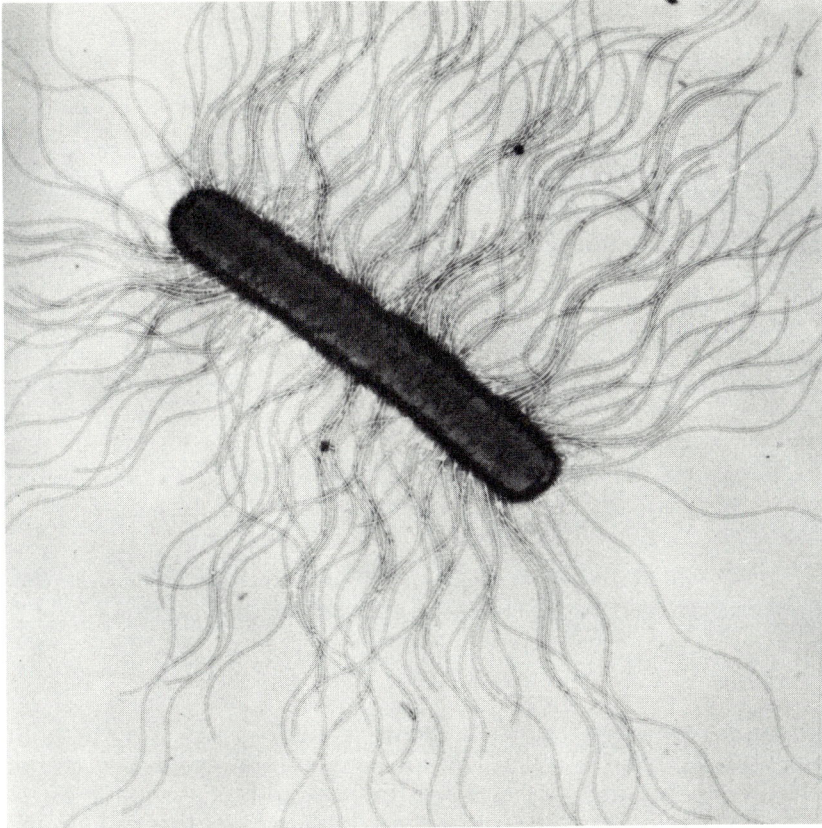

(c)

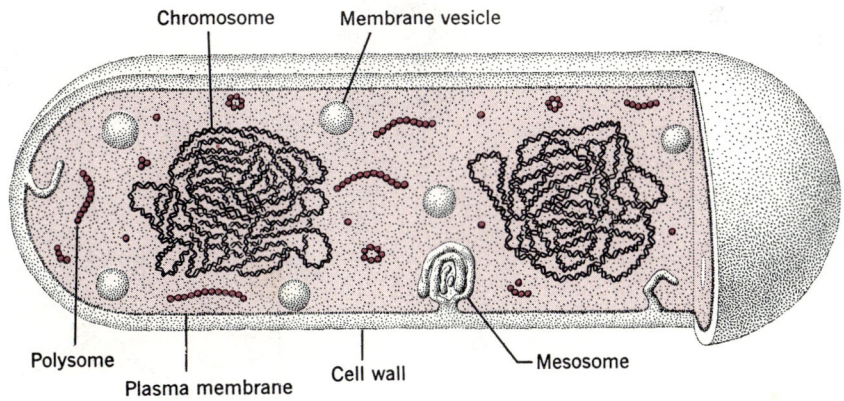

Figure 6-6   A model of a generalized bacterial cell. The chromatin is not surrounded by a nuclear membrane. The cytoplasm contains ribosomes and various membranous vesicles.

153   Types of Cell and Tissue

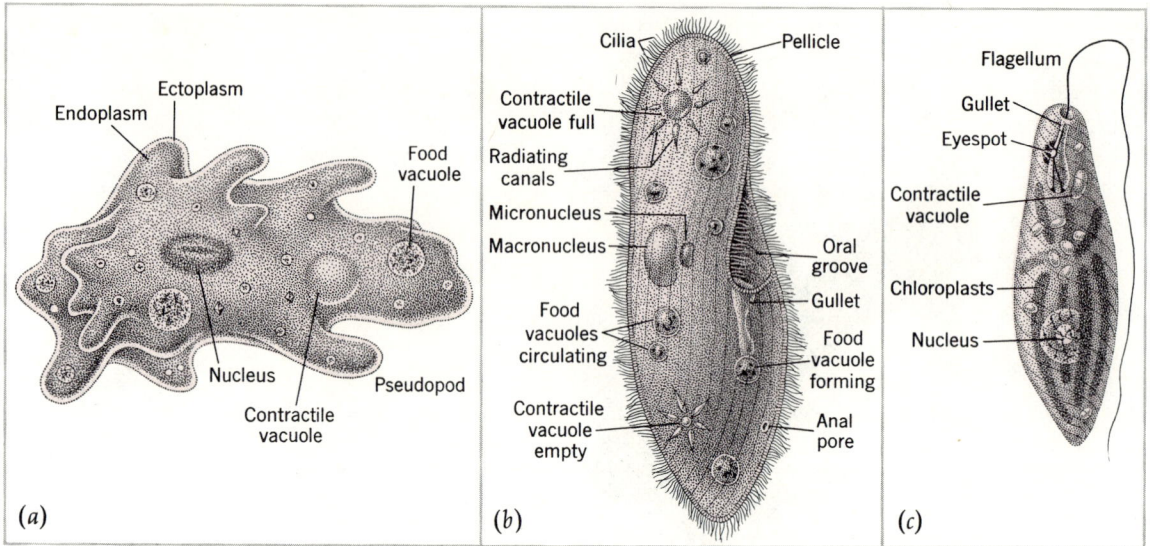

Figure 6-7 (a) *An* Ameba, *moving by pseudopodal activity to the right.* (b) *The ciliate* Paramecium. (c) Euglena *is a typical flagellate.*

forming a cell with a hard covering, called a **cyst.** The nuclei are eventually liberated as individual flagellate cells that swim about and soon fuse in pairs. Each fused pair, now called a **zygote** (p. 191), grows into the adult giant cell to complete the cycle. Because of its size and the easy physical separation of the nucleated rhizoid from the rest of the

Figure 6-8 *Volvox. Each hollow colony has several smaller daughter colonies within it, which in turn have still smaller daughter colonies inside them. Magnification 110×.*

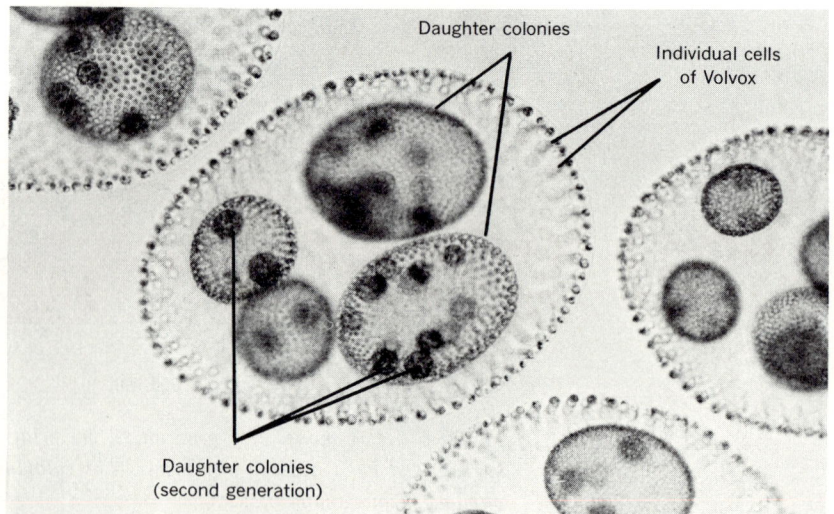

cytoplasm, this organism has been used in important studies of how the nucleus and cytoplasm affect each other during early development. We shall describe this work in Chapter 12.

Cells and Tissues of Multicellular Organisms

At first glance the cells that make up the tissues of our body, and those of other animals and of plants, seem very different from one another. The lens of the eye appears to bear no structural resemblance at all to a muscle; a bone does not look like a nerve. If we look more closely, however, we see that the cells that make up these various parts all share the subcellular organelles and functions described in Chapter 5. That is, they all have membranes, nuclei, mitochondria, ribosomes, and so on. They all absorb oxygen from their environment and use it to obtain energy by oxidizing nutrient materials. They all synthesize proteins and other macromolecular substances from smaller subunits. They exhibit many common enzymic reactions. Yet these cell types are obviously very different. Each is endowed with unique structural and functional properties, and is thus highly specialized. At some point in their development there has been a dramatic division of labor among the cells of a multicellular plant or animal.

In multicellular organisms cells that are similar in structure and function combine to form a **tissue.** The cells of a tissue are usually bound together with varying amounts of some sort of **interstitial** (intercellular) substance to form an organized, compact group. There are some types of cell, like the endothelial cells that make up the blood vessels and the interstitial **fibroblasts,** that are found in essentially all organs and tissues. Cells that are typical of each tissue type, such as hepatic cells in the liver, contractile cells in muscle, or secretory cells in the kidney, are

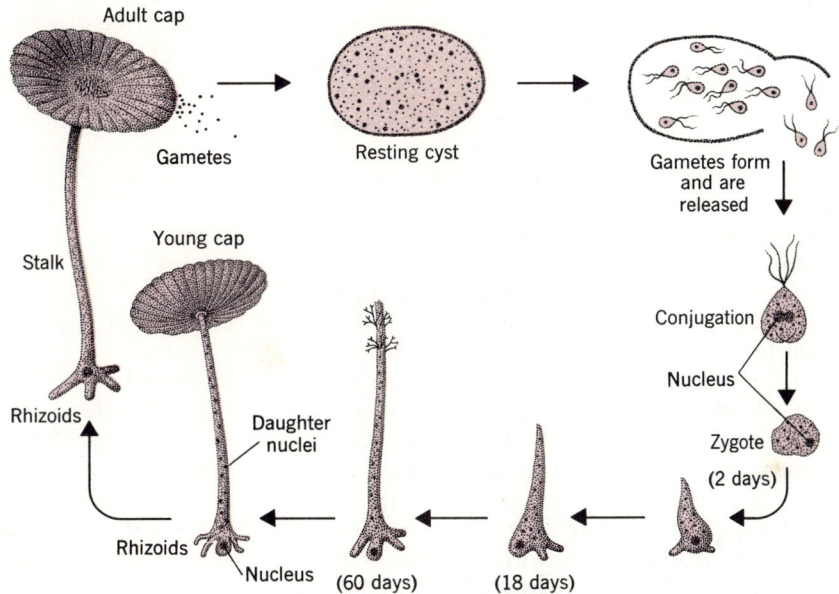

Figure 6-9 Life cycle of Acetabularia.
Acetabularia is a green alga that, in its adult form, is a single giant cell, in the shape of a stalk up to several centimeters long with an umbrella-like cap (Drawn after Brachet,

Adult cap

Gametes

Resting cyst

Gametes form and are released

Conjugation

Nucleus

Zygote

(2 days)

Young cap

Stalk

Rhizoids

Daughter nuclei

Rhizoids

Nucleus

(60 days)

(18 days)

referred to as the **parenchyma** of that organ or tissue. Blood cells appear to contradict this statement, but **plasma** (the fluid portion of the blood) can also be regarded as interstitial material (Chapter 15).

Various tissues are usually bound to one another and coordinated in their activities to form **organs.** Groups of organs function together in a highly integrated and organized manner as an **organ system.** The various organ systems (e.g., the digestive and nervous systems in higher animals), interacting together, collectively make up a whole organism.

Multicellular plants and animals are complex organisms composed of a variety of tissues and organs. An adult animal's organs and the cells that comprise them are—in the ordinary course of events—more or less stable throughout its life. Organs attain their mature size and then stop growing. They retain their shape and their functional relationships. Their cells may be **nonmitotic** (as in the heart and nervous system), or they may continue to divide in order to replace sloughed cells (as in many epithelial tissues), but a liver cell never loses its identity by changing into a nerve cell; a muscle cell is always recognizably different from a skin cell or any other type. Plant cells are not so distinctive or stable as animal cells. The different cell types are more alike in structural characteristics, and a given cell may change from one type to another during the course of its life.

How do these various cells know when, where, and in what quantities to develop and maintain themselves? What accounts for their intricate structural and functional coordination? How do they manage to synchronize their various activities with one another to maintain a normal functioning plant or animal? These are questions we attempt to answer in the chapters on the nervous system (Chapter 17), hormones (Chapter 16), and embryonic development (Chapters 11 and 12). In a sense they are all part of the same problem: How do the cells of a multicellular organism exchange information and interact with one another? An answer to this problem may lead to an understanding of why, at a higher organizational level, metazoa have proved so successful from an evolutionary point of view.

Plant Cells and Tissues

Plant tissues can be classified into two major categories: **meristematic** tissue and **permanent** tissue (Fig. 6-10). Meristematic tissues are made of embryonic, unspecialized cells that undergo cell division and give rise to all other plant tissues. The cells are usually small and cubical, contain a single nucleus, have a dense, metabolically active cytoplasm with few or no vacuoles, and are thin-walled and tightly packed, with few if any intercellular spaces. Regions of meristimatic tissue at the growing tips of roots and stems are called **apical meristems;** growth in diameter of stems and roots is made possible by **lateral meristem tissue,** known as **cambium.**

Permanent tissues, unlike meristematic tissues, are usually fully specialized and do not generally divide or change into other kinds of tissue, although there are exceptions. Permanent tissues are often subdivided into three classes: **fundamental, protective,** and **conductive.** Each of these in turn contains several different tissue types. We will explain what these various tissues are, and how together they form a plant, in Chapter 13.

Animal Tissue Types

The various specialized cells or tissues of complex animals such as man can be classified into five fundamental types: **epithelial, connective, muscle, nervous,** and **blood.**

Epithelial Tissues

Epithelial tissues consist of cells that are characteristically closely packed, with little interstitial material. They occur as a single or multiple (stratified) layer of cells that line the inner and outer body surfaces, serving many functions. They provide **protection** against injury, excess dehydration, and invasion by

Region of differentiation

Region of maximal cell division

(a)

Bark   Xylem   Phloem   Vascular cambium

Wood ray

Phloem ray   (b)

microorganisms. They are involved in the **absorption** (digestive and respiratory tracts) of materials from the external environment and the **excretion** (lungs, skin, kidney tubules) of waste products. As glands, they **secrete** important specific substances (e.g., hormones and digestive enzymes) used in other parts of the body. As **components of sense organs** such as the eye they are highly specialized and extremely sensitive to certain kinds of stimuli. They may be conveniently subclassified into four kinds according to shape and specialized structure of their cells.

Squamous. Squamous epithelial cells (Fig. 6-11a and g, p. 158) are thin, flattened, and tile-like, occurring in single or stratified layers. They line such cavities as the mouth, esophagus, and vagina, and make up the outer layers or epidermis of the skin.

Cuboidal. Cuboidal epithelial cells (Fig. 6-11b and e) as their name implies, are as tall as they are wide. Cells with this shape make up glands, including the thyroid, and the lining of the kidney tubules. They also form epithelial tissues that produce eggs and

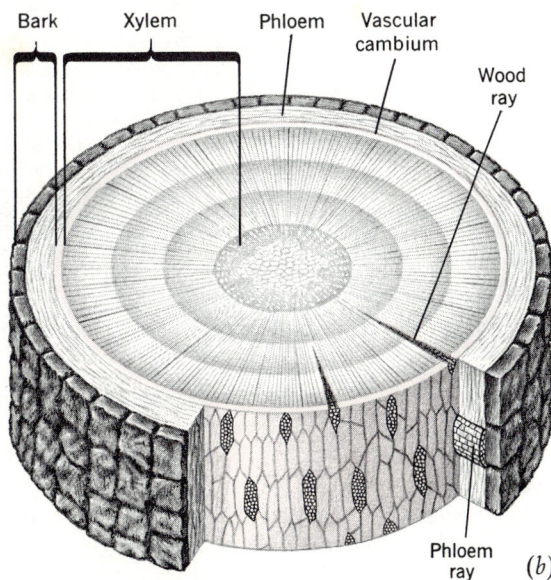

Figure 6-10   Plant tissues (a) Apical meristem in a corn root tip. Magnification 300×. (b) A woody stem, showing vascular cambium, xylem, and phloem.

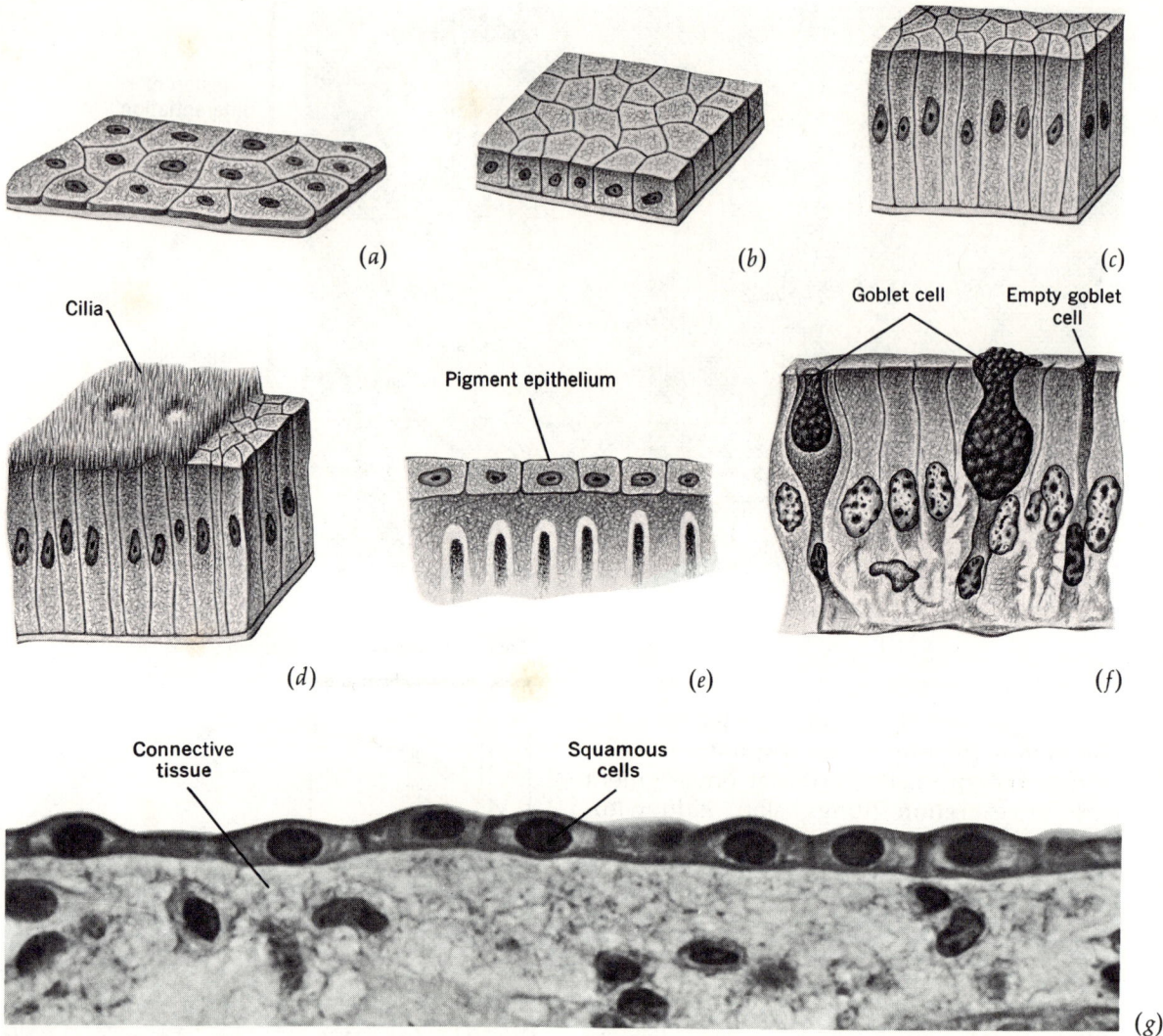

(a)

(b)

(c)

Cilia

Pigment epithelium

Goblet cell   Empty goblet cell

(d)

(e)

(f)

Connective tissue

Squamous cells

(g)

sperm in the ovaries and testes.

Columnar. Columnar epithelial cells (Fig. 6-11c and f) occur in one or more layers and are cylindrical, with their height markedly exceeding their width. Cells of this type form, for example, the lining of the stomach and intestines.

Ciliated. Ciliated epithelial cells (Fig. 6-11d and h) are also columnar, but in addition possess rhythmically beating cilia on their free surface (the part exposed to the cavity they line). They line most of the respi-

ratory passages and the egg-carrying ducts of the female reproductive tract. By the coordinated beating motion of their cilia they function in the respiratory tract to remove dust and other foreign particles, and in the oviducts to direct the movement of the liberated egg (see Chapter 11).

No matter what the shape of their cells, all epithelia form layers that provide a continuous protective barrier between the underlying cells and the external medium. Anything entering or leaving the body or organ that the

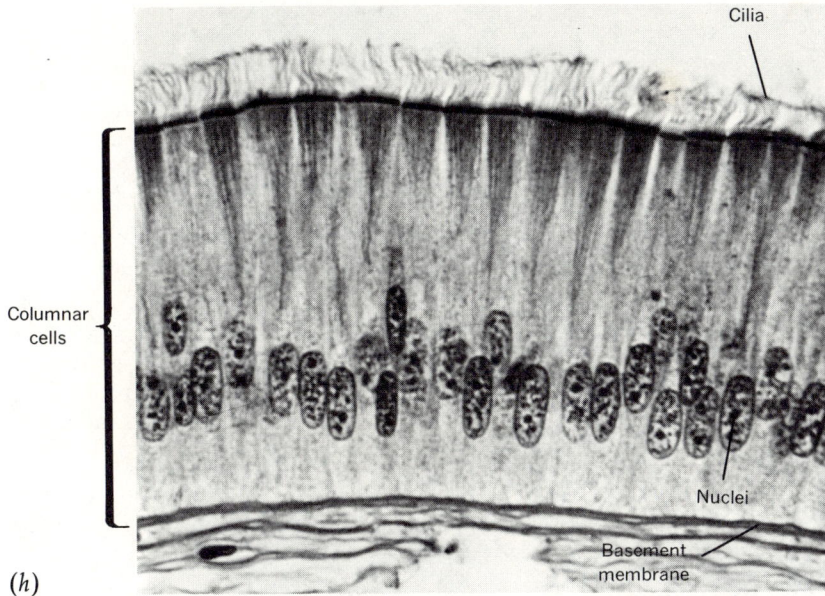

(h)

Figure 6-11 Types of epithelial tissue: (a) squamous; (b) cuboidal; (c) columnar; (d) columnar with cilia; (e) sensory epithelium (eye retina); (f) glandular. (g) Epithelium of the collecting ducts of a dog kidney; a single layer of squamous cells. These cells may also be cuboidal. Magnification 600×. (h) Ciliated columnar epithelium from the gut of a mussel. Note the distinct, dark basement membrane. Magnification 500×.

epithelium surrounds must cross that barrier, which means that the permeability of the various epithelial cells is important in regulating the exchange of materials between different parts of the body and between the body and the external environment.

In view of this special role it is understandable that the cells of an epithelial layer are bound tightly together with little or no intercellular space (Fig. 6-12). In the electron microscope some cells can be seen to be bound together by specialized adhesive structures called **desmosomes** (Fig. 6-12). In other regions the surfaces of neighboring cells are pressed tightly together, so that their unit membranes become fused into **tight junctions.** In addition to these adhesive specializations, all epithelia are characterized by a **basement membrane,** an extracellular fibrous layer that separates the epithelial cells from those of the underlying tissue.

Connective Tissue

Connective tissue cells function primarily to support the body and to bind or connect its parts. They also provide a mechanical framework (the skeleton) for locomotion in higher animals. They are characterized by

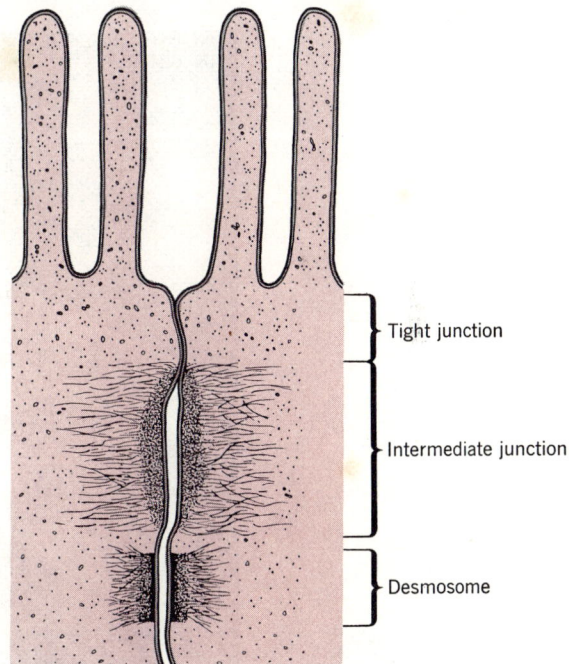

Figure 6-12 Diagram of neighboring epithelial cells forming various kinds of specialized junctions where their membranes come into contact. These include desmosomes, intermediate junctions, and tight junctions.

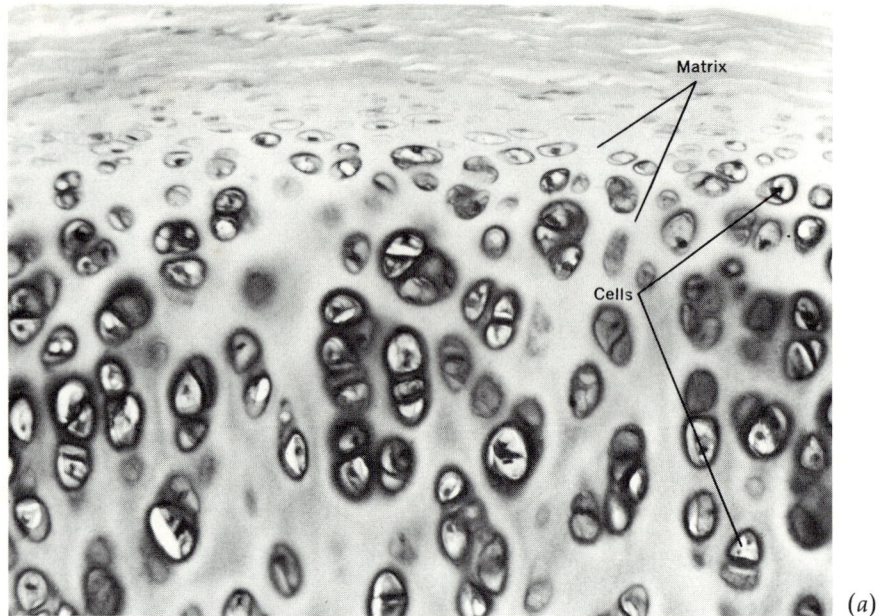

(a)

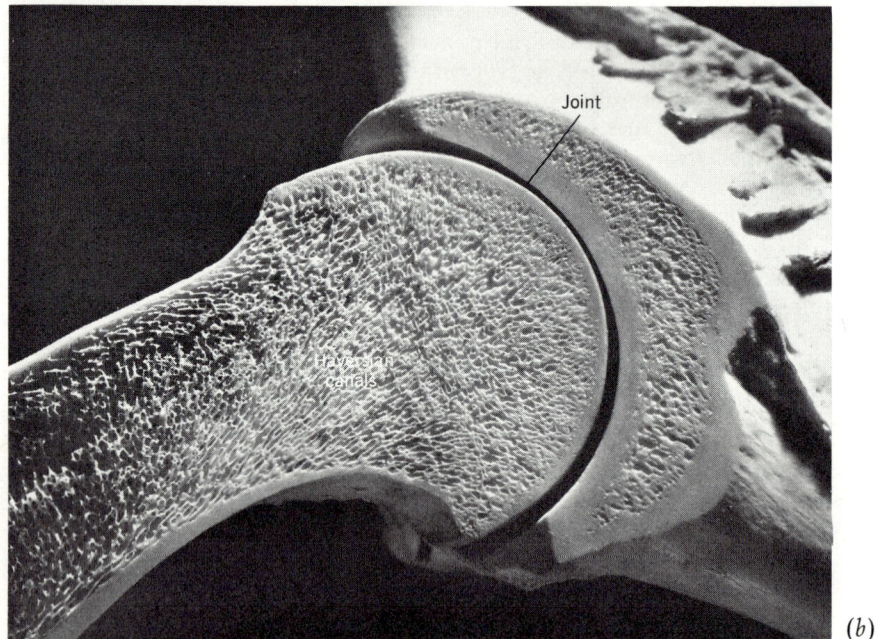

(b)

Figure 6-13   *The three major types of connective tissue are (a) cartilage, in which cells are embedded in a tough matrix (Magnification 300×); (b) bone, penetrated by a system of openings called Haversian canals; and (c) fibrous material consisting of bundles or alternating layers of collagen, as in the cornea (Magnification 12,000×).*

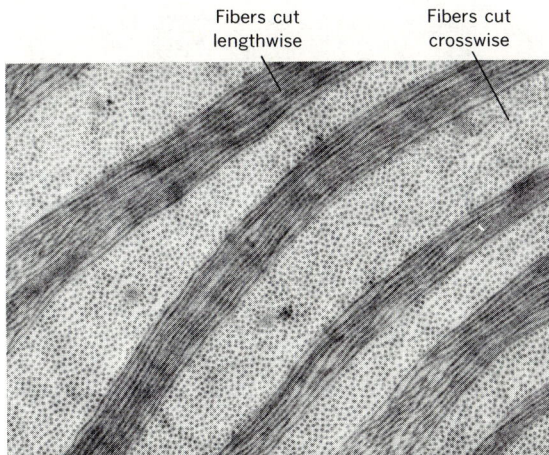

Fibers cut lengthwise   Fibers cut crosswise

(c)

relatively huge deposits of intercellular material, called **matrix,** which is secreted by the cells. In most instances the matrix represents the bulk of the tissue and is responsible for its supporting and binding qualities. Connective tissues are conveniently grouped into three classes: **fibrous** (including **fat** tissue), **cartilage,** and **bone.**

Fibrous Connective Tissue. Fibrous tissue is the material of the body that holds together organs and tissues and sheaths muscle and nerve fibers (Fig. 6-13c). The major cell type of the fibrous connective tissue is the **fibrocyte,** an irregular or spindle-shaped cell that synthesizes, secretes, and is embedded in a matrix consisting largely of interlacing protein fibers (mainly **collagen** and **elastin,** Chapter 3). When these fibers are closely packed and oriented in the same direction they form **tendons,** which connect muscle to bone or other muscle, or **ligaments,** which fasten bones together. Less compact connective tissue supports, surrounds, and connects most of the other organs and tissues of the body. Fibrous connective tissue attaches the skin to the underlying tissues; it forms a framework for the liver and bone marrow; it constitutes the tough sclera and cornea surrounding the eye; it forms the cell sheets—called **mesenteries**— that suspend the internal organs in their

proper positions; it functions as packing material in the spaces between organs; and it forms a thin sheath around every blood vessel.

Cartilage. Cartilage (Fig. 6-13a) is primarily a supporting connective tissue typified by an extensive and unusual rubbery matrix that is both firm and elastic. Cartilage cells themselves are generally spherical, and lie singly or in small groups in cavities within the matrix.

Cartilage is the temporary skeleton of the embryo of vertebrates (animals with backbones) and the template for the development of most bone; it gradually converts into bone during embryo development and the growth period after birth. Some cartilage, however, persists in adult animals, as in the joints at the end of long bones, at the end of the ribs, in some of the respiratory passages, and in the ears and nose.

Bone. Bone (Fig. 6-13b) is uniquely hard and rigid connective tissue whose dense matrix is chiefly made of complex inorganic salts of calcium and phosphate (as much as 65 percent of the dry weight of adult human bone), and to a lesser extent of organic matter. Although its most obvious role is the skeletal support of the body, it also protects the brain and other organs and houses the **bone marrow,** the site of formation of red blood cells. In addition, bone acts as a reserve of calcium and phosphate for use by other cells in the body.

Microscopic examination of hard, compact bone reveals the presence of canals, known as **Haversian canals** (Fig. 6-14, p. 162), each containing an artery, vein, and nerve and surrounded by concentric layers of bone matrix and concentric rows of bone-forming cells. The bone cells are connected with one another and with the Haversian canals by a system of minute canals that penetrate the matrix in all directions, permitting the transfer and exchange of nutrients and metabolic wastes.

There are essentially two types of bone cell: those that deposit the bone matrix and those that are responsible for its dissolution and absorption. Their combined action helps de-

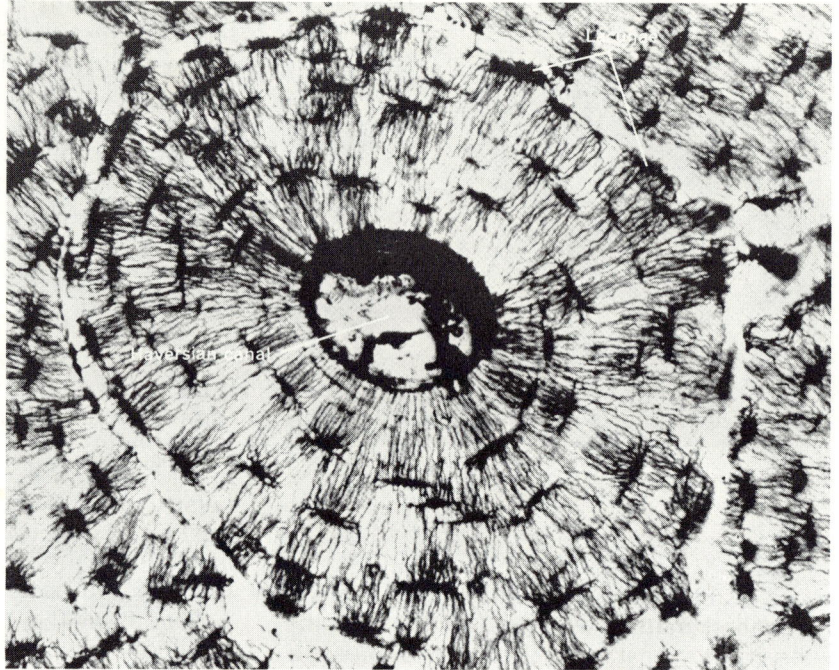

Figure 6-14 A Haversian system in the human femur. Magnification 400×.

termine the characteristic shape and size of a bone. With increasing age, the composition of bone matrix tends to become richer in inorganic salts and correspondingly poorer in organic material. The consequent increase in brittleness accounts for the more frequent occurrence of bone fracture in elderly persons.

Muscle Tissue

The main function of muscle is to contract, and thereby to perform mechanical work. We shall discuss this functional aspect in detail in Chapter 18. Muscle tissue typically consists of slender, elongated cells, usually grouped into bundles ranging in length from a few microns to as much as 4 cm. Humans, like most higher animals, have three distinct types of muscle in their bodies: **skeletal, cardiac,** and **smooth.**

Skeletal Muscle. Skeletal muscle is also called **voluntary muscle** because its action can be consciously controlled. Skeletal muscle makes up 40 percent of the weight of the body, and is completely and directly controlled by the nervous system. It is attached to the bones or tendons and is responsible for the movement of the skeleton and therefore of the body. It contracts and relaxes the most rapidly of the three muscle types. Skeletal muscle consists of long, cylindrical, tapering multinucleate cells called **fibers.** Muscle fibers are formed during embryonic life by the fusion of hundreds of single premuscle cells. Such tissues, with many nuclei in a common cytoplasm, are called **syncytia.** Each muscle bundle is surrounded by a thin envelope of fibrous connective tissue that binds many muscle fibers together. Large numbers of fibers are thus enveloped in connective tissue sheath, which is continuous with the tendons, attaching skeletal muscle to bone. The many nuclei in each fiber lie just under the plasma membrane. Within the fiber are long strands of contractile material, **myo-fibrils.** Myofibrils are in turn made of bundles of **myofilaments,** which are giant, chain-like molecules of contractile proteins called **myosin** and **actin.** Where these molecules overlap at regular intervals they produce alternating light and dark regions, giving a distinctive cross-banding or cross-striated appearance to the muscle fiber (Fig. 6-15). For this reason both skeletal and cardiac muscle are referred to as **striated** muscle (Chapters 15 and 18).

Cardiac Heart Muscle. The striated muscle that makes up the walls of the vertebrate heart contains myofibrils similar to those of skeletal muscle, but heart muscle differs in some important ways from skeletal muscle. We noted that a fiber of skeletal muscle contains the nuclei of many hundreds of cells that have fused into a syncytium. In heart muscle the cells do not fuse; they merely **associate.** Each cell remains a discrete unit, with one or occasionally two nuclei surrounded by an intact plasma membrane. Cardiac cells are generally shaped like irregular cylinders, often branched, with flat ends. The cells are joined end to end into long, branching chains, which form the cardiac fibers. The membrane at the end of each cell is specialized as an adhesive structure that is visible as a darkly staining region in the microscope. These junctional regions are called **intercalated discs** (intercalated means, literally, between cells). It has recently been shown that heart muscle cells are not only tightly adherent to each other, but are also in **electrical communication.** That is, ions (or even dyes) are able to pass from one cell to another through specially permeable sites in the plasma membrane called **nexal junctions.** The significance of this discovery will become clear when we discuss the functional aspects of heart muscle and the general problem of electrical activity of cells (Chapter 15).

The rhythmic contraction and relaxation of heart muscle is not under conscious control. The heart beats (and will continue to do so even outside the body of the animal) as a result of periodic electrical stimuli arising from specialized heart muscle cells, called

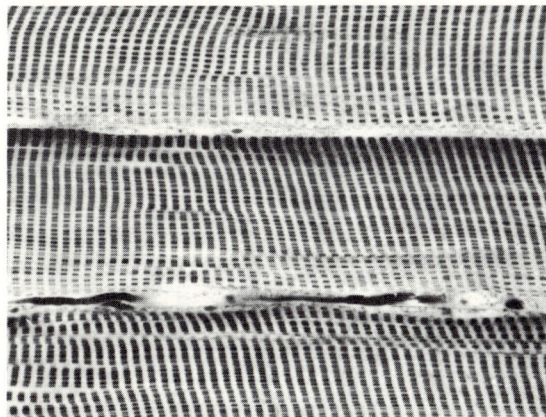

Figure 6-15 · Photomicrograph through a light microscope of three muscle fibers from a dog. The alternating dark and light segments aligned side by side give skeletal muscle its striated appearance. Magnification 700×.

pacemaker cells, that form a network throughout the working muscle of the heart wall. This network is referred to as the **sino-ventricular conduction system.** In the normal mammalian heart the main driving stimulus comes from a small mass of these pacemaker cells, termed the **sino-atrial node,** located in the back (dorsal) wall of the heart. The rate of "firing" or pulsation of the sino-atrial node can be increased or decreased by influences from the involuntary nervous system.

Smooth Muscle. Smooth muscle cells are found mainly in the walls of internal organs (digestive tract, respiratory passages, arteries, and veins). They are spindle-shaped cells with long, tapering, pointed ends and a centrally located nucleus. Unlike skeletal and cardiac muscles, they are slow to contract, are capable of prolonged contraction, and show no cross-striation (Fig. 6-16, p. 164). The contraction and relaxation of smooth muscle cells are controlled both by the nervous system and by their chemical environment; they are not under voluntary (conscious) control.

Nerve Tissue

The basic cellular unit of the nervous system is the nerve cell **(neuron),** a cell highly specialized for carrying electrical signals.

Figure 6-16 (a) Smooth
muscle cells. (b) Electron
micrograph of part of a
smooth muscle cell;
myofibrils are present but
are indistinct and show no
sign of cross-striation.
Magnification 20,000×.   (a)

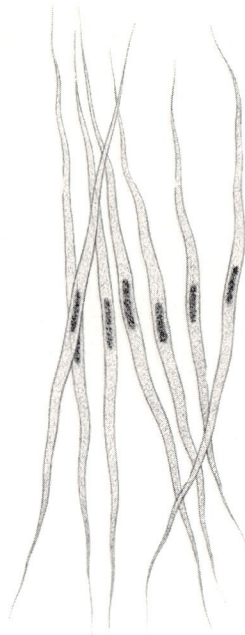

Each neuron consists of a **cell body** made up of a nucleus and surrounding cytoplasm. The cytoplasm often extends into two kinds of fiber called **processes;** these are the **dendrites** and **axons.** In one type of nerve cell, the **motor neuron** (Chapter 15), the dendrites are fairly short and branched, whereas the axons, which usually arise from the opposite side of the cell body, are generally single, long, and slender extensions or fibers with branches near the terminal portion (Fig. 6-17).

Nerve cells connect at a junction known as a **synapse,** where the terminal branches of an axon and the dendrites of another neuron come together. Nerves stimulate muscles at specialized junctions called **motor end plates.** It is this nerve-muscle combination that allows animals to move rapidly (see Chapters 17, 18).

Blood Tissue

Several different kinds of cell floating in a complex fluid called **plasma** make up the tissue known as **whole blood.** The blood cells

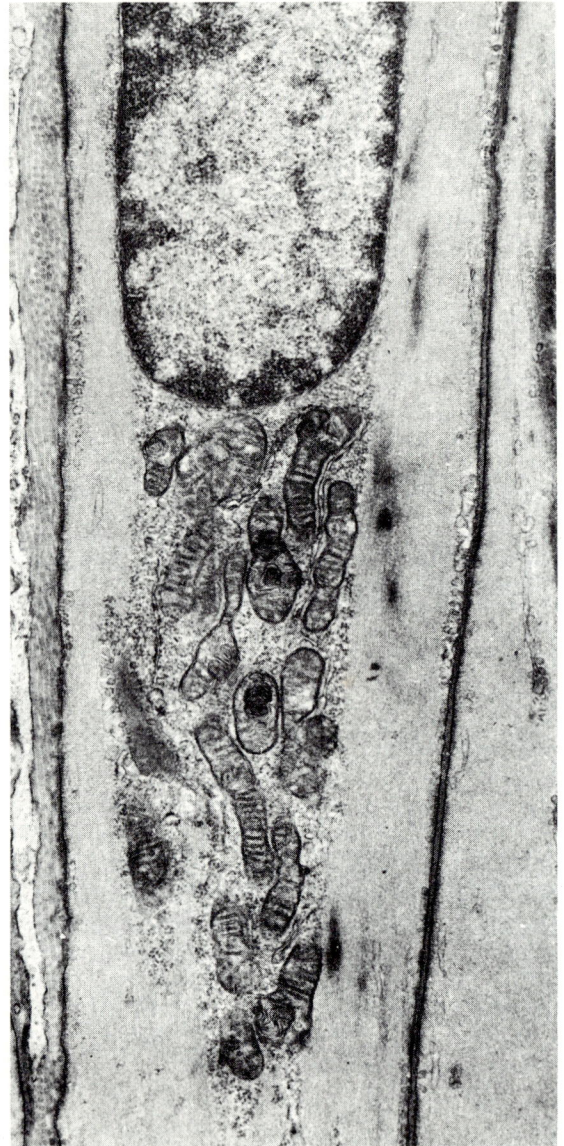

(b)

include: red blood cells, or **erythrocytes** (Fig. 6-18) which (in mammals) are nonnucleated discs filled with the oxygen-carrying pigment **hemoglobin** (p. 75); white blood cells, or **leucocytes,** of which there are several types, all concerned mainly with fighting disease; and **blood platelets,** which are derived from special bone-marrow cells and are involved in

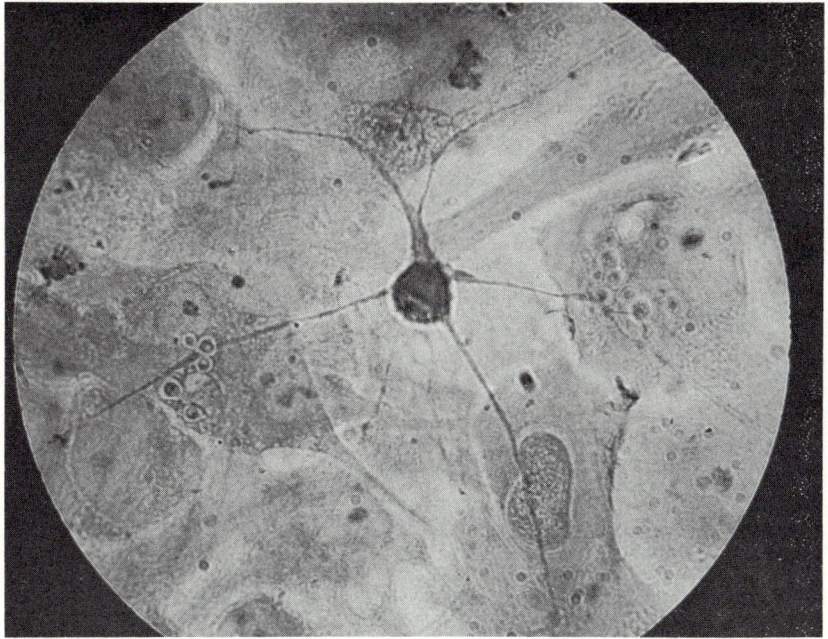

Figure 6-17 A single nerve cell from the brain of a chick embryo after 15 days in tissue culture. Magnification 850×.

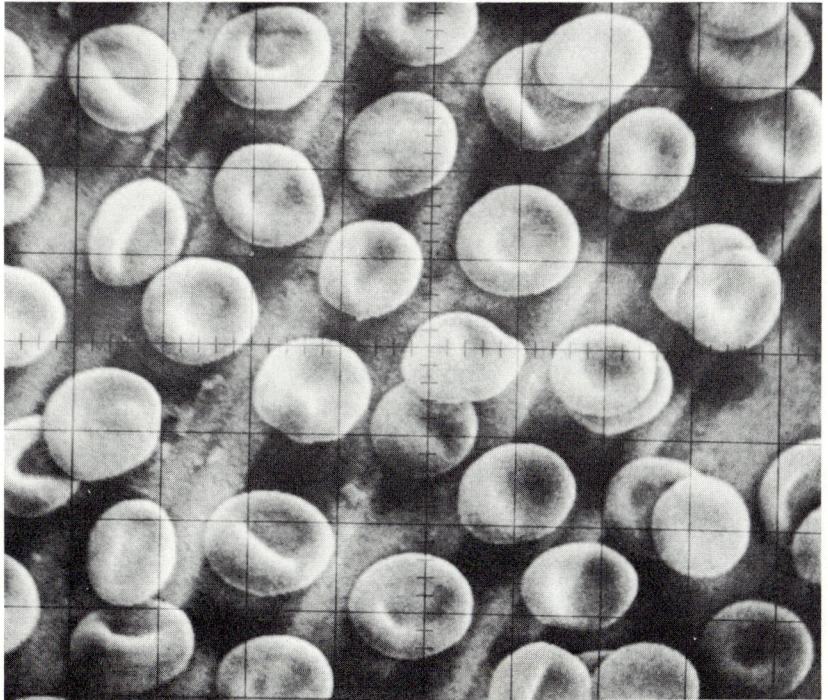

Figure 6-18 Human red blood cells. Magnification 2000×.

blood-clotting. We shall discuss these aspects of blood in greater detail in Chapter 15, on circulation.

The Problem of Differentiation

We have seen that multicellular animals are composed of many cells, and noted also that their most striking characteristic is that these cells are very different from one another. A muscle cell is filled with highly organized myofibrils; a nerve cell has none. A red blood cell synthesizes an enormous quantity of hemoglobin, which it stores in its cytoplasm before its nucleus disintegrates; a fibroblast cell synthesizes collagen, which it secretes into an extracellular matrix. But a fibroblast contains no hemoglobin; an erythrocyte does not synthesize collagen. Yet we know that the organism that contains these different cells developed from a fertilized egg, a single cell. We know that the zygote divided into many cells, and that those cells were not all alike. That process of cells becoming different is called **differentiation;** how it happens is one of the central questions of modern embryology—indeed, of biology. Before we can ask that question, we need to know more about how cells divide and about how information is transferred from cell to cell and from generation to generation. Therefore we take up these matters in Chapters 7 to 11 and return to the problem of cell differentiation in Chapter 12.

Reading List

Cairns, J., "The Bacterial Chromosome," *Scientific American* (January 1966), pp. 36–44.

Fraser, R. D. B., "Keratins," *Scientific American* (August 1969), pp. 86–96.

Loewenstein, W. R., "Intercellular Communication," *Scientific American* (May 1970), pp. 78–86.

Loewy, A. G., and P. Siekevitz, *Cell Structure and Function.* Holt, Rinehart & Winston, New York, 1969.

McElroy, W. D., and C. P. Swanson, *Modern Cell Biology.* Prentice-Hall, Englewood Cliffs, N.J., 1968.

Racker, E., "The Membrane of the Mitochondrion," *Scientific American* (February 1968), pp. 32–39.

Ross, R., and P. Bornstein, "Elastic Fibers in the Body," *Scientific American* (June 1971), pp. 44–52.

Swanson, C. P., *The Cell.* Prentice-Hall, Englewood Cliffs, N.J., 1969.

Wood, W. B., and R. S. Edgar, "Building a Bacterial Virus," *Scientific American* (July 1967), pp. 60–74.

chapter seven | *Cell Reproduction*

"Omnis cellula e cellula." (*All cells arise by the division of previously existing cells.*)

RUDOLPH VIRCHOW, 1858, *CELLULAR PATHOLOGY*

We have seen that a cell is a marvelously complex structure containing many different components. Cells reproduce themselves by a process of cell division called **mitosis,** in which a cell first duplicates all of its parts, distributes those parts equally to two separate regions of the cell, and then pinches in two in such a way that each half has an identical set of components.

If we examine a population of cells isolated in culture dish—bacterial cells, unicellular organisms, or tissue cells—we can observe cell division, or at least the results of cell division, merely by counting the cells in the population at regular intervals. Starting with only a few cells in a dish of appropriate nutrient medium (one in which all of the nutritional needs of the cells are satisfied), we will find that the number of cells in the dish increases (often after a brief lag) and that it does so according to a geometric progression: 2, 4, 8, 16, 32 . . . (Fig. 7-1, p. 168). That is, it takes the same length of time for the population to go from two to four as it does a little later to double from 16 to 32, implying that every newly formed cell immediately divides into two more cells. A progression that doubles at each step may also be written 2^1, 2^2, 2^3, 2^4, 2^5 . . . , where the superscript number, or **exponent** determines the number of times the base, in this case 2, is multiplied by itself. A population doubling at such a geometric rate is said to be in the **exponential phase** of growth. A population in a limited space (in a dish, for example)

167

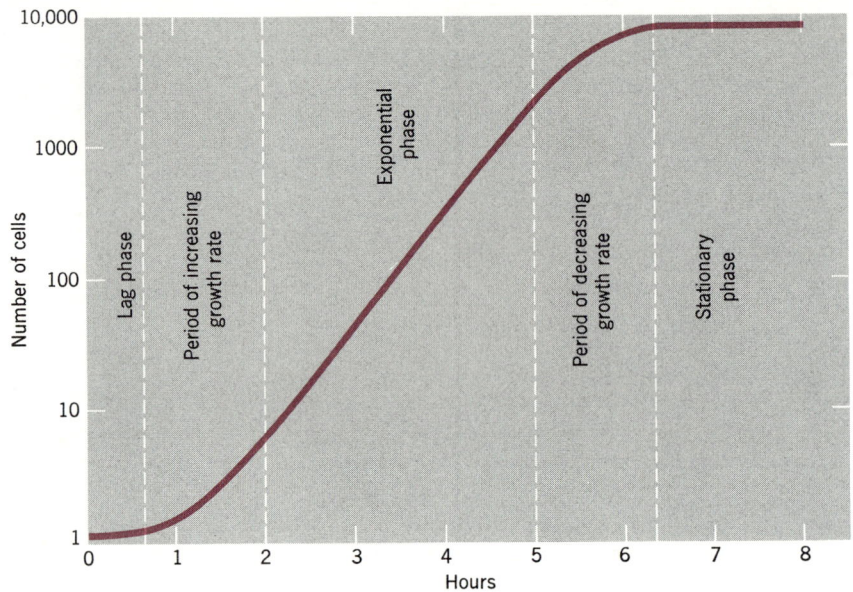

*Figure 7-1 A growth curve for a flask of bacteria. No divisions take place for the first 30 to 40 minutes after the medium is inoculated with cells. This is the **lag phase** of the curve. Over the next hour or so cells begin dividing with increasing frequency, soon reaching a constant (**exponential**) rate of division. In this phase a plot of the exponents would form a straight line. As the medium becomes exhausted, growth slows, and finally net growth ceases, meaning either that divisions have stopped or the number of divisions is just balanced by the rate of cell death.*

will continue to grow exponentially until it becomes overcrowded or the cells run out of some nutrient; then divisions begin to be spaced further apart and may finally stop altogether. Now we say the population is in a **stationary,** or **nonmitotic,** phase.

The significance of the exponential phase of growth, of course, is that each doubling of the population represents the doubling, or division, of each cell in the dish. For some cells the time between doublings, called the **generation time,** may be one or two days; in some bacteria it is as little as 18 minutes. In any case a doubling means that each cell has duplicated all of its parts and divided into two cells. We shall see in Chapter 20 that cells in nature do not normally grow at the same rate as they do under optimal laboratory conditions; they are usually confronted with re-

straints on their rate of reproduction, nutritional deficiencies, or agents that inhibit or regulate division. However, the fact that many types of cell exhibit exponential growth tells us two things. First, the processes that underlie cell division are repeating, or **cyclical,** processes. Each newly divided cell immediately begins to duplicate each of its parts. When it has done so it divides into two cells. Each of those **daughter cells** immediately begins to replicate each of *its* component parts; when it has done so it divides into two cells, and so on. Second, the capacity for exponential growth suggests that unicellular organisms that reproduce by mitosis may be considered immortal. An individual ameba or bacterium that divides doesn't die; it merely contributes its parts to its daughters, and each of these does the same.

The Cell Cycle

The terms growth and cell division are not synonymous. We have seen that a population of cells grows when its numbers increase. But when does a *cell* grow? Clearly, it grows during the time between divisions. If we measure the volume of cells in well-fed cultures, we find that the volume of a cell just before division is, on the average, twice the volume of either of its new daughter cells. This should come as no surprise. We have said that with each turn of the cell through its life cycle all of its elements—nuclear and cytoplasmic—double. The chromosomes are replicated, as are ribosomes, lysosomes, membranous structures, and mitochondria.

How are all these synthetic or replicative activities coordinated? What tells a cell when all of its parts have been duplicated, so that it can then (and only then) parcel them out equally to the two prospective daughter cells? One of the major generalizations of cell biology in the last decade is that all of these activities are controlled by the chromosomes as they duplicate themselves. As a cell goes from one mitotic doubling to the next, it progresses through a series of phases, all keyed to the synthesis of a new complement of DNA. We refer to the repeated sequence of synthesis and division as the **cell cycle.**

The phases of the cell cycle are normally identified by the letters M, G_1, S, and G_2 (Fig. 7-2). The brief period labeled M (mitosis) includes the events that physically divide a cell into two cells. S stands for synthesis, the period during which a new copy of the all-important DNA of the cells' chromatin is synthesized. When the events of cell division were first being worked out, it was thought that S and M were separated by periods during which the cell rested. These were referred to as gaps (G) in synthetic activity, G_1 before M and G_2 following M. We now know that the cell manufactures essentially all of its protein, RNA, and other components during the G_1 and G_2 phases. In most protista and animal tissues M phase lasts only an hour or two (in higher plants usually 3 to 4 hours). But because this is the period when the most

dramatic events associated with division can be seen under the microscope, and especially because this is the only time that the chromosomes of the cell become visible, all of the early work on cell division was concentrated on describing and explaining what happened during this period. The entire remainder of the cycle is referred to as **interphase**—the phase between divisions.

The Events of Mitosis

During the mitotic phase several more or less distinct processes occur together. We reserve the term mitosis for those processes that result in the division of the nucleus and chromosomes. Separation of the rest of the cytoplasm, with all of its organelles, is called **cytokinesis.** The mitotic period is customarily subdivided into four stages: **prophase, meta-**

Figure 7-2   Phases of the cell cycle. Note that mitosis (M phase) actually lasts only a small fraction of the entire cycle—perhaps one hour out of 20—but that it is during this brief period that the visible events of cell division take place.

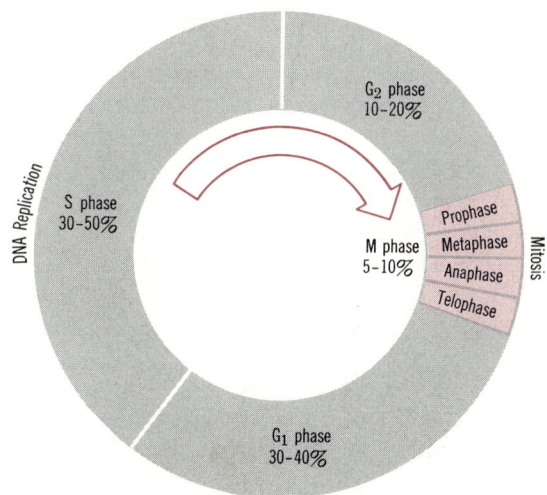

169   *Cell Reproduction*

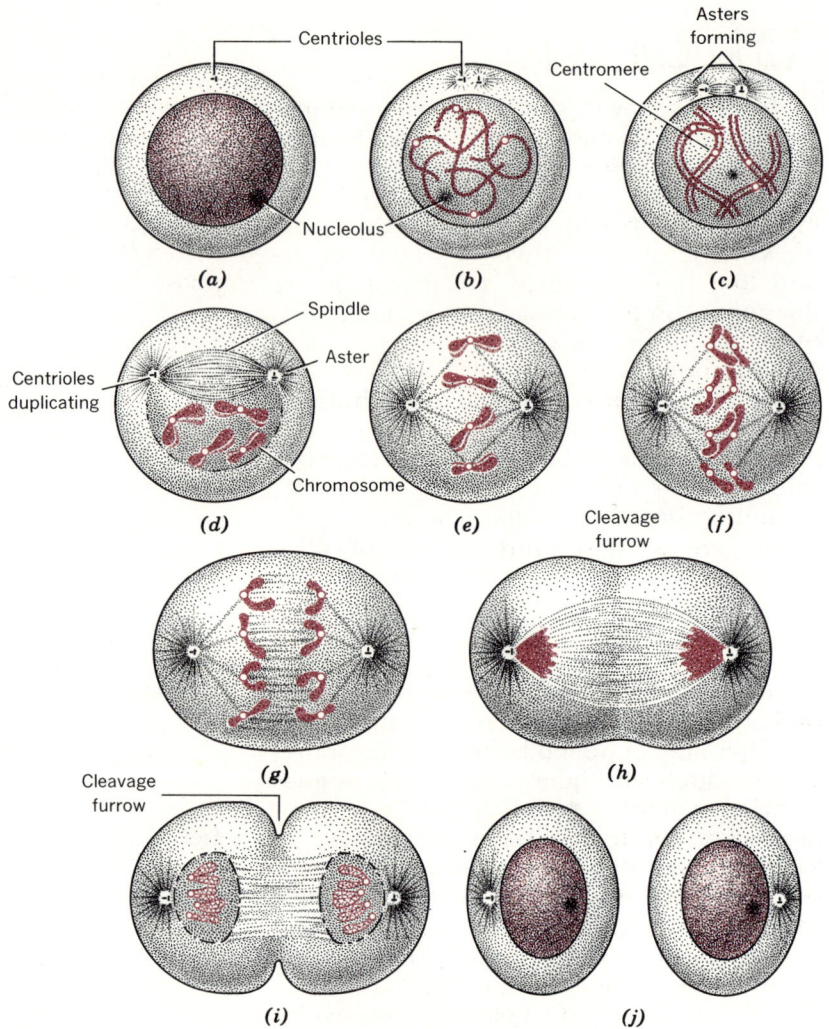

Figure 7-3 Mitotic division. (a) Interphase. The chromosomes are not seen as distinct structures; nucleolus is visible; nuclear membrane is intact. (b) Early prophase. The centrioles are moving apart; the chromosomes condense as visible threads; the nucleolus is becoming less distinct. (c) Middle prophase. The centrioles have moved farther apart and begin to organize a spindle. Each chromosome is visible as two chromatids attached by a centromere. (d) Late prophase. The centrioles are nearly at opposite sides of the nucleus and the spindle is nearly complete. The nuclear membrane is disappearing; the chromosomes are moving toward the equator; and the nucleolus is no longer visible.

(e) Metaphase. The nuclear membrane has disappeared; the centromeres are attached to spindle fibres at the equator. (f) Early anaphase. The centromeres have divided and are moving toward opposite poles. (g) Late anaphase. The two sets of new, single-stranded chromosomes are nearing the poles and cytokinesis is beginning. (h, i) Telophase. New nuclear membranes begin to form; the chromosomes lengthen and become thinner and less visible; the nucleolus reappears. The centrioles have replicated and cell division is nearly complete. (j) Daughter cells in interphase. The nuclear membranes are complete, the chromosomes no longer visible, and the process of cell division is complete.

phase, anaphase, and **telophase.**

We have noted (Chapter 5) that when the nucleus of the interphase cell is treated with basic dyes that stain nucleic acids, the chromatin stains only as diffuse threads, hardly visible in the light microscope. This is because the chromatin in the interphase nucleus is present as a diffuse, tangled network of fibers 200 to 300 Å in diameter. When such interphase chromatin fibers are spread out, dried, and observed with the electron microscope, they have a beaded appearance (Fig. 5-23). The best evidence is that each of these beaded strands that form the chromosomes of our cells contains a single enormous molecule of DNA, in the shape of a double helix (p. 80) 25 Å thick and over 1 in. long, wrapped

in a thick coat of protein. Linked end to end, the 46 chromosomes in each human cell would make a 5-ft strand of DNA.

Prophase

The earliest indication that mitosis has begun in a cell is the movement of its two centrioles (Fig. 7-3a, b) toward opposite sides of the nucleus (in animal cells; plant cells have instead two clear regions called **polar caps** outside the nuclear membrane that serve the same function as centrioles), and the increase in distinctness of the chromatin as it condenses into visible threads that become progressively shorter, thicker, and more readily stainable. This condensation (Fig. 7-4)

Figure 7-4 Condensation of chromosomes. (a) A single chromatid in the interphase cell consists of a DNA double helix wrapped in a protein sheath, loosely folded into a torturous strand. (b) During S-phase, replication proceeds, probably from either end. (c) At metaphase, the sister fibers fold up as daughter chromatids held together a centromere.

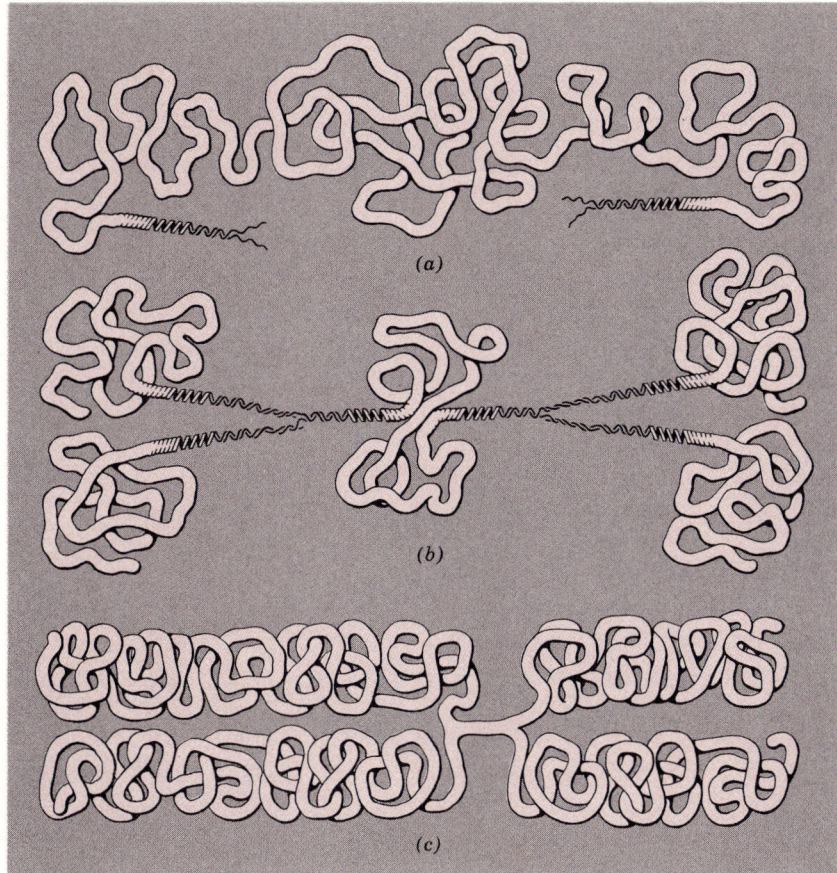

results from each of the two chromatin strands coiling along its entire length, and each of those coiled "ropes" again coiling to form **supercoils.** By late prophase the chromatin has condensed into a definite number of short, rodlike individual chromosomes, each consisting of two separate strands of chromatin called **chromatids** joined at one point by a small region called a **centromere** (Fig. 7-4c). The chromatids are the duplicated DNA of the cell, one of which was formed by replication of the other during the preceding S-phase.

As the chromatids condense from the chromatin threads, the nuclear membrane and nucleoli gradually disintegrate and a new structure, the **mitotic spindle apparatus,** is formed by the centrioles (Fig. 7-3c, d). As the

Figure 7-5 Comparison of chromatin arrangement in embryonic cells in (a) interphase and (b) metaphase.

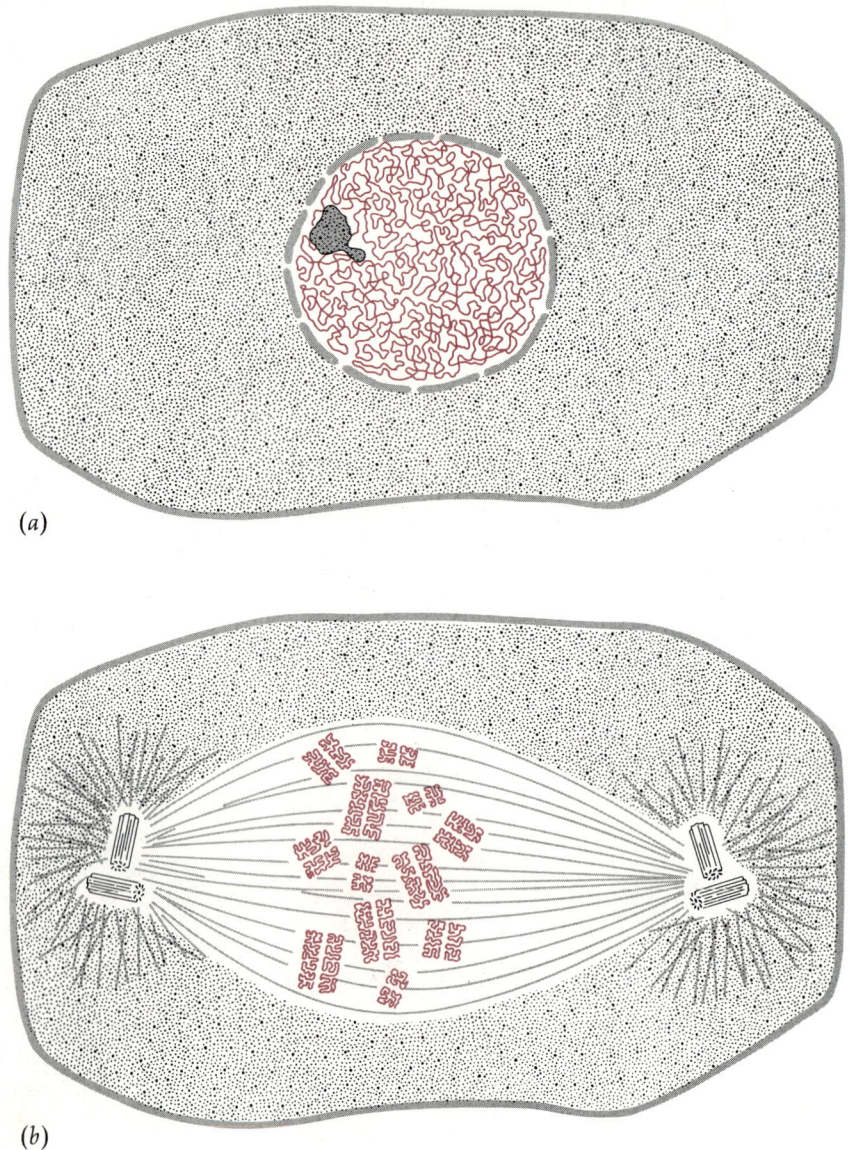

(a)

(b)

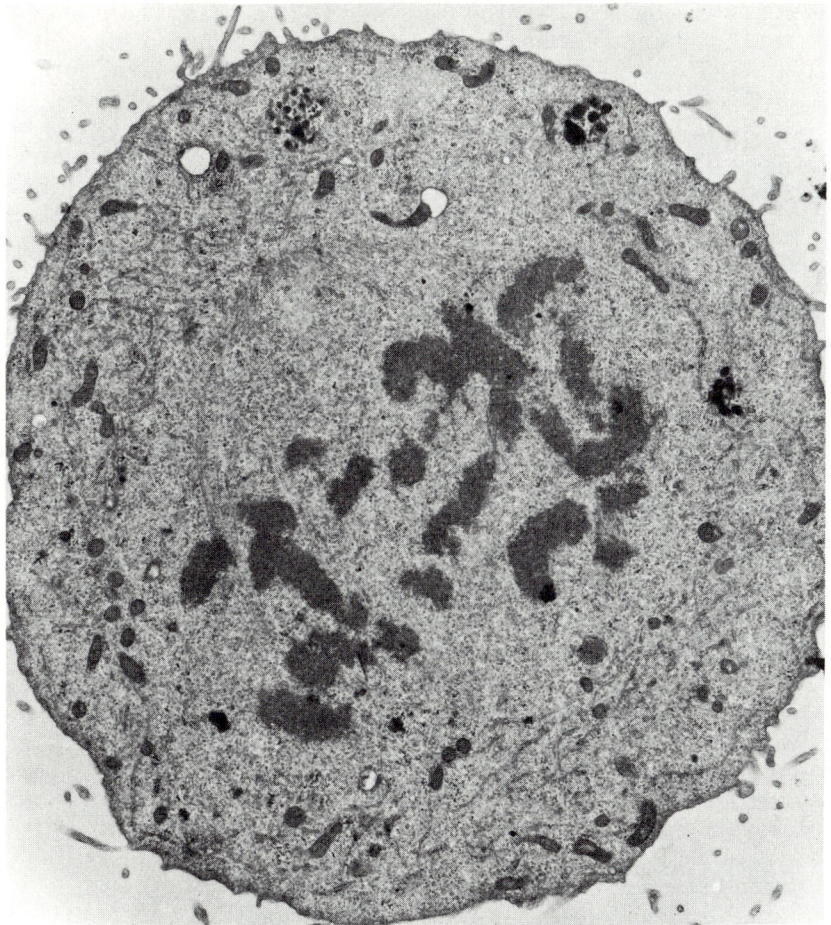

Figure 7-6 A human cancer cell in tissue culture at metaphase or early anaphase. The chromosomes have condensed in the middle of the cell; spindle fibers are visible; the nuclear membrane has disappeared. Magnification 5000×.

two centrioles move apart they send out a system of thin **fibrils,** or microtubules (p. 140), radiating in all directions (Fig. 7-3*d*, e). Three kinds of fibrils are formed (Fig. 7-5*b*): (1) curved **spindle fibers** that are continuous from one centirole to the other, forming a football-shaped basket around the nuclear region; (2) spindle fibers that pass from each centriole to connect to the centromere of each pair of chromatids; and (3) fibrils, called **astral rays,** that radiate out away from the centrioles to form a tuft, termed an **aster,** at each pole of the spindle. Toward the end of prophase (Fig. 7-3*d*) the chromosomes begin to move toward the midline or **equator** of the spindle region, apparently pulled by the spindle fibers attached to their centromeres.

Prophase ends and metaphase begins when the nuclear membrane disappears completely and the chromosomes are moving toward alignment in a flat plane crossing the equator of the spindle (Fig. 7-3e).

Metaphase

In the second stage of mitosis the double-stranded chromosomes become aligned in the equatorial plane. In this position they are called the **metaphase plate** (Fig. 7-6). Each chromosome is attached by its centromere to a spindle fiber, and the centromeres still join the two chromatids of each chromosome. As metaphase ends centromeres divide in such a way that each chromatid of a pair becomes a separate, single-stranded chromosome attached by a spindle fiber to one pole of the spindle.

Anaphase

At the beginning of anaphase (Fig. 7-3f) the two new single-stranded daughter chromosomes begin to move apart. As the spindle fibers attaching them to the centrioles shorten, the members of each pair are pulled to opposite ends of the cell. Thus, as anaphase ends (Fig. 7-3g), the cell contains two identical groups of chromosomes, one at each pole of the spindle apparatus. In Figure 7-5 the organization of the chromatin in an interphase and metaphase cell are compared.

Telophase

The final mitotic stage (Fig. 7-3h and i) is essentially the reverse of prophase. The spindle apparatus quickly disappears; nucleoli reappear in each new nuclear region; a nuclear membrane forms around each group of daughter chromosomes; and these chromosomes gradually uncoil to regain the diffuse appearance characteristic of the interphase cell (Fig. 7-3j). In summary, then, if mitotic prophase begins in a human cell with 46 double-stranded chromosomes, telophase ends with two new nuclei each containing 46 single-stranded chromosomes. Mitosis has separated the replicated chromosomes in such a way that each new daughter nucleus has the same number and kinds of chromosomes as the original parent cell.

Of what advantage to a cell is this sequence of coiling and uncoiling its chromatin with each division? The answer to this question leads directly to the significance of the whole elaborate process of mitosis. We shall see in Chapter 10 that each triplet of nucleotides in the DNA helix represents a word, or **codon,** in the genetic code, and that groups of codons constitute genes or hereditary determinants. Much evidence from both plant and animal genetics indicates that the genes are lined up along the chromatin

strand. Thus when each chromatin strand duplicates itself in the interphase S period, each gene throughout its length is copied in order, and the copy is an exact replica of the original.

Mitosis, then, is clearly a mechanism for distributing identical chromosomes to the two daughter cells in such a way that one receives the original and the other the copy, but both receive exactly the same genetic information. As long as no mistakes are made in copying, mitosis results in a population of cells whose members are all genetic duplicates of some ancestral cell. For multicellular organisms this means that every cell of the body must have the same genetic information as every other cell, the same as that in the original fertilized egg. The advantage of the coiling and condensation of the chromosomes at prophase should then be clear. The mitotic chromosomes are distinct bodies only a few microns in length. In this tightly packed form they can easily be moved about without becoming hopelessly tangled. But why should the chromosomes become uncoiled at telophase? The answer is that only in the extended, uncoiled configuration can the DNA helix be replicated and RNA messages be transcribed (see Chapter 4).

Cytokinesis

In most cells cytokinesis (division of cytoplasm, Fig. 7-7) is a separate process synchronized with mitosis. It begins late in anaphase or in telophase. Cytoplasmic division is markedly different in animal and plant cells. In animal cells it is achieved by a process initiated by the appearance of a shallow groove, or **cleavage furrow** in the cell membrane in the same plane as the spindle's equatorial plane (Fig. 7-3g). This peripheral indentation gradually deepens toward the center of the cell, breaking any spindle fibers that remain, until the cytoplasm is divided into two cells; imagine a balloon constricting into two compartments by the gradually tightening of a string around its middle. Each of the two new cells, which are not neces-

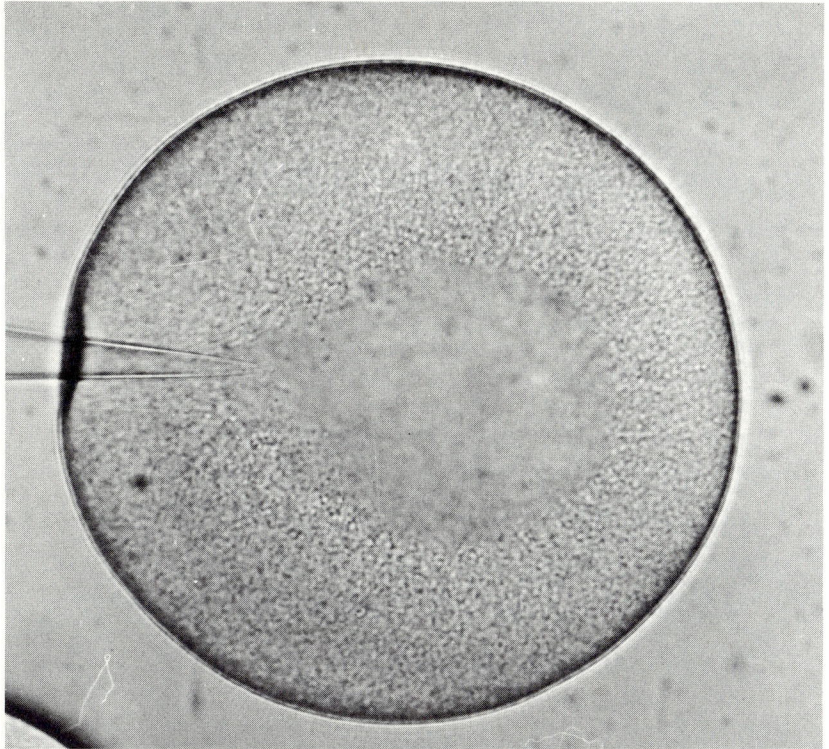

(a)

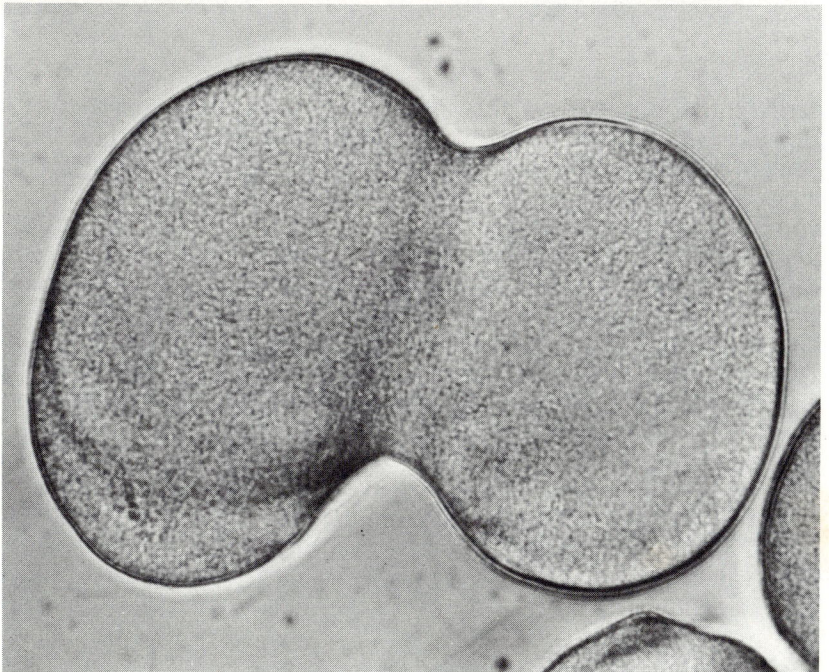

(b)

Figure 7-7 Cytokinesis (cell cleavage) usually is synchronized with mitosis, but can take place without the mitotic apparatus. In (a) a solution of sucrose, injected into a sea urchin egg just before the first cleavage after fertilization, dissolves the spindle apparatus and allows it to be sucked out of the cell. (b) Cleavage proceeds despite this operation. Magnification 700×.

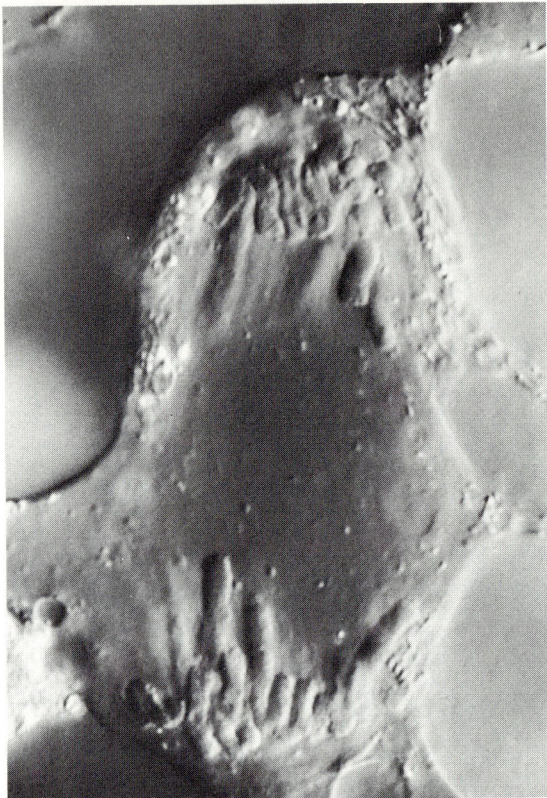

(a)

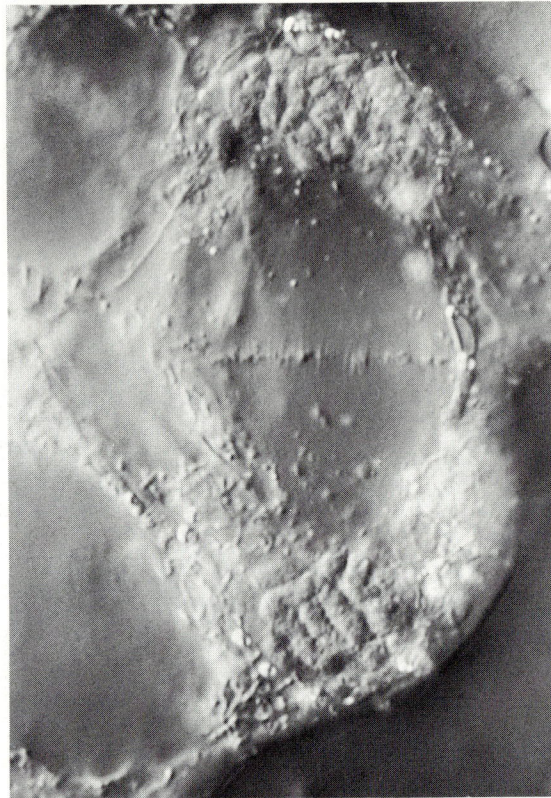

(b)

sarily equal in size, contains one of the two daughter nuclei.

Mitosis does not directly cause the constriction of the cleavage furrow, although the spindle establishes the plane of cleavage. If at late metaphase we push the spindle out of its normal position in the cell, we find that the cleavage furrow always corresponds to the plane of the metaphase plate. Once the cleavage furrow is established, even if all of the chromosomes and spindle apparatus are removed from the cell, constriction will continue and the cell will divide (Fig. 7-7).

In plant cells the cytoplasm is partitioned by the formation of a special membrane called a **cell plate.** It forms in the equatorial plane, deep in the center of the cytoplasm, and gradually grows outward toward the surface until the original cell is divided (Fig. 7-8).

Recent evidence indicates that the cell plate is formed by the endoplasmic reticulum. The two daughter cells soon form new cell membranes on their respective sides of the plate, followed soon afterward by the appearance of a cell wall. The cell plate is sandwiched between the cell walls of the cells on either side of it, and is now called a **middle lamella.** The daughter cells remain together.

The major differences between plant and animal cells during mitosis and cytoplasmic division are: (1) Animal cells possess centrioles that serve as poles for the spindle apparatus, whereas plant cells develop polar caps instead. (2) The division of the cytoplasm into two cells occurs in animal tissues by a process of cleavage that proceeds from the cell periphery toward the center, whereas in plants a cell plate grows from the center

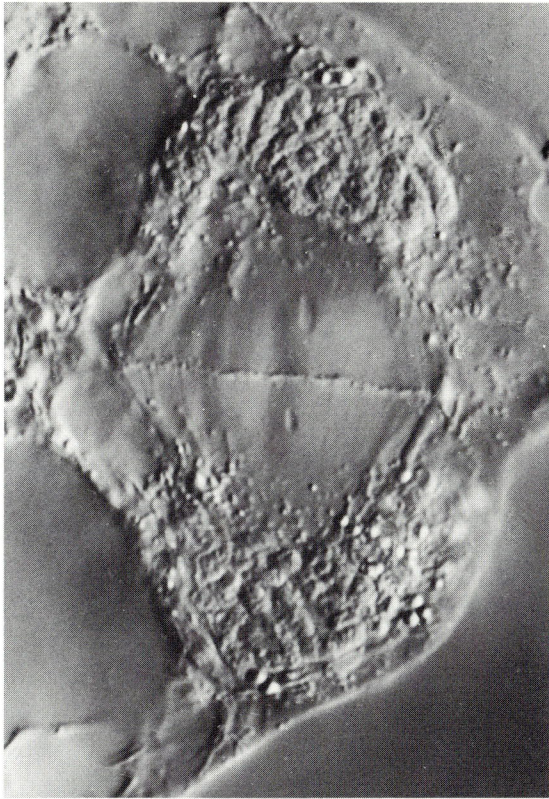

(c)

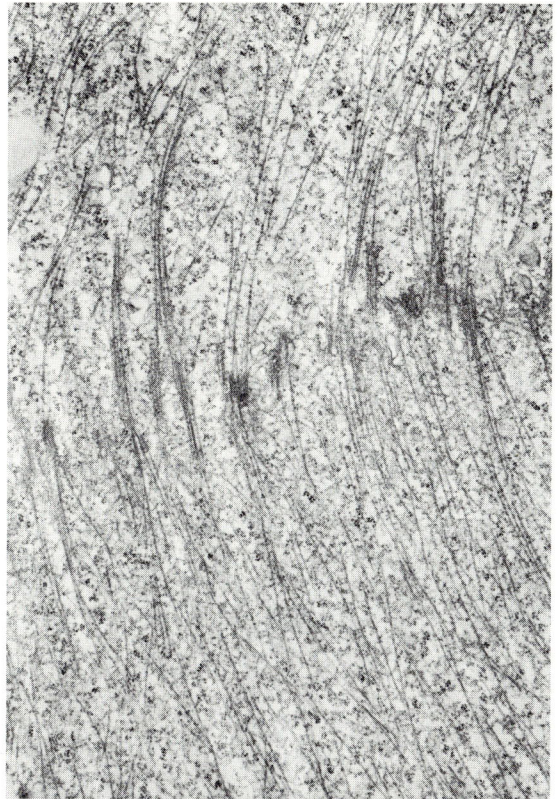

(d)

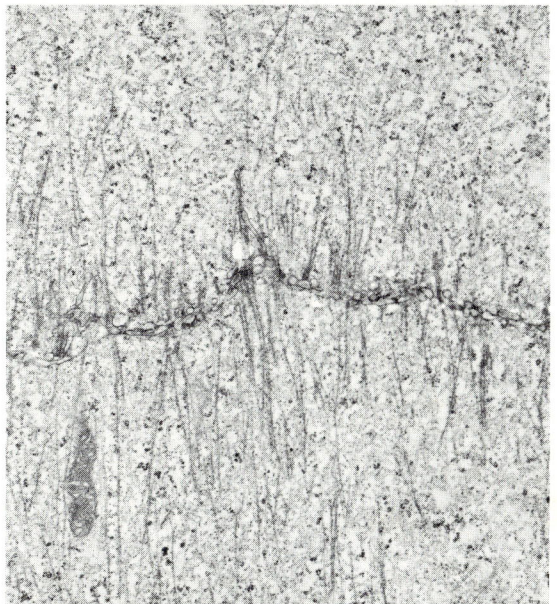

(e)

Figure 7-8 Formation of the cell plate in
Haemanthus, the blood lily. The living cell was
photographed with a microscope equipped with
special Nomarski optics, which give objects a
translucent, three-dimensional appearance. (a) Late
anaphase. The chromosomes are distributed in
opposite poles of the cell, top and bottom. (b)
Telophase. The chromosomes are tightly clustered at
opposite poles and the cell plate has started to form
as a grainy horizontal line. (c) Late telophase. The
cell plate is continuous. Magnification 4000×. (d),
(e) Comparable telophase stages as seen with the
electron microscope. Magnification 26,000×.

of the cell toward the surface to become the middle lamella. (3) In animals daughter cells often move apart; in plants they do not.

The Building of a New Cell in Interphase

To understand how a cell accomplishes the almost incredible task of duplicating all of its components as it progresses through interphase, and how these synthetic activities are keyed to the duplication of the chromosomes, we must examine more closely the mechanisms of synthesis of DNA and protein. We shall show that it is the DNA that is responsible both for its own duplication and for controlling the production of protein and other cell components; let us begin by asking, "How does DNA replicate?"

S Phase: DNA Replication

We know that the interphase chromatin fiber contains a single long DNA double helix; that each chain is composed of alternating sugar (deoxyribose) and phosphate groups; and that these chains are held together in their helical coils by hydrogen bonds between adenine and thymine from opposite chains. (If this picture isn't clear in your mind, refer back to Chapter 3, pp. 77–81.) We also know that the genetic information that is passed on at each cell division is built into each DNA helix in terms of the linear order, or **sequence,** of the nucleotide bases, so that each group of three nucleotide bases forms a code word or codon, and a group of codons acting together forms a gene (Chapter 10). The basic question of genetic replication, then, is **how** the nucleotide bases that form the new chromatin strand are put together in exactly the same sequence as those in the original.

After postulating their DNA double helix model (see pp. 98–102 and Chapter 10), Watson and Crick suggested that if the two

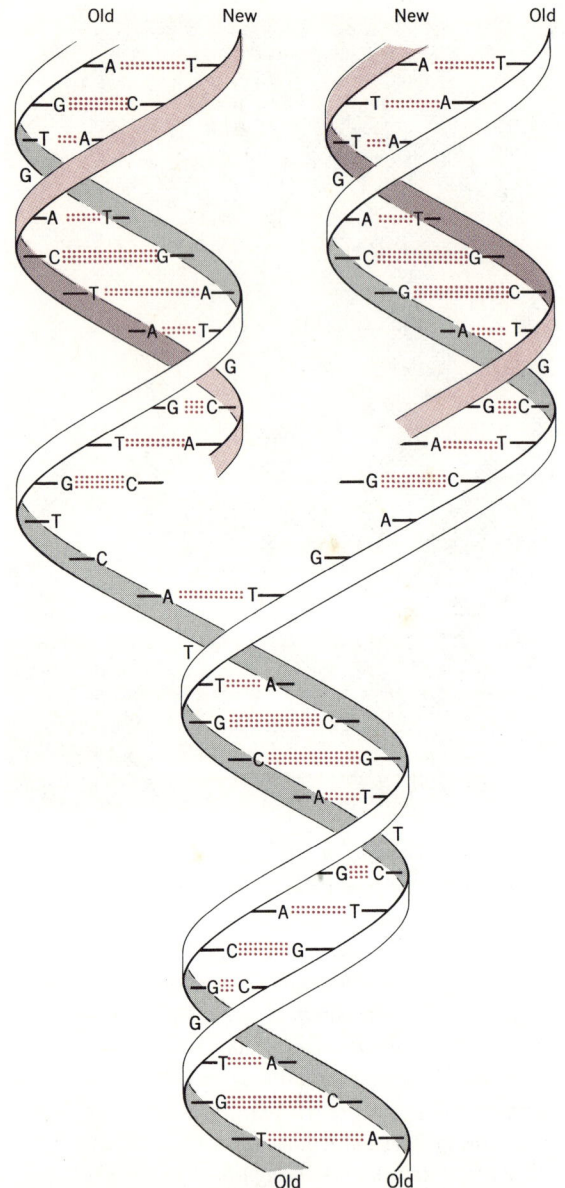

Figure 7-9   Model of DNA in the process of replication.

chains of the helix were separated by breaking the hydrogen bonds between the base pairs, each single chain could provide all the information needed to synthesize a new partner. Because an adenine (A) must always pair with a thymine (T), and a guanine (G) always pairs with a cytosine (C), the sequence of nucleotide bases in one chain determines precisely the sequence in its complementary chain. A sequence of ATGC in one strand, for example, would require the sequence TACG in the complementary strand. That is, if the two chains of a double helix were separated, each one could act as a model, or template, for synthesizing a new partner, and the result would be two complete new double helices, identical to the original (Fig. 7-9; also see Figs. 7-10 and 7-11 for further elaboration).

Shortly after Watson and Crick suggested this hypothesis, Arthur Kornberg and his colleagues at Washington University discovered an enzyme they called **DNA polymerase,** which they extracted from the bacterium *Escherichia coli.* In the presence of the four nucleotide bases, plus ATP as a source of energy, the enzyme was unable to function. However, if a single chain of DNA was added to the mixture as a **primer,** new lengths of DNA helix were formed. If the four nucleotide bases were labeled with radioactive isotopes, Kornberg could show that newly formed chains were likewise labeled, and,

Figure 7-10   Autoradiograph showing the circular chromosome of the bacterium E. Coli *in the process of replication. The distance around loop B from X to Y is about 500μ. In the inset the chromosome is shown diagrammatically. The original double DNA strands are drawn as two solid lines from the initiation point (X) around loop C to the growing point (Y). The newly replicated strands are shown as broken lines around loops A and B. Replication is about two-thirds completed.*

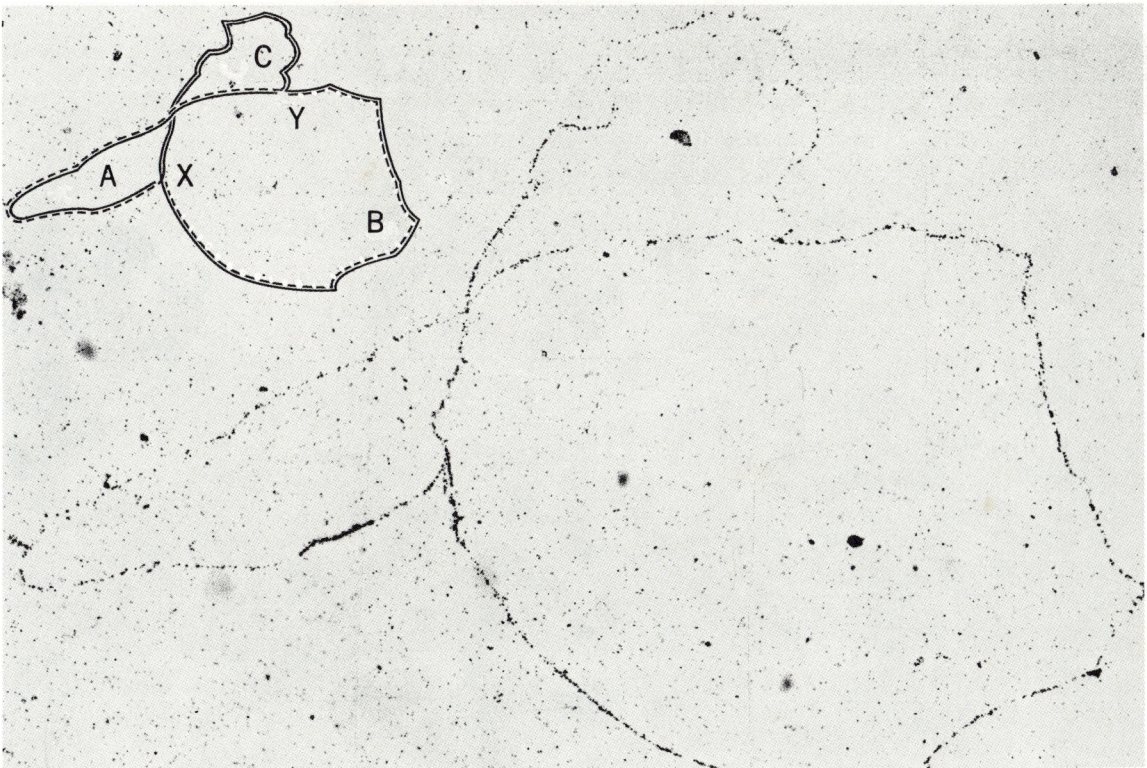

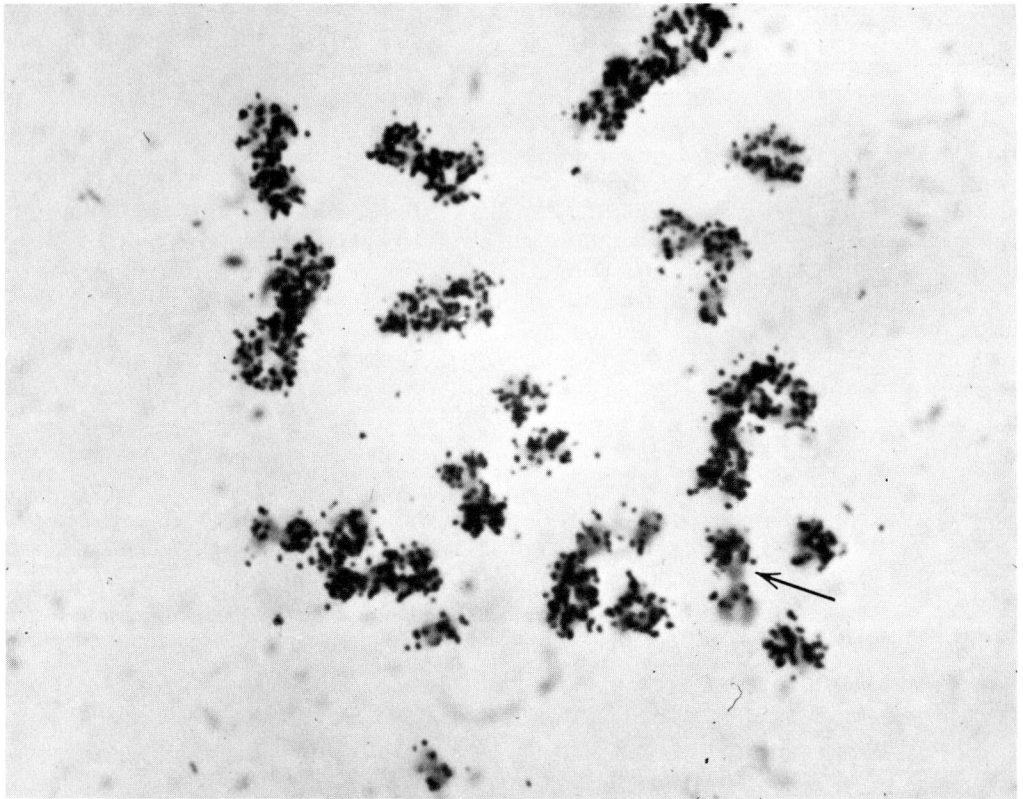

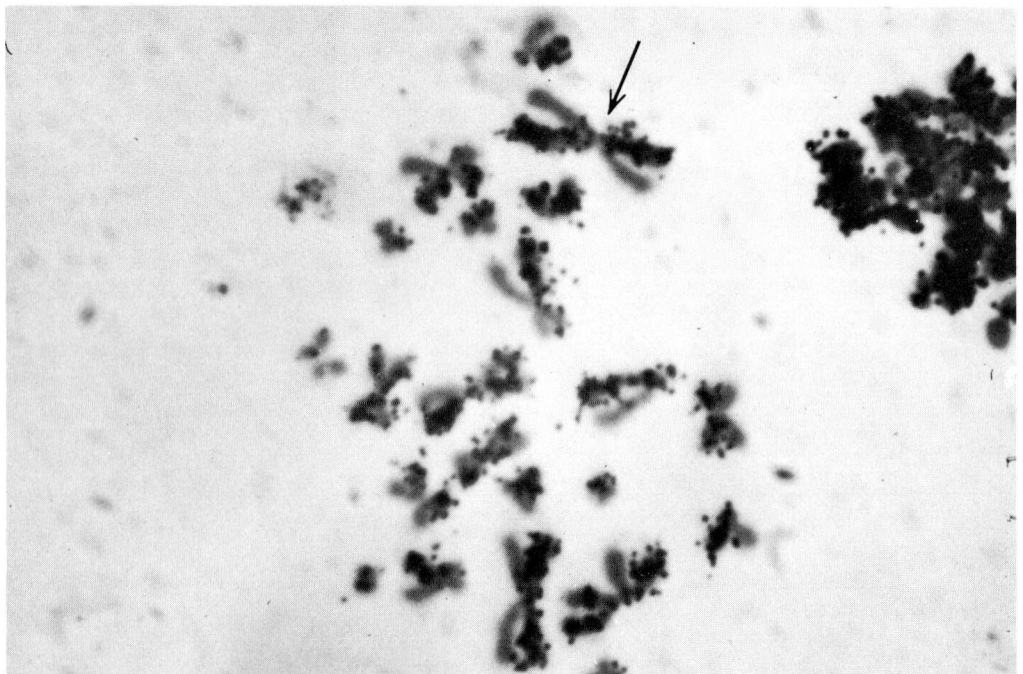

The Cell as the Unit of Life

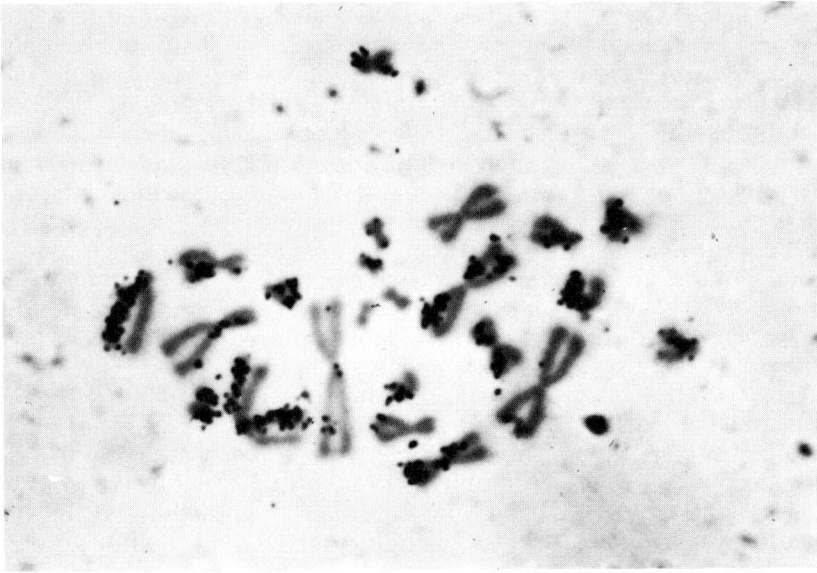

(c)

Figure 7-11 An experiment to show how mammalian chromosomes are duplicated. Hamster cells growing in tissue culture were exposed for 30 minutes to labeled thymidine. (a) First metaphase following DNA labeling. All chromatids are labeled equally except one arm of the X-chromosome (arrow), which synthesizes DNA later than the rest. (b) Metaphase chromosomes of the second division after labeling. In each pair only one of the chromatids is labeled. At the arrow the labeled chromatid crosses over the unlabeled one. (c) Metaphase chromosomes of the third division after labeling. Roughly half of the chromatids are free of label; the other half have one chromatid labeled and one unlabeled.

moreover, that they had exactly the same sequence of nucleotides as the original primer chain. Whatever the sequence of the primer, the new DNA was always identical. Clearly, the polymerase was functioning by making complementary copies of the primer chain, using it as a template. Since that time several DNA polymerases have been found in the nuclei of all cells examined, both animal and plant. Some are now known to be DNA **repair enzymes** (including the one Kornberg originally found). But at least one is involved in DNA replication.

We should point out that it isn't the base pairing, G and C or A and T, that requires an enzyme. Formation of hydrogen bonds is a spontaneous process that occurs without enzymatic interaction. The enzyme functions instead in hooking the mononucleotides together into a chain with covalent phosphodiester bonds (Chapter 3). To get an idea of how efficiently the polymerase can perform this task, we can calculate that during S phase in *E. coli* the new DNA strand grows at the rate of 2100 nucleotides per second.

If this picture of DNA replication, in which the two strands of a double helix separate and each is used as a template for a new complementary strand, is correct, then it should be possible with the electron microscope to see a growing point as a Y-shaped configuration in a strand of duplicating DNA, like that modeled in Figure 7-9. With very elegant techniques, John Cairns has been able to show just such a structure in the single chromosome of *E. coli*. In fact, because all bacteria have circular molecules of DNA, they have two Y-shaped regions, one representing the growing point and the other the initiation point of a cycle of DNA replication; that is, the beginning of the chromosome (Fig. 7-10).

Evidence that this kind of copying mechanism occurs in mammalian cells comes from experiments done by David Prescott and his associates at the University of Colorado. They obtained a sample of the nucleotide thymidine in which some of the hydrogen atoms had been replaced by a radioactive isotope of hydrogen called tritium (H^3) (p. 41). This labeled nucleotide was then added to cells from hamster embryos growing in tissue culture, so that it could be taken up by the cells and used in the replication of chromosomal DNA. After 30 minutes' exposure to the labeled nucleotide, the cultures were "washed" with fresh medium containing nonradioactive thymidine, and were then allowed to continue growing.

The fate of the labeled thymidine could be traced when any of the treated cells entered metaphase at some later time. If such cells were placed in contact with a photographic emulsion (in the dark, of course), the emulsion formed tiny black dots or "grains" wherever it was hit by electrons emitted from the radioactive tritium atoms built into the DNA of the chromosomes. (This technique is **radioautography,** one of the valuable uses of radioactive isotopes in biological research.)

In Prescott's experiments, if a radioautograph were made on cells that had entered metaphase shortly after treatment with the label (i.e., in the first division after treatment), all of the chromosomes were labeled (Fig. 7-11a). However, if the cells were allowed to go through two divisions after labeling before they were autoradiographed, photographic grains appeared over only one of the chromatids in each chromosome. The other chromatid was free of the radioactive nucleotide (Fig. 7-11b). If the cells were autoradiographed at metaphase of the third division after labeling, about half of the chromosomes had no label on either chromatid and the other half had only one chromatid labeled (Fig. 7-11c).

These results can best be interpreted as shown in Figure 7-12. The first S-phase replication occurred while the cells were in the medium containing the radioactive label (i.e., shortly before metaphase). Thus the new chain of DNA forming the double helix of each chromatid contained the label, and produced grains along its whole length in the emulsion. In the next cycle, however, the two halves of each double helix separated again, and each replicated in a medium that did not contain the label; the new chains were free of radioactive thymidine. Those that were formed on the original unlabeled chain formed a double helix containing no label. Those replicated on a labeled chain did not themselves contain the label, but the double helix—the chromatid—still exhibited the radioactivity of the one labeled chain (Fig. 7-12, second metaphase). In the third generation each nonradioactive chromatid duplicated itself to form the two chromatids of an entire chromosome, and each labeled strand was passed on to one of the daughter chromatids.

From these data, and many others equally convincing, we can conclude that the original DNA replication hypothesis suggested by Watson and Crick was correct. During the S-phase the entire chromosome complement of the cell is replicated by this mechanism, a process that normally takes 30 to 50 percent of the total generation time of the cell. During this time also, protein synthesis occurs, so that by the end of the S period two complete, protein-clad chromatids have been fabricated.

Doubling the
Cytoplasmic
Components:
G_1 and G_2

A newly divided cell in a population that is dividing does not immediately begin to synthesize DNA. Instead there is a pause, or gap (G_1), that may last 30 to 40 percent of the cell cycle. This G_1 period is a time of active synthesis of protein and RNA, partly to prepare for DNA synthesis. Just how active this synthesis may be is emphasized by look-

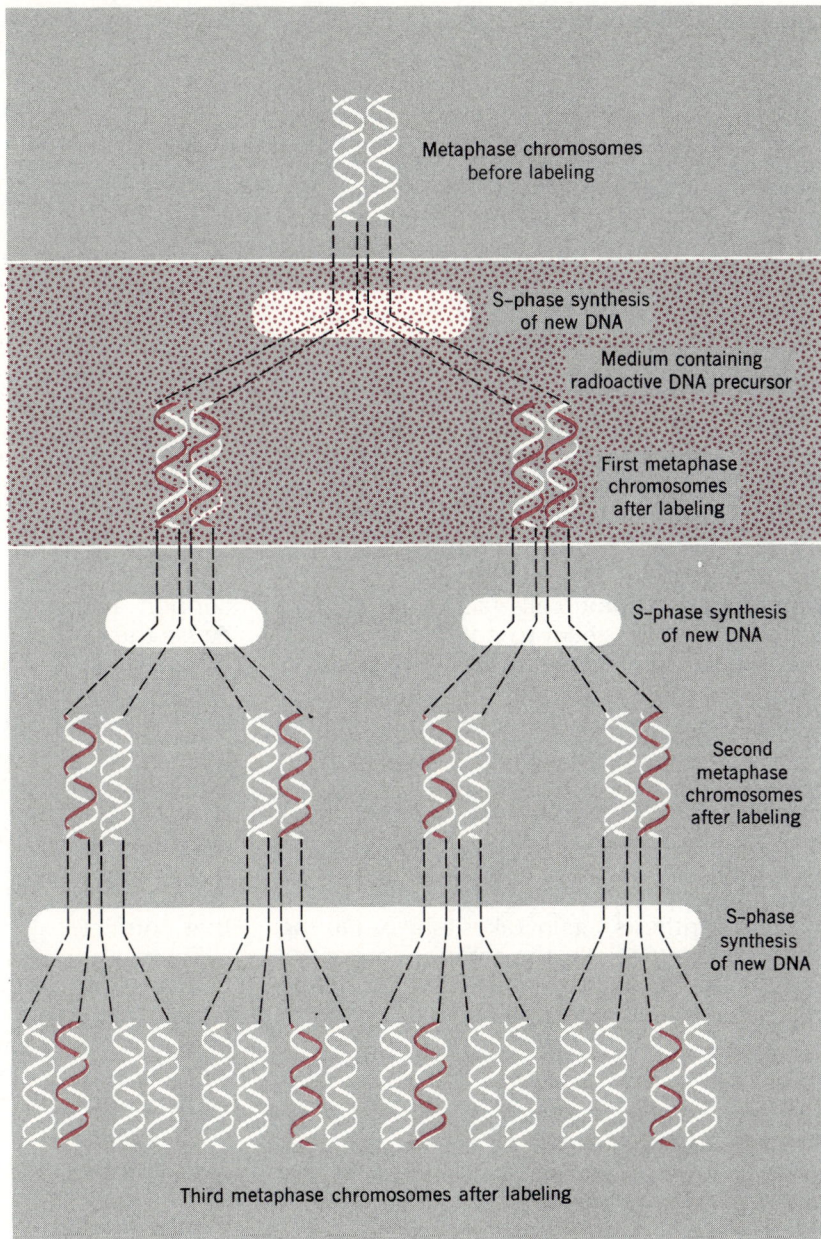

Figure 7-12 Diagram to explain how chromosomes are duplicated. Chromosomes are shown as a pair of chromatids, each in turn made of a DNA double helix. The stippled zone represents treatment with tritium-labeled thymidine. As the chromosomes duplicate in this medium, the newly copied strands become labeled (black). During the next two replications, the labeled strands are transmitted intact to one or the other of the chromatids, which appear as radioactive in an autoradiograph.

Metaphase chromosomes before labeling

S-phase synthesis of new DNA

Medium containing radioactive DNA precursor

First metaphase chromosomes after labeling

S-phase synthesis of new DNA

Second metaphase chromosomes after labeling

S-phase synthesis of new DNA

Third metaphase chromosomes after labeling

ing more closely at a single cell component.

Recall from p. 104 that a ribosome is a ribonucleoprotein particle about 200 Å in diameter, with a molecular weight of 1 to 2 million daltons. It may be made of 10,000 precursor amino acids and mononucleotides. A bacterium such as E. coli contains about 90,000 ribosomes per cell, and, in exponential growth, duplicates itself every 20 minutes. That is, a cell of E. coli can synthesize 4500 ribosomes per minute, and will do so hour after hour for an indefinite period, as long as the appropriate nutrients are supplied.

The length of the G_1 phase may vary enormously in different cells or under different conditions. In general it may be said that variation in the cell generation time is associated mainly with a longer or shorter duration of G_1, whereas S, G_2, and M tend to remain more constant. In a study of epithelial cells in different parts of the gut, for example, the average length of the cell cycle ranged between 17 hours and 181 hours, but the duration of $S + G_2 + M$ remained constant at about 10 hours; that is, G_1 varied between 7 and 171 hours.

It is also during G_1 that some **initiator** substance is probably synthesized, which triggers the beginning of S-phase. One hypothesis is that this substance functions by causing an interaction between the chromosomal DNA and the enzyme DNA polymerase. On the other hand, in some cells factors other than specific initiators seem to control whether a cell divides. In amebas, for example, division occurs as soon as the cell mass has exactly doubled; division can be delayed indefinitely simply by cutting off bits of cytoplasm from time to time.

Even when a cell progresses through G_1 and S, completing DNA replication, it is still not ready for mitosis. Normally, there is a second gap period (G_2), during which active protein synthesis again takes place. During this period, especially, the components of the mitotic apparatus are synthesized and made ready for assembly at prophase.

The Significance of Mitosis

In the process of cell division two cells are created from one. They are chromosomally identical and therefore contain precisely the same genetic information. Clearly, the primary act in this process is replication of the chromosomal DNA during the interphase S period. Mitosis itself, although visibly the most dramatic part of the process, actually represents only an elaborate mechanism for ensuring the equal distribution of the replicated DNA to the two daughter cells.

But why should cells divide at all? In all organisms we know that protoplasm has an inevitable tendency to increase itself, and there is clearly an evolutionary advantage for individual organisms to increase in size. Yet, largely because of physical limitations on rates of diffusion in aqueous systems, there seems to be an upper limit on the mass that a given unit of protoplasm can attain. Because the cell membrane acts as a diffusional surface for the protoplasmic mass within it, maintenance of a cell size on the order of 10 to 100μ permits relatively enormous masses of protoplasm to retain an appropriate ratio of surface area to mass (see Fig. 14-1). For example, a blob of protoplasm weighing more than a few grams could not possibly support an active metabolism internally because the diffusion of oxygen into the center of such a system would be much too slow. Yet 200 pounds of protoplasm, divided into the 100 trillion cells that make up the human body, is highly successful. Thus organisms seem to be confronted with two conflicting requirements: the tendency to increase in size on one hand, and the restriction on maximum unit size on the other. The device of cell division seems to be the resolution of that conflict.

There is apparently another justification for cell division. Cytoplasm seems to undergo certain aging processes whereby its organelles tend to wear out. Cell division allows a nucleus, or perhaps the genetic material, to gather around it fresh cytoplasm on a recurrent basis and still not exceed the limiting ratio of surface to mass. Evidence for this idea derives from the fact that most cells from mature organisms, animal and plant, tend to be nonmitotic. However, even though not-actively dividing, some of these cells retain the *capacity* to divide.

If we categorize cells (whether unicellular organisms or tissue cells) on the basis of their ability to divide, we generally find three classes. One type, like a protozoan colony in exponential growth phase, continually passes through the cell cycle. Among multicellular organisms, the epithelial cells of the gut of mammals are of this type; they never stop dividing throughout the life of the organism. Thus this tissue always contains cells in all phases of the life cycle. A second class of cells, which may be termed **true nondividing cells,** are those that have differentiated to the point at which they are incapable of initiating DNA synthesis or division. An extreme example of this class is the mammalian erythrocyte, which has lost its nucleus. The third class of cells, which includes most of the tissues of our bodies and those of most other metazoa, exist in organisms under normal conditions in a state of G_1 **arrest;** that is, they are blocked in G_1 more or less permanently; they do not pass to S. These may be termed **resting** cells. The G_1 block may, under normal circumstances, last throughout the life of an organism, but at any time, on the proper stimulus, such arrested cells can resume their passage through the cell cycle, initiating S, passing through G_2, and dividing. Examples of this class are liver cells, which rarely show mitosis normally, but begin to divide rapidly if a piece of liver is removed; or cells in the mouse salivary gland, which normally are in a resting state but can be stimulated to begin dividing by injection of certain drugs. The fact that most of the tissues of our bodies are in mitotic arrest can be demonstrated when fragments are removed and the cells separated and placed in tissue culture. In many cases cells that had not undergone a division for years in the tissue now begin to divide with a mitotic interval of only a day or two. In fact, in some desert plants living cells may survive in mitotic arrest for 100 years and still be able to divide when stimulated.

A cell dividing by mitosis normally replicates its chromosomes exactly. Through this process, the characteristics of the cell are passed down without change from one cell generation to the next. In most organisms, however, the information that is passed from generation to generation does not remain exactly the same. How this slight variation occurs is a major concern of the field of genetics, to which we turn in the next chapter.

Reading List

Cairns, J., "The Bacterial Chromosome," *Scientific American* (January 1966), pp. 36–44.

Dounce, A. L., "Nuclear Gels and Chromosomal Structure," *American Scientist* (January/February 1971), **59** pp. 74–83.

German, J., "Studying Human Chromosomes Today," *American Scientist* (March/April 1970), **58,** pp. 182–201.

Goldstein, L., *Cell Biology: A Book of Readings.* William C. Brown, Dubuque, Iowa, 1966.

Goodenough, V. W., and R. P. Levine, "The Genetic Activity of Mitochondria and Chloroplasts," *Scientific American* (November 1970), pp. 22–29.

Kornberg, A., "The Synthesis of DNA," *Scientific American* (October 1968), pp. 64–78.

Loewy, A. G., and P. Siekevitz, *Cell Structure and Function* (2nd ed.). Holt, Rinehart & Winston, New York, 1969.

Watson, J. D., *The Double Helix.* Signet Books, New York, 1968.

Watson, J. D., *Molecular Biology of the Gene* (2nd ed.). W. A. Benjamin, New York, 1970.

Wilson, G. B., *Cell Division and the Mitotic Cycle.* Reinhold, New York, 1966.

"Heredity means the process whereby men, mice, flies, plants, fungi, bacteria, viruses, and all forms of living things reproduce themselves, or at least something unmistakably like themselves."

W. HAYES, 1968, *THE GENETICS OF BACTERIA AND THEIR VIRUSES*

part four | *Heredity: The Transfer of Information*

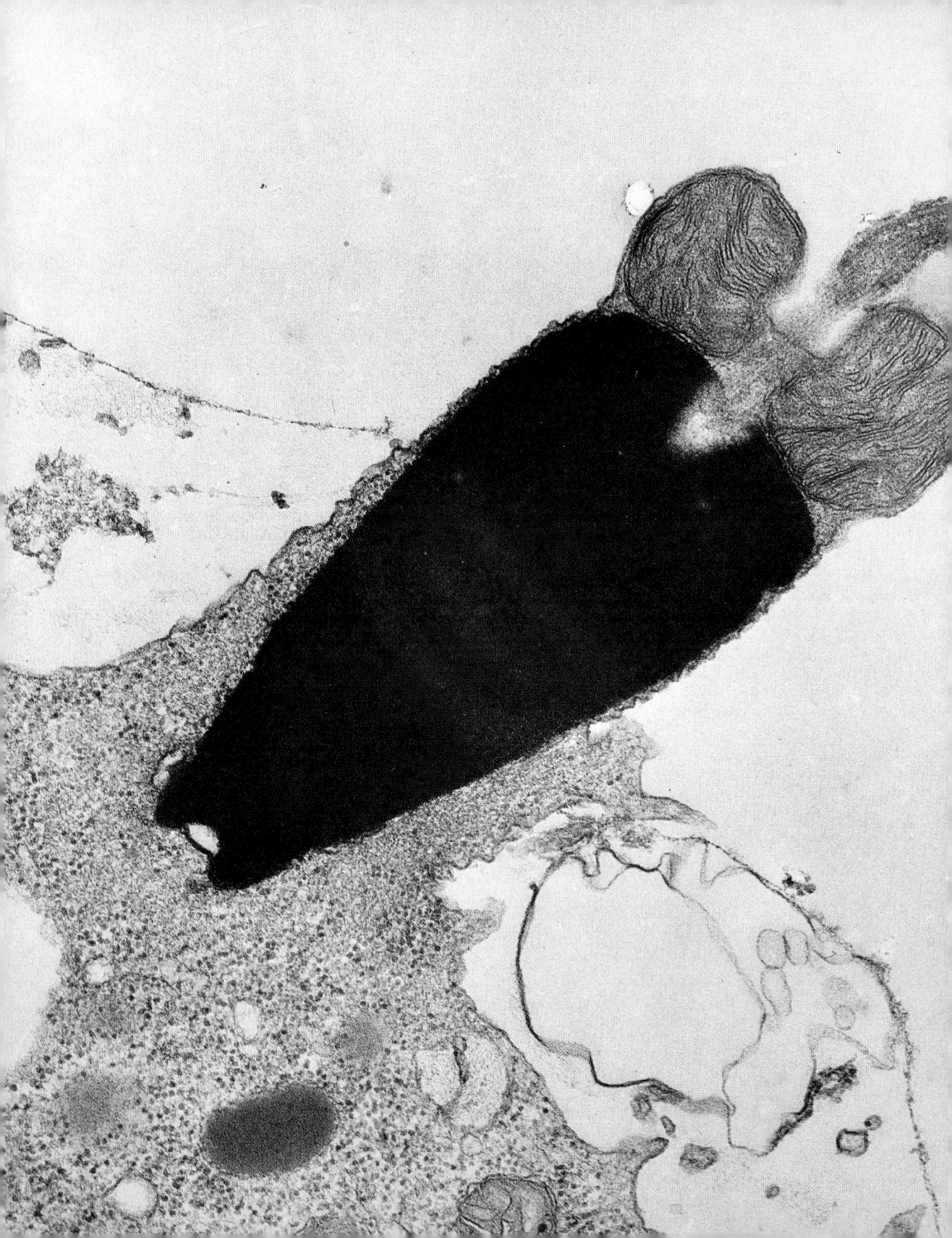

The Mechanics of Heredity

"Scientists announced today they have created a gene, the unit of heredity that controls all life processes" (from an article describing the work of Dr. H. G. Korhana and his colleagues).

THE BALTIMORE SUN, JUNE 3, 1970

In Chapter 7 we saw that a unicellular organism that reproduces by mitosis produces essentially identical copies of itself with each division. Mitotic division is a form of **asexual** reproduction that is **semiconservative;** this means that, with the exception of mistakes in copying the DNA sequence during the S-phase of a cell cycle, each new daughter cell carries the same information as cells of the previous generation. It is common observation that—like amebas— frogs only give rise to frogs and oak trees to oak trees, but there is a profound difference. In both frogs and oak trees, as in all other plants and animals including humans, the offspring are similar but not identical to the parent. Frogs and oak trees and humans reproduce by **sexual** reproduction (in case you didn't know). From a genetic point of view, the main difference between asexual and sexual reproduction is that the latter provides new combinations of hereditary characteristics with each generation; that is, it acts as a source of **genetic variability.**

Recall that a gene is a length of chromosomal DNA that can be transcribed into messenger RNA and then translated into an amino acid sequence to form an enzyme or a polypeptide subunit of a structural protein (Chapter 3). With semiconservative mitotic division a copying error might occur only once in every 100,000 or 200,000 replications (i.e., divisions). This is an amazing degree of accuracy, but for a population of millions or billions of cells dividing

Sperm chromatin entering sea urchin egg. Magnification 42,000×.

every few hours it still allows for a goodly number of miscopies. This process of mis-copying the DNA of a chromosome is called **mutation;** the organism that results is a **mutant.**

Consider a hypothetical unicellular organism with four chromosomes in its nucleus. On chromosome I it has a gene that controls the synthesis of a pigment that makes the cell dark, and on chromosome III it has another gene that allows the formation of a protein in the cell coat that causes it to be rough and nubbly. The initial population of cells, then, is dark and rough-surfaced. This is called the **wild-type population,** the population as it normally occurs in nature. Now, suppose that at some point in time the pigment-producing gene is miscopied in a cell division. The daughter cell that carries the original chromatid of chromosome I will still be dark and will continue to give rise to dark daughter cells, but the mutant cell with the miscopied chromatid will be unable to synthesize pigment and will be pale. When it divides, if it copies its DNA accurately, its daughter cells will likewise be unable to make pigment, and thus a mutant pale **strain** (descendant line) will arise, assuming, of course, that the lack of pigment is not harmful to the cells so that they cannot survive. If some time later another mutation occurs in the original population, this time on chromosome III, so that the cell coat is smooth instead of rough, there will now be found in the population three strains or **variants:** the original wild-type population, dark-rough; the pigment mutant, pale-rough; and the coat mutant, dark-smooth.

As long as these new gene mutations do not influence the survival of the cells, the three different sets of chromosomes will continue to be produced in the population in roughly constant proportions. The position of a gene on the chromosome is termed its **locus** (plural, **loci**). In our hypothetical cell population genes at two different loci are available in two different forms, the original and the mutant. We refer to different forms of a gene at a given locus as **alleles** of that gene; genes for the smooth coat and the rough coat are alleles of one another. The particular set of chromosomes an organism has, that is, the specific genetic information encoded into its chromosomes, is termed the **genotype** of the organism. The **expression** of the genetic information, in terms of the appearance or behavior of the organism (in our case whether cells are dark or pale, rough or smooth), we call the **phenotype** of the organism.

Thus mutation introduces some variation into an asexually reproducing population, but this is a slow process. Although we now have dark-rough, pale-rough, and dark-smooth cells, there is no way to produce a pale-smooth cell except by another mutation. That is, there is no way of putting the different characteristics of a parent cell together in new combinations. Moreover, since random mutations are usually harmful to an organism (as if you were to change one connection in a computer at random), the mutant usually does not survive to pass on its allelic form of the gene. This means that the introduction of variation into a population by mutation alone is even slower than a particular mutation rate would suggest.

In contrast, although women never give birth to oak trees or frogs, it is not at all uncommon for a blond, blue-eyed mother to give birth to a child with brown eyes and dark hair. We know from common experience that this results from the random **segregation** of characteristics from two parents whose genes have been joined and recombined through sexual reproduction. *In genetic terms, sexual reproduction means the formation of a new individual by the fusion and mixing of two sets of genetic information.*

Among unicellular organisms that reproduce by mitosis, each cell is a reproductive unit. Replication of the organism's DNA, and the cytoplasmic components that support it, is sufficient for the continuation of the species. But higher organisms are multicellular, and their cells become highly specialized into different organs and tissues (see Chapter 6). Division of a mouse liver cell many times over does not produce a mouse, it yields only a mass of liver tissue. These specialized, non-reproductive cells that make up most of the body of an individual are called **somatic** cells.

Among higher organisms only certain cells are reserved as reproductive units; these are the sex cells, or **gametes.** In most species there are two different kinds: male gametes are small (and therefore called **microgametes**) and usually motile; female gametes (or eggs, or **ova**) are larger and filled with nutritive material. In animals the male gametes are called **sperm.** Among plants the pollen grains contain the microgametes.

Sexual reproduction involves the fusion of a sperm and an egg in such a way that the genetic information carried in the two cells can mix. This fusion is termed **fertilization.** The product of the fusion, the fertilized egg, is called a **zygote.** The zygote and all the somatic cells that form from it carry a complete set of chromosomes from the two parent individuals. These chromosomes occur in pairs (called **homologous pairs**), one from each parent. A full complement of chromosomes is referred to as the **diploid** (double), or **2n,** number; the single set of chromosomes that comes from each parent is called the **haploid (n)** number. For example, the diploid number in human beings is 46; the haploid number, the number of chromosomes that comes from each parent, is 23; and there are 23 homologous pairs. These terms are discussed more fully on p. 196.

Mendel: The Beginning of the Science of Genetics

The discovery of how genes combine (in the formation of a zygote), and sort out from one generation to the next, we owe to the work of an Austrian priest named Gregor Mendel. Mendel joined an Augustinian monastery in Moravia (now part of Czechoslovakia) in 1847 when he was 21 years old. He remained a monk (eventually becoming abbot of the monastery) until his death 40 years later. In the mid-nineteenth century monasteries occupied an important place in the cultural life of their communities; Mendel's monastery included scholars of high repute, philosophers, natural scientists, composers, men of letters. No atmosphere could have been more stimulating to a young man interested in natural history.

Mendel served mainly as a substitute teacher in a secondary school (he was unable to be fully accredited because he twice failed the necessary examination); he taught mathematics, Latin, Greek, and German. He devoted his spare time, which was apparently ample, to a wide range of interests: he was an amateur meteorologist, for a while in charge of a meteorological station maintaining records of sunspots and tornadoes; he did studies on bees, and had a collection of 50 hives under observation; he was an active member of the horticultural section of the Moravian Agricultural Society, and won prizes for producing new sorts of apples and pears. Mendel's primary work, however, to which he devoted 10 years of serious study, was on the inheritance of traits in the common garden pea, a sexually reproducing plant. In this work, which he published in 1866, he formulated the basic principles of hereditary transmission and laid the groundwork for a meaningful experimental genetic science. Ironically, these studies went unrecognized for 35 years, and Mendel died unknown and embittered.

Mendel's Experiments

The flower of a pea plant is so constructed that it is easy to prevent fertilization of a plant by stray foreign pollen; thus a **controlled cross** (cross-fertilization) is possible. Mendel's famous plant-breeding experiments consisted simply of mating, or crossing, different varieties of garden pea and systematically recording the distribution of parental traits among the offspring. Observing and recording the distribution of a few separate traits in generation after generation of offspring derived from selected crosses of pea plants, Mendel discerned an underlying pattern in his data. In devising an explanation

table 8-1
Mendel's Results in the F_2 Generation of Garden Peas

| TRAIT | NUMBER OF PLANTS | | RATIO OF DOMINANT TO RECESSIVE |
	DOMINANT	RECESSIVE	
Seed color	6022 yellow	2001 green	3.01:1
Stem length	787 long	277 short	2.84:1
Seed shape	5474 round	1850 wrinkled	2.96:1
Flower color	705 red	224 white	3.15:1
Pod shape	882 smooth	299 wrinkled	2.95:1
Pod color	428 green	152 yellow	2.82:1
Flower location[a]	651 axial	207 terminal	3.14:1

[a] **Axial** flowers are distributed along the length of a stem; **terminal** flowers are located at the end of a stem.

to fit this data he was inevitably led to certain conclusions that today are the foundations of the modern theory of inheritance.

Before carrying out his experiments, Mendel made certain that he was dealing with "pure, inbred lines" of plants with respect to each trait. This he could determine by mating the plants of a single variety for several generations to make sure that they would breed true; that is, that the trait would always appear in all the offspring. Then, having established the purity of the lines, Mendel made numerous crosses of varieties of peas that differed in one contrasting characteristic: a variety with yellow seeds and one with green seeds; one that had a long stem with another that had a short stem; and so on. He studied seven traits in all (Table 8-1).

When plants of inbred lines that differed in only one trait were mated (he called this first cross the **parental** or **P_1 generation**), the first generation of offspring (known as the **first filial** or **F_1 generation**) always possessed **only one** of the parental traits. For example, when Mendel made a parental cross of yellow-seeded plants with green-seeded ones, he never obtained offspring with chartreuse or mixed yellow and green seeds. This cross produced only yellow-seeded plants; the green-seed trait seemed to have disappeared.

The trait that was present both in one parent and in the F_1 generation Mendel called the **dominant** trait. The characteristic of the other parent, which seemed to disappear in F_1, was termed **recessive.** Yellow seed is therefore dominant over green seed in the garden pea. Of the seven pairs of contrasting (allelic) traits examined, Mendel found that in all cases one of the allelic traits was dominant and the other recessive.

Mendel continued his experiments by crossing members of this F_1 generation, called **hybrids,** which now showed only the dominant characteristic (yellow seed). He found that the recessive trait (green seed) reappeared in about 25 percent of the progeny (called the **second filial** or **F_2 generation**). The other 75 percent were again dominant, a ratio of about 3 to 1. The actual figures for Mendel's experiments are shown in Table 8-1. Note that this ratio of 3 to 1 holds very closely for every trait.

Mendel's Theory of Segregation

In order to explain his results Mendel postulated that there was some sort of hereditary unit responsible for transmitting inherited traits from parents to offspring. Such a hereditary factor presumably would maintain its identity—neither contaminating nor blending with other hereditary units—although at times it would fail to express its given trait. From his results Mendel constructed the following set of postulates, which are now known as **Mendel's First Law** or the **Law of Segregation.**

1. **Each heritable trait is controlled or determined by hereditary units that exist in pairs.** We now call these units genes. We have noted that alternative or contrasting forms of the same gene are called alleles; the genes for a yellow-seeded pea plant and a green-seeded pea plant are alleles of one another. Two or more alleles may exist for a given trait. At any one time in a diploid organism two alleles may occupy any pair of loci, one at each of the corresponding positions of a homologous pair of chromosomes (e.g., yellow and green peas). When both genes are exactly alike for a given trait, that is, when the allelic loci are occupied by identical genes, the organism is said to be **homozygous** for that characteristic. When the two genes are different, the plant or animal is called **heterozygous,** or **hybrid.** In the P_1 generation the yellow-seeded and green-seeded pea plants were homozygous for seed color, whereas all the F_1 offspring were heterozygous, or hybrid, for the same characteristic.

2. **In the formation of gametes the two genes of each pair are separated (segregated) from one another, so that each gamete receives only one gene of a given pair.** Therefore each gamete has only one kind of gene for any particular trait; in the next generation one of the two factors for any trait is provided by each parent.

3. **When the two members of an allelic pair of genes are different (i.e., in the heterozygous condition) only the trait determined by the dominant factor will be expressed.** Although the recessive factor is present in a heterozygous individual, it will not manifest itself. For example, in the F_1 generation of a cross between homo-

zygous yellow-seeded plants and homozygous green-seeded ones all the plants had only yellow seeds even though they contained a dominant gene for yellow color and a recessive gene for green seed color. These yellow-seeded F_1 plants **appeared** not to be different from their yellow-seeded parents, but in genetic constitution they were very different. Recall that the hereditary makeup of an organism is its genotype. The observable traits under consideration—its appearance—is the organism's phenotype. The yellow-seeded parents and all the F_1 plants shared the same phenotype (i.e., yellow seeds). In their genotypes, however, the plants differed: parental plants were homozygous for the gene producing yellow seeds; the F_1 plants were heterozygous, having the dominant yellow gene on one chromosome and the recessive green gene as the allelic member of the pair.

Mendel's postulates can be illustrated by a convenient system of shorthand now widely used by geneticists. In the P_1 cross of yellow-seeded and green-seeded pea plants a common practice is to use a capital **Y** to designate the dominant gene, yellow, and lower-case **y** for the recessive or nonyellow gene, in this case green. The crossings and resultant progeny can be represented as shown in Figure 8-1 (p. 194).

It should be emphasized at this point that the expected distribution of inherited traits, called the **probability ratio,** is based on statistics or probability. It does not necessarily mean that when there are four offspring in the F_2 generation three of them will always be yellow-seeded and one green-seeded. All four **might** be yellow-seeded or, less probably, all four **might** be green-seeded. With a limited number of offspring any ratio could appear. If many progeny are produced, however, the expected ratios will be more closely approached. The probability ratio of 3:1 merely states that in the offspring arising from the crossing of two identical heterozygous individuals (e.g., Yy) there are three chances out of four that the dominant trait (yellow)

P₁ Generation			
Phenotype	Yellow–seeded		Green–seeded
Genotype	YY		yy
Gametes	Y	X	y
F₁ Generation			
Phenotype	100% Yellow–seeded		Yellow–seeded
Genotype	Yy		Yy
Gametes	Y y	X	y y
F₂ Generation			
Phenotype	75% Yellow–seeded		25% Green–seeded
Genotype	YY, Yy, Yy		yy

Figure 8-1 Summary of Mendel's results from the mating of two varieties of garden peas. In the first mating (P₁ generation) a dominant yellow-seeded variety (Y, shown in color) is crossed with a recessive strain that has green seeds (y, shaded black). The × means "crossed with." Because every zygote resulting from the P₁ cross must have a Yy genotype, all of the offspring (F₁) are yellow-seeded. Segregation of the traits in the F₁ cross, however, leads to a 3:1 probability in the next generation (F₂) of plants being yellow-seeded.

will appear in any offspring, and there is only one chance out of four that the recessive trait (green) will manifest itself.

It can be seen from Mendel's original data in Table 8-1 that he recorded the distribution of traits in hundreds and even thousands of offspring. His data in the F₂ generation showed a good approximation of the 3:1 ratio. Had he recorded these traits in only 10 or 15 progeny instead of several hundred, he might not have been able to discern the true distribution ratios in the offspring and therefore might have been led to the wrong conclusions.

Look again at Figure 8-1. Remember that according to Mendel's postulate 2 each gamete of an organism receives only one gene of a pair. In the P₁ generation among plants

that are homozygous for yellow (YY) every gamete must receive a Y gene; among the plants homozygous for green (yy) all the gametes must receive a y gene. Fertilization of Y egg by a y pollen cell (or a y egg by a Y pollen cell) always yields a Yy F₁ individual. But when these F₁ plants form gametes, each new gamete receives one gene of the allelic pair, either or both of which may be Y or y. The results of the random unions in F₂ can be represented in a **checkerboard,** or **Punnett, square,** a chart first used by the English geneticist R. C. Punnett. In this case the Punnett square is:

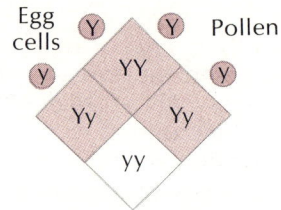

That is, on the average the **genotypic** combinations will be 1YY, 2 Yy, 1 yy. But because Y is dominant, and both YY and Yy will have yellow seeds, the **phenotypic** ratio will be 3:1, yellow to green (as shown by the shading in the square).

Mendel's Theory of Independent Assortment

Having provided a reasonable set of postulates that could adequately explain and predict results of **monohybrid** mating (involving a single pair of characteristics), Mendel then studied the simultaneous inheritance in peas of two traits (*dihybrid* crosses). He postulated a second law, the **Law of Independent Assortment,** which states: **In crosses with two or more pairs of independent traits, each pair of genes is assorted at random and inherited independently of the other pair.** For example, Mendel examined the distribution in the F₁ and F₂ generations of seed color and shape, starting with a P₁ cross of a variety of peas having yellow (YY), round (RR) seeds and another having green (yy), wrinkled (rr) seeds. He had already shown that the hereditary factor for the yellow seed color was dominant to that for green and that round seed shape was dominant to wrinkled (see Table 8-1). As expected, all the F_1 generation plants displayed only yellow, round seeds.

In the F_2 generation all four possible types of plants appeared: yellow round, yellow wrinkled, green round, and green wrinkled. The results of these dihybrid experiments are summarized in Figure 8-2. The gametes of the P_1 generation are all either YR or yr; therefore the F_1 plants are all YyRr, and have yellow round seeds. When these F_1 plants form gametes and each gamete receives one gene of each allelic pair, there are four possibilities: YR, Yr, yR, and yr. Again assuming random

fertilization, and further assuming that the genes of the different traits will be passed on independently of one another (i.e., will **assort independently**), the different genotypic possibilities are shown in the Punnett square. These possibilities will lead to a phenotypic ratio of 9:3:3:1.

We now know that the Law of Independent Assortment applies only if each pair of genes is located on a different pair of homologous chromosomes; that is, it is the chromosomes that assort independently. If two different genes are on the same chromosome, they do not assort independently but are distributed in a combined fashion. Mendel was extremely fortunate in his selection of pea-plant characteristics for genetic study; apparently in all his dihybrid crosses he happened to

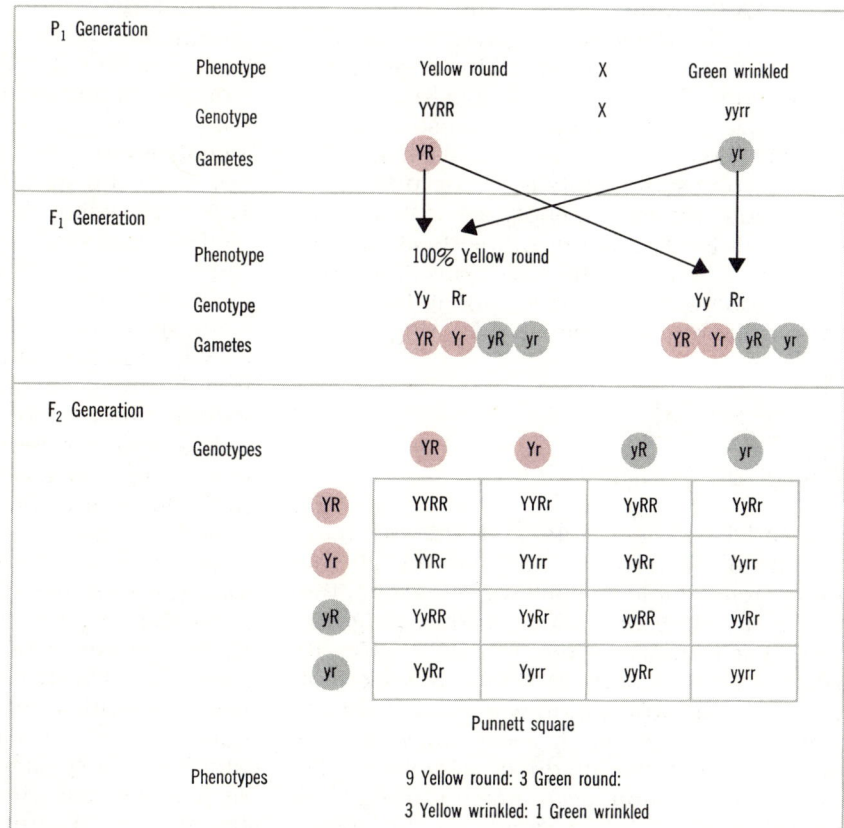

Figure 8-2 Representation of dihybrid cross (inheritance of two separate traits) between a pea plant having yellow, round seeds (YYRR) and one having green, wrinkled seeds (yyrr) as the parental generation.

select conspicuously different traits that were controlled by genes located on different pairs of chromosomes.

We have noted that the independent assortment of chromosomes in sexual reproduction seems to be a device for producing genetic variability among organisms. To understand the power of chromosomal segregation in generating diversity, consider the number of genetic combinations that could be produced by a hypothetical man and woman with identical **genomes** (genotypes). Assume for purposes of simplicity that each parent is heterozygous for only one gene on each of its 23 pairs of chromosomes. For one pair of genes (A and a) offspring with any of three genotypes are produced: AA, Aa and aa. For a second pair of allelic genes (B and b) three more genotypes are generated: BB, Bb and bb. But in combination with the first three, these yield nine possible combinations: AABB, AABb, AaBB, AaBb, and so on. For each additional pair of genes the number of combinations is multiplied by three, yielding a series: 3, 9, 27, 81, 243, 729, etc., or 3^n, where n is the number of chromosome pairs. For 23 pairs of genes, one on each pair of chromosomes, the figure is 3^{23}, or more than 80 billion different genotypes. Obviously, even with the limitations of identical parents and only one pair of heterozygous genes per chromosome, essentially no possibility exists for producing two identical offspring.

The Material Basis of Heredity

Mendel's experiments and the accurate conclusions he drew from them are all the more remarkable when we consider the backward state of biological knowledge in 1866. For instance, it was not until 1876, 10 years after the reports of Mendel's work, that the first clear observation was made of fertilization in any animal. It was realized shortly afterward that the union of gamete nuclei during fertilization accounts for the equal contribution of both parents in the trans- mission of inherited factors to the offspring. The process of **meiosis** (p. 197), a means of distributing half the number of chromosomes to each of the newly formed gametes, was discovered about the same time. But it was not until the 1920s that the evidence finally accumulated to indicate conclusively that the hereditary substance of cells is located in the chromosomes.

As we shall see later (Chapter 23), the physical basis of heredity is also the physical basis of evolution, for changes in the chromosomal DNA sequence (i.e., mutations) produce the variations that account for biological evolution.

The Chromosome Complement

We saw in Chapter 7 that the fundamental element of a chromosome is a long, thin chromatin thread made of a DNA double helix (Chapter 3) coiled and supercoiled, and wrapped in protein to form a chromatid. Each chromosome has a single specialized region, a **centromere**, which is the point of connection with the spindle fibers during division. The centromere is typically a permanent, well-defined structure, and its position—near the middle or end of the chromosome—is characteristic; the shape of each chromosome in an organism can usually be recognized, and each chromosome can be identified, by the position of the centromere and various other knobs and constrictions.

The diploid number (2n) of chromosomes per nucleus is usually constant for all the individuals of a species and varies from one species to another. Human beings have 46, foxes 34, rats 42, garden peas 14, corn 20, fruit flies (*Drosophila*) 8, and so on. The chromosomes in each cell of an organism (with the exception of its sex cells) are present in pairs, and the two members of a pair are usually identical in size or shape. It is often convenient to speak of the chromosome number of a species in terms of the number of pairs present; thus humans have 23 pairs, the fruit fly 4, and so on. The number of chromosome pairs, of course, equals the haploid number (n). Although the members of a pair are alike, different pairs are usually readily distinguishable by size and shape, and are symbolized by numbers. The 46 chromosomes of a human cell at metaphase are shown in Figure 8-3 as they appear pressed ("squashed") down on a microscope slide and stained to form a chromosome squash. In Figure 8-4 (p. 198) the chromosomes are drawn to match those in an enlarged photograph of the chromosome squash. They have been arranged in pairs numbered from 1 (the longest) to 22 (the shortest). The unpaired chromosomes, labeled X and Y, are the sex chromosomes. These are discussed in Chapter 9. A presentation of the chromosomes of an individual like that in Figure 8-4 is called a **karyotype**.

Meiosis and Gamete Formation

Because fertilization consists of the fusion of a sperm nucleus with an egg nucleus, it is obvious that some process must exist for

Figure 8-3   Human male karyotype squash. Chromosome 14 (arrow) will be important to the discussion in Chapter 9.

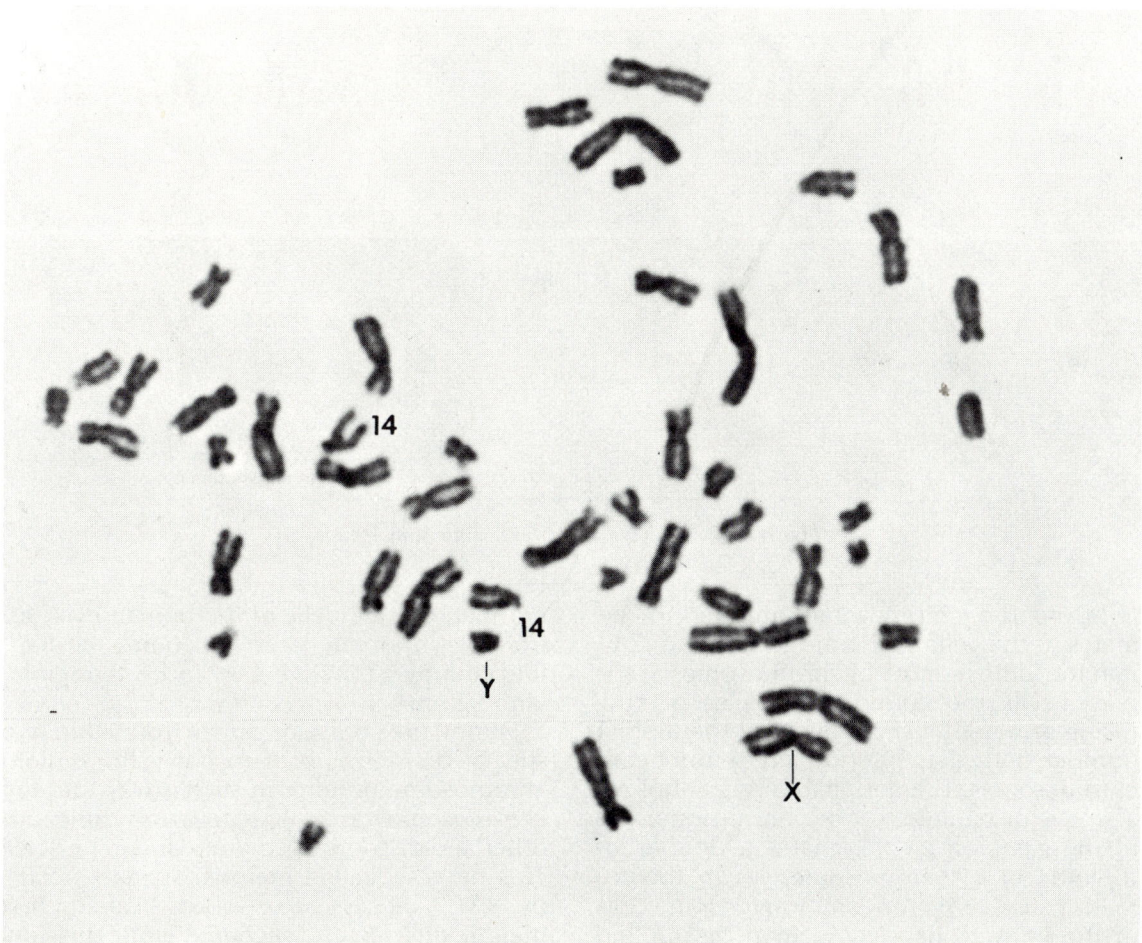

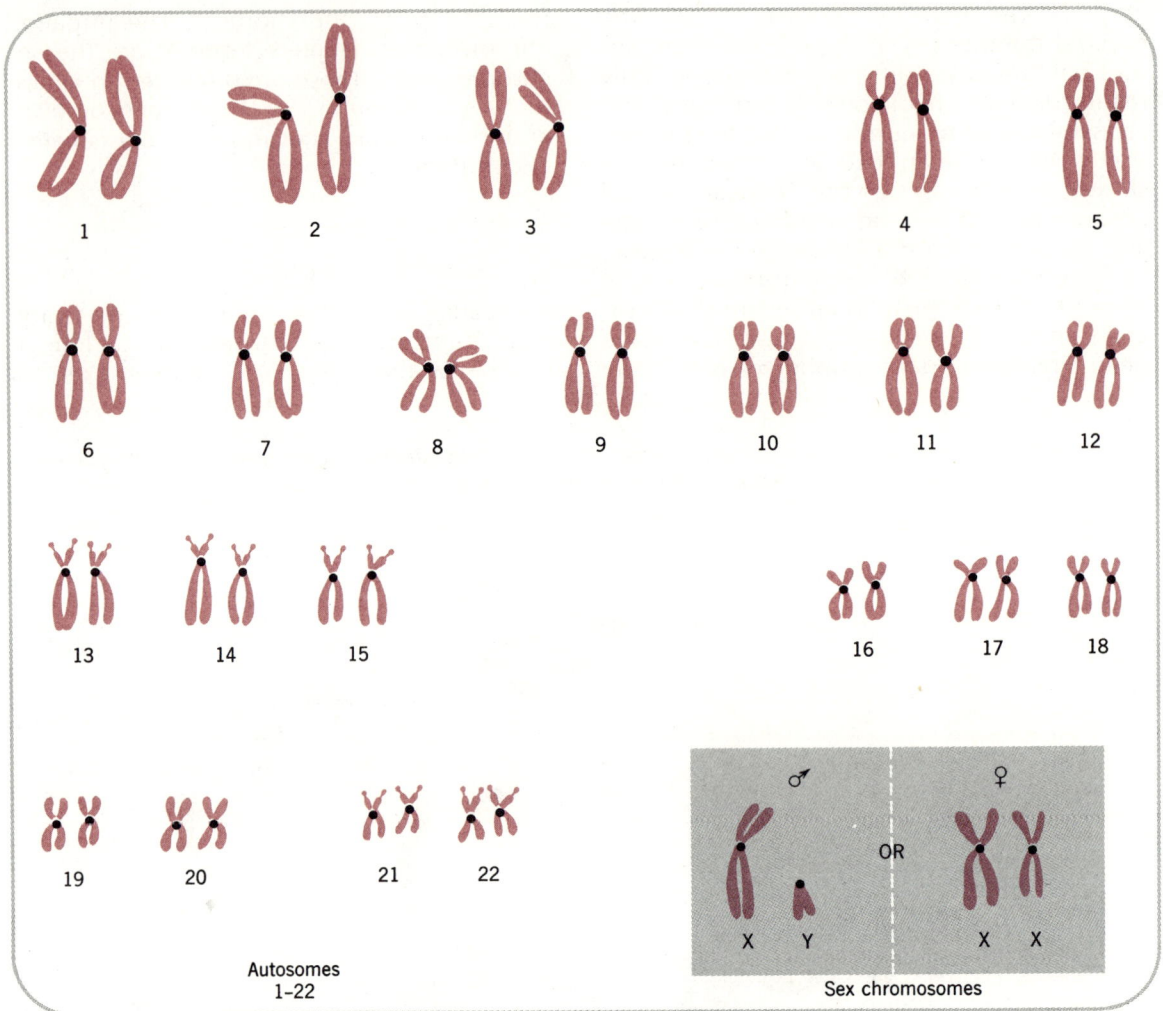

Figure 8-4 *Human karyotype, male and female.*

reducing the chromosome number in gametes. If the cells of the progeny are to contain the same number of chromosomes as the parent cells (the diploid number), each of the fusing gametes must contain half the diploid number, or the haploid (unpaired) number of chromosomes. The designation n symbolizes a single, or haploid, set of chromosomes, as in the gametes; 2n designates a double, or diploid, set of chromosomes, as in the somatic cells. It is not uncommon for some cells of the body to have more than the diploid number. Certain cells of the human liver, for example, normally have four times the haploid number; they are said to be **tetraploid,** with 92 chromosomes (4n).

During the course of gamete formation two special divisions occur to halve the diploid chromosome number in such a way that the chromosomes are separated from one another and distributed to four daughter nuclei. This process, called **meiosis,** actually occurs by two successive divisions called **first meiotic division** and **second meiotic division.**

Both meiotic divisions consist of a sequence of events that is conveniently described by the same terms used in mitosis: prophase, metaphase, anaphase, and telophase. However, note the differences by comparing each phase as shown in Figure 7-4 with those in Figure 8-5 (p. 200), which illustrates the process of meiosis. In mitosis one nuclear division accounts for the equal distribution of already duplicated chromosomes, whereas in meiosis two nuclear divisions distribute the duplicated chromosomes to four daughter nuclei, each containing the haploid number of chromosomes. Meiosis therefore decreases the number of chromosomes by half, segregating the members of homologous pairs into separate cells, in contrast to mitosis, which preserves the original chromosome number in succeeding generations of cells.

First Meiotic Division

The prophase stage of the first meiotic division resembles the prophase of mitosis in the usual thickening and rearrangement of the threads of the chromatin network into rod-like chromosomes. The critical difference is that in the first meiotic division homologous chromosomes pair together but do not fuse, a process known as **synapsis.** One member of each pair of homologous chromosomes in a prospective sex cell, as in all cells, is of paternal origin, the other of maternal origin (Fig. 8-1).

At about the time of synapsis it can be observed that each chromosome consists of two chromatids (chromosomal DNA having replicated in the previous S phase) so that each pair of chromosomes has four chromatids; the entire unit is called a **tetrad.** Toward the end of prophase the typical spindle apparatus is formed and the nucleoli and nuclear membranes disappear.

In the metaphase stage the entire complement of chromosomes orients itself on the equator of the spindle. Anaphase is characterized by the movement of the chromosomes away from one another toward the opposite poles of the cell. Thus the homologous chromosomes of each pair, **but not the daughter chromatids,** are separated. This differs from the anaphase of mitosis, when the daughter chromatids separate.

The telophase stage, like that of mitosis, is typified by a retransformation of the two sets of chromosomes into two haploid daughter nuclei and separation of two daughter cells, each of which then enters a period of **interkinesis,** which is an interphase **without DNA synthesis** (because the chromosomes are already replicated in the form of two chromatids).

Second Meiotic Division

The two daughter nuclei then undergo a second meiotic division, which is essentially similar to mitosis, resulting in the separation of daughter chromatids from one another to yield four haploid nuclei. Like mitosis, in the second meiotic division the chromosomes contract, thicken, and become more distinct (prophase); align themselves on the equatorial plane of the spindle (metaphase); the daughter chromatids separate and are now called daughter chromosomes (anaphase); and eventually (in telophase) four nuclei (each containing the haploid number of chromosomes) form, followed by cytoplasmic division into four daughter cells.

Gametogenesis

Meiosis takes place during the process of gamete formation, called **gametogenesis.** In males gametogenesis results in the formation of sperm and has the more specialized name of **spermatogenesis.** In females gametogenesis is the process of egg (ovum) formation and is called **oögenesis** (pronounced oh-oh-genesis). The steps in spermatogenesis and oögenesis are summarized in Figure 8-6 (p. 201).

Spermatogenesis

In mammals primitive unspecialized cells called **spermatogonia** line the seminiferous tubules of the testes (Chapter 11) and ultimately give rise to the sperm. With the onset of sexual maturity and throughout most of a male's life some spermatogonia undergo spermatogenesis, whereas others continue to divide mitotically to produce new spermato-

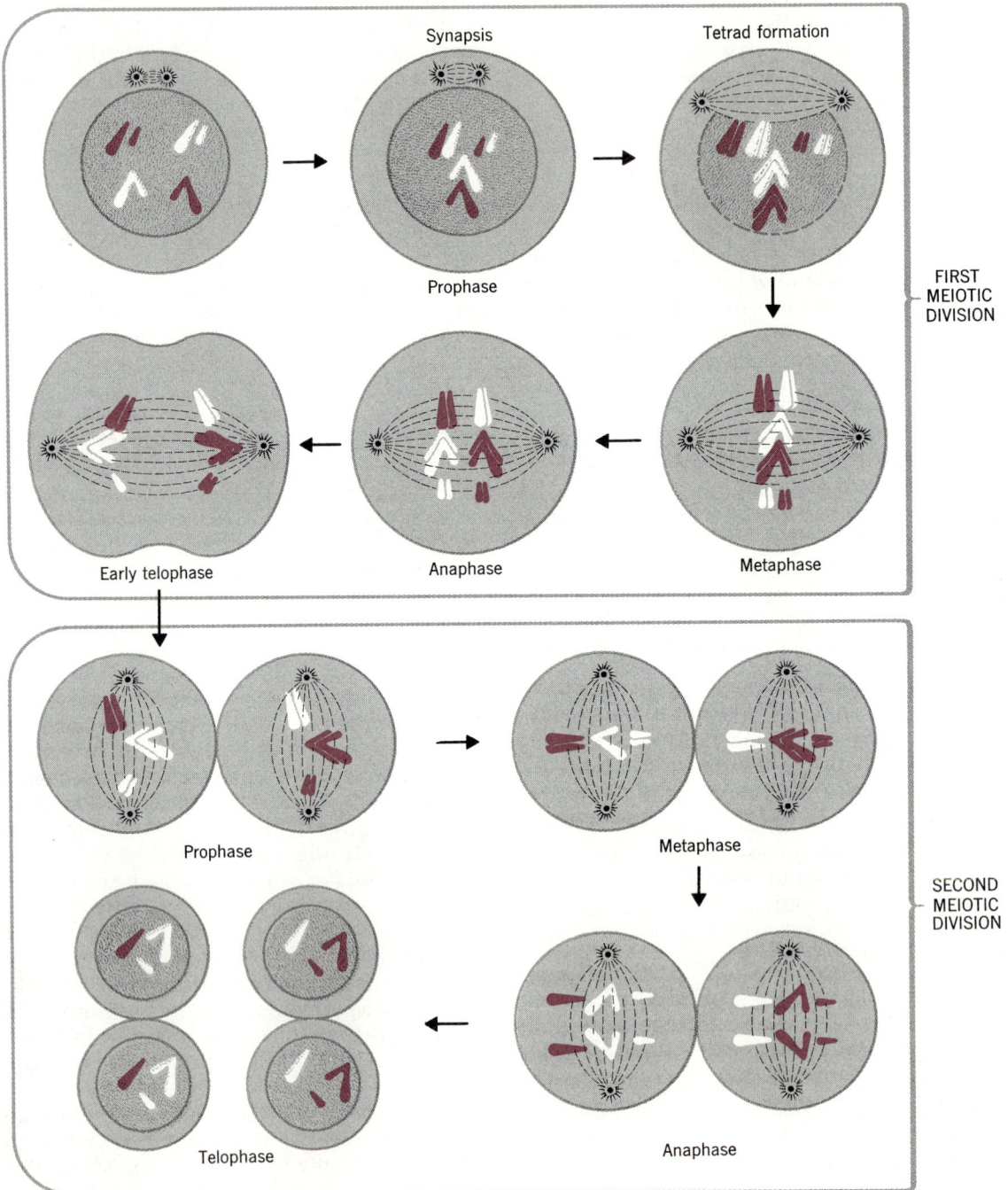

Figure 8-5 *Diagrammatic summary of meiosis in animal reproductive tissue, showing the first and second meiotic divisions, each consisting in turn of prophase, metaphase, anaphase, and telophase stages.* Note that synapsis and tetrad formation occur only in the first meiotic division, and that each mother cell ultimately yields four daughter cells having only half the number of chromosomes.

gonia. In the process of spermatogenesis the spermatogonia first form cells called **primary spermatocytes** and then undergo meiosis to yield four spherical haploid cells or **spermatids.** Each spermatid eventually develops into a functional sperm.

Sperm are produced in the testes at a rate of hundreds of thousands per hour from early boyhood throughout life. An average man produces 200 million sperm or more in a single ejaculate. Because humans have 23 pairs of chromosomes that are assorted at random in meiosis, a man can produce 2^{23} different kinds of sperm cell—that is, more than 8 million different combinations of chromosomes; any healthy male normally produces, on the average, two or three of each different kind daily.

Oögenesis

In mammalian females the corresponding primitive unspecialized sex cells, called **oögonia,** located in the surface layers of the ovaries, ultimately give rise to the egg cells, or ova. With the beginning of sexual maturation, and throughout the fertile lifetime of a human female, one (or more) oögonium per month undergoes oögenesis (Chapter 11).

The oögonium first forms a **primary oöcyte.** The first meiotic division experienced by the primary oöcyte distributes the chromosomes normally but results in a strikingly unequal distribution of the cytoplasm and stored food material, or **yolk.** Thus it forms a large cell called a **secondary oöcyte,** containing nearly all the cytoplasm and yolk, and a small at-

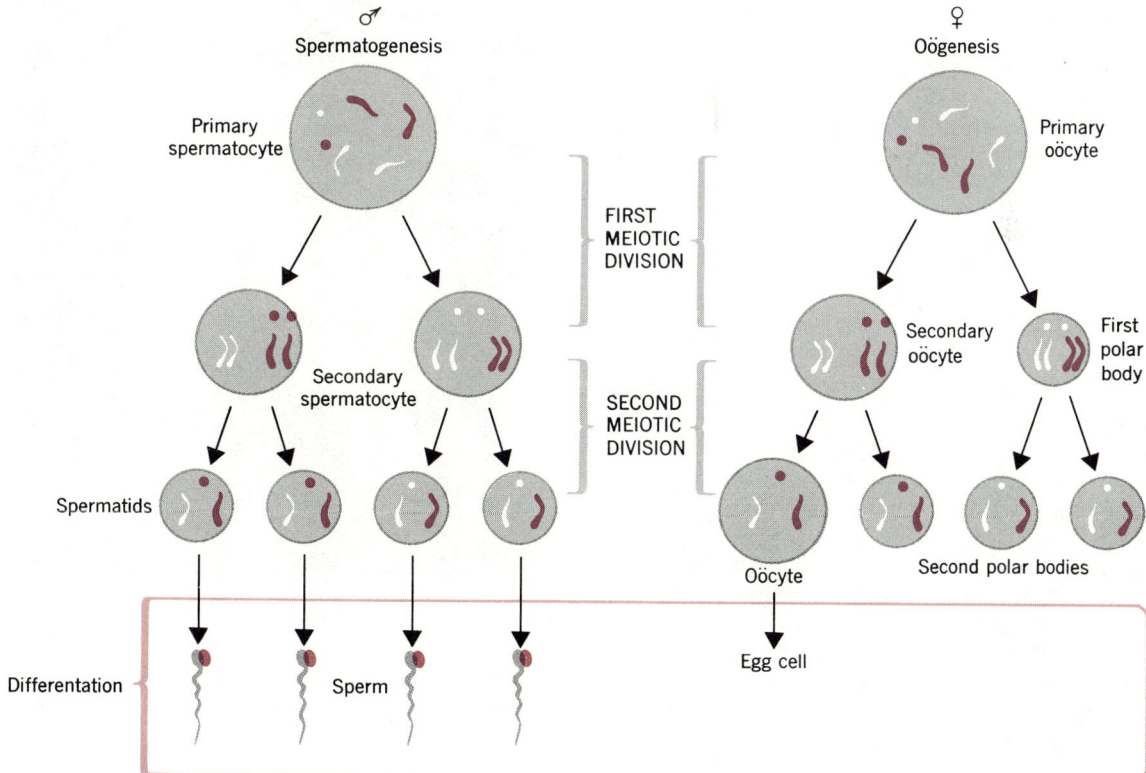

Figure 8-6 Diagrammatic summary of the processes of spermatogenesis and oögenesis in mammalian reproductive tissue.

tached cell called the **first polar body.** The secondary oöcyte then undergoes a second meiotic division to produce a large **oötid** and a **second polar body,** both of which are haploid. During this time the first polar body may have degenerated or it may have divided to form two second polar bodies. All the polar bodies, which often cluster together on the surface of a maturing egg, eventually disintegrate and disappear. With time the oötid becomes a mature ovum.

Whereas males constantly produce new gametes, females are born with several thousand potential ova already present in the ovaries; these represent the total supply throughout life. The ova in a newborn girl infant have already begun their first meiotic division. At puberty a female begins to release eggs from the ovaries, usually one each month. Occasionally two eggs are released at the same time, and if each is fertilized, both develop into infants. Such infants, known as **dizygotic** (two-egg), or **fraternal,** twins (Fig. 8-7*b*) bear no closer genetic resemblance than any brothers and sisters. **Monozygotic** (single-egg), or **identical,** twins (Fig. 8-7*a*) may develop if a single newly fertilized egg undergoes a mitotic division before its normal development begins. In this case the two infants that result have identical sets of chromosomes.

(a)

(b)

Fertilization and Subsequent Development

Fertilization is accomplished by the penetration of the sperm into the egg and the union of their haploid nuclei. The fertilized egg possesses the diploid number of chromosomes, a haploid set from the mother and a haploid set from the father. It subsequently undergoes successive mitotic divisions to form, ultimately, all the body cells, each with the identical number and kinds of chromosomes. The original chromosomes contributed by the parents do not maintain their identity as maternal and paternal chromosomes as such; in subsequent gamete forma-

Figure 8-7 *The greater similarity of features between monozygotic twins (a) as compared with fraternal twins (b) from two different eggs shows the influence of the identical genetic constitution of the former.*

tions by the new generation segregation of any given pair of chromosomes will be entirely independent of that of any other pair. We shall examine these processes in detail in Chapter 11.

Relationship of Chromosomes to Genetic Traits

So far in this chapter we have seen that a hereditary trait in a diploid organism is essentially determined and controlled by a pair of genes. Each member of a gene pair is located on one member of a particular pair of chromosomes. Any trait determined by a dominant gene will express itself, whether the dominant gene is present singly or in duplicate. Both genes for the corresponding recessive trait, however, must be present if the trait is to manifest itself. The example we selected earlier of Mendel's monohybrid crosses (Fig. 8-1) can be reexamined in the light of chromosome theory; Mendel's hereditary units are now designated as specific gene pairs on particular pairs of chromosomes.

Although Mendel's postulates form a conceptual framework within which to view genetic theory, as we proceed it will become evident that some traits are regulated by sev-

eral sets of genes instead of only one, that there are exceptions to the phenomena of dominance and recessiveness, and that the inheritance of many traits is more complex than Mendel originally perceived.

The Back-Cross, or Test-Cross

The geneticist is frequently confronted with the problem of determining the genotype of an individual from its phenotype. If he knows that the trait in question is a recessive characteristic, the problem is simple enough. The genotype for any recessive trait that appears must obviously be homozygous. (If this point is not obvious to you, reread pp. 191–193.)

To distinguish whether the phenotype for a dominant trait results from a homozygous or heterozygous genotype, a **back-cross,** or **test-cross,** is performed (Fig. 8-8). An organism that shows the dominant trait is mated with one that exhibits the corresponding recessive trait. Let us take an example among the mammals. In the inheritance of coat texture in guinea pigs, a **rough** coat (S), characterized by growth of fur in whorls to give the animal a roughened appearance, is dominant over **smooth** coat (s). To determine whether a rough-coated guinea pig is homozygous or heterozygous with regard to coat texture, it is crossed with a smooth-coated animal and

Figure 8-8 Representation of the back-cross, or test-cross, for determining genotypes of one or more pairs of genes.

Unknown						
Phenotype		Rough-coated				
Possible genotypes		SS	or	Ss		
Results of back-cross	SS x ss		or	Ss x ss		
Gametes	S	s		S	s	s
Progeny	100% Ss			50% Ss, 50% ss		
	All rough-coated			50% Rough-coated and 50% Smooth-coated		

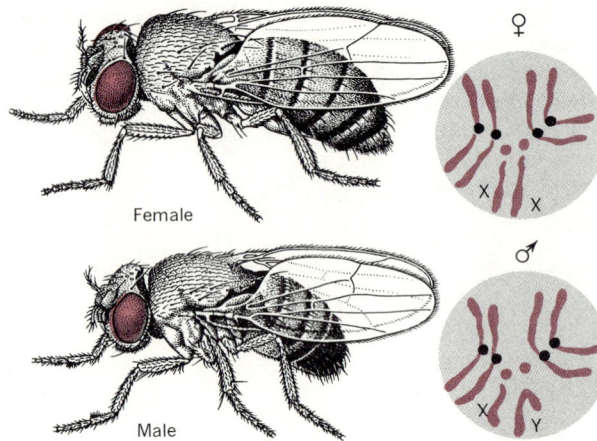

Figure 8-9 The chromosome complement of Drosophila, male and female. In addition to its autosomes the female has two X chromosomes, whereas the male has an X and a Y.

Female

Male

the coat texture in the resulting progeny is examined. As shown in Figure 8-8, if the guinea pig in question were homozygous for a given trait, it would be expected that all the offspring would display only rough coats. Alternatively, if the guinea pig were heterozygous, then 50 percent of the progeny would be likely to have rough coats and 50 percent smooth coats.

Linkage and Crossing-Over

Linkage of Genes

Genes are arranged in a linear order, and, because there are thousands of genes but only a few pairs of chromosomes for any organism, many different genes occur on the same chromosome. But it is whole chromosomes that segregate independently in meiosis. Therefore Mendel's Law of Independent Assortment must apply only to genes on different pairs of chromosomes, not to genes on the same pair. When genes occur on the same pair of chromosomes they will not assort independently of one another but together because they are connected, or **linked.**

By way of example we may look at a situation in the fruit fly, *Drosophila* (Fig. 8-9), in which the inherited gray body color (B) is dominant to black body color (b), and a long wing (V) is dominant to a stunted, wrinkled wing known as a **vestigial** wing (v). If the two pairs of genes responsible for these traits are located on different pairs of chromosomes, then in the F_2 generation of a cross between a homozygous gray-bodied, long-winged (BBVV) fly and a black-bodied, vestigial-winged (bbvv) one we should expect the typical 9:3:3:1 ratio of a dihybrid cross, based on the Mendelian principle of random assortment (Fig. 8-10).

If, on the other hand, these two pairs of genes are linked (i.e., located on the same pair of homologous chromosomes), then in the F_2 generation of the same parental cross we should obtain, as indicated in Figure 8-11 (p. 206), a ratio of phenotypes for these two inherited traits different from that predicted by the Mendelian principle of independent assortment. In fact, in this case the two gene pairs behave as if they were a single gene pair because of the linkage, yielding a phenotypic ratio typical of a monohybrid cross, 3:1.

Crossing-Over

In the cross just described there was complete linkage between two genes on the same chromosome. Actually, complete linkage is relatively rare; that it occurs regularly among male fruit flies is the exception, not the rule. Much more usual is a phenomenon referred to as **crossing-over,** which involves an exchange of genes between homologous chromosomes.

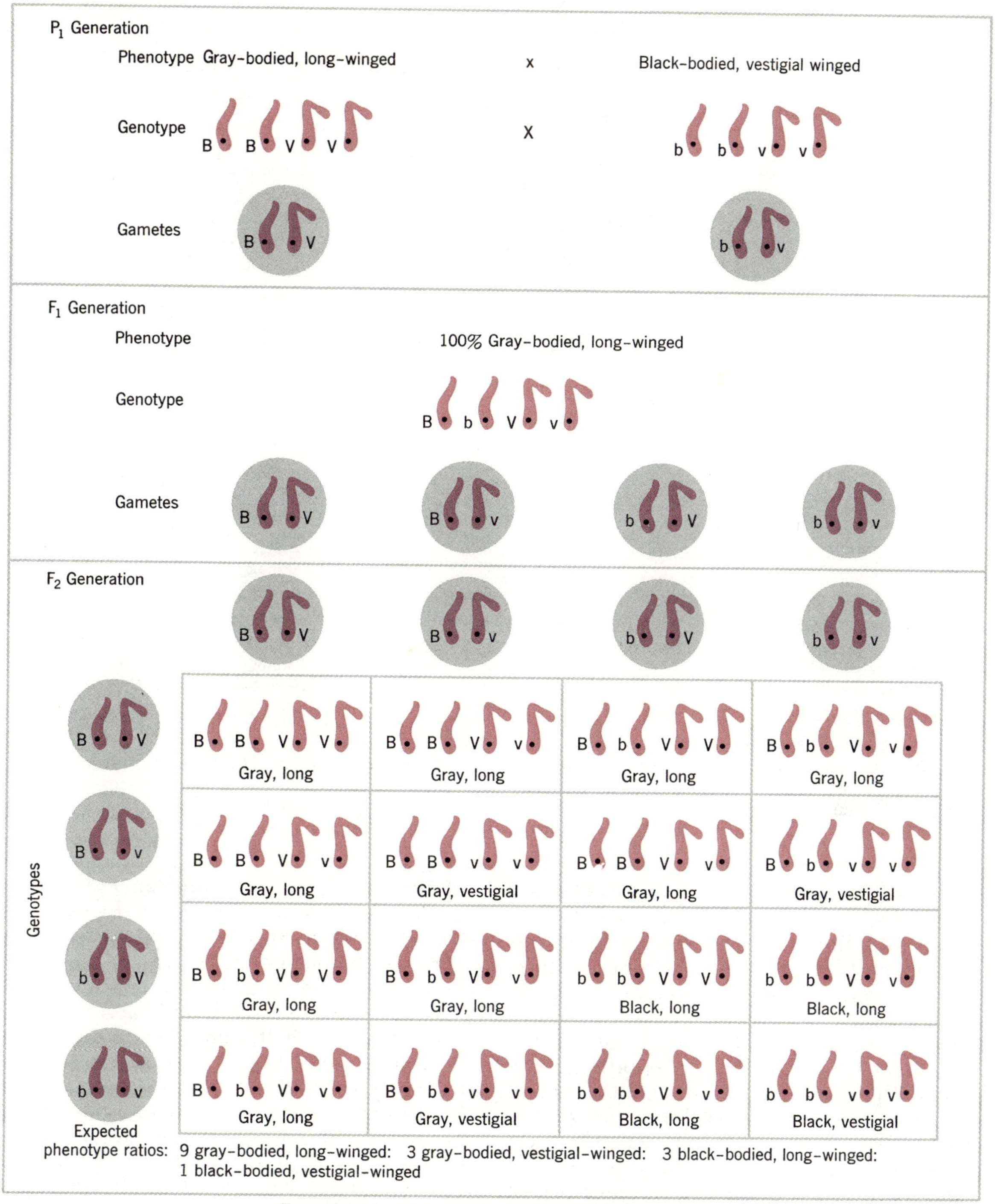

Figure 8-10 *Representation of a dihybrid cross involving two pairs of genes on different pairs of chromosomes (i.e., unlinked genes).*

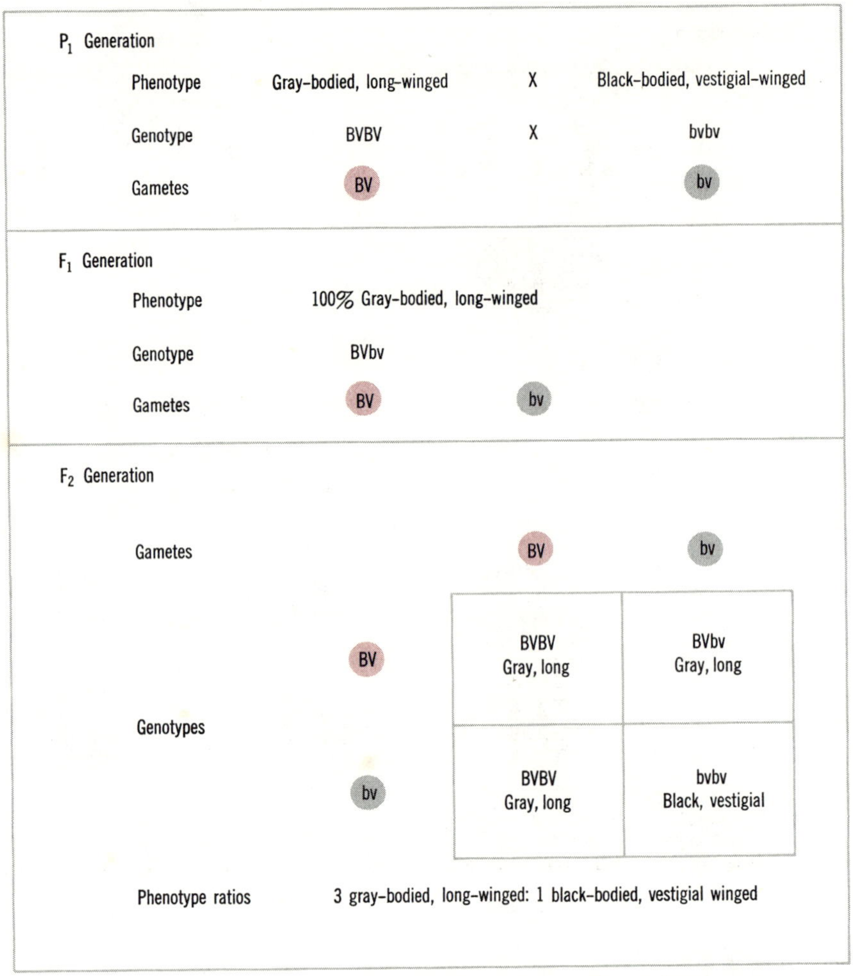

P₁ Generation

Phenotype	Gray-bodied, long-winged	X	Black-bodied, vestigial–winged
Genotype	BVBV	X	bvbv
Gametes	BV		bv

F₁ Generation

Phenotype	100% Gray-bodied, long-winged	
Genotype	BVbv	
Gametes	BV	bv

F₂ Generation

Gametes — BV — bv

	BV	bv
BV	BVBV Gray, long	BVbv Gray, long
bv	BVBV Gray, long	bvbv Black, vestigial

Phenotype ratios: 3 gray-bodied, long-winged: 1 black–bodied, vestigial winged

Figure 8-11 Representation of the same dihybrid cross as in Figure 8-10, except that both pairs of genes are located on the same pair of homologous chromosomes (i.e., linked chromosomes).

During prophase of the first meiotic division, when the chromatids align side by side in synapsis, there is often a considerable coiling of the chromatids around one another. Occasionally two of the chromatids break at a corresponding place along their length and rejoin in such a way that a portion of one chromatid is now joined with a portion of the other. This phenomenon, the coiling, breaking, and rejoining of the chromatids of homologous chromosomes, is crossing-over (Fig. 8-12). Suppose one of the chromosomes in a synaptic pair carries a gene A near one end and a gene B near the other, and its sister chromosome carries genes a and b at the allelic loci. If one chromatid of each of these homologous chromosomes breaks at a point between the two genes, the two broken chromatids may exchange parts before repair of the breaks occurs. The result of this breakage-fusion mechanism is one chromatid bearing genes A and b and another with genes a and B. If the crossing-over had not occurred, two of the four gametes produced by meiosis would have carried genes A and B and two a and b; there would have been no Ab or aB gametes. After crossing-over, however, each of the four gametes carries different genes: AB, Ab, aB, or ab. Such new combinations of linked genes are called **recombinations.**

A case of crossing-over can be illustrated

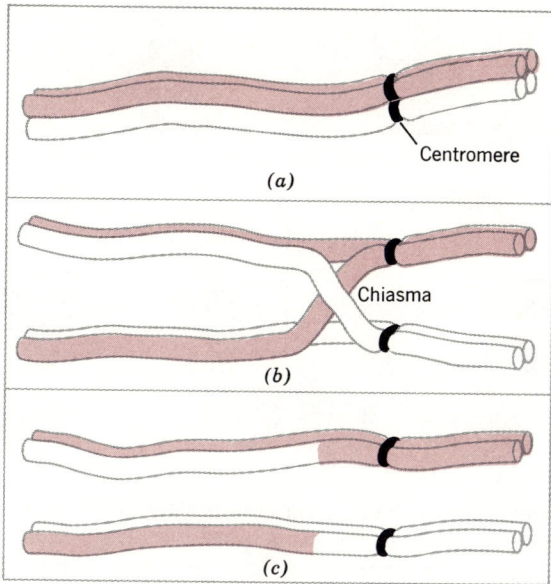

Figure 8-12 *Simplified schematic representation of the crossing-over phenomenon between nonhomologous chromatids.*

for the genes of *Drosophilia* used in the last example. A heterozygous, gray-bodied, long-winged fly was crossed with a homozygous, black-bodied, vestigial-winged fly. Because body color and wing type are linked, we would expect the 1:1 results shown in Figure 8-13. Instead, it was consistently observed that the following percentages of offspring appeared:

Gray-bodied, long-winged	41.5 percent
Black-bodied, vestigial-winged	41.5 percent
Gray-bodied, vestigial-winged	8.5 percent
Black-bodied, long-winged	8.5 percent

The simplest interpretation is that the organism produces four different gametes, as in Figure 8-14 (p. 208), in unequal frequency, instead of the theoretically expected two-gamete types. In effect, in a small but definite percentage of gametes, some crossing-over

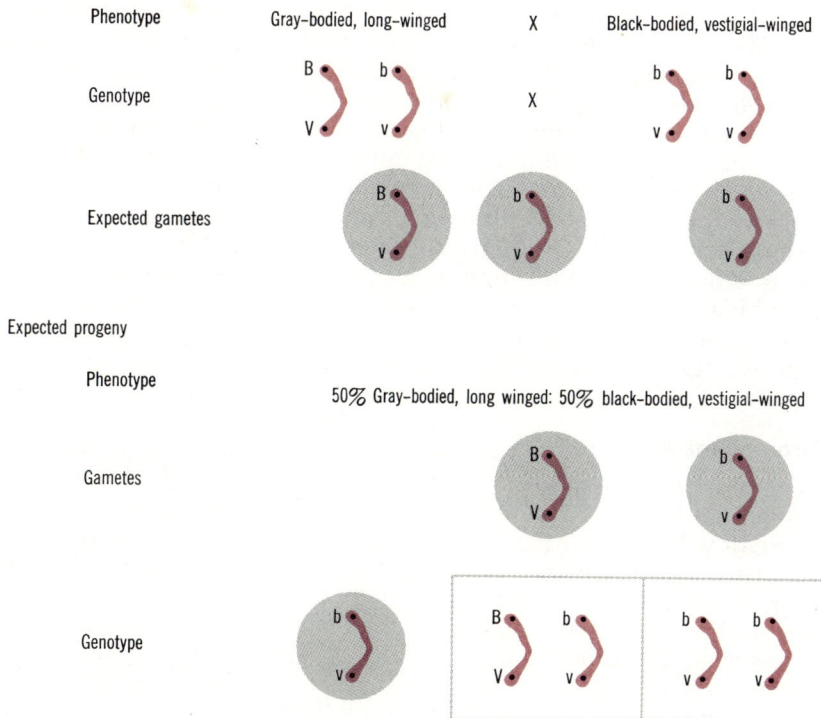

Figure 8-13 *Theoretically expected results of a dihybrid cross involving linked genes.*

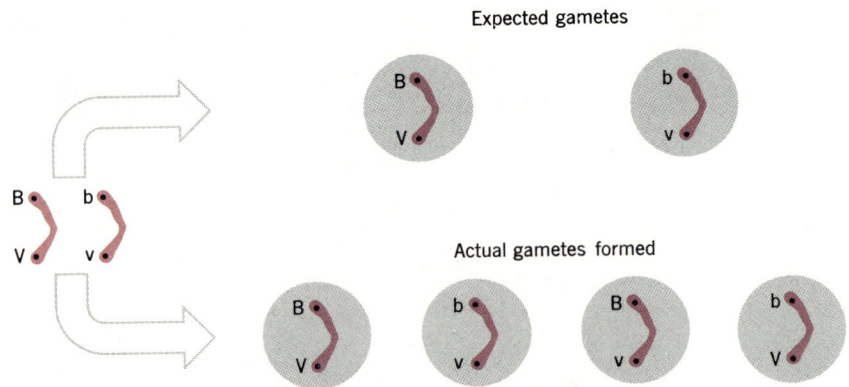

Expected gametes

Actual gametes formed

Figure 8-14 Occurrence of expected gametes and those actually formed.

between homologous chromosomes has occurred.

It appears from the example in Figure 8-13 that 17 percent of the progeny (8.5 percent gray-bodied, vestigial-winged, and 8.5 percent black-bodied, long-winged) must be derived from gametes whose chromosomes underwent crossing-over between the two genes being investigated. The remaining 83 percent of the offspring arose from gametes that did not experience an exchange of chromosome segments for these two genes.

It should be emphasized that crossing-over is a normal and widespread biological phenomenon and is not to be considered an aberrant or abnormal process. It occurs in virtually all higher plants and animals and in lower forms also. Crossing-over represents still another mode for increasing genetic recombination; from an evolutionary viewpoint it is therefore another one of nature's ways of obtaining greater variety and mixing of genes in offspring.

Chromosome Mapping

We know that breakage of a chromosome during meiosis can occur at any point along its length with about equal probability. Then the chance of a cross-over between any two genes must be directly proportional to the distance between them; the farther apart two genes are on a chromosome, the greater is the chance of crossing-over because there are more points between them at which a break

may occur. Moreover, no matter how many times a given cross is made in peas, *Drosophila*, mammals, or fungi, the frequency of crossing-over between two genes of a linked pair is always the same. Therefore geneticists have found that they can use the frequency of crossing-over as a tool for mapping the positions of genes on chromosomes with reference to one another.

The percentage of crossing-over yields no information about the absolute distance between two genes. We cannot estimate such distances in microns, for example. Cross-over frequency gives only relative distances. By convention, one unit of map distance on a chromosome is the distance within which crossing-over occurs 1 percent of the time. In the *Drosophila* test-cross described in the last section crossing-over occurred 17 percent of the time between genes B and V; we would say that these two genes are located 17 map units apart on the chromosome.

Suppose we know that three genes, A, B, and C, occur on the same chromosome, and that the cross-over frequency between A and B is 30 percent and between B and C is 12 percent. We would then conclude that A and B are 30 map units apart and B and C 12 map units apart. But how do we determine the order of these genes? The order could be A-B-C in which case the map distance between A and C would be 42 units (30 + 12). Or it could be A-C-B, so that A and C would be only 18 units apart (30 − 12). Obviously the way to decide is to do the appropriate crosses to determine the cross-over frequency between A and C. Using this procedure, by determining the frequency of cross-over between any gene and at least two other genes, it is possible to build up a map showing the linear sequence of many genes on a chromosome (Fig. 8-15).

One of the main implications of cross-over frequency and chromosome mapping is that genes are arranged in linear sequence along chromosomes, like a string of beads. We can even define a gene, purely on the basis of this kind of strictly genetic result, as **a point (or locus) on a chromosome that controls one or more characteristics of the organism.** If two traits are always linked and recombination never occurs between them, then they must be controlled by the same locus on the

chromosome—that is, by the same gene. If, on the other hand, cross-over does occur between them, even rarely, then they must be controlled by different loci—that is, different genes. According to this view a gene is the smallest unit of recombination.

It should be apparent by now that this concept of a linear sequence of genes on the chromosome, derived strictly from genetic cross-breeding experiments, bears a strong similarity to the view of the gene as related to the linear sequence of the nucleotides that make up a DNA helix (Chapter 3).

The Cytology of Genes

If hereditary traits controlled by genes on nonhomologous chromosomes segregate and sort out independently of one another,

Figure 8-15 Maps of the X chromosome and two others of the chromosomes of Drosophilia melanogaster. Only a few of the many known genes are shown. The figures indicate their position in cross-over map units from the "zero end" of the chromosome.

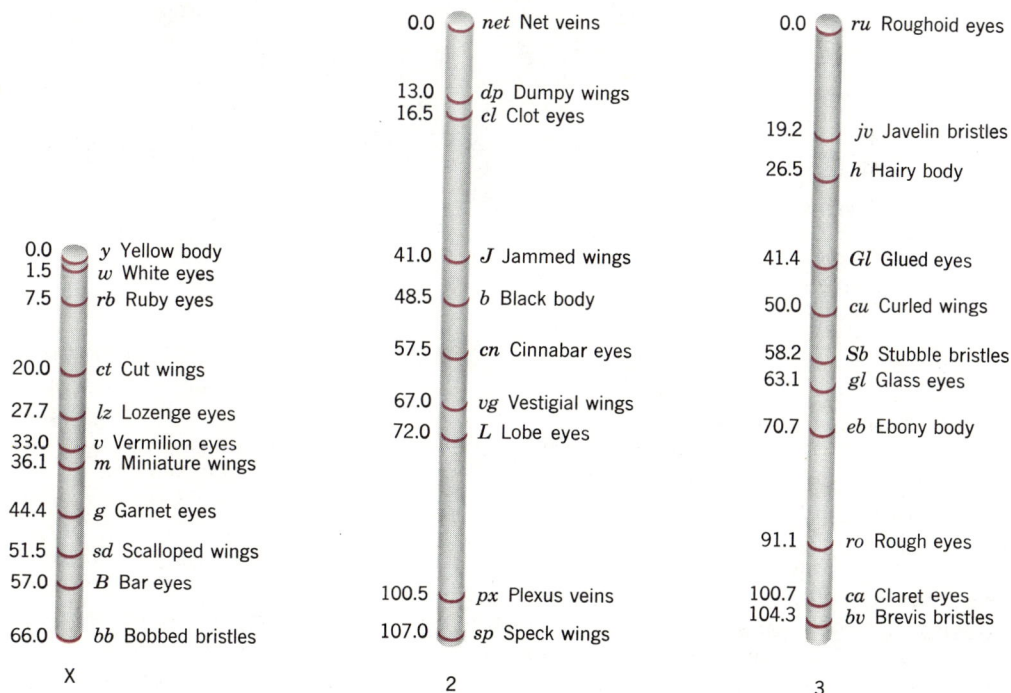

	0.0 *net* Net veins	0.0 *ru* Roughoid eyes
	13.0 *dp* Dumpy wings	
	16.5 *cl* Clot eyes	19.2 *jv* Javelin bristles
		26.5 *h* Hairy body
0.0 *y* Yellow body	41.0 *J* Jammed wings	41.4 *Gl* Glued eyes
1.5 *w* White eyes		
7.5 *rb* Ruby eyes	48.5 *b* Black body	50.0 *cu* Curled wings
	57.5 *cn* Cinnabar eyes	58.2 *Sb* Stubble bristles
20.0 *ct* Cut wings		63.1 *gl* Glass eyes
	67.0 *vg* Vestigial wings	
27.7 *lz* Lozenge eyes	72.0 *L* Lobe eyes	70.7 *eb* Ebony body
33.0 *v* Vermilion eyes		
36.1 *m* Miniature wings		
44.4 *g* Garnet eyes		
51.5 *sd* Scalloped wings		91.1 *ro* Rough eyes
57.0 *B* Bar eyes	100.5 *px* Plexus veins	100.7 *ca* Claret eyes
		104.3 *bv* Brevis bristles
66.0 *bb* Bobbed bristles	107.0 *sp* Speck wings	
X	2	3

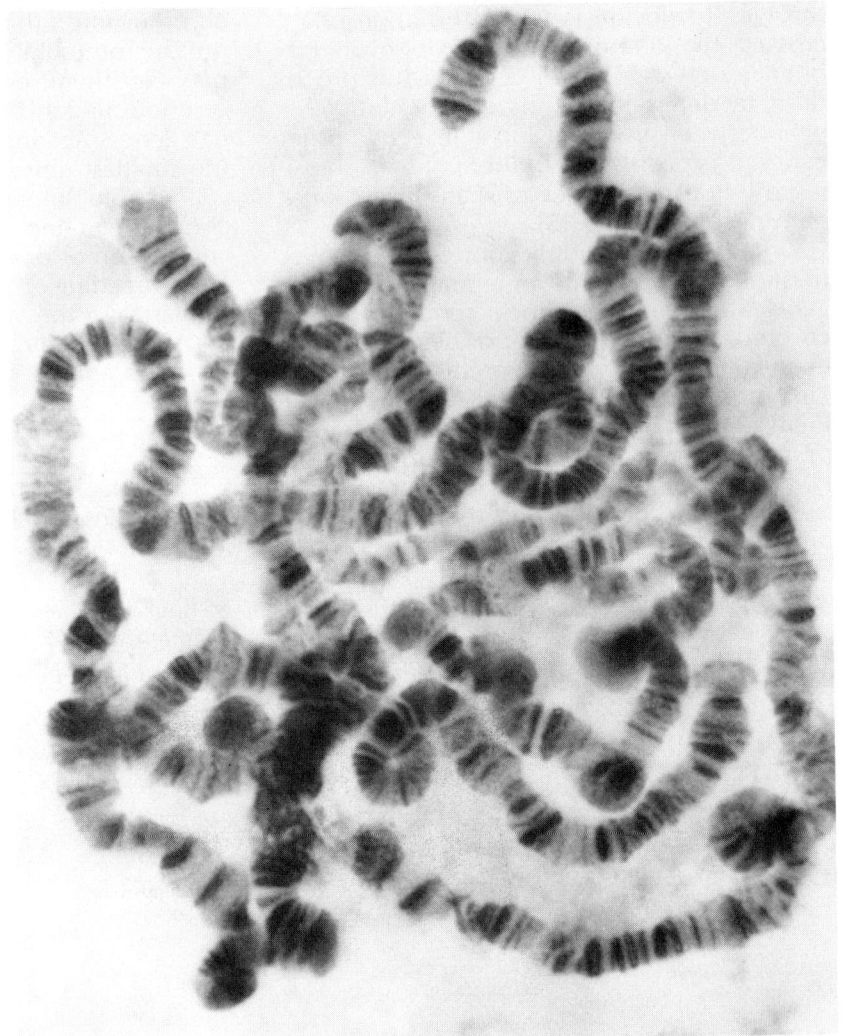

Figure 8-16 Drosophila *salivary gland giant chromosomes.*

whereas traits controlled by linked genes do not, then even without chromosome mapping it should be possible to divide a number of traits of an organism into groups in such a way that all of the members of one group (1) segregate independently of the traits in any of the other groups, and (2) exhibit cross-over recombinations with one another but not with members of other groups. In all species in which substantial numbers of hereditary traits have been studied such **linkage groups** have been found; and in all cases the number of linkage groups corresponds to the number of chromosome pairs observable under the microscope: there are four pairs of chromosomes and four linkage groups in *Drosophila melanogaster; Drosophila pseudobscura* (a closely related species) has five linkage groups and five pairs of chromosomes; corn has 10 pairs of chromosomes and 10 linkage groups; and so on.

Mendel knew nothing about chromosomes; he constructed his postulates exclusively from the results of his cross-breeding experiments. But shortly after his work, unrecognized for many years, was rediscovered in 1900, Columbia University geneticist Walter S. Sutton pointed out that Mendel's hereditary factors behaved in ways that were remarkably similar to the newly discovered chromosomes. Genes were supposed to occur in pairs; so did chromosomes. Allelic pairs of genes separated at meiosis; chromosomes were observed to do so. In 1910 Thomas Hunt Morgan (also at Columbia) recognized the first linkage groups on the chromosomes that determine sex in *Drosophila;* this added another strong piece of evidence that genes were in fact physically aligned on the chromosomes.

In the last 40 years this evidence has become incontrovertible, largely as a result of the discovery of **giant chromosomes** that occur in the salivary glands of the larvae of many flies like *Drosophila,* and in the oöcytes of many other animals, including man. Metaphase chromosomes in most mitotic and meiotic cells are generally small and difficult to work with. Details of their structure, other than the position of the centromere and a few other knobs and bumps (Fig. 8-3), usually cannot be seen. In contrast, salivary gland giant chromosomes may be 200 times larger than normal chromosomes. As in Figure 8-16, they show a striking banded pattern. In these chromosomes the original chromatin thread remains uncoiled and duplicates itself hundreds of times. As noted in Chapter 7, each thread has a beaded appearance (Fig. 5-23) with irregularly spaced, nodule-like enlargements called **chromomeres.** The massing together of many chromomeres in register is what gives a giant chromosome its banded structure.

Because differences in width, spacing, and appearance of these bands can be recognized under the microscope at ordinary magnifications, cytogeneticists who are familiar with them can identify many regions on each of the four chromosome pairs of *Drosophila.* By studying chromosome abnormalities it has been possible to determine the location of individual genes in relation to the bands. Geneticists initially thought that each band would represent one gene; we now know

that in *Drosophila* each band controls a linked family of related traits (or the synthesis of a series of related proteins). There is recent evidence that the replication of chromosomal DNA occurs in discrete steps, one step equivalent to one chromomere. Moreover, a few mutations have been described that have been related to deletions or specific alterations in a single chromomeric band. Thus cytological studies of this sort have provided a way of mapping the arrangement of genes on the chromosomes that is independent of cross-over frequency (Fig. 8-17). Such cytological maps have in general corroborated the earlier maps that had been based on the percentage of crossing-over, at least with regard to the sequence of genes on a chromosome.

The patterns of transmission of hereditary traits from one generation to the next that we see in pea plants and fruit flies and guinea pigs appear also to be present in human genetics. This fact is of some importance both from a theoretical and a practical point of view. It means, first, that the mechanisms underlying genetic relationships in man are the same as those that apply in other orga-

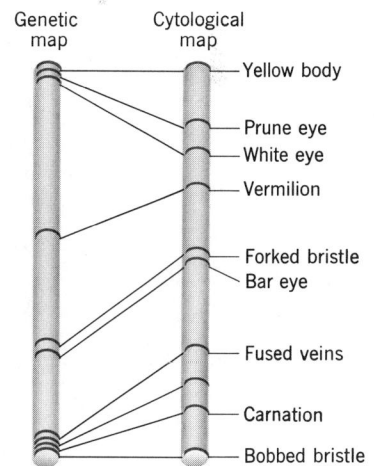

Genetic map   Cytological map

— Yellow body

— Prune eye
— White eye

— Vermilion

— Forked bristle
— Bar eye

— Fused veins

— Carnation

— Bobbed bristle

Figure 8-17   Comparison of cytological and cross-over maps of a portion of the Drosophila X *chromosome.*

211   *The Mechanics of Heredity*

nisms. If this proves valid, then we should be able to use what we have learned from the genetics of animals and plants in ways that can be of major medical significance. We know, for example, that many diseases and abnormalities are inherited. Can we learn enough to prevent or correct these maladies? To try to answer this question, let us focus our attention in the next chapter on the genetics of man.

Reading List

Burns, G. W., *The Science of Genetics—An Introduction to Heredity.* Macmillan, New York, 1969.

Crew, F. A. E., *The Foundations of Genetics.* Pergamon Press, New York, 1966.

German, J., "Studying Human Chromosomes Today," *American Scientist* (March–April 1970), **58,** pp. 182–201.

Grobstein, C., *The Strategy of Life.* W. H. Freeman, San Francisco, 1965.

McKusick, V. A., *Human Genetics.* Prentice-Hall, Englewood Cliffs, N.J., 1965.

McKusick, V. A., "The Mapping of Human Chromosomes," *Scientific American* (April 1971), pp. 104–113.

Mettler, L. E., and T. G. Gregg, *Population Genetics and Evolution.* Prentice-Hall, Englewood Cliffs, N.J., 1969.

Moody, P. A., *Genetics of Man.* W. W. Norton, New York, 1967. Chapters 2, 3, 4, 14.

Peters, J. A. (ed.), *Classic Papers in Genetics.* Prentice-Hall, Englewood Cliffs, N.J., 1959. Papers by Mendel, Sutton, Bateson and Punnett, Morgan, Creighton and McClintock, and Painter.

Sigurbjörnsson, B., "Induced Mutations in Plants," *Scientific American* (January 1971), pp. 86–95.

Srb, A. N., R. D. Owen, and R. S. Edgar, *General Genetics* (2nd ed.). W. H. Freeman, San Francisco, 1965.

Stahl, F. W., *The Mechanics of Inheritance.* Prentice-Hall, Englewood Cliffs, N.J., 1964.

Stern, H., and D. L. Nanney, *The Biology of Cells.* Wiley, New York, 1965. Chapters 2, 12, 14.

Watson, J. D., *Molecular Biology of the Gene* (2nd ed.). W. A. Benjamin, New York, 1970. Chapters 1, 7, 10.

Wills, C., "Genetic Load," *Scientific American* (March 1970), pp. 98–111.

Inheritance in Man

"*An individual is a product of his heredity and his development. His heredity is the substance he receives from his parents—his biological inheritance. The essence of this substance is the set of instructions that it contains. An ovum and a sperm, the hereditary substance of man, unite to form a fertilized ovum, the zygote. The zygote contains all the instructions required to produce another human being.*"

J. A. MOORE, 1963, *HEREDITY AND DEVELOPMENT*

Although the basic principles laid down by Mendel are the very foundation of genetics, subsequent experiments by other investigators have extended and modified Mendel's conclusions. These researchers have revealed that the relationships among genes and their influence in determining phenotypes may be different and more complex in many instances than had been supposed. Observed traits are not often so clearly dominant or recessive as in Mendel's pea plants; some traits appear to blend or show intermediate phenotypes. Many genetic traits in man, such as skin color or intelligence, are not controlled by a single gene but are subject to the effects of many genes.

Linkage Groups and Mutation

One exception to Mendel's conclusions that we have already discussed is linkage groups. Only genes on different chromosomes sort independently; genes on the same chromosome are distributed together except when recombinations occur through crossing-over (pp. 204–208).

A second exception is mutation. One of the three scientists who rediscovered Mendel's principles around the beginning of this century was Hugo DeVries, a Dutch botanist who worked mainly on heredity in the evening primrose. He found that the transfer of traits from one generation to the next in the primrose was generally just as orderly and statistically predictable as in the garden pea. Occasionally, however, a plant appeared in one of his strains with a characteristic that was not present in either of its purebred parents or in any of its previous lineage.

DeVries guessed that such new traits appeared as a result of some change in a gene, and that the characteristic that was the manifestation of that altered gene was passed along like any other hereditary trait. DeVries referred to these heredity changes as mutations (p. 190). He listed some 2000 such changes in the primrose. We now know that most of those were not actually mutations, but **recombinants;** a recombinant is an organism that carries a genetic recombination. It is one of the paradoxes of science that a profound and valid insight can sometimes be gained on the basis of incorrect observation.

Hermann J. Muller, three decades later, reasoned that if natural mutations represent miscopies of the DNA sequence, short-wavelength radiations such as X-rays, which can break chemical bonds, should be able to produce similar alterations in gene structure. Indeed, he found that irradiating *Drosophila* sperm with high doses of X-rays resulted in a frequency of mutations several thousand times greater than would be found in the normal population.

It was soon discovered that other forms of radioactive emission, ultraviolet light, and some chemicals could also act as **mutagens,** agents that produce mutations. Moreover, it has since been demonstrated that even small doses of radiation have similar effects, and that all organisms are susceptible. These discoveries have been of enormous value to geneticists in analyzing the mechanisms of gene function and evolutionary change. But they have been of even greater value in alerting us all to the potential danger of radioactive isotopes. Radiation, we now know, can be damaging to organisms exposed to it, and to the future generations of those organisms. We do not yet know whether any detectable dose of radiation is small enough to produce no change in genetic makeup. Therefore many scientists and others are seriously concerned about any activity that increases exposure to radiation; not only fallout from testing of nuclear bombs, but by-products from nuclear power plants, atomic engines, medical X-ray equipment, and the like. Decisions that we make now and will be making in the next few years concerning these issues can have profound effects on humanity for many generations to come.

Recent work on mutation has shown that at least in some species—*Drosophila,* corn, and certain bacteria for example—genes called **mutators** are present. These are mutant alleles that can cause an increase in mutation frequency of other genes by as much as a hundredfold. This discovery is so new that its significance for genetics and evolution is not yet fully understood. We do not know whether mutator genes exist in man.

The mean mutation rate among humans has been estimated at about 4×10^{-5} (4/100,000). This means that about four gametes in every 100,000 carries a newly mutated gene. Out of the 200 million sperm an average man produces on any given day, about 8000 would have new mutants—the vast majority of which would produce nonviable or defective offspring. About 6 percent of all newborns have been found to have defects partially or wholly of genetic origin. Perhaps half of these (3 percent) result from new mutations. Until recently the main man-made source of damage to the genes was the increasing exposure of the population to radiation. Now, in addition, many scientists have become concerned about the variety of mutagenic chemicals beir introduced into our environment as pesticides, industrial by-products, agents that leach out of plastic containers, air contaminants, and so forth. Clearly our desire for technological progress, coupled with our lack of understanding of many basic biological mechanisms, can produce undesirable and irreversible effects on humanity—and perhaps have already done so (see Chapter 21).

Incomplete Dominance

In Mendel's experiments genes were clearly dominant or recessive, and the heterozygote Aa was indistinguishable in the phenotype from the homozygote AA. We now know that this "rule of dominance" is the exception rather than the rule; usually the heterozygous phenotype is intermediate between the two parents, displaying a mixture of traits determined by both alleles. This phenomenon is called **incomplete dominance.** Incomplete dominance is classically illustrated by a cross between red-flowered and white-flowered snapdragons, shown in Figure 9-1. The resulting F_1 generation is made up neither of red-flowered nor of white-flowered but entirely of pink-flowered plants. The F_2 generation resulting from a cross between two pink F_1 plants has plants with white, pink, and red flowers in the ratio 1:2:1. These data agree completely with Mendel's first law of segregation, except that the condition of incomplete dominance prevails.

An example of incomplete dominance in humans is the gene for sickle-cell anemia. In the homozygous condition the gene causes malformation and improper functioning of the red blood cells, which tend to clump and rupture easily, producing painful circulatory obstructions and severe anemia. The heterozygote exhibits an intermediate phenotype; the red blood cells show the characteristic sickle shape, but only very mild symptoms of the disease occur. Sickling must be dominant, because it is expressed in the heterozygous condition, but incompletely so, as demonstrated by the difference in expression between the homozygote and heterozygote.

Dominance is best understood by considering the enzymes that are usually the primary products of genes (Chapter 4). In the snapdragon, for example, suppose the **wild-type gene**—the unmutated allele that occurs most often in the normal population as it exists in nature—produces an enzyme necessary for the production of red pigment in a flower. Then a mutation of the gene may, in a homozygote, result in a reduced flower color or no color at all. This loss of pigment in the mutant may result from one of three causes: (1) the mutant gene may produce no functional enzyme; (2) it may produce a less effective enzyme than the wild-type gene; or (3) it may produce a product that actively

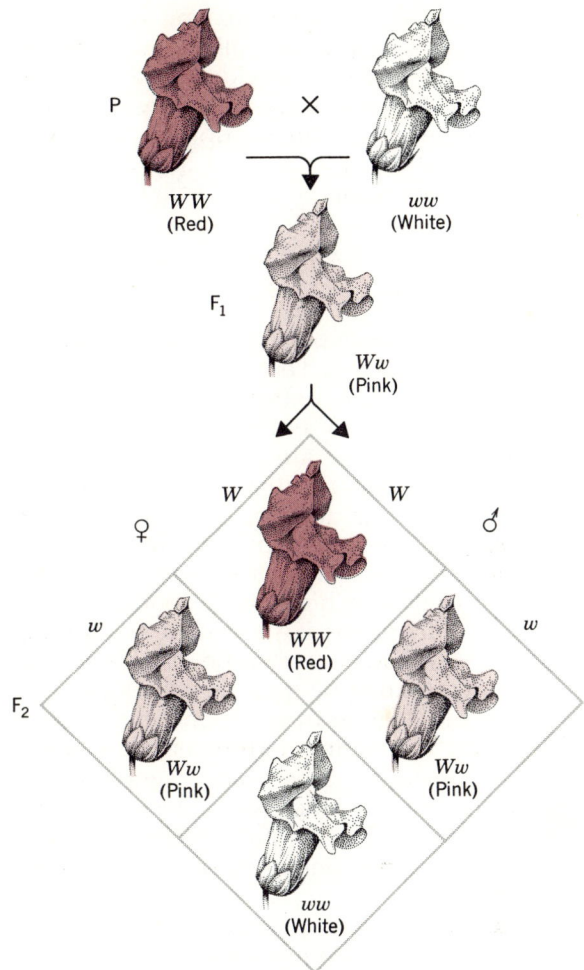

Figure 9-1 A cross between a red (WW) and a white (ww) purebred snapdragon. Unlike the simple Mendelian crosses described earlier (Fig. 8-1), the hybrid (Ww) is an intermediate pink color. The results of the F_2 generation, shown in the Punnett square, demonstrate that the colors sort themselves out in accordance with Mendel's 1:2:1 ratio, indicating that the traits did not blend but that the pink color is an example of incomplete dominance.

blocks pigment production. In the first two cases the gene will be recessive or may show only a small effect in the heterozygous condition. This is because the heterozygote will have one dose of normal enzyme, which may be sufficient to produce all or most of the pigment necessary to provide the flower with full color. In an actively inhibitory mutation the mutant gene will be dominant; it will express itself completely or nearly completely in the heterozygote.

Multiple Genes and Multiple Alleles

Unlike the examples of classic Mendelian inheritance discussed earlier, in which a given phenotypic trait is controlled by a single gene that can occur in one of two alternative forms, most traits are affected by several genes. Very often a particular characteristic is determined in an **additive** fashion by several or many pairs of independent genes at different loci on the same or different chromosomes, each of which alone produces only a small effect. This is spoken of as **multiple gene inheritance.** It can be expected that traits determined by multiple genes will display many degrees of variation between two extremes, depending on the genetic makeup of an individual, rather than a simple dominant-recessive relationship. This holds for such human characteristics as height, weight, intelligence, skin and eye color, and special abilities.

Moreover, many genes can exist in more than two allelic forms. A given trait may be controlled by only one gene locus on one chromosome pair, but that locus may be occupied by the gene in any one of several forms, a series of **multiple alleles.** Although at any one time in a diploid organism only two alleles can occupy the homologous loci in a chromosome pair, the dominance of a phenotypic trait will clearly depend on which two allelic forms of the multiple series happen to be represented. Genetic studies with *Drosophila* have demonstrated that there are at least a dozen different alleles of one eye-color gene (Fig. 9-2). In man the inheritance of blood types designated A, B, AB, and O is a well-established case of multiple alleles, in this instance three.

Human Genetics

The same principles of genetics that apply to *Drosophila*, peas, and all other organisms also govern the transmission of heritable traits in man. Most fundamental problems in human genetics have been investigated in other organisms that are more suitable to the experimental approach, because the study of heredity in man is encumbered by several serious hindrances and difficulties. Human beings have comparatively few offspring, and their generation time, averaging 25 to 30 years, is too long, at least as far as the experimental geneticist is concerned. In addition,

Figure 9-2 *Some eye colors in* Drosophila.

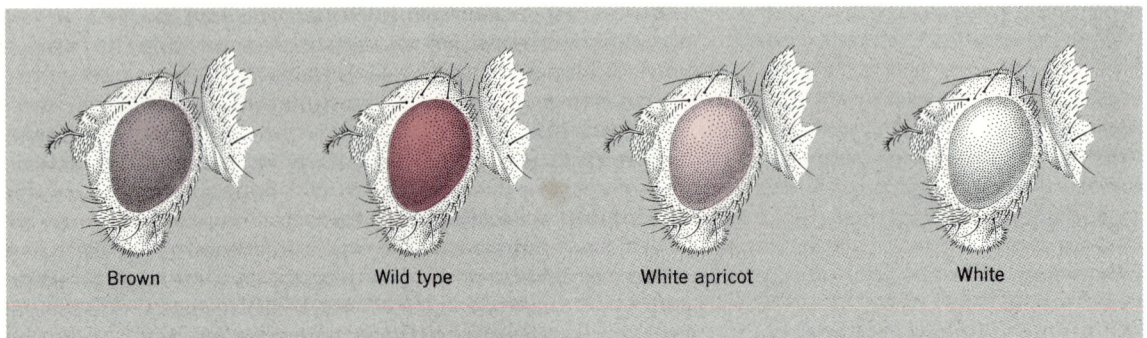

Brown   Wild type   White apricot   White

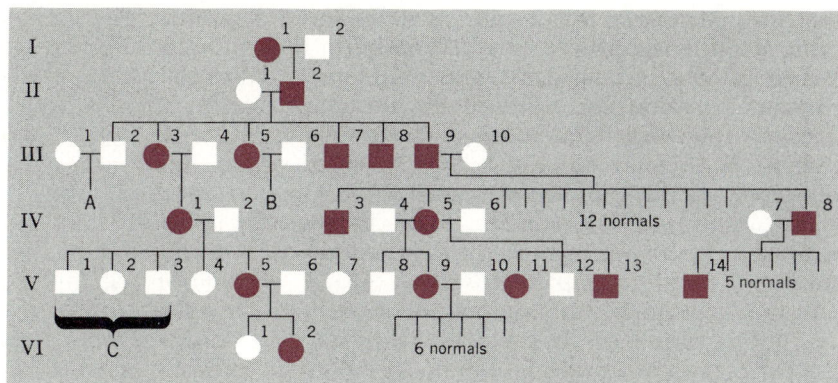

Figure 9-3   Inheritance of Huntington's chorea. Squares represent males; circles females. Colored symbols indicate individuals affected by the disease. A horizontal line connecting a circle to a square represents a mating; a vertical line extending downward connects to the offspring produced by that mating. Each horizontal row or tier of circles and squares is a generation. The generations are identified by Roman numerals, and each individual in a generation is given an Arabic numeral (shown slightly above and to the right of the square or circle). Some of the individuals in generations V and VI who were diagnosed as normal as children or young adults may have developed the disease later, after this study was terminated.

for obvious reasons, man would not lend himself well to planned matings, to genotype standardization, or to maintenance in a uniform environment.

On the other hand, human beings offer certain distinct advantages as experimental material for genetic investigations. Their long life span, for example, has permitted the study of hereditary abnormalities that develop late in life, phenomena known in man but unknown in most other animals.

Various techniques have established conclusively the existence of numerous gene-controlled traits in man. Early approaches to the study of inheritance in man were based on pedigree histories of families (next section), usually with emphasis on conspicuous or unusual traits. Recent studies of identical twins and other multiple births (as a means of attaining some degree of genotype standardization) have yielded new data on the inheritance of human characteristics, especially relating to intelligence, special aptitudes, temperament, and certain diseases.

Human Pedigrees

The classic and most common genetic technique applied to humans involves tracing the inheritance of a trait through generations of an ancestral line. Such a line is called a family tree, or more properly, a **family pedigree.** Figure 9-3 is a pedigree of a Vermont family showing the distribution of **Huntington's chorea,** a disease in which mental deterioration is accompanied by uncontrollable, involuntary muscular movements. Although the disease always leads to insanity and death, it frequently is not manifested until adulthood (usually age 30 to 60)—that is, until after marriage and childbearing.

According to the conventions for such pedigrees, circles represent females, squares represent males. The unshaded squares and circles designate people who are normal or unaffected by the characteristic being studied, in this case Huntington's chorea; the colored squares and circles indicate affected individuals.

In generation I of this pedigree a woman (I-1) who later developed Huntington's chorea married a man (I-2) who was unaffected. They had one son (II-2) who showed the trait. He married a normal woman (II-1). Individuals 2, 3, 5, 7, 8, and 9 of generation III represent the six children of that couple; individuals 1, 4, 6, and 10 are their spouses. Note that five of the six children (III-3, III-5, III-7, III-8, III-9) manifested the disease. Couple A in generation III had nine normal children and 53 normal grandchildren. Couple B had 11 children, three with Huntington's chorea. Individual IV-1, who later turned out to have the disease, married a normal wife. Of their six children only one (V-5) was affected. Individuals V-1, V-2, and V-3 (labeled C) each married and produced a total of 19 children, all normal.

What genetic meaning can we derive from such a pedigree? First, it is obvious that no individual shows the trait unless one of his parents is affected; but a mating between an affected and a normal individual can produce both normal and affected progeny. Second, the trait never skips a generation. We saw in Chapter 8 that recessive characteristics tend to skip generations (as heterozygotes), whereas dominant traits do not. Third, the trait is about equally distributed between male and female progeny. It is therefore not sex-linked (see p. 226). We may conclude that this disease is controlled by a dominant gene, and that each of the affected individuals is a heterozygous dominant.

A Cytologic Demonstration of Mendelian Heredity in Man

In any small sample of blood taken from a person's vein some white blood cells (leukocytes) will be found in the metaphase stage of mitosis. Such cells can be pressed down on a microscope slide to form a chromosome squash (Fig. 8-3) and stained with dyes specific for DNA so that the metaphase chromosomes can be observed and photographed. The individual chromosomes may be matched in pairs of homologues and arranged in order of descending length to form a karyotype of that individual.

Victor McKusick of Johns Hopkins University has made a karyotypic analysis of the Amish community of Lancaster County, Pa., a community that has become relatively inbred and genetically stable because the religious traditions of the sect prohibit marriage outside the group. In about 10 percent of the Amish people McKusick found that both homologues of chromosome number 14 were characterized by a knob, or **satellite,** on one end, which was conspicuously larger than that of a normal individual (Fig. 9-4). By following the transmission of this so-called marker chromosome, McKusick was able to demonstrate the operation of Mendel's laws of segregation and independent assortment in man.

Mendel's Law of Segregation of Alleles is illustrated in Figure 9-5a (p. 220). The somatic chromosomes of the parents, shown in metaphase, undergo two meiotic divisions in the production of eggs and sperm. The heterozygous parent (Fig. 9-5a) produces two types of haploid gamete: those with marker chromosome 14 and those with normal chromosome 14. When the heterozygous parent is married to a normal person, normal gametes have an equal probability of joining with a marker or a normal gamete from the heterozygotic mate, and therefore heterozygous offspring should be produced in a ratio of about 1:1. Among 82 offspring of matings between heterozygotes and normals, McKusick reported that 47 showed heterozygous karyotypes. Among matings between two heterozygotes (Fig. 9-5b), because each parent can produce either class of gamete, three types of offspring can result: homozygous for marker chromosome 14, heterozygous for marker chromosome 14, and homozygous for normal chromosome 14, in the ratio of 1:2:1. Among the Amish population McKusick studied, he observed such an F_1 cross only once. This couple had eight children, among whom the exact Mendelian ratio (2:4:2) was realized!

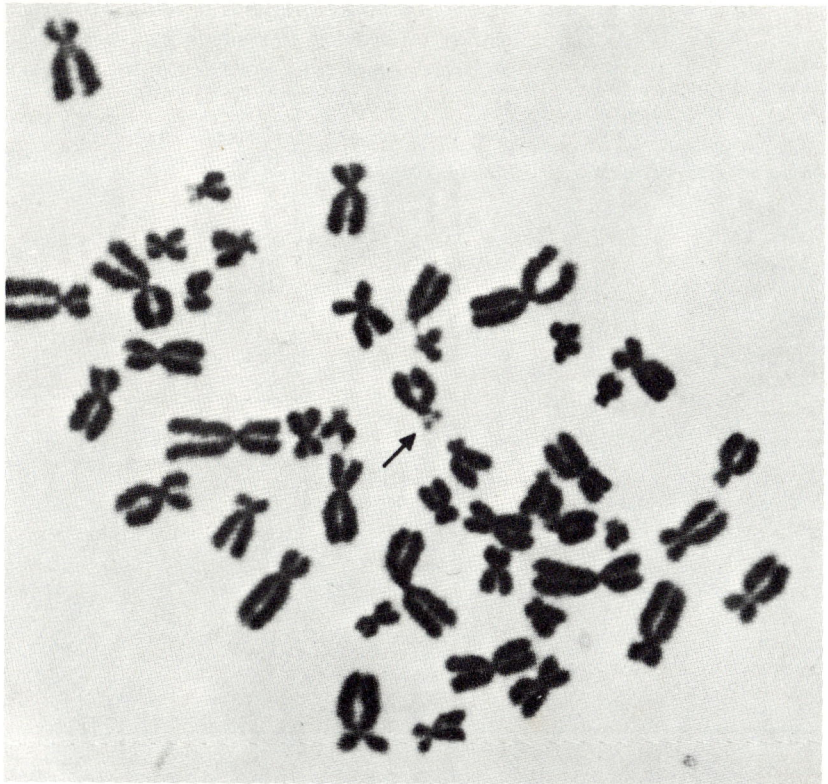

Figure 9-4 Metaphase chromosomes of an Amish man. Chromosome number 14 (arrow) exhibits a giant satellite (knob) on each homologue. Compare this figure with Figure 8-3.

Mendel's Law of Independent Assortment, which refers to traits on nonhomologous chromosomes, that is, **nonalleles,** can also be illustrated by the Amish marker chromosome if the hereditary passage of that chromosome is followed in relation to another cytologically identifiable chromosome such as the male Y chromosome. If a male marker heterozygote mates with a homozygous normal female, four types of sperm will be available to fertilize a single class of egg: marker X, marker Y, nonmarker X, and nonmarker Y (Fig. 9-6, p. 221). Therefore four classes of offspring should result, all in equal proportions: female heterozygote, male heterozygote, female homozygote, male homozygote. Among 41 children actually born from such matings, McKusick did indeed find all four classes in approximately equal numbers.

The Genetics of Sex

The Sex Chromosomes

Just before the turn of the century cytologists noted differences between the chromosome squashes of males and females, and began to suspect that they were related to sex determination. In 1905 the story was worked out by Miss Nettie Stevens and Edmund B. Wilson. In grasshoppers the female has an even number of chromosomes; the male has an odd number, one less than the female. One of the chromosomes that is paired in the female has only a single representative in the male. Miss Stevens termed that the X chromosome.

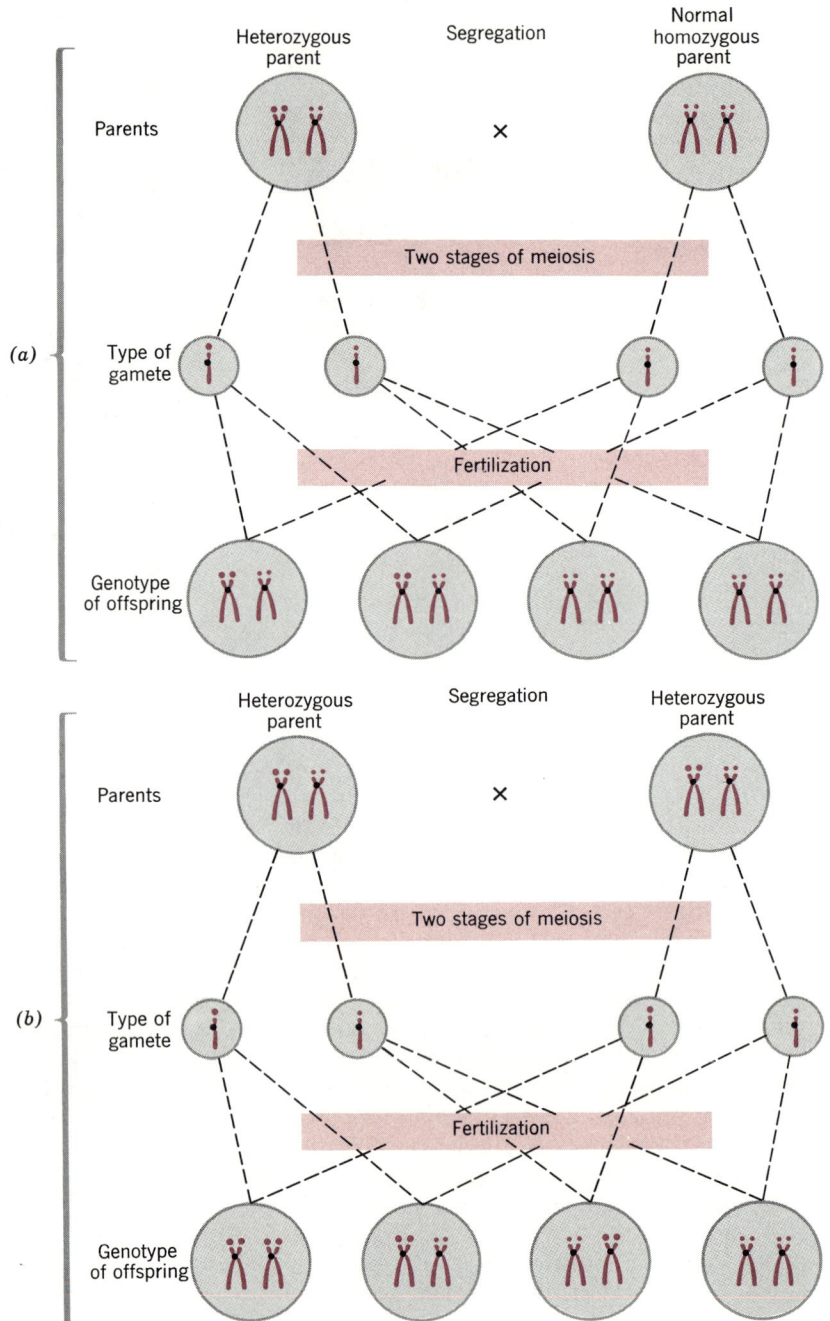

Figure 9-5   (a) Segregation of alleles, as demonstrated with the Amish marker chromosome (carrying a giant satellite) in the mating of a heterozygote and homozygote. (b) Segregation in a cross between two heterozygotes (F_1 intercross).

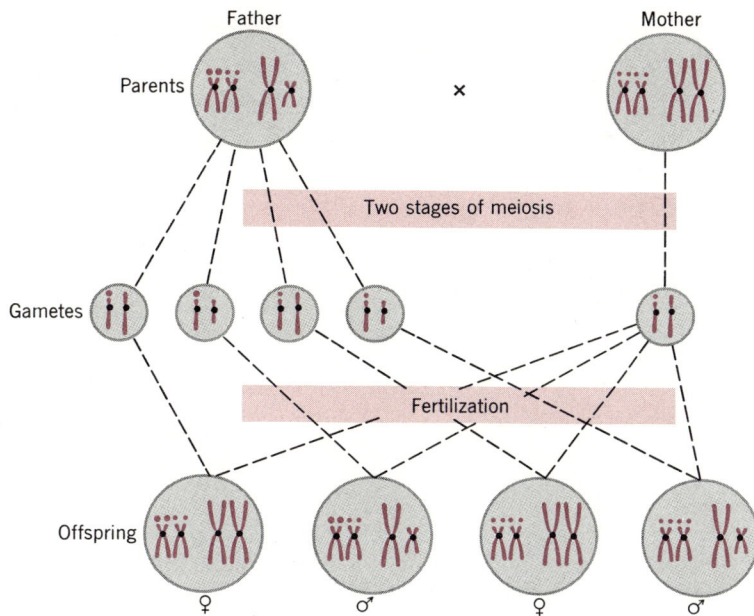

Figure 9-6 Independent assortment of marker chromosome 14 and the sex chromosomes in the mating of a heterozygous Amish man and a homozygous woman.

When eggs are formed in a female the members of each pair of chromosomes separate; one homologue goes into each egg, so that every egg carries an X chromosome. When a male forms sperm, however, half of the sperm cells receive an X chromosome but the other half do not; that is, two classes of sperm are formed. When an egg happens to be fertilized by an X bearing sperm, the zygote is female (XX); if it is fertilized by a sperm lacking the X chromosome, the zygote is male (X0) having X derived only from the egg. The sex of an individual appears, therefore, to be determined by a chance event, the type of sperm that happens to fertilize the egg from which the individual is derived. In other animals—Drosophila and most mammals, for example—the male does have a homologue for the X chromosome, but that chromosome is smaller than the X or different in shape (Fig. 8-8). Miss Stevens first noticed such an unlike partner of the X in the common mealworm, and dubbed it the Y chromosome. We now refer to X and Y as **sex chromosomes,** in contrast to the other chromosomes, which are known as **autosomes.**

In man the diploid number of chromosomes is 46 (Fig. 8-4): 22 pairs of autosomes plus a pair of sex chromosomes (XX in the female and XY in the male). In the male the Y chromosome is only a fraction of the size of its X homologue. In humans, just as in Miss Stevens' grasshoppers, each egg carries an X chromosome, but half of the sperm receive an X and the other half a Y chromosome (Fig. 9-7, p. 222). Union of an X-bearing sperm with an egg yields a female child (XX); fertilization by a Y-bearing sperm gives rise to a boy (XY). Both sexes always receive an X chromosome from the mother.

Sex Determination

From what has already been said, it appears that the sex of an individual is determined by its sex chromosomes. But is a man a male by default? Does the presence of two X chromosomes determine a female and one a male, or does the Y chromosome play an active role in determining maleness? In humans the answers to these questions have come largely from the study of sex anomalies.

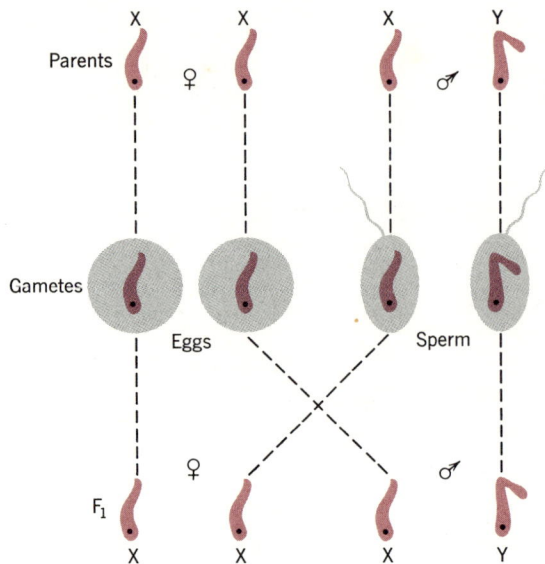

Figure 9-7 How the sperm cell determines the sex of a child. During meiosis in the mother, the egg (and each polar body) receives an X chromosome. The father's sperm will contain either an X or a Y chromosome in equal numbers. Whether the zygote becomes female (XX) or male (XY) will depend on which type of sperm fertilizes the egg.

In the late 1940s in Ontario Murray Barr noticed that the interphase nucleus of many cells contained a darkly staining body (Fig. 9-8). This mass of chromatin was unique because it stained so strongly in the interphase cell when the rest of the chromosomal material was too finely dispersed to be visible in a light microscope. It was also notable because it was present only in females; cells from normal males never contain a **Barr body.** This observation provided a means for distinguishing male tissue from female tissue, even when the X and Y chromosomes themselves were not visible, as in metaphase.

When the Barr technique was applied to sex anomalies it was discovered that certain persons had sex chromatin inappropriate to their phenotype. For example, individuals with an anomaly called **Turner's syndrome** (Fig. 9-9, p. 224) have female genitalia (va-

gina, clitoris, etc.) and breasts, but their cells do not exhibit Barr bodies. They have only 45 chromosomes, with an X0 sex chromosome. Persons with **Klinefelter's syndrome** (Fig. 9-10, p. 225) have a Barr body in their cells but develop male genitals. The karyotype of these individuals shows 47 chromosomes, with two X chromosomes and a Y. The sex chromosome constitution is said to be XXY.

A **syndrome** is not a single disease, but refers to a group of symptoms or disorders that usually occur together. In cases like Turner's and Klinefelter's syndromes the broad range of defects results from the massive chromosomal abnormality.

Figure 9-8 Interphase nuclei from cells of a normal man (a) and woman (b). The darkly stained spots or Barr bodies just inside the nuclear membrane (arrows) in the female represent the condensed chromatin of one X chromosome. (c) Nucleus of an oral mucosal cell of a mentally retarded boy with chromosome complement 49, XXXXY. Note that it has three Barr bodies, suggesting that the number of Barr bodies is one less than the number of X chromosomes.

(a)

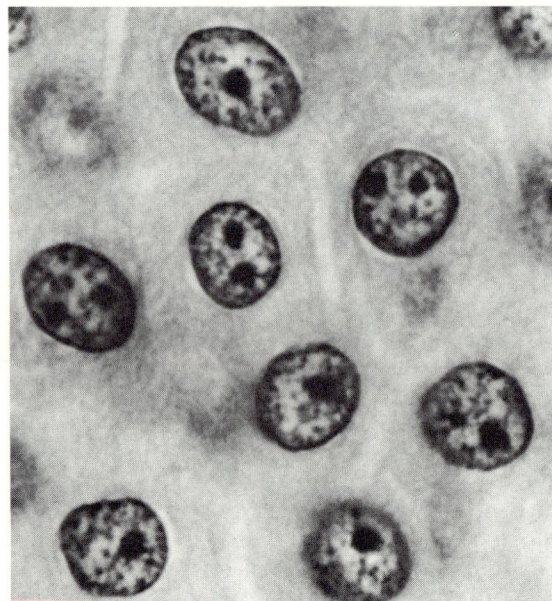

Other sex-chromosome anomalies have been discovered in recent years. One of these, termed **triple X,** occurs in persons who are phenotypically female. Their nuclei exhibit two Barr bodies, and their metaphase cells show three X chromosomes and no Y. Another, also with two Barr bodies, occurs in males, with the sex chromosome constitution XXXY. There are even persons with four (Fig. 9-8c) or five X chromosomes (who show three or four Barr bodies, respectively). Many of these patients are mentally retarded and have been detected by sex-chromatin surveys in institutions for the retarded. The brain is such a delicately balanced mechanism that chromosomal aberrations easily affect it. This is illustrated in a subtle way in so-called XYY males. These are often tall, aggressive men with an extra Y chromosome, who are frequently found in prisons or other penal institutions because they are incapable of normal social adjustment.

How can a person develop an extra X chromosome? Gains or losses of a single chromosome occasionally occur during meiosis, by a process known as **nondisjunction.** Recall that in the first meiotic division the homologous chromosomes undergo synapsis (i.e., they pair off), form tetrads, and then each pair of chromatids separates (Chapter 8). In the second meiotic division the chromatid pairs are in turn separated and distributed as a haploid set to each gamete. Sometimes in this process the two homologous chromosomes fail to line up properly on the metaphase plate, or for other reasons fail to attach to the proper spindle fiber. If in such a case the two homologues (e.g., both X chromosomes in a female) remain together, they may be drawn to the same pole and will thus end up in the same gamete. Half of the gametes will have a complete haploid chromosome set plus one extra X. The other half will have only autosomes and no X. When an egg carrying an extra X is fertilized by an X-bearing sperm, the result is an XXX chromosome. If it is fertilized by a Y-bearing sperm, the result is XXY, which leads to Klinefelter's syndrome. When an egg that has not received an X chromosome is fertilized by an X-bearing sperm, the resulting X0 individual suffers from Turner's syndrome. If an egg with no X chro-

(b)

(c)

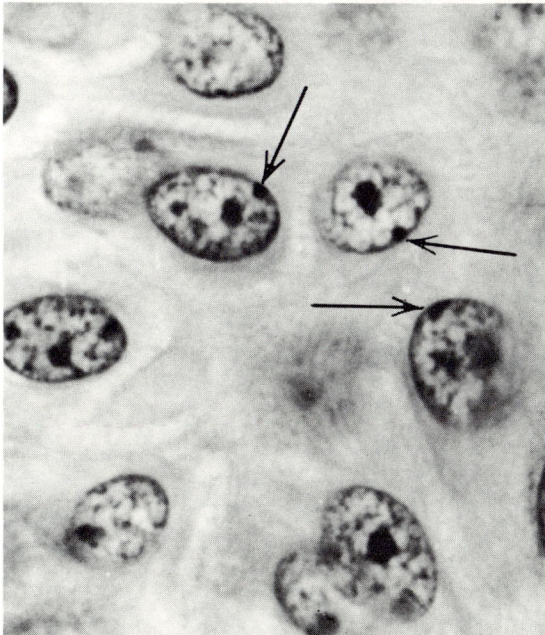

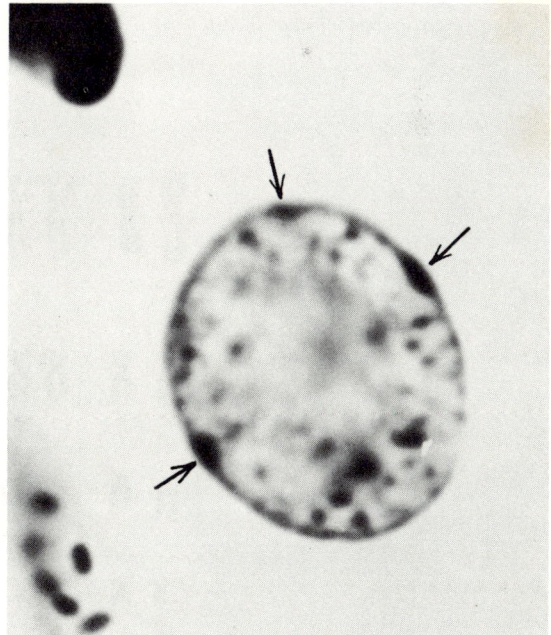

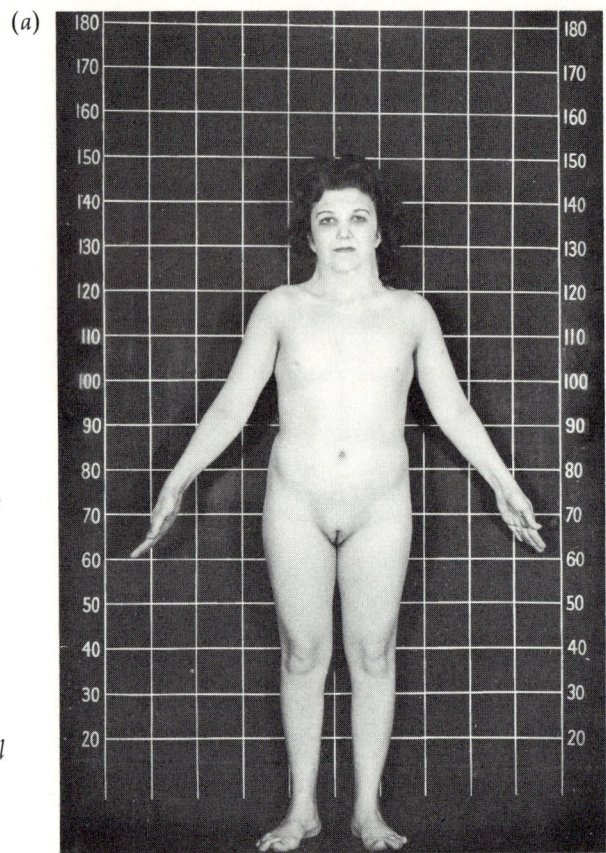

(a)

Figure 9-9   *Turner's syndrome. (a) Persons suffering this anomaly have female genitals, short stature (the scale is in centimeters), webbed neck, low-set ears, a broad chest with widely spaced nipples and poorly developed breasts, a small uterus, and no ovaries. This patient has undergone surgery to correct a birth defect, a constriction of the aorta. This condition occurs with high frequency in individuals with Turner's syndrome, and is usually fatal if not corrected early in life. It is another manifestation of the genetic imbalance. (b) The karyotype of this patient, showing 45 chromsomes and X0 sex chromosomes. Compare with the normal karyotype shown in Figure 8-4. The chromosomes have been grouped into categories (A–G) according to a common system of classification.*

(b)

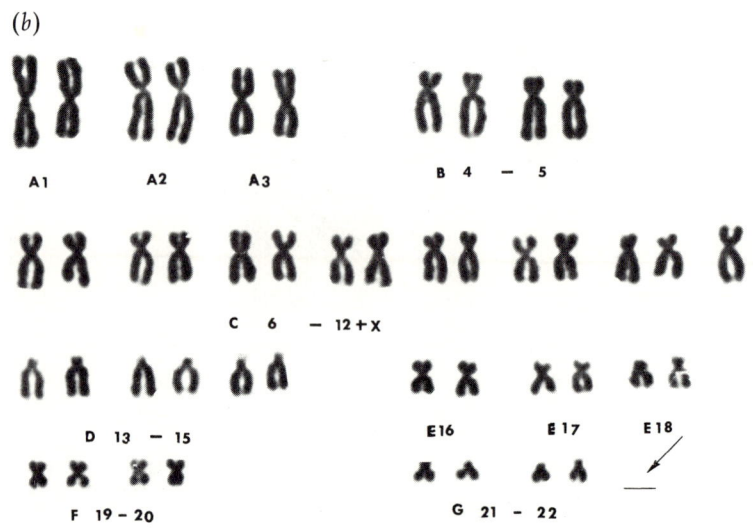

(a)

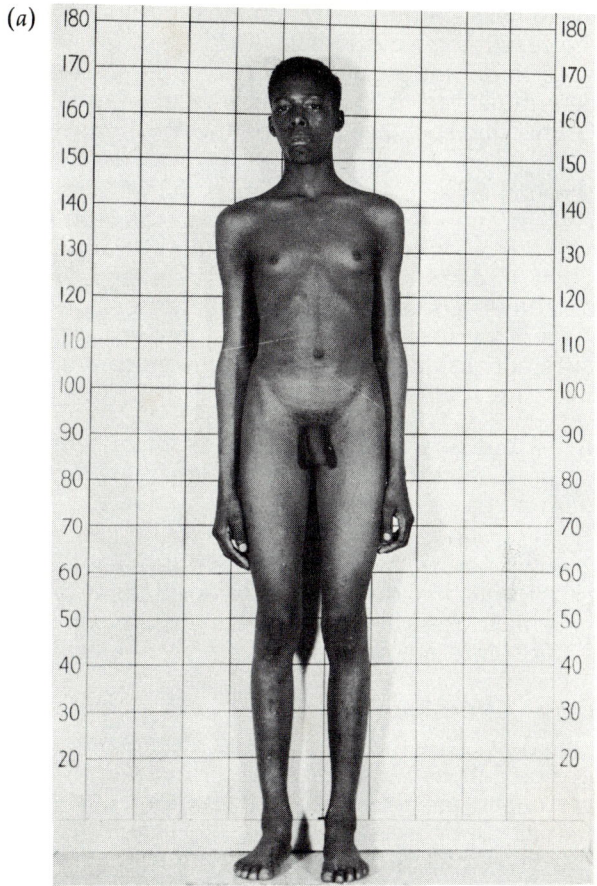

Figure 9-10 Klinefelter's syndrome. (a) These patients have male genitals but may develop small, female-like breasts. The nuclei of their interphase cells contain a Barr body, and the karyotype (b) shows a total of 46 chromosomes with an XXY sex-chromosome make-up.

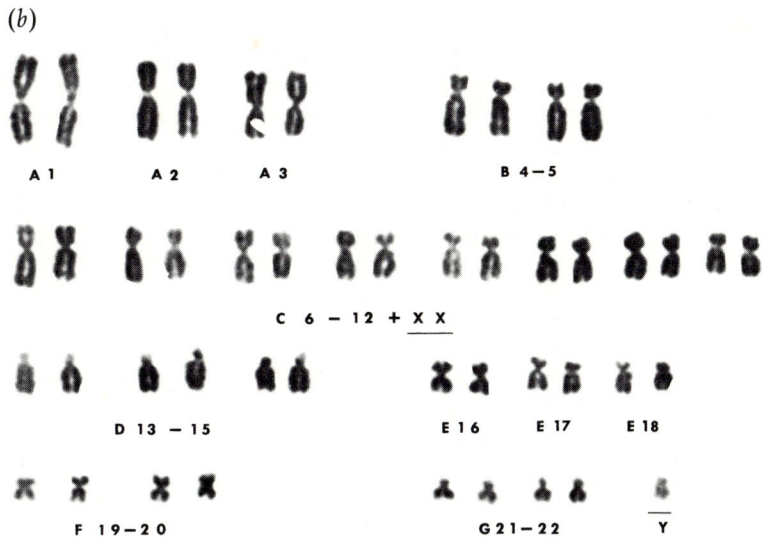

(b)

A 1   A 2   A 3   B 4−5

C 6 − 12 + X X

D 13 − 15   E 16   E 17   E 18

F 19−20   G 21−22   Y

mosome is fertilized by a Y-bearing sperm, the result is always lethal.

An important new discovery relating to sex determination is that the malarial drug Quinacrine specifically stains the Y chromosome, even allowing identification of the Y chromatin in sperm and interphase cells. Thus it should soon become possible routinely to recognize and separate the Y-bearing or male sperm from X-bearing sperm, and so be able to control the sex of offspring.

From these studies of sex anomalies and differential staining of the sex chromosomes, two principles emerge:

1. All but one of the X chromosomes form Barr bodies, no matter how many there are in a cell.
2. A person develops phenotypically as a male only if he carries a Y chromosome. Without a Y the individual develops as a female, regardless of how many X chromosomes are present.

Origin of the Barr Body

In describing the structure of chromosomes during mitosis (Chapter 7) we noted that the chromatin at metaphase is in a supercoiled, highly condensed state; we suggested that this state of condensation accounts for the visibility of the chromosomes during the brief mitotic phase. During interphase the chromosomes uncoil and become extended into single, delicate strands of chromatin, each the thickness of a DNA double helix with its nucleoprotein coat. There are, however, exceptions to this behavior. Certain regions of some chromosomes remain condensed during interphase, and in this condition are referred to as **heterochromatin.** Heterochromatin is known to be genetically inactive; that is, none of its genes is being used as templates for the synthesis of RNA (Chapter 4).

These observations suggested to Mary Lyon, a British geneticist, the idea that the Barr body might be an X chromosome in a heterochromatin condition. The **Lyon hypothesis,** as it is now commonly termed, is based on the following observations: (1) In chromosome squashes from female cells in metaphase one X chromosome stains more darkly than the other because of its heterochromatic or condensed state. (2) In cultures of cells from women thymidine labeled with tritium (H^3) can be incorporated into newly synthesized DNA of all cells in the S phase. Any difference in timing of DNA synthesis by the two X chromosomes can then be observed by autoradiographs of cells dividing in these cultures (Fig. 7-11). Such autoradiographs show that the heterochromatic X chromosome synthesizes DNA late in the process of mitosis, after most of the other chromosomes have completed their replication. (3) Anomalous individuals with three X chromosomes always exhibit two heterochromatic X chromosomes (identifiable by staining), two late-labeling chromosomes (as shown by autoradiography), and two Barr bodies; persons with four X chromosomes have three of each, and so on. Thus it is clearly the heterochromatic, late-labeling X chromosomes that form the Barr bodies, while one X remains noncondensed and genetically active.

Sex-Linked Inheritance

In addition to genes related to sex determination, it is clear that the sex chromosomes carry other genes, apparently unrelated to sex. Genes carried on the sex chromosomes are called **sex-linked genes.** The discovery of sex-linked traits was made during early experiments with *Drosophila* in 1910. In that year Thomas Hunt Morgan published a short paper in *Science* that began, "In a pedigree culture of *Drosophila* which had been running for nearly a year through a considerable number of generations, a male appeared with white eyes. The normal flies have brilliant red eyes." When that white-eyed male mutant was mated with homozygous red-eyed females, the cross yielded an F_1 generation of 100 percent red-eyed flies (Fig. 9-11). The F_2 generation showed the familiar ratio of approximately 3 red to 1 white, indicating that the gene (W) for red eyes was dominant, the mutant allele (w) for white eyes recessive.

The most striking feature of the F_2 generation, however, was the fact that all the fe-

P₁ Generation

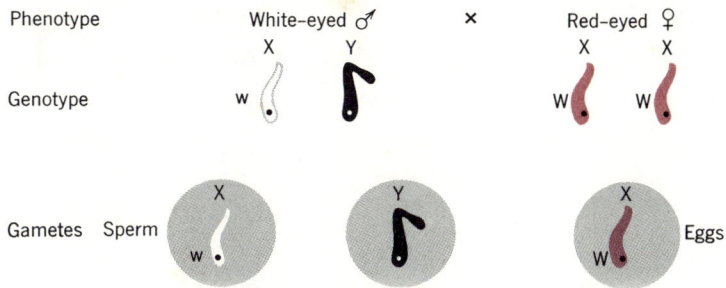

Phenotype   White-eyed ♂   ×   Red-eyed ♀

Genotype

Gametes Sperm   Eggs

F₁ Generation

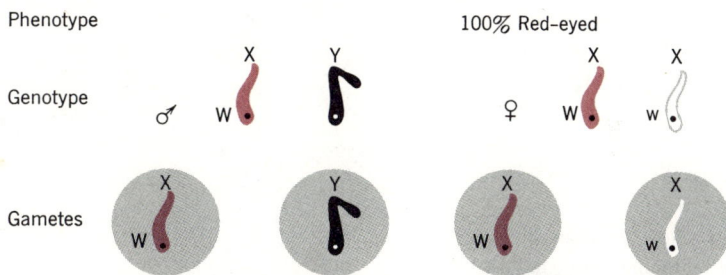

Phenotype   100% Red-eyed

Genotype   ♂   ♀

Gametes

F₂ Generation

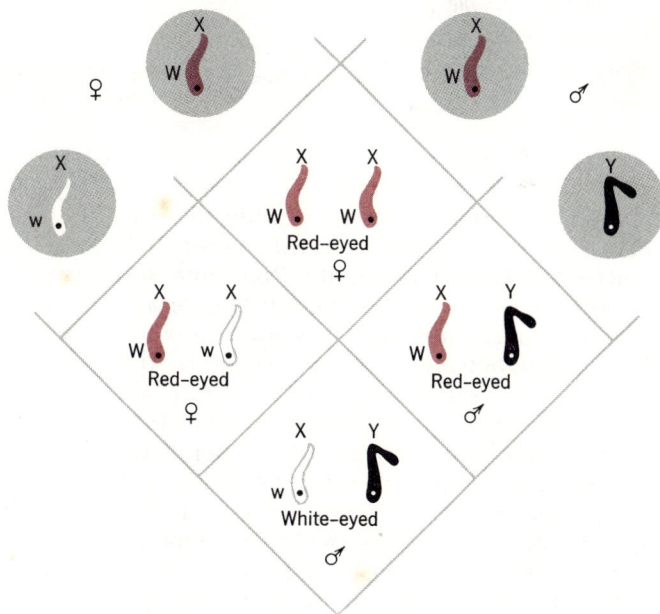

♀   ♂

Red-eyed
♀

Red-eyed   Red-eyed
♀   ♂

White-eyed
♂

Figure 9-11   A cross between a wild-type, red-eyed (W) female and a white-eyed (w) male illustrates that when a recessive gene is sex-linked—that is, is carried on an X chromosome—half of the females and half of the males in the F₂ generation carry the gene, but only the males exhibit the phenotypic trait.

Phenotype 3 red-eyed: 1 white-eyed (100% red-eyed females; 50% red-eyed and 50% white-eyed males)

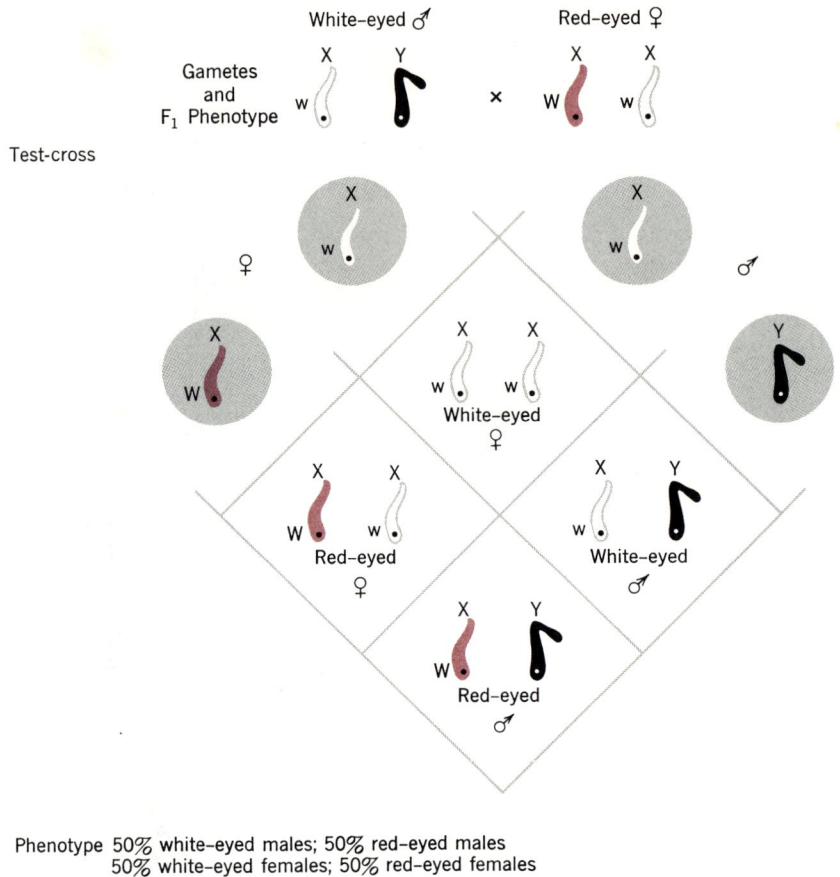

White-eyed ♂   Red-eyed ♀

Gametes
and
F₁ Phenotype

Test-cross

White-eyed
♀

Red-eyed
♀

White-eyed
♂

Red-eyed
♂

Figure 9-12   A back-cross between the mutant white-eyed (w) male and one of his heterozygous red-eyed (Ww) daughters to show that half the males and half the females among the offspring are white-eyed.

Phenotype 50% white–eyed males; 50% red–eyed males
50% white–eyed females; 50% red–eyed females

males were red-eyed, whereas only half the males were red-eyed and the other half white-eyed (Fig. 9-11). Until that time the inheritance of all traits studied in *Drosophila* and other organisms had shown no unequal distribution between the two sexes of the progeny. From these and other crosses it was finally concluded that the results could best be explained by assuming that the genes for eye color are carried on the X chromosome.

Although the trait for white eyes is recessive, the presence of a gene for this characteristic in the male will be expressed in the phenotype because the male has only one X chromosome. The Y chromosome lacks the allele of this gene, and a single recessive sex-linked gene expresses itself in the absence of the dominant allele. A male transmits his X chromosome to all his daughters but to none of his sons, and his Y chromosome to all his sons but to none of his daughters. A female derives one X chromosome from her mother and one from her father, whereas a male receives his single X chromosome only from his mother. As shown in Figure 9-11, the white-eyed males in the F₂ generation have received their gene for this trait from their normal appearing but heterozygous red-eyed mother of the F₁ generation.

It is also possible to produce some white-eyed females by crossing a heterozygous red-eyed female with a white-eyed male, in effect performing a back-cross (Chapter 8). This produces a 1:1 ratio between white-eyed males and white-eyed females (Fig. 9-12).

In human beings such conditions as red-green color blindness (inability to distinguish between red and green), hemophilia (the bleeder's disease, pp. 404–408), and certain other abnormal and normal traits are among the many known sex-linked traits. Their mode of transmission in inheritance is identical with that of white-eye in *Drosophila*. Males suffering from recessive sex-linked diseases but having apparently normal parents must have inherited the condition from their mothers (who would have to be heterozygous for these traits) and not from their fathers. The ability of single recessive sex-linked genes to express themselves in the male, and the segregation of the X chromosomes in gametogenesis, account for the higher incidence of red-green color blindness in males than in females. In the United States about 8 percent of the males are red-green color blind, whereas less than 1 percent of the females display the same condition. In humans there are now known to be at least 70 traits controlled by genes on the X chromosomes. Most of these produce diseases; most are X-linked recessives.

As soon as it became apparent that there were genes on the X chromosome unrelated to sex determination, a new question arose. If each gene controls the production of a specific product—an enzyme, an antihemophilia factor, or some structural protein, for example—then does a female with two X chromosomes have twice as much gene product as a male with only one X? If we consider that each chromosome produces a given "dose" of a variety of gene products, how does a female contend with twice the dose of a male? We could imagine that to avoid disruptive dosage effects the X chromosome might have been largely stripped of genetic information during evolution. We know, however, that this is not the case. The X chromosome seems to carry at least as much genetic information as an autosome of comparable length. The Lyon hypothesis (p. 226) has its greatest significance as an answer to this question of dosage. The explanation that occurred to Mary Lyon and to several other workers is that in female embryos one X chromosome in every cell becomes genetically inactive and forms the Barr body of an interphase nucleus. The Lyon hypothesis further suggests (1) that it is a random matter in any single cell whether the X chromosome derived from the mother or the one from the father becomes inactive; and (2) that once one X chromosome attains a heterochromatic condition, that same X chromosome (maternal or paternal) remains inactive in all mitotic descendants of that cell. Thus a female has no more active X chromosomes than a male and therefore has no imbalance in gene products.

Inherited
*Physical and Mental
Human Traits*

What is the difference between a blond and a brunet? Between a mouse and a man? The answer, of course, is that they have several—or many—different genes. The patterns of inheritance, and in some cases the mechanisms of inheritance, are known for many physical features of man, including the detailed structure of the face; color and texture of the hair, skin, nails, and teeth; characteristics of the eyes, including color, shape, and various defects of vision; the structure and proportions of the skeleton and muscles that determine stature; and a host of enzymes and biochemical products.

Genes and Intelligence

There is good evidence that there is a large component of heritability in intelligence. We noted earlier that many chromosomal abnormalities result in mental retardation. Yet many people have concluded that the effects of environmental differences are more important than genetic differences in determining intelligence, at least within the "normal" range. In no other field has the "nature-nurture" question been so hotly debated.

There are two main problems in determining if intelligence is inherited. First, intelligence is difficult to define; there are no exact and objective means for measuring the native intellectual endowment with which an individual is born. We could ask whether height or blood type is inherited and have no difficulty in determining, by accepted methods, the phenotype of the individuals. But there is nothing comparable to a centimeter scale or an immunochemical test for measuring intelligence. Much ingenuity has entered into the devising of intelligence tests, but even the best are only partially successful in distinguishing what is innate (i.e., genetic) from what is acquired. Most tests employ language; this immediately introduces the effects of different environments and education. Nonverbal tests attempt to avoid the difficulty, but are only partly successful. We are left having to define intelligence as "what intelligence tests measure," but the correlation between what intelligence tests measure and what we intuitively consider native intellectual capacity may not be very close.

Aside from difficulties in defining and measuring intelligence, the second major problem in determining its genetic input is the degree of interaction between native capacity and environmental factors. We commonly think of a trait such as body height as highly susceptible to environmental modification. Throughout history, for example, most individuals with the genetic capacity to exceed six feet in height actually have been much shorter as a result of improper diet and childhood diseases. Such influences are much easier to control for and take into account than those involved in the complex interactions that produce measurable intelligence; thus mass programs of intelligence testing, in which attempts are made to compare different populations (separated along socioeconomic, ethnic, educational, geographic, or other lines) may give relatively accurate information regarding certain abilities of the persons tested, but they tell little or nothing about differences in the innate intellectual endowment of the individuals.

How, then, can we determine what roles hereditary components play in intelligence? The approach that seems to have been most successful so far is the **twin method,** which makes use of measurements of monozygotic and dizygotic twins. The theory is that if a genetic component is involved in the production of a given trait, then the **average difference** between monozygotic twins should be less than the difference between dizygotic twins. To put it more simply, identical twins should be more consistently alike in the measured trait than fraternal twins. Many studies indicate that this is the case with intelligence. For example, in one investigation the Stanford-Binet intelligence test was given to 50 monozygotic pairs of twins and 52 like-sexed dizygotic pairs. The average intrapair difference for the identical twins was 3.1 IQ points; for the dizygotic pairs it was 8.5 IQ points; the fraternal twins differed almost three times as much as the identical twins. If we assume that dizygotic like-sexed twins raised together have an environment that is as similar as the environment of monozygotic twins raised together, then it seems reasonable to ascribe the greater similarity of IQ scores for identical twins to their greater genetic similarity.

It is convenient to express the intrapair differences in terms of a statistic known as a **correlation coefficient.** If two groups of measurements for a given trait always agree exactly, the correlation coefficient is 1; if there is no agreement at all between one set of measurements and the other, the correlation coefficient is 0. In the study just described the correlation coefficient for IQ scores in the monozygotic pairs was .93, whereas for dizygotic pairs it was .66. A number of studies have yielded similar results (Fig. 9-13); intelligence test scores of identical twins are generally much more alike (with correlation coefficient near .90) than those of any other pairs of individuals, no matter how they are raised or what their environments are. Thus the evidence seems conclusive that there is a genetic as well as an environmental component in performance on intelligence tests. It is also safe to conclude that many genes must be involved in determining such a complex trait; that is, intelligence is **polygenic.**

Genes and disease

Genetic factors are involved in all diseases, in the same sense that they are involved in

Figure 9-13 chart:

CATEGORY		Correlation coefficient	Groups included
Unrelated persons	Reared apart	(~0 to .30)	4
	Reared together	(~.20 to .30)	5
Fosterparent–child		(~.20 to .40)	3
Parent–child		(~.25 to .80)	12
Siblings	Reared apart	(~.35 to .45)	2
	Reared together	(~.30 to .80)	35
Twins – Two-egg	Opposite sex	(~.35 to .60)	9
	Like sex	(~.45 to .90)	11
Twins – One-egg	Reared apart	(~.65 to .90)	4
	Reared together	(~.80 to .95)	14

Correlation coefficient scale: 0 .10 .20 .30 .40 .50 .60 .70 .80 .90

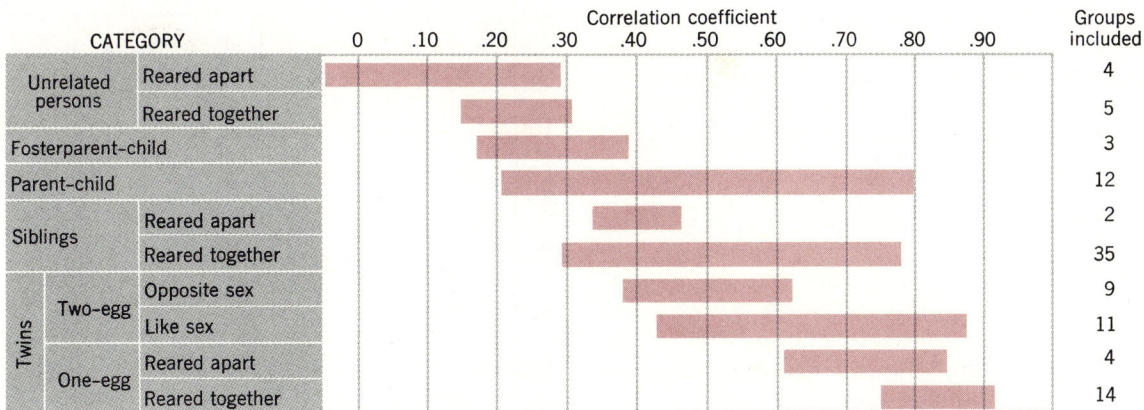

Figure 9-13 Correlation coefficients for intelligence-test scores. Most of the data were derived from Stanford-Binet IQ test. This chart shows that the greater the degree of genetic relationship, the more similar the individuals are in intelligence. Note the low foster-parent-to-child correlation, and the difference between fraternal and identical twins

every phenotypic trait. It is difficult to imagine a disease in which hereditary factors played no role, just as it is difficult to imagine one in which environmental factors played no role. An English physician, Sir Archibald Garrod, was among the first to suggest that some human diseases might be caused by the hereditary lack of a specific enzyme. He referred to such diseases as "inborn errors of metabolism." All diseases can, however, be viewed as falling on a spectrum of **genetic** versus **environmental** (external, or **exogenous**) causal factors (Fig. 9-14).

Some diseases, like **galactosemia** and **phenylketonuria** (PKU) are near the extreme genetic end of the spectrum; they are known to result from specific single gene defects. In galactosemia the specific enzyme that metabolizes galactose in the energy cycle is missing, causing that sugar to accumulate in the blood. PKU results from the lack of the enzyme that normally breaks down the amino acid phenylalanine. Without the enzyme, phenylalanine (or its derivative, phenylketone) accumulates in the bloodstream and appears in the urine. These prod-

Figure 9-14 A group of diseases, selected to represent a spectrum of relative importance of genetic and environmental factors.

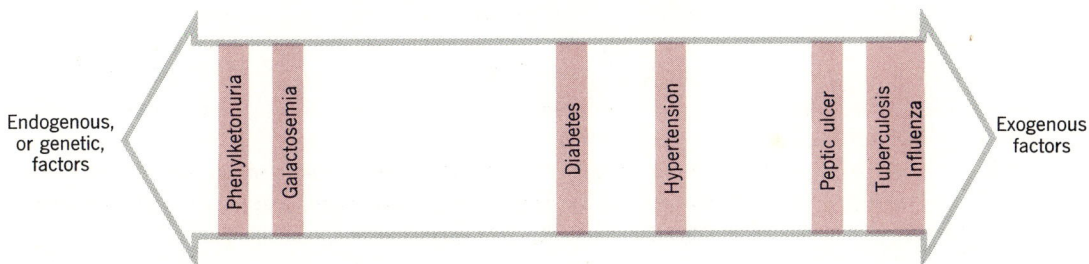

Spectrum diagram — Endogenous, or genetic, factors ← Phenylketonuria | Galactosemia | Diabetes | Hypertension | Peptic ulcer | Tuberculosis | Influenza → Exogenous factors

ucts are harmless at normal concentrations and essential for protein synthesis; at excessive concentrations, however, they are harmful, particularly to cells of the brain. In the United States about one infant in every 15,000 is born homozygous for this recessive gene. The disease is relatively easy to detect by a simple blood or urine test. If it is diagnosed early enough in infancy, mental retardation can be prevented by a diet low in phenylalanine because this essential amino acid cannot be synthesized (Table 4-1). Even galactosemia and phenylketonuria cannot be considered purely genetic, because the effects of both diseases can be alleviated by affected individuals special diets; thus exogenous factors play a significant role.

On the other hand, diseases like tuberculosis or influenza, resulting from infectious agents, lie near the exogenous end of the spectrum. But, again, these cannot be considered entirely environmental because studies (like those on intelligence) comparing

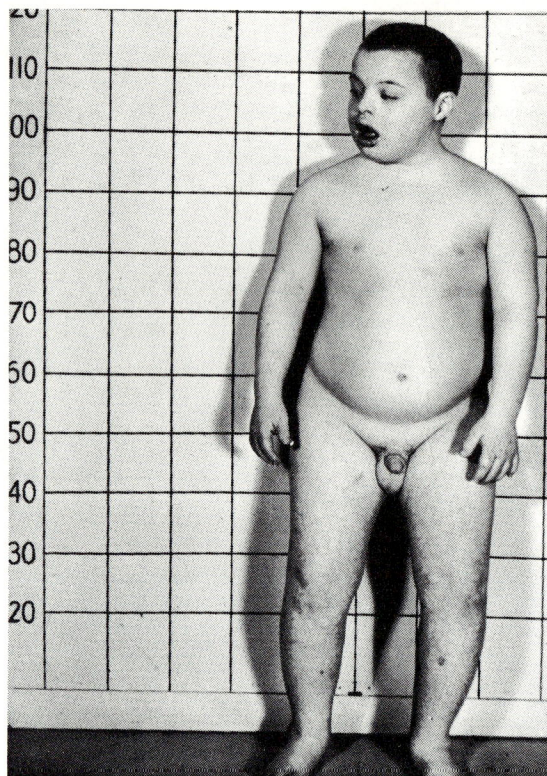

(a)

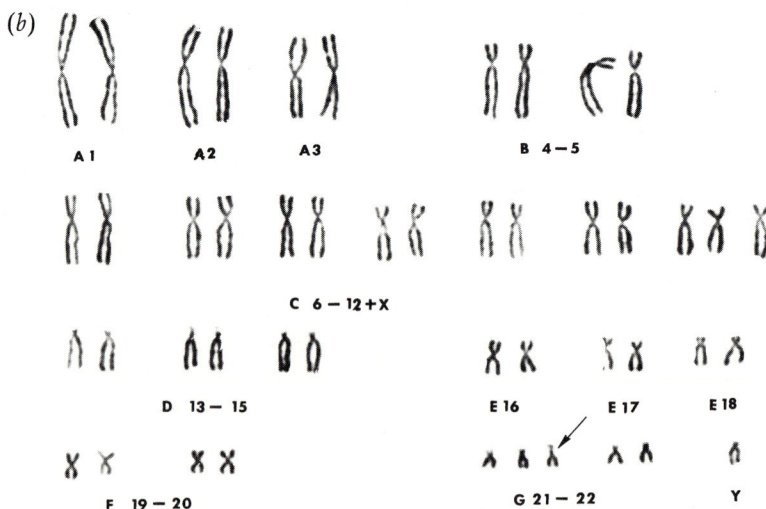

(b)

Figure 9-15 (a) A patient with Down's syndrome or mongolism. Note the short stature, stocky body, and peculiar facial features. (b) A karyotype of such a patient, showing that chromosome 21 is triploid.

identical and fraternal twins and others comparing ethnic groups indicate that a person's genetic constitution can determine how susceptible he is to infection and how severely he is affected by a disease.

Perhaps the best known and most commonly recognized genetic defect is **mongolism** (Fig. 9-15). The term derives from the facial shape and "mongoloid" eyefold characteristic of these individuals. Mongolism is more properly referred to as **Down's syndrome,** after a physician, Langdon Down, who first described the condition in 1866. The syndrome includes not only the characteristic eyefolds, but also mental retardation, a short, stocky body, a thick neck, stubby hands and feet, and often malformations of other organs, especially the heart.

Mongolism was, in fact, the first autosomal anomaly to be described in man. The basis for this abnormality appears to be a nondisjunction comparable to that described for the sex chromosomes. All patients with Down's syndrome have what is known as **trisomy 21;** that is, each cell of the body contains three of chromosome 21 (Fig. 9-15*b*) rather than two, as in normal diploids. No individual has ever been found with only one chromosome 21; presumably gametes that fail to receive one homologue of chromosome 21 cannot survive. It is probably significant that chromosome 21 is one of the smallest in the human complement; trisomies of two larger chromosomes (13 and 18) are known, but these lead to an early death. Apparently larger chromosomes contain so much genetic information that the effects of overdosage are always lethal.

Down's syndrome is by no means rare. It occurs once in every 500 or 600 births. In fact, it is the most frequent single definable entity that causes severe mental deficiency. Like a number of other defects involving gross chromosomal abnormalities, it is more likely to occur among infants born to women over 40 than among children of younger women (Fig. 9-16). Although it is not clear why this is so, the most reasonable hypothesis is that as oöcytes or oögonia age in the ovary, they become more susceptible to damage, and

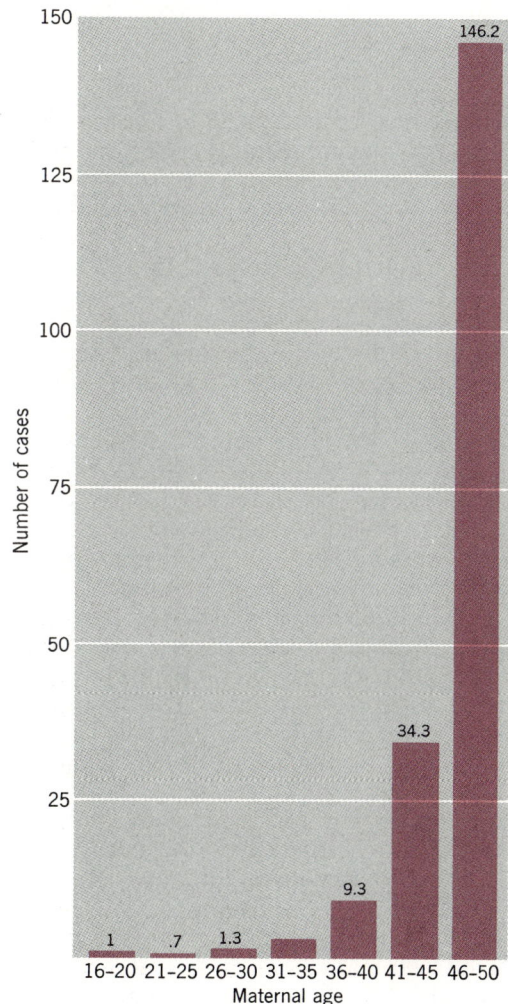

Figure 9-16   Frequencies of Down's syndrome in infants, as related to age of the mother. The figures shown are for number of cases per 2240 births at each age level. The risk of having a mongoloid child increases steadily after age 25. Although only about 2 percent of all births are mongoloids, approximately 40 percent of these mongoloid children are born to women over 40.

therefore exhibit a greater frequency of abnormalities.

Knowledge of the genetic contribution to disease can help a physician in diagnosing a problem and often allows him to perform tests for early signs of an ailment and to provide treatment before serious consequences result. It also allows him to provide a married couple with **genetic counseling.** If a husband or wife carries a harmful mutant gene, it is often possible to calculate the probability that the couple will have a defective child. If that probability is great, many couples choose to adopt children rather than to have their own.

A Brief Review

We have now considered many of the ideas important to an understanding of how heritable characters are transmitted from parents to children. A brief review at this stage may be valuable.

A new individual is formed when a sperm fertilizes an egg and the nuclear contents of the two gametes are brought together. Development of the new individual is controlled by a series of hereditary regulators called genes. Genes are carried on chromosomes in linear array, each gene being a sequence of nucleotides in the chromosomal DNA molecule. Genes may occur in alternative forms, alleles, that produce alternative traits. Most somatic, or body, cells contain homologous pairs of chromosomes, one member of each pair contributed by each parent. Such cells are termed diploid. Alleles occupy corresponding locations, or loci, on homologous chromosomes. If more than two alternative forms of a gene can appear at a given locus, the gene constitutes a series of multiple alleles.

New alleles are formed occasionally by mutation, which may occur in genes either spontaneously (that is, for unknown reasons) or under the influence of agents such as radioactive emissions or specific chemicals called mutagens. Human genes mutate at a spontaneous rate of about four mutations per 100,000 gametes formed. Because mutations represent random changes in the DNA sequence, they are almost always harmful; most are so harmful that the gamete or individual carrying one does not survive. Some are neutral; a few are beneficial. One allele may be dominant over another, in which case the recessive allele is expressed only if it occurs on both homologues, a homozygous condition. The recessive allele in a heterozygous individual is not usually expressed in a phenotype, but is passed on as an independent unit at the next generation.

During development of an organism from zygote to adult, chromosomes are passed from cell to cell by mitosis. In this process each daughter cell receives an exact copy of each chromosome in the diploid complement, with the exception of copying errors. Hereditary transmission of genes from one generation to the next begins with gametogenesis, the formation of haploid sperm and egg cells, containing only one of each homologous chromosome. The process by which homologous chromosomes are distributed to different gamete cells is meiosis. Because gametes contain only one homologue and one allele of each gene, they are haploid cells. At meiosis the homologues are assorted randomly in the gametes; therefore genes on different chromosomes also assort independently and at random. Genes on the same chromosome do not assort independently but form linkage groups that may be distributed together. More often, however, during meiosis alleles may move from one homologue to another by the process of crossing-over, producing new arrangements of genes called recombinations. The frequency of crossing-over is a measure of the linear distance between two genes on a chromosome.

A mutant gene that actively blocks some other gene formation but is only partly successful behaves phenotypically as an imcomplete dominant. Many traits are controlled by more than one gene and are therefore subject to multiple gene or polygenic inheritance.

Heredity in humans follows the same laws and is controlled by the same basic mechanisms as in other animals. This is strikingly demonstrated by studying human pedigrees, by the genetics of sex determination, and by a variety of heritable abnormalities. In addition to 44 autosomes, women have two X chromosomes, men have an X and a Y chromosome. One of the X chromosomes in females becomes inactive by condensing into heterochromatin, and forms the Barr body. This process occurs in the early embryo. It explains why the dose of X gene products in females and males remains in balance. Most of the physical and mental traits that characterize humans (and other organisms), including normal characteristics and diseases, result from an interaction between genetic factors and environmental agents. Two methods of assessing the relative importance of genes and exogenous factors in determining any given trait are studies of family pedigrees and comparisons between monozygotic and dizygotic twins.

Reading List

See reading list for Chapter 8. Also:

Moore, J. A., *Heredity and Development.* Oxford University Press, New York, 1963.

Whitehouse, H. L. K., *Towards an Understanding of the Mechanism of Heredity* (2nd ed.). Edward Arnold, London, 1969.

The Mechanism of Gene Action

"Making and breaking of hydrogen bonds seems to be all there is to understanding the workings of the hereditary substance."

GUNTHER S. STENT, 1966, *IN PHAGE AND THE ORIGINS
OF MOLECULAR BIOLOGY*

We have said on several occasions that genetic information is stored in the long molecules of DNA that comprise the chromosomes. But how do we know this? What do we mean by genetic information? What is the basis for the capacity of a molecule to store information? What is the molecular code in which genetic information is written?

From classical genetics (Chapter 8) we saw that the existence of crossing-over of homologous chromatids during meiosis provides a means of locating genes along chromosomes. Because crossing-over occurs randomly throughout the chromosome length, the farther apart two genes are, the greater is the probability that a break will occur between them to cause genetic recombination. By studying cross-over frequencies in a large number of crosses, it is possible to show that all genes on chromosomes are located in linear sequence, never branched or helter-skelter. Chromosomes, we discovered, are linear both in shape and in gene arrangement (refer back to Figs. 8-15 and 8-16).

*Genetics of
Microorganisms*

Heritable traits studied in classical genetics—eye color, coat color, wing shape, or any other visible feature—are the end products of one or many chemical reactions. When it became apparent, early in the 1940s, that the synthesis of enzymes might be directly controlled by genes, a new approach to genetic analysis was made possible. Instead of picking a genetic trait and working out its chemistry, early molecular geneticists could begin with chemical reactions known to be controlled by specific enzymes and determine how genetic changes affected those reactions.

This kind of approach proved especially important when applied to microorganisms, which provide several major advantages for genetic analysis. First, they are usually haploid, so that the expression of a mutation does not depend on whether it is dominant or recessive; in haploid organisms mutant genes generally express themselves almost immediately. Second, microorganisms multiply very rapidly. Instead of having life cycles of days or weeks duration—which are very short for most multicellular forms—many bacteria produce a new generation every half hour or less. Finally, whereas most of the mutant genes studied by the early Mendelian geneticists arose spontaneously and were rare; it is easy to produce large numbers of mutations in microorganisms with the use of mutagenic agents. Treatment of bacteria with a chemical mutagen such as **nitrosoguanidine,** for example, can produce nonlethal mutations in as many as 1 percent of bacterial genes.

A major breakthrough in the use of bacteria for genetic studies came in 1944, with the realization that mutations could be obtained that affected the ability of a microorganism to synthesize an essential compound. For example, a common bacterium, *Escherichia coli,* found in the human colon, ordinarily grows well with glucose as its only carbon source. That is, it is an autotroph (p. 591); from this one compound it is normally capable of synthesizing all the other organic compounds that it needs (amino acids, vitamins, lipids, nucleotides, etc.).

When *E. coli* bacteria are treated with mutagens, however, strains can be produced that grow only if their normal medium is supplemented with one or more compounds (called **metabolites,** or **growth factors**) in addition to glucose. To test for their presence a suspected mutant may be grown on two different media, one in which the metabolite (e.g., arginine) is present and one in which it is absent (Fig. 10-1). If a mutation has occurred that prevents the bacterium from synthesizing its own arginine, it will grow only if that amino acid is present. Such a strain is called **arginine-requiring** (the notation for this kind of mutation is arg⁻). The use of this approach very quickly led to the isolation of a large number of gene mutations that affected the enzymatic synthesis of specific molecules.

These mutants proved to be powerful genetic tools in two ways. First, they could be used as conventional genetic markers to reveal the chromosomal arrangement of the bacterial genes. Second, they provided a method for demonstrating genetic recombination in bacteria. In *E. coli,* for example, it had not been realized that cell fusion and genetic recombination existed. Simple microscopic examination of the cells gave no clues that such a sexual process might be taking place. To detect recombination it was necessary to design an experiment in which only recombinant cells (cells in which a recombination had occurred) would be able to multiply. This was done by using parental strains with specific mutant growth requirements (like those isolated as in Fig. 10-1), which could not multiply in a minimal medium containing only glucose and inorganic salts. Two strains of bacteria, each with different growth requirements, would be mixed together.

One strain (thre⁻, leu⁻) was unable to grow without the amino acids threonine and leucine added to its medium. The other (met⁻, bio⁻) required methionine and the vitamin biotin. Neither of these strains could grow on a minimal medium, but after they had been mixed together for a period a small number of cells were found that could grow on minimal medium without any growth factors. This meant that somehow these cells had acquired nonmutated copies of each of their mutant genes, and indicated that *E. coli* has a sexual phase during which the chromosomes of two different cells are brought together. Crossing-over can then place "good" copies of all necessary genes in one chromosome.

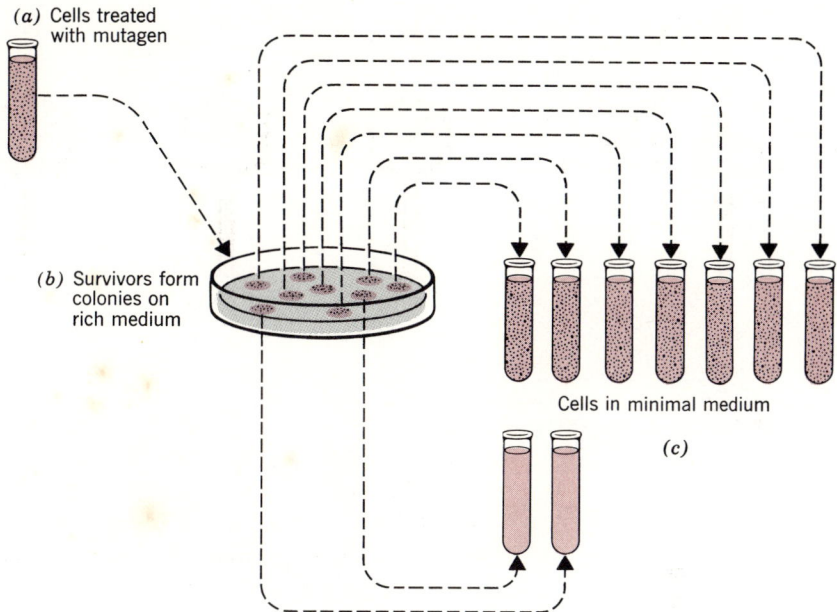

(a) Cells treated with mutagen

(b) Survivors form colonies on rich medium

Cells in minimal medium

(c)

Figure 10-1   *The method of isolating mutant strains of* E. coli *that have specific metabolic requirements. A suspension of cells in a test tube (a) is treated with a mutagen. (b) Cells are then placed on a rich medium containing all 20 amino acids and a long list of vitamins, pyrimidines, purines, carbohydrates, and other compounds that might be required for growth. Cells that are not lethally affected by the mutagen multiply to form colonies of several thousand cells each. (c) When cells from each of these colonies are transferred to tubes with a minimal medium containing only glucose and inorganic salts (upper row), most are still able to grow. The transferred cells from a few of the colonies, however, fail to grow (bottom row). These are mutant cells in which the ability to synthesize some required compound has been affected. The identity of the required growth factor can be determined by adding the compounds present in the rich medium, one at a time.*

The Role of DNA in Viral Infection

At about the same time that these experiments on bacterial recombination were being done Max Delbruck and Salvadore Luria were beginning to investigate another kind of mutation, resistance to bacterial viruses. A French physician, Felix d'Herelle, had discovered a group of viruses that attacked bacterial cells; and showed that only 20 minutes after the infection of a bacterial cell by a single virus particle the cell would burst open and 200 or more new virions (virus particles) would be released. It was this phenomenal rate of reproduction that attracted Delbruck and Luria.

Every known type of bacterium is susceptible to infection by at least one virus, and many are host to several different viruses. Delbruck and Luria began their studies with a series of related bacteriophages that attacked *E. coli* (Fig. 10-2, p. 240); one such group was numbered T1 through T7 (T stood

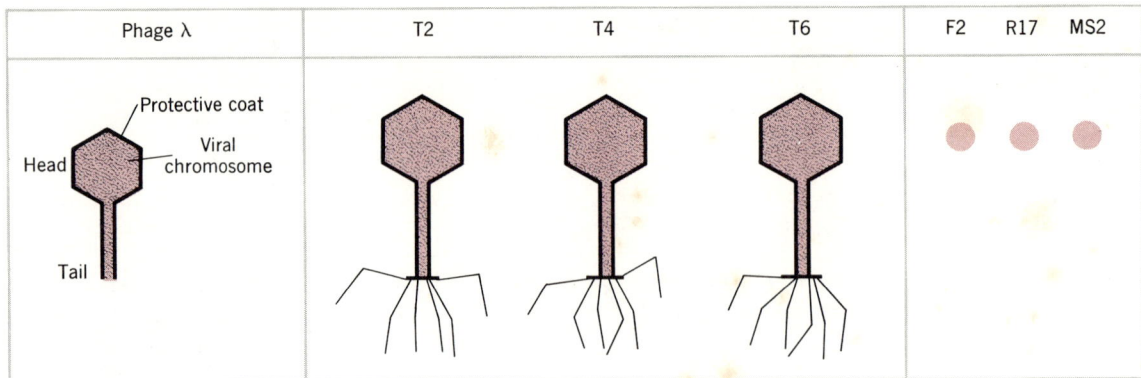

Figure 10-2 *Several different viruses known to infect* E. coli *are drawn to the same scale. T2 is a particle about 1000 Å long with a molecular weight of about 2.5 × 10⁸ daltons. It is one of a group of genetically related viruses of identical appearance in the electron microscope. These are identified as T2, T4, and T6, and are referred to as the* **T-even phages.** *Similar in appearance, but lacking attachment fibers on its tail, is phage lambda (λ). The smallest* E. coli *viruses, F2, R17, and MS2, appear as spherical particles about 200 Å in diameter, having a molecular weight of only about 4 million daltons.*

simply for type). Most of their work, however, was done on T2 and T4, which proved to be very similar in certain respects and were therefore referred to as the **T-even** phages. (The model shown in Fig. 6-4a is a T-even phage.)

Viruses are so small that they cannot be seen in a light microscope. The T-even phages are about .1μ long; the largest known virus, the one that causes smallpox, is about .25μ, which is just at the limits of visibility with the light microscope. Thus it was not until the studies of Delbruck and Luria were well under way that viruses were first seen in the electron microscope. When phage-infected *E. coli* cells were viewed with the electron microscope, it was found that the viral particles totally disappeared shortly after infection, and for about half the infectious cycle no virions could be found in the bacterial cells. Then, gradually, increasing numbers of virus particles began to appear.

The growth cycle starts when a phage particle collides with a bacterial cell and the phage tail attaches to the cell wall. An enzyme in the tail then breaks down a small portion of the cell wall, creating a hole through which the viral chromosome can be "injected" into the cell, leaving the empty protein coat on the outside (Fig. 10-3). In effect, the protein coat is merely a container and injector for the viral DNA. It is only the DNA that is required in a bacterial cell to cause it to synthesize new phage DNA and new phage protein. The viral chromosome duplicates and the daughter chromosomes duplicate again to form, eventually, up to several hundred new chromosomes. Each of these then becomes surrounded by a newly synthesized protective coat, to form a large number of new phage virions (Fig. 10-4). The growth cycle is completed when the bacterial cell breaks open **(lyses)** and releases the newly formed progeny virions into the surrounding medium.

It is easy to demonstrate live phage particles in a solution. *E. coli* is normally grown in a layer on the surface of a nutrient culture

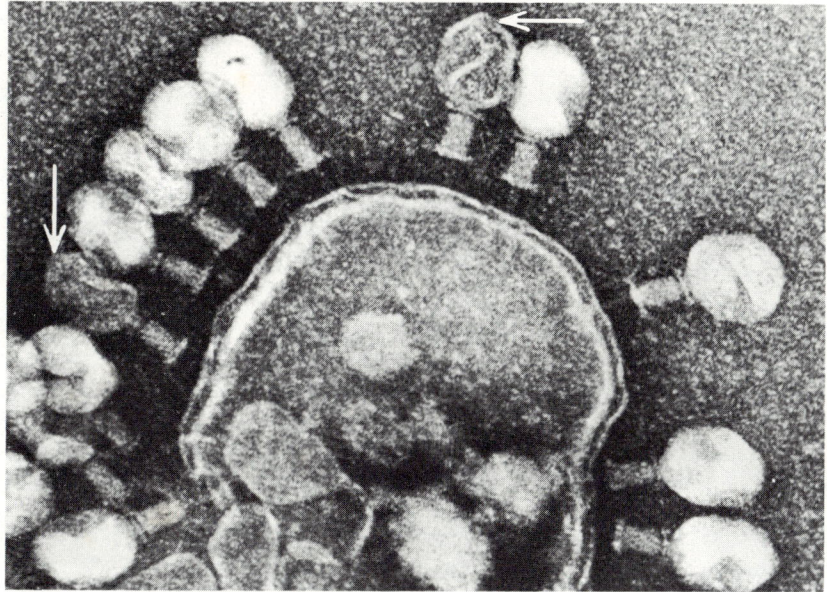

Figure 10-3   A cell of E. coli infected with T4. Many virions are attached at the surface. Note (arrows) the virions that have injected their DNA into the bacterium and consist of empty protein coats. Inside the cell several viral particles are being synthesized.

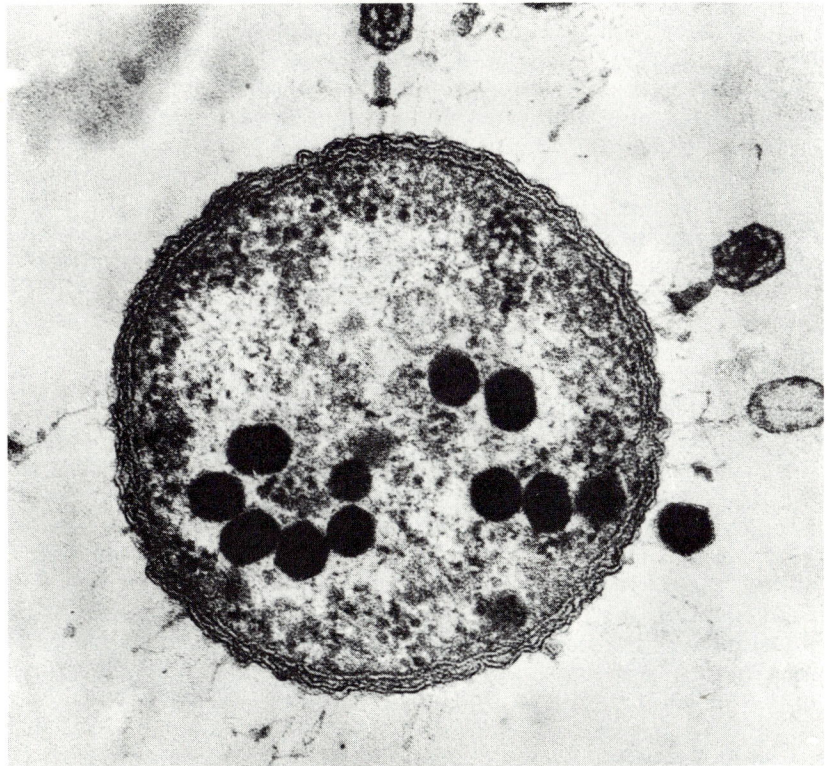

Figure 10-4   A cross-section of an E. coli cell infected with bacteriophage T2. The cell wall and plasma membrane are clearly visible, with three phage particles, one empty and two still partially filled, attached to the cell surface. The phage at the top shows the long tail fibers in contact with the cell wall. The clear area in the cell is composed of the bacterial and phage nucleic acid pool, and contains 11 condensed phage DNA cores.

dish, where it divides rapidly. A solution suspected of having live viruses in it may be added to the surface of such a dish. If no virions are present, the bacteria will form a uniform thick layer on the surface. But if even one virus particle is present, it will attach to a bacterium, inject its DNA, and multiply to form several hundred new phage particles. These will be released into the medium 15 to 60 minutes after infection. Each of these several hundred progeny particles then attaches to a new nearby bacterium and multiplies. After several such cycles of attachment, multiplication, and release by **lysis,** all the bacteria in the immediate region of the original virus particle are killed. These regions of dead bacteria appear as circular holes, called **plaques,** in the layer of healthy E. coli (Fig. 10-5).

When the T-even phages were analyzed chemically, they were found to consist entirely of DNA and protein. Both DNA and protein attached to bacterial cells; which one carried the information on how to make new viral particles? This question was answered in 1952 by Alfred D. Hershey and Martha Chase. Recall that protein contains sulfur as —SH groups in cysteine, but none of the common amino acids contains phosphorus. In contrast, DNA is rich in phosphorus but contains no sulfur groups (Chapter 3). Hershey and Chase therefore grew E. coli on a medium that contained radioactive isotopes of both phosphorus and sulfur (P^{32} and S^{35}) and then infected the bacterial cells in this medium with T2 bacteriophage (Fig. 10-6). When the viruses multiplied inside these labeled bacteria, the DNA of the virus was labeled with P^{32}. The protein coats of the new virions also were labeled, but with S^{35}. The experimenters then used these labeled virions to infect fresh E. coli growing on a normal, nonradioactive medium. At intervals after the virus was added, the cells were violently stirred in a food blender to rip the phage protein from the surface of the bacterial cell. This did not interrupt the process of phage reproduction, however. The P^{32}-containing DNA of the infecting phage had already entered the bacteria and was not removed. Thus a normal

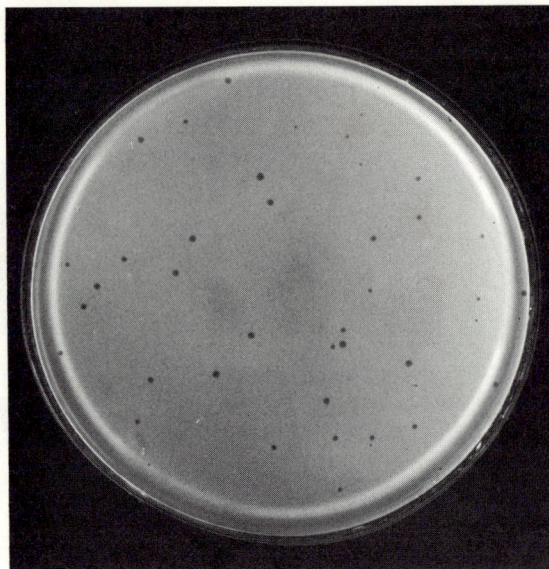

Figure 10-5   Phage T4 plaques on a layer of E. coli. *Each black spot represents a region of lysed bacteria.*

number of new virions were produced; these contained most of the labeled phosphorus but none of the labeled sulfur. This experiment showed that only the phage DNA entered the cell, and that it carried the information for making new virus particles. Phage protein could not have been involved.

Transduction and Transformation

The discovery of two additional phenomena related to viral infection of bacteria lent strength to the concept that genetic information is carried on DNA. We saw earlier that bacterial genes are transferred during a sexual phase. It was soon found that they can also be passively carried from one bacterium to another by phage particles in a process known as **transduction;** this happens when a virus is formed that accidentally contains a small piece of its host chromosome rather than its own viral chromosome. Such a virion

is called a **transducing phage.** When it is released and attaches to another bacterial cell,
its fragment of bacterial chromosome is injected into the new host cell and can engage
in crossing-over with the host chromosome.
If the transducing phage was derived in a
strain of bacteria genetically different from
the strain later infected, a genetically altered
bacterium may result (Fig. 10-7, p. 244). Thus
a methionine-requiring (met⁻) strain can be
transduced to a met⁺ form by genetic recombination with a met⁺ fragment carried to it
by a phage particle from a previous met⁺
host. Transduction was discovered by Norton

Figure 10-6   *The Hershey-Chase experiment, to determine whether the viral
chromosome or the protein coat is the component required for viral duplication.
(a) Virions are prepared with protein coats labeled with S^{35} (black), and
chromosomes containing P^{32} in the DNA (red). (b) Bacteria are infected with
these labeled viruses. (c) As soon as the attached virion has injected its DNA,
the particles are agitated violently in a food blender, which tears the empty
viral coat from the surface of the bacteria. (d) The labeled viral chromosomes
are duplicated and encapsulated in newly synthesized protein coats. (e) After
lysis, some of the released virions have DNA labeled with P^{32}, but none have
S^{35} in their protein coats (dotted lines). Thus the component that was used
for viral duplication must have been the original viral DNA, not the protein.*

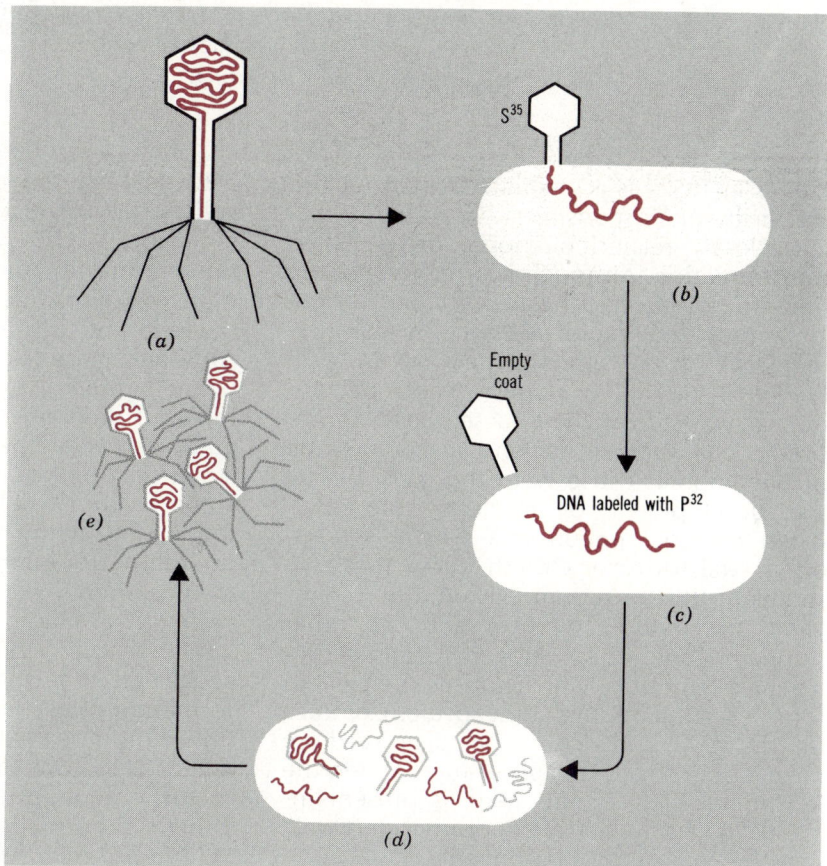

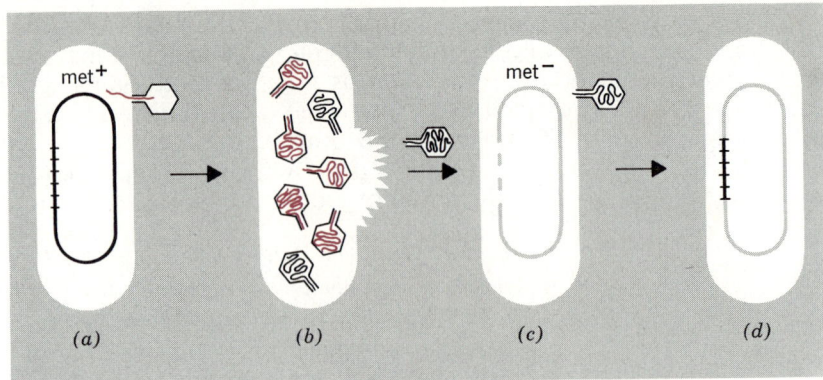

Figure 10-7 *Transduction is the process whereby genetic material from one bacterium is carried to another by means of a phage particle. In (a) a wild-type strain of E. coli (met⁺) is infected by a phage particle, which soon (b) produces a large number of complete new virions and a few defective phages that carry a small fragment of the bacterial chromosome instead of the viral DNA. If one of these transducer particles (c) infects a bacterial cell that carries a mutant chromosome (met⁻) it may (d) insert the nonmutated bacterial gene into the new host by crossing-over, thereby correcting the deficiency of the host bacterium.*

Zinder in 1952, as a graduate student at the University of Wisconsin.

A closely related phenomenon is called **transformation.** Although not realized at the time, the discovery of transformation in the 1920s provided crucial evidence for identifying DNA as the genetic material. In 1928 Frederick Griffith, a public health bacteriologist, was studying the bacterial cell *Diplococcus pneumoniae,* which causes one type of pneumonia. Those were the days, before the development of modern antibiotics, when bacterial pneumonia was a serious, often fatal, disease. Griffith knew that two strains of this bacterium existed, one pathogenic (i.e., causes disease) and the other nonpathogenic. The pathogenic strain carries a **gene S (smooth),** which causes the cells to form a thick carbohydrate capsule. Colonies of this strain growing on an agar medium have a smooth, glassy appearance. When the **R (rough)** allele of this gene is present instead, the bacteria have a normal cell wall, and the cells are nonpathogenic.

Griffith wanted to find out if pathogenic cells, killed by heat to render them harmless, could be used to vaccinate persons against pneumonia. In one experiment he injected mice with a mixture of heat-killed pathogenic bacteria and the live nonpathogenic strain. Although these bacterial cells should all have been harmless, to Griffith's amazement the mice all died of pneumonia. When Griffith autopsied these dead mice he found live S strain (i.e., pathogenic) bacteria. Apparently, some of the R cells had been transformed into S cells by the heat-killed S type bacteria.

Griffith found that when descendants of these newly pathogenic cells were themselves heat-killed, they too could confer virulence on a nonpathogenic strain. This suggested that the phenomenon of transformation was hereditary and that somehow chromosomal material from the heat-killed cells was acting as a **transforming factor,** entering the live cells and conferring genetic information on them (Fig. 10-8). Within the next few years this interpretation was confirmed and it was found that purified cell fractions could produce the same effect. In a paper that is now considered an important landmark, Oswald T. Avery and two colleagues showed that DNA isolated in highly purified form from S cells was just as effective as a transforming factor as whole heat-killed cells. In contrast, other S-cell components—proteins or carbohydrates for example—and R-cell DNA were all without effect in transforming R cells into pathogenic S cells. Furthermore, whole cell extracts treated with the enzyme deoxyribonuclease (which specifically degrades DNA to its nucleotide building blocks) lost their transforming activity, whereas ribonuclease (which degrades RNA) or proteolytic (protein-digesting) enzymes had no such effect; that is, only when DNA was broken down was the transforming capacity lost.

Although these early transformation experiments involved only changes in coat chem-

Figure 10-8 *Transformation is the process whereby chromatin or DNA from one strain of bacterial cells carries genetic information to a different mutant strain. (a) The pathogenic pneumococcus strain S produces a carbohydrate capsule outside the cell coat, while nonvirulent R cells do not. (b) If S cells are killed by heating, the cell wall becomes leaky and small fragments of their chromosomes are released; these can be taken up by live R cells. When the R cells divide, an S chromosome fragment may be inserted into one of the daughter R chromosomes by crossing-over. (c) If the S fragment carries the gene for capsule formation, the resulting cell is transformed into an encapsulated S type.*

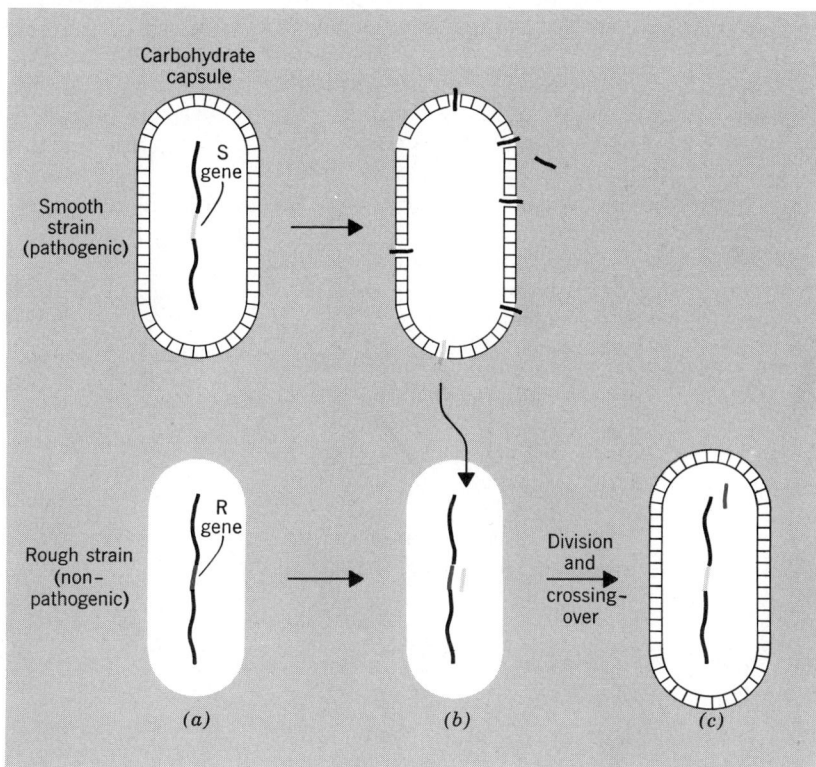

istry of a single bacterium, it is now clear that all genes in almost any bacterial strain can be transformed by incorporation of a length of foreign DNA. Not only did these experiments prove that genetic information is carried by DNA—with the aid of a viral particle in transduction and without it in transformation—they also provided a means for locating genes on the viral or bacterial chromosome. Because only small chromosomal fragments can be transferred in this way, the process reveals which genes are located close together on the bacterial chromosome.

The Structure of DNA

What properties must the prime carrier of genetic information have to perform its role?

1. There must be a means of coding information into it in a relatively permanent way; and the molecule must be large and complex enough to contain a great deal of information.
2. The molecule must be able to replicate itself, and do so with great precision, for we know hereditary characters do this.
3. On the other hand, the copying mechanism must be susceptible of some error to allow for mutations.
4. There must be a mechanism for translating the stored genetic information into structures and processes within cells.

We have seen in Chapter 3 that DNA is, in fact, big enough, complex enough, and of the right structure (a sequence of nucleotide bases) to exhibit all of these properties. The two men primarily responsible for working out this structure were Francis Crick and James Watson. Their discovery represents one of the most dramatic and important advances in the biological sciences.

The Double Helix

In the early 1950s a young American virologist, James D. Watson, went to Cambridge, England—having just received his Ph.D. at

Indiana University—on a research fellowship to study problems of molecular structure. There, at the Cavendish lab, he met Francis Crick, a physicist working on his Ph.D. in X-ray crystallography. Although each was occupied on a different research problem, they both became interested in DNA and soon began to work together to solve the problem of its molecular structure. They actually did no experiments, in the usual sense; instead they tried to gather all the known information concerning DNA and from it to build a meaningful model. The story of the months of activity that led up to their eventual success in this attempt was later set down in a remarkable autobiographical book by Watson called *The Double Helix*[1]. Some excerpts from this book will illustrate both the structural model and the intensely creative act that was its discovery.

Watson, alone in a strange country, did not find his chosen task a simple one, especially at first:

"Most of my time I spent walking the streets or reading journal articles from the early days of genetics. Sometimes I daydreamed about discovering the secret of the gene, but not once did I have the faintest trace of a respectable idea. It was thus difficult to avoid the disquieting thought that I was not accomplishing anything."

Watson and Crick soon became familiar with the information already available on the subject. First, it was known that the DNA molecule was very large and had a long, thin shape. Next, the composition of DNA from four nucleotides was understood, and the structure of those nucleotides had been worked out. Third, Maurice Wilkins at King's College, London, had been studying DNA by X-ray diffraction techniques, and had demonstrated crystal-like regularities in its structure. The molecule was known to have a regularly repeating unit 34 Å long, and in fact

[1] *The Double Helix*, James D. Watson, Atheneum, New York, 1968.

these studies even had shown a hint of a spiral structure of some sort. Fourth, the American physical chemist Linus Pauling had shown that many protein chains were arranged in the shape of an alpha helix (see pp. 68–72), with hydrogen bonds between amino acids holding together successive turns of the helix. Pauling had suggested that DNA might have a similar structure. Finally, Erwin Chargaff had analyzed the purine and pyrimidine content of the DNA of several different organisms including man, and had found that the nucleotide bases do not occur in the same proportions. The base composition of DNA varies from one species to another, although within a species every individual—and each organ within an individual—has DNA with the same base composition. For example, human DNA contains 30 percent of adenine (A) and 30 percent of thymine (T), but only 20 percent each of guanine (G) and cytosine (C); some crab DNA contains 47 percent each of adenine and thymine and only 3 percent guanine and cytosine. Chargaff found consistently, in most DNA's, that A = T, and G = C.

Watson and Crick attempted to construct a model of DNA that would fit these facts. For some time they had attempted to work out a two-stranded helical model in which the two strands were identical, each nucleotide pairing with one of its own kind in the sister strand, A with A, G with G, and so on, but this proved incorrect. The actual act of discovery is best described in Watson's own words (from *The Double Helix*, pp. 120–127).

"My scheme was torn to shreds by the following noon. Against me was the awkward chemical fact that I had chosen the wrong tautomeric forms [forms that differ only by the position of one hydrogen atom] of guanine and thymine [Fig. 10-9]. Before the disturbing truth came out, I had eaten a hurried breakfast at the

Figure 10-9   The enol and keto forms of guanine and thymine. Because these molecules differ only by the shift of a hydrogen atom between adjacent positions (shown shaded) they are referred to as **tautomeric** *forms.*

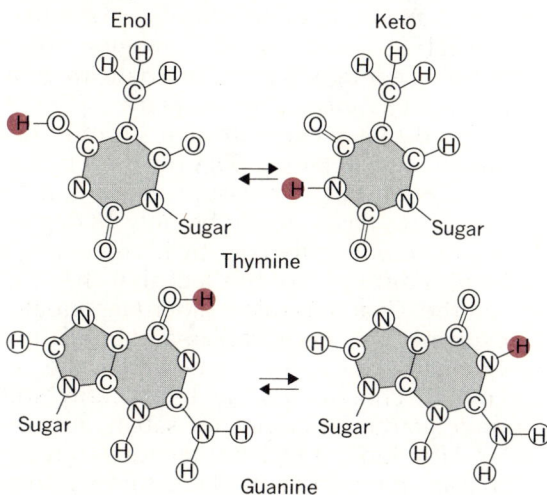

Whim, then momentarily gone back to Clare [the departmental secretary] to reply to a letter from Max Delbruck which reported that my manuscript on bacterial genetics looked unsound to the Cal Tech genetists. Nevertheless, he would accede to my request that he send it to the *Proceedings of the National Academy*. In this way, I would still be young when I committed the folly of publishing a silly idea. Then I could sober up before my career was permanently fixed on a reckless course. . . .

"The letter was not in the post for more than an hour before I knew that my claim [to have devised a beautiful new structure for DNA] was nonsense. I no sooner got to the office and began explaining my scheme than the American crystallographer Jerry Donohue protested that the idea would not work. The tautomeric forms I had copied out of Davidson's book were, in Jerry's opinion, incorrectly assigned. My immediate retort that several other texts also pictured guanine and thymine in the enol form cut no ice with Jerry. Happily he let out that for years organic chemists had been arbitrarily favoring particular tautomeric forms over their alternatives on only the flimsiest of grounds. In fact, organic-chemistry textbooks were littered with pictures of highly improbable tautomeric forms. The guanine picture I was thrusting toward his face was almost certainly bogus. All his chemical intuition told him that it would occur in the keto form. He was just as sure that thymine was also wrongly assigned an enol configuration. Again he strongly favored the keto alternative.

"Though my immediate reaction was to hope that Jerry was blowing hot air, I did not dismiss his criticism. Next to Linus Pauling himself, Jerry knew more about hydrogen bonds than anyone else in the world. . . . Thoroughly worried, I went back to my desk hoping that some gimmick might emerge to salvage the like-with-like idea. But it was obvious that the new assignments were its death blow. Shifting the hydrogen atoms to their keto locations made the size differences between the purines and pyrimidines even more important than would be the case if enol forms existed. Only by the most special pleading could I imagine the polynucleotide backbone bending enough to accommodate irregular base sequences. Even this possibility vanished when Francis [Crick] came in. He immediately realized that a like-with-like structure would give a 34 Å crystallographic repeat [see p. 80] only if each chain had a complete rotation every 68 Å. But this would mean that the rotation angle between successive bases would be only 18 degrees, a value Francis believed was absolutely ruled out by his recent fiddling with the models. Also Francis did not like the fact that the structure gave no explanation for the Chargaff rules (adenine equals thymine, guanine equals cytosine). I, however, maintained my lukewarm response to Chargaff's data. So I welcomed the arrival of lunchtime, when Francis' cheerful prattle temporarily shifted my thoughts to why undergraduates could not satisfy *au pair* girls.

"After lunch I was not anxious to return to work, for I was afraid that in trying to fit the keto forms into some new scheme I would run into a stone wall . . . so I spent the rest of the afternoon cutting accurate representations of the bases out of stiff cardboard. But by the time they were ready I realized that the answer must be put off till the next day. After dinner I was to join a group from Pop's at the theater.

"When I got to our still empty office the following morning, I quickly cleared away the papers from my desk top so that I would have a large, flat surface on which to form pairs of bases held together by hydrogen bonds. Though I initially went back to my like-with-like prejudices, I saw all too well that they led nowhere. When Jerry came in I looked up, saw that it was not Francis, and began shifting the bases in and out of various other pairing possibilities. Suddenly I became aware that an adenine-thymine pair held together by two hydrogen bonds was identical in shape to a guanine-cytosine pair held together by at least two hydrogen bonds [Fig. 10-10]. All the hydrogen bonds seemed to form naturally; no fudging was required to make the two types of base pairs identical

Figure 10-10   The adenine-thymine and guanine-cytosine base pairs used to construct the double helix (hydrogen bonds are dotted). The formation of a third hydrogen bond between guanine and cytosine (as shown) was considered, but rejected because a cyrstallographic study of guanine hinted that it would be very weak. Now this conjecture is known to be wrong. The three strong hydrogen bonds drawn between guanine and cytosine represent the correct structure.

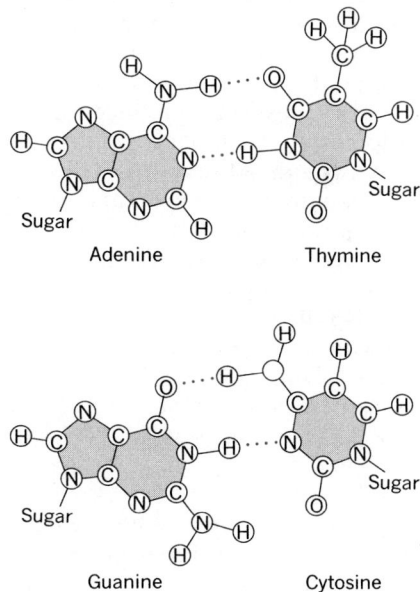

Adenine   Thymine

Guanine   Cytosine

in shape. Quickly I called Jerry over to ask him whether this time he had any objections to my new base pairs.

"When he said no, my morale skyrocketed, for I suspected that we now had the answer to the riddle of why the number of purine residues exactly equaled the number of pyrimidine residues. Two irregular sequences of bases could be regularly packed in the center of a helix if a purine always hydrogen-bonded to a pyrimidine. Furthermore, the hydrogen-bonding requirement meant that adenine would always pair with thymine, while guanine could pair only with cytosine. Chargaff's rules then suddenly stood out as a consequence of a double-helical structure for DNA. Even more exciting, this type of double helix suggested a replication scheme much more satisfactory than my briefly considered like-with-like pairing. Always pairing adenine with thymine and guanine with cytosine meant that the base sequences of the two intertwined chains were complementary to each other. Given the base sequence of one chain, that of its partner was automatically determined. Conceptually, it was thus very easy to visualize how a single chain could be the template for the synthesis of a chain with the complementary sequence.

"Upon his arrival Francis did not get more than halfway through the door before I let loose that the answer to everything was in our hands. Though as a matter of principle he maintained skepticism for a few moments, the similarly shaped A-T and G-C pairs had their expected impact. His quickly pushing the bases together in a number of different ways did not reveal any other way to satisfy Chargaff's rules. A few minutes later he spotted the fact that the two glycosidic bonds (joining base and sugar) of each base pair were systematically related by a diad [i.e., two-nucleotide] axis perpendicular to the helical axis. Thus, both pairs could be flipflopped over and still have their glycosidic bonds facing in the same direction. This had the important consequence that a given chain could contain both purines and pyrimidines [see Fig. 3-28]. At the same time, it strongly suggested that the backbones of the two chains must run in opposite directions. The question then became whether the A-T and G-C base pairs would easily fit the backbone configuration devised during the previous two weeks. At first glance this looked like a good bet, since I had left free in the center a large vacant area for the bases. However, we both knew that we would not be home until a complete model was built in which all the stereochemical contacts [i.e., the geometric arrangement of chemical bonds] were satisfactory. There was also the obvious fact that the implications of its existence were far too important to risk crying wolf. Thus I felt slightly queasy when at lunch Francis winged into the Eagle to tell everyone within hearing distance that we had found the secret of life.

"Francis' preoccupation with DNA quickly became full-time. The first afternoon following the discovery that A-T and G-C base pairs had similar shapes, he went back to his thesis measurements, but his effort was ineffectual. Constantly he would pop up from his chair, worriedly look at the cardboard models, fiddle with other combinations, and then, the period of momentary uncertainty over, look

The Watson-Crick model satisfies all four of the requirements listed on p. 246. It shows clearly how the DNA molecule is sufficiently complex to be able to carry genetic information as a sequence of nucleotides, because any sequence of the four base pairs (AT, TA, CG, GC) is possible. Even the smallest DNA molecules of simple viruses range in the millions of daltons; for example, the chromosome of T2 bacteriophage has a molecular weight of 130 million. Because the four nucleotides average only 660 in molecular weight, the base pairs in even a small chromosome number in the thousands. If the single tangled thread of DNA in each of the 46 human chromosomes (Fig. 10-11) were

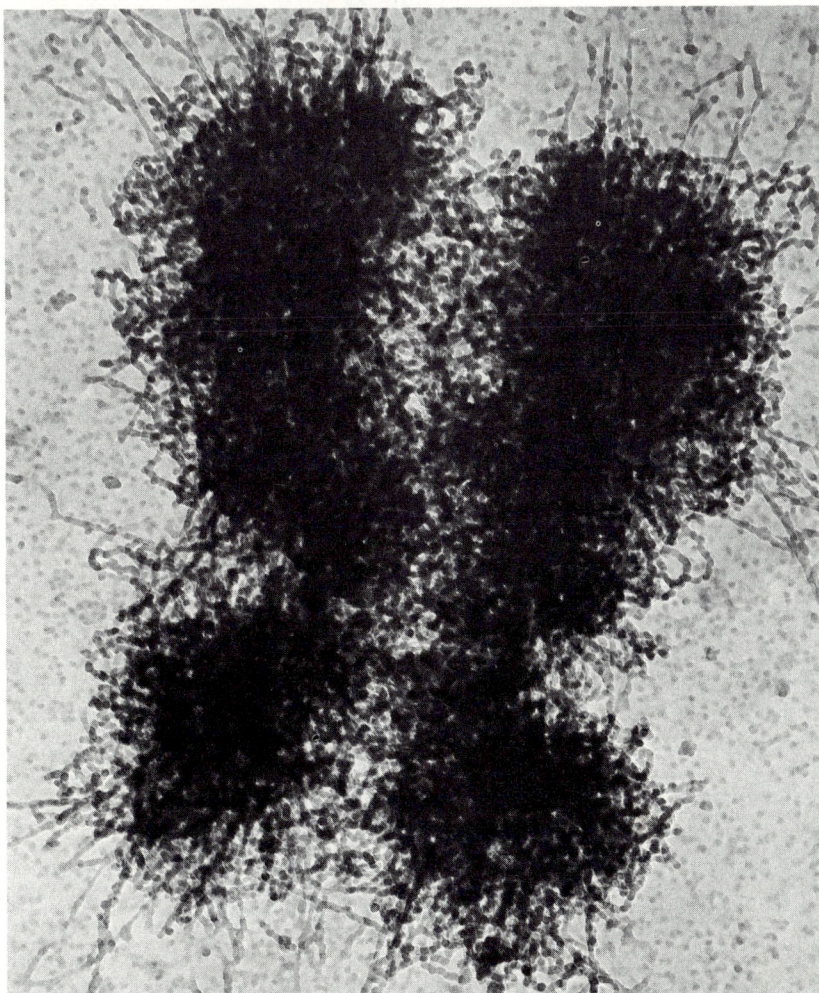

Figure 10-11   Two human metaphase chromosomes, showing the skein of chromatin fibers. Magnification 37,500×.

stretched in a straight line and placed end to end, they would measure about 5 ft long. This is equivalent to about 5 billion base pairs, which is enough to contain the information in a library of a thousand average-sized books.

Implicit in the model also is its copying mechanism. Paired bases break apart at the connecting hydrogen bonds, and the two original complementary strands separate. Each strand then forms a new complementary strand, using itself as a template and the enzyme DNA polymerase to link a new chain of nucleotides together (Chapter 7). If an A is present in the original strand only a T will bind to it in the new helix; a G will pair only with a C, and so on; each original strand forms a complementary strand, a copy of the other. Base pairing permits very accurate replication. Because of its structure and the way it is hydrogen bonded, the probability of an adenine pairing with a thymine is 100,000 times greater than with a cytosine. Two exact replicas of the original double helix are usually produced; however, the very complexity of the molecule assures an occasional copying error.

The Watson-Crick model answered most of the questions about how DNA carries genetic information, and how that information could be replicated, but the mechanism whereby this information could be translated into cellular structures—proteins, carbohydrates, lipids—was less obvious. We shall consider one of these problems—the coding of proteins—in the next section.

The Coding of Proteins

Beadle and Tatum (Chapter 4) showed that genes control the manufacture of proteins, and it was quickly realized that proteins, as enzymes, were in turn responsible for the synthesis of all other cellular components. Enzymes were obviously also needed to link amino acids into peptide chains, but it was clear that enzymes alone could not deter-

mine the **order** of those amino acids unless there were as many different ordering enzymes as there are amino acid combinations. And because all enzymes are themselves proteins, still more enzymes would be necessary to synthesize the enzymes, and so on. What was required logically was another linear structure that could line up the amino acids in some predetermined order, and then a single enzyme for forming peptide bonds between any amino acid pair. Thus during the 1940s the concept of a **template** became accepted.

After the publication of the Watson-Crick model, many scientists in the United States and Europe devoted their attention to the problem of the biochemical steps between the genes on the chromosomes and a particular sequence of amino acids in a protein molecule. It was known that most of the DNA in a cell is concentrated in the nucleus and most of the protein synthesis occurs in the cytoplasm, so it seemed unlikely that the chromosomal DNA was itself the protein-forming template. In fact it had already been shown that protein synthesis can take place in an enucleated cell. Clearly some intermediary "message" had to exist to carry the genetic information from the chromosomes to the cytoplasm.

The most obvious clue in solving this mystery was that the amount of protein synthesized in a cell seemed to be directly related to its content of a second kind of nucleic acid, ribonucleic acid (RNA, pp. 80–81). It was known that cells containing much RNA manufacture a great deal of protein, whereas very little protein is made in RNA-poor cells. It was soon shown that proteins could be synthesized in the test tube if only amino acids, an energy source, and the **microsome fraction** of the cytoplasm (p. 103) were included in the reaction mixture; this fraction is composed mainly of ribosomes, which are rich in RNA.

RNA, moreover, is concentrated mainly in the cytoplasm, but it is synthesized primarily in the nucleus. It is chemically very similar to DNA; the main differences are that the sugar of RNA is ribose rather than deoxyribose (p. 77), and thymine, one of the pyrimidines of DNA, is replaced in RNA by uracil (Fig. 3-31). If an active cell is given uracil labeled with radioactive atoms of tritium, after a brief time most of the label will be found in newly synthesized RNA in the nucleus. A little later much of that labeled RNA will have left the nucleus and entered the cytoplasm (Fig. 10-12).

In 1959 investigators in three separate laboratories independently discovered an enzyme called **RNA polymerase,** which catalyzes the formation of RNA on a DNA template. Such enzymes have since been found in virtually every cell examined. They have been shown to make lengths of RNA that are exactly com- plementary in their sequence of nucleotides to specific segments of one strand of a chromosomal double helix. Moreover, the enzyme itself is only broadly specific; that is, it will work on any DNA template. RNA polymerase purified from the bacterium *E. coli,* for example, will form RNA fibers on the DNA of a toad that are similar to those made by the toad's own cells. Thus the fundamental mechanism for the synthesis of RNA is similar to that of DNA (pp. 178–182). In both cases the immediate precursors are nucleoside triphosphates; in both cases the sequence of the newly formed strand is dictated by the base sequence in the original template DNA strand. In the normal cellular

Figure 10-12 Autoradiograph of a cell labeled with uridine, which is taken up exclusively in RNA. (The uridine contains radioactive tritium (H³) as the label.) After labeling, the cell is fixed and coated with a photographic emulsion. Wherever the label is incorporated into a newly formed molecule of RNA, radioactive disintegrations of the tritium produce dark grains in the overlying emulsion. (a) Cell fixed immediately after 15 minutes exposure to the radioactive uridine. Most of the grains are over the nucleus. (b) A cell treated with label as in (a) but then allowed to stay for 88 minutes more in nonradioactive nucleotides. Essentially all the labeled RNA has moved from the nucleus to the cytoplasm.

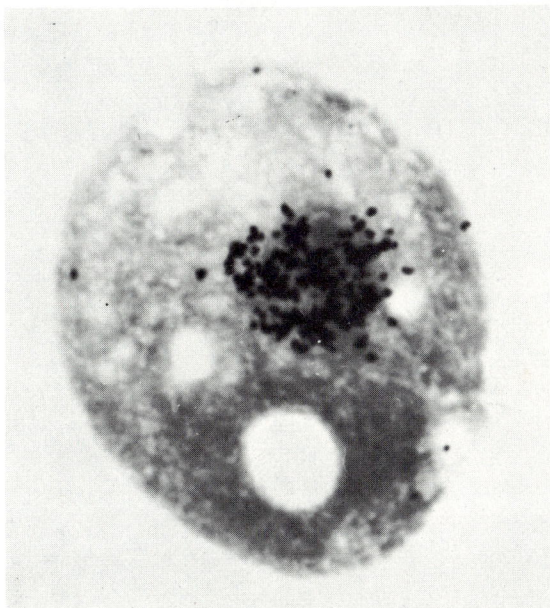

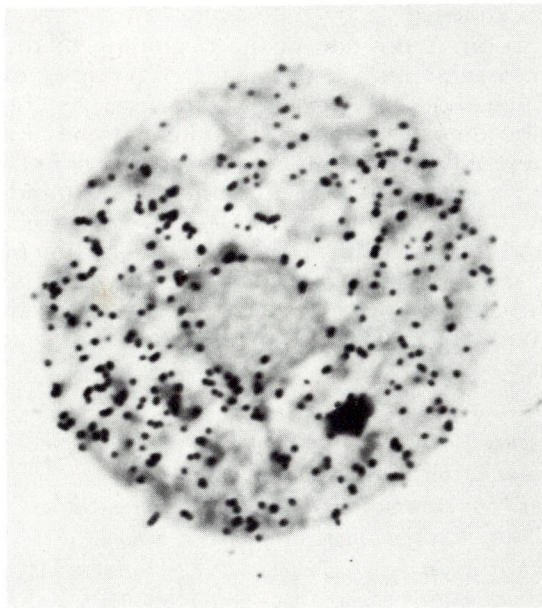

(a)

(b)

process of RNA synthesis in each gene only one of the strands of the DNA double helix is copied.

Recall from Chapter 4 that there are basically three kinds of RNA, each transcribed from different genes on the chromosomal DNA; and that it is these molecules that act as the intermediate links between genetic information coded into the DNA double helix and the synthesis of cytoplasmic proteins. Ribosomal RNA (rRNA) comprises about 50 percent of the two major subunits of the ribosomal particles, on which protein synthesis occurs (Fig. 4-14). Messenger RNA (mRNA) carries genetic information from the gene, where it is transcribed, to the ribosome. Messenger RNA molecules vary in size from a few hundred nucleotides up to 10,000, depending on the size of the gene they represent. Transfer RNA (tRNA) carries specific amino acids to the ribosome and attaches them to the correct site on a strand of mRNA being "translated" into protein. There are about 60 different tRNA molecules.

The role of RNA in protein synthesis was summarized in Figure 4-18. A messenger RNA molecule, formed on one strand of DNA in the gene, attaches by one end to a ribosome. A cloverleaf-shaped molecule of tRNA, with its attached "activated" amino acid, locates its correct position at the beginning of the message template by means of a triplet of nucleotides in the RNA sequence. As the ribosome moves along the mRNA strand, the next tRNA molecule moves into place, finding a second amino acid to the mRNA strand. At this point the first tRNA detaches itself, and the energy released by that detachment is used to form a peptide bond between the two neighboring amino acids. Once released, the tRNA becomes available again to attach

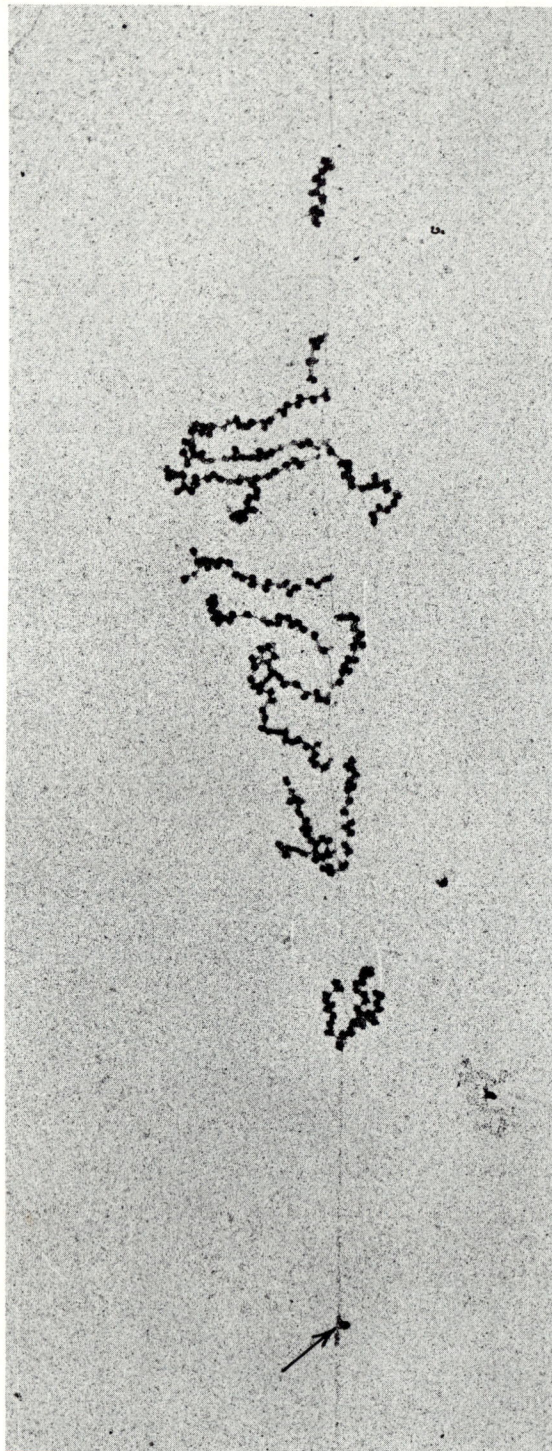

Figure 10-13 A long strand of chromatin DNA from E. coli is shown with attached fibers of mRNA, on which are aligned ribosomal particles. The spot at the bottom (arrow) is a molecule of RNA polyerase, presumably about to begin forming a new mRNA chain. Magnification 64,500×.

to another amino acid of the same kind. Meanwhile, a third tRNA-amino-acid complex has attached, the ribosome moves to the next position, and a **tripeptide** is formed. In this way, as the ribosome moves along the message, a polypeptide chain grows in which the sequence of amino acids is an exact representation of the mRNA nucleotide triplets; which is, of course, a complementary copy of the homologous sequence in the gene.

In the last few years most of these steps have been described in both microorganisms and animal cells. In preparations from *E. coli*, for example, strands of DNA can be seen with attached fibers of newly forming messenger RNA, to which, in turn, ribosomes have adhered (Fig. 10-13). Similarly, fibers of nucleolar genes (which code for rRNA) from amphibians can be seen with their attached, newly forming RNA strands. This gene is repeated, lengthwise, about 450 times along the DNA strand (Fig. 10-14). Several kinds of RNA have been isolated and purified from animal cells. The gene for at least one of these molecules, the small 5s component of amphibian ribosomes, has also been chemically identified (Fig. 10-15, p. 256). Even its location on the *Drosophila* chromosome has been demonstrated by an ingenious technique.

We described in Chapter 4 (p. 101) the molecular hybridization technique with which two lengths of single-stranded nucleic acid can be tested for complementarity. Complementary RNA as well as DNA will bind to a length of purified DNA from chromatin attached to a nitrocellulose filter (Fig. 4-12). In 1970 Joseph Gall and his colleagues at Yale University refined this technique for application to the giant salivary gland chromosomes of *Drosophila* (Fig. 8-16), in which the banding pattern of individual genes or groups of genes is clearly visible in the light

Figure 10-14   *Electron micrograph of the nucleolar genes from an amphibian egg cell. Each long, thin strand of DNA is composed of active regions in which about 100 fibers of rRNA are being transcribed as an RNA polymerase molecule moves along the gene; these alternate with inactive regions called* **spacers.** *Two such repeating sequences along the DNA strand are marked off by arrows. Magnification 20,000×.*

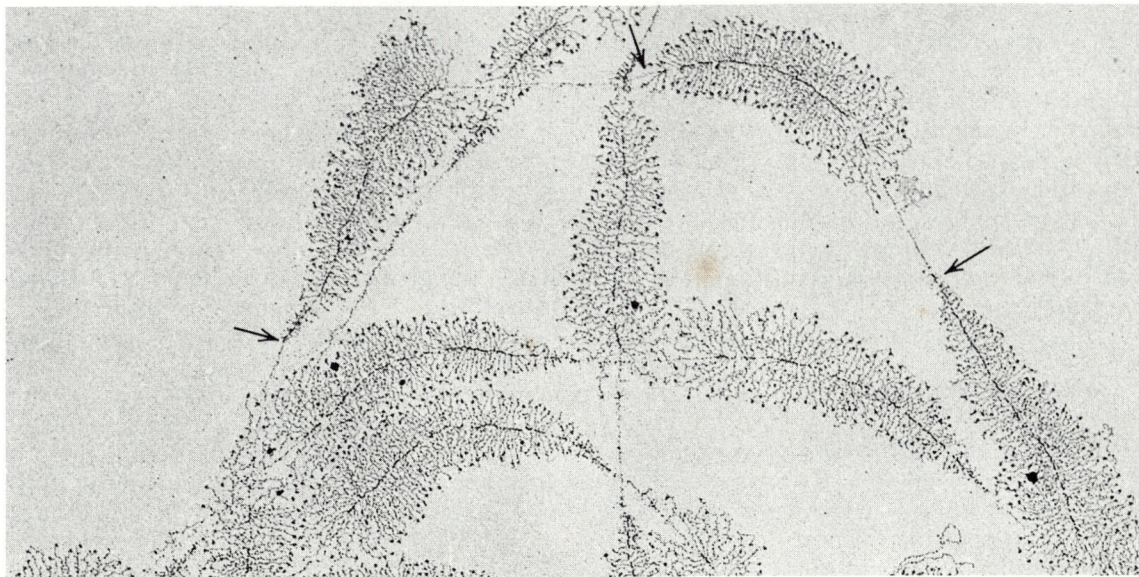

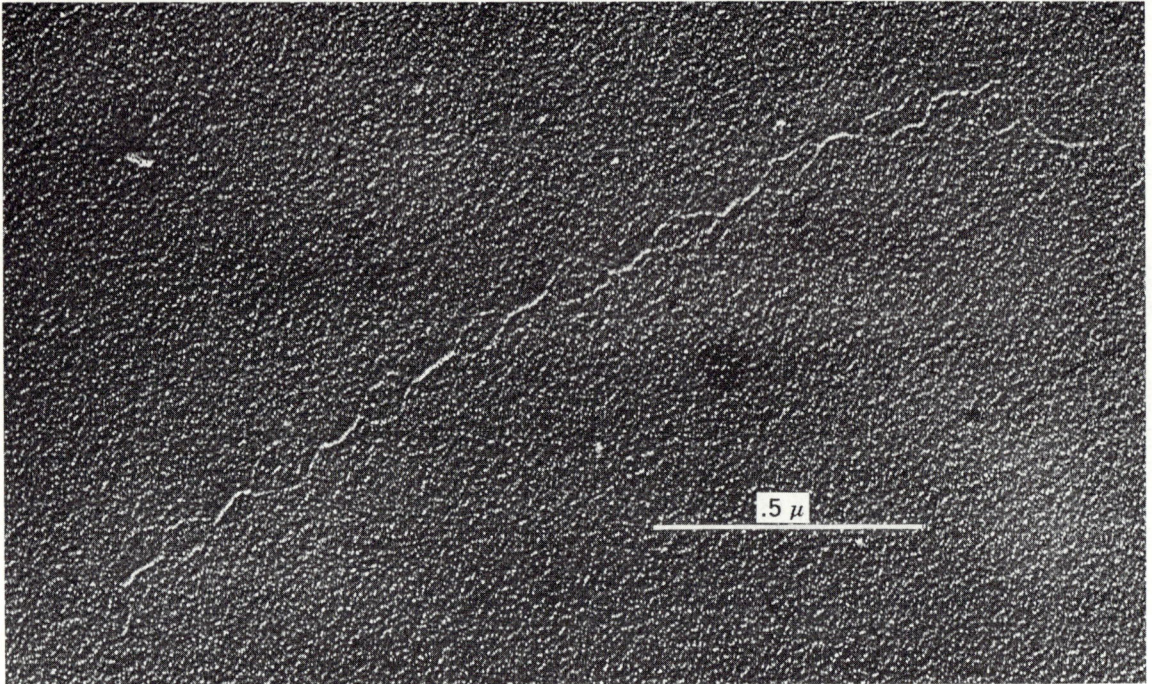

Figure 10-15 *An electron micrograph of a partly denatured molecule of amplified (i.e., repeated) rDNA. Each repeating unit, consisting of a straight strand and a loop, is about .2μ long, which is equivalent to a length of about .5 × 10⁶ daltons of DNA. Each contains a nucleotide sequence that codes for one 5S RNA molecule. Magnification 75,000×.*

microscope. With this technique an RNA molecule can be tested for its capacity to hybridize with specific regions of the chromosome. Figure 10-16 indicates that the 5S RNA of ribosomes hybridizes exclusively with a specific band in chromosome 2 of *Drosophila melanogaster*. This means that that particular RNA is complementary to the DNA only in that location—and is presumably synthesized there.

The Genetic Code

The general role that the three classes of RNA play in protein synthesis was fairly well established by 1960. During the next decade a growing volume of work by many investigators culminated in "cracking" the genetic code itself.

Because there are 20 biologically important amino acids and only four different nucleotides, the first questions to be answered were, How is a nucleotide sequence represented by an amino acid sequence? How is the four-letter "language" of nucleic acids translated into the 20-letter language of proteins? If each nucleotide were to code for one amino acid, only four amino acids could be accounted for. If two nucleotides together were to code for one amino acid, there would still only be 16 different combinations, still not enough. Therefore, on such mathematical considerations alone, it was obvious that at least a triplet of nucleotides in sequence must be required to specify an amino acid.

This would provide for 64 possible combinations, which is more than enough.

Molecular geneticists were working on the code problem at about the same time that other biochemists were experiencing their initial success in sequencing proteins and other macromolecules (see pp. 65–68). It was known, for example, that the tobacco necrosis virus contains an RNA "chromosome" (instead of DNA) made up to 1200 nucleotides, and that this molecule codes for the coat protein of the virus. When it was found that each of the protein subunits had 400 amino acids, the relationship $1200/400 = 3$ nucleotides per amino acid could hardly go unnoticed.

Early studies had also shown that ribosomes isolated from eukaryotic cells or from bacteria could make proteins in a test tube if only ATP, amino acids, and cell extracts containing RNA were present in the reaction mixture. Moreover, the template RNA could be from any source. Viral, bacterial, or tissue RNA could all be read as messenger RNA by ribosomes of *E. coli*, for example. Thus the genetic code seemed to be a universal language.

Synthetic Messengers

If *E. coli* ribosomes could read a "foreign" message and translate it into protein, perhaps they could read a totally synthetic message. Severo Ochoa of New York University had synthesized an RNA molecule made of only a single repeating uracil ribonucleotide. In 1961 Marshall Nirenberg of the U.S. Public Health Service and a colleague, Heinrich Matthei, incubated Ochoa's **polyuridilic acid** (poly U) with purified *E. coli* ribosomes in a series of 20 test tubes. Each of the tubes also contained tRNA, ATP, the necessary enzymes, and all 20 amino acids; in each tube, however, a different amino acid was radioactively labeled. In 19 of the test tubes no polypeptides were made, but in the one in which phenylalanine had been labeled, labeled polypeptide chains were formed of repeating units of phenylalanine; poly U had coded exclusively for polyphenylalanine. If the code

is indeed a triplet, then UUU is the code word, or **codon,** for phenylalanine.

By repeating this experiment with poly C, poly A, and poly G, Nirenberg and his colleagues soon worked out other codons. CCC, for example, codes for proline, AAA for lysine, GGG for glycine. They found that a ribonucleotide synthesized mainly from uracil plus a small amount of guanine would produce a peptide made largely of phenyl-

Figure 10-16   Autoradiographs of chromosome 2 of Drosophila melanogaster. Purified preparations of the small component of ribosomal RNA, the 5S subunit, radioactively labeled with tritiated uridine, were allowed to hybridize with the chromosomes. The clusters of black grains at position 56EF indicate that the RNA has bound to that portion of the DNA and nowhere else.

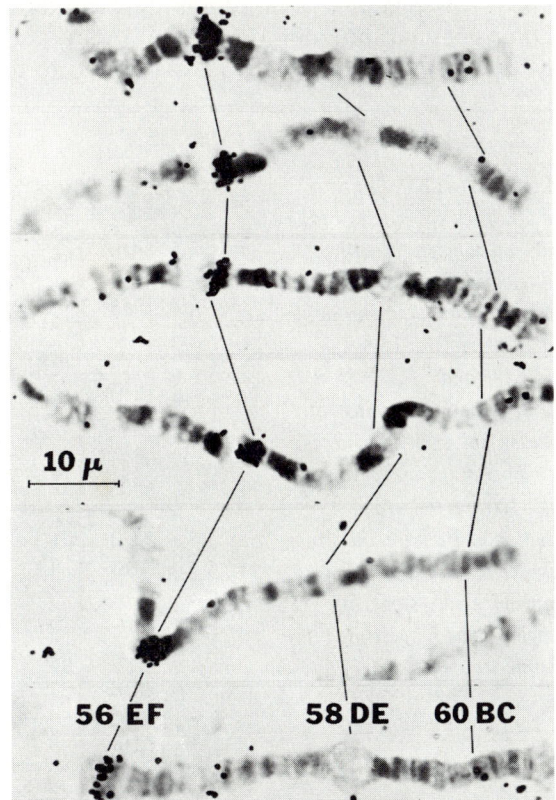

10 μ

56 EF   58 DE   60 BC

alanine plus some valine. Because the phenyl-alanine must have been specified by UUU sequences, the codon for valine must have been GUU, or UUG, or UGU. At first, from this kind of experiment, the investigators had no way to distinguish the order of the bases. For UUU this presents no problem. But a triplet of three different nucleotides—A, U, and C, for example—can be arranged in six different ways: AUC, ACU, UCA, UAC, CAU, CUA, each of which might represent a different codon.

This problem was solved largely by H. Gobind Khorana and his colleagues at the University of Wisconsin. They worked out a method for chemically synthesizing lengths of DNA or RNA in which two or three nucleotides were repeated in known sequence, forming long chains. Thus, for example, they could produce DNA strands of alternating cytosine-thymine residues, CTCTCTCT . . . , which would be transcribed by RNA polymerase to produce complementary mRNA chains of guanine-adenine, GAGAGAGA. . . . Khorana found that when it was added to a cell-free protein-synthesizing system like the *E. coli* ribosomal fraction used by Nirenberg, such a message would produce poly-peptides of arginine and glutamic acid. This indicated that one of these must be specified by the codon GAG, and the other by AGA, as these are the only triplets available in such a sequence. By the use of these methods, Nirenberg, Khorana, Ochoa, and others soon succeeded in working out the base sequence of the codons for all of the amino acids (Fig. 10-17). Khorana has recently synthesized nucleotide chains of known repeating sequences of all four nucleotides, thereby re-producing exactly a known gene of a yeast cell (see p. 189).

The Code is Universal

Although the dictionary of genetic codons shown in Figure 10-17 was determined primarily with experiments using protein-synthesizing systems from *E. coli* and other microorganisms, there is ample evidence that the genetic code is the same for all types of

Figure 10-17 The dictionary of genetic code words consists of 64 nucleotide triplets and their corresponding amino acids. Three triplets, UAA, UAG, and UGA do not code for amino acids, but cause polypeptide synthesis to terminate; these three act like periods in a sentence. Because the other 61 codons code for the 20 amino acids, some are "synonyms," specifying the same amino acid.

SECOND LETTER

FIRST LETTER	U	C	A	G	THIRD LETTER
U	UUU } PHE UUC UUA } LEU UUG	UCU UCC } SER UCA UCG	UAU } TYR UAC UAA TERM. UAG TERM.	UGU } CYS UGC UGA TERM. UGG TRP	U C A G
C	CUU CUC } LEU CUA CUG	CCU CCC } PRO CCA CCG	CAU } HIS CAC CAA } GLN CAG	CGU CGC } ARG CGA CGG	U C A G
A	AUU AUC } ILE AUA AUG MET	ACU ACC } THR ACA ACG	AAU } ASN AAC AAA } LYS AAG	AGU } SER AGC AGA } ARG AGG	U C A G
G	GUU GUC } VAL GUA GUG	GCU GCC } ALA GCA GCG	GAU } ASP GAC GAA } GLU GAG	GGU GGC } GLY GGA GGG	U C A G

plant and animal, including man.

We mentioned in Chapter 3 that human hemoglobin is made of two kinds of polypeptide chains, alpha chains and beta chains. The alpha chains consist of 141 amino acids each, the beta chains of 146. It was shown some years ago that the hemoglobin of patients suffering from sickle-cell anemia differs from normal hemoglobin in only one amino acid in the beta chain (Fig. 10-18), where a valine is substituted for glutamic acid.

Since then over 100 mutant hemoglobins have been found. Most of these also result in only a single amino acid replacement in either the alpha or beta chain. When these mutations are examined, most of them can be explained on the basis that only a single nucleotide in one codon has been altered; the result of the mutation is to convert the codon into a triplet specifying a different amino acid. To return to the example of sickle-cell anemia, a glutamic acid is replaced by a valine. According to Fig. 10-17, GAA specifies glutamic acid; GUA codes for valine. The difference between the mRNA that codes for the two is the replacement of an adenine by a uracil. The gene defect that represents the mutation that causes sickling must therefore be a substitution for thymine by an adenine in the chromosomal DNA.

Analysis of a large number of mutations in many vertebrates and invertebrates reveals a complete consistency between the code words established with bacterial protein-synthesizing systems *in vitro* and those derived from animal mutations that have occurred throughout the course of evolutionary history. The genetic code has been shown to be identical in forms as varied as man, *E. coli*, the tobacco plant, and the silkworm.

Conclusion

The main source of information (although not the only information, as we shall see in Chapter 12) passed from one generation to the next is that carried on the haploid chromosome set of the egg and sperm. We have seen that this genetic information is coded into the sequence of nucleotides that comprises a strand of DNA, and is transcribed in complementary fashion by RNA polymerase in the formation of a molecule of mRNA. In the cytoplasm the sequence of nucleotides in that strand of mRNA is translated by an attached ribosome into a corresponding sequence of amino acids on the basis of the three-for-one code. Using this information, Watson has argued recently that organisms are complex but not infinite in their complexity.

Thus, if an average-size protein contains 300 amino acids, then the gene coding for that protein requires 900 nucleotide pairs (i.e., 900 nucleotides in sequence along each strand of the DNA double helix). Because each nucleotide pair has a molecular weight of 660, an average gene must have a molecular weight of about 600,000 (i.e., 660×900). This means that if we know the molecular weight of the total DNA in a cell, we can estimate the maximum number of different proteins that that cell can possess (i.e., total DNA divided by 600,000). For example, the haploid molecular weight of the DNA in *E. coli* is about 2 billion daltons. Dividing that figure by the weight of an average-size gene (2 billion divided by $600,000 = 3000$) corresponds to a maximum of about 3000 average size different proteins. Since most proteins are enzymes, and we know that on the average it takes one or two enzymes to form each macromolecule of every kind (lipids, carbohydrates, nucleic acids, etc.), we can predict with some confidence that a bacterium such as *E. coli* can contain no more than a few

Hbᴬ --his-val-leu-leu-thre-pro-glu-glu-lys--

Hbˢ --his-val-leu-leu-thre-pro-val-glu-lys--

Figure 10-18   Corresponding portions of the beta chains of normal hemoglobin (Hbᴬ) and that of a person suffering from sickle-cell amenia (Hbˢ). The sickle-cell mutation causes anemia by substituting a nonpolar amino acid (value) for a negatively charged one (glutamic acid). This changes the solubility of the hemoglobin molecule, and decreases its oxygen-binding capacity.

```
pCATCGATCGCGCGATAGCGCGATCGCGATCGCGATCGAGATCTCGAGCGATCGAGCGTAGCGCGATATAGCAGAGATC
ACGCTAGCGATCGAATCTCGAGAGCTAGCGATCGAATCGCCGCGATAGCGATCGAGACTTAGCGCGATAGCAGAGCTT
AGCTAGCTAGCGCGAGAGAGCTCTATAGAGCTATATCGAGAGCTATATCGCGATAGCTAGCTAGCGAGAGATCTAGAG
ATCGATCGAGATCTCGAGATCGAGATATAGCGCGATATAGCGCGATATAGCGAGATCGAATATCGCATAGCGATAGCTAGCT
AGCGCGATAGCGCTATATAGCGCGAGATATAGACTAGCTAGCTAGATCTAGCTATTAGCCTAGACTAGATCTACATCG
AGCTCTCGAGATCTTATATAGCGCGATAGCTATAGCGATATAGCTAGGCTAGCTAGCTAGCTAGAGATCGAGCTATAT
ACGAGATAGCGATATATGCGAGATATGCGATAGCTATATATCGAGATGCTAGCTAGCTAGCTCTAGACTCTCTAGATA
GACTCGATCGATCGATCTCTCGGAGACTCTAGCTATGCGATCGATCCGAGATCGCGATCGCATCGAAGCTAGCTAGCT
CTGACTAGCTAGCTATATCGCGATAGCGATCGAGATCGAAGCTAGCTAGCTAAAGCTCTAGAGCTAATAGAGACTAGT
GCGATAGCAGAGCTCGCGATCGAGATCTCTAGAGACTCTCGAGAGCTCGAGCTAGTCGGATCGAGCTAGAGATCGAGC
AACCGGTTAAAGGCCCTTAAGGCCCTTTAAGGCTCCTGGAACGCTAATGCAGGCTAGCGCTAGCCTTAGG
TTAGAGCTAGGACTCTAGACTCTCGACTCATAGAGCGCGCTATATAGCGCGAATATCGCGCGAATAAATTGGCCATTA
AAACTCGGAATCTCGAGGCCGAATTAAGGCGCGAGACTGACTGACTGACCGCGCTATATAGCGCGAGATCGCGATAGC
AGCTCTAGAGACTTCGAGACTCTAGCGCTAGCTCGATCGATCGAGATATAGAGCTCTCTGAGCGCGATATAGCGA
TAGCTCGATCGATCGCGAGATATAGCGCGCGATATAGCGCGAGAATTTGAAGAGACTTTAAAGCCGAGATCGACTCGG
CGAGCTCGAGCGCTATAGCGCGATCGCGAGAGATCTCGCGCTCGCGATATAGCGCTATATAGCGCGATCGCGAGCGTC
AGCTAGCTAGCGCGATATAGCGCGAATATAGCGATAGCGAATTCGAGAGACCTAGCGAGAGCTACGAGACTAGA
TGCTAGCGCTAGCGCGATATAGCGCGAGATCTCTAGCGCATAGCGAGATCGAGCTATCTCTCAGAGAGCTATAGAGGT
AGCGCTATAGCGCTATAGCGCGATATATAGCGCGATATCTCTATATAGAGAGACACATATAGACAGATAGATATATAT
GCTATATAGAGACACACATATAGAGCGCGCTAGAGCTATAGCGCCGAATAAGAGACACATAGACTAGCTAGCGATAGA
CAGATCGCGATATCGAGAGCTCTAGAGCTCTCTATAGAGAGACACATAGAGCGCGATAGAGCGACGAGCAAGCGAGCT
TAGAGCGCGATATCGCGCGATATACAACAGAGACATATAGAGAGCGCGTGTGCTTAGAGATGATACAGAGATAGAGTC
GCAGATAGACACAGATACAGATACAGATACAGATAGATAGAGCGCTAGCGTGTGCGACATGAAGACTAGCGATGCTGA
AGAGCTCTAGAGCTAGCGCTCGCGATCTCGAGAGCGAGAGCTCTCGAGAGCTCTCGAGACATAGACAGCTCGATCGATGCGAGGAACGA
TGGCGATATAGCGCGCGATATAGCGCGATATAGCACACAGATTATCGCGCTTAAGGCGGAATTGCCGTGCAGAAGTG
CTATATACTATATACACATATAGAGACACATAGAGAGCGCGGGGAATTAACCTTGGGAAACCTTGAAACCTTGTGCAC
ACTAGAGCCTAGCTAGCGCGATATAGCGCATAGAGCAGCGCGAGAGCGCGTGGCAGCTAGCTCGATCGGAGCGGATGCA
GCGCGATATAGCCGCGATATAGCGCGGAGACACATAGAGCGCGTGTGTACAGATACAGATAGACATGACT
CTCGCGGACAACCAGAATTAGGCGCTCGGACACATAGCCGGAGATACACATAGACAGATACAGATACAGATACAGC
TGCGCGATAGCGCGATATCGCGCTAGAGACAGATACAGATCGCGTAGCGCGATAGCGCGATATGCCGGATAGACATTG
GACACATAGACAGTAGACAGATATAGACACAGATAGACACCAGGTTGTGCTCGAGAGACACAGAGATAGATGCGTG
AGACCATAGAACACAGATATGAGACACATAGAGAGCGCTCGCGAGACATGACCAGATAGAGCGTAGACAGATAGCAGT
TAGACACAGATATGCGCGCTAGAGACACAAGAGCTCTCGGACAGATACAGATAGAGACACAGATAGACAGGAGATTCG
CTCGGACACAGATAGAGCGCTAGAGCGCTCGCGCTTAGACAGAGCTCGAGAGAGCTCTCGAGACACAGTAGGACAGATC
GCTGAGGACAACCAGGGATTTAGAGACACAAAGTTGGAACCGGATTAATGCCGGAACATAGCGATGACAGAATTGAGA
AGACCATAAGGAACCAGATATGGAACAGAGATAGACAGATAGACACGAGATAGAGACAGATAGACAGATACGCTGGAT
TTAGACCAGATAGAGACAAGAGAGACCCATTGGCCAAGGTCCGGAACTTAGGAACGAATTGGCCTGACAGATAGACGA
GCCTGGAACCAAGGCCTTGGAACCAGAAGACAATTAGACAGGAATTCCGTAGACGAATGGACAGGCTCGGACTAGACG
CCAGATATTCCGGAATTAGGCCGGAATTGCCGGCTTAGAGACACAAGGCCTGGAACGAATTCGGGAATGGGAACCAGT
AAGGCCTGGAAGCGCCTTGAAGGCCTGGAAGCCTCTAGAGCGCCTTAAGGCCGGAATTGGCCTTGGAACGGGAACTGA
GGCCTGGAACCAAGGCCTGGACCAATTAGAGAGAACCAGAGATTCCGGGAACCAATTAAGGCCTTGAACAGGCTGGAC
TTGGCCTTGGAATTCGGAACCAATTAGGGCCTGGAACCAAGGCCTTGAACGCCTTAGGACCTCGGTACAAGGCTGAGT
CATTGCCGGATACCAGGCCTGGAACCTTCGGACAGGCCTGACAGGTCGGACATGCTGACAGGCTGACAGACAGTAGGCG
GGACCAATTGCCGGCTTAGACCAGGCCTGCCAGAGGACCTAGGCTTGAACGGTCCGGTGACAGGCTGACGTAGGCTGA
TGGCTTAGGACCATTGGCTCGCGCTTAGACAAGGCCTGACAGGCCTGGAACATAGGCCTGGAACCAGGCTGGACAGAT
TGCGGATAGAGACCAATGGACAGATTAGGACCCAGGATTCGCGTAACCAGATTCGGACAATTCGGACTAGGACATGACG
AATTAGGAACCAAGGAATTCGCCCTTGGAACCAGGCTCCGGAACGGGTCCTGGAACAATGGAACGATAGACTGACAGT
GCTTAGGCCTTGACCAAGGCCTGGACAAGGCTTGGACCCAGATTCGGACAATGCGGTGGTAGACAGA
CCTGGAACCAGAATTCCGGAACCAATTGCCTGGACAGGGTCGGACCATTAAGGCCTGACCATAGGACATAGATTGTAT
AATTCGGACCAAGGTTAACCAGGAATTCCGGTAAGGCCTGGACAAGGCTTACGTGACGATTCGGACTAGACATGAGAC
CAGGATTCGGAATTAACCAGGGCCTTGGAACCAAGGCCTTGGAAGGTCGGAATTAGGACAGGATAGACATGACGTGA
TGCCTGGACCAGGATTCGCGCTTAGGAACCAGGCCTGGAACCAGAGGCCTGGACCAGGAACCATTGGAACTGACGACG
CTGGAACCAGAGGCCTGGACACAAGGCCTGGACAACAAGGCCTGACCAGGCCTGACAGGATGAGACGATGACGTCGAT
GCTTAGGAGCCTTGAGAGAGCTCTCGAGAGCTCTCGAGAGACACATAGCGCTCTCGAGACAGCTCGACAGGCTGAGCA
ACTTAGACCAAGGCCTGGACCCAAGGATTCCGGATTACCAAGGCTTGGACCAGGCTGGACAAGGTCGGATGACAGATG
CGGATTAGGCCTGGACCAGGCTGGACAGGCTGACAGGCTGACAGGCTTGAGGCTGACCAGGCTGGAACGATGATAGGT
TGGCCTGGAACCAATGCCGGATAGGCTGGACCAAGGCCTGGACAAGTTCGGACCAATGCGTAGGAACGATGAGACAGC
GCAATGCGGATGGAACCAGATTCGGACCAAGGCCTTGGAACCAATGGCTTGAACCAGGCTTGAACCAGGCTGACAGGCA
AATTCGGAATTCGGAATTCCGGACCAATTCGGACCAATTCCGGAAATGGCTTAGAACGGAATGCTGAACATGACAGTG
GCTTAGGCTGGAACCAAGGATTCGGAATGGCCTGGAACCAGGCTGGACCAGGCTTGAACGAATTCGGACAGTGACAGT
CGGATTAGGCTGGAACCAAGGCCTGGACCAAGGCCTGGACAGGCTGACAGCTGACAGCTGACATTACAGTCGACGAAG
AGCTGGACCAAGGCCTGGAACCAAGGCCTGGACAACAAGGCTATAAGGACCAAGGCCTGGACAGGCTGACAGATGCTC
TGGACCAGGCCTGGAACCAAGGCCTGGACCAAGGCCTGGAACCAGGCCTGGACAAGGCCTGGAACCAAGGCTGACAGT
GACCAATGCTGGACCAGGCCTTGGAACCAAGGCCTGGACAACCAAGGCTCCGGATACAAGGCCTGACAAGCTGACAGT
ATGAACCAAGGCCTGGACCCAAAGAGAGGCCTCTCGGGACACAGGCCTTGGAATTACCAAGGCCTGGACCTGGACTGAG
AAGGCCTGGAACCAAGGCCTGGACCAGGCCTGGAACCAAGGCCTGGAACCAAGGCCTGGGAACCAATGCGCTGACAGA
TTCCGGATTACCAAGGCCTGGACCAAGGCTGGAACCAGGCCTGGAACCAGGAATTCGGAATTCGGATACCAAGGCCTG
GGCCTGGAACCAGGCCTGGACCAATGGAACCAATGGCTGGACCAGGCCTGACAACCAATGGACAGGATACGATGCGGT
CCTGGAACCAAGGCCTGGACCAAGGCCTGGACCAGGCCTGGACCAGGCTGGACCAGGCTGACCAGGCTGACAGTGAGA
TGAGGCCTGGAACCAAGGCCTGGACCAAGGCCTGGACACAAGATACAGATAGGCTGACAGGCCTGACAGGCTGACGGT
AAGGCTGGAACCAAGGCCTGGACCAGGCCTGGACCAGGCTGGACAGGCTGACAGCTGACAAGCTGACAGTCGAGATGA
GCCTGGACCAAGGCCTGGACCAAGGCCTGGAAACGGTAGGACAAGGCTGACAAGGCCTGACAGGCCTGACAGTCGTGG
CCGGACCAATTGCCGGATTACCAAGGCCTGGACCAGGCTGGACCAGGCTGGACCAGGCTGACAGGCTGACAGCTGAGA
```

Figure 10-19   A hypothetical base sequence for the chromosome of a bacteriophage such as ΦX174, one of the smallest viruses known. Although this is an imaginary sequence, it is of the right length, and contains about the right proportions of the four nucleotides, to be representative of the viral genome. The chromosome of an E. coli cell would have a base sequence requiring about 2000 of these pages. About a million pages would be necessary to show the base sequence of the DNA in a human cell.

thousand different molecules of biological importance. This, in fact, represents a vast quantity of information (Fig. 10-19). However, biochemists are already familiar with metabolic pathways in cells (Chapter 4) that involve perhaps 1000 small molecules and metabolites. Watson concludes that this represents a substantial fraction, perhaps one-third of the total. This conclusion is most heartening, for it suggests that within the next few decades we may approach a situation when we can describe every reaction that takes place in a living bacterial cell.

If these considerations should lead to a sense of self-satisfaction, consider the awesome fact that an *E. coli* cell in a simple growth medium reads off its 2 million nucleotide pairs and synthesizes all of its several thousand component molecules *every 20 minutes!* Human cells, containing 1000 times as much DNA as *E. coli* (and therefore 1000 times as many codons), can perform the same feat in 24 hours when growing in tissue culture or in the embryo.

It is this fantastic capability of an organism to translate a set of instructions coded into the DNA of a fertilized egg into a fly, or a chick, or a man that we turn to next, as we consider the problem of reproduction.

Reading List

Benzer, S., "The Fine Structure of the Gene," *Scientific American* (January 1962), pp. 70–84.

Britten, R. J., and D. E. Kohne, "Repeated Segments of DNA," *Scientific American* (April 1970), pp. 24–30.

Crick, F. H. C., "The Genetic Code," *Scientific American* (October 1962), pp. 66–74.

Hartman, P. E., and S. R. Suskind, *Gene Action* (2nd ed.). Prentice-Hall, Englewood Cliffs, N.J., 1969.

Kornberg, A., "The Synthesis of DNA," *Scientific American* (October 1968), pp. 64–78.

Lehninger, A. L., *Biochemistry*. Worth, New York, 1970.

Loewy, A., and P. Siekevitz, *Cell Structure and Function*. Holt, Rinehart & Winston, New York, 1969.

Losick, R., and P. W. Robbins, "The Receptor Site for a Bacterial Virus," *Scientific American* (November 1969), pp. 120–124.

Luria, S. E., "The Recognition of DNA in Bacteria," *Scientific American* (January 1970), pp. 88–102.

Nomura, M., "Ribosomes," *Scientific American* (October 1969), pp. 28–35.

Ptashne, M., and W. Gilbert, "Genetic Repressors," *Scientific American* (June 1970), pp. 36–44.

Sistrom, W., *Microbial Life*. Holt, Rinehart & Winston, New York, 1969.

Spiegelman, S. S., "Hybrid Nucleic Acids," *Scientific American* (May 1964), pp. 48–56.

Stahl, F., *The Mechanics of Inheritance* (2nd ed.). Prentice-Hall, Englewood Cliffs, N.J., 1969.

Stanier, R. Y., M. Douderhoff, and E. A. Adelberg, *The Microbial World* (3rd ed.). Prentice-Hall, Englewood Cliffs, N.J., 1970.

Stent, G. S., *Molecular Biology of Bacterial Viruses*. Freeman, San Francisco, 1963.

Swanson, C. P., T. Merz and W. J. Young, *Cytogenetics*. Prentice-Hall, Englewood Cliffs, N.J., 1967.

Temin, H. M., "RNA-Directed DNA Synthesis," *Scientific American* (January 1972), pp. 24–33.

Tomasz, A., "Cellular Factors in Genetic Transformation," *Scientific American* (January 1969), pp. 38–44.

Watson, J. D., *The Double Helix*. Atheneum, New York, 1968.

Watson, J. D., *Molecular Biology of the Gene*. W. A. Benjamin, New York, 1970.

Whitehouse, H. L. K., *Towards an Understanding of the Mechanism of Heredity* (2nd ed.). Edward Arnold, London, 1969.

Woese, C. R., *The Genetic Code*. Harper & Row, New York, 1967.

Yanofsky, C., "Gene Structure and Protein Structure." In R. H. Haynes and P. C. Hanawalt (eds.), *The Molecular Basics of Life*. Freeman, San Francisco, 1968.

part five

Development: The Expression of Genes

chapter eleven | *Human Reproduction*

"The purpose of this awe-inspiring impulsion [sex] is nothing more, nor less, than the enrichment of life. Reproduction purely considered gets on a great deal better without anything so chancy as mating. What sex contributes to it is the precious gift of variation, as a result of commingling. And as variety is the spice of life, it has come to be—thanks to the invitation of sex which creatures accept with such eagerness—one of life's chief characteristics."

DONALD CULROSS PEATTIE, 1935, AN ALMANAC FOR MODERNS

Reproduction is the main theme of life. All other aspects of living systems that we discuss—energy utilization, digestion, circulation, respiration, growth, coordination—may be viewed as processes that enable organisms to survive in order to reproduce. In this sense each adult individual is an elaborate device for producing eggs and sperm and for bringing them together in the process of fertilization. "An adult," it has been said, "is merely an egg's way of making another egg."

Each of us began our individual lives as just such a single cell, a fertilized egg (Fig. 11-1, p. 266). In nine months of development in our mother's uterus, each of us replicated our diploid gene complement and cleaved into two cells, then four, then eight (Fig. 11-2, p. 266). Soon we had produced a tiny ball of a few dozen cells, from which we began to mold the complex set of functional organs and tissues and grew to the 50,000 billion cells that form a baby (Fig. 11-3, p. 267). Each of us increased in weight from a few tenths of a microgram to perhaps 4 kg (8 lb) in the process.

After a bacterial cell in a culture replicates the sequence of 2 million nucleotide pairs in its DNA and synthesizes a new set of 90,000 ribosomes and many thousands of other organelles, it divides, forming two cells from one. The two daughter cells represent a new generation; the process by which they were formed we call **asexual reproduction.** Our bacterial cell has clearly accomplished the goal

Oöcytes of the South African clawed toad. Magnification 8×.

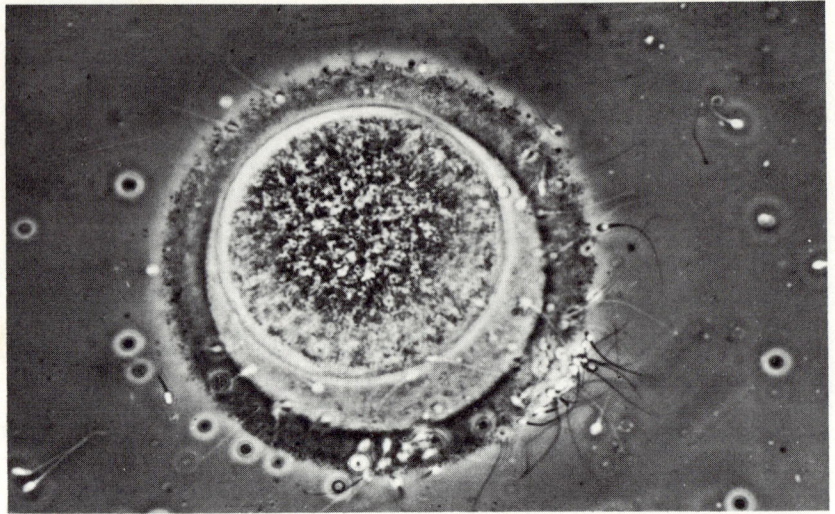

Figure 11-1 *A human egg with its surrounding layer of supporting cells, being penetrated by sperm. Magnification 500×.*

Figure 11-2 (below) *Early cleavage stages in human development. (a) This two-cell stage was recovered from the Fallopian tube about two days after ovulation. (b) A blastocyst containing 58 cells, four days after ovulation.*

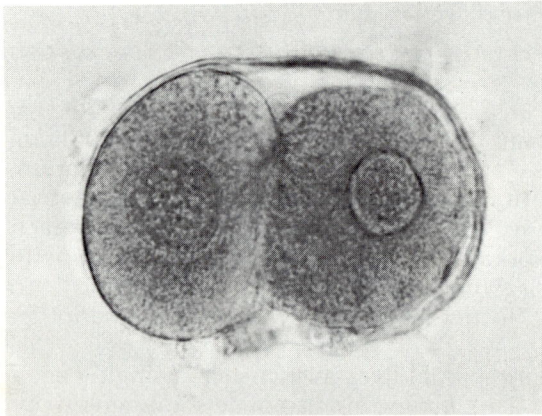

(a)

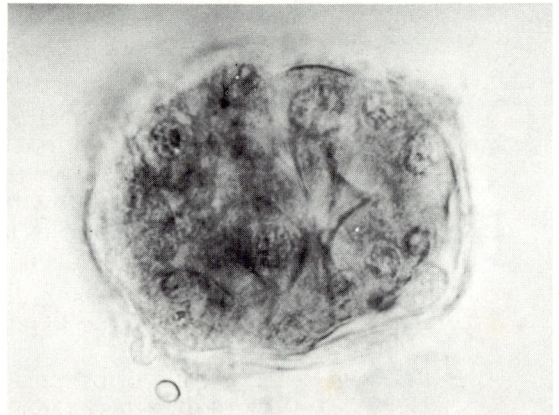

(b)

of a living system to increase—to make more of itself. From an evolutionary point of view, however, this is not enough. Genetic information has been transferred to the new generation exactly, but there has been too little opportunity for the input of new information or for the mixing of the original information in new combinations.

It is now apparent that the greatest significance of sex is that (1) it **maintains a diploid condition,** and (2) it **permits genetic recombination.** Both of these results tend to increase

the genetic variability of populations (Chapter 20). Diploidy prevents the elimination of genes that may be disadvantageous under one set of conditions but are potentially beneficial if the environment changes; it allows those traits to be carried as recessives in heterozygotes. Recombination, of course, permits the mixing and reassorting of genes from two parental genomes in new ways within an individual, again allowing selection to choose the most effective combinations. And, in fact, we saw in Chapter 10 that bacteria, like almost all other forms, exhibit a sexual phase that permits recombination at least every few generations.

Sexuality, then, is simply a device for bringing together the haploid gametes, produced in meiosis, of two different individuals. The incredibly complex process of development of a sexually mature adult, the hormonal interactions in our bodies that endow us with sexual characteristics, the instinctive urge for sexual contact that we all experience, and our all-but-universal participation in the sex act are all merely means for ensuring the bringing together of the male and female gametes, sperm and egg, in the production of a new individual—a zygote.

Just as the rules of heredity and the DNA code are universal among all living things, plant and animal, so (with few exceptions) is the phenomenon of sexuality. We have made the point before, and we will again, that man shares certain fundamental biological attributes with all other organisms. Sexuality is one of these. Thus in any discussion of reproduction and development it makes little difference, from a theoretical point of view, whether examples of the anatomy and physiology of reproduction are taken from chickens or frogs or men. The organism with which we have the most experience and about which young people are most curious is, of course, ourselves. Thus we devote this chapter to a discussion primarily of human reproduction, not because human beings are special or different in our reproductive behavior, but because we are as representative as any other species; moreover, we have at least as much information about the reproductive biology of man as about any other species, derived from two centuries or more of medical experience.

This approach seems to us eminently reasonable and pedagogically valid, but it is not without certain minor difficulties. For example, an illustration of a pair of frogs copulating (Fig. 11-4, p. 268), an act called **amplexus,** is considered by all as acceptable and informative. A similar illustration of a healthy

Figure 11-3   Human developmental sequence, showing the shape of the embryo at several stages. Note that the stages are not all drawn to the same scale.

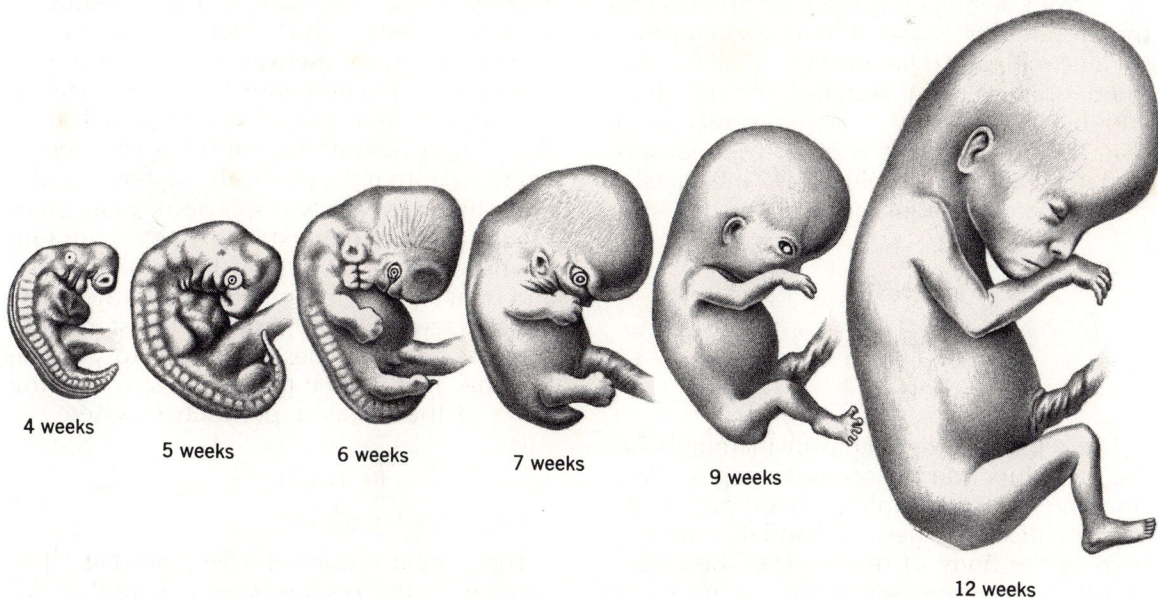

4 weeks

5 weeks

6 weeks

7 weeks

9 weeks

12 weeks

Figure 11-4 External fertilization in the frog is accomplished in the process known as amplexus, in which the male mounts the female, grasps her tightly around the body, and sheds sperm on the eggs as she lays them.

young man and woman in the analogous act, called **coitus,** would probably prevent this book from being published. Our point is that the distinction is based purely on current societal mores and has no biological validity. In our opinion the anatomical, physiological, and behavioral aspects of reproduction in man should be taught as objectively, dispassionately, and accurately as are the facts of heredity or circulation or digestion.

The Human Reproductive System

The reproductive system in mammals has the basic function of accomplishing fertilization and the added role of ensuring the development of the new individual as an embryo in the body of the female. The organs in which the gametes are formed are called **gonads.** The male gonads produce sperm and are called **testes** (singular, **testis**); the female gonads, which produce the eggs, are called **ovaries.** The gametes are formed by **gametogenesis** (p. 199), which includes the fundamental process of **meiosis.** The fusion of two gametes, the phenomenon known as **fertilization,** results in a diploid zygote, which then divides repeatedly by mitosis. The reproductive system is the only system of the body that exists in two morphologically and functionally different states, thus accounting for two different types of individuals, male and female. It includes the gonads, the ducts or passageways that permit the transfer of the gametes, and, in males, the copulatory organ, or penis, that is used to deposit sperm in the ducts of the female reproductive system.

The Female Reproductive Organs

The genital system of a woman (Fig. 11-5) consists of the ovaries, oviducts, uterus, vagina, and external genitalia.

The Ovaries

Each of the two adult ovaries is 4 to 5 cm long, has the shape of a large almond, and is located in the pelvic region. During embryonic development of a female, small groups of cells in the ovary become arranged into structures known as **Graafian,** or **ovarian follicles,** each consisting of a large oöcyte surrounded by a single layer of follicle cells. A newborn female infant has several thousand primordial ovarian follicles in each of her ovaries. Some dozen years later, when sexual development (called **puberty**) begins, a **follicle-stimulating hormone (FSH)** from the pituitary gland influences the successive maturation (about once every 28 days) of the ovarian follicles. In a maturing follicle a single egg cell is formed and is finally liberated from the ovary by the rupturing of the follicle, a phenomenon known as **ovulation.** The mature egg is probably the largest cell produced by the body, averaging about .14 mm in diameter. Because a woman normally ovulates only one egg each month during the 30 to 40 years of her reproductive life, it follows that only a few hundred oöcytes ever mature and leave the ovaries. The details of this process and its hormonal control are discussed on pp. 270–275.

The Oviducts

Close to each ovary is a large, funnel-shaped, fringed opening of a 6-in.-long tube called an **oviduct, Fallopian tube,** or **uterine tube** (Fig. 11-5). At its opposite end each oviduct opens into the upper part of the uterus. At the time of ovulation the egg is swept into the funnel opening and is moved toward the uterus. In this stage of development the egg cell is not yet mature, and is in fact only in the early steps of oögenesis (Chapter 8), usually in the secondary oöcyte stage. During the course of its passage along the oviduct the second meiotic division occurs, to yield ultimately a mature egg with a haploid number of chromosomes. It is usually in the uterine tube that the egg is reached by a sperm and fertilization takes place.

Figure 11-5   Diagram of a section of the female body showing the reproductive system of a nonpregnant woman.

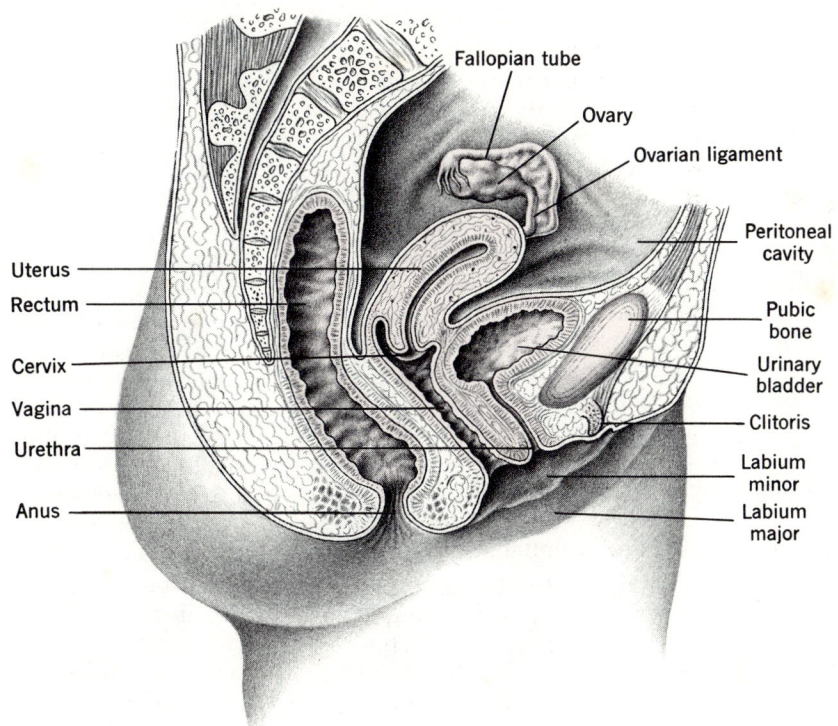

The Uterus

The organ to which the oviducts connect is a muscular, somewhat flattened, pear-shaped uterus (Fig. 11-5), about the size of a clenched fist in an adult nonpregnant woman. It is located between the bladder and rectum. The uterine cavity (i.e., the space inside the organ) is somewhat triangular in shape and is comparatively small because its walls are so thick. These walls are made of a thick coat of smooth muscle fibers and are lined on the inner surface by a mucous membrane called the **endometrium.** The lower end of the uterus, the **cervix** (which is smaller and narrower than the main body of the organ), opens into the vagina. During pregnancy the uterus increases to about 200 times its usual size.

The Vagina

The copulatory organ of the female is a sheathlike tube composed mostly of smooth muscle and lined with mucous membrane. It leads from the cervix to the outside of the body and in sexual intercourse receives the erect penis. At the birth of a child the vagina also serves as part of the canal through which the infant passes to the outside. In young girls the lower, external end of the vagina is partially closed or constricted by a fold of membrane called a **hymen;** ordinarily a sufficient opening is present to allow the escape of menstrual flow. The hymen may be ruptured during the first act of coitus. In many girls, however, it may have been broken long before intercourse by a tampon (an absorbent cotton cylinder inserted into the vagina to control menstrual discharge), by a physician, or through participation in sports. Thus the presence or absence of the hymen is not necessarily related to virginity, contrary to common belief.

The Female External Genitalia

The external genital structures of a woman are the vestibule, the major and minor labia (lips), and the clitoris (Fig. 11-6). The **vestibule** is the common opening chamber of the vagina and the urethra (p. 432). The vestibule itself is enclosed by two pairs of fleshy, lip-

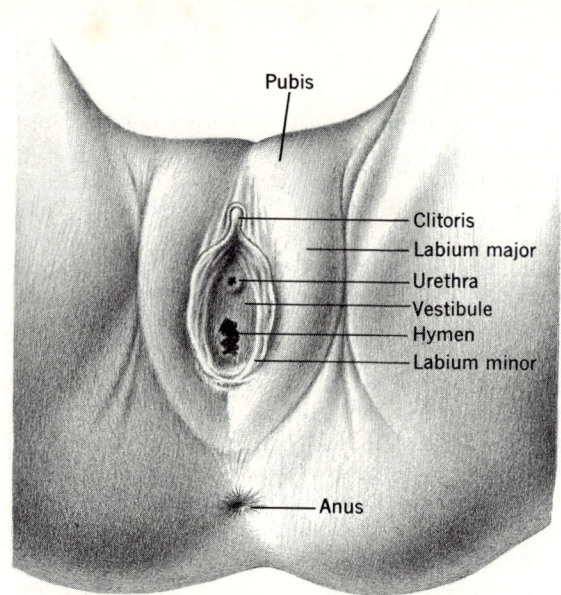

Figure 11-6   *The external genital organs of the female.*

like skin structures, called the **major** and **minor labia.** The **clitoris** is a small mass of erectile tissue that is analogous to the penis in the male but does not include the urethra. Like the penis, it becomes engorged with blood during sexual excitement, and serves as an excitatory organ in coitus.

The Female Reproductive Cycle

Puberty in girls begins when a special region in the brain, the hypothalamus (Chapter 17), produces special chemical messengers called **releasing factors,** which stimulate the anterior pituitary gland (p. 451) to begin secreting two hormones with potent effects on the gonads. These hormones are two of the so-called **gonadotropic** hormones, **FSH** (follicle-stimulating hormone) and **LH (luteinizing hormone),** which cause maturation of the ovaries. We discuss the interactions of the various endocrine glands in some detail in Chapter 16; suffice it to say here that under the influence of these gonadotropins the ovaries begin secreting the female sex hormones **estrogen** and **progesterone.** These hormones stimulate development of the **secondary sexual characteristics** (Fig. 11-7):

Figure 11-7   Development of secondary
sex characteristics in women.

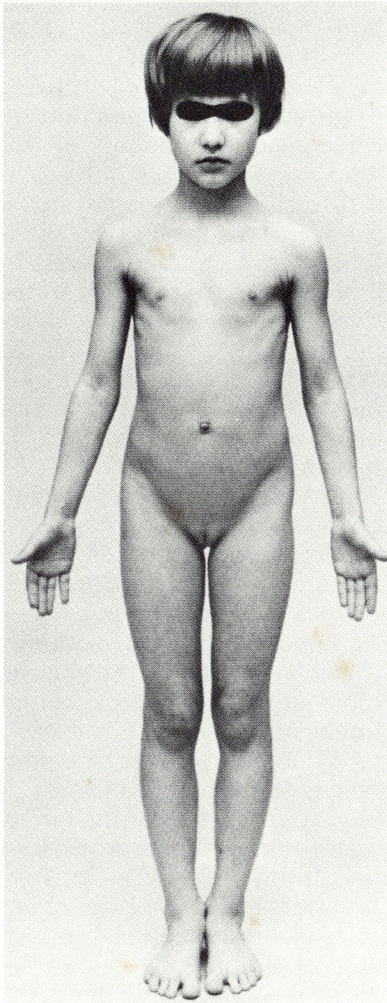

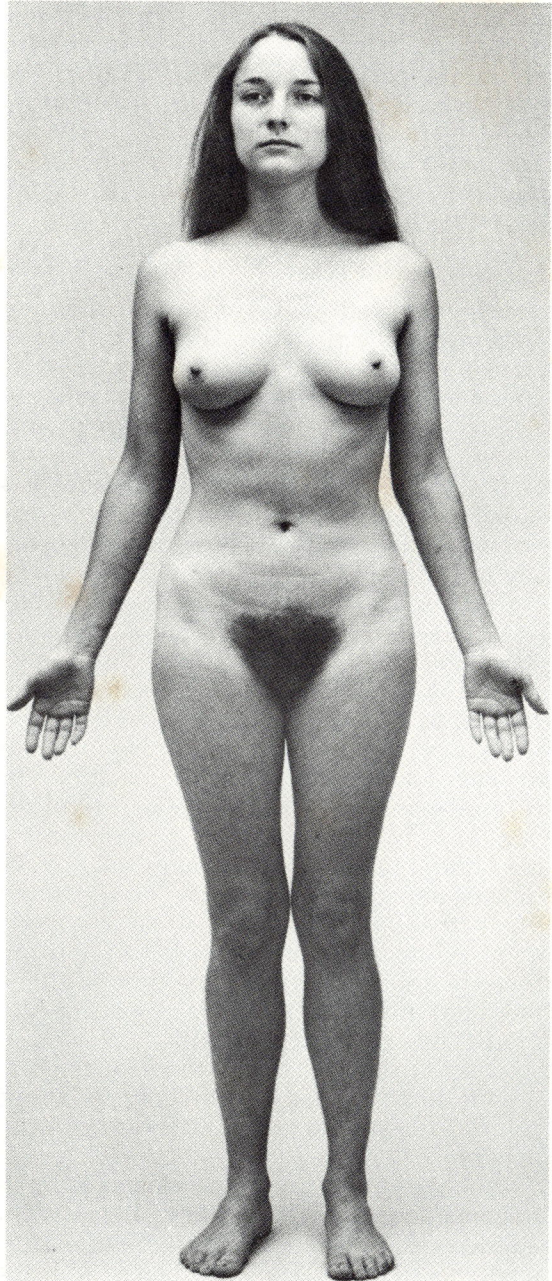

(a)   9 years   20 years   (b)

271   Human Reproduction

growth of pubic and body hair; broadening of the hips to allow an opening in the pelvic arch large enough to provide passage of a fetus at birth; growth of the breasts and development in them of milk-forming mammary-gland tissue; and increase in size of the uterus and vagina. The period of puberty is completed with the first **menstrual** (monthly) bleeding. At this time, generally between the age of 11 and 14, reproduction becomes physically possible (though not necessarily desirable) for a girl.

The Menstrual Cycle

Beginning with puberty, and continuing until **menopause** (when ovulation ceases) at age 45 to 50, the uterine endometrium undergoes a cyclic change in structure in response to rhythmic variations in the secretion of hormones by the ovaries. At the end of each cycle, part of the thickened uterine lining is sloughed off and appears as a bloody vaginal discharge, the **menstrual flow,** which generally lasts three to five days.

The menstrual cycle represents an elaborate interaction between the anterior pituitary gland, the ovaries, and the uterus, which results in the production approximately every 28 days of a mature, fertilizable ovum at the same time that the uterine wall has become thickened and engorged in preparation for accepting and nourishing the egg if fertilization does indeed take place. It is customary to consider the first day of menstrual flow as the beginning of the cycle (Fig. 11-8). During the three to five days of menstrual discharge, about 35 ml (1 oz) of blood is lost and the thickened surface layers of the uterine endometrium are sloughed away, leaving the uterus lined for a few days with a raw wound surface. At the end of this flow phase, the uterine lining, called the **mucosa,** is about 1 mm thick, and no ripe follicles exist in the ovaries.

The next phase of the cycle, called the **follicular,** or **proliferative,** phase, begins with an increase in secretion of FSH by the pituitary, which stimulates some of the follicles in the ovaries to grow (Fig. 11-9, p. 274). The growing follicles are also glandular, and

begin secreting estrogen. The estrogen stimulates the uterine endometrium to grow and thicken. Blood vessels become engorged. Many cells throughout the tissue begin dividing. At the height of its proliferative phase the mucosa attains a thickness of about 5 mm. This phase (Fig. 11-8) lasts about 9 to 10 days after cessation of the menstrual flow (or about 14 days after the start of menstruation).

As these growth activities are occurring in the uterus, one follicle in one ovary responds dramatically to the continued production of gonadotropins by the pituitary and begins to mature or ripen. Why only one follicle at a time matures, and why it alternates each month from the right ovary to the left, remain mysteries. Maturation of the follicle consists of growth and proliferation of the follicular cells, accompanied by an accumulation of fluid near the center of the follicle that causes it to distend and move outward toward the surface of the ovary (Fig. 11-9a). The mature follicle grows to about the size of a pea, until it projects from the ovary surface like a small, swelling cyst. As the follicle enlarges, the production of estrogen increases to a point at which it has an inhibiting effect on further production of FSH and a stimulating effect on the anterior pituitary to produce luteinizing hormone (LH). This is the hormone that causes ovulation. The follicle bursts and discharges the egg.

The development of the follicle and its rupture at ovulation occur approximately 10 days after the cessation of menstrual flow (or 14 days after its start), marking the end of the follicular phase of the menstrual cycle. Almost immediately following ovulation, the **luteal** phase begins. The cavity of the ruptured follicle, under the influence of LH, begins to fill with a mass of developing yellowish cells, the **corpus luteum,** which produces the hormone **progesterone.** Progesterone promotes even greater growth and proliferation of the uterine lining, especially of its glandular tissue, thus providing a favorable environment for the implantation of a zygote. At the same time progesterone inhibits the development of any other ovarian follicles and stimulates further development

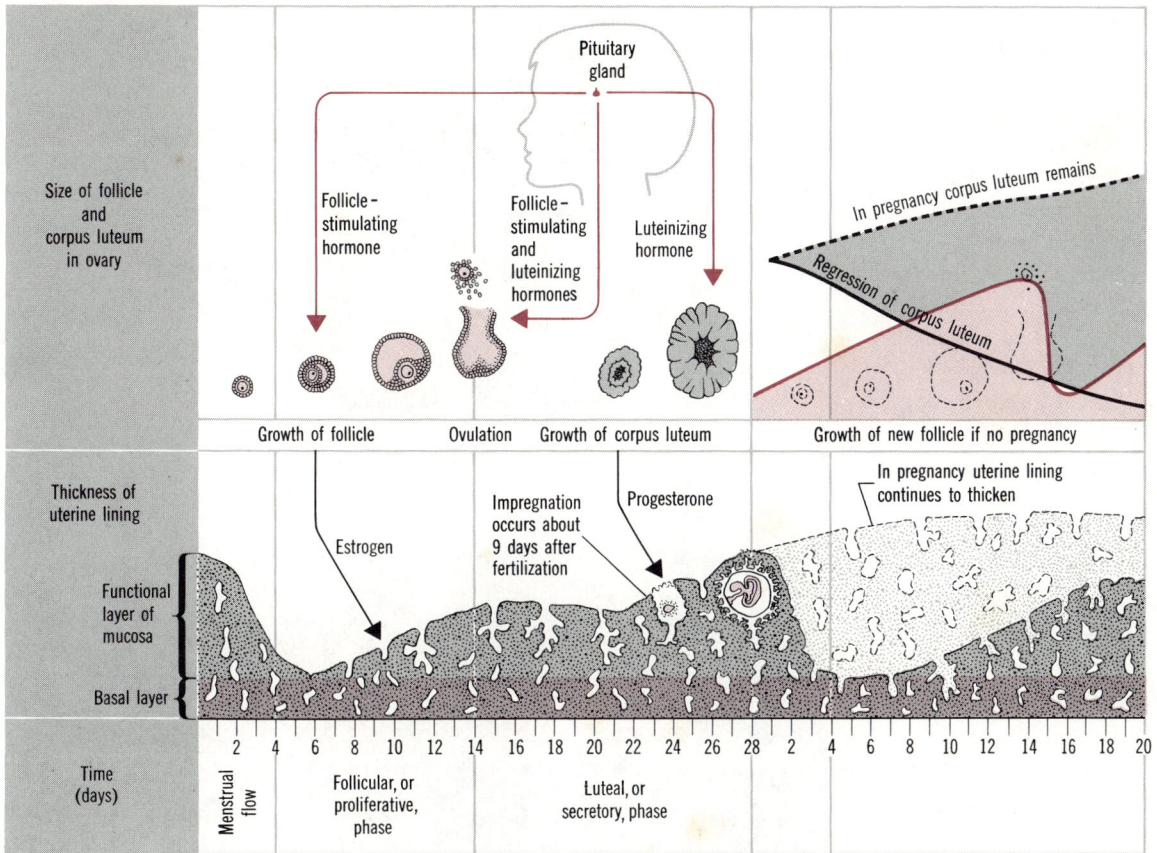

Figure 11-8   In this diagram, which correlates the events in the ovary and uterine lining (endometrium) during the 28-day human menstrual cycle, one complete cycle and part of another are indicated. Ovarian developments include the growth of a follicle under the influence of FSH, its rupture at ovulation, formation of the corpus luteum under the influence of LH, and its regression. The developing follicle secretes estrogen, which causes the proliferation of the endometrium during the follicular phase. If fertilization does not occur, progesterone, secreted by the corpus luteum, induces the next endometrial sloughing and the beginning of a new cycle. If fertilization and implantation occur, the corpus luteum remains and continues to secrete progesterone and the uterine lining continues to thicken.

of the mammary glands in preparation for milk formation that takes place after birth.

During the luteal phase the uterine lining becomes fully developed, awaiting implantation. However, if the newly ovulated egg passes through the Fallopian tube and reaches the uterus without being fertilized, it cannot implant. If implantation has not occurred by about the twenty-seventh day of the cycle, the corpus luteum begins to degenerate and stops secreting progesterone. When the blood levels of progesterone fall, the thickened lining of the uterus cannot be maintained. It soon begins to degenerate and

Figure 11-9 (a) Diagram showing the development of egg and follicle in the ovary. Starting at the arrow, the stages progress clockwise as a primary follicle containing a primordial germ cell enlarges to form a mature follicle with an ovum surrounded by a fluid-filled cavity. (b) Photograph of a section of an adult cat ovary. Magnification 100×.

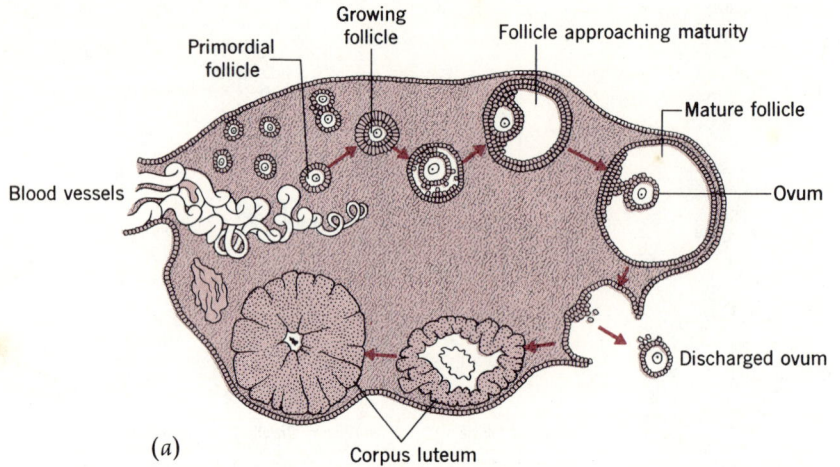

(a)

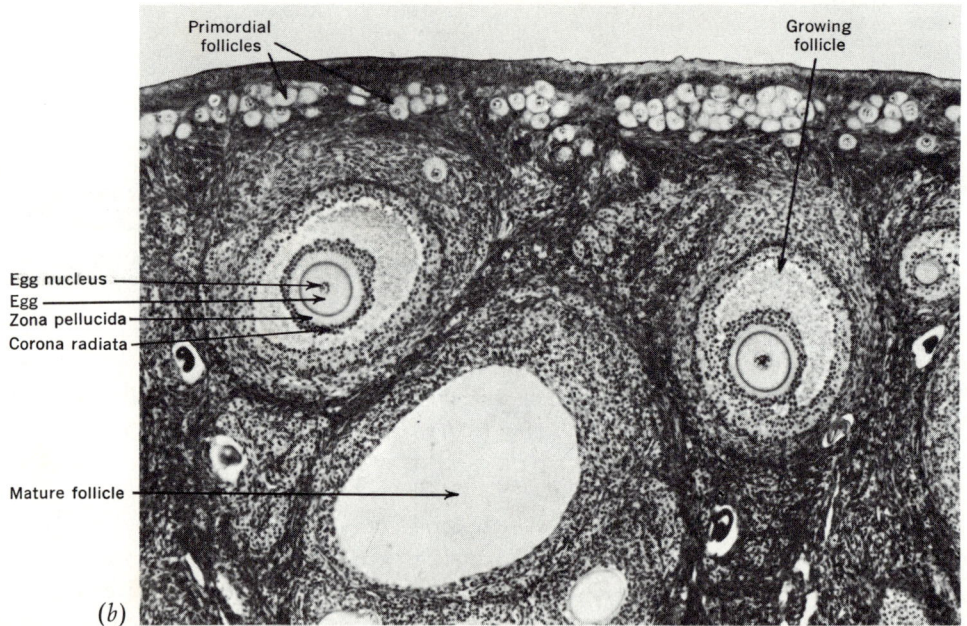

(b)

slough away, marking the beginning of the new menstruation. The absence of progesterone that results from atrophy of the corpus luteum and the low level of estrogen combine to free the hypothalamus from inhibition and allow its releasing factors to stimulate the pituitary. FSH is secreted, triggering a new round of follicular growth, and a new cycle ensues. Although we normally speak of a 28-day cycle, and this is the average for most women, the time between periods can vary from 25 to 35 days, and may be quite irregular.

There can be little doubt that the hormones active in the female sexual cycle have effects on behavior. Just before and after ovulation, when estrogen levels are at their peak, many women become increasingly receptive and find intercourse especially gratifying. As the ratio of estrogen to progesterone declines in the nonpregnant cycle, many women become progressively more restless and irritable, experiencing what is known as **premenstrual tension.**

Similar hormonally induced cycles of reproductive behavior in females of all non-human mammals result from so-called **estrus** cycles. In fact, as we mentioned earlier, the females of most species will accept a male in copulation only during that brief period of the cycle when successful implantation is possible. At this time the sex urge is at its height; the female is said to be "in heat," or in estrus. Many mammals breed seasonally, and have only a few estrus periods each breeding season, although rats and mice may undergo estrus cycles every 5 to 7 days year round. If fertilization does not occur, in most mammals the thickened uterine lining is reabsorbed gradually without bleeding. In a few mammals (the dog, for example) some bleeding does occur, but this is just before estrus; that is, near ovulation in midcycle, and is therefore not equivalent to menstruation.

Puberty

In boys of age 13 to 14 (that is, about two years later than in girls), an unknown timing event causes the same hypothalamic region in the brain that brings on puberty in girls to stimulate the anterior pituitary to begin secreting the gonadotropic hormones FSH and LH; these hormones in the two sexes are identical. The response of the male testis to FSH is to begin maturation of the **seminiferous tubules,** where formation of the sperm takes place. LH causes the development of certain secretory cells in the gonad; these **interstitial cells** begin secreting the male sex hormone **testosterone.**

The appearance of testosterone in the bloodstream stimulates a complex of changes associated with the development of male secondary sex characteristics (Fig. 11-10): growth of a beard and pubic hair; "cracking" and deepening of the voice; development of larger and stronger muscles; and generally a rapid increase in body growth. Additional

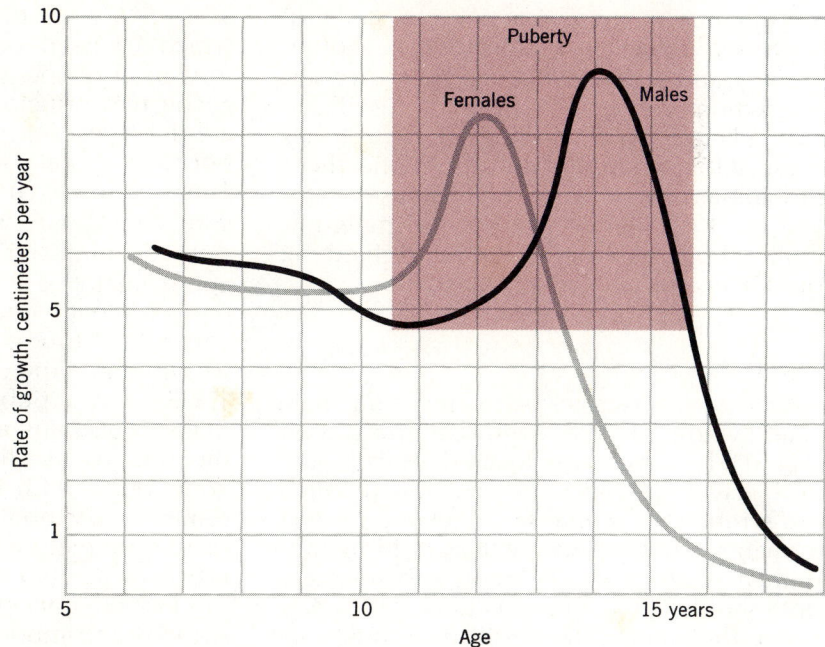

Figure 11-10   A graph showing the relative growth rates of boys and girls. At about the time of puberty, the growth rate of a young man may nearly double for a period of 2 to 3 years.

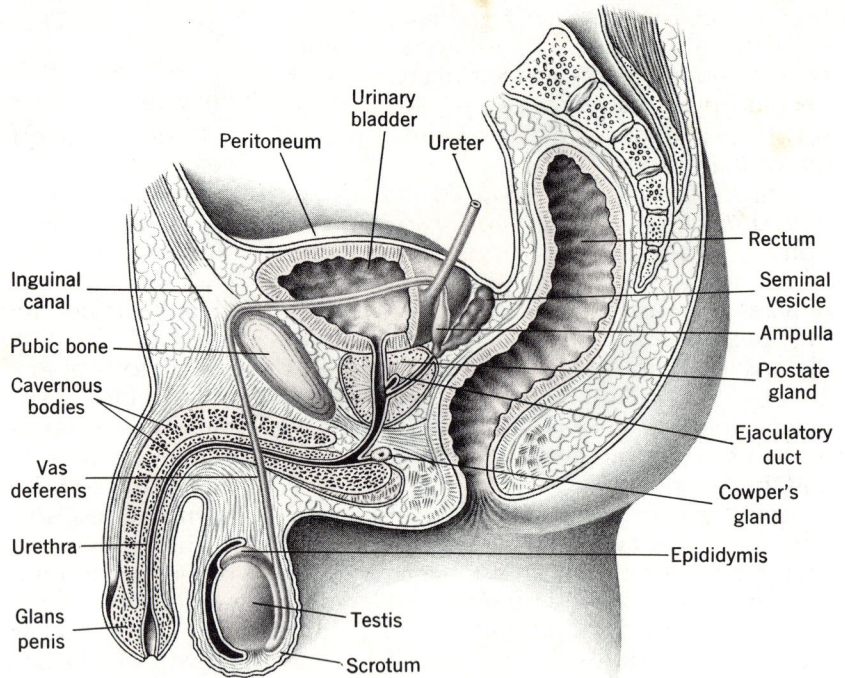

Figure 11-11 Diagram of a section of the body, to show the male reproductive system.

Labels (clockwise from top): Peritoneum, Urinary bladder, Ureter, Rectum, Seminal vesicle, Ampulla, Prostate gland, Ejaculatory duct, Cowper's gland, Epididymis, Testis, Scrotum, Glans penis, Urethra, Vas deferens, Cavernous bodies, Pubic bone, Inguinal canal

effects appear as behavioral changes in the adolescent male; he may alternate from restless, strenuous physical activity to sluggish, depressed indolence. His disposition is often grumpy, and occasionally even aggressive. His interaction with girls is often confused and contradictory. With continued high levels (normal for an adult) of male hormone, these physiological and behavioral responses stabilize, and the "adolescent behavior" pattern is replaced by a more integrated and maturing personality, usually by age 17 to 18.

The Testes

Each of the two testes in the adult male is an ovoid body about the size of a walnut (Fig. 11-11). Both are located outside the body cavity in a skin-covered pouch called the **scrotum,** or **scrotal sac.** They are formed in man, as in most mammals, within the abdomen close to the kidneys, and lie there until shortly before or soon after birth. Thereupon they normally descend through an opening in the abdominal wall, called the **inguinal canal,** into the scrotal sac. The scrotum has a slightly lower temperature (about 3° to 5°C less) than the abdominal cavity, which has been found to be essential for the formation of sperm.

The two principal functions of the testes are the formation of sperm and the production of the male sex hormone testosterone. The wall of each testis consists largely of connective tissue, which forms wall-like partitions dividing the interior of the testis into approximately 250 small compartments. Each compartment contains from one to three tiny convoluted tubes, the **seminiferous tubules,** within which the sperm are formed (Fig. 11-12). Starting at puberty under the stimulation of gonadotropins, and continuing throughout the reproductive life of the male, spermatogonia (Fig. 11-13) lining the interior of each seminiferous tubule undergo spermatogenesis to form mature sperm (p. 199). Other special cells located between the seminiferous tubules and collectively called **interstitial tissue** have the important function of producing and secreting testosterone.

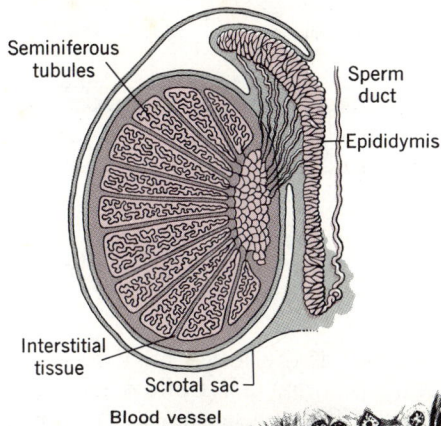

Figure 11-12 *An enlarged section of the human testis.*

Seminiferous tubules

Sperm duct

Epididymis

Interstitial tissue

Scrotal sac

Figure 11-13 *The human testis, showing several seminiferous tubules cut in cross-section to illustrate spermatocytes in various stages and interstitial cells. Magnification 200×.*

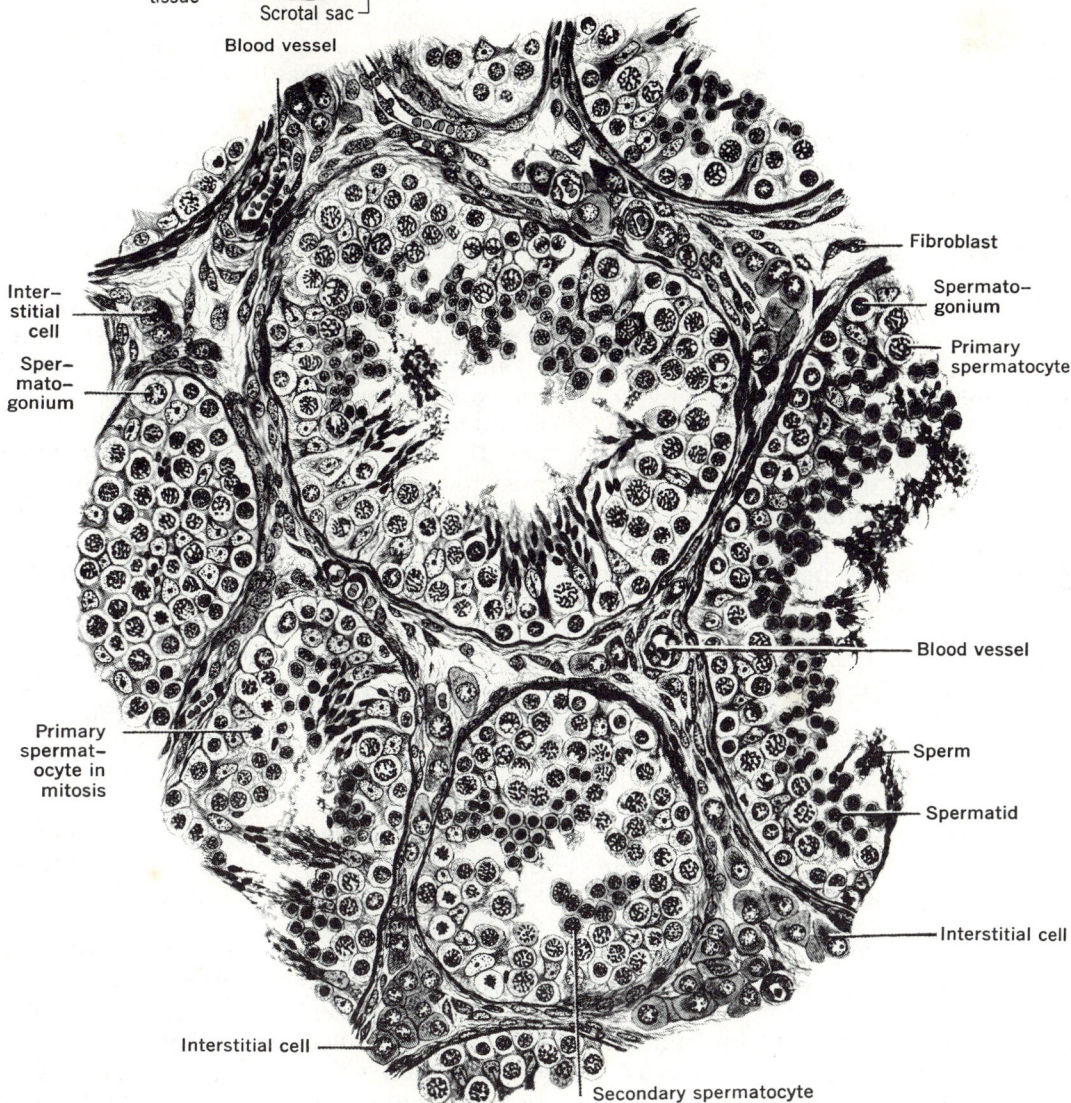

Blood vessel

Inter-stitial cell

Sper-mato-gonium

Primary spermat-ocyte in mitosis

Interstitial cell

Secondary spermatocyte

Fibroblast

Spermato-gonium

Primary spermatocyte

Blood vessel

Sperm

Spermatid

Interstitial cell

277 *Human Reproduction*

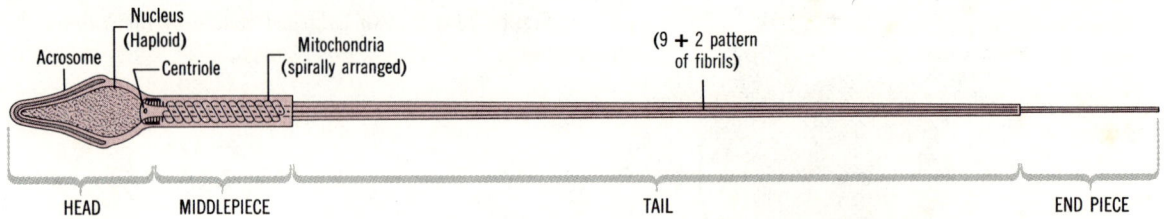

Figure 11-14   Diagram of a mature human sperm.

A mature human sperm averages about .05 mm in length, but has only about 1/100,000 of the volume of an egg. The sperm consists of three principal regions: head, middle piece, and tail (Fig. 11-14). The **head** is made up mostly of the haploid nucleus, a small amount of golgi material, and a structure called an **acrosome** located just beneath the cell membrane (Fig. 11-15). The acrosome is essential in two ways for the sperm's penetration of an egg. On contact of the sperm with the zona pellucida, the acrosome releases hydrolyzing enzymes that digest a portion of the zona pellucida, allowing for penetration. The **middle piece** possesses a pair of centrioles and a rich concentration of mitochondria arranged in a loose spiral around a central core of fibrils that extends the length of the tail. The mitochondria apparently provide the energy for the mobility of the sperm. The tail is the longest part of the sperm (about .045 mm) and accounts for the mobility of the gamete. It has the typical structure of a flagellum, with a 9 + 2 pattern of longitudinal fibrils (Fig. 11-16, p. 280, and Fig. 5-26).

Accessory Ducts and Glands

Within the testes the sperm pass from the seminiferous tubules into a single large convoluted tube called an **epididymis** (Fig. 11-12). One epididymis lies along the side and top of each testis and stores sperm before their ejaculation in the sexual act. When uncoiled it measures 15 to 20 ft in length. The epididymis leads into the **sperm duct,** or **vas deferens,** which passes from the testis through the inguinal canal into the abdominal cavity and connects with the urethra.

The paired **seminal vesicles** (Fig. 11-11), located at the lower back surface of the urinary bladder, are hollow glands about 2 in. long. They secrete into the sperm ducts a thick, viscous seminal fluid containing some nutrients. The seminal fluid, the sperm arriving from the testis, and lesser secretions by the prostate and Cowper's glands (described below) are collectively called **semen.**

The two sperm ducts drain into the urethra, which originates from the urinary bladder and is surrounded at that point by the **prostate gland.** This gland secretes, by way of numerous ducts leading into the urethra, an alkaline fluid that is believed to activate or increase the motility of the sperm and to neutralize the acid condition of the urethra and the female reproductive tract. In older men the prostate gland frequently becomes enlarged, constricting the urethra and making urination difficult; surgical removal of the gland is often used as a remedy for this condition. Slightly below the prostate gland is a pair of small glands, each about the size and shape of a small pea, called **Cowper's glands.** The mucous secretions of Cowper's glands drain into the urethra, serving as a lubricant and buffer.

The Penis

The main body of the penis consists of three longitudinal, cylindrical columns of spongelike **erectile tissue** (Fig. 11-11). The urethra passes through one of the erectile tissue columns and empties at the conical tip of the penis, called the **glans penis.** This is covered completely or in part by a double fold of skin called a **foreskin** (Fig. 11-11). **Circumcision** is the surgical removal of a portion of the foreskin. During sexual excitement

Acrosome

Nucleus

Centriole

Mitochondria

Flagellar central filaments

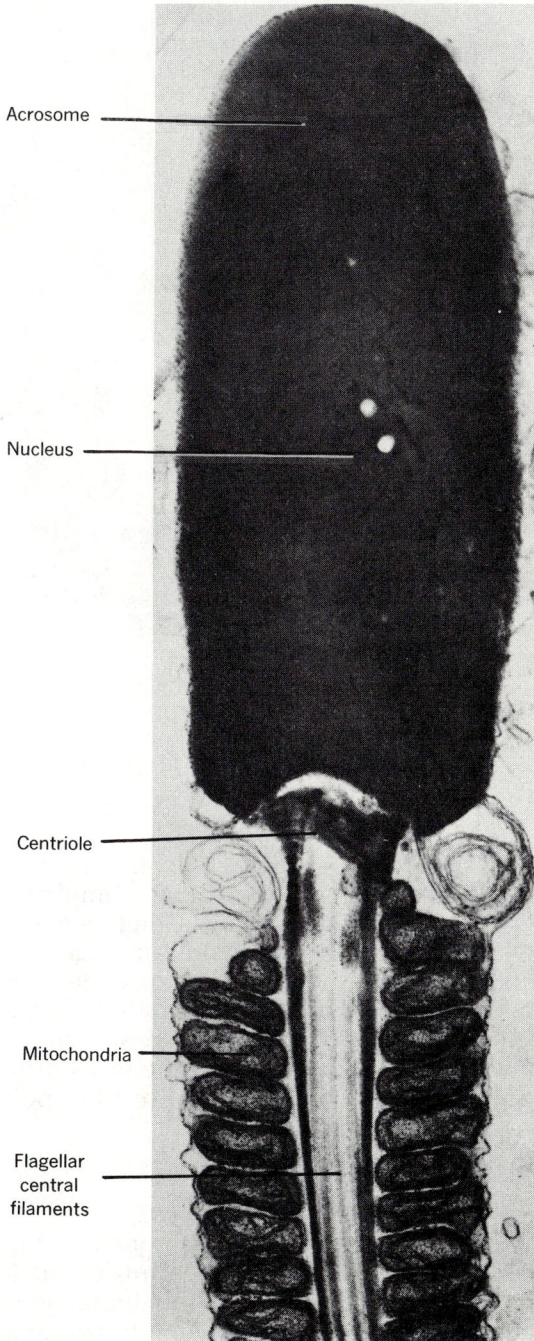

the spongelike spaces, or **cavernous bodies,** of the erectile tissue become engorged with blood under pressure because the tissue's arteries dilate and its veins constrict. The resulting turgidity causes the penis to become enlarged, hard, and erect.

Sexual Behavior

Until recently, human sexual behavior has been obscured by taboos, rituals, and social inhibitions. Men have written poems and created great works of art, accepted torture and even death in the name of the powerful attractive force between male and female.

Male birds display to females only during the breeding season (Chapter 19); similarly most male and female mammals are largely indifferent to one another except during the period of estrus. That is, overt sexual behavior is likely to occur only when fertilization is possible. In this regard man is unique among animals; humans can be sexually receptive at all times, without regard for seasonal considerations. Given the proper impetus, normal adult men could perform successful coitus daily without harm, from puberty to the **climacteric,** the time at which many aging men experience a decrease in sexual drive. The sex drive of women is somewhat different from that of men; however, women are to varying degrees receptive to male advances throughout their cycle, regardless of the amount of sex hormones released into the blood, or of the menstrual flow, which makes coitus untidy, or of cultural mores that may dictate when sexual relations are permissible. Even after the cessation of the ovarian cycle (**menopause),** which occurs in most women in their late forties or early fifties, many women continue to experience sexual interest.

Sexual Excitation, Foreplay

In most human societies sexual excitation is brought about by an assortment of sensory stimulations collectively referred to as **fore-**

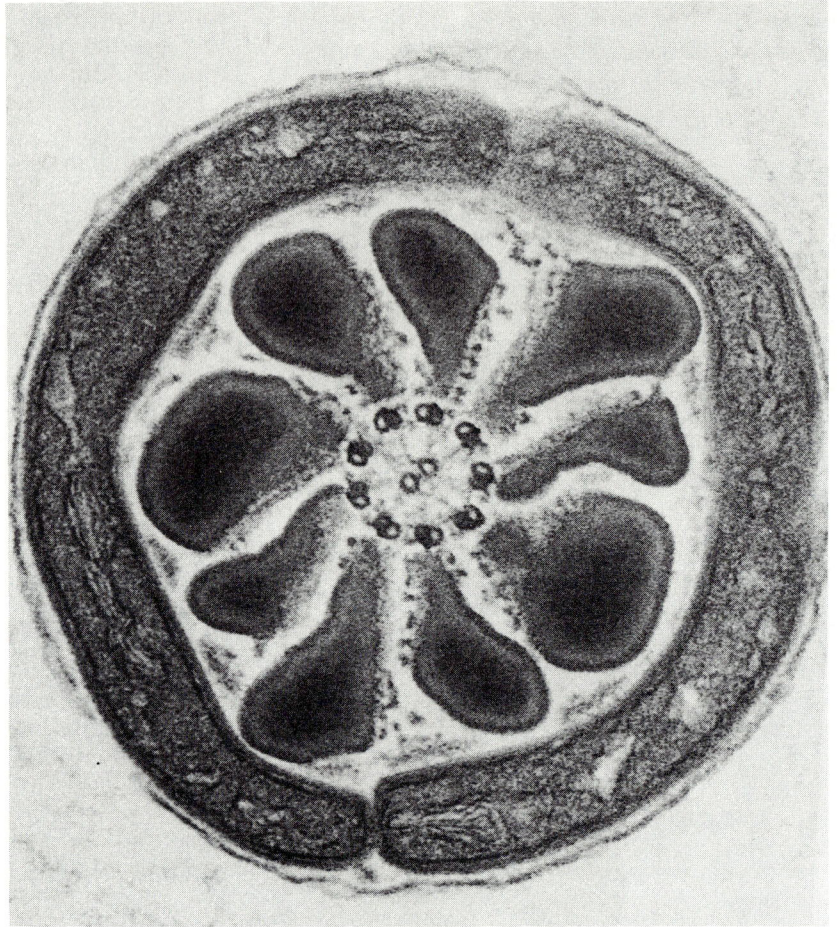

Figure 11-16   Electron micrograph of a cross-section of a sperm tail at the level of the midpiece. Magnification 83,000×.

play. Foreplay, or lovemaking, serves largely to arouse desire to such a level that each partner is ready for coition. The wide variety of sensory receptors that are stimulated by interaction with the partner—optic receptors, touch receptors of the skin, olfactory (smell) receptors—all form sensory input for sexual stimulation. In the male the penis enlarges as the cavernous bodies (Fig. 11-11) become engorged with blood, and soon becomes erect and firm. The heartbeat of an aroused male jumps from a normal rate of about 70 beats per minute to 110 to 180 beats per minute; blood pressure and respiratory rate also increase. Women, who usually become aroused more slowly than men, experience similar increases in heart rate, blood pressure,

and respiration. Nipples become enlarged and acutely sensitive to tactile stimulation. The vaginal walls and lips become engorged with blood, and the clitoris undergoes an erection analogous to that of the male penis, becoming enlarged and firm. With further stimulation, glands in the cervix secrete a mucous lubricant that coats the walls of the vagina, permitting insertion of the penis without discomfort.

Sexual Intercourse

Sexual intercourse (**coitus**) consists of the coordinated body activities of male and female that make possible the bringing together of the sperm and ovum. In mammals it involves insertion of the erect penis into the vagina of the female, followed by a series of deep thrusting movements. In males the stimulation of sensory nerve endings in the penis as a result of friction against the walls of the vagina culminates in a short period of heightened paroxysmal sexual excitement, called **orgasm,** accompanied almost simultaneously by **ejaculation,** the forceful ejection of semen from the penis. Ejaculation is the result of several coordinated involuntary muscular contractions, including those of the walls of the epididymis, sperm duct, and urethra, propelling the sperm from their place of storage in the epididymis, through the ducts, and from the urethra. At the same time the seminal vesicles, and to a lesser extent the prostate and Cowper's glands, release their secretions, which mix with the sperm to form the semen. The expulsion of the semen completes the act of ejaculation.

In females the friction of the erect penis on the clitoris and vaginal wall during coitus ordinarily culminates in an orgasm that is accompanied by involuntary rhythmic contractions of the vaginal and uterine walls. No special discharge or ejaculation of fluid occurs during the orgasm of the female.

The volume of semen ejaculated during coitus is normally 2 to 5 ml; a single ejaculate contains about 300 million sperm. The sperm are usually released in the upper region of the vagina near the cervix, and they are moved principally by muscular contractions of the female reproductive tract into the uterus and then into the oviducts. Until the sperm are actually ejaculated they are inactive, but at orgasm, when the sperm mix with the glandular secretions that comprise the seminal fluid, they are activated and begin strenuous swimming movements with their flagellar tails.

If sexual excitement and foreplay are prolonged excessively without the release afforded by ejaculation, sperm may accumulate in the epididymus and vas deferens of a man and produce discomfort or even aching in the groin or testes. This condition is harmless and usually soon passes, or may be relieved by self-induced emission of sperm. It is common in adolescent males for such an accumulation of sperm to be discharged spontaneously during sleep, a normal process termed **nocturnal emission.**

Since the urge for sexual gratification is one of the most powerful psychic drives, it is perhaps not surprising that sexual release is sometimes sought through means other than coitus. For example, **masturbation** is the production of orgasm by self-manipulation of the penis or clitoris. According to recent evidence masturbation is practiced by 90 percent of males and only a slightly smaller percentage of females at some time in their lives. In the past the practice has been erroneously related to insanity, impotence, acne, warts, and a variety of other major and minor disabilities. It is now recognized, however, that masturbation represents a normal part of sexual development in both males and females; there is no evidence whatsoever of any harmful effects of the practice in men or women.

During preadolescence or adolescence many individuals, both male and female, experience sexual urges toward members of their own sex. Normally, with further development, these feelings are replaced by fully heterosexual behavior. In some persons, however, the strong sexual attraction toward members of their own sex persists into adulthood. Such individuals are called **homosexuals.** Homosexuals are apparently normal physically. Their homosexuality is considered to be the result mainly of emotional rather than biological causes, as the condition generally does not respond to hormone treatments. Homosexuality is not necessarily accompanied by behavior associated with the opposite sex; for example, in men the condition is not uncommon among athletes and otherwise aggressively masculine individuals.

Early Human Development

Fertilization

The egg is usually capable of fertilization for a day or two after ovulation; thus this interval is often called the **fertile period,** the

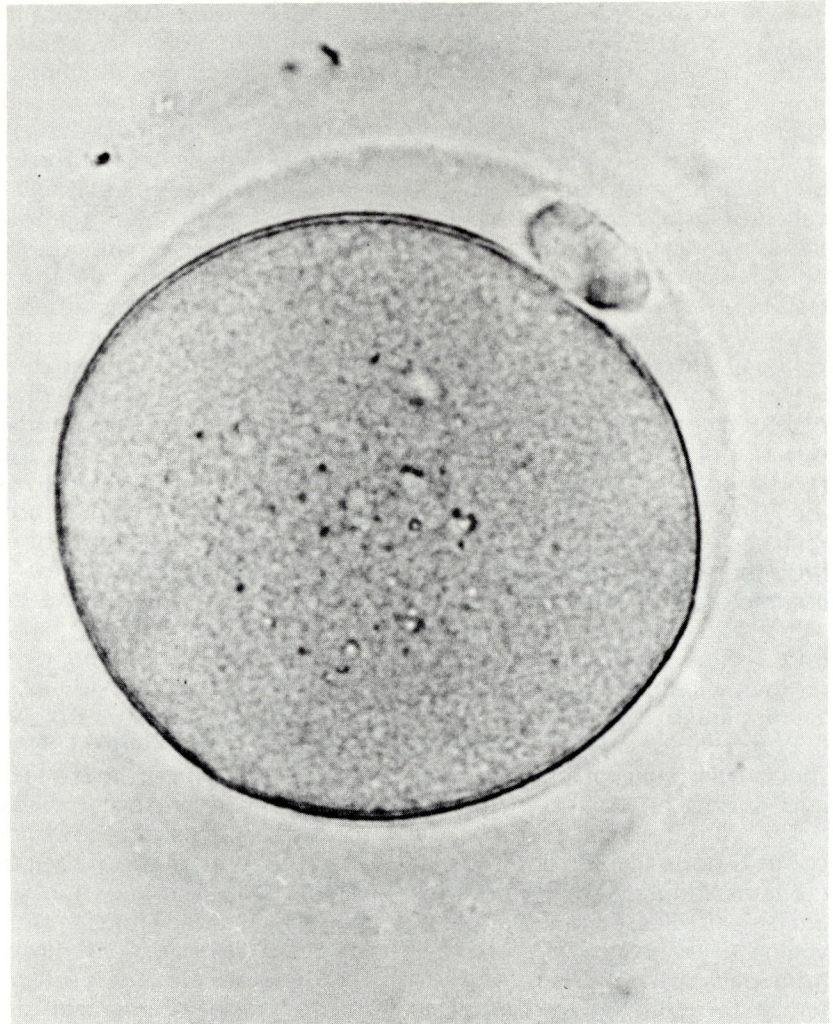

Figure 11-17 Human eggs maintained in a culture medium during fertilization and early cleavage. (a) Unfertilized; the first polar body, formed at meiosis, has been extruded and is visible at the upper right. The whitish halo around the egg is the zona pellucida. Magnification 700×. (b) After sperm entry the head separates from the tail and swells to form the male pronucleus. The male and female pronuclei can be seen in the center of the egg, which was fertilized about 8 hours before this photograph was taken. Extruded polar bodies are also visible at the lower right. Magnification 700×.

(a)

time when fertilization is likely to occur. Sperm introduced into the female reproductive tract are capable of fertilizing an egg for about the same period of time.

Fertilization ordinarily takes place in the upper third of the oviduct. Before fertilization the egg cell is surrounded by a gelatinous layer, the **zona pellucida,** and a layer of follicle cells called the **corona radiata** (Fig. 11-17). For fertilization to take place the corona radiata must be penetrated or dissolved. This is accomplished by enzymatic hydrolysis of the material cementing the cells of the corona radiata together; the acrosomes of the sperm that come in contact with this cell layer release hydrolyzing enzymes called **hyaluronidase** and **acrosomal protease.** Once the corona radiata cells have been parted, a single sperm penetrates the zona pellucida and enters the egg; subsequent changes in the egg membrane prevent the entry of other sperm. As mentioned on p. 269, oögenesis at this time has not been completed; only the secondary oöcyte stage has been reached.

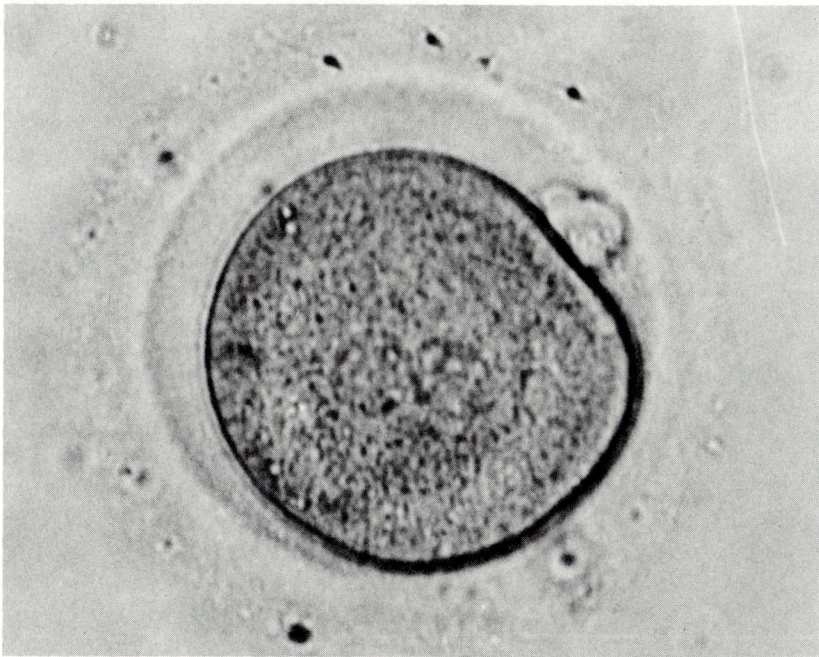

(b)

The entrance of the sperm into the egg stimulates the egg to undergo its second meiotic division and subsequent maturation so that fusion of the egg and sperm nuclei can occur. After the two nuclei meet and fuse, the resulting zygote begins the first in a long series of mitotic divisions. This dividing structure moves down the oviduct, and in 10 days has become several hundred cells. It is finally implanted in the uterus and completely enveloped in the thick endometrium. These developments are considered in the next section.

Fertilization not only permits the haploid nuclei of the two gametes, called **pronuclei,** to unite, thereby restoring the diploid chromosome number (p. 191), it also **activates** the egg cell, causing it to begin cell divisions and development. The mechanism by which fertilization transforms a quiescent egg cell into one undergoing numerous, rapid cell divisions and development into an embryo is not known. The fact that a wide variety of artificial chemical and physical agents act similarly strongly suggests that the egg already possesses most, if not all, of the capabilities to develop. Activation of the egg by fertilization is therefore the triggering of a series of genetically predestined, well-ordered reactions in space and time that progressively transform the zygote and its descendents into an embryo and eventually a new individual. We consider some of the mechanisms underlying this process, as worked out in experimental animals, in Chapter 12.

Cleavage and Implantation

After fertilization has occurred in the oviduct the zygote begins to develop immediately. We know from studies of human eggs, fertilized *in vitro* (in a culture medium), that it takes about 8 hours for the two pronuclei to unite (Fig. 11-17b), and the first division (Fig. 11-2) takes place about 24 hours after fertilization. The second and third cleavages

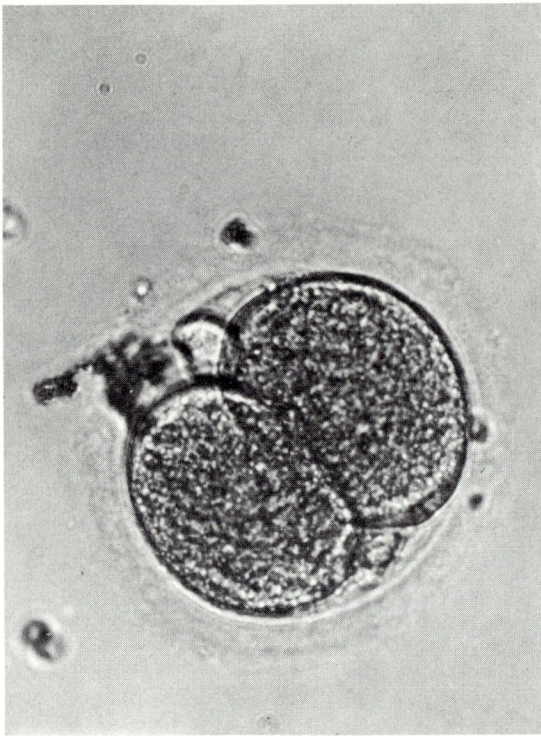

(a)

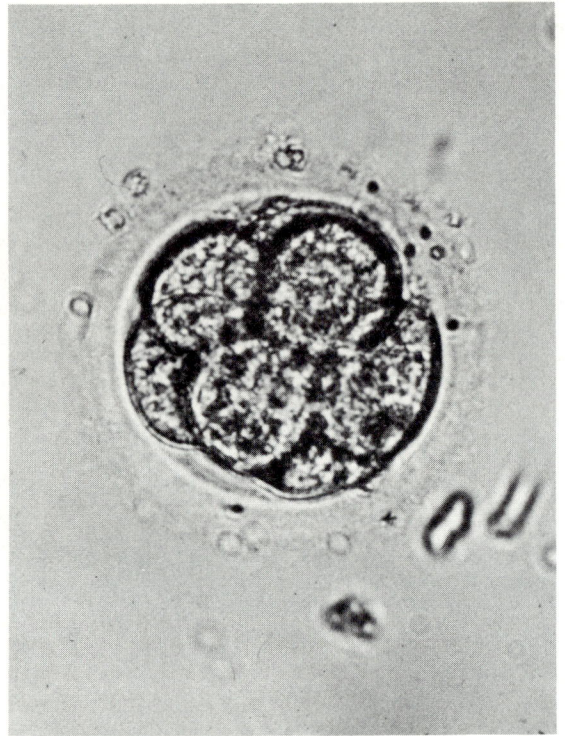

(b)

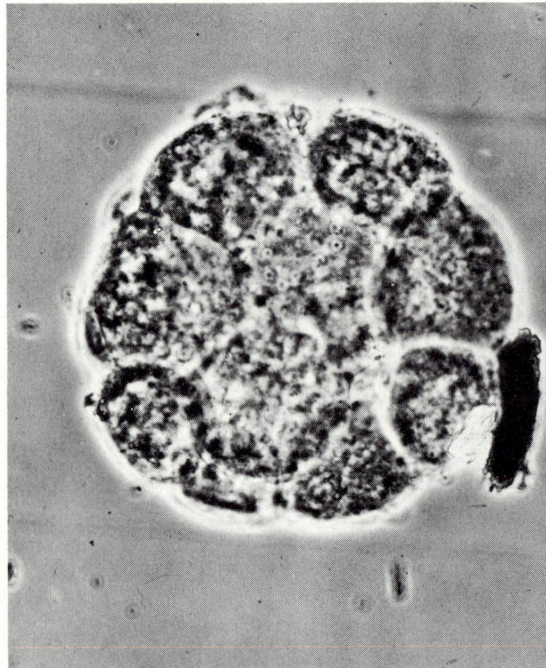

(c)

Figure 11-18 Human cleavage-stage embryos, developing in a culture medium. (a) Two-cell stage, observed 40 hours after in vitro *fertilization. Sperm can be seen adhering to the zona pellucida, which still surrounds the embryo. (b) Six-cell stage, 54 hours after in vitro insemination. (c) Twelve-cell stage, 72 hours after fertilization. The zona pellucida and corona have disintegrated.*

embryo stage, a **blastocyst**—a hollow, fluid-filled, spherical body with an inner mass of cells at one side (Fig. 11-20, pp. 286–287). The enveloping sphere of the blastocyst, a single layer of cells, is called a **trophoblast;** the inner cell mass, which will develop into the embryo, is made up of **formative** cells. The trophoblast is an embryonic structure, but it does not enter into the formation of the embryo itself. Instead, it remains external to the developing embryo, giving rise to the **chorion** and **amnion,** two extraembryonic membranes that function in its care and maintenance (p. 291).

(Fig. 11-18) take place during the next 24 to 28 hours (Fig. 11-19) as the egg passes through the oviduct.

Once the embryo is in the uterus, where it arrives at about the fourth day, the size of the cells increases and the total number of cells in the embryo increases rapidly to approximately 100. These form a solid, mulberrylike mass of cells called a **morula.** A cavity filled with fluid soon begins to form in its midst, to yield the next mammalian

Soon after entering the uterus the blastocyst begins to **implant** in the endometrium (Fig. 11-20b and c). Cells of the trophoblast, usually the portion immediately overlying the inner mass of formative cells, embed themselves in the uterine lining. Shortly after the trophoblast makes contact with the uterine endometrium, the trophoblast cells begin to

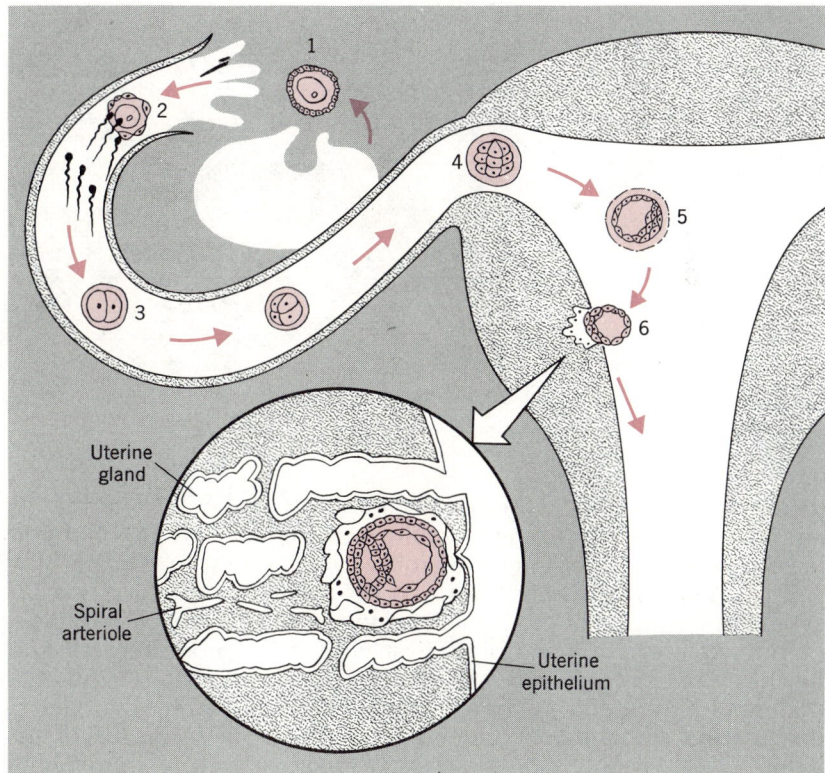

Figure 11-19 Diagram showing the fate of the oöcyte after ovulation (1). Fertilization (2) takes place in the upper oviduct. As the egg travels down the oviduct it goes through the early cleavage stages (3) to form a morula (4). Once in the uterine cavity the embryo enlarges to form a blastocyst (5) and implants in the uterine wall (6 and inset).

Uterine gland

Spiral arteriole

Uterine epithelium

Figure 11-20 (a) A blastocyst of a monkey, silhouetted by light from below, to show the inner cell mass of embryonic formative cells (arrow) and the individual cells of the trophoblast. Magnification 300×. (b) A section through a similar monkey blastocyst, just after it has attached to the uterine wall. The inner cell mass lies close to the uterine endometrium. Magnification 330×. (c) The same monkey blastocyst as in (b), shown at low magnification, to illustrate its attachment to the endometrium; about nine days after ovulation Magnification 85×.

(a)

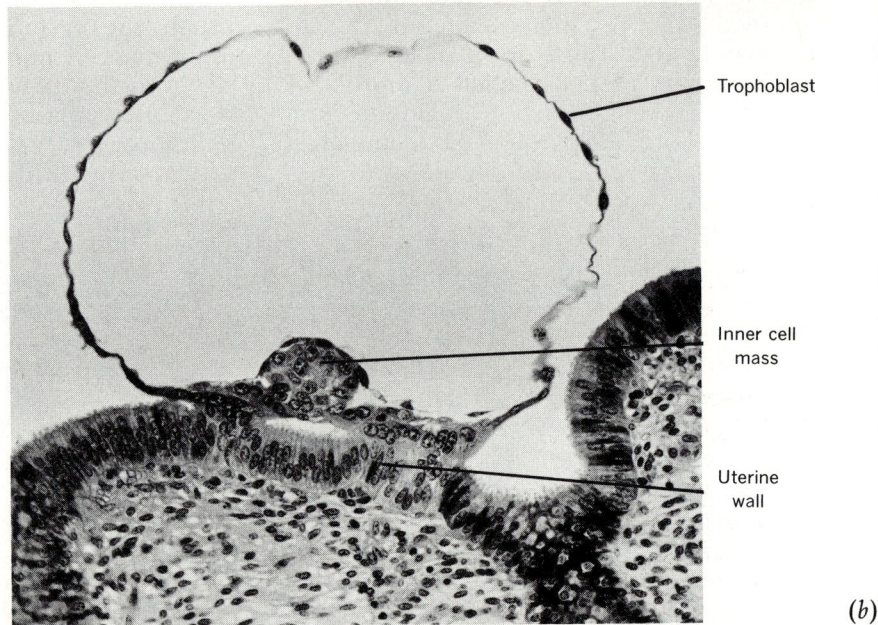

Trophoblast

Inner cell mass

Uterine wall

(b)

divide rapidly. As they do so their cell boundaries break down, forming a multinucleate mass of tissue surrounding the embryo. This is called a **syncytial trophoblast.** It is the syncytial trophoblast that erodes its way into the uterine wall, carrying the blastocyst with it. By the ninth or tenth day after fertilization the entire blastocyst lies completely beneath the uterine epithelium, nourished by the richly supplied mucosa (Fig. 11-21, p. 288).

Hormonal Effects of Implantation

We saw earlier that in the normal menstrual cycle (Fig. 11-8) after ovulation the pituitary stops secreting LH; the corpus luteum in the ovary begins to degenerate and stop secreting progesterone; and the uterine lining begins to slough off in menstruation. Clearly, this whole sequence of events must be prevented if pregnancy is to ensue. The answer to how this is done lies in the remarkable properties

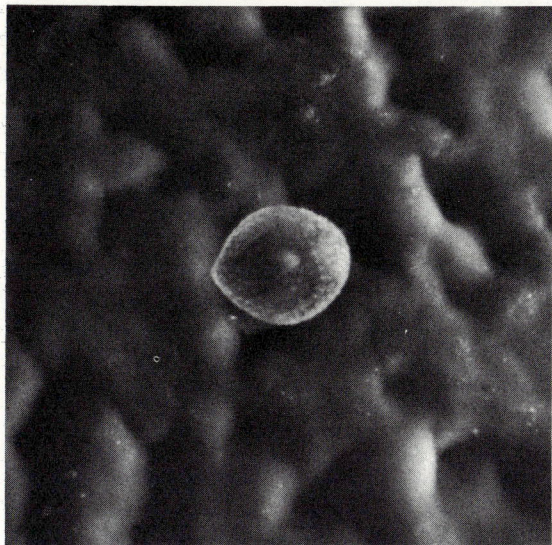

(c)

of the syncytial trophoblast. As the tropho-blast invades the uterine wall it forms a thick spongy layer called a **chorion** (p. 291) which will later become part of the **placenta,** the organ through which the embryo obtains its nourishment from the mother. Immediately upon implantation, the chorion cells begin secreting a hormone very similar to LH, called **chorionic gonadotropin.** This gonadotropin takes the place of the LH secreted by the pituitary and thus preserves the corpus luteum, maintains its secretion of progester-one, and permits the uterine lining to remain intact and grow throughout the entire preg-nancy.

So much chorionic gonadotropin is nor-mally produced in a pregnant woman that much of it is excreted in the urine. Because it is a relatively easy substance to identify, many commonly used tests for pregnancy are based on this phenomenon.

A variety of illnesses in the mother, or any serious misdirection of development in the embryo, may result in reduction of progester-one output, often causing breakdown of the endometrium and a **miscarriage** (loss of the embryo), with abnormal or heavy menstrual flow. Miscarriage most often occurs during the first month of embryonic life; at this stage the embryo is so small that the entire preg-nancy and loss of the fetus may go unrecognized.

Formation of the Three Primary Germ Layers

All the structures of the body are ultimately formed from one or more of three **primary germ layers.** While implantation is occurring, the inner cell mass continues to develop. The formative cells have arranged themselves into a flattened **embryonic disc,** or **blastodisc** (Fig. 11-22, p. 289), which will develop into the embryo. By the tenth day the embryonic disc consists of two germ layers, the **ectoderm** (which will form the skin and nervous sys-tem) and the **endoderm** (from which the digestive system will develop). A space called the **amniotic cavity** has formed next to the ectoderm layer. On the opposite side of the embryonic disc from the amniotic cavity, a single layer of cells forms a sac, called the **yolk sac,** in the cavity within the blastocyst.

During the next 10 days several important developments take place. The trophoblast develops complex villilike structures over its entire surface, which grow deeper into the uterine wall. The trophoblast and its villi de-velop blood vessels and are collectively called the chorion (Fig. 11-23, p. 290). In time the endodermal lining of the yolk sac develops into the epithelial lining of most of the diges-tive tract. A fingerlike projection of the end-oderm extends laterally to form a rudimentary **allantois** (Fig. 11-23b), another of the extra-embryonic structures. The embryonic disc is now attached to the chorion only at its pos-terior end by a mass of cells known as a **body stalk** (see Fig. 11-23a, b) which will later form the **umbilical cord** (Fig. 11-23d).

The **mesoderm** (Fig. 11-22c), the third germ layer, arises from cells in the midline of the embryonic disc between the endoderm and ectoderm; these cells condense to form a longitudinal thickened strip, or axis, called a **primitive streak.** Mesoderm cells migrate through the primitive streak and move out

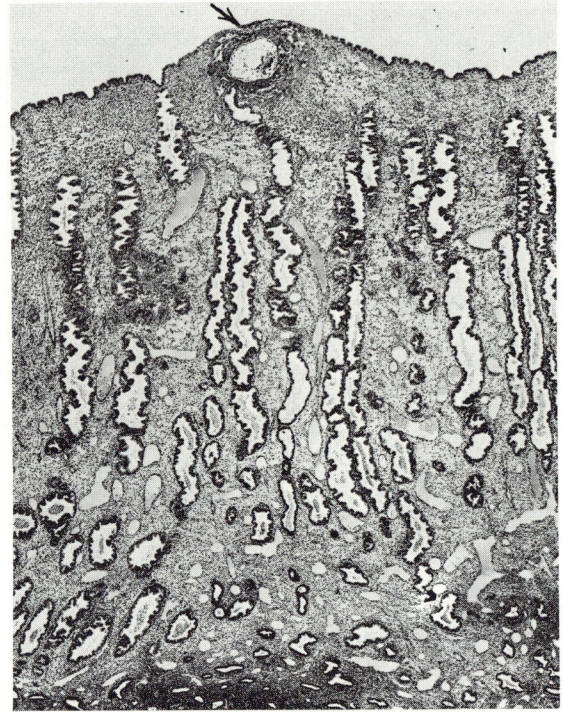

(a)

Figure 11-21   A human egg, 11 days after
fertilization, implanted in the uterus. (a) At low
magnification (18×) the entire endometrium is
shown, illustrating the swollen secretory glands and
veins of the uterus that carry nourishment to the
embryo. (b) The same section as in (a), magnified
100×, to show the embryonic disc that has formed
from the inner cell mass, the early amniotic cavity,
and the syncytial trophoblast.

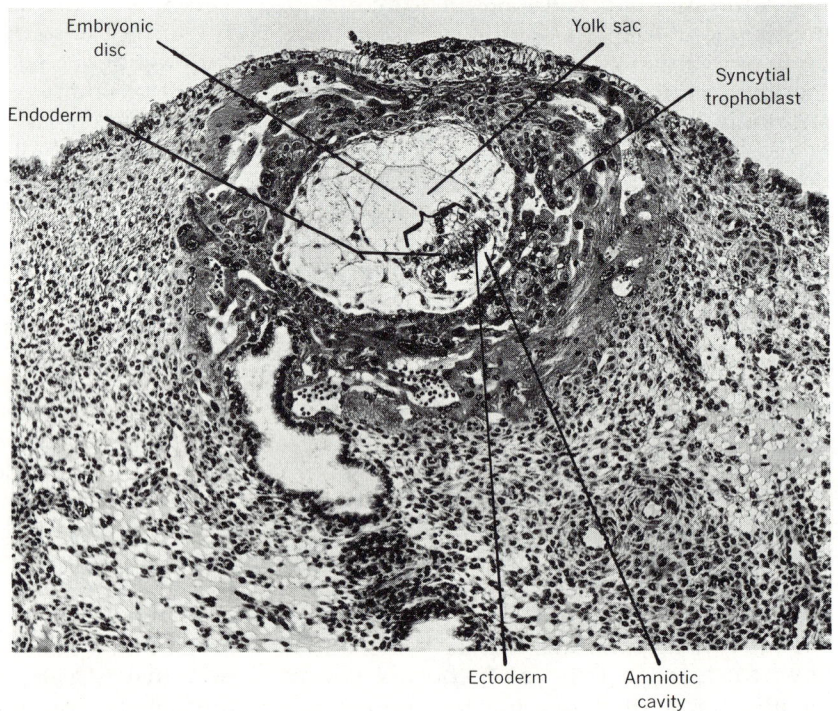

(b)

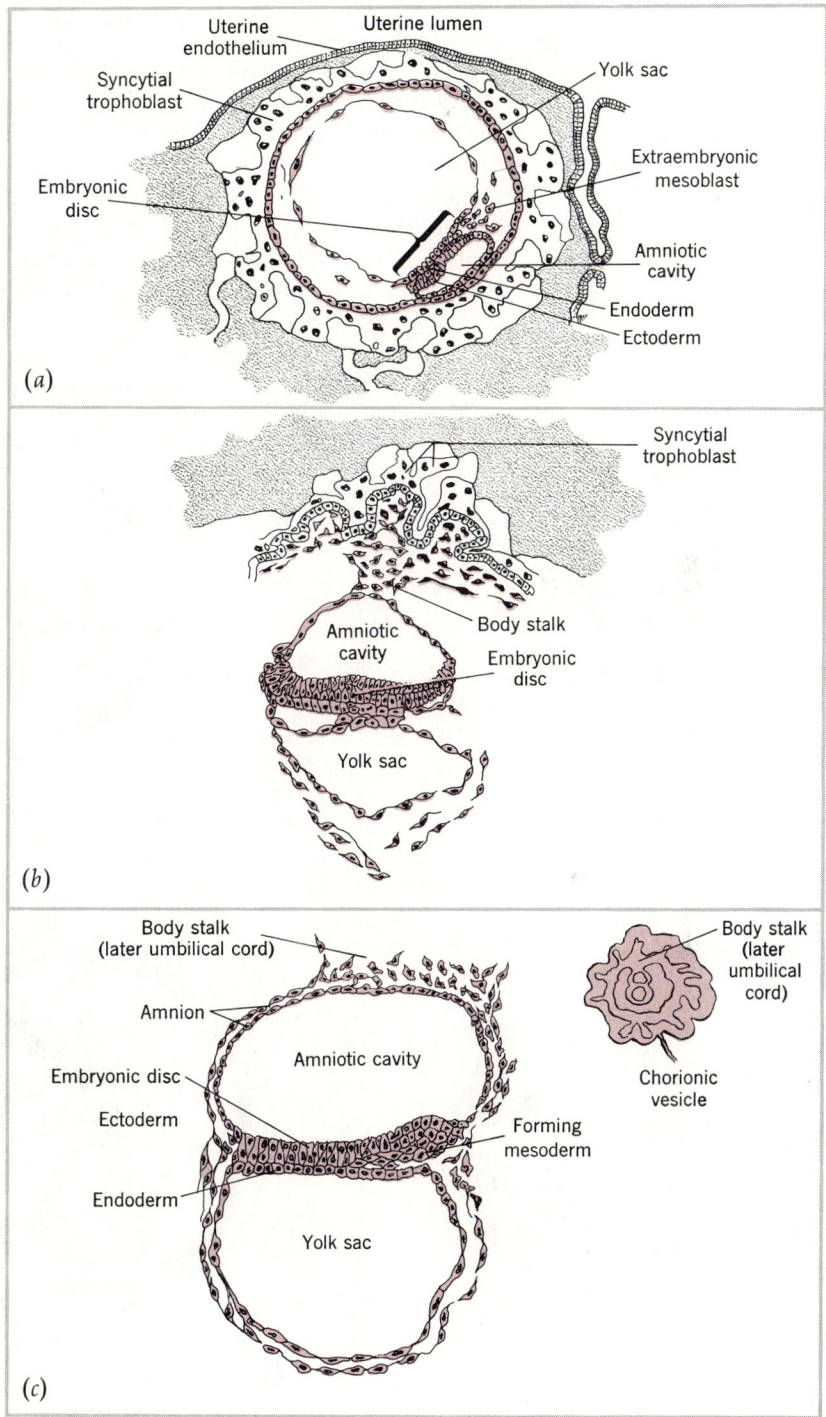

Figure 11-22 Diagrams of the implanted embryo at (a) 12 days; (b) 15 days; and (c) about 16 days after fertilization, to show the development of the embryonic disc, the amnion and yolk sac, and the body stalk.

Figure 11-23   *Development of the extraembryonic membranes. (a) As the blastocyst implants in the uterine wall, two spaces develop in the inner cell mass—the amniotic cavity and the yolk sac. The embryo separates from the trophoblast except in a narrow region called the body stalk. (b) As the embryo enlarges, villi and blood vessels appear in the trophoblast, now called the chorion. An extension of endoderm grows out along the body stalk to form the allantois, which becomes a sac for storing waste materials that the embryo secretes. (c) A diagrammatic section of the mother's body shows the spatial relationship of the embryo and chorion to the uterus. (d) The body stalk gradually narrows and lengthens to form the umbilical cord, through which run the placental blood vessels that carry nourishment to the embryo, and the allantoic stalk.*

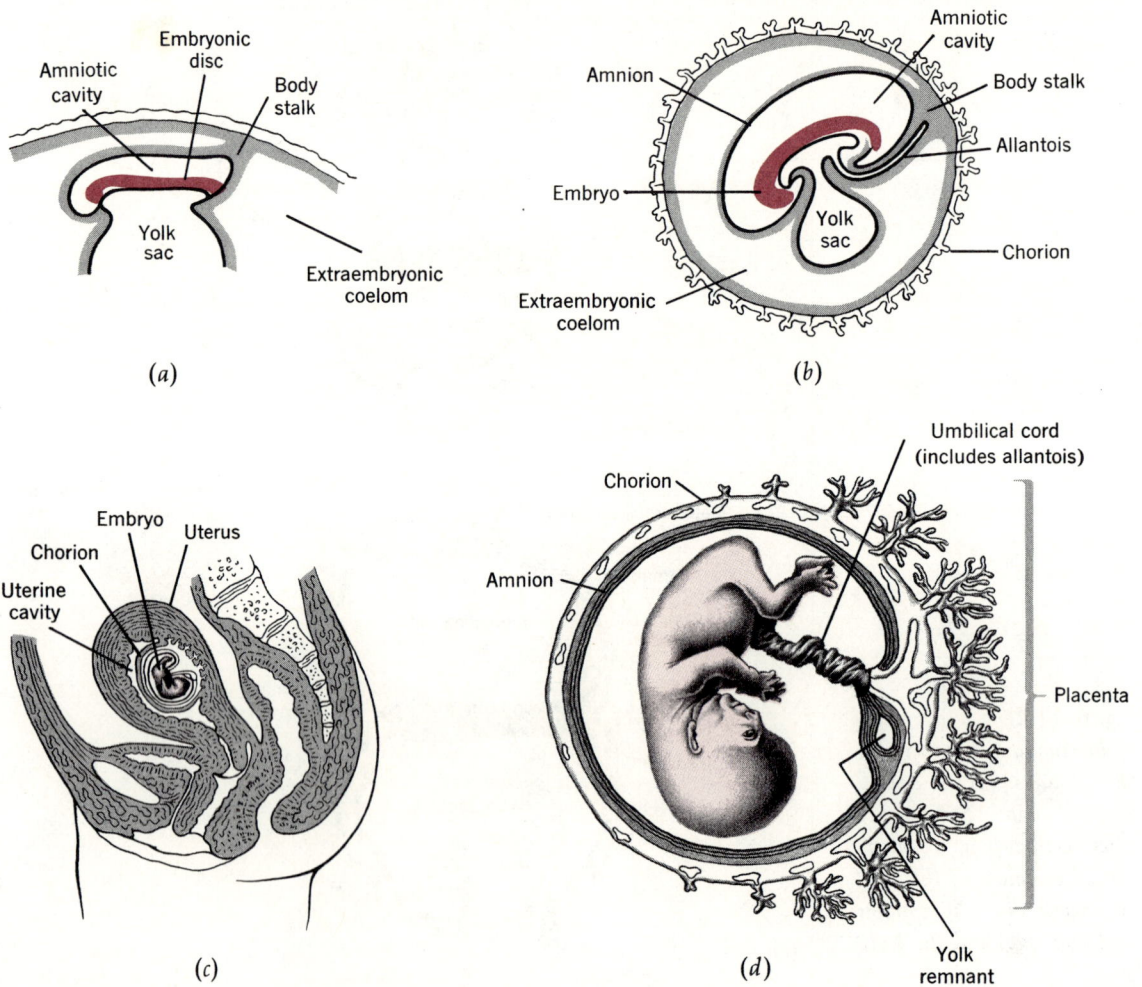

(a)

(b)

(c)

(d)

laterally between the ectoderm and endoderm to contribute to every organ and part of the developing embryo.

No organ is exclusively derived from any one of the three primary germ layers. In many instances the basic functional portion of an organ may originate from a single germ layer, but products of the other two germ layers may contribute. For example, the intestinal lining, which is of endodermal origin, is the principal functional structure of the digestive tract, acting in the secretion of digestive juices and absorption of digested materials. However, the muscles, connective tissue, blood vessels, nerves, and outer lining, which make up the bulk of the intestines, are derived from mesoderm and ectoderm.

Extraembryonic Structures

In the course of its development the embryo produces certain tissues or structures that are not part of itself but function in its care and maintenance. These parts are the extraembryonic membranes, and include the amnion, chorion, placenta, yolk sac, allantois, and umbilical cord (Fig. 11-23).

Amnion

The amnion is a membrane that originates from the ectoderm, in the early stages of embryonic development, as a wall enclosing the amniotic cavity (Fig. 11-23a). It comes to surround the entire embryo and umbilical cord. The amniotic cavity is filled with a clear, lymphlike fluid called amniotic fluid, in which the embryo is suspended. This fluid envelope is a protective, shock-absorbing cushion allowing for growth and freedom of movement of the embryo. It also contains cells that have sloughed away from the embryo, or **fetus.** Samples of these cells can be obtained at later fetal stages by piercing the mother's abdomen and uterus with a long, hollow needle and drawing off a small quantity of amniotic fluid; the cells are then stained and their chromosomes examined under the microscope for abnormalities and identity of sex.

At birth the amniotic sac, under pressure of the contracting uterus, aids in dilating the cervix in preparation for the passage of the fully formed infant outward through the vagina. At about this time in the birth process the amnion usually ruptures, releasing amniotic fluid that lubricates the birth canal.

Chorion, Placenta, and Umbilicus

The chorion, which is a composite layer of ectoderm and mesoderm cells, originates from the trophoblast, and is therefore the outermost extraembryonic membrane, enclosing the amnion, embryo, and umbilical cord (Fig. 11-23b, c).

Soon after villi have developed over the outer surface of the chorion (p. 287) and penetrated into the uterine mucosa, they begin to atrophy, except those on the side of the blastocyst bearing the inner cell mass or future embryo. This area of attachment of the embryo to the uterine wall develops into the placenta, which is formed from the villi of the chorion and the maternal tissue of the uterine lining in which they are embedded (Fig. 11-23c). It is connected to the embryo by the body stalk, which gradually elongates into a tube called the **umbilical cord,** or **umbilicus.** The mature placenta at birth is usually a disclike structure 6 to 7 in. in diameter and 1 in. thick. By that time the umbilical cord is a tubular, spirally twisted structure about 2 ft long and $\frac{3}{4}$ in. in diameter (Fig. 11-23b). The **navel** is a site on the abdomen of all individuals that represents the point of attachment of the umbilical cord.

During the course of pregnancy the spaces surrounding the villi and the developing placenta become filled with maternal blood, bathing the villi as soil water bathes the roots of a plant. The villi themselves contain numerous capillaries that receive and return blood via arteries and a vein from the embryo through the umbilical cord. It should be emphasized that the maternal blood circulation and the blood circulation in the embryo and placenta are separate and distinct. Normally only the slightest intermixing of these two bloods occurs. The exchange of material in the placenta takes place between the two bloods by diffusion through the walls of the capillaries and villi. Metabolic waste materials such as carbon dioxide and urea diffuse from villi capillaries to the surrounding maternal blood, whereas oxygen, nutritive substances, and hormones are transferred in the opposite direction. The placenta therefore performs the important functions of an excretory, respiratory, and nutritive organ during embryonic life.

The placenta also permits the transfer of antibodies from maternal to embryonic circulation in the last months of gestation. In this way a newborn infant temporarily acquires some of his mother's immunity to tide him over about the first six months of life, until he produces his own antibodies in response to various foreign agents in his environment.

Finally, the placenta serves as an endocrine gland, furnishing, among other hormones, sufficient quantities of progesterone for the maintenance and successful completion of embryonic development after the fourth or fifth month of pregnancy, when progesterone secretion by the corpus luteum declines.

Allantois

The allantois is a sac attached to the abdominal region of the embryo, arising as a small outpocketing of the forming gut. It is located between the chorion and the amnion. In egg-laying vertebrates such as reptiles and birds, it fuses with the chorion to enclose the embryo; it is well ramified with blood vessels that take up oxygen and give off carbon dioxide through the porous eggshell. Thus it serves as an embryonic respiratory device until the time of hatching.

In mammals the allantois is small and no longer functions in respiration. It does, however, possess blood vessels, and when it becomes incorporated into the umbilical cord its blood vessels become the umbilical arteries and vein connecting the blood circulation of the embryo proper with that of the placenta (Fig. 11-23c).

The period of development of the young in the female reproductive system is called **pregnancy** or **gestation;** it begins at fertilization and ends at birth. In humans the normal period of gestation is approximately 266 days from ovulation, or 280 days from the last menstruation.

Early human embryos are available only rarely for experimental investigation. Most have been obtained when, as a result of some illness, a woman's uterus must be surgically removed—an operation called a **hysterectomy.** Over the last 50 years or so such embryos, at all stages of development, have been collected and studied. Perhaps the most remarkable thing about the development of the human embryo is how similar it is to the process in other mammals and even other vertebrates (Fig. 11-24). During early phases the most rapid changes take place (see Fig. 11-3).

First Week

The fertilized human egg undergoes cleavage in the oviduct. Toward the end of the week the blastocyst implants in the uterine wall and begins to obtain nourishment from the maternal bloodstream.

Second Week

The blastocyst embeds deeper in the endometrium; the amniotic cavity and embryonic disc develop. By the fourteenth day a primitive streak has formed at the caudal (tail) part of the disc, and mesoderm is spreading between the ectoderm and endoderm. The size of the embryo is about 1.5 mm.

Third Week

The embryonic disc has broadened into a three-layered, pear-shaped plate that has folded into the tubular body of the embryo. A primitive tubular heart has formed but is not yet beating. The embryo is about 2 mm long.

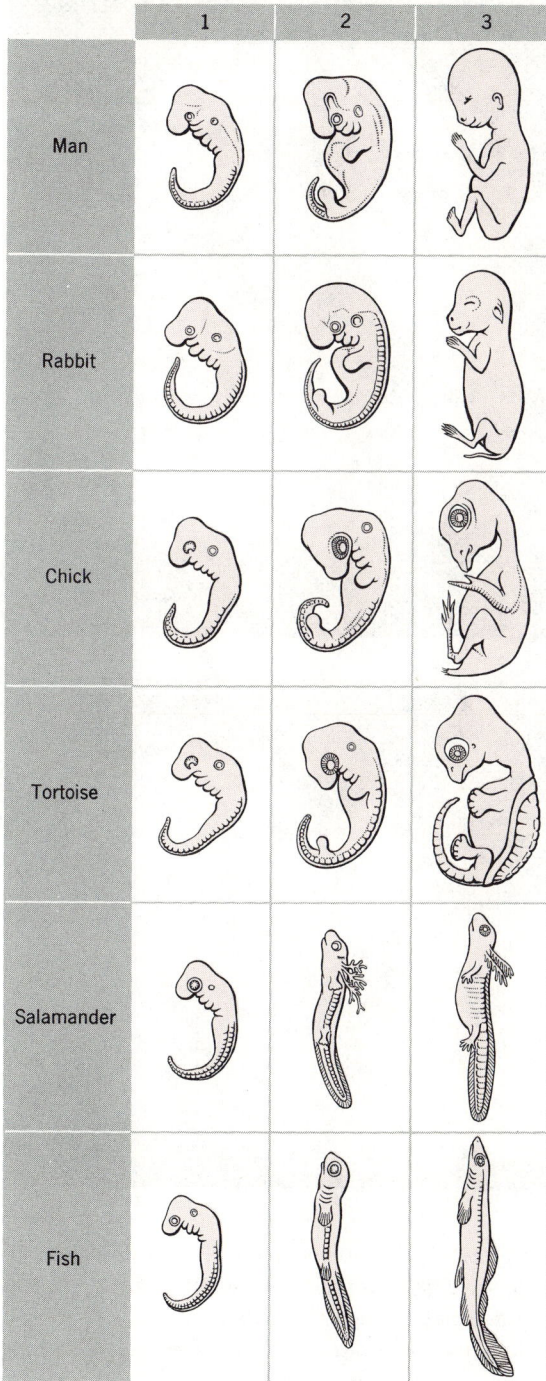

Figure 11-24   A comparison of vertebrate embryos at three equivalent stages of development.

	1	2	3
Man			
Rabbit			
Chick			
Tortoise			
Salamander			
Fish			

Fourth Week

By the end of the first month the embryo (Fig. 11-25a, p. 294), protected and suspended in amniotic fluid, is less than ½ in. in length and has the beginnings of brain, eyes, stomach, kidneys, and heart. The heart is beating (at approximately 60 times per minute) and the embryo has already increased in weight 10,000 times over the egg from which it originated. The primitive umbilical cord has formed and the embryo at this stage displays gill pouches and a tail-like appendage. The entire exterior surface of the enclosing chorion is fringed with rootlike villi, which anchor to the maternal tissue and draw nourishment from the maternal blood. Some of the villi will soon be incorporated into the placenta, whereas the remainder will atrophy and disappear.

Second Month

Between the fifth and eighth weeks the principal parts of the face and neck develop. The limbs begin to appear, first as **buds** or paddlelike protruberances (Fig. 11-25b), and soon develop and differentiate into arms and legs. The tail becomes most prominent about the sixth week and subsequently retrogresses and disappears.

By the end of the second month the embryo possesses most of the features and internal organs of the future adult, and from this stage until birth it is usually called a **fetus** rather than an embryo. The arms, hands, and fingers are formed by the seventh week (Fig. 11-25c). The slower-growing legs display recognizable knees, ankles, and toes. The fetus at the end of two months is no longer than 2 in. and weighs about ⅟25 oz. The nervous and muscular systems have developed to the extent that the fetus can move its arms and turn its body slightly. The eyes and ears are well formed.

Third Month

During the third month the limbs become longer, nails begin to make their appearance, and the external sex organs differentiate sufficiently so that male and female can be distinguished. Some movements of the body and

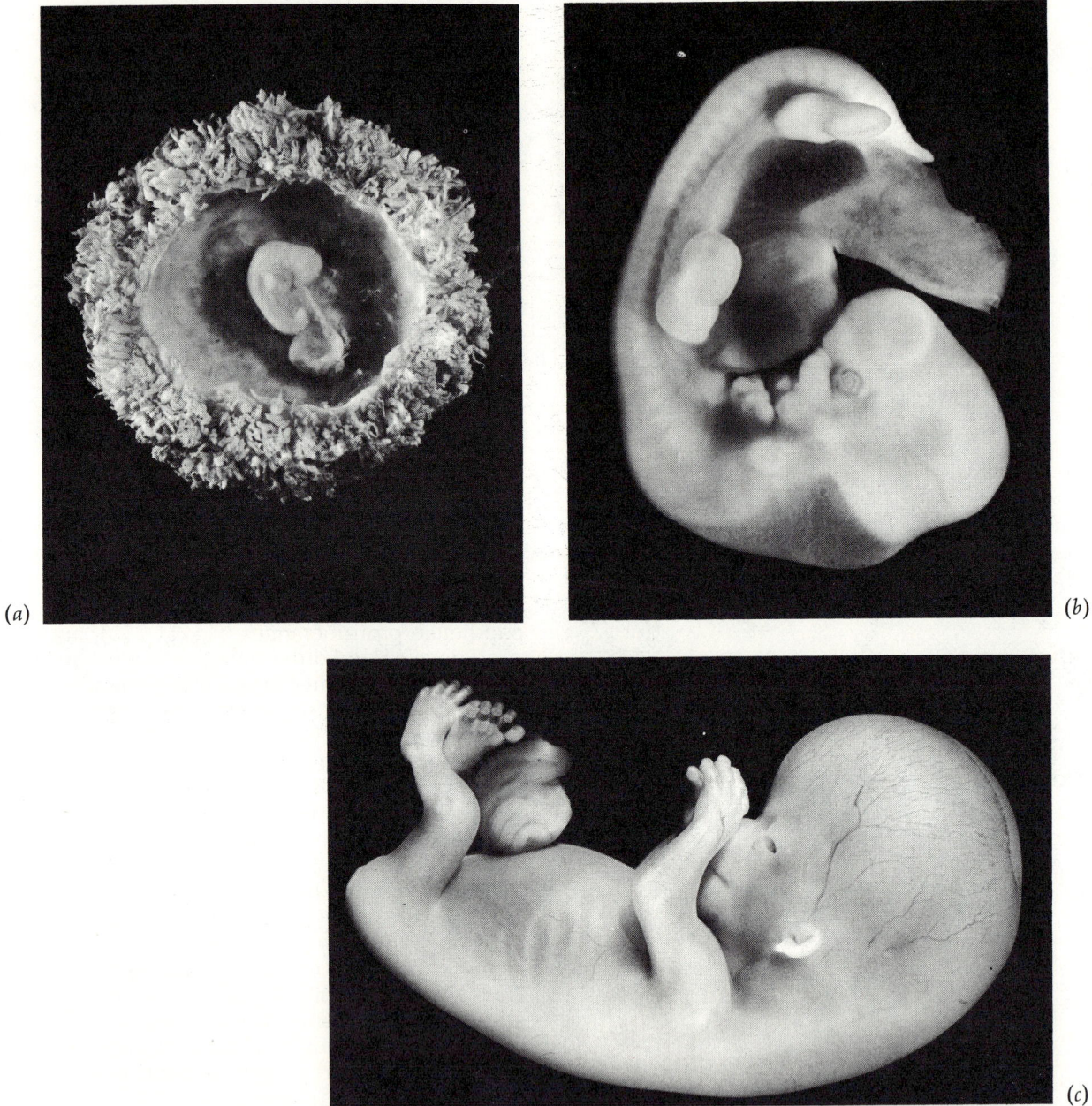

(a)

(b)

(c)

Figure 11-25 Further development of the human embryo. (a) a 28-day embryo with chorion opened to show embryo within aminiotic cavity. Magnification 3×. (b) A 34-day embryo, actual length 11.6 mm. Magnification 10×. (c) A 56-day embryo, actual length 37 mm. Magnification 5×.

limbs occur, but the fetus' total length of approximately 3 in. is so small that the mother does not yet feel its presence. Most of the subsequent development of the fetus in the remaining six months or so of gestation has to do chiefly with an increase in size, and to a lesser extent with the final steps in the formation of some organs.

Four to Six Months

In the fourth month hair appears on the head and body and the facial features become more distinct. By this time the placenta is firmly established and the entire uterine cavity is occupied by the fetus. After the fourth month the uterus, which is increasing in size, pushes up toward the abdominal cavity, displacing several of the internal organs. In the fifth month the fetus is about 1 ft long, and by the sixth month its movements in the uterus are vigorous and extensive; its presence is very clearly felt by the mother. It experiences intervals of sleep and of wakefulness similar to that of a newborn infant. At the end of the seventh month the body of the fetus has attained more nearly adult proportions.

Seven to Nine Months

In the final two or three months of gestation (Fig. 11-26) the fetus gains most of its birth weight and is increasingly able to survive if born prematurely. Approximately 5 lb are added to its weight during this period, and

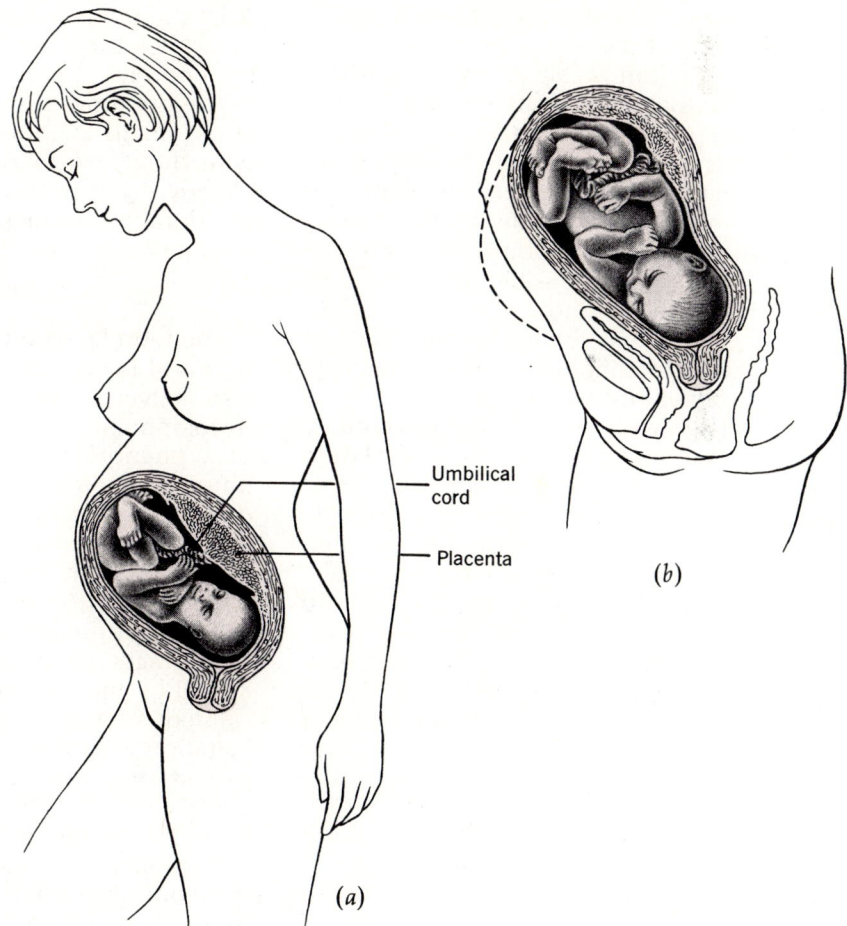

Umbilical cord

Placenta

(b)

(a)

Figure 11-26 The position of the fetus during pregnancy. (a) After seven months. (b) Full term, just before birth; broken line shows the configuration of the compressed abdomen during the birth process.

its body becomes smooth and plump with the deposition of subcutaneous fat. Most of the antibodies the fetus receives from its mother are transmitted at this time.

During the course of gestation the uterus has increased in size by approximately 200 times, from a weight of about 50 g and a capacity of approximately 5 ml in the non-pregnant state to a weight of about 1000 g by the end of the gestation period. Toward the latter part of pregnancy the human fetus normally assumes a position with its head directed downward (Fig. 11-26b), in preparation for its birth.

Childbirth

The process by which the fetus is expelled from the body of the mother to terminate the period of gestation is called **birth,** or **parturition.** The factors that initiate and control the remarkable process of birth are still obscure. Several hormonal changes, particularly a decreased secretion of progesterone relative to that of estrogen, occur in the mother's body before the onset of childbirth; they are undoubtedly in part responsible for the softening of the ligaments and certain other structures of the pelvic joints, rendering them more pliable for childbirth and also increasing the contractility of the uterus.

Birth begins with an onset of involuntary contractions of the uterine walls, called **labor.** These contractions are at first weak and of short duration, but become progressively stronger and more prolonged. The contractions of the uterus force the fetus, especially the head, against the cervix, causing it to dilate; **dilation** normally takes several hours. The still-intact amniotic sac with its enclosed fluid enveloping the fetus begins to bulge into the vagina. It usually ruptures shortly afterward, liberating the quart or so of amniotic fluid, which drains out through the vagina. Meanwhile the vagina itself has become more pliable and distended. The birth process is culminated by powerful contractions of the uterus, aided by voluntary contractions of the abdominal muscles, which

are sufficient to expel the fetus, normally head first, through the maximally enlarged cervix and vagina to the exterior (Fig. 11-27).

The umbilical cord, which still connects the newborn infant to the uterus by way of the placenta, is tied off by a physician soon after birth and cut close to the baby's body. Within the next 15 to 30 minutes further contraction of the uterine wall serves to loosen and expel the placenta and other attached extraembryonic structures, collectively called the **afterbirth,** from the mother's body. When the placenta, which includes part of the uterine wall, is torn away, a hemorrhaging wound is opened on the wall of the uterus. The bleeding is ordinarily controlled by the same contractions of the muscular uterine walls that expel the newborn; these contractions constrict the blood vessels, impeding the flow of blood. Clotting finally terminates the bleeding entirely.

The duration of the birth process averages about 16 to 20 hours for women pregnant with their first child. For women who have already given birth to one or more children the duration averages about 12 hours.

Lactation

The term **lactation** includes not only initiation and maintenance of milk secretion, but also delivery of milk to the young. Mammary development and function are highly complex phenomena. They are controlled and influenced directly and indirectly by a variety of hormones (Chapter 16), including estrogen, progesterone, adrenal corticoids, lactogenic hormone, FSH, LH, and oxytocin.

According to our present knowledge, the following primary events occur. During the later stages of pregnancy progesterone secreted by the placenta aids in further development of the mammary glands for future lactation. The relatively high levels of estrogen at this time, also secreted by the placenta, inhibit secretion of the lactogenic hormone from the anterior pituitary gland; the expulsion of the placenta during parturition is responsible for a sudden decrease in the mother's estrogen level, resulting in an increased secretion of the lactogenic hormone and subsequent synthesis of milk. Soon after birth the ejection of milk from the mammary glands is stimulated by the suckling action of the newborn at the nipples of the mother's breasts. This suckling stimulus apparently gives rise to nerve impulses to the hypothalamus, resulting in the release from the posterior pituitary gland of oxytocin; the latter stimulates the ejection of milk from the mammary glands. Oxytocin is also responsible for stimulating contraction of the uterus; it thus aids in restoring the loss of blood from the torn area of the uterus by constricting the blood vessels.

The first milk secreted after parturition is a watery fluid, called **colostrum,** with a high content of the mother's antibodies. The antibodies can be absorbed through the infant's intestine and help immunize against common infectious diseases during the first six months or so of life.

Figure 11-27   Birth, the beginning of independent life.

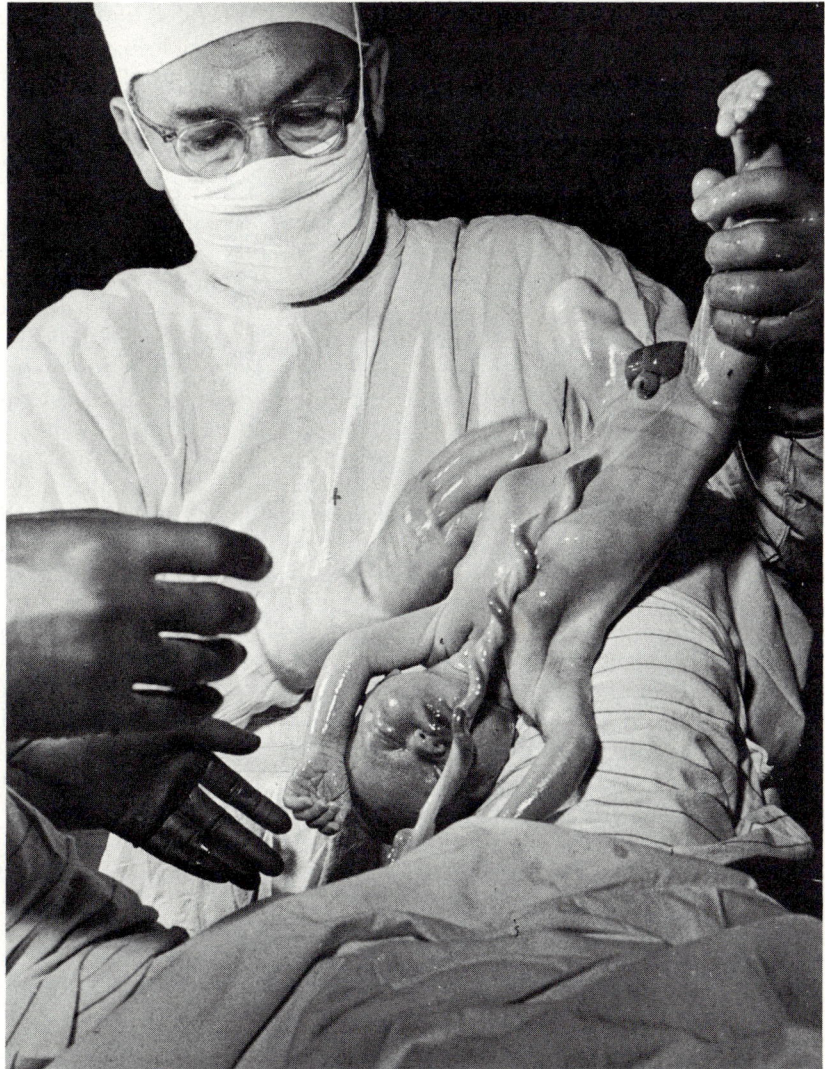

Suckling, or **nursing,** is essential for the continued formation of milk by the breast and its secretion; milk will continue to be secreted as long as the infant suckles, ordinarily for eight or nine months. Milk secretion progressively decreases and finally ceases, and the breasts decrease in size, when the nursing stops.

Birth Control

We have emphasized so far the similarities in reproduction between man and other animals. An important difference involves the frequency of the human monthly ovulatory cycles and the fact that women are sexually receptive continuously without regard for seasonal considerations. Because ovulation may occur less than a month after childbirth, many women are capable of having a new baby every 11 or 12 months throughout their reproductive lives (although families of more than 20 surviving children are rare indeed). We will see in Chapter 20 that the reproductive rate of all other animals is subject to certain natural checks that prevent uninhibited population growth of any species at the expense of others. Man seems as yet to be under no such restraints; the human population explosion, with its biologically disastrous consequences, is a problem of wide concern. Moreover, a child that is unwanted and uncared for by its parents is a needless social tragedy. Thus most couples, for a variety of reasons—physical, economic, social, ethical— seek to prevent unwanted pregnancies.

Any method that keeps sperm from reaching the egg will prevent pregnancy; this is called a **contraceptive** measure. In theory, complete sexual abstinence is the most direct and infallible approach to preventing the conception of unwanted children. This method, however, does not take into account the strength of the sex drive in most individuals, and has rarely proved to be workable as a means of regulating reproduction except when accompanied by strong religious vows and isolation.

Based on the observations that ovulation usually occurs about 14 days before the next menstrual period, and that eggs and sperm are viable for only one to three days, some techniques call for abstinence only during the five days before and the five days after the day midway between menstrual periods. Careful application of this plan—called the **rhythm method**—reduces, but does not seem to eliminate fully the probability of fertilization, because many women experience irregular menstrual cycles at least occasionally. One recent study found that the rhythm method permitted a 24 percent rate of unwanted pregnancies.

A variety of **sperm-blocking devices** are available, some of very ancient origin, that are designed to permit sexual intercourse but to block the entry of the sperm into the uterus. Currently used devices include the **condom,** a thin rubber sheath worn over the penis during coitus; the **diaphragm,** a rubber disc placed in the vagina to cover the cervical opening to the uterus; and various **spermicidal jellies** or creams that are deposited in the vagina to kill sperm. These methods are variously successful, depending mainly on the care and expertise with which they are utilized, but on the average 10 to 20 percent of couples using these methods fail to prevent unwanted pregnancies.

Physicians have tried placing various devices in the human uterus to prevent implantation for at least a century, but without much success in the past because the uterus, like other body tissues, becomes infected or tends to expel foreign matter. This problem has been solved with the use of nontoxic plastics. Today the commonly used **intrauterine device (IUD)** probably functions by blocking implantation of the fertilized egg, causing the equivalent of a very early miscarriage. The device may be up to 95 percent effective (it occasionally may be expelled) and can be removed at any time—also by a physician—with normal pregnancies following.

After implantation of the embryo, its removal from the uterine wall is termed **abortion.** A great deal of research is now being done to seek chemical abortants that will block implantation at its earliest stages. After implantation is well under way, abortion can be induced routinely by a physician.

The most popular current method of birth control is the **oral contraceptive pill,** which prevents ovulation through hormonal effects. Recall that after fertilization the normal menstrual cycle is interrupted by the secretion of chorionic gonadotrophin, which prevents degeneration of the corpus luteum and thereby maintains the blood level of its hormone progesterone. A high progesterone-to-estrogen ratio maintains the uterine endometrium and also prevents further ovulation. Birth control pills contain a balance of progesterone and estrogen (or synthetic analogs of those molecules), which is designed to prevent egg production via the same mechanism. There are several types of contraceptive pills available, containing various hormone combinations. In general, when used regularly, these are highly effective in controlling conception; failure rates average .1 to 1 unwanted pregnancies per 100 couples per year.

Many women experience undesirable side effects such as nausea and a tendency to gain weight when on the pill. A more serious difficulty, although rare, is a tendency for a few women to suffer **thrombosis,** a condition in which blood clots form in the blood vessels, blocking circulation. There is also a very small increase in the probability of multiple births (i.e., twins or triplets) after the use of contraceptive pills is stopped. As with any other pill, oral contraceptives should be used only under a physician's supervision.

No doubt as we learn more about the physiology of reproductive hormones, today's oral contraceptives will be improved and wholly new approaches will be discovered. It seems safe to predict that this particular product of our scientific age will have profound effects on humanity for many generations to come.

Reading List

Allen, F. D., *Essentials of Human Embryology* (2nd ed.). Oxford University Press, London, 1969.

Arey, L. B., *Developmental Anatomy* (7th ed.). W. B. Saunders, Philadelphia, 1965.

Bodemer, C. W., *Modern Embryology.* Holt, Rinehart & Winston, New York, 1968.

Corner, G. W., *Ourselves Unborn.* Yale University Press, New Haven, 1944.

Edwards, R. G., and R. E. Fowler, "Human Embryos in the Laboratory," *Scientific American* (December 1970), pp. 44–57.

Fertig, D. S., and V. W., Edmonds, "The Physiology of the House Mouse," *Scientific American* (October 1969), pp. 103–110.

Hardin, G., *Biology, Its Principles and Implications* (2nd ed.). Freeman, San Francisco, 1966.

Harrison, R. J., and W. Montagna, *Man.* Appleton-Century-Crofts, New York, 1969.

Havemann, E., *Birth Control.* Time-Life Books, Time, Inc., New York, 1967.

Jones, K. L., L. W. Shainberg, and C. O. Byer, *Sex.* Harper & Row, New York, 1969.

Masters, W. H., and V. E. Johnson, *Human Sexual Response.* Little, Brown, Boston, 1966.

Tietze, C., and S. Kewit, "Abortion," *Scientific American* (January 1969), pp. 21–27.

chapter twelve *Principles of Development*

"A single cell out of the millions of diversely differentiated cells which compose the body, becomes specialized as a sexual cell; it is thrown off from the organism and is capable of reproducing all the peculiarities of the parent body, in the new individual which springs from it by cell division and the complex process of differentiation."

AUGUST WEISMANN, 1889, *ESSAYS UPON HEREDITY AND KINDRED BIOLOGICAL PROBLEMS*

Consider the amazing quality of the process we described in Chapter 11. A single cell of one organism joins with that of a mate. The zygote divides, grows, undergoes mysterious contortions, and—miraculously—a baby, or a frog, results. It has skin and eyes, a beating heart; it has limbs and brain and a voice to cry or croak. The egg has none of these, and whereas an egg is a single very small cell, a baby, or a frog, is made of billions or trillions of cells and weighs many grams or kilograms.

Thus, merely by comparing a fertilized egg with the finished organism, we can infer many of the processes that must have occurred in the change from one to the other. The first of these is **growth,** meaning a permanent increase in mass. A human egg is about .15 mm in diameter and weighs perhaps 3 millionths of a gram. The egg of a whale or a mouse is not much different in size. In all three cases the egg's growth represents an increase in weight of billions of times and an increase of thousands of billions in numbers of cells. What initiates this growth? What determines the rate of growth? During development, different parts or organs of an animal or plant grow at different rates (Fig. 12-1, p. 302). How are these differential growth rates regulated? And what causes a group of cells to stop growing, as in the experiment shown in Figure 12-2?

The second process of development we must consider is **differentiation.** A multicellular organism is characterized not only by its many

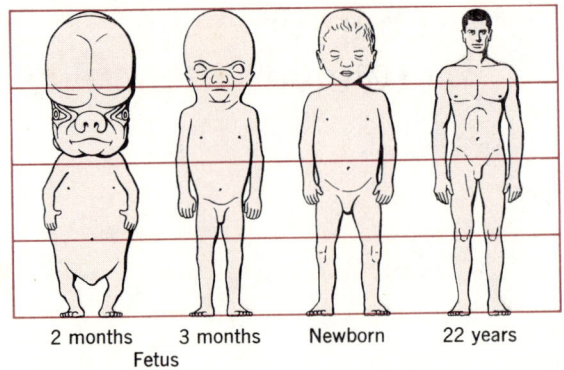

Figure 12-1 *Diagram showing the changing proportions of the human body that result from different rates of growth in different parts during the fetal and postnatal period. Note that the legs elongate at a faster relative rate than does the body, whereas the head grows much more slowly from two months of gestation through adulthood.*

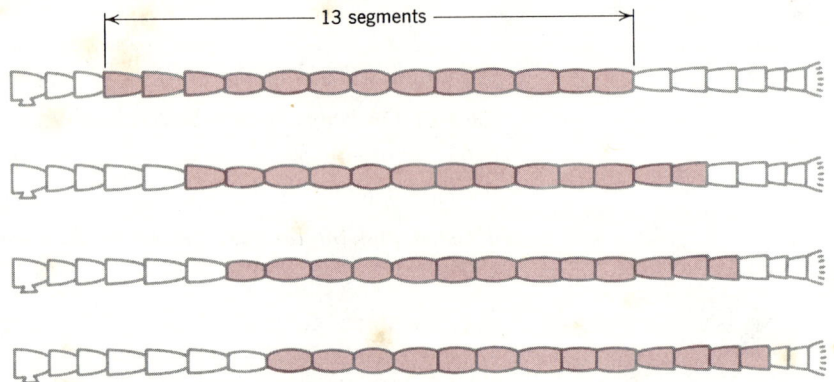

Figure 12-2 *An example of precise growth control. All individuals of the marine segmented worm Clymenella have exactly 22 segments. If several segments are cut off, at either the head or tail end, or both, exactly the correct number of segments regenerates to restore the original number. For example, if a total of nine segments is removed, leaving 13 segments intact, the same number that was cut off at each end is restored.*

cells, but even more by the fact that those cells are of many different types. Retinal cells synthesize rhodopsin; nerve cells are specialized to conduct electrical signals; muscle cells are filled with the contractile proteins actin and myosin. For the first few divisions of an egg in many species, all of the daughter cells, called **blastomeres,** seem to be identical. In the sea urchin or sand dollar (see Fig. 12-13), for example, any one of the four blastomeres resulting from the first two cleavages can be isolated and will develop into a perfectly normal larva, called a **pluteus.** Such larvae are about one-fourth normal size but contain all

the usual different cells and tissues that would have formed from the entire egg (Fig. 12-3). Clearly the zygote contains the instructions, coded into its genome, necessary for constructing a new individual, and these are replicated and parceled out to each blastomere in the mitotic divisions.

After two or three more cleavages, for example at the 16-cell stage, a single blastomere is capable of forming only a partial embryo in which some of the tissues that normally would have been contributed by the other 15 blastomeres are lacking. As cleavage and development proceed, each cell eventually has

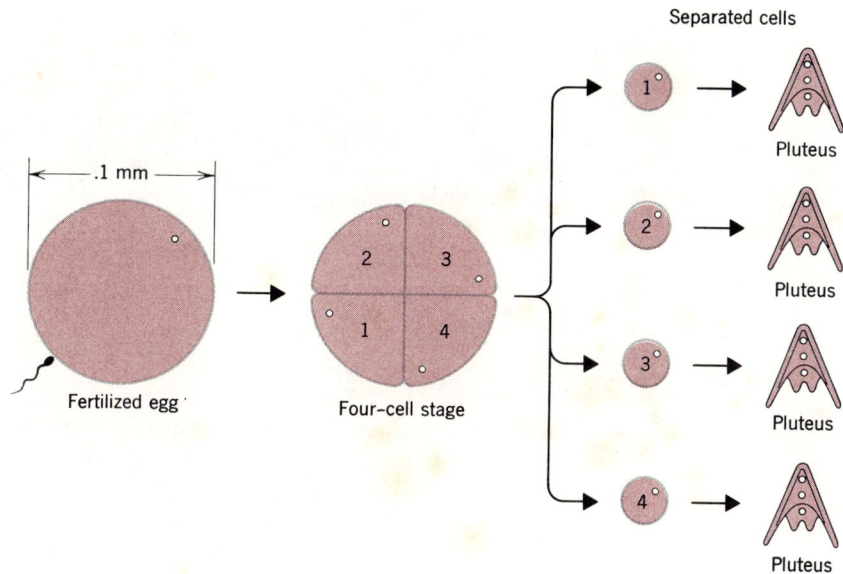

Figure 12-3 An experiment with a sea urchin egg is diagrammed to show that each of the first four cleavage blastomeres has the same developmental capabilities. Each develops into a complete pluteus larva with all the normal structures but only one-fourth normal size.

the ability to form only one or a few types of tissue, at which point we say it is **proto-differentiated;** that is, its fate is now determined in the normal course of events.

Differentiation is the process whereby the daughter cells of a zygote become different. But if the early blastomeres are all identical, and all the cells of an embryo are produced by mitotic division of those blastomeres so that each chromosome is faithfully replicated, why should cells ever become different? How is the course of differentiation regulated in such a precise sequence that the stages of embryonic development of each organism are exactly the same, generation after generation?

An egg that grows and differentiates will form an embryo only if a third developmental process also occurs. This is **morphogenesis,** a term that derives from Greek words meaning "the origin of form." Morphogenesis is the process whereby cells become spatially organized into functional organs and tissues with proper shapes and sizes (Fig. 12-4, p. 304). It includes the organized movements of individual cells and groups of cells, the folding and rolling cell sheets into tubes or balls—called **morphogenetic movements**—and the regulation of different mitotic rates

in different parts of a tissue to produce swelling or curvature from **differential growth.** In plant embryos each cell is surrounded by a rigid cell wall and cells cannot move or migrate, so that shape changes must result exclusively from differential cell growth. Both morphogenetic movement and differential growth are responsible for forming the vertebrate eye (from a hollow ball of ectoderm) or the heart (from a spongy sheet of mesoderm cells). Retinal cells would be of no value to an organism, no matter how much rhodopsin they might synthesize, if they were scattered throughout a shapeless blob of cells. Only if they are organized into a concave sheet on the inner wall of the eyeball, positioned exactly the proper distance from the lens to receive light focused on them, do they form a retina (Fig. 12-5, p. 305). Muscle cells, actively synthesizing actin and myosin, do not form a useful organ for mobility unless they are aligned side by side into elongated fibers and are attached at each end to a properly shaped and jointed bone (Chapter 18).

A host of questions arise in the mind of any awestruck viewer of embryonic development. How do cells move among and upon one another? What guides their movements

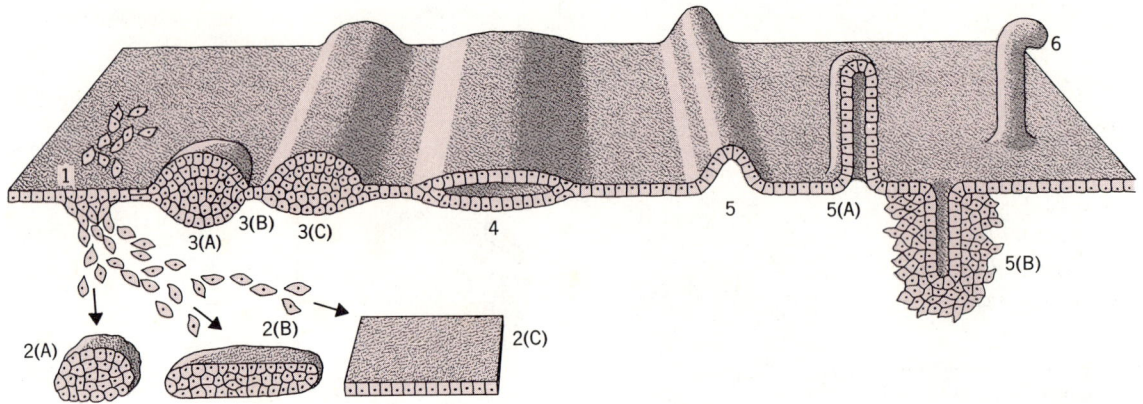

Figure 12-4 *A diagram illustrating various morphogenetic processes. (1) Cell migration either on or from a solid cell sheet. (2) Cell aggregation forming (2A) masses, (2B) rods, or (2C) sheets. (3) Localized growth resulting in various kinds of enlargements (3A, 3C) and (3B) constrictions. (4) Fusion or splitting of cell sheets. (5) Folding, including (5A) evagination, to form hollow balls or tubes, and (5B) inpocketing, or invagination, into a mass of tissue. (6) Bending resulting from unequal growth.*

to a distant location? What signals indicate "stop here"? What forces cause a sheet of cells to fold into a tube or bring about the outpocketing—called **evagination**—that deforms the sheet into a hollow ball?

A developing system need not be an embryo. A regenerating salamander limb is a developing system; so is the metamorphosis of a tadpole into a frog. Many animals and plants exhibit life cycles during which they progress from one stage to another; if during that progression an organism undergoes the three fundamental processes of development—growth, differentiation, and morphogenesis—even though we may not refer to it as embryonic, it may be considered as a developing system.

We saw in Chapters 4 and 10 that the synthetic processes of a cell are controlled by information coded into the cell genome. Differentiation requires different cells to manufacture different products; how can the cells of an embryo ever become different if they begin as identical daughters of a mitotic division? The answer to that question, which we shall detail in this chapter, is, first, that the

cleavage blastomeres are never actually identical. Although their chromosomes may be exact duplicates, the cells contain different parts of the original egg cytoplasm and membrane. These subtle differences can cause some genes to be activated in some blastomeres and not others. Thus different cells begin synthesizing different products very early in development. These differences in turn, can result in cell behavior that leads cells along progressively divergent paths of development.

We know, for example, that the cells of an embryo move among themselves and take up specific locations as a result of the properties of their surface membranes. If a cell begins to synthesize new membrane components, it may move to a different position as a consequence of this early differentiation. The resulting change in position may place the cell in a new environment, which may, in turn, set off a further differentiation. Thus in general the process of development takes place because early small differences among cells are progressively magnified by interactions between the genome and environment of

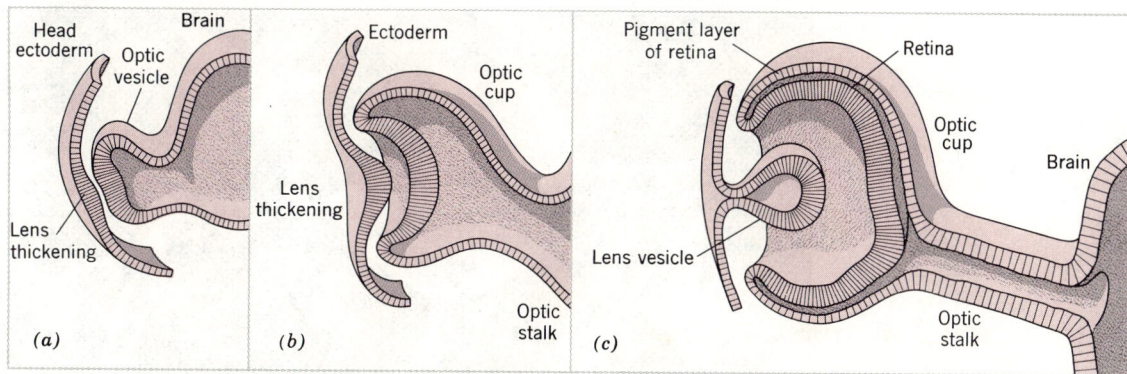

Figure 12-5   Development of the vertebrate eye, to show its morphogenesis from (a) a simple vesicle (ball) by inpocketing to form (b) a cup; and reorganization of two cell layers (c) to produce the lens and double-layered retina.

each cell. As more and more different sets of genes are activated in different cells, their properties and behavior diverge further, leading eventually to the formation of the many highly differentiated cell types of the adult.

Some Developing Systems

Cap Formation in Acetabularia

A bacterial cell in a rich medium replicates its DNA and uses the genetic information coded there to synthesize an entire stock of new components before dividing into two cells. An ameba does the same thing. In both cases we refer to the process as reproduction, but not as development. Nonetheless, there is ample evidence that the developmental processes that produce a multicellular embryo are just as much under the control of information on the chromosomes as are the synthetic processes that occur in the reproduction, say, of *E. coli.*

One of the best bits of evidence that de-

velopment is controlled by activities of the nucleus comes from experiments by Max Hämmerling on a peculiar plant called *Acetabularia* (Fig. 12-6, p. 306), a green marine alga, thin and delicate in structure, with a shape like a tiny umbrella. Even though the whole plant may be several centimeters high, it has no crosswalls. The stalk and cap enclose one continuous pool of cytoplasm, with one giant nucleus in a rootlike structure at the base called a **rhizoid.** At one stage in its life cycle (see Fig. 6-9), however, the nucleus breaks down into many small nuclei that spread throughout the stalk and cap. Around each daughter nucleus a wall is produced to form **cysts,** or **spores.** These all migrate to the cap and are released. Eventually the cysts germinate, releasing into the water motile, flagellated gametes; these undergo a process called **conjugation,** in which they exchange chromosomal material to form a zygote. The zygote then differentiates a stalk, rhizoid, and cap, and grows from a fraction of a millimeter to several centimeters in length. Clearly the plant has undergone the three fundamental processes of development—growth, differentiation and morphogenesis—even though it is not multicellular and is not considered an embryo at any stage.

Hämmerling found that if he cut a stalk in

Figure 12-6 The alga Acetabularia mediterranea. *Each umbrella-like structure is one giant cell several centimeters long.*

half (Fig. 12-7) the upper half would survive for some weeks but would not regenerate a new base. The lower half, containing the nucleus, would grow a new tip and then produce the radial spokes that make up a new cap, just as in the course of normal development. Moreover, the nucleated stalk could regenerate a new cap repeatedly. Hämmerling also found that it was easy to remove just the nucleus from *Acetabularia* by cutting off the rhizoid in which it was located. The **anucleate** plant could survive this operation for as long as several months. If the cap were removed after the base healed, one new cap would regenerate; but if the cap were removed a second time, no further regeneration would occur, and the stalk would eventually die. To explain these results Hämmerling suggested that some material produced by the

nucleus (presumably mRNA) is required for the alga to regenerate a new cap. The anucleate stalk had enough of this material in its cytoplasm to permit the regeneration of one cap, but without a nucleus it could produce no more. This tells us that nuclear products are required for regeneration (or development), but provides no evidence as to whether the nucleus directs those processes.

Fortunately, there are several species of *Acetabularia* that can be distinguished by the shape of their caps, and it is possible to graft the stem and cap of one species onto the base of another (Fig. 12-8). Such **interspecific grafts** may be made between *A. mediterranea* and *A. crenulata*, for example. When healing is complete after the grafting, the cap is removed from the grafted stalk, and after a few weeks a new cap is regenerated. At first

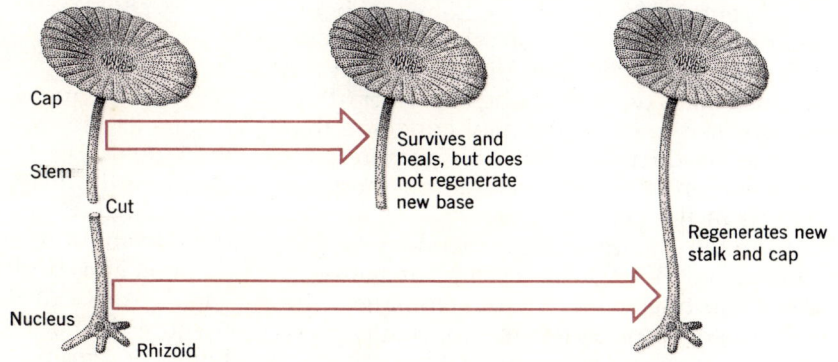

Figure 12-7 *Experiment demonstrating regeneration of the nucleated half of Acetabularia.*

Figure 12-8 *Transplantation experiments using two species of Acetabularia, identifiable by the shape of their cap (here colored differently also), show that the character of the cap that is regenerated is controlled by the nucleus.*

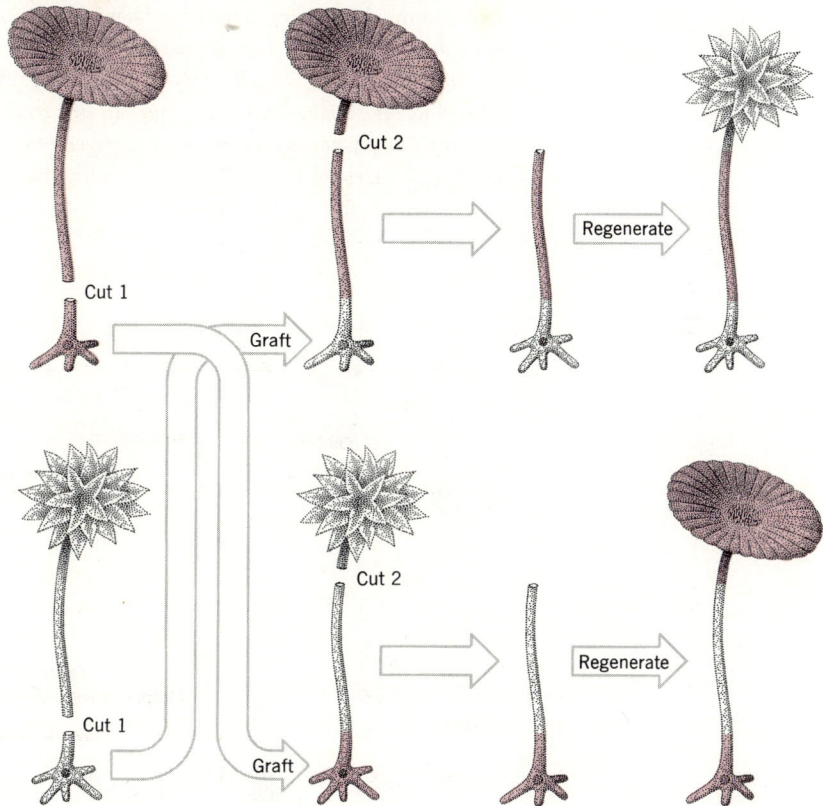

the regenerating cap shows signs of being appropriate to the stem on which it is growing; but these signs soon disappear, and the new cap takes on entirely the character of the nucleus in the base. We may conclude that nuclear products are not only required for regeneration, they also direct the morphogenesis of the cap.

Although Hämmerling's experiments argue convincingly for nuclear control of synthesis and morphogenesis in *Acetabularia*, the component processes are not really very different from those required to produce a new ameba or bacterial cell since only one nucleus is involved; it produces structures appropriate to its genotype in the cytoplasm of its domain. In the next section we shall examine evidence for similar gene control of development in multicellular systems.

Cellular Slime Molds: a Differentiative Model

The slime mold *Dictyostelium discoideum* begins its life cycle (Fig. 12-9) as a group of individual free-living amebas that inhabit moist earth. In the laboratory we can study these amebas growing on moist agar. They migrate about on the agar feeding actively on bacteria and dividing rapidly. When the food supply runs out they stop dividing and undergo a dramatic change in both shape and behavior. From irregularly shaped, typical ameboid cells (Fig. 12-10a), they become smooth, elongated, streamlined cells (Fig. 12-10b). At first they move independently, but soon begin flowing in streams, each ameba sticking to its neighbors in front and behind, migrating toward centrally located **aggrega-**

Figure 12-9   The life cycle of the cellular slime mold Dictyostelium discoideum. *Free-living amebas swarm together into aggregation centers to form a multicellular, slug-like pseudoplasmodium. During the migration of the pseudoplasmodium, its cells differentiate into prospective stalk, spore, and basal cells. Soon it stops migrating, points its tip in the air, and grows into a fruiting body (sporangium) with a differentiated stalk rising out of the basal plant and carrying a droplet of fluid with spore cells suspended in it. The spores are eventually released and from each an individual ameba germinates, completing the cycle.*

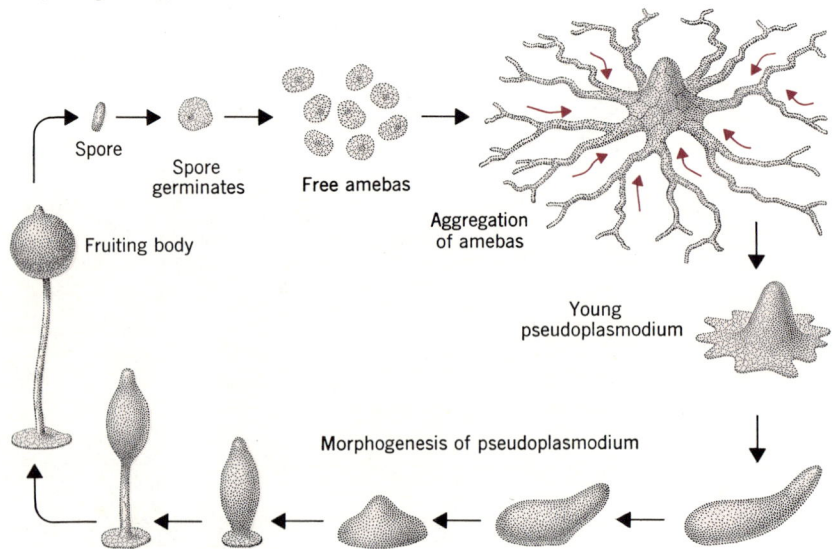

Figure 12-10   Amebas of D. discoideum (a) in the feeding stage; (b) in the aggregation stage. (c, p. 310) Aggregation centers with streams of ameba swarming in. (d) Pseudoplasmodia gliding over the surface of an agar-coated dish, leaving a trail of slime and sloughed-off dead cells behind. (e to k) The culmination stage in the life cycle of the cellular slime mold begins when the pseudoplasmodium stops migrating, tilts its forward tip upward, and pushes under to form a hat-shaped mass (e, f). The mass pushes off the agar on the tip of a rapidly forming, thin, hard stalk (g to i) to produce the mature fruiting body, or **sorocarp** *(j, k).*

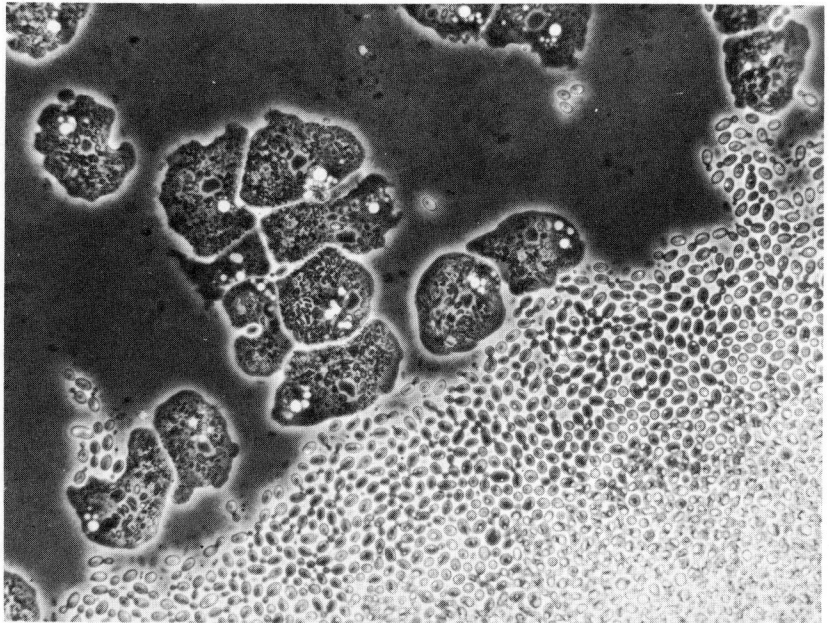

(a)

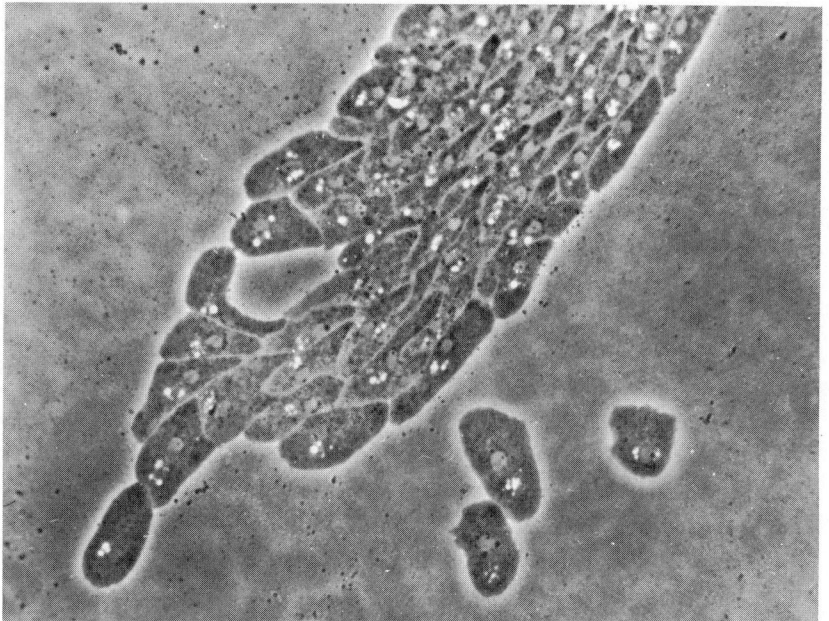

(b)

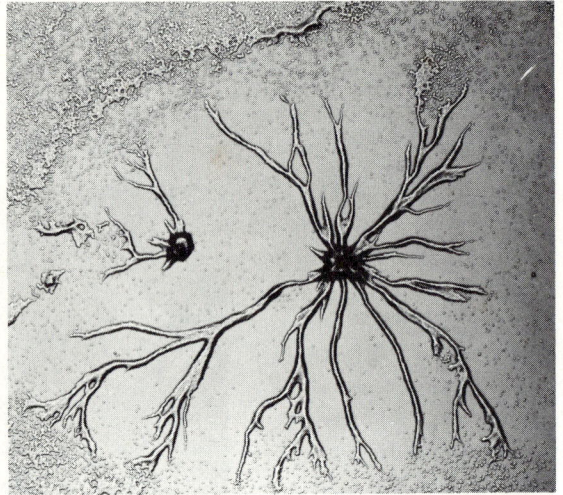

(c)

Figure 12-10 (cont.)

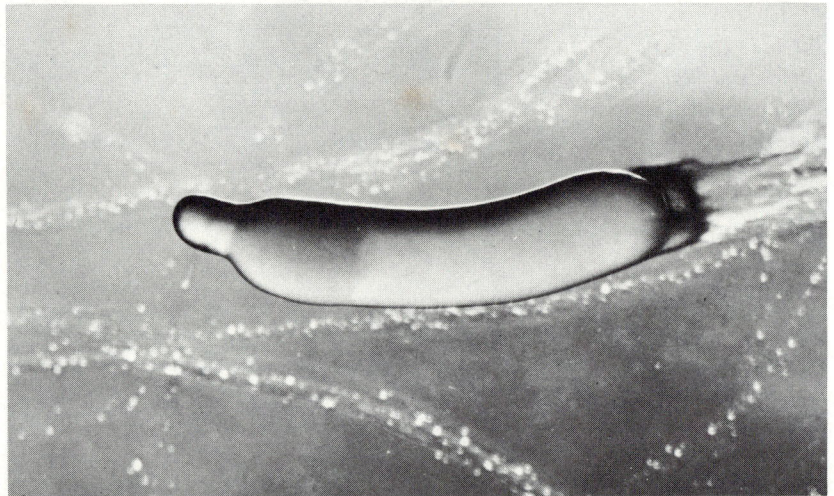

(d)

(e)   (f)

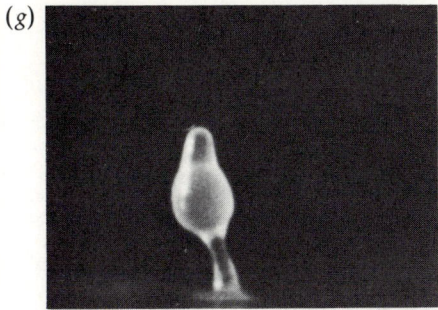

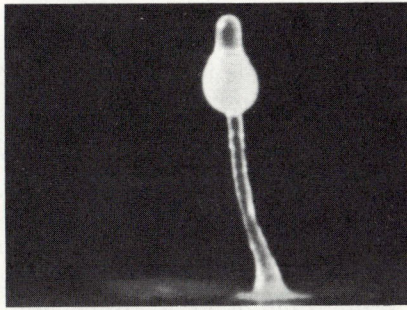

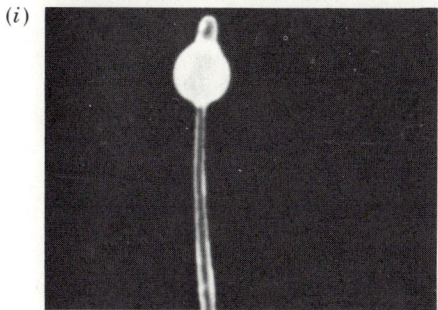

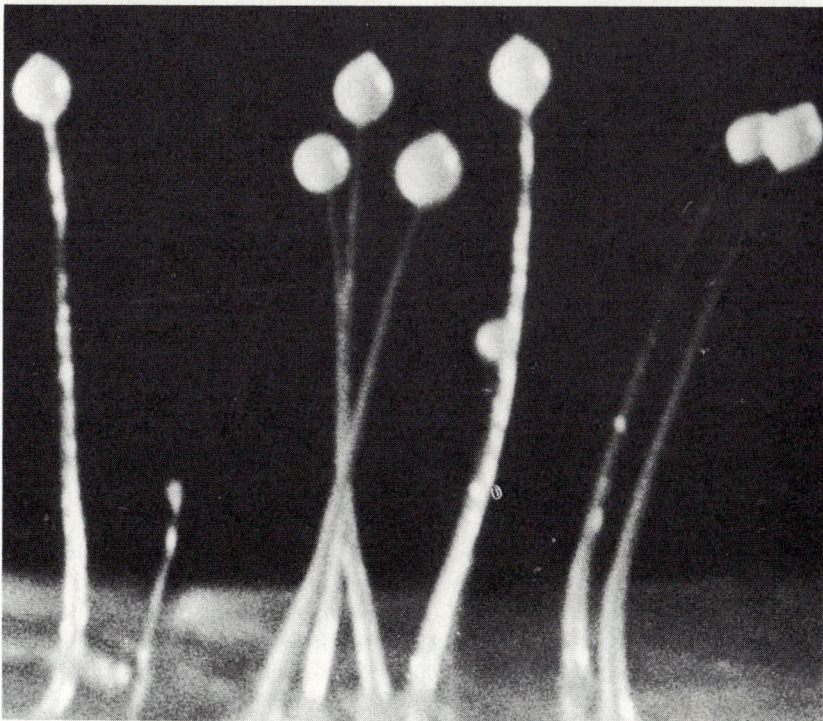

(g)　(h)　(i)　(j)　(k)

tion centers (Fig. 12-10c). Aggregation centers are groups of cells that secrete a hormone-like substance called **acrasin.** Recent evidence indicates that acrasin is **cyclic AMP** (a circular form of adenosine monophosphate), an important nucleotide that seems to play a role in hormone action in a wide variety of tissues in vertebrates and invertebrates alike.

The swarming amebas mass together in the aggregation centers, piling up into cylindrical, sausagelike structures several millimeters long, called **pseudoplasmodia.** As a cylinder grows longer, it leans over onto one side and glides off over the agar, leaving a trail of slime behind it and resembling nothing so much as a tiny garden slug. (Fig. 12-10*d*). What began as a group of independent cells has now produced a multicellular organism with distinctly new properties.

If, at the free-living, feeding ameba stage, we sweep a bunch of cells together into a pile, they quickly disperse, showing no sign of adhesiveness. In contrast, if a migrating pseudoplasmodium is minced or gently squashed, it is possible to dissociate the amebas; within minutes these dissociated cells reaggregate and reconstitute the slug. Clearly the adhesive properties of the individual cells at the two stages are very different.

An intact pseudoplasmodium also has properties that are not demonstrated by individual amebas. Migrating pseudoplasmodia, for example, are **phototropic;** that is, they move toward light, orienting in the direction of even very faint light sources. Individual amebas show no such tropism; they do not respond to light of any intensity. The multicellular mass has another important property lacking in the individual amebas: its tendency to permit or promote protodifferentiation. Recall (p. 303) that this is the point at which a cell must differentiate along a given route.

Migration of the pseudoplasmodium may continue for several hours or days, lasting longer, generally, in a more humid atmosphere. Why it stops we do not know, but when it does a complete change in the structure takes place. The front tip of the pseudo-

plasmodium stops moving first and points upward, while the rear sections continue to move forward, bunching underneath (Fig. 12-10*e,f*) to form a hat-shaped mass. The cells that originally formed the tip of the pseudoplasmodium become swollen, develop vacuoles, and begin to synthesize a celluloselike product. They move into the center of the mass, organize themselves into a thin cylinder, and there deposit a layer of the cellulose, which hardens into a stiff stalk. The stalk grows in height as additional cells are added to it. Similar cells form a flat circular base at the foot of the stalk.

As the stalk forms, the cells that make up the remainder of the pseudoplasmodium are carried upward off the agar surface. As it is lifted into the air each cell shrinks into a small sphere by excreting its excess cytoplasmic water and forms a covering of clear, slimy fluid and a hard outer cell wall. Each cell sphere thus differentiates into a spore, several thousand of which are suspended in a drop of clear, mucous fluid atop the newly formed stalk. This whole structure, generally 2 to 3 mm tall, is called a **mature, fruiting body,** or **sorocarp** (Fig. 12-10*k*). Ultimately the spores are released. If they fall upon a suitable moist substrate, each spore releases a free-living, feeding ameba, and the cycle can begin again.

Regulation

In the completed fruiting body three types of cell are present: stalk, base, and spore. How did these three cell types differentiate? Did three different kinds of ameba aggregate to form the pseudoplasmodium? Might the cells have become genetically different during the migratory phase? These possibilities can be ruled out by a simple experiment. If a plasmodium is dissociated, an individual cell may be transferred to an agar plate with an ample food supply. After a while it stops trying to reaggregate and begins feeding and dividing. Curiously, it does not now aggregate with its newly formed brothers, but instead the cells continue dividing and feeding.

These cells are called a **clone,** a group of cells all derived from a single common ancestor. If the original mother cell had represented a genetic strain of one of three cell types— stalk, for example—this should be manifested in the behavior of the descendants. In fact, however, when the mass of amebas has devoured all the food, it aggregates to form a pseudoplasmodium that migrates in the usual fashion and culminates in a normal sorocarp made of the usual numbers of each of the three cell types. Thus each ameba seems to have the same potential (i.e., genetic information) as every other, and can transmit that information to its descendants.

It can be shown experimentally that the cells at the forward tip of a migrating pseudoplasmodium normally form the stalk, the rear end produces the base, and the spores differentiate from the large middle section (Fig. 12-11a). Suppose, then, that we cut a migrating slug in half. The front piece contains about half of the cells that would have formed spores, all of the prestalk cells, and none of the prebase cells. The rear piece has all of the cells that normally would have differentiated into the base, half the prespore cells, and none of the prestalk cells. If this cut is made under conditions of high humidity, the two pieces stop moving only briefly while the cut end smooths over. Each piece then moves off in its own direction and will continue migrating for several hours before it culminates. When it does so, each fragment develops into a small but perfectly proportioned fruiting body, with the usual percentage of all three cell types (Fig. 12-11b). This capacity of a developing system to compensate for experimental or accidental alterations and to progress normally is called **regulation;** most embryos have some regulative ability.

If humidity is not kept high during such an experiment, in some cases at least, each fragment of the pseudoplasmodium promptly elevates its forward end and culminates on the spot (Fig. 12-11c). In this case the result is different; the structure of the fruiting body much more nearly reflects the proportion of protodifferentiated cells in the two frag-

Figure 12-11   Experiments with a migrating pseudoplasmodium. (a) The location of the three different cell types can be demonstrated by staining, and a diagram showing the developmental fate of each region can be drawn. (b) A pseudoplasmodium is cut into forward and rear halves, and conditions are adjusted to allow the slug to continue migrating for several hours after the operation before culmination. Both halves regulate to form a small but normally proportioned sorocarp. (c) A pseudoplasmodium is cut at the same position as in (b), but under conditions when culmination takes place shortly thereafter. Malproportioned fruiting bodies form, showing only slight regulation.

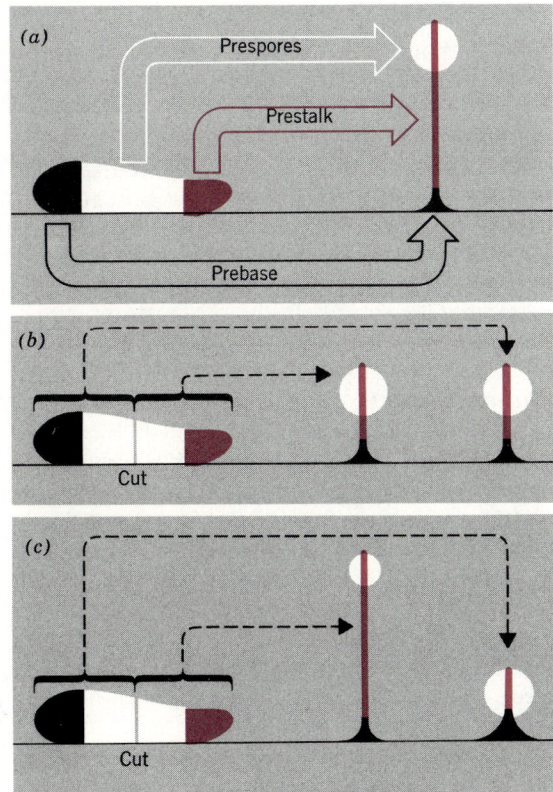

ments. The front piece forms a sorocarp with a nearly normal-size stalk and a drop of spores about half normal size; but it also forms a small group of basal cells. Similarly, the rear piece forms a large base plate from which arises a tiny but distinct stalk carrying a half-normal-size spore drop.

These results tell us that the cells of the front piece retain the potential for forming a base, and those of the rear fragment for making a stalk, although they normally would never have to do so. The evidence also indicates that there is some form of communication among the various cells. If a group of cells that normally would not have formed a stalk does so after amputation, then some of them must have received some kind of signal redirecting their activities to compensate for the loss. We shall see more evidence of **intercellular communication** in other regulative developing systems.

Reversibility of Differentiation

Such experiments with *Dictyostelium* allow us to draw three conclusions that, as we shall see, are generalizable to most developing systems. First, **developing cellular systems have amazing capabilities to compensate for accidental or experimental modifications—that is, to regulate them.** Like *Dictyostelium*, for example, most animal embryos can be operated on and have pieces removed or extra pieces added. If the operation is performed early enough in development, the embryos of many species exhibit regulation, forming perfectly proportioned individuals at later stages. For example, a Polish biologist, André Tarkowski, first showed that two whole mouse embryos, each at the four-cell or eight-cell stage, can be fused to form a single giant morula. This ball of cells is maintained in culture until it reaches the blastocyst stage, at which time it is implanted into the uterus of a foster-mother mouse. There it will develop into a large but normally proportioned fetus, and be born to grow into a healthy adult.

Such animals grown from cells with different genetic constitutions are referred to as

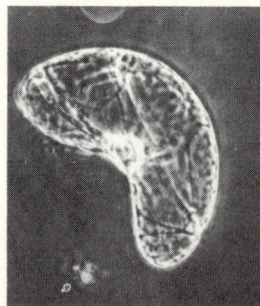
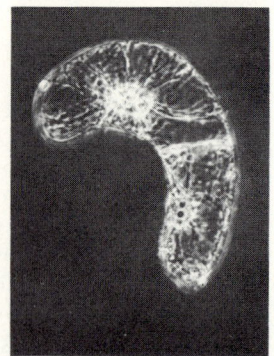
(a)
(b)
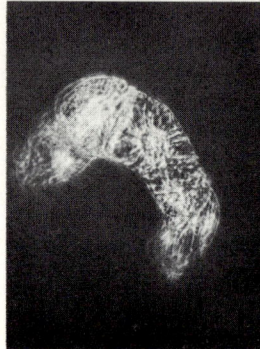
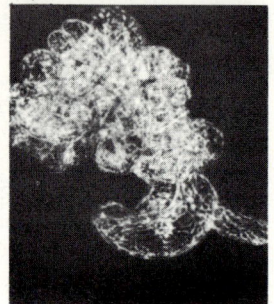
(c)
(d)

allophenic. They need not start with equal numbers of blastomeres from the two parental strains. The combination of cells makes no apparent difference; up to the eight-cell stage any combination can be compensated for by the embryo's regulative powers, and a normal animal results. As is true for the slime mold pseudoplasmodium, however, later in the course of differentiation the system has less capacity for regulation. Allophenic mice can be produced only rarely by combining embryos at the 32-cell stage, and combinations of blastomeres from older embryos never develop successfully.

This brings us to our second conclusion: **the process of differentiation involves the progressive repression, or switching off, of developmental capacities, but not their irreversible loss.** This can be demonstrated readily in the higher plants. Single cells can be teased apart from carrot roots or tumors of tobacco plants. These cells can be grown into clones in tissue culture and maintained as

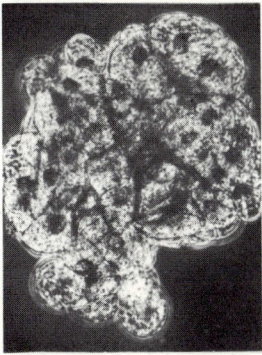

(e)

(f)

(g)

(h)

(i)

Figure 12-12 Development of a tobacco plant from a single cultured cell. On an appropriate agar medium the cell undergoes division (a to e) to form a small callus (tumor) mass (f). After further growth roots and shoots are differentiated (g, h). Young plants can then be transferred to soil, where they continue to grow and flower (i).

actively dividing groups of single cells indefinitely. If certain changes are made in the culture conditions, however, the dividing cells remain together to form small cell masses. These masses, like plant embryos, soon differentiate roots and shoots, and grow into normal plants indistinguishable from those derived in the usual manner from a seed (Fig. 12-12). Obviously the original cell could not have experienced any irreversible loss of genetic information and still have been capable of producing an entire normal plant.

This leads irresistibly to a third conclusion: **in multicellular organisms not all genes in a cell function at the same time.** A prestalk cell in a slime-mold slug actually has the capability (i.e., the genes) to synthesize products characteristic of all of the three cell types. As a stalk cell, however, it produces stalk cellulose but does not secrete the slimy fluid typical of a spore; that is, in the stalk cell the genes for the latter secretion are not active. Similarly, a carrot root cell does not normally produce the products characteristic of a leaf or stem, despite the fact that it retains the capacity to do so. Something must dictate that a cell of a given type synthesizes only certain characteristic products as a result of the activity of a given set of genes, while other genes are inactive. In other words, differentiation appears to represent **differential gene action.** In developing systems particular genes may be activated, function for a limited period, and then become inactive. Moreover, whether active briefly or for more extended periods, different sets of genes are transcribed in different cells. To examine the extensive evidence for this concept, we must look to embryos of several animal species.

Embryonic Development

Spatial Organization of Eggs

An egg liberated from the ovary and ready to be fertilized already has a long history of growth and preparation. Large amounts of RNA, especially messenger RNA and ribosomal RNA, have been made from the egg's diploid maternal genome long before meiosis has reduced the egg to the haploid number of chromosomes. Because there is a relatively enormous demand for ribosomal RNA during growth of the egg, the oöcyte faces the severe problem of supplying enough. Thus in many eggs the ribosomal gene is copied many times over at the beginning of the egg growth phase (a process called **amplification**); this means the DNA that codes for rRNA is copied. In the South African clawed toad *Xenopus laevis,* for example, there are about 450 copies of the ribosomal gene normally present and lined up on one of the 16 chromosomes of the diploid set. During growth of the oöcyte, however, another half-million gene copies are made, and these are liberated from the chromosomes as about 1000 nucleoli, all manufacturing rRNA that is stored for later development.

A striking characteristic of all unfertilized eggs is that the materials that fill them are not homogeneously distributed. All eggs have a **polarity,** like the north and south poles of the earth. The "north" pole of the egg is called the **animal pole;** the "south" pole, where most of the yolk of the egg is usually concentrated, is the **vegetal pole.** An imaginary line passing through the two poles defines the **animal-vegetal** axis of an egg. The unequal distribution of material in the egg,

as we shall see, is crucial for development. As the egg divides into two cells, then four and eight, materials that were restricted to one part of the egg will be parceled out to some cells but not to others. Thus, although the chromosomes are faithfully replicated, the early cleavage blastomeres will be different as a result of the inhomogeneous organizations of the egg.

Cleavage in Echinoderm Eggs

Among the common marine animals found along the east and west coasts of the United States are the sea urchin, sand dollar, and starfish (Fig. 12-13). These are all members of the phylum Echinodermata (Chapter 24), whose eggs are favorite subjects for embryologists; because of their transparency and because fertilization and development take place externally (i.e., normally in the sea water), many of the processes of embryo formation can easily be viewed in the laboratory. Photographs of serial stages in cleavage and development of the sand dollar, *Echinarachnius parma* (Fig. 12-14, pp. 318-319), for example, clearly illustrate many of the events that are common to all animal embryos. These photographs also indicate the rapidity of development in these forms, the whole process going from fertilization to pluteus larva in only three days.

Following fertilization the egg begins to cleave, divisions occurring every hour or so, at precisely timed intervals. For the first several hours cleavages increase in frequency; within 7 hours after fertilization eight divisions have taken place, producing 64 cells arranged as a solid ball called a **morula.** The cells of the morula continue dividing four or five more times, and also separate in the center of the ball, leaving a cavity called a **blastocele.** The embryo at this stage is called a **blastula** (equivalent to the blastocyst in mammals) and is composed of 1000 to 2000 cells. Note that it is about the same size as the original egg (Fig. 12-14); cleavage has

(a)

(c)

Figure 12-13   Common marine echinoderms, the sea urchin, sand dollar, and starfish, whose eggs are excellent for embryological study because of their transparency.

(b)

Figure 12-14   (pp. 318–319) Developmental stages of the sand dollar. The times given are hours after fertilization. (a) Sand dollar sperm. (b) The unfertilized egg is surrounded by a transparent jelly coat that has pigment granules embedded in its surface. (c) About one and a half hours after fertilization, the first cleavage divides the egg in two. The fertilization membrane is clearly visible surrounding the two blastomeres. Inside each cell spindles are forming for the next cleavage. (d) Four-cell stage, after three hours. (e) Eight-cell, after four hours. (f) 16 cells, after five hours. The arrow points to the four micromeres. (g) 16-cell embryo, after five and a half hours, looking down at the egg from the animal pole. (h) Morula at six and a half hours. (i) Formation of a blastocele produces an early blastula at seven hours. (j) Blastula at eight hours. (k) Late blastula, hatching out of its membrane, at 12 hours. (l) The vegetal plate begins to push into the blastocele, forming an early gastrula, at 25 hours. (m) Late gastrula at 31 hours. (n) Prism-stage pluteus larva at 48 hours. (o) Early pluteus, with short arms, at 50 hours. (p) Late pluteus, with long arms, at 72 hours.

distributed the relatively enormous mass of egg cytoplasm into about a thousand cells of normal size. There has been little, if any, growth or synthesis during this period; thus each blastula cell is roughly one-thousandth of the original egg material. If, as we noted earlier, materials in the unfertilized egg are

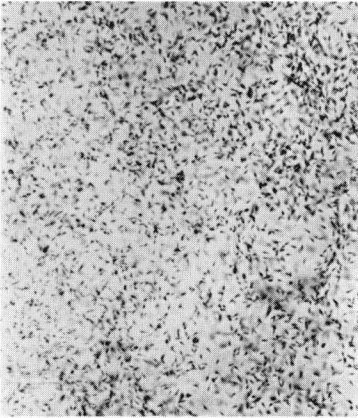
(a)

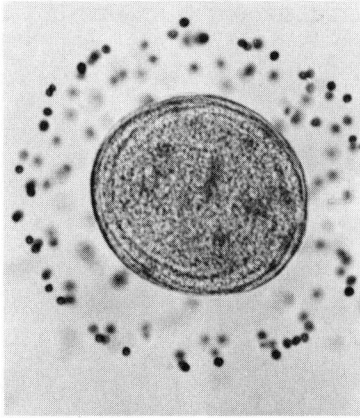
(b)

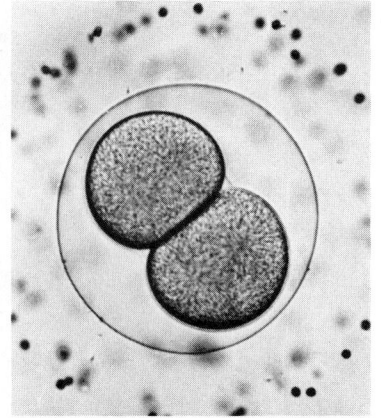
(c)

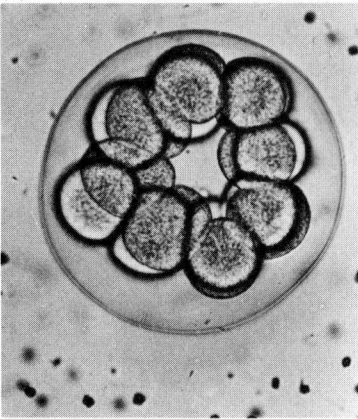
(g)

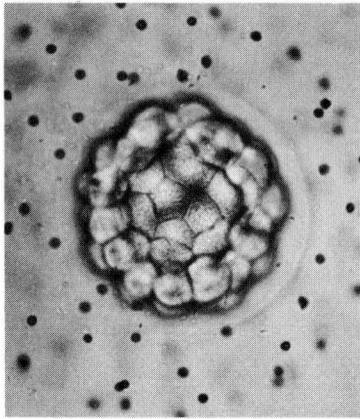
(h)

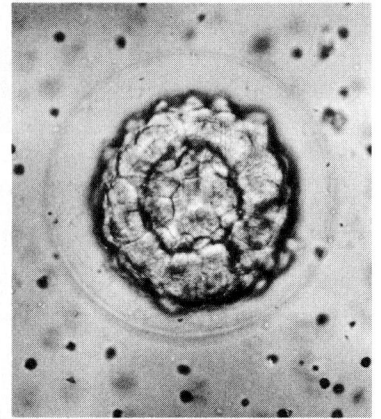
(i)

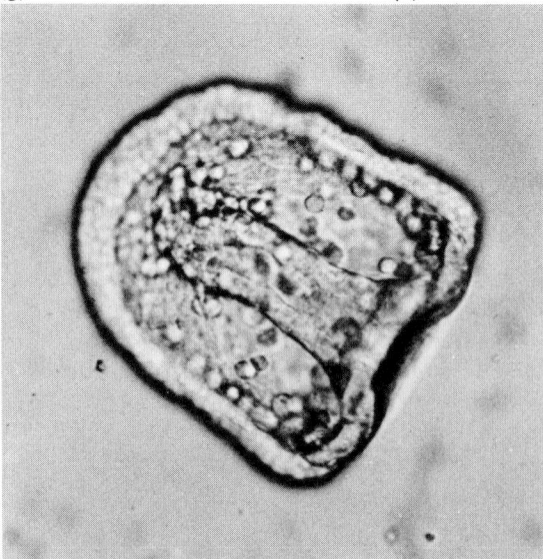
(m)

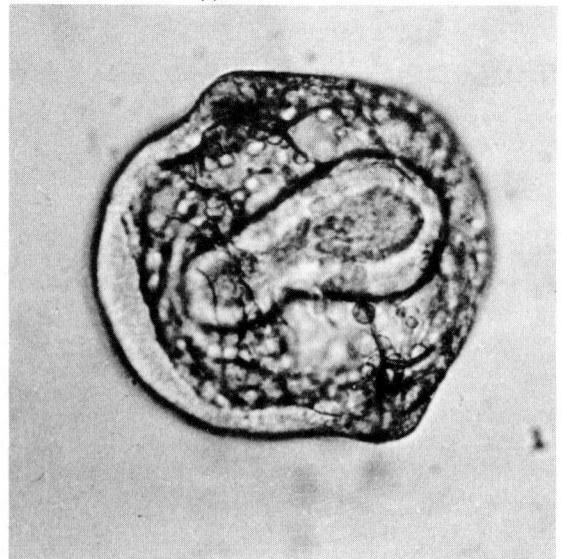
(n)

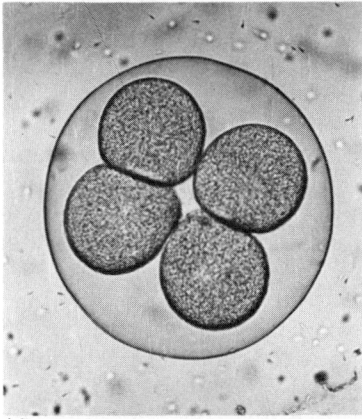

(d)

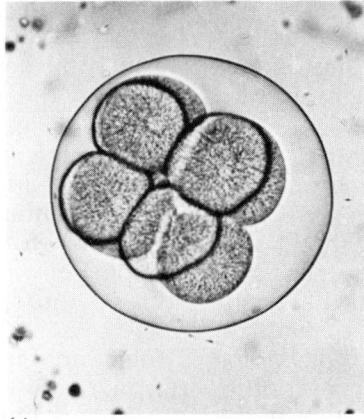

(e)

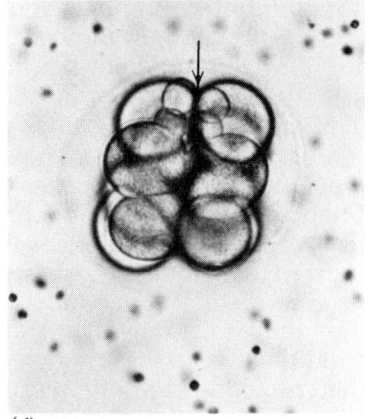

(f)

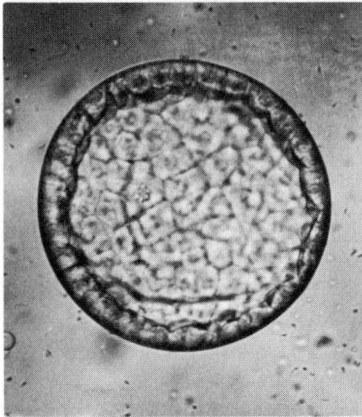

(j)

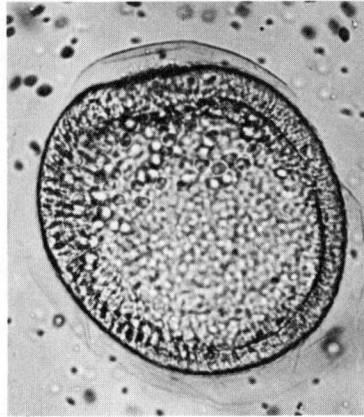

(k)

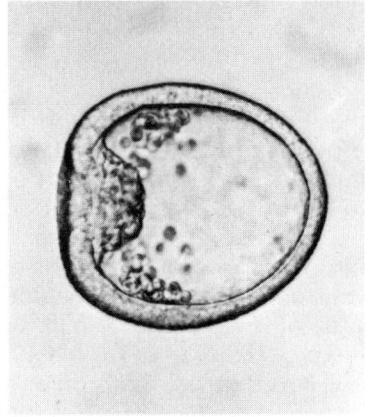

(l)

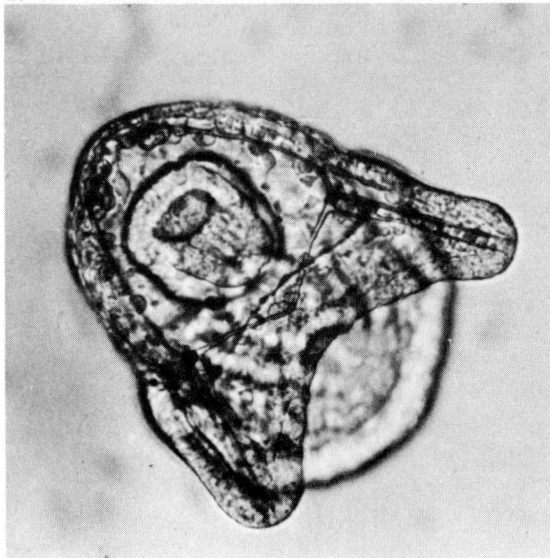

(o)

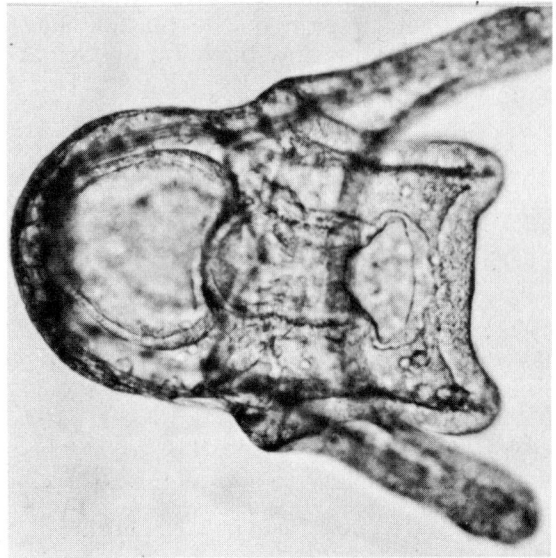

(p)

not distributed homogeneously throughout the cytoplasm, then the cells of the blastula are inevitably different because blastomeres in different locations must have received different cytoplasmic components.

Most multicellular animals (those above the level of the coelenterates or hydroids; see Chapter 24) are made of three body layers (Fig. 12-15): (1) an outer covering that is specialized to include the skin, sense organs, and nervous system; (2) an inner layer that forms the tubular digestive tract and its derivatives; and (3) between these, a mass of bulky tissue that provides support, motility, and other "service" functions in the form of cartilage, bone, muscle, and vascular system, and so on.

The blastula, however, is a single-layered hollow ball of cells. The echinoderm embryo produces a three-layered configuration from this ball in a very straightforward way by a process known as **gastrulation.** Gastrulation includes the buckling in of the blastula wall to form two layers, an endoderm and ectoderm; and the migration of individual cells into the space between those layers to constitute the mesoderm. Imagine pushing in the wall of a soft tennis ball with your index finger (Fig. 12-16a). What happens? The one-layered ball is converted to a two-layered structure. This is precisely the arrangement shown in Fig. 12-14, where cells in the vegetal region of the blastula have thickened and are now buckling (invaginat-

ing) into the blastocele cavity. Within a few hours they have formed an **endodermal tube,** which pushes upward to meet the opposite wall of the embryo, now referred to as a **gastrula** (Figs. 12-14m and 12-16c). The space within this endodermal tube (where your finger would be in the tennis ball) is called an **archenteron;** this becomes a digestive canal when the endodermal tube meets the opposite wall of the gastrula and a hole breaks through to form the mouth opening. The opening formed by the invagination of the floor of the blastula is called a **blastopore,** which later becomes an anal opening.

As invagination forms the endodermal tube, individual cells migrate from the region of the vegetal pole into the blastocele cavity. These are **primary mesenchyme cells,** from which all mesodermal structures of the animal will form. The cells of the outer wall of the gastrula remain as the ectoderm layer. Once gastrulation is complete the gastrula quickly transforms into a prism-shaped larva, which in its definitive form is called a **pluteus** (Fig. 12-14o, p). In this condition it swims and feeds actively until it settles down on the surface of a convenient rock and metamorphoses into the adult form.

Cleavage Patterns Control Differentiation

We have now identified two factors that appear to be important in development: (1) the genetic material containing the funda-

Figure 12-15   *The basic three-layered body plan of animals. The outer covering (1), which forms the skin and nervous system, is derived from the embryonic ectoderm. An inner tube (2), produced from the endoderm, forms the digestive system. Between these is a space (3), the coelom, which is generally filled with the viscera and the muscular and bony tissues derived mainly from mesoderm.*

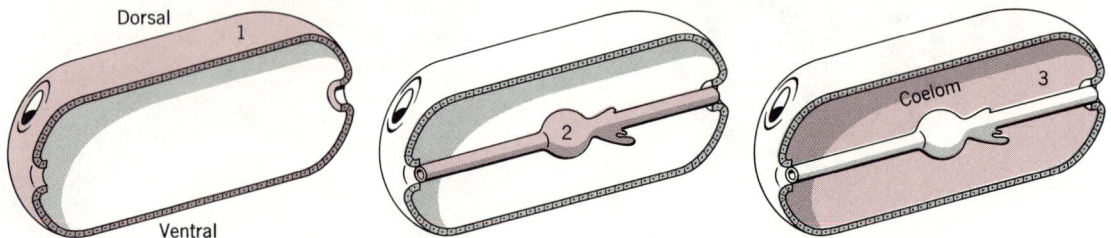

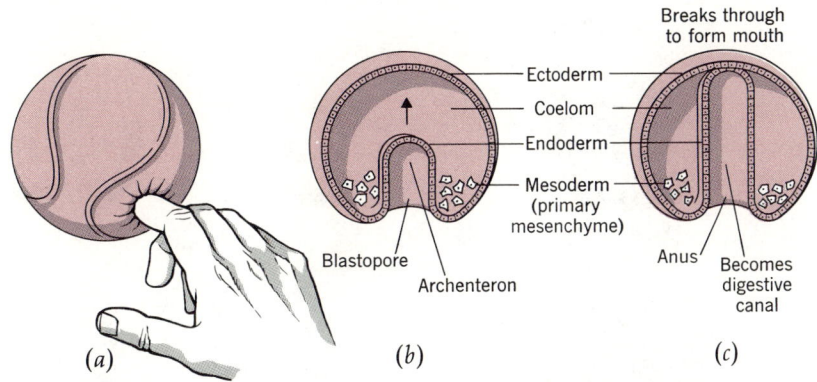

Figure 12-16   Diagrams showing the mechanism of gastrulation whereby an embryo forms the primary germ layers.

(a)

(b)

(c)

Breaks through to form mouth

Ectoderm
Coelom
Endoderm
Mesoderm (primary mesenchyme)
Blastopore
Archenteron
Anus
Becomes digestive canal

mental developmental instructions is duplicated at each cleavage and distributed in equal fashion to each blastomere; (2) non-nuclear factors—cytoplasmic granules, preformed ribosomes, materials associated with the egg membrane—appear to be unequally distributed throughout the egg and are therefore not equally parceled out to the blastomeres. Their distribution to particular cells of the blastula or gastrula depend on a specific pattern of cleavage.

Even a casual examination of Figure 12-14 indicates that the cleavage pattern of the sand dollar egg is exact and follows a precise sequence. This is apparent when several hundred eggs in a dish are fertilized and their development is observed; the precision with which every egg divides along the same plane as its neighbors, and at the same time, is a most convincing demonstration that there is no element of randomness in the process.

The way the egg contents are distributed during cleavage can be seen most readily in a species of sea urchin (called *Arbacia*), related to the sand dollar, which, in addition to the usual cytoplasmic granules and particles, contains a group of red pigment granules that form a flat band across the egg below the equator (Fig. 12-17a, p. 322). The pigment band provides a natural marker for a specific region of the unfertilized egg; by observing this band during development it is possible to see what structures differentiate from this region. In fact, with the aid of various stains and other markers, the developmental fate of all of the regions of the echinoderm egg has been examined.

The first cleavage along the animal-vegetal axis bisects the egg and pigment band. The second cleavage is also vertical, but perpendicular to the first (Fig. 12-17b). Both of these cleavage planes are said to be **meridianal** because they pass through the north and south poles of the egg as a meridian of longitude passes through the earth. In both cases the **spindle axis** (a line from one centriole to the other) is horizontal, the plane of cleavage perpendicular to the spindle (Chapter 7).

The four blastomeres now divide (Fig. 12-17c), but this time the mitotic spindles are parallel to the animal-vegetal axis. Simultaneously in each of these four cells a cleavage furrow appears in the plane of the equator, yielding eight blastomeres (Fig. 12-17d). The red pigment band is now restricted to the four cells in the vegetal half, the first obvious difference among the blastomeres. At the fourth cleavage (Fig. 12-17e) there is a radical departure from the roughly equal divisions seen so far. The upper tier of cells divides meridianally to form a ring of eight cells, but in the lower tier the spindles arrange themselves obliquely so that four small cells divided from the lower inner sides of the larger cells. The small cells are called **micromeres** and the larger cells are **macromeres.** The cells of the upper tier, intermediate in size, are called **mesomeres.** At the 32-cell stage (Fig. 12-17f) the macromeres are divided by an equatorial cleavage into upper

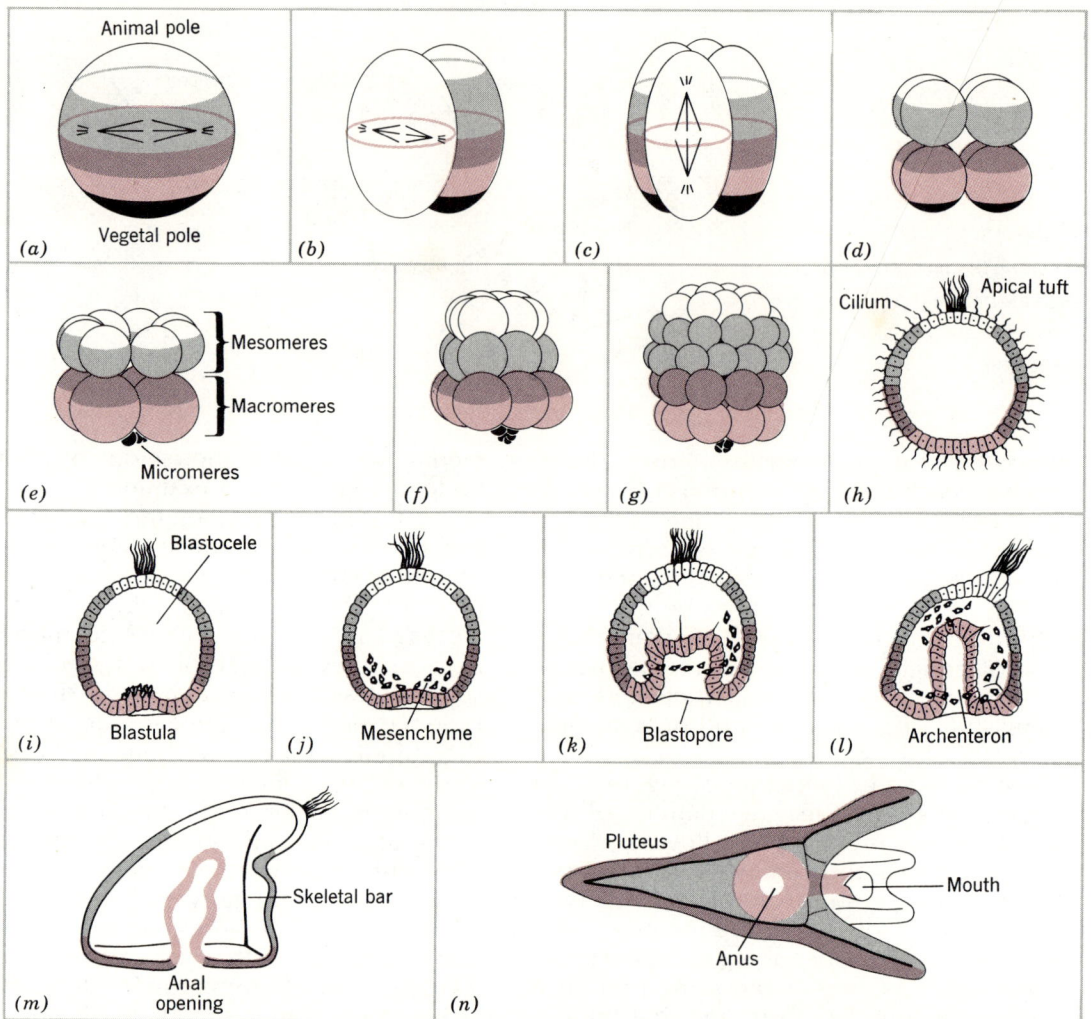

Figure 12-17 Distribution of egg contents during cleavage and gastrulation of the sea urchin egg. The egg is diagrammed with a series of five color-keyed layers that can be followed in stepwise fashion into the pluteus larva at the bottom. The reddish-gray band just below the equator of the egg represents the band of red cytoplasmic pigment granules that the egg contains. The other bands indicate regions identified with experimentally applied stains or other markers.

and lower tiers of cells; the red pigment is now restricted to the upper tier of these half-macromeres.

Other materials are similarly limited in their distribution. The egg contents that were originally at the extreme vegetal pole of the sphere (called **pole plasm** and shown in black), for example, at the 64-cell stage (Fig. 12-17g) are found only in the micromeres. During gastrulation derivatives of the micromeres alone, with their content of vegetal pole plasm, migrate into the blastocele cavity to form mesenchyme (Fig. 12-17i, j) and it is these cells that produce the skeletal material of the pluteus larva (Fig. 12-17m). Cells containing the red pigment granules form a ring of ectoderm surrounding the anal opening; cells that derive from the lower tier of half-macromeres (just above the micromeres) invaginate and form the endoderm. The region of the egg above the pigment band that produces the mesomeres produces the rest of the ectoderm of the gastrula and pluteus larva.

These observations indicate that certain regions of the egg are normally associated with particular developmental fates, but they do not demonstrate that specific cytoplasmic factors determine differentiation. Is such evidence available?

We indicated earlier that any of the first four blastomeres of the sea urchin egg would form a complete pluteus (Fig. 12-3). An analogous experiment can be performed by microsurgery on an unfertilized egg, with similar results (Fig. 12-18a), as long as the egg is cut along the animal-vegetal axis as the first two cleavage planes would do. Dividing the egg in the equatorial plane gives a very different result (Fig. 12-18b). Each half may be fertilized and undergo cleavage, but neither produces a normal pluteus. The animal half forms a small blastula that lacks endoderm or mesenchyme and therefore never gastrulates. The vegetal half forms a malproportioned larva with a relatively enormous digestive tube, which, however, is deficient in

Figure 12-18 (a) An experiment demonstrating the equivalence of egg halves separated by a meridianal cut. An unfertilized sea urchin egg is cut in two through the animal-vegetal axis (A-V). Each half may be fertilized and develops into a normal embryo of half size. (b) Egg halves separated by an equatorial cut are also fertilizable. However, neither half develops normally. The animal hemisphere produces a blastula that lacks endoderm and never gastrulates (called a **dauerblastula**). The vegetal half forms a malproportioned and incomplete embryo deficient in ciliated ectoderm.

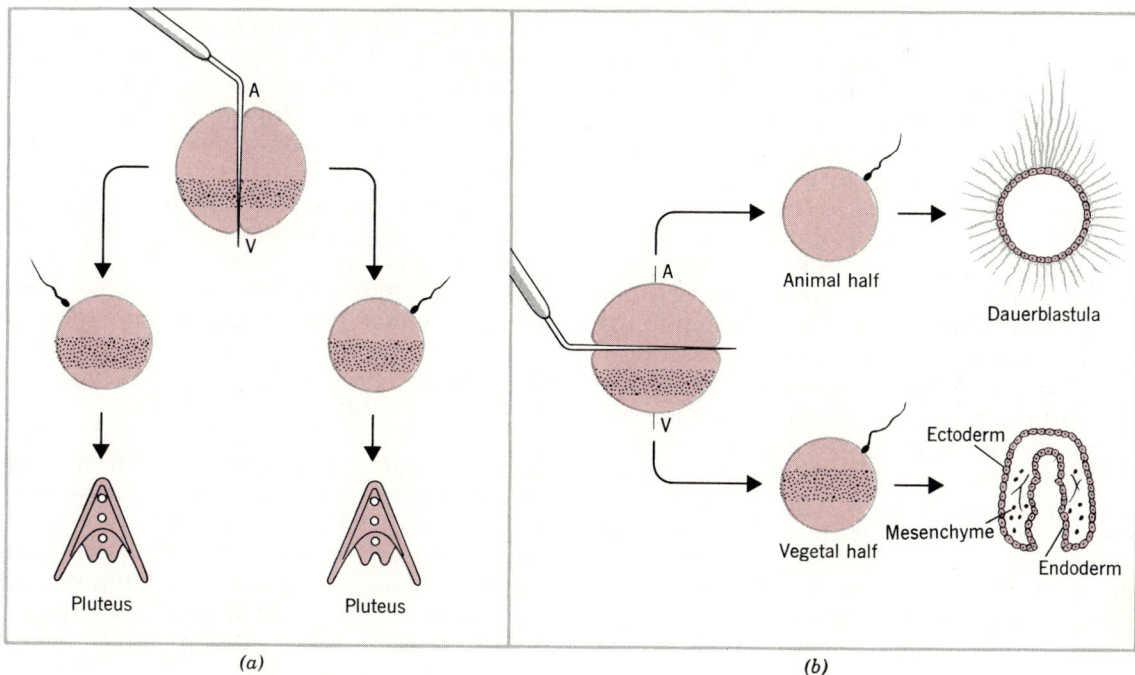

(a)

(b)

ectoderm; these embryos fail to develop to the pluteus stage.

These experiments confirm observations on the prospective fate of the whole egg, again indicating that the animal half forms ectoderm but no mesoderm or endoderm, and the vegetal half forms mainly mesoderm and endoderm but only a small amount of ectoderm. Recall that in the normal development of the echinoderm embryo the third cleavage is horizontal, and it is at this eight-cell stage that the first difference among the blastomeres can be recognized. If the animal and vegetal halves of such an eight-cell embryo are separated at this point and each set of four cells is allowed to cleave further and develop in isolation, precisely the same result is obtained as for the surgically divided egg. This suggests that the differentiative fate of a blastomere is not merely associated with the distribution of material in the egg, but is causally dependent on it. If differentiation, as we said on p. 315, means differential gene action, then these results suggest that certain egg components repress some genes and switch others on, that is, **derepress** them.

That none of the genetic information is irreversibly lost from a cell through differentiation can be demonstrated readily in sea urchin embryos by allowing them to develop in the presence of certain chemical agents that modify differentiation. Treating an intact embryo with a solution of lithium chloride, for example, causes it to behave much like a vegetal-half embryo; this substance inhibits cells of the animal hemisphere and produces malformed larvae in which endoderm and mesoderm predominate. We speak of lithium as a **vegetalizing** agent. Other compounds, such as sodium thiocyanate, have the opposite effect, depressing endoderm and mesoderm to yield blastulas composed exclusively of ectoderm. Such agents can be used to balance a lack in a particular set of blastomeres. An animal-half egg produced by microsurgery (Fig. 12-18*b*), developing in ordinary sea water, would produce an endodermless blastula. Treated with the proper concentration of lithium chloride, however, it will form a complete and normal pluteus,

indicating that the ectodermal protodifferentiation of some cells, at least, is reversible and that the genes for producing endoderm and mesoderm are still present and capable of being activated.

Nuclear Equivalence: The Vertebrate Embryo

So far we have seen that differentiation requires only reversible genetic changes in forms as diverse as the cellular slime molds, higher plants, and echinoderms. We saw earlier (pp. 312–314), for example, that a single cell could be isolated from a slime mold pseudoplasmodium or a carrot root, and that from that single progenitor a clone of cells could form and produce a new, complete individual. That is, all cell nuclei of these organisms are fundamentally alike, or **equivalent,** even after differentiation.

Vertebrate embryos are more complex in many ways. A large number of vertebrate cells—mammals, chick, and amphibian—have been cloned, several under conditions in which their differentiated state has been retained. Muscle fibers able to contract (Fig. 12-19), for example, and cells that synthesize collagen or the mucoproteins of cartilage have been grown from single differentiated cells; so also have clones of cells that produce thyroid hormone, insulin, and other products of differentiated cells. Most such cells do tend to lose their differentiated function after many generations in culture, but no such cell line has ever been capable of giving rise to an entire organism, or even to a different differentiated cell type, as slime mold or carrot root cells are. Does this mean that higher forms have had to resort to irreversible changes to support differentiation?

The pattern of cleavage (Fig. 12-20, p. 327) and gastrulation (Fig. 12-21, p. 328) in a vertebrate—a frog or salamander—is clearly more complicated than that in a sea urchin, although the processes achieve the same goals of distributing egg cytoplasm differentially to a large number of small cells and bringing those cells into new spatial relations. An in-

Figure 12-19 *Clonal development of differentiated muscle in tissue culture. (a) A single myoblast cell taken from the leg muscle of an 11-day-old chick embryo. (b) On the fourth day of culture the cell has produced a clone of about 36 cells. (c, p. 326) After two weeks the clone contains many thousands of cells that have fused into long, multinucleate* **muscle straps**. *(d) At higher magnification the cross-striations in the well-organized myofibrils are readily seen. At this stage the muscle is functional; that is, it contracts when stimulated.*

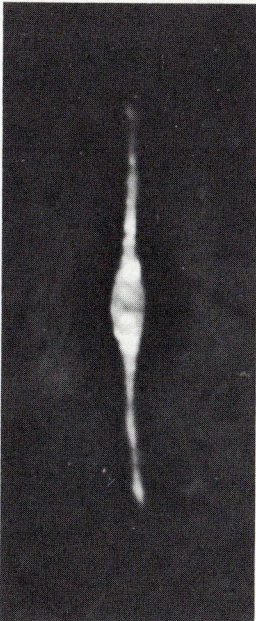

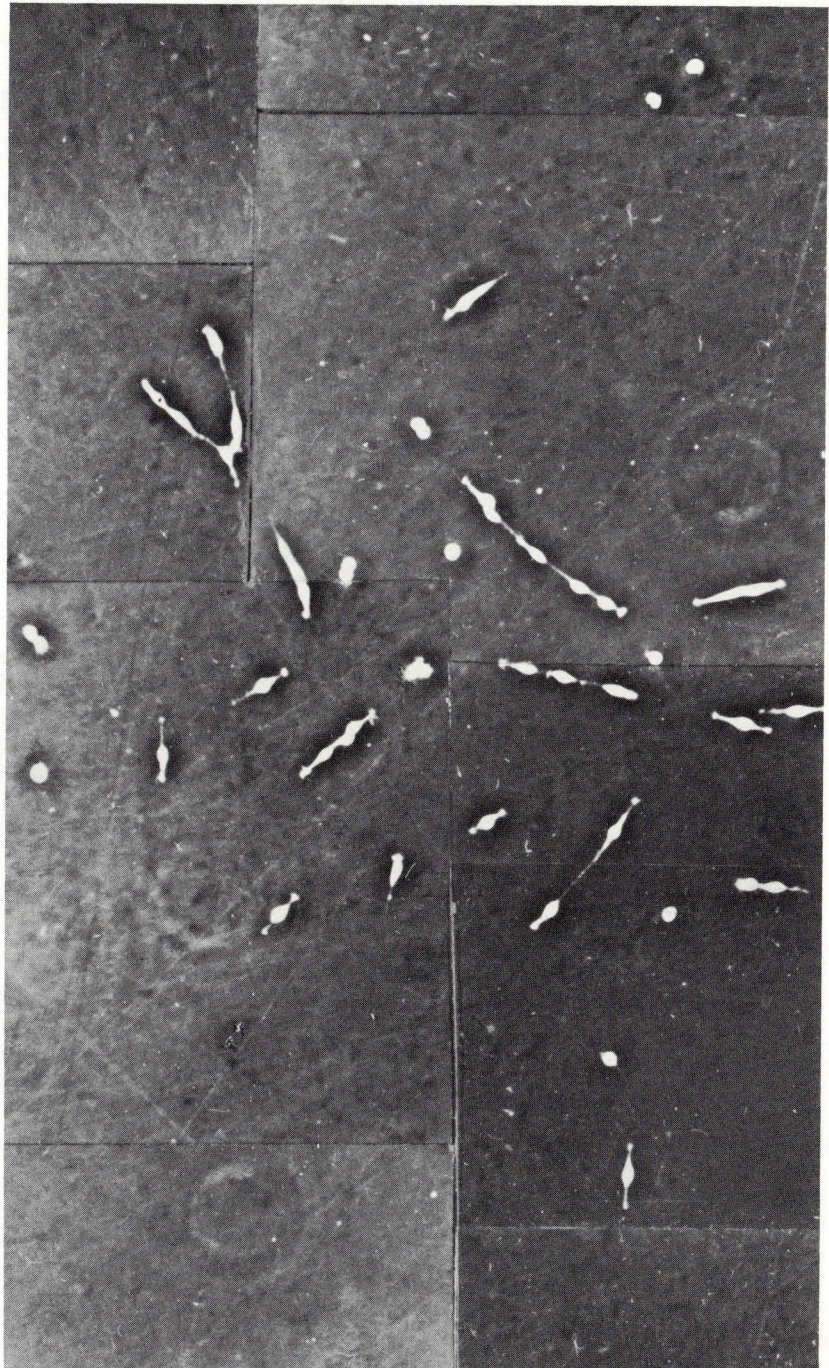

(a)

(b)

Figure 12-19 (cont.)

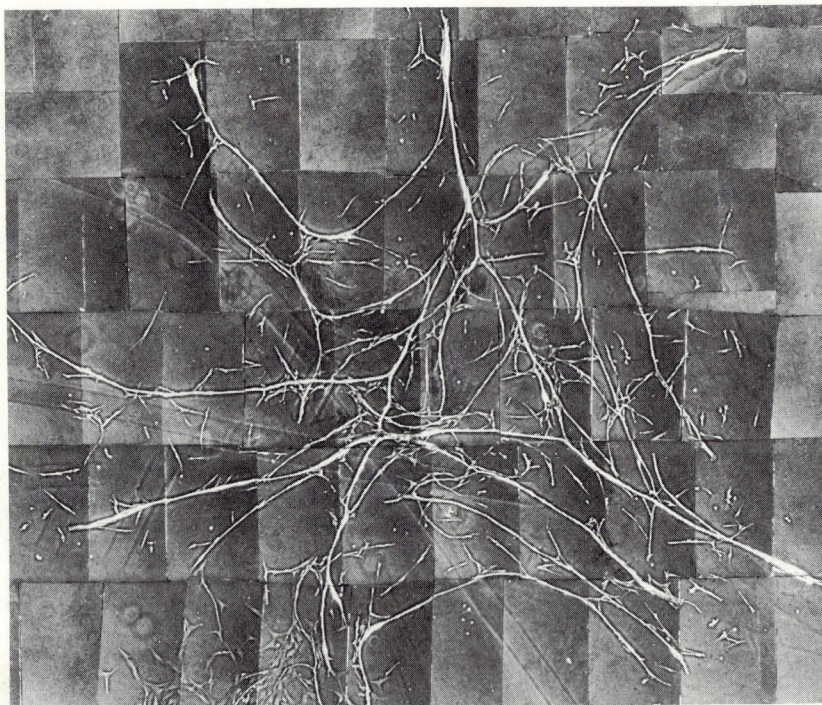

(c)

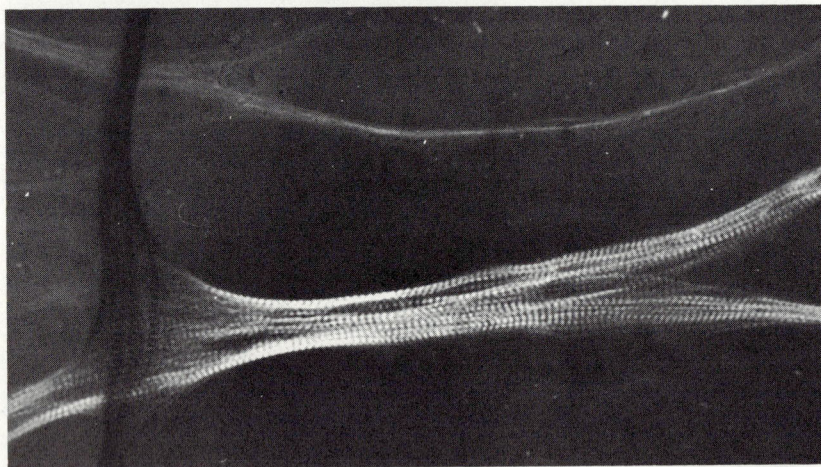

(d)

genious experiment in 1928, by the pioneer embryologist Hans Spemann, suggests that in a salamander embryo, just as in the sea urchin, the nuclei of different cleavage blastomeres are identical. Using a hair from a blond baby girl (the finest material available for this purpose), Spemann constricted a newly fertilized newt egg into a dumbbell shape in such a way that the zygote nucleus was forced to one side (Fig. 12-22, p. 329). The nucleated half proceeded to cleave normally; no divisions occurred in the non-nucleated part. After the third cleavage the constriction was loosened slightly, permitting one of the 16-cell cleavage nuclei to slip through the cytoplasmic bridge to the non-nucleated half, which then began to divide. Eventually, Spemann found, complete larvae were produced from each half of the egg. Thus, up to the 16-cell stage, no irreversible changes had taken place in the nucleus.

The most elegant and convincing demonstration of the equivalence of nuclei in differentiated amphibian cells, however, comes from **nuclear transplantation** studies. In the 1950s Robert Briggs and Tom King perfected a method for transferring the nucleus of a cell from frog embryos at various stages of development into a frog egg whose own nucleus had been removed. They found that the nucleus of a cell from a late blastula, for example, could be transferred undamaged and the recipient egg would cleave and develop normally into a healthy, fertile frog. Similar results have since been obtained by other workers with the South African clawed toad *Xenopus.* Even from embryos as late as the gastrula stage, nuclei from at least some well-differentiated endoderm cells can be injected into enucleated eggs and will support development of a complete and normal

frog with the usual complement of mesoderm and ectoderm. Thus our earlier conclusion seems equally applicable to development in vertebrates: irreversible nuclear changes are apparently unnecessary for differentiation. Although the evidence is not yet definitive, the processes of development seem more to depend on continuing interactions between the genome and specific cytoplasmic components.

We described earlier the animal-vegetal differences in contents of the sea urchin egg that seem to control development. These differences become manifest at the third cleavage, which divides the egg horizontally (see Fig. 12-17). The amphibian egg is even more sensitive, normal development being dependent on the correctness of the very first cleavage plane. Not only must the plane of that division be vertical through the animal-vegetal axis, it must also separate the two blastomeres in such a way that each contains part of a special region called the **gray crescent** (Fig. 12-20). When this occurs, as it normally does, the two cleavage blastomeres may be physically separated and each will produce a normal, complete embryo (Fig.

Figure 12-20   Cleavage in the amphibian egg. (a) *The animal hemisphere of a frog or salamander egg is black or dark brown; the vegetal half is milky white. After fertilization the pigment at the junction of the dark and light zones shifts upward on the side opposite the point of sperm penetration, leaving a zone of diluted color, called a* **grey crescent,** *that determines the future tail end of the embryo.* (b) *The first cleavage always bisects the grey crescent, dividing the embryo into left and right halves. Further sequential perpendicular cleavages* (c, d, e) *divide the egg into 16 cells. Because of the relatively great amount of yolk in the vegetal hemisphere, division is slower and the vegetal cells are much larger. Rapid division in irregular planes continues in the pigmented half to yield* (f), *a morula of several hundred cells. The cells soon separate to produce the blastocele of the early blastula* (g).

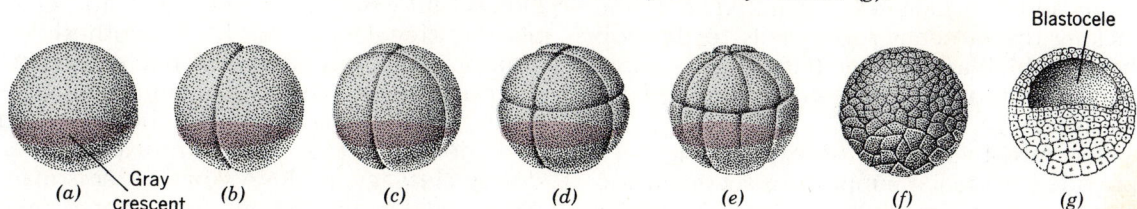

(a)   Gray crescent   (b)   (c)   (d)   (e)   (f)   (g)

Blastocele

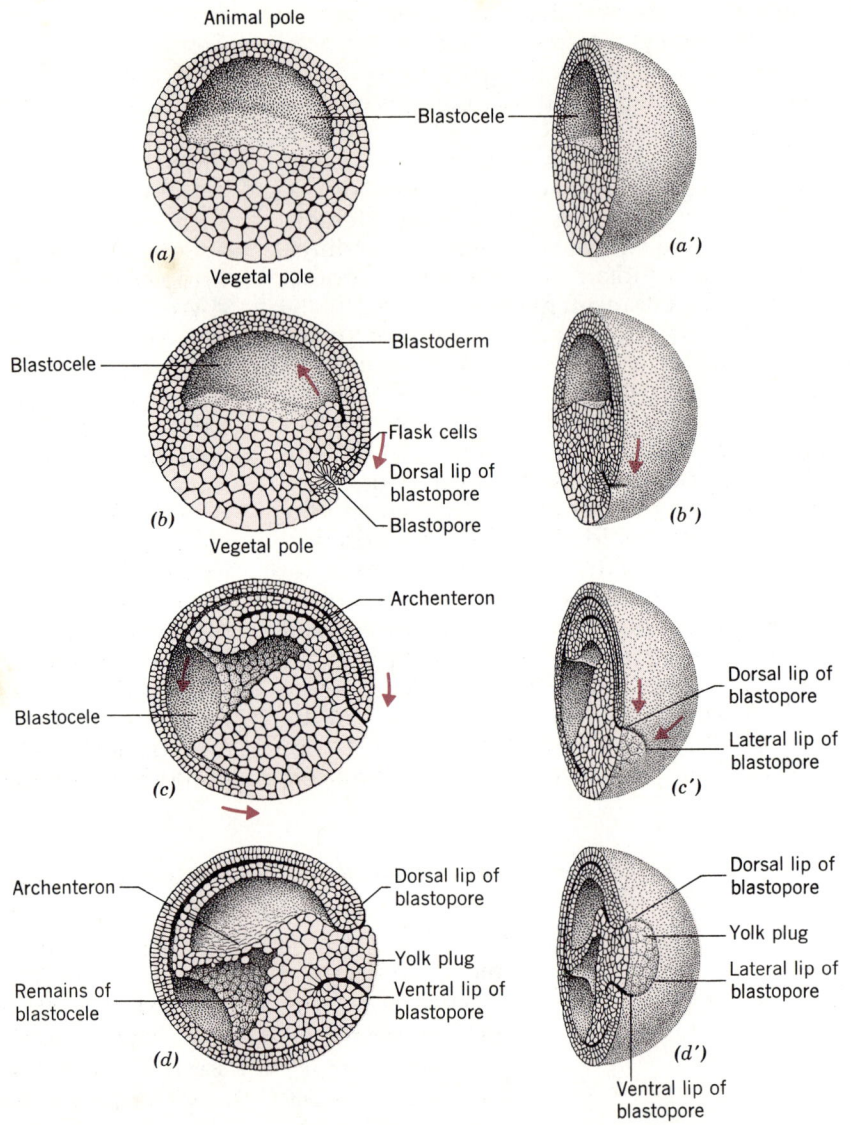

Animal pole

Blastocele

(a)

Vegetal pole

(a')

Blastocele

Blastoderm

Flask cells

Dorsal lip of blastopore

Blastopore

(b)

Vegetal pole

(b')

Archenteron

Blastocele

(c)

Dorsal lip of blastopore

Lateral lip of blastopore

(c')

Archenteron

Dorsal lip of blastopore

Yolk plug

Ventral lip of blastopore

Remains of blastocele

(d)

Dorsal lip of blastopore

Yolk plug

Lateral lip of blastopore

(d')

Ventral lip of blastopore

Figure 12-21   Four stages in amphibian gastrulation. The left half of the embryo is cut away to indicate internal changes. In a' to d' the half-embryos are viewed from the back side. (a, a') late blastula; (b, b') beginning of gastrulation, showing the dorsal lip of blastopore just forming as cells begin to invaginate. (c, c') midgastrula. (d, d') late gastrula. The blastocele is gradually obliterated and replaced by the archenteron. The position of the yolk plug now marks the posterior end of the embryo and future anal opening.

12-23a, p. 329). If cleavage takes place so that the gray crescent lies entirely in one blastomere and the cells are separated, the one lacking the crescent region fails to develop beyond the blastula stage (Fig. 12-23b); some component of the crescent is required for gastrulation.

Recently two kinds of experiment have provided evidence supporting a conclusion that the cytoplasm in which a nucleus resides determines the functions of that nucleus. One example, reported by John Gurdon and Donald Brown, is concerned with the cytoplasmic regulation of RNA synthesis in *Xenopus* embryos. We mentioned earlier that there is intense synthesis of ribosomal RNA during growth of an egg in the ovary. At ovulation, however, this synthesis ceases; during cleavage no rRNA whatsoever is manufactured by any cell in the embryo, and

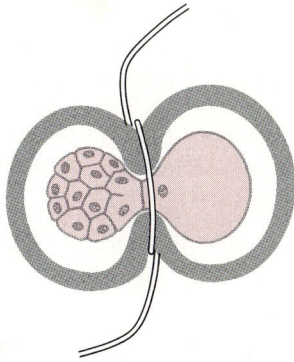

Figure 12-22 *Constriction of a fertilized newt egg permits cleavage only in the nucleated half until a nucleus passes through the cytoplasmic bridge to the other side. The previously nonnucleated side then cleaves. Complete larvae are formed from each half.*

synthesis resumes only after gastrulation. Gurdon and Brown found that an endoderm nucleus taken from a late gastrula, which was actively synthesizing rRNA, would stop within minutes after transplantation to egg cytoplasm. Moreover, synthesis of ribosomal RNA would resume in the descendants of that nucleus only after the embryo had completed cleavage and gastrulation.

However, the rRNA synthesis in such a nucleus did not stop if it were transplanted to an immature oöcyte—one whose own nucleus was still engaged in manufacturing

ribosomes. This study clearly demonstrates that the synthesis of rRNA is regulated by the type of cytoplasm surrounding a nucleus.

The second source of evidence that gene action depends on cytoplasmic components comes from recent experiments with the **cell hybridization** technique. In this method cells of different types are caused to fuse so that their two nuclei lie in a single mass of cytoplasm surrounded by a single merged plasma membrane. Such fused cells are called **heterokaryons.** Occasionally the two separate nuclei also fuse, yielding a cell with a single

Figure 12-23 *The significance of the egg contents in the region of the grey crescent is demonstrated by separating the two blastomeres formed at first cleavage. When each blastomere contains a share of the crescent, (a), each is capable of forming a complete embryo. If the division separates the cells in such a way that one blastomere contains all of the crescent, (b), that cell forms a normal embryo, while the one that lacks crescent material fails to gastrulate.*

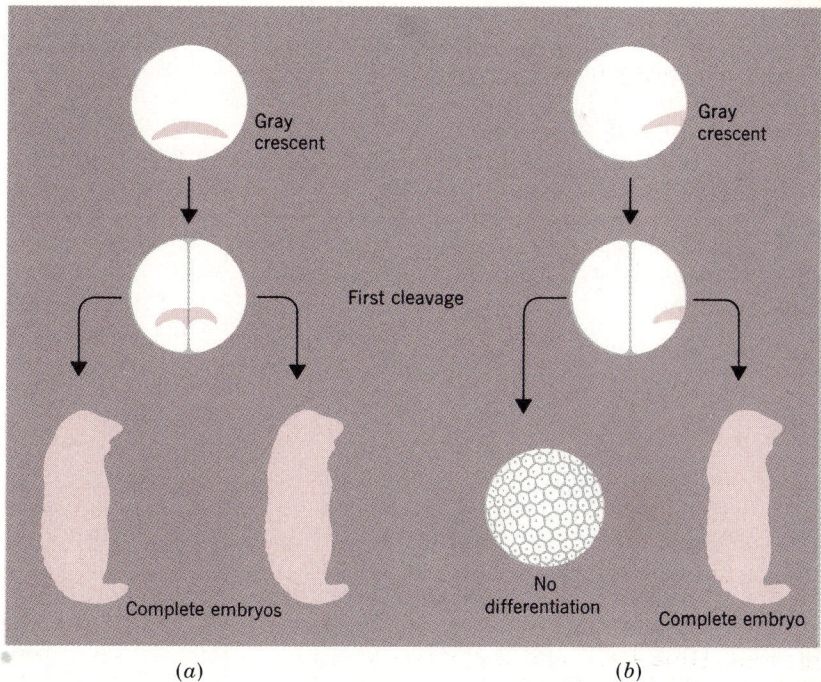

Gray crescent

Gray crescent

First cleavage

Complete embryos

No differentiation

Complete embryo

(a)

(b)

nucleus containing chromosomes from both parent cells. Such **somatic cell hybrids** may divide rapidly, but generally some of the chromosomes of one of the parent strains degenerate. Hybrids of mouse-human, mouse-chick, and even mammal-insect cell combinations have been produced, some of which throw light on the question of cytoplasmic regulation of genes. For example, human red blood cells of the type known as HA synthesize a specific protein called **interferon.** Chicken red blood cells do not produce interferon, although other cells in chickens do. If human HA cells and chicken erythrocytes are fused, however, the hybrid produces interferon consisting of both the human and the chicken type. The interferon gene in the red blood cell nuclei of chickens, which normally is never functional, is somehow activated in the HA cell cytoplasm.

Morphogenetic Mechanics

An embryo then, is a highly organized group of cells that progress through an orderly sequence of movements, localized mitotic activity, and shape changes; as we have seen, the embryo undergoes morphogenesis as well as differentiation. Somehow the continuous interaction between genome and cytoplasm must account for the movements and spatial relationships of the cells that comprise the parts of the embryo, as well as for the synthesis of specific products of differentiation.

What Determines the Cleavage Pattern?

We concluded earlier that, at least up to gastrulation, development is controlled by the cleavage pattern of the egg, which determines in which parcel of cytoplasm each cleavage nucleus will reside and therefore which genes in that nucleus will be derepressed. Look back at Fig. 12-17; the most striking aspects of the sequence of divisions

are their timing and precise organization. Because the cleavage furrow always divides a cell along the plane of the metaphase plate, that is, perpendicular to the spindle axis (p. 174), whatever factors control the position of the spindle thereby also determine the cleavage plane.

Cytoplasmic Factors Position the Spindle

In most blastomeres of a variety of organisms the mitotic spindle forms in the center of the yolk-free region of the cytoplasm. This accounts for the difference in cell size between the animal and vegetal blastomeres of a yolky egg like that of the frog; because most of the yolk is located in the vegetal hemisphere, the animal blastomeres will be smaller. However, it is not the yolk itself that determines the spindle location. Eggs can be gently centrifuged to pull all of the yolk to one side or the other with respect to the animal-vegetal axis. This has no effect on the normal position of the cleavage spindle or plane; the egg divides independently of the location of the yolk.

However, the control factors are indeed cytoplasmic. They can be demonstrated by an experiment first performed with sea urchin eggs in 1928 by the Swedish embryologist Sven Horstadius (Fig. 12-24). By vigorously shaking eggs just before cleavage was expected, Horstadius found that the forming spindle apparatus would degenerate and cleavage would fail to occur. At the next scheduled division, however, the spindle axis was determined not by the number of blastomeres present, but by the **time elapsed since fertilization.**

For example, if the sea urchin eggs were shaken just before the third cleavage (Fig. 12-24b), which normally would have been horizontal, dividing four mesomeres from four macromeres, that cleavage would fail. At the next cleavage the spindles would be oriented in the oblique direction characteristic of the fourth cleavage, despite the existence of four blastomeres instead of eight. Similar treatment just before the second cleavage (Fig. 12-24c) resulted in the next cleavage being horizontal (as in the normal third cleavage) although two blastomeres were present instead of four. The factors that determine this temporal sequence move about in the egg with strong centrifugation as the granules and inclusions such as mitochondria, lysosomes, pigment, lipid droplets, etc. become stratified (Fig. 12-25, p. 332). Such stratified eggs develop into complete (but oddly pigmented) plutei. When they are cut in half along the animal-vegetal axis, however, only the half that contains the layer of granules above the pigment cleaves normally. Halves lacking this granular layer fail to divide at all, or do so very slowly.

Figure 12-24 *Experiments with sea urchin eggs show that cytoplasmic components cause the cleavage spindles to progress through a sequence of orientations independent of cytokinesis. In (a) the successive positions of the spindles and location of the blastomeres is shown for normally developing eggs. At the temperature used it takes five hours for the eggs to reach the 16-cell stage. In (b) eggs are shaken just before the third cleavage (arrow), preventing the division of mesomeres from macromeres in the equatorial plane. At the next cleavage, about four hours after fertilization, the spindles have the same orientation as would be normal for the macromeres. In (c) eggs are shaken just before the second cleavage (arrow), preventing cleavage into four cells. Despite this, at the next division the spindles are oriented vertically, characteristic of the third cleavage.*

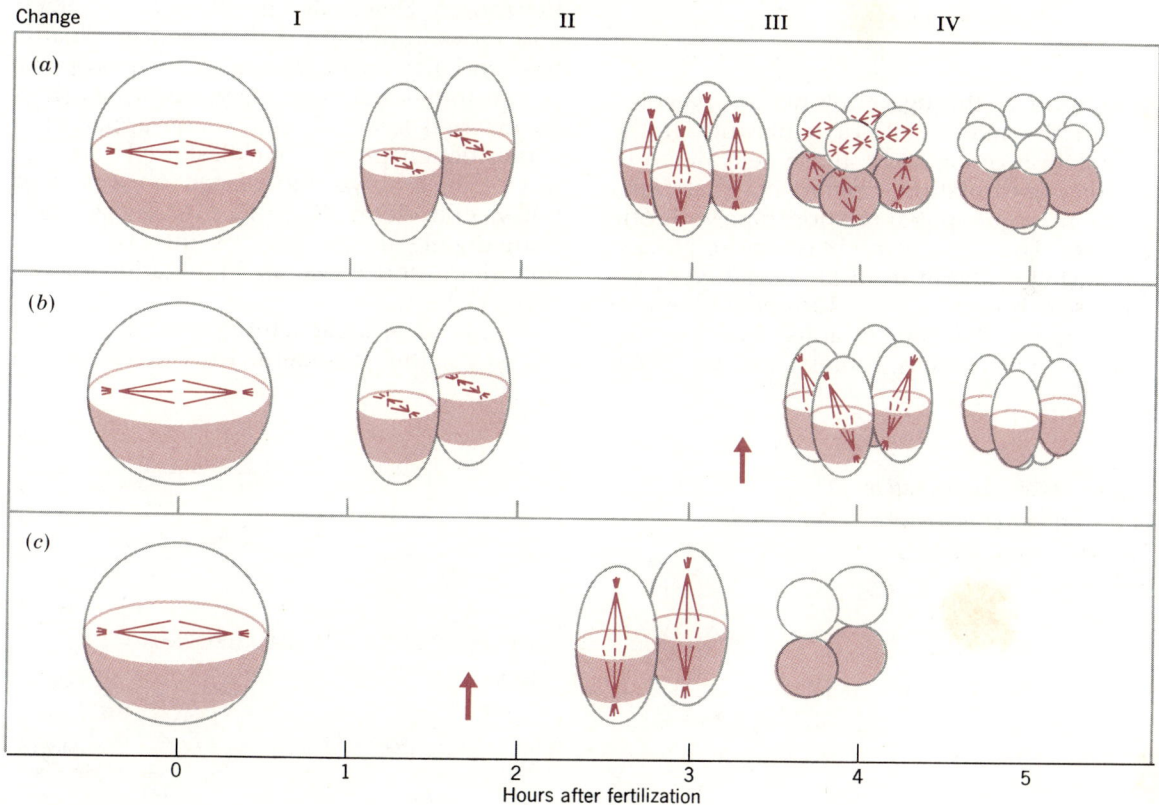

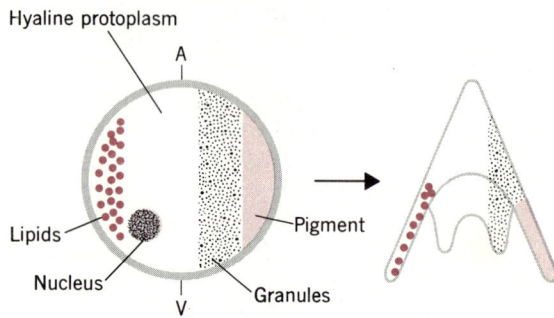

Figure 12-25 *Exposing sea urchin eggs to centrifugal forces of 10,000 times gravity or more stratifies the contents of the egg into layers that are independently oriented in relation to the animal-vegetal axis. Such eggs develop into normal plutei; the animal-vegetal polarity of the egg is not altered.*

*Spatial
Rearrangements
of Cells*

As soon as the egg contents are parceled out to the many cells of a blastula at the completion of cleavage, these cells begin to undergo gastrulation movements that bring them into new spatial relationships with one another. The cells of the blastula are already somewhat different from one another; as we shall see, however, their further differentiation requires that they establish new associations with other cells, somehow a cell's loca-

tion among the neighboring cells with which it is in contact influences its differentiation. What forces cause these gastrulatory movements?

The Cellular Basis of Morphogenesis

Look back at Figure 12-21b. The first sign of gastrulation in a frog egg is a small indentation that forms just below the equator of the blastula as some of the cells in that region suddenly begin to move from the surface into the yolky interior. This cleft marks the **dorsal lip** of the blastopore (Fig. 12-21) through which cells migrate to form the roof of the archenteron. The early invagination is brought about by a few surface cells that elongate into the interior, pushing aside other cells, each cell stretching itself into the shape of a graceful, narrow-necked flask (Fig. 12-21b) as it applies tension to draw the surface inward. The combined force of these invaginating cells indents the surface and forms the blastopore. Eventually, the flask cells detach and move further into the interior. Clearly, the tendency to move inward is a property of the individual cells; they are not pushed by external forces. This is made apparent if a few dozen prospective flask cells are cut out of an embryo and placed on a bed of yolky endoderm (Fig. 12-26). These cells immediately move into the interior of the endoderm and cause an indentation of the surface there.

This kind of invagination is similar to that by which the mesoderm layer is formed in

Figure 12-26 *Invagination of a fragment of blastopore lip tissue into a bed of endoderm.*

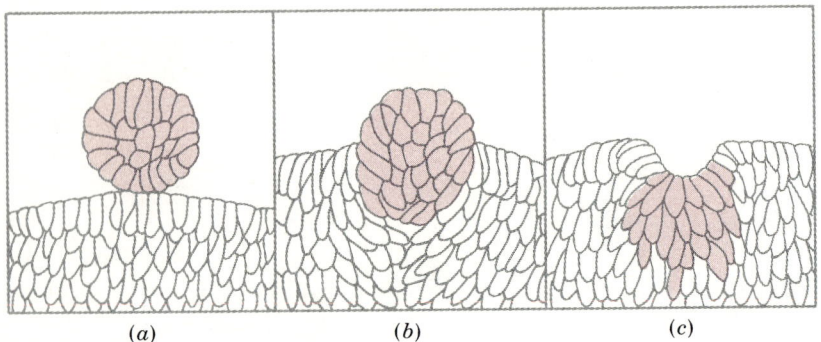

(a) (b) (c)

embryos of birds and mammals, which, you will recall from Chapter 11, arise from a flat plate of cleavage cells, the embryonic disc, or blastodisc (Fig. 12-27). In these embryos the blastodisc (or **blastoderm**) becomes arranged into two layers of cells, called **epiblast** and **hypoblast** (Fig. 12-28, p. 334). As these layers are forming, the cells begin to move within each layer, to pile up at what will be the back of the embryo, forming in the midline a thickened region called a **primitive streak** (Fig. 12-29, p. 335). Gastrulation begins when epiblast cells invaginate through the primitive streak into the space between epiblast cells and hypoblast to form the mesoderm (Fig. 12-30, p. 335). The epiblast cells that remain after this invagination are ectoderm. The hypoblast forms the endoderm of the three-layered embryo.

Cell Movements and the Cellular Surface

Cells appear to have properties, probably located in their plasma membranes, that determine whether they will adhere to a particular group of neighbors and establish a more or less permanent relationship with them. Such specific adhesive properties can be readily demonstrated in embryonic chick or mouse tissues. The tissues of a chick embryo after five days of incubation (at which time a well-formed heart, brain, limbs, etc., are present) can be dissociated into their component cells. If separate suspensions of heart and cartilage cells, for example, are combined (Fig. 12-31a, p. 336), random mixtures of the two cell types form into aggregates (Fig. 12-31b), but within a few hours the cells begin to sort out. After a day or two in culture, all of the heart cells are aggregated into a compact mass enclosed by an equally homogeneous mass of cartilage tissue (Fig. 12-31c, d).

Experiments by Malcolm Steinberg have shown that 11 combinations of six types of tissue cells always become organized into an inner and outer homogeneous cell mass. He has found that a series may be arranged in which each tissue segregates internally to all those below and externally to all those above. Steinberg argues convincingly that a cell's rank in this hierarchy is determined by its

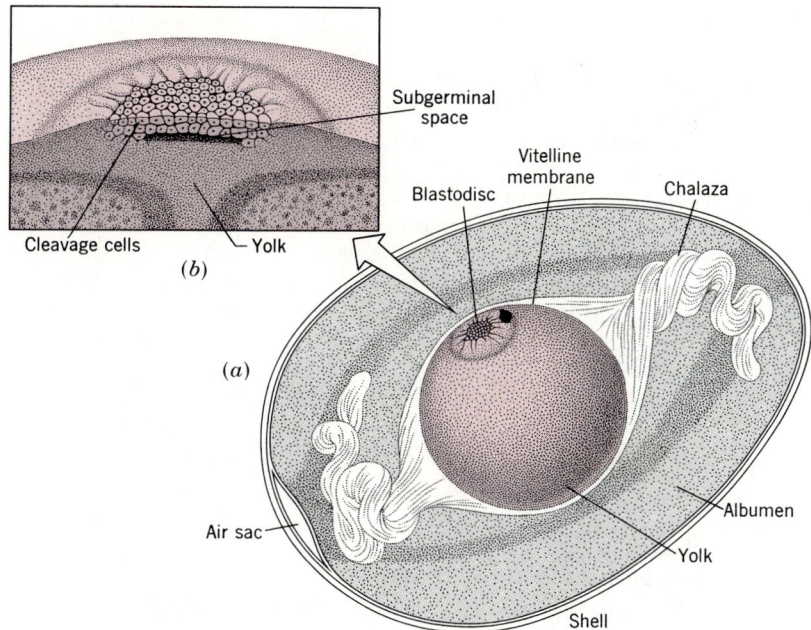

Figure 12-27   (a) A hen's egg, showing the relationship of the cleavage stage blastodisc to the yolk. (b) An enlarged section through the blastodisc showing the cleavage cells and the subgerminal space separating the cells and yolk.

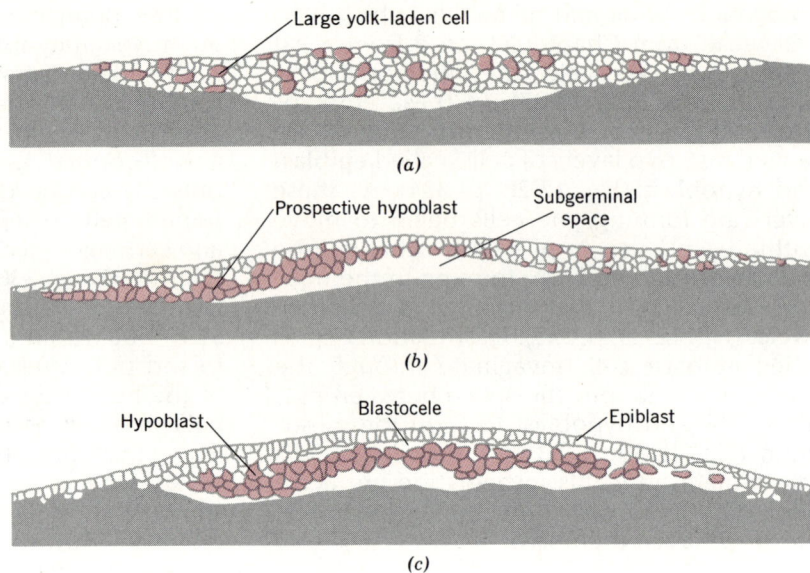

Large yolk-laden cell

(a)

Prospective hypoblast Subgerminal space

(b)

Hypoblast Blastocele Epiblast

(c)

Figure 12-28 Sections through chick blastoderms of successive ages showing the origin of the hypoblast.

specific adhesiveness to all of the other cell types. The results of these experiments—and the observed movements of gastrulation—are best explained by assuming that cells become reshuffled and alter their spatial relationships with one another as different adhesive types sort out and gradually group together. The adhesive capacities of cells depend on the molecular structure of the cell surface, which is determined, in turn, by the transcription of specific genes and synthesis of membrane components. Thus we begin to see emerging a set of causal relationships: DNA: mRNA: ribosomes: cell surface components: cell adhesiveness: cell movement: morphogenesis. Now we must bring the analysis back full circle and show that the spatial relationships of cells do have an influence on differential gene activation.

Cell Relationships and Development

If a small group of cells near the vegetal pole of a frog gastrula is stained with a non-toxic dye, the stained cells will be found in the belly skin of the embryo after it hatches out as an early tadpole. That is, in the normal course of development, these cells form skin. On the other hand, if the same group of cells is stained and then transplanted to a position near the animal pole, instead of producing skin it contributes to the brain, differentiating into perfectly normal nerve cells. That is, the grafted cells develop in conformity to their new position.

If the same experiment is performed a day or two later, however, at the end of gastrulation, the results are different. Prospective epidermis (skin) placed in the neural region ignores its new location and differentiates as epidermis. Thus during the course of gastrulation the developmental capacities of the cells become restricted—the cells become protodifferentiated, or determined (p. 303). Moreover, it can be demonstrated that the state of determination is the result of gastrulatory movements that permit cells to interact in new ways. One such interaction is **embryonic induction,** in which materials that pass from one tissue layer to another cause the recipient tissue to take a new differentiative pathway.

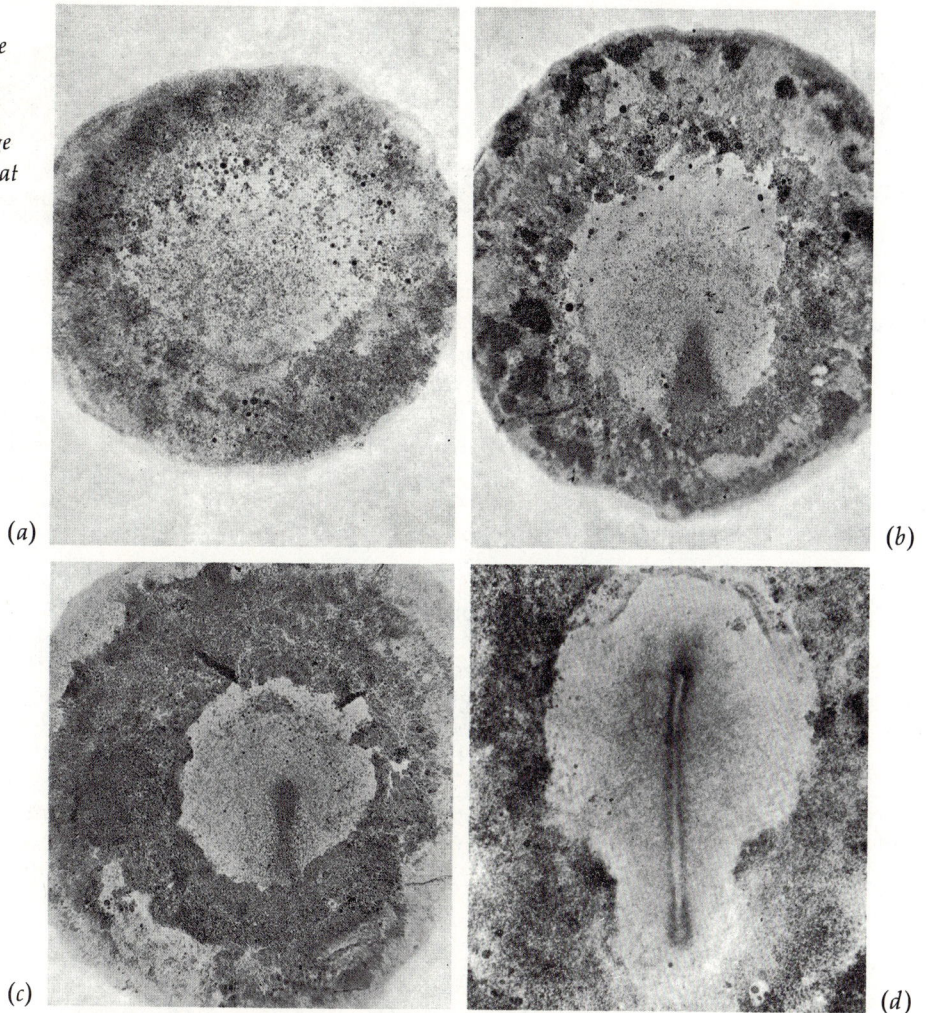

Figure 12-29 Early developmental stages of the chick embryo. The blastoderm is viewed atop the yolk, from above. Stage (a) shows the blastoderm at the time the egg is laid. Stages (b to d) illustrate successive stages in the formation of the primitive streak during the first 18 hours or so of incubation.

(a)

(b)

(c)

(d)

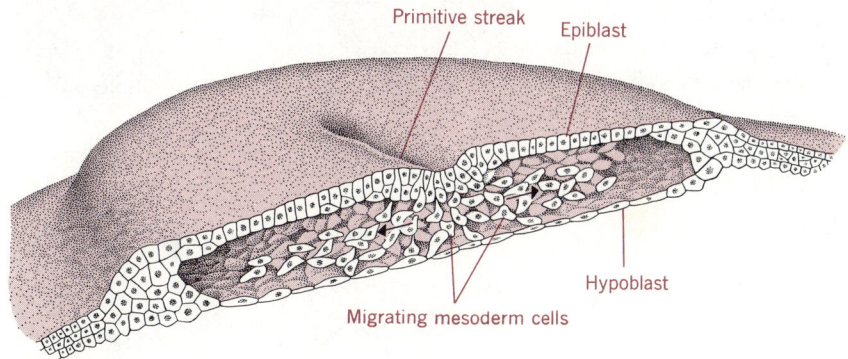

Figure 12-30 Front half of the chick blastoderm at stage d (Fig. 12-29), cut transversely to show invagination of cells through the primitive streak to form mesoderm.

Primitive streak   Epiblast

Hypoblast

Migrating mesoderm cells

335   Principles of Development

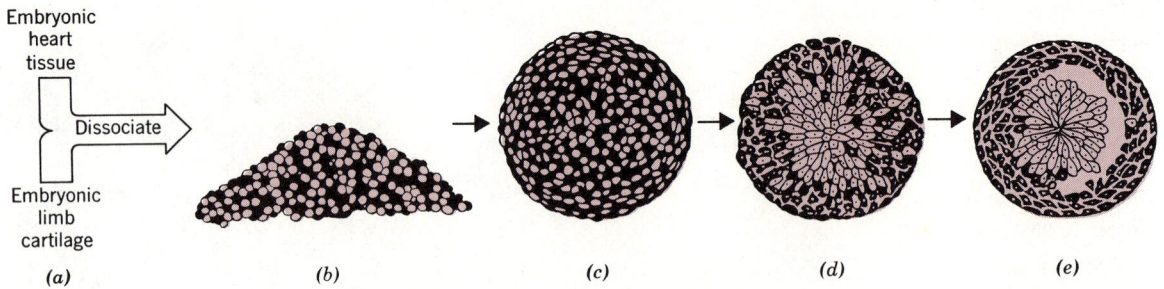

Figure 12-31 (a) Embryonic tissues such as heart and limb cartilage can be dissociated into their component cells with trypsin. (b) When they are mixed the cells reaggregate in random combinations (c). Soon, however, (d) the different cell types begin to sort out on the basis of their adhesive characteristics, separating into homogeneous tissues (e), usually arranged with the more adhesive cell type inside, surrounded by a mass of less adhesive cells.

Embryonic Induction

Shortly after gastrulation is completed in the frog, the ectoderm in the animal hemisphere that has come to be underlain by the cells of the archenteron roof begins to condense into a thickened plate. This is called a **neural plate** and is bounded by ridges known as **neural folds** (Fig. 12-32). The neural folds roll together to form a closed tube at the dorsal midline of the embryo. That tube is the neural tube, which gives rise to the animal's brain and spinal cord. Formation of the neural plate and its subsequent development is dependent on contact of that layer of the ectoderm with the mesoderm cells that form the roof of the archenteron. If invagination of the mesoderm is prevented, or if it is separated from the overlying ectoderm by an interposed piece of tissue, the ectoderm does not develop into nervous tissue but only skin. We speak of this mesoderm as an **inducer** of neural differentiation of the over-

Figure 12-32 Development of the neural tube and body form in the amphibian. (a, b, and c) Dorsal views of gastrula and early neurula stages, to show rolling together of the neural folds. (d, e, f) Views from the right of later stages.

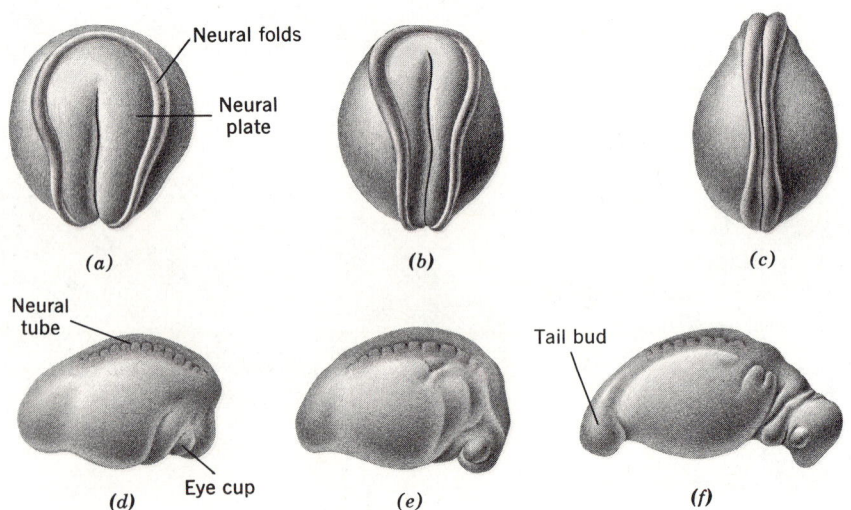

lying ectoderm, and of the interaction that must take place as **inductive interaction.**

Many such interactions in embryos have been demonstrated. The lens of the eye is induced to form from head ectoderm by the optic cup that underlies it; heart is induced in the mesoderm by the endoderm with which it is in contact; kidney tubules are induced in mesenchyme by epithelial cells that invade it. Unfortunately, the molecular nature of the inducer is not understood in any of these interactions. It may be that specific gene regulators are secreted by the inducing tissue. Another recent suggestion derives from the finding that many early embryonic tissues are **electrically coupled.** Recall that when cells form low-resistance nexal junctions (p. 126) with neighbors, ions and small molecules, as well as more complex substances, can pass freely from the cytoplasm of one cell to that of a coupled neighbor. It is possible that such cell coupling plays an important role in inductive interactions; at present we can only speculate about this.

Differentiation of Specific Cell Functions

There is one further characteristic in the process of differentiation that we have not yet considered: the development of function in organs. Gland cells secrete; muscle cells contract; nerve cells conduct. The aspect of embryology termed **developmental physiology** concerns itself with the development of function.

Heart Formation: A Model of Organogenesis

Consider, for example, the development of a beating heart in a chick embryo. At the primitive streak stage (Fig. 12-29d) the chick embryo shows no sign of a heart, limbs, eyes, or any other differentiated tissues. Over the next 18 hours or so of incubation, however, the embryo undergoes a series of morphogenetic movements whereby the three germ layers fold and move upon one another (Fig. 12-33, p. 338) to form a pouchlike foregut from which a digestive canal will develop, a neural tube, and a primitive tubular heart. Within an hour or two after the tubular heart first forms it begins spontaneously and rhythmically to contract.

Fragments of embryos such as those in Figure 12-33a, cut out and placed in culture, round up and differentiate into small balls, or vesicles, of tissue. Within two days of culture such vesicles develop an outer covering of skin and contain a variety of tissue types within, depending on where the fragment was in the embryo. Some of those vesicles contain masses of actively beating heart tissue, although there is no indication of heart tissue in Figure 12-33a. The cells that normally form the heart are located in the mesoderm, in a pair of areas called **heart-forming regions;** this tissue is termed **cardiogenic mesoderm.** The location of the cardiogenic mesoderm in an embryo at the stage of that in the figure can be mapped by transplanting fragments from a radioactively labeled donor embryo to an unlabeled recipient (Fig. 12-34, p. 339), where it participates in normal development of the host's heart tissues. Later, after the host has developed to the tubular-heart stage, its distribution of labeled cells is determined by autoradiography. The bilateral position of the two heart-forming regions is seen in Figure 12-34c.

How is the flat sheet of cardiogenic mesoderm transformed into a tubular heart 10 to 12 hours later? The same autoradiographic mapping method shows that the mesoderm condenses, deforms, and folds, but it does not lose its integrity as a sheet of cells. As the ectoderm and mesoderm begin their morphogenetic folding, forming the head fold and foregut, the mesoderm folds in between them toward the midline (Fig. 12-35a to c, p. 339) to form two curved channels facing each other, known as **myocardial troughs.** Starting at the anterior end, these

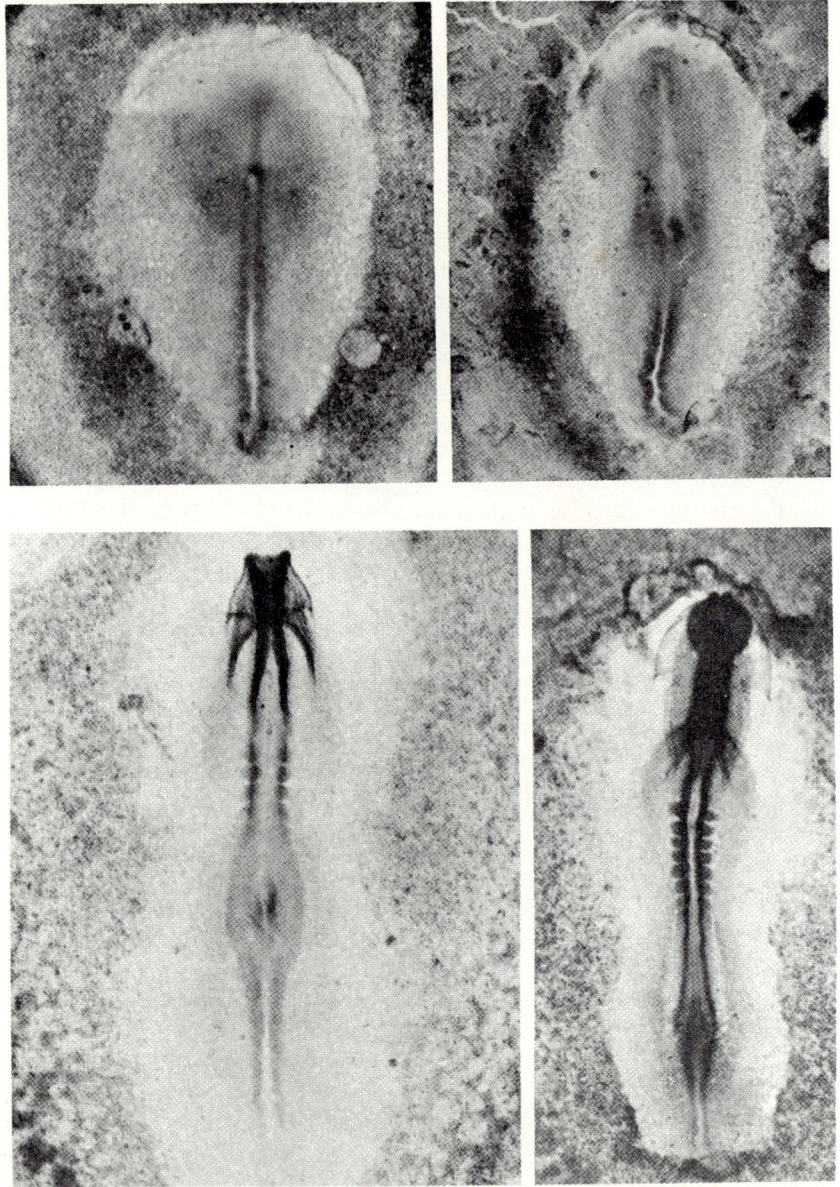

Figure 12-33 Successive developmental stages of the chick embryo from about 18 to 30 hours of incubation, showing formation of the headfold, neural tube, and heart.

troughs fuse in the middle to form the primitive curved cardiac tube (Fig. 12-35d to f).

Cells of the cardiogenic mesoderm in Fig. 12-33a show none of the microscopic or physiologic properties of heart tissue. Twelve hours later they have molded themselves into a tube of myocardium and are actively, rhythmically beating. To perform this feat these cells must differentiate at least the following three kinds of machinery: (1) Organized myofibrils made of actin, myosin, and other associated proteins, to act as a contractile mechanism (Chapter 18). (2) A spontaneously active, rhythmic source of electric

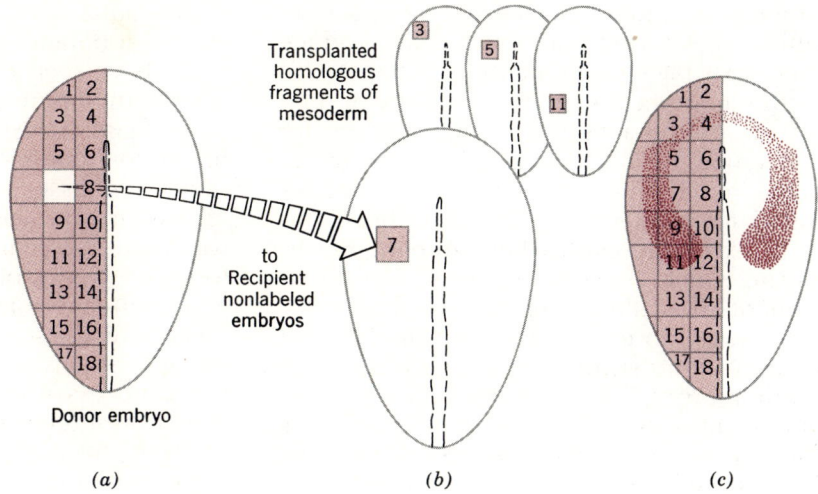

Figure 12-34 The transplantation mapping technique used for locating cardiogenic mesoderm in the stage 5 chick embryo. Fragments of endoderm-mesoderm are transplanted from (a) an embryo grown in H^3 thymidine to (b) nonlabeled recipient embryos. After development of the host to stage 12 (when it has a well-formed tubular heart), the position of labeled cells in the heart is determined by autoradiographic analysis. (c) Regions shaded on the embryo contributed cells to the stage 12 heart.

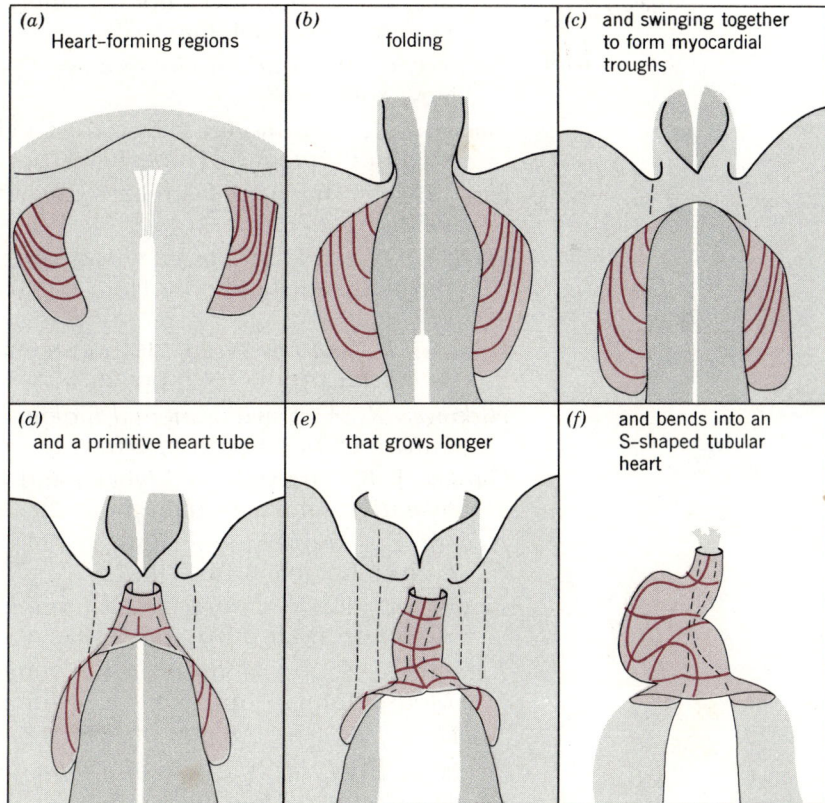

Figure 12-35 Heart formation in the chick embryo. Only the mesodermal sheet is shown. (a) The heart map is superimposed on the folding mesoderm layer and in (b, c) has been appropriately distorted in shape to fit the forming myocardial troughs and (d, e, f) heart tube.

stimulus to cause the periodic contraction. Unlike skeletal muscle, heart cells produce their own pacemaker stimulus (Chapter 15) by the spontaneous electrical activity of their plasma membranes. (3) A mechanism for converting the electrical activity of the membrane into a stimulus that acts as an effective trigger for contraction of the myofibrils. This is the **excitation-contraction coupling mechanism,** which consists of a specialized network of intracellular membranous structures.

To watch an embryonic heart give its first weak beat under the microscope is always a startling proof that these developmental requirements have been met. It is even more startling to see these properties exhibited by isolated cells. Single cells may be removed from a newly formed heart; these cells adhere to the surface of a culture dish and there exhibit spontaneous rhythmic pulsations.

We have seen that through the processes of growth, differentiation, and morphogenesis an egg is transformed into a highly organized, functioning individual, with a variety of different tissues, organs, and organ systems for carrying out specialized functions. So far we have concentrated mainly on animal forms. However, plants exhibit wholly analogous—often similar—processes of development, which also result in the production of various organs and tissues with different functions essential to the plants well-being. It will be interesting now to compare some of these processes, anatomical structures, and organ functions as we turn our attention to the plants.

Reading List

Balinsky, B. I., *An Introduction to Embryology* (3rd ed.). Saunders, Philadelphia, 1970.

Bodemer, C. W., *Modern Embryology.* Holt, Rinehart & Winston, New York, 1968.

Bonner, J. T., "Hormones in Social Amoebae and Mammals," *Scientific American* (June 1969), pp. 78–86.

Ebert, J. D., "The First Heartbeats," *Scientific American* (March 1959), pp. 87–96.

Ebert, J. D., and I. M. Sussex, *Interacting Systems in Development* (2nd ed.). Holt, Rinehart & Winston, New York, 1970.

Ephrussi, B., and M. Weiss, "Hybrid Somatic Cells," *Scientific American* (April 1969), pp. 26–35.

Flickinger, R. A., *Developmental Biology: A Book of Readings.* Brown, Dubuque, Iowa, 1966.

Gurdon, J. B., "Transplanted Nuclei and Cell Differentiation," *Scientific American* (December 1968), pp. 24-35.

Hadorn, E., "Transdetermination in Cells," *Scientific American* (November 1968) pp. 110–120.

Loewenstein, W. R., "Intercellular Communication," *Scientific American* (May 1970), pp. 78–86.

Rafferty, K. A., Jr., *Methods in Experimental Embryology of the Mouse.* Johns Hopkins Press, Baltimore, 1970.

Saunders, J. W., Jr., *Animal Morphogenesis.* Macmillan, N.Y., 1968.

Saunders, J. W., Jr., *Patterns and Principles of Animal Development.* Macmillan, N.Y., 1970.

Saxen, L., and J. Rapola., Congenital Defects. Holt, Rinehart & Winston, New York, 1969.

Torrey, T. W., Morphogenesis of the Vertebrates. Wiley, New York, 1967.

Weiss, P., Principles of Development. Holt, Rinehart & Winston, New York, 1939.

Wessells, N. K., and W. J. Rutter, "Differentiation Phases in Cells," *Scientific American* (March 1969), pp. 36–44.

Wessells, N. K., "How Living Cells Change Shape," *Scientific American* (October 1971), pp. 76–82.

Willier, B. H., and J. M. Oppenheimer, Foundations of Experimental Embryology. Prentice-Hall, Englewood Cliffs, N.J., 1964.

"Have you thanked a green plant today?"

SLOGAN POPULAR WITH ENVIRONMENTALISTS

part
six *Plants*

Form, Function, and Reproduction

"I flatter myself that I have accidentally hit upon a method of restoring air which has been injured by the burning of candles, and that I have discovered at least one of the restoratives which nature employs for this purpose. It is vegetation. In what manner this process in nature operates, to produce so remarkable an effect, I do not pretend to have discovered.

"One might have imagined that, since common air is necessary to vegetable, as well as to animal life, both plants and animals had affected it in the same manner, and I own I had that expectation, when I first put a sprig of mint into a glass-jar, standing inverted in a vessel of water; but when it had continued growing there for some months, I found that the air would neither extinguish a candle, nor was it at all inconvenient to a mouse, which I put into it."

JOSEPH PRIESTLY, 1772, "OBSERVATIONS ON DIFFERENT KINDS OF AIR"

We have seen that all living organisms, including plants, have similar requirements for energy, nutrients, water, and a proper environment. The most important, indeed, the unique feature of green plants is their ability to tap light as an energy source through photosynthesis (pp. 361–366). The energy stored and the oxygen produced by photosynthetic organisms provide the basis for the survival of all other forms of life on our planet. Although nonflowering plants such as algae, mosses, liverworts, ferns, and conifers are photosynthesizers, we consider them only briefly in this chapter. The **flowering plants,** or **angiosperms,** represent the most advanced evolutionary development in the plant kingdom, as man does in the animal kingdom, and we emphasize this extensive and most familiar group of plants in this chapter. We must, however, begin with a very brief discussion of the plant kingdom.

Thallophytes

Digitalis (*foxglove*).

The plant kingdom is divided into two major groups (for a detailed classification, see Appendix 2). The subkingdom **Thallophyta** includes the lower, or primitive, plants. It is in large part an artificial group, consisting of seven phyla of algae and three phyla of fungi (the bacteria, the slime fungi, and the true fungi) and including numerous unrelated plants that range from microscopic to large multicellular forms.

Although the group is artificial, the thallophytes share several common traits, the most notable of which is a simple plant body known as a **thallus.** This form is relatively un-differentiated; that is, it has no true roots, leaves, or stems; the vascular tissues are absent. Whether a species reproduces sexually or asexually, the reproductive structures of most thallophytes are unicellular. Finally, the thallophytes, in contrast to the members of the other plant subkingdom (the embryophytes), form zygotes that do not develop into multicellular embryos until after they leave the female sex organs.

Embryophytes

Although the second plant subkingdom, **Embryophyta,** contains only two phyla, the **Bryophyta** and **Tracheophyta,** these phyla include perhaps half a million species representing three-fourths of the plant kingdom. The two phyla are believed to have evolved independently from green algae; despite their independent origins, they share several common characteristics.

First, nearly all members of the Embryophyta are essentially terrestrial, or land, plants. It is true that a few higher plants of this group are aquatic types (e.g., water lilies), but there is every indication that they evolved from terrestrial forms, just as the seagoing mammals such as whales and seals surely originated from terrestrial mammalian ancestors.

Second, all develop their embryos in the female sex organs, from which they derive protection and nutrition, a situation markedly resembling that of mammals and contrasting with that of the lower plants. Third, all plants belonging to the Embryophyta reproduce sexually. They possess multicellular **sporangia** (p. 347) and multicellular sex organs, the **archegonium,** representing the female sex organ and the **antheridium (anther),** representing the male sex organ.

Other traits of the embryophytes include male and female gametes, chlorophyll and carotinoids (p. 363), and a waterproof cuticle covering most of the aerial parts of the plant. Finally, all display a definite **alternation of generations** involving a multicellular **gameto-** phyte generation (the plant that produces haploid gametes) and a multicellular **sporophyte generation** (the plant in which meiosis occurs).

The bryophytes are small plants that lack conducting tissue and are dependent on water as a medium for the motile gametes. They include the mosses and liverworts, in which the spore-producing diploid stage of the life cycle is attached to and dependent on the haploid gametophyte for nourishment.

The tracheophytes, or **vascular plants,** are characterized by two important features: the presence of **vascular tissue** (xylem and phloem, p. 351); and a sporophyte that is larger than the gametophyte and entirely independent of it except during the early stages of development. Included in the tracheophytes are the club mosses, horsetails, ferns, and seed plants. All members of the tracheophytes experience alternation of generations, although at times it may seem masked, especially in the higher forms, because the gametophyte is very much reduced in size and may remain contained in the sporophyte generation.

The Seed Plants

Within the phylum Tracheophyta are those plants with which we are most familiar, the **gymnosperms (conifers)** and the **angiosperms (flowering plants).** Together these two groups are often called the seed plants. The key characteristics that distinguish the seed plants from all others and account for their phenomenal success as terrestrial forms are the **production of seeds** and the **formation of pollen tubes** to convey the male gamete to the ovule.

Production of Seeds

The seed is an extremely efficient stage in the development of a new plant, and has conferred important advantages for existence in a terrestrial environment. Seeds are generally well adapted for wide dissemination and are viable for long periods of time, in some cases for hundreds of years. They are resistant to dessication, extremes of temperature, and numerous toxic substances. All seeds are equipped with a built-in supply of food, assuring that a newly germinating plant will be adequately developed by the time it becomes self-supporting.

Spores, like animal gametes, are tiny cells that are products of meiosis. They remain temporarily within the parental cell wall, called a **sporangium.** They are eventually liberated and, under favorable conditions, resume growth. The gymnosperms and angiosperms are universally **heterosporous;** that is, they produce two types of spore, a microspore and a megaspore.

The **megaspore** is not released from the sporangium (called a **megasporangium**), but remains to develop into a female gametophyte. The megasporangium, surrounded by certain maternal sporophyte tissue (the **integument**) and containing the female gametophyte, is called an **ovule.** After fertilization by the male gametophyte (which has developed from a **microspore**) an embryonic sporophyte is formed. It ceases development temporarily to enter a resting phase. Meanwhile, the integument matures into a seed coat. This entire structure, the embryo in the resting stage, the enclosing megasporangium, and the surrounding matured integument, is now collectively known as a seed. Under suitable conditions the young sporophyte embryo in the seed will resume development, germinating and emerging into an older and recognizable sporophyte with true roots, stems, and leaves (Fig. 13-1).

The Pollen Tube

Pollen-tube formation by the male gametophyte, the second unique characteristic of the seed plants, has also contributed to their phenomenal success. **Microspores** mature into the male gametophytes known as **pollen grains,** which are carried to the female plant in various ways, including wind, birds, and animals, liberating the seed plants to a large degree from dependence on free water for

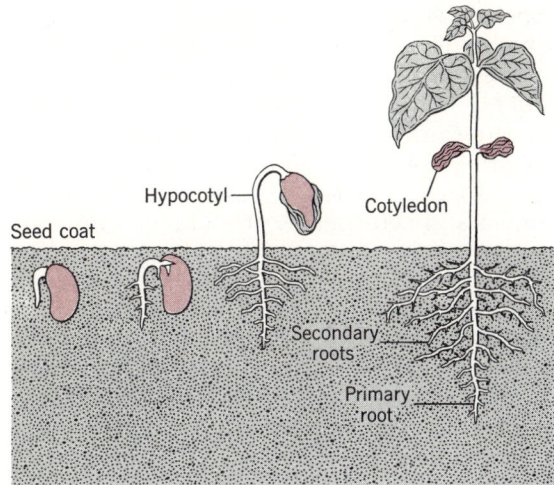

Figure 13-1   Some stages in the germination of the common bean Phaseolus vulgaris.

fertilization. When conditions are suitable, each microspore produces a pollen tube that grows toward the nearby ovules to transport the male gamete. This process is discussed further on p. 349 and 357.

The Gymnosperms

The gymnosperms as a class possess certain characteristic features. Unlike the angiosperms (discussed in a later section), their ovules and resulting seeds are not borne in an enclosed structure; they are said to be naked (gymnospermae means naked seeds). The sporophyte generation has true roots, stems, and leaves; the gametophyte is dependent and extremely small. In most cases the sporophytes have woody stems and evergreen, scalelike or needlelike leaves. This group is economically important as a source of timber, turpentine, and resin.

The gymnosperms, which are descended from an ancient line extending back into the Upper Devonian Period, include seven orders with more than 60 genera and 700 species. Three of the orders, Cycadofilicales, Bennettitales, and Cordaitales, are known only as fossils. Ginkgoales has only a single living species, **Ginkgo biloba** (maidenhair tree), which is characterized by broad, lobed leaves and features similar to those of another living order, Cycadales. The other two living orders are the Coniferales and the Gnetales.

Coniferales (the conifers), the major living order of gymnosperms, consists of some 50 genera and 600 species. It includes the well-known pines, spruces, firs, larches, junipers, cedars, yews, and redwoods. All characteristically possess seed-carrying **cones,** or **strobili,** which are either male, and carry microspores, or female, and carry megaspores. Depending on the species, some plants bear both types of cone (i.e., are **monoecious**) or only one type (i.e., are **dioecious**). All produce nonmotile sperm that are conveyed to the egg by pollen tubes. In the great majority of conifers the leaves are scalelike or needlelike and remain on the plant throughout the year.

The order Cycadales, the **cycads,** once abundant during the Upper Mesozoic Period, is probably the most primitive of the living gymnosperms. It now has only about 100 species of palmlike tropical and subtropical plants. Most representatives of the group possess certain features (e.g., motile sperm, vascular bundle arrangement, and leaf structure) similar to those of the ferns (class Filicineae, another group of vascular plants) in addition to characteristics of the gymnosperms. The order Gnetales, in addition to its gymnosperm characteristics, shows several angiosperm features, such as true vessels in the secondary wood (p. 351) and, in several genera, a female gametophyte resembling that of angiosperms. In these respects it is more like the angiosperms than any other gymnosperm. The suggestion has been made that an ancient line of the Gnetales may have given rise to the angiosperms, but there is no supporting fossil record.

Gymnosperm Reproduction

We may consider the familiar pine (order Coniferales, genus *Pinus*) as representative of the gymnosperms in its reproductive process. It has a life history that includes the well-known pine tree sporophyte generation, with its large, branched stem and extensive root system. The male, or **staminate,** cones bear only microsporangia, consist of spirally arranged **microsporophylls** (a **sporophyll** is a spore-bearing leaf, in this case a scale), and are found in clusters at the ends of branches.

Two microsporangia containing haploid microspores **(pollen)** are carried on the undersurface of each microsporophyll. Each pollen grain, at the time of pollination, is an immature male gametophyte.

The female, or **ovulate,** cones, which are larger but fewer in number than the staminate cones, consist of spirally arranged **megasporophylls** on whose upper surface two ovules are located. Each ovule is made of a single integument, with a single opening called a **micropyle** (through which pollen tubes may enter) surrounding the megasporangium, which is now called a **nucellus.**

Each megasporangium produces four haploid megaspores; this marks the beginning of the gametophyte generation. Three of the megaspores degenerate and the fourth develops into a small female gametophyte consisting of two or three archegonia embedded in a mass of haploid vegetative storage tissue called the **endosperm.**

Pollination, the transport of pollen grains to the female cones, is accomplished by the wind. Some of the airborne pollen grains come to rest in a sticky fluid near the micropyle. As the fluid dries, the pollen grains are drawn down through the micropyle until they come into contact with the nucellus itself. Pollen tubes begin to grow from the pollen grains, but fertilization does not occur until about a year later. As the pollen tube slowly develops, a **generative cell** in the microspore divides to form a **stalk cell,** which controls the growth of the pollen tube and a **body cell.** The latter enters the pollen tube and eventually divides into two nonmotile male nuclei, thus completing development of the fully matured male gametophyte, which is still dependent on the sporophyte for its nutrition.

Meanwhile the female gametophyte has also been developing slowly, and is mature by the time of fertilization, when the tip of the pollen tube reaches one of the archegonia of the female gametophyte and one of its male nuclei fuses with the egg cell to form the zygote—the beginning of the new sporophyte generation.

During the cell division that follows, only a single embryo develops from each female gametophyte; this finally consists of several leaf-like structures called **cotyledons,** the **embryo axis,** called the **epicotyl** (which will become the young stem of the plant), and a **hypocotyl** (which will become the primary root of the plant).

The Angiosperms

The angiosperms are the most advanced class of the entire plant kingdom. They include approximately 300,000 different species divided into two subclasses, the **Dicotyledoneae (dicotyledons, or dicots)** and the **Monocotyledoneae (monocotyledons, or monocots).** The angiosperms exceed in number of species all other plant groups together, making up the vast majority of the vascular plants. They are also of far greater economic importance to man than any other plants. As agricultural crops they represent an important source of man's food supply and in this respect have played a central role in the development of human society. They also provide us with textiles, medicines, and oils.

Although they share many features with the gymnosperms, the angiosperms are distinguished by the following important characteristics.

1. All angiosperms possess **flowers,** which are essentially highly modified strobili composed of microsporophylls called **stamens** and megasporophylls called **carpels.** They are often borne together, in contrast to the staminate and ovulate strobili of the gymnosperms. A flower, like a strobilus, is therefore essentially a stem with leaves modified for carrying on reproduction (p. 357).
2. The ovules and seeds of angiosperms are contained within an enclosed structure, the ovary; by contrast the gymnosperms bear naked seeds on the surface of the megasporophylls. Thus in angiosperms the pollen tube must grow through the tissues of the carpel before reaching the ovule.
3. With but a few exceptions vessels are present in the xylem of all angiosperms.
4. Angiosperms undergo a double fertilization, as we shall soon see.
5. Whereas pollination (p. 357) in gymnosperms is accomplished by wind, in angiosperms it is accomplished by insects and birds as well as by wind.
6. Although the angiosperms include woody plants that persist year after year **(perennials)** they also include many soft-tissued plants (**herbaceous** plants) that live for one or two years (called **annuals** or **biennials,** respectively), in contrast to the gymnosperms, which are all woody perennials.

The angiosperms, like the gymnosperms, have reduced, dependent gametophytes, large, conspicuous, and almost entirely independent sporophytes, heterospory, pollen-tube formation, seeds, and true roots, stems, and leaves.

The subclass **Dicotyledoneae** consists of about 225,000 species having two cotyledons in their embryos; the subclass **Monocotyledoneae** consists of some 75,000 species with only one cotyledon. Other important differences between the two subclasses are that the dicotyledonous plants usually show vascular tissues in a cylinder or bundle arranged in a single circle, have a cambium, and show net-veined leaves (Fig. 13-2a, p. 350). Monocotyledons, on the other hand, usually have scattered vascular bundles (Fig. 13-3, p. 350), no cambium, and leaves with parallel veins (Fig. 13-2b). The flower parts of dicotyledons often occur in multiples of four or five (e.g., four sepals, four petals, four stamens, four carpels), whereas those of monocotyledons are usually found in multiples of three. From an evolutionary point of view the monocotyledons, which include grasses, grains, lilies, irises, and orchids, are considered to be more advanced than the dicotyledons, which resemble the gymnosperms in their posses-

(a)

(b)

Figure 13-2 (a) The net-veined leaf of a dicotyledon; (b) the parallel-veined leaves of a monocotyledon.

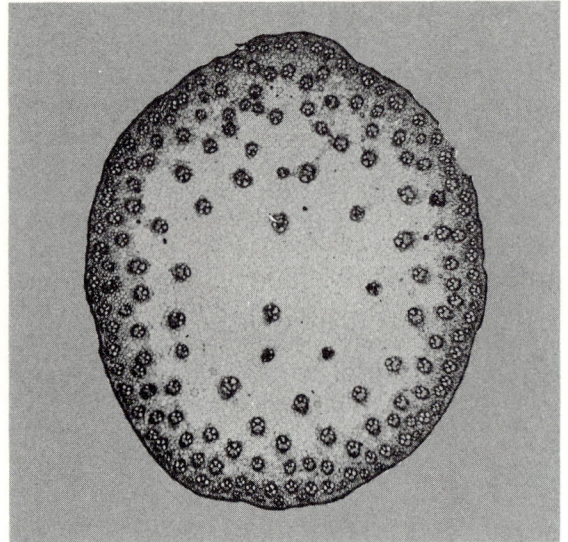

Figure 13-3 Photograph of a corn stem in cross-section, showing the numerous individual vascular bundles.

sion of a cambium and the circular arrangement of vascular tissues. Presumably the monocotyledons arose from some ancient line of dicotyledons.

Anatomy of the Angiosperms

We can now turn to a more detailed examination of the flowering plants. The basic structural features of the plant cell have already been described (pp. 140–143). Like that of an animal, the life of a flowering plant starts with a fertilized egg cell; the stem, branches, roots, and other parts arise through mitosis and differentiation.

The Stem

The plant stems of angiosperms are classified as either woody or herbaceous. **Woody** stems are usually covered with a layer of cork tissue and are tough and thick. They are characteristic of the plants called perennials.

Herbaceous stems are generally soft, green, and thin, and are typical of plants that live for one growing season (annuals) or two growing seasons (biennials). The woody plants are considered to be the more primitive, having appeared before the herbaceous ones in the course of evolution.

Stem Structure

In a typical woody flowering plant (Fig. 13-4) the mature tissues of the stem are arranged in concentric cylinders around a central core of thin-walled, loosely packed, colorless **parenchyma** cells (Chapter 5) known as the **pith.** The pith is encircled by a layer of highly specialized water-conducting tissues. These **conductive,** or **vascular, tissues** are responsible for transporting various materials, including water, dissolved minerals, carbohydrates, and so on, to differ-

ent parts of the plant. They consist of two complex tissue types, **xylem,** or wood, and **phloem,** each of which includes several kinds of cells, and a single layer of meristematic cells called the **vascular cambium.** The vascular tissue is continuous from the roots throughout the stem, branches, and leaves.

Xylem. The xylem in both gymnosperms and angiosperms is involved in support of the stem and in the transport of water and dissolved mineral salts from the roots, where they are absorbed, to the other parts of the plant. Xylem is described as either **primary** or **secondary,** depending on the part of the plant in which it develops. The secondary xylem, which is the major type, is formed by the vascular cambium. The primary xylem develops from the apical meristem.

The two chief conducting units of the xylem are the **tracheids** and **vessels.** Mature

Figure 13-4   Diagram of a cross-section of a woody stem.

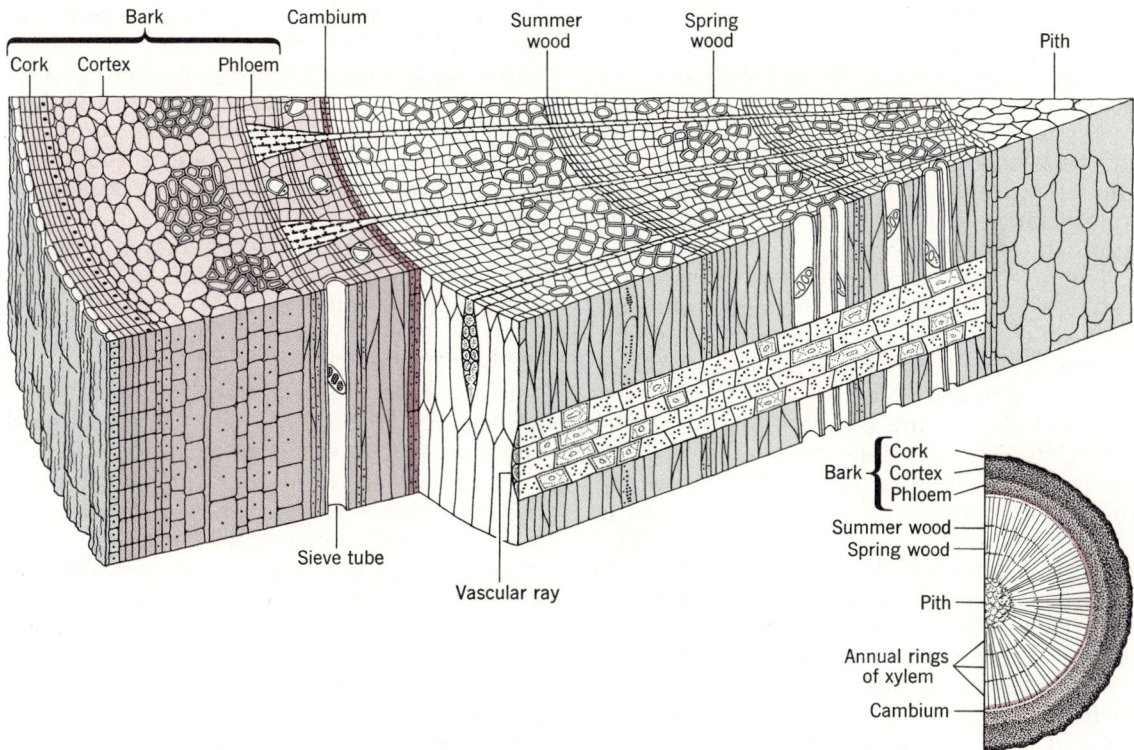

tracheids are elongated, tapering cells whose protoplasts have died. The large empty spaces in the tracheids, formerly occupied by the protoplasts, make it well suited for water conduction. Their cell walls, which aid in support of the plant, often exhibit thickened spiral rings of cellulose and small thin or pitted areas connecting them to other tracheids and cells.

Further specialization of the tracheids for water conduction during the evolutionary course of plant development has resulted in the enlargement of these cells, and perforations, usually in their end walls, to form long, vertical tube systems known as **vessels,** at times several yards in length. These contain various kinds of wall thickenings and pits like those of the tracheids from which they evolved. The vessels, too, function primarily in water conduction and secondarily as a supporting tissue. Tracheids are the chief conducting elements of the gymnosperms; the more highly evolved vessels are characteristically present in the more advanced flowering plants, which possess both structures.

Phloem. Surrounding the xylem is the vascular cambium, and this in turn is surrounded by the phloem, which is the chief passage for the transfer, or **translocation** (p. 360), of organic compounds from the sites of their manufacture in the leaves and stem to regions of use and storage.

The fundamental structural and functional cell type in the phloem is the **sieve tube cell,** just as the trancheid is the important cell type in the xylem. Parenchyma cells and fibers, which function in structure and support, are also included in the phloem, as they are in the xylem. The sieve tube cell is an elongated, living cell containing protoplasm. It loses its nucleus at maturity. The vertical living sieve tubes constitute a continuous vertical cytoplasmic connection that aids in the transport of food to the various parts of the plant.

A specialized type of parenchyma cell called a **companion cell** is structurally and functionally associated with the sieve tube cells in the angiosperms but not in less advanced plants. Companion cells are elongated, nucleated cells somewhat shorter than the sieve tube cells. They are connected by numerous cell wall pits to the sieve tube cells with which they are associated, and are believed to assist the sieve tube cells in the transport of food.

Cortex. A layer of cortex surrounds the phloem, and outside of this is a layer of cork. Together the phloem, cortex, and cork make up the **bark** of a woody plant. The cortex is made largely of parenchyma cells, and may contain **collenchyma** cells. Collenchyma is a supporting tissue whose cells are not so elongated as fiber cells and are characterized by cell walls that are usually markedly thickened at the corners. It is relatively soft and plastic. It occurs commonly in the elongating parts of stems and to a lesser extent in parts of leaves and in soft, mature plant parts.

The cortex may also contain **chlorenchyma,** which are chloroplast-bearing cells that are generally similar to parenchyma cells, but are able to photosynthesize and help provide the plant with energy.

In young woody plants and in most herbaceous plants the cortex is covered with a continuous layer of living epidermal cells extending over the surface of the entire plant. In the older parts of woody stems, however, the outer layer of the cortex eventually produces a single layer of meristematic cells, the **cork cambium,** which then produces an overlying layer of cork. When this happens, the epidermis is sloughed off and replaced by the cork tissue.

Stem Growth

In most higher plants growth in stem length takes place at the tip, where the **apical meristem,** or **meristematic zone,** is located (Fig. 13-5). Each cell division of the meristematic tissue produces two daughter cells, of which one remains meristematic and the other forms one of the specialized stem cells. As a result, the meristematic cells of the stem remain in the apical portion, which moves upward, whereas progressively older cells are found in correspondingly lower regions of the stem.

In their conversion to specialized cells the

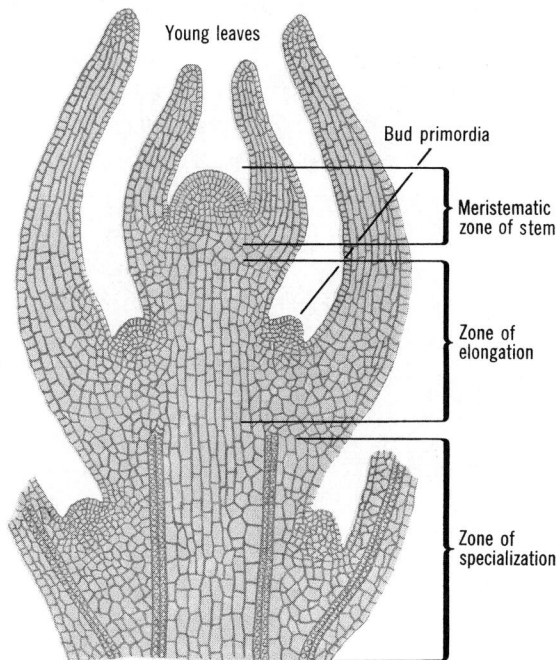

Figure 13-5   Zones of longitudinal growth and differentiation in the stem tip.

daughter cell next to the xylem may develop and mature into a secondary xylem cell, whereas the outer daughter cell remains meristematic. The latter subsequently divides and the outer of the two daughter cells may develop and mature into a secondary phloem cell; this time the inner cell remains meristematic. In this manner the cambium layer gives rise to secondary xylem and phloem (although not necessarily in an alternating order); cambium is always maintained between the other two types of conductive tissue. Generally a greater quantity of secondary xylem than secondary phloem is formed during a growing season, so that the xylem makes up the bulk of a stem. Except for the peripheral layers close to the vascular cambium, most of the trunk of an old tree consists of dead xylem.

Most monocotyledonous plants, nearly all of which are herbaceous, possess no cambium tissue and therefore display no secondary growth. All their tissues are primary in origin, having developed from apical meristematic cells. Their xylem and phloem are organized into individual vascular bundles dispersed through the stem (Fig. 13-3).

Annual Rings. In the spring, when abundant water is available and the temperature favorable, stem growth is most rapid. The size of cells produced, especially of the xylem, is also greatest at this time, accounting for a relatively porous wood. In the progressively drier weather of summer and fall cell size and rate of division decrease; division essentially ceases with the coming of winter, and resumes again the following spring.

In cross-sections the difference between the denser wood of the summer and fall and the large-celled, porous wood of the following spring are seen as alternating rings of summer and spring wood. These are called **annual rings** (Fig. 13-4). The youngest, or most recently formed, annual ring is located immediately inside the cambium.

The Bud. A plant stem produces **shoots** (branches), **leaves,** and **flowers** from specific kinds of apical meristem. Generally such meristems remain **dormant** (inactive) during a portion of the year. A dormant meristem is

daughter cells first become greatly enlarged and elongated, increasing to at least 10 to 20 times the size of the meristematic cells and accounting almost completely for increase in stem length. A process of maturation or differentiation ultimately converts some of the cells into epidermis, parenchyma, primary xylem, and primary phloem. The sequence of changes can be visualized by examining a longitudinal section of the stem (Fig. 13-5). Starting at the apical portion of the stem and moving downward, we see first the meristematic zone, the zone of elongation, and the zone of specialization. The zones are not sharply delineated, but gradually blend into one another. The stem below the zone of specialization is fully formed and mature, but in woody plants growth in girth continues in this region because of continuing cell division in the cambium layer.

When a cambium cell divides, the inner

called a **bud.** Leaf buds and flower buds usually develop seasonally, although some species, such as night-blooming jasmine or day lilies, also exhibit a 24-hour cycle of bud development. Shoot buds (Fig. 13-5) may remain inactive for extended periods, or they may develop into lateral branches with their own apical meristems.

The extent to which these buds develop or remain dormant determines the **habit,** or over-all appearance, of a plant. If the bud primordia remain inactive, the plant will grow vertically; if they develop, the plant will take on a bushy, many-branched form. Commonly, the development of the lateral buds is inhibited by hormones formed in the central growing apex, and as a result most plants show **apical dominance**—their main stems show the greatest development.

The Root

The Structure of a Young Root

The arrangement of tissues in the roots of flowering plants is essentially like that of the stem, with just a few differences.

Unlike most stems, roots have no pith. Instead, xylem occupies the central core, with several lateral extensions (Figure 13-6). In a cross-section the xylem looks like an X or a star, with phloem tissue occupying the spaces between the arms. The vascular cambium (when it is present, as in the woody plants) separates xylem from phloem and accounts for growth in root diameter. Surrounding the core of xylem and phloem are two layers each usually only a single cell in thickness. The first layer is called the **pericycle,** and is the point at which lateral root formation originates. The second layer is called the **endodermis,** and consists largely of cells whose walls are impregnated with a waxy material called **suberin.**

Enclosing the endodermis is the cortex, which is in turn surrounded by a single cell layer of **epidermis.** In a region along the root that roughly corresponds with the zone of maturation (Fig. 13-7), the epidermal cells exhibit delicate projections of cytoplasm called **root hairs.** It is here that absorption of

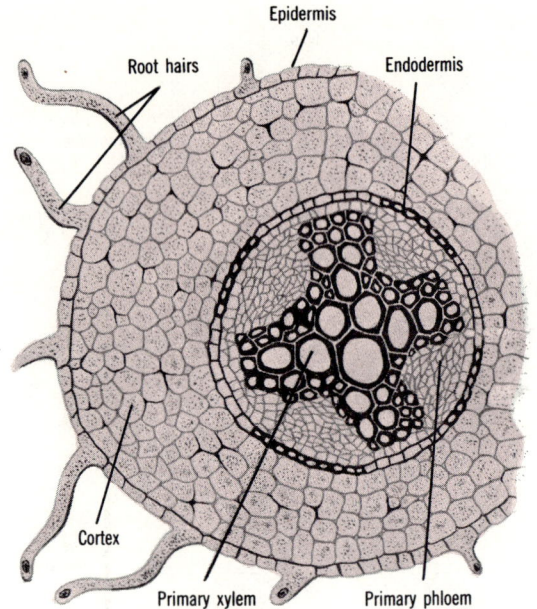

Figure 13-6 *Cross-section of a young root.*

Figure 13-7 *Longitudinal section of a root tip.*

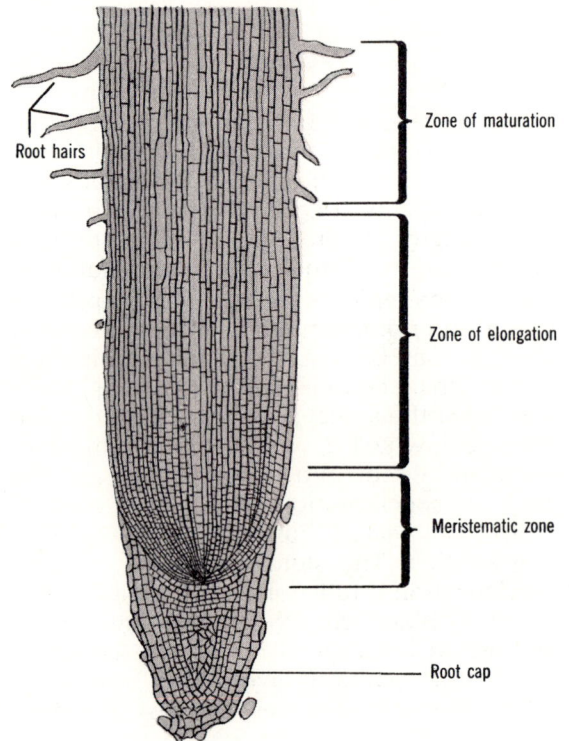

water and dissolved mineral nutrients takes place, for the root hairs are often present in high concentrations (thousands per square centimeter of root surface), and thus provide a tremendously increased area for absorption of water and dissolved salts. For example, the 14 billion root hairs of a single rye plant represent a total area of 4000 square ft. The other principal roles of the roots, anchoring the plant and serving as food-storage organs, are secondary to the role of absorption.

The Structure of the Mature Root

As the cambium in the young root begins to divide, secondary xylem and phloem are laid down. Almost simultaneously the pericycle immediately underlying the endodermis generates a cork cambium that produces cork between it and the endodermis. When this new cork layer completely surrounds the roots, all the layers external to it (i.e., endodermis, cortex, and epidermis) are sloughed off. As more secondary xylem and phloem are laid down, the mature root comes to resemble the stem.

Root Growth

Growth in length of the root is very much like that of the stem, but with at least one added feature. In addition to the usual meristematic zone at the tip, a zone of elongation, and a zone of maturation or specialization, there is a thimble-shaped mass of cells called a **root cap** (Fig. 13-7) covering the meristematic zone. Its obvious function is to protect the meristematic tip as the root pushes out among the soil particles in the course of growth. The zone of maturation may extend for several centimeters, depending on the root and its rate of growth. As in the stem, the increase in root length is largely the result of increase in cell number and in cell size in the zone of elongation.

The Leaf

The primary role of the leaf is to carry on photosynthesis, for which it is well suited in terms of structure and tissue arrangement. The anatomy of the leaf allows for maximal exposure of chlorophyll-containing cells to light and makes possible a large surface area for the necessary exchange of gases between the cells and their environment.

The typical leaf of a dicotyledonous plant consists of a broad, flattened **blade** (the part that we call the leaf), which is usually attached to the plant stem by a **petiole,** the leaf stem. The petiole contains vascular tissue that is continuous with the **midrib** (the principal vascular bundle of the blade), or, depending on the plant species, with several large **veins.** The veins may form different patterns in different species, including a netlike design, parallel veins, and a palmlike effect (see Fig. 13-2).

Leaf Structure

In a cross-section (Fig. 13-8, p. 356) the upper surface of a typical dicotyledonous leaf is covered with a single layer of colorless epidermal cells. On their outer side the epidermal cells are protected by a waxy layer called the **cuticle,** consisting of **cutin,** a substance that is nearly impervious to water and to such gases as oxygen, nitrogen, and carbon dioxide.

Immediately below the upper epidermis is a densely packed layer, called the **palisade layer,** of columnar parenchyma cells containing numerous chloroplasts. Beneath the palisade layer is a loosely packed layer of chloroplast-containing parenchyma cells with numerous intercellular spaces, collectively referred to as the **spongy layer.** The lower surface of the leaf, like the upper surface, is bounded by a layer of colorless epidermal cells that in most angiosperms are characteristically interspersed with numerous small, specialized openings, or **stomata** (singular, **stoma**). Stomata may also occur (usually to a more limited extent) on the upper surface, depending on the species. The number of stomata varies greatly, and may range from as low as 1400 per square centimeter to more than 100,000, again depending on the species.

Each stoma is surrounded by two highly specialized and modified epidermal cells, the **guard cells** (Figs. 13-8 and 13-9, p. 356). These are typically sausage shaped, usually contain chloroplasts, and can change their shape, thus regulating the stomatal pore size. The

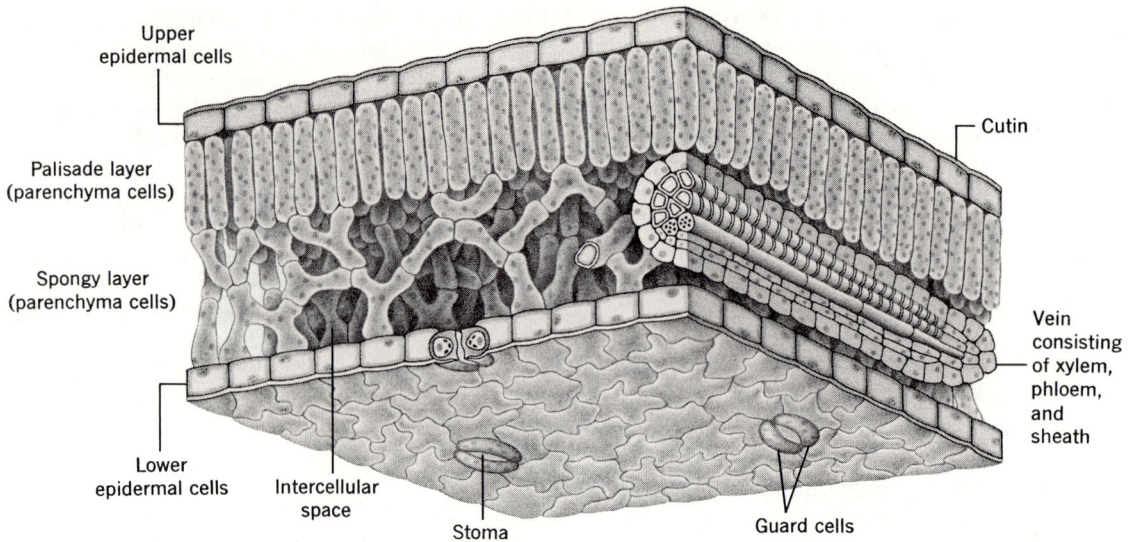

Figure 13-8   *Cross-section through a typical dicotyledonous leaf.*

Labels (clockwise from upper left): Upper epidermal cells; Cutin; Vein consisting of xylem, phloem, and sheath; Guard cells; Stoma; Intercellular space; Lower epidermal cells; Spongy layer (parenchyma cells); Palisade layer (parenchyma cells)

mechanisms controlling stomatal size and the influence of stomatal size on photosynthesis and water relations are discussed in the physiology section (pp. 359–366).

Leaf Growth

In woody plants buds form in the early fall from small masses of meristematic tissue. They open the following spring and quickly develop into mature leaves and a growing stem. The transformation of a new, tiny leaf to a mature one principally involves the enlargement of most of the cells in the blade; the contribution to growth by cell division is relatively minor. The growth of a leaf is limited; it reaches a particular size, functions usually for one season, and then drops off.

The Flower

The flower is a modified stem or branch bearing reproductive structures. Some of the flower parts have presumably originated from the stem itself and others from modified leaves.

Figure 13-9   *Cell-wall thickness under conditions of increased turgor determines the shape of the guard cell and therefore the size of the stomatal aperture.*

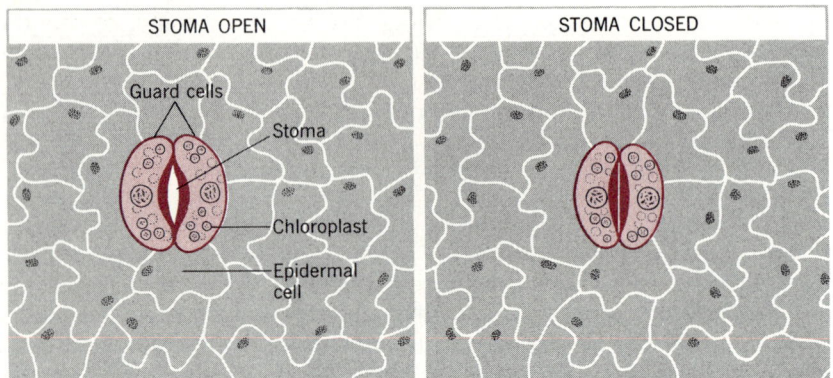

STOMA OPEN — Guard cells; Stoma; Chloroplast; Epidermal cell

STOMA CLOSED

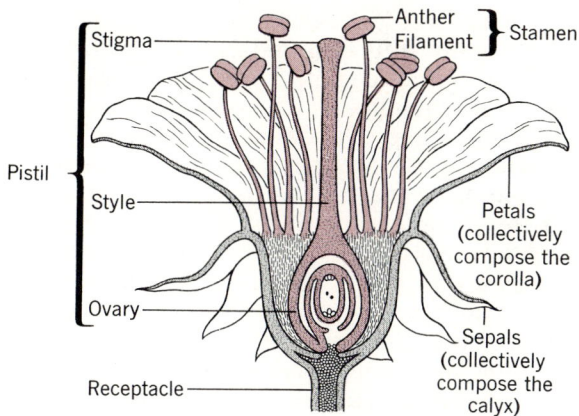

Figure 13-10   Structure of a typical flower.

A typical flower (such as a rose or petunia) is made of the tip of a stem (called a **receptacle**) and the four types of floral part that grow from it: sepals, petals, stamens, and pistil (Fig. 13-10). The **sepals,** collectively called the **calyx,** are the outermost of the floral organs and usually enclose the other flower parts in the bud. They are generally small, green, leaflike structures.

The **petals,** collectively called a **corolla,** form a second group, or **whorl,** of structures and are often the most conspicuously colored of the flower parts. By virtue of their bright colors, sweetish nectars, and distinctive scents, the petals tend to attract insects, important in pollination.

Within the corolla are the **stamens,** each usually consisting of a slender stalk, or **filament,** bearing at its tip an enlarged portion, the **anther** (antheridium), which produces pollen. The stamens correspond to the microsporophylls of the gymnosperms. Surrounded by the stamens at the center of the flower is the **pistil,** composed of an enlarged lower portion, the **ovary** (consisting of one or more carpels containing the ovules), a slender structure, the **style,** which rises from the ovary, and the expanded tip of the style, called a **stigma.**

A good deal of variety occurs among the angiosperms with respect to flower structure, shape, arrangement, size, color, and number of members for each of the flower parts. Although most species of angiosperms possess complete flowers (sepals, petals, stamens, and pistil), there are other species that lack in one or several of the flower parts; the sepals, or petals, or both, may be absent. This is true, for example, of grains and grasses. In other species, such as corn, the stamens are restricted to some flowers and the pistils to others, so that the species has separate male and female flowers on each plant. Some species, the willow and the date palm, for example, produce staminate and the pistillate flowers on separate plants.

Reproduction in Angiosperms

The Reproductive Structures

Haploid (1n) pollen grains, or microspores, are produced in the anthers by meiosis. Liberation of the pollen grains and their transmittal by wind and insects to the stigma of flowers of the same plant **(self-pollination)** or other plants of the same species **(cross-pollination)** mark the beginning of the process of sexual reproduction. At the time of liberation the pollen grains contain a **tube nucleus** and a **generative nucleus,** each of which is haploid (1n).

After being deposited on the stigma, a pollen grain absorbs food and water and sends out a **pollen tube** (Fig. 13-11, p. 358); this grows downward through the style toward the ovary. The tube nucleus regulates the growth of the pollen tube. It is usually during the period of early growth of the pollen tube that the generative nucleus divides to produce two male, or sperm, nuclei.

The ovules in angiosperms are attached by separate short stalks to the body wall of the ovary. The outer covering of the ovule has two **integument** layers covering the underlying **megasporangium,** or **nucellus.** In the megasporangium a single cell undergoes meiosis to form four haploid **megaspores,** three of which disintegrate. The remaining megaspore produces an **embryo sac** in which the embryo plant will develop. At maturity

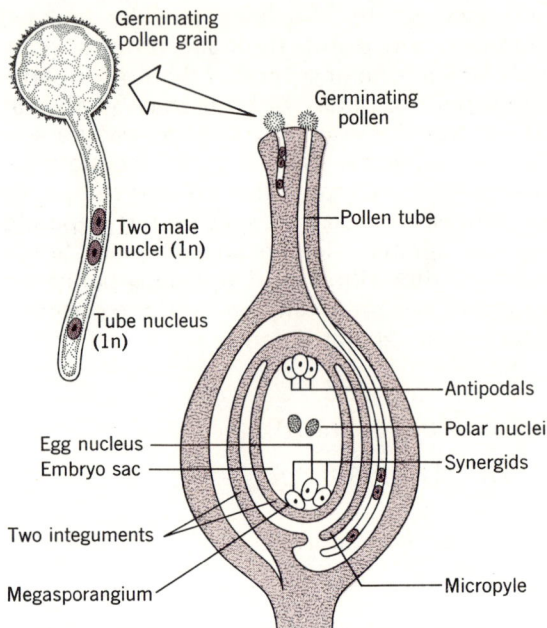

Figure 13-11   Germinating pollen grain and mature embryo sac of an angiosperm.

the embryo sac typically contains eight haploid nuclei. Three of the nuclei, an **egg nucleus** and two so-called **synergid nuclei,** are present at the end of the embryo sac near the **micropyle** (a tiny pore extending through the integuments); two **polar nuclei** are located near the center of the embryo sac; and three nuclei, called **antipodals** are present at the end of the embryo sac opposite the micropyle.

The Process

Before fertilization can occur the pollen tube must grow downward through the style into the ovary, through the micropyle, and finally penetrate into the embryo sac to discharge its male nuclei. One of the two male nuclei fuses with the egg to form a diploid zygote, marking the beginning of the sporophyte generation. The other male nucleus meanwhile fuses with the two polar nuclei to form a triploid (3n) nucleus. This triploid nucleus will eventually develop into triploid **endosperm tissue,** which serves as the food-

storage tissue for the seed.

Soon after this double fertilization, a phenomenon found only in angiosperms, the remaining five nuclei of the embryo sac (the antipodals and synergids) disintegrate. The diploid zygote divides and differentiates into an embryo. Meanwhile, the triploid endosperm divides successively to form the multicellular triploid endosperm tissue. During their development the embryo and endosperm receive food materials from the parent plant. The foods (for the most part oils, starches, and proteins) that accumulate in the endosperm are utilized by the embryo as the seed subsequently germinates into a young plant. As the embryo and endosperm grow and develop, the integuments become hardened and modified into a **seed coat,** thus collectively transforming the enlarging ovule into a developing seed.

At the same time that the seeds are forming, the ovary also increases in size and the other flowering parts (stamens, petals, and sepals) deteriorate and disappear. Thus the enlarging ovary with its developing seeds eventually becomes the **fruit.**

Angiosperm seeds may differ widely in appearance from one species to another, but all have the same fundamental parts (Fig. 13-12a). There are always one or two **cotyledons,** temporary leaf structures that function in the digestion, absorption, and storage of food from the endosperm, making it available to the embryo before and at the time of seed germination. There is always an **epicotyl** (the part of the main embryo axis above the point of attachment of the cotyledons), and a **hypocotyl** (the portion of the axis below the cotyledons, at whose lower end is a **radicle,** or **rudimentary root.**

The seeds of many species normally experience a period of rest, or **dormancy,** usually during the winter, when growth and development of the embryo are at a standstill. The biochemical or physiological mechanisms responsible for seed dormancy are not well understood, but it is during this stage of seed dormancy that an embryonic plant is best able to withstand unfavorable environmental conditions. Favorable external conditions

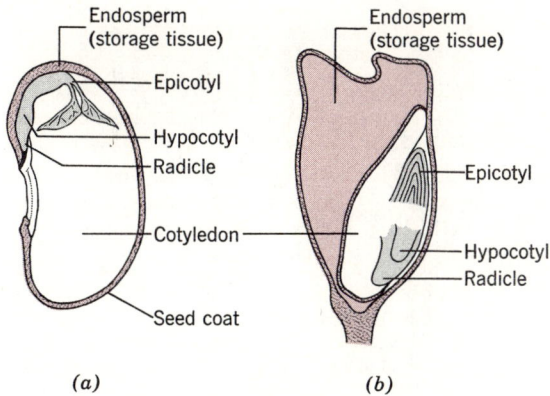

Endosperm
(storage tissue)

Epicotyl

Hypocotyl
Radicle

Cotyledon

Seed coat

(a)

Endosperm
(storage tissue)

Epicotyl

Hypocotyl
Radicle

(b)

Figure 13-12 Structure of (a) the common bean seed (a dicotyledon) and (b) a corn kernel (a fruit, a monocotyledon).

such as available water and suitable temperatures result in the resumption of growth and development of the embryo, a process known as **seed germination** (Fig. 13-1).

Physiology and Biochemistry of Flowering Plants

Stomatal Size and Exchange of Gases

The intercellular spaces of a leaf connect with the outside air only by way of the stomata. Thus the main function of the stomata is to permit the exchange of gases (carbon dioxide, oxygen, and water vapor) between the leaf and its environment. The size of the stomatal opening therefore influences the photosynthetic rate as well as the loss of water from plants (by **transpiration,** p. 360) by controlling the rate of gas exchange.

During the evening, and often during the warmest part of a summer day, the stomata are closed, resulting in a greatly decreased exchange of gases. During the first half of the day and in the late afternoon the stomata are usually open widest, allowing the maximum rate of gas exchange and therefore the highest rates of photosynthesis and transpiration.

The size of the stomatal aperture is primarily determined at any given time by the **turgor** (swelling) of the guard cells. The guard cell wall bordering on the stomatal pore is considerably thicker than the other portions of the cell wall (Fig. 13-9); an increase in turgor causes a greater expansion of the thinner portion of the cell wall than of the thicker portion, causing the latter to become concave. This effect is illustrated by inflating a cylindrical balloon that has been made thicker on one side by the use of rubber patching; it will assume a sausage shape. The same process in the guard cells forms the stomatal aperture.

The mechanisms that control the opening and closing of the stomata are complicated; we shall not discuss them here except to point out that light, CO_2 concentration, wind velocity, relative humidity, and temperature all influence the turgor of the guard cells and thus the size of the stomatal opening. For example, on exposure to light the osmotic concentration of the guard cells usually increases, promoting a net movement of water into the guard cells, an increased turgor, and an opening of the stomata.

Water Relationships

The major constituent of all active living cells on our planet is water. One of the important developments that has better adapted higher plants to a terrestrial existence has been the evolution of a vascular system, making possible the efficient transfer of water from the absorbing root to the upper parts of the plant.

Absorption of Water and Salts

The delicate root hairs and the adjacent epidermis are the actual absorbing structures of a plant, and water passes from them through the cortex and endodermis to the xylem, where it moves up the stem. Mineral absorption by roots is selective and essentially independent of water absorption; each mineral nutrient enters the root at its own rate. The root requires energy to function in this selective fashion, and it obtains this energy through respiration (pp. 83–95). The nutrients consumed in respiration are obtained from the above-ground photosynthetic portions of the plant, primarily the leaves, by **translocation** (the process by which organic materials in solution and suspension, synthesized for the most part in the leaves and to some extent in the stem and roots, are transferred to other parts of the plant, usually downward and chiefly by way of the sieve tubes of the phloem). The needed oxygen is obtained from the soil, which explains why plants must have adequately aerated soil for satisfactory growth.

The important factors involved in the absorption of salts and water and the ascent of sap in plants are summarized in Figure 13-13.

Transpiration

As plants absorb water from the soil, water is continuously evaporated from the upper parts of the plants (principally from the leaves) to the surrounding air, a process known as **transpiration.** Transpiration occurs almost entirely by way of the stomata. Water evaporates from the cells of the palisade and spongy layers into the intercellular air spaces of a leaf, where it collects in the gaseous form of water vapor. The water vapor concentration, also termed **vapor pressure** or **water potential,** is generally higher in the intercellular air spaces than in the outside atmosphere. A diffusion gradient is created, favoring a net movement of water outward through the stomata to the external air. (We discuss the significance of this gradient in the next section.) Because of the waxy, essentially waterproof cutin covering the epidermis, transpiration takes place only as long as the stomata remain open. Under normal circumstances of adequate water supply the rate of transpiration shows a typical daily pattern: it is greatest during the day and declines considerably, and often ceases, at night.

Transpiration is responsible for immense losses of water from higher plants and therefore indirectly from the soil. A single corn

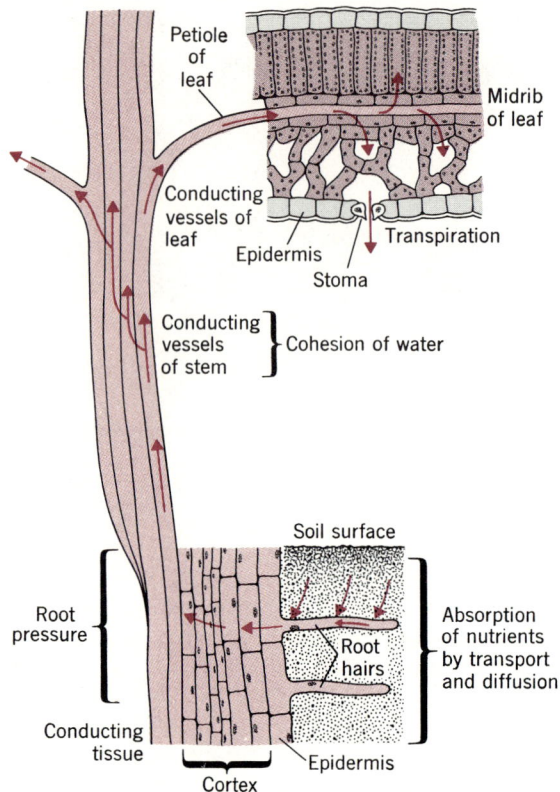

Figure 13-13 *Schematic representation of the entry of salts and water into a plant, and the ascent of sap by root pressure, transpiration, and cohesion.*

plant, for example, during its normal life span of three months or so, transpires as much as 50 gal of water, corresponding to a loss of more than 300,000 gal of water by an acre of corn plants. Consider such water losses in terms of great expanses of vegetation such as large forests; the process of transpiration may profoundly influence regional climates by affecting the moisture content of the air and therefore the temperature and rainfall of a given area.

When the rate of transpiration exceeds the rate of water absorption by the roots (often because of an inadequate water supply in the soil), **wilting** occurs, and may be followed by death of the plant. When the leaves begin to wilt the guard cells lose their turgidity, the stomata close, and transpiration ceases. In addition, such factors as water supply, wind velocity, humidity of the external air, and temperature influence the rate of transpiration.

The tremendous water loss from plants through transpiration does provide several benefits to plants. Because evaporation has a cooling effect, transpiration lowers the temperature of leaves on hot summer days. The process also aids in the upward movement of dissolved salts through the xylem once the nutrient salts have entered the roots, a mechanism that is described in the next section.

Ascent of Plant Sap

The rise of **sap**—water and dissolved substances—via the xylem to the tops of the very highest trees has been ascribed in part to transpiration. According to the **cohesion-tension theory,** transpiration creates a **pulling force** between the leaf cells and the water-conducting vascular tissue. The water in the xylem of leaves is continuous with that in the xylem of the trunk and root, particularly because water molecules are attracted to one another by strong cohesive forces, so that as water leaves the leaf by transpiration, water from the lower parts of the plant is drawn in to replace it. This assumed continuity of water between leaf and root and the pulling force of transpiration are presumed to be responsible for the lifting of a water column in the xylem of plants. These pulling forces have actually been measured in actively transpiring plants (and in model systems), and shown to be sufficient to raise water to the tops of the tallest trees.

Another force, called **root pressure,** is believed to be of secondary importance but also seems to contribute to the ascent of plant sap. When the stem of a plant is severed close to the ground, sap oozes from the stump, a visible demonstration of root pressure. Measurements of root pressure (Fig. 13-14) show that appreciable pressure sometimes can be developed, depending on the species, but never to magnitudes that alone could raise water as far as the tops of trees.

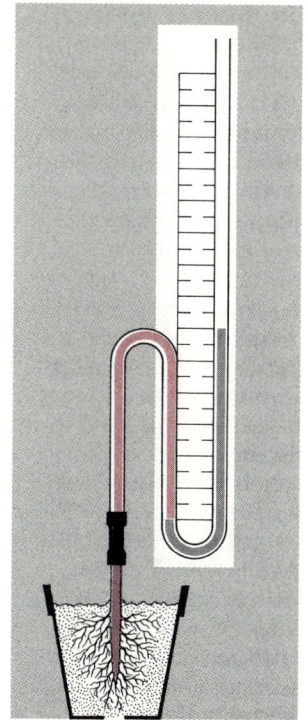

Figure 13-14 *The difference in height of mercury in the two arms of a manometer is a measure of root pressure.*

Photosynthesis

The over-all energy pattern of the biological world consists of (1) a transformation of the light energy of the sun by photosynthesis to the chemical form represented by the carbon-hydrogen bonds of organic compounds; and (2) its ultimate utilization and final dissipation by the activities of living cells. Sunlight therefore is the primary source of energy for nearly all living things.

Photosynthesis (see pp. 363–366) is the biological process by which green plants are able to utilize the energy of sunlight to convert carbon dioxide and water to energy-rich organic compounds such as carbohydrates and proteins. Molecular oxygen is liberated as one of the products.

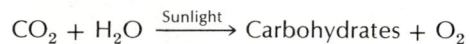

$$CO_2 + H_2O \xrightarrow{\text{Sunlight}} \text{Carbohydrates} + O_2$$

In all plants, with the exception of the blue-green algae, photosynthesis is confined to the chlorophyll-containing chloroplasts (p. 142). Most of the chloroplast-containing tissue of higher plants is located in the leaves. Studies with isolated chloroplasts have demonstrated that the entire process of photosynthesis can take place in the chloroplasts themselves.

About 80 percent of the total photosynthesis on our planet occurs in the oceans. Each year an estimated 150 billion tons of carbon dioxide and 60 billion tons of water enter into the photosynthetic process of our planet to produce about 110 billion tons of molecular oxygen and 100 billion tons of organic matter. Sizable fractions of this organic material are used in respiration and other life activities of the plants themselves; only a minute fraction is actually available for consumption as food by animals. By far the greatest portion of this material is decomposed, or decayed, to carbon dioxide, water, and mineral salts through the action of microorganisms.

The Nature of Light

White light is made up of a mixture of light of different colors (Fig. 13-15); each of the colors corresponds to a different wavelength of light. Visible light—light that we can detect with our eyes—is really only a small portion of the total range, or **spectrum,** of radiant energy.

In the visible part of the electromagnetic spectrum the longest wavelengths (about 700 nanometers) appear reddish, the shortest (about 400 nm), violet. Between these extremes are orange, yellow, green, blue, in descending order of length. Beyond the red end of the visible spectrum are the longer, invisible **infrared,** or heat, waves and the even longer **radio waves.** Beyond the violet end of the visible spectrum are the even shorter, invisible **ultraviolet** waves followed successively by **X-rays** and the **gamma rays** emitted by radioactive elements.

Figure 13-15 *Diagram of the radiant energy spectrum, showing the position of visible light and its resolution by a prism into its different component light wavelengths, or colors.*

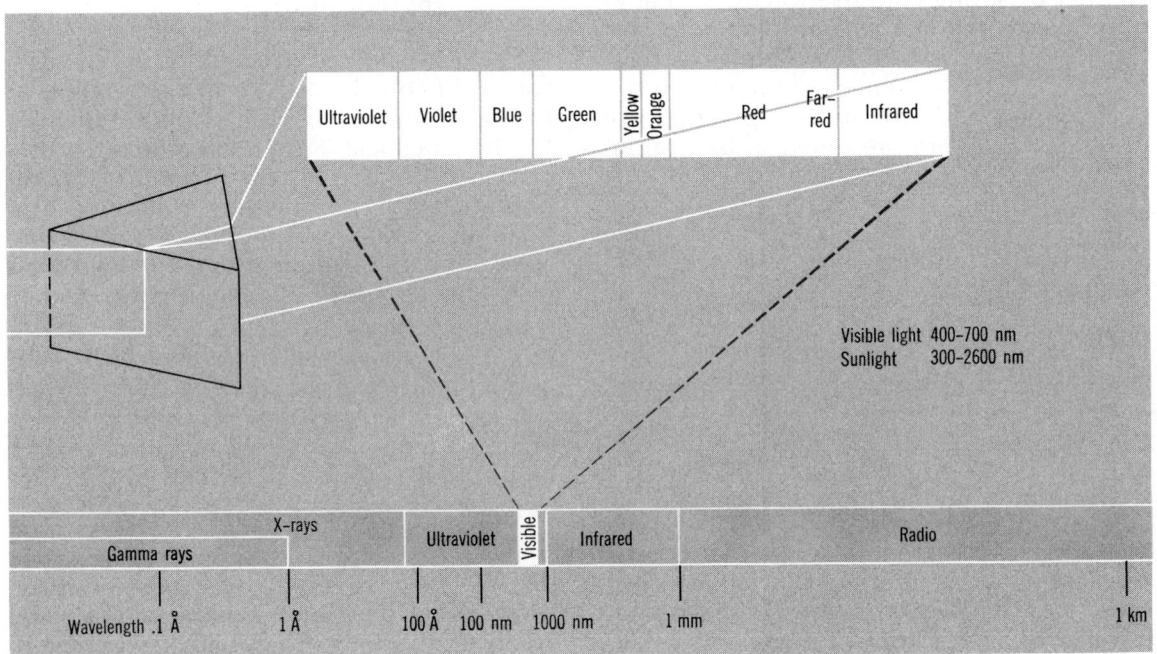

Radiant energy is propagated in packets of energy called **photons,** or **quanta.** The shorter the wavelength, the more energy associated with each photon. For example, an ultraviolet photon with a wavelength of 200 nm has five times the energy of a photon of infrared radiation with a wavelength of 1000 nm.

The Photosynthetic Pigments

The green pigments known as chlorophylls participate directly in photosynthesis by absorbing certain wavelengths of visible radiant energy; the most strongly absorbed wavelengths are red and blue.

Although the yellow and orange carotenoid pigments (carotenes and xanthophylls) universally present in chloroplasts, and some of the unique pigments associated with certain algae (e.g., phycocyanin of the blue-green algae and phycoerythrin of the red algae), have been implicated as light absorbers in the photosynthetic process, they are unable to substitute completely for chlorophyll. No plants have yet been found that can carry on photosynthesis without possessing at least one of the chlorophyll pigments.

Products of Photosynthesis

Molecular oxygen and carbohydrates are the major products of photosynthesis. The oxygen diffuses out of the photosynthesizing leaf cells into the intercellular spaces and then through the stomatal openings into the surrounding air; a portion may be used directly by the plant cells for respiration. During the day the average rate of photosynthesis by most plants far exceeds the rate of aerobic respiration, so that there is a net release of molecular oxygen to the atmosphere.

The most common carbohydrates that accumulate in leaf cells as a result of the photosynthetic process are starch and certain other polysaccharides. Frequently, nonphotosynthesizing colorless plant cells such as those of the potato tuber will accumulate carbohydrates, largely in the form of starch granules suspended in the cytoplasm. The starch is synthesized from sugars that were formed and transported from the leaves.

Biochemistry of Photosynthesis

The biochemical approach, especially at the enzymatic level, has proved to be a powerful tool in unraveling several basic features and detailed mechanisms of photosynthesis. In the early 1900s it was found that the photosynthetic process consisted of two successive events, a **light phase,** a series of rapid reactions dependent on light, followed by a slower series of non-light-requiring reactions called a **dark phase.** We now know, as a result of extensive biochemical investigations, that the photosynthetic reactions constituting the light phase provide both the reducing power (as NADPH; see p. 88) and energy (as ATP), and that the subsequently occurring dark phase consists of several enzymatic reactions by which carbon dioxide is transformed to carbohydrates.

The Light Phase. In 1937 the English biologist Robin Hill observed that cell-free suspensions of chloroplasts liberated oxygen in the presence of light, but without carbon dioxide. His work suggested that the molecular oxygen released in photosynthesis originated from water, because carbon dioxide, which is the other usual source of oxygen, was not needed. In order for this biochemical reaction (now known as the **Hill reaction**) to occur appreciably, a suitable receptor for hydrogens or electrons such as ferric ions (Fe^{+++}) must be added.

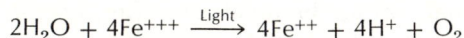

$$2H_2O + 4Fe^{+++} \xrightarrow{\text{Light}} 4Fe^{++} + 4H^+ + O_2$$

In the 1950s it was found that the coenzyme NADP could also serve as an acceptor of hydrogens or electrons.

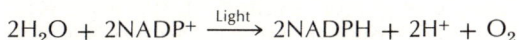

$$2H_2O + 2NADP^+ \xrightarrow{\text{Light}} 2NADPH + 2H^+ + O_2$$

This process does not occur in a single step, as the equation might suggest. Instead, it consists of a series of interconnected photosynthetic reactions. In 1954, in another highly significant discovery, it was found that, on exposure to light, isolated chloroplasts were also able to phosphorylate ADP to ATP. Since that time numerous studies have led to the following still tentative picture of the events comprising the light phase of photosynthesis.

Electron Flow and Energy Capture in Photosynthesis. The early or light phase of photosynthesis can currently best be explained on the basis of two kinds of functional systems called **photosystem I** and **photosystem II** operating in a connected sequential fashion. The photosynthetic pigments (primarily chlorophyll) of higher plants are not uniformly distributed in the chloroplast, but are organized, on the lamellar system of the organelle (p. 142), into many functional groups consisting fundamentally of two types corresponding to photosystems I and II. Each system possesses its own characteristic organization of light-absorbing pigments. Photosystem I responds to longer wavelengths of light and consists of several hundred molecules of chlorophyll *a,* carotenoids, and a **photoreactive center** (believed to be a specialized chlorophyll molecule) that is associated with a particular electron-transport chain. Photosystem II, activated by shorter wavelengths of light, is made up of several hundred molecules of both chlorophyll *a* and chlorophyll *b, c* or *d,* depending on the species, probably other light-absorbing pigments, and a photoreactive center connected to its own characteristic electron transport chain.

Photosynthesis is initiated by the absorption, or **capture,** of the energy of light by chlorophyll in the chloroplasts. This is the primary photochemical act of photosynthesis (all other subsequent reactions of the photosynthetic process can take place in the dark). As a result, an electron of each chlorophyll molecule of the illuminated chloroplasts is converted from its normal energy level to a higher-energy, or **excited** state. An excited chlorophyll may transfer its added energy to a neighboring chlorophyll molecule, thus exciting it, and in this way the extra energy can migrate in its own photosystem until it reaches the photoreactive center chlorophyll and causes the latter to emit a **high-energy electron.** This electron is then taken up by the electron transport pathways associated with both photosystems. Once in the electron transport chain, the high-energy electrons follow a path similar to that found in mitochondria: each of the carriers undergoes an oxidation-reduction reaction in sequence as the electrons drop from a relatively high energy level to a lower level. As we shall soon see, it is during the course of this electron flow that both reducing power (as NADPH) and useful energy (as ATP) are produced to be ultimately used for the subsequent dark phase conversion of carbon dioxide to carbohydrates.

Both photosystems are believed to function in photosynthesis in an integrated sequence with one another as shown in Figure 13-16. When photosystem I is activated by light, the high-energy electrons that it generates are passed along from its photoreactive center to an iron-containing protein called **ferredoxin.** The electrons of ferredoxin in its reduced state are then enzymatically transferred to NADP, reducing it to NADPH, with the needed hydrogen ion being supplied by water.

The electrons thus lost by photosystem I may be replenished in two ways. The first is by a process that is postulated to occur by way of photosystem II. The high-energy electrons produced by light from the photoreactive center of photosystem II, as in the case of photosystem I, also flow downhill (in an energy sense) along an electron transfer pathway containing cytochromes and other carriers to the photosystem I chlorophyll molecule that has lost its electron to ferredoxin.

The occurence of ATP formation from ADP and inorganic phosphate during electron flow through this chain is called **photophosphorylation** and resembles the oxidative phosphorylation that accompanies the respiratory (cytochrome) chain in mitochondria (Chapter 4). The electrons lost from photosystem II are restored by a transfer of electrons from water by a process that also results in the production of oxygen. Thus in the above postulated connected sequence of photosystems I and II, the path of electron flow may begin with water and photosystem II, proceed through photosystem I and end at the point of synthesis of NADPH. These steps collectively form the so-called **noncyclic electron transport pathway.**

The second process for replenishing the depleted electrons of the photoreactive center of photosystem I occurs by a flow of the high-energy electrons, not to NADP, but back to the photoreactive center chlorophyll via certain cytochrome carriers, to constitute the **cyclic electron transport pathway** which is also accompanied by ATP production (i.e., photophosphorylation).

In summary, the light phase of photosynthesis is currently conceived to consist of two types of interconnected light reactions, one of which is primarily responsible for the formation of NADPH and the other for the production of oxygen. Both systems appear to be involved in the formation of ATP. The production of NADPH is always the result of noncyclic electron transport, but ATP can also be formed from either noncyclic electron transport (accompanied by noncyclic photophosphorylation) or from cyclic electron transport (accompanied by cyclic photophosphorylation). The relative amount of energy transduced by each of the two paths is under the over-all metabolic control of the chloroplast. The light energy thus harnessed can be used in the subsequent dark-phase transformation of carbon dioxide to carbohydrates.

The Dark Phase. A sequence of enzymatic reactions that appears to be a principal route for the conversion of carbon dioxide to carbohydrates and other cellular components was discovered during the 1950s by Melvin Calvin and his associates at the University of California. For this work Calvin was honored with the Nobel Prize in 1962. Radioactive carbon dioxide was an important tool for the study of this aspect of photosynthesis. Green plants were exposed to radioactive carbon dioxide in the light and labeled products subsequently formed in the cell were identified. This technique provided important clues as to the intermediates and reactions constituting the pathway of the dark phase.

When the green alga *Chlorella* was exposed to light in the presence of labeled carbon dioxide, appreciable photosynthesis

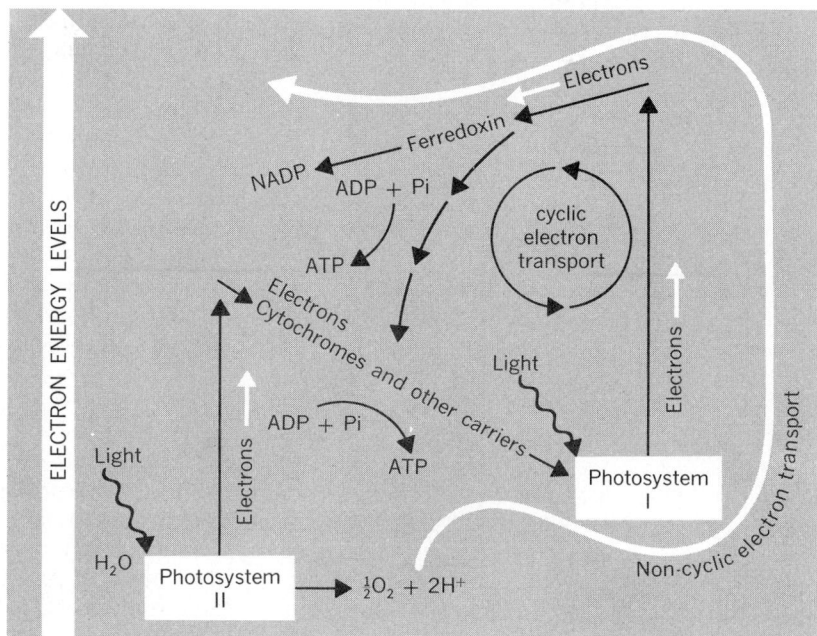

Figure 13-16 A diagrammatic summary of photosystems I and II of the light phase of photosynthesis.

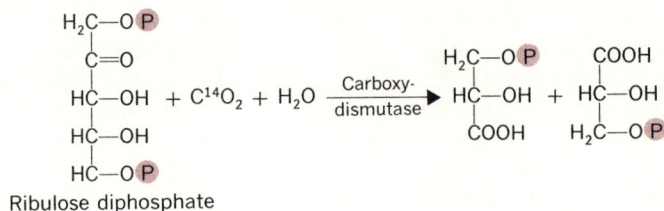

Figure 13-17 *Splitting of ribulose diphosphate by the addition of CO_2 to give two molecules of phosphoglyceric acid.*

occurred. A high concentration of radioactivity first appeared in phosphoglyceric acid, the same substance that serves as an important intermediate in the glycolysis pathway (p. 87). It is not difficult to conceive that by a reversal of the glycolysis pathway, starting with glyceric acid (and NADPH + ATP furnished by the light phase of photosynthesis), carbohydrates such as glucose could be formed. The following enzymatic steps illustrate how this appears to happen.

1. A key step involves the addition of carbon dioxide to a five-carbon compound called **ribulose diphosphate,** by the action of the enzyme carboxy dismutase, to form two molecules of **phosphoglyceric acid** (Fig. 13-17).
2. The phosphoglyceric acid arising from this reaction is converted to the corresponding

phosphoglyceraldehyde by a reaction that requires energy (ATP) and reducing power (NADPH) originally provided by the light phase of photosynthesis (Figure 13-18). It apparently proceeds by reversal of the corresponding step in glycolysis.
3. The phosphoglyceraldehyde thus produced is really the end product of photosynthesis, and has one or more of three possible fates: it may be used as a source of respiratory fuel; or be converted into glucose and eventually into di- and polysaccharides; or undergo a series of enzymatic steps referred to as the **pentose phosphate pathway,** which maintains the supply of ribulose diphosphate, thus illustrating the cyclic nature of the dark phase of photosynthesis (Fig. 13-18).

Several years ago two workers from Australia, M. D. Hatch and C. R. Slack, discovered an alternative route for carbon dioxide conversion that operates in some plants, including corn and a number of grasses. In this route, now referred to as the **Hatch-Slack pathway,** CO_2 combines with a three-carbon compound to form the four-carbon **oxalacetic acid,** which is also an intermediate in the Kreb's cycle (p. 91). This sequence is now the object of intense investigation in laboratories throughout the world.

Plant Hormones

Plant hormones, like animal hormones, are produced in one part of the organism and transmitted to another part. They act in extremely small quantities and influence the physiological processes of organs and tissues. Unlike animals, plants do not produce their hormones in special glands, but in rapidly

Figure 13-18 *Summary of the dark phase of photosynthesis.*

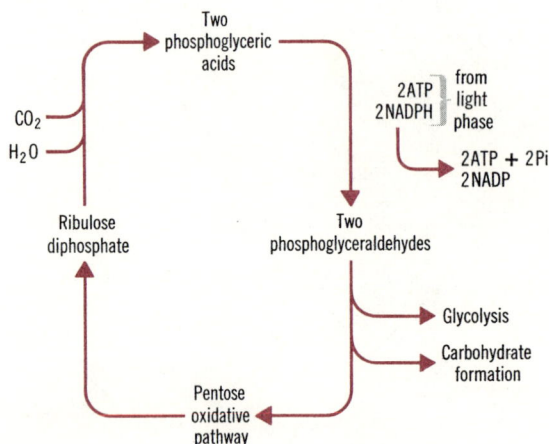

growing embryonic tissues or in cells of leaves.

The plant hormones that function in the regulation of plant growth are collectively known as the **growth hormones,** or **growth regulators.** We discuss two types, the **auxins** and the **gibberellins,** and briefly note two others, the **kinins** and **abscissic acid.**

Auxins

The auxins are synthesized by young, physiologically active cells, especially by the meristematic or growing apices of stems and roots, including buds, young leaves, and developing flowers and fruits. They are usually transported away from the parts of the plant where they are produced toward the zone of elongation and the more basal parts. Although several auxins have been isolated from the plant tissues and chemically identified, the most widely distributed of these, and probably the most important, is **indoleacetic acid** (Fig. 13-19).

Indoleacetic acid **(IAA)** influences several cellular processes including the initiation, enhancement, and regulation of cell division. It also stimulates cell elongation in growing stems and roots, apparently by affecting the deposition of cellulose and increasing the plasticity of cell walls, thus allowing for a greater absorption of water by the plant cells. If the tip of a young growing stem is severed, the elongation of cells in the remaining portion slows dramatically, and the growth of the stem soon ceases. If the tip is replaced, stem growth is largely restored for at least several hours. The hormone is also required for cell division to occur in the tissue culture of plant cells. IAA can suppress the develop-

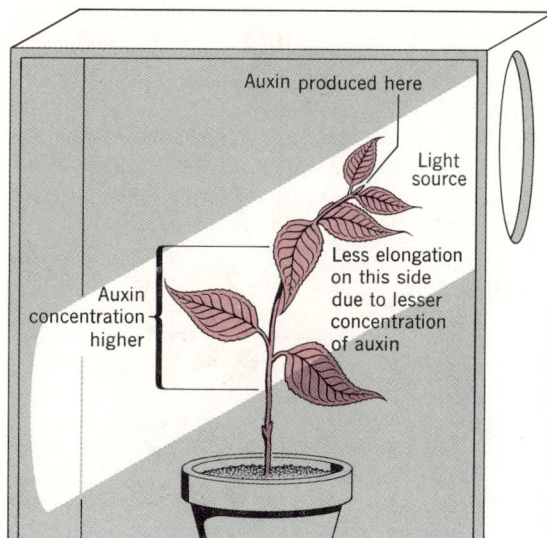

Figure 13-20   *Demonstration of phototropism. The side of the stem exposed to the light has less auxin than the opposite side, accounting for the stem curvature.*

ment of lateral bud primordia in the growing stem (Fig. 13-5), and this is a controlling factor in the development of apical dominance.

Tropisms. The direction of growth movements or curvature of plants in response to certain environmental factors are called **tropisms.** They are attributed to a differential distribution of auxins in the plant tissues concerned. The growth of a plant stem toward a light source, as in Figure 13-20, is called **phototropism.** It apparently results form uneven distribution of auxin in the stem; the side of the stem away from the light possesses significantly more auxin than the side of the stem exposed to the light and therefore has a greater rate of growth, leading to the observed curvature.

Similarly, the eventual upward curvature of the stem after a plant has been placed on its side in the dark is ascribed to a greater auxin concentration on the lower side than on the upper (Fig. 13-21, p. 368). The resulting greater rate of growth on the lower side of the stem accounts for the observed upward curvature, or **geotropism.** The opposite cur-

Figure 13-19   *Indoleacetic acid.*

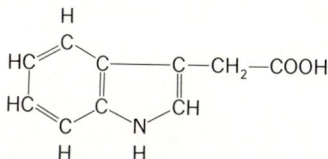

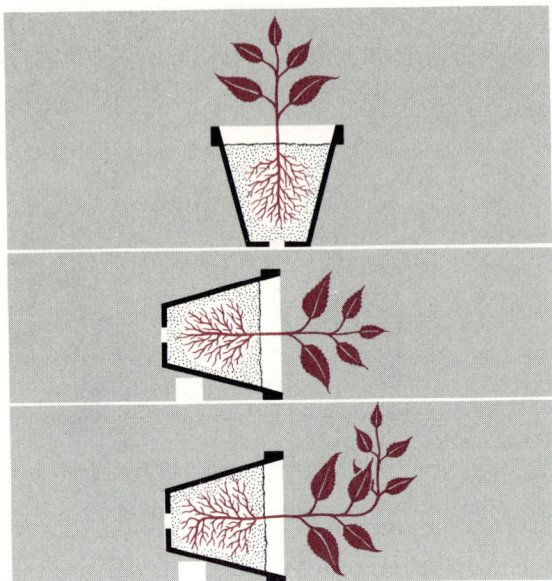

Figure 13-21 *Demonstration of geotropism. Concentrations of auxin on the lower side of the stem stimulate its growth; similar concentrations on the lower side of the root inhibit root growth.*

vature of the root—its growth downward—presumably results from an inhibitory effect of auxin on the lower side of the roots; roots are inhibited by high auxin concentrations that still can stimulate stem growth.

Abscission. A decrease in auxins is responsible in part for the fall of leaves, flowers, and fruits, a process known as **abscission.** In autumn the diminishing auxin supply from the leaf blade to the leaf petiole results in the formation at the base of the petiole of a zone of parenchyma known as an **abscission layer.** Its thin cell walls finally separate, and the structure of the petiole is so weakened that the leaves drop from the plant. The process may also be stimulated by the hormone abscissic acid.

At times it is desirable to delay the fall of premature fruit, as in apple and orange trees. Spraying of young fruits with a solution of an appropriate auxinlike synthetic growth regulator prevents their drop until they are ready for harvest. In contrast, by spraying

with a suitable chemical that has the opposite effect of auxin, leaves and fruit can be made to fall off prematurely when desirable (as in preparation for the mechanical harvesting of cotton).

Gibberellins

Gibberellins are a group of organic substances originally discovered and isolated by Japanese research workers from an ascomycete fungus, *Gibberella fujikuroi,* that infects rice seedlings. The fact that the rice seedlings infected by the fungus tended to be extremely elongated was one of the initial observations that led to the eventual discovery of the gibberellins. Gibberellic acid (Fig. 13-22), representative of this group of substances, has a powerful growth-regulating effect when applied to some higher plants. It causes dwarfed plants to grow much higher, the heights directly in proportion to the concentration of the hormone applied (Fig. 13-23).

Gibberellins also influence a wide variety of other processes in plants and in some cases have been shown, for example, to stimulate flower initiation, inhibit lateral root initiation, and promote cell expansion. Since their discovery in fungi, gibberellins have been found in a wide variety of higher plants, and there is little doubt that they are important natural growth regulators.

Kinins

Kinins are a class of growth-promoting substances, found in coconut milk and yeast extract, that greatly enhance the growth of numerous plant cells in tissue culture. One member of this class is **kinetin,** a compound

Figure 13-22 *Gibberellic acid.*

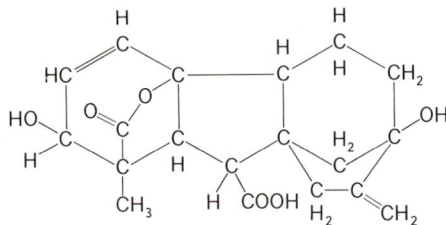

Figure 13-23   Dwarf peas (Pisum sativum) *show the effect of gibberellic acid, which was added to each plant in varying amounts seven days before the photograph was taken.*

related to the purines occurring in nucleic acids. As far as we know, kinetin is not a naturally occurring substance, but is found in aged or somewhat decomposed nucleic acid preparations or in fresh nucleic acid preparations that have been boiled. When kinetin is applied to plant cells such as those of roots grown in tissue culture, it appears to act with auxin to produce a striking increase in growth or division of the cells. This effect possibly reflects the presence in whole plants of a naturally occurring kinin that normally interacts with auxin.

Abscisic acid is a recently isolated hormone, occurring in a wide variety of plants, that promotes absicission of leaves and fruits. It also induces dormancy of buds, inhibits seed germination, and plays some part in flowering. In general, auxins, gibberellins, and kinins stimulate growth, whereas abscisic acid retards it.

Photoperiodism

A number of experimental studies clearly point to the production of a diffusible chemical agent or hormone, tentatively called **florigen,** that somehow initiates flowering. All attempts so far to isolate and identify this chemical substance and its mechanism of action have been futile. We do know, however, that there is a close relationship between the length of daily exposure of a plant to light and the onset of flowering, a phenomenon known as **photoperiodism.** Moreover, experimental results suggest that plants exposed to a suitable photoperiod produce in their leaves the unidentified substance called florigen, which is then transmitted to the flower buds, where it initiates flowering.

On the basis of the length of their photoperiod, all angiosperms can be divided into three groups: (1) **short-day plants** (e.g., asters, dahlias, poinsettias, violets), which produce flowers only when their daily photoperiod is shorter than a certain critical period that varies with the species (cockleburrs, e.g., flower only when the day length is about 15 hours or less); (2) **long-day plants,** which produce flowers only when their photoperiod is longer than a certain critical period (e.g., spinach flowers only when the day length is 13 hours or more); and (3) **indeterminate,** or **day-neutral, plants** (e.g., sunflower, tomato, dandelion, carnation), which flower independently of the photoperiod. Short-day plants usually flower in early spring or late summer or fall; long-day plants flower in late spring and early summer. Both types can be induced to flower at any time by subjecting them to the proper artificial light periods in a greenhouse; such practices are widely used in the commercial flower-raising industry.

In some plants, at least, the photoperiodic response (i.e., initiation of flowering) in reality is not controlled by the length of the light period so much as by the length of the **dark period.** For instance, it has been shown that interruption of the dark period, even by a momentary flash of light, prevents initiation of flowering. Soybean plants, for example, soon flower when grown under conditions of 16 hours of darkness and eight hours of light. Flowering is suppressed, however, if the routine 16-hour periods of darkness are interrupted by a moderately intense light for a few seconds. Thus short-day plants are in reality long-night plants, requiring a minimal uninterrupted dark period; long-day plants are short-night plants, requiring a dark period whose duration does not exceed a particular maximum.

The fact that flowering can be suppressed by a small amount of light implies involvement of a light-sensitive pigment. We now know from studies of the flowering response at various wavelengths that interruption of the long dark period of a short-day plant by red light (in the wavelength region of 660 nm) is most effective in inhibiting the flowering response. Interestingly, if the plants are subsequently exposed to a flash of longer-wavelength red light, called **far-red,** (about 730 nm) after a flash of red, flowering is stimulated (Fig. 13-24).

The pigment responsible for this effect is called **phytochrome,** and exists in two forms: a red-absorbing form, and a far-red-absorbing form, which are interchanged by light. When plants are exposed to red light or to sunlight, the red-absorbing pigment (designated P_{660}) is converted to the far-red-absorbing form (P_{730}). Darkness (or exposure to far-red light) favors the reverse reaction, as shown below.

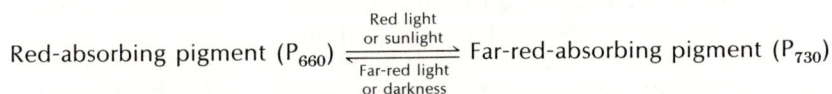

$$\text{Red-absorbing pigment } (P_{660}) \underset{\substack{\text{Far-red light} \\ \text{or darkness}}}{\overset{\substack{\text{Red light} \\ \text{or sunlight}}}{\rightleftharpoons}} \text{Far-red-absorbing pigment } (P_{730})$$

Figure 13-24 *Flowering response in the kalanchoe plant: (a) grown under conditions of an uninterrupted dark period; (b) same conditions as in (a), but with a one-minute interruption with red light at midnight; (c) same conditions as in (b), but the red-light interruption was followed by one minute of exposure to far-red light, which nullifies the action of red light.*

(a)   (b)   (c)

It is the longer-wavelength red in sunlight that is the controlling factor necessary for flowering. For flowering to occur in short-day plants, then, the period of darkness must be sufficiently long to decrease the far-red-absorbing pigment (P_{730}) to a low level by conversion to the red-absorbing form (P_{660}), and/or by destruction that occurs in the dark, and to maintain this low level for some hours.

A similar antagonistic action of red and far-red light is also observed in such phenomena as leaf expansion, stem elongation, and seed germination, suggesting phytochrome involvement.

Reading List

Clayton, R. K., *Light and Living Matter. Vol. 2: The Biological Part.* McGraw-Hill, New York, 1971.

Cutter, E. G., *Plant Anatomy: Experiment and Interpretation.* Addison-Wesley, Reading, Mass., 1969.

Echlin, P., "Pollen," *Scientific American* (April 1969), pp. 80–93.

Esau, K., *Plant Anatomy* (2nd ed.). Wiley, New York, 1965.

Galston, A. W., and P. J. Davies, *Control Mechanisms in Plant Development.* Prentice-Hall, Englewood Cliffs, N.J., 1970.

Greulach, V. A., and J. E. Adams, *Plants—An Introduction to Modern Botany* (2nd ed.). Wiley, New York, 1967.

Hendricks, S. B., "How Light Interacts with Living Matter," *Scientific American* (September 1969), pp. 174–186.

Ledbetter, M. C., and K. R. Porter, *Introduction to the Fine Structure of Plant Cells.* Springer, New York, 1970.

Levine, R. P., "The Mechanism of Photosynthesis," *Scientific American* (December 1969), pp. 58–70.

Rabinowitch, E., and Govindjee, *Photosynthesis.* Wiley, New York, 1969.

Raven, P. H., and H. Curtis, *Biology of Plants.* Worth, New York, 1970.

Salisbury, F. B., and C. Ross, *Plant Physiology.* Wadsworth, Belmont, Calif., 1969.

Torrey, J. G., *Development in Flowering Plants.* Macmillan, New York, 1967.

Van Overbeek, J., "The Control of Plant Growth," *Scientific American* (July 1969), pp. 75–81.

Weier, T. E., C. R. Stocking, and M. G. Barbour, *Botany, An Introduction to Plant Biology* (4th ed.). Wiley, New York, 1970.

"What has the organism gained by the constancy of temperature, constancy of hydrogen ion concentration, constancy of water, constancy of sugar, constancy of oxygen, constancy of calcium, and the rest? . . . How often have I watched the ripples on the surface of a still lake made by a passing boat, noted their regularity and admired the patterns formed when two such ripple-systems meet . . . but the lake must be perfectly calm. . . . To look for high intellectual development in a milieu whose properties have not become stabilized is to seek . . . ripple patterns on the surface of the stormy Atlantic. . . . For the mammals all, homeostasis was survival; for man, emancipation."

W. GREG WALTER, QUOTING SIR JOSEPH BARCROFT, 1953,
THE LIVING BRAIN

part seven
Functions of a Complex Animal: Man

Maintenance of the Body: Digestion, Nutrition, and Protection

"He may live without books—what is knowledge but grieving? He may live without hope—what is hope but deceiving? He may live without love—what is passion but pining? But where is the man who can live without dining?"

OWEN MEREDITH, 1873, LUCILE

Let us begin this chapter by recalling that man is a multicellular organism, a collection of cells. The primary requirement of any organism is to supply each of its cells with the materials it requires. For most cells this includes water, a group of essential nutrients such as salts, a supply of amino acids, a source of carbon chains, and oxygen for the energy-conversion reactions. A second, equally important requirement is that cells must be able to eliminate the waste products from these reactions, especially excess carbon dioxide, nitrogenous compounds produced in the breakdown of amino acids, and excess water.

For a single-celled or small colonial organism that lives in a watery environment, the problem of transporting materials into and out of itself is not difficult to solve. Nutrients and gases dissolved in liquid can simply diffuse or be actively transported across the cell membrane (Chapter 5). The inside of the animal or plant is simply the volume enclosed by its cell membrane. But as organisms have become larger and more complex the problem of servicing each cell has gotten more complicated. Nearly all metazoan organisms have had to overcome this common problem of how to deal with the new space formed when they became multicellular; that is, with the volume that is outside the cells but inside the organism (Fig. 14-1, p. 376). This is the space we call the **internal environment.**

"The Broad Jumper," an 1884 photograph by Thomas Eakins.

375

1 4-cm cube
(96 cm²)

8 2-cm cubes
(192 cm²)

64 1-cm cubes
(384 cm²)

(a)   (b)   (c)

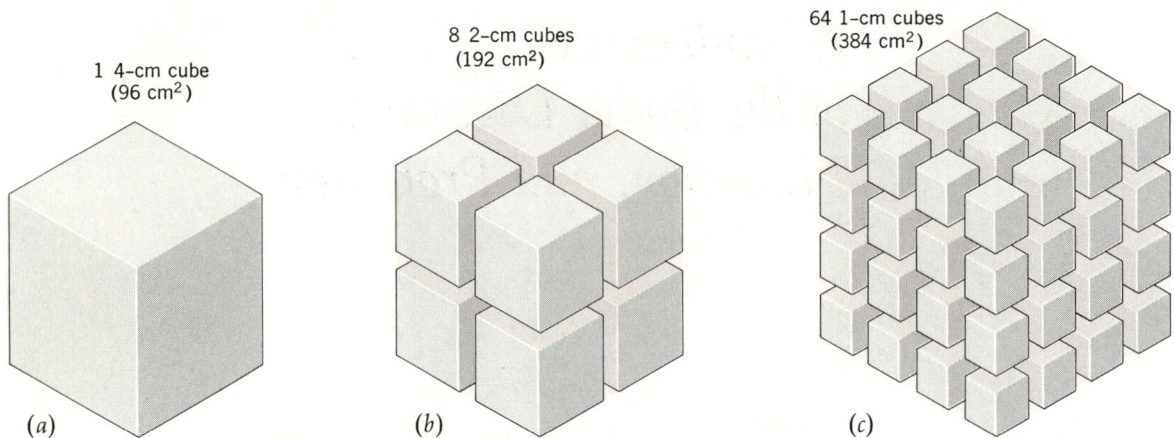

Figure 14-1   The total volume enclosed within cubes is the same in (a), (b), and (c). But the total surface area of (b) is twice as great as (a), and that of (c) is four times greater than (a). Moreover, all of the surface in (a) is exposed to the environment. In (b) one-half, and in (c) three-fourths of the surface (the colored area) faces the internal environment.

The Human Digestive System

Most organic matter exists as components of organisms or as large, complex molecules that are often insoluble in water. **Digestion** is the process whereby enzymes break down nutrient materials until they are sufficiently small and soluble to pass into a cell and be used in its various metabolic reactions for the release of energy. The process may be summarized as follows:

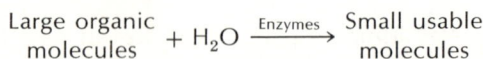

$$\text{Large organic molecules} + H_2O \xrightarrow{\text{Enzymes}} \text{Small usable molecules}$$

In protozoa, sponges, and most plants food material is digested by enzymes that the organism secretes into a vacuole inside the cell (Fig. 14-2). In metazoan animals digestion takes place outside the cells; digestive enzymes are secreted by cells into a specialized extracellular space called a **digestive cavity.** In the long course of evolution man has been endowed with a highly specialized and complex system of organs and tissues known as a **digestive** or **alimentary system,** which converts food into a physical and chemical state suitable for proper absorption into the circulatory system and utilization by the innumerable cells of the body. The **alimentary tract,** or **canal,** in man can be conveniently divided into the mouth, pharynx, esophagus, stomach, small intestine and associated glands (pancreas and liver), large intestine, and rectum. The various parts of man's digestive system are shown in Figure 14-3 (p. 378), and the principal digestive enzymes are outlined in Table 14-1 (p. 380).

The Mouth

Food enters the digestive tract in the mouth. The teeth and the action of the tongue reduce food into a soft pulpy mass in preparation for swallowing. This process is called **mastication.** In mammals such as man a tooth is typically divided into three regions: the **crown,** which protrudes into the mouth cavity above the gum; the **neck,** which is surrounded by the gum; and the **root,** which is firmly embedded in the jawbone (Fig. 14-4, p. 379). The outer layer of the crown is an

Figure 14-2 An ameba flows around food particles to ingest them by phagocytosis. Enzymes are secreted into the newly formed food vacuole within the cell to digest the particles.

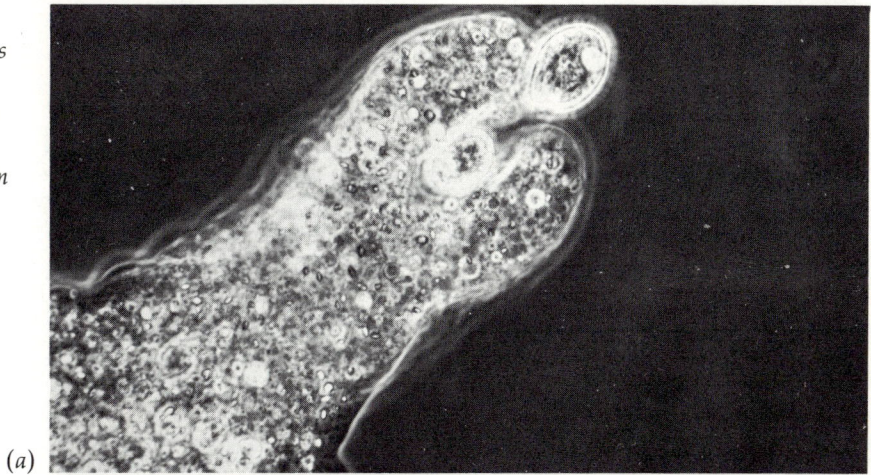

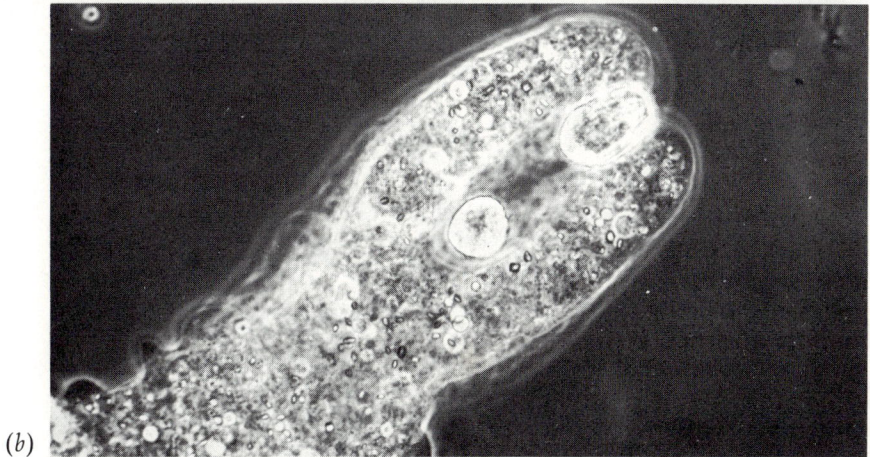

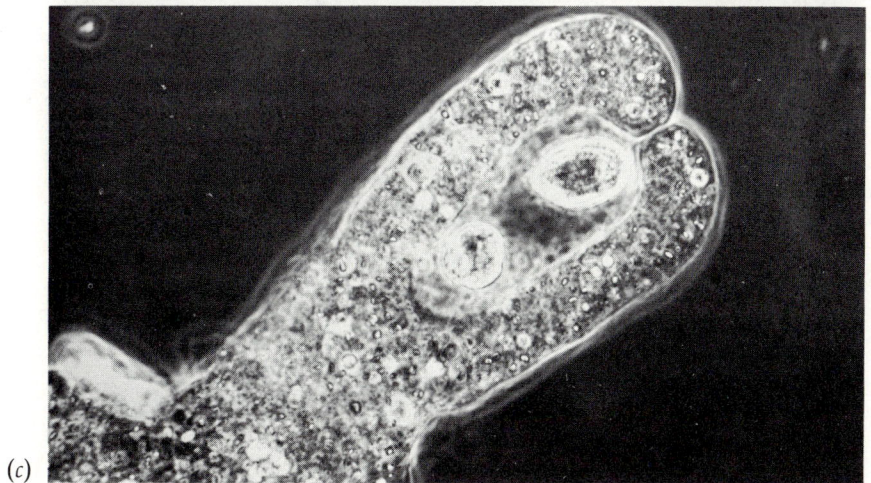

(a)

(b)

(c)

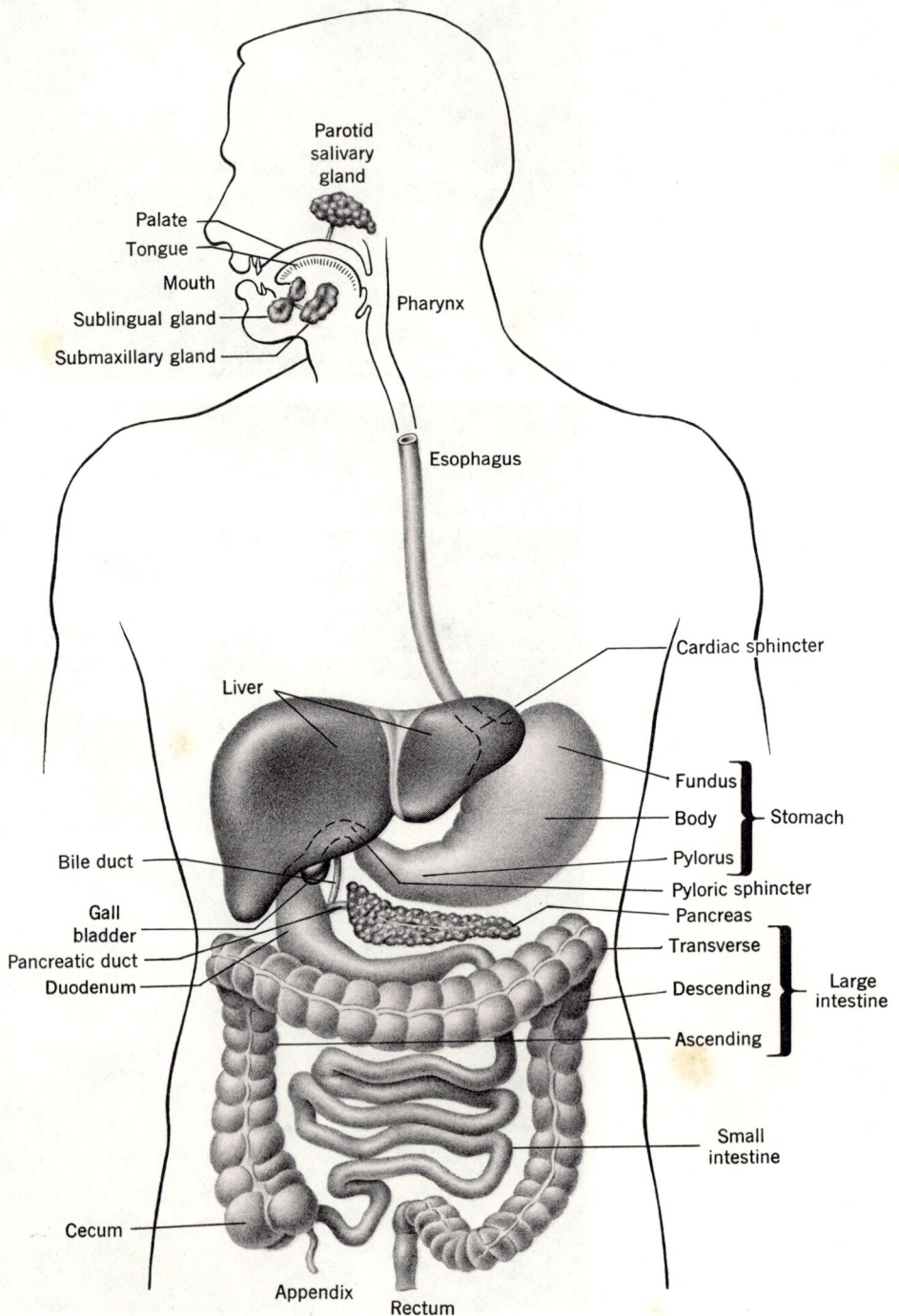

Figure 14-3 *The digestive system of man.*

Parotid salivary gland

Palate

Tongue

Mouth

Sublingual gland

Submaxillary gland

Pharynx

Esophagus

Cardiac sphincter

Liver

Fundus

Body] Stomach

Pylorus

Pyloric sphincter

Bile duct

Pancreas

Gall bladder

Transverse

Pancreatic duct

Descending] Large intestine

Duodenum

Ascending

Small intestine

Cecum

Appendix

Rectum

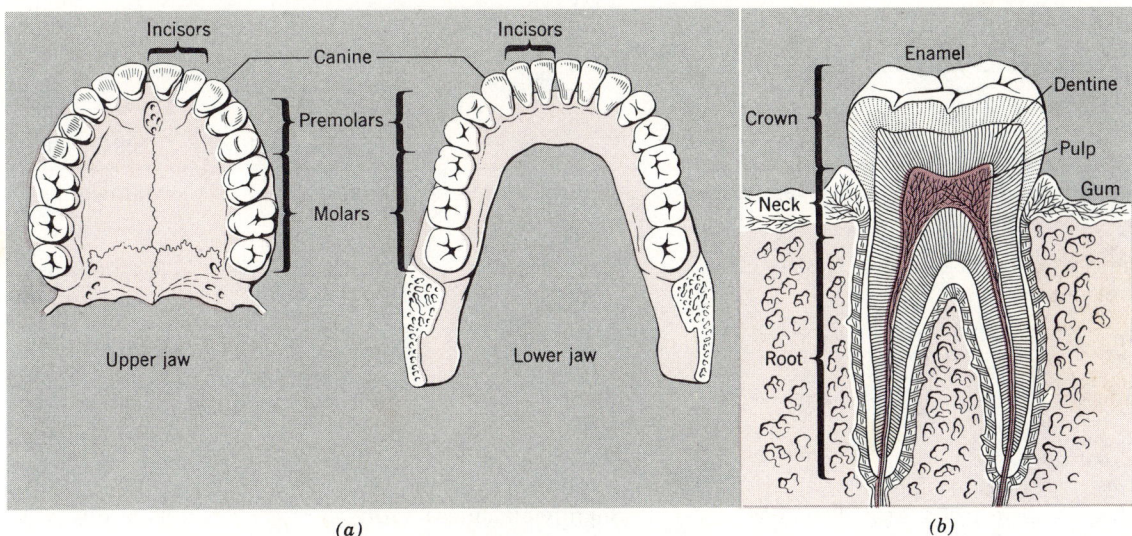

Figure 14-4   The teeth of man. (a) A view of the upper and lower jaws; each half of each jaw has eight teeth, a total of 32. (b) A section through a molar tooth.

exceedingly hard substance, called **enamel,** that consists largely of crystals of calcium apatite, a complex of calcium phosphate. The bulk of the tooth lies beneath the enamel. A thick layer called **dentine,** whose composition is somewhat similar to bone, surrounds the interior of the tooth, called the **pulp,** which is made up of connective tissue, blood vessels, and nerves. The tooth is fastened to the jawbone by a bone-like cement on the outer surface of the dentine. Adult human beings have 32 teeth; each half of the upper and lower jaw has (from front to rear) two chisel-shaped teeth **(incisors)** for cutting food, one pointed, conical tooth **(canine)** for tearing food, and two **premolars** and three **molars** with grinding surfaces for crushing and grinding food. The rearmost molars are called **wisdom teeth,** and are usually nonfunctional.

A **hard** and a **soft palate** make up the roof of the mouth. The hard palate, located immediately behind the upper lip, consists of several bones. The soft palate is behind the hard palate and is made of muscle in the shape of an arch that separates the nasal cavities from the throat, or pharynx.

The **tongue** projects from the floor of the mouth. Its major digestive functions are to manipulate food for chewing and to shape the chewed food into a spherical mass that it pushes into the pharynx to initiate swallowing. It is covered with an epithelial lining and is made of several sets of skeletal muscle (p. 502) oriented in different directions, which accounts for its ability to move in different planes. The tongue is richly supplied with sensory nerves embedded in its surface, which serve the sense organs of taste and touch (Chapter 17).

In the mouth food is mixed with **saliva,** which is secreted from three pairs of **salivary glands** called the **parotid,** the **sublingual,** and the **submaxillary glands.** These are clusters of cells whose ducts open into the mouth cavity. Saliva consists of thin, watery and thicker, viscous mucous fluids that serve principally to moisten and lubricate ingested food. In humans and most other mammals it also contains a starch-digesting enzyme called

table 14-1

Principal Digestive Enzymes in Man

ENZYME	GLAND	OPTIMUM pH	SUBSTRATE	PRODUCT
Salivary amylase	Salivary glands	Slightly acid (pH 6.4–6.8)	Starch	Maltose
Pepsin (secreted as pepsinogen)	Gastric glands	Very acid (pH 2)	Proteins	Polypeptides and peptides
Amylase	Pancreas	Slightly alkaline (pH 7–8)	Starch	Maltose
Lipase	Pancreas	Slightly alkaline (pH 7–8)	Fats	Glycerol and fatty acids
Trypsin (secreted as trypsinogen)	Pancreas	Slightly alkaline (ph 7–8)	Proteins, polypeptides, peptides	Peptides and free amino acids
Chymotrypsin (secreted as chymotrypsinogen)	Pancreas	Slightly alkaline (pH 7–8)		
Carboxypeptidase (secreted as procarboxypeptidase)	Pancreas	Slightly alkaline (pH 7–8)		
Enterokinase	Intestinal glands	Slightly alkaline (pH 7.5–8.5)	Trypsinogen	Trypsin
Peptidases	Intestinal glands	Slightly alkaline (pH 7.5–8.5)	Peptides	Free amino acids
Maltase	Intestinal glands	Slightly alkaline (pH 7.5–8.5)	Maltose	Glucose
Sucrase	Intestinal glands	Slightly alkaline (pH 7.5–8.5)	Sucrose	Glucose and fructose
Lactase	Intestinal glands	Slightly alkaline (pH 7.5–8.5)	Lactose	Glucose and galactose
Lipase	Intestinal glands	Slightly alkaline (pH 7.5–8.5)	Fats	Fatty acids and glycerol
Nucleases	Intestinal glands	Slightly alkaline (pH 7.5–8.5)	Nucleic acids (RNA and DNA)	Nucleotides, nucleosides, and bases

salivary amylase, which catalyzes the hydrolysis of starch into the disaccharide maltose (Chapter 3). The parotid glands lie below and in front of the ear and secrete only the thin, watery type of saliva. In the virus infection mumps these glands become painfully inflamed and swollen. The sublingual glands are located beneath the tongue, toward the front of the mouth cavity, and secrete only the mucous type of saliva. The submaxillary glands are also situated beneath the tongue, but toward the rear of the mouth cavity at the angle of the lower jaw. They secrete both types of saliva. Human saliva also contains mucous secreted by small glands scattered over the surface of the mouth cavity. About 1200 to 1500 ml (about 1½ qt) of saliva are secreted by the average adult in 24 hours.

The Pharynx, or Throat

The region of the alimentary canal between the mouth and the food pipe (esophagus) is called the **pharynx,** or throat. It is also the crossing-over point between the digestive and respiratory tracts, serving as a common passageway for the transmission of food from mouth to esophagus and air to trachea and lungs. The general structure of the pharynx is described on pp. 424–425.

The pharynx plays an important role in swallowing, an involuntary phenomenon in which the muscular walls of the pharynx contract, squeezing a mass of food and forcing it downward into the esophagus. At the same time other involuntary mechanical movements take place to assure that the food will not move into the nose or larynx, back into the mouth, or into the trachea and lungs: the soft palate is raised, preventing the food from entering the nose; a trapdoor-like flap at the back of the mouth, called the **epiglottis,** closes over the entrance to the air passageways; and the base of the tongue and the muscular walls surrounding the entrance to the pharynx block the movement of the food mass back into the mouth.

The Esophagus

The **esophagus** in man is a collapsible muscular tube about 10 in. long extending down from the pharynx between the lungs, behind the heart and trachea, and through the diaphragm directly to the stomach. Food is carried through the esophagus by **peristalsis,** a series of rhythmical, ringlike contractions of the muscular walls that travel downward, sweeping before them any food contained in the tube (Fig. 14-5). Peristalsis also occurs in the stomach and intestines and is, in fact, a characteristic action of most hollow muscular organs of the body.

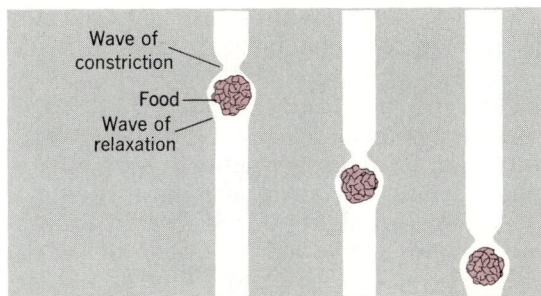

Figure 14-5   Peristalsis in the digestive tube. A wave of muscular contraction (constriction) pushes a food mass ahead of it.

The site at which the esophagus joins and empties into the stomach is called the **cardiac sphincter.** This muscular valve is ordinarily contracted to close off the stomach from the esophagus and to prevent regurgitation of the stomach contents into the esophagus. As part of the swallowing reflex the cardiac sphincter relaxes and opens, allowing food to be swept into the stomach.

The Stomach

The stomach is a dilated muscular structure of the alimentary canal that is connected at its upper end to the esophagus and at its lower end to the small intestine. It stores food, furthers the mechanical breakup of food through its powerful churning movements, and aids digestion through the action of digestive enzymes that it secretes.

The stomach is situated largely in the upper left side of the abdominal cavity beneath the diaphragm, and extends at its lower part toward the middle of the abdomen under the liver (Fig. 14-3). At the junction of the stomach and the small intestine is a circular valve of smooth muscle called the **pyloric sphincter,** whose action is similar to that of the cardiac sphincter.

The size of the stomach varies according to the size of the individual and the extent to which the stomach is distended by food.

When it is filled with food its walls may be considerably distended, accommodating about $2\frac{1}{2}$ to 4 qt in the average adult. This relatively large capacity accounts for its function as a reservoir and makes it possible to satisfy our food requirements in two or three meals a day instead of the more frequent meals that would be needed, say, by an individual whose stomach has been surgically reduced in size or removed. When empty, the stomach has a tubular shape resembling the letter J and is about the size of a large sausage; its walls are partially collapsed and its inner epithelial lining takes the form of longitudinal folds.

Stomach Function

Our modern knowledge of stomach function is based on the work of an American Army surgeon, William Beaumont. In 1833 Beaumont studied the effects of dietary and emotional factors on the secretion of gastric juice in a patient, Alexis St. Martin. St. Martin had been shot and the wound never healed properly, resulting in a windowlike opening from the surface of the abdomen to the stomach; this afforded a remarkable opportunity to investigate the functioning of the stomach. In the course of his studies with this unusual patient Beaumont demonstrated the acidity and digestive properties of gastric juice.

The muscular walls of the stomach contain three layers of smooth muscle fibers, which are responsible for peristalsis and the powerful grinding action of the stomach. The inner surface of the stomach is lined with columnar epithelial cells in which are scattered an estimated 10 million to 35 million tubular **gastric glands.** These pour their secretions, called **gastric juices,** through small ducts into the stomach cavity. The gastric glands of the stomach contain three structurally and functionally different types of cell: one type produces mucin, a sugar-protein responsible for the viscous nature of mucus; another the digestive enzymes (primarily pepsin); and a third, remarkably enough, hydrochloric acid.

Mucin forms a coating over the stomach lining (made famous by television commercials) and has an important protective capacity. Mucin is steadily renewed and only slowly digested; this assures that the digestive enzymes are mechanically separated from the tissues of the digestive tract. Because it buffers or binds acid and inhibits the enzymatic action of pepsin, mucin plays an essential role in helping to prevent the digestion of the protein that makes up the walls of the stomach and intestines. Occasionally a malfunction occurs and a portion of the stomach or intestinal wall is eaten away; this is called an **ulcer.**

Pepsin, like most of the digestive enzymes, is secreted in an enzymatically inactive form, called **pepsinogen,** by special glandular cells of the stomach. Pepsinogen is activated to pepsin when exposed to the acid medium of the stomach or to already activated pepsin; for reasons unknown, it is maximally active at very acid pH levels. This enzyme catalyzes the hydrolysis of proteins to smaller fragments, or polypeptides.

Hydrochloric acid is probably the most unusual component of the gastric juices. It is secreted by specialized glandular cells in a remarkably concentrated form, having a pH of about 1, which represents a hydrogen ion concentration 1 million times greater than that of blood plasma. Its most important function in the digestive process appears to be furnishing suitable acid conditions both for the activation of pepsinogen to pepsin and optimal digestive activity of the pepsin. The inordinately low pH is also responsible for the precipitation (denaturation) of many soluble proteins, such as casein, the protein of milk. Because denatured protein degrades slowly, the soluble proteins remain for a longer time in the stomach and are subjected to a longer period of digestion by pepsin. The strong acidity also kills many microorganisms, preventing bacterial invasion and putrefaction of food.

The consistency of food is apparently a major factor determining how soon it will leave the stomach. Fluids pass through the stomach rapidly, often in 20 minutes or less. Solid foods take much longer, and may remain in the stomach for three or four hours.

When food has attained a consistency appropriate for discharge into the small intestine, it is evacuated by progressively stronger descending peristaltic contractions of the stomach wall that move the **chyme** (partly digested food that is now in a semifluid state) through the pyloric sphincter into the small intestine.

Small Intestine

The small intestine in the average adult is a tube about 21 ft long and approximately 1 in. in diameter. It is the principal digestive organ of the alimentary tract. It is arranged in coils and loops that fill a large part of the abdominal cavity, and is attached to the body wall through most of its length by a membrane called the **mesentery.** This membrane also supports the blood vessels, lymphatic vessels, and nerves that service the intestinal walls. The first 12 in. or so of the small intestine leading from the stomach is called the **duodenum;** this segment receives secretions from the pancreas and liver. Like most of the alimentary canal, the small intestine is essentially constructed of three layers of tissue: an inner epithelial layer, or **mucosa,** with an underlying coat of connective tissue in which are embedded the main blood vessels of the digestive tract; a middle **muscular layer;** and an outer **connective tissue layer.**

Several types of movements occur in the small intestine as a result of its muscular activity. Peristaltic contraction waves propel material along the intestinal tube in somewhat the same manner as that already described for the esophagus. Other contractions mix the contents with digestive juices. Partly digested food normally travels through the small intestine in about eight hours.

The numerous digestive glands embedded in the wall of the small intestine (Fig. 14-6, p. 384) are made of several types of secretory cell. Together they secrete a mixture of digestive fluids at a daily rate of approximately 3000 ml (about 3 qt) in the average adult. Intestinal juice is slightly alkaline and contains large quantities of mucus as well as several different enzymes. The role of the mucus in lubrication and protection during the digestion process is the same as in the stomach.

At least two types of mechanism, neural (Chapter 17) and hormonal (Chapter 16) are responsible for the integration and control of digestive juice secretion. The factors that regulate the flow of each of the digestive juices are summarized in Table 14-2 (p. 385).

Villi

The inner epithelial lining of the small intestine provides an enormous surface for the absorption of nutrients, not only because of its great length but most importantly because of innumerable fingerlike projections called **villi** (Fig. 14-6). So numerous are the tiny villi that the inner surface of the small intestine has a velvety appearance. It has been estimated that the small intestine of man possesses approximately 5 million villi, making up a surface of about 10 square meters—or more than five times the skin surface. Each villus includes a blood capillary, a small lymph vessel called a **lacteal** (Chapter 15), and an outer covering of a single cell layer of specialized columnar epithelial cells that are continuous with the intestinal lining and embedded intestinal glands.

The villi are highly specialized absorptive organs whose epithelial cells are entirely responsible for the absorption of water and the products of digestion from the small intestine. They exhibit properties of selective absorption, and there is little doubt that their action also includes an active, energy-requiring transport mechanism. The villi appear to be in ceaseless motion, moving from side to side and lengthening and shortening. Presumably these movements mix the intestinal fluids in the immediate neighborhood of the villi, thus helping the process of digestion and absorption. Lying between the villi are microscopic pockets and indentations, the **intestinal glands,** which produce mucus and digestive enzymes.

The Pancreas

Separate accessory organs, the pancreas and liver, also pour their juices into the small

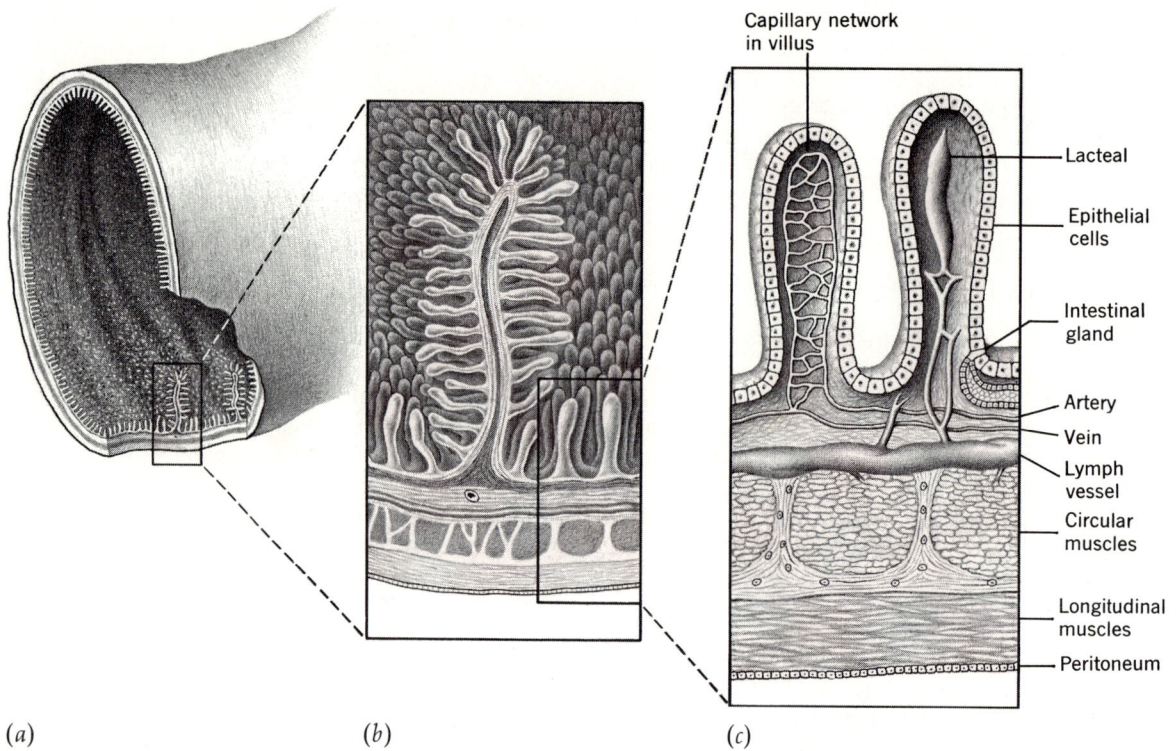

(a) (b) (c)

Capillary network in villus

Lacteal

Epithelial cells

Intestinal gland

Artery

Vein

Lymph vessel

Circular muscles

Longitudinal muscles

Peritoneum

Figure 14-6   A section of the human small intestine. The major stages of digestion take place and the products of digestion are absorbed here. The absorptive area of the long tubular organ is increased by extensive infolding of its internal surface (b), from which project millions of small, hairlike projections called villi (c), which make the internal surface of the intestine seem like velvet. A microscopic section of such a villus (d) shows its cellular structure and the location of the secretory goblet cells.

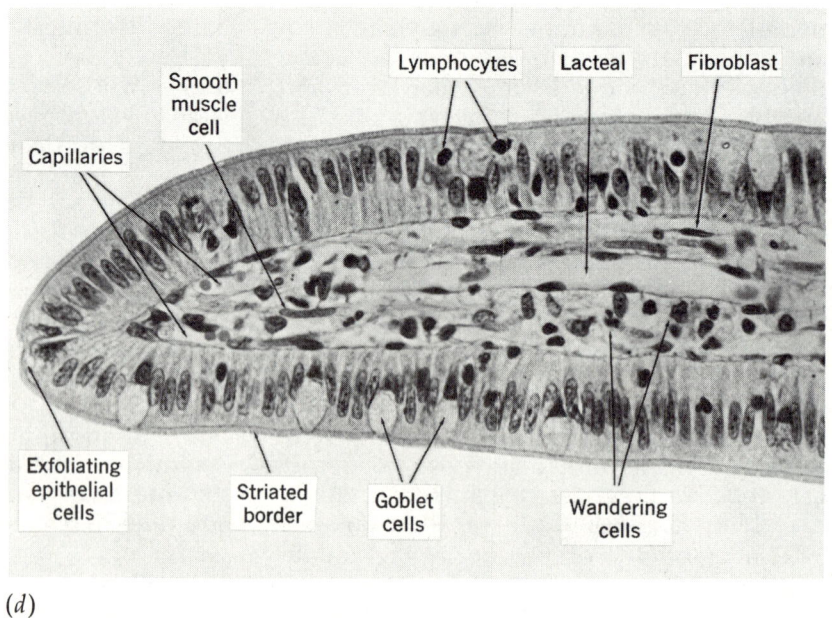

Lymphocytes Lacteal Fibroblast

Smooth muscle cell

Capillaries

Exfoliating epithelial cells

Striated border

Goblet cells

Wandering cells

(d)

table 14-2

Summary of Neural and Hormonal Factors Affecting the Flow of Digestive Juices

DIGESTIVE JUICE	FACTORS AFFECTING SECRETION	PRINCIPAL SITE OF FORMATION	PRINCIPAL SITE OF ACTION	PRINCIPAL PHENOMENA EFFECTED
Saliva	Simple reflex (mechanical factors in mouth) Conditioned reflex (thought, sight, taste, and smell)		Salivary glands	Secretion of saliva
Gastric juice	Simple reflex (mechanical factors) Conditioned reflex (thought, sight, taste, and smell)		Gastric glands	Secretion of gastric juice
	Polypeptides	From partially digested proteins in pyloric region of stomach	Gastric glands	Secretion of gastric juice
	Gastrin	Pyloric region of stomach	Gastric glands	Secretion of gastric juice
Pancreatic juice	Reflex (probably simple reflex)		Pancreas	Secretion of pancreatic juice rich in digestive enzymes
	Pancreozymin	Duodenum	Pancreas	Secretion of pancreatic juice rich in digestive enzymes
	Secretin	Duodenum	Pancreas	Secretion of large volume of pancreatic juice low in digestive enzymes
Bile	Reflex (probably simple reflex)		Gall bladder	Contraction and emptying
	Cholecystoknin	Duodenum	Gall bladder	Contraction and emptying
Intestinal juice	Mechanism as yet unknown; presence of food in intestine stimulatory			

intestine. The pancreas is a diffuse, fish-shaped organ ranging from 6 to 9 in. in length, 1 in. or so in width, and slightly less in thickness. It is situated partially behind the stomach within the curvature of the small intestine (Figure 14-3), and has two roles: it secretes **pancreatic juice** for digestive purposes by way of a duct that leads to the small intestine; and it serves as an endocrine gland by virtue of specialized scattered clusters of cells **(islets of Langerhans)** that secrete two different hormones, **insulin** and **glucagon,** into the bloodstream (Chapter 16). The cells that produce pancreatic juice make up the bulk of the pancreas. They are drained by a system of microscopic ducts. These tiny ducts unite to form progressively larger tubes that eventually fuse into one main **pancreatic duct** that enters the wall of the duodenum about 3 or 4 in. below the pyloric sphincter. The average volume of pancreatic secretion is about 500 to 800 ml (about a pint) a day.

Pancreatic juice is alkaline (pH 7 to 8), and thus partially neutralizes the acid contents emerging from the stomach and provides a suitable pH for the action of the digestive enzymes in the intestine, which, like most other enzymes, are maximally active near pH 7.

Pancreatic juice contains several important enzymes concerned with the digestion of carbohydrates, fats, and proteins. The carbohydrate-digesting enzyme is **amylase.** It is similar in enzymatic action and in several of its properties to salivary amylase, catalyzing the hydrolysis of starch to maltose. Pancreatic amylase is extremely active and is the most important enzyme in the alimentary canal in the digestion of starch (Fig. 14-7).

Lipase is the principal enzyme responsible for the hydrolysis of fats. Pancreatic lipase is apparently secreted in an inactive precursor form that is converted, by an unknown mechanism in the small intestine, to an effective fat-digesting enzyme.

Figure 14-7   Enzymatic digestion of starch. Amylase in the saliva and pancreatic juice hydrolyzes the bonds between every other pair of glucose units, yielding the dissacharide maltose. Maltose is digested to glucose by maltase, secreted by the intestinal glands.

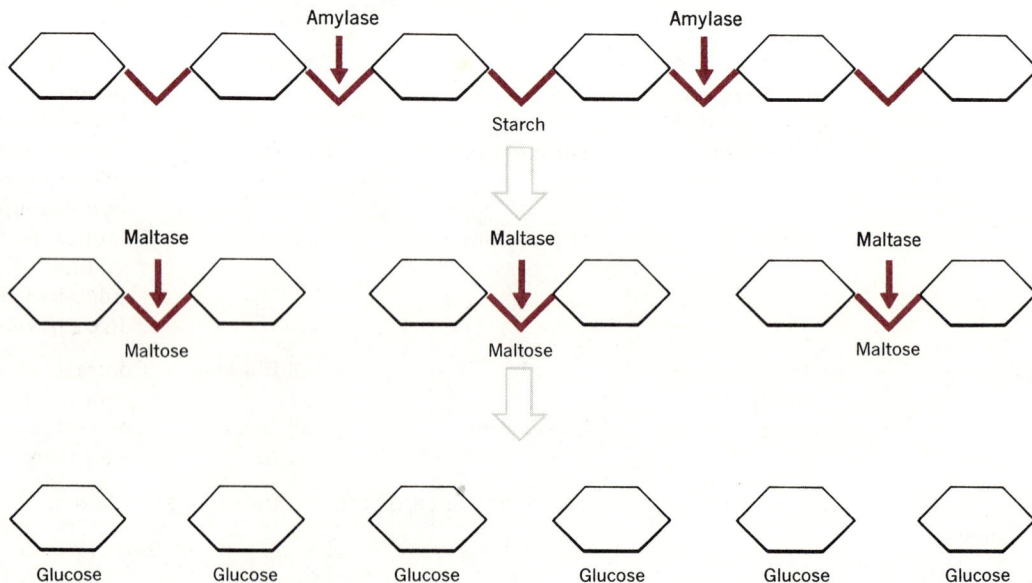

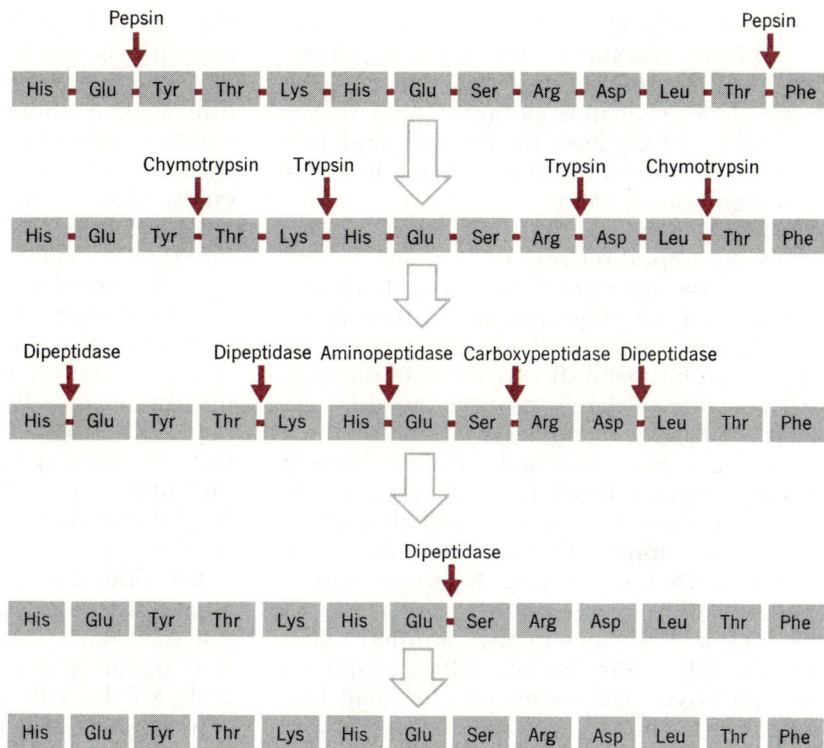

Figure 14-8 Enzymatic digestion of protein. In the stomach the peptide bonds holding tyrosine (Tyr) and phenylalanine (Phe) in the protein chain are hydrolyzed. Then the food moves into the small intestine, where trypsin and chymotrypsin break other bonds. Terminal bonds at the amino and carboxyl ends of peptide chains are split by aminopeptidase and carboxypeptidase respectively. Small two-peptide fragments of the original protein are broken into single amino acids by dipeptidases.

The pancreas also secretes several important protein-digesting enzymes, in inactive precursor states, that are activated only when they reach the small intestine so they cannot digest the pancreatic tissue. These include **trypsinogen, chymotrypsinogen,** and **procarboxypeptidase,** which are transformed to the corresponding protein-digesting enzymes **trypsin, chymotrypsin,** and **dipeptidase** (Fig. 14-8).

The Liver

The liver, the largest gland of the body, performs a variety of important functions. Its major digestive function is the formation of bile, a complex, clear, yellow- or orange-colored fluid. In normal adults the daily production of bile probably ranges between 500 and 1000 ml.

Bile is continuously produced by the cells of the liver; it is carried by a system of **bile ducts** to the **gall bladder,** which drains into the duodenum via a common opening with the pancreatic duct. The gall bladder itself is a pear-shaped muscular sac, of approximately 50 ml capacity, that is used to store bile until needed. In some instances a complex insoluble material appears in the gall bladder and forms hard structures called **gallstones,** which usually are 80 percent or more cholesterol.

Bile serves in the excretion of certain waste products and in the process of digestion. In excretion it acts as a vehicle for the removal of excess calcium, certain breakdown products of hemoglobin from worn-out red blood cells, and other wastes. It participates in lipid digestion (although it contains no known fat-digesting enzymes) and lipid absorption by promoting emulsification and solution of lipids, including the fat-soluble vitamins and the fatty acids liberated in fat digestion. By their emulsifying action the bile salts (p. 63) lower the surface tension of fat droplets, thus promoting the division of fatty materials into smaller globules. As a result, the total surface area of the fats and oils exposed to the action of pancreatic lipase is increased considerably.

This aid in the digestion of fats is secondary to the important role of bile salts in enhancing the absorption by the villi of the products of fat digestion and other lipid-soluble substances such as the fat-soluble vitamins. It has been shown, for example, that without bile a large portion of the fat-digestion products and other fat-soluble substances are not absorbed but are instead eliminated in the feces. The exact mechanism by which bile salts enhance lipid absorption is not clear; apparently they become associated with the lipids during the absorption of the latter through the epithelial cell layer of the villi, and are eventually returned again into the bile. Thus the bile salts are utilized over and over again with only a small loss appearing in the feces. As we shall see on pp. 391–392, the liver has other important functions as well.

Absorption of Nutrients from the Digestive Tract

In general the epithelial cells of the intestine are impermeable, or nearly so, to the large molecules of carbohydrates (starch), proteins, and fats. Digestion of these molecules produces smaller products, including simple sugars, glycerine, fatty acids, and amino acids, which pass much more freely across the intestinal mucosa.

For all practical purposes virtually all absorption in the digestive tract takes place through the villi of the small intestine, with the exception of water, which is mainly absorbed from the large intestine. Some absorption of water, alcohol, and small quantities of mineral salts also occurs in the stomach; certain drugs and hormones can be absorbed from the mouth. Very few digestible substances reach the large intestine.

The absorption of small molecules, including the final products of digestion, across the epithelial cells of the villi of the small intestine is not accomplished by diffusion alone; active transport (p. 125) plays a major role also. It should be noted that the absorption process is highly selective; some substances are absorbed rapidly; other molecules, despite their small size, are hardly absorbed at all. For example, magnesium sulfate (Epsom salts), whose molecules are considerably smaller than those of simple sugars and most amino acids, is for all practical purposes not absorbed from the digestive tract. A concentration of unabsorbed salt in the intestine causes the intestinal contents to be hypertonic to the body fluids (see Fig. 5-6), thus drawing water into the lumen of the intestine. This process is one of the reasons for the effectiveness of Epsom salts taken as a laxative.

Considerable absorption of the simple sugars and disaccharides apparently takes place by active transport. The digestion products of fats enter the lymph vessels, or lacteals, of the villi rather than the capillaries. (The name lacteal, from a Latin word meaning "of milk," refers to the milky appearance the lymph vessels assume during absorption; this is caused by the suspension of fat droplets in the lymph.) The small-chain fatty acid molecules and possibly glycerin, which is water soluble, are chiefly absorbed by the capillaries of the villi rather than the lacteals. As in sugar absorption, the absorption of the products of fat digestion also depends on the metabolism of the intestinal epithelial cells, and probably involves active transport mechanisms.

Amino acids, the major end products of protein digestion, are rapidly and almost exclusively absorbed in the small intestine, entering the blood stream directly by the blood capillaries of the villi. Different amino acids are known to be absorbed at different rates, but the exact mechanisms of amino-acid absorption are not known. Appreciable amounts of peptides and, under special circumstances, proteins (e.g., milk proteins and egg whites, particularly in infants) may also pass through the epithelial cells of the small intestine and appear in the blood.

The Large Intestine

The large intestine, or **colon,** is essentially an inverted, U-shaped tube in the abdominal cavity. It is considerably shorter than the small intestine (5 to 6 ft long in the average adult); but its diameter is approximately $2\frac{1}{2}$ in., noticeably larger than that of the small intestine. The small intestine connects with the large intestine at the lower right side of the abdomen 2 to 3 in. above the beginning of the colon, leaving a "blind" portion (the **cecum**) of the large intestine below the T-shaped junction. At the end of the cecum is a small blind structure called the **appendix.** In herbivores (animals subsisting solely on a diet of grasses and other plants) such as horses and cows the cecum is large and is an important site for the digestion of cellulose by its rich and specialized bacterial population. In man the cecum and appendix have no important function. Painful inflammation of the appendix in man as a result of infection is called **appendicitis** and is usually treated by surgical removal of the appendix.

The opening of the small intestine into the colon is guarded by a ring, or sphincter, of smooth muscle that controls the passage of material into the large intestine and prevents a reverse flow back into the small intestine; thus it protects the latter from the rich bacterial population in the large intestine.

The large intestine performs several important functions. One of the most significant is the absorption of water and salts from the fluid from the small intestine. The water that is absorbed includes not only that ingested in food and drink but also the large volume contributed by the secretions of the salivary glands, stomach, pancreas, liver, and intestinal glands.

A too-rapid passage of fluid through the colon does not permit adequate reabsorption of water and results in the condition known as **diarrhea.** It is usually caused by excessive muscular activity of the walls of the large intestine, often as a result of a physical irritation, infection, drug action, or emotional disturbance. If permitted to go unchecked it will lead to dehydration and excessive loss of salts, which can be fatal, especially in infants. At the other extreme, an unusually slow passage of material through the large intestine may be responsible for excessive removal of water, resulting in **constipation,** which is characterized by a relatively dryer and therefore harder mass of undigested residues.

A second important role of the large intestine is performed by its varied bacterial population. The ability of many intestinal bacteria to synthesize certain vitamins, amino acids, and other compounds that we absorb in part from the colon helps us meet some of our nutritional requirements; symptoms of vitamin deficiencies appear when these bacteria are destroyed, for example, by excessive or prolonged use of antibiotics. Some bacterial strains in the large intestine are also responsible for the breakdown of some of the small quantities of undigested or partially degraded proteins that reach the colon. They liberate such foul-smelling substances as indole, skatole, and hydrogen sulfide, which account for the characteristic odor of feces.

The last 6 to 8 in. of the intestinal tract are called the **rectum.** The opening from the rectum to the outside of the body is called the **anus** and is guarded by two muscular sphincters—an internal one of smooth muscle and an outer one of striated muscle. As the rectum is filled and compacted with feces it distends until it is sufficiently stimulated to give rise to a **defecation reflex.** Under ordinary circumstances this reflex can be consciously inhibited in individuals other than small infants. Defecation consists of a powerful peristaltic contraction of the terminal portion of the colon and rectum, accompanied by a voluntary contraction of the abdominal muscles and a relaxation of the anal sphincters. The feces that are eliminated consist of a compacted mixture of undigested food residues, remains of bile pigments, minerals, epithelial cells of the intestinal mucosa, and bacteria. They also contain an appreciable quantity of materials (fats, nitrogenous substances, and minerals) eliminated or excreted from the blood. Bacteria make up 10 to 50 percent of the feces. The indigestible food substances, consisting mostly of cellulose of plant materials, are called roughage; they serve to stimulate further the secretory and muscular activity of the intestinal wall.

Energy Requirements of Man

Basal Metabolic Rate

The minimal amount of energy required by the human body during the course of a single day is called the **basal metabolic rate (BMR).** By definition it is equal to the minimal energy an individual expends per unit of time; it is just the amount of energy necessary to keep an individual alive. Determination of the basal metabolic rate can be made directly by recording how much heat the body gives off, for heat production is a principal means of energy dissipation. However this requires very complicated and sophisticated equipment; the basal metabolic rate is far more conveniently determined by the measuring oxygen consumption and carbon dioxide production (Fig. 14-9). The ratio of these two values ($CO_2 : O_2$) is called the **respiratory quotient (RQ).** In the aerobic respiration of carbohydrates (Chapter 4), glucose, for example, for each molecule of glucose oxidized six molecules of oxygen are used and six molecules of carbon dioxide are produced.

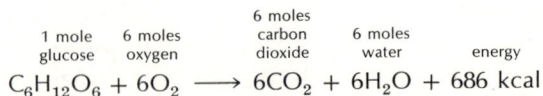

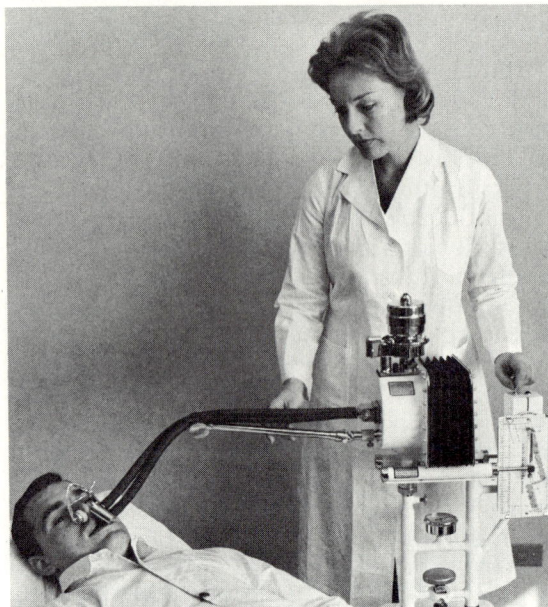

Figure 14-9   Basal metabolic rate is determined in a quietly resting individual by measuring oxygen uptake and CO_2 production rates.

$$\underset{\substack{\text{1 mole}\\\text{glucose}}}{C_6H_{12}O_6} + \underset{\substack{\text{6 moles}\\\text{oxygen}}}{6O_2} \longrightarrow \underset{\substack{\text{6 moles}\\\text{carbon}\\\text{dioxide}}}{6CO_2} + \underset{\substack{\text{6 moles}\\\text{water}}}{6H_2O} + \underset{\text{energy}}{686 \text{ kcal}}$$

The RQ is therefore 1 for carbohydrates.

Fats have more carbon-hydrogen bonds per carbon atom than carbohydrates. Accordingly, more oxygen is consumed in the course of fat oxidation than carbon dioxide produced. The RQ for fats is significantly less than 1, averaging close to .7. Proteins or amino acids are also in a more reduced chemical state than carbohydrates; their RQ averages about .8. Because the RQ is different for each of the major energy sources, its value can be used as an indicator of which one (fats, carbohydrates, or proteins) is being used during metabolism.

The equation shown above tells us that the oxidation of a mole of glucose (180 g) makes available 686 kcal, or roughly 4 kcal per gram. The basal metabolic rate of an individual is 1500 kcal per day, or about 80 kcal per hour. Therefore at rest, just to keep ourselves alive, we are all burning the equivalent of 20 g of sugar, a heaping tablespoon per hour.

The basal metabolic rate shows wide variations from individual to individual, depending on such factors as age, hormonal balance, sex, diet, inheritance, climate, and weight. It is low during sleep. With only moderate activity, say housework or a leisurely walk, energy expenditure jumps to 200 to 300 kcal per hour. For a well-trained athlete, the maximum continuous rate of energy utilization is about 800 kcal per hour, 10 times more than the minimum rate.

Energy Storage

It is interesting to note that energy storage in living systems is achieved by the carbon-hydrogen bonds of lipids and carbohydrates. Most, but not all, animals store their energy largely as fats, whereas the chief energy reservoir of plants is carbohydrates (starch). The energy stored in fats is more than twice that contained by an equal weight of carbohydrate. In muscle tissue some energy storage is also attained by means of the energy-rich phosphate bonds (Chapter 4) of compounds such as creatine phosphate.

Role of the Nutrients

Carbohydrates as a Food Source

For most heterotrophs carbohydrates are the major source of both energy and carbon. Autotrophic organisms (p. 591) require carbohydrates, but they produce these by photosynthesis. In higher animals, including man, 50 percent or more of the total food ingested may be carbohydrates. Most of it is usually in the form of large polysaccharide molecules, chiefly starches.

The liver exercises a central role, together with the kidneys and the endocrine glands, in maintaining the internal environment of an organism by regulating the concentrations of the numerous substances in the blood. In mammals the liver is the organ normally richest in carbohydrate stores, which occur principally as an animal starch called **glycogen** (Fig. 3-11). Glycogen makes up 2 to 8 percent of the fresh weight of the liver in various mammals, but it is usually entirely depleted if an animal has not eaten in 24 hours.

Liver glycogen is in a dynamic state in the sense that it is constantly being synthesized from glucose in the blood and degraded. The quantity of glycogen stored in the liver depends on an individual's diet, the amount of exercise he gets, and the levels of certain hormones in his blood, mainly insulin, glucagon, and glucocorticoids (see Chapter 16).

The normal glucose level of blood in human beings is about 100 mg per 100 ml of blood; the main sources of blood glucose are sugar absorbed from the digestive tract and glycogen stored in the liver. Because carbohydrates, fats, and proteins are all interconvertible by cells of the liver, the level of liver glycogen is influenced by the intake of all energy-rich nutrients in the diet.

The second principal site of storage of glycogen in mammals is skeletal muscle. Although the .5 to 1 percent concentration of muscle glycogen is less than in liver, it is nevertheless an appreciable total amount because of the large percentage of muscle (about 40 percent) in body weight. Muscle glycogen is more or less influenced by the same factors that affect liver glycogen levels, except that it is not as readily depleted by fasting, nor does it make a significant contribution to the blood glucose level. It is the immediate substrate for muscle respiration (Chapter 18).

Fats as a Food Source

Fats, like carbohydrates, serve as both a source of energy and of carbon for synthesizing other compounds; they are furnished chiefly in the human diet as triglycerides (p. 62) of both animal and vegetable origin. Fats are enzymatically hydrolyzed in the small intestine by digestive lipases (Table 14-1) to yield glycerol and fatty acids.

In mammals fats represent 10 percent or more of the body weight and serve as a principal reservoir of stored energy. They are present in varying quantities in all tissues as liquid droplets in the cytoplasm. The lipid concentration in human blood is about 500 mg per 100 ml. The amount, distribution, and storage of fat are determined in part by the basal metabolic rate, hormonal pattern, type of exercise, genetic makeup of the individual, and diet. Most mammalian fat is localized in the cells immediately below the skin, forming a padded layer called **adipose tissue.** There it functions not only as an energy store, but also in a secondary role as an insulating and protective layer against excessive heat loss and mechanical injury. This is particularly significant in maintaining a constant body temperature in mammals living in the sea (e.g., the whale), where the water is generally colder than body temperature.

The biochemical steps in the metabolism of fats, including pathways of breakdown and synthesis of fatty acids and interrelationships to carbohydrates and proteins, were indicated in Chapter 4. Most of the fatty acids of the body can be synthesized from simpler molecules in man and other mammals in sufficient quantity. However, at least three known unsaturated fats must be provided in the diet of experimental animals such as rats and chickens to maintain them in a healthy state and apparently also in man; accordingly these are called **essential fatty acids.**

The major site in the body of fat metabolism—synthesis and breakdown of fatty acid components—is the liver. Present evidence indicates that fats are transported by the bloodstream as triglycerides and as dissolved glycerin and liberated fatty acids bound to the albumin protein fraction of the blood.

When the caloric intake is less than the energy expenditure of an organism there is a net decrease in the lipid content of the body. As glycogen is depleted, fats are increasingly utilized. Stores of glycogen in the liver are completely consumed within 12 to 24 hours after the last intake of food; thereafter the principal energy source is the stored fat. The rate of fat degradation is very high under these conditions, especially in the absence of adequate carbohydrates and proteins in the diet. It may result in the accumulation of pyruvic and citric acids, which are toxic in high concentrations and are a major cause of death by starvation. In the terminal stages of starvation the structural proteins of the body (such as muscle) are used as the final energy and carbon source. Death follows shortly.

When the caloric intake of mammals exceeds their immediate needs, the extra energy is stored as fat molecules, largely in the adipose fat. When the caloric intake is the same as the energy expenditure of an organism, the quantity of body fat remains constant, but it is in a state of dynamic balance between the constant deposit of fats and their breakdown and removal. This also seems to be true for most of the other components of the body.

table 14-3
Probable Essential Amino Acids for Man

Arginine	Methionine
Histidine	Phenylalanine
Isoleucine	Threonine
Leucine	Tryptophan
Lysine	Valine

Proteins as a Food Source

Animals, unlike most plants and microorganisms, are unable to utilize the organic forms of nitrogen such as nitrate, nitrite, or ammonia as a source of nitrogen to synthesize proteins and nucleic acids. Instead, they require a source of amino groups, principally in the form of protein or amino acids; the primary source of amino nitrogen is plants. The unique and essential roles of proteins (and of amino acids) in the diet are twofold: to furnish the essential amino acids (Table 14-3) that cannot be synthesized in sufficient quantities by the biochemical machinery of the animal itself; and to restore the steady loss of nitrogen that is excreted in the course of metabolism. Like carbohydrates and fats, proteins act as both an energy and a carbon source.

Protein is hydrolyzed in the digestive tract into its component amino acids through the action of a number of digestive enzymes (Table 14-1). The resultant amino acids are transported by the bloodstream to the body tissues, where they are utilized for new protein synthesis (Chapter 4). In living cells proteins, like fats and carbohydrates, are in many instances in a dynamic state; they are constantly being synthesized and degraded.

Unlike carbohydrates and lipids, which can be stored as glycogen or fat deposits, amino acids or proteins in mammals cannot be stored. Whether they arise from the diet or from an unusually rapid metabolic breakdown of body proteins, excess amino acids are rapidly broken down. Excess amino groups are ultimately excreted, whereas remaining carbon chains usually enter the pathways of carbohydrate and lipid metabolism.

The principal waste product of nitrogen metabolism in mammals and certain other animals is **urea.** It is the final product of a series of biochemical reactions taking place largely in the liver. Urea is subsequently excreted by the kidneys as a component of **urine** (Chapter 15).

Vitamins

Vitamins are organic compounds that are needed in extremely small quantities in the diet or in nutrient media for the well-being and normal functioning of an organism. They are not confined to any particular class of organic substance; they may be organic acids, amino acids, esters, alcohols, steroids, and so on. A substance may be a vitamin for one species or organism and not for another, depending on whether it is required nutritionally. Some of the main vitamins for man are described in Table 14-5, (p. 394).

The need for vitamins in the diet reflects an inability to synthesize this compound from other dietary and metabolic substances, often because one or more enzymes necessary for the formation of a vitamin are absent. In animals a prolonged dietary deficiency of any vitamin results in characteristic diseases and eventual death. The fact that the vitamins must be regularly supplied in the diet also implies that they are continually being broken down and excreted in the course of metabolism. This has been verified by studies employing isotopically labeled compounds.

Vitamins are generally divided into two broad groups based on their solubility properties. Most dissolve relatively easily in water and are called **water-soluble** vitamins. Those that are insoluble in water can be dissolved in fats or in solvents that dissolve fats, and are accordingly called **fat-soluble** vitamins.

A pattern that has emerged from the vast welter of details concerning the action of vitamins is of fundamental significance: most vitamins serve as components of specific co-enzymes; and these substances function in identical fashion in all living cells, be it microbe or man. This fact has provided us with further evidence for the fundamental likeness and unity of all forms of life and thus supports the evolutionary concept that all living things are related.

Inorganic Nutrients

Approximately 95 percent of the dry weight of protoplasm is made of four elements—carbon, hydrogen, oxygen, and nitrogen—occurring principally as carbohydrates, proteins, fats, and nucleic acids; minerals account for only 5 percent of the body weight. About 15 mineral elements have been shown to be necessary for normal growth and development (Table 14-4); the omission of any of them from the diet or nutrient medium results in disease or death.

The inorganic nutrients, or **mineral salts,** required by living things are divided into two broad classes: the **macronutrient,** or major, elements (needed in relatively large amounts) and the **micronutrient,** or **trace,** elements (required in only very small quantities). About 4 percent of the dry weight of protoplasm is

table 14-4
Essential Minerals

MACRONUTRIENT (MAJOR ELEMENT)	MICRONUTRIENT (TRACE ELEMENT)	
Potassium (K)	Iron (Fe)	
Sodium (Na)	Copper (Cu)	
Chlorine (Cl)	Zinc (Zn)	
Phosphorus (P)	Manganese (Mn)	
Calcium (Ca)	Cobalt (Co)	Required by higher animals only
Magnesium (Mg)	Iodine (I)	
Sulfur (S)	Selenium	
	Vanadium (V)	Required by higher plants only
	Boron (B)	
	Molybdenum (Mo)	Thought to be required by higher plants only

table 14-5
Some Vitamins Required by Man

VITAMIN	SOME DEFICIENCY SYMPTOMS	IMPORTANT SOURCES	CHEMICAL FUNCTION
Fat-Soluble			
Retinol (vitamin A)	Dry, brittle epithelia; night blindness; malformed rods	Green and yellow vegetables and fruit; dairy products; egg yolk; fish-liver oil	Component of rhodopsin
Calciferol (vitamin D)	Rickets; very low blood calcium level, soft bones, distorted skeleton; poor muscular development	Egg yolk, milk, fish oils	Calcium absorption
Tocopherol (vitamin E)	Male sterility in rats (and perhaps other animals); muscular distrophy in some animals; abnormal red blood cells in infants; death of rat and chicken embryos	Meat, egg yolk, green vegetables, seed oils	Unknown
Vitamin K	Slow blood clotting and hemorrhage	Green vegetables	Synthesis of the blood clotting factor prothrombin
Water-Soluble			
Thiamine (vitamin B_1)	Beriberi; muscle atrophy, paralysis; mental confusion; congestive heart failure	Whole-grain cereals; yeast; nuts; liver; pork	Coenzyme in oxidation of pyruvic acid
Riboflavin (vitamin B_2)	Vascularization of the cornea; conjunctivitis and disturbances of vision; sores on the mouth and tongue; disorders of liver and nerves	Milk, cheese; eggs; yeast; liver; wheat germ; leafy vegetables	Electron carrier in cytochrome system
Pyridoxine (vitamin B_6)	Convulsions; dermatitis; impairment of antibody synthesis	Whole grains; fresh meat; eggs; liver; fresh vegetables	Coenzyme in transamination reaction
Pantothenic acid	Impairment of adrenal cortex function; numbness and pain in toes and feet; impairment of antibody synthesis	Almost all foods, especially fresh vegetables and meat, whole grains, eggs	Part of coenzyme A in Krebs cycle

table 14-5

Some Vitamins Required by Man (continued)

VITAMIN	SOME DEFICIENCY SYMPTOMS	IMPORTANT SOURCES	CHEMICAL FUNCTION
Water-Soluble (continued)			
Biotin	Clinical symptoms in man are extremely rare, but can be produced by great excess of raw egg white in diet. Symptoms are dermatitis; conjunctivitis	Liver, yeast, fresh vegetables	Coenzyme in fatty acid synthesis
Nicotinamide	Pellagra; dermatitis; diarrhea; irritability; abdominal pain; numbness; mental disturbance	Meat, yeast, whole wheat	Part of coenzyme NAD
Folic acid	Anemia; impairment of antibody synthesis	Leafy vegetables, liver	Nucleic-acid and amino-acid synthesis
Cobalamin (vitamin B_{12})	Pernicious anemia	Liver and other meats	Red blood cell formation
Ascorbic acid (vitamin C)	Scurvy; bleeding gums, loose teeth; anemia; painful and swollen joints; delayed healing of wounds	Citrus fruits, tomatoes	Collagen formation

made up of the major elements; less than 1 percent is accounted for by the trace elements. The fact that the trace elements are needed in such extremely small quantities does not necessarily mean that they are less important than the macronutrient elements, but suggests that they function as specific components of enzyme systems.

The Problem of Diet

Throughout most of man's existence his numbers have been limited by the supply of food. Even now, despite the recent "green revolution" (see Chapter 20), two-thirds of mankind is still hungry and malnourished most of the time. As a result of recent major advances in agricultural technology, most persons living in North America, Europe, Australia, and Japan have an adequate food supply, and the populations of developing countries have been spared massive famines (except in Biafra in 1969–1970). Nonetheless, human nutrition on a global scale is still in a sorry state. Malnutrition, particularly protein deficiency, exacts an enormous toll from the physical and mental development of the young in poorer countries—and of perhaps 20 percent of the population of the affluent countries, including the United States (Fig. 14-10, p. 396). Protein is the key to human health and vigor. Protein may be consumed

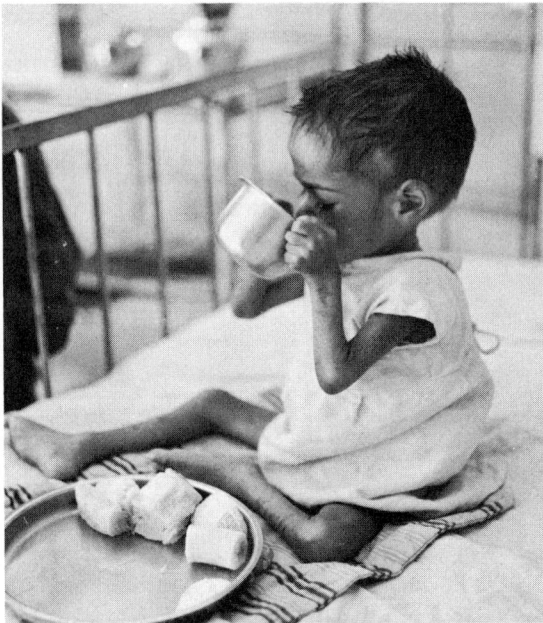

Figure 14-10 A malnourished Indian child. Her bloated belly and spindly limbs are the common result of Kwashiorkor, a severe protein deficiency resulting from starvation.

Figure 14-11 There are more than 30 million overweight Americans. Obesity results from an imbalance between caloric intake and expenditure.

directly from grain products by way of bread and cereals, or indirectly in the form of eggs, meat, milk, and milk products, from livestock fed on grain.

More than 52 percent of man's total dietary energy intake is supplied by cereal crops (wheat, rice, corn, soybean, etc.), which occupy more than 70 percent of the world's farm land. Eleven percent is supplied by livestock products (meat, milk, and eggs); 10 percent by potatoes and other tubers; 10 percent by fruits and vegetables; 9 percent by animal fats and vegetable oils; 7 percent by sugar, and 1 percent by fish. However, the world distribution of diet varies greatly. Each of the 2 billion people living in the poor countries consumes about 360 lb of grain per year—about 1 lb per person per day. Nearly all grain produced in these countries must be consumed directly to meet minimal energy

needs, with little remaining to feed to livestock for conversion into meat and milk. In contrast, the average North American consumes more than 1600 lb of grain per year, of which he eats only about 150 lb directly as bread, breakfast cereal, and so on. The rest is consumed indirectly in the form of meat, milk, and eggs. That is, he enjoys the luxury of the highly inefficient animal conversion of grain into tastier and somewhat more nutritious proteins. Thus the average North American currently makes about four times as great a demand on the earth's agricultural capacity as a person living in one of the poor countries. In effect, only about one-third of the world's people get enough of the proper nutrients, and in doing so they consume three-fourths of the world's available food supply.

It is therefore understandable, although

nonetheless tragic, that the problems of nutrition in our western culture have shifted from undernutrition and deficiency disease to "overnutrition" and obesity (Fig. 14-11). **Obesity** is a condition of excess fat caused by ingesting more calories than the body uses. The extra calories not expended by the numerous activities of the body are stored in the form of fat. Approximately 30 million persons in the United States are overweight. Obesity carries with it a more frequent occurrence of diabetes than in persons of normal weight; degenerative disease of the heart, blood vessels, and kidneys; and greater risks during surgery and pregnancy. What are the factors that cause an individual to take in more calories than he can use? The answer is not simple. In addition to genetic constitution, emotional problems apparently have an important bearing on excessive calorie intake, as does a disturbance in the appetite-regulating mechanism.

The Skin

In introducing this chapter we distinguished between the internal environment and the external environment, the former

being inside the body but outside the cells, the latter outside the body. In a very real sense the space within the digestive system is not part of the internal environment, but is actually "outside" the body, in the same sense that the space inside a cup is outside the material of which the cup is made, or the hole in a doughnut is outside the doughnut itself. This can be seen by comparing the digestive cavities of simpler animals with that of man (Fig. 14-12). The lumen of the digestive system is clearly continuous with the external environment. Moreover, the epithelium that lines the digestive tract is continuous with the outside epithelium, or **skin.**

Functions

The skin, or **integument,** is the outermost covering of the body and performs a number of diverse functions. First, it has a highly important role in **protection.** It protects the underlying tissues and organs against mechanical injury, invasion by bacteria and other infective agents, excessive loss of moisture, chemical damage, extreme temperature changes that occur in the external environment, and, finally, from overexposure to ultraviolet irradiation from the sun.

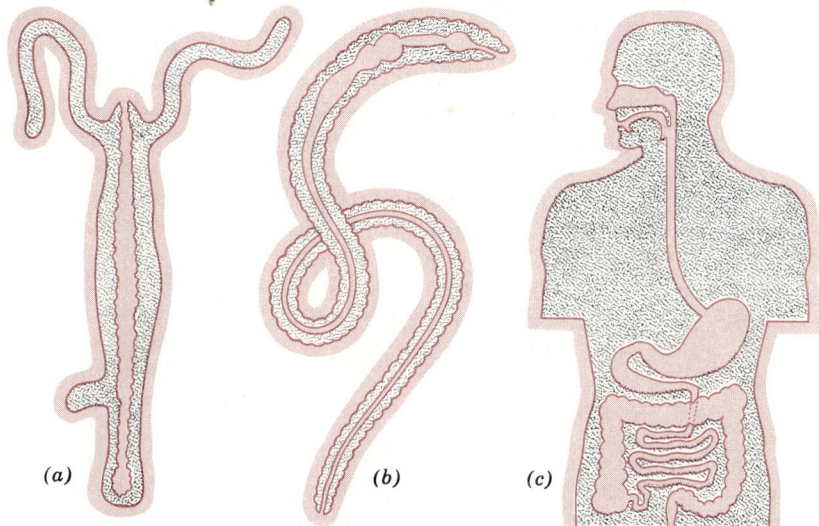

Figure 14-12 Diagrams of the digestive cavities of (a) hydra, (b) a worm, and (c) man. In all cases the lumen of the digestive system is continuous with the outside space and the epithelial lining of the mouth, gut, and anus is continuous with the skin.

(a)   (b)   (c)

The skin (with the adipose layer under it) also is involved in the **regulation of heat loss.** The diameters of the arteries and capillaries to the skin, and therefore the volume of blood flowing through it, are controlled by the autonomic nervous system (Chapter 17) according to the needs of the organism: dilation of these blood vessels brings more blood to the skin, resulting in a loss of heat by radiation from the surface of the body; constriction of the blood vessels reduces the heat loss. Increased perspiration by the sweat glands also promotes greater heat loss, or cooling, by evaporation.

The skin has an **excretory** function that supplements the role of the kidneys. In addition to its large water content (about 99 percent), sweat contains a variety of substances filtered from the blood (including organic salts, urea, uric acid, ammonia, and creatinine) whose composition resembles that of urine.

In addition, by housing a vast number of nerve endings or receptors concerned with the senses of touch, pressure, temperature, and pain (Chapter 17), the skin serves in a **sensory** capacity; it is one of several means by which man is informed of activities in his immediate external environment.

The skin also has a **secretory** role. In mammals, including man, oil is secreted to the surface by special glands. The oil prevents the skin and hair from drying out and cracking and protects against excessive ultraviolet irradiation from the sun. The milk-secreting (mammary) glands of women, as specialized derivatives of the skin, have an obviously important secretory function.

Finally, the skin also has a **nutritional** function, for it contains the steroid dehydrocholesterol, which is transformed to vitamin D on exposure to ultraviolet light in the sun's rays. The vitamin is then absorbed into the system through the skin.

Structure

The skin is made up of two principal parts: the outer layers of epithelial cells, called the **epidermis,** and the inner, thicker **dermis** (Chapter 6) made up of loose connective tissue containing the blood vessels, nerves, and other specialized structures (Fig. 14-13). The dermis is attached to the underlying tissue by a layer of connective tissue called the **subcutaneous layer,** which contains varying amounts of fat.

In human adults the skin covers a total area of some 20 square feet and weighs about 7 lb. It is from $\frac{1}{8}$ to $\frac{1}{32}$ in. thick, and is laced with nerve receptors in which the sensations of touch, pain, and temperature are initiated. The skin and central nervous system both originate from the outer embryonic layer of cells (**ectoderm,** Chapter 12) in the embryo.

Epidermis

The epidermis has no blood vessels of its own; it depends on the rich blood supply in the dermis. Only the deepest layer of the epidermis undergoes mitosis. The new cells produced are pushed outward toward the surface of the body, away from the necessary blood supply; they become progressively more flattened, and soon die. In the course of these events the outer cell layers of the epidermis attain a tough, horny, water-resistant quality because of the deposition in these dying cells of a hard, protein substance, **keratin,** the same material found in the fingernails and toenails. These outer, horny cells of the epidermis are continually worn away, but they are steadily replaced at a more or less equal rate. Dandruff is an obvious example of the shedding, or flaking, process in man, particularly because the hair tends to retain the flakes.

The fingernails and toenails are masses of modified horny cells that grow from the epithelial cells lying under the white crescent at the proximal (beginning) end of each nail.

Dermis

The bulk of the dermis is a matrix of connective tissue fibers among which are scattered an abundance of blood vessels, lymphatics, nerve endings, sense organs, fat cells, smooth muscle, special glands, and other structures. The dermis also contains numerous elastic fibers. With age, the loss in elasticity of the dermis and a decrease in subcutaneous fat account for the appearance of wrinkles.

Skin Glands

There are two general types of gland present in the human skin: the **sebaceous** glands secrete oil; the **sweat** glands produce sweat, or perspiration. Although both kinds of gland occur well within the dermis, they are actually derivatives of the epidermis, having formed as ingrowths from its innermost, multiplying layer of the epidermis. The oil secretions of the tiny sebaceous glands are beneficial for both the hair and skin. (There are at least two sebaceous glands associated with each hair.) The oil they secrete passes out through an opening in the epidermis to the surface of the skin. This oil keeps the hair supple and the skin soft and pliant, and is partially responsible for an animal's individual, characteristic scent or odor.

Sweat glands are small, coiled, tubular glands whose ducts empty onto the surface of the body (Fig. 14-13). They are distributed over the entire surface of the body but are especially profuse in the palms, soles, forehead, and armpits. Their function in excretion and in regulation of heat loss has already been indicated. The mammary glands are essentially highly specialized and modified sweat glands. Other modified sweat glands in the external ear secrete the waxy, pigmented material that accumulates in the ear instead of the usual aqueous sweat.

Hair

Hair is characteristic of mammals. In man it is distributed over the entire body except the palms and soles, and is more heavily con-

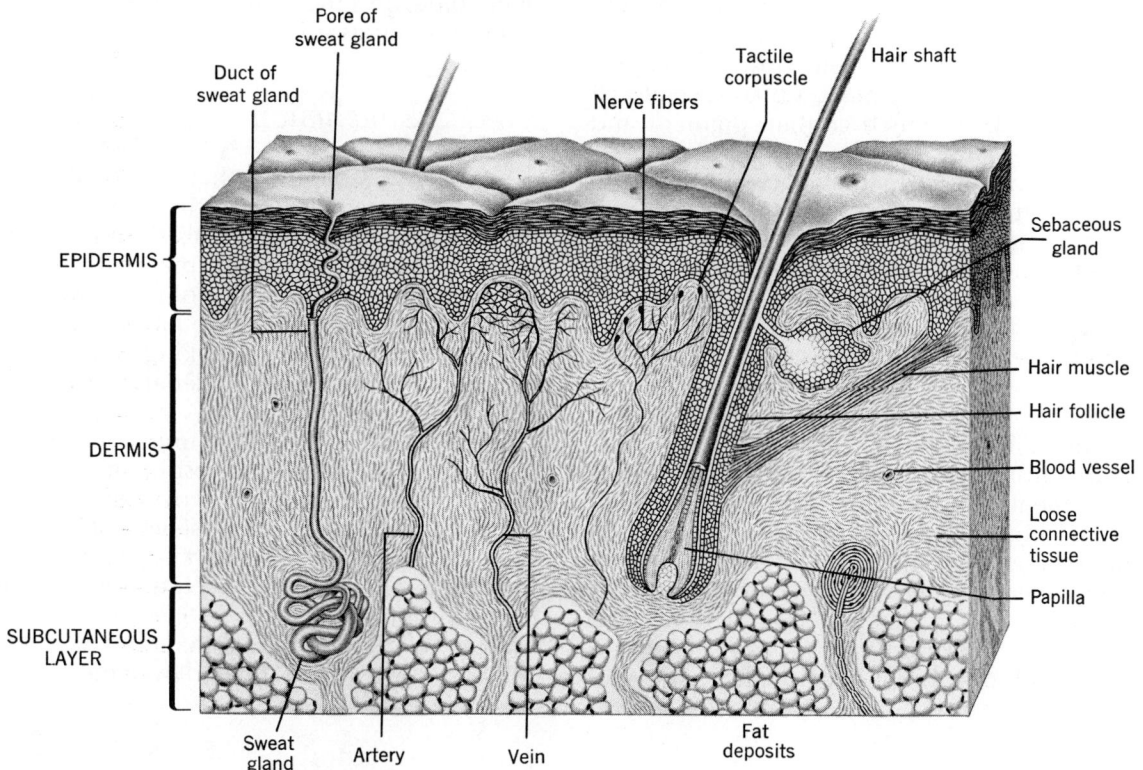

Figure 14-13 *Diagram of a slice of human skin, illustrating its microanatomy. Magnification about 15×.*

centrated in some areas than others. Although hairs are anchored deep in the dermis they are essentially derived from inpocketings of the epidermis. A hair typically consists of two portions: the **shaft,** the visible portion projecting above the surface of the skin, and the **root,** the portion extending into the dermis (Fig. 14-13). The hollow tube sunk into the dermis within which the root is located is called a **hair follicle.** To each hair follicle is attached one or more delicate bundles of smooth muscles whose contraction is under the control of the autonomic nervous system. When these muscles contract, the hair is pulled into a more upright position, raising the skin immediately around it to produce the familiar "goose pimples."

An expanded, hollow, bulbous end of the root at the bottom of the hair follicle encloses some capillaries and nerves collectively called the **papilla.** The epithelial cells nearest the papilla in this bulbous end of the root are the only living cells of the hair; these cells are constantly dividing, pushing new cells upward that soon die and eventually become the cells of the shaft.

A cross-section of a hair shows a complex structure; the hair is made up of several cell layers, some of which contain pigment and air spaces. Hairs that have a circular cross-section tend to be straight, whereas those that are flat or oval are wavy or curly.

In most mammals hair serves as insulation; this function has essentially been lost in man. It may also have specialized uses, as the eyebrows, eyelashes, and hairs of the nose and ears serve to keep out dust and insects.

Color

The color of the skin and hair results primarily from the presence in the epidermis of a pigment known as **melanin.** Melanin pigments are fine granules ranging in color from pale yellow through various shades of brown. The pigment is derived from the amino acid tyrosine and is formed in cells called **melanophores,** or **melanoblasts,** part of the innermost multiplying epithelial cells of the epidermis. Because the melanin-synthesizing cells, like the other cells of this layer, proliferate to build up all the overlying epidermal layers, melanin is distributed throughout the epidermis and hair. The greater the deposits and distribution of melanin in the epidermis—characteristics controlled by both genetic and environmental factors—the darker the skin. Light influences the amount of pigment produced; suntanning, for example, is a result of increased multiplication of the melanophores under the stimulus of ultraviolet light. In **albinos**—either animal or human—melanin-forming cells are entirely absent from the tissues.

Skin color also depends in part on the blood supply to the dermis. When the capillaries of the skin, especially in the outer portion of the dermis, are dilated and the blood flow is rapid, the skin becomes reddish in color. When the vessels are constricted the skin is pale. If the blood flow is unusually slow, most of its oxygen is given up in the tissues. Blood devoid of its oxygen is relatively blue, and the skin under these circumstances acquires a bluish tint.

Conclusion

We saw in Chapter 1 that one of the defining characteristics of all living systems is the ability to take in materials from the environment and to use those materials as fuel for energy and for synthesis of new structural elements. In this chapter we have defined as digestion the process of making soluble and breaking down nutritive materials so they can be metabolized.

To get energy from the soluble compounds that are the products of digestion, most organisms must have a supply of oxygen and they must be able to rid themselves of carbon dioxide and the other materials produced by metabolism. These are the functions of the organ systems that have evolved for gas exchange, circulation, and excretion. We turn now to these closely related systems.

Reading List

Best, C. H., and N. G. Taylor, *The Living Body* (5th ed.). Holt, Rinehart & Winston, New York, 1968.

Boerma, A. H., "A World Agricultural Plan," *Scientific American* (August 1970), pp. 54–69.

Bowen, H. J. M., *Trace Elements in Biochemistry*. Academic Press, New York, 1966.

Brown, L. R., "Human Food Production as a Process in the Biosphere," *Scientific American,* (September 1970), pp. 160–173.

Davenport, H. W., "Why the Stomach Does Not Digest Itself," *Scientific American* (January 1972), pp. 86–93.

Grollman, S., *The Human Body*. Macmillan, New York, 1968.

Harrison, R. J., & W. Montagna, *Man*. Appleton-Century-Crofts, New York, 1969.

Holt, S. J., "The Food Resources of the Ocean," *Scientific American* (September 1969), pp. 178–197.

Lehninger, A. L., *Biochemistry*. Worth, New York, 1970.

Montagna, W., "The Skin," *Scientific American* (February 1965), pp. 56–66.

Marples, M. J., "Life on the Human Skin," *Scientific American* (January 1969), pp. 108–115.

Phillipson, J., *Ecological Energetics,* Edward Arnold Publishers, London, 1966.

Schmidt-Nielsen, K., *Animal Physiology* (2nd ed.). Prentice-Hall, Englewood Cliffs, N.J., 1964, pp. 1–36.

Whittaker, R. H., *Communities and Ecosystems*. Macmillan, New York, 1970.

White, E. H., *Chemical Background for the Biological Sciences* (2nd ed.). Prentice-Hall, Englewood Cliffs, N.J., 1970.

chapter
fifteen

Maintenance
of the Body: Circulation,
Respiration, and Excretion

*"As we are about to discuss the motion, action, and use of the heart
and arteries, it is imperative on us first to state what has been
thought of these things by others in their writings, and what has been
held by the vulgar, and by tradition, in order that what is true may
be confirmed, and what is false set right by dissection, multiplied
experience, and accurate observation."*

WILLIAM HARVEY, 1628, *AN ANATOMICAL DISQUISITION ON THE
MOTION OF THE HEART AND BLOOD IN ANIMALS*

Circulation

In man and other complex multicellular animals a direct interchange of nutrients and metabolic waste products between organism and environment is impossible, because most cells are far from
the body surface. We therefore possess a **circulatory,** or **vascular,
system,** an interconnecting network of tubes **(vessels)** in which a
complex, cell-containing fluid, the **blood,** is circulated as the means
for this exchange. Blood is pumped by the heart, a unique organ
that we shall consider a little later, into the **arteries** (the major
vessels carrying blood away from the heart), which lead to the
capillaries (microscopic blood vessels made of a single layer of
endothelial cells), and then to the **veins,** which return the blood
to the heart (Fig. 15-1, p. 404). The **lymphatic system** is considered
a secondary structure of the circulatory system. The principal role
of the blood and circulatory system is to maintain a suitable environment for the 300 trillion cells of the human body by delivering
nutrients and regulatory substances and by removing the products
of metabolism. **It is only during the passage of blood through the
capillaries that this interchange occurs.** These vessels are the important physiological units of the circulatory system and not merely
connecting networks between the arteries and the veins.

403

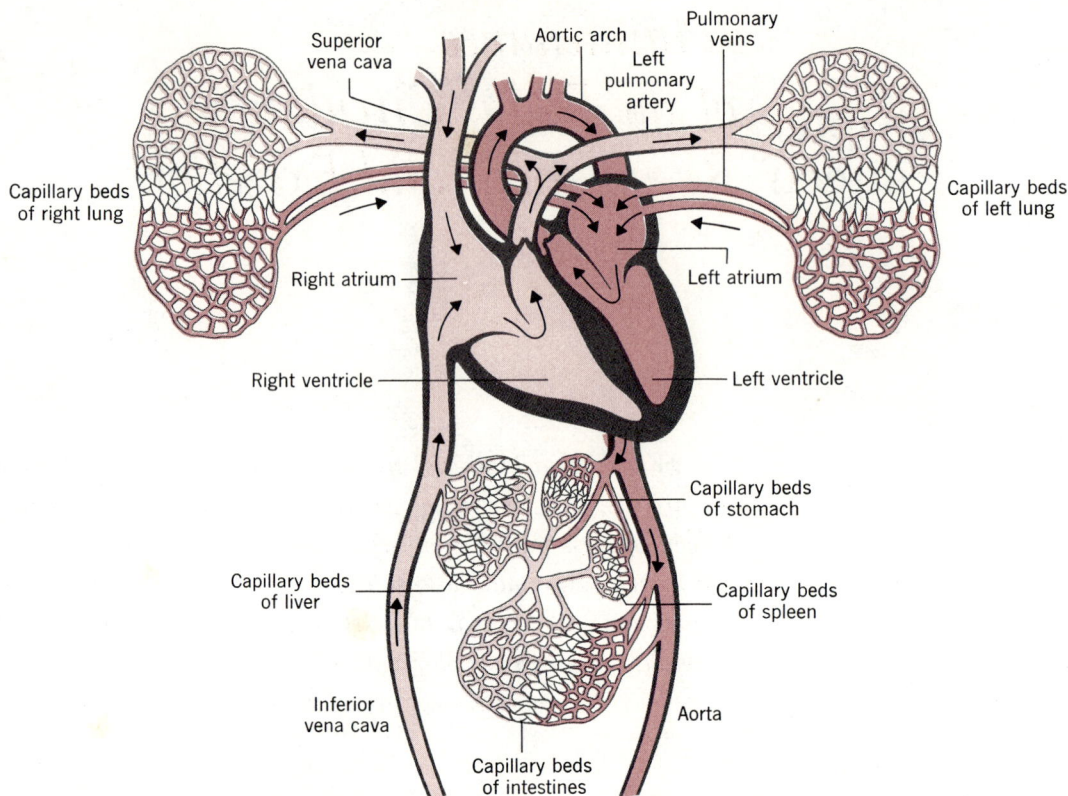

Figure 15-1   *A schematic view of the pathway of the blood through the two separate pumps of the heart. Aerated blood (shown in deep color) from the lungs returns to the left atrium and is pumped by the left ventricle through the aorta to the viscera, brain, and muscles of the body. Deoxygenated blood (shown in pale color) returns to the right atrium, and is pumped by the right ventricle through the pulmonary arteries to the lungs for reaeration.*

Composition and Functions of Blood

Blood has the following important functions: (1) It supplies the cells and tissues with oxygen and nutrients for growth, repair, and other life activities. (2) It transports the waste products of cellular metabolism, including carbon dioxide, to the excretory organs, where they are eliminated. (3) Its **white blood cells, antibodies,** and other protective substances are an important defense against disease and injury. (4) It helps regulate and equalize body temperature by evenly and rapidly distributing heat that is stored to a large extent in the blood and lymph. (5) It bathes the various body cells and tissues in fluid, thus regulating the water, acid, base, and salt content of all tissues, including itself, and exercising a significant effect on osmotic concentrations throughout the body. (6) It transports the secretions of the endocrine glands, the hormones (Chapter 16), which are so necessary for the regulation and coordination of the activities of the various cells and tissues. (7) It maintains its own composition and halts its own escape by a **blood-clotting mechanism.**

Blood is made of a fluid called **plasma,** which contains a variety of dissolved and suspended inorganic and organic substances, and three kinds of cell collectively called **formed elements:** red blood cells (erythrocytes), white blood cells (leucocytes), and blood platelets. Because these cells are present, blood is considered a tissue (p. 155). In higher animals such as man blood is about 8 percent of the body weight; an individual weighing 165 lb has about 6 qt of blood. Blood that has recently passed through the lungs and is therefore well oxygenated (as it usually is in the arteries) is scarlet; less oxygenated blood is dark red and has a bluish color (as seen in the veins). Blood is five to six times more viscous than water (supporting the old adage) and has a pH of 7.39.

Plasma, or **serum,** contains about 90 percent water, 7 to 8 percent plasma proteins, and 2 to 3 percent of dissolved substances such as salts (mainly sodium, calcium, potassium, and chloride), glucose, vitamins, lipids, hormones, and urea. Certain nerve cells in the brain are particularly sensitive to the blood glucose level, which is normally about .1 percent. If that level falls to .04 percent or less, convulsions and death result.

The blood of an adult contains about 5 million red blood cells *per milliliter.* Mature **red blood cells** are disc-shaped, about 8μ in diameter, and have no nucleus (Fig. 15-2). Each erythrocyte is packed with perhaps 300 million molecules of the oxygen-binding pigment **hemoglobin.** Under normal conditions erythrocytes are continuously being destroyed and new ones produced at the rate of about 140 million per minute; the average

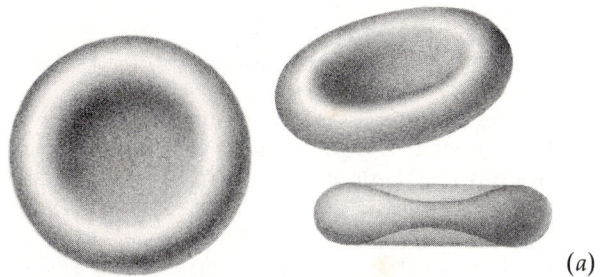

(a)

Figure 15-2   (a) *Mammalian erythrocytes are disc-shaped cells, thinner in the center than at the edge, with an average diameter of 7–8μ. Red blood cells are not self-propelled. They are flexible, and are often bent and twisted out of shape during their passage through the capillaries. (b) Four cells— the dark shapes in the middle, two on the left and two on the right—are shown in this electron micrograph. These mature cells are little more than masses of hemoglobin enclosed in membranes. Magnification 8500.*

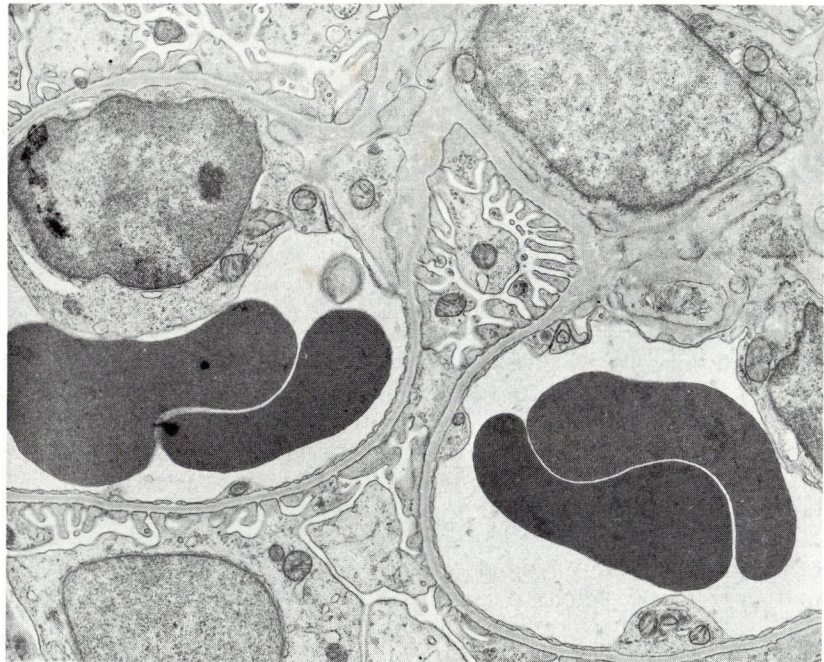

(b)

life of a red blood cell in the body is about 120 days. In an adult the formation of red blood cells takes place in the red bone marrow of the long bones and flat bones (e.g., the limbs and ribs). Before entering the bloodstream, erythrocytes appear as undifferentiated cells **(erythroblasts),** each with a nucleus but without hemoglobin. As these cells mature they synthesize vast quantities of messenger RNA, which, combined with ribosomes, produces the enormous hemoglobin content characteristic of mature cells. Toward the end of the maturation process 90 percent of the total synthesis occurring in each cell is devoted to hemoglobin. At this stage, for unknown reasons, the nucleus of each cell degenerates and the cell escapes from the bone marrow into the bloodstream as a mature erythrocyte.

The main function of **white blood cells** is to protect the body against invading microorganisms. Leucocytes differ considerably from red blood cells (Fig. 15-3a). They are devoid of hemoglobin, colorless, contain a nucleus, and are generally larger (8 to 15μ in diameter) than the erythrocytes. White blood cells are irregularly shaped, extremely active (Fig. 15-3b), and move under their own power by ameboid motion. By forming pseudopods they engulf and destroy foreign particles, including bacteria, in the process of phagocytosis (p. 126 and Fig. 15-3c), and thus are sometimes called **phagocytes.**

In a region of the body where bacteria have intruded—for example, at a wound—the capillaries dilate in response to the invading microorganisms and bring more blood to the area; the result is a reddish appearance of the tissue commonly called an **inflammation.** The white blood cells are somehow attracted to the source of irritation, especially during bacterial invasion. A dense aggregation of leucocytes collects and prevents the infestation from spreading by surrounding the threatened area and engulfing the invaders. The germs are ingested alive, in much the same manner that an ameba engulfs food particles. These accumulated white blood cells, together with destroyed tissue cells and erythrocytes, are collectively called **pus.**

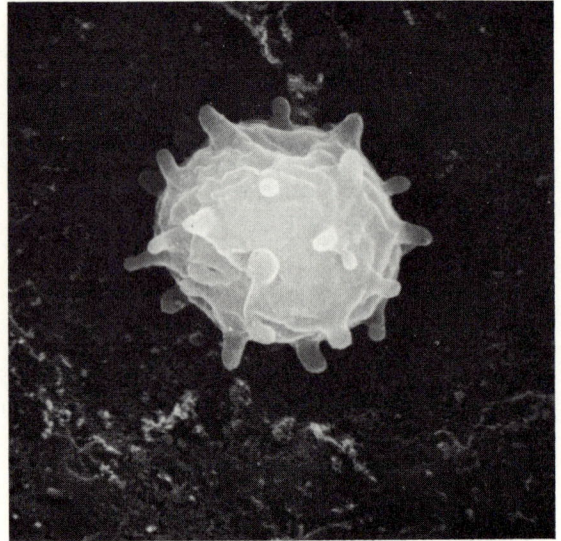

(a)

Blood platelets are small (2 to 4μ in diameter) oval fragments of specialized bone marrow cells. They usually lack nuclei. Their main function is in triggering the clotting of blood after an injury. Contrary to popular belief, blood does not clot as a result of exposure to air; rather, the formation of a blood clot involves the conversion of a soluble plasma protein, **fibrinogen,** to an insoluble protein, **fibrin,** in a network of fibers that enmesh the formed elements. A specific enzyme, **thrombin,** catalyzes this conversion.

Thrombin is ordinarily present in the blood in an inactive state known as **prothrombin,** but is activated by a process that begins when broken blood platelets release a complex group of blood-clotting substances. Thus any injury severe enough to disrupt the fragile platelets sets off a series of steps leading to the formation of a fibrin clot. If a small fragment of clot material breaks free in the blood stream it can cause damage to a rapidly metabolizing organ like the heart or brain by clogging blood flow to a localized region. Blood clots are especially dangerous in persons with hardening of the arteries, or **atherosclerosis,** a disease in which the blood vessels

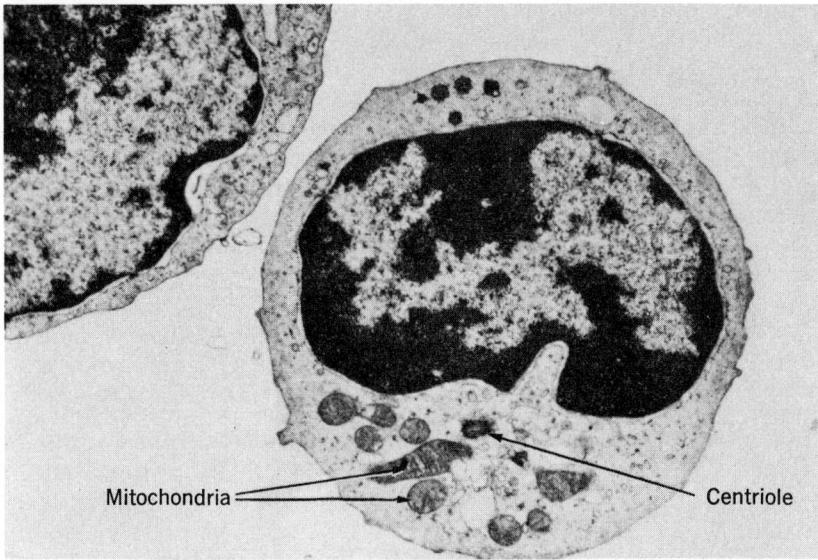

(b)

Mitochondria

Centriole

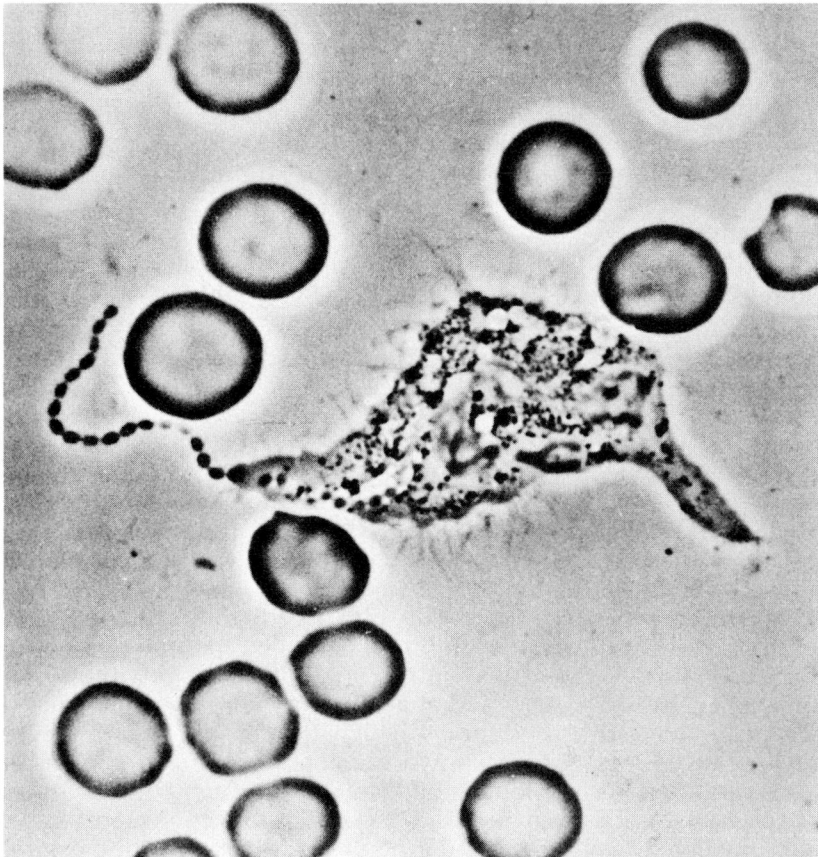

(c)

Figure 15-3 (a) This micrograph reveals the irregular surface features of leucocytes in contrast to the smooth surface of the doughnut-shaped, smaller erythrocytes. Magnification about 1500×. (b) Electron micrograph of a guinea pig lymphocyte. Magnification 3500×. (c) A human leucocyte phagocytizing a chain of bacteria. To get an idea of relative size, remember that the disc-shaped red blood cells are about 8μ in diameter. Magnification 2000×.

are reduced in diameter by the accretion of layers of fatty material.

Because each step in the chain of events leading to blood coagulation depends on the reaction before it, clotting can be prevented in such cases by blocking any one of the steps involved. Substances that inhibit or prevent coagulation are called **anticoagulants;** one of the most commonly used is **heparin,** which inhibits the conversion of prothrombin to thrombin.

If blood is to perform its functions it must circulate continuously through the body. Interruption of the blood flow to the brain, for example, causes a loss of consciousness within five seconds, irreparable destruction of the mental processes in a matter of minutes, and death shortly afterward. In other tissues also the failure of blood to flow may result in extensive damage often followed by death.

The Role of Blood in Immunity

The term **immunity** refers to the specific reactions by which invading microorganisms and their toxic products or other substances foreign to an animal are inactivated or destroyed. **Immunology** is the field of study concerned with these specific reactions.

Certain white blood cells produce specific proteins called **antibodies** in response to the introduction into the body of microorganisms and other foreign biological materials designated as **antigens.**

An antigen is defined as any substance that is capable of inducing the synthesis of a specific antibody. In general, antigens are natural products having molecular weights greater than 5000. They are foreign to a given animal because of their chemical structure. With few exceptions, virtually all proteins not produced by an animal are antigens, and many polysaccharides are also antigenic. The fact that some relatively simple nonprotein or noncarbohydrate compounds seem to act as antigens has been tentatively attributed to the union of their chemical substance with protein in an animal's tissues, which thus becomes capable of inducing antibody production.

Our present concept of antigens is that they possess particular chemical groupings in their molecular structure, called **determinant groups,** or **haptens,** equivalent to active sites on an enzyme, that are responsible for the specific reaction between antigen and antibody and, when chemically combined with a protein, for the induction of antibody formation.

Antibodies

An antibody is a specific protein that occurs in the body fluids as a direct result of the introduction of a specific antigen, and is characterized by its affinity for that antigen. All antibodies belong to a class of serum proteins called **immunoglobulins.** They may contain carbohydrates and perhaps lipids. Most antibodies in man have molecular weights of about 160,000 and consist of four polypeptide chains, of which two are so-called **heavy** chains and two are **light** chains. Although the majority of antibodies are found in the gamma-globulin fraction of the blood serum (p. 405), other antibodies also occur in other globulin fractions.

It seems clear that antibodies are synthesized from amino acids and are not formed by modifications of already existing globulin proteins. One major theory proposes that globulin synthesis is somehow altered by the presence of the antigen, which in an unknown manner directs the production of the new globulin molecules or antibodies. The number of antibody molecules produced is usually greater, in some instances by several million, than the number of antigen molecules.

In general, two broad groups of antibodies are recognized: so-called **natural** antibodies, which are present in the blood without apparent antigenic stimulation, and **acquired** antibodies, which appear only after exposure to a known antigen. The anti-A and anti-B blood group antibodies (p. 409), which are involved in blood type reactions, are examples of the natural type. The antibodies that are developed during the course of immunization are acquired.

The Antibody-Antigen Reactions

The combination, or **binding,** of antibodies to antigens is manifested in several different ways. One of the most common is the formation of a visible **aggregate,** or **precipitate.** It is usually called **agglutination** when antigens such as large, visible particles or whole cells (e.g., bacteria or red blood cells) are aggregated. It is called **precipitation** when soluble antigen molecules become insoluble. Frequently when antigenic bacteria have combined with an antibody they are more effectively phagocytized by the white blood cells. Another manifestation of antibody-antigen reaction is a disruption, or **lysis,** of cells. This is often observed with red blood cells (resulting in a release of hemoglobin) and with certain types of bacteria, involving a rupture or fragmentation of the cells.

Blood Groups

The complex nature of blood is very well illustrated by the phenomenon of blood types. In 1900 Karl Landsteiner, a young Viennese pathologist, discovered that human blood was not the same in all individuals. On the basis of the frequent occurrence of clumping (agglutination) of red blood cells when the blood of two different individuals was mixed, he found that human blood can be classified into four well-defined groups described in the following section. Before Landsteiner's Nobel-Prize-winning research, many deaths resulted from the mixing of incompatible blood during transfusion. His work led to a better understanding and the successful use of blood transfusion.

Blood Group Antigens and Antibodies

Blood grouping is founded on the presence or absence of two specific antigens located on the surface of red blood cells. These antigens, designated as A and B, are present on the erythrocytes of each individual's blood in one of three possible combinations (A, or B, or A and B), or are completely absent. When an individual has only one of the antigens, the blood group is called **A** or **B,** depending on which is present. When both are present, the blood is known as **AB;** when both are absent, the classification is designated by the letter **O** to indicate zero. The blood of all humans falls into one of these groups.

On the other hand, blood plasma contains one, both, or neither of two antibodies, which are complementary to the blood cell A and B antigens. The serum of blood type A always contains the specific antibody designated as **anti-B;** blood type B has the antibody called **anti-A;** type AB has neither; and type O has both anti-A and anti-B. Because anti-A reacts specifically with antigen A and anti-B with antigen B, if type A and B bloods are mixed together, the anti-A from the plasma of group B causes an agglutination of the type A or type AB erythrocytes and the anti-B from the plasma of type A blood does the same to type B or type AB cells. The four different blood groups with respect to both the antigen and antibody are: A_{anti-B}, B_{anti-A}, AB, and $O_{anti-A, anti-B}$.

Blood Transfusion

If a group A individual should inadvertently receive a transfusion of type B blood, his anti-B plasma will react with the incoming erythrocyte B antigen, resulting in the clumping of the donor's red blood cells. The A antigen of a B donor's blood would be expected to react similarly with the anti-A of a B-type patient, with the same effect. Such agglutination can lead to serious complications and death. Aside from the possibility of clogging the smaller blood vessels and thus impairing circulation in vital areas, the subsequent breakdown of the clumped cells may cause kidney damage through obstruction of the kidney tubules. Obviously bloods of types A and B should never be mixed; the main concern in blood transfusion is the clumping of the donor's red blood cells in the vessels of the recipient. The recipient's erythrocytes generally escape clumping, because the donor's plasma antibodies are sufficiently diluted during transfusion.

Individuals belonging to group AB are **uni-**

versal recipients because they contain neither anti-A nor anti-B and will therefore produce no clumping of any donor's red blood cells. Individuals of the O group are regarded as **universal donors** because they contain neither A nor B antigens, which means that their red blood cells will not be agglutinated no matter what the composition of the receiver's blood.

Rh Factor

Antigens other than A and B are also present on the red blood cells. The **Rh factor** (first found in the Rheusus monkey, thus accounting for the designation Rh) is an antigen found on the red blood cells and is responsible for their clumping under certain conditions. Individuals who possess the antigen are called **Rh-positive,** and make up about 85 percent of the American population; the remaining 15 percent are **Rh-negative.** The introduction of Rh-positive blood into the bloodstream of an Rh-negative individual stimulates the production of specific antibodies by the latter within approximately two weeks.

This phenomenon may have serious consequences in the developing fetus of an Rh-negative mother and an Rh-positive father, for it can result in a blood condition fatal to the fetus or to the newborn child. If the blood of the fetus is Rh-positive, inherited from the father, formation of the antibody to Rh-positive erythrocytes is usually stimulated in the mother's Rh-negative blood because there is a slight mixing of the fetal and maternal bloods. Therefore during subsequent pregnancies any mixing of the mother's blood, now containing the Rh-antibodies, with that of an Rh-positive fetus will cause extensive agglutination and destruction of the fetal red blood cells.

The Heart

The heart, by means of its pumping action, propels blood throughout the entire human body continuously, 24 hours a day, year after year, during the average 70-year lifetime of an individual. The heart is actually two pumps, fused together and beating in unison. The right side of the heart, which is concerned with **pulmonary circulation,** receives blood returning from all parts of the body (except the lungs) and sends it to the lungs; aerated blood returns from the lungs to the left side of the heart, which in turn propels it out to all other parts of the body. This is called **systemic** circulation (Fig. 15-1).

The remarkable and almost legendary characteristics of the heart can be illustrated by many statistics and analogies. It is usually regarded as synonymous with life itself, the last heartbeat marking the onset of death. As a pump weighing approximately 11 oz—the size of an average grapefruit—the adult human heart continuously circulates 6 qt of blood through a network of vessels that, if laid end to end as a single tube, would extend some 60,000 miles, 2.5 times the earth's circumference. In a resting individual the heart pumps approximately 5 qt of blood per minute, 75 gal per hour. It beats at a steady tempo of one to two beats per second, more than 100,000 times a day, 36 million times a year. In an average lifetime of 70 years it beats approximately 2.5 billion times, pumping nearly 18 million barrels, or more than 400,000 tons, of blood and doing enough work to raise a weight of 45 tons to a height of 5 miles.

Structure and Action of the Heart

The heart is almost entirely cardiac muscle. This muscle tissue consists of long, striated muscle fibers (Chapters 6 and 18) that differ in important characteristics from skeletal muscle. It used to be thought (and many textbooks still state) that the long, branching muscle fibers of the heart are syncytial, that is, contain the nuclei of many cells that have fused together; skeletal muscle fibers do form in this way (Fig. 12-19). But in the last 15 years or so examination of heart tissue under the electron microscope has shown that cardiac fibers are chains of many cells—each with one or occasionally two nuclei—joined end to end (Fig. 15-4a, p. 412). In each heart cell the nucleus lies deep in the center, surrounded by striated **myofibrils** (Fig. 15-4b) similar to those found in skeletal muscle. Interspersed among the myofibrils are rich concentrations of large mitochondria, called **sarcosomes,** which provide the energy required for the continued repeated contractions. The cells may be simple cylinders, or they may be branched, but each cell abuts on its neighbor at each end across a specialized structure called an **intercalated disc** (Fig. 15-4c,d). Recall that the apposed surfaces of cells that appear to be in contact are actually separated by a gap of about 200 Å (Chapter 6). Within the intercalated disc region the surfaces of adjoining heart cells establish much closer junctions at which the cell membranes remain separated by a space no greater than about 20 Å, just resolvable at the highest magnification of the electron microscope (Fig. 15-4e, p. 413). It is through these close junctions, called **nexuses,** that electric current can pass from cell to cell, allowing a contractile stimulus (action potential; see p. 413) to be conducted down the entire length of a fiber. Several fibers may be bound together by connective tissue into bundles and sheets of muscular tissue.

The main pumping chambers of the heart are two **ventricles,** one on each side, which are separated by a thick muscular dividing wall called a **septum.** Each ventricle is connected to its own antechamber, called an **atrium,** or **auricle** (Fig. 15-5, p. 414), which is a collecting point for delivery of blood to the ventricles. Blood low in oxygen and high in carbon dioxide enters the right atrium of the heart via the two largest veins of the body, the **superior** and **inferior vena cava.** It then passes into the right ventricle and is propelled by way of the pulmonary artery to the lungs, where carbon dioxide is discharged and oxygen is taken up. The freshly oxygenated blood returns from the lungs through the pulmonary vein into the left atrium. It then flows into the left ventricle, which pumps the blood out through the largest artery, the aorta, and ultimately to almost all parts of the body.

The opening from the right atrium to the right ventricle is guarded by the **tricuspid valve,** made up of three flaps of connective tissue; the blood is permitted to flow from atrium to ventricle but not in the reverse direction. The corresponding valve between the left atrium and the left ventricle is similar in structure and action, except that it has two flaps instead of three and is accordingly called the **bicuspid valve.** Each ventricle is also equipped with an outlet valve **(semilunar valve),** opening away from the heart into the pulmonary artery and the aorta.

When the ventricles fill with blood they begin to contract. The pressure causes the inlet valves between the atria and the ventricles to snap shut, producing the typical sounds heard through the physician's stethoscope. Almost at the same time the outlet valves open, permitting the discharge of blood to the pulmonary artery and the aorta. The subsequent relaxation of the heart and the consequent fall in pressure in the ventricles are followed by the closing of the outlet valves (also accompanied by characteristic sounds) and the opening of the inlet valves, again permitting blood to enter from the atria. Defective or leaky valves produce additional sounds or "murmurs" as they close, which may furnish important clues about the nature of the defect.

Although the average volume of each ventricle is about the same, the muscular wall of the left ventricle is considerably thicker than that of the right; the left ventricle must be a more powerful pump to propel the blood on the longer route through the body. Blood from the right ventricle passes only to the lungs, a much shorter route, and needs a pressure of only about 20 percent of that on the left.

Control of the Heartbeat

The average pulsation rate of the heart differs considerably among different members of the animal kingdom. For example, the heart of a horse or a whale beats 30 to 40 times per minute, that of a mouse 300 to 500 times per minute, and the human heart about 70 times per minute. The rate is slightly lower during sleep and is appreciably increased by physical or emotional excitement.

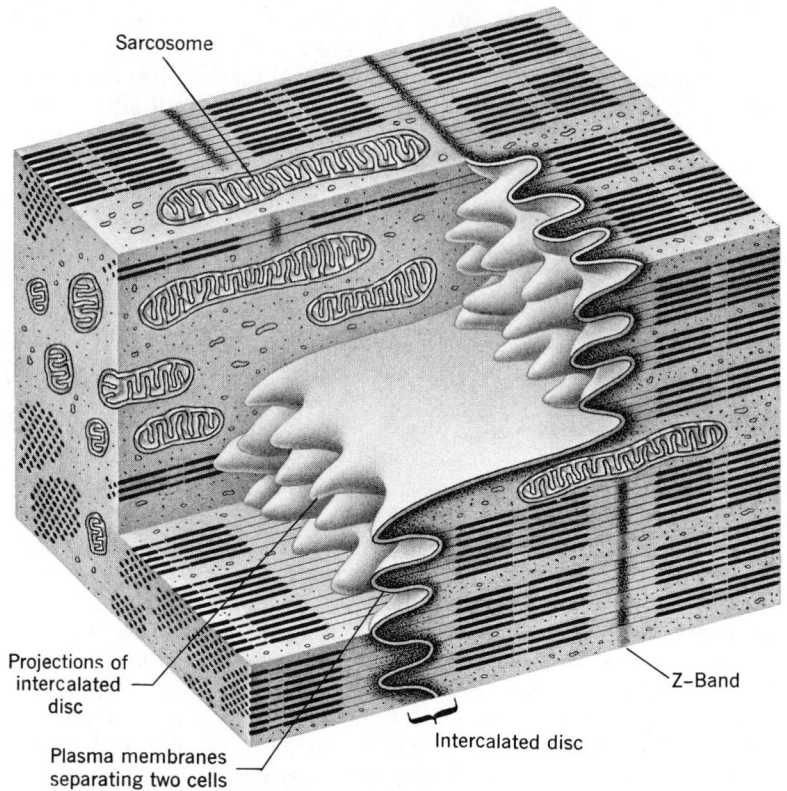

Figure 15-4   Ultrastructure of human cardiac
muscle. (a) Diagram of a
heart muscle cell to show
myofibrils, an end view of
an intercalated disc, and
other intracellular structures.

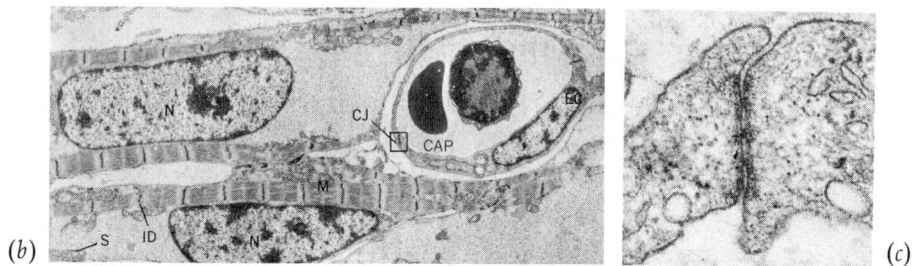

Sarcosome

*Projections of
intercalated
disc*

*Plasma membranes
separating two cells*

Intercalated disc

Z–Band

(a)

(b)

(c)

Figure 15-4 (cont'd.)   (b) Parts of two cells from a human fetus, showing
striated myofibrils (M), the two cell nuclei (N), and muscle mitochondria or
sarcosomes (S). The circular structure in the upper right is a capillary (CAP)
of the coronary system formed by two thin endothelial cells (EC) joined at
their edges by close junctions (CJ), enlarged in the inset at right (Fig. 15-4c).
In the lower left corner a tiny intercalated disc (ID) crosses a single myofibril.
Magnification 5000×. (c) Endothelial close junction (enlarged from box in b),
showing that at some points the cell membranes appear actually to fuse.
Magnification 65,000×.

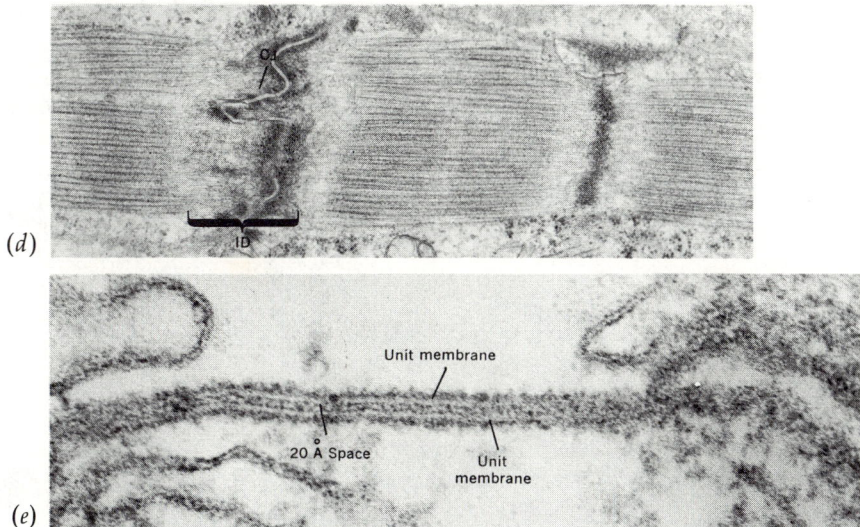

(d)

(e)

Figure 15-4 (cont'd.)   (d) Enlarged eight times over Fig. 15-4b, the intercalated disc is seen to consist of the interdigitated membranes of the adjoining cells crossing a myofibril. Within the folded membranes of the intercalated disc a small region of close junction is seen. Magnification 40,000×. (e) A close junction (Cj) enlarged to near the limits of electron microscopy shows the structure of the unit membranes of the two apposed cells, separated by a gap of about 20 Å. Magnification 300,000×.

In 1628 William Harvey cut the heart out of an eel, and—keeping it moist with saliva—found that it would continue beating rhythmically outside the body. This was the first piece of experimental information indicating that the ability of the heart to maintain a rhythmic beat is an inherent property of the heart itself. Unlike skeletal muscle, contractions of the heart do not depend on stimuli from the nervous system. We now know that the hearts of all vertebrates, including mammals, contain specialized muscle cells called **pacemaker** cells. These are capable of generating and transmitting an electrochemical disturbance called an **action potential,** similar to a nerve impulse, which can be conducted from cell to cell and stimulates each cell to contract. Each heartbeat originates from a small mass of these specialized cells, called a **pacemaker node** or **sino-atrial (S-A) node,** situated in the rear upper right wall of the right atrium (Fig. 15-5, p. 414).

From the S-A node the action potential travels across the muscular walls of the two atria, stimulating the muscle fibers to contract in a progressive and coordinated manner, starting first in the right atrial wall and spreading rapidly to the left atrium. These impulses also reach another knot of nodal tissue, located between the ventricles, called the **atrio-ventricular (A-V) node.** The impulses stimulate this node to transmit similar impulses rapidly (via a specialized network of conducting fibers) to every muscle fiber of the two ventricles, resulting in their coordinated and almost simultaneous contraction. Relaxation of the atrial walls begins at about the time the ventricles start to contract.

Each heartbeat, then, consists of progressive and coordinated contraction (**systole,**

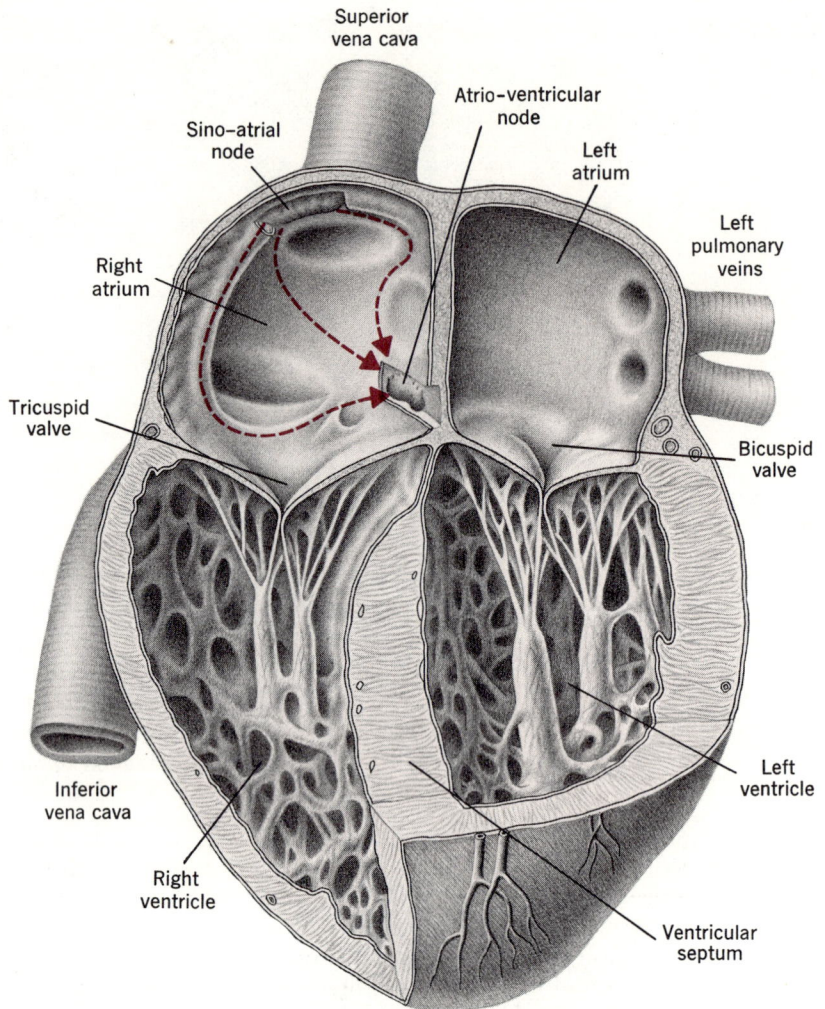

Figure 15-5 The human heart, with the walls cut away to show the internal structure.

pronounced sis'-tol-lee) of the heart, expelling blood into the arteries, followed by its relaxation (**diastole,** di-as'-tol-lee) and resultant refilling of the heart chambers by blood from the veins.

Although the heart will continue to beat rhythmically under the control of the S-A node alone, the heartbeat rate is in part modified by the nervous system. The S-A node has nerve endings from both the sympathetic and parasympathetic nervous systems (Chapter 17); impulses arriving from the parasympathetic nervous system slow down the beat, whereas impulses arriving from the sympathetic nervous system accelerate it. The two sets of nerves can be traced back ultimately to a specific nerve center in the medulla oblongata region of the brain (p. 476). This heart-rate center is in turn influenced by impulses from other parts of the body, including those from higher centers of the brain. Thus anger, fear, and other emotional states may exert a profound effect on the heartbeat rate.

Obviously the complete story of heartbeat control is complex, the result of the interplay of several factors. High temperatures, low blood pH (as a result of higher-than-normal concentrations of CO_2) and certain hormones such as adrenaline and the thyroid hormone also increase the heartbeat rate, apparently by acting directly on the S-A node.

The amount of blood pumped by each contraction of the heart is called the **stroke volume.** Each stroke volume is normally about 75 ml of blood, which means that about 5 qt of blood, almost the total amount of blood in the body, is pumped through the heart every minute. Not all the blood passes through the heart in one minute, however. The blood on shorter routes, such as those nourishing the cardiac muscle, circulates through the heart a number of times, whereas the blood on longer routes, such as those to the legs, takes much longer to return. Under physical or emotional stress the volume of blood pumped by the heart in a given time may be increased several fold by an accelerated heart rate and by an increase in the volume of blood pumped during each ventricular systole. Within limits, the greater the stretching of cardiac muscle fibers before the contraction, the greater the force of the contraction and therefore the greater the amount of blood ejected during systole. The extent of cardiac muscle stretching in turn depends on the amount of blood that has flowed into the ventricle by the end of the diastole. In brief, we have a self-regulatory mechanism whereby the volume of blood entering the heart plays an important role in determining the contraction strength of the heartbeat and therefore its output. If the heart muscle is stretched too far by a flow of excess blood into the ventricle because of leaky valves, the contractions may weaken and even cease, resulting in death.

The Riddle of Heart Attack and Atherosclerosis

About 10 percent of the total blood pumped out of the heart passes directly into the muscles of the heart via the two coronary arteries branching off the aorta (Fig. 15-1). After serving every muscle fiber of the heart, the blood flows, via the coronary veins, back to the right atrium of the heart.

Any impairment in the delivery of oxygenated blood to a portion of the heart muscle will seriously disrupt the metabolism of the cardiac muscle fibers concerned, causing appreciable damage in a very short time. The sudden blocking of one of the coronary arteries or its large branches by a blood clot results in a **heart attack.** The rapid deterioration of the oxygen-starved cardiac muscle fibers of heart-attack patients results in the release of several enzymes such as transaminase and lactic dehydrogenase from the tissue into the bloodstream. The striking increase of these enzymes and other symptoms (chest pains, breathlessness, collapse, etc.) are important clues in the diagnosis of a heart attack. During recovery the necrotic (dead) muscle fibers are slowly replaced by scar tissue; the damaged region is referred to as an **infarct.**

A heart attack may be fatal if sufficient heart tissue is deprived of oxygenated blood; heart attack is the leading killer in the United States, causing about one-sixth of all American deaths and approximately one-fourth of all deaths among men over 35. A major factor in a large majority of heart attacks is **atherosclerosis,** the most common form of **hardening of the arteries.** In atherosclerosis the inside diameter, or **lumen** of the arteries become narrowed by fatty deposits that gradually harden into tough, fibrous plaques. The presence of these obstructions beneath the lining of the arteries, especially in the coronary arteries and their branches, somehow sets the stage for blood-clot formation and possible resultant heart attack.

Atherosclerosis is extremely widespread among adult males of our western culture. Some estimates indicate that approximately 40 percent of all men over 40 have this condition in one or more main branches of their coronary arteries. The causes are still obscure, although most evidence indicates a disturbance in lipid metabolism. Arterial plaques themselves are rich in fats, and especially in the fatty substance **cholesterol** (Chapter 3), the raw material for making the sex hormones and other hormones of similar chemical structure known as steroids. We know that high cholesterol levels in the blood are correlated with an increased incidence of atherosclerosis; however, a complex interplay of numerous factors, including heredity and diet, seems to be involved in this disease. Female sex hormones apparently provide some protection; atherosclerosis occurs to a considerably lesser extent in women. It is also known that elevated blood pressure and excess weight definitely have unfavorable effects.

The Blood Vessels

Arteries and Arterioles

Arteries transport blood away from the heart. They vary considerably in size, from the aorta (1 in. in diameter), which subdivides into smaller and smaller arteries, to those of almost microscopic dimensions, the **arterioles** (about .2 mm in diameter). Arteries are characterized by thick, muscular, strong, elastic walls made up of essentially three layers of tissue: an outer layer of thick, fibrous connective tissue; a middle layer of smooth muscle and elastic tissue; and an inner layer (endothelium) of elastic membrane and epithelial cells. (Fig. 15-6).

The smooth muscles in the arterial wall are connected to both sympathetic and parasympathetic nerves, one set controlling constriction and the other dilation of the arteries. Thus the opening of the artery may vary from very narrow to very wide, depending on the degree of contraction of the muscle fibers. The size of the lumen of an artery is responsible in part for the rate at which blood flows through it, and the distribution of blood to the different parts of the body according to their needs is mainly controlled by the diameter of the arterioles that lead to the capillaries in those areas. Contraction of the smooth muscle in the walls of the arterioles in one area constricts the blood vessels, forcing blood out of that tissue, while relaxation of the smooth muscles produces dilation of the arterioles in another region. The state of contraction or relaxation of the arteriolar muscles is regulated by two sets of nerves (sympathetic and parasympathetic), by the amount of CO_2 and other chemical factors produced locally in each tissue, and by the action of specific hormones such as adrenaline and vasopressin.

The first branches from the aorta are the two coronary arteries that curl around the surface of the heart. The aorta then forms an ascending arch from which there are three major arterial branches: (1) the **innominate**

Figure 15-6 Structure of arteries, veins, and capillaries.

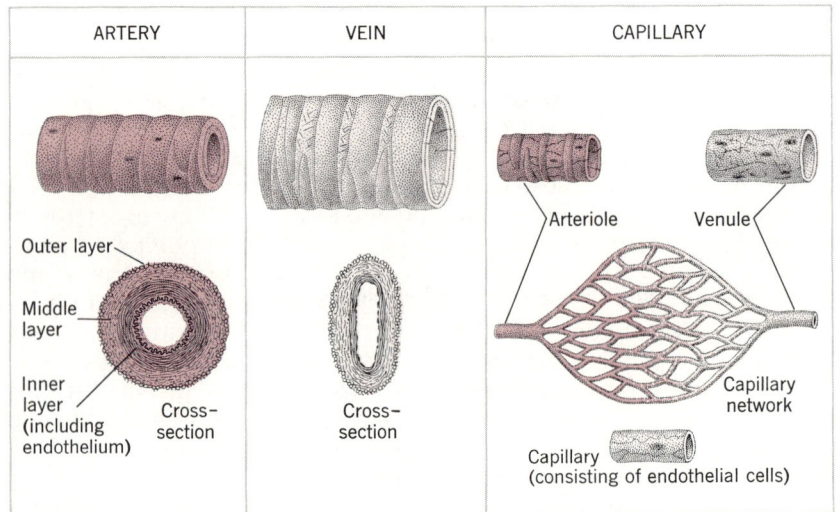

ARTERY   VEIN   CAPILLARY

Outer layer
Middle layer
Inner layer (including endothelium)
Cross-section
Cross-section
Arteriole
Venule
Capillary network
Capillary (consisting of endothelial cells)

artery, which divides almost immediately into the **right subclavian** artery, carrying blood to the right shoulder and arm, and the **right common carotid** artery, which goes up through the right side of the neck and provides blood to the right side of the head; (2) the **left common carotid,** which supplies blood to the left side of the head; and (3) the **left subclavian** artery, which conveys blood to the left shoulder and arm (Fig. 15-7).

The aorta descends through the chest cavity and abdomen, sending out some small branches (not shown in Fig. 15-7) that transport blood to the chest walls and a number of large arteries that pass to the viscera. The first of these large arteries is the **coeliac** artery, which subdivides into smaller arteries carrying blood to the liver, stomach, pancreas, spleen, and a portion of the small intestine. The next is the **superior mesenteric** artery, leading to the remaining portion of the small intestine and to a segment of the large in-

Figure 15-7 *Major arteries and veins in the human. Arteries (shown in deep color) carry blood away from the heart to the organs of the body. Veins (in pale color) return blood to the heart.*

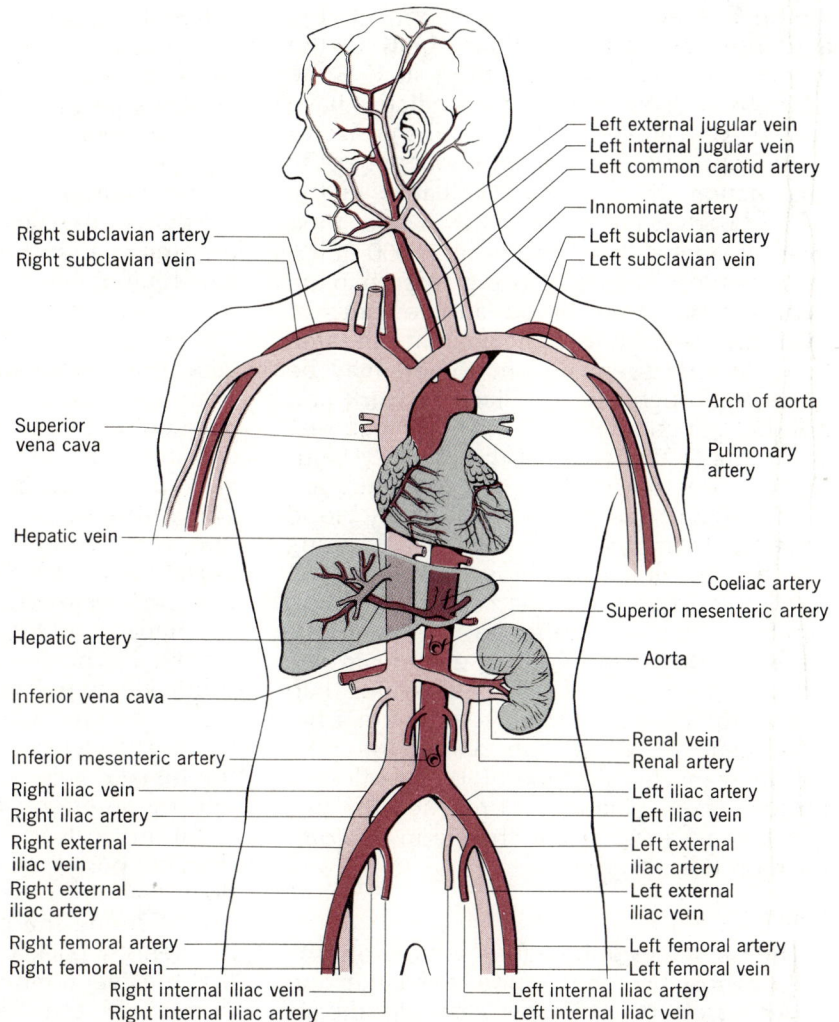

Right subclavian artery
Right subclavian vein

Left external jugular vein
Left internal jugular vein
Left common carotid artery
Innominate artery
Left subclavian artery
Left subclavian vein

Superior vena cava

Arch of aorta
Pulmonary artery

Hepatic vein

Hepatic artery

Coeliac artery
Superior mesenteric artery
Aorta

Inferior vena cava

Inferior mesenteric artery
Right iliac vein
Right iliac artery
Right external iliac vein
Right external iliac artery
Right femoral artery
Right femoral vein
Right internal iliac vein
Right internal iliac artery

Renal vein
Renal artery
Left iliac artery
Left iliac vein
Left external iliac artery
Left external iliac vein
Left femoral artery
Left femoral vein
Left internal iliac artery
Left internal iliac vein

testine. Then there are two **renal** arteries, one to each kidney. The last major arterial branch from the aorta in the abdominal region is the **inferior mesenteric** artery, which provides blood to a section of the large intestine and to the rectum. The aorta finally forks in the lower abdomen into the **right** and **left iliac** arteries; these extend into the legs to become the **femoral** arteries.

Capillaries

The arterioles lead into a branching series of microscopic vessels, the capillaries, which are made up of a single layer of endothelial cells. Capillaries are present in all tissues of the body; they are most highly concentrated as dense nets, called **capillary beds** in the most metabolically active regions such as the brain, heart, liver, and so on. Capillaries have an average diameter of about 7μ—the size of a red blood cell. Their total number defies the imagination: if all of the capillaries in the human body were placed end to end, their over-all length would be about 50,000 miles. Because their total capacity is greater than the total volume of the blood, all capillaries of the body are obviously not open at the same time. At any given moment blood may be flowing through one capillary network and not through another, with the shift occurring according to the needs of the tissues. Extensive data demonstrate that the concentration of oxygen and nutrients in the capillary blood is greater than in surrounding cells, resulting in a net movement of these substances out of the blood and into the cells and tissues. Similarly, the concentration of metabolic waste products and carbon dioxide in the surrounding cells and tissues is greater than that of the capillary blood, resulting in a net movement of these substances into the bloodstream. By the time blood has flowed through the capillaries and reached the first small veins, a significant change in its composition has occurred.

Veins

The capillaries unite into larger vessels, the veins, whose branches join to form larger and larger vessels that ultimately carry blood back to the right side of the heart. The veins have essentially the same three tissue layers as the arteries, but with certain modifications (Fig. 15-6): the middle layer in veins is thinner and far less developed than in arteries of comparable size; veins have a smaller quantity of muscle than arteries, but a larger diameter; they have no connections with the nervous system and no inner elastic membrane. Some veins have internal valves that prevent the backflow of blood.

The venous circulation to a large extent parallels the arterial system (Fig. 15-7). Blood drains from each side of the head by a pair of large veins, the **right** and **left internal** and **external jugular** veins, the venous counterparts of the carotid arteries. Each set of jugular veins unites with the corresponding **subclavian** veins, which in turn join to form the **superior vena cava,** transporting blood to the right atrium.

The **femoral** veins from the legs lead into the **iliac** veins, carrying blood from the lower extremities into the large **inferior vena cava,** which ultimately drains into the right atrium. In its ascent through the abdomen the inferior vena cava receives blood from a **renal** vein from each kidney and the **hepatic** vein from the liver.

The blood leaving the major organs of the digestive tract (the stomach, small intestine, and large intestine) does not drain directly into the inferior vena cava; rather, it flows into veins connecting to the large **hepatic portal** vein, which leads to the liver. (Note that the liver also receives aerated blood from the hepatic artery, a branch of the coeliac artery). In the liver the hepatic portal vein subdivides into smaller veins and ultimately into capillaries, which then unite into veins again. The veins leaving the liver join to form the **hepatic** vein, which then drains directly into the inferior vena cava. This is a unique situation; instead of an artery, it is a vein (hepatic portal) that leads to capillaries that drain back into a vein (hepatic).

It is during the passage of blood through the densely packed capillary bed of the liver that the liver removes and stores a significant portion of the freshly absorbed digestive products (amino acids, simple sugars, and vitamins). The subsequent release of these stored materials to the capillaries by the liver under appropriate conditions results in an enrichment of blood leaving the liver.

Blood Pressure

If an artery is cut blood gushes out with considerable force, indicating that it is under pressure in the circulatory system. The first reported measurement of blood pressure was made in 1733 by an English biologist named Stephen Hales, who connected a vertical glass tube to one of the carotid arteries of a horse. The blood reached a height of 9 ft 6 in. in the glass tubing and its level fluctuated rhythmically, rising with each ventricular systole and falling with each diastole. It is now more common to measure blood pressure with a device called a **sphygmomanometer,** which contains a vertical column of mercury instead of blood or water; mercury is 13 times heavier than water and therefore requires a column only a few inches high to counterbalance the blood pressure. By scientific convention, blood pressure is generally stated in terms of millimeters of mercury.

Blood pressure varies appreciably in different parts of the vascular system. The highest pressure is encountered in the aorta, close to the heart, where it normally ranges to 140 mm of mercury that is, a vertical column $5\frac{1}{2}$ in. high. It progressively decreases in the arteries and arterioles more distant from the heart, becoming lower in the capillaries and still lower in the veins. It is lowest in the inferior and superior vena cava veins, reaching 0 mm of mercury just before it enters the right atrium. The sharpest drop in blood pressure occurs between the arterioles and small veins, where the friction of the blood is greatest as it passes through the capillaries. Blood pressure variations are roughly indicated by the rate and force with which blood escapes from cut vessels—rapid, forceful spurts from an artery, a slow flow from capillaries and veins.

Arterial blood pressure undergoes a rhythmic change, reaching a maximum during ventricular systole when blood is pumped into the arteries (accordingly called **systolic pressure**), and decreasing during ventricular diastole (called **diastolic pressure**). The normal systolic pressure in man ranges from about 120 to 140 mm of mercury, whereas disastolic pressures range from about 75 to 90. These vary considerably, however, and are affected by numerous factors including age, sex, heredity, and physical and emotional states.

The Pulse

As each ventricular systole causes blood to spurt into the aorta the elastic walls of the aorta are temporarily distended. This slightly expanded area is transmitted as a wave along the arterial walls that passes to all arteries and arterioles. This phenomenon is called the **pulse wave,** or simply the **pulse.** The pulse is most conveniently detected by placing a finger over an artery near the body surface, say at the wrist. The pulse does not signal the arrival of blood at the wrist, but rather the transmission along the elastic arterial walls of the impact of blood suddenly entering the aorta under high pressure. The pulse travels independently of blood flow or pressure, 10 to 15 times more rapidly than blood itself. Its velocity is approximately the same throughout the arterial system, in contrast to the progressively decreased blood pressure in arteries farther away from the heart.

The Lymphatic System

Lymph is a fluid derived primarily from blood plasma that has filtered out through the capillary walls into the spaces surrounding all cells. It has about half the protein concentrations that blood plasma has. Concentrations of other substances in the lymph also differ from plasma, depending on additions and withdrawals made by the surrounding cells.

Although some lymph may reenter the bloodstream directly by the capillaries, most of it passes instead into a separate network of vessels collectively called the **lymphatics,** or **lymphatic system.** The lymphatics are

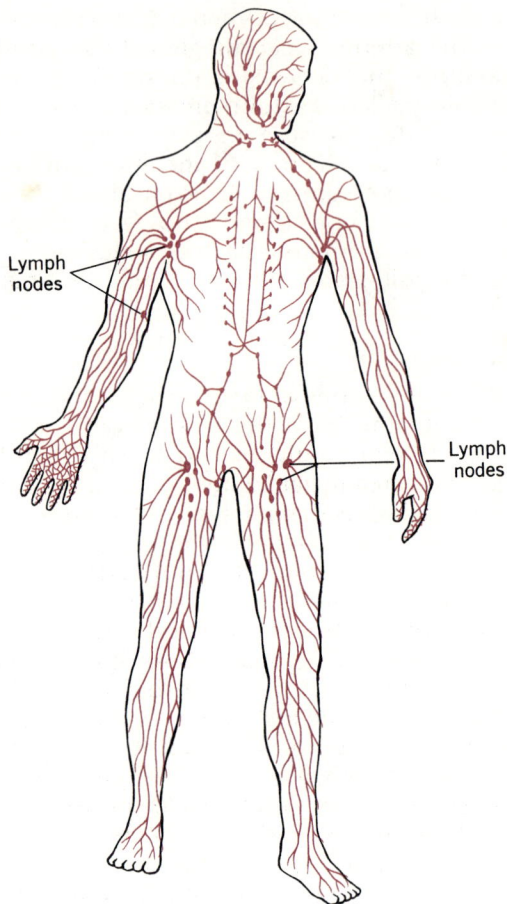

Figure 15-8 *The lymphatic system in man.*

directly to the venous bloodstream.

At many points in the body where the smaller lymph vessels unite to form larger ones there are lumpy masses or aggregations of cells known as **lymph nodes** (Fig. 15-8). These play an important role in the body's defense against disease. From the lymph nodes several kinds of lymph cell, or **lymphocytes,** are derived. These function mainly in producing special globular proteins, called **antibodies,** which combine with foreign substances such as invading bacteria and viruses to make them more susceptible to destruction by the white blood cells, or phagocytes (p. 406). Antibodies are one of the body's main defenses against disease-producing organisms.

The lymphatic system performs a number of other important functions. It accounts for the return of most intercellular fluid to the bloodstream, although the blood capillaries also take up some of this fluid. Also, because the microscopic lymph vessels have greater permeability to large protein molecules, fats, and so on, than the capillaries have, the lymphatic system returns and transports these substances to the bloodstream; many of the products of fat digestion are absorbed from the small intestine by the lymphatic network in the villi. Finally, the lymph nodes tend to remove not only microorganisms from the body fluids, but also any foreign particles that may have broken loose from other tissues and entered the lymph stream. In some instances the trapping of bacteria in the lymph nodes may result in an infection and swelling of the nodes, or, in the case of freely circulating cancer cells, the trapping and subsequent proliferation of malignant tissues in the nodes.

The Spleen

The spleen is probably best classified as an organ belonging to the circulatory system. It is about the size of a fist, reddish-brown in color, and situated in the left side of the abdomen behind the stomach. It contains a large amount of smooth muscle, and its internal structure is permeated by small areas of lymphoid tissue and many cavities con-

found in all parts of the body (Fig. 15-8) and range from the size of capillaries to the size of large veins. Their wall structure is like that of the capillaries and veins, and the larger lymphatic vessels possess valves that prevent backflow. One important difference, however, is that the lymphatics do not have a continuous closed circulation; the microscopic lymph vessels that correspond to capillaries terminate in a dead end. Lymph is largely taken up by these tiny vessels and moved slowly in one direction to the larger lymph vessels, which ultimately converge into two large trunks. Each trunk then drains into a subclavian vein, thus returning the fluid

nected directly to the blood vascular system.

The spleen appears to serve as a unique blood reservoir, containing blood of a high red cell concentration in its cavities. Hemorrhage, physical or emotional stress, low oxygen pressure in the blood, carbon monoxide poisoning, or high environmental temperatures cause the smooth muscles of the spleen to contract, squeezing the stored blood directly into the bloodstream. In addition, the spleen produces lymphocytes, destroys worn-out red blood cells (as does the liver), and contributes to the production of red blood cells in the embryo. The spleen is not essential under normal conditions; it can be removed without causing any serious effects. It is probable, however, that under certain stress conditions such as those just mentioned it could be critical in determining survival of the organism.

Respiration

Gas Exchange

To burn nutrients as fuels and obtain energy from them, cells require dissolved oxygen for oxidation reactions, and they produce carbon dioxide by those reactions. The term **respiration** refers fundamentally to this exchange of gases between a living cell and the air. Unfortunately for the beginning student, however, we use "respiration" and "respire" in at least three different but related ways, each referring to a different level of the overall process. In Chapters 3 and 4 we described how oxygen is used as an electron acceptor in "respiratory metabolism." We might think of this as respiration at a **molecular level.**

Recall that the molecules of a gas, free or in solution, experience constant spontaneous movement (Fig. 2-1). We have referred to this random movement as diffusion, and have demonstrated that it results in a net movement of molecules from a region of high concentration to one of lower concentration. In all organisms dissolved oxygen moves into cells and CO_2 moves out of them by diffusion across the cell membranes. In unicellular ani-

mals and small aquatic organisms—sponges, coelenterates, some worms, for example—no specialized structures are needed; the exchange of oxygen and CO_2 between the organism and its water environment is carried out directly and adequately by diffusion (Fig. 15-9, p. 422). The gas exchange that takes place across the cell membrane may be considered as respiration at the **cell level.**

As a general rule, the greatest distance an adequate supply of oxygen can diffuse through tissue to maintain a reasonable level of metabolism is about .5 mm. Therefore animals dependent on simple diffusion are limited in size or thickness to about 1 mm. As organisms evolved multicellularity and an internal environment (Fig. 14-1), and as they became larger, the problem of gas exchange became more complicated. All higher animals have evolved means of exchanging oxygen and carbon dioxide between the innermost cells of an organism and its external environment. Most have evolved channels through which air or water, containing dissolved oxygen and CO_2 can pass deep into the organism. This process, too, is called respiration. The organs and tissues of the body that function together as a system to accomplish the exchange of gases are called the **respiratory system;** Thus the process of breathing (inhaling and exhaling) is included as another meaning of respiration—at the **organismic level.**

A common gas-exchange organ among aquatic animals is the gill. Gills present a large surface to the water in which they are bathed; oxygen in the water diffuses into their surface cells. The gills are supplied with circulating blood, which transports oxygen to the interior tissues of the body. The same blood picks up carbon dioxide from the tissues and transports it out to the gills, from which CO_2 passes into the water by diffusion (Fig. 15-10, p. 423). It is important to note that all such systems include three distinct, essential parts: a **circulating fluid** that acts as a gas-transport system; an extensive **epithelial surface** across which gases can diffuse between the internal environment (circulating fluid) and the external environment (the

Figure 15-9 In the aquatic environment of a pond or ocean, as in a well-balanced aquarium, oxygen is released and CO_2 is consumed during photosynthesis in green plants. Oxygen is consumed and CO_2 is produced during cellular respiration. Every cell in a small animal such as the thin planarian worm is close enough to the body surface for gas exchange to take place by simple diffusion across the surface, just as in the protista.

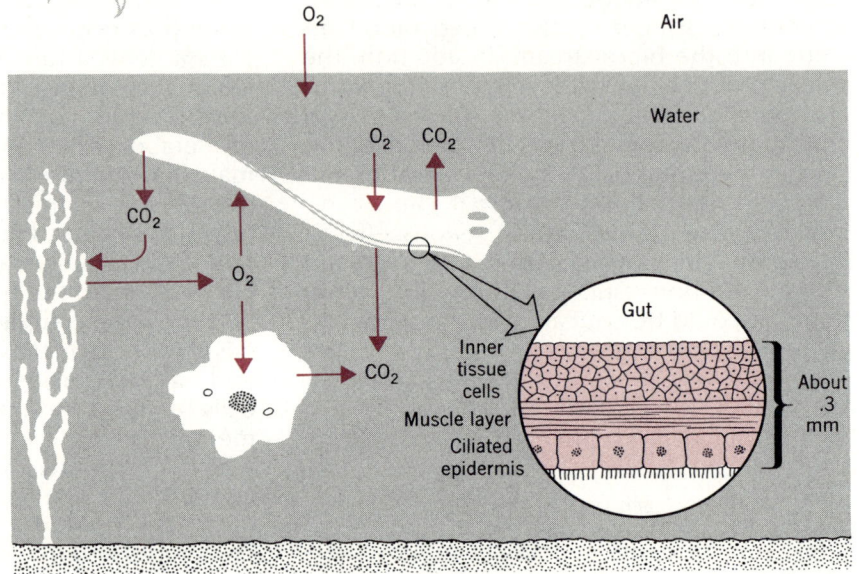

water); and a mechanism to provide a **continuous oxygen supply,** in this case from fresh water over the gill surface.

The problem of gas exchange among terrestrial (land-living) animals is complicated by the fact that an organ like a gill, covered with a delicate layer of cells, tends to dry out if exposed to air instead of water. Among some species, such as the insects and spiders, this problem has been solved by an elaborate network of tubes, the **tracheae** (Fig. 15-11), through which air is pumped into the internal tissues of the organism.

All of the terrestrial vertebrates (amphibians, reptiles, birds, and mammals) have solved the gas-exchange problem by developing lungs, which are, in effect, gills folded inside the body. Instead of exposing the respiratory surface to a watery environment, these organisms have evolved a specialized respiratory epithelium, in close contact with an oxygen-rich supply of blood capillaries, which is kept moist by mucous secretions.

The surface area of the respiratory system must be extensive to meet the requirements of an organism for obtaining an adequate supply of oxygen and disposing of waste carbon dioxide. The human body at rest, for example, must obtain 250 ml of oxygen from the surrounding atmosphere each minute, and may require up to 10 times this much during physical exertion. To obtain this much oxygen it must present to the atmosphere a large surface containing many blood vessels for gaseous exchange. This surface, the **pulmonary alveoli** of the lungs, has a total surface area of more than 100 square yd—some 50 times greater than the body surface.

The Respiratory System of Man

In man, as in all other forms, respiration has three phases: the exchange of gases between the blood and external air; the transport by the blood of dissolved gases from the lungs to other cells and tissues of the body; and the exchange of gases between the blood and body cells. The third phase fulfills the ultimate function of the respiratory system: it provides oxygen to body cells and tissues and rids them of waste carbon dioxide.

In man the respiratory system includes the lungs and the air passageways that lead to them (Fig. 15-12, p. 424).

Figure 15-10 The gills of a fish serve to transport gases between the surrounding water and the interior tissues of the body.

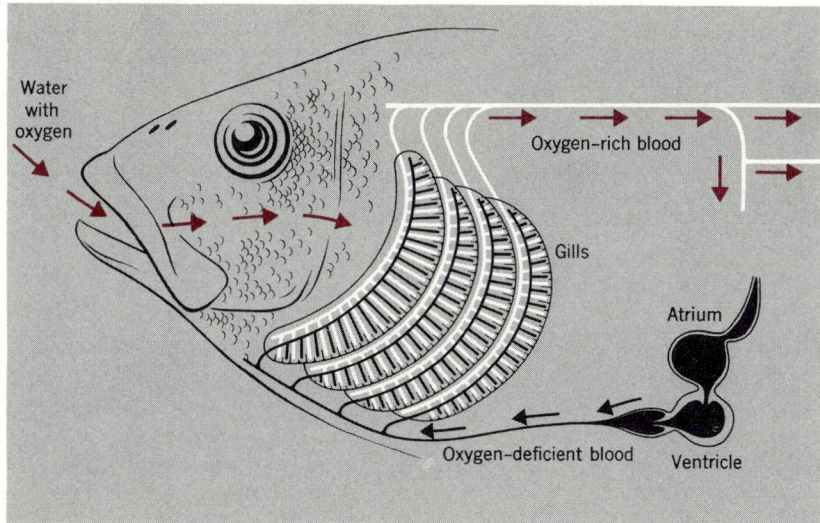

Water with oxygen

Oxygen-rich blood

Gills

Atrium

Ventricle

Oxygen-deficient blood

Figure 15-11 Tracheal system of an insect.

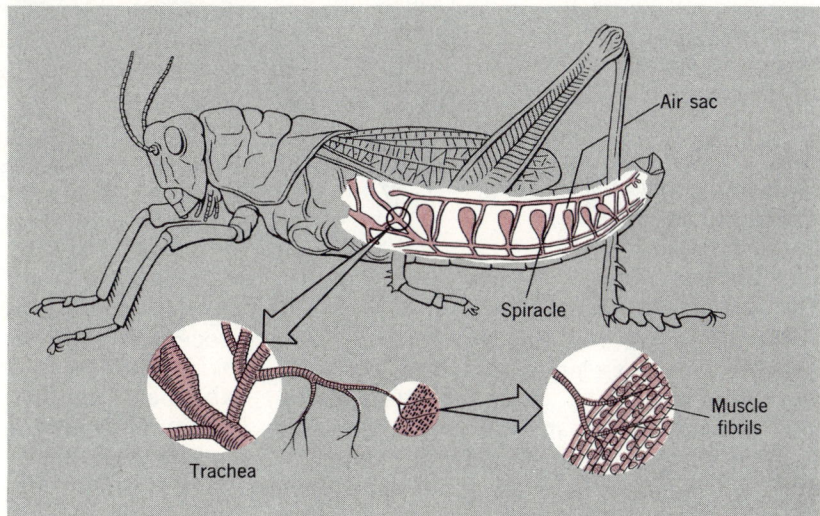

Air sac

Spiracle

Trachea

Muscle fibrils

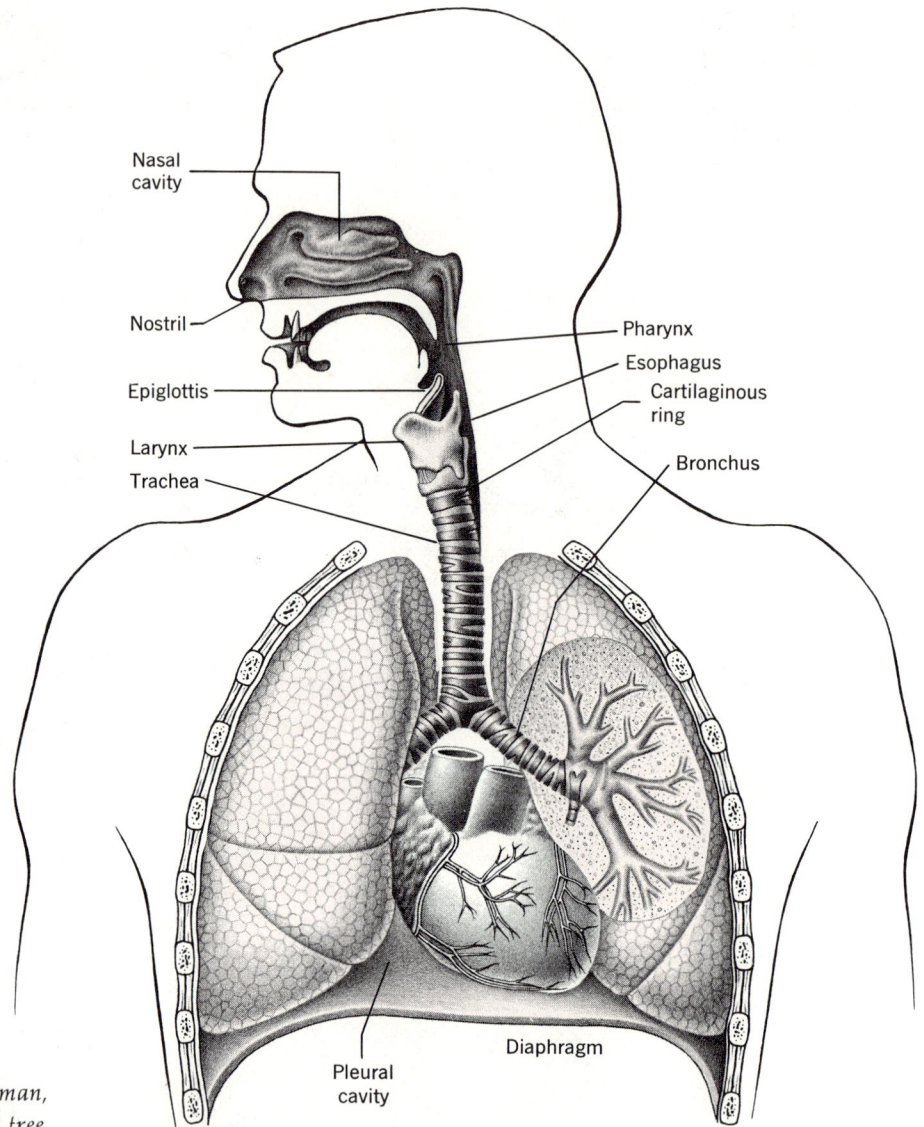

Figure 15-12   The respiratory system of man, showing the bronchial tree.

Labels on figure:
Nasal cavity
Nostril
Epiglottis
Larynx
Trachea
Pharynx
Esophagus
Cartilaginous ring
Bronchus
Pleural cavity
Diaphragm

The Passageways

The air passageways are the nostrils, nasal cavities, pharynx, larynx, trachea, bronchi, and bronchioles. Air enters by the **nostrils** and during its passage through the **nasal cavities** is warmed, moistened, and filtered of dust and foreign particles by the mucous membrane covering the bony ridges, called **turbinates,** that project into the nasal cavities.

Odors in the air register in specialized smell **receptor cells** in the mucous membrane, which are located in a small upper recess of the nasal cavities. The cavities lead to the pharynx by two internal openings.

The **pharynx,** or throat, is a muscular passage through which food passes on its way from mouth to esophagus and air passes on its way from the nasal cavities to the trachea (windpipe). The **larynx,** or voice box, located in the neck just below the pharynx, is made of several cartilage sections. Inside the larynx is a cavity called the **glottis;** it is lined with mucous membrane that is continuous with that of the pharynx and trachea. The larynx is responsible for voice or sound production; stretched across the glottis are two thin-edged, fibrous bands, the **vocal chords.** Vibrations of the vocal chords are produced by the passage of air through the voice box; if the vocal chords are drawn short by the chord muscles, like a tightly drawn violin string, they produce high notes; low-pitched notes are produced by longer, more relaxed vocal chords.

During its passage to the esophagus, food must pass over the opening of the larynx; to keep food out of the wrong channel, the entrance to the larynx is guarded by a hinge-like cartilage section, the **epiglottis,** which automatically closes during the act of swallowing.

The **trachea,** or windpipe, is a tubular structure about 4 to 5 in. long, approximately 1 in. in diameter. Its inner surface is lined with ciliated epithelial cells whose cilia beat upward, sweeping small foreign bodies away from the air passageway. The trachea is in front of the esophagus and extends from the larynx downward through the neck just into the chest (or **pleural**) cavity, or **thorax,** where it divides into two tubes called the **right** and **left bronchi.** The trachea wall and the bronchial tubes contain a series of C-shaped cartilage rings that prevent them from collapsing, thus maintaining an open passageway for air (Fig. 15-12). The bronchial tubes subdivide into smaller and smaller branches; at a diameter of 1 mm or less the cartilage rings are entirely lacking, and the tubes are now called **bronchioles.** Their walls, made up essentially of circularly arranged smooth muscle, are also lined on the inner surfaces with epithelial tissue. The muscle of the bronchioles expands and contracts as a reflex response to signals from both the sympathetic and parasympathetic nervous systems. Excessive bronchiolar constriction is responsible for the respiratory distress that is commonly a charac-

teristic of asthmatic attacks.

The trachea, the two connecting bronchi, and their many branches resemble an inverted miniature tree, and for this reason are often referred to collectively as the **bronchial tree** (Fig. 15-12). No interchange of gases can occur across its relatively thick walls; its primary function is to serve as a passageway by which air reaches the area of the lungs where this exchange does occur.

The Lungs

As they proceed deeper and deeper into the lungs, the bronchioles continue to divide and subdivide, each finally opening into one of the many air sacs that are the functional units of the lung. Each air sac looks like a cluster of grapes (Fig. 15-13) and consists of several (usually four to six) microscopic outpocketings of the bronchiole called **alveoli.** The total number of alveoli in both lungs is estimated at nearly 1 billion. It is at the alveo-

Figure 15-13   A microscopic portion of the human lung, showing a bronchiole leading into two air sacs. Each air sac consists of several alveoli, tiny spherical bubbles .2–.3 mm in diameter, formed of a single layer of thin squamous epithelial cells and surrounded by a rich network of blood capillaries to absorb oxygen and transfer CO_2.

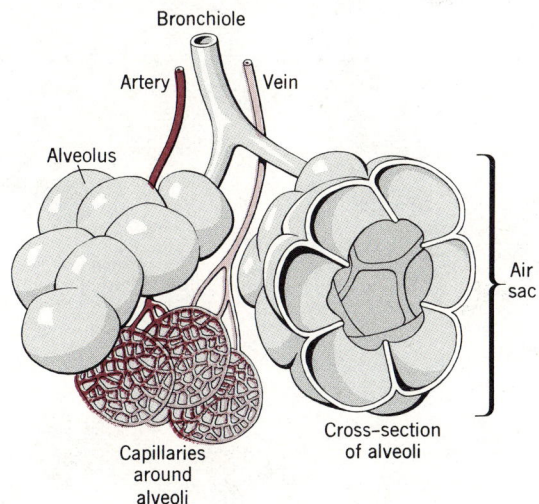

Bronchiole
Artery
Vein
Alveolus
Air sac
Cross–section of alveoli
Capillaries around alveoli

lar surfaces that the interchange of gases between air and blood occurs; the walls of the alveoli are simply a single layer of epithelial cells that overlie a rich network of capillaries and are admirably suited to function as a respiratory membrane (Fig. 15-14).

Although blood is pumped through the lungs at a rapid rate (approximately 5 L per minute), several factors assure adequate gas exchange during this relatively short period. First, within the rich network of capillaries surrounding the alveoli blood is distributed in extremely thin layers and is therefore exposed to a large alveolar surface (about 1000 sq ft). Second, the blood cells pass through the lung capillaries largely in single file, so that each comes in close proximity to the alveolar air. Finally, the blood in the lungs is separated from the alveolar air only by the extremely thin membranes of the capillaries and alveoli (estimated to be no more than 4μ in total thickness).

Each lung, then, has a bronchial tree with its many air sacs and alveolar units, together with associated structures such as blood vessels and nerves. Both lungs and the interior surface of the chest cavity are covered with a thin membrane called a **pleura.** Under normal circumstances, during expansion of the lungs in the process of breathing, the part of the pleura that covers them and the part of the pleura that lines the internal chest surface come in contact. There is usually just enough lubricating mucous fluid between the two pleuras to avoid friction and irritation during contact. When the pleuras are inflamed the

Figure 15-14 *An electron micrograph of the alveolus and adjacent capillary from the lung of a mouse. Note how thin the epithelial layer is between the capillary and the air space. Magnification 1600×.*

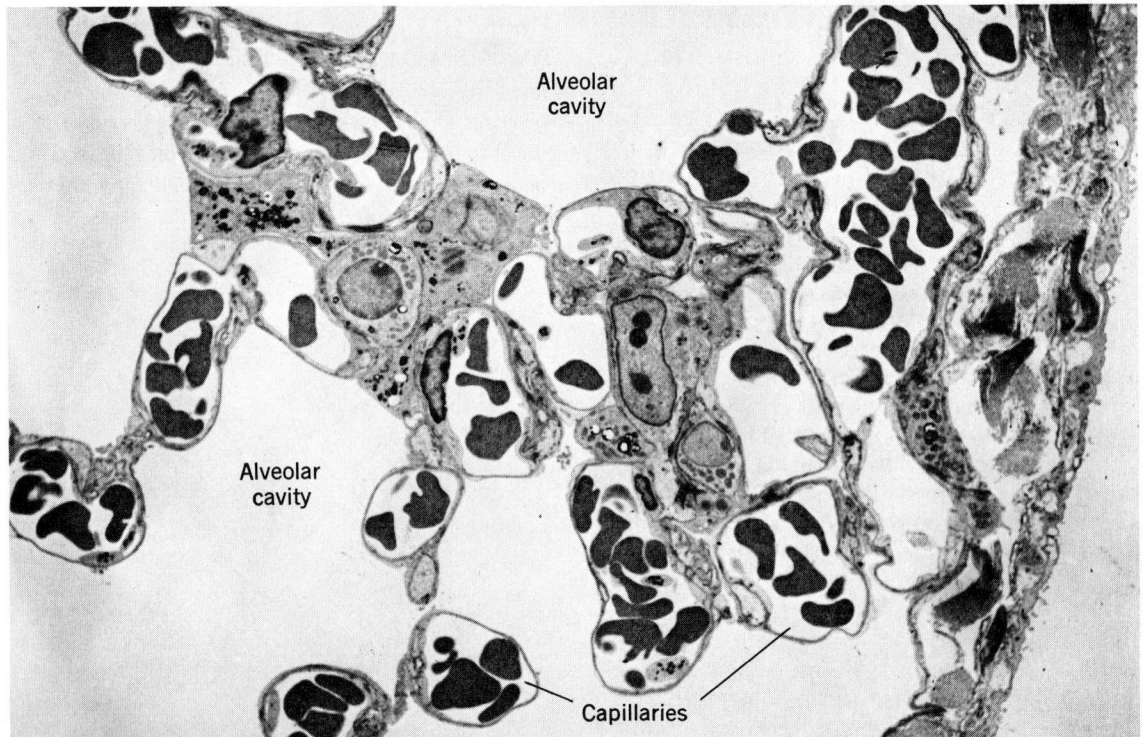

condition is known as **pleurisy,** and is frequently accompanied by the accumulation of excess fluid between the two membranes.

The enormous total surface area of the alveoli normally allows for ample gas exchange. Under certain conditions, however, this area can be severely reduced. For example, during some viral or bacterial infections, some of the alveoli become filled with a fluid of lymph and mucus, reducing the surface area exposed to air. This condition of fluid-filled alveoli is termed **pneumonia.** In critical cases a patient may turn blue from oxygen deficiency (recall that deoxygenated blood is darker in color than that containing an adequate amount of oxygen). Such a patient may be allowed to breathe pure oxygen under an oxygen tent until his body—with the aid of antibiotics—can reduce the infection and the fluid can be reabsorbed from his lungs.

Another disease of the lungs in which the gas-exchange area is reduced is **emphysema.** In this disease the alveolar walls actually break down (Fig. 15-15) as a result of irritation and death of epithelial cells. Unlike pneumonia, this condition develops slowly and is rarely the direct cause of death. However, the gradual loss of gas-exchange surface area forces the heart to pump increasing quantities of blood to the lungs in order to satisfy the needs of the body, and this added strain on the heart can lead to heart failure. In recent years the frequency of emphysema has increased at an alarming rate. There is now good evidence to show that the disease is aggravated or even caused by breathing air polluted with certain contaminants; industrial wastes, automobile exhaust, and cigarette smoke seem to be involved.

Mechanics of Breathing

The closed chest cavity that houses the lungs is bounded by the ribs, spinal column, breastbone, body wall, and, at the bottom, by a dome-shaped sheet of skeletal muscle, the **diaphragm.** Breathing is a mechanical process in two phases, the taking in of air, called **inspiration,** and the letting out of air, or **expiration** (Fig. 15-16, p. 428). Under resting conditions the two phases occur alternately and

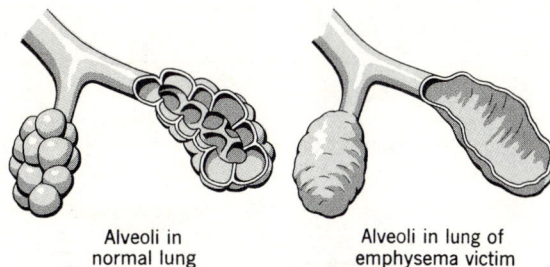

Alveoli in
normal lung

Alveoli in lung of
emphysema victim

Figure 15-15   The structure of the alveoli in a normal lung and the lung of a patient with severe emphysema. In this disease the alveolar walls break down, reducing the surface area for gas exchange.

rhythmically, 12 to 20 times per minute.

The lungs themselves do not actively expand and contract; they respond passively to the contraction and relaxation of the diaphragm and muscles of the chest wall. During an inspiration the muscle between the ribs contract, lifting the ribs up and out; the dome-shaped diaphragm is pulled down, increasing the volume of the pleural cavity and forcing the lungs to expand. About 500 ml of air is drawn in and exhaled with each breath. As a result of exercise or conscious thought an individual can breathe more deeply merely by contracting all these muscles more strongly. With great effort an average adult can flush his lungs with 3 to 5 L of air at each breath. This is known as the **vital capacity** of the lungs, and can be increased to about 5 L with athletic training. Even at the point of maximum expiration about 1200 ml of air (the **residual volume**) remains in the lungs.

Regulation of Breathing

The breathing movements are entirely under the control of the nervous system. The process is essentially involuntary; it is performed automatically and usually not at the level of awareness, although it can to a certain extent be placed under the control of the will—we can voluntarily speed up or slow down or even halt the rate and depth of breathing movements for a limited period of time.

In all instances, however, whether the

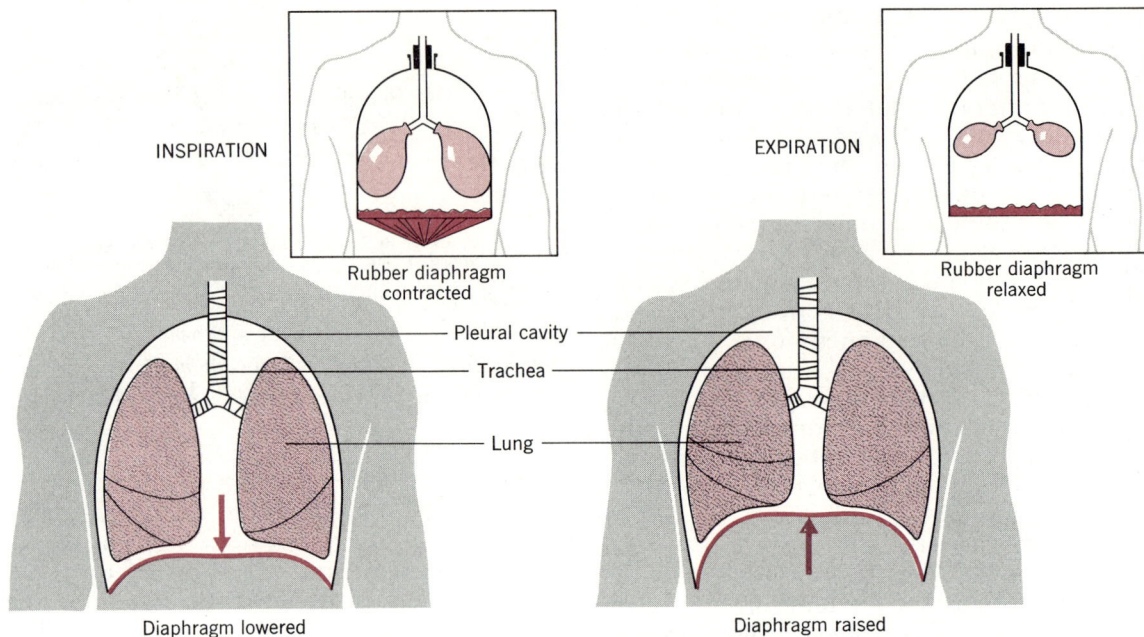

INSPIRATION

Rubber diaphragm contracted

EXPIRATION

Rubber diaphragm relaxed

Pleural cavity

Trachea

Lung

Diaphragm lowered

Diaphragm raised

Figure 15-16   The mechanics of breathing can be illustrated by comparing the respiratory system with a model made of two balloons (representing the lungs) attached to a Y-shaped glass tube (the bronchial tree), all inside a glass bell jar (the pleural cavity) with a rubber diaphragm stretched across the bottom (the diaphragm). When the volume of the pleural cavity is increased by pulling the diaphragm down, the lungs expand as air is drawn in through the trachea and bronchial tree.

breathing movements are voluntary or involuntary, the contraction and relaxation of the respiratory muscles depend on the regular and alternating discharge of nerve impulses from a group of nerve cells, collectively called the **respiratory center,** in the medulla oblongata of the brain (Chapter 17). The respiratory center in turn consists of two subdivisions: an **inspiratory center** and an **expiratory center.** The inspiratory center sends out nerve impulses that cause the contraction of the diaphragm and the elevation of the ribs. When the expanding lungs reach a certain size, specialized stretch-sensitive nerve endings in the lung tissues are stimulated to send out nerve impulses that inhibit the inspiratory center. That center stops discharging impulses to the respiratory muscles,

which consequently relax and the expiration phase of breathing is initiated. As the lungs deflate, the stretch receptors are no longer stimulated, and the inhibitory effect of the expiratory center ceases. Within a few seconds the impulses from the inspiratory center are resumed, and the breathing cycle starts again.

Voluntary control of breathing is mediated through the same nervous-system mechanisms as involuntary breathing; during voluntary breathing impulses are transmitted from the cerebral cortex to the respiratory center. Strong emotions, fright, excitement, and other mental or emotional factors also have a striking influence on the breathing process. Coughing and choking are reflexes of protective value that are usually caused by stimulation of nerve endings in the lining of the air passageways of chemicals or foreign particles; these reflexes generally result in expulsion of the irritating substances.

It would be reasonable to assume that the factor that triggers the inspiratory center might be an oxygen deficiency in the blood. Experiments have shown that this is not usually the case. Rather, the amount of **carbon dioxide** in the blood is probably the most important factor that influences the respiratory center and therefore the rate and depth of the breathing process; the slightest elevation in the carbon dioxide concentration in the blood stimulates the respiratory center in the brain, resulting in an increased rate and depth of the breathing cycle.

Increased **acidity** of the blood has a similar stimulating effect on the respiratory center. For example, the elevated concentrations of both carbon dioxide and lactic acid that result from muscular exercise are responsible for the accompanying increase in the rate and the depth of breathing.

Under ordinary circumstances, then, the normal variations in oxygen content are of little importance in the regulation of breathing. Only large decreases in the concentration of oxygen in the blood affect the breathing rate; for example, at high altitudes (10,000 ft or more) the oxygen in the atmosphere, and therefore in the blood, is greatly reduced. Special receptors located in the walls of the aorta and the carotid arteries respond to major decreases in blood oxygen level and stimulate the respiratory center of the brain to increase the breathing rate. At even higher altitudes, where the amount of oxygen in the atmosphere is so low that more rapid breathing can no longer compensate for the insufficiency of oxygen, **altitude sickness** results.

Transport of Oxygen and Carbon Dioxide

Air normally consists of 20 percent oxygen, just under 80 percent nitrogen, .04 percent carbon dioxide, and traces of other gases. When blood is exposed to air it **equilibrates** with it; that is, it absorbs oxygen at the same fractional level as it exists in the air (20 percent, or 20 ml of oxygen per 100 ml of blood). An oxygen-saturated solution of water (or of serum alone) carries only about .3 ml of oxygen per 100 ml of solution, not nearly enough to supply the actively metabolizing tissues of the body.

Because of its structure, hemoglobin in the red blood cells has the remarkable ability to take up and carry molecular oxygen; this accounts for the capacity of the blood to transport 60 times as much oxygen (20 percent) as the oxygen-carrying capacity of the blood plasma alone would permit. A hemoglobin molecule consists of four heme groups (p. 75), each containing an iron atom. The four are attached to the protein globin. When hemoglobin has an oxygen molecule attached to each heme group it is called **oxyhemoglobin.**

Alveolar air does not contain 20 percent oxygen; that deep in the lung oxygen is reduced to about 13 percent of the air. The blood in the capillaries surrounding the alveoli has the same amount of oxygen (about 13 ml per 100), having come into equilibrium with the alveolar air. At that oxygen level 98 percent of the hemoglobin is oxygen saturated; that is, it is in the form of **oxyhemoglobin.** Deep in the tissues of the body, where the oxygen level is lower still (1 to 5 percent), oxygen is liberated from the oxyhemoglobin and diffuses out from the blood through the thin capillary walls and eventually into the cells. This is made possible by the important fact that the combination of oxygen and hemoglobin in the red blood cells to yield oxyhemoglobin is reversible.

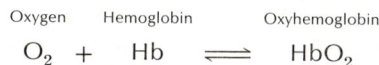

Oxygen	Hemoglobin		Oxyhemoglobin
O_2	$+$ Hb	$\rightleftharpoons$	HbO_2

That is, hemoglobin picks up oxygen when it is equilibrated with a gas mixture rich in oxygen and releases it in a low-oxygen environment. Hemoglobin would serve no useful purpose in the transportation of oxygen if it formed a union with oxygen that was not easily dissociable.

The relationship between level of oxygen and the uptake and release of oxygen by the blood is shown by an **oxygen-hemoglobin dissociation curve** (Fig. 15-17). Increasing percentage of oxyhemoglobin is represented along the vertical axis, and increasing level of oxygen to which the blood is exposed along the horizontal axis. The colored curve in the figure tells us that hemoglobin is almost completely oxygenated (95 percent or more) by an oxygen level of 13 percent, which is the same as that in the lungs. This means that for all practical purposes hemoglobin can be as completely oxygenated by 13 percent oxygen as it would by any higher oxygen level, say if we were breathing in pure oxygen. The S-shape of oxygen from the hemoglobin dissociation curve illustrates both the high oxygen-carrying capacity of hemoglobin and its ability to unload oxygen very rapidly where it is needed most by the cells and tissues of the body.

The curve also tells us that when the oxygen pressure falls below 5 percent (as it does in many of the cells and tissues of the body) the affinity of oxygen for hemoglobin decreases very sharply. This is indicated by the steeper slope of the curve at the lower oxygen levels; it means that oxyhemoglobin releases its oxygen very rapidly below about 5 percent, resulting in the liberation of relatively large quantities of oxygen in areas where it is most needed.

The movement of gases between the blood and the cells of the body is, of course, the reverse of movement in the lungs because the level of oxygen in the tissues is low and that of carbon dioxide high, compared with that of the blood.

The carbon dioxide concentration significantly affects the capacity of hemoglobin to combine with oxygen. Increasing levels of carbon dioxide from metabolically active tissue shift the oxygen-hemoglobin dissociation curve to the right, as is shown by the black curve in Figure 15-17; this means that the quantity of oxygen that hemoglobin will hold at any given oxygen level decreases with increasing carbon dioxide concentrations. Thus more oxygen is released to cells and tissues

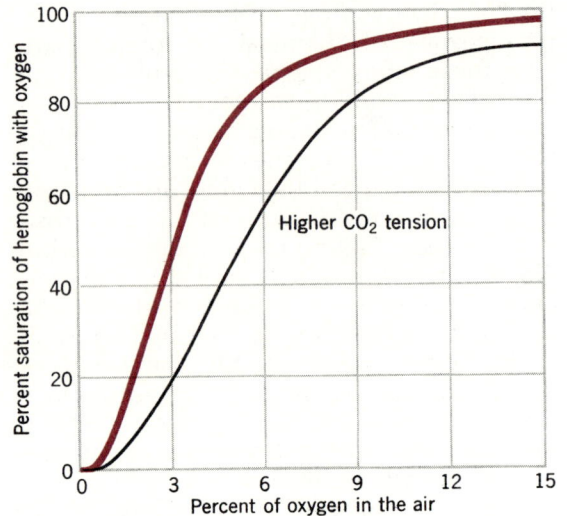

Figure 15-17 *The oxygen-hemoglobin dissociation curve, showing that at oxygen levels above about 5 percent most of the hemoglobin is in the form of oxyhemoglobin. At lower oxygen levels, oxygen is released and the percentage of oxyhemoglobin falls. An increased CO_2 concentration shifts the curve to the right. This means that at any given O_2 level more oxygen is released from the blood to the tissues. For example, at 6 percent O_2 with little CO_2, about 83 percent of the hemoglobin is saturated with oxygen (colored curve). Raising the CO_2 level of the blood without changing oxygen level causes O_2 to be released until only 59 percent of the Hb has O_2 bound to it (black curve).*

that need it most—those that are producing large amounts of carbon dioxide. Conversely, the uptake of oxygen and the formation of oxyhemoglobin in the blood coursing through the lungs greatly reduce the blood's carbon dioxide carrying capacity, facilitating the release of carbon dioxide to the alveoli.

Carbon dioxide diffuses out of the body cells, largely in simple aqueous solution, through the capillaries and into the blood plasma. As each 100 ml of blood passes through the tissues it takes up about 4 ml of CO_2, which dissolves in the plasma. In solution carbon dioxide reacts chemically with water at an extremely slow rate to form carbonic acid, H_2CO_3. When CO_2 enters the red blood cells, however, this reaction is rapidly catalyzed by the enzyme **carbonic anhydrase.** The carbonic acid thus formed is quickly converted to sodium and potassium bicarbonates ($NaHCO_3$ and $KHCO_3$), which are not acid. Carbonic anhydrase, like all other enzymes, also catalyzes the reverse reaction, in this case the splitting of carbonic acid to water and carbon dioxide; thus it also accounts for the observed rapid release of 4 ml of carbon dioxide per 100 ml of blood during its passage through the lungs.

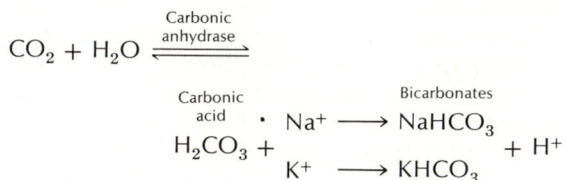

$$CO_2 + H_2O \underset{\text{Carbonic anhydrase}}{\xrightleftharpoons{}}$$

$$\underset{\substack{\text{Carbonic} \\ \text{acid}}}{H_2CO_3} + \begin{array}{l} Na^+ \longrightarrow NaHCO_3 \\ K^+ \longrightarrow KHCO_3 \end{array} \overset{\text{Bicarbonates}}{} + H^+$$

Notice that in the above reaction hydrogen ions are liberated when bicarbonates are formed. If it were not for the buffering action of the blood proteins (of which 75 percent are hemoglobin), the pH of the blood would quickly drop to a fatal level. An average adult male produces about 400 L of carbon dioxide each day; if it all remained in the blood as carbonic acid, the effect would be equivalent to an accumulation of .5 L of concentrated sulfuric acid. However, because of the ability (buffering capacity) of blood proteins to bind with the hydrogen ions of carbonic acid to form sodium and potassium bicarbonates, there is little or no change in the pH of the blood. Instead, the carbon dioxide is harmlessly conveyed (as bicarbonates) to the lungs, where it is eventually exhaled from the body as carbon dioxide.

Although the body successfully copes with carbon dioxide, it is less successful with carbon monoxide (CO), for this deadly gas, which is present in automobile exhaust fumes, binds strongly to the iron of hemoglobin. Because carbon monoxide has a much greater affinity for hemoglobin iron than oxygen has, it can displace molecular oxygen from oxyhemoglobin. The presence of .5 percent carbon monoxide in the air will render more than half the bloodstream's hemoglobin useless in its oxygen-carrying role for at least several hours after exposure.

The movement of gases between the blood and the cells of the body is, of course, the reverse of movement in the lungs, because the amount of oxygen in the tissues is low and that of carbon dioxide high, compared with that of blood.

Excretion

Excretion is the term applied to the process by which an organism rids itself of metabolic wastes, which may include by-products of carbohydrate, fat, amino-acid, protein, and nucleic-acid metabolism, various salts, and excess carbon dioxide and water. **Secretion** is the discharge, from particular cells or tissues, of substances that in many cases are utilized in other parts of the body; digestive enzymes and hormones are examples. Finally, **defecation** means the elimination from the body of undigested material that has passed through the digestive tract (see Chapter 14).

The Excretory System

In most unicellular organisms metabolic waste products are disposed of by diffusion into the external aqueous environment. As organisms increase in size and complexity, diffusion is no longer adequate; specialized excretory organs must assume this function.

The major organs of excretion in man are the **kidneys,** which dispose of excess water (also an end product of oxidative metabolism), nitrogenous wastes such as **urea,** the breakdown product of proteins and amino acids, salts, and other materials, in the form of **urine.** Urine is formed in the kidneys and is an aqueous solution (95 percent water) of all these excreted products of the body. Because of their excretory function the kidneys are also largely responsible for maintaining the delicate balance of pH, water, salts, and other substances in the bloodstream and therefore, indirectly, in all the tissue fluids of the body. In effect, then, the kidneys have a primary role in stabilizing the internal environment of the body.

The kidneys and associated organs that produce and eliminate urine are collectively called the **urinary system.** In man the urinary system accounts for the excretion of about 75 percent of the metabolic wastes of the body. Although the kidneys are the key organs for the regulation of water balance in the body, the lungs, skin, and digestive tract also function in this respect. The average quantity of water lost per day by a normal adult human, for example, is usually about 1500 ml through the kidneys, 500 ml through the skin, 300 ml through the lungs, and 150 ml through defecation. These average figures vary considerably, depending on the inherent makeup of the individual, his water and salt intake, physical activity, hormonal balance, and other factors, including environmental conditions. In addition, the lungs excrete carbon dioxide and water; the sweat glands eliminate salts, water, and some nitrogenous wastes in the form of **perspiration,** whose composition somewhat resembles that of urine; finally, the liver, by disposing of certain hemoglobin breakdown products via the bile, has an excretory capacity, as does the lining of the colon, which excretes certain heavy metals into the feces.

Gross Structure of the Urinary System

Each of the two kidneys in humans has the general shape of a lima bean, measuring in adults about $4\frac{1}{2}$ in. long, 2 to 3 in. wide, and 1 in. thick. They are located at the back of the abdominal cavity, one on each side of the spine, just above the waist (Fig. 15-18). Both are anchored by connective tissue to surrounding structures to maintain their positions. In longitudinal section the kidney shows two general regions, an outer, darker region, the **cortex,** and a considerably thicker inner region, the **medulla.**

The indentation at the side of each kidney is the site at which the main blood vessels and nerves connect to this organ. It is also the site at which a **ureter,** a tube about 10 to 12 in. long, emerges from each kidney and extends downward to the **urinary bladder.** The upper portion of the ureter enlarges at its junction with a large, funnel-shaped space in the kidney called the **renal pelvis.** At the opposite end, where each ureter enters the urinary bladder, is a valvelike fold of epithelial tissue that prevents the backflow of urine from the bladder.

The urinary bladder is a muscular bag whose muscular layer is made up of three coats of smooth muscle fibers. Because of the elasticity of its walls it is capable not only of contraction but also of considerable distention, with a capacity that varies greatly among individuals.

Urine, which is continually being formed in the kidneys, drains into the renal pelvis and trickles down through the ureters into the bladder. Regular waves of muscular contraction and relaxation along the lengths of the ureters facilitate this movement. The continuous production of urine soon results in an accumulation in the urinary bladder, which increases rapidly in size. When the urine volume reaches approximately 300 ml, the bladder is distended to a point at which stretch-sensitive receptors in its wall are stimulated, resulting in awareness of pressure or fullness in the bladder.

Urine is transported from the bladder to the outside of the body by a duct called **urethra.** The opening of the urethra near its junction with the bladder is controlled by two sets of **sphincter** muscles. In infants the urine is voided, involuntarily and without restraint, by reflex contraction of the bladder and relaxation of both sphincters. Older children and adults learn to inhibit reflexes of the bladder and sphincter muscles, and urination can therefore be restrained until an appropriate time.

In males the urethra not only transports urine out of the body but is also the terminal passageway of the reproductive tract (through the penis; see Chapter 11), carrying semen to the outside of the body. In females the urethra opens to the outside of the body just in front of the vaginal orifice and is exclusively excretory in function.

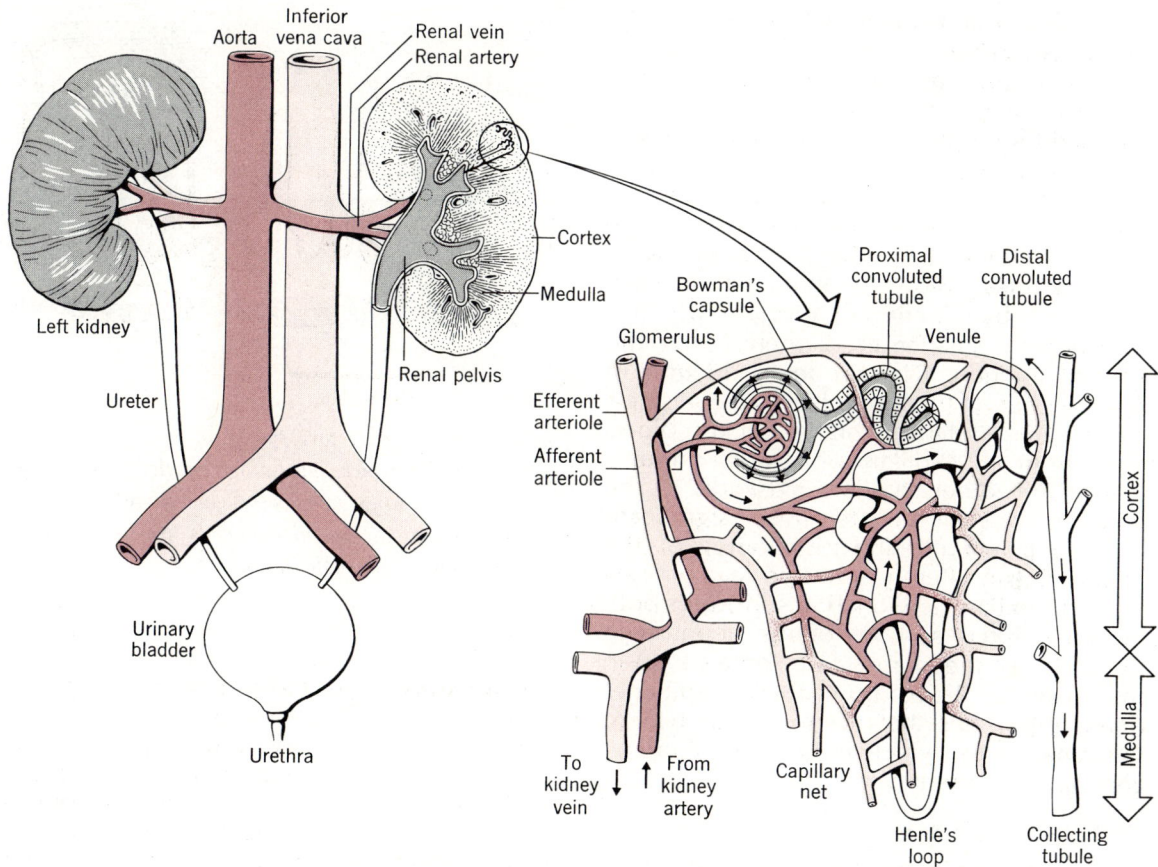

Figure 15-18   The urinary system of man, showing a section cut through the kidney and a detailed view of a nephron.

The Nephron

The functional unit of the kidney is an elongated, highly specialized tubule called a **nephron;** each human kidney contains approximately 1 million nephrons. Each nephron is about 3 cm in length, has thin walls of epithelial cells, and consists of four regions (see the detailed diagram in Fig. 15-18).

One end of the human nephron, lying in the cortex of the kidney, is a double-walled, cuplike or funnellike structure called **Bowman's capsule.** Bowman's capsule surrounds a thick cluster of capillaries called a **glomerulus** (plural: **glomeruli**). A winding, convoluted section of the nephron tubule extends from

Bowman's capsule toward the medulla region of the kidney; this is the **proximal convoluted portion.** The next part of the nephron, called **Henle's loop,** has a relatively straight portion of tubule that extends into the medulla and loops back in a hairpin turn to the cortex. The final section, the **distal convoluted portion,** is also a winding and convoluted structure located close to the proximal convoluted portion. It joins, with the distal convoluted tubules of several other nephrons, to larger ducts called **collecting tubules.** These ultimately conduct the urine formed in the nephrons to the renal pelvis of the kidney.

Arterial blood enters the kidneys by the renal arteries, which subdivide into successively smaller and smaller arteries and finally to **afferent arterioles.** Each afferent arteriole leads into the capillaries of a glomerulus. Approximately one-fourth of the blood pumped by the heart in each circuit around the body enters the kidneys through the renal arteries and passes through the glomeruli (except for a small fraction that services the connective tissue and other accessory tissues of the kidney). Blood is drained from each glomerulus by the **efferent arteriole,** which is considerably smaller in diameter than the afferent arteriole. The sudden reduction in diameter before the blood leaves the glomerulus creates a relatively higher backup blood pressure in the capillaries of the glomerulus than in the other capillary networks of the body. As we shall see later, this is significant in the efficient filtering functioning of the nephron. The efferent arteriole leads into a second capillary network that surrounds both the convoluted portions of the nephron in the cortex of the kidney and Henle's loop in the medulla. Blood from these capillaries drains into veins that eventually unite into the renal veins, which return the blood to general circulation.

Evolution of Nephron Structure

The structure of the nephron is essentially the same, with some variations, in the kidneys of all vertebrates. It has even been identified in the fossil imprints of the oldest known vertebrates. However, the nephron of vertebrates has undergone a series of changes in the course of evolution from the invertebrate condition. In many invertebrates, the excretory structure is a simple absorbing tubule called a **nephridium** (Fig. 15-19), which is instructive to examine as a simplified model of the later nephron.

Among the lower vertebrates, for example in fish, a primitive kind of kidney called the **pronephros** is found. It consists of three to five kidney tubules near the heart. The kidney of most adult fishes and amphibians, however, is called a **mesonephros,** and contains a more complicated type of tubule. Finally, adult reptiles, birds, and mammals have an even more advanced type of kidney called a

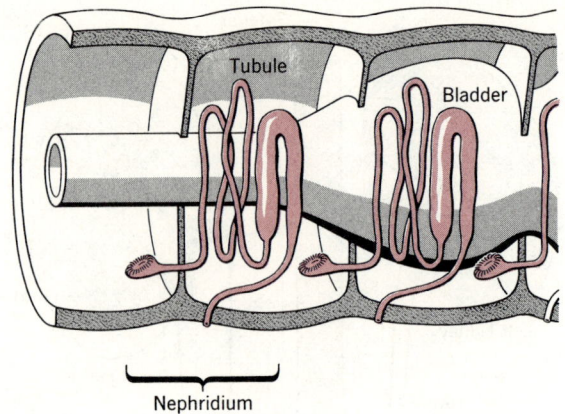

Figure 15-19 *Excretory system of an earthworm. The nephridium picks up fluid from the body cavity and gut containing both wastes and useful substances, but the useful materials are reclaimed as the fluid moves through the tubule.*

metanephros, the most advanced type of kidney found in vertebrates.

In the human embryo a pronephros forms in the third week of embryonic development and begins to degenerate during the fourth, without ever functioning as an excretory organ. A mesonephros begins to form during the fourth week, develops for a month, and then degenerates; during its brief existence it functions as a true kidney. A metanephros begins to appear during the fifth week of development and remains as the functional kidney for life. It is as if a developing embryo recapitulates the evolutionary history of its vertebrate relations.

Urine Formation in the Nephron

Urine formation is the result of three processes that occur in each nephron: **pressure filtration** through the glomerulus; **reabsorption** of the filtrate as it passes through the various portions of the nephron tubule; and **tubular excretion.**

The first step in the formation of urine is the separation of a cell-free, protein-free fluid, called a **glomerular,** or **pressure filtrate,** by a simple mechanical process of squeezing the circulating blood out under pressure through the capillary walls of the glomerulus. This pressure filtrate, which collects in Bowman's capsule, normally contains only dissolved materials to which the capillary and tubular walls are permeable; its composition is the same as that of the blood, except for the absence of cellular elements (red blood cells, leucocytes, and platelets), plasma proteins, and fats. Water is its main component; in addition the glomerular filtrate contains sugars, amino acids, salts, nitrogenous wastes, and other dissolved substances in approximately the same concentrations as in the blood.

It has been estimated that 1200 ml of blood passes through the 2 million glomeruli in the kidneys each minute, and of this approximately 125 ml of fluid and dissolved substances are filtered out each minute, for an estimated total of 180 L per day. Obviously the amount of urine excreted could not equal the amount of glomerular filtrate produced, or the body would quickly be dehydrated. Instead, most of the filtrate must be reabsorbed into the bloodstream.

The glomerular filtrate passes along the proximal convoluted portion of the renal tubule from Bowman's capsule; in this stage about 85 percent of its water and some of its dissolved materials are reabsorbed through the walls of the tubule into the blood of the surrounding capillaries. The reabsorption process is highly selective; water, sodium and chloride ions, most of the bicarbonate, and all the glucose of the glomerular filtrate is reabsorbed back into the bloodstream, while other substances, such as urea and other products of protein metabolism, remain. During its subsequent passage along Henle's loop and the distal convoluted tubule most of the remaining filtrate is further selectively reabsorbed, so that only about 1 percent of the volume of the original glomerular filtrate, now considerably different in composition, is finally excreted as urine. The final concentration of urea in urine is 60 to 80 times as great as that in the plasma or glomerular filtrate. Of the 180 L of glomerular filtrate produced each day by an average adult, only

about 1 or 2 L are excreted as urine.

The reabsorption of water from the filtrate in the proximal convoluted tubule occurs by ordinary diffusion. However, an active transport mechanism (Chapter 5) is necessary in both the proximal and distal tubules for the selective reabsorption of other substances (e.g., glucose, amino acids, sodium, calcium, and potassium ions) because these molecules are generally present in greater concentration in the blood than in the filtrate. Because it involves the transport of substances against a diffusion gradient, metabolic energy is required.

The tubule cells obviously display remarkable characteristics of selectivity and specificity, reabsorbing most efficiently substances that the body vitally needs. If the concentration of one of these substances in the blood exceeds a certain level, the active transport mechanism may not be adequate for complete reabsorption of that material in the time that it takes for the filtrate to pass along the tubule. As a result the substance will appear in the urine; this is a symptom often used by doctors to diagnose kidney disease.

The energy-requiring active transport mechanism also accounts for the third and final step in urine formation: several substances are transported by the tubules from the blood into the filtrate in the process called tubular excretion. The active transport process in this instance is more or less similar to that in reabsorption, except that it goes in the other direction. In tubular excretion materials such as ammonia, potassium, and hydrogen ions are extracted from the blood and actively passed into the tubule. This means that the cells of the walls of the tubule are responsible for a two-way traffic between the filtrate and the blood: most of the water and essential food materials are restored to the bloodstream; and additional substances, mostly wastes, are removed from the blood and added to the concentrated filtrate that soon becomes the urine.

The excretion of hydrogen ions and ammonia in varying amounts by the cells of both the proximal and distal convoluted tubules regulate the pH of the urine. This is an important mechanism in the control of the acid-base balance of the blood tissues and fluids. The kidneys, in fact, rid the body of excess acid in several ways. They neutralize most of the excess acid by the formation of ammonium and phosphate salts before excreting it, and also excrete a small amount of free hydrogen ions.

Hormonal Effects on Kidney Function

The distal convoluted tubule, and probably Henle's loop, are subject to several hormone control mechanisms. These are part of a regulatory system that, by affecting the reabsorption function of these tubules, serves as a means for adjusting excesses or deficiencies in the internal environment.

The **antidiuretic** hormone secreted by the posterior pituitary gland (Chapter 16) promotes reabsorption of water by the distal convoluted tubules and the collecting tubules by changing their permeability. How it does this is an unsolved riddle. What is known is that a decreased water supply to the body is reflected by an increased osmotic concentration of the blood, which in turn results in an increased secretion by the pituitary of antidiuretic hormone. By stimulating tubular reabsorption of water through permeability changes, the hormone retains water

in the body and decreases the volume of urine formed. On the other hand, increased water intake by the body leads to a decreased secretion of the hormone. The over-all effect, therefore, is a self-regulating mechanism for maintaining the water balance of the body. Individuals suffering from an insufficient secretion of the antidiuretic hormone, a condition called **diabetes insipidus,** have a less efficient reabsorption of water; the result is a urine volume increased as high as 20 L per day (compared to a maximal output of 2.5 L in normal individuals). Alcohol inhibits the secretion of the antidiuretic hormone and thus causes increased urine volume, but the diuretic effects of beer drinking can be attributed more to the water consumed than to its alcohol content (only about 4 percent).

Urea Formation

Next to water, the major component of urine is **urea,** constituting 60 to 90 percent of all nitrogenous material in the urine of man. Urea is not formed in the kidney but mainly in the liver as a by-product of protein and amino acid metabolism. For example, when amino acids are converted to fats or carbohydrates rather than to protein, their amino groups are removed. The process is referred to as **deamination** and is catalyzed by the enzyme **deaminase** (Chapter 3):

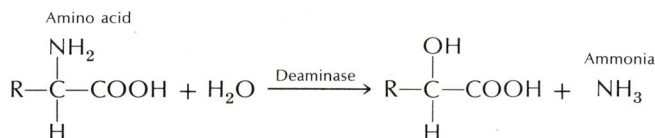

$$\underset{\text{Amino acid}}{R-\overset{\overset{\displaystyle NH_2}{|}}{\underset{\underset{\displaystyle H}{|}}{C}}-COOH} + H_2O \xrightarrow{\text{Deaminase}} R-\overset{\overset{\displaystyle OH}{|}}{\underset{\underset{\displaystyle H}{|}}{C}}-COOH + \underset{\text{Ammonia}}{NH_3}$$

As shown, the reaction converts the amino group to ammonia, a product so highly toxic that it must be rapidly converted to a less poisonous substance—in man, urea. The ammonia liberated by deamination is combined

with the amino group of aspartic acid, and with carbon dioxide in the presence of ATP, by a separate cyclic sequence of enzymatic reactions called the **urea cycle.** The over-all reaction is shown in the following equation:

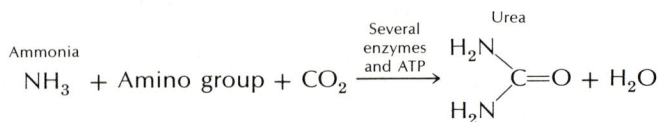

$$\underset{\text{Ammonia}}{NH_3} + \text{Amino group} + CO_2 \xrightarrow[\text{and ATP}]{\text{Several enzymes}} \underset{\text{Urea}}{\overset{H_2N}{\underset{H_2N}{\diagdown}}C=O} + H_2O$$

The urea, which is in solution, is transported via the circulatory system to the kidney, where it is excreted as a component of urine.

Composition of Human Urine

The normal volume of urine voided by the average adult may range from as little as 750 ml to as much as 2500 ml each 24 hours; the average is usually 1500 ml. The pH of urine may vary between 4.6 and 8.

Freshly voided urine is usually clear and straw colored or amber because of the presence of several pigments, collectively called **urochrome,** made of a number of products of hemoglobin breakdown in the liver and intestinal tract. Urine ordinarily consists of 95 percent water and 5 percent dissolved solids. Of the 50 g or so of solids per liter of urine, somewhat more than half are organic components and the remainder are inorganic salts. **Urea** accounts for about half of the solids in urine (20 to 25 g per liter). **Creatinine,** an organic by-product of muscle metabolism, makes up about 1.5 g per liter; **uric acid,** the third principal organic component, is derived in part from the diet and from nucleic-acid metabolism of the body tissues, and averages about .5 g per liter of urine. Small quantities of other organic substances are also present, including several other breakdown products of hemoglobin, water-soluble vitamins, and certain hormones. Tests based on the presence of certain sex hormones in the urine of women are used in the early diagnosis of pregnancy (Chapter 11).

Sodium and potassium chlorides, the major inorganic salts of the diet, are also the major salts in human urine. They are excreted in varying amounts, depending on the quantities ingested. The average concentration of sodium chloride in urine is about 9 g per liter, that of potassium chloride approximately 2.5 g per liter. Progressively smaller quantities of sulfate, phosphate, ammonia, calcium, and magnesium are also present.

Various abnormal constituents may appear in the urine depending on the condition of an individual. Glucose, which is normally completely reabsorbed in the proximal convoluted tubules, may be present in the urine for several different reasons; the most common is **diabetes mellitus,** a disorder that results from an insufficiency of the hormone insulin (Chapter 16). In this condition the blood sugar level becomes so high that the normal rate of active reabsorption in the renal tubule cannot completely remove the excessive quantities of glucose; glucose in the urine is therefore diagnostic indication of the disease. Diabetics characteristically excrete a great volume of urine because of the resulting higher osmotic concentration of the filtrate, which decreases the rate of water reabsorption from the tubules. Disorders of renal tubule function and extreme emotional states may also be responsible for the presence of glucose in the urine. Other sugars, such as certain pentoses, galactose, fructose, and lactose, may also appear in the urine, depending on extent of dietary intake, genetic defects, and other factors; for example, in the hereditary disease **galactosemia** (Chapter 9).

Evolution of Nitrogen Excretion

In virtually all animals either ammonia, urea, or uric acid accounts for two-thirds or more of the total nitrogen excreted. The chief chemical state in which metabolic nitrogenous waste products are excreted by members of the animal kingdom depends on the species and the evolutionary adaptations it has made to the availability of water in its environment, especially during embryonic development.

Animals are divided into three groups on the basis of their main nitrogenous excretory product: **ammonotelic** (ammonia), **ureotelic** (urea), and **uricotelic** (uric acid). Ammonia is the major and immediate nitrogenous product of protein and amino acid metabolism for all organisms. Most animals living in water are ammonotelic, for ammonia, which is easily diffusible, is largely excreted without having to be metabolized further to the less toxic form of urea or uric acid. This is because aquatic and marine animals and fish have at their disposal a vast environmental reservoir of water into which they can release waste ammonia without fear of self-intoxication.

The uricotelic condition is best explained as an evolutionary adaptation to a limited environmental supply of water. Under such conditions ammonia cannot be disposed of fast enough from the body of an organism to avoid toxic or fatal effects. Its biological conversion to uric acid, an insoluble and comparatively harmless material that is also an end product of purine metabolism, represents a biochemical means of surmounting the problem. Most birds and reptiles, and certain insects, are uricotelic.

Animals that have evolved in an environment in which water availability is intermediate between the extremes experienced by ammonotelics and uricotelics tend to be ureotelic; most mammals are ureotelic. The mammalian embryo has its mother's water supply at its disposal, in contrast to uricoteles such as birds and reptiles, whose embryos develop separately from the maternal tissue in eggs of limited water content.

A chick embryo in the course of its development is first ammonotelic, then ureotelic, and finally uricotelic, suggesting that this may have been the course of the biochemical evolution of nitrogen excretion.

Kidney Function: A Summary

The kidney exerts a major influence in stabilizing the body's internal environment by maintaining the composition of the blood, and therefore of the body and tissue fluids, through (1) removal of waste products of nitrogen metabolism; (2) elimination of excess inorganic salts; (3) regulation of the acid-base balance of the body; (4) elimination of excess water; and (5) excretion of substances present in the blood in excessive amounts. Should the kidneys abruptly stop functioning, death would soon follow.

Conclusion

We have described three separate organ systems of circulation, respiration, and excretion, but it should be obvious by now that the three are so interrelated in their functions that such distinctions are blurred.

The circulatory system pumps the blood throughout the body; but we have seen that the composition of the blood is determined, in large part, by the respiratory and excretory systems. Moreover, the major function of the respiratory system, gas exchange, is accomplished via the circulatory system, carrying oxygen from the lungs to the tissues in the form of oxyhemoglobin and carbon dioxide from the cells back to the lungs as dissolved bicarbonate ions. Similarly, the prime excretory product, urea, is transported from its site of manufacture in the liver to the kidney by distribution through the bloodstream.

In general the three systems function together to accomplish a common task, that of regulating the composition of the watery internal environment that bathes the cells of the body.

Reading List

Adolph, E. F., "The Heart's Pacemaker," *Scientific American* (March 1967), pp. 32–37.

Best, C. H., and N. G. Taylor, *The Living Body* (5th ed.). Holt, Rinehart & Winston, New York, 1968.

Buchsbaum, R., *Animals Without Backbones* (revised ed.). University of Chicago Press, Chicago, 1948.

Comroe, J. H., "The Lung," *Scientific American* (February 1966), pp. 56–68.

Ingram, M., and K. Preston, Jr., "Automatic Analysis of Blood Cells," *Scientific American* (November 1970), pp. 72–82.

Macalpine, I., and R. Hunter, "Porphyria and King George III," *Scientific American* (July 1969), pp. 38–57.

Macey, R., *Human Physiology*. Prentice-Hall, Englewood Cliffs, N.J., 1968.

Schmidt-Nielsen, K., *Animal Physiology* (3rd ed.). Prentice-Hall, Englewood Cliffs, N.J., 1970.

Schmidt-Nielsen, K., "How Birds Breathe," *Scientific American* (December 1971), pp. 72–79.

Smith, H. W., *From Fish to Philosopher*. Doubleday, Garden City, N.Y., 1959.

Windle, W. F., "Brain Damage by Asphyxia at Birth," *Scientific American* (October 1969), pp. 76–87.

Winter, P. M., and E. Lowenstein, "Acute Respiratory Failure," *Scientific American* (November 1969), pp. 23–29.

Wood, J. E., "The Venous System," *Scientific American* (January 1968), pp. 86–96.

chapter sixteen

Coordination: The Endocrine System

"All the vital mechanisms, varied as they are, have only one object, that of preserving constant the conditions of life in the internal environment."

CLAUDE BERNARD, 1855, MEMOIRS

The human body, like that of most advanced multicellular animals, is comparable to a vast city with its population of thousands of billions of cells living and functioning together harmoniously in a smoothly running, well-coordinated organization or community. As in any large city or other organized system, one or more kinds of control and communication are essential to assure an integration of activity.

Integration means forming into a whole, the attainment of unity out of diversity. By its very nature integration is more than merely a summing-up process. As a result of integration new and often unpredictable features appear, especially in biological phenomena, that at times are difficult to explain on the basis of our knowledge of the individual parts. In the integrative process the whole has become equal to more than the sum of its parts, for a new dimension has been added.

One of the unique and remarkable abilities of many animals is to maintain a stability, or constancy, of their internal environment despite extreme changes in external conditions. This physiological phenomenon is known as **homeostasis** (p. 9). The relative constancy of the internal environment depends on a complex interplay of numerous dynamic processes, ranging from the molecular to the gross levels, that are sensitive to external as well as internal conditions. Maintenance and alteration of these dynamic processes are

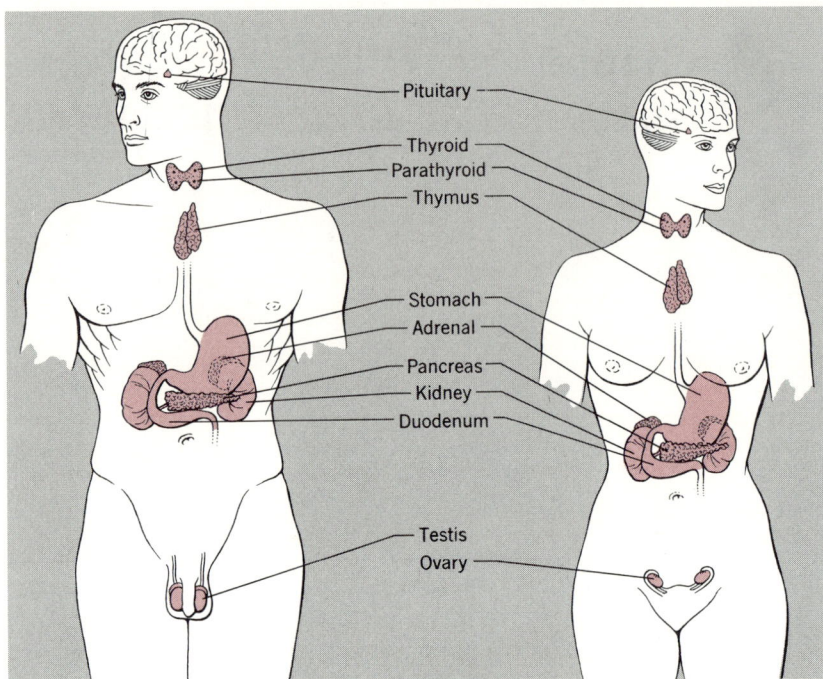

Figure 16-1 Location of the main glands of the endocrine system in the human male and female.

Labels in figure: Pituitary, Thyroid, Parathyroid, Thymus, Stomach, Adrenal, Pancreas, Kidney, Duodenum, Testis, Ovary

mediated in large part through the combined action of the **nervous** and **endocrine systems.**

The endocrine system consists collectively of 10 specialized glands, located in different parts of the body (Fig. 16-1). The glands have no ducts; they liberate their secretions, known as **hormones,** directly into the bloodstream. For this reason they are sometimes called the **ductless glands.**

Chemical Nature of Hormones

Hormones are organic substances that are carried by the blood to the parts of the body where they exert their specific effects. In simplest terms they are chemical messengers; in a broader sense they are important regulating substances that control virtually every aspect of the metabolism of living cells. Usually they do not initiate a process, but they may either stimulate or inhibit it. For example, sugars may be metabolized in the absence of the hormone insulin, but the reaction does not proceed at the normal rate. In general hormones fall into two chemical classes. They are either **steroids** (p. 63) or **amino acid derivatives** (p. 64) which include proteins, peptides, and modified amino acids.

Hormones are effective in remarkably small quantities; for example, the injection of a few micrograms of adrenaline into a dog causes an increase in the rate of heartbeat. There is still little conclusive evidence about the mechanisms by which hormones produce their effects, despite tremendous research efforts in this area over the years. The fact that hormones, like vitamins and trace elements, function in extremely small quantities appears to implicate them in a catalytic role, possibly as components of enzyme systems. Recent experiments indicate that some hormones function by increasing the transcription rate of certain genes (Chapter 10) and thus increase the production of messenger RNA specifying the synthesis of specific enzymes. For example, injection of adrenal cortical steroid hormones (see p. 450) into rats has been found to cause a large increase in the enzyme tryptophan pyrrolase in the liver. Other hormones apparently change the permeability or transport activity of cell membranes, thereby making enzyme substrates or coenzymes more readily available. However, the detailed mechanism of hormone action at the molecular level is still one of the major unsolved riddles of biology.

Endocrine function is based on an intricate complex of interactions. Some hormones appear to be highly specific in their action (e.g., the follicle-stimulating hormone and the luteinizing hormone, p. 454). Others, such as the thyroid hormone, insulin, and the adrenal cortical hormones, affect a wide variety of cells. Moreover, the interrelationships between the nervous system and the endocrine system are very involved.

Although the endocrine system and nervous system share the control of body activities, there are several obvious differences in how they act. In general the action of hormones occurs more slowly and is longer lasting than that of the nervous system, which affects high-speed, short-duration responses. The endocrine system uses the blood to carry hormones to the target tissues; the nervous system uses a sequence of neurons as a system of communication that finally secretes chemicals (e.g., acetylcholine and sympathin) to elicit a response.

A closer comparison of the nervous and endocrine systems, however, indicates that they share more similarities than heretofore believed (see Chapter 17). In some respects the nerve cell resembles an endocrine cell: Both secrete certain chemicals that act on a target organ or tissue, but in contrast to the endocrine glands, nerve cells are in direct contact with their target organ. Some endocrine glands, for example the adrenal medulla (p. 448), secrete the same hormones or chemicals (noradrenaline or sympathin) as the nerve cells. The adrenal medulla is actually a modified portion of the sympathetic nervous system, and several pituitary secretions are in reality secretions of modified nerve cells or are indirectly regulated by nervous mechanisms residing in the hypothalamus region of the brain (see Chapter 17).

The secretory activity of most of the endocrine glands themselves seems to be regulated by certain substances carried by the blood. In some instances these chemical factors are hormones from other glands. In other cases they are relatively simple substances such as sugars (e.g., glucose in the blood stimulates the flow of insulin from the pancreas) or acids (e.g., the secretion of secretin by the small intestine is stimulated by the acid chyme entering from the stomach). In a few glands (e.g., the adrenal medulla) secretory activities are regulated directly by the visceral nervous system.

The Thyroid Gland

The thyroid gland consists of two lateral lobes and a connecting isthmus of tissue (Fig. 16-2a and b, p. 444). In the human adult it has an average weight of about an ounce (25 to 30 g). It is located in the midpoint of the neck slightly below the larynx. The gland contains numerous secretory cells responsible for the production of the thyroid hormones, which temporarily accumulate (both in the free state and as part of a large glycoprotein **thyroglobulin**) in the many microscopic cavities, or vesicles, in the gland until they enter the bloodstream. The thyroid hormones are not transported by the blood in a free form but combine largely with one of the blood proteins, alpha globulin, which serves as a carrier.

The thyroid gland has the remarkable ability to store and accumulate iodine, which is a constituent of the thyroid hormones. At least one-quarter of the total body iodine is concentrated in the gland itself. Experiments with radioactive iodine (I^{131}) have demonstrated that within a few hours after injection of sodium iodide as much as 70 percent has accumulated in the thyroid gland. It is now known that the thyroid hormones consist of

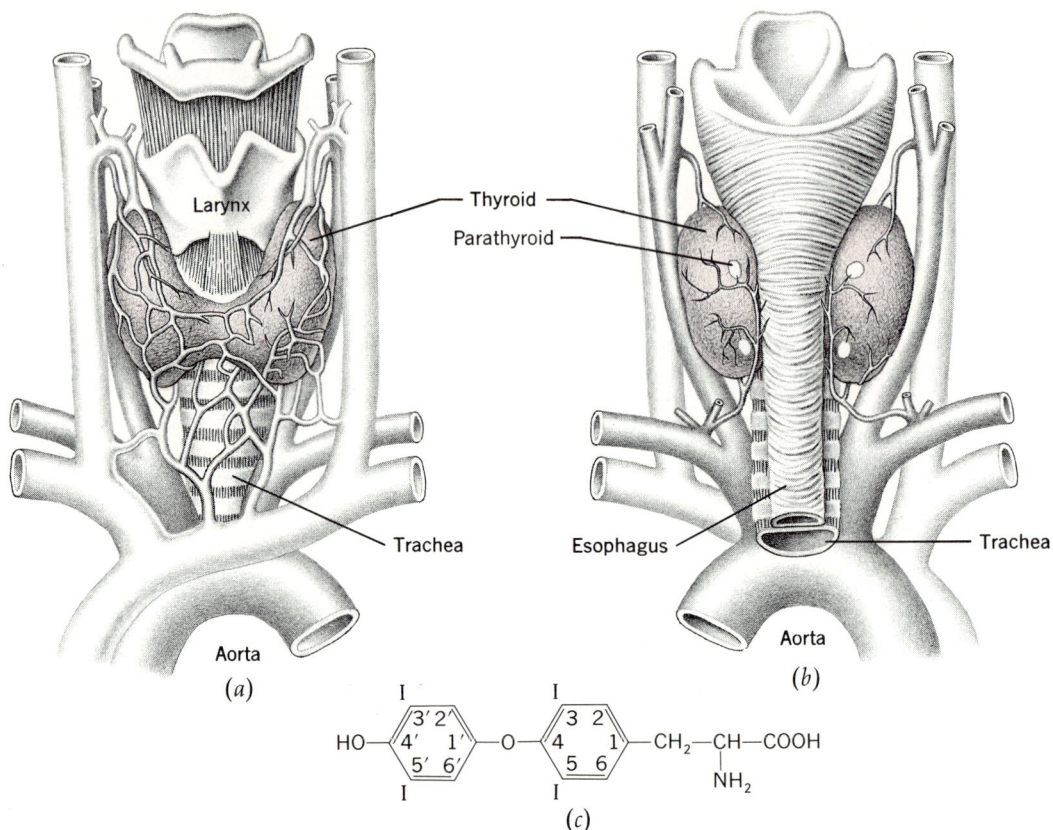

Figure 16-2 *The thyroid and parathyroid glands, situated in the throat just below the larynx. (a) Front view; (b) back view, showing the location of the parathyroids. (c) The chemical structure of thyroxin shows that it is an amino acid containing two phenyl groups, combined with four iodine (I) atoms. Iodine present in the bloodstream is taken up selectively by the thyroid gland to produce thyroxin. In the absence of iodine, thyroxin cannot be produced, and the resultant continuous flow of TSH from the pituitary may cause a compensatory overgrowth of the thyroid, or* **goiter.**

at least four iodinated derivatives of the essential amino acid **L-tyrosine.** The most abundant of the four is called **thyroxin** (Fig. 16-2c), which until recently was believed to be the only physiologically active component or hormone secreted by the gland. It is now known that the other three forms, which have fewer iodine atoms, are also important.

Thyroid hormones stimulate the rate of oxygen consumption (oxidative metabolism) and resulting heat production of all cells and tissues of the body. The thyroid gland also has a second important function, a regulatory role (by its effect on cellular metabolism) in the general processes of growth and development of cells and tissues. These include sexual development, maturation of bones and teeth, mental development, energy metabolism, and so on.

The mechanisms of action of the thyroid

hormones at the molecular level is still not established. We do know that when thyroxin is added to isolated tissues and mitochondria it has the ability to prevent the formation of ATP that normally accompanies the terminal steps of aerobic respiration (Chapter 4). Thus, instead of being stored as ATP, a portion of the energy derived from oxidative metabolism would be wasted as heat. Whether thyroxin has the same effect on the production of ATP in an intact animal is not yet clear, but excessive heat production is known to accompany the occurrence of an overactive thyroid.

Control of Thyroid Secretion

The secretion of thyroid hormones is controlled by a thyroid-stimulating hormone (known as **TSH,** or **thyrotropic hormone**) secreted by the anterior pituitary gland (p. 451). If the level of thyroxin in the blood falls, an increased amount of TSH is produced by the pituitary, whereas a rising level of blood thyroxin inhibits the production of TSH and thus decreases the stimulation of the thyroid. Such a relationship represents another example of a **negative feedback system** (Chapter 1), in which the end product regulates the system that produces it.

Thyroid Malfunction

As with all other hormones, nearly all our knowledge of the functions of the thyroid secretion is derived from experiments with laboratory animals in which the thyroid gland has been removed or excessive quantities of the hormone have been administered. A second important source of information has been clinical observation of patients.

An insufficient secretion by the thyroid gland is known as hypothyroidism; excessive secretion is called hyperthyroidism. **Hypothyroidism** may result from a variety of factors, including atrophy of the gland, deficient dietary supply of iodine, or inadequate de-

velopment of the glandular tissues. Hypothyroidism at birth gives rise to a dwarfed condition called **cretinism,** which is typified by retarded physical, sexual, and mental development and a significantly lowered metabolic rate. Early administration of the thyroid hormones by mouth or injection is necessary to prevent permanent effects. The occurrence of hypothyroidism in adults produces a condition called **myxedema** characterized by a lowered metabolic rate, thickness and puffiness of the skin, coarseness and brittleness of the hair and fingernails, and a general physical and mental lethargy. There is usually a weight gain, loss of hair, slowed pulse rate, reduced blood pressure, and decreased body temperature. The administration of adequate doses of thyroid extracts quickly restores normal function.

Any enlargement of the thyroid gland is designated by the term **goiter.** One of the causes of simple goiter is a dietary deficiency of iodine; at one time goiter was prevalent in geographic areas where the soil and water were deficient in this element. The wider distribution of foods grown in different areas and the addition of iodine to certain foods (e.g., table salt) and drinking water have reduced its incidence.

Hyperthyroidism is accompanied by physiological effects and symptoms that are the opposite of those observed in hypothyroidism. One of the most severe conditions of hyperthyroidism is called **Graves' disease,** or **exophthalmic goiter** (exophthalmic means "bulging eyes"). This condition is characterized by an enlargement of the entire gland, elevated metabolic rate, weight loss, profuse perspiration, high pulse rate, and a typical protrusion of the eyeballs (Fig. 16-3, p. 446). The condition may be treated by surgical removal of a portion of the thyroid gland; administration of radioactive iodine, which accumulates in the thyroid and destroys some of its tissues; or use of certain thyroid-inhibiting drugs that interfere with the synthesis of thyroid hormones. Hyperthyroidism may also be caused by tumorous growth of the thyroid gland tissue.

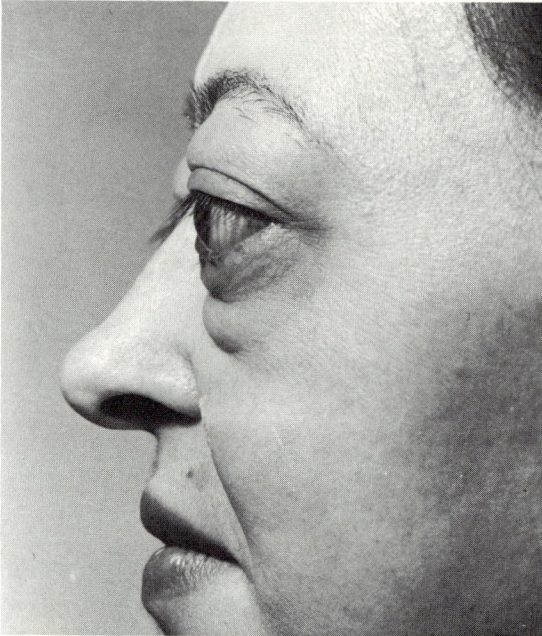

Figure 16-3   A patient with exophthalmic goiter. Note the typical protrusion of the eyeballs.

The Parathyroid Glands

The parathyroid glands in man include two pairs of small oval glands made of densely packed cells (each gland averaging about 5 mm, or $\frac{1}{4}$ in., in diameter) attached to the thyroid gland (Fig. 16-2b). Despite their close anatomical relationship to the thyroid gland, the chemical function of the parathyroid hormone is totally unrelated to that of thyroxin.

The parathyroid hormone is a protein (probably a straight-chain polypeptide) that regulates the level of calcium and phosphorus in the blood. By affecting the levels of these substances in the blood, the hormone indirectly regulates their levels in all other cells and body fluids. Too much parathyroid hormone, caused by a tumor of the glands, results in an increased blood calcium level, lowered blood phosphate concentration, and an increased excretion of calcium by the kidneys.

The exact mechanism by which parathyroid hormone exerts its influence is not clear. In excess it apparently causes an increase in the calcium concentration of the blood at the expense of the major reserves of calcium in the bones. The hormone somehow stimulates the action of bone-dissolving cells (Chapter 18). The resulting demineralization of bone also releases phosphate, which is liberated together with calcium. Under these conditions the removal of excess phosphate from the blood by the kidneys is also stimulated by parathyroid hormone. The net result of the action of the hormone in stimulating both bone demineralization and phosphate excretion by the kidneys is to increase the level of blood calcium. In hyperparathyroidism the maintenance of a high blood calcium level at the expense of bone often leads to deposits of calcium (**calcification**) in the soft tissues of the body.

Too little parathyroid hormone causes a marked decrease in blood calcium concentration. It is followed within a few days by increased irritability or excitability in nerves and muscles (a condition known as **tetany**), convulsions, and death. Administration of calcium salts by mouth or by injection temporarily relieves the condition, as do injections of parathyroid gland extracts.

The Pancreas and Islets of Langerhans

The pancreas in human adults is an elongated gland lying near the stomach. It functions both as a digestive gland (Chapter 14) and an endocrine gland. It consists primarily of glandular cells that secrete a mixture of several digestive enzymes that drain through a duct into the small intestine. The pancreas also contains numerous scattered, rounded microscopic clusters, or islets, of tissue called the **islets of Langerhans.** These are not connected with any ducts. They do, however, perform important endocrine functions.

The islets of Langerhans have several cell types. **Beta (β)** cells form the majority of the islets and are responsible for secretion of the hormone insulin. About 20 percent are **alpha (α)** cells, which secrete another hormone, glucagon. Recall that insulin is a protein of 51 amino acid residues, arranged in two linked peptide chains with a molecular weight of 5000 (Fig. 16-4). Glucagon is also a relatively small protein molecule, with a molecular weight of approximately 4000, and consists of a single straight chain of 29 amino acid residues.

Insulin and glucagon (together with certain other hormones from the thyroid, adrenal, and pituitary glands) play a central role in regulating the carbohydrate metabolism of the body. Their mechanisms of action, however, are still relatively obscure. Because both pancreatic hormones affect carbohydrate metabolism in the liver and other cells of the body, they have a major function in maintaining blood glucose at a constant level, about 100 mg per 100 ml of blood in mammals.

Insulin decreases the concentration of blood glucose by promoting (1) its use in cell respiration and fatty acid synthesis; and (2) its deposition in the storage form of glycogen, particularly in skeletal muscle tissue and probably in the liver. **Diabetes** is a disease characterized by a low level of insulin in the blood, which results in inadequate use of glucose and derangement of certain other enzymic reactions. This condition is reflected in several symptoms: the blood glucose level is elevated, the liver glycogen content is lowered, and fatty acid synthesis is decreased. Despite the high concentrations of blood glucose, the cells of the body respond as if none were present because insulin is not available. The liver also reacts as if glucose were lacking in the bloodstream, and continues to transform its glycogen to blood glucose. The high levels of blood glucose soon exceed the level that the kidneys can reabsorb and the sugar spills over into the urine.

Under these circumstances the cells resort to fat as an energy source, but the oxidation of fatty acids does not proceed to completion. As a consequence, there is an accumulation in the blood and urine of keto-acid fragments, or **ketone bodies,** of fatty acid metabolism, a condition known as **ketosis.** If left untreated, ketosis eventually results in death. Other changes accompanying diabetes include a reduced rate of respiration and lowered body protein; the latter condition results in part from the metabolic breakdown and conversion of proteins to glucose and

Figure 16-4   A model of beef insulin, made with children's plastic beads. Compare this with Figure 3-18.

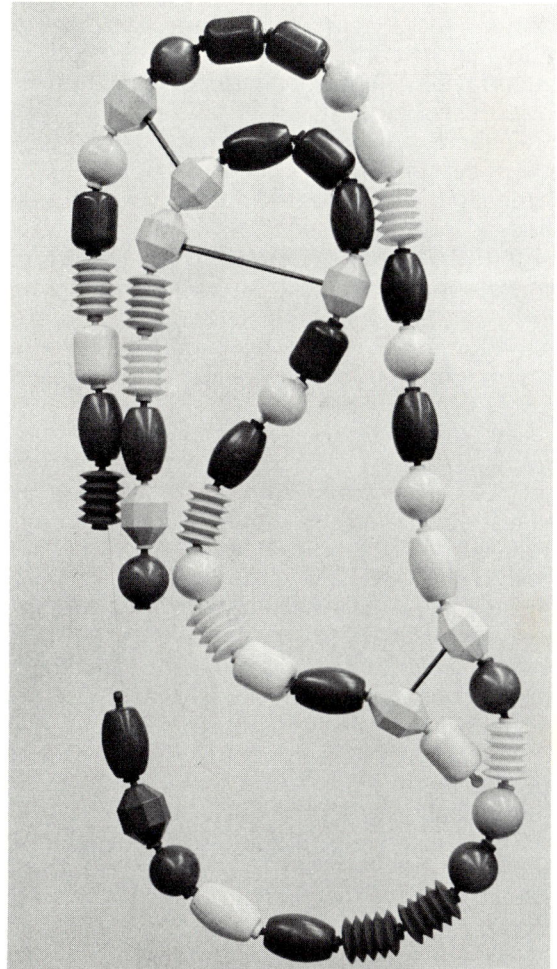

from a decrease in protein synthesis.

How does insulin stimulate the conversion of glucose to glycogen? One theory proposes that insulin stimulates the movement of glucose across the cell membrane into the cell by somehow increasing the latter's permeability to this sugar. Although there is appreciable evidence in favor of this view, there is also experimental support for a second major theory, which contends that insulin influences the early stages of glucose metabolism, possibly by affecting **hexokinase,** the enzyme responsible for the conversion of glucose by ATP to glucose-6-phosphate (Chapter 4).

Elevated glucose levels in the blood apparently serve as the stimulus for insulin secretion by the β-cells of the pancreas. A lowering in blood glucose concentration is accompanied by a decrease or cessation of insulin secretion.

Diabetes and starvation share several biochemical similarities. Glucose is not metabolized at a normal rate in either condition. Instead, fatty acids serve as the energy source, with the eventual accumulation in the blood of ketone bodies, leading to ketosis. However, the diabetic has excessively high concentrations of blood glucose, whereas the starving individual has normal or subnormal levels of blood glucose.

Therapy for diabetics normally involves periodic injections of insulin. Because this hormone is a protein, it would be destroyed by digestion if taken by mouth. Recently drugs such as tolbutamide have been discovered that lower the blood sugar level in diabetics; these can be taken orally, and may replace insulin treatment. In either case the treatment causes an increased deposition of glycogen in the liver and a generally increased utilization of glucose in metabolism.

Glucagon has the opposite effect from that of insulin; it tends to increase the level of blood glucose by promoting the conversion of liver glycogen to glucose. Glucagon is believed to enhance the activity of the enzyme **phosphorylase,** which is responsible for the breakdown of glycogen to glucose-1-phosphate. Adrenaline and several hormones of the adrenal cortex also increase the blood glucose level by stimulating the breakdown of liver glycogen. Their action accounts for the observation that surgical removal of the pancreas gives rise only to symptoms that reflect an insulin deficiency, even though glucagon is also lacking.

The Adrenal Glands

Each adrenal gland measures 1 to 2 in. in length; one is attached to the top of each kidney. Each gland has two parts that are embryonically, histologically, and functionally different: an external reddish-brown **cortex,** which surrounds an internal grayish **medulla** (Fig. 16-5). So far it has not been possible to explain why the cortex and medulla are shaped into a single organ. Both parts have an endocrine function, but they are otherwise totally unrelated.

Hormones of the Adrenal Medulla

The tissue of the adrenal medulla is in many ways related to nervous tissue. Its embryonic origin is from neural crest cells (Chapter 12), and the two compounds it releases are identical to those released by sympathetic nerve endings (Chapter 17). Moreover, the secretory cells of the adrenal medulla in some ways resemble sympathetic neurons.

The closely packed cells of the adrenal medulla are of two distinct cell types: one that secretes **adrenaline** (or **epinephrine**), and another that is responsible for the production of **noradrenaline** (or **norepinephrine**) (Fig. 16-6, p. 450). In man 70 to 90 percent of the hormonal activity of the adrenal medulla is the secretion of adrenaline. The release of the two hormones from the secretory cells in the medulla is controlled by two different centers in the hypothalmic region of the brain. In general, both exert the same physiological effects as those that are brought about by the stimulation of the sympathetic nervous system: they tend to rally the body in situations that involve stress or emergency.

The two hormones have related but distinctly different physiological effects. Adrenaline functions as an emergency hormone for the body. It elevates the glucose level in the bloodstream by aiding enzymatic conversion of liver glycogen to glucose. The rise in blood glucose is accompanied by an increase in oxygen consumption, body temperature, and heat production. Adrenaline also tends to increase the blood flow in skeletal muscle, heart, and other viscera by dilating the blood vessels. It is responsible in part for the relaxation of smooth muscle of the digestive tract and bronchial tubes of the lung, but causes contraction of certain other muscles. Its over-all effect, then, is to direct the circulation of blood where necessary during exertion or increased activity. Adrenaline also augments blood pressure and cardiac output

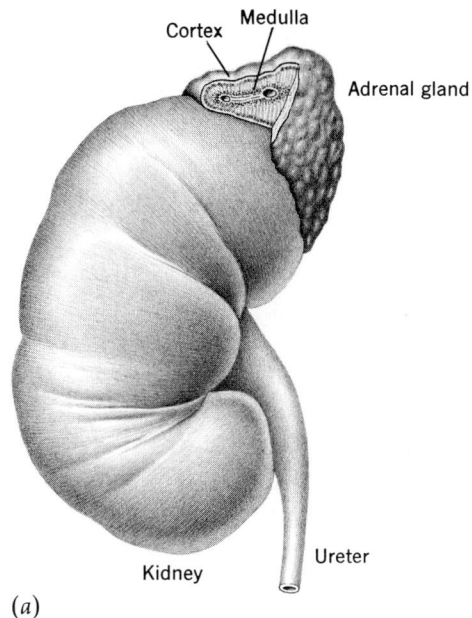

(a)

Figure 16-5 (a) The adrenal gland, shown in position on the kidney, cut to illustrate its internal structure. (b) A cross-section of the adrenal, showing details of the cortex and medulla. ×30

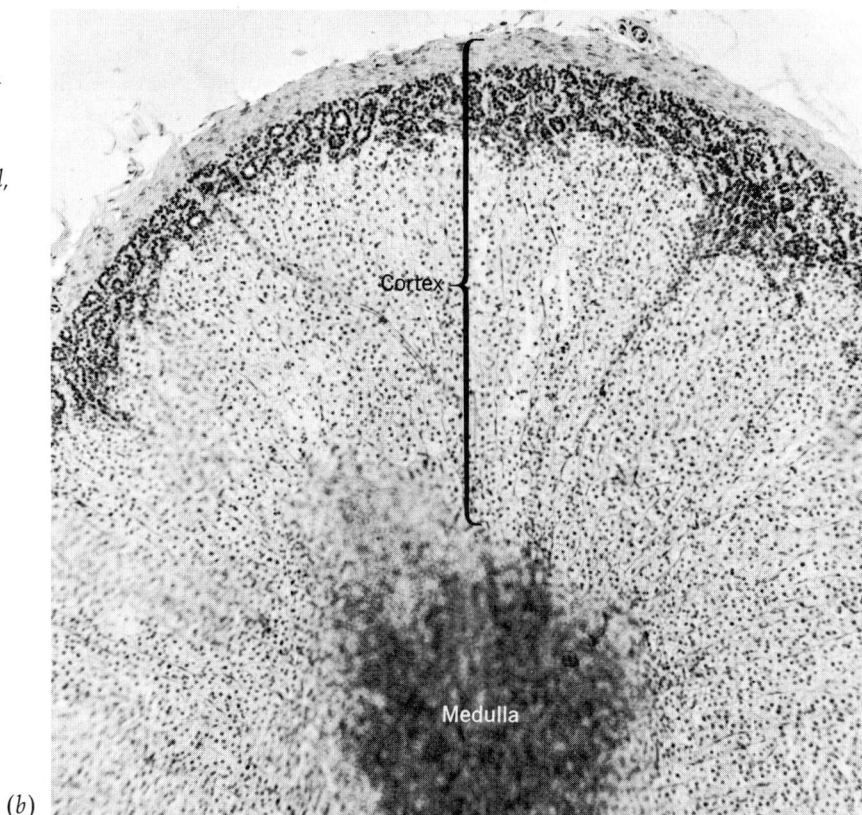

(b)

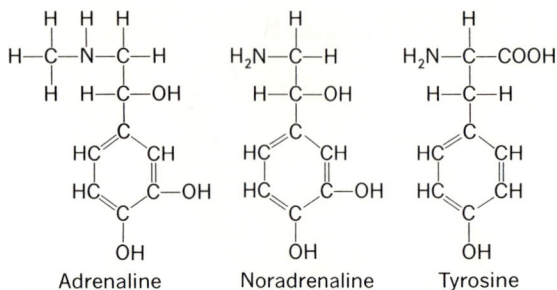

Figure 16-6 *Structure of adrenaline and noradrenaline, as compared with the amino acid tyrosine from which they are derived.*

(the amount of blood pumped by the heart) and steps up the rate of heartbeat.

By contrast, the effects of noradrenaline are less extreme. Noradrenaline has no appreciable influence on carbohydrate metabolism or oxygen consumption and exerts little or no action on cardiac output. It produces milder inhibitory effects on smooth musculature and fails to relax the muscles of the bronchial tubes. It does raise the blood pressure, but, unlike adrenaline, it constricts instead of dilates the blood vessels. Both hormones are metabolized and inactivated by enzymatic processes and finally excreted in the urine in a modified form.

The adrenal medulla, unlike the thyroid and parathyroid glands, pancreas, and adrenal cortex, is not essential for life, for it can be removed surgically without causing drastic effects.

Hormones of the Adrenal Cortex

The adrenal cortex is essential for life. It is made up of several zones that secrete different steroid hormones. These are **cortisone** (Fig. 16-7), **corticosterone, deoxycorticosterone, aldosterone, cortisol,** and **dehydrocorticosterone,** all related in structure and all derived from the steroid cholesterol (Fig. 3-14). The various functions of the adrenal cortical hormones fall into four main groups: carbo-

hydrate and protein metabolism; mineral, salt, and water metabolism; sexual development; and inflammation and allergy protection.

Carbohydrate and Protein Metabolism

Cortisone, cortisol, and corticosterone exert their influence primarily on carbohydrate metabolism; they are therefore often called **glucocorticoid** hormones. They increase the blood glucose level at the expense of liver glycogen, and indirectly enhance the conversion of proteins to carbohydrates.

Mineral, Salt, and Water Metabolism

The cortical hormones that have a regulatory effect on the relative concentrations of mineral ions (especially sodium and potassium) in the body fluids, and therefore on the water content of the tissues, are called **mineral corticoids.** Aldosterone and deoxycorticosterone are mineral corticoids that are highly potent in this respect.

An insufficiency of mineral corticoids results in increased excretion into the urine of sodium ions, chloride ions, and water, accompanied by a fall in blood sodium, chloride, and bicarbonate. A concomittant rise in potassium ions at the expense of the body cells and tissue fluids also occurs. The loss of bicarbonate reflects itself in a lowered pH, or **acidosis,** of the blood. These are the symptoms of an illness called **Addison's disease;** if untreated, it is usually fatal. Administration of the mineral corticoid hormones stimulates the convoluted tubules of the kidney (Chap-

Figure 16-7 *Structure of cortisone; compare it with that of cholesterol (Fig. 3-16).*

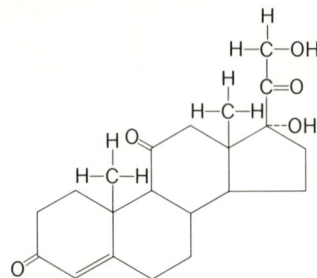

ter 15) to decrease reabsorption of potassium and increase that of sodium; water is thus retained and blood volume and blood pressure return to normal.

Sexual Characteristics

Under certain circumstances the adrenal cortex appears to secrete steroids that are similar in their effects to male hormones. Adrenal cortical tumors secrete large amounts of such steroids; in females this may give rise to masculine characteristics (Chapter 11) as illustrated by a bearded lady in a circus. Although female sex hormones have also been detected in the adrenal cortex, it is believed that they are simply intermediates in the formation of the adrenal cortical steroid hormones.

Inflammation and Allergy

Several cortical hormones have a marked influence in preventing the appearance of inflammation (Chapter 15) and accompanying breakdown of connective tissue. Cortisol, and to a lesser extent cortisone, are effective in this respect. Some cortical hormones (e.g., cortisol) counteract the symptoms of allergic reactions. They are also employed to control the symptoms of some arthritic conditions and rheumatic fever.

Control of Adrenal Cortex Secretions

The development and function of the adrenal cortex is under the control of a hormone called **adrenocorticotropic hormone (ACTH),** secreted by the anterior pituitary gland (see next section).

Oversecretion by the adrenal cortex as a result of excessive secretion by the pituitary of ACTH gives rise to symptoms including high blood pressure, excess blood salt, swelling of the tissues with water, demineralization of the bones, and loss of sexual function. Such oversecretion may also result from a tumor in the cortical cells; symptoms then depend on the kinds of hormone secreted by the tumor. If cortisol is predominant, changes in metabolism occur, and these are accompanied by weakness, wasting of muscle, and certain types of obesity.

The Pituitary Gland

The **pituitary gland,** or **hypophysis,** is a two-lobed body about the size of a large pea, directly attached by a small stalk to the hypothalamus portion of the brain at the undersurface of the cerebrum (Fig. 16-1). Each of the two lobes has a different endocrine function and origin: the **anterior** (front) lobe is derived from the same epithelial tissue as that of the mouth, the **posterior** (rear) lobe (also called the **neurohypophysis**) forms from brain tissue and bears structural resemblances to nervous tissue. The lobes are essentially two different endocrine glands, in the same manner as adrenal medulla and adrenal cortex and the thyroid and parathyroid glands.

The Anterior Pituitary Gland

The anterior pituitary is often called the **master gland** of the endocrine system because of its important effects in regulating and maintaining the development and function of other endocrine glands. It has a key part in integrating growth and metabolism and in the development and functioning of the reproductive system, including the secondary sexual characteristics. The central role of the anterior pituitary gland is summarized in Figure 16-8 (p. 452).

The anterior pituitary in humans secretes at least six distinct hormones. All are proteins except for the adrenocorticotropic hormone (ACTH), which is a polypeptide. All serve **trophic** functions (from the Greek word **trophos,** to feed; by convention, the suffix -trophic has been simplified to -tropic); that is, they stimulate or sustain other glands or tissues. The nature and function of these six hormones are described in the following sections.

Thyrotropic (Thyroid-Stimulating) Hormone (TSH)

TSH controls various aspects of thyroid gland function, including development and

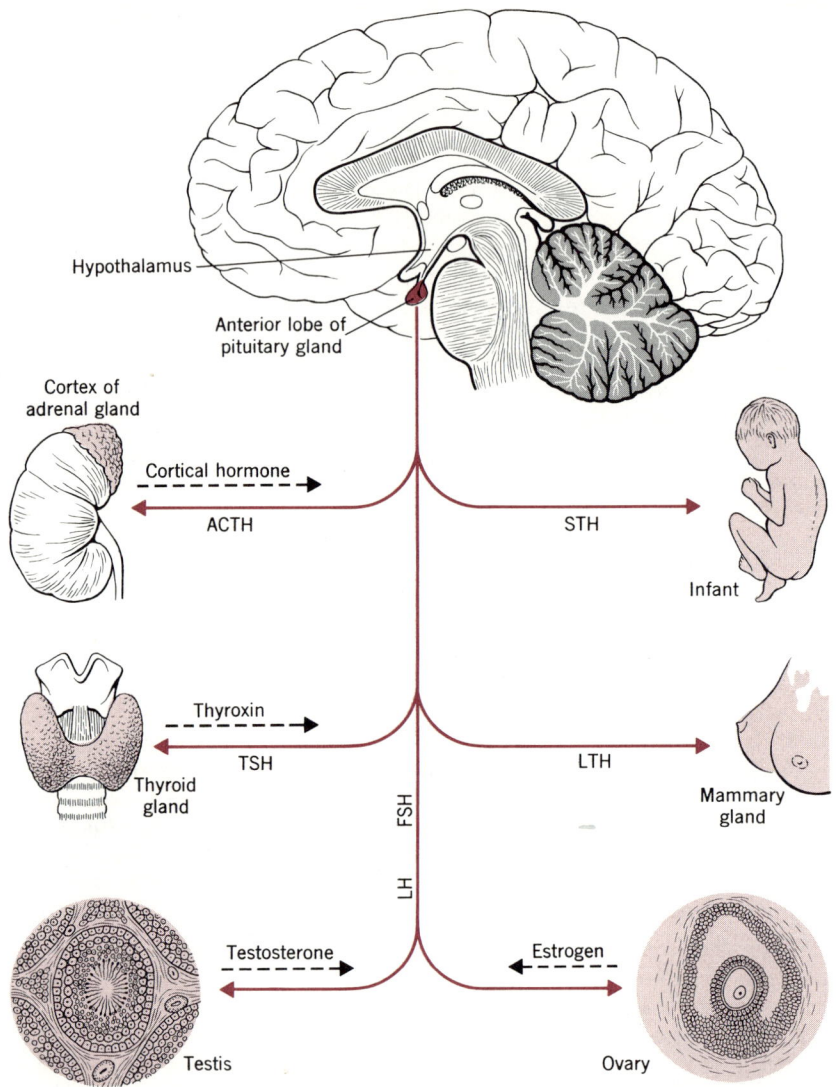

Figure 16-8 The central role of the anterior pituitary is emphasized in this diagram, which indicates some of its major interactions with other hormone-producing glands and tissues. Under the influence of the hypothalamus, which lies directly above the pituitary, the anterior lobe produces tropic hormones that regulate the secretions of the thyroid, the adrenal cortex, and the sex glands, as well as body growth. (These stimulatory effects are shown by solid arrows.) Most of the target glands produce substances (their own hormones) that feed back as regulators of the output of tropic hormones by the pituitary. (Feedback effects are shown by broken lines.)

maintenance. It also stimulates iodine accumulation, the conversion of iodine into the thyroid hormones, and their release from the gland into circulation. We have already noted the operation of a negative feedback mechanism in regulating TSH secretion. It is thought that certain nerve secretions from specific areas of the hypothalamus (p. 477) are transmitted to the pituitary and secreted into the bloodstream to regulate the secretion of TSH.

Adrenocorticotropic Hormone (ACTH)

ACTH regulates the development, maintenance, and secretions of the adrenal cortex. ACTH principally stimulates glucocorticoid secretion and has only a small effect on mineral corticoid production. The secretion of ACTH by the anterior pituitary is apparently stimulated by a decreased glucocorticoid concentration in the blood. The ACTH in return stimulates adrenal cortical secretion.

Thus a classical negative feedback relationship exists between the pituitary and the adrenal cortex, whereby the increasing glucocorticoid level in the blood reciprocally inhibits further ACTH secretion by the pituitary.

The secretion of ACTH and adrenal cortical hormones are mutually affected by a negative feedback mechanism. In addition, there is evidence of nerve secretions from certain hypothalamus areas of the brain that influence the secretion of ACTH. Such a mechanism suggests a basis for the release of ACTH by several different stimuli including emotional stress, extreme temperatures, poisons, drugs, and various other substances.

The ACTH molecule is a polypeptide containing 39 amino acids. Fifteen of these amino acids can be removed, leaving a basic chain of 24 amino acids that still retains most of the hormonal activity.

Growth Hormone

The growth hormone (also called **somatotropin** or **STH**), stimulates body weight and the rate of growth of the skeleton. A striking retardation of growth of young experimental animals as a result of surgical removal of the pituitary gland was one of the earliest findings relating to the function of the pituitary.

Undersecretion of growth hormone during the years of skeletal growth of an individual results in **dwarfism.** An adult dwarf may attain a height no taller than 3 or 4 ft, and is usually sexually immature. Dwarfism, in contrast to cretinism (arising from thyroid insufficiency), is not accompanied by physical deformity or mental retardation. Oversecretion of growth hormone results either in **gigantism** (individuals growing as tall as 8 ft or more) or **acromegaly** (overgrowth of the bones and tissues of the hands, feet, jaws, and face—Fig. 16-9). The latter condition results from excess hormone in an adult after full skeletal growth is completed.

The Three Gonadotropic Hormones

Early observations of atrophy and impaired development of the reproductive system following the degeneration or surgical removal of the pituitary gland indicated that the gland was essential for normal sexual development. Over the years three distinct hormones necessary for sexual development and activity have been discovered; all are secreted by the anterior pituitary gland, and all are proteins. They are referred to as the **gonadotropic** hormones because they control the growth and development of the gonads. Secretion of the gonadotropic hormones is apparently controlled by feedback mechanisms (secretion is inhibited, e.g., by a certain level of sex hormones in the blood) and by secretions called **releasing factors** from the hypothalamus.

The gonadotropic hormones are secreted in slight amounts until just before the onset of puberty (Chapter 11). They are then produced in progressively larger concentrations, which initiate the onset of sexual development characteristic of puberty. In females they

Figure 16-9   An individual with acromegaly, the result of oversecretion of growth hormone by the anterior pituitary gland beginning in adulthood.

subsequently account for cyclic changes that recur in the ovaries and uterus throughout the years of sexual maturity (pp. 270–275). In males they regulate the formation of sperm and the production of the male sex hormone testosterone. The details of the interrelationships between the gonadotropic hormones and the sex hormones produced by the ovaries and testes are discussed in a later section of this chapter.

Follicle-Stimulating Hormone (FSH). In females, starting at puberty and continuing throughout the years of sexual maturity, FSH promotes certain changes during a portion of each menstrual cycle. It stimulates the development and maturation of an ovarian follicle, which produces the egg cell and acts as a gland to secrete estrogen. In males, starting at puberty, FSH induces spermatogenesis in the testes (see Chapter 11).

Luteinizing Hormone (LH). In females LH is involved in further development of the egg cell and in its release from the ovarian follicle (Chapter 11). The hormone also stimulates the ovary to produce the hormone progesterone (p. 272). In males LH stimulates formation and secretion of the male sex hormone testosterone by specialized cells in the testes.

Lactogenic, or Luteotropic, Hormone (LTH). LTH apparently functions only in females. It was originally described as the hormone necessary for the initiation of milk secretion in the mammary glands of mothers shortly after giving birth; it is now known that, along with estrogen, LTH promotes development of the mammary glands. Together with LH, the hormone also serves to maintain the secretion of progesterone that is vital for the maintenance of pregnancy. The secretion of LH (unlike that of FSH) seems to be stimulated by female sex hormones. The factors causing the release of LTH are poorly defined.

The Posterior Pituitary Gland

The posterior pituitary gland, or **neurohypophysis,** releases two hormones, oxytocin and vasopressin. Neither of these is actually synthesized in the pituitary; instead, they are formed in the hypothalamus, at the base of the brain, and flow along nerves to the posterior pituitary, where they are stored.

The Nobel-prize-winning work of Vincent du Vigneaud and his colleagues at Cornell Medical School on the identification and synthesis of oxytocin and vasopressin has made it possible to study the physiological action of each as a pure chemical compound. Each hormone is composed of eight amino acids; six of these are the same in both. The two amino acids that differ, however, cause the two compounds to have very different physiological effects. **Oxytocin** (Fig. 16-10) generally promotes the contraction of smooth muscles of the body—for example, the smooth muscles of the urinary bladder and intestines—but the greatest response is elicited in the uterine muscle, especially during the late stages of pregnancy. The hormone is frequently administered by injection during and after birth to stimulate contraction of the uterus and constriction of blood vessels, thus facilitating the birth process and impeding the loss of blood.

The normal ejection of milk from the mammary gland of a lactating female is initiated by mechanical stimulation of the nipple by feeding infant; the sucking of the nipple gives rise to nerve impulses that are transmitted to the hypothalamus, causing oxytocin to be released from the pituitary. Milk ejection is therefore a response to the release of oxytocin from the neurohypophysis, and is different from milk secretion, which is a response to the lactogenic hormone secreted by the anterior pituitary.

Vasopressin is often called the **antidiuretic hormone** because of its striking influence in decreasing loss or excretion of water from the body by way of the urine. Changes in the osmotic concentration of the blood and body fluids stimulate particularly sensitive neurons

Figure 16-10   Structure of the polypeptide oxytocin.

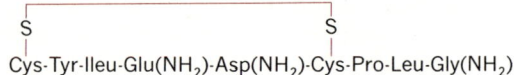

$$S \qquad\qquad\qquad\qquad S$$
Cys-Tyr-Ileu-Glu(NH$_2$)-Asp(NH$_2$)-Cys-Pro-Leu-Gly(NH$_2$)

in the hypothalamus, resulting in the release of vasopressin from the neurohypophysis. We noted earlier (p. 436) that an undersecretion of vasopressin results in diabetes insipidus; (not to be confused with diabetes mellitus, which is caused by insufficient insulin). This disease is characterized by the daily excretion of huge volumes of urine. It can usually be controlled by injections of vasopressin.

A secondary function of vasopressin is to cause constriction of the musculature of the arterioles and capillaries, including the coronary and pulmonary vessels, thus causing some rise in blood pressure.

The Sex Hormones

In Chapter 11 we discussed the functions of the gonads in reproduction. We know that in addition to producing eggs and sperm, the ovaries and testes are also the primary sources of the sex hormones that control the maturation and function of the reproductive system and certain other tissues of the body. The sex hormones stimulate and regulate the mature development of the gonadal ducts and auxiliary glands, which assures the proper passage of the sex cells, and initiate the development of the secondary sexual characteristics. The gonads and their hormonal secretions are not necessary for the life of an organism, although they are obviously essential for continuation of a species.

Female Sex Hormones

The principal function of the female sex hormones is the maintenance of the reproductive tract for (1) the normal passage of the egg cell and its fertilization, and (2) the subsequent development of the fertilized egg into a new individual. The ovaries secrete two types of female sex hormone: estrogen and progesterone, both of which are steroids.

Estrogen is a collective name for four closely related steroid hormones primarily secreted by the maturing ovarian follicle. An-

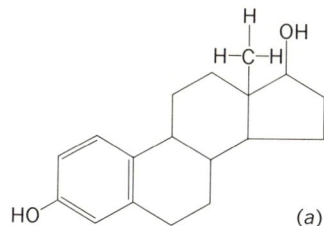

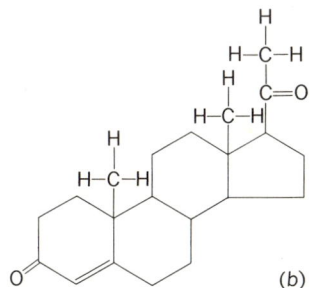

Figure 16-11
Structure of the sex hormones (a) estradiol and (b) progesterone.

other important source of estrogen is the placenta, which develops from tissues arising from both the embryonic membranes and maternal uterus. The most common form of estrogen is **estradiol,** whose structure is shown in Figure 16-11a. Starting shortly before puberty and continuing throughout the sexaul maturity of a female, the secretion of estrogen from the developing follicle is periodically stimulated by FSH from the pituitary gland. Estrogen is responsible for mature growth, development, and maintenance of the female reproductive tract, the secondary sex organs (breasts, vagina, uterus), and other secondary sexual characteristics. It very likely also contributes to sexual desire. Estrogen promotes repair of the uterine lining following menstruation in preparation for possible implantation of a fertilized egg. It also tends, in general, to decrease secretion of the gonadotrophic hormones of the pituitary.

Progesterone (Fig. 16-11b) is the second principal ovarian sex hormone. It is secreted during the latter half of the menstrual cycle in women, chiefly by a temporary endocrine tissue, the corpus luteum, which develops in the ruptured follicle of the ovary shortly after the release of the egg cell. Progesterone further promotes the development of the uterine lining (already initiated by estrogen during the first half of the menstrual cycle) in preparation for implantation of the fertilized egg. At this phase of the female menstrual cycle (when estrogen is being secreted at a minimum level), progesterone takes over the general functions of estrogen in maintaining the reproductive tract and the secondary sexual characteristics. Thus, together with estrogen, progesterone is responsible for the sequence of changes of the repeating female menstrual cycle (Chapter 11).

One additional hormone is produced by the mammalian corpus luteum. It is called **relaxin,** and is a polypeptide instead of a steroid. It specifically causes relaxation of the pubic ligament of the pelvic girdle, thus increasing the size of the birth canal during delivery of a child.

Oral Contraceptives

One of the greatest contributions to humanity that has come from the field of endocrinology is the development of an effective birth control method in the form of an orally ingestible pill. The pill's basic hormone is **progestin,** a synthetic compound similar to progesterone. Most oral contraceptives also contain estrogen. Contraceptive pills are designed primarily to prevent ovulation (pp. 270–275). Recall that the pituitary hormones FSH and LH are necessary for the production and release of a mature egg from the ovary, and that the estrogen and progesterone normally produced by the follicles in the ovary act as feedback inhibitors of secretion of FSH and LH. When a woman takes a birth control pill daily, starting on the fifth day after the beginning of menstruation, the progestin and estrogen provided by the pill inhibit the secretion of FSH and LH by the pituitary. Without FSH and LH ovulation does not take place, so no egg is present in the uterine tubes to be fertilized by the sperm released during intercourse. This inhibition of egg production is very much like the suppression of ovulation that normally takes place while a woman is pregnant; under the influence of the high levels of natural estrogen and progesterone, no other eggs are ovulated until after the pregnancy is terminated.

Male Sex Hormones

The steroid **testosterone,** the principal male sex hormone, is secreted by the interstitial cells in the testis. It is responsible for the maturation, at puberty, and maintenance of the various ducts and accessory glands of the male reproductive tract, as well as for the male secondary sexual characteristics (p. 275) and sexual desire. Testis removal, or **castration,** in the human male before the onset of puberty results in the failure of the secondary sexual characteristics to appear.

Other Hormones

A variety of other hormones or hormonal effects are also known; those concerned with the integration of the digestive process have been discussed in Chapter 14.

The **pineal apparatus** is a small body lying deep in the brain that serves a neuroendocrine function in regulating some of the body's intrinsic daily biological rhythms (Chapter 19).

It is known that damaged tissues release a substance called **histamine,** which relaxes the muscles in the walls of blood vessels and increases their permeability. The localized effects of histamine are beneficial. It facilitates movement of white blood cells and antibodies into a damaged area and helps prevent and fight infection. Increasing levels of histamine in the blood, however, can have unpleasant effects—particularly as regards allergies. In hay fever, for example, antigen-antibody reactions (pp. 408–410) occur in the nasal mucosa; these reactions damage cells and cause the local release of histamine. The histamine makes the walls of the capillaries more permeable, which results in a copious flow of fluids both from the vessels themselves and from the mucosal glands. This "runny nose" is the body's attempt to flush out irritants from the mucosa, and, although uncomfortable, is beneficial. The histamine released into the bloodstream may also have a distant, hormonelike effect in causing constriction of the walls of the bronchioles, which may be so severe as to make breathing difficult, as in the condition termed **asthma.** Excessive release of histamines can be slowed by drugs commonly called **antihistamines.**

Endocrine Interactions in Other Organisms

So far we have dealt only with the hormones of man. There is ample evidence, however, that all multicellular organisms are coordinated chemically by the effects on target organs of substances released by cells of another type located in a different part of the body. In 1849 A. A. Berthold, a German physician, castrated a group of young roosters in the first scientific experiment in the field of endocrinology. He found that castrated birds became typical capons; the combs, wattles, plumage, aggressiveness, crow, and sexual urge typical of the mature cock failed to develop. If at the time of the operation, however, the removed testes were reimplanted in the same animal, but at a different site, the birds developed normally. It was from these experiments on birds that Berthold and his co-workers developed the concept that substances produced by certain tissues could be carried by the bloodstream and exert specific effects on other tissues.

It is now clear that every group of animals produces hormones. In amphibians hormones are especially active during development. Thyroid hormones in particular have a remarkable action in accelerating the rate of metamorphosis to the adult stage. Tadpoles treated with thyroxin, for example, undergo differentiation and development to the adult frog stage much sooner than they would otherwise. In insects hormones cause color changes in the body, stimulate development of eggs in the female, and control growth and metamorphosis through the larval, pupal, and adult stages (Fig. 16-12, p. 458).

If the brain of a mature caterpillar is re-moved before the caterpillar spins its cocoon, pupation does not take place. This is not the result simply of surgical shock; if the brain tissue is reimplanted in some other part of the caterpillar's body, the pupal stage does occur. This experiment suggests that the insect brain produces a hormone necessary for pupation. In fact, about two dozen special **neurosecretory** cells have been found in the caterpillar brain that secrete a **brain hormone.** This hormone does not stimulate pupation directly, but instead acts on a pair of glands, called **prothoracic glands,** in the caterpillar thorax. These glands secrete a second hormone, a steroid called **ecdysone,** which directly initiates molting and pupal formation.

There is a third hormone, called **juvenile** hormone, involved in this process of insect development. Studies by Sir Vincent Wigglesworth on the South American bloodsucker *Rhodnius* have shown that this juvenile hormone is secreted by a pair of endocrine glands, the **corpora allata,** located near the brain. Juvenile hormone promotes larval development but prevents metamorphosis. If the corpora allata are removed from an immature silkworm, it spins a cocoon and pupates at its next molt, to form a miniature moth. On the other hand, if the corpora allata of a young caterpillar are implanted into the body of a mature larva, metamorphosis does not occur; molting results simply in the production of an extra-large caterpillar (Fig. 16-12). Normal pupation thus seems to occur when the output of juvenile hormone decreases spontaneously in the mature caterpillar.

The development of the typical structures of the larva, pupa, and adult requires the sequential activation of different sets of genes in the insect cells. The giant chromosomes of certain insects such as *Drosophila* and *Chironomus* have enlarged regions of "puffs" that occur at different bands on the chromosome during the course of development (Chapter 8). Each time a larva prepares to molt, its chromosomes exhibit a definite sequence of puffing. If the hormone ecdysone is injected into a larva just after a molt, when its own level of the hormone is low,

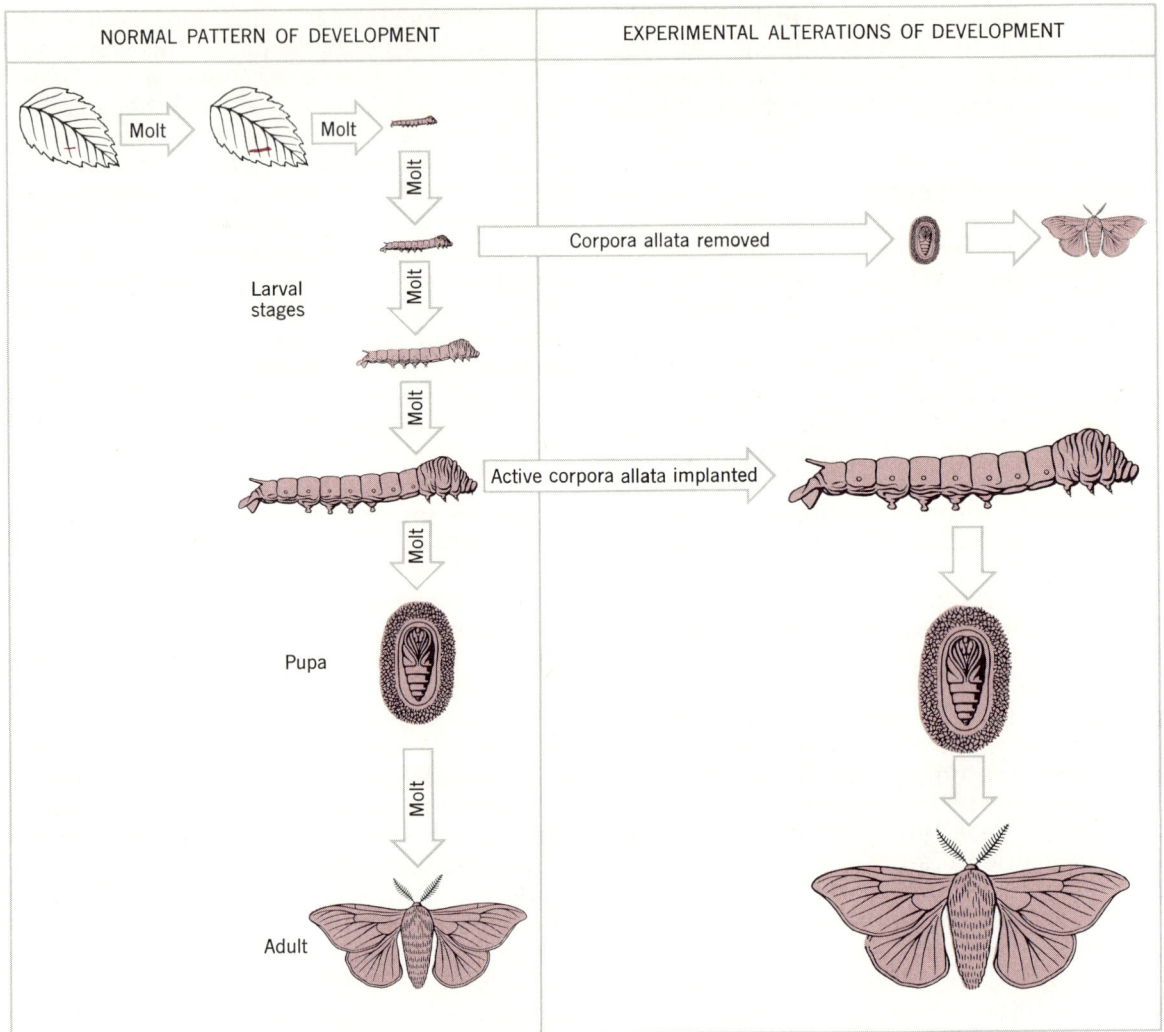

Figure 16-12   *Development of the silkworm* Bombyx mori. *Insects grow by periodically shedding a rigid exoskeleton in the process known as* **molting,** *which occurs several times during larval development. Just before the final molt, the mature caterpillar spins a silk cocoon and enters a dormant stage called the* **pupa,** *during which the marked transformation in body structure, termed* **metamorphosis,** *takes place. At the final molt the organism emerges as an adult moth. This developmental sequence can be shortened by removing the corpora allata from a young caterpillar, and can be extended for several extra molts by surgically introducing active corpora allata into a mature caterpillar. As long as substantial amounts of juvenile hormone are being secreted, pupation does not take place.*

the normal premolt puffing sequence begins again. It is in these puffs that especially active RNA synthesis occurs—that is, it is the operons in the puffed regions that are activated. In fact, a messenger RNA has been extracted from these ecdysone-treated chromosomes that serves as a template for the synthesis of a specific enzyme associated with pupation; it is not found in the cells of untreated larvae.

Hormone Action

It seems that enough evidence is accumulating from several different systems to allow a generalization regarding one class of hormone function. We have just seen that ecdysone acts by stimulating specific RNA synthesis. We noted earlier (p. 442) that cortisone causes a similar increase in messenger RNA synthesis in mammalian liver cells, leading to an enhanced production of certain enzymes. A number of other hormones are known to act in similar ways: among the mammals, estrogen, testosterone, insulin, ACTH, thyroxin, and growth hormone; even the plant hormones gibberellic acid and indoleacetic acid (Chapter 13) seem to work by increasing specific RNA synthesis. Thus one class of hormones seem to be specific gene regulators.

Other hormones seem to have their primary action by affecting the cell membrane of the target cell. Recently E. W. Sutherland and his colleagues made a discovery of great importance to our understanding of this mechanism of hormone action; Sutherland received the Nobel Prize in 1971 for his investigations of cyclic AMP and adenyl cyclase. While trying to learn how adrenaline from the adrenal medulla stimulates the liver to release more glucose into the bloodstream, these investigators found that the first response of the liver cells to the hormone was to increase their content of a compound called **cyclic adenosine monophosphate,** or **cyclic AMP.** This increase in cyclic AMP in the

cells led, in turn, to the activation of the enzyme glycogen phosphorylase (Chapter 4), which promotes the breakdown of glycogen to glucose. Sutherland found that cyclic AMP was formed in the cells by an enzyme called adenyl cyclase, which is built into the cell membrane. Thus adrenaline acts as an extracellular messenger from the adrenal medulla to the adenyl cyclase molecule in the liver cell surface, while cyclic AMP is an intracellular messenger between the cell membrane and the appropriate intracellular enzymes.

We now know that adenyl cyclase is present in almost all cell membranes, and more than a dozen hormones have been discovered that act in a similar fashion—that is, by stimulating this surface enzyme and the build-up of cyclic AMP in their target cells. But if many hormones act in the same way, how can we explain their remarkable specificity? Why does TSH circulating in the bloodstream activate adenyl cyclase in thyroid cells—thereby stimulating those cells to produce thyroxin—whereas adrenaline, which activates the same enzyme in liver cells, has little or no effect on the thyroid?

The most probable answer to this question, which, however, has not yet been verified experimentally, is that each kind of target cell has different hormone receptor sites along with the adenyl cyclase in its cell membrane. Thyroid cells presumably have receptors that bind thyrotropic hormone but not cortisone or adrenaline. Liver and muscle cells, on the other hand, must have adrenaline receptors, but not sites for TSH.

A more difficult question is why the increased cyclic AMP that results from triggering adenyl cyclase has such different effects inside each different cell type. It has recently been shown that cyclic AMP can act as a regulator of a vast number of enzymatic reactions. Why this compound should act on enzymes necessary to produce thyroxin in thyroid cells, without apparently influencing the thyroid's glucose level, but still cause profound changes in glucose metabolism in liver cells, is still not known.

Conclusion

It should be apparent, by now, how immensely complex and interwoven are the life functions of living things. That our bodies, and those of all other organisms, function in an orderly fashion despite this complexity, is the result of intricate, interreacting, coordinating mechanisms. The endocrine system is one of the major mechanisms for providing that coordination; hormones are chemical messengers that signal activities of the gland producing them and evoke functions in the target tissue.

The other main biological coordinating system sends its messages as electrical signals along nerve fibers: the nervous system. It is to that system that we turn our attention now.

Reading List

Beerman, W., and V. Clever, "Chromosome Puffs," *Scientific American* (April 1964), pp. 50–58.

Bonner, J. T., "Hormones in Social Ameba and Mammals," *Scientific American* (June 1969), pp. 78–88.

Davidson, E. H., "Hormones and Genes," *Scientific American* (June 1965), pp. 36–45.

Etkin, W., "How a Tadpole Becomes a Frog," *Scientific American* (May 1966), pp. 76–88.

Gorbman, A., and A. A. Bern, A Textbook of Comparative Endocrinology. Wiley, New York, 1962.

Jones, K. L., L. W. Shainberg, and C. O. Byer, Sex. Harper & Row, New York, 1969.

Levey, R. H., "The Thymus Hormone," *Scientific American* (July 1964), pp. 66–77.

Levine, S., "Stress and Behavior," *Scientific American* (January 1971), pp. 26–31.

Mayr, O., "The Origins of Feedback Control," *Scientific American* (October 1970), pp. 110–118.

Peakell, D. B., "Pesticides and the Reproduction of Birds," *Scientific American* (April 1970), pp. 73–78.

Schneiderman, H. A., and L. I. Gilbert, "Control of Growth and Development in Insects," *Science,* (1964), **143,** pp. 325–333.

Turner, C. D., General Endocrinology. Saunders, Philadelphia, 1966.

Williams, C. M., "Third-generation Pesticides," *Scientific American* (July 1967), pp. 13–17.

Wurtman, R. J., and J. Axelrod, "The Pineal Gland," *Scientific American* (July 1965), pp. 50–63.

Coordination: The Nervous System

*"Now that the success of molecular genetics has made it an academic
discipline, one can expect that in the coming years students of the
nervous system, rather than geneticists, will form the avant garde of
biological research."*

G. S. STENT, 1960, PHAGE AND THE ORIGINS OF MOLECULAR BIOLOGY

In the introduction to Chapter 16 we noted that there are, in
principle, two mechanisms by which messages may be transmitted
from one part of an organism to another. One method depends
on the release of specific chemical agents—hormones—at some
point; these can diffuse or circulate to other sites, and there cause
responses. The other method is conversion of the desired message
into some sort of electrochemical code that can be conducted from
one end of a cell to another, or from cell to cell. All animals have
nervous systems made of cells specialized to do just that.

Nearly all our responses at both the conscious and unconscious
levels, including physical movement, thoughts, emotions, memories,
and personality traits are made possible through the functioning of
the nervous system. Aside from its general role in controlling and
integrating the activities of different parts of the body, the nervous
system has a broad and unique function: it is our principal, if not
only, direct means of contact with the external environment. Not
only our responses to objects and events in the world surrounding
us, but our entire awareness of their existence, is necessarily mediated by the nervous system, accounting for every thought, sensation,
and movement. A defect or impairment in any portion of the nervous system can therefore limit our contacts with the outside world:
destruction of the eyes or their nerves (the optic nerves) leading
to the brain means that we can no longer see what is happening;

461

defects in the nerve cells leading from the specialized taste buds in the tongue can result in a loss of the ability to taste certain substances.

We know that there are numerous physical phenomena that cannot be detected directly by means of our sense organs and nervous system; radio waves, X-rays, cosmic rays, and atomic radiation are examples. We are only aware that they exist because we have been sufficiently clever to transform some of their effects by various devices into phenomena (such as heat, visible light, sound) that we can detect or decipher through the action of the nervous system.

Our complete dependence on our nervous system for knowledge of the world around us raises an interesting philosophical question. Is it possible that the universe might be very different from what we picture it to be because of limitations or aberrations in our nervous systems? The answer is probably no. Our perception of the external world cannot be too far from reality, merely because it is hard to imagine, if it were, how we could have survived and evolved in the inexorable process of natural selection for the last 2 billion years.

In all vertebrates, including man, the nervous system is made up of a connected vast and intricate branching network of nerve cells and fibers, collectively called the **peripheral nervous system,** which extends to every part of the body; and a central structure consisting of the brain and spinal cord, collectively called the **central nervous system (CNS),** which serves as a central exchange or switchboard. It should be noted that the central and peripheral nervous systems are not independent and separate entities. From one point of view the peripheral nervous system is an extension of fibers from the central nervous system, like the branches of a tree from its trunk. All messages, or nerve impulses, transmitted by the peripheral nervous system must eventually pass along a nerve pathway that always includes one or more of the neurons that are part of the central nervous system.

Nerve Function

Any qualitative or quantitative change in the environment that can be detected by an organism is defined as a **stimulus.** The ability to respond to a stimulus, a characteristic called **responsiveness,** or **irritability,** is a fundamental feature of living systems. The **response,** which is essentially similar in all organisms, consists of three successive steps: **receipt** of the stimulus, **conduction** of the resultant signal, and **reaction** to the signal. For example, in a unicellular organism such as an ameba the reaction to a foreign non-edible object is usually a movement away from it. In this case the response occurs entirely in the protoplasm of a single cell. Among some of the protista, for example in the familiar ciliate *Paramecium,* there are specialized areas in the protoplasm for detecting changes in the environment, other structures for transmitting the signals to various parts of the cell, and other areas for effecting a response.

In the course of evolution of the multicellular animals the ability of an organism to respond to particular stimuli has become highly developed in specialized tissues and organs. These include: (1) certain specialized **nervous tissues** or closely associated cells or organs, called **receptors** (e.g., the eyes, ears, taste buds), each of which is sensitive to a particular kind of stimulus in the environment; (2) the **nerve cells** themselves, which propagate, transmit, and integrate the resultant signal, or **nerve impulse,** initiated by the stimulus at the receptor; and (3) the **effector organs,** which are muscles or glands that react to a nerve impulse by being stimulated or repressed in their activities. The nervous system thus links receptors to effectors by conducting and integrating nerve impulses from the former to the latter.

The most primitive nervous system among early animals must have been similar to that in existing coelenterates (see Chapter 24) such as the jellyfish or *Hydra* (Fig. 17-1a), which consist simply of a group of similar nervelike cells joined end to end to form a

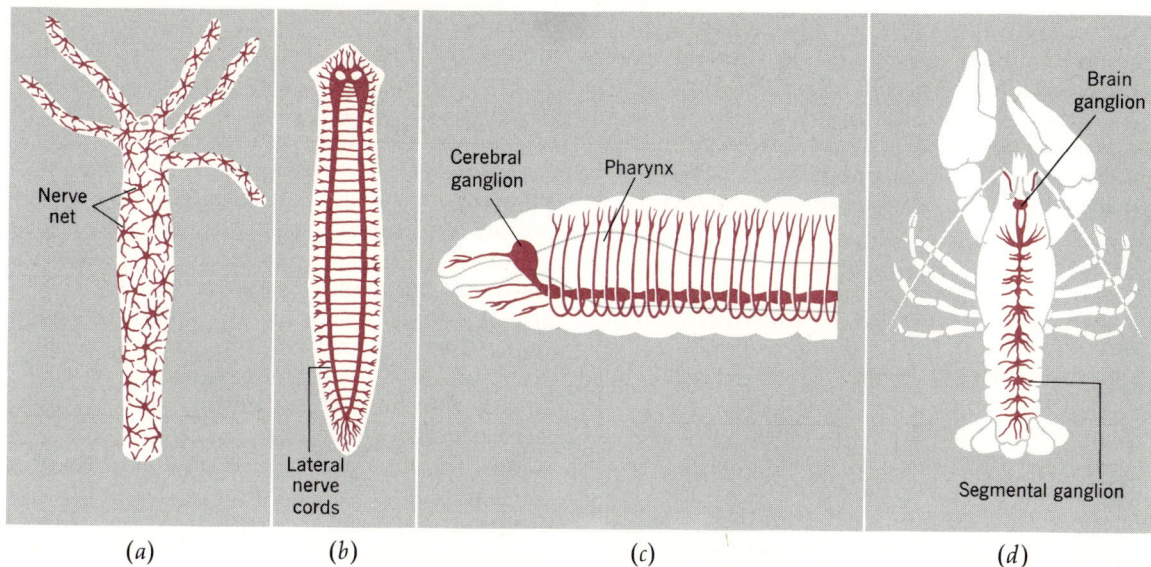

Figure 17-1   Simplified diagrams of the nervous systems of Hydra, Planaria, a segmented worm, and a crayfish. (a) The nerve net of Hydra shows a primitive form of differentiation into receptors, conductors, and effectors, but there is no central nervous system. (b) In Planaria two new developments are seen: some of the nerve net is condensed into two **lateral nerve cords,** and an aggregation of nerve cells at one end suggests the beginnings of a brain. (c) The **lateral nerve fibers** of the earthworm are able to transmit signals more rapidly to and away from the **cerebral ganglia** that form the worm's brain. (d) In the arthropods the brain ganglia are connected to a chain of **segmental ganglia** that permit many activities to be controlled locally by reflex arcs.

network called a **nerve net** throughout the organism. Nerve impulse conduction is slow and occurs in all directions away from the stimulus. Nonetheless, some functional differentiation exists among the "nerve" cells. Some ("receptors") are more sensitive than others to particular chemical or mechanical changes in the environment; that is, they are more readily able to convert those stimuli into electrochemical excitation. Other cells seem better adapted to transmitting the excitation to different parts of the body. Still others ("effectors") are more capable of responding to the excitation by moving or contracting.

The successively more advanced inverte-brates show progressively more specialization in structure and function of the nervous system (Fig. 17-1). Some cells of the nerve net condense into a cable of nerve fibers on each side of the body, and at one end aggregates of nerve cells form enlarged portions of the nerve called **ganglia** (Fig. 17-1c, d), which are the forerunners of the brain. These specializations are the true beginnings of a central nervous system in which incoming signals can be integrated and coordinated; they permit the differentiation between **afferent** (or **sensory**) neurons (those that carry signals from receptors *to* the central nervous system) and **efferent** (or **motor**) neurons (which transmit signals *from* the CNS to the effectors).

Man, the most complex and advanced of the multicellular animals, possesses a nervous system made up of between 10 billion and 100 billion specialized neurons. These cells are organized into a vast, complex communication network like a telephone system that coordinates the activities of a great factory.

The Neuron

As the organization of the nervous system evolved, so did the structure of the nerve cells themselves. We have already seen (Fig. 6-17) that a typical neuron is a cell with an enlarged area called a **cell body,** containing the usual nuclear and cytoplasmic structures, and one or more specialized cytoplasmic extensions called **fibers,** or **processes.** The cell bodies and fibers of neurons vary so widely in size and shape that it is almost impossible to describe a generalized neuron (Fig. 17-2 indicates this variety). Cell bodies range in diameter from 4 to 25 μ, whereas the processes and their branches may extend in length from a few microns to as long as 3 ft. There is still some disagreement about the terminology for neurons, especially for their fibers.

Efferent neurons (Fig. 17-2c and p. 465) possess a long, slender, cytoplasmic process known as an **axon,** which branches near one end, and one or more cytoplasmic processes called **dendrites,** usually arising from the opposite part of the cell body. As a general rule, dendrites transmit impulses toward the cell body, whereas axons transmit impulses away from the cell body. Many afferent neurons (Fig. 17-2d), however, have only a single process that divides close to the cell body into two main branches. Still other neurons, called **connecting,** or **associational,** neurons (Fig. 17-2b) consist of a cell body and many short, branching cytoplasmic processes. Connecting neurons occur exclusively in the spinal cord and brain, and serve as intermediate links or cross-connections between numerous sensory and motor neurons. By greatly increasing the possible pathways for nerve impulse transmission, they vastly expand the capabilities of the nervous system.

Neurons never occur as isolated single cells or isolated clusters of cells, but are always linked to other nerve cells as part of the vast and branching nervous system. Neurons may be linked to one another through their axons and dendrites in **linear** or **branching** sequence.

The nerve fibers in turn are organized as a **nerve,** made up of several or many fibers bound together within a common sheath of connective tissue. In vertebrates the fibers of many neurons are wrapped by the membranes of specialized cells called Schwann cells (Chapter 5), which form a whitish, stratified lipid sheath—**myelin sheath** (Figs. 17-2 and 17-3, p. 466). This is interrupted at regular intervals along its length by constrictions, or **nodes.** The myelin layer has several functions: (1) it insulates the densely packed fibers of the brain and spinal cord from one another; (2) the Schwann cells that form the myelin sheath (Fig. 5-8; 17-3b to d) supply materials to the nerve cells; (3) the sheath increases the rate of conduction of the impulse along the axon (myelinated neurons conduct impulses 10 to 20 times faster than nonmyelinated cells). Disintegration of the myelin sheath in the CNS, and its replacement by scar tissue, occurs in the crippling disease known as **multiple sclerosis.**

Neurons are a major means of communication among the body parts, serving to integrate their many activities. They perform this highly specialized function by receiving various stimuli to which they respond by transmitting the messages called nerve impulses to different tissues and organs. All messages carried by the nervous system of mammals are transmitted within a fraction of a second. They must always pass through the brain or spinal cord, which form the central nervous system in all vertebrates, just as all telephone messages entering or leaving a factory must be relayed by way of the switchboard. The flow of information in the nervous system goes not only from the CNS to all parts of the body, but also via different nerve circuits from all parts of the body to the brain and spinal cord. Messages from the brain and spinal cord in turn control the body's response to this information.

In many instances we are unaware of the

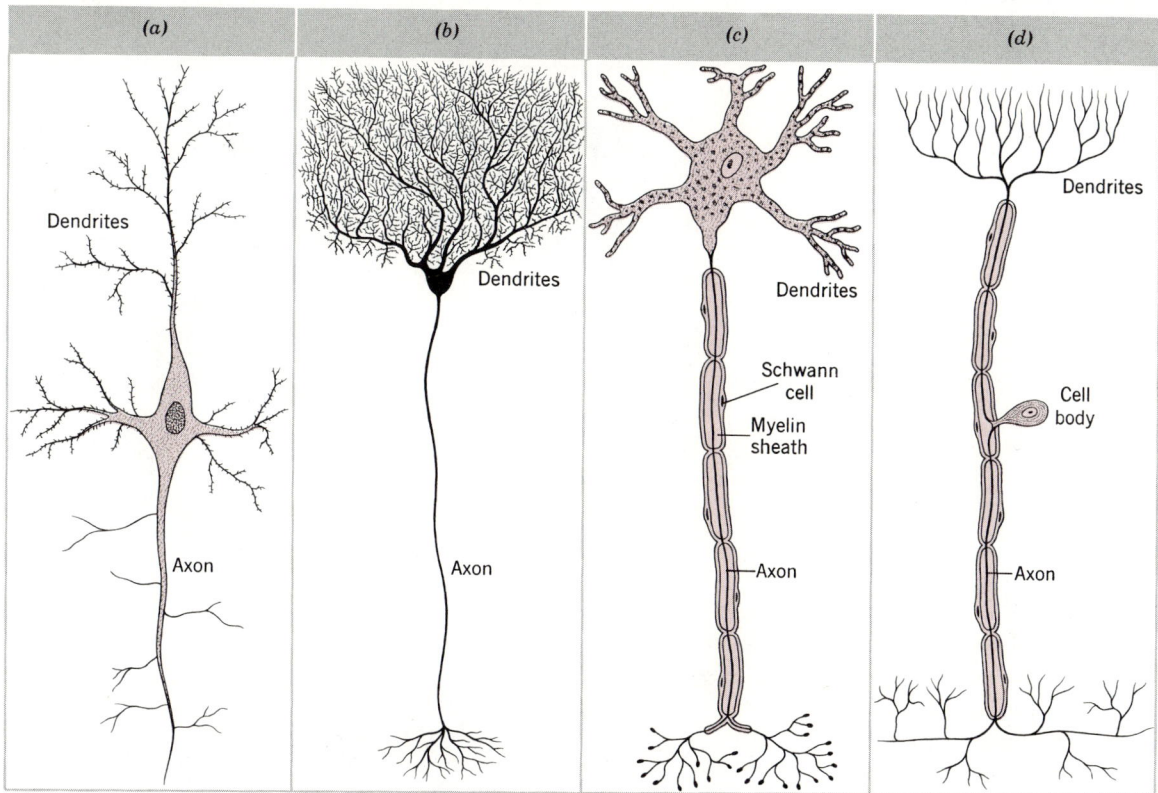

Figure 17-2   *A variety of neuron types.* (a) *A* **bipolar** *cell, showing that the dendrites, unlike the axon, often have a "spiny" appearance.* (b) *A* **connecting** *cell in the brain with profusely branched dendrites.* (c) *A* **motor neuron** *that transmits signals from the CNS to an effector such as a muscle. The axon of this cell is surrounded by a myelin sheath, which increases the speed of conduction of the signal.* (d) *A* **bipolar sensory neuron** *with one process carrying impulses from a receptor to the cell body and the other running from the cell body to the CNS.*

precise, detailed information that is transmitted (such as how much sugar, oxygen, and carbon dioxide are present in the blood). In other instances we experience nerve messages as emotions and interpretations at the conscious level as a result of the integrative activity of different parts of the brain. It is worth noting that the level of complexity of these two processes is entirely different. We have a relatively good working knowledge of the physiology of the nervous system with regard to such phenomena as the effects of

carbon dioxide concentration on the blood, but we are as yet entirely unable to explain or correlate the physiology of the neuron with thought, emotions, or interpretation.

The Nerve Impulse

Messages are transmitted via nerve cells from one part of the body to another by means of a wave of electrochemical activity called a nerve impulse. What is this electrochemical signal? It should be noted that a

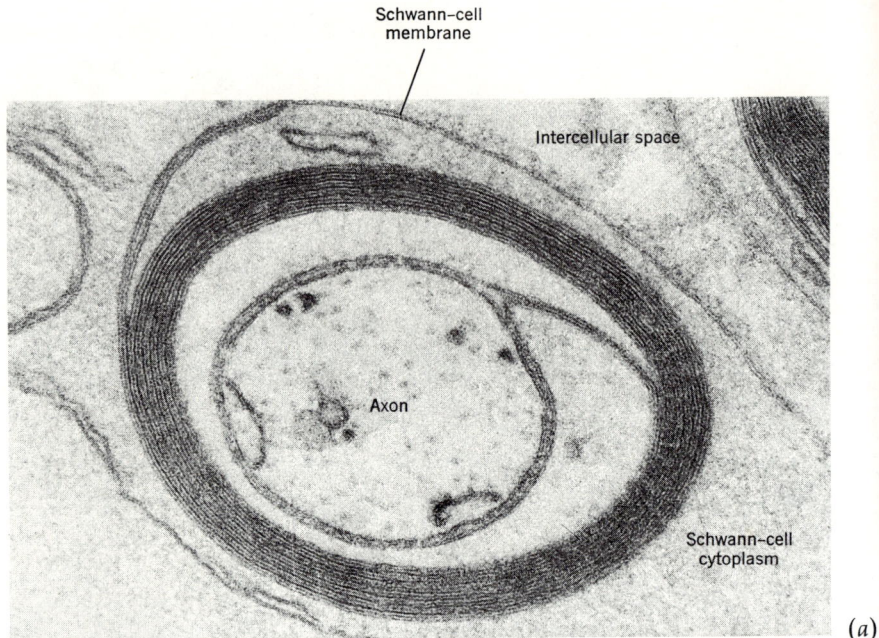

Figure 17-3 (a) Electron micrograph of a section through a myelinated nerve fiber, showing that the myelin sheath consists of wrap after wrap of the cell membrane of Schwann cells around the axon. Magnification 70,000×. (b) to (d) show the development of the myelin sheath.

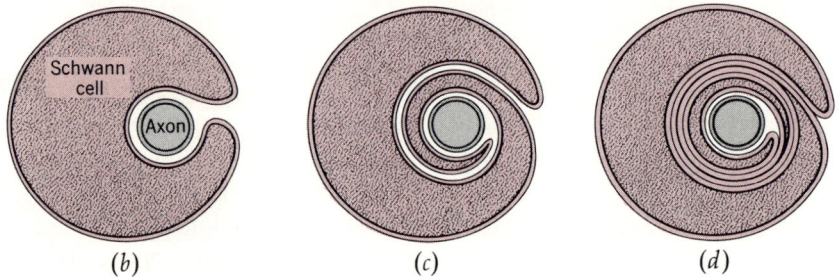

nerve impulse is not identical with an electrical impulse or an electrical current in a wire. Its fastest observed rate of travel is at best about 100 m per second, or about 4 miles per minute in a nerve fiber, in contrast to the 100 million m per second that can be achieved by an electric current (stream of electrons) through wire. A second very important difference is that the nerve impulse is generated, or **self-propagated,** by the neurons that transmit it. In some respects a nerve impulse resembles a burning spark traveling along the length of a fuse. The physical-chemical change instigating the electric impulse in one section of a nerve cell causes a similar change in the adjacent section, and so on until the impulse reaches the end of the fiber.

A major portion of our knowledge of the nature of the nerve impulse has developed from studies with certain giant nerve cells of a marine mollusk, the squid (Chapter 24), found in the waters of the North Atlantic. Their giant axons, which are as large as 1 mm in diameter (at least 50 times larger than the biggest axons in our own bodies), are so big that electrodes can be positioned inside the fiber to record electrical changes between the inside and outside (Fig. 17-4), and are even large enough to permit chemical analysis of the cytoplasm in the cells. Recent investi-

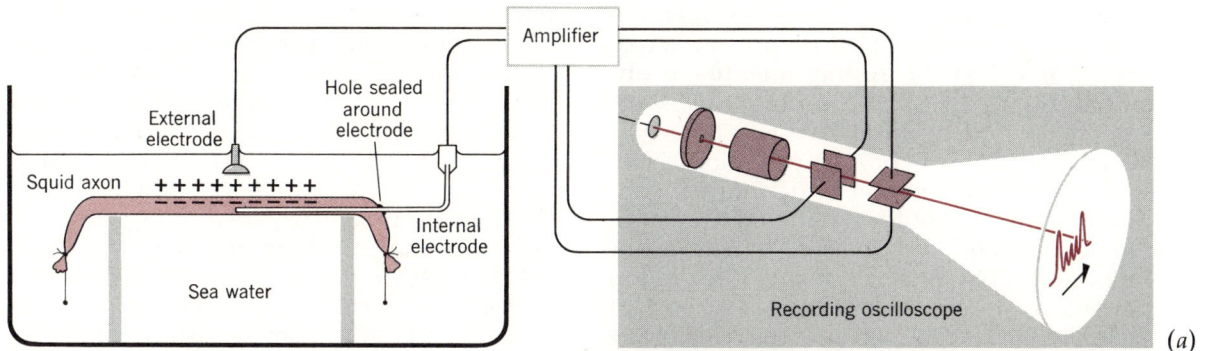

Figure 17-4 (a) Diagram of an apparatus for recording impulses in a length of squid giant axon. The nerve membrane acts like a battery, maintaining a difference in electrical potential between the inside and outside of the fiber. With two electrodes, one inside the fiber and the other in the surrounding fluid, changes in that potential can be amplified and recorded. (b) Photomicrograph of a recording electrode consisting of a glass tube filled with salt solution inside a giant axon of a squid. The recording electrode in this setup is about 150μ in diameter. The cylindrical axon is about 900μ.

gations have shown that nerve-conduction mechanisms in man are fundamentally the same as those observed in the squid, despite the difference in size of the neurons.

Resting and Active Membrane Potentials and Action Currents

The answer to how a nerve impulse is propagated along the entire length of a nerve lies in the structure and chemistry of the neuron membrane and in the composition of the fluid surrounding it. We can think of a nerve axon as a long cylinder of cell membrane filled with cytoplasm that differs in chemical composition from the surrounding extracellular fluid. We have already seen (Chapter 5) that cell membranes exhibit differential permeability; that is, they are more permeable to some substances than to others. If there are more negative charges (anions)

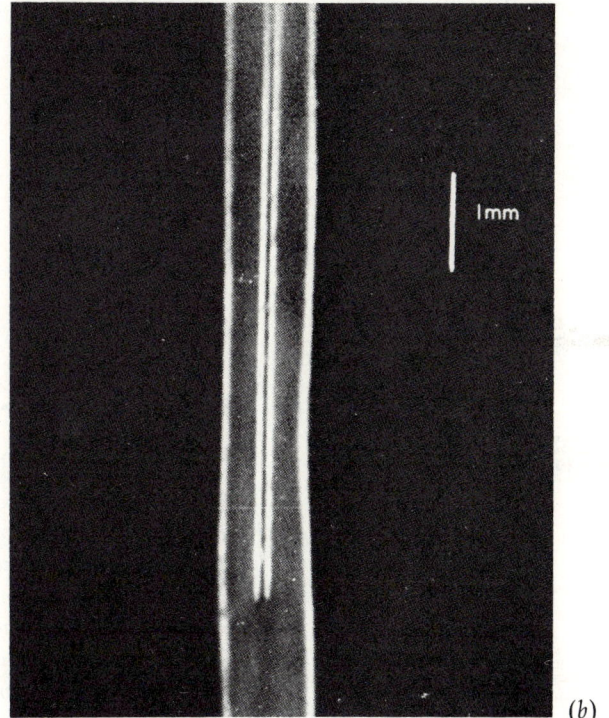

stored inside the cell than outside, then the membrane will behave like a battery, maintaining a difference in electrical potential between the inside and outside.

Microelectrodes in the cytoplasm of squid giant axons have shown that the outer surface of an axon, when not transmitting a nerve impulse, is electrically more positive than the inside. This difference in charge between the inner and outer surface of a nonconducting neuron is called the **resting membrane potential** (Fig. 17-5). A nerve impulse represents a brief, rapid change in that resting potential, associated with very brief changes in permeability of the membrane. Under normal conditions a nerve impulse is initiated by a stimulus at one end of a neuron, which results in a remarkable localized change in the ion permeability at that point. For just an instant (actually about .001 second) the membrane at the site of the stimulus becomes highly permeable to the positive ions to which it is normally rather impermeable, as if holes were suddenly punched in the membrane. In that instant positive ions on the outside flow rapidly across the membrane; the resting potential disappears and is replaced by a new potential, called the **active membrane potential;** the outside of the cell is now more negative than the inside, just the reverse of what it was before. Only a small portion of the neuron (usually a few millimeters) is in the active state at any one time, and the cell membrane quickly returns to its original resting potential once the nerve impulse has passed.

The rapid flow of charged ions across the active membrane is an electric current, which we call an **action current.** But because all electric currents must flow around a complete circuit, action currents move around cyclic pathways that pass inward from the outer membrane surface at the active membrane potential region, then along the inside membrane surface to the resting potential region just ahead of the impulse, and finally out through the membrane and back through the external fluid to the active membrane potential region to complete the circuit (Fig. 17-5). The flow of the action current through the resting potential membrane region just ahead of the nerve impulse causes the membrane potential in that region to change from the resting to the active state. The action current then advances into the next region along the membrane, and the entire process repeats itself. The moving nerve impulse is actually the migration of the change in potential from the resting to the active state in successive regions of the cell membrane.

Of the numerous ions present in the nerve cell and the surrounding fluid, only a few are of great importance in the nerve impulse mechanism. Chemical analysis has demonstrated that sodium ions (Na^+) are 10 times

Figure 17-5   *Diagram of an axon showing the electrical charge distribution between the inside and the outside. (a) The resting membrane potential is maintained because the membrane is impermeable to organic anions inside the cell and differentially permeable to Na^+ and K^+. (b) When the membrane changes permeability, localized action currents flow across it and produce a small region of active membrane in which the potential is reversed.*

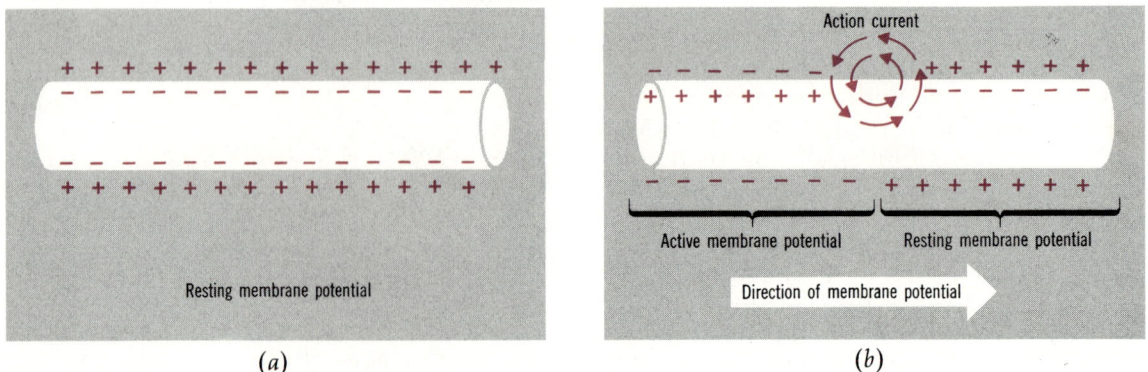

Resting membrane potential

(a)

Action current

Active membrane potential   Resting membrane potential

Direction of membrane potential

(b)

higher in concentration outside than inside the cell membrane, whereas potassium ions (K+) are 20 times more concentrated inside than outside. In addition, neuronal cytoplasm contains large negative organic ions, that is, organic molecules containing many amine groups; these molecules, called **polyelectrolytes** are in much greater concentration inside the cell than outside.

In the resting state the cell membrane is much more permeable to potassium ions than to sodium and organic ions; that is, it permits only a small movement of sodium into the cell, accounting for the high concentration on the outside surface. For the same reason the negative organic ions remain concentrated inside the cell. This separation of charges (negative inside, positive outside) is what generates the resting membrane potential (negative inside). In contrast, during the active state the membrane becomes suddenly very permeable to sodium ions as specific sodium "channels" open up. The consequent rapid net movement of the positive sodium ions inward (leaving the negative chloride ions behind) not only abolishes the resting potential (called **depolarization**), but makes the inside momentarily more positive than the outside. Soon after passage of the nerve impulse, the resting membrane potential is restored **(repolarization).** The sodium channels close; the membrane rapidly becomes impermeable to sodium again; and a small quantity of potassium diffuses out of the fiber to reinstate the internal negatively charged condition of the resting state. The neuron is now ready to conduct another impulse.

Small quantities of sodium and potassium, respectively, thus enter and leave the nerve fiber each time a nerve impulse is propagated. Although a nerve can conduct many impulses before its chemical composition is greatly changed, sodium must eventually be moved back out and potassium back in. The energy required to return these ions uphill against their respective concentration gradients (in contrast to downhill transfer occurring during impulse propagation) is generated by metabolic processes in the nerve cell. Such a **metabolic pump** is a dramatic example of an active transport mechanism fueled by the energy-rich ATP (Chapter 4) that is produced during cellular respiration. Without a constant supply of oxygen for the metabolic synthesis of ATP, nerve cells quickly lose their ability to conduct impulses.

The time interval required for restoration of the resting membrane potential before another nerve impulse can be transmitted by the neuron is called the **refractory period.** In mammalian nerves the observed refractory period ranges from 1 to 5 milliseconds (thousandths of a second), which means that approximately 200 to 1000 impulses per second can be fired along a nerve.

In order to initiate a nerve impulse, a stimulus of a certain minimum intensity **(threshold level)** must be applied; if this threshold of stimulus strength is reached or exceeded, a nerve impulse is started. Once started, the nerve impulse travels at a speed that is entirely independent of the intensity of the stimulus. Its rate depends only on the physical and metabolic state of the nerve itself. This relationship is called the **all-or-none law.** The threshold stimulus that initiates a nerve impulse is comparable to pulling the trigger of a gun or lighting a fuse. Sufficient force to pull the trigger must be applied for the gun to fire; once that force is reached or exceeded, pulling the trigger harder will not make the bullet travel any faster.

Moreover, once a nerve impulse is started, whether from a sound receptor in the ear, a retinal rod in the eye, or in an axon deep in the brain, it is essentially the same as any other nerve impulse that occurs in any other part of the nervous system. Our ability to distinguish both qualitative and quantitative effects, given only identical nerve impulses, can be attributed to several factors, including the number of nerve endings stimulated, the nerve pathways traveled, and the frequency of nerve impulses (the number of impulses per second and their spacing in time). The most important factor is the sorting out and integration of this information by the brain, which functions as a giant complex control panel with different regions specializing in receiving different kinds of information.

The Synapse

Neurons never occur as single, isolated units, but are always part of a linear or branched sequence of other nerve cells. Consecutive neurons are always so arranged that the axon endings are juxtaposed with the dendrites of the next. In principle there are two ways an impulse may cross from one cell to another: electrically or chemically. That is, if two cell membranes are touching, the action currents can cross the junctional boundary, assuming that the electrical resistance between the two cells is low (Fig. 17-6a); this is called **electrotonic conduction,** and is the mode of impulse transmission among many nerve cells and for cardiac and smooth muscle. In these tissues cells are joined across special close junctional structures called **nexuses** (Fig. 5-14). Ions and other molecules diffuse freely through these points, where the adjoining membranes come to within a few angstroms of each other, and therefore they can carry current from cell to cell.

Among most cells of the body, however, including most neurons, neighboring cell membranes never come so close. Instead there is a gap, an intercellular space about 200 Å wide. The nerve impulse cannot cross such a gap. The points at which the axon and dendrites of two nerves come together across such an intercellular gap is called a **synapse** (Fig. 17-6b). At the synapse the nerve impulse is converted to a chemical signal in the form of a **transmitter substance** that can diffuse across the intercellular gap. The transmitter substance is synthesized and stored in tiny membrane-bound sacs, called **synaptic vesicles,** in the ends of the axon that leads to the synapse (the **presynaptic axon**). A nerve impulse in the presynaptic neuron triggers the release of transmitter substance from some of these vesicles into the intercellular gap. It diffuses across the gap and binds to special receptor sites on the outer surface of the **postsynaptic cell,** altering the permeability of its membrane and thereby producing a small change in the membrane potential. In most cases there are hundreds or thousands of presynaptic nerve endings converging on each postsynaptic cell body (Fig. 17-6).

Two kinds of chemical synapse are known to exist: excitatory and inhibitory (Fig. 17-6b). At an **excitatory synapse** the transmitter produces an increase in permeability to sodium ions, of the postsynaptic membrane, which results in a depolarizing potential change; the membrane potential becomes momentarily less negative. This swing of voltage toward the firing threshold of the cell is called an **excitatory postsynaptic potential (EPSP).** If enough transmitter substance diffuses across the gap for the resulting EPSP to reach threshold, it acts as a generator potential, and fires an all-or-none action potential in the postsynaptic cell.

At an **inhibitory synapse** the transmitter increases the membrane permeability to potassium or chloride ions, resulting in a hyperpolarizing potential change; that is, away from threshold. This is called an **inhibitory postsynaptic potential (IPSP),** which counteracts excitatory stimuli. After producing a postsynaptic potential, extra transmitter material is rapidly destroyed by an enzyme, so the synapse is ready to function again. This whole process of synaptic transmission takes about half a millisecond.

The Reflex and the Reflex Arc

Most of our body activities and reactions to environmental changes, both internal and external, are automatic, or **involuntary.** Such responses, which occur outside of our awareness, are immediate and rapid and are called **reflexes.** Reflexes are usually classified in two broad subdivisions, the simple, or unconditioned, reflex, and the conditioned reflex.

A **simple** reflex is an inborn, unlearned response to a stimulus or change in environment. A familiar example is the knee jerk, in which the leg is involuntarily and momentarily extended as the result of a sharp tap below the kneecap. Certain stretch-sensitive receptor nerve endings in the knee are stimulated, starting one or more successive nerve impulses that quickly travel by a particular

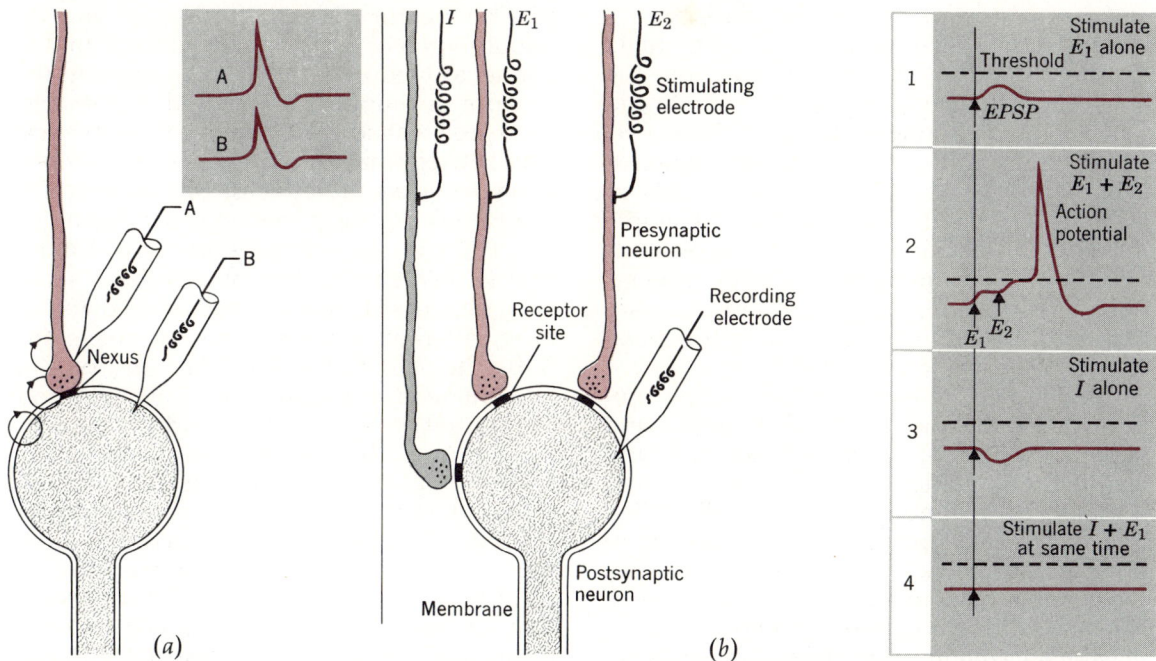

Figure 17-6 (a) At an electrical synapse the intercellular gap and the electrical resistance between the two cells are both reduced. Action currents sweeping down the presynaptic fiber (recorded in trace A) therefore can cross directly into the postsynaptic cell, causing it to depolarize and fire (trace B) with essentially no synaptic delay. (b) Chemical synapses may by excitatory (shown in color) or inhibitory (gray). When transmitter molecules diffuse across the synaptic gap to receptor sites on the postsynaptic cell membrane, a small depolarization of the postsynaptic cell occurs. This excitatory postsynaptic potential is seen as a small change in the voltage recorded from inside the postsynaptic cell (Panel 1). Sequential impulses from two excitatory endings (E_1 and E_2, Panel 2) combine to depolarize the membrane, causing the postsynaptic cell to fire an action potential. An inhibitory fiber changes the membrane potential in the opposite direction (Panel 3), or counteracts an EPSP arriving at the same time from an excitatory nerve. (c) Afferent presynaptic fibers converging on the cell body of a spinal motor neuron illustrate the number of synapses normally present on each postsynaptic cell.

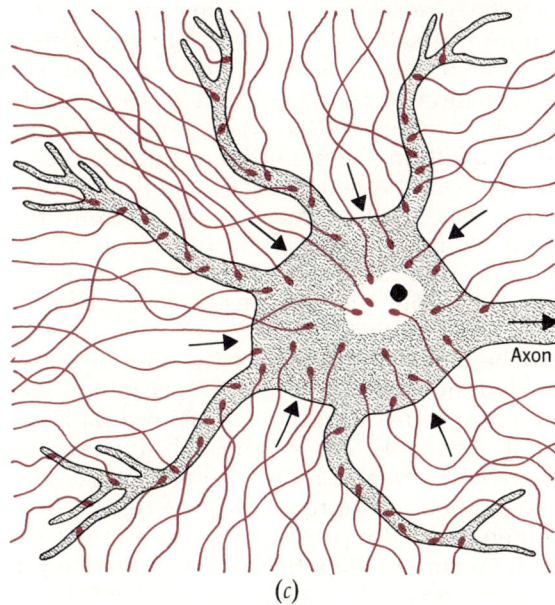

pathway of neurons (including a portion of the CNS) to specific muscles of the thigh, causing them to contract and straighten the knee. Similarly, the quick closing of an eyelid when an object suddenly approaches the eye, or the rapid withdrawal of a hand when it is burned or pricked (even before there is an awareness of pain) are illustrations of simple reflexes. Innumerable simple reflexes continually occur in the body, controlling secretions of our glands, breathing, muscular activity of large portions of the digestive tract, rate of heartbeat, and so on.

Conditioned reflexes, unlike simple reflexes, are not inborn but are acquired and dependent on past experience, training, and learning. A conditioned reflex is an acquired response to a stimulus that originally failed to evoke the reaction. It was first clearly demonstrated in the early part of this century by the eminent Russian physiologist Ivan Pavlov. He showed that conditioned reflexes can be developed experimentally to an extraordinary degree. Under normal circumstances the secretion of saliva when food is ingested is a simple reflex, a response initiated by the stimulation of the taste buds in the mouth. Working with dogs, Pavlov demonstrated that if an additional stimulus were furnished regularly, such as the ringing of a bell whenever food was given to the animals, in time the new stimulus (the bell) would result in the secretion of saliva without either the sight or smell of food. This was the conditioned reflex (see Chapter 19).

The anatomical basis of the simple reflex, called the **reflex arc** (Fig. 17-7), is an arrangement of at least three neurons in a pathway that invariably passes through the central nervous system. A reflex, therefore, may also be defined as the response that results from a nerve impulse passing through a reflex arc.

In a simple reflex (Fig. 17-7) the nerve impulse is normally initiated in the dendrites of a **sensory neuron** in the receptor organ or tissue. The impulses travel along the sensory neuron until they reach the terminal branches of the axon at the synapse in the spinal cord. Impulses are initiated in a second neuron

(the **connector neuron**) in the arc, and the impulse proceeds along the length of this connector neuron to its axon endings. At this synapse an impulse is initiated in the third nerve cell, called the efferent, or **motor neuron.** The terminal axon branches of the motor neuron end in the effector organ; the reflex or response made by the effector organ as a result of the arriving nerve impulse is either a muscular contraction or glandular secretion, depending on the effector involved. In brief, afferent neurons transmit nerve impulses toward the central nervous system; efferent neurons transmit impulses away from the central nervous system; and the connector neuron is the link between them.

The foregoing account of a single isolated reflex arc is an oversimplified description intended to indicate the structural and functional basis of a simple reflex. Unfortunately, it tends to impart the erroneous impression that (1) only a single isolated reflex arc is the result of a single stimulus; and (2) an organism is a relatively passive protoplasmic mass whose responses are governed solely by environmental stimuli.

With regard to the first point, reflex actions of higher animals almost always involve an appreciable number of reflex arcs, not just one. For example, should a person receive a sudden, unexpected burn on his fingertip, he would not only jerk his hand as a result of a simple reflex, but would also respond in other ways. He would probably turn his head to the source of danger, cry out, and very likely experience a series of sensations and reactions including pain, fear, and emotional tension. Obviously numerous muscles—including those of the arm, shoulders, neck, head, trunk, tongue, larynx, and respiratory tract—and several mental processes have participated in his total reaction. What started out as a simple stimulus (i.e., a slight, highly localized burn) has evoked an extensive reflex behavior involving the integration of the rest of the nervous system.

This multiplicity of reactions is explained by the **integrative** and **coordinating** properties of the reflex arc. Axons of sensory neu-

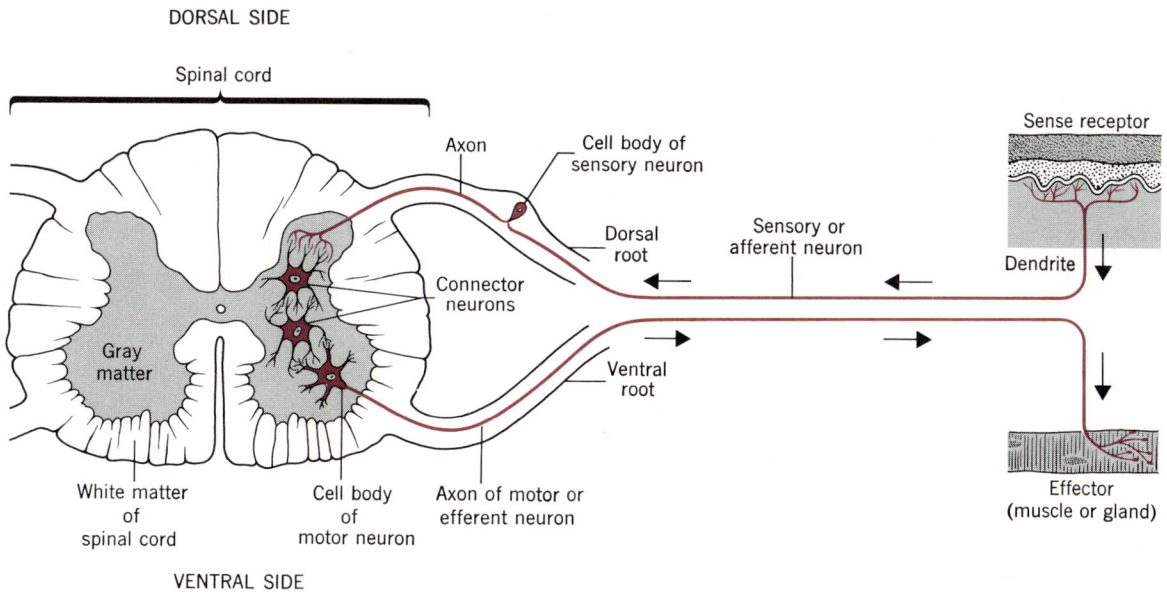

DORSAL SIDE

Spinal cord

Axon

Cell body of
sensory neuron

Dorsal
root

Sensory or
afferent neuron

Sense receptor

Connector
neurons

Dendrite

Gray
matter

Ventral
root

White matter
of
spinal cord

Cell body
of
motor neuron

Axon of motor or
efferent neuron

Effector
(muscle or gland)

VENTRAL SIDE

*Figure 17-7   The anatomy of a reflex arc. Impulses from a receptor cell travel
along the sensory neuron to the spinal cord. The cell body of the sensory
neuron is located in a dorsal root ganglion lying just outside the spinal cord.
The sensory axon enters the cord and synapses with one or more connecting
neurons in the gray matter. The connectors in turn form synapses with the
dendrites of a motor neuron, the axon of which transmits signals out the
ventral root to an effector (muscle or gland).*

rons in many cases have several branches, each of which may form a synapse with other connector neurons in the central nervous system. Each sensory neuron therefore has potential connections with many effectors and can evoke activity in many parts of the body, eliciting a number of simultaneous responses and accounting for the ability of a single, highly localized stimulus such as a pinprick to elicit several responses or reflexes. Similarly, a single motor neuron may have synapses with numerous other neurons, accounting for the ability of different stimuli in different parts of the body to elicit the same response; for example, an injury to any part of the body may evoke the same use of muscles of the larynx to emit a cry of pain.

With regard to the matter of organisms as passively controlled by stimuli, it seems that an organism itself has appreciable control over what constitutes stimulation. The body's receptors are continually being bombarded by innumerable stimuli, yet only a relatively limited number of responses or reflexes occur, indicating some control or inhibitory mechanism that is still not clear. This control is exercised mainly at the synapse.

The term **reflex center** refers to an aggregate of neurons (often including synapses) in the spinal cord or brain that controls the activity of a particular group of effectors. For example, the reflex center for breathing in the medulla oblongata region of the brain consists of a group of neurons that controls the impulses conducted to muscles of the chest responsible for breathing action. The cells of the reflex center are in turn regulated by appropriate afferent impulses. Much of the brain and spinal cord are made up of reflex centers affecting body activities.

Neurochemistry

Our information on brain biochemistry, particularly with reference to mental and emotional conditions, is still at a primitive stage, although a major research effort is being made in this field. Recent studies of brain function suggest that emotions, moods, and behavior may be significantly influenced by certain compounds normally found in the brain. Two such substance have been identified: **norepinephrine** and **serotonin,** a derivative of the amino acid tryptophan. Many facts about severe mental diseases and deficiencies suggest the existence of specific metabolic disturbances, possibly related to genetic phenomena. Two types of drug have been developed for treatment or alleviation of emotional illnesses: **tranquilizers** have a calming effect; **antidepressants** elevate mood and restore drive. The physiological or biochemical basis for mental diseases and for the relief-giving effects of these drugs is, however, unknown.

How does the autonomic nervous system stimulate or inhibit the visceral organs? The pioneering experiments of Otto Loewi some 40 years ago furnished the first important clues as to how the parasympathetic nervous system exerts its effect and earned Loewi the honor of the Nobel Prize. He showed that the effects of the visceral nerves on the pulsation rate of the vertebrate heart result directly from the action of certain chemical substances liberated by the nerve endings by the arrival of a nerve impulse. Loewi removed the beating hearts of two frogs, filled their chambers with a dilute chemical solution of inorganic salts, and inhibited the beat of one heart by stimulating its attached parasympathetic nerve, the **vagus** nerve. He transferred some of the fluid from the inhibited heart to the second heart; the rate of beat of the second heart was also inhibited (although its vagus nerve had not been stimulated). Loewi proposed that the axon branches of the vagus nerve terminating in the heart released an inhibitory substance when the nerve impulse arrived at the axon ending. This inhibitory substance in fact has been identified in more recent research as **acetylcholine.**

Acetylcholine is liberated at the terminals of postganglionic parasympathetic neurons that innervate the internal organs. It is also very likely the transmitter substance for synapses between neurons outside the central nervous system. If acetylcholine released at the parasympathetic nerve endings were permitted to accumulate at the synapse, the connecting neuron or effector organ would be continuously stimulated or inhibited. It does not accumulate, however, because of a ubiquitous enzyme, **cholinesterase,** which very rapidly destroys acetylcholine by hydrolysis.

By contrast, the axons of the postganglionic fibers of the sympathetic nervous system (with a few exceptions) secrete, instead of acetylcholine, a substance called **sympathin,** which consists primarily of norepinephrine (p. 450). This different secretion accounts in large part for the opposite action of the parasympathetic and sympathetic systems. There is apparently no specific enzyme to destroy the action of the sympathin released at the synapse; and its influence lasts longer than acetylcholine. Eventually, however, it is destroyed by oxidation reactions.

Central Nervous System

The central nervous system performs a key function in the vastly intricate nervous system. It serves as a central clearing house or switchboard, controlling, directing, and integrating all messages (i.e., nerve impulses) transmitted by the nervous system.

The brain and spinal cord, vital and delicate organs of the human body, are well fortified against injury by two types of protective covering and by an external cushion of fluid. The outermost covering that encases the brain is the skull; the spinal cord is largely enclosed in the vertebral column. The inner covering of both the brain and the spinal cord has three distinct layers, membranes called the

meninges. (Infection of the meninges constitutes the disease called **meningitis.**) The meninges are made up for the most part of connective tissue and separated from one another by spaces. The space between the inner and middle meninges is filled with **cerebrospinal fluid,** which serves as a protective cushion for the brain and spinal cord. Cerebrospinal fluid resembles lymph in composition and is produced largely by the filtration of blood from the mass of capillaries in the brain.

The Spinal Cord

The spinal cord in an average adult human is an oval-shaped hollow cylinder, 17 to 18 in. long, which tapers slightly as it extends downward from the brain to the pelvic region. In cross-section the core displays an inner butterfly- or H-shaped core of gray matter consisting of neuron cell bodies and nonmyelinated fiber processes; the outer portion is composed of white matter made up of bundles, or **tracts,** of myelinated fibers (Fig. 17-7). The dorsal horns of the butterfly region contain mostly the cell bodies of connector neurons. The ventral horns contain cell bodies of the efferent neurons of the spinal nerves (Fig. 17-8).

Within the white matter are bundles of nerve fibers that transmit impulses upward to the brain **(ascending tracts)** and others that send impulses downward **(descending tracts).** All tracts, ascending and descending, **cross over** from one side of the spinal cord to the other somewhere along their pathways to and from the brain; the right side of the brain receives impulses arriving from the left side of the body and in general regulates the activities of that side, and the corresponding relationship applies to the left side of the brain and the right side of the body. The significance of this curious phenomenon has not yet been satisfactorily explained.

The two main functions of the spinal cord are (1) as a pathway for the conduction of impulses between the peripheral nervous system and the brain, and (2) as a center for a great many reflexes.

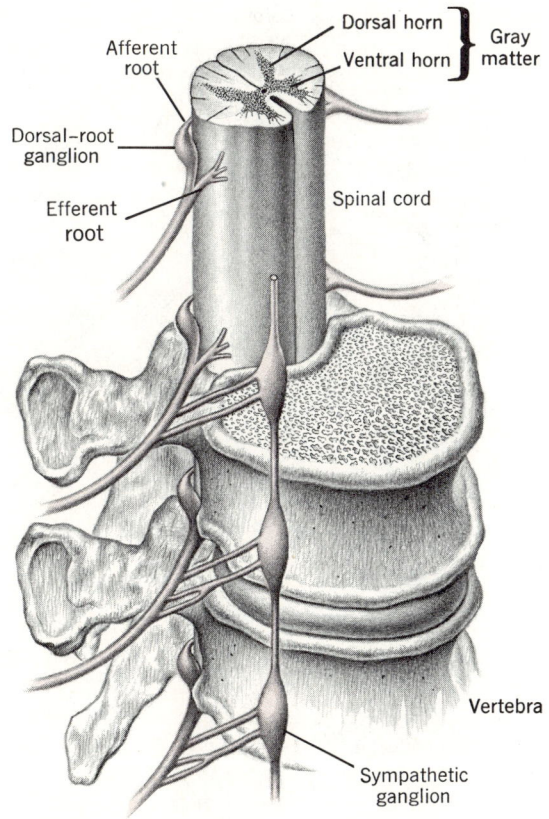

Figure 17-8   A segment of the spinal cord. Spinal nerves enter and exit through spaces between the vertebrae, afferent nerves on the dorsal side and efferent motor nerves on the ventral side. The gray area shaped like a butterfly in the center of the cord is made of nerve cell bodies.

The Brain

The adult human brain, a most complex and highly developed organ, has an average weight of about 3 lb. Its different parts have been classified in several ways, but here we describe six main regions (Fig. 17-9 p. 476): medulla oblongata, cerebellum, midbrain, thalamus, hypothalamus, and cerebrum, which is composed mainly of the all-important cerebral cortex. The medulla oblongata and midbrain are often collectively called the **brain stem** because of their location and shape.

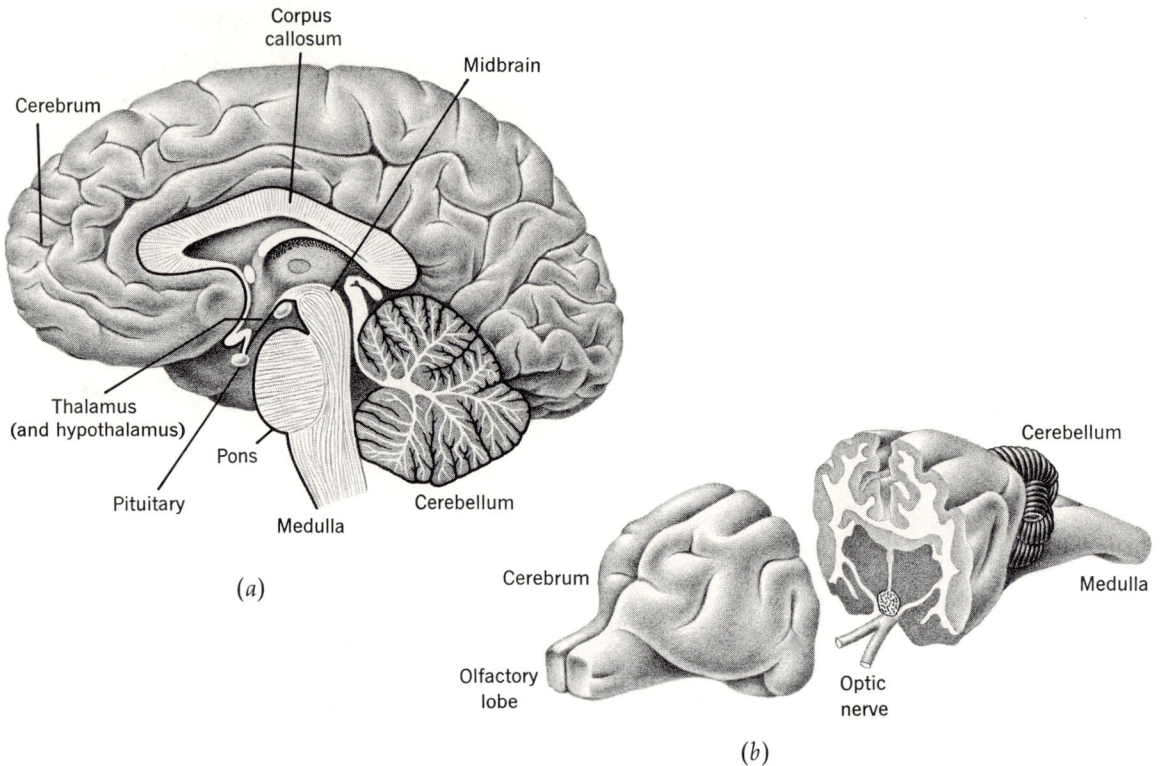

Figure 17-9 (a) A midline section of the human brain. (b) Cat brain for comparison.

The way in which the brain is organized in general indicates the degree or level at which each part functions. The lower anatomical regions of the brain deal with automatic functions (e.g., heartbeat rate and respiration), whereas successively higher regions are concerned with correspondingly more integrative and advanced activities, such as coordinated muscular activity and reasoning or abstract thought. It is generally believed that the lower regions of the brain, which are similar in most mammals, are the most primitive portions. The evolution of higher mammals has been accompanied by an increased development in size and function of the front (anterior) portion of the brain, the cerebrum. Although the more primitive parts of the brain (i.e., the brain stem) still carry out their original autonomic functions, in man they are

subject to various degrees of control and regulation by the higher centers.

Medulla

As an enlarged extension of the spinal cord, the **medulla oblongata** is the lowest part of the brain. It is about 1 in. long and consists mostly of ascending and descending tracts of myelinated fibers (white matter), with some gray matter in its interior. It has an enlarged cavity (an expanded portion of the spinal canal) within it that is called a **ventricle,** one of four in the brain. The masses of nerve cell bodies that make up the gray matter scattered in the interior of the medulla contain several vital reflex centers, including those that control the rate of heartbeat, breathing, and constriction and dilation of blood vessels. Other reflex centers in the medulla include those responsible for vomiting, coughing, sneezing, hiccuping, and swallowing.

Cerebellum

Above and behind the medulla is the **cerebellum,** which is made up for the most part of two large hemispheres. Although it is the second largest portion of the brain (exceeded in size only by the cerebrum), it has only 10 percent of the brain's weight. Bundles of fibers that extend from the cerebrum to other parts of the brain or spinal cord are called **projection tracts;** each hemisphere of the cerebellum has an interior composed of such white (i.e., myelinated) tracts and an exterior (cortex) of gray matter (mainly neuron cell bodies) lying in numerous folds and convolutions.

The cerebellum does not itself directly control body activities. Instead, impulses from its gray matter somehow coordinate the activities of several other brain centers that regulate and integrate certain body functions, particularly skeletal muscle activity. It is responsible for normal movements, like those we use when picking up an object, or walking, that are smooth, timed, steady, precise, and graded in terms of force, extent, and rate. Injury, disease, or experimental removal of the cerebellum in an animal results in a disorder characterized by movements that are jerky, shaky, and poorly regulated.

Recent evidence indicates that the cerebellum also has to do with the integration of the sensations of touch, hearing, and sight. The various functions of the cerebellum in coordinating specific body activities have in fact been localized or mapped in distinctly defined areas of its cortex.

Another small mass of white matter at the top end of the medulla is the **pons,** which means bridge. It consists largely of projection tracts that conduct impulses among the cerebellum, medulla oblongata, and cerebrum.

Midbrain

Just above the pons atop the medulla is the **midbrain,** a mass of projection tracts made up mostly of white matter surrounding a central cavity. These tracts serve largely as conduction pathways between the spinal cord and the medulla, pons, and cerebellum. The back portion of the midbrain also possesses a prominent mass of gray matter, which collaborates with the cerebellum in controlling muscular coordination, and four rounded protruberances in which lie certain auditory and visual reflex centers. these areas mediate such reflexes as constriction of the pupil of the eye in response to strong light or an animal pricking up its ear in response to sound.

Thalamus and Hypothalamus

The **thalamus** and **hypothalamus** lie above the midbrain and below the cerebrum. As the location suggests, the thalamus serves as a relay station, receiving nearly all impulses arriving from the different sensory areas of the body before passing them on to the cerebrum, where they give rise to conscious sensations.

The hypothalamus is the principal region of the brain that has visceral reflex centers for controlling and integrating the metabolism and various activities of the internal organs and tissues. It regulates body temperature, smooth muscle activity, water balance, appetite, blood pressure, and possibly carbohydrate and fat metabolism. For example, reflex centers in the forward part of the hypothalamus prevent overheating by increasing blood circulation to the skin (accelerating heat loss by perspiration) and by causing a faster rate of breathing. Fever is thought to be associated with the release from damaged body cells of a substance that affects the temperature-regulating centers of the hypothalamus.

Nerve centers in the hypothalamus also participate in producing sleep and maintaining the waking state; they determine the sexual drive; and are involved in such basic sensations as hunger, thirst, fear, and rage. Studies have been made with electrodes inserted into the hypothalamus of animals, or of people undergoing brain surgery. Mild electrical stimulation of a specific area produces feelings of intense hunger, even in subjects that have just eaten. Stimulation of other regions arouses fear or violent rage and aggressive behavior entirely unrelated to external circumstances. No matter how contented or secure a cat is, for example, when the "rage center" in the hypothalamus is stimulated the resulting **agonistic** (threatening or fighting) behavior overrides all other external stimuli. When the stimulator is turned off, the cat immediately returns to what it was doing before.

Finally, as we saw in Chapter 16, the hypothalamus is also a center of control for anterior pituitary function by its production of releasing factors.

The Cerebrum

The **cerebrum,** by far the largest part of the human brain, has more than half the 10 billion neurons of the entire nervous system. It is the largest brain region of apes and monkeys, but is a less prominent structure in the brains of lower vertebrates (Fig. 17-10). It is the unique organ of the human species, for its various activities account for the basic differences between man and all other existing animals and those that preceded him in the evolutionary sequence. Within the cerebrum reside the most advanced functions of the nervous system, including memory, intelligence, insight, personality, and judgment, and the most highly developed centers for various sensations, including sight, hearing, smell, taste, and so on.

The surface layer of the cerebrum, called the **cortex,** is tremendously expanded in man so that it covers most of the other brain structures, extending to a depth of $\frac{1}{10}$ to $\frac{2}{10}$ in. and covering an area of about 400 square in. It is composed of gray matter, some 2 billion cell bodies in a vast interconnecting mass arranged in folds or convolutions like the surface of a walnut kernel. The cerebrum is divided by a deep groove, running from front to back, into two halves called **hemispheres,** which are not completely separated from one another by the groove but are connected by the underlying white matter. This bridge of white matter, called the **corpus callosum,** is composed for the most part of nerve tracts connecting different parts of the cortex to

one another and to other parts of the brain and spinal cord.

Our knowledge of cerebral function has been obtained over the years in a variety of ways, including examination of the effects of removing or destroying different cerebral areas in experimental animals, observation of the reactions evoked by stimulating cerebral regions in humans and animals, study of the symptoms of patients with known brain damage, and use of a relatively recent technique, **(electroencephalography)** for measuring the summed electrical activity of the **brain action potentials,** in brains of intact, living individuals. Microscopic investigations have established that there are more than 100 different structural areas in the cortex. Many studies have collectively established a number of important facts concerning functions of the cortex and other brain regions.

It has been possible to map particular cerebral locations, called **sensory areas,** associated with the reception of certain sensory impulses arriving from the peripheral nervous system. Sensory regions of the surface of the cerebral cortex such as the **visual** area (responsible for sight), the **auditory** area (hearing), the **olfactory** area (smell), the **gustatory** area (taste), and **somatosensory** areas (touch, pain, pressure, perception of body position, etc.) have been located and mapped as in Figures 17-11 and 17-12 (pp. 480, 481). Destruction of or damage to any specific sensory area will cause a loss in ability to experience that particular sensation, even if all other parts of the cerebral cortex and that sense organ are functioning normally. Destruction of the visual sensory areas, for example, would cause blindness just as certainly as the loss of both eyes would.

Experiments have also led to the mapping on the cerebral cortex of distinct areas responsible for the voluntary contraction of different skeletal muscles (Fig. 17-11). They are called the **motor areas** of the cortex because nerve impulses from these localized regions eventually pass by way of motor neurons to specific muscles. A mild electric shock to a given motor area will cause a spasmodic contraction of the muscle groups it affects. There is evidence that if impulses from the

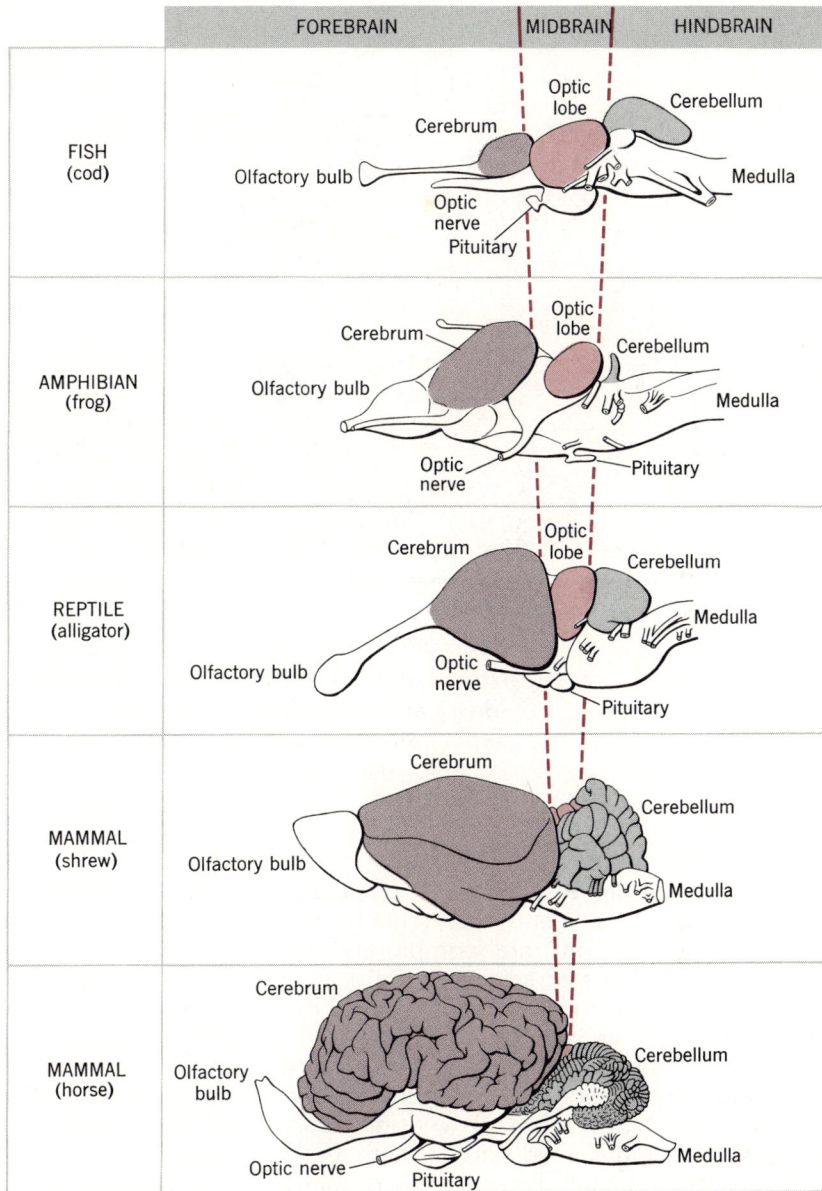

FISH (cod)

Optic lobe · Cerebrum · Cerebellum · Olfactory bulb · Optic nerve · Pituitary · Medulla

AMPHIBIAN (frog)

Cerebrum · Optic lobe · Cerebellum · Olfactory bulb · Optic nerve · Pituitary · Medulla

REPTILE (alligator)

Cerebrum · Optic lobe · Cerebellum · Olfactory bulb · Optic nerve · Pituitary · Medulla

MAMMAL (shrew)

Cerebrum · Cerebellum · Olfactory bulb · Medulla

MAMMAL (horse)

Cerebrum · Cerebellum · Olfactory bulb · Optic nerve · Pituitary · Medulla

Figure 17-10 During evolution from the fish through the amphibian, reptile and mammal, the relative sizes of the forebrain, midbrain, and hindbrain have changed enormously. The midbrain, with its optic lobes, became relatively smaller than the cerebellum of the hindbrain, but both were overshadowed by the tremendous expansion of the cerebral cortex and other parts of the forebrain.

cortical motor area are transmitted through a pathway other than their usual one, they exert an inhibitory instead of stimulating effect on muscular movement.

Of some interest is the fact that the size of the motor area for each group of muscles appears to be proportional to the complexity of the movements performed and to the ex- tent of use, rather than to the size or bulk of the muscles. For example, the motor areas concerned with the fingers and the facial muscles are larger than those for the muscles of the legs (Fig. 17-11). Man's outstanding manual dexterity—his remarkable ability to use his hands—is dependent on the develop- ment of the appropriate area of his cerebrum.

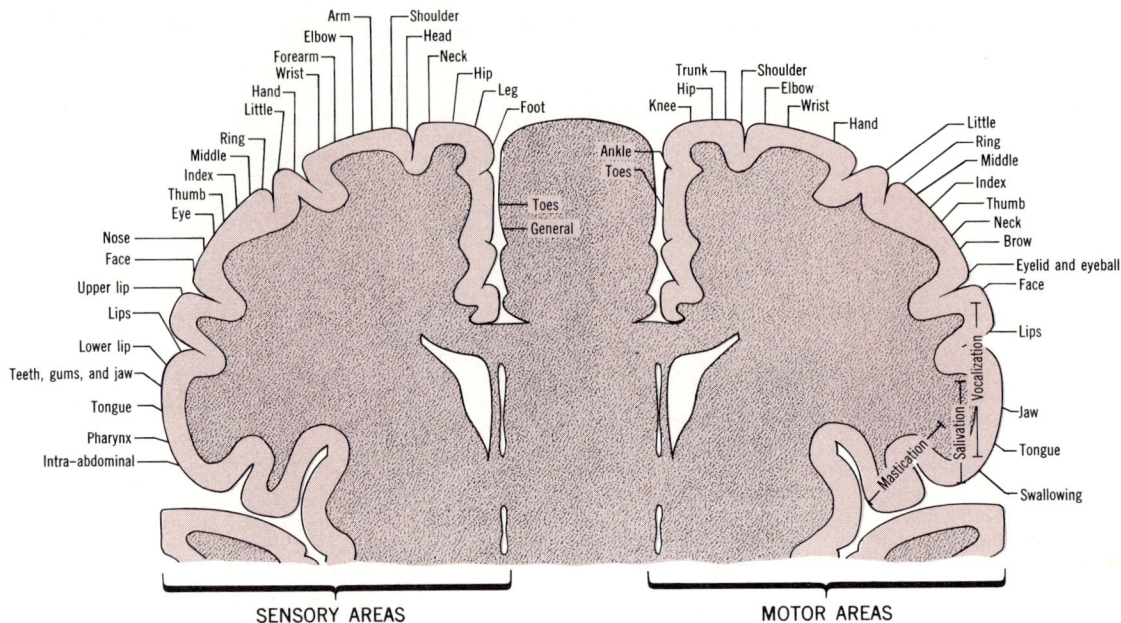

Figure 17-11 *Map of the somatic sensory and motor functions in the human cerebral cortex. The left side of the figure represents a section through the sensory area; the right side is a section through the motor area. Note that the cerebral areas are proportional to the complexity of movements and extent of use rather than to the size of the muscles or sense organs.*

Although we know that certain areas of the cerebral cortex are responsible for specific roles (Fig. 17-12), it seems highly likely that many other functions are completely dependent on the cortex as a whole. Because of the interconnections among regions of the cortex as well as similar connections to different regions of the brain and spinal cord, any conscious sensation, even when ascribed to a particular cortical area, is very probably the result of widespread cortical activity as well as of the integrated functioning of many parts of the nervous system. The cortex, moreover, does not merely register separate sensations as they occur, but in addition evaluates, compares, and puts them together into meaningful concepts. When we view an object, for example, we get a sense of light and dark, size, shape, and color, among other impressions. Even more important, we interpret these properties not as separate and iso-

lated entities, but as a unified, integrated, whole object—a pencil or an automobile.

Furthermore, there are still large areas of the human cerebral cortex, known as **association** areas, that have defied all attempts at localization of function. The association areas, which in man represent a major portion of the cortical surface, are believed to be responsible for the highly advanced traits of memory, intelligence, learning, foresight, imagination, emotions, verbalization of concepts, and various other mental processes. All attempts to solve the great riddle of the association areas and to explain intellectual and mental phenomena in terms of neural mechanisms have so far met with failure. At most we can only speculate from limited evidence.

There is also some recent evidence that the highest levels of integration in mental processes do not reside solely in the neurons of the cerebral cortex; there is an intricate and poorly understood interchange of nerve impulses between nerve cells of the cerebral cortex and those of a region, called the **reticular system,** extending through the central portion of the brain stem. The reticular system is primarily concerned with alerting, or "awakening," the cerebral cortex to receive and interpret incoming sensory signals; in addition it contributes to the regulation of all motor activities of the body. It is roughly comparable to a traffic control center, helping to direct the flow of messages to the brain and alerting the conscious centers to activity.

Electrical Brain Waves

In 1929 it was discovered that the cerebral cortex is almost constantly emitting small rhythmical changes in electrical potential that are commonly called **brain waves.** These can be detected and measured by means of electrodes applied to the scalp. Brain waves seem to result from the inherent electrical activity of the nerve cells of the cerebral cortex; they apparently are not caused by nerve impulses arriving at the cerebrum from the lower levels of the nervous system.

Most normal individuals show several regular types of brain wave. These include **alpha waves,** which are best recorded over the cortical visual sensory area when the eyes are shut and when the visual field is uniform; **beta waves,** which are of lower intensity; and a third group of considerably greater strength, **delta waves,** which appear during sleep.

In general, visual activity, muscular activity, and mental activity are responsible for drastic changes in the frequency, regularity, and strength of the brain waves.

The instrument for recording brain waves, the **electroencephalograph** (or **EEG**) is a valuable diagnostic tool in distinguishing several kinds of brain disease and damage that characteristically alter the brain-wave pattern; it has been especially useful in diagnosing a condition known as **epilepsy.** Epilepsy results from certain types of brain damage or from a genetic defect and is characterized by irregular periods of convulsions and spasms that show up in a recognizable brain-wave pattern.

It is worth reemphasizing that the neural mechanisms that account for learning, memory, intelligence, and all other associations made by the brain are still obscure. Man's

Figure 17-12 Localization of functional areas in the human cerebral cortex.

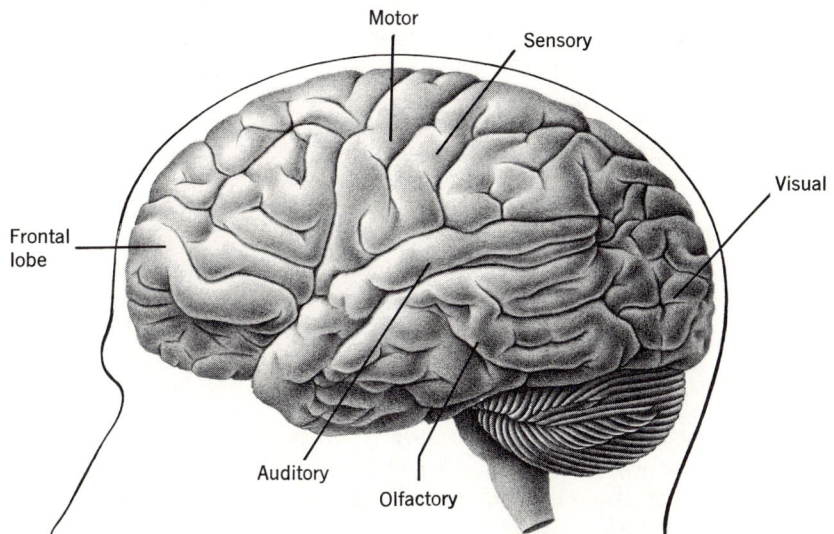

expectation is that his brain, which is transforming the world, is clever enough to understand itself. The hope is that someday in the not-too-distant future our growing knowledge of biochemistry, genetics, biophysics, mathematics, and information theory (the manner in which information is transmitted, coded, and integrated) will ultimately unravel the complex mechanisms that account for brain function.

The Peripheral Nervous System

The peripheral nervous system consists of 12 pairs of cranial nerves and 31 pairs of symmetrical spinal nerves and their many branches. The **cranial nerves** originate in different sections of the brain and extend to various parts of the head, the neck, and the internal organs, or viscera, of the chest and abdomen. The **spinal nerves** originate in pairs at various locations along the length of the spinal cord and form branches that extend into the legs, arms, and trunk of the body.

The peripheral nervous system is in turn subdivided, on the basis of structure and function, into two groups, the somatic nervous system and the visceral nervous system (also called the **autonomic nervous system**). The **somatic nervous system** is made up of nerves that come from the sense organs in the skin to the brain and spinal cord and nerves that innervate the skeletal muscles, skin, and certain other parts of the body; it is therefore responsible for the somatic reflex movement of various parts of the body as well as for consciously controlled movements, both of which we have discussed earlier (pp. 472–479).

The **visceral nervous system** innervates cardiac muscle, smooth muscle, and glands, and therefore governs and controls the functions of the viscera (heart, guts, glands, etc.) that are carried out automatically and without awareness. These include rate of heartbeat; contraction of the smooth muscle of the digestive tract, blood vessels, urinary bladder,

and other internal organs; and secretions by the digestive glands and sweat glands. The visceral nervous system thus has an indispensable role in maintaining the constancy of the internal environment, that is, in **homeostasis** (Chapter 1).

The Spinal and Cranial Nerves

Within the peripheral nervous system the dendrites and axons of most neurons are bound together with connective tissue to constitute larger nerves, or **nerve trunks.** **Sensory** nerves and **motor** nerves are made up only of fibers of sensory neurons and motor neurons, respectively. **Mixed** nerves possess the fibers of both types. All 31 pairs of spinal nerves in man are mixed nerves, whereas some cranial nerves are mixed and others are exclusively either sensory or motor.

The spinal nerves are attached at more or less regular intervals along the length of the spinal cord. Just before it joins the spinal cord, each spinal nerve divides into two branches or roots (Fig. 17-7). The **dorsal root** contains only sensory neurons and connects with the dorsal (back) portion of the spinal cord. The **ventral root** contains only motor neurons and connects with the ventral (front) portion. Because the dorsal root of a spinal nerve is made up only of sensory neurons, it is the point of entry of each impulse into the spinal cord. Within the cord impulses are carried up and down along the longitudinal tracts to join reflex arcs, and to higher centers in the brain. Axons of the motor neurons leave the spinal cord to become the ventral root of the spinal nerve, joining the dorsal root to make up the main trunk of the spinal nerve; this progressively subdivides into branches that innervate various parts of the body.

The role of the dorsal and ventral roots in conducting impulses to and from the spinal cord can be demonstrated by severing one of them and observing the effects. When the dorsal root is cut, no impulses reach the spinal cord from the part of the body innervated by the severed spinal nerve; as a result, that part of the body experiences a complete loss of sensation. If the cut end of the dorsal root attached to the spinal cord is stimulated, or if impulses are sent from the brain, a response will take place if the ventral root is still intact. When the ventral root is cut instead, complete paralysis of the muscles innervated by that nerve ensues because no impulses can reach the muscle from the cord. The senses of pressure, temperature, pain, and so on are unaffected, however, because the dorsal root is still intact, allowing passage of sensory impulses to the central nervous system. If the cut end of the ventral root not attached to the spinal cord is stimulated, the muscles innervated by the cut nerve will respond. When the spinal nerve, which is a union of the dorsal and ventral roots, is cut, the portion of the body affected will suffer both a loss of sensation and a paralysis of the muscles.

The Visceral Nervous System

The visceral, or autonomic, nervous system is subdivided into sympathetic and parasympathetic nervous systems (Fig. 17-13, p. 484). The cell bodies of the **sympathetic** nerves are located in the spinal cord only in the area of the chest (thorax) and abdomen (lumbar region); these autonomic nerves are therefore also referred to as the **thoracolumbar** system.

The **parasympathetic** nerves arise from the ends of the cord, that is, some from cranial nerves at the base of the brain and some from the lowest **(sacral)** part of the cord, which is why the parasympathetic system is also called the **cranio-sacral** system.

Unlike the somatic reflex arc, in which a single motor neuron extends between the spinal cord and any given muscle in the periphery, autonomic pathways all have two motor neurons. An autonomic neuron arising from the spinal cord leads first to a mass of cell bodies, called a **ganglion,** outside the cord. There it synapses with a second neuron,

which leads to the end organ. The first neuron is called the **preganglionic,** or **presynaptic,** fiber; the second is the **postganglionic,** or **postsynaptic,** nerve. The ganglia of the sympathetic nervous system lie relatively close to the spinal cord, forming two parallel chains with connecting fibers between them (the sympathetic trunks), one row on each side of the cord (Fig. 17-13). In the parasympathetic nervous system the ganglia of the postsynaptic neurons are further away from the spinal cord, tending to be near or in the organs they innervate.

One of the noteworthy features of the anatomy of the sympathetic nervous system is that each preganglionic neuron has synapses with several postganglionic neurons that lead to a number of scattered tissues. By contrast, each parasympathetic preganglionic neuron has synapses with only one postganglionic neuron that supplies only one structure. These structural facts account for the observation that sympathetic stimulation in all cases elicits responses by several organs, whereas parasympathetic stimulation often evokes a response from only one organ. Nevertheless, nearly all internal receptor organs and tissues are innervated by both the sympathetic and parasympathetic divisions of the visceral nervous system.

The visceral nervous system plays an important role in the adaptation of the body to its immediate needs and in the maintenance of a stable or relatively constant internal body environment. It increases or decreases the activity of each internal organ in an integrated fashion. Visceral reflex arcs are usually activated by several different stimuli, including stretch, pressure, and chemical factors. Impulses thus initiated move to groups of neurons in the brain and spinal cord, which serve as integrating centers for the activities of the sympathetic and parasympathetic systems. These centers regulate such body functions as breathing, water balance, heat control, and sexual responses.

In most body activities, including heartbeat rate, blood pressure, and activation of the sweat glands and adrenal glands, nerve impulses arriving via the sympathetic nervous

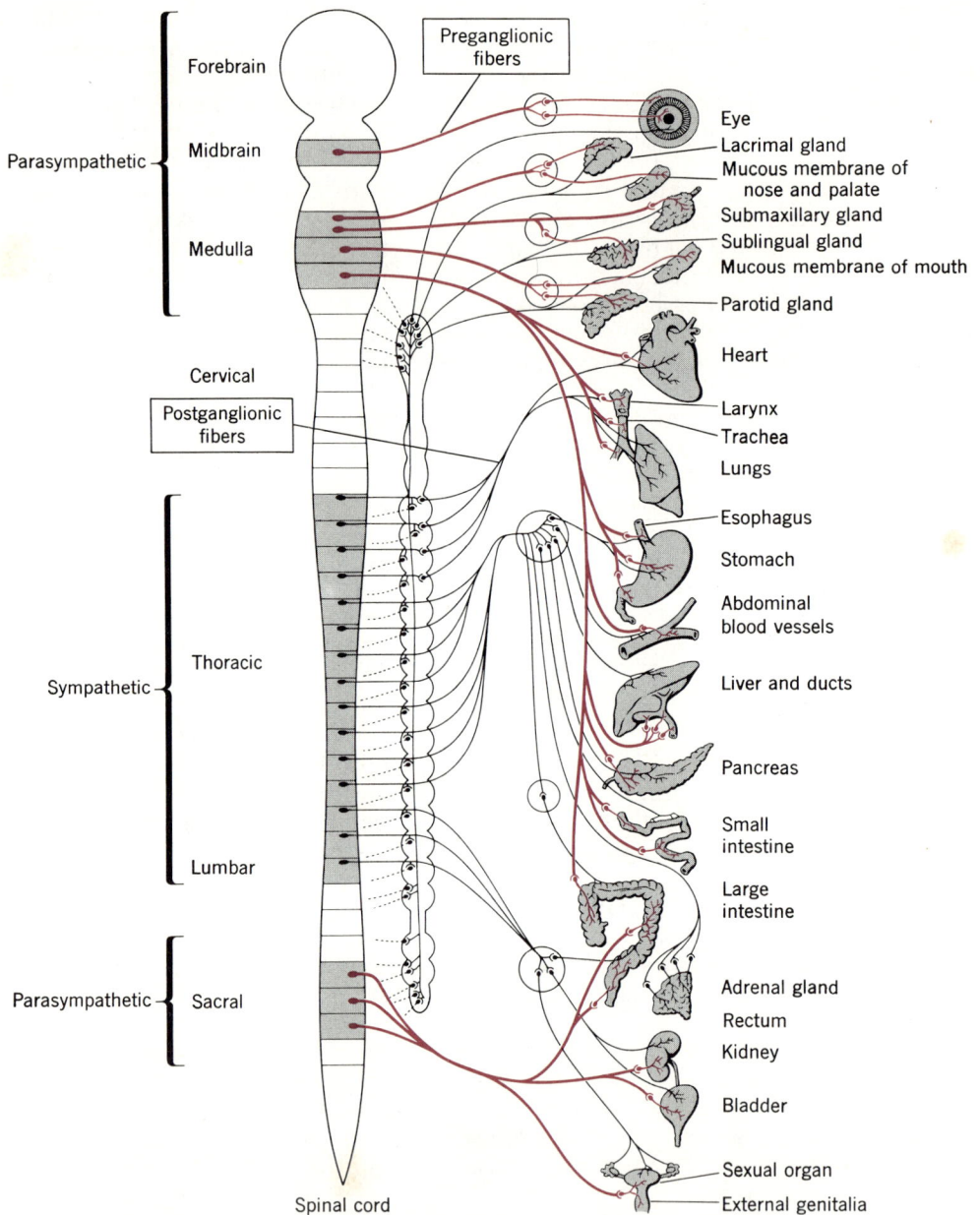

Figure 17-13 *The autonomic nervous system, consisting of the sympathetic and parasympathetic systems. The presynaptic neurons of the parasympathetic system leave the CNS from the medulla of the brain and from the sacral spinal cord. The sympathetic system originates in the thoracic and lumbar regions. Most internal organs are innervated by both systems, one of which stimulates activity in the organ while the other inhibits.*

system have a stimulating effect; nerve impulses going to these organs by way of the parasympathetic system depress or inhibit these activities. Thus the sympathetic nervous system can act in an emergency role, enabling the body to cope with situations of stress by promoting maximal energy production. The parasympathetic division of the autonomic nervous system, on the other hand, is primarily concerned in making adjustments that bring about or maintain a constancy of internal environment. For certain other body activities, however, including the action of smooth muscle in the walls of the digestive tract and secretion by the salivary glands, the effects are precisely the opposite: the sympathetic system inhibits and the parasympathetic system stimulates.

Sense Organs

A **sense organ,** or **receptor,** may be defined as specialized nervous tissue in contact with a nerve cell that is sensitive to a specific stimulus or change in the environment. Receptors are essentially the dendrites of sensory neurons or highly specialized sensitive cells in close association with them. They perform the vital functions of informing the body at both the conscious and unconscious levels of changes in its external and internal environments, thereby enabling the body to respond by protecting itself, improving its chances of survival.

All receptors act as **transducers,** structures that transform one kind of energy into another. They convert the stimulus to which they respond into an electrical change, a depolarization leading to a nerve impulse that is called a **generator potential.** Generator potentials are in many ways like the pacemaker potentials of heart cells (p. 413); the cell membrane is depolarized sufficiently to elicit an action potential.

Several important features of receptor organs should be emphasized. First, specific receptors are particularly sensitive to specific stimuli—the eyes to light, the ears to sound. They can also respond to other stimuli if the stimuli are of sufficient strength or intensity; for example, pressure on the eyeball causes a sensation of light, and irritation caused by disease produces a ringing in the ears. Second, the specific sensitivity arises from (1) specialized structure of the receptor cell (e.g., light-absorbing pigments in the photoreceptors); and (2) auxiliary structures that efficiently channel the specific stimuli to the receptor (e.g., the lens and cornea focus light rays on the photoreceptor cells of the eye). Third, the kind of sensation that is elicited depends only on what nerve pathways are activated, not on how they are activated. Under normal conditions, specific stimuli induce a local excitatory state in specific receptors, which in turn initiate nerve impulses that travel along the connecting nerves.

All nerve impulses, whether they give rise to conscious to unconscious responses, are essentially alike regardless of the stimulus that initiates them. The nerve impulses are transmitted to the central nervous system, where they are sorted and deciphered. The quality of sensation depends on where nerve impulses arrive in the brain. As an illustration, the region of the cerebrum receiving impulses from an olfactory receptor interprets them as a specific odor or aroma. If an olfactory receptor is artificially stimulated with an electrode, the same nerve pathways to the brain will be activated and the brain will interpret the arriving impulses as an odor.

Finally, although we usually take it for granted that we see with our eyes and hear with our ears and smell with our noses, in fact this is not the case. Seeing, hearing, and smelling all take place in the brain. The brain does not receive a picture or a sound or an odor, but only a series of nerve impulses. Patients undergoing brain surgery, for example, see vivid visual images or hear snatches of tunes when specific areas of their cortex are stimulated electrically. **Perception,** then, consists of organization, analysis, and interpretation of stimuli, in addition to their reception.

The commonly described five senses— touch, smell, taste, hearing, and vision—in reality represent an incomplete list. At least

11 distinct sensations or senses can be recognized in man: touch, smell taste, hearing, vision, warmth, cold, equilibrium, pain, proprioception (awareness of position and movement), and visceral sensations (such as hunger, nausea, and sexual sensations). Some of the specialized sensory receptors can be grouped according to the nature of the stimulus into three main subdivisions: chemical receptors (those for taste and smell), which are sensitive to chemical stimuli; mechanical receptors (those for touch and pressure, pain, temperature, sound, and motion), which are sensitive to mechanical stimuli; and photoreceptors (the eyes), which are sensitive to visible light.

Chemical Receptors

Taste

The receptors for taste in higher vertebrates such as man are called **taste buds.** They consist of barrel-shaped clusters of specialized, slender receptor cells located on the surface of the tongue. Each cell has a fine, hairlike projection extending toward the small pore of its taste bud, which opens on the surface of the tongue (Fig. 17-14). Incoming substances must be dissolved in water or saliva before entering the microscopic pores of the taste buds if they are to stimulate particular receptor cells. Several sensory-neuron dendrite branches are distributed among the receptor cells of each taste bud; nerve impulses are initiated in the nearby sensory neurons and eventually give rise to the sensation of taste.

In man there are four fundamental sensations of taste—**sweet, salty, sour** (acid), and **bitter**—for which there are separate taste buds located on different parts of the tongue. Taste buds sensitive to sweetness and saltiness tend to be concentrated at the tip of the tongue; those for sourness are mostly distributed along the sides; and those for bitterness are found along the back of the tongue.

Although all taste buds are structurally alike, they are physiologically different in specificity and sensitivity. Bitterness (as tes-

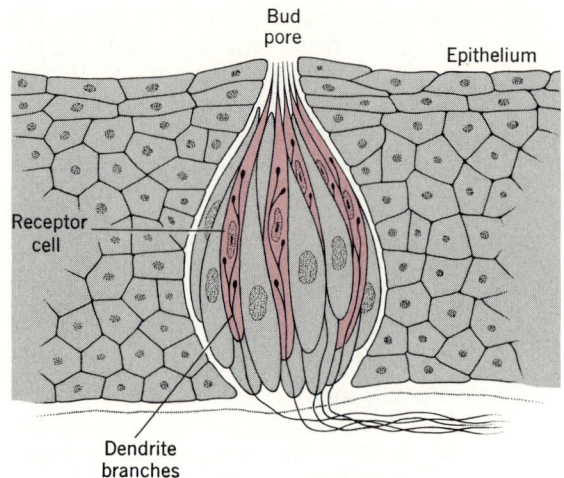

Figure 17-14   Structure of a taste bud, based on electron micrographs of the human tongue.

ted by man's sensitivity to the drug quinine) can be detected in a dilution of one part in 2 hundred million; sweetness in one part in 200; sourness (hydrochloric acid) in one part in 130,000; and saltiness in one part in 400.[1] The taste for sourness appears to be the most specific of the four taste sensations. We seem to experience other taste sensations because of a blending of these fundamental sensations with one another and with the sensations of smell (see the following section) and because of the stimulation of other nerve endings in the mouth.

Smell

The receptors for smell are specialized ciliated cells embedded in the mucous membranes that line the main upper passages of the nose and nasal cavities (Fig. 17-15). If any material is to arouse a sensation of smell, it

[1]Most people have no conception of what constitutes one part per million (ppm) or per billion (ppb). Two examples may help: an inch is one part per million in 16 miles or one part per billion in 16,000 miles; a minute is one part per million in two years or one ppb in 2000 years.

must reach the nasal passages in gaseous form; it must therefore be **volatile,** giving off particles of molecular size that are carried in the air by diffusion and air current to the smell receptors in the nasal cavities. Here they must be dissolved in the mucous secretions before they can stimulate the appropriate receptors. There is an almost infinite variety of odors; and there seems to be only poor correlations between the chemical or physical properties of materials and the sensations of smell that they elicit.

The sensitivity of the smell receptors is in general several thousand times greater than that of the taste receptors. In many animals the sense of smell is of central importance for survival in terms of detecting enemies and seeking out food or mates (Chapter 19); an appreciable portion of the brain in such animals is devoted to this function. Man, with a relatively poorly developed sense of smell, can nevertheless detect certain substances in the air in dilutions of one part in 50 billion.

Interestingly enough, the smell receptors are also apparently easily fatigued; odors that are at first very strong are not sensed at all after a few minutes.

Mechanical Receptors

Touch and Pressure Receptors

The sensation of touch is elicited by a light, momentary contact such as a light brushing of the skin. The sensation of pressure is usually experienced on sustained and more intense contact. The sense of touch derives exclusively from specialized nerve endings near the surface of the skin; the sense of pressure arises from different receptors in deeper parts of the skin (Fig. 17-16, p. 488).

The touch receptors are not uniformly distributed over the surface of the body, but are more concentrated in particular areas such as the fingertips and lips. Certain receptors sensitive to pressure or tension are found in the

Figure 17-15 (a) and (b) The nasal passages and cavities of man, showing the location of the receptors for smell in the olfactory epithelium.

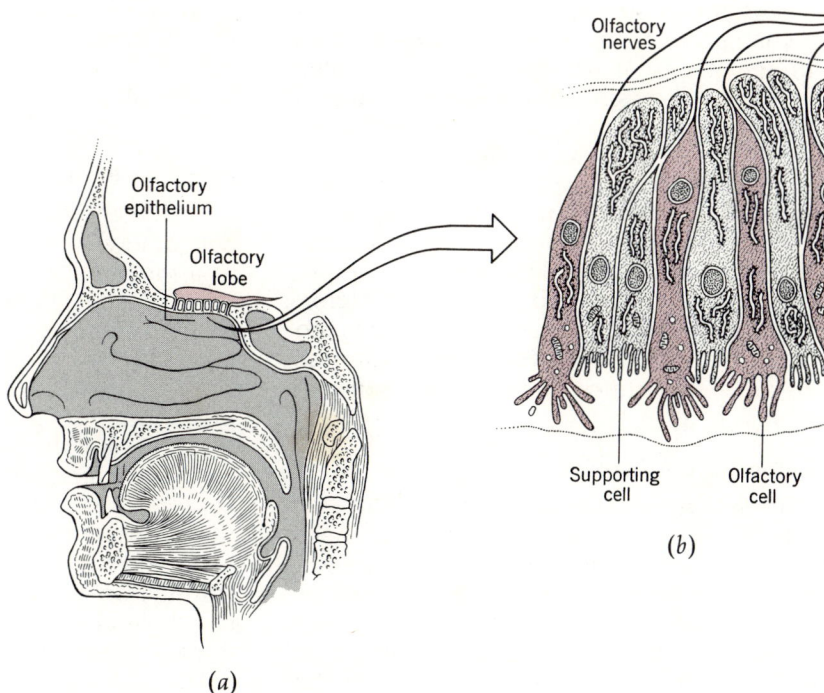

Olfactory epithelium

Olfactory lobe

Olfactory nerves

Supporting cell

Olfactory cell

(b)

(a)

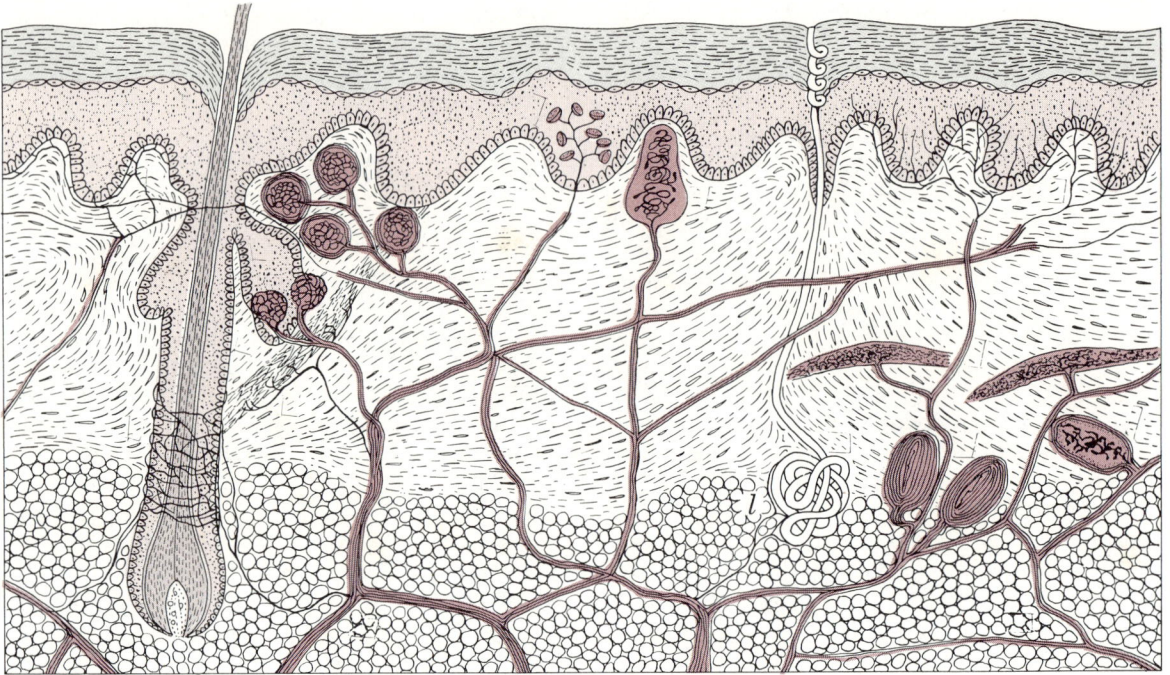

Figure 17-16   A variety of mechanical receptors—touch, pressure, and pain.

carotid arteries (Chapter 15); these respond to changes in blood pressure. At least three different types of pressure receptor, collectively known as **proprioceptors,** are located in the skeletal muscles, tendons, and joints. The stretching of a tendon or the contraction or stretching of a muscle stimulates the proprioceptors; stimulation gives rise to awareness of position and movement of body parts. Our sense of physical balance is a complex phenomenon attributed to several factors, including the proprioceptors, pressure receptors in the soles of our feet, the equilibrium organs of the inner ear, and our sense of vision.

Temperature Receptors

The sensations of heat and cold result from stimulation of two different kinds of nerve ending. Cold receptors fire more rapidly as skin temperature falls; warmth receptors fire more rapidly as it rises. When the skin is explored millimeter by millimeter with a cold or hot pointed probe, spots are found that respond with only one sensation or the other. On the forearm, cold spots average about 90 per square inch and warm spots 10 to 12 per square inch.

Pain Receptors

The receptors for pain are simply free, naked nerve dendrites. They respond not only to a mechanical stimulus but to any type of stimulation—chemical, thermal, or otherwise—as long as it is of sufficient intensity. Compared to the receptors for touch, pressure, and temperature, those for pain are the most numerous, occurring in all tissues of the body except the brain. The brain, ironically enough, is insensitive to pain. Of all the receptors, only those for pain are located in the deeper tissues and organs of the body as well as at the surface; thus pain may arise from stimulation of pain receptors in the skin, in the skeletal muscles, tendons, joints, and internal organs such as intestines and kidneys.

The presence of pain receptors obviously

favors survival of an organism by informing it of environmental changes potentially dangerous to health and life. At times, for reasons that are not entirely clear, the cerebrum misinterprets the source of pain when it arises from the stimulation of deeper structures in the body such as the skeletal muscles and internal organs. In these instances the brain refers the stimulus to an external surface area, a phenomenon known as **referred pain.** For example, pains originating in the heart at times are referred to the inner surface of the left arm or to the chest, a symptom called **angina pectoris.**

Visceral Sensations

The mechanisms of initiation of certain visceral sensations such as thirst, hunger, and appetite are not very well understood. For lack of a better place, they are tentatively classified in this book under mechanical receptors. These sensations originate in the internal organs, but no specific receptors have yet been identified. Thirst is associated with a reduced water content of the body. Hunger is a localized sensation associated with lack of food; it is apparently initiated by rhythmic contractions of the muscles in the walls of the stomach. Appetite, although somewhat similar to hunger, is a more generalized and pleasant sensation and represents an enjoyable desire for food.

The Ear

The ear (Fig. 17-17), the receptor organ of the auditory apparatus, includes the ears, the auditory nerves, and the auditory areas of the cerebrum. The ear is highly sensitive to sound waves, which are best defined as vibrations of any physical medium—gas (air), liquid, or solid—in which they are propagated.

The ear of man has three regions: outer, middle, and inner. The outer and middle ears are basically auxiliary structures that receive, amplify, and transmit sound waves; receptors for sound waves are actually in the inner ear.

Outer Ear. The outer ear (Fig. 17-17) consists of three parts: (1) A trumpet-shaped flap of elastic cartilage and skin called the **pinna.** This is the facial feature usually referred to as the ear. In many mammals the pinna may

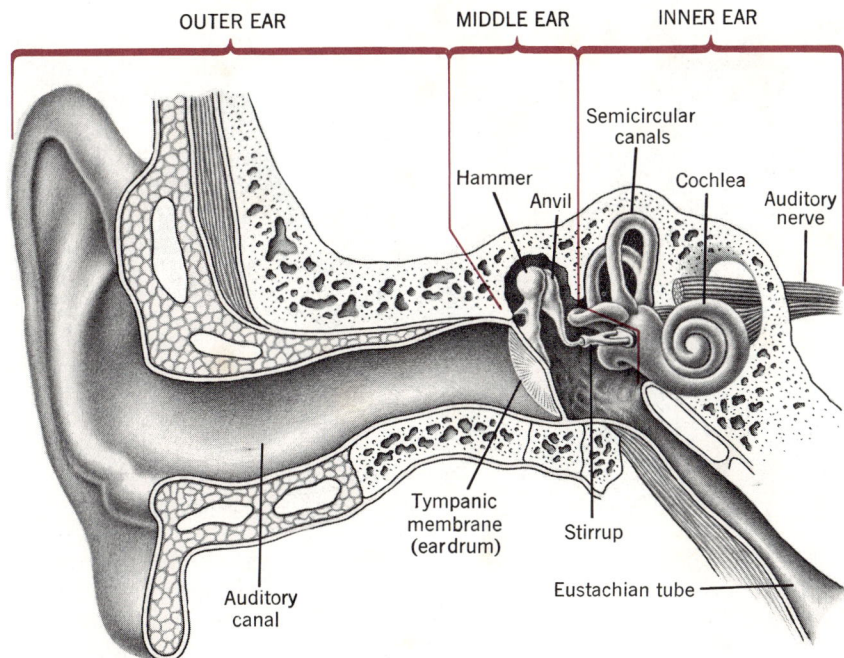

Figure 17-17 The human ear consists of the outer, middle, and inner ears. The actual sound receptors are on the organ of Corti in the cochlea of the inner ear.

OUTER EAR   MIDDLE EAR   INNER EAR

Semicircular canals

Hammer   Anvil   Cochlea   Auditory nerve

Tympanic membrane (eardrum)   Stirrup

Eustachian tube

Auditory canal

be moved—"pricked up"—by well-developed muscles so as to collect sound waves. (2) A funnel-shaped passage, or tube, about 1 in. long, called the **outer auditory canal,** leading from the pinna into one of the bones of the head. (3) A thin, semitransparent, elliptical, flexible membrane called the **eardrum,** or **tympanic membrane** (about 1 cm in diameter), stretching across the inner end of the auditory canal and separating it from the middle ear.

Middle Ear. The middle ear, which connects the outer ear to the inner ear, is a small, hollow, air-filled cavity in one of the bones of the skull. It is lined with an epithelial membrane and contains three tiny bones joined together, commonly called the **hammer, anvil,** and **stirrup** because of their shapes (Fig. 17-18). The handle end of the hammer bone is attached to the inner surface

Figure 17-18 Sound path through the mammalian ear. The cochlea is represented as if it were uncoiled into a straight tube.

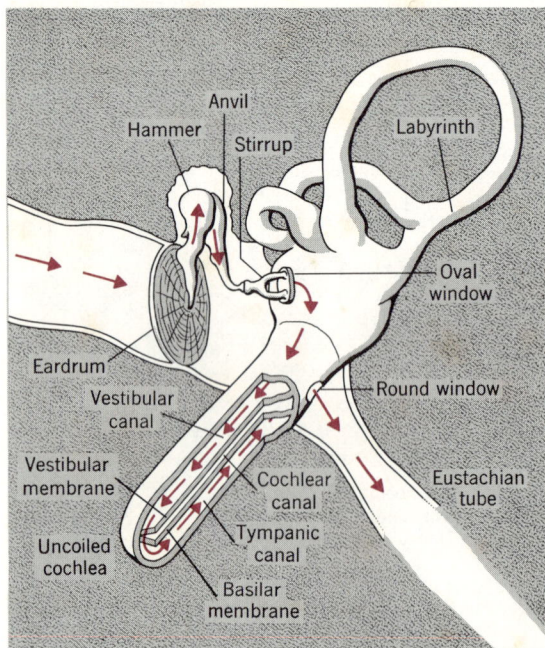

of the tympanic membrane. Its opposite end is connected by a small joint to the anvil, which in turn is joined to the stirrup. The footplate of the stirrup fits into a so-called **oval window,** a membrane-covered opening leading to the inner ear. A second opening, which is also covered by a thin membrane, connects the middle and inner ears and is called a **round window.**

The **Eustachian tube** connects the middle ear to the throat. It is a collapsible tube that acts as an air passage between these two areas and thereby equalizes pressure on both sides of the tympanic membrane. Sudden pressure changes on the tympanic membrane brought about by loud explosions or rapid ascent or descent in an elevator or airplane might cause the membrane to rupture; such an event is prevented by swallowing or yawning, a process that opens the normally closed Eustachian tube and balances the pressure on the external surface of the tympanic membrane. Unfortunately, the Eustachian tube can also be a passageway for the spread of infections from the nose and throat to the middle ear.

Inner Ear. The inner ear is a membranous labyrinth of passageways and cavities in the bones of the skull. Of the different sense organs in the inner ear only one, the **cochlea,** is concerned with the sense of hearing. The others, the **saccule, utricle,** and **semicircular canals,** have to do with the sensation of physical equilibrium and orientation. These are described in a later section.

The cochlea, as the name might suggest, has a shape resembling that of a snail shell. Internally the cochlea is divided into three membranous channels running the full length of the organ. The organization and arrangement of these channels are best visualized by hypothetically unwinding the spirals of the cochlea, as shown in Figure 17-18. The uppermost channel, known as the **vestibular canal,** is attached at its base end to the oval window, into whose membrane the stirrup bone of the middle ear is inserted. At the outer end, or apex, of the cochlea the vestibular canal joins the lowermost channel, called the **tympanic canal.** The base end of the tympanic canal terminates in a membrane-covered round window leading to the middle ear. Both the tympanic and vestibular channels are filled with a clear fluid called **perilymph.** The third and smallest channel is called the **cochlear canal,** which rests between the other two and is filled with a clear fluid called **endolymph.** It is separated from the overlying vestibular channel by a membrane called the **vestibular membrane,** and from the underlying tympanic channel by a ledge-like projection of the bony cochlear wall plus a membrane, the **basilar membrane.**

The actual receptors for hearing are present in the cochlear canal as several rows of approximately 24,000 specialized **hair cells.** They contain numerous cilia projecting into the endolymph from the free end of each cell. The rows of hair cells, together with supporting cells and surrounding dendrites, constitute the structure called the **organ of Corti.** It rests on the basilar membrane in the cochlear canal.

Physiology of Hearing. The process of hearing depends on a succession of events that starts with the entry of sound waves into the external auditory canal (Fig. 17-18), causing the tympanic membrane to vibrate. The vibrations of the tympanic membrane are mechanically transmitted (and magnified) by the hammer, anvil, and stirrup bones of the middle ear to the oval window membrane of the vestibular canal in the cochlea of the middle ear. The vibrations of the oval membrane set up movement, or **pressure waves,** in the perilymph of the vestibular and tympanic channels. These pressure waves are in turn transmitted, via the basilar and vestibular membranes, to the endolymph of the cochlear channel, mechanically stimulating the hair cells of the organ of Corti. Impulses are initiated in the surrounding dendrites and ultimately transmitted by way of the auditory nerve to the auditory sensory area of the cerebral cortex, resulting in the sensation of hearing. Fluids are for all practical purposes incompressible, and pressure waves in the perilymph are compensated for and finally dissipated by corresponding inward and outward bulgings of the round window membrane at the base of the tympanic channel.

Several aspects of hearing are not yet understood with certainty. Sound as perceived by the human ear has three properties: **pitch,** or **frequency,** which is the number of cycles or vibrations per second of a pure tone; **loudness,** or the amplitude of the cycles or vibrations; and **quality,** or **timbre,** which is dependent on overtones produced along with the fundamental tone. According to the usually accepted theory we are able to distinguish between differences in pitch or frequency because some hair cells are stimulated more strongly than others because of certain properties of the basilar membrane. This membrane is approximately .04 mm wide at the oval window end, and broadens gradually to .5 mm at the other end, becoming progressively less rigid as the widest portion is approached. Although vibrations in the perilymph cause the entire basilar membrane to vibrate, the varying flexibility along its length enables it to respond maximally to waves of high frequency at the narrow, stiff end and to waves of low frequency at the other end. As the basilar membrane vibrates, it pulls and pushes the hairs of the organ of Corti, causing nerve impulses to be initiated in different locations. The result is that nerve impulses are sent to correspondingly specific sites in the cerebral cortex, presumably accounting for our ability to discriminate between differences in pitch.

The range of audible frequency varies considerable among individuals depending on age and other factors. The human ear can detect sounds ranging in frequency from about 20 to 20,000 cycles per second (cps). It is most sensitive to sounds between 1000 and 2000 cps. Most sounds made by a mouse are usually of frequencies greater than 20,000 cps, and are therefore inaudible to the human ear; a cat, however, can hear noises of such high frequencies.

Deafness. Any impairment or defect in the ability to hear sounds normally heard by an average individual is known as **deafness.** The common disorders of hearing are usually classified into three types. **Conduction deafness** results from such defects in the sound-transmitting mechanisms as an obstruction in the external auditory canal, damage to the eardrum, or damage or stiffness of the middle ear bones from excessive deposits of calcium salts. **Perception deafness** often results from damage of the cochlea, usually of receptor hair cells in the organ of Corti. Individuals exposed in their work to continual loud noises, for example airplane pilots and artillery men, often are deaf to particular frequencies of sound, apparently because of destruction of specific groups of hairs in the organ of Corti. A similar type of deafness has recently begun to appear among teenagers who spend substantial periods listening to excessively loud, electronically amplified music. Deafness caused by damage to the auditory nerve is also included in perception deafness. **Central deafness** is usually from a physical or psychological disorder in the auditory sensory region of the cerebral cortex or in the brain tracts leading to it.

Organs of Equilibrium in the Inner Ear

The saccule, utricle, and semicircular canals play a major role in our sense of equilibrium and awareness of position (Fig. 17-18). The utricle and saccule are sacs that are only about $\frac{1}{8}$ in. in their longest diameter. In these sacs are special receptor hair cells and nerve endings, projecting into an overlying gelatinous mass called the **endolymph.** The mass contains many small, bonelike fragments, called **otoliths,** made of calcium carbonate and protein. Any changes in position of the head or body cause the otoliths to stimulate the hair cells; this initiates nerve impulses that travel to the brain and ultimately evoke muscular reflexes that tend to restore the body to its normal position. Stimulation of the visual receptors and proprioceptors induces the same type of reflexive response. The nerve endings from the various organs of equilibrium in the inner ear ultimately lead into the same cranial nerve as the auditory nerve extending from the cochlea.

The three semicircular canals are hollow tubes containing endolymph fluid. One end of each canal is enlarged into a swelling called an **ampulla.** In each ampulla is a small elevation of receptor hair cells and nerve endings that project into the endolymph. Each semicircular canal is in a plane approximately perpendicular to the other two, so any movement of the head causes movement of endolymph in one or more of them. The receptor cells are stimulated, initiating nerve impulses that are transmitted to the brain. The position of the head and body are therefore signalled to the brain by impulses coming from various combinations of stimulated ampullae, giving rise to reflexes that tend to right the body.

Photoreceptors (the Eyes)

Light is a form of electromagnetic radiation. Substances that absorb light energy at certain wavelengths and not others are called **pigments;** their color depends on the wavelength of the light they reflect into our eyes; but they gain energy from the light that they absorb. Light is made visible to us because our eyes contain materials called **photopigments,** which are capable of absorbing light energy at wavelengths between about 400 and 700 millimicrons ($m\mu$), the portion of the spectrum we call **visible light** (p. 362).

The first step in seeing occurs when light strikes the photopigments in the receptor cells of the eye. The energy absorbed initiates chemical reactions in the receptor cell membranes that set off action potentials; the production of these nerve impulses is the second step in visual functions. The third step begins with the arrival of the impulses at the visual sensory areas of the cerebral cortex, where they are integrated and analyzed to yield the sensation of vision.

The Human Eye

The adult human eye, measuring approximately 1 in. in diameter, consists of a hollow sphere called an **eyeball,** approximately five-sixths of which is enclosed in the bony eye socket of the skull. In addition there are accessory structures, including eyebrows, eyelids, the tear gland apparatus, the **conjunctiva** (a delicate protective membrane covering the exposed part of the eyeball and inner surface of the eyelid), and six small **ocular muscles** attaching each eye to its socket. The ocular muscles account for the ability of the eye to rotate in various directions. The eyeball itself has essentially three coats, or layers (Fig. 17-19a).

Outer Layer. The outermost coat, called the **fibrous tunic,** is divided into two regions: the cornea and the sclera. The **cornea** (itself covered by a corneal epithelial layer and conjunctiva) is the transparent portion covering the exposed part of the eyeball and representing about one-sixth of the surface. It is actually five separate layers of cells.

Why the cornea is transparent to light has never been satisfactorily explained. Most of the energy metabolism of this structure depends on aerobic metabolism of glucose. Because the cornea possesses no direct blood supply (nor does the lens), the diffusion of O_2 from the atmosphere and nutrients from the aqueous humor (p. 495) is essential. The **sclera,** a firm, white, dense, semirigid membrane, is continuous with the cornea; it covers the remainder of the eye, giving shape to the eyeball and protecting its inner parts. The ocular muscles extend from the sclera to the eyesocket.

Middle Layer. The middle layer of the eyeball is called the **vascular tunic** and has three regions: the choroid coat, the ciliary body, and the iris. The **choroid coat** is a thin, dark membrane containing many blood vessels

Figure 17-19 A cross-section through the human eye. (a) The parts of the eyeball. (b) Detailed structure of the retina, diagrammed at higher magnification.

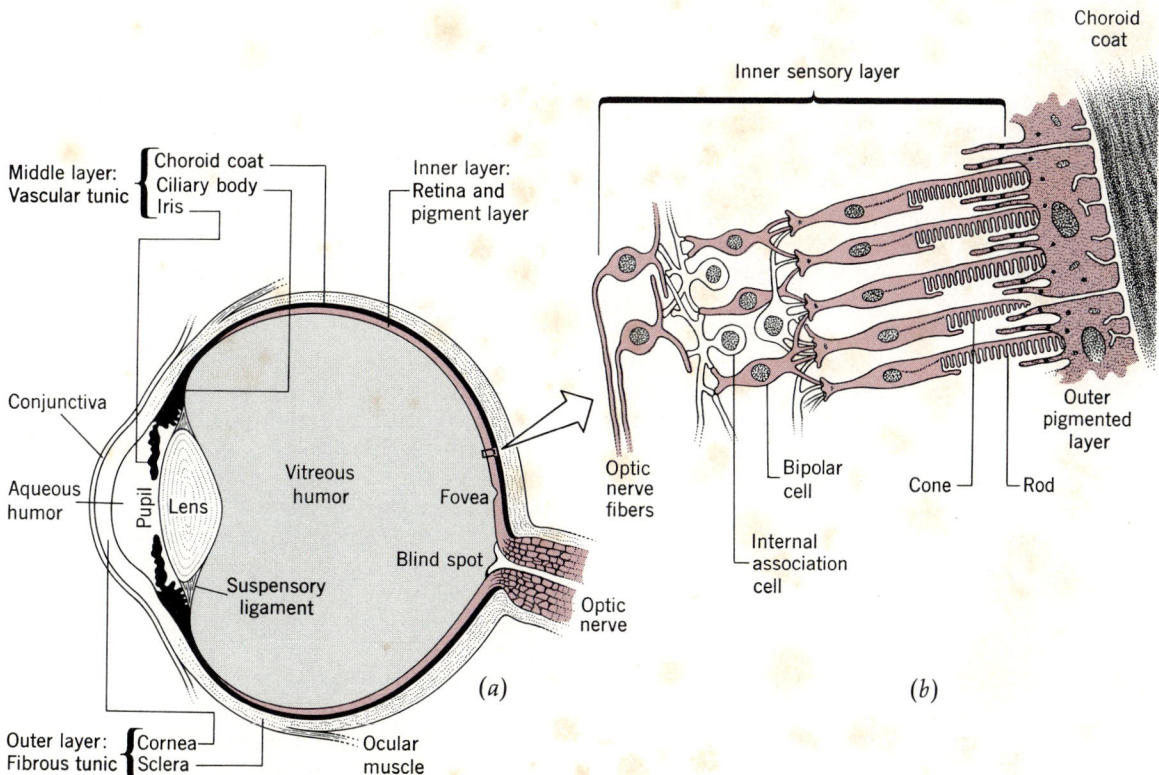

and a large quantity of pigment. It closely adheres to the undersurface of the sclera and therefore covers about five-sixths of the eyeball. The choroid coat is modified into the **ciliary body** near its junction with the iris. The ciliary body has several parts, including a **ciliary muscle,** which is responsible for changing the shape of the lens.

The **iris** is a thin, pigmented, doughnut-shaped structure consisting of circular and radial smooth muscle fibers arranged about a central opening called the **pupil.** The iris is suspended between the cornea and lens and is attached at its outer margin to the ciliary body. It plays a major role in determining the amount of light entering the eye by regulating the size of the pupil by the reflex response of its muscles. The circular fibers of the iris are supplied by the parasympathetic fibers of thf visceral nervous system, and act to narrow the pupil; the radiating fibers are connected to the sympathetic fibers and enlarge the pupil. The muscles of the iris also aid in the formation of clear images on the retina; constriction of the pupil by contraction of the circular muscle fibers normally accompanies the viewing of nearby objects. In this manner divergent rays from the viewed object are prevented from entering the eye through the periphery of the cornea or lens. The entry of such rays would cause a blurred image because they could not be brought into focus properly on the retina.

Inner Coat. The third and inner coat of the eyeball, present only on the back half of the eyeball, is called the **retina.** It is the photosensitive structure of the eye. The retina has the highest rate of oxygen consumption (per unit weight) of any tissue in the body. It contains the actual receptors for vision and consists of two major layers, an **outer pigmented layer** of epithelial cells in contact with the choroid coat, and an **inner sensory layer.**

Two kinds of receptor cell, **rods** and **cones,** so called because of their shapes, form the sensory layer. The rods are characteristically long and narrow; the cones are short and thick (Fig. 17-19*b*). The rods and cones are modified neurons with special segments that are highly sensitive to visual light stimuli. Nerve impulses originate in the rods and cones and are passed, via connector neurons, to the optic nerve and then to the visual sensory areas of the cerebral cortex.

At about the exact geometric center of the surface of the retina is a tiny depression, about 1 mm in diameter, called the **fovea.** The fovea is the region of keenest vision, the area on which the image of an object is normally focused. It accounts for acute visual perception of details and color. It is structurally unique in that its light-sensitive layer consists only of densely packed cones. Vision in the retina surrounding the fovea is far less sensitive to detail and color and becomes progressively more ill-defined toward the more peripheral areas. This correlates with the distribution in the retina of an estimated 7 million cones; the cones are most highly concentrated at the fovea, becoming correspondingly less numerous with increasing distance from the center of the retina.

You can test the difference between the central and surrounding **acuity** (sensitivity to detail) of the retina without putting down this book. Turn to a new page and, as you do, fix your gaze immediately and steadily on the first word you pick out in the center of the page. If you do not move your eyes you will find that you will be able to read only a few words surrounding the one you are fixed on; all the rest of the page, though visible, will be indistinct and unreadable. It is not a matter of focus. To read this page, you are scanning it line by line; your eyes are moving so as to bring the words in sequence into the center of your visual field. Images are then received on the one tiny patch of the retina that has fine enough grain to resolve the detail for word recognition.

By contrast, the estimated 120 million rods of the eye, which are primarily responsible for night vision or vision in dim light, tend to increase in concentration with increasing distance from the fovea. Vision in dim light is therefore best attained by viewing an object indirectly, that is, slightly sideways, so as to focus the image toward the periphery of the retina where the rods are more plentiful.

Although the rods are considerably more sensitive than the cones to lower light intensity, they fail to elicit a color sensation; this response arises exclusively in the cones. Rods produce visual reactions only in terms of black and white or shades of gray.

The Lens. Within the eyeball itself, just behind the pupil, lies an elastic, pale yellow **lens,** a transparent, biconvex structure about $\frac{1}{3}$ in. in diameter. It is clear and glassy in appearance and is composed of an outer layer of epithelial cells enveloping an elastic protein coat, or capsule, that contains a clear, viscous fluid. The thick, gel-like fluid is 25 percent protein and at least 10 percent lipid. Defective lens metabolism often manifests itself by an opacity called **cataract,** the result of a change in protein metabolism that results in the formation of fibrous opaque masses in the lens.

The lens is suspended in position by a series of fibers, collectively called the **suspensory ligament,** that connect the entire circumference of the lens to the ciliary body (Fig. 17-19a). The portion of the ciliary body that contains a mass of smooth muscle fibers is the **ciliary muscle;** contraction of the ciliary muscle causes the lens to change its shape, and therefore determines the extent to which the lens focuses light. When an image striking the retina is out of focus, an involuntary reflex causes the ciliary muscle to contract or relax and to alter the curvature of the lens just enough to bring the image into sharp focus. Relaxing the ciliary muscle allows the lens to flatten, focusing on distant objects; contracting the ciliary muscle increases the curvature of the lens, focusing the eye on nearby objects. The ability of the eye to bring objects at various distances into focus is known as **accommodation.**

The Chambers of the Eye and Their Media. The hollow sphere of the eye contains two main chambers (Fig. 17-19a). In the front of the eyeball, between the cornea and the lens, is the first main chamber of the eye. It is filled with a clear, watery fluid called the **aqueous humor.** The second and larger main chamber, between the lens and the retina, is occupied by a transparent, jelly-like material called the **vitreous humor.** To reach the retina, light rays must pass successively through the cornea, aqueous humor, lens, and vitreous humor, which collectively serve in the transmission and focusing of light on the light-sensitive rods.

How We See

A man-made camera shares many similarities with the mammalian eye (Fig. 17-20, p. 496). Both use a lens to focus light from objects onto a photographic plate or retina. One important mechanical difference is that a camera (like the eyes of some animals) is focused by changing the distance between the lens and the film; the mammalian eye is focused by adjusting the shape of the lens. The diaphragm (the light-regulating mechanism of a man-made camera) is functionally equivalent to the iris. The light-sensitive material in both the eye and the camera is chemical; usually silver bromide crystals on a film in the camera, and rhodopsin in the eye.

Stimulation of the light-sensitive rods and cones of the retina gives rise to a barrage of nerve impulses that are conducted from the eye, via the optic nerve, to the visual areas of the cerebral cortex. Here the features of viewed objects—degree of light and dark, color, form, and motion—are recorded, integrated, and interpreted to yield the over-all sensation of vision. The image of an object that falls on the retina is upside down and reversed from side to side, just like an image on a photographic plate. Yet we do not see the object in that way; in the process of interpretation in the brain the image is again inverted and reversed from side to side, causing the object to appear as it actually is.

Defects in Seeing. For various reasons the eyes may be unable to focus light properly on the retina; this causes a blurred image. Visual acuity is usually measured by the ability of an individual to see letters of an established size that can be viewed by a normal eye at a distance of 20 ft. In the commonly used **Snellen test,** normal visual acuity is rated as 20/20, which means that the eye sees the standard size letters as they are seen by a

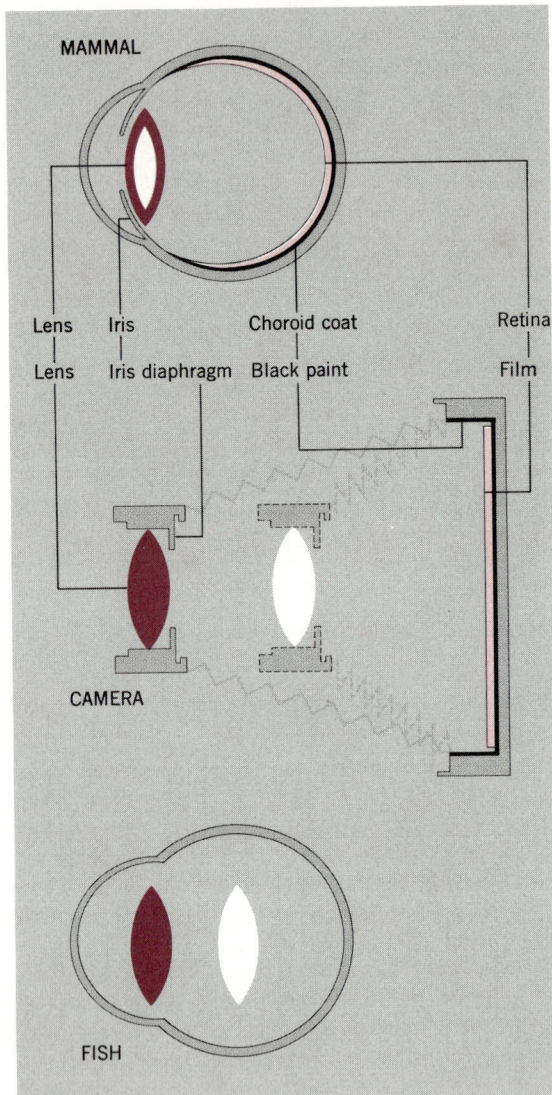

MAMMAL

Lens Iris   Choroid coat   Retina
Lens Iris diaphragm Black paint   Film

CAMERA

FISH

Figure 17-20 A comparison of the eyes of a mammal and a fish with a simple camera. Each is shown with the lens focused on a distant object (white) and a near object (color). The eyes of amphibians, snakes, and most mollusks work like the eyes of a fish or the lens of a camera; that is, focus is adjusted by moving the lens closer to or farther from the light-sensitive surface (retina or photographic film). Birds and mammals focus by changing the curvature of the lens.

normal eye at 20 ft. Ratings of 20/40 or 20/80, for example, mean that the eye must be at 20 ft to see clearly letters that can be seen by a normal eye at 40 or 80 ft. A visual acuity of 20/40 is 50 percent of normal vision, 20/80 is 25 percent of normal.

Among the most common defects in vision are myopia (nearsightedness), hyperopia (farsightedness), and astigmatism. In **myopia** the light rays are focused at a point in front of rather than on the retina (Fig. 17-21), usually because the eyeball is longer than normal. It can be corrected by placing a concave, or **divergent,** lens in front of the eye, so that the extent of bending, or **refraction,** of light rays entering the eye is sufficiently decreased so that they focus on the retina. In **hyperopia,** which is just the opposite, the light rays are focused at a point that would be beyond the retina, usually because the eyeball is shorter than normal. It can be corrected by use of a convex, or **convergent,** lens, which causes a greater convergence of light rays entering the eye. As most people reach middle age they tend to become farsighted because the elasticity of the lens progressively decreases, so that it is less able to bulge or assume maximum curvature. In **astigmatism** the irregularities of the curvature of the cornea, or more usually of the lens, cause part of the retinal image to be out of focus. It can be corrected by lenses that compensate for the irregular curvature.

The Physiology of Vision. How is light falling on the retina converted to an electrical impulse in the optic nerve? We know that the change that takes place in the rods and cones to translate the physical energy of light into nerve impulses involves a **photochemical** step; that is, light waves set up chemical changes in the rods or cones, which in turn give rise to the generator potential. The biochemical steps in this visual stimulation involve (1) an initial light-dependent chemical reaction (i.e., the photochemical reaction) in which one or more specific pigments absorb light and are chemically transformed in the process; (2) initiation of the nerve impulse, presumably by one or more of the pigment products of this photochemical reaction; and

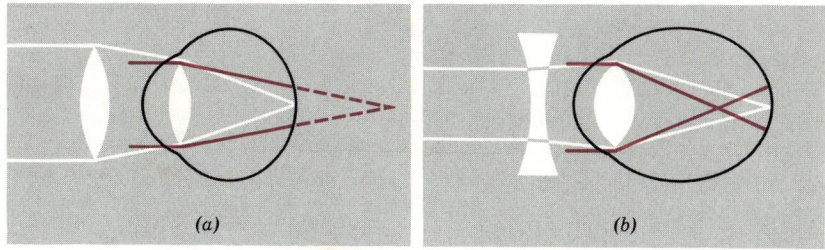

Figure 17-21   The path of light-rays in the eye with (a) hyperopia
(farsightedness) and (b) myopia (nearsightedness) is shown uncorrected (in
color) and corrected (in white) with properly ground glasses.

(3) regeneration of the pigment to its original state so that it can be used again in subsequent visual stimulation. Step 1 is considered the only photochemical reaction in the process.

The rod cells contain a unique reddish pigment called **rhodopsin,** or **visual purple,** which is a conjugated protein—a complex of a specific protein known as **opsin** and the aldehyde form of vitamin A (p. 394), called **retinal.** George Wald and his co-workers at Harvard have shown that when rhodopsin absorbs a quantity of light, the retinal changes from a bent (cis) shape to a straight chain (trans) form. In the process the complex is bleached (that is, it loses its color, and is therefore called **visual yellow**) and the protein undergoes a series of physical rearrangements. We are not completely certain which of the photochemical products of rhodopsin serves as the stimulus that initiates nerve impulses, but there is now evidence that at some point the protein formed in the reaction binds to the infolded membrane layers of the rods, called **lamellae** (Fig. 17-22 p. 498), and alters membrane permeability, thereby producing the generator potential. This entire sequence of events, from the absorption of light by retinal to the relay of a nerve signal down a ganglion cell, takes place in less than 2 milliseconds (two thousandths of a second).

A process called **dark adaptation** involves the reverse action, the conversion of visual yellow to visual purple; this is a relatively slower and more complex process involving several enzymatic reactions. During the process of conversion an individual is relatively insensitive to light—at least until the normal level of rhodopsin is restored. We have all experienced the common difficulty of entering a darkened movie house from a bright street or lobby; in a few minutes our ability to see in the dark is restored. The retinal that appears upon bleaching of rhodopsin cannot recombine directly with opsin to form active rhodopsin, but must first undergo several enzymatic reactions including reduction, isomerization, and oxidation.

In man maximal sensitivity to semidarkness is obtained after a period of about three-quarters of an hour in the dark. A deficiency of vitamin A in the diet results in defective vision in dim light but not in bright light, a condition known as **night blindness.**

Far less is known about the physiological or biochemical mechanism of action of the cones, the receptors responsible for color and detailed vision. The cones are considerably less sensitive to light than the rods and require 50 to 100 times greater light intensity in order to be stimulated. Experiments with retinas from humans and monkeys have demonstrated the existence of three kinds of primary color receptor—a blue-sensitive cone, a green-sensitive cone, and a red-sensitive cone—that are collectively responsible for color vision. The evidence also indicates that the green-, red-, and probably the blue-sensitive pigments consist of the same

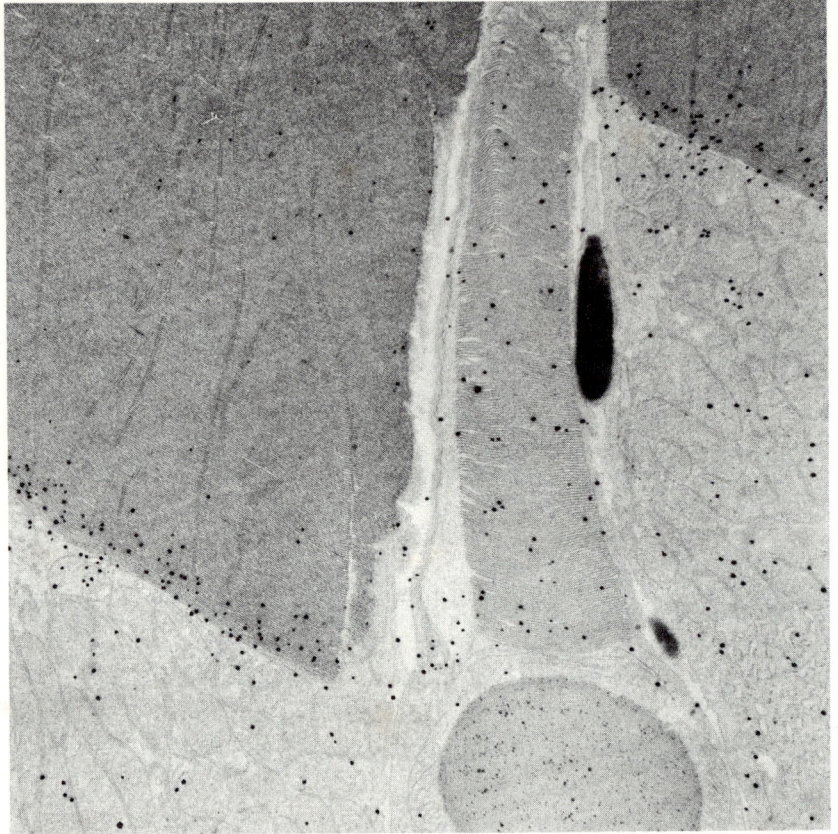

Figure 17-22 Electron micrograph of a rod cell from a human retina. The outer segment and stalk develop as a highly specialized cilium. Visual pigment is packed between the closely spaced membranous lamellae, where products of the photochemical reaction can exert their effects on a large area of membrane to produce a generator potential. Magnification 9000×.

retinal as in rhodopsin but bound to different proteins or opsins, thus accounting for their different light-absorbing properties.

Stereopsis. In many mammals, including man, and in certain birds of prey, the eyes are directed forward. The two fields of vision although very similar, are not identical; the two eyes view an object from different angles, and the images overlap to a large extent. The two slightly different images produce nerve impulses that are integrated by the brain into a single composite visual image, a response known as **stereoscopic** or **binocular** vision. The ability to judge distance or depth is largely ascribed to binocular vision. Other factors also contribute to the estimation of depth and distance, including relative size of objects, distinctness of detail, and so on.

Conclusion

It may seem that we included a great deal of detailed information about the nervous system in this chapter. Our knowledge of how the nervous system works is, however, really very primitive. We know the anatomy of its various parts only in gross terms; and our ignorance about the relationship between nerve organization and such functions as behavior, emotion, and intellectual activity is even more profound. We shall explore these areas further in Chapter 19 to see what progress has been made.

In general, we have seen that the nervous system, like the endocrine system (Chapter 16) provides organisms with a means of coordinating body functions, and that, also like the endocrines, nerves perform this task by distributing chemicals to receptive locations. Hormones are transported broadly throughout the body by the bloodstream. Neural transmitter substances are applied at highly localized points by nerve endings. Hormones are much more versatile than nerves; we have examined some of the wide variety of chemical effects they can have. But because messages are sent over nerves by "direct wire" to specific locations, the nervous system performs its coordinating activities in a much more rapid and precise way than do hormones. Nerves, however, have only a limited repertoire of effects. A nerve can stimulate or inhibit another nerve across a synapse; it can cause a gland to secrete; or it can stimulate a muscle to contract. It is to this last activity, muscular contraction, that we turn next.

Reading List

Case, J., *Sensory Mechanisms*. Macmillan, New York, 1966.

Dean, G., "The Multiple Sclerosis Problem," *Scientific American* (July 1970), pp. 40–56.

Dethier, V. G., *To Know a Fly*. Holden-Day, San Francisco, 1963.

Di Cara, L. V., "Learning in the Autonomic Nervous System," *Scientific American* (January 1970), pp. 30–39.

Dowling, J. E., "Night Blindness," *Scientific American* (October 1966), pp. 78–87.

Haber, R. N., "How We Remember What We See," *Scientific American* (May 1970), pp. 104–115.

Hailman, J. P., "How an Instinct Is Learned," *Scientific American* (December 1969), pp. 98–108.

Heimer, L., "Pathways in the Brain," *Scientific American* (July 1971), pp. 48–60.

Kandel, E. R., "Nerve Cells and Behavior," *Scientific American* (July 1970), pp. 57–70.

Katz, B., *Nerve, Muscle and Synapse*. McGraw-Hill, New York, 1966.

Luria, A. R., "The Functional Organization of the Brain," *Scientific American* (March 1970), pp. 66–79.

MacNichol, E. F., "Three-Pigment Color Vision," *Scientific American* (December 1964), pp. 48–56.

Marler, P. R., and W. J. Hamilton, *Mechanisms of Animal Behavior*. Wiley, New York, 1966.

Michael, C. R., "Retinal Processing of Visual Images," *Scientific American* (May 1969), pp. 104–114.

Pribram, K. H., "The Neurophysiology of Remembering," *Scientific American* (January 1969), pp. 73–87.

Thomas, A., S. Chess, and H. G. Birch, "The Origin of Personality," *Scientific American* (August 1970), pp. 102–109.

Young, J. Z., *Doubt and Certainty in Science. A Biologist's Reflections on the Brain*. Oxford University Press, Fairlawn, N. J., 1960.

Young, R. W., "Visual Cells," *Scientific American* (October 1970), pp. 80–91.

chapter eighteen | *Muscular Locomotion*

"It is only by the use of our muscles that we are able to act on our environment—to exert forces and to move objects, including ourselves, around the world. The muscles are biological machines which convert chemical energy, derived ultimately from the reaction between food and oxygen, into force and mechanical work. . . . As our knowledge of the mechanism by which they operate is advancing by leaps and bounds, it seems likely that muscle may be the first tissue whose function is completely understood in terms of ordinary physics and chemistry."

D. R. WILKIE, 1968, MUSCLE

In Chapter 17 we discussed the sensory and conductor components of the nervous system as one way in which an organism coordinates its activities. However, if an organism is to exhibit behavior of any kind, it must not only be able to respond to stimuli or generate its own, and to conduct those stimuli throughout its body, it must also be able to act. That is, it must be able to effect a change in shape or position of some part of itself or of itself in relation to its environment. An **effector** is an organ that accomplishes such a change.

Basically there are only two kinds of effector that can be influenced by nervous innervation: those that secrete materials and those that bring about motion. Secretory effectors are glands; we have already discussed their activities (Chapter 16) and their control by the autonomic nervous system (Chapter 17). We know relatively little about how nerve impulses cause a gland to secrete its products or stop it from doing so. But the activity of animals that is usually most obvious to us involves movement; most of what we call behavior involves motion of some kind, and we do know a great deal about how organisms effect this activity.

Motion is a fundamental property of essentially all living systems. It may involve the movement of structures in a cell, or of the whole cytoplasm. The latter, called **cytoplasmic streaming,** is common in plant cells and in a wide variety of eggs after fertilization; cytoplasmic material streams up and down the nerve axons of vertebrates; and cytoplasm moves actively in many protista. Cytoplasmic streaming distributes materials throughout a cell. It is also responsible for the **ameboid** movement (Fig. 18-1 and Chapter 5) exhibited by many single-celled organisms, metazoan plasma cells, and most cells in tissue culture. We know that cytoplasmic proteins organized into microtubules (Chapter 5) are able to contract (as in the spindle apparatus of a dividing cell, Fig. 7-5) and that the contraction is somehow related to cytoplasmic movements. But we do not yet understand the molecular mechanism of this action or where the force is exerted. Other small organisms move by means of cilia or flagellae (Chapter 5).

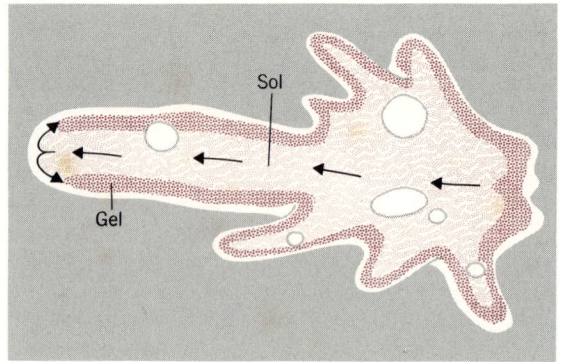

Figure 18-1   *Ameboid movement. The cytoplasm of ameboid cells is generally gelatinous around the cell periphery and liquified near the center. Such cells move by pushing out temporary bulbous cytoplasmic projections (**pseudopods**) into which additional cytoplasm flows until the whole animal has moved forward. As the liquid cytoplasm (sol) reaches the tip of a pseudopod, it flows to either side like a fountain and is reconverted into the gel state. At the same time, the gel at the posterior end becomes a sol and flows forward.*

Muscles as Effectors

The most effective means of motion, used by all multicellular animals except sponges, is **muscular contraction.** Muscles are tissues made of cells that are highly specialized for contractility, in which most of the cytoplasm is filled with threadlike **myofibrils** essentially identical to those in heart muscle (chapter 15). Myofibrils are able to shorten rapidly, by a substantial fraction of their length, at an appropriate stimulus. It is interesting that muscular contraction must have developed early in animal evolution; the structure and physiology of muscle, from the lowliest worms and insects to the mammals, including man, are amazingly similar. In fact, the proteins extractable from the contractile microtubules of amebas have chemical properties resembling those that make up the myofibrillar proteins of vertebrates.

All vertebrates, including man, have three distinctly different types of muscle: smooth muscle, cardiac muscle, and skeletal muscle (Chapter 6). Together these represent about 40 percent of the weight of a person.

Smooth muscle, which is generally regulated by the visceral nervous system, (pp. 482–485) exists in most of the organ systems of the body. It is a major component of the walls of the internal organs such as the digestive tract, urinary bladder, and nearly all tubular structures, including those of the circulatory system, respiratory system, urinary system, and reproductive system. Smooth muscle, like cardiac muscle, frequently exhibits an inherent rhythmic contraction independent of its innervation. For example, pieces of intestinal or uterine smooth muscle will contract and relax rhythmically for hours when placed in a suitable nutrient solution.

Cardiac muscle is found only in the heart. Its action propels blood through the circulatory system. Like skeletal muscle, it is striated (Chapter 15), but like smooth muscle its rhythmic beat is stimulated by specialized pacemaker muscle cells in the tissue, regulated by the autonomic nervous system; that is, its actions are involuntary.

Skeletal muscle, as the name implies, is primarily associated with the skeleton and comprises the large bulk of the body musculature. Although most skeletal muscles are attached from bone to bone, a few, such as those of the face, are connected from one part of the skin to another, and several are attached to cartilage or to other muscles. All operate not as individual muscles but in groups under the integrated control of the central nervous system. The coordinated movements of groups of skeletal muscles are a conscious action, responsible for movements of the bones, body location, and posture. Skeletal muscles are also a major source of heat production in the body; because they are fairly inefficient machines, 75 to 80 percent of their metabolic energy appears not as contractile work but as heat.

The Skeletal Musculature

Skeletal muscles show a wide variation in size, shape, and mode of attachment to bones and other structures. Each muscle has a main portion, called the **belly,** and two ends by which it is anchored. By convention the end of a muscle attached nearer to the central portion of the body is called its **origin,** which remains relatively fixed when the muscle contracts; the other, more distant point of attachment is called the **insertion** (Fig. 18–2).

Muscles perform work by their ability to contract. They never push; rather, they pull by shortening. They apply power at the insertion, just as the shortening of the strings produces movements by pulling on a marionette's body parts. In effect, the bones act as levers, the joints as fulcrums for these levers, and the contracting muscles as the source of power for the movement of the bones (Figs. 18-2 and 18-3, p. 504).

As a rule, muscles act in groups or teams rather than singly, usually exerting opposite, or **antagonistic,** effects. This is well illustrated in the bending of the arm at the elbow, which is caused by the contraction of the **biceps** (Fig. 18-2). When the biceps contracts, it bends **(flexes)** the arm, and in this action the biceps is called a **flexor.** For bending to occur, however, an opposite-acting muscle, the **triceps,** must relax. To execute the reverse movement, that is, to straighten, or **extend,** the arm, the triceps (called an **extensor**) must contract and the biceps relax. Saying that the flexor and extensor are antagonistic, however, does not mean that when one begins to contract the other immediately and completely relaxes. This would produce jerky, all-or-none

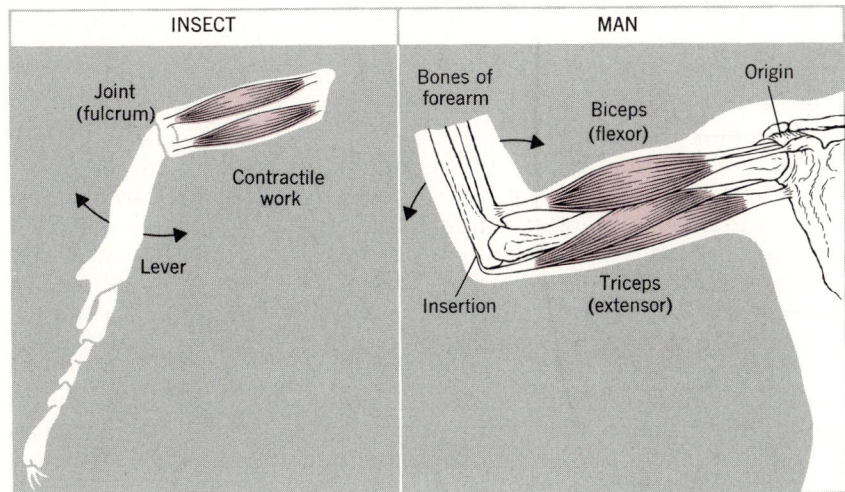

Figure 18-2 Limb movement in man and insect. Antagonistic pairs of muscles cause the movement of the exoskeleton of insects and the endoskeleton of vertebrates. When one muscle (e.g., the biceps) contracts, the arm is flexed (bent) at the elbow; simultaneously, the antagonistic muscle (in this case, triceps) automatically relaxes.

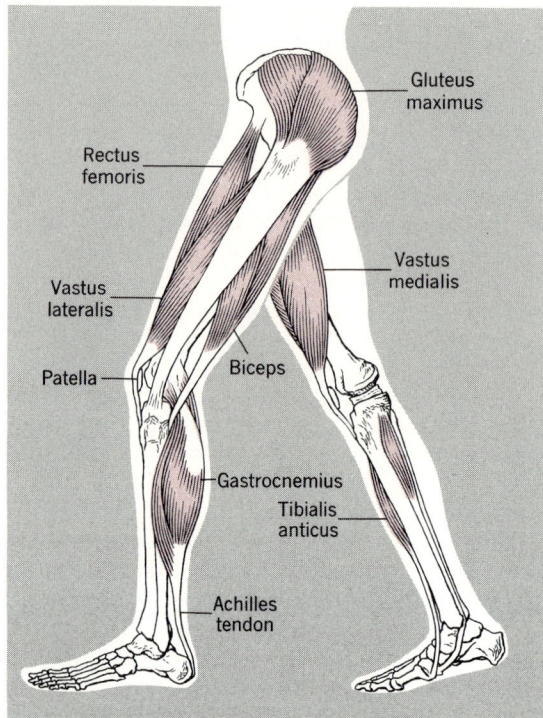

Figure 18-3 The muscles of the leg that are concerned with walking.

(those of the forearm) to it. The origin of the muscle, therefore, usually remains comparatively stationary as the muscle contracts, whereas the insertion experiences relatively more movement.

Cellular and Subcellular Structure

Skeletal muscle is composed of multinucleated cells called **muscle fibers** (Chapter 6). Each fiber contains the cytoplasm of hundreds of single **myoblasts** or muscle-forming cells that fused during embryonic life. Each muscle fiber is encased in a delicate connective-tissue sheath, called a **sarcolemma,** which is part of a larger connective-tissue membrane enclosing different sized bundles of muscle fibers (Fig. 18-4). Groups of such muscle-fiber bundles are in turn wrapped together and invested in a common outer connective-tissue envelope, to make up the skeletal muscle (Fig. 18-4). Through this complex system of connective tissue, the nerves and blood vessels are distributed to each part of the muscle. For most skeletal muscles the connective tissues usually extend from each end of the muscle to form a tough, white, fibrous cord, called a **tendon,** by which the muscle is attached to the bone.

The muscle fiber itself contains a gellike cytoplasm called **sarcoplasm,** which is filled with numerous longitudinal **myofibrils** ranging from 1 to 3μ in diameter. The structure of the myofibrils gives the effect of alternating light and dark bands along their length. The many nuclei scattered throughout a fiber are generally located immediately under the sarcolemma. The mitochondria are termed **sarcosomes.**

Skeletal muscle fibers range in diameter from 10 to 100μ and may extend through the entire length of a muscle, joining with the tendons at each end. The myofibrils are the actual contractile elements of the muscle fiber, and it is the side-by-side arrangement of the light and dark bands of the many myofibrils that gives muscle fiber its characteristically striated appearance (Fig. 18-5, p. 506).

movements. A moment's thought will show that the smooth, graded movements we normally use result from the simultaneous controlled action of both flexor and extensor, with just a little more power being given to one or the other depending on whether the desired motion involves bending or straightening the arm.

The flexion and extension of the arm illustrates another generalization concerning muscle arrangement and function. The origin and belly of many skeletal muscles are usually located on one side of a joint, the insertion on the other; thus muscle contraction moves body parts at the joint by pulling from one bone across the joint to the other bone. One of the bones remains relatively stationary (in this instance, that of the upper arm), acting as an anchor point for the contracting muscle as it pulls the other bone or bones

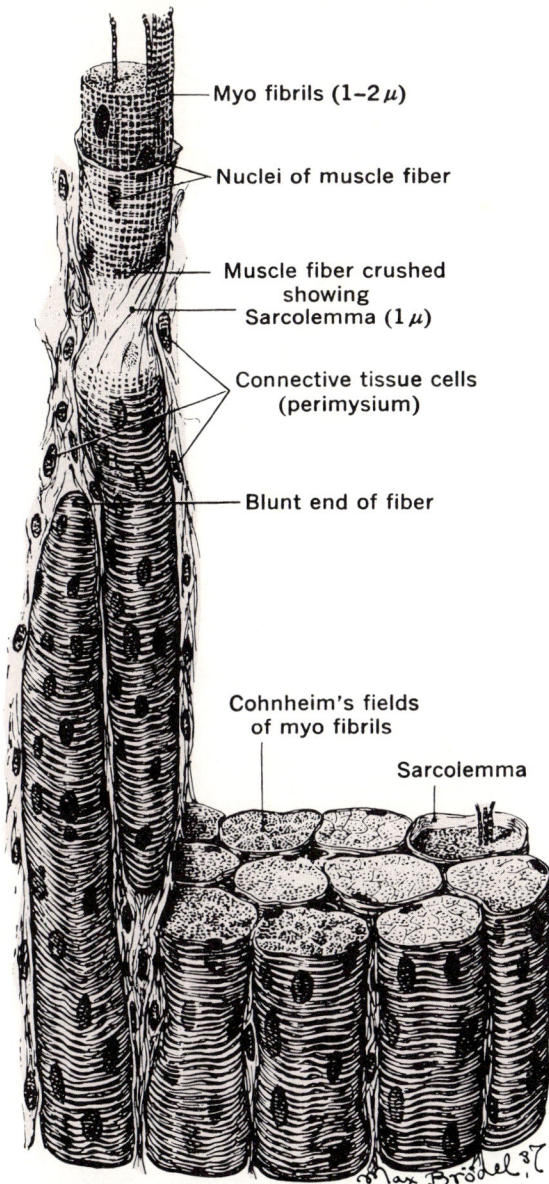

Myo fibrils (1–2 μ)

Nuclei of muscle fiber

Muscle fiber crushed
showing
Sarcolemma (1 μ)

Connective tissue cells
(perimysium)

Blunt end of fiber

Cohnheim's fields
of myo fibrils

Sarcolemma

Figure 18-4   The gross structure of muscle fibers.

Electron microscope studies of cross-sections of myofibrils have established that each myofibril is made of an ordered arrangement of two types of filament. The **thick filaments** (about 150 Å in diameter) are made of the protein **myosin;** the **thin filaments** (about 60 Å) are made of a second protein, **actin.** Neither of these fibrous proteins is in itself able to shorten. Yet electron micro-scopic examination of muscles (Fig. 18-6, p. 507) shows that the myofilaments do not extend the entire length of the fibrils, but are stacked in repeating arrays, called **sarcomeres,** that are arranged end to end throughout the length of a myofibril. Each sarcomere is 2 to 3μ long, and is bounded at each end by a dense black line (the Z band in Fig. 18-5d and Fig. 18-6). In Figure 18-5d, e, you can see the alternating thin and thick filaments arranged in a regular pattern between two Z bands.

Myofibril Contraction

Albert Szent-Gyorgyi at the Marine Biological Laboratory at Woods Hole showed that if actin and myosin are separately extracted from muscle, purified, and mixed at high ionic concentration, they form a viscous, opalescent, colloidal solution. If such a solution is slowly squirted through a hypodermic syringe into dilute saline, the extruded material gels into a long fiber, remarkably tough, made of a loose complex termed **actomyosin.** Such an actomyosin fiber exposed to calcium ions and ATP will contract, and in fact will actually lift a small weight. Neither actin nor myosin alone will do so. If the individual myofilaments themselves are unable to shorten, how do the myofibrils contract?

We have seen that myosin is located in the thick filaments of a fiber that make up the A band in Figure 18-5d; actin forms the thin filaments located in the light I band. Electron micrographs of muscles in relaxed and contracted states show that the widths of the two bands change after contraction: the I bands and H zones become narrower, but the A bands remain the same, with the result that the Z lines move closer together (Fig. 18-7, p. 508). During contraction the filaments telescope together by sliding past each other. The mechanism of this sliding can be seen in very high magnification electron micrographs of the cross-bridges that link the thick and thin filaments (Fig. 18-7c, d). These bridges are part of the enlarged end of each myosin molecule (Fig. 18-5i), and are movable. That is, the bridges can act like hooks or levers, exert-

505   *Muscular Locomotion*

Figure 18-5   *The organization of skeletal muscle from the gross to the molecular level. Any typical muscle, such as the deltoid of the shoulder (a), is made up of muscle bundles called* **fascicles** *(b), each in turn made of tens or hundreds of muscle fibers (c). The muscle fiber, normally 10 to 30μ in diameter, is the cellular unit of muscle. It is enclosed by a plasma membrane* **(sarcolemma)** *and contains* **sarcosomes** *(mitochondria) and a special system of membranous tubes called the* **sarcoplasmic reticulum.** *These structures are found between and surrounding the closely packed, striated myofibrils (d). Each myofibril includes a series of interdigitated (interlocked) myofilaments (e), which are made of the contractile proteins* **actin** *(the thin filaments, f) and* **myosin** *(the thick filaments, g). Contraction of the myofibrils (and therefore of the muscle) is brought about when the thick and thin filaments are caused to slide lengthwise past one another.*

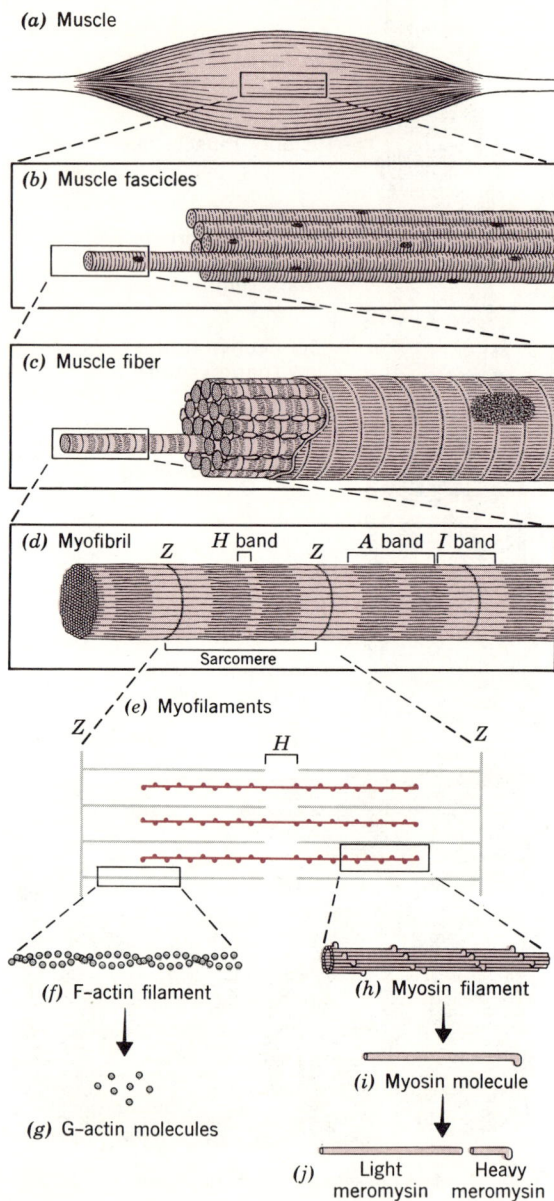

(a) Muscle

(b) Muscle fascicles

(c) Muscle fiber

(d) Myofibril   Z   H band   Z   A band   I band

Sarcomere

(e) Myofilaments

Z   H   Z

(f) F-actin filament

(g) G-actin molecules

(h) Myosin filament

(i) Myosin molecule

(j)   Light meromysin   Heavy meromysin

ing tension on the neighboring actin filaments (Fig. 18-7b).

Another extremely important characteristic of the remarkable protein myosin is that it contains enzyme activity. It is an ATP-ase, and cleaves the high-energy bond of ATP to liberate energy (Chapter 4). If each cross-bridge swings back and forth, sliding the neighboring actin filament along like a ratchet mechanism, the exertion of force in the process requires energy.

Hugh Huxley of University College, London, and A. F. Huxley of Cambridge (not kin at all) were the biophysicists who discovered and unraveled this **sliding filament** mechanism of muscle contraction. Hugh Huxley assumed that each oscillation of a cross-bridge requires the energy of one ATP molecule. By counting the number of cross-bridges per unit length, and noting the rate of contraction of the muscle he used, he was able to calculate that each bridge would have to go through 50 to 100 oscillations per second. Thus 50 to 100 molecules of ATP should be used up per cross-bridge per second.

If the total number of cross-bridges in a fiber is estimated, and this value is used to predict the total ATP consumption per second by that fiber, the resulting figure is remarkably close to the rate, known from other studies, at which myosin is able to catalyze the removal of the terminal ATP phosphate group. The Huxley calculations also permit predictions of energy utilization of muscle,

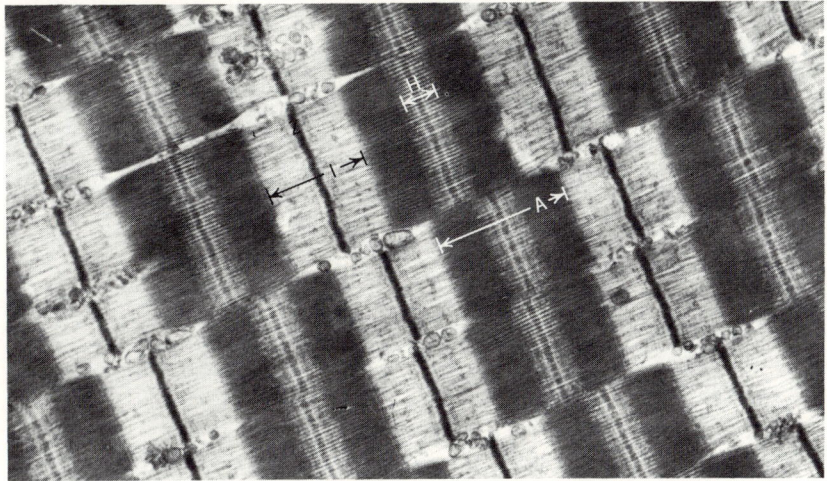

Figure 18-6
Ultrastructure of rabbit skeletal muscle. Myofibrils run diagonally across the micrograph from upper left to lower right. Narrow dark Z bands separate each sarcomere. The wide dark gray bands in the middle of each sarcomere are A bands, with a lighter H zone in the center. The wide light areas are called I bands. Magnification 20,000×.

which agree well with calorimetric investigations. Thus the sliding filament mechanism seems to explain essentially all the known observations.

Physiology of a Muscle Twitch

Let us follow the events of a single muscle contraction. As it leads to a skeletal muscle, a motor-nerve fiber divides into numerous small branches. Each branch terminates in a specifically organized flattened muscle-fiber structure, called a **motor end plate**, located at the membrane surface of the muscle fiber. The junction between the motor neuron terminal and the motor end plate is called a **neuromuscular junction** (Fig. 18-8, p. 509) and resembles in several aspects the synapse between neurons. The arrival of the motor neuron impulse at the motor end plate causes the muscle fiber to undergo a single contraction, or **twitch** (Fig. 18-9, p. 509).

A resting muscle fiber, like a resting nerve cell, is electrically polarized at its membrane surface. The inner surface is negatively charged compared to the outer surface, to the extent that a potential difference of about 80 millivolts (mV) exists across the membrane. The arrival of the nerve impulse at the motor end plate produces acetylcholine, initiating a wave of depolarization along the muscle fiber similar to that described for nerves (Chapter 17). If the stimulus is great enough (i.e., if a sufficient number of transmitter vesicles cross the synapse), a generator potential is produced that reaches the fiber's threshold, and an action potential sweeps down the length of the fiber. The response is all-or-none; the fiber either fires or it does not. A stimulus greater than threshold produces no greater contraction than a barely threshold stimulus in each fiber.

No visible change takes place in the fiber during the phase in which the action potential sweeps down it. This period, from 3 to 10 milliseconds (msec) is therefore termed the **latent** phase. During this time the action potential penetrates deep into the interior of the fiber via inpocketings of the cell membrane called the **T-tubular** system. The tubes of the T system terminate at the Z lines of the myofibrils; that is, at the boundaries of each sarcomere.

The sarcoplasmic reticulum is well developed in these regions, and has specialized chambers, or **cisterns,** in which calcium is stored. The arrival of the action potential triggers the release of some of this calcium from the sarcoplasmic reticulum. These Ca^{++} ions, in combination with the ATP present in the cytoplasm, trigger the cross-bridge ratcheting

Figure 18-7 (a) The arrangement of thick and thin myofilaments in a myofibril in the relaxed state (above) when the I band and H zone are wide; and contracted (below), when the I band and H zone are both narrowed by the sliding together of the filaments. Note the increased number of cross-bridges in the contracted state. (b) The ratchetlike action of a cross-bridge moves neighboring filaments lengthwise relative to one another. (c and d) Electron micrographs at very high magnification (400,000×) of actin and myosin myofilaments, showing the cross-bridges in different "ratchet" positions.

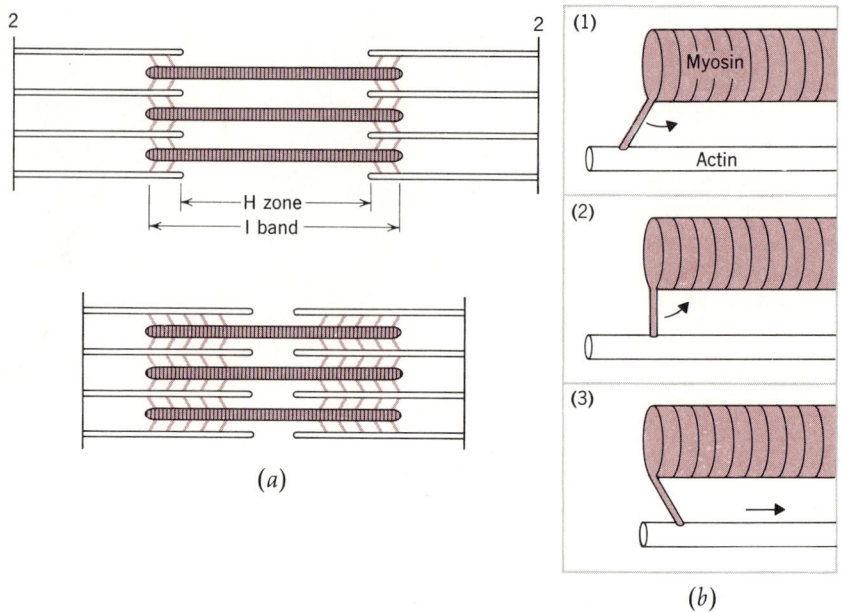

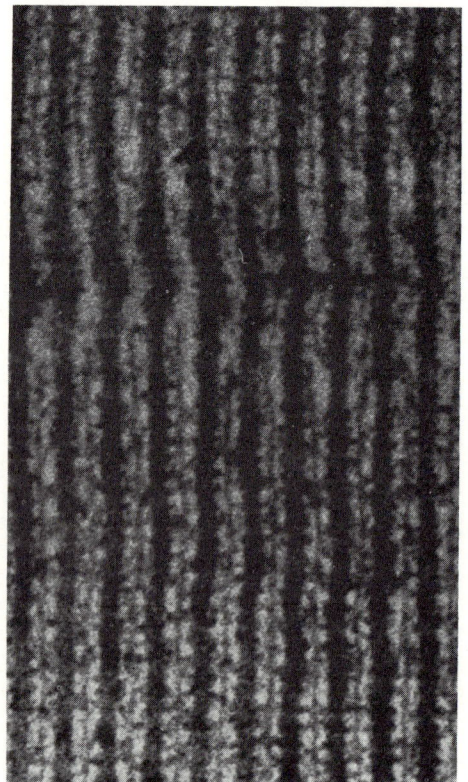

Figure 18-8 Conduction from a nerve to a muscle takes place across a special intercellular space called a **neuromuscular junction,** *or* **motor end plate.** *Transmitter molecules stored in synaptic vesicles in the motor nerve endings are liberated into the synaptic cleft by a nerve impulse. They diffuse across the gap and interact with receptor molecules in the muscle membrane, resulting in a generator potential.*

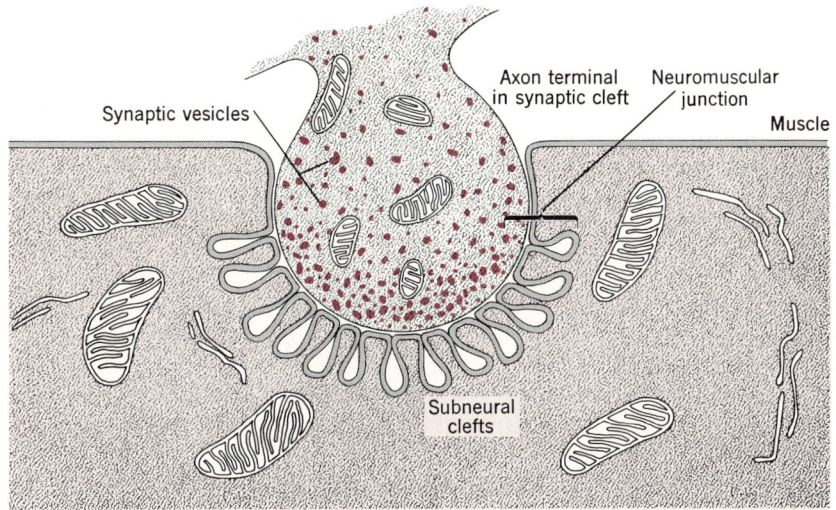

Synaptic vesicles

Axon terminal in synaptic cleft

Neuromuscular junction

Muscle

Subneural clefts

Figure 18-9 (below) (a) A single muscle twitch can be studied by recording the contraction on a moving drum called a **kymograph.** *The upper end of a frog leg muscle is held firmly in a clamp and the lower end is connected to a sensitive lever, tipped with a penpoint in contact with a rotating drum covered with graph paper. (b) A single stimulus results in a curve that shows the duration of each of the three phases—***latent, contraction,*** and* **relaxation** *phase—of the twitch, as well as the strength of contraction (indicated by the height of the contraction phase).*

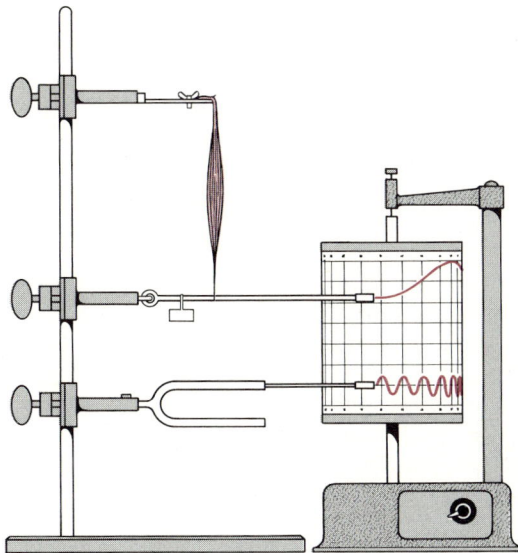

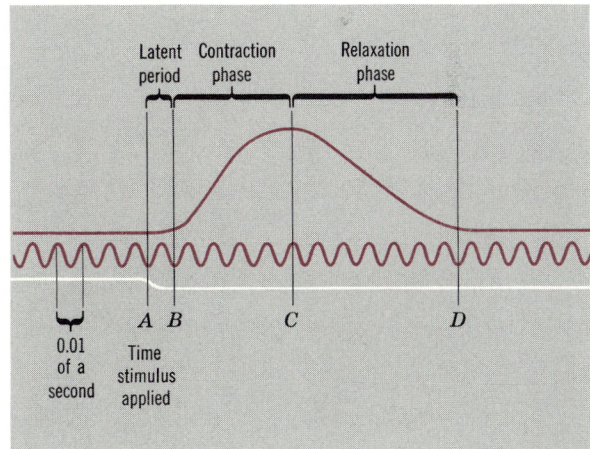

Latent period

Contraction phase

Relaxation phase

0.01 of a second

A B

Time stimulus applied

C

D

(a)

(b)

mechanism and the fibers contract. This **contraction phase** (Fig. 18-9*b*), during which the thick and thin filaments slide together, requires about 50 msec.

Immediately after contraction, the released Ca^{++} is soaked up by the sarcoplasmic reticulum membranes again, and the muscle relaxes. The **relaxation phase** also takes 50 to 60 msec. If a fiber is stimulated again before it relaxes, however, it holds its contraction or contracts even further. That is, the second stimulus adds onto the first in a process called **summation.** This means that a fiber can be maintained in a contracted state as long as it continues to be stimulated every 10 or 20 msec. Such a sustained contraction is called a **tetanus.** As we normally use our muscles to produce smooth, sustained movements of a limb or digit, the individual fibers are usually held in a degree of tetanus by continued, summated, nervous stimulation. We rarely employ single muscle twitches for purposes other than an eye-blink.

Unfortunately, the term tetanus is also used to refer to a bacterial infection commonly called **lockjaw.** The disease is characterized by a complete spasmic tetanus, or locking, of the jaw muscles, making it impossible for a patient to relax these muscles at will. It is the result of a powerful nerve poison, or toxin, produced by the infecting bacteria.

Biochemistry of a Muscle Twitch

We have seen that ATP is the immediate source of energy for muscle contraction. When actomyosin threads were made to contract in Szent-Gyorgyi's experiments, the contraction was accompanied by the hydrolysis of ATP to yield ADP and inorganic phosphate (and energy). This has also been demonstrated with intact myofibrils. It has been established that myosin itself acts as the enzyme that catalyzes the breakdown of ATP to liberate its stored energy.

ATP as the immediate energy source for muscle contraction is derived from the metabolism of muscle glycogen, phosphocrea-tine, and lactic acid. Because the quantity of ATP present in skeletal muscle at any one time, although adequate to meet the requirements of resting muscle, is not sufficient to supply the needs of a contracting muscle for more than a brief period, additional ATP sources must be available. One is the high-energy-containing compound **phosphocreatine,** which is present in muscle in approximately fivefold greater amounts than ATP. Although it cannot serve as an immediate source of energy for contraction, phosphocreatine is able to transfer its high-energy phosphate to ADP (in the presence of the enzyme **creatine kinase**) to form ATP:

$$\text{Phosphocreatine} + \text{ADP} \xrightarrow{\text{Creatine kinase}} \text{creatine} + \text{ATP}$$

Thus phosphocreatine is an energy reservoir that contributes ATP for muscle contraction.

Additional ATP for actively contracting muscle is also generated by glycolysis, which is responsible for the stepwise breakdown of glycogen to lactic acid for a net yield of three ATP molecules per molecule of glucose consumed (Chapter 4). The lactic acid produced is carried by the blood from the muscle to the liver, where most of it is eventually resynthesized back to glycogen. This synthesis occurs at the expense of liver ATP, which is necessary to push lactic acid up the energy scale to glycogen (the ATP is provided by the aerobic oxidation of about one-sixth of the lactic acid to carbon dioxide and water via aerobic respiration, i.e., the Krebs cycle and terminal respiratory chain, Chapter 4). Liver glycogen is ultimately released as glucose to the bloodstream and is resynthesized to glycogen in the muscle, to be used in due course for muscle contraction. Finally, some ATP for muscular contraction is also undoubtedly provided by aerobic oxidation in the muscle tissue itself of a portion of the lactic acid to carbon dioxide and water.

In summary, the immediate energy source for muscle contraction is ATP, which is present only in small quantities at any given time in the muscle. ATP must therefore be generated to account for contraction, and is formed in muscle by at least three important processes: (1) enzymatic transfer of high-energy phosphate from phosphocreatine to ADP, (2) glycolysis; and (3) aerobic respiration.

During strenuous physical exercise oxygen is not delivered fast enough to the muscles to metabolize glycogen completely to carbon dioxide, water, and ATP. Large quantities of lactic acid accumulate and spill over into the bloodstream. Nevertheless, the muscles are still able to sustain maximal activity under these essentially anaerobic conditions via the glycolytic pathway. Under such circumstances the muscle builds up a so-called **oxygen debt,** which is paid back by heavy, prolonged breathing when the strenuous exercise ceases. The oxygen thus supplied results in the production by the liver of sufficient ATP to transform the remaining lactic acid to glycogen (and eventually to restore the normal phosphocreatine concentration in muscle).

In effect, some of the energy for muscular work is derived through oxidation performed in the liver. For example, a sprinter runs the 100-yard dash in 10 seconds. Approximately 6 L of oxygen should be necessary to produce enough ATP by complete oxidation of glycogen to carbon dioxide and water for the energy used during the dash. Actually, however, the sprinter only consumes about 1 L of oxygen during the 10 seconds of the race, but continues to breathe heavily for some time after he has stopped running. Thus he obtains the additional 5 L of oxygen that represent the oxygen debt necessary to convert the remaining lactic acid to glycogen and to restore the normal phosphocreatine level of his muscles.

The Skeleton

In order for muscular contraction to be effective in causing movement, muscles must exert force or provide leverage against a movable framework.

The term **skeleton** usually refers to the hard, supportive connective tissues around or within which an organism is built. The skeleton includes all the bones of the body, the joints formed by the attachment of bones to one another, and associated cartilage and connective tissue, or **ligaments,** that connect bone to bone. Not only does the skeleton support and protect the softer and more delicate organs of the body, it also serves as the mechanical framework for locomotion. In addition, it functions in other important roles, such as the all-important body reservoir of calcium and phosphorus and as the site of formation of red blood cells and certain white blood cells by particular tissues of the bone marrow (Chapter 15).

The mechanical principles involved in skeletal movement of all vertebrates are the same; and in fact are similar to those employed by the arthropod invertebrates that have chitinous exoskeletons (Fig. 18-2). Therefore the skeleton of man can serve as a representative example as well as that of any other form.

Human Skeleton

For convenience, the 206 bones of the human skeleton (Fig. 18-10, p. 512) may be subdivided into an **axial skeleton,** representing the main axes of the body; and the **appendicular skeleton,** consisting of the bones in the shoulders, hips, arms, and legs. The individual bones of the body are usually classified in four general types, according to shape: **long** bones, as in the arms and legs; **short** bones, as in the wrists and ankles; **flat** bones, as in the skull and rib basket; and **irregular** bones, as in the vertebral column.

The Axial Skeleton

The axial skeleton in man has some 80 bones and consists of the skull, vertebral column, ribs, and sternum. The **skull** itself, which is formed from 28 irregularly shaped bones, including those of the middle ear, consists of the **brain case,** or **cranium,** and the bones of the face. All the skull bones, with the exception of the lower jawbone, are so united with one another as to be virtually immovable. Several bones of the cranium are not fused

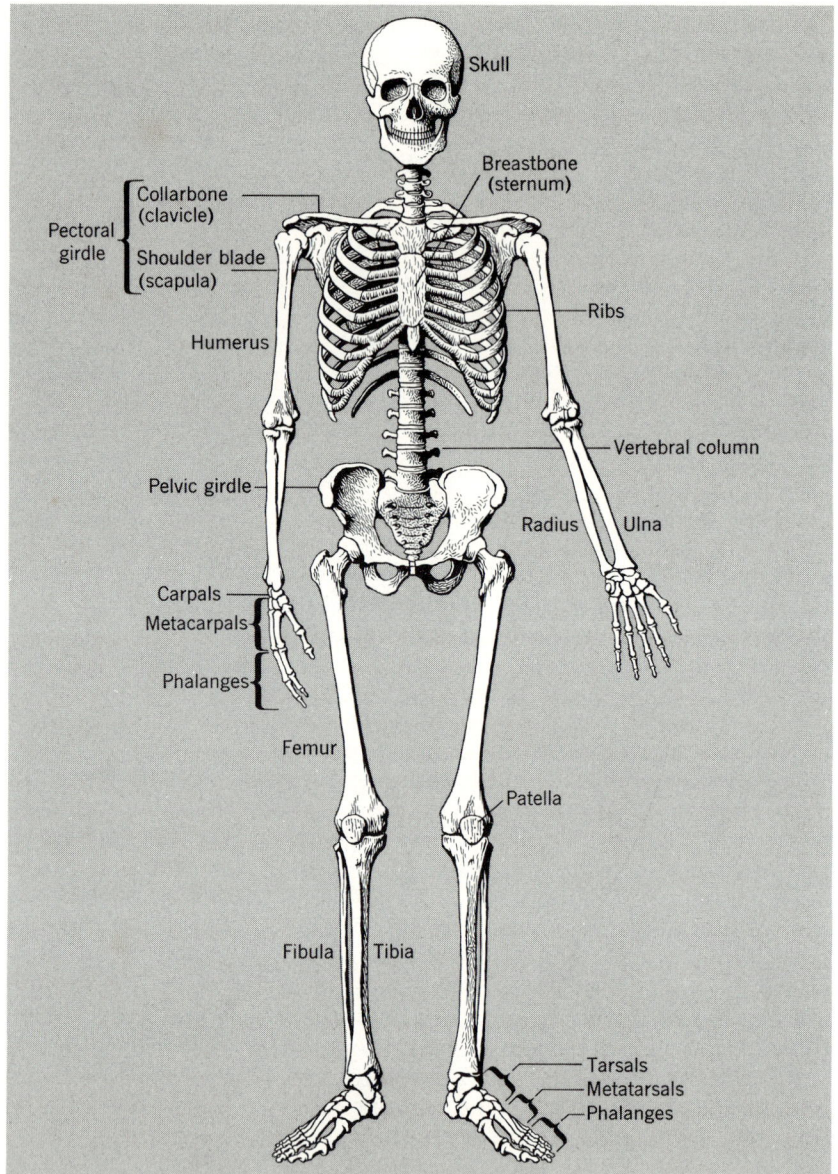

Figure 18-10   The human skeleton.

by the time of birth, allowing for greater plasticity of the head and therefore greater ease of passage through the narrow birth canal. Fusion of these bones is almost entirely completed within the first two years of life.

The **vertebral column,** also called the **spinal column** or **backbone,** is the longitudinal axis of the body on which the head is balanced. It consists of 33 irregular bones, the **vertebrae,** so connected **(articulated)** to one another as to allow forward, backward, and sideways movement. Some regions of the column are more flexible than others. In man the vertebral column is curved, a characteristic that is associated with the greater carrying strength and balance of an upright animal.

Although the vertebrae differ in size and shape at different sites along the backbone, they share certain common structural features. The main portion of a typical vertebra is called the **centrum,** from whose back **(dorsal)** surface arises an arch of bone called the **neural arch.** This surrounds and protects the spinal cord. Bony projections from the neural arch lock each vertebra to the one above and the one below. In addition, the vertebrae are securely tied together by numerous ligaments. A small pad of tough cartilage called a **disc** lies between each of the vertebrae. The spinal nerves emerge from pairs of small openings between the vertebrae.

The upper seven **cervical** vertebrae make up the skeletal framework of the neck. These are followed by 12 **thoracic** vertebrae, five **lumbar** vertebrae, and below these the **sacrum** and **coccyx.** In man the sacrum consists of five **sacral vertebrae** fused into a single solid bone. It is joined to one of the bones of the pelvic girdle, which in turn attaches the legs to the spine. The coccyx in man is also a single bone that has resulted from the fusion of four tiny vertebrae. In many vertebrates it gives support to the tail, but in man the coccyx is now a useless vestigial structure.

The bony **chest cage** consists of the 12 thoracic vertebrae and 12 pairs of **ribs,** which are joined at the back to the thoracic vertebrae and in front to the **breastbone,** or **sternum.** Ten of the 12 rib pairs are directly fused or indirectly attached by means of cartilage to the breastbone. The eleventh and twelfth pairs of ribs are attached only to the backbone, and for this reason are called **floating ribs.**

The Appendicular Skeleton

The appendicular skeleton consists of 126 bones, organized into the **appendages** (arms and legs) and the pectoral and pelvic girdles that attach these appendages to the axial skeleton. Each half of the **pectoral girdle** is made up of a **collarbone (clavicle)** and a **shoulder blade (scapula).** Each clavicle forms a joint to the breastbone at one end; at its other end it articulates with a scapula. The latter in turn connects to the ribs, not by a joint, but by muscles and ligaments, an arrangement that accounts for the flexibility and freedom of movement of the shoulders and arms.

A socket formed by the union of clavicle and scapula is the site to which the arm is attached. The bone of the upper arm is called the **humerus,** and is jointed to the shoulder socket by means of ligaments. The two bones of the forearm are the **radius** and the narrower and smaller **ulna.** The eight small bones of the wrist are called **carpals;** the hand consists of five slender **metacarpals** of the palm and 14 finger bones, or **phalanges**—three phalanges for each finger, except for the thumb, which has only two.

The **pelvic girdle** consists of two **hip bones** firmly bound to the sacrum to form a stable foundation that supports the trunk and attaches the legs to it. Each hip bone in reality has three separate bones that have fused together. The plevic girdle of women is broader and more flaring, with a wider central opening in contrast to the narrower corresponding structure in males; this breadth makes childbearing easier.

The bone of the upper leg, or thigh, is called the **femur;** it is the longest and heaviest bone in the body. Its end nearest to the torso fits into a deep socket in one of the hip bones. The lower leg, or shank, has two bones, the **tibia,** or **shin bone,** and the narrower and smaller **fibula.** The tibia is articulated to the femur by a hingelike attachment, the **knee joint.** This is protected by a small, separate bone called the **kneecap,** or **patella,** which has no counterpart in the arm.

The structure of the foot is somewhat similar to that of the hand, but with certain modifications that adapt it for supporting weight. The arch of the foot is constructed of some of the seven ankle bones, or **tarsals,** and the five **metatarsals,** which correspond to the metacarpals of the palm. Each foot, like each hand, has 14 phalanges, three for each toe, except for the big toe, which has only two (Fig. 18-11, p. 514).

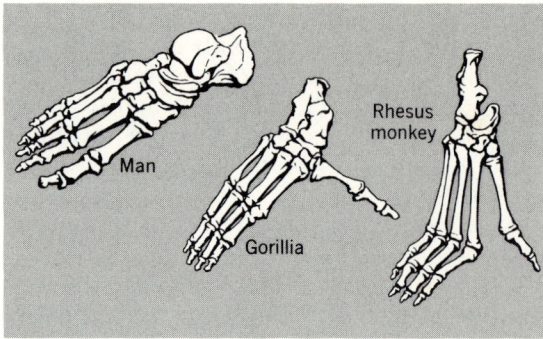

Figure 18-11   *The skeletal elements of the foot of three primates.*

The Joints

Bones are joined to one another in a number of ingenious ways. The connection, or articulation, of one bone to another is called a **joint.** Although movement of bone depends on the activity of the attached skeletal muscle, the type or the degree of freedom of bone movement will be largely determined by the nature of the joint.

In general, joints may be classified in three principal types, depending on the degree of movement they permit: (1) those that allow virtually no movement, as in the bones of the cranium, which are dovetailed, or **sutured,** to one another by small bony projections inter-locking like the teeth of a zipper; (2) those that permit only very slight movements between the bones, as in many of the vertebrae of the spine; and (3) those that permit, in varying degrees, free movement between the bones. The last include several types such as the shoulders and hip joints, which enjoy the widest range of movements, and the elbow and knee joints, which have movement in one plane (Fig. 18-12).

Different structural devices at different joints determine the degree and kind of movement. At one extreme, in joints where free movement prevails, a thin layer of cartilage covers the articulating ends of each bone, and the joint is encased in a fibrous capsule that forms a lubricant-filled cavity. Several strong ligaments (made up of tough connective tissue) attach the bones of such joints.

Small, fluid-filled sacs called **bursae** are often found wherever pressure is exerted between moving parts of the skeletal system (Fig. 18-12). They serve as cushions to relieve this pressure. When they become inflamed, the condition is known as **bursitis.**

At the other extreme, in joints where little or no movement occurs, the space between the two articulating ends of the bones may be filled either with cartilage or dense connective tissue, as in the cranial bones before their ultimate fusion.

Figure 18-12   *The knee and shoulder joints of man.*

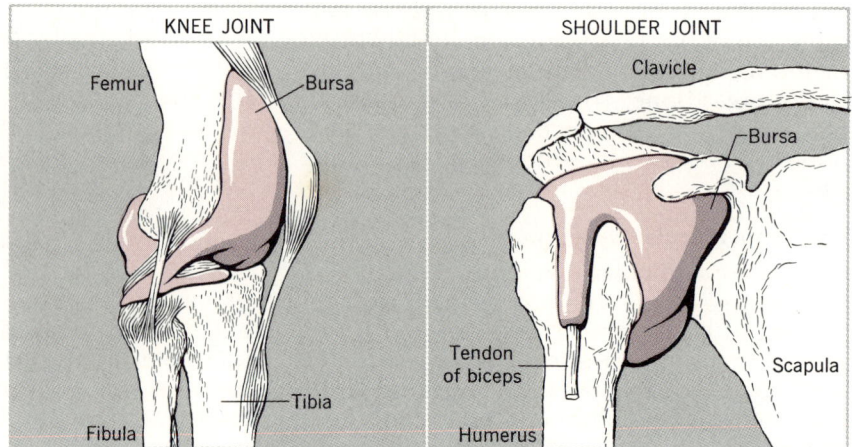

The skeleton of an embryo when first formed is made up principally of cartilage shaped like the bones that will later form (see Fig. 18-14, p. 516). The cartilage is slowly destroyed and its place taken by bone, a process known as **ossification.** The adult skeleton is predominantly bony.

The typical gross anatomy and growth of a bone can be illustrated by a long bone such as the thigh bone, or femur (Fig. 18-13). The main **shaft** of a long bone consists of a hard, outer, cylindrical **bony matrix** enclosing a central cavity occupied by **marrow.** Each end of the bone is called an **epiphysis.** In general an epiphysis has a more spongy type of bone matrix than the rest of the bone, and its matrix is interspersed with the marrow tissue responsible for blood cell formation. Increase in bone length occurs by a continual formation of new cartilage at the epiphysial ends, followed by subsequent ossification. Growth in bone diameter takes place as a result of two general processes: (1) destruction of bone cells in the interior of the shaft, thus increasing the size of the marrow cavity; and (2) the simultaneous deposition of new bone tissue around the outside of the bone.

Living bone is not a static substance, but is in a constant state of flux; old bone matrix is destroyed and new material is formed. The steady breakdown of bone is brought about by special bone-destroying cells under hormonal control—specifically, of the parathyroid hormone (Chapter 16). Their activity is directly correlated with the requirements of the body: a deficiency of calcium in other tissues of the body causes an increased activity of the bone-destroying cells and a subsequent release of calcium ions into the blood; conversely, an increase in the calcium level of the body leads to increased formation of bone matrix. Thus the skeleton serves as a reservoir for calcium, an essential nutrient element for a variety of functions, including blood clotting and general cell metabolism. In addition, bone is a storage site for phosphate, whose all-important role in metabo-

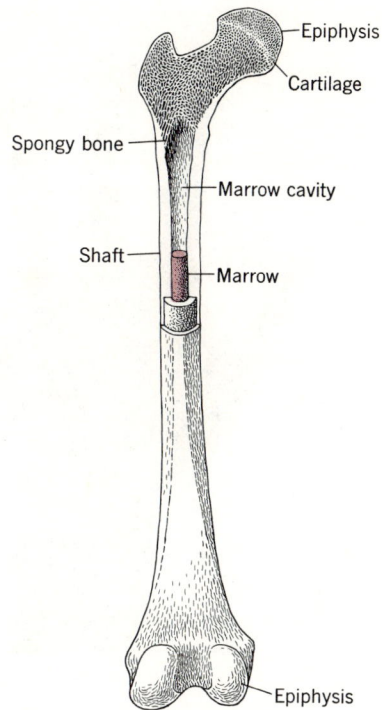

Figure 18-13   The anatomy of a long bone.

lism and energy transfer is described in Chapter 4.

Radioactive strontium[90] is a product of atomic fallout in our present age and has a strong affinity for bone tissue. When it enters the body, usually by way of the digestive and respiratory tracts, strontium[90] is concentrated in the bony matrix, much as calcium is. Because of its radioactivity it has a serious effect on the blood-forming tissue in the marrow as well as other parts of the body, producing damage and inducing cancerous malignancies.

Conclusion

We take it for granted that we see with our eyes, hear with our ears, and move with the muscles of our arms and legs. In a real sense, however, this is not the case. Seeing, hearing, and moving are all functions of the brain and nervous system. Moreover, the brain doesn't

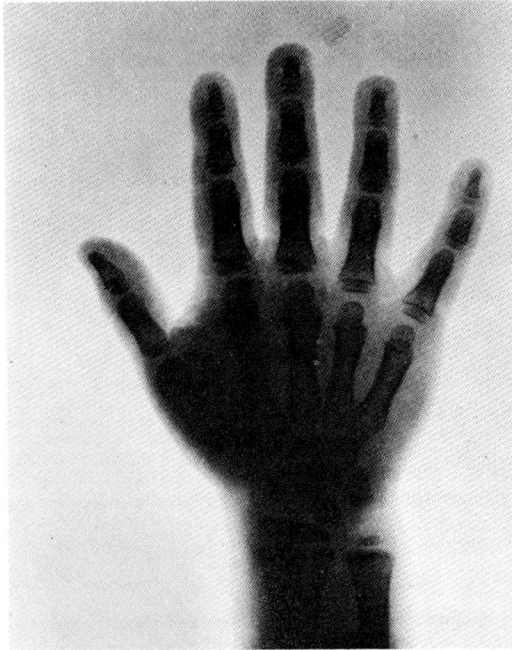
(a)

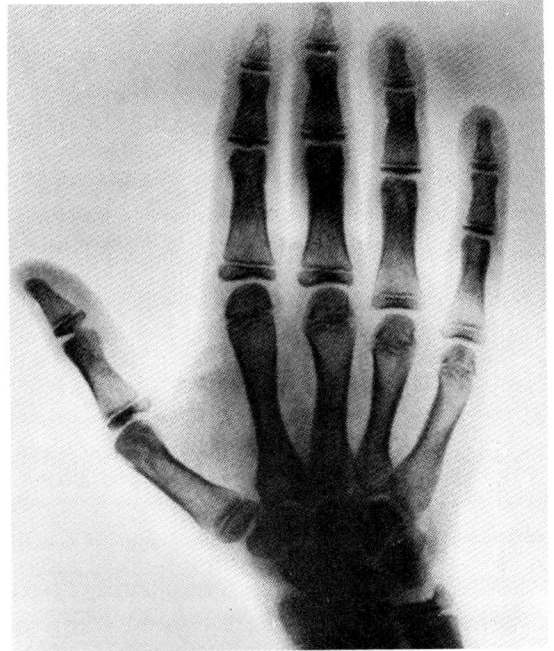
(b)

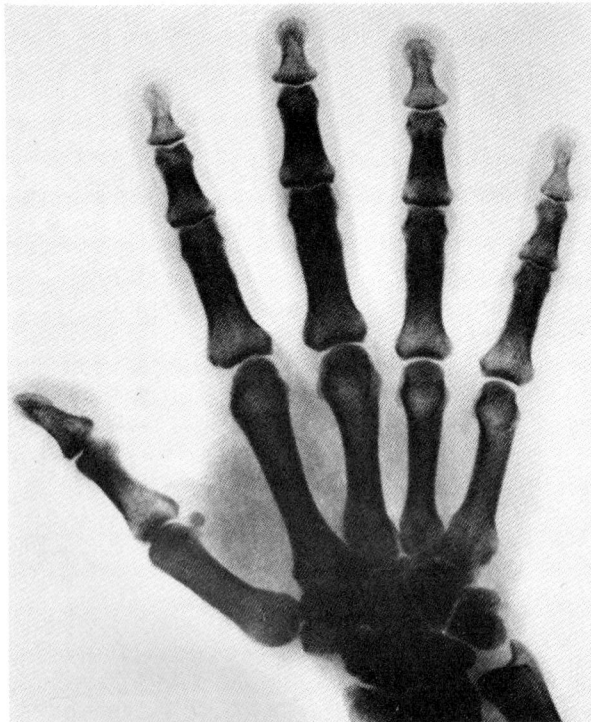
(c)

Figure 18-14 X-ray films of the hands of (a) a child, (b) an adolescent, and (c) an adult. Note the large spaces and cartilaginous pads between each bone in the child's hand. These are gradually filled in and ossified by adulthood.

receive a visual image or a sound, but only patterns of nerve impulses. Similarly, the brain itself does not move or contract; it only initiates more of the same nerve impulses, in different patterns.

It is interesting to note that a muscle contraction and a nerve impulse begin with the same primary event, the depolarization of the cell membrane, so that both perception and action are based on a common phenomenon: the action potential.

What organisms do, and how they do it, are dependent on the interactions among their receptors, their nervous systems, and their effectors. The sum total of an organism's activities, its movements, responses, emotions, mental processes, we refer to as its **behavior,** to which we direct ourselves next.

Reading List

Bendall, J. R., *Muscles, Molecules and Movement.* Heinemann, Ltd., London, 1969.

Gans, C., "How Snakes Move," *Scientific American* (June 1970), pp. 82–90.

Harrison, R. J., and W. Montagna, *Man.* Appleton-Century-Crofts, New York, 1969.

Hoyle, G., "How Is Muscle Turned On and Off?" *Scientific American* (June 1970), pp. 84–93.

Huxley, H. E., "The Mechanism of Muscular Contraction," *Science* (1969), **164,** pp. 1356–1366.

Katz, B., *Nerve, Muscle and Synapse.* McGraw-Hill, New York, 1966.

Lehninger, A. L., *Biochemistry.* Worth, New York, 1970.

Merton, P. A., "How We Control the Contraction of Our Muscles," *Scientific American* (May 1972), pp. 30–37.

Napier, J. R., "The Antiquity of Human Walking," *Scientific American* (April 1967), pp. 56–66.

Ross, R., and P. Bornstein, "Elastic Fibers in the Body," *Scientific American* (June 1971), pp. 44–59.

Smith, D. S., "The Flight Muscles of Insects," *Scientific American* (June 1965), pp. 76–90.

Tucker, V. A., "The Energetics of Bird Flight," *Scientific American* (May 1969), pp. 70–80.

Wilkie, D. R., *Muscle.* Edward Arnold, London, 1968.

**part
eight** *The Interactions of Organisms*

chapter nineteen

The Biology of Behavior

"The functioning of the brain plainly depends on a system of physical organization of its billions of nerve cells—a wiring plan, so to speak. The cells in the brain's various functional centers are linked together by an elaborate circuitry of pathways and interconnections forming a network of communication. In order to arrive at a detailed understanding of how the brain works we need a clear knowledge of this wiring diagram." LENNART HEIMER, 1971, *PATHWAYS IN THE BRAIN*

Our description of sensors, conductors, and effectors in Chapters 17 and 18 provides a background for a consideration of behavior. In order to "behave," an organism must be able to respond to signals, internal or external. We shall see that an animal's behavior is directly related to the complexity of the structures it has developed for handling signals—that is, its **information-handling capacity.** Among multicellular animals, this capacity resides largely in their nervous system.

In each of the preceding sections of this book we have tried to describe and explain biological concepts, using man as an example whenever possible. We feel that it makes perfectly good biological sense—in most cases—to do this. There is an added incentive in learning about ourselves; even a nonbiologist is interested in how his own body works, and why it sometimes fails to work. And a human nerve conducts impulses in a fashion identical in principle with nerves of a frog or a mollusk. The biochemistry of the synthesis of DNA or the breakdown of ATP is essentially the same in man as in all other species, animal or plant. To paraphrase Gertrude Stein's notorious line, a muscle is a muscle is a muscle. When we come to problems of animal behavior, however, we find a somewhat different situation. Behavior is so complicated in most organisms, which is to say that we know so little about how it is regulated, that it is essentially impossible to deal with human behavior *except* by comparison with the activities of other, "simpler" forms.

Honeybee on honeysuckle.

521

The nervous system of a man has between 10 billion and 100 billion cells; a rat's brain may contain 2 to 3 billion. There are perhaps 3 million sensory neurons entering the human central nervous system. Each of these sensory axons fires in response to one type of stimulus at one point in or on the body. There are 10 to 50 billion **interneurons** (i.e., connectors); these connect with one another and with the motor neurons, of which there are about 150,000. The motor neurons carry all of the output from the nervous system. Even specialized parts of the vertebrate nervous system have an enormous number of cells. The retina of the eye has about 130 million receptor cells and sends more than a million nerve fibers to the brain. A single segment of spinal cord, which may control only a few muscles, operates through several thousand motor neurons, which receive instructions from perhaps 10 times that many sensory elements.

Such vast populations of cells present an impossible situation to a biologist wishing to understand how the nervous system works in terms of its cellular activities, or one trying to trace neural connections in order to construct a "wiring diagram." Many biologists have therefore begun to study intensively the behavior of invertebrate animals that have few nerve cells but exhibit complex forms of activity. The claw-bearing limb of a crab, for example, shows impressive coordination and a wide range of movements, made possible by six movable points and a pair of antagonistic muscles operating each joint. Yet the crab controls this limb with only about two dozen motor nerve cells. The assumption, of course, is that an understanding of the circuitry that underlies behavior in such a simple system can lead to useful conclusions about the organization of much more complicated ones. So far that assumption seems to be justified.

There are two main categories of behavior: instinctive and learned. **Instinctive** or **innate,** behavior consists of relatively complex patterns of activity that are inherited and are relatively unaffected by experience. **Learned** behavior, on the other hand, to a much greater extent can be modified by experience. Much current research at the cellular, or neurophysiological, level asks the questions "Are innate patterns of behavior based on predetermined configurations of neural circuitry that are constant among all members of a species? If so, how does learning affect behavior? Do neural connections exhibit plasticity; that is, are they modifiable by the patterns of impulses they conduct?"

In addition to biologists who study behavior at the cellular level, another group of behavioral scientists—most with psychological backgrounds—is concerned with how animals learn, how much of their behavior is innate, and how stereotyped patterns of activity develop and are inherited. These workers generally train laboratory animals, such as rats, pigeons, and monkeys, to carry out a variety of tasks. They then attempt to define some of the principles of behavior, especially of learning, using these animals in a laboratory situation where a wide variety of variables can be controlled. One of the modern leaders of the **behaviorists** is B. F. Skinner, an experimental psychologist at Harvard University.

A different approach to behavior is taken by **ethologists** like Konrad Lorenz or Niko Tinbergen, who are concerned not so much with modifications of behavior produced by learning as with analysis of patterns of instinctive behavior among related species, to see how stereotyped behavior patterns evolve. **Ethology** is the science of behavioral characteristics of species.

We mentioned in Chapter 1 that science generally progresses through three conceptual levels. (1) At the **descriptive** level scientists ask, merely, "What does the system (organism, cell, molecule) do? What does it look like? How is it built? What is its range of possibilities?" (2) At the **experimental** level scientists attempt to test hypotheses, to establish consistent rules or principles that are obeyed by the system. They attempt to establish from observations and experiments general laws that have predictive value. And (3)

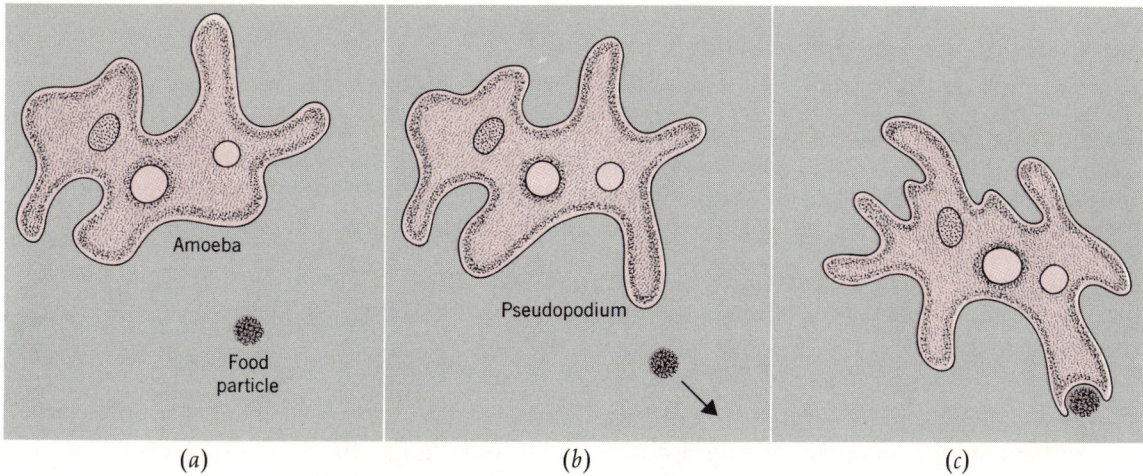

Figure 19-1 (a) An ameba can sense a food particle at some distance. (b) Its initial response is to send out a pseudopod of the appropriate shape to engulf its intended prey. (c) As the food particle moves away, the ameba pursues it, and finally surrounds it.

at the **analytical** level they attempt to understand the system and its functions in terms of simpler, more fundamental systems; that is, in terms of mechanism.

When Mendel spent a decade observing the inheritance of traits in garden peas and doing experimental crosses to establish the principles of heredity, he was functioning at levels (1) and (2). Modern molecular genetics, which explains the transfer of hereditary traits in terms of gene function and the sequence of nucleotides in the DNA molecule, exemplifies the third level. It is worth noting that all three levels of analysis are useful and necessary, and that one level never replaces another. Some understanding at the first level is prerequisite for analysis at the second or third, but, in practice, a scientist generally functions at two or all three conceptual levels at the same time.

Behavior can be treated in the same ways. Some investigators are interested in making generalizations from the observed behavior of organisms in a variety of circumstances, and thereby discovering the laws that govern their behavior. Others are interested in analyzing the ways in which behavior is related to the functions of the nervous system, or how behavioral characteristics develop, or how they are inherited. Needless to say, these studies tend to be complementary and to deepen mutual insights.

Behavior among Protists

At its most basic level, behavior includes four fundamental components: **irritability, habituation, avoidance,** and **learning.** These qualities are shared by all organisms, and in fact are among the criteria for distinguishing living from nonliving objects (Chapter 1).

Irritability is the ability to respond to a stimulus. If a morsel of food is placed in a clean dish of water near an ameba (Fig. 19-1), the protozoan senses the presence of the food long before it comes in contact with it— at least its own body length (perhaps 200μ)

away. Its first visible response is to send out an appropriately shaped pseudopod in the direction of its prey: a small, nonmotile bacterial cell will be approached by a narrow **filopod;** a larger or more vigorously moving object will elicit a larger pseudopod. The stimulus an ameba senses may be chemical, like diffusible substances emanating from a food particle; or mechanical, like currents in the water; or both. If the prey moves away, the ameba will follow as long as it remains close enough to continue receiving stimuli from it. The cell is clearly responsive to the food.

Amebas also respond to light. If a micro-spot of light shines on an advancing pseudo-pod, the pseudopod is withdrawn. This is a typical **avoidance,** or **withdrawal,** response. If the entire cell is flooded with bright light, it contracts suddenly into a round sphere. If the light is left on at the same intensity for a few minutes, however, the ameba soon sends out a pseudopod—at first tentatively, as if to test its surroundings—and then gradually resumes its normal activities. This ability to ignore a prolonged stimulus and resume a previous behavior pattern in the face of it is known as **habituation.**

Primitive forms of **learning** can also be demonstrated in the protists. If we dip a clean platinum wire into the center of a culture of *Paramecium,* we will find that the ciliates avoid the wire; they swim away. If the wire is smeared lightly with bacteria before it is lowered into the water, the paramecia approach and cluster around the wire to feed on the bacteria.

The wire then may be dipped in and out of the water repeatedly, in a series of 15 or 20 "training trials." If, after a waiting period of up to several hours, the thoroughly cleaned wire is again dipped into the culture, the paramecia will approach and cling to it. In general, the more training trials that are given, the greater the number of paramecia that cling to the bare wire. The paramecia have "learned" that the wire is a source of food. Learning, then, may be defined as the process by which behavior is altered by past experience.

Reflexes and Pacemakers

The behavioral repertoire of protozoa, although surprisingly varied, can generally be understood in terms of the four components described above. The activities of even the simplest multicellular forms, however, are much more complex. A coelenterate such as a sea anemone, for example, seems at first glance to exhibit only the simplest behavior. Sea anemones live a **sessile** life—that is, they are firmly attached to rocks—and they wave their tentacles lazily to feed in the tidal currents (Fig. 19-2). Their nervous system consists only of a diffuse nerve net; they have no aggregations of nerve cells into ganglia or a brain. To a casual observer, the behavior of an anemone seems, at best, passive; the animal appears to react only when there is an obvious external stimulus. If the mouth region near the tentacles is prodded with a glass rod, the body withdraws from the stimulus. Such a single mechanical stimulus evokes a response because the sensory receptors fire a burst of impulses after even the briefest stimulus. If the animal is poked sufficiently hard or prodded several times, the tentacles are pulled in and the animal rounds itself into a compact ball. The same response can be elicited with a mild electric shock.

With carefully regulated electrical stimuli, however, an interesting new set of behavioral properties can be demonstrated. An animal may be stimulated with a shock of such low intensity that it does not respond. If repeated single shocks are given, each at a slightly greater intensity than the last, eventually an intensity will be reached that just elicits the withdrawal response. This is called the **threshold level** for that response. For a given animal under constant conditions, the threshold for withdrawal remains much the same; say 5 volts. At any time, a single shock at an intensity below 5 V, a so-called **subthreshold stimulus,** fails to produce a response, whereas 5 V or more reliably elicits withdrawal. But a series of stimuli, say, 10 per second—even at a subthreshold intensity—

Figure 19-2   A sea anemone, (a) feeding and (b) contracted in an extreme withdrawal reaction.

produces a response. This form of **summation** is exactly analogous to what happens when a mammalian muscle is repetitively fired by a motor nerve (p. 472). In behavioral terms it is called **facilitation.**

The behavior exhibited by some anemones, controlled only by a nerve net, is almost unbelievably complex. There is an anemone from European waters that normally lives on the shell of a large species of marine snail called a whelk. Occasional anemones will attach to the shell of a different species, or to a rock, but if a whelk shell becomes available, the anemone always moves to it. There are four steps in the transfer of an anemone to a whelk shell.

1. Some of the anemone's tentacles come into contact with the shell. They immediately attach. The stalk of the animal then extends and the unattached tentacles begin to wave more actively. As each tentacle touches the shell, it attaches. Soon the entire ring of tentacles is adhering to the shell, and the anemone is fastened by its tentacles to the whelk shell and by a **pedal disc** at its bottom end to its old attachment.
2. Peristaltic waves of contraction begin to move down the column of the animal, starting near the oral disc and running toward the foot. The column becomes longer and thinner, and the pedal disc frees itself from the mucous envelope by which it was attached to the old surface.
3. Now the column bends to swing the pedal disc over onto the whelk shell in cartwheel fashion. The pedal disc swells like a mushroom, and the tentacles shift their hold on the shell, moving over toward one edge.
4. Finally the pedal disc attaches to the new surface, the tentacles loosen their hold on the shell, and the anemone resumes a normal stance.

This behavior seems remarkably complex. Yet the nerve net of such a coelenterate contains about 100,000 neurons, is about the same number as in the central nervous system of a honeybee, whose behavior (as we shall see) is still vastly more complicated. Presumably the difference relates to the difference in neural organization—a diffuse nerve net as compared to a well-organized CNS.

The behavior we have discussed so far—the pursuit of a tasty morsel of food by an ameba or the withdrawal response of a poked anemone—can generally be understood in terms of a preprogrammed set of actions triggered in an organism by an external signal. However complex the pattern of activity, it is nonetheless passive, a reflex action, initiated by a stimulus outside the organism. For many years controversy has raged among behavioral scientists as to whether, in fact, all behavior can be described in similar terms. Are any actions of any organisms truly self-initiated (**endogenous**)? If all external and proprioceptive stimuli were blocked, would an animal "behave" at all?

This argument has been laid to rest with the recent discovery of pacemakers in the central nervous systems of a wide variety of animals. A **neural pacemaker** is a cell that produces a pattern of spontaneous nerve impulses in the complete absence of neural stimulation. The mechanism of this spontaneous activity appears to be similar to that we described in the pacemaker cells of the heart (Chapter 15). Intracellular electrical recordings from neural pacemakers have demonstrated that these cells do not have stable resting potentials. Instead, the potential across the membrane starts at about 80 millivolts (with the inside of the cell negative to the outside) and drifts fairly rapidly toward zero, until the firing threshold of the cell is reached, and the cell generates an action potential. The membrane then repolarizes back to 80 mV, and the potential immediately starts to drift down again, resulting in a rhythmic pattern of spontaneous firing.

Some behavior is clearly generated in animals by such pacemakers. For example, jellyfish swim by contracting their umbrellas rhythmically. Each contraction is initiated by pacemaker nerve impulses that arise spontaneously in any one of eight clusters of nerve cells spaced evenly around the circumference of the umbrella. The impulse started at one cluster sweeps across the umbrella, causing the muscles to contract. Jellyfish continue to swim as long as they have at least one such cluster to set the pace; the other seven can be removed without stopping the beat.

We have learned most about the properties of neural pacemakers from studies of the nervous system of a giant marine snail called *Aplysia*, the "sea hare" (Fig. 19-3). The nervous systems of these marine mollusks are particularly useful to study because they contain relatively few cells and the cells are unusually large. Some reach as much as .5 to .7 mm in diameter, and are therefore especially convenient for study with intracellular microelectrodes. The abdominal ganglion of *Aplysia* has only about 1800 cells. Eric Kandel and his colleagues at New York University School of Medicine, and Ladislaw Tauc in Paris, have identified 30 of these cells as unique individuals and mapped a number of their central connections. Twenty-four of these 30 cells exhibit spontaneous activity.

Similarly, the lobster cardiac ganglion contains nine cells. All of these cells are capable of endogenous activity in the absence of neural input, and they normally interact with one another to initiate a specifically patterned burst of impulses at regular intervals.

Figure 19-3   The sea hare, Aplysia. *These animals range from 6 to 10 in. long.*

It is these rhythmic stimuli that drive the lobster's **neurogenic** heart. (In crustacea and insects, the heartbeat is initiated in the nervous system, unlike the **myogenic** hearts of vertebrates and most other invertebrates, where the generator potential arises in specialized cardiac muscle cells) (Chapter 15).

The impulses produced by neural pacemaker cells such as those in the abdominal and cardiac ganglia are truly endogenous; they persist even when the cells are isolated from all sensory input. However, the pattern of output of each cell, or of the whole ganglion, can be **modulated** (varied or regulated) by neural input from accelerator or inhibitor fibers from the CNS and by hormonal influences.

Visual Perception and Behavior

Whether a given behavioral pattern is stimulated endogenously or is reflexive, every animal activity may be influenced by sensory input. Many aspects of behavior are in fact related directly to particular sensory capabilities.

Are Bees Color-Blind?

In 1910 a paper was published suggesting that honeybees are color-blind. Yet many zoologists believed that the evolution of color in plants had the adaptive advantage of attracting bees for pollination. If bees are color-blind, why should flowers be colored? A young German zoologist, Karl von Frisch, set up an ingenious series of experiments to test color vision in bees. He placed a card table near a beehive. On the table was a square of blue cardboard with a dish of sugar water sitting on it. Bees soon arrived at the table, some took sugar water, and carried it back to the hive. After many such visits, Von Frisch removed the cardboard dish and placed two fresh squares on the table, one red and one blue. The great majority of the bees coming to the table continued to land on the blue square.

The experiment showed definitely that bees can distinguish between red and blue cardboard. But did it show that bees have color vision? Von Frisch realized that totally color-blind men can also distinguish between two such cards. Red light is absorbed very poorly by rhodopsin and blue light is absorbed well. Even if cards are prepared that reflect exactly the same amount of light, red appears as a dark gray; blue looks much lighter. Von Frisch then demonstrated that bees could learn to select the blue card from among a whole series of gray cards, ranging from almost white to almost black. They could also be trained to come to orange, yellow, green, violet, or purple cards, but not to scarlet. They cannot distinguish between scarlet and a certain shade of dark gray; bees are "scarlet-blind." They can, however, recognize ultraviolet projected through a prism. To bees, ultraviolet is a color.

If bees cannot see scarlet, why have some plants evolved scarlet flowers? The answer is simple, and fits the hypothesis. Scarlet flowers are all polinated by birds.

Like bees, most animals can distinguish at least some colors, and this ability is often important in their behavior. Studies of the three-spined stickleback, a small freshwater fish, have shown that fighting behavior among males is elicited by the red color of their underbellies. Tinbergen, who spent many years studying the instinctive behavior of sticklebacks in the laboratory and in nature, found that male fish would rush to the sides of their aquarium tanks and assume a threatening posture every time a red truck passed in front of the window.

Pigeon Vision

Elaborate and accurate methods are now available for measuring the sensory abilities of a variety of laboratory animals. Learned behavior patterns are used as indicators of the sense being tested. For example, how might we determine how sensitive a pigeon's vision is?

The pigeon is kept in a special experimental cage. Every day the light in the cage is made dimmer, so the animal gradually adapts to living in total darkness. At one end of the cage are two keys that the pigeon can peck; they are located under a small screen on which a spot of light can be shown. The pigeon can be trained to peck a key when it sees a spot of light and to peck another key when the light goes off: one or the other of the keys can open a food hopper. For example, the apparatus can be arranged so that to get food a pigeon must see a light spot, peck key A to make the light disappear, and peck key B to open the food hopper and make the light reappear. Pigeons learn such a sequence readily.

The final step in the training sequence is to arrange the setup so that a peck at key A no longer turns the light off, but decreases it by one step in intensity; a peck at key B makes it more intense by one step and opens the food hopper. Now the pigeon sees the light and pecks key A repeatedly until, to its eye, the light goes off. The animal then immediately pecks key B to open the hopper and make the light reappear. Any time the pigeon stops pecking key A and pecks key B, even though the light is actually on at a low intensity, it tells the experimenter that the intensity of light has fallen below its visual threshold; the pigeon behaves as though the light were turned off because it no longer sees it. It then pecks key B, which increases the light intensity to the point at which it is again visible.

With this method it is possible to measure the visual threshold of a pigeon's eye in constant light or the change in threshold that occurs when the pigeon is suddenly placed in darkness. The resulting **dark-adaptation curve** is remarkably similar to those determined in human subjects simply by presenting visual stimuli and asking whether they can be seen. In similar fashion, the sensitivity of a pigeon's eye to different wavelengths of light can be determined by presenting light spots of different colors.

Visual Patterns

In many vertebrates vision is a direct stimulus to action. Certain visual patterns often trigger specific behavioral responses that may be reflexive and highly stereotyped. Many predatory fish actually cannot help striking at a moving object of appropriate size. Frogs behave much the same way. In fact, a frog, which normally eats live flies, will starve to death surrounded by perfectly edible but anesthetized ones. However, if such a motionless fly is moved across the frog's visual field on the end of a thread, the animal will immediately strike at it.

Electrical recordings can be made from single axons in the optic nerve of a frog. A tiny spot of light projected on the retina is a stimulus. Different fibers generate an action potential in response to the light spot on a specific area of the retina—the **receptive field** for that fiber.

Turn back to Figure 17-19, and recall that light entering the eye stimulates the rods and cones, which produce generator potentials in the bipolar cells to which they synapse. Nerve impulses in these cells are carried to ganglion cells in the front of the retina, and then via fibers that course through the optic nerve to the brain. At least four classes of ganglion cell have been found in frogs' eyes. Some respond only if there is a sharp boundary between light and dark in their receptive fields. Some of these fire only if that boundary is curved and the convex side of the boundary is dark; others only if the boundary is moving.

What small, moving dark spots would normally stimulate a frog's receptive fields? The answer, of course, is flies and other insects. The fibers that leave these ganglion cells have been traced to the frog's midbrain, but no further. However, it seems reasonable to speculate that signals processed in this part of the brain may be sent, more or less directly, to the efferent motor fibers that control the muscles involved in the striking reflex.

Behavior in response to a much more complex visual pattern is seen in bees. When von Frisch was studying color vision in honeybees, he noticed that after a dish of sugar water was placed on a table, many hours might pass before the first bee arrived to investigate. But once that first bee had landed and departed, many more quickly followed. Within a short time there would be tens or hundreds of bees at the table, all from the same hive. It seemed reasonable to von Frisch that the first bee acted as a scout, and somehow communicated to its mates that a new source of food had been found.

On investigation, he discovered that when a scout returns from the table to the hive, it first gives much of the sugar water to its mates, causing great excitement in the hive. It then begins a characteristic and amazing **wagging dance** (Fig. 19-4), which conveys an incredible amount of information. The dancing platform is usually the vertical surface of the honeycomb inside the hive, although occasionally the dance is performed on a horizontal landing area in front of the hive. First the scout runs for a short distance in a

Figure 19-4   The wagging dance of a honeybee scout communicates a great deal of information to her hivemates about the location of a food source. On a vertical surface the bee orients herself by gravity, a point straight overhead being the reference point for the position of the sun. The straight, wagging run of the dance takes place at the same angle to the vertical (X) as the angle between the food source and the sun. The distance of the food source from the hive by the frequency with which she repeats the dance pattern.

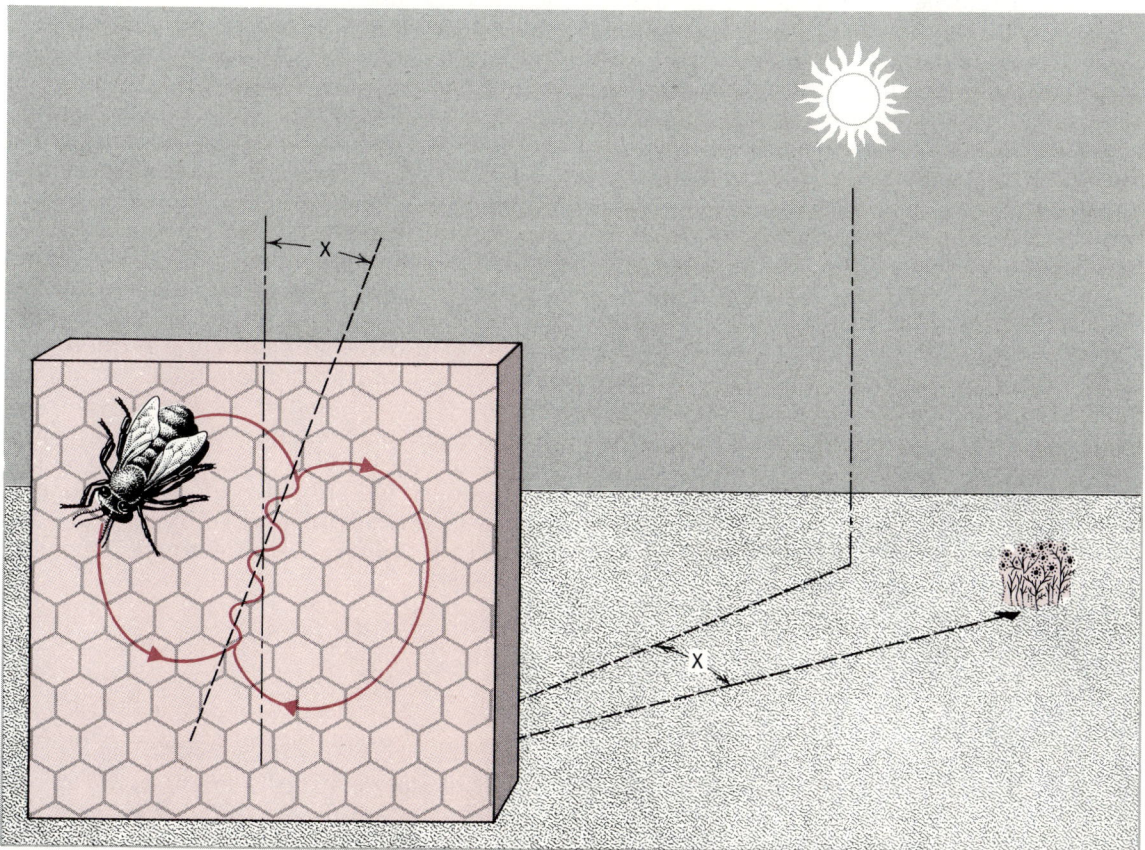

straight line, wagging its abdomen from side to side. It then turns in a clockwise semicircle and repeats the wagging, straight-line movement. The next semicircle is counterclockwise.

In this dance the scout tells both the direction and the distance of the food source from the hive. The distance the scout has traveled to get to the food is correlated with the rate at which the dance is performed; the closer the food, the faster the movement. If the food is 100 m away, a complete turn occurs every 1.5 seconds. A food source at 1000 m elicits one turn about every 3 seconds. By timing 3885 dances, von Frisch was able to construct a curve relating the rate of performance of the wagging dance to distance of the food from the hive (Fig. 19-5).

The direction of the food source from the hive is communicated by the orientation of the straight-line part of the wagging dance. If the dance is performed on the horizontal landing area, the straight-line portion of the dance points directly at the food. Inside the hive, on a vertical surface, the bee orients with respect to gravity. The straight part of the dance is performed at an angle to an imaginary vertical plumbline, and that angle is the same as the angle between the sun and the direction of the food on the flight out (angle X, Fig. 19-4). If the food is to be reached by flying directly into the sun, the straight part of the dance is oriented straight up. If the food can be found with the sun 30° to the bee's left side, the dancer orients its wagging line at 11 o'clock, 30° from the vertical. Apparently the visual pattern this creates in the eye of the observing bees can be very accurately read. Von Frisch reports that after observing such a dance most bees fly out of the hive and head immediately in a line within 3° of the correct compasss direction, to reach food one-fifth of a mile away.

Occasionally a scout continues her dance for hours after returning to the hive. In these cases von Frisch noted that the wagging directional line of the dance gradually shifts orientation to take into account the movement of the sun, just as the hour hand of a clock does. Thus the angle indicated after the first few minutes is no longer the same as the one the bee itself has flown with respect to the sun, but is continually corrected for the slow movement of the sun in the sky. We know now that the dancer is able to make this correction by means of an internal **biological clock,** called a **circadian rhythm** (p. 541).

There is increasing evidence that pattern recognition, even among humans, is a serial operation in which the brain examines a pattern feature by feature. It is possible to determine a person's eye movements by recording the reflection of a spot of light off his cornea in a dimmed room. When a subject views a pattern projected on a screen for the first time, his eye usually scans over it, following repeatedly a more or less fixed path (Fig. 19-6, p. 532). This pattern of eye movements—a **scanpath**—may be different from person to person viewing the same pattern, but for each subject it remains remarkably consistent for a given visual image. That is, it appears that a person recognizes a visual image, in part at least, by remembering the scanpath he first used to view it. These findings suggest that the eye movements involved in perception do not merely move the pattern over the retina, but are an integral part of the mental processing, or memory, on which recognition is based. The way we perceive a pattern seems to involve an alternating sequence of sensory and motor memory traces, recording alternately some feature of the pattern and the eye movement required to reach it from the last feature. When we "recognize" a visual pattern, we are reproducing the successive eye movements and verifying that each leads to the remembered feature.

Nonvisual Perception

The Sound Sense

In a wide variety of animals specific patterns of behavior are produced by auditory stimuli. Many insects call to one another by

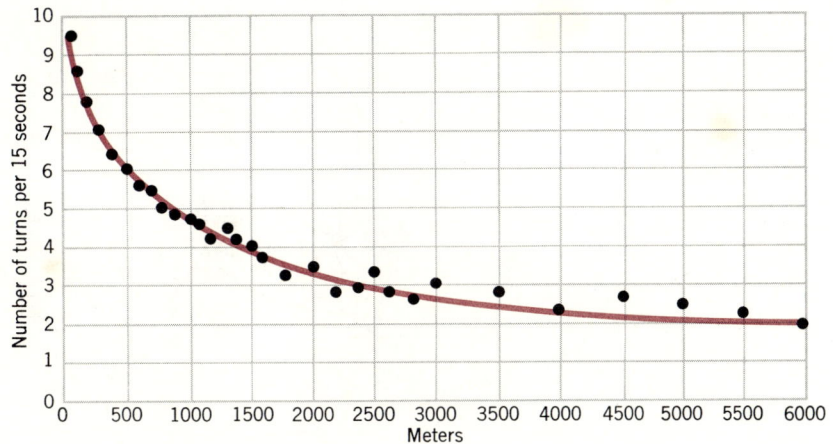

Figure 19-5 The relation between the frequency of honeybee wagging dance and the distance of the food from the hive, as determined by von Frisch.

rubbing their legs or wing parts together or against their bodies. Male mosquitoes, for example, detect and locate females by the buzzing whine of their wing beat. The male antenna is designed to be set vibrating by sounds at the frequency of the female wing beat (300–350 Hz; Herz, abbreviated Hz, means cycles per second). The male beats its wings at a much higher frequency (about 500 Hz); thus the male antenna is "tuned" to the female sound, and fails to sense other males. Similarly, crickets produce their familiar sounds by rasping together specialized wing parts; these calls are important in attracting females of the same species and stimulating their reproductive behavior and in warning away other males. In most cases the calls of insects are so species-specific that closely related forms, which may be anatomically indistinguishable, have very different calls.

The calls of frogs serve similar functions to those of insects, and may be equally specific. Those of birds are certainly more complex, but in the main serve similar purposes. Despite the poetic notion that birds sing with joy, the evidence seems to indicate that bird songs function mainly (1) as a species-recognition signal; (2) as a display to attract females to the male and increase their sexual drive; and (3) to establish or lay claim to a defensible territory.

A wide diversity of animals, from crickets to howler monkeys and fish to birds, establish **territories.** A territory is an area to which an animal (usually the male) lays claim and that it will defend from intrusion by another male of the same species. The role of the bird song in territorial defense is illustrated by experiments by William C. Dilger of Cornell University. Dilger tied a stuffed model of a male woodthrush to a loudspeaker wired to a hidden tape recorder. When Dilger played the recorded song of a male thrush, the male bird that had claimed the territory responded in characteristic fashion. If the recording was played at low volume, the defending male attacked the stuffed bird. If the volume was high, the defending bird retreated. Moreover, the same stuffed bird would be completely ignored if the recorded song of some other bird species was played.

We often tend to think of audible signals as those that can be sensed by our own ears. In fact, many animal sounds are in ranges we cannot hear. Infant rats and mice, for example, constantly squeak at sound frequencies beyond our range of detection. Perhaps the best-known example of the use of high-frequency sound by other animals is that of **echolocation** in bats. Almost two centuries ago Lazzaro Spallanzani observed that bats in a completely darkened room could fly about without hitting anything. Thinking that they might have extremely acute vision, Spallanzani put black hoods over their heads. If the

531 *The Biology of Behavior*

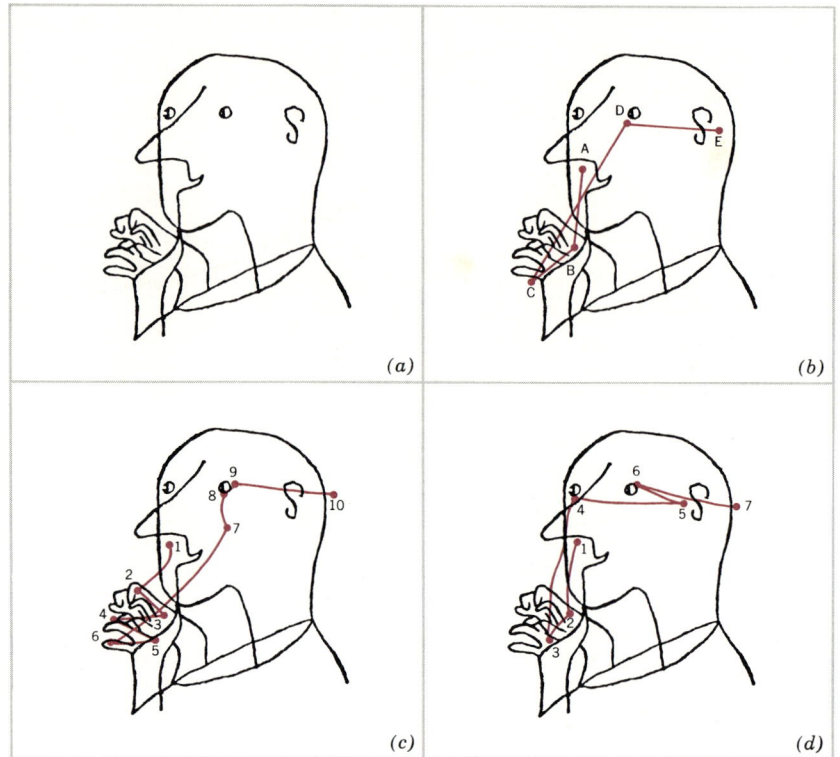

Figure 19-6 The scanpath (colored line) represents the pattern of eye movements when a subject views a large visual pattern from close at hand. The line drawing (a) is from an etching by Paul Klee. The scanpath of one typical subject (b) is shown diagrammatically always to start under the figure's nose, at a point near the center of the composition. Two actual scanpaths (c, d) traced during a series of 20-second viewing trials are remarkably consistent, even though (c) represents the subject's first 20-second exposure to the pattern and (d) is from a subsequent trial in which the subject was asked to recognize the pattern from among nine other similar drawings.

hoods masked only the bats' eyes, the animals suffered no loss of ability to fly in the dark. However, he found that hoods that also covered the bats' ears left them apparently blind.

In 1938 Donald R. Griffin demonstrated that flying bats produce bursts of high-frequency sounds, at frequencies as high as 100,000 Hz. These sounds do not travel far—only about 20 ft—but their extremely short wavelength makes them well suited for identification when reflected back from a small object. On the basis of their highly sophisticated form of echolocation, or "sonar," bats are able to fly through a dark room strung with wires only 200μ thick, or locate an insect only a few millimeters in diameter at a distance of 15 ft and catch it on the wing. Despite the fact that bats of some species often fly in flocks of thousands, each bat in such a flock is able to distinguish echoes of its own sounds, and behave accordingly.

Figure 19-7 *A housefly extends its proboscis by a reflex action whenever the taste receptors in its feet are stimulated. You can test this reflex by holding a fly immobile and touching a drop of weak sugar solution to its feet. Watch what happens.*

Chemical Senses

Chemoreception among mammals is represented by our sense of taste and of smell. A human can distinguish four basic flavors, p. 486) and perhaps 10,000 different odors. Yet, by comparison with other organisms, our chemoreceptors are dull indeed. A fly, for example, can detect many different types of sugar, all of which are sensed merely as sweet on our tongues. Moreover, a hungry fly is 10 million times more sensitive to sugar than we are. If such a fly—whose taste receptors are in its feet—contacts a drop of water containing .1 micrograms (a tenth of a millionth of a gram) of sugar, its **proboscis**—a tubelike sucking organ—extends by an automatic reflex (Fig. 19-7).

Pheromones

Many kinds of behavior other than feeding can be influenced by the chemical senses. A variety of animals excrete hormone-like substances that affect the activities of other members of their own species. Such compounds are called **pheromones** (Figs. 19-8 and 19-9, p. 534).

Pheromones are chemical messengers of two general types, usually produced in special glands. One type, a **triggering** pheromone, sets off immediate specific patterns of

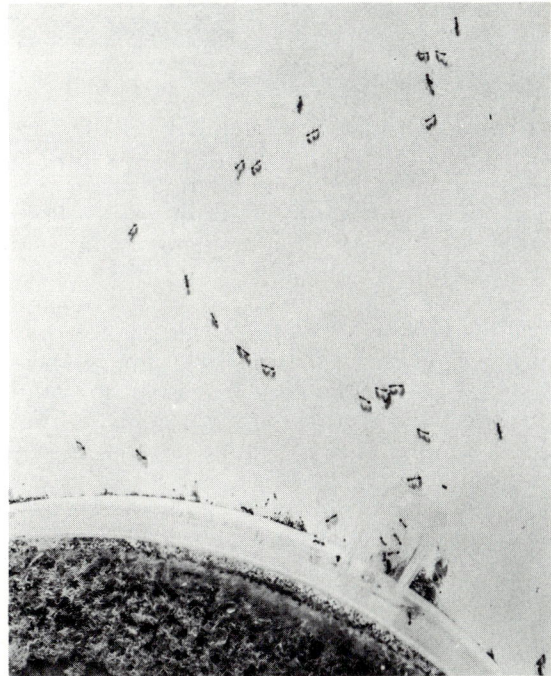

Figure 19-8 *If the trail pheromone of ants is smeared in a curved line to make an artificial trail, ants emerging from the nest will follow it closely for a brief period. The pheromone evaporates and the trail soon disappears unless the ants are stimulated to secrete more by finding food at the end of the trail.*

(a)

Figure 19-9   A number of mammals use pheromones for recognizing members of their community or marking their territories. (a) The flying phalanger secretes a pheromone from a gland on its forehead. (b) The common golden hamster has a similar gland on its flank that it rubs against objects in its territory. (c) The female marmoset secretes a "marker" substance from glands in her genital region. (d) Maxwell's duiker, a small West African antelope, has a large scent gland underneath each eye. It rubs the pheromone against objects in its habitat or presses its gland against the cheek of potential mates.

behavior in responsive members of the species; the other type, often called a **primer** pheromone, alters the endocrine and reproductive system of recipient animals without necessarily causing immediate behavioral changes.

In the first category are sex attractants and trail markers (Fig. 19-8) released by a variety of insects. Female moths, for example, have a chemical sex attractant that is so powerful that males are attracted to a female from distances of 2 miles or more by the release of as little as .01 μg (a hundredth of a millionth of a gram) into the air. Evidence suggests that a male moth is able to detect and respond to only a few molecules in contact with its receptors. Among mammals such triggering pheromones are involved in many aspects of behavior. Most animals hunt by smell. A wounded fish releases a chemical that sets off a fright reaction in other fish. This is an example of an **alarm** pheromone. By chemoreception—in this case their sense of smell—most mammals other than man distinguish friend from enemy, male from female, infant from adult. They identify their mates, their parents, their offspring, by odor. A male knows when a female is in heat and when she is not by sense of smell. Many animals mark their territories by pheromones (Fig. 19-9).

The primer pheromones may be considered essentially as hormones released into the environment. Their main function is to produce relatively long-term physiological alterations. For example the estrous cycles of female mice in a laboratory colony can be triggered by the odor of a male mouse. In contrast, pregnancy in a female mouse may be interrupted by the odor of a strange male. This effect of smell may play an important role in limiting birthrate in crowded colonies.

(b)

(c)

(d)

Another type of primer pheromone is found in social insects such as ants, termites, and bees. In a beehive the queen bee produces a pheromone that is ingested by all members of the hive and inhibits the other females—the workers—from producing eggs. The entire caste system of termite communities is maintained in similar fashion by an elaborate system of pheromones.

Innate Behavior

Instincts

Instincts are complex behavioral patterns that are genetically determined. Instinctive behavior is therefore alike among all members of a species and is relatively little influenced by experience or learning. Classical examples of this may be seen in the nest-building behavior of birds. Members of one species of the lovebird *Agapornis,* for instance, use their sharp bills to cut long strips of paper, bark, or leaves with which to build their nests. The birds then tuck several strips of the material into the feathers on their rumps (Fig. 19-10) and fly to their nests with these strips dangling behind them. Birds of a closely related species cut similar strips but carry them, one by one, in their bills back to the nest. The behavior of genetic hybrids between these two species shows clearly the effects produced by the genes of the two parents. F_1 Hybrids cut strips and try again and again to tuck them into the tail feathers, but without success. Sometimes the tucking movements are incomplete, sometimes the bird fails to let go of the strip at the right time. Some of these hybrids were studied over a period of three years. During that time the futile tucking activities decreased considerably, but the birds never entirely gave up trying to tuck.

The behavior of *Agapornis* is an example of an **innate,** or **stereotyped,** activity, but after a time the genetically determined behavioral pattern or tendency is clearly modified in the *Agapornis* hybrids by the experience of futility. This nicely illustrates the fact that it is not possible to categorize animal actions neatly as either entirely genetically determined or entirely learned. Rather than thinking of behavior as *either* learned or innate, it is probably more suitable to consider every activity as a mix of a genetic and an experiential input. We can visualize a behavioral continuum with reflexive and highly stereotyped behavior at one end and experiences like our own in learning to play baseball or a piano sonata at the other—but with all activities having both genetic and learned determinants.

Ethologists study instincts by attempting to dissect patterns of behavior into sequences of stimuli and specific responses. As an ex-

Figure 19-10   A female lovebird Agapornis *tucking strips of paper into her rump feathers. This bird had cut each strip from the heavy construction paper on which she is standing.*

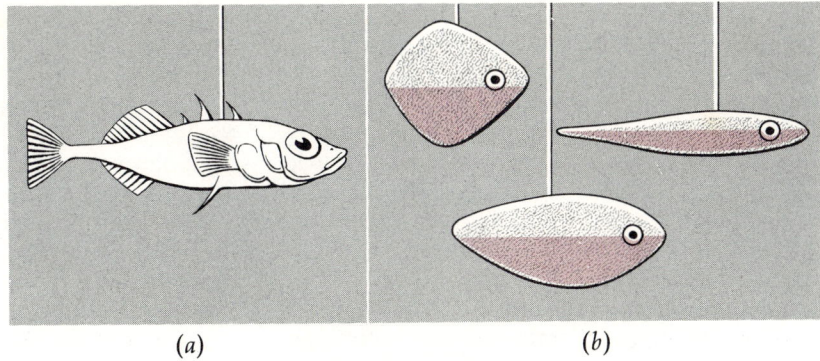

Figure 19-11 (a) An accurate model of a male stickleback, lacking only the red coloration of the belly, does not elicit territorial defense behavior from males. (b) Models of a variety of shapes, featuring only an eye and a red belly, are immediately attacked.

(a)   (b)

ample, consider the courting behavior of the three-spined stickleback, whose defensive response to color we mentioned on p. 527. Sticklebacks live in the sea or brackish water for most of the year, but in the spring, under the influence of hormonal changes brought on by the increasing length of days, they migrate up rivers into shallow freshwater streams.

When the male fish arrive in the streams, they have changed coloration from rather dull brown to brightly colored, with a greenish back and bright red underbelly. Each male takes up a position in the stream and establishes a territory that he defends from invasion by other males. Female sticklebacks or other species of fish can swim into the territory without challenge; another male stickleback is immediately attacked.

Tinbergen has demonstrated that it is the red color of the belly that elicits defensive behavior in the stickleback. A wooden model, shaped and painted exactly like a male stickleback (Fig. 19-11a), but without the red belly, is allowed into the territory without protest. The same model with a red coloration is immediately attacked. In fact, models of a variety of shapes (Fig. 19-11b) bearing little resemblance to the real fish, elicit territorial defense if only they feature an eye and a red belly. The specific stimulus for the defensive behavior in this case is the red belly. Such stimuli that trigger specific patterns of stereotyped behavior are called **releasers.** The red

belly is the releaser for territorial defense.

After establishing a territory, the male stickleback digs a shallow pit in the sand and constructs a breeding tunnel, or nest, out of threads of algae and a mucous material it secretes. While the male sticklebacks build their nests, the females become swollen with unfertilized eggs. They cruise about in schools, passing through the territories of different males. Each male, if ready to receive a female, reacts to the group by beginning a **zigzag dance,** consisting of a series of darting movements. The releaser effect of females in triggering this dance can be duplicated by a model bearing little relation in shape to the females; all they must have are an eye and a swollen belly (Fig. 19-12).

In response to the zigzag dance a female

Figure 19-12 Model (a) is an accurate copy of a female stickleback; (b) bears little resemblance to a real fish except for the eye and the bulging abdomen that are similar to the swollen belly of an egg-carrying female. Model (b) acts as a releaser for courting behavior in males; model (a) does not.

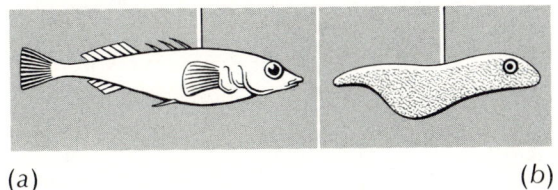

(a)   (b)

enters the nest. This prompts the male to begin a series of thrusting movements with his snout, prodding the female at the base of the tail, stimulating her to spawn her eggs into the nest. Spawning can be induced by prodding the female's tail with a glass rod, but only after she has entered the nest. Finally, the eggs provide a chemical stimulus that releases ejaculation by the male, resulting in fertilization of the eggs.

This is an example of how a chain of releaser-response interactions between a pair of animals can produce a complicated behavioral sequence. Descriptions of such behavioral patterns, however, are likely to be misleading unless it is clearly understood that a releaser will not work every time. When a male does a zigzag dance it is very likely that a female will respond with her courting behavior, and it is very unlikely that she will do so without this stimulus. But a free-living animal is barraged at every moment with sensory inputs from all sides and must integrate many variables into its CNS. The behavior of an organism, even in the most stereotyped activities, is rarely automatic. Stickleback courtship may be broken off at any time as one fish or the other swims away. The female may fail to enter the nest; the male may then return to its zigzag dance. The normal sequence of behavior represents merely the most probable set of interactions, not a rigid, unvarying protocol.

The interaction between instinctive behavior and learning is well illustrated in recent studies by J. P. Hailman of the pecking behavior of young laughing gull chicks (Fig. 19-13).

When a parent laughing gull returns to its nest from a fishing foray, it lowers its head in front of its chick. The chick aims a well-coordinated peck at the red bill of its parent, grasping the bill in its own with a downward stroking motion. After a few such **begging pecks,** the parent regurgitates a mass of partly digested food onto the floor of the nest, which the chick then tears apart and eats with a **feeding peck,** which takes a different form.

If chicks are allowed to peck at a model of a laughing gull head, immediately after hatching about 30 percent of the pecks strike the red beak. This is true even if eggs collected in the field are hatched in a dark incubator to prevent the chicks from receiving any visual stimuli before being tested.

Chicks improved their accuracy greatly one or two days after hatching. If chicks were removed from their nest after two days of normal feeding with the parent, more than 75 percent of the begging pecks hit the model beak. Is this improvement in accuracy related to practice (i.e., learning), or merely to maturational changes such as improved motor coordination?

To answer this question Hailman reared newly hatched chicks in dark brooders to prevent visual experience. One group of these chicks was force-fed in the brooder to prevent the chicks from pecking at all. A second group received no food for two days, the chicks living adequately on their reserves of yolk. The third group had been removed from their eggs just before hatching; these chicks were maintained in an incubator without food, to test whether the hatching movements themselves were required for proper pecking. On various days after hatching the chicks of all groups were tested for pecking accuracy. It was found that all three groups increased in accuracy somewhat, but none reached the normal control level of 75 percent hits. Thus the denial of visual experience after hatching did have a strong effect.

We can draw two conclusions from these experiments: even immediately after hatching

Figure 19-13 The normal feeding behavior of a laughing gull chick involves two separate types of pecking movement. As the parent lowers its head (a) the chick aims a well-coordinated "begging" peck at the parent's red beak, grasping it with a downward stroking motion (b). The parent then regurgitates a partly digested fish (c) and the chick tears it apart and eats it with a feeding peck (d).

(a)

(b)

(c)

(d)

The Biology of Behavior

there is something about a parent's head that elicits a typical begging peck from a completely inexperienced chick; and the accuracy and coordination of these pecks increases with age, at least partly as a result of practice.

What particular aspect of the parent's head acts as the releaser to the chick's behavior? Models were built with various parts systematically eliminated or changed (Fig. 19-14). Chicks were again hatched in dark incubators and maintained there for 24 hours before being tested. They were then replaced in nests to be fed by parents, and were tested again a week later. These experiments demonstrated that newly hatched chicks respond equally well to any shape that has a red, beaklike structure. Older chicks, however, pecked with the greatest frequency at the most realistic model. The number of pecks decreased progressively as more and more features of the model were altered (Fig. 19-14).

These experiments suggest that in a newly hatched chick instinctive behavioral patterns, although present, may be in a relatively primitive state. The specific stimulus is simple; the stereotyped response is not yet well coordinated. During the first few days after hatching, a process of **perceptual sharpening** occurs. The releaser grows more complex and is better differentiated; the behavioral pattern becomes more coordinated and integrated, at least in part as a result of experience.

An especially dramatic example of the importance of releasers in behavioral development is the phenomenon of **imprinting,** discovered by the ethologist Konrad Lorenz. Shortly after hatching, ducklings, chicks, goslings, and certain other birds go through a **critical period** during which they exhibit a standardized **following response** to the first moving object they encounter. Usually this is the parent, and results in the young running behind their mother. It normally depends on several releaser stimuli associated with the parent bird: its size and shape, noises it makes, but above all, its movement.

Birds reared in a laboratory, however, may become imprinted on any moving object: an animal of a different species, a man, even an animated toy. Henceforth the offspring will accept that animal, person, or object as its

Figure 19-14   Features of the parent's head that elicit pecking behavior were tested with models presented to newly hatched (white bars) and seven-day old (black bars) laughing gull chicks. The newly hatched birds were stimulated to pecking with about equal frequency by any shape that included a red, beaklike structure. None of the other features was important. Older chicks, in contrast, were sensitive to head shape, and to the amount of black and white area represented.

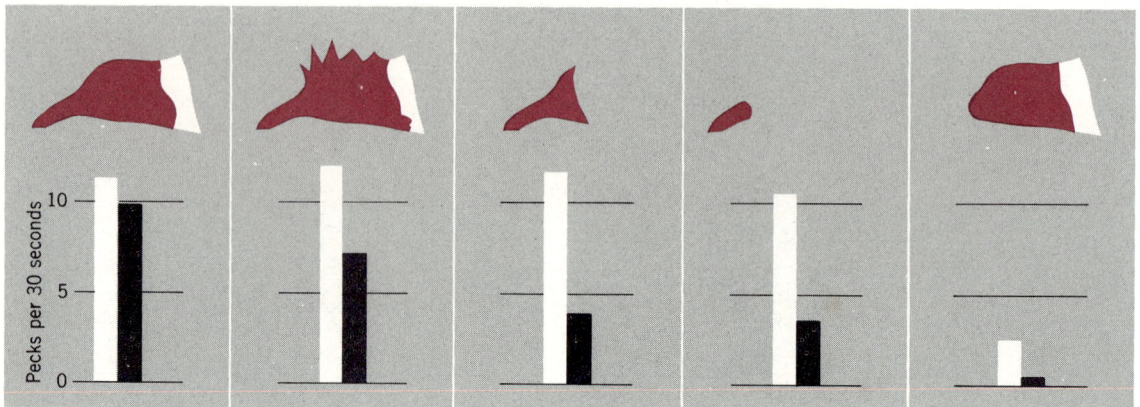

"mother." The fixation on such surrogate mothers appears to be just as strong and permanent as on a natural parent. Moreover, because the early imprinting normally forms the basis for later mate selection, birds raised to maturity with surrogate mothers may reject potential mates of their own species in favor of the surrogate, even if the latter is an inanimate object.

Biological Rhythms

Essentially all organisms, plant and animal, exhibit rhythmic behavior of some kind. Birds migrate with amazing regularity, south in the winter and north again the following spring, in an **annual cycle.** Many women menstruate regularly every 28 to 30 days, maintaining a **monthly cycle.** And it is a commonplace to note that the activity of animals and plants is different at different times of the day: birds and butterflies are active during the daylight hours, whereas moths and many small mammals confine their activity to the night. Essentially every activity in every living thing, from food gathering to egg laying, exhibits some kind of rhythm that bears a relationship to the **day-night cycle.** For years scientists observed such daily rhythms of organisms with the preconceived notion that cycles were imposed on living things by the physical environment. Statements such as "birds begin to sing when the sun rises," or "crickets chirp at dusk," are more or less accurate, but they imply that organisms are passive and that activity is forced on them or triggered by changes in the environment. Only in the last 20 years or so has it become generally recognized among behavioral biologists that organisms are not passive responders to time changes around them. They have internal, accurate time-measuring systems, or **clocks.** The environment may lead an organism periodically to reset its clock, but time-keeping itself is an innate, fundamental biological process.

If we set a pendulum swinging, its steady oscillations establish a rhythm (Fig. 19-15). The repeating unit of a rhythm, that is, one full swing from left to right and back to left

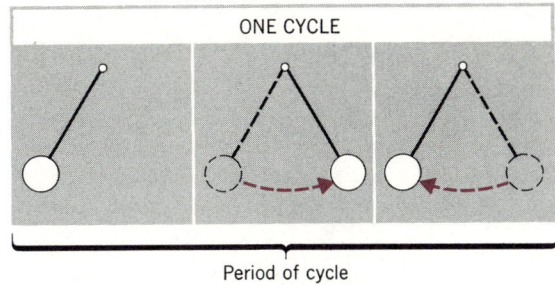

Figure 19-15   *The steady oscillations of a pendulum establish its rhythm. One full swing and back is a cycle; the length of time to complete one cycle is the period of the rhythm.*

again, is a **cycle.** The length of time required to complete one cycle is called the **period** of the rhythm. The period of the earth rotating on its axis as it revolves around the sun is 24 hours. (Astronomers, however, use a period called the **sidereal day,** based on the period of the earth's rotation as measured by fixed stars; it is 23 hours, 56 minutes, and 4 seconds.)

Daily rhythms have been found in all major groups of organisms, but they rarely have a period of exactly 24 hours. Among the protists, for example, rhythms in *Euglena, Paramecium,* and the luminescent marine alga *Gonyaulax* have been well studied. J. Woodward Hastings has shown, for example, that three separate daily rhythms operate independently in the single cell of *Gonyaulax:* (1) a **luminescence** cycle, in which the firefly-like flashing of the organism reaches a peak of brightness in the middle of the night; (2) **photosynthesis,** which reaches its peak at midday; and (3) **mitosis,** which occurs only in the hours just before dawn.

Several fungi, including *Neurospora,* also have conspicuous daily cycles. In many higher plants leaf movement, petal movement, and nectar secretion occur in daily rhythms. Among animals, daily cycles of various activities have been discovered in coelenterates, annelids, mollusks, arthropods, insects, reptiles, birds, and a wide variety of mammals. A daily rhythm has been extensively studied in the emergence of adult *Drosophila* from the pupal case. Human births are also statistically rhythmic, in the sense that they occur most often in the early morning hours than any other time of day.

Daily cycles occur not only at the level of overt behavior. Many physiological or cellular activities also show daily cycles: body temperature, metabolic rate, electrical activity of the central nervous system, concentration of various blood constituents, liver glycogen levels, synthesis of DNA in regenerating liver. Even the spontaneous rate of beating of heart cells isolated in tissue culture fluctuates according to a regular daily rhythm.

How do we know that such daily rhythms do not represent merely a passive response of an organism to daily environmental changes (light, temperature, cosmic rays, etc.), and that they actually are endogenous to an organism? The evidence is simple. When an organism is placed in a laboratory situation under constant conditions (for example, continuous dark, constant temperature and humidity), its daily rhythms persist, but the period of cyclic activity usually shifts to something other than 24 hours. The onset or peak of the measured activity does not occur at the same time every day, but takes place a constant period of time earlier or later on each successive day. The activity of a flying squirrel running in a wheel, for example, has been recorded under conditions of constant darkness and temperature (Fig. 19-16). In continuous darkness the squirrel began running 24 minutes earlier each day than it had the day before; its period was 23 hours, 36 minutes. There are no known environmental factors that have such a period, so the rhythm must have been internally generated.

The period of such an endogenous rhythm is called the **free-running period.** Because most free-running periods thus measured do not differ from 24 hours by more than an hour or two on either side, they are called **circadian periods** (meaning "about one day"). Biological rhythms that exhibit circadian periods under constant conditions are referred to as **circadian rhythms.**

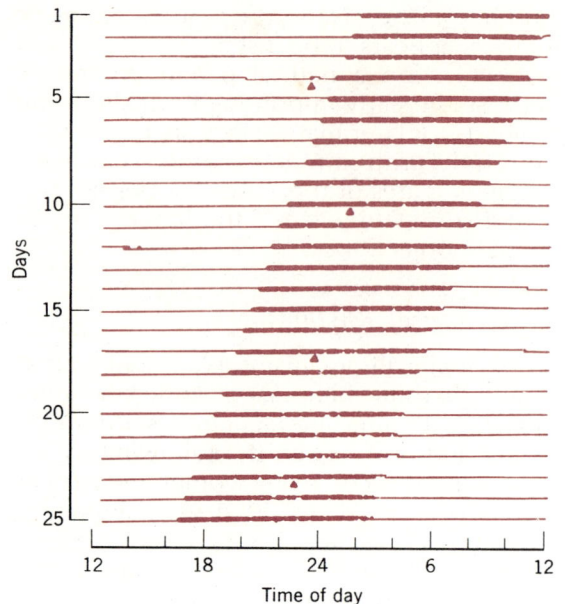

Figure 19-16   *A continuous record of 25 days of running-wheel activity of a flying squirrel maintained in a cage in the laboratory in constant darkness and temperature. Reading from the top down, each line represents a 24-hour day; day 1 represents the first day in continuous darkness. Black bars indicate intense activity. Note the persistence of the rhythm under constant conditions, and its precision. The period of the endogenous cycle, as measured by the time of onset of intense activity, is 23 hours and 36 minutes. That is, the squirrel began to run on the wheel exactly 24 minutes earlier each day, without any sign of the actual day-night cycle.*

The argument that such circadian rhythms are indeed endogenous is strengthened by the fact that different individuals in a species have different free-running periods. That is, other flying squirrels in the same room, at the same time as the one whose record is shown in Figure 19-16 exhibit periods ranging from 23 to 24.3 hours. Individual sparrows, also under similar conditions, show a range of periods of "perching activity" of 24.3 to 26.2 hours. It is difficult to imagine how

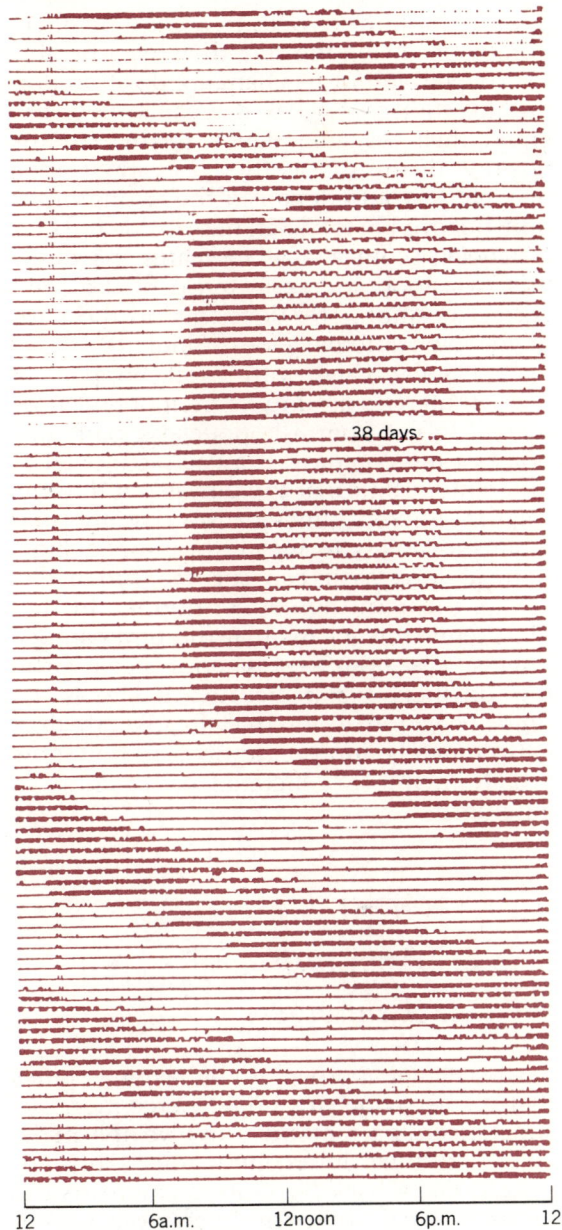

Figure 19-17 Entrainment of the perching rhythm of a house sparrow to a light cycle of three hours of light and 21 hours of darkness. The data have been handled as in Fig. 19-16. For the first 18 days (first 18 bars) the bird was free-running, with a period of about 26 hours in constant darkness. On the nineteenth day the light cycle was imposed (lights on at 8 A.M.). The dense black bars indicate almost continuous perching while the lights were on. Note that the bird was entrained by the light cycle, beginning his perching activity about 30 minutes before the lights went on each day. The bird remained entrained for 78 days of continuous light-cycle periodicity. When replaced in constant darkness (lower portion of the figure) the bird gradually (over about 10 days) returned to his previous free-running period, which was maintained for 40 days until the experiment was terminated.

any environmental rhythm with a period of 24 hours could induce such a wide range of periodicity in different animals. Moreover, for most of the organisms so far studied, circadian rhythms have been found to be innate, heritable characteristics. Mice, lizards, and insects raised for several generations under constant environmental conditions still exhibit circadian rhythms much like those of their parents. In fact, it has been possible to select for genetic stocks of *Drosophila* with specific free-running periods.

Although many circadian rhythms are clearly endogenous, they usually may be **entrained** (that is, synchronized to a period of 24 hours) by an appropriate stimulus delivered every day at the same time. Periodic light cycles are usually most effective in this regard (Fig. 19-17); Organisms keep synchronized with real time by daily resetting their clocks according to the light cycles. The ability of an organism to adjust its inner rhythm to that of the outside world seems to be an obvious advantage. In only a few cases, however, is the adaptive significance of biological rhythms apparent. Some plants, for example, secrete nectar at specific times of the day. As

a result, bees, which have their own circadian periods, visit these flowers at the "right" time, ensuring maximal cross-pollination of the flowers.

Another obvious advantage is navigational.

In order for birds or bees to determine direction by observing the sun, they must be able to compensate for its changing position during the course of a day or year. If bees oriented their wagging dance relative to the position of the sun as they last saw it, their hivemates would depart in the wrong direction, but even in total darkness inside the hive a dancing scout always indicates the direction of a food source in terms of the sun's correct position. She does this on the basis of her own internal clock. Similarly, a bird in a circular cage with food hoppers at regular intervals around its periphery can be trained to eat from a hopper at a particular point of the compass. If a cage is moved to different localities, or the bird is put into the cage at different times of the day, the bird retains its ability to locate the proper direction; as long as it can see the sun it will always eat from the correct hopper. In simplest terms this means that the bird can distinguish east from west when it sees the sun low on the horizon, because it knows the time of day from its internal circadian rhythm. Unfortunately, in no case do we know how a daily rhythm is generated or regulated in an animal.

A Cellular Approach to Behavior

The cellular approach to behavior assumes that all behavior—including perception, instincts, learning, memory—involve only nerve cells and their interconnections. This approach dates back 50 years or more to the morphological studies of the nervous system by the great Spanish anatomist Santiago Ramón y Cajal, who held that the central nervous system was constructed from discrete cellular units, the neurons. The way to understand the brain was to analyze its functional architecture—its wiring diagram. In the 1940's Roger Sperry revealed the importance of specific neuronal interconnections in be-

havior in a series of studies of regeneration of neural connections in fish and amphibians. Sperry's work indicated that the coordination of visual perception and motor activities in these animals depended on a highly specific set of neuronal interconnections between the retina and the optic center of the brain. Moreover, these connections seemed to be constant from one individual to the next, and appeared to be unaffected by experience.

But if connections between neurons in the CNS are rigidly determined according to a specific wiring diagram, then how is behavior modified? How does learning occur? This apparent paradox had already been forseen by Cajal, who proposed the idea of **plasticity.** Even though the anatomical connections between neurons may develop according to a predetermined plan, the effectiveness of those connections is not necessarily constant. It is the *effectiveness* of synapses and other conductive properties of neurons that is altered by experience. For example, a train of rapidly repeated impulses, called a **tetanic stimulus,** applied to a cat nerve produces a larger response in a postsynaptic neuron than a single impulse. However, immediately after the tetanic stimulus a single impulse produces a response two to three times larger than usual. That is, the tetanic stimulus results in an increased effectiveness of the synapse that may last many minutes (Fig. 19-18). This is the phenomenon known as **post-tetanic potentiation.**

Can such plasticity of nerve cells and synapses be related in any way to a recognizable form of behavior? A simple form of learning is **habituation,** which we mentioned in describing the behavior of amebas. Any behavioral response that decreases, or **extinguishes,** when a stimulus is presented repeatedly may be said to exhibit habituation. A sudden loud noise presented unexpectedly immediately draws our attention and may even cause an increase in heartbeat and respiratory rate. If the same noise is repeated continuously over an extended period, our attention and physiological responses gradually diminish. The response is restored after a period of rest from the stimulus; if the noise stops and is fol-

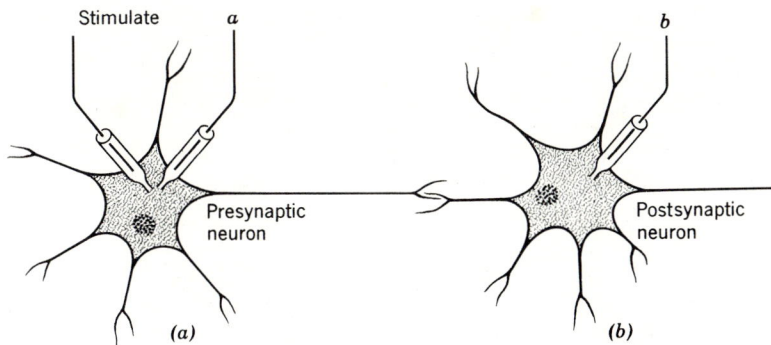

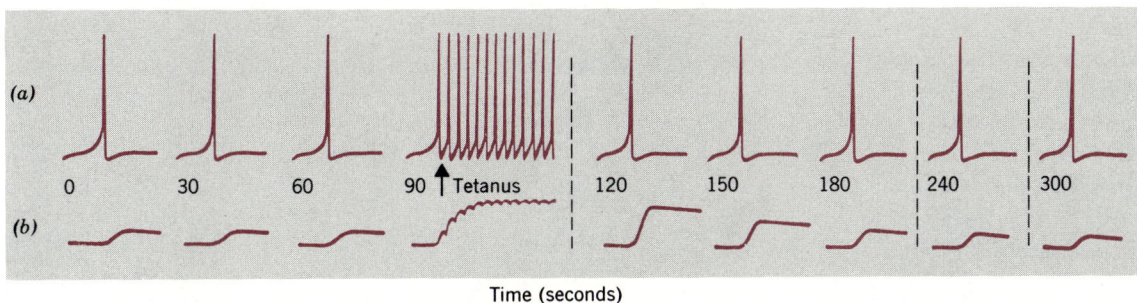

Figure 19-18   An experiment showing posttetanic potentiation—a form of plasticity at a synapse. Stimulation of the presynaptic neuron produces an action potential in it (trace a) and a response in the postsynaptic cell. The single shocks at 0, 30, and 60 seconds illustrate the constant size of the postsynaptic potential. High-frequency stimulation yields an additive, or **summated,** response of the postsynaptic cell, after which the cell is much more sensitive to single shocks than before. This increased effectiveness of the synapse gradually decays over a period of several minutes.

lowed by an hour of silence, the same noise then will produce an equally strong response. A habituated response can also be restored by altering the stimulus in certain ways, for example, making the noise suddenly louder. This is termed **dishabituation.**

The sea hare *Aplysia* (Fig. 19-3) shows a defensive withdrawal response that clearly involves habituation. The snail's gill is normally partly extended, waving in the water (Fig. 19-19a, p. 546). A gentle touch on the gill causes it to be convulsively withdrawn into its cavity under the mantle shelf (Fig. 19-19b). The defensive purpose of this reflex is obvi-

ous; it protects the gill, which is an important and delicate organ, from possible harm. However, if the gill is touched repeatedly, each consecutive touch elicits a weaker and weaker withdrawal response; after nine or ten touches the animal essentially ignores the stimulus. Gill withdrawal is analogous to the reflexive jerk of a man's hand away from a hot iron, but the nervous system of *Aplysia* is simple enough to examine experimentally (p. 526). Recall that the abdominal ganglion of *Aplysia*, which contains the cell bodies of all of the neurons related to the gill-withdrawal reflex, both motor and sensory,

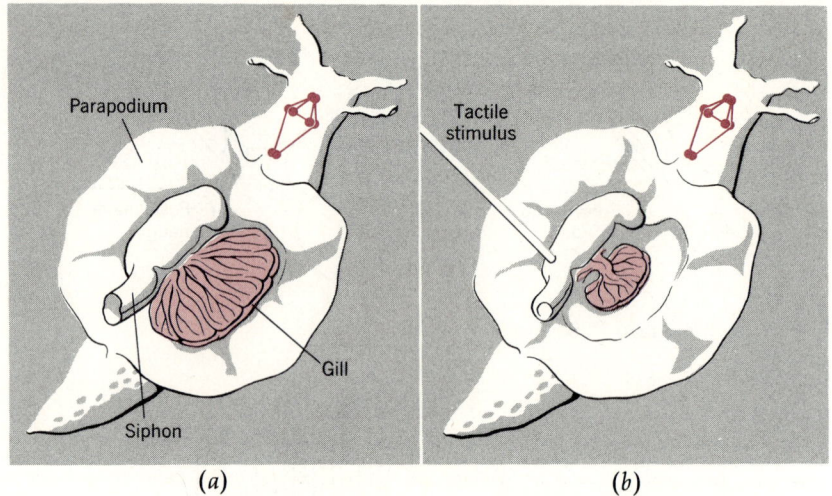

Figure 19-19 (a) The gill of Aplysia is normally extended between the parapodia. (b) When the gill or siphon is touched, the gill retreats into its cavity in a typical defensive reflex. The cell bodies of both the motor and sensory neurons innervating the gill are located in the paired abdominal ganglion (color). (c) About 30 of the 1800 cells of the abdominal ganglion have been classified and identified as recognizable from one animal to another by stimulating cells one at a time and recording with a microelectrode. Five cells (shown in color) in one part of the ganglion have been identified as the motor neurons that control the gill-withdrawal reflex.

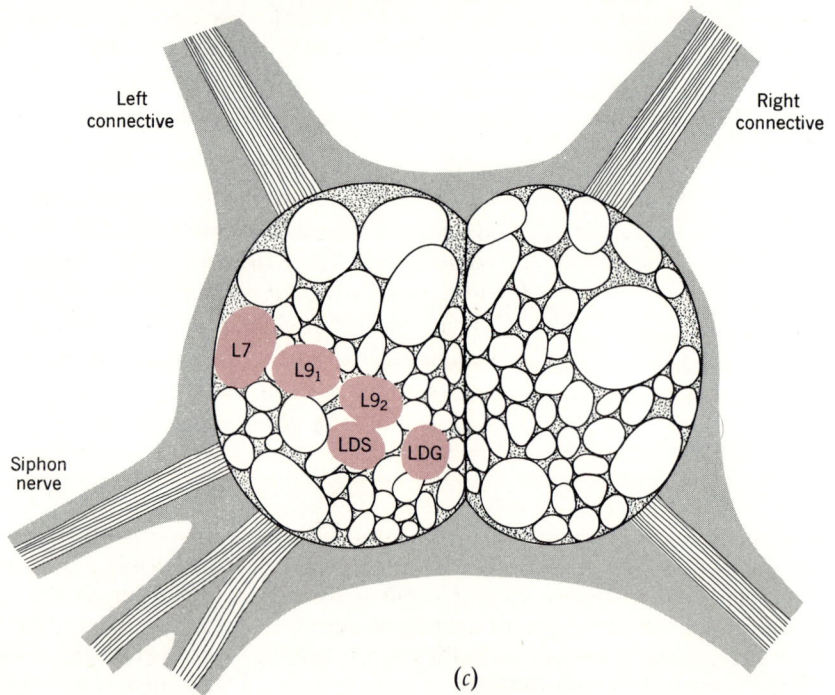

has a total of only about 1800 cells.

Eric Kandel and his colleagues have recently completed an analysis of the neuronal circuit for gill withdrawal, and have been able to identify the site and nature of the functional change involved in its habituation. By stimulating different cells in the abdominal ganglion, one at a time, with a microelectrode, they could observe which cells produced movements of the gill. By this means they identified five motor cells, clustered together in the ganglion, that produced contractions of the gill (Fig. 19-19c). By recording from the same five cells they were able to show that all five received distinct excitatory postsynaptic potentials (EPSP's) from sensory cells when the gill or siphon was lightly stroked.

With this anatomical background, they were then able to explore the mechanism of habituation. In theory, the decrease in response to repeated pokes could result from: (1) fatigue in the withdrawal muscles; (2) fatigue in the sensory tactile receptors in the gill; (3) decrease in conductivity of the motor neurons; or (4) inhibition or decreased effectiveness of synaptic transmission.

The first two possibilities were easily ruled out by stimulating the five motor neurons directly. Even when the withdrawal reflex was almost totally extinguished, direct firing of the motor neurons resulted in a gill retraction just as strong as in a fresh animal. Moreover, recordings from sensory cells in animals in the habituated state revealed that the sensory neurons were firing just as rapidly as in a rested animal. Therefore neither muscle fatigue nor fatigue of the sensory receptors could account for habituation of the gill reflex.

By directly measuring the membrane resistance and conductive properties of the motor neurons in fresh and in habituated snails, Kandel was also able to discount the third possibility. The motor nerves of a habituated animal do not themselves become ineffective. On the other hand, the EPSP produced in the motor neurons when the gill was touched underwent characteristic changes as the withdrawal response habituated. The EPSP diminished gradually in size, and the frequency of action potentials fired by the EPSP decreased correspondingly. Thus habituation of this response—like posttetanic potentiation in the cat spinal cord—results from a long-lasting decrease in effectiveness of a specific synapse. In the case of *Aplysia* gill withdrawal, the modification occurs in the excitatory synapse between the sensory and motor neurons, and results from a decrease in the release of transmitter substance (p. 470) from the presynaptic sensory terminals after repeated tactile stimuli.

Learning and Memory

Learning is the process of modifying behavior by experience. It is one of biology's most intriguing problems. As we suggested in the last section, learning undoubtedly has a cellular basis, but most research on the subject has been devoted to attempts at understanding and regulating the behavior of animals (and people) rather than to cell physiology. Learning has been subdivided into a large number of categories; we restrict ourselves here to a few of the more widely recognized forms.

Habituation

We have already discussed habituation among protists and mollusks. It is perhaps the simplest form of learning, involving only the decline in response to a frequently repeated stimulus. It differs from sensory adaptation or fatigue mainly in its long duration and in the fact that associative regions of the CNS are usually involved.

Response Conditioning

The **conditioned-reflex** experiments by the Russian physiologist Ivan Pavlov are perhaps the best-known studies in animal behavior. He presented a hungry dog with food, which consistently elicited a reflex of salivation (Fig. 19-20, p. 548).

Pavlov then rang a bell each time he presented the food in order to associate the **unconditioned response** (salivation) with a neutral stimulus (the bell). After a few trials he was able to demonstrate that the dog salivated at the sound of the bell, without being presented with food. This is known as a **conditioned response,** and is a form of **associative learning.**

Once a conditioned response has been established, it can be extinguished by prolonged repetition. That is, if the bell is sounded repeatedly, but the dog is never rewarded with food, the salivation response gradually disappears. On the other hand, an

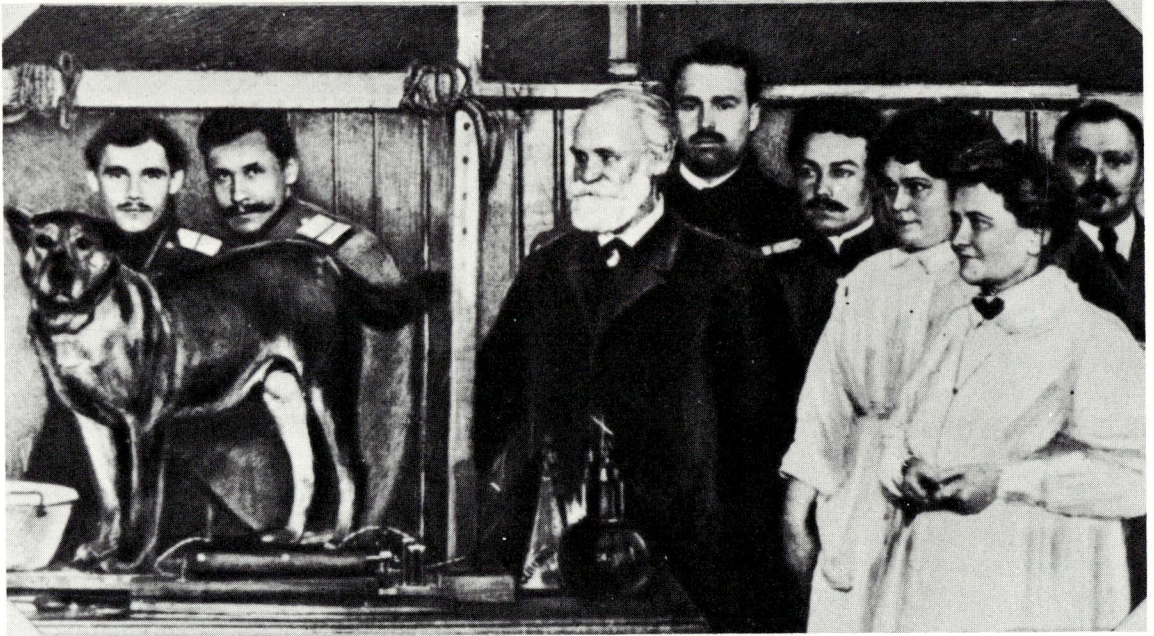

Figure 19-20   Pavlov and his staff demonstrating the conditioned reflex phenomenon. This photograph was taken in about 1910.

occasional reward of food with the bell still acts to strengthen, or **reinforce,** the desired response.

Operant Conditioning

A second widely used method for producing and studying learning is called **operant conditioning.** We have already seen operant conditioning in the experiment (p. 528) in which a pigeon was trained to provide information on its visual threshold. In its simplest form, operant conditioning is merely the repeated rewarding of a specific, restricted element of behavior. For example, a rat is placed in a soundproof box (Fig. 19-21a) that has a bar on one wall that the animal can press and a food hopper filled with food that opens with each press. The bar is wired to a **cumulative recorder,** which plots a graph of the total number of times the bar is pushed in a certain period of time. (Such an apparatus

is called a **Skinner box** after its inventor, B. F. Skinner of Harvard University.) When the animal is put in the box, it begins to do all sorts of things; it may turn around, thrust its head into a corner, stand still for a moment, perhaps lick its fur. None of this behavior seems to be reflexive; none of it seems to be in direct response to a given stimulus. Skinner called these more or less random testing movements **operant behavior.** The animal "operates" on its surroundings. Operants also might be called **voluntary** movements.

One of the operants the rat is likely to perform, sooner or later, is to press the bar (Fig. 19-21b). If this produces no effect, the animal will repeat the action occasionally, but no more often than it performs any other action. If pressing the bar opens the food hopper, however, the rat will see the food and pick up a piece before the hopper closes. There is then an excellent chance that it will immediately press the bar again. If each press is followed by food, the rat will continue to press, gradually attaining a high rate of pressing. In this case the reward of food has **reinforced** the pressing behavior. The reinforcement has changed an element of behavior from a low-frequency, random act to a high-frequency, purposive one. This is operant conditioning, the conditioning of voluntary actions. It is also known as **instru-mental** conditioning.

Operant conditioning tends not to be forgotten, but it can be readily extinguished. If a pigeon that has been conditioned by the same method to peck at a high rate on a key is placed in an inactive Skinner box—one in which a peck fails to open the food hopper—the pigeon will continue to peck at a high rate for a while; gradually, however, the frequency of pecking will fall to the level found before reinforcement.

A well-conditioned pigeon will peck at about 6000 to 8000 times per hour, and will continue to peck at that rate for an hour or more after reinforcement is stopped. Over the next several hours pecking frequency may fall off to only a few hundred pecks an hour. This decrease in frequency of the behavior that is no longer being reinforced is called **extinction.**

Pigeons undergoing extinction frequently exhibit "emotional" responses, turning round and round or beating their wings in apparent frustration; men in similar circumstances show signs of annoyance or even rage. Conditioned pigeons removed from a Skinner box with no opportunity for extinction may be returned much later, even after several years, and will begin pecking immediately with no diminution in rate. Extinction is clearly not the same as forgetting.

By appropriate schedules of reinforcement, animals can be conditioned to carry out remarkably complex behavior patterns. For example, a rat may be taught the following sequence of actions: it raises up on its hind legs to pull a chain, causing a marble to fall from a rack; picks up the marble with its forepaws; carries it to a chute in the bottom of its cage; and drops the marble into the chute. This sequence is gradually built up by reinforcing each small segment of the pattern. Although the complete performance seems highly improbable for a rat, each isolated act is closely related to its normal activities—manipulating nesting materials, handling and storing food, retrieving the young, and so on. Animals cannot be trained to perform actions for which they lack sensory or motor equipment or that conflict with other,

Figure 19-21   Operant conditioning: rat in a Skinner box (a) licking its fur; (b) pressing a bar.

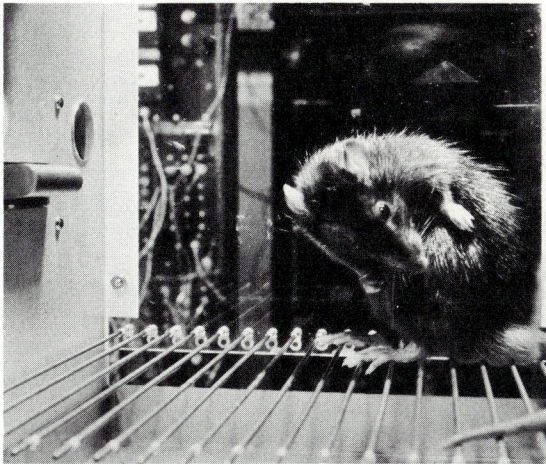

(a)

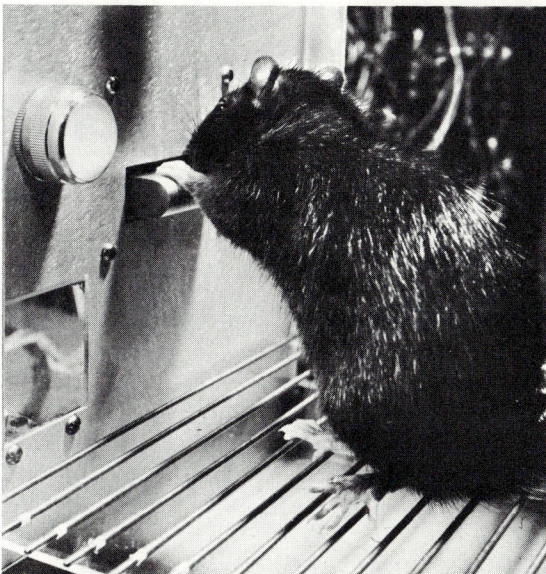

(b)

more powerful, behavior patterns. For example, a raccoon can be conditioned to pick up small objects such as coins and drop them into a box, but never two at a time. When a racoon has an object in each "hand," it reacts to an instinctive pattern of washing behavior; it can never be conditioned to override that innate behavior pattern and drop the coins in the box without first "washing" them.

What Is a Reinforcer?

For a hungry or thirsty animal, food or drink are highly effective rewards, and are commonly employed as reinforcers in the laboratory because they are convenient to manipulate. Sexual contact is reinforcing, as are the courtship displays that precede mating in many animals. That is, a cock will peck a key for the chance to see a potential mate. Even curiosity can be a positive reinforcer. A monkey kept in a closed box will press a lever or perform other tasks for the reward of having a window unlatched for a view outside. Chimpanzees will spend hours doing simple puzzles, and will work for the opportunity to do so; therefore, puzzle solving is a reinforcer for chimpanzee behavior.

These are all **positive** reinforcers. They all satisfy some need, or **drive,** of an animal; they are rewards. Behavior can also be reinforced by removing an uncomfortable stimulus: turning off a loud noise, stopping an electric shock, warming a cold animal. The noise, shock, and cold in these cases are **negative reinforcers;** animals will work to avoid them.

We can readily measure the relative strengths of various reinforcers in given animals. A fighting cock will peck a key to look at itself in a mirror. Such a bird may be placed in a Skinner box with three keys, the first for food reinforcement, the second for a drink of water, and the third for a mirror. After a brief period of training, the cock will peck, on the average, 100 times a day for a look in the mirror, about the same number for food, and about 800 times a day for water. A rat in a similar situation devotes essentially all its lever pressing of food and water and ignores the mirror key.

Rats can be taught elaborate tricks by reinforcing their behavior with an electrical stimulus to the proper locus of the brain. If an electrode is permanently implanted in a particular location in the brain, which James Olds has termed the **pleasure center,** a very weak stimulus will act as a very strong positive reinforcer. Such a rat may be placed in a Skinner box and its implanted electrode connected to a stimulator by a thin wire, allowing the animal free movement in the box (Fig. 19-22). The stimulator is activated by a lever, and delivers a half-second pulse. Thus whenever the rat presses the lever it stimulates its own brain. With the electrode properly placed, rats will press the lever as fast as possible, up to 5000 times an hour, for many hours. They pause only briefly and rarely, to obtain food and water.

A few experiments have been reported in which similar electrodes were implanted into the homologous area in human patients undergoing brain surgery. Humans push a key just as readily to stimulate their brains as rats. The experience they report is a feeling of joy, serenity, and sometimes a pleasant, vague sensation in the pelvic region. It has been postulated that drugs such as heroin and morphine may be addictive in part because they specifically activate the pleasure center and reduce the peripheral input of negative reinforcers.

Punishment is a negative reinforcer; an unwanted response results in either a painful stimulus or the withdrawal of a positive reinforcer. It is surprising that humanity places such reliance on punishment in child-rearing and social affairs, and on concepts such as military deterrence in international relations, considering that all the evidence suggests that punitive acts are rather ineffectual means of changing behavior. Mild punishment must follow immediately after an undesired operant and must be continued for the life of the animal. If punishment is discontinued, even for a few trials, the initial behavior usually reappears. Extremely painful punishments can permanently suppress a piece of behavior, but such strong stimuli also produce undesired side effects; they usually act as an

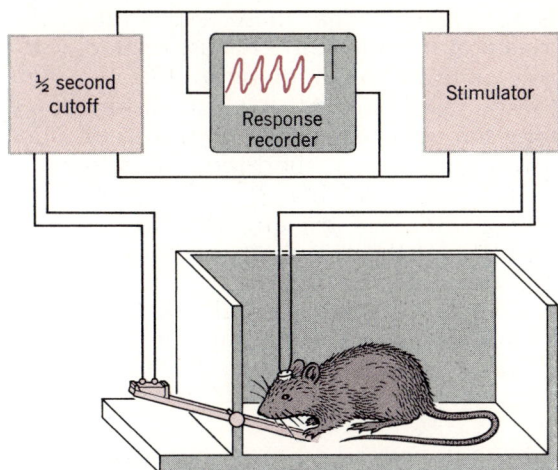

Figure 19-22 A rat can be allowed to stimulate its own brain through implanted electrodes. Depending on where the electrodes are located, this behavior may act as a powerful reinforcer for lever pressing.

unconditioned stimulus for the release of epinephrine from the adrenal medulla, for changes in heartbeat rate and blood pressure, and for negative emotions of fear and anxiety. An animal that has received severe punishment repeatedly will also be conditioned to negative emotional responses to the training box or experimenter. Monkeys forced to perform tasks under the threat of electric shock develop stomach ulcers; control monkeys given an equal number of shocks at random throughout the day remain healthy.

Memory

Learning must change some part of the nervous system in a relatively permanent way. It is clear that throughout evolution the development of the capacity to learn has paralleled the development of the cortex (Chapter 17). When it was discovered that vivid recollections of past events could be caused by electrical stimulation of the cortex, the theory was put forth that specific items of memory, called **engrams**, were filed away in separate

places in the brain. Because the sensory and motor cortex of the brain are subdivided into areas with different functions, it seemed reasonable to assume that associative behavior would work in the same way. However, different pieces of brain can be removed surgically from a trained rat without eliminating different specific memories. There is only very general forgetting, related to the amount of tissue removal. Moreover, multiple cuts, which presumably sever millions of nerve connections, can be made through the cortex of a well-trained rat without having any measurable effect on the animal's remembrance of things learned. Thus it seems very unlikely that memory traces are represented by specific chains of nerve cells and synapses that are somehow altered in their function.

Insight into this problem has come from studies of memory in mollusks. J. Z. Young and Brian B. Boycott, two British physiologists, have trained octopi to attack and eat crabs, but to avoid a crab when it is accompanied by a white card. If certain parts of an octopus's brain are removed, it no longer makes this discrimination. If such octopi are retrained on the same schedule as before (say, with a trial every 2 hours), they never relearn the association of white card and no attack. If training is speeded up, with trials every few minutes, the octopi can learn the association after several trials, but when they are retested 2 hours later they seem to have forgotten.

The octopus, then, seems to have two memory systems; a **short-term** memory retains training for a few minutes, and a separate, **long-term** system stores engrams for longer periods. It is now well established that mammals also exhibit long- and short-term memories. Rats or humans receiving electric shocks (as in electroshock therapy) usually fail to recall events or training sessions during the few minutes before the shock; this does not affect their memory of events in the more distant past.

These observations all suggest that memory is acquired in two steps. The first depends on the integrated firing of chains of nerve cells, perhaps impulses circulating (or **reverberating**) in closed chains of neurons. These may last only a relatively brief period—perhaps some minutes—and may be readily disrupted by any treatment (like a massive electric shock) that discharges a large number of cells in the brain. While the **transitory** memory lasts, however, a second, more stable **memory trace** is being laid down in the brain; this is no longer easily disrupted. Occasional epilepsy patients have been treated by removal of the hippocampus and temporal lobe (see Fig. 17-9) from the brain. Such patients find no interference with old memories, but are no longer able to establish new ones. They may carry on a rational conversation in which new information is discussed, but they completely forget it in minutes. The hippocampal system seems to be required for the establishment of new long-term memory traces.

Evidence seems to have been accumulating recently that long-term memory is encoded at the molecular level, in RNA molecules; some dramatic evidence has been presented in support of this idea. RNA, remember, is related to protein synthesis, and the amount of RNA in cells is generally proportional to how actively they are making new proteins.

In Sweden Holger Hyden trained rats to perform complicated tasks and found that the RNA in their neurons increased appreciably during the process. Some investigators have reported that the effects of training can be transferred from one animal to another by injecting RNA from the brain of a trained animal into an untrained **(naive)** subject, after which the untrained animal seems to know tasks that it has never been trained to perform, or learns them in fewer trials than an uninjected control animal. More recent work suggests that this effect may be an artifact of the experimental treatment, resulting, perhaps, from an inadequate statistical sample, but in any case unrelated to the RNA injected. It is found, moreover, that the antibiotic actinomycin D, which inhibits RNA synthesis, does not seem to interfere with learning when it is injected into animals. It

is not clear whether these results refute the idea that memory and RNA are related. Curiosly, puromycin, an inhibitor of protein synthesis, does prevent the establishment of long-term memories in goldfish and mice, but it has no effect on short-term memory.

This field—the chemistry of memory—is a young and very exciting area of research. Considering the complexity of the system and the vast number of questions still to be answered, we can safely predict that it will remain an active field for many decades.

Generalization

We spoke earlier of a behavioral continuum, with highly stereotyped, innate behavior at one end and progressively more experimental or learned behavior toward the other. At the extreme learning end of that continuum would be behavior involving **generalization,** the ability to see relationships. **Intelligence,** which is often tested in animals as if it were the ability to learn tasks, is probably more suitably defined as the capacity to generalize.

A hungry animal that sees food on the other side of a screen may scratch helplessly or attempt to dig under. It usually becomes agitated and, eventually, by running back and forth may find its way around the screen. Some puppies will immediately run around a fence the second time they are confronted with this problem. If the fence is made longer, some puppies will run only as far as the length of the original fence, whereas others will follow the fence to its end; the latter group has demonstrated the capacity to generalize. Similarly, a monkey readily learns to find a piece of food under a lid marked X, when the other two lids are marked O, or under a yellow lid when the other two are red. He will then look first under a round lid when the other two are square, or under a green one when the other two are purple, recognizing that the food will be in the one that is different. In both cases the animals have clearly understood not just the specific problem itself, but the **nature** of the problem.

The immense differences between the mental capacity of man and that of even the highest primates are nowhere more evident than in their capacity for generalization, because it is this ability that lies at the base of logical reasoning. Man shares with other animals instinctive behavioral patterns and physically induced motivational states. It is, however, the things that he learns, remembers, and integrates that comprise the raw materials for his rational thought, and—at least in part—for his emotions. Man has a temperament and is **self-conscious;** that is, he is aware of his own personality and mental processes. He dreams when he sleeps, as do some other animals, but he alone continues to dream and imagine when awake. Man searches endlessly for his free will. He, uniquely, needs to communicate to others his inner experiences. To do so he creates the ultimate form of generalization, a language.

Social Behavior

Many animals live in groups or communities and direct much of their behavior to other members of those communities. Such species are often referred to as **social** animals. In general, however, social behavior refers simply to the behavioral interactions among individuals of the same species. It is these interactions that represent the behavioral "cement" that keeps a group together and functioning. Animal interactions range from occasional encounters in a loosely organized species to those in enormously complex and interdependent societies represented by a beehive or a city.

In this brief section we make no attempt to survey the subject matter of the fields of sociology, social psychology, or political science, all of which, after all, are manifestations of the social behavior of man. Instead we restrict ourselves to some of the forms of social behavior seen most commonly among animals, with only passing references to their relevance to the human condition.

Mating Behavior

The primary social interaction among all animals (including man) is sexual. It is essential for perpetuation of the species that males and females **synchronize** their reproductive behavior; that is, that they be together and in a physiological and endocrinological state of readiness for reproduction at the same time. In many animals mating and nesting behavior is timed in part by external stimuli, such as day length or season, ensuring that eggs will hatch or young be born during the mild spring and summer months. Often, however, irrespective of the season, only the presence of a male and his display of specific patterns of **courting** behavior, acts as the releaser for mating responses in a female. Among birds, courting (Fig. 19-23, p. 554) involves a constant interplay between visual stimuli and endocrine responses.

The relation between endocrine responses and the visual stimuli of courting has been clearly demonstrated by the studies of Daniel Lehrman on ringdoves. These domesticated birds may be kept under carefully controlled laboratory conditions. Lehrman found that a solitary male or female will show no sign of mating or nesting movements. However, if a male and female are paired, they will carry out repeated cycles of reproductive behavior, each lasting six to seven weeks and each accompanied by profound physiological changes. During the first week the time is spent in courtship. The male bows, coos, and struts. A nest site is selected, and the two birds cooperate in gathering materials for the nest; they may mate toward the end of the week. In the course of this week the oviducts of the female increase to about five times their average weight. This same fivefold increase occurs in a female separated from a male by a glass partition; however, if she is paired with a castrated male that refuses to participate in the courting or nest-building behavior, the female's oviducts do not enlarge.

The physiological change and behavioral responses clearly result from reciprocal interactions. If a pair of birds is placed in a cage

Figure 19-23 *Courtship behavior in cormorants.* *(a) Food-sharing; (b) "necking"; (c) preening and wing-drying; and (d) mating.*

containing a nest with eggs, the nest is completely ignored. Courting and nest-building ensue. When the time comes for the birds to build their nest, they may select another site or even build it right on top of the already present eggs. However, if a pair of birds is allowed to court for several days and a nest with eggs is introduced into the cage, the birds stop their activities and promptly sit on the newly introduced nest. In fact, if isolated birds are treated with a series of injections with the hormone progesterone for a week, they will sit on eggs in a ready-made nest with no preliminaries; presumably the courtship behavior results in the secretion of high levels of progesterone. This interplay between hormone action and behavior synchronizes the complex activities required for reproduction and the rearing of young.

Courting behavior can be very complex. The herring gulls studied by Tinbergen normally live on the sand dunes along the Dutch coast of the North Sea. All of the breeding and raising of young takes place in special breeding areas, occupied only in spring and summer. Early in April, at some unknown signal, the flock flies off together and lands on the breeding ground. Herring gulls are **monogamous;** pairs quickly find one another and establish a territory that may be from about 2 to 50 yd in diameter.

Many of the birds on the breeding ground are youngsters who haven't previously paired. These birds settle in groups and spend much of their time, at first, dozing or preening. Occasionally a female walks toward a male, with her neck pulled in and her head held horizontal, pointing forward. She walks around the male several times, tossing her head and uttering a special melodic cry. The male may react by raising up to his full height. If another male is nearby, the male who is propositioned will attack the other and drive it off. The male and female strut about together, and then make some of the movements characteristic of nest building. The female soon begins strutting back and forth before the male. Every now and then she takes his bill in her own with a motion similar to the begging peck of a nestling. After a while the male regurgitates some half-digested food, which she avidly pecks from his mouth.

The males and females go through these ceremonies many times. In the beginning a female makes advances toward several males, but gradually she spends more and more time with one. Finally they leave the group as a pair and establish a territory. After the pair is formed the female may begin to entice the male as before, but now he never regurgitates food. Instead, he begs for food in return with tossing movements of the head and begging pecks. After a while, the male moves behind the female with head stretched out, moving in a rhythmic fashion. He gives a series of harsh cries, jumps on the female's back, and they copulate.

For mammals courting displays are by no means as elaborate as among birds, but perhaps they are equally important. In many mammals that are nocturnal in habit, visual stimuli may be less significant than chemical and tactile means of synchronization. We have already mentioned the importance of odor among many mammals for recognizing sexually receptive partners. Deer have a series of calls and nudges that function to stimulate a female to stand for the male and allow copulation.

The behavior of many primates involves mutual grooming, licking, even a type of oral contact—kissing—before mating. Some primates, baboons and rhesus monkeys, for example, have what is known as "sex skin" localized around the anal area. The sex skin and lips of the vagina become swollen and brightly colored midway through the estrous cycle, at about the time of ovulation. Male chimpanzees rarely make sexual advances until this time, and they diligently inspect the genital area of an interesting female with much sniffing and licking as a prelude to coitus. The physiological readiness of the female is one of the most potent releasers of interest in the male. A female chimpanzee will mate with a number of different males; the males show no sign of jealousy or possessiveness. Sometimes mating is initiated by the male, who may display himself with the hair on his head and arms erected, swinging from branch to branch. If the female is receptive, she will go into a coital crouch almost immediately. At other times, the female solicits coitus by crouching before a male (As does the female gorilla in Fig. 19-24, p. 556).

In humans, also, a period of sexual arousal is common before the coital act is completed. Women—alone among mammals—are sexually receptive at essentially all times during their estrous cycle. Thus sexual behavior among humans, more than any other mammals, may be divorced from reproduction. In all animals, including man, precoital activities are important physiological steps in preparing the sex partners for successful mating (Chapter 11).

Another important function of mating rituals is **species isolation.** Courtship behavior, no matter how simple or elaborate, is always highly species specific. Premating behavior of members of one species fails to elicit mating activity in closely related species. Numerous examples exist of closely related species that are **sympatric;** that is, they share common habitats, yet never interbreed. Even if interspecies hybrids can be produced in the laboratory, the species remain isolated because of the courtship behavior of one species is not a releaser of sexual activity in members of the other.

Animal Groups

Essentially all animals form groups, at least from time to time. It is extremely rare that any individual lives in isolation from others of its species for more than a brief period. Many animals such as gorillas and many other primates form aggregations based primarily on **family** relationships. Many species—some birds, bats, locusts, fish, antelope, bison, and gnats, to name a few—live always in **flocks** or **herds.** For some, such as bees, termites (Fig. 19-25), prairie dogs, and man, complex group interactions have evolved into the formation of **communities** or **societies,** which are relatively permanent; that is, they retain their organization and structure over many generations and have a division of labor. Individuals play separate and mutually inter-dependent roles.

Group life affords a variety of adaptive advantages. Aside from merely bringing the sexes together for reproduction, grouping in all species ensures a greater likelihood that the young will be cared for. Groups of animals are less likely than individuals to be attacked by predators. For some small, warm-blooded animals that nest together, such as field mice, the group provides warmth.

The stimuli that hold congregations of animals together are often not difficult to see. Mating provides its own reward. The bonds that unite parent and offspring are varied, but clearly reciprocal. The larvae of social wasps, for example, are fed by workers, but they exude droplets of fluid that their nurses drink. Licking is a positive reinforcer for many mammals, who repeatedly lick their nipples

Figure 19-24 Mating behavior in primates like the gorillas pictured here is usually preceded by grooming and other pleasurable activities.

Figure 19-25   Termites are organized in a highly structured society by a rigid system of occupational castes. Members of the different castes have body parts specialized for distinct functions. Compare the elongated head and powerful pinching mandibles of the soldier (center) with the more generalized structure of the three workers.

and genital areas during pregnancy and birth and lick their young all during their infancy. Similarly, nursing is a positive sensation to all female mammals, including women, and so provides its own inducement for a mother to suckle her offspring. The specific stimuli that compel larger groups—flocks, schools, societies—to remain together are only poorly understood; they are complex, and they are reciprocal among individuals.

Territoriality

A frequent aspect of the organization of animal groups is **territoriality,** in which, as we've already mentioned, an individual or pair defends a definite area from intrusion by other individuals (especially males of the same species). Territories are established in some form among species as widely diverse as crickets, dragonflies, lizards, howler monkeys, red deer, beaver, fur seals, fish, and many birds.

Several types of territory are known. One type is an area within which mating, nesting, and feeding all occur. In another type a pair mates and builds its nest, but feeding occurs in a nearby feeding ground where animals congregate amicably. Territories vary greatly in size; a territory may be an area of a plain, a small wooded region, or a few feet at the bottom of a pond. A golden eagle may de-

fend a territory of 30 to 40 square miles, a song sparrow's territory is a few hundred square meters (Fig. 19-26a, p. 558); a nesting penguin's is about $\frac{1}{2}$ meter square (Fig. 19-26b), just large enough for a nest site.

Whatever the type or size, territories seem to function mainly in spacing individuals in a way that minimizes the more severe aspects of competition and individual antagonisms and improves social stability. Territoriality also has significant evolutionary significance, because males that are unable to secure a territory are generally not able to reproduce.

Social Hierarchies

Many animal groups, whether a tribe of baboons or a flock of hens, are often organized according to a scale of **social dominance,** in which a senior male or female may act as the leader of the group. A new flock of hens, for example, will establish, in the first few days, a series of dominance relationships that rank its individuals. These relationships are based on the first few hostile encounters between each pair of birds. Generally one bird dominates all the others; she can peck any other hen without being pecked in return. A second hen can peck all birds in the flock but the first; a third can peck all hens but the first two; and so on. Thus a flock

Figure 19-26 (a)
Territories of song sparrows
on Mandarte Island.
Because of the small size
of the island (about a mile
long), each territory is
somewhat smaller than
occurs on the mainland.
The 53 territories cover
the entire available
habitat. Area shown as
white does not contain
vegetation suitable for song
sparrows; each territory
contains only as many
birds as it will support. (b)
Nesting site of Adelie
penguins.

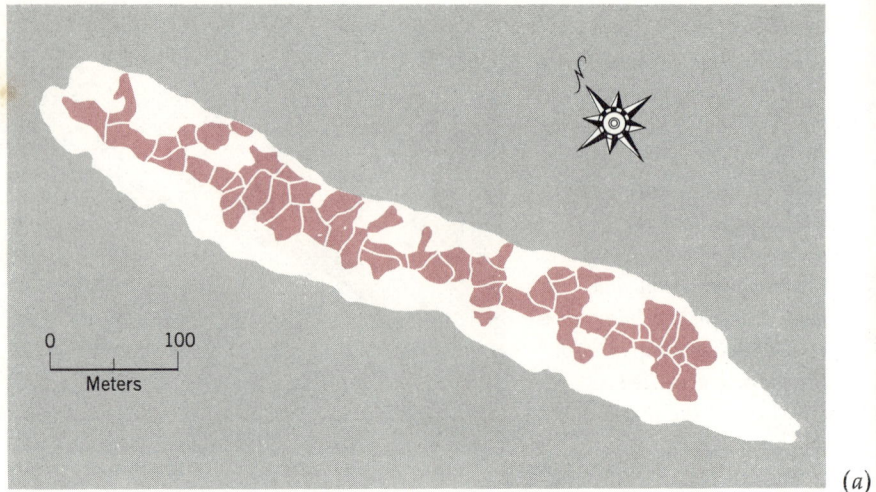

(a)

(b)

quickly establishes a **pecking order.** Hens that rank high in the pecking order have privileges such as first chance at the food trough and best position on the roost and in the nest boxes.

Once established, a social hierarchy tends to give order and stability to a group. Often a mere raising or lowering of the head is sufficient to acknowledge the dominance or subordinance of one hen to another; life then proceeds in relative harmony. There is little tension or fighting. A flock that is disrupted by the removal of hens or addition of new ones must reestablish a new pecking order. Frequent and sometimes bloody battles occur; the birds eat less, gain less weight, and lay fewer eggs—a fact recognized by every poultry farmer.

The Relevance of Behavioral Studies

It is probably less necessary than in many other subjects to point out here the relevance of the topic to our own activities. Animal behavior, like animal function, morphology, or development, is a proper part of biological science, but it is one that impinges very much on the more human subjects such as psychology, sociology, or political science. It should be clear that to understand the behavioral activities and interactions of organisms, vertebrate or invertebrate, "low" or "high," it is necessary to comprehend the biological bases of these phenomena. A clear example of this can be seen in our concept of aggression.

It was once thought that animals in nature are extremely aggressive toward one another, exhibiting much **agonistic** (threatening or combative) behavior and fighting to the death at every encounter. By and large this idea was based on inaccurate and incomplete observations, and on a misinterpretation of Darwin's concept of the survival of the fittest (which to him was an evolutionary construct, not a description of individual behavior).

However, it led many social thinkers to conclude that aggressiveness among humans is a legacy from our evolutionary past. War and murder, it was said, are merely human expressions of our animal instinct toward aggression.

In fact, a large number of recent, well-controlled studies, both in the field and laboratory, have shown that animals almost never kill members of their own species, and, except for predator-prey relationships, rarely kill those of other species. Although an animal will resort to aggressive behavior to obtain a territory or establish dominance, these relationships actually serve to diminish fighting between individuals of the same species. Most of this aggression is highly ritualized, involving prancing, fearsome noises, gestures, and threats, but these actions rarely lead to bloody battles.

Male iguanas of the Galapagos Islands fight in defense of their territories by pushing their heads against one another; one eventually drops to its belly and admits defeat, and the fighting ceases. Stags of the Indian black buck, which have long, vicious horns, follow a careful ceremony in fighting (Fig. 19-27, p. 560). They attack only when they are facing one another; the horns are used only for dueling, not for goring. Many other horned animals fight in this way. Male gorillas occasionally display combative activity and temper. They will stand at arm's length beating their chests and making terrifying faces at each other. They stamp the ground, jump, threaten, and feint toward their opponent (Fig. 19-28, p. 560). Only rarely, however, do they actually come to blows, and even less often in such brief fights is one of the combatants badly injured. Usually, after a time, one animal gives a signal of **appeasement,** indicating defeat, and the fight is over.

Aggressive behavior among animals, then, is almost always related either to predation or to territorial defense; and in the latter case is often highly ritualized in its expression. In this sense aggression is instinctive. However, at least in mammals, recent evidence has shown that aggression is highly susceptible to modification by experience. The psychologist Harry Harlow and his colleagues at the

Figure 19-27 A territory is defended by many animals by means of semiritualized contests. The male black bucks of India have needle-sharp horns, but rarely gore one another in these duels. The victor acquires food and breeding rights.

Figure 19-28 Gorilla in an agonistic posture.

University of Wisconsin have raised baby monkeys under a variety of conditions. Monkeys growing up with inadequate maternal attention, under hostile or even abusive circumstances, show extremely aggressive behavior toward other monkeys. In addition, they tend to be hypersexual, devoting an unusual amount of time to a variety of sexual activities; and they are difficult to train.

On the other hand, experiments with rats and mice have demonstrated that "natural" aggressions can be diminished by early experience. Adult male laboratory rats will usually attack and kill mice. However, rats raised as infants in mouse litters, or paired with young mice for the first month after weaning, do not kill mice when they become adults. Moreover, in a converse experiment it was found, unexpectedly, that mice reared by rat mothers were *less* aggressive toward other male mice than control animals. Obviously we must reject any hypothesis that states that aggression among mammals is a genetically determined, instinctive response that cannot be modified by experience. We may hope that we will learn enough, soon enough, to apply these concepts to human behavior.

Reading List

Atkinson, R. C., and R. M. Shiffrin, "The Control of Short-Term Memory," *Scientific American* (August 1971), pp. 92–102.

Bitterman, M. E., "The Evolution of Intelligence," *Scientific American,* (January 1965), pp. 92–99.

Carr, A., *So Excellent a Fishe.* The Natural History Press, Garden City, N.Y., 1967.

Davis, D. E., *Integral Animal Behavior.* MacMillan, New York, 1966.

DiCara, L. V., "Learning in the Autonomic Nervous System," *Scientific American* (January 1970), pp. 30–39.

Gibson, E. J., "The Development of Perception as an Adaptive Process," *American Scientist* (January–February 1970), **58,** pp. 98–107.

Haber, R. N., "How We Remember What We See," *Scientific American* (May 1970), pp. 104–115.

Hailman, J. P., "How an Instinct Is Learned," *Scientific American* (December 1969), pp. 98–106.

Heimer, L., "Pathways in the Brain," *Scientific American* (July 1971), pp. 48–60.

Kagan, J., "The Determinants of Attention in the Infant," *American Scientist* (May–June 1970), **58,** pp. 298–306.

Kandel, E. R., "Nerve Cells and Behavior," *Scientific American* (July 1970), pp. 57–71.

Kennedy, D. G., "Small Systems of Nerve Cells," *Scientific American* (May 1967), pp. 44–52.

Kennedy, D. G., "Nerve Cells and Behavior," *American Scientist* (January–February 1971), **59,** pp. 36–42.

Klopfer, P. H., "Sensory Physiology and Esthetics," *American Scientist* (July–August 1970), **58,** pp. 339–403.

Konishi, M., "Ethology and Neurobiology," *American Scientist* (January–February 1971), **59,** pp. 56–63.

Levine, S., "Stress and Behavior," *Scientific American* (January 1971) pp. 26–31.

Lorenz, K., *On Aggression.* Bantam Books, New York, 1967.

Luria, A. R., "The Functional Organization of the Brain," *Scientific American* (March 1970), pp. 66–78.

McGill, T. E., *Readings in Animal Behavior.* Holt, Rhinehart & Winston, New York, 1965.

Marler, P., "Birdsong and Speech Development: Could There Be Parallels?" *American Scientist,* (November–December 1970), **58,** pp. 669–673.

Marler, P. R., and W. J. Hamilton, III, *Mechanisms of Animal Behavior.* Wiley, New York, 1966.

Menaker, M., "Biological Clocks," *Bioscience* (August 1969), **19,** pp. 681–692.

Noton, D., and L. Stark, "Scanpaths in Eye Movements during Pattern Perception," *Science,* (1971), **171,** pp. 308–311.

Ralls, K., "Mammalian Scent Marking," *Science* (1971), **171,** pp. 443–449.

Roeder, K. D., "Episodes in Insect Brains," *American Scientist* (July–August 1970), **58,** pp. 378–389.

Schaller, G. B., The Mountain Gorilla: Ecology and Behavior. University of Chicago Press, Chicago, 1963.

Thomas, A., S. Chase, and H. G. Birch, "The Origin of Personality, *Scientific American* (August 1970), pp. 103–109.

Tinbergen, N., Herring Gull's World. Doubleday, Garden City, N.Y., 1967.

Van der Kloot, W. G., Behavior. Holt, Rinehart & Winston, New York, 1968.

Von Frisch, K., Dancing Bees. Harcourt Brace Jovanovich, New York, 1965.

Willows, A. O. D., Giant Brain Cells in Mollusks," *Scientific American* (February 1971), pp. 68–75.

chapter twenty

Population Biology

"In Indian Bengal, I was very touched to see little girls carrying babies around, and commented to a Hindu doctor on this concern for their brothers and sisters. He looked at me to see if I was joking. . . . I then found that the average age for marriage for girls in this district was 12 years, and that this was a decided improvement on the average a generation ago, when it was 10."

RITCHIE CALDER, 1962, *POPULATION PERSPECTIVE.*

Populations and Social Interactions

In Chapter 5 we introduced the cell as the basic biological unit. We soon saw that cells interact with one another to form multicellular organisms in much the same way that molecules form supramolecular aggregates, which in turn combine to form cells. Organisms, too, interact, as we tried to show in Chapter 19, and we have explored how some of their interactions work. A **population** is an aggregation of organisms of one species and as such shares many of the properties we have already discussed at other levels of organization. Populations develop and grow. They possess means of obtaining energy and distributing it. They secrete waste products. Populations exhibit coordination of their parts; and they have homeostatic mechanisms to defend against environmental change. Populations evolve, age, and die.

What are the interactions that members of a population experience? What advantages do those social phenomena confer? The primary social interaction in a population, you will recall, is **mating.** The result is the sexual exchange of genetic information between members of a species, and, often, cooperative care of the young. Both are clearly in the interests of the individual.

Territoriality, a social phenomenon we discussed on p. 557, means that resources in short supply, such as food or space, are parcelled out to a select few. The rest of the population fails to breed. The alternative to this behavior is open competition for resources, which could lead to much disruptive violence and the possibility that no

Figure 20-1 A mixed herd of wildebeests and zebras grazing together in Tanzania. Cattle egrets accompany the herd. An elaborate and sensitive alarm system functions among the three species for the protection of all.

individual would obtain enough to survive.

Social behavior also confers the advantages of numbers. Animals may increase their **effective size** by operating in bands in defense against predators (Fig. 20-1), or as predators against larger, stronger prey. Small birds in a group drive away a large hawk or owl; a pack of wolves can run down a much larger animal than any individual would attack. Larger effective size is also a protection against the physical environment. Beehives and termite mounds are maintained at almost constant temperature by the activities of their communities; birds and mammals keep warm by huddling in groups; a school of goldfish can remove toxic substances from water to which a single fish would succumb.

A group is not only larger than an individual, it can obtain more **information** from a larger area. Whereas a single organism must constantly interrupt its feeding or other activities to guard against approaching predators, those in a group can spend more time feeding or resting because only a few need to be on guard at any one time (Fig. 20-2). The **division of labor** that a group or society permits is one of its definitive characteristics and greatest advantages. To the extent that these advantages operate, social behavior will evolve by natural selection because an individual acting in a group will leave more offspring than a solitary member of a species.

Once a communicative social life is established, animals reap another benefit, perhaps the most important one: they learn from one another; that is, they **transfer experience.** Greenfinches and great tits are both common birds in England. The greenfinch is solitary; the great tit lives much of the time in gregarious flocks. If seeds are filled with aspirin to make them distasteful, and placed on a colored dish, individual birds of both species soon learn to avoid all seeds on that dish and to eat only from a white dish. When pairs of each species are placed in adjacent compartments of a cage in full view of each other, however, the normally gregarious tits learn, faster than an isolated bird learns, not to eat the seeds. Each bird apparently is able to benefit from witnessing the other's mistakes. Two greenfinches, in contrast, never learn to avoid the distasteful seeds. When one sees the other trying the seeds, he ignores his own previous experience and tries them again himself. This confuses the other, which repeats the cycle.

The ability to benefit from the experience of others is probably the most important by-product, in evolutionary terms, of social behavior. It leads to the ability to discriminate among various signals of other individuals, which results eventually in the elaboration of more complex communication systems.

If groups of animals enjoy distinct advantages, they also run the risk of at least one major disadvantage. A population is more than merely an aggregation of organisms; it is an aggregation *in space.* Populations are characterized not only by numbers, but by **density;** that is, numbers per unit area (or per unit volume). A population of 1000 animals distributed over a square mile has very different properties and allows for different interactions than does the same population con-

Figure 20-2 *Gannets flock by the thousands on Bonaventure Island, Gaspé.*

densed into a smaller area. The disadvantage is the danger of too great numbers, of **overpopulation.** If a population grows too fast or too large for its **habitat** (the range it normally occupies) the effects can be disastrous.

The suggestion of overcrowding among animal populations seems to contradict the widely held view of a **balance of nature,** the idea that prey and predators always balance one another. It has long been assumed that population size is regulated by a set of natural negative-feedback controls, of which the main factors are predators, starvation, disease, and accidents. More recent thinking has suggested that these ideas do not apply to all populations and circumstances.

Nature's balance clearly does work, and in fact is highly efficient; but it is not perfect, and it does not rely simply on predator-prey relations. Some species overproduce consistently; their numbers increase and then abruptly decline in fairly regular cycles. Minnesota jackrabbit populations rise and fall through cycles of several years. During the times when they are dying in great numbers, there is usually plenty of food in the habitat and no sign of starvation; there is no evidence of excessive predation, nor of epidemic disease. There are, moreover, a number of animals that effectively have no predators and are not readily subject to disease; the lion, the eagle, and the skua (a North Atlantic sea bird) are notable examples. Yet these forms maintain limited populations under circumstances in which even casual observation shows that there is plenty of food.

About 40 years ago, a pair of deer was put on a small island of about 150 acres in Chesapeake Bay. The deer were kept well supplied with food, and soon a colony grew to a density of about one deer per acre. Then the animals began to die unexpectedly, despite adequate food and care.

When the dead deer were examined, they exhibited a characteristic set of symptoms: liver disease, hypertension (high blood pressure), atherosclerosis (hardening of the arteries), and degeneration of the adrenal glands, thyroid, and kidneys. These symptoms, called the **adrenal stress syndrome,** are typical of the acute stress that results from prolonged overactivity of the pituitary and adrenal glands (Chapter 16). Interestingly, the Minnesota jackrabbits showed similar symptoms when they died in their regular cycle. Studies in the Philadelphia Zoo indicated that a wide variety of wild animals (woodchucks, antelope, monkeys, etc.) suffered up to tenfold increases in fatal heart disease and atherosclerosis under crowded conditions. In 1950 John Christian concluded a now-famous study of crowding among mice with the working hypothesis that population-wide death occurs under conditions of high population density, by "exhaustion of the adreno-pituitary system" resulting from increased stress.

Much other evidence since has indicated that the single factor of too many individuals in a given space, with no other privation whatever, can lead to adrenal stress syndrome and a dramatic increase in death rate. The apparent applicability of these findings to some of humanity's problems is too obvious to miss, and we shall return to this point later (p. 572).

Origin and Growth of Populations

The Growth Curve

One way a new population can arise is by **physical isolation.** In the laboratory a single bacterium or protozoan (or cultured tissue cell) may be placed in a dish with appropriate nutrient medium. Soon the cell divides (Chapter 7), producing two daughter cells. Each daughter increases in size to that of the original **progenitor cell,** and reproduces to form another pair, and so on, doubling in every generation in a geometric progression 1, 2, 4, 8, 16, 32 . . . .

If we plot the total number of organisms at the end of each generation in a **growth curve** (Fig. 20-3), we see that the population increases slowly at first but that *the rate of growth increases with each generation.* Thus in the second doubling the population of bacteria increases only by two in 20 minutes (from two to four), but at the end of the ninth doubling, for example, in a similar period of time (20 minutes), the population increases by more than 500. Such progressions are expressed algebraically in terms of exponents, or logarithms; hence, this kind of geometrically accelerating growth is called **exponential,** or **logarithmic** growth. If it goes unchecked, exponential growth can lead to fantastic numbers. At a continued rate of one doubling every 20 minutes, in only 36 hours our population of bacteria would grow enough to form a layer a foot thick over the entire earth.

A single pair of sexually reproducing animals can produce a population in similar fashion if isolated in a suitable habitat. The pair of deer isolated on the Chesapeake Bay island formed a colony of about 150 animals

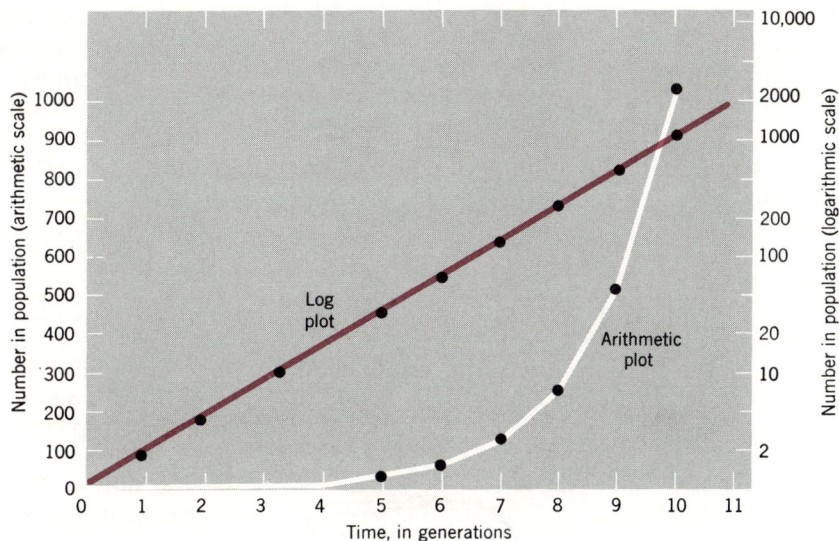

Figure 20-3 An exponential population growth curve in which the total population at the end of each generation is plotted. The curve in white shows an arithmetic plot; the population size, shown on the vertical ordinate at the left, is arranged in linear fashion. The straight-line curve (in color) is a log plot; the number of organisms in the total population has been converted to a logarithm (ordinate on the right).

in less than three decades. From 1879 to 1881, 435 striped bass from the Atlantic Ocean were planted in San Francisco Bay. In 1899, 20 years later, the commercial catch of these fish amounted to 1,234,000 lb. Clearly, the exponential growth of both the deer and the bass must have been essentially unchecked. In cases like these the original animals from which a new population is formed are called **founders.** The reproductive capacity of an individual is called its **biotic potential.** A small, localized population of a single species is referred to as a **deme.**

New populations can also form as a result of mutation, leading to **genetic isolation.** For this to occur, the new deme must be unable to exchange genes (that is, to mate) with the originally homogeneous parent population. There are a variety of ways this happens; we shall discuss them in Chapter 23 when we deal with evolution and the formation of new species.

Whether a new population forms by physical or genetic isolation, if it finds itself in a favorable habitat it tends to increase at an exponential rate. In Figure 20-3 the slow increase in numbers early in the growth of the population and the much faster increase after

several generations is obvious in the arithmetic plot. But the gradually rising curve can be converted to a straight line by the log plot. The value of this mathematical conversion is that the slope of that line is a direct measure of the **average doubling time** of the population, and more nearly represents the biotic potential of each individual. A nearly flat slope means that the population is doubling slowly; a steeper slope indicates more rapid reproduction.

The straight log line shown in Figure 20-3, applied to a dish of bacteria, indicates that from the very first division the doubling time was 20 minutes, and that it remained constant at that period for the 11 doublings recorded. In nature this would be unusual. More often organisms find themselves in an environment that is less than optimal, and the doubling time is therefore longer. The first division of a bacterium might take 40 minutes instead of 20, the second 38 minutes, the third 35 minutes, and so on, the doubling time gradually decreasing down to 20 minutes as the organisms accommodate to their new environment. Such a period of longer but gradually diminishing doubling times is called the **lag phase** of the growth curve (Fig.

20-4), and precedes the **log phase** in many natural situations.

Note that the lag phase is not the same as the gradually increasing slope of the accumulated population curve plotted arithmetically (Fig. 20-3). The arithmetic curve shows the increase in numbers with a *constant* doubling time; the transition from lag to log phase shows the increase in the rate of that increase with a *changing* doubling time.

The rate of growth of every new population eventually slows down, and it enters the final phase of the growth curve, the **equilibrium phase.** Now death raises its ugly head, as

chemical, physical, or biological factors of the environment become **limiting.** Our bacterial population could not actually climb out of the dish and begin spreading over the earth. Long before that happened the colony would succumb to one or all of three obvious restrictive factors: it would run out of nutrients in the medium; the layer of cells would become so dense that oxygen from the atmosphere could not diffuse through it fast enough (remember the diffusion limitation of .5 mm discussed on p. 421); or it would produce metabolic by-products such as CO_2, lactic acid, and so on, so rapidly that toxic concentrations would accumulate in the medium. That is, if left to itself the population would either starve, asphyxiate itself, or drown in its own excreta. In any case some of the organisms would fail to survive long enough to reproduce, and the constant increase in size of the population would diminish.

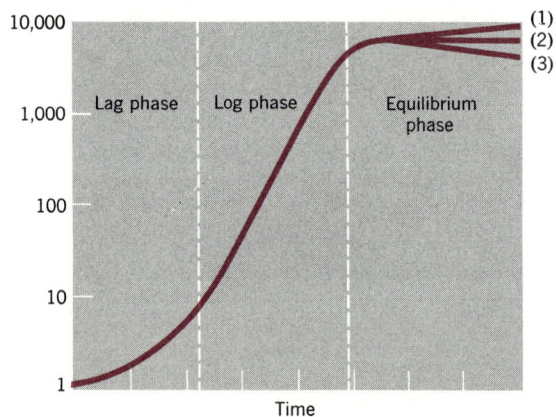

Figure 20-4   The three phases of a growth curve are plotted logarithmically. During the lag phase the rate of doubling of the population gradually increases up to some maximum level. During the log phase, in which the average rate of doubling remains constant, the population increases exponentially, with a slope determined by the doubling time. Eventually the population reaches a point at which maximal growth is limited by an environmental restriction, and some organisms fail to survive long enough to reproduce. In the equilibrium phase the rate of mortality and reproduction may be equal, in which case the size of the population remains constant (curve 2). If mortality does not quite balance the reproductive rate, the population may continue to increase slowly (curve 1). With a higher death rate, the size of the population declines (curve 3).

Mortality and Fertility

The rate of death, or **mortality rate,** determines the slope of the growth curve during the equilibrium phase. A bacterium has only two possible fates: it can divide into two cells, or it can die. From the point of view of the individual bacterial cell, either of these acts is terminal. The log phase represents the condition in which essentially all cells reproduce; few or none die. The population doubling time is identical with the cell doubling time (or **generation time;** see Chapter 7).

As the environment begins to decay in its capacity to support growth, some cells die. If the mortality rate reaches a constant level that is lower than the reproductive rate, the population will continue to increase in size (Fig. 20-4, curve 1), but at a much reduced rate. The time it takes the population to increase by two will be much greater than the doubling time of the individual. If, on the average, one out of two cells dies with each division, the death rate equals the reproductive rate, and the size of the population reaches a constant level. The equilibrium phase then represents a **zero-growth condition** (Fig. 20-4, curve 2). On the other hand, when more cells die than are produced in a unit of time, the equilibrium-phase curve will have a negative slope and the size of the population will decline (curve 3).

A population of deer, or any other sexually reproducing organisms, progresses through the phases of an S-shaped, or **sigmoid,** growth curve similar to that for bacteria in Figure 19-4, but with some differences. A pair of deer "doubles" itself when the female has produced four offspring, not two (assuming that equal numbers of males and females are born). A sexual animal is not restricted to the choice of *either* reproducing *or* dying; most, in fact, do both. Finally, a sexual organism is not limited to reproducing itself always in units of two (as normally occurs in mitotic reproduction). A pair of deer may bear only

a single fawn each year for two or three years; a woman during her reproductive life can have 15 children or more (Fig. 20-5); a male and female clam produce several million fertilized eggs each season, each one of which, if it escapes predation, may form an adult.

There are several ways of measuring reproductive capacity in sexual animals. The **reproductive potential** of a female, her ability to bear young, is called **fecundity.** Fecundity is generally much larger than would be predicted by the natural birth rate (or **natality**), which refers to the actual number of young born (or hatched) per female. The **net reproductive rate** is the number of female offspring that replace each female of the previous generation. In a stable-equilibrium population in zero-growth conditions, the net reproductive rate must be 1; each female is replaced by one female in the next generation.

Figure 20-5 Although some families, like this one, still have many children, relatively few modern women reproduce at or near their level of maximum biological capacity.

The rate of growth of a population of sexual organisms thus reflects the balance between natality and mortality. Because the reproductive period of most vertebrate and invertebrate females represents an extended portion of their lives, the generations usually overlap. In such cases the population has an **age structure,** which is described by the fraction of the total population in each age group. Each age group has a specific birth and death rate, and these usually differ from one age group to another. A limited subpopulation, defined in terms of species, sex, and age, is called a **cohort;** we speak, for example, of the cohort of women between the ages of 20 and 24 years.

Everyone who is born must die; in the long run the number of births and deaths in a given population must be equal. But the birth or death **rate** (number per year) can vary widely. Suppose each of 100 women, at age 20, has one female baby. In long-range terms, assuming no infant mortality, that population has merely replaced itself, and we would see no growth. However, any census taken after that wave of births would count more than 100 persons in the population, depending on how many of the mothers survived, and for how long. If the life expectancy of all of these women were 30 years, then counts made 10 years after the birthwave would show a population not much larger than 100, made up mainly of 10-year-old children. If the life expectancy of the women were 70 years, however, these counts would show a population more nearly approximating 200, with about equal numbers of mothers and children. This hypothetical example illustrates two points: (1) the birth rate and death rate in a population must be equal for a long time—at least as long as the average lifetime of an individual in the population—before the growth rate becomes constant. Fluctuation in fertility or mortality result in a nonequilibrium population. (2) How great the difference in rate of birth and death (that is, how rapidly the population increases or decreases) depends on the age structure of the population.

Under ideal conditions, a death rate would occur only among the oldest individuals in a population, a result of the downhill physio-logical changes of old age. In actuality, however, the death rate is usually distributed among the different age groups of a given population in a manner that depends on the size and composition of the population as well as on environmental factors. In many species newborn offspring are especially susceptible to disease or predation; therefore in most populations the death rate is highest among the youngest *and* oldest members.

The death rate distribution among the various age groups of any given population may be illustrated by plotting the number of survivors (per thousand individuals born) against maximum life span. The resulting **survival curve** in effect shows the percentage of the population still alive at any time during the total life span of the species. The survival curves for a number of different organisms can be directly compared if the life span is expressed on a percentage basis, as shown in Figure 20-6. Under conditions when the theoretical minimum death rate prevails and results primarily from the ravages of old age, we would expect the survival curve to be a horizontal line for virtually the entire life span and then drop rapidly to zero near the end of the maximal life span (as is the case with the rat flea). At the other extreme there are populations (e.g., many birds) in which many members perish early in the life span as a result of predation, and fewer and fewer individuals survive for the maximal lifetime. In most natural populations the survival curves fall between these two extremes.

As a general rule, the proportions of various age groups in any given population can serve as a useful guide for predicting the future growth of the population. Populations with a large percentage of young individuals would be expected to expand, whereas those with a greater proportion of older members tend to decline. A population in which the age structure is about equally distributed tends to remain constant in size. Age structure is as accurate a predictor of future population size among humans as with other species. The birth rate is highest for women around 20 years, lower before and after; human death rates are lowest between 12 and 20 years, highest in the first year and in old age. This results in a large percentage of young adults and in the currently observed rapid population expansion.

Mechanisms of Population Control

We have already noted many of the mechanisms whereby population size is controlled. Some organisms, such as small desert animals living under extremely harsh conditions, are at the mercy of their physical environment. Their populations grow only during the short wet season, and the size of a population often depends on their intrinsic rate of increase (fecundity) and the length of the growing season (Chapter 21).

Larger organisms, especially those living in more favorable climates, may be relatively independent of their physical environments; the challenges polar bears face in the arctic regions are starvation and predation by hunters, not the extreme cold. Under these conditions the size of a population is determined either by interactions among its members or by interactions of its members with those of other species.

Cross-Species Interactions

Interactions of one species with another often will take on the forms of **competition, predator-prey,** or **host-parasite** interactions.

When two species occupy a habitat and use any of the same resources (food, water, space, light, etc.), each of the two demes will grow to the point at which mortality balances reproduction. If any of the common resources is essential, both populations will inevitably be smaller than if either deme occupied the habitat alone. This is an operational definition of the existence of **competition** among populations.

Joseph Connell has studied interspecific competition among two types of barnacle, *Balanus balanoides* (which we shall label B) and *Chthamalus stellatus* (C). Their interaction provides a good example of such competition.

A small founder population of B and one of C entered a favorable environment at about the same time. Initially both were scarce in the new region and their growth rate was exponential. As their combined populations increased, however, their rates of growth began to slow down and eventually one (C) entered the equilibrium phase of the growth curve, but B continued to grow.

If any of the resources that two species share is truly essential (as attachment space is to barnacles), the continued growth of one reduces still further the ratio of birth rate and

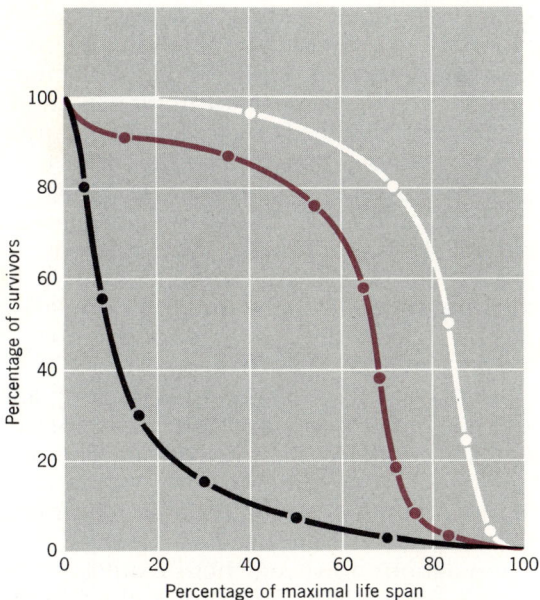

Figure 20-6 Survival curves for three species: rat fleas (the white curve); white American males in 1940 (colored curve); and European song thrush (black curve), showing the percentage of the population still alive at any time during the total life span of each species. Under nearly optimal conditions, very few rat fleas die until "old age," when population loss is precipitous. Among humans there is a relatively high death rate among infants and children, but most mortality occurs in the older population.

mortality of the other to such an extent that the other population begins to decline. Thus B increased while C decreased; B developed a **competitive advantage** over C and will thrive while C becomes rare. In fact, Connell observed individual barnacles actually prying type C off the rocks or growing over them. If B maintains this competitive advantage, C will be completely eliminated from the habitat. This is indeed what usually happens in a wide variety of species, and explains why it is rare to find two competitive species sharing a given habitat; organisms generally exhibit **competitive exclusion.**

A second natural means of regulating population growth is **predation.** Despite our comments on p. 565, the explosive increase of some species when their predators are removed shows that their growth had been controlled by those predators. Before 1907 the deer herd on the Kaibab Plateau in Arizona numbered about 4000. In 1907 a bounty was placed on mountain lions and wolves, the natural predators of deer. Within 15 years the predators were essentially exterminated and the deer population had increased more than tenfold. Ironically, the resultant over-browsing of the vegetation in the area led to massive deaths of deer and yielded a population of about half of what could have been maintained. Insects and plants that reach pest proportions are usually those introduced into a new habitat without predators or parasites. Prickly pear cactus (Fig. 20-7) was brought from South America into Queensland, Australia, early in the 1900s and soon spread over millions of acres, driving cattle off the land. When a caterpillar that normally lived on the cactus in South America was introduced into Queensland, it quickly destroyed most of the cactus population. We do not know whether these are special cases, but it is obvious that some populations are regulated by predators.

Parasites, unlike predators, cannot afford to kill their hosts, for they are then also eliminated. The fungus *Endothia parasitica* is parasitic on the bark of oriental chestnut trees in China. It rarely causes damage to its host. In 1904, however, the fungus was accidentally carried to the United States. Within 30 years the majestic American chestnut, which had

been predominant in forests of eastern North America, was virtually extinct. We are witnessing a similar eradication of the Dutch elm from our forests as a result of the ravages of a related fungus. Thus under equilibrium conditions a parasite-host relationship can exist without causing substantial harm to the host population; in a nonequilibrium state, the host population may be eliminated rather than regulated, to the detriment of both host and parasite.

Intraspecific Interactions

Interactions among members of the same species are social interactions. We have already mentioned several of these that are important in regulating population size. Mating is the source of population growth. Territoriality provides a mechanism for preventing certain members of the population from breeding—those that cannot defend a territory successfully. Density can be a population regulator when animals begin dying of adrenal-stress syndrome as soon as the population density exceeds some critical level. Thus no animal starves, but the population never outgrows the available food supply; it has a homeostatic mechanism.

Birds feeding on insects in the spring and summer are a good example of a **socially regulated homeostatic mechanism.** The production of insects in the early spring is so abundant that it could feed an enormous population. If it were necessary to feed such a population, however, there would not be enough for the later critical period when the need for feeding young birds is greatest and the production of insects is diminishing. To balance food need against food production, the birds must restrict the size of their population in advance. It is the threat of starvation tomorrow, not hunger today, that controls the population density. Among birds the mechanism for accomplishing this is usually territoriality. Instead of competing directly for food, the birds compete for pieces of ground. If the standard territory is large enough to feed a family, the entire deme is

Figure 20-7 (a) Without predators, the prickly pear cactus spread thickly over the grazing lands of Queensland. In 1926, a caterpillar that normally lives on the cactus was introduced. (b) The same area as above, three years later, showing almost complete destruction of the cactus.

safe from the danger of overtaxing the food supply.

Another social interaction that serves as a mechanism of population control involves certain group activities that allow a population to sense its size and to reduce reproduction when size becomes too great. The evening flights of huge numbers of starlings (Fig. 20-8), for example, the dawn chorus of birds and frogs, and the swarming of gnats and whirligig beetles are all thought to be activities of this kind, which perhaps by hormonal feedback mechanisms, may regulate reproductive activity.

Growth of the Human Population

We have presented a good deal of evidence so far that populations in nature are self-regulating. The population size or density appears to be controlled through a variety of automatic feedback mechanisms. Primitive man, in contrast, seems not to have been subject to such controls. He was, of course, limited to the food he could get by hunting;

but the major system for restricting his numbers involved social mechanisms such as tribal traditions and taboos: prohibiting sexual intercourse for mothers while they were still nursing their babies; practicing compulsory abortion and infanticide; offering human sacrifices; and waging wars between rival tribes. These customs, deliberate or not, could keep the population density balanced with the feeding capacity of a given hunting range.

Then, perhaps 8000 to 10,000 years ago, the agricultural revolution removed the limitation on food resources. There was no longer reason to restrict the size of the tribe; on the contrary, power and wealth accrued to tribes that increased. The old checks on population growth were gradually discarded; the rate of reproduction became a matter of individual choice rather than of tribal or community control. It has remained so ever since, and the human population now shows a tendency to expand without limit. Lacking the built-in homeostatic mechanism that regulates the density of animal populations, man apparently cannot look to natural processes to restrain his growth. He must do it by his own deliberate, socially applied efforts.

The first apelike predecessors of man prob-

Figure 20-8   Mass wheeling flights by starling flocks on fall evenings may include several hundred thousand birds. These maneuvers represent a communal activity that provides the flock with an indication of population density. If that density gets too great in relation to the food supply, the flock automatically reduces its reproductive rate to improve the balance.

Figure 20-9 (a) *World population growth from 1600 to 1970. The points before 1800 are estimates; the data are projected to the year 2000.* (b) *The rate of growth is calculated in percentage per year, averaged over 50- or 10-year periods.* (c) *This shows the times the population would have multiplied itself in 25 years if the rate of growth calculated for that period had continued.*

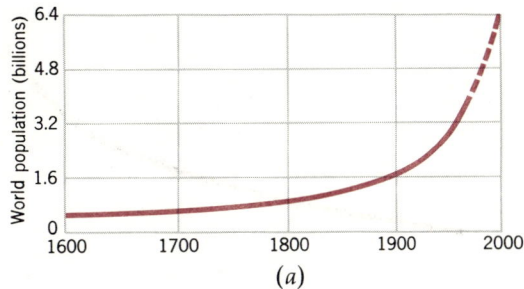

(a)

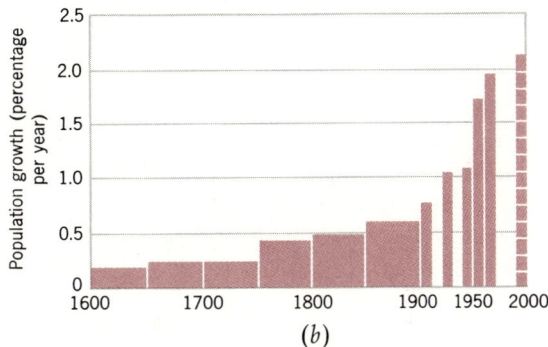

(b)

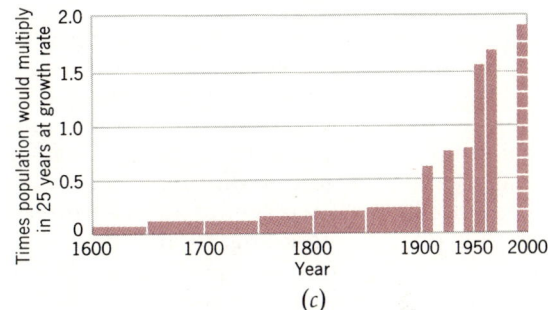

(c)

ably evolved more than 2 million years ago, but our own species, *Homo sapiens,* originated no more than 100,000 years ago. Obviously estimates of the numbers of men in prehistoric times cannot be very accurate, but a well-accepted guess is that by 8000 B.C. the total human population was about 5 million. With the rise of the first civilizations in Mesopotamia and the Nile Valley 4000 years later, the population of the earth had increased 16 times (i.e., had doubled four times) to a total of 80 to 90 million.

During the next 4000 years, to the time of Christ, world population reached an estimated 300 million, and climbed to about 1 billion by 1840. At the time of World War II (1940) the world population was over 2 billion; in 1970 it was between 3.5 and 4 billion. Thus from "creation" to about the middle 1800s the planet's population grew to 1 billion people; the second billion was added in less than 100 years, and the third billion in about 25 years (Fig. 20-9).

Demographers (people who study distribution and changes in population) calculate that the world's population is now increasing by about two people every second, about 75 million per year (the figure in 1970), or about a billion per decade. A reasonable prediction is that we will reach 6 billion, barring a disaster, by the year 2000—when most readers of this book will have families of their own.

In the 1800s the doubling time of humanity was close to 200 years. It was inconceivable to the average person of that day that man would ever suffer a population problem. Nonetheless, in 1798 an English clergyman, Thomas Malthus, had pointed out that in some "back settlements where the sole em-

ployment is agriculture, and vicious customs and unwholesome occupations are little known, the population has been found to double itself in fifteen years."[1] Malthus postulated that a worldwide doubling time of 25 years might soon exist, and noted that at such a rate the population would rapidly outstrip its capacity to grow food for itself. His ideas were clearly oversimplified and ahead of his

[1] From Malthus, Thomas R., *Essay on the Principle of Population* (7th ed.) J. M. Dent & Sons Ltd., London, 1816. Quoted in *Population in Perspective,* Young, L. B. (ed.). Oxford University Press, New York, 1968, p. 10.

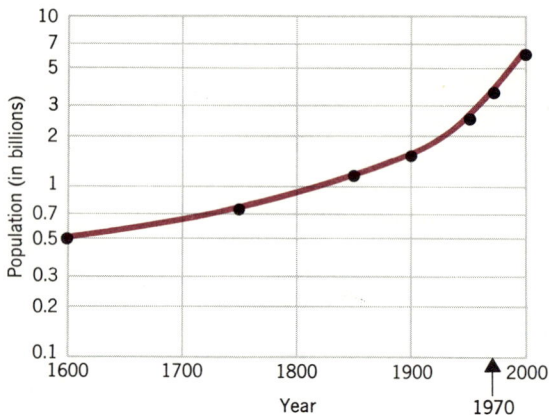

Figure 20-10   Growth of the world's population plotted logarithmically. Note that the slope of the line, representing the population doubling time, increases gradually. Only recently does growth seem to be entering a log phase.

time, but not by much. Our doubling time is now approaching 35 years.

If we replot the human growth curve (Fig. 20-9) as a log plot (Fig. 20-10), it is obvious that until recently we have been in a lag phase, with a low rate of doubling; but in the last 30 to 40 years we seem to have entered the log phase. According to the best estimates, the present average annual rate of population growth is between 17 and 20 per thousand (shown as 1.7 to 2 percent in Fig. 20-9). These figures are not birth rates. They represent the *difference between birth rate and death rate;* they mean that in 1970 (e.g.) the birth rate and death rate balanced out slightly in favor of births, so that for every 1000 persons alive in the world at the beginning of that year, there were 1017 to 1020 by the end of the year.

This appears at first to be a negligible change. Let us, however, make a population **projection,** by simply assuming that this rate of growth will remain constant worldwide until the year 2400 (for our arithmetic we use the more conservative figure of 17 per thousand, and start with the population census of 1964). What Malthus called the "fantastic

power" of a geometric progression is shown in Table 20-1. In the physical sciences any sudden large acceleration is called an explosion. Demographers have adopted this metaphor to describe the current rapid increase in growth of humanity as a **population explosion.**

At a rate of growth of 1.7 percent both the total population and its density increase about fivefold every century. In 1964 the population per square mile of the earth's surface (excluding the oceans and polar regions, was 61.4. By the year 2400 it would be about 100,000. This last figure may be compared with some current population densities: in 1960 Manhattan Island (New York County) had a density of about 77,000 persons per square mile, the highest for any political unit in the United States. By comparison, the density in the well-populated state of Massachusetts was only 650 per square mile, and in the United States as a whole it was 50.5.

Regulation of Birth Rate and Death Rate

How has such a rate of growth come to be? Why has our doubling time got so much shorter in the last hundred years or so? The reason is no mystery: mortality has fallen sharply and fertility has not. Worldwide, until 300 years ago, the birth rate and death rate were both very high, but the two balanced out almost exactly. A common death rate might have been about 50 per thousand in any given year, and that was also not unusual as a birth rate.

Of course, both tended (and still tend) to vary considerably from year to year and from place to place. In years when crops were good deaths would decrease slightly and population growth would be 5 or 10 thousand. Years of food scarcity tended to have higher death rates. Food shortage caused some starvation, but more importantly (for a demographer as well as for the people involved), it produced widespread malnutrition and opened a population to death from vari-

table 20-1

Projection of World Population Assuming Annual Increase of 17 per 1000

YEAR	POPULATION IN BILLIONS	POPULATION PER SQUARE MILE
1964	3	61
1975	4	74
2000	6	113
2025	9	178
2050	14	265
2075	21	405
2100	33	620
2200	178	3394
2300	974	18579
2400	5330	101672

ous infectious diseases. In one particularly bleak period the famous Black Death epidemic occurred. An epidemic of bubonic plague began in Constantinople in 1347 and spread throughout Europe during the following years. By 1352 the Black Death had killed one-quarter of the population of Europe (Fig. 20-11).

In the late medieval period the average life expectancy in Europe was about 27 years. During the seventeenth and eighteenth centuries, probably because of improved diet and hygienic conditions, it increased to about 31, and by the mid-nineteenth century in England, France, and the Scandinavian countries, it had advanced to about 41. During this period the birth rate remained fairly stable (until the latter part of the nineteenth century), resulting in a marked increase, in any given year, of births over deaths. For example, during the 1800s the average annual growth rate in Sweden, Norway, and Denmark was about 12 per thousand—nearly five times what it had been in 1750, and sufficient to triple the population during the century. In fact, it was at least partly in response to the increased population pressure that people began to emigrate from Europe by the millions in the latter part of the nineteenth century. Between 1846 and 1932 an estimated 27 million people emigrated from Europe, mostly to the United States, Australia, and South Africa.

Figure 20-11 Woodcut from an illustrated manuscript printed in the 1360s, showing burning of clothes during the Black Plague.

During the same period in the United States the birth rate fell drastically, from about 55 per thousand in 1800 to 22 per thousand in 1930. But a combination of factors, here as in Europe, caused mortality to fall even faster. These were all consequences of the scientific-industrial revolution, and included a rise in level of nutrition; great improvements in sewage and water supply systems; medical advances in the prevention of disease through inoculation against infectious agents (Fig. 20-12); and medical advances in the cure of diseases, mainly through the discovery of antibiotics.

In the short span since the founding fathers landed on Plymouth Rock, life expectancy in the United States has increased from about 30 years to 70 years; and in the period from

Figure 20-12 Compulsory vaccination during a smallpox scare, Jersey City, 1880's.

1900 to 1970 the total death rate, all ages included, dropped from 17 to 9 per thousand.

Today about half of all the people in the world have been born since 1945. The figure would be even more than half, but the death rate of the born-before-1945 group has been greatly retarded by science and medicine. Outside the economically developed regions death rates remained high until after World War II. After the mid-1940s the pronounced reduction in mortality in these countries has been a major cause of the tremendous recent acceleration in world growth rate. This decrease in death rate has resulted mainly from massive programs of inoculation against infectious diseases in India, Asia, Africa, and the Sino-Soviet continent, from a reduction in malaria in the tropical countries through DDT spraying, and from the widespread availability of antibiotics.

One of the most potent population effects of improvements in health and hygiene of a society is a phenomenal reduction in infant mortality. The survival curve for man (Fig. 20-6) reflects a relatively high infant mortality, a low death rate until about 50, and then an increasing rate. In 1965 the infant mortality in the United States was 24.1; that is, out of every 1000 live births, 24.1 babies died before the age of 1 year. This may be compared with the rate in 1900, which was close to 200 per 1000. Compared with those of other modern countries, however, the infant mortality in the United States is disgraceful (Fig. 20-13). For a country that considers itself to be among the most advanced medically, it is unacceptable to be ranked fifteenth in infant deaths, after the industrialized countries of Europe and Asia, with a rate more than twice that of Sweden, for example. The U. S. rate dropped from 24 to 21 deaths per 1000 infants in 1970, but this change did not appreciably alter our rank among nations. The rate among American nonwhites, in fact, compares more closely with those of some less developed nations (Poland, Jamaica). It is still lower than in some African and South American countries, whose infant mortality rates are numbered in the hundreds.

As these figures show, truly dramatic re-

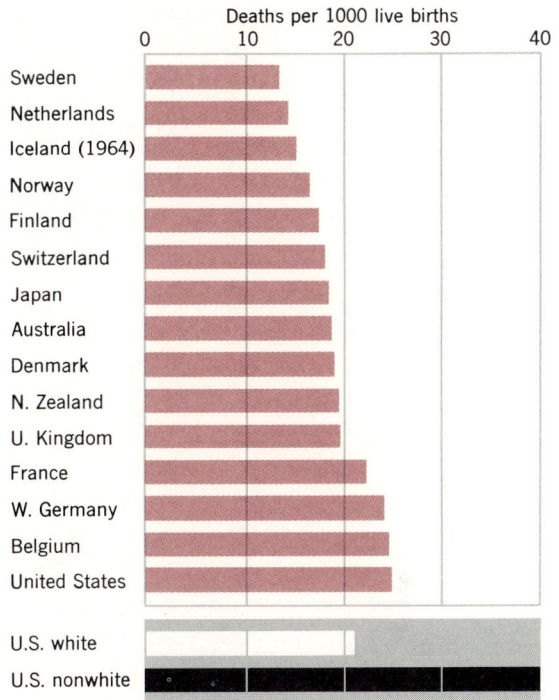

Figure 20-13   Infant mortality rates in 1965 of 15 economically developed nations. The United States is at the bottom of the list.

ductions in infant and adult mortality have attended the growth of science and technology. It is equally apparent, however, that this reduction has not yet affected some of the major undeveloped regions of the world, nor has it reached socially deprived individuals in the economically developed nations.

Economic development and the growth of science and industry clearly result in a drop in fertility as well as mortality, and at least in the United States this drop is continuing (Fig. 20-14, p. 580). In 1972, for example, the number of children in first grade across the nation was 5 percent smaller than it was in 1967. This drop followed a general trend that has continued with only brief upswings since the early 1800s. In 1910, for example, birth rate in the United States was about 30 per thousand. It dipped following the great de-

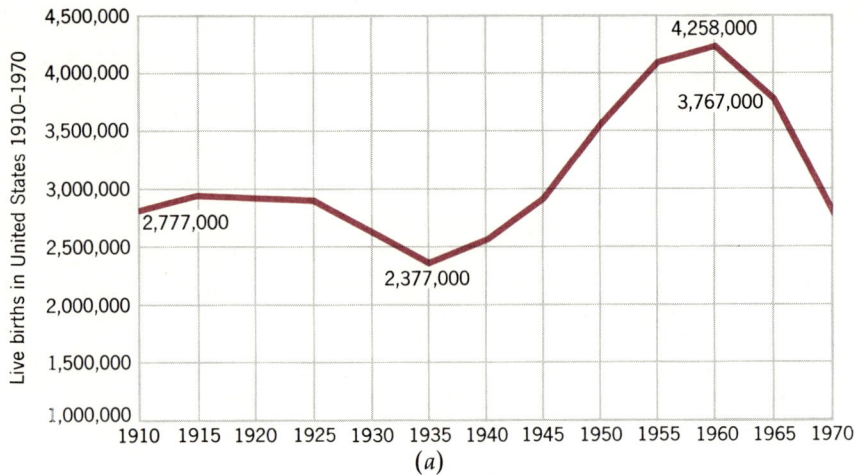

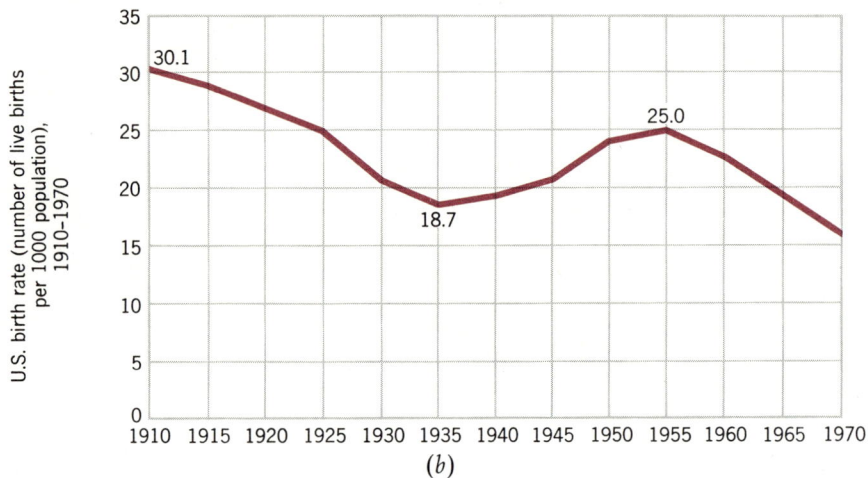

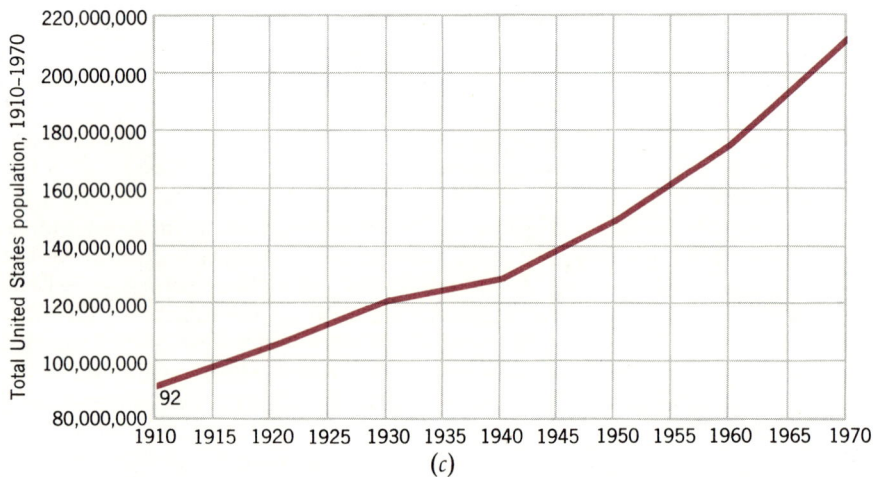

Figure 20-14   Population data for the United States from 1910 to 1970. The number of (a) live births, (b) birthrate, and (c) total population are shown.

pression of 1931 and rose briefly during World War II. But the trend down continues; in 1970 natality was 17.4 per thousand (Fig. 20-14); with a total U. S. population in 1970 of 206 million, that means one person born every $11\frac{1}{2}$ seconds (Fig. 20-15).

Men die from causes that are not usually under individual control: disease, war, accident, old age. Death rates rise or fall depending on such things as health measures, nutrition, and sanitation. To bear young, however, requires a conscious individual act: the act of sexual intercourse. As a result, birth rates rise or fall for reasons that trace back to the psychology, sociology, and emotions of individuals. If large numbers of people want to have children, or want not to, their motivation generally has a major impact on the birth rate.

Changes in fertility generally relate, too, to social factors such as age and frequency of marriage, birth-control practices, and the economics of child-raising. In preindustrial societies, for example, with high rates of infant mortality, large families are the norm. Children are needed to help work the land, and they begin to do so early; it is assumed that not all babies will survive; and the security of the parents in old age depends on having grown children to support them. The social cost of a large family in modern society, however, is very different. Compared to their peasant ancestors, modern children are expensive. They do not work; they go to school and must be supported. Social stability and governmental programs of social security reduce the dependence of parents on their children. A peasant woman expended essentially all her energies in child-bearing and maternal tasks; modern woman demands liberation from the fetters of maternity and seeks fulfillment elsewhere, in education and a career.

We can summarize the historical growth patterns of societies in terms of the **theory of demographic transition,** which places human societies in three categories. (1) **Pre-industrial societies** with high fertility and mortality have a low natural growth rate. (2)

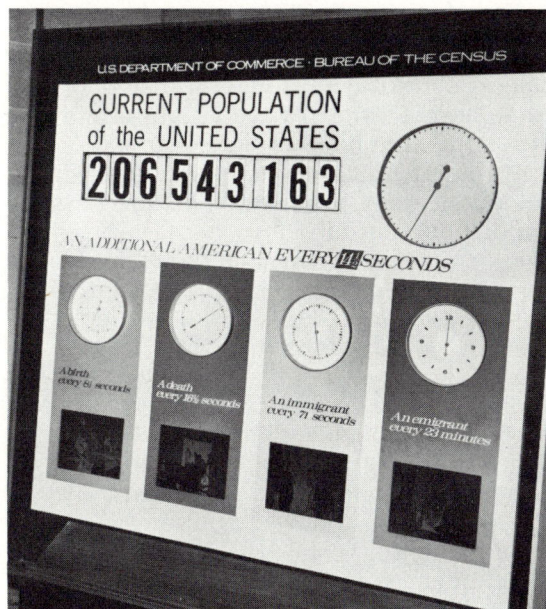

Figure 20-15   The U. S. Census Clock as of 8 A.M. on January 1, 1971. At a growth rate of .9 percent and a population of 206 million, one person was being added every 14.5 seconds. Although the rate of increase has declined slightly since then, the U. S. population is projected to exceed 300 million within the next 25 years.

Societies in **technological transition** continue to have high birth rates, but mortality declines, resulting in high natural increase. (3) **Advanced** societies have both fertility and mortality stabilized at lower levels, and therefore grow less rapidly or not at all.

Population Control

It is obvious that the world population is drastically out of balance, and is headed on a collision course (as we shall see in the next chapter) with the facts of ecology. Humanity is increasing at the rate of more than a million per week, yet we cannot adequately provide for those already on earth. More than half the world's population is malnourished, poorly housed, ill-clad, improperly educated or not educated at all. Medical care, social services, physical security are inadequate or nonexistent for many. Clearly the need for world population control is no longer a matter of debate. How, then, shall we go about solving this worldwide biological problem?

According to David Ehrenfeld, three widely held concepts seem to block progress in population control. These are: (1) The original doctrine of Malthus that the main danger in population growth is the potential exhaustion of world food supplies. (2) The idea that if birth control information and devices are readily available, people will automatically control their own family size to maintain a birth rate in balance with the death rate. (3) The idea that the groups most in need of population control are the underprivileged classes of industrialized countries and the inhabitants of the underdeveloped nations.

Although there is some truth in all these ideas, each has been distorted in ways that are antithetical to further human population control. It is true that malnutrition and famine exist in many parts of the world, and that these conditions are associated with high population density. However, when examined, these problems are generally traceable to local social, economic, and political conditions that interfere with distribution of food. It has been estimated that with present agricultural techniques the world can provide enough food to feed about 6 billion people an adequate diet. Thus it is not a proved fact that the human population is yet outgrowing its food supply. We have witnessed in recent years the latest "Green Revolution"; the use of new genetic strains of wheat has increased the annual production of that crop as much as sevenfold in many countries.

Ultimately, of course, Malthus may be proved right; it is hard to imagine that our present population, doubled several times over, could be fed. But at least for the next generation or two it seems reasonable to assume that improvements in agricultural technology can keep pace with world food needs, and we will continue to see food products rotting in some parts of the world while malnutrition exists in others. Thus, by placing too much emphasis on the threat of worldwide famine as a reason for population control, we run the risk of appearing to diminish the urgency of the need. It is too easy, especially for the public in a technologically advanced nation, to say, "Oh, we have plenty of time to face that problem," and do nothing. Considering the environmental and psychosocial damage that we are already doing to ourselves as a result of our excessive numbers, it seems wiser to develop and accept a justification for the practice of population control that emphasizes present dangers and current ills.

The second idea, that every society will reduce its own numbers if birth control information and equipment is made available, is also dangerously oversimplified. Judith Blake has analyzed data regarding desired family size from surveys of large numbers of couples, and has concluded that most American couples want more than three children; and that they already know about, and are using, reasonably effective contraceptive methods. The reason they generate a birth rate in excess of the over-all death rate is that, on the average, they *want* larger families.

This view is confirmed by a 1970 study in which it was concluded that if an ideal contraceptive were available—that is, one that was easy to take and was required only once a month or once a year, completely safe and reliable, nominal in cost, and associated with no side effects—for the American public it would decrease family size by only about 20 percent.

We are not suggesting that existing birth control and family planning programs are unimportant; clearly they are vital and should be expanded. However, it is of even greater importance that changes in public attitudes be brought about concerning ideal family size and the acceptability of having few or no children. Men and women who choose to remain single or childless, and thus deviate from traditional reproductive and family patterns, should be encouraged to do so, and social institutions should be modified to allow them to lead normal, healthy, and rewarding lives.

Progress is being made in these areas. The rise of the Women's Liberation movement is a current sign; a demonstrable increase in number of women in professions is another. Clearly the downward trend in birth rate in the United States in recent decades indicates that there has been a growing acceptance for, and means of, self-imposed control of fertility; the widespread use of contraceptive pills and intra-uterine devices (Chapter 11) attests to that. In countries where abortion has been legalized, as in Sweden and Japan, that method in addition to contraception has resulted in dramatic reductions in birth rates. It is especially noteworthy, therefore, that in New York State, in the six months after abortion was legalized in 1970, the number of abortions performed exceeded the number of live births.

The necessity for changing attitudes is especially pertinent in underdeveloped societies. Until a situation is seen in rational terms, no techniques or devices for altering it are relevant, however effective they may be. The point can be illustrated by citing a health survey of a village in Iran, made by the inhabitants themselves. According to this self-estimate, the state of health in the village was good. In fact, however, the incidence of many diseases was so high that they were not seen as abnormalities. In the view of the vil-

lagers all persons have two eyes, most ulcerated with trachoma. Both two-eyedness and trachoma are a part of nature; neither can be changed by human action. Only after such a community comes to recognize that there is a problem to be solved, a sickness to be cured, can a program of modern medical treatment be effective. Similarly, so long as societies look on children as facts of nature, gifts of God, and instruments of security in old age, any program of population control is doomed to failure (Fig. 20-16).

The third idea, that the problems of population growth arise primarily among underprivileged members of society, must also be reexamined. It is true that the poor in most modern societies have growth rates and family sizes that are slightly (but not much) larger than the middle and upper economic classes. But in most industrialized countries, groups classified as underprivileged by the standards of that country represent a relatively small percentage of the total population. Accordingly, changes in growth rates of these groups have proportionately little impact on the total national rate of increase.

It is also true that the birth rate in most of the underdeveloped nations still ranges be-

Figure 20-16 Street scene in India, illustrating a government-sponsored program of education to promote population control.

tween 40 and 60 per thousand, while mortality in recent years has continued to fall, in many cases below 20 per thousand. Growth rates of 2 to 3 percent are not uncommon, in contrast to 1 to 2 percent in the technologically advanced nations. This situation will inevitably lead in the next 15 to 20 years to appalling levels of human misery. In a city like Calcutta, today, there are 600,000 people who sleep, eat, and live in the streets, without shelter or means of subsistence. Similar situations exist, with only minor variations in severity, in most parts of the underdeveloped world. It is more than a mere statistic to state that half of the world's population today suffers some degree of malnutrition.

In terms of man's biological future, however, in terms of the carrying capacity of the planet Earth, a growth rate of 1 percent in the United States is far more disastrous than 2 percent or even 3 percent in India or Pakistan. Each resident of North America makes four times as a great a demand on the world's agricultural resources as someone living in one of the underdeveloped nations. Every American burns 10 times more fossil fuel, uses 20 times more fertilizer and 60 times more steel, than his Indian or African counterpart. In terms of total energy U.S. consumption averages about 10,000 watts per person per day, compared with about 100 watts in most of the rest of the world. Severalfold increases in the populations of the nonindustrialized nations thus will lead to further human suffering, but this will continue to be felt mainly on a local or national basis. Similar increases in the size of modern societies, however, can lead to a worldwide exhaustion of nonrenewable resources, which can be disastrous on a planetary scale.

Is There an Optimal Population Size?

Until only very recently in human affairs an expanding population has been equated with progress. "Increase and multiply," the Bible tells us. As we shall see, the number of surviving offspring is the measure of fitness in natural selection and the prime determiner of evolution. If number is the criterion, the human species has made great progress.

In all other populations of organisms, as we have seen, three kinds of restriction operate to change logarithmic growth into an equilibrium zero-growth condition. These are environmental limits (space, food, and other resources, species interactions), (processes predation, disease, etc.), and psychosocial (territoriality, stress syndrome, etc.).

During the 8000 years before the Christian era the 60-fold increase in the human population was supported by genuine increases in the means of subsistence. A shift from animal to plant food and the development of agriculture were ample to support the growing numbers. The scientific-industrial revolution and, especially, recent developments in agricultural technology and genetics have further expanded our ability to produce foods. But can this increase of efficiency keep pace with the population—and if so, to what limit? Can the present population enlarge by two? by 10? Or is man now pressing so hard on his food supply that much further growth in numbers is impossible? "One is made a little nervous," notes demographer Edward Deevey, "by the thought of so many hungry mouths."

The answers to these questions are not easy to find. In terms of mere subsistence, Harrison Brown allows (reluctantly) that a total of 100 billion people could perhaps be supported on Earth (that is, 25 times its present population), and Deevey agrees that such a number in theory is possible, given adequate technological improvements. But to do this, man would have to displace all other animals and utilize vegetation with an efficiency about 10 times greater than now possible. At the present rate of multiplication, allowing for no further acceleration, humanity will reach that figure of 100 billion in somewhat less than 200 years (Table 20-1). Even assuming that we are able to invent ways of feeding those vast numbers, what will happen then, when men perceive themselves to be overcrowded?

Other authorities in the field consider the figure of 100 billion humans to be grossly overoptimistic. Geneticist H. R. Hulett, for example, has calculated that if all the cropland in the world were as productive as that in the United States today, total world food production would supply only about 1.2 billion people with a normal American diet.

Food and other plant and animal products (lumber, paper, fibers, etc.) are considered **renewable** resources. The basic limitation on their utilization is the photosynthetic process. When we turn to human exploitation of **nonrenewable** resources, such as fossil fuels and minerals, the picture is bleak indeed. World consumption of energy in 1967 was equivalent to about 6 billion tons of coal; that of the United States was equivalent to about 2 billion tons. Thus no more than three times the American population, perhaps 600 million people, could have used energy at the rate we did. In one sense, coal, oil, and gas should also be considered renewable resources, since they, too, are derived ultimately from photosynthesis. However, if all organic plant matter present on the entire land surface of the earth were burned in one year, it would still provide energy at U. S. rates of consumption for at most 4 billion people—and of course, there would be nothing left over for food, or for burning next year.

Steel consumption of the world in 1967 was 443 million tons; in the United States, about 128 million. About 700 million people could have used steel at the rate we did. Perhaps 500 million could have used aluminum at our rate. Essentially all the mineral resources show similar ratios. The world is now capable of providing resources and materials at a rate sufficient to maintain fewer than a billion people at a standard of consumption and affluence equivalent to our own.

Obviously there is no conclusive answer to the question of an optimal population size. We have not considered in this chapter the effects of man's activities on his environment, or on the biological process of evolution, both topics to which we return in the next chapters. But whether the earth can support a population of 100 billion, or 10 billion—or whether we have already exceeded the optimal level by threefold—is not really the primary issue. We must first answer one question: Is there a number that we on earth cannot, or should not, exceed? If there is, then the issue at hand is whether we determine that number by the use of our highest human faculties or whether we have the question decided for us by the subhuman forces of starvation, predation, and disease.

Reading List

Brown, L. R., "Human Food Production as a Process in the Biosphere," *Scientific American* (September 1970), pp. 161–170.

Cavalli-Sforza, L. L. "'Genetic Drift' in an Italian Population," *Scientific American* (August 1969), pp. 30–39.

Deevey, E. S., Jr., "The Human Population," *Scientific American* (September 1960) pp. 195–204.

Djerassi, C., "Birth Control after 1984," *Science* (1970), **169,** pp. 941–951.

Ehrenfeld, D. W., *Biological Conservation*. Holt, Rinehart & Winston, New York, 1970.

Hauser, P. M., "The Census of 1970," *Scientific American* (July 1971), pp. 17–25.

Heer, D. M., *Society and Population*. Prentice-Hall, Englewood Cliffs, N.J., 1968.

Hocking, B., *Biology or Oblivion*. Schenkman, Cambridge, Mass., 1965.

Hulett, H. R., "Optimum World Population," *Bioscience* (February 1, 1970), **20,** pp. 160-161.

Joyce, J. A., "The People Problem," *Vista* (January–February 1971), pp. 16 ff.

MacArthur, R. H., and J. H. Connell, *The Biology of Populations.* Wiley, New York, 1966.

Paddock, W. C., "How Green is the Green Revolution?" *Bioscience* (August 15, 1970), **20,** pp. 897–902.

Petersen, W., *Population* (2nd ed.). Macmillan, New York, 1969.

Shepard, P., and D. McKinley, *The Subversive Science.* Houghton Mifflin, New York, 1969.

Singer, S. F., "Human Energy Production as a Process in the Biosphere," *Scientific American* (September 1970), pp. 175–190.

Sladen, B. K., and F. B. Bang, *Biology of Populations.* Elsevier, New York, 1969.

Taylor, C. E., "Population Trends in an Indian Village," *Scientific American* (July 1970), pp. 106–115.

Tietze, C., and S. Lewit, "Abortion," *Scientific American* (January 1969), pp. 21–30.

Wynne-Edwards, V. C., "Population Control in Animals," *Scientific American* (February 1964), pp. 68-74.

Young, L. B., *Population in Perspective.* Oxford University Press, New York, 1968.

chapter twenty-one

Organisms and Their Environment: Ecology

"As I look about me in this forest, I see scarcely a square inch devoid of a life covering of some sort. . . . What I see is all life, the whole vast skein, interwoven, interdependent, stationary or moving, furred, feathered, scaled, flowered. . . . Life is but one great unity, a unity that includes bluets and twinflowers as well as you and me."

DALE REX COMAN, 1966, *PLEASANT RIVER*

Ecology and the Biosphere

It is nearly impossible these days to read a newspaper or magazine or view a television screen for an evening without encountering the term ecology or references to ecological concepts. Ecology derives from the Greek word *oikos,* which means house, or environment; and in a general way the term deals with an organism's relationship with its environment. But "environment" in this sense means much more than it does in ordinary usage. No organism is an isolated independent entity. All forms of life experience a vast and complicated interplay with their surroundings as they satisfy the requirements for their maintenance and growth—food, shelter, moisture, respiratory gases, mates, and so on. Ecology is the study of all the interactions that meet these environmental needs; it is the study of the interrelationships that exist among animals, plants, and microbial life, and between these groups of organisms and the physical environments with which they interact.

Ecological relationships of necessity include a set of nonliving components (called **abiotic** components), which include inorganic elements and compounds such as water, CO_2, oxygen, nitrogen, and salts, and an array of relatively small-molecular-weight organic compounds, as well as such other physical characteristics as heat energy and light, winds, currents, pressure, moisture. It is in this nonliving matrix that the **biotic** components—plants, animals, and prokaryotes—interact. This assemblage of physical and biological entities comprises an **ecological system,** or **ecosystem.** An ecosystem, then, is a place or environment; it is a group of animal and plant communities; and it is a set of relationships. A pond or a forest is an ecosystem; a region containing field, pond, forest, and city is also an ecosystem. Ecological systems are characterized not merely by the fact that one component is dependent on another, but also by the interdependencies of each on all the others and on the integrity of the whole.

Generally, men live in houses, whereas animals live in "nature." We humans regulate our artificial environments to a large extent: we control our temperature, light cycles, humidity. We bring filtered air, chlorinated water, and several forms of energy, all through convenient conduits, into our rooms. Animals in field and forest interact frequently and in direct ways with their biotic environments; an urbanized human encounters species other than his own much more rarely, and often in ways that minimize his awareness of the relationship. We tend, therefore, to forget that the air we breathe, the meat and vegetables we eat, the clothes we wear, and the waste we produce—no matter how sanitized and plastic-wrapped—are all part of the great planetary ecosystem within which we exist, called the **biosphere.**

Properties of the Biosphere

The biosphere is that thin film surrounding the earth in which life exists. It is the region extending a few thousand feet above and below the earth's surface that receives an ample supply of energy from the sun, that contains water, and in which the three states of matter—solid, liquid, and gaseous—are in direct contact with one another. The term biosphere was introduced by an Austrian geologist, Eduard Suess, to denote the envelope around the earth occupied by living things.[1] The total biosphere has several

[1] See Hutchinson, G. E., "The Biosphere," *Scientific American* (September 1970), p. 45.

different subdivisions, each with a particular kind of physical climate and characteristic set of organisms. These subdivisions are called **biomes;** an ocean or a desert is a biome, as is a seashore or a tundra region.

The biosphere did not always exist. During at least the first 2 billion years of the Earth's 4.5-billion-year history, the atmosphere was probably free of oxygen. The modern biosphere apparently had its beginnings 2 to 3 billion years ago in the primitive oceans with the evolution of photosynthetic bacteria-like organisms that could convert solar energy into the chemical bond energy of organic compounds, and did so by splitting water molecules to release free oxygen (Fig. 21-1). The beginning would have been slow. Molecular oxygen released by marine plant cells accumulated for hundreds of millions of years, gradually building an atmosphere that could be used for oxidative respiration, and one that filtered out most of the deadly ultraviolet in the sun's rays. Thus the land was opened to exploitation by living systems perhaps a half-billion years ago.

Energy Flow and Food Chains

The sun releases a tremendous, continuous flow of energy, about one 50-millionth of which reaches the earth as sunlight; this fraction is called the **solar flux.** About half of this is absorbed by the upper atmosphere or is reflected back out to space. Most of the energy that actually strikes the earth's surface is converted into heat energy. Only a fraction of 1 percent is absorbed by plant life and used in photosynthesis. This fraction, however, small as it is, may be represented in a field or forest by the manufacture of several thousand grams of organic matter per square meter per year. Worldwide, it is equivalent to the annual production of 150 to 200 billion tons of plant material, which provides food for every animal on earth including man.

All living things, in any biome or ecosystem, are sustained by this energy from the sun; radiant energy in the form of sunlight

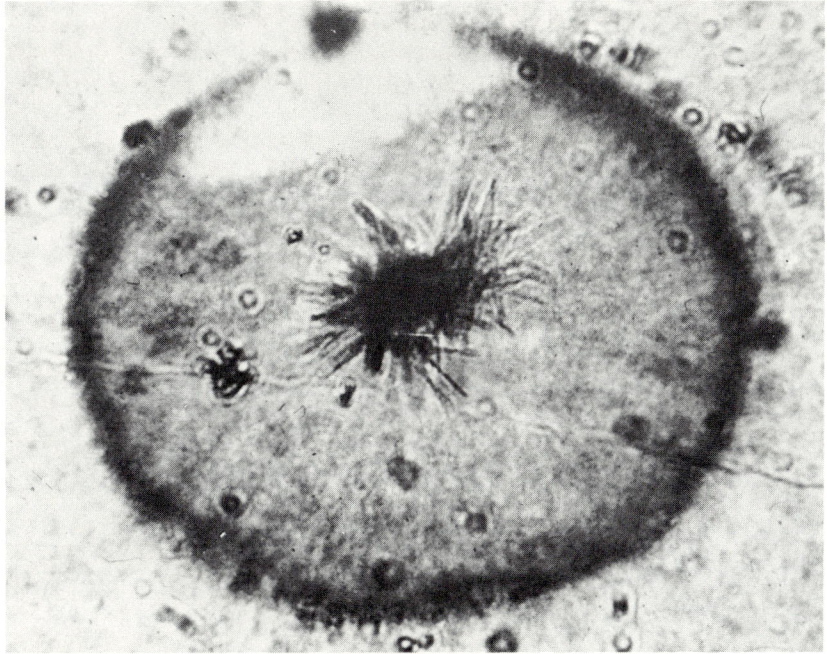

Figure 21-1 A fossilized blue-green alga, estimated to be approximately 2 billion years old, found in the Gunflint geological formation in Canada by S. A. Tyler and E. S. Barghoorn of Harvard University. The Gunflint algae are the oldest known photosynthetic organisms on earth. They must have contributed to the original oxygenation of the earth's atmosphere, and thus helped to form the biosphere.

is the ultimate source of energy for all ecosystems. Energy is absorbed by green plants and used in the process of photosynthesis; low-energy components such as CO_2, H_2O, nitrogen, and phosphorus, are converted to more complex organic molecules and supermolecules (Chapter 4). We refer to this process of converting light energy to organic material as **energy fixation;** when we speak of a green plant **fixing energy,** we mean using it to manufacture more plant material by photosynthesis.

In ecological parlance photosynthetic organisms are called **producers.** These are the chlorophyll-bearing plants; the algae of pond and ocean, the trees of the forest, the grass of the field and marsh. The term producer is misleading, however; as we have noted, plants are devices for converting one form of energy into another (radiant energy into chemical bond energy). Nonetheless, photosynthetic organisms are the primary manufacturers of organic compounds; it is in this context that they are producers.

Producers are **autotrophic** (i.e., "self-feeding"; see p. 95). All other organisms are **heterotrophic** ("feeding on others"), and are therefore referred to as **consumers.** Animals that derive their nutrition from eating plants directly are **herbivores,** or **primary consumers.** **Carnivores** are **secondary consumers;** they eat herbivores and thus derive their energy indirectly from the producers. This transfer of energy from one organism to another in a series of **trophic levels** of eating and being eaten (Fig. 21-2, p. 592) is called a **food chain.**

In Figure 21-3, (p. 593) the flow of energy through the food chain of a typical aquatic ecosystem (a river) is diagrammed. About half the light energy that reaches green plants and algae is absorbed by their chlorophyll. Most of this energy is lost by transpiration (evaporation of water) and by the organisms' own respiratory metabolism. Only a fraction—in this case about 1 percent—of the 50 percent that is absorbed is converted into organic compounds to produce more plant substance (i.e., for growth). This fraction is called **net production.**

Figure 21-2   Big fish eat little fish. By Peter Brueghel, 1556.

If the plants are eaten by small aquatic animals, some of the energy that has been fixed (perhaps 1 to 5 percent) will be transferred, and if fish eat these animals the **efficiency** of energy transfer will only be 5 to 10 percent. Thus at every trophic level, from producer to carnivore, a major portion of the original energy is lost; the flow of energy through the ecosystem is **unidirectional** and **noncyclic.**

But another major group of heterotrophs in all ecosystems does not eat food in the usual sense that herbivores and carnivores do. These are the **decomposers**—mainly bacteria and fungi—which obtain nutrients by releasing enzymes into dead plant and animal material and absorbing the digested products. The released enzymes, called **exoenzymes,** degrade the giant molecules of protoplasm, the large-molecular-weight proteins, carbohydrates, nucleic acids, and so on, into inorganic and small organic components. Components that the decomposers do not use themselves are absorbed into the ground or dissolved in the surrounding water and are

thus made available for reutilization by producers. These nutrient molecules do not flow through an ecosystem, as energy does, but are used and reused over and over again in cyclic fashion. Thus it becomes apparent that a food chain comprises two primary processes that proceed simultaneously: the unidirectional flow of energy, and the cyclic movement of nutrient elements.

Food Chains

Food chains exhibit various degrees of complexity. The simplest would involve only two links: green plants that carry on photosynthesis and microorganisms that obtain their nutrients from these plants by degrading or decaying them after their death:

Photosynthetic organisms $\longrightarrow$

Microorganisms.

In this notation the arrow refers to the flow of energy from one trophic level to the next.

More usual food chains consist of many links. For example, in the oceans microscopic

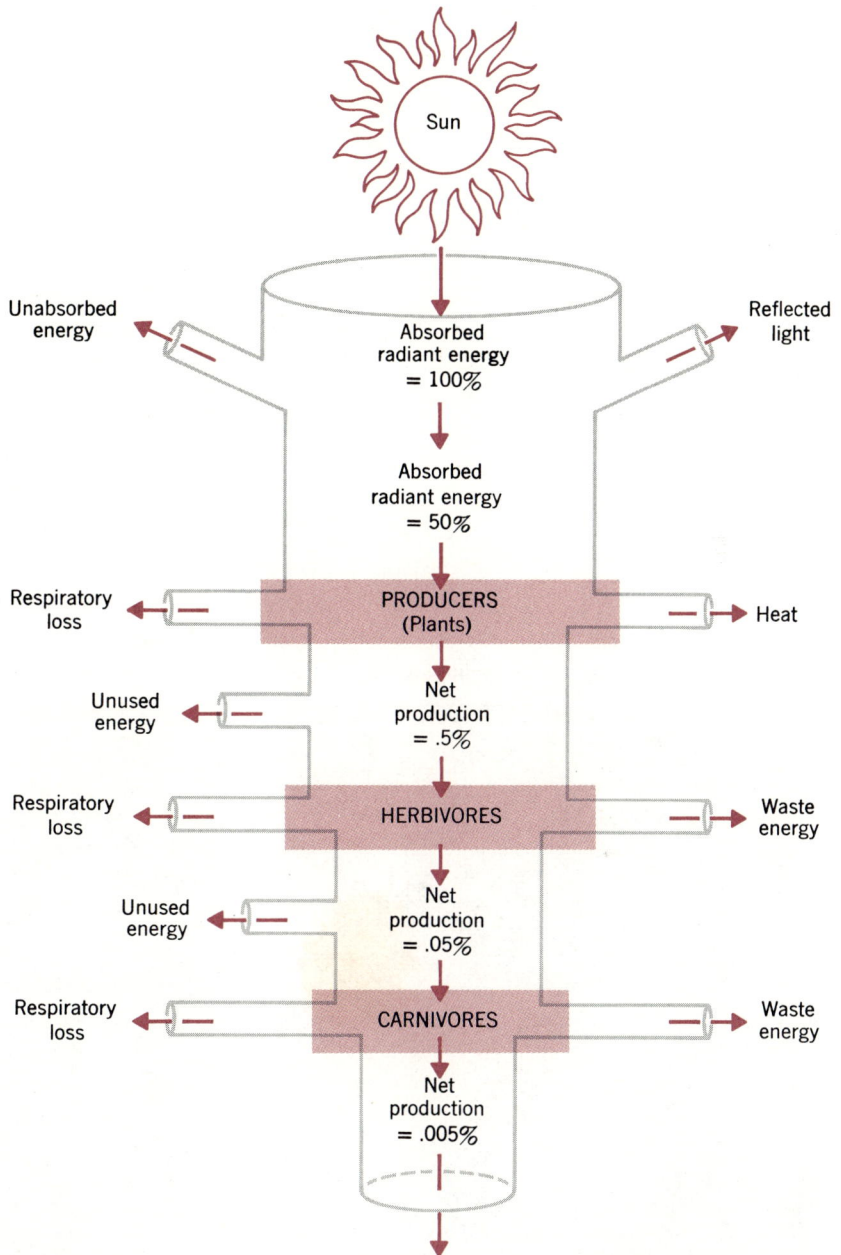

Figure 21-3   The flow of energy through the food chain of a typical ecosystem (a river). The colored boxes represent the different trophic levels; the conduits and arrows indicate energy flow. At each step in the food chain, only a fraction of the total energy content is passed on.

floating plants called **phytoplankton** are primarily responsible for the production of organic matter by photosynthesis. These tiny autotrophic plants are consumed by unicellular and multicellular **animal plankton,** or **zoöplankton** (microscopic floating animal forms: tiny shrimp and other crustaceans, and the larvae of starfish, sea urchins, fish, etc.) (Fig. 21-4). These plankton are then ingested by the slightly larger carnivorous forms, which in turn may be eaten by worms, shrimps, lobsters, crabs, and so on. The smaller of these animals may then be devoured by small fish, which are eaten by larger fish. The larger fish then serve as food for various mammals and birds, whose eventual death and subsequent decomposition by microorganisms of decay terminate the food chain:

Microscopic plants $\longrightarrow$ Herbivorous animal plankton $\longrightarrow$ Carnivorous plankton $\longrightarrow$ Small fish $\longrightarrow$ Large fish $\longrightarrow$ Birds and mammals $\longrightarrow$ Decomposers

For simplicity we have indicated food chains as straight pathways. They are, in fact, far more complicated, and are often accompanied by various branching and parallel sequences. In the food chain above, for example, the herbivorous plankton may also be eaten directly by fish, which eventually perish and decay, thus constituting short, branching

Figure 21-4   (a) *Living phytoplankton consist mainly of single-celled algae called* **diatoms.** *Each diatom synthesizes a transparent case of silicon from minerals in the sea around it.* (b) *A variety of tiny creatures, each only a few millimeters long, comprise the animal plankton. These include crustacean and molluscan larvae as well as copepods and worms.*

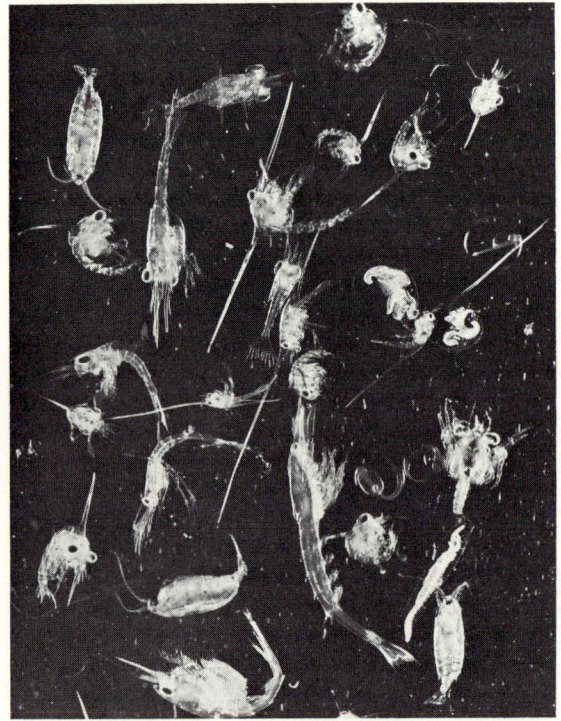

(a)

(b)

food chain connected to the larger one. Thus it should be noted that in reality **food webs** occur, with all sorts of short-circuits and connections occurring in a system.

On land, food chains extending from higher green plants through several animal links and ending with the microorganisms of decay are readily observed. In our own agricultural economy grasses are eaten by cattle, which are in turn eaten by man. The final link in this sequence, like that in all other food chains, is decomposition by microorganisms of our human remains.

Food Pyramids and the Biomass

We have noted that energy is lost along a food chain at every trophic level; 100 lb of grass in a field does not produce 100 lb of meat "on the hoof." To measure how much animal product is actually formed from a given amount of plant material, we cannot simply weigh whole plants and animals. Most living tissues, both animal and vegetable, are made up mainly of water. About 80 percent of the weight of beef muscle is water—for example, the water of blood trapped in the tissue and the water that comprises the bulk of the cytoplasm. This is weight that is not produced or synthesized by the cow; it is simply absorbed.

An easy way to distinguish total weight of a tissue from the inorganic and organic substance of which it is made is to evaporate the water by placing the tissue in an ordinary oven for several hours, The weight of the residue, called the **dry weight,** gives a fairly accurate representation of the amount of energy required to synthesize that tissue. The energy losses and the figures for net production shown in Figure 21-3, for example, are calculated on the basis of dry weight.

Because energy is lost at every trophic level, food chains may also be represented as pyramids that indicate either the number of organisms at each trophic level or the total dry weight of all the organisms at that level (called the **biomass**). The broad base of the pyramid represents the plant producers. The

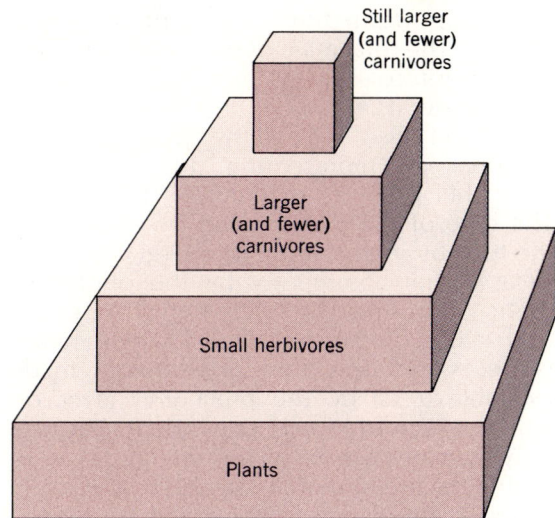

Figure 21-5 *An ecological pyramid. The size of each layer represents the biomass, or total energy, contained in that trophic level. The broad base is usually the photosynthetic producers, with each successively higher trophic level consisting of progressively larger but fewer consumers.*

next higher trophic level in the pyramid is composed of small animals that feed on the many plants of the lowest pyramid layer. The smaller animals are the food source for the larger ones that make up the next trophic layer in the food pyramid, and the successively higher trophic levels in the pyramid are typically made up of progressively larger but fewer animals.

In general in terrestrial pyramids the total biomass for any trophic level is always less than that of the layer below, which is why the structure has a pyramid shape (Fig. 21-5). The food pyramid is a graphic device that represents the distribution of the protoplasmic mass of an ecosystem. The proportion of protoplasmic dry weight is greatest at the base and diminishes with each succeeding higher trophic stratum because a substantial portion is expended as an energy source for a variety of activities and thus never appears in the weights of the larger animals. The total mass of protoplasm at each level therefore is always less than that of the layer below it.

We can also regard the pyramid organization as representing the inevitable "downhill pathway of energy" (Chapter 2). The base of the pyramid can be considered the total biological input of energy furnished to a natural community by the photosynthetic organisms. The total energy present in each succeeding trophic layer of the pyramid is necessarily less than the energy made available to it by the preceding layer on which it is nutritionally dependent, because only a small fraction of the energy taken in during the lifetime of an organism is actually stored in the tissues; a good part of the energy has been used for the various life activities of the animal other than synthesis of various substances.

Types of Pyramid

Several different pyramids or ecosystems often exist in any geographical area. For example, a pyramid may occur in a pond: the usual fundamental level of aquatic photosynthetic microorganisms is followed by successively ascending layers of bacteria and protozoa, semimicroscopic multicellular animals, small fish, and progressively larger fish, until the peak of the pyramid is reached (Fig. 21-6a).

On the land surrounding this pond there may be another natural community (Fig. 21-6b) represented by photosynthetic plants, small herbivores, small carnivores, larger and larger carnivores, and **omnivores** (devourers of both plants and animals), culminating, for example, in a pyramid peak of bears (who actually obtain most of their food from fruits) and mountain lions. Simultaneously, on the very same land, another natural community of organisms may exist that is represented by a pyramid of photosynthetic organisms, bacteria and unicellular animals, semimicroscopic multicellular animals, small visible multicellular animals such as earthworms and various insects, small birds, and larger birds. The peak of this pyramid (Fig. 21-6c) is occupied by large predatory birds such as hawks and eagles.

Pyramids of varying degrees of complexity exist. Some interlock, sharing the same lower layers but separating into individual peaks. Most interact at several levels. Insects that hover near a pond are eaten by both birds and fish; fish in turn are shared as food by

Figure 21-6   Three natural ecological pyramids illustrating (a) a pond ecosystem, and (b, c) two different terrestrial food chains occupying the same geographic locale.

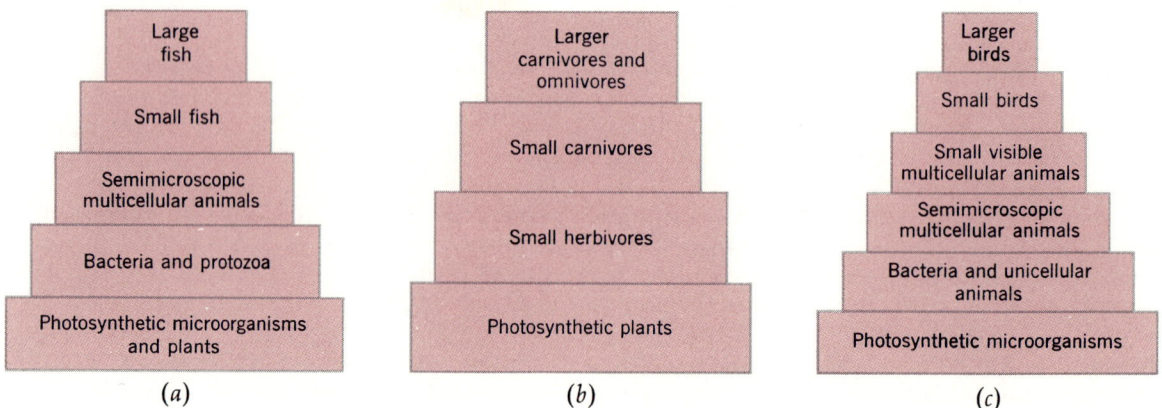

(a) Large fish / Small fish / Semimicroscopic multicellular animals / Bacteria and protozoa / Photosynthetic microorganisms and plants

(b) Larger carnivores and omnivores / Small carnivores / Small herbivores / Photosynthetic plants

(c) Larger birds / Small birds / Small visible multicellular animals / Semimicroscopic multicellular animals / Bacteria and unicellular animals / Photosynthetic microorganisms

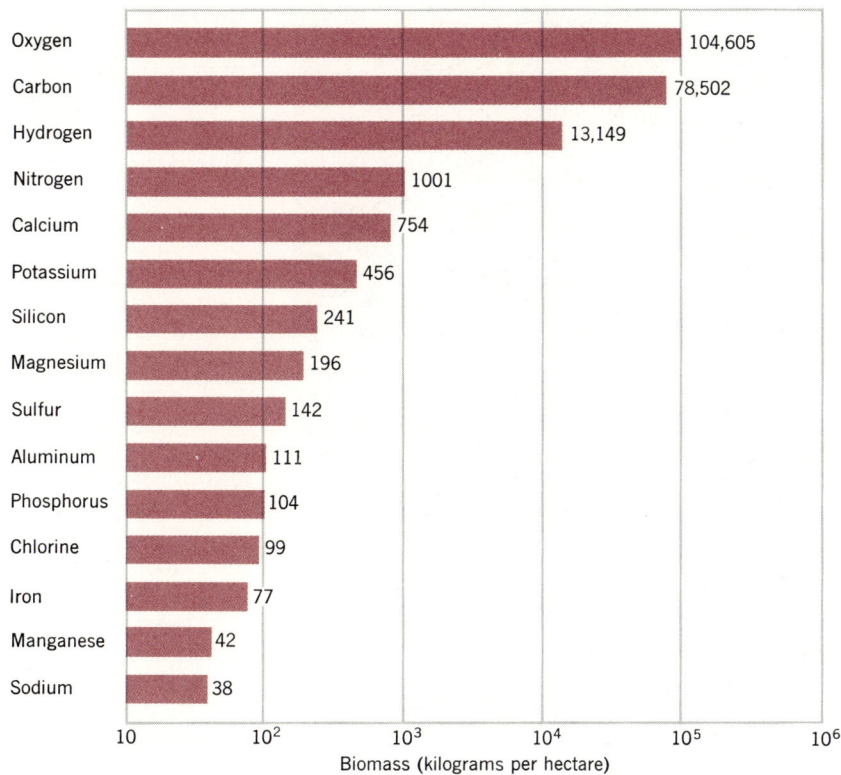

Element	Biomass (kg/ha)
Oxygen	104,605
Carbon	78,502
Hydrogen	13,149
Nitrogen	1001
Calcium	754
Potassium	456
Silicon	241
Magnesium	196
Sulfur	142
Aluminum	111
Phosphorus	104
Chlorine	99
Iron	77
Manganese	42
Sodium	38

Biomass (kilograms per hectare)

Figure 21-7   Composition of the biosphere, dominated by oxygen, carbon, and hydrogen, is indicated by the bars in this logarithmic chart. The units are kilograms per hectare of land surface. (A hectare is a square area 100 m on a side, or about 2.5 acres.)

many larger birds and mammals, as are small rodents. Thus, although we can speak of the ecosystem of a pond, or of a field, in actuality organisms in a given geographic locale generally overlap in their food chains to such a degree as to form a single large ecosystem.

The Cyclic Use of Matter in the Biosphere

There are more than 100 chemical elements on earth. Yet ecologists define the biosphere mainly in terms of the interaction of only six of these: hydrogen, carbon, nitrogen, oxygen, phosphorus, and sulfur. From the periodic table the atomic numbers (Appendix 1) of these six are 1, 6, 7, 8, 15, and 16; from this it should be apparent that all life derives from the exceptional reactivity of six of the 16 lightest elements. A glance at a partial list (Fig. 21-7) of the elements of the biosphere shows why hydrogen, oxygen, carbon, and nitrogen dominate the chemistry of the biosphere. These four make up all but a tiny fraction of the biomass and more than 99 percent of the world's plant life.

Life is mainly aqueous; most organisms are 80 to 90 percent water. Water, of course, is continuously recycled near the earth's surface by runoff, evaporation, and condensation. It flows over the land in rivers and springs, carrying materials from the soil to the sea. It enters the atmosphere by evaporation, and returns as rain to the land (Fig. 21-8, p. 598). In this planetary water are dissolved the elements of life, circulating endlessly and entering the cycle of the biosphere whenever they are absorbed by a plant and fixed as part of an organic compound via the process of photosynthesis.

Figure 21-8   A radar image of a 600-square-mile area of land near Sandy Hook, Ky. Rainfall and runoff of the tributaries of the Ohio river have produced the "etched" appearance of the land.

The Carbon Cycle

The complex process of photosynthesis, which we have examined in Chapters 4 and 13, can be summarized by a single simple reaction:

$$CO_2 + H_2O + energy \longrightarrow CH_2O + O_2$$

The formaldehyde molecule CH_2O is the simplest organic compound; the term energy indicates that light energy is converted to chemical bond energy and stored in the reaction. Thus carbon and water are used, organic compounds and oxygen are produced. Respiration, of course, has the opposite effect. Organic molecules are burned with oxygen to liberate CO_2 and water (Chapter 4).

The average concentration of carbon dioxide in the earth's atmosphere is about 320 ppm. If we measure the CO_2 in the air over a forest periodically over 24 hours, we will find revealing changes. When the sun rises photosynthesis begins. As leaves and grass convert CO_2 into organic compounds, the concentration of CO_2 in the air falls. By noon, measurements of 300 to 310 ppm may be recorded. As the temperature rises, so does the rate of respiration, and the net consumption of CO_2 slowly declines. At sunset photosynthesis stops, but respiration continues throughout the night, with the result that the CO_2 concentration in still air immediately over the trees may reach levels of 400 ppm.

A fair estimate is that terrestrial plants fix a total of about 40 billion tons of carbon per year into organic compounds. Photosynthesis in the world's oceans, mainly by phytoplankton, fixes somewhat less CO_2, perhaps 20 billion tons. The carbon fixed by photosynthesis sooner or later is returned to the atmosphere by decomposition of dead organic matter: leaves and trees fall to the ground and are oxidized in the soil; phytoplankton and animal plankton die, sink to the bottom, and there decompose.

We have an approximate idea of the rate at which the carbon fixed from the atmosphere into plant material is returned there by decomposition. Measurements of the content of the radioactive isotope carbon[14] in decaying forest organic matter suggest that this cycle takes from several decades to several hundred years. The total amount of carbon at or near the earth's surface is on the order of 20,000,000 billion (20×10^{15}) tons. Most of that, 99.9 percent, consists of inorganic deposits, chiefly calcium carbonates (Fig. 21-9), and a few hundredths of a percent as fossil deposits (coal and oil). A tiny fraction of a percent is in constant circulation through the carbon cycle of the biosphere. This fraction includes the 700 billion tons in the form of CO_2 in the atmosphere, a like amount that is present in all of the living animals and plants on earth, and perhaps 10 times that quantity locked up in dead organic matter on the land and in the sea.

Most of the exchange of carbon between

Figure 21-9   The White Cliffs of Dover consist of almost pure calcium carbonate, representing the skeletons of phytoplankton that settled to the bottom of the sea, over a period of millions of years, more than 70 million years ago. The worldwide deposits of limestone and other carbon-containing sediments are the largest repository of carbon.

earth and atmosphere occurs across the surface of the oceans. The rate of this exchange has recently been estimated by measuring the disappearance from the atmosphere of C^{14} that was produced in nuclear weapons tests in the early 1960s. These measurements indicate, first, that the atmosphere over the entire earth is rapidly mixed from one geographic location to another; and that the total amount of carbon dioxide in the air (about 700 billion tons) exchanges with a like amount in the ocean every five to 10 years.

In sum, then, about 40 billion tons of carbon per year is fixed from the atmosphere and returned to it by plants and animals on land in a cycle that lasts on the order of 100 years; perhaps twice that amount is exchanged annually between the ocean and air in a cycle that lasts five to 10 years. Clearly, if these carbon cycles did not exist, we would deplete the atmosphere of CO_2 in only a few short years. The over-all carbon cycle is diagrammed in Figure 21-10 (p. 600).

For about the last 100 years man has unknowingly (until very recently) been conducting a global geochemical experiment by burning large amounts of fossil fuel and thereby returning to the atmosphere carbon that was fixed by photosynthesis millions of years ago. At current rates of consumption of coal and oil, we are releasing about 5 billion tons of fossil carbon into the atmosphere per year. About two-thirds of this reenters the carbon cycle or dissolves in the ocean, but the remainder adds to the atmospheric content.

In the last century the average carbon dioxide content of the atmosphere has risen from 290 ppm to 320 ppm (Fig. 21-11, p. 600), 20 percent of that rise occurring in the last decade. With the present rate of acceleration in burning of fossil fuels, the amount of CO_2 in the atmosphere will undoubtedly continue to climb. Present estimates are that it will reach 380 to 400 ppm in the next 30 years (Fig. 21-11). What will the result be on the biosphere? What happens in the next 100 years?

The Oxygen Cycle

The oxygen of the atmosphere is also in a state of dynamic equilibrium. Its 20 percent concentration is essentially the result of an equal rate of its removal from the atmosphere by oxidative processes of biological systems (largely respiration), balanced by an equal rate of replenishment by photosynthesis. Therefore our diagram for the carbon cycle (Fig. 21-10), with attention focused on the oxygen atom instead of carbon, also represents the oxygen cycle in nature. It is estimated that if the atmospheric oxygen were suddenly depleted, it would take about 3000 years to restore the oxygen concentration to its original level by photosynthesis.

The interrelation between the oxygen and

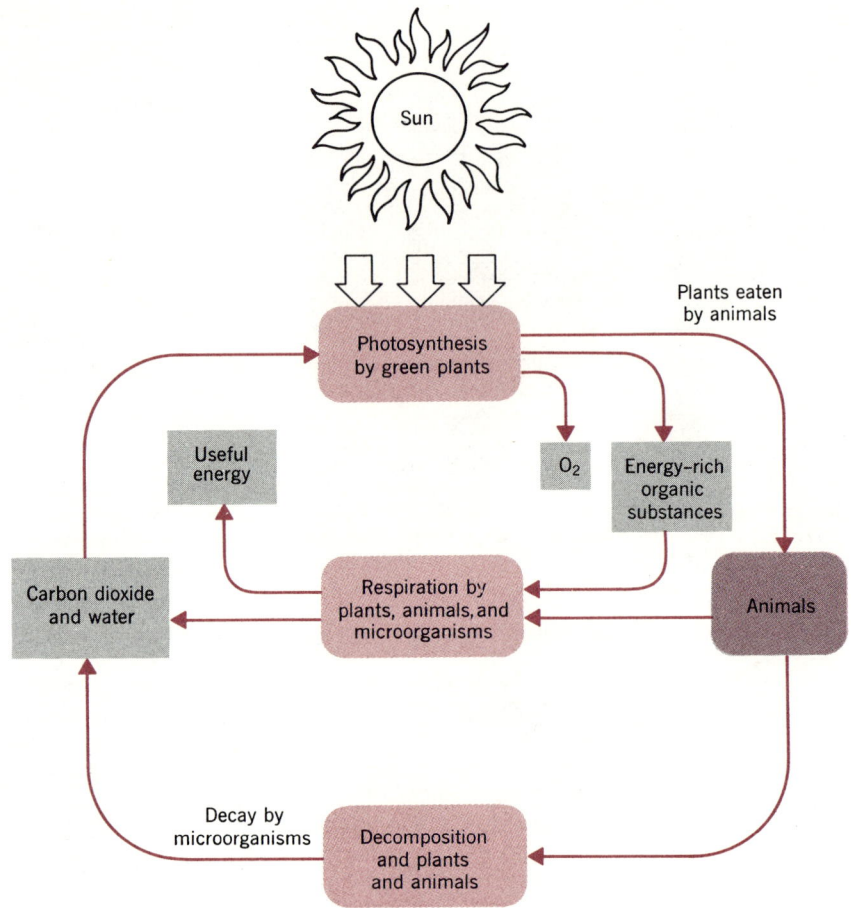

Figure 21-10   The carbon cycle.

Figure 21-11   The increase in atmospheric carbon dioxide since 1900, as measured at Mauna Loa observatory in Hawaii, is shown, with a projection to the year 2000. The sawtooth curve superimposed from 1960–1970 indicates the annual oscillations that result from seasonal variations in rate of photosynthesis.

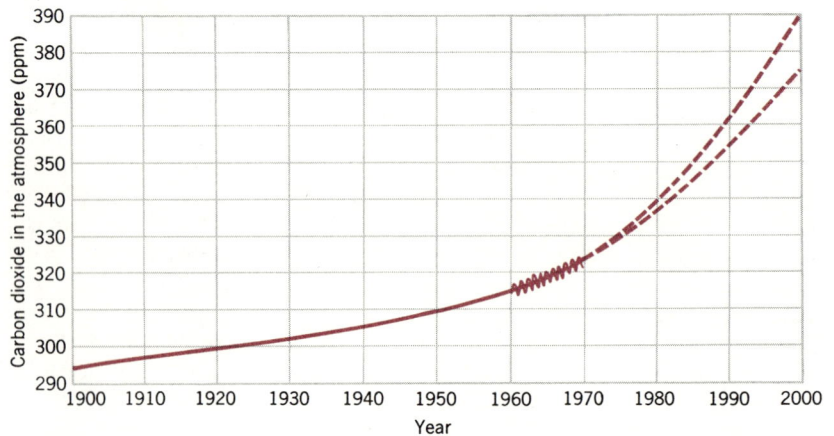

carbon cycles is also apparent from the biochemistry involved. From the basic photosynthetic reaction $CO_2 + H_2O + energy \longrightarrow CH_2O + O_2$, it is not obvious whether the free molecular oxygen comes from CO_2 or H_2O, or whether the three atoms of oxygen that go into the reaction are pooled in random fashion. Experiments at Berkeley by Samuel Ruben and Martin Kamen, using the heavy isotope oxygen[18] as a tracer, demonstrated that molecular oxygen is derived from the splitting of water. The oxygen that appears in photosynthesized carbohydrates is the oxygen that was in the carbon dioxide; the two elements are used together.

The Nitrogen Cycle

The nitrogen cycle includes four separate processes: nitrogen fixation, nitrogen assimilation, nitrification, and denitrification. Although men and other animals live in an atmosphere that is 79 percent nitrogen, their supply of food is limited more by the availability of fixed nitrogen than of any other plant nutrient. Nitrogen as a gas cannot be used by most organisms. It must first be incorporated, or fixed, into compounds that can be utilized, such as ammonia (NH_3) or nitrate (NO_3), the principal forms of nitrogen in green plants and microorganisms.

Nitrogen gas is transformed to other chemical forms by two processes. First, electrical discharges arising from electrical storms in the atmosphere form small amounts of nitrogen oxides by displacing electrons from molecular nitrogen. Second, and much more important, is the biological process called **nitrogen fixation.** A few groups of bacteria and algae are capable of transforming molecular N_2 to ammonia; these species represent the most critical link in the nitrogen cycle. Among the most common are those of the genus *Rhizobium*.

Nitrogen-fixing organisms enjoy their special capacity to split molecular nitrogen because they contain an enzyme called **nitrogenase.** Although the list of bacteria and algae known to be capable of fixing free nitrogen is substantial, and although these mi-croorganisms are widespread over the earth's surface, it has been estimated that the total amount of nitrogenase in the world is probably no more than a few kilograms.

Some of the nitrogen-fixing bacteria live in the roots of certain higher plants called **legumes** (e.g., peas and beans). The invasion of these plants by the bacteria causes large root swellings, or **nodules** (Fig. 21-12, p. 602).

Green plants convert nitrates to amino acids via a sequence of enzymatic reactions. This process is called **nitrate assimilation.** Plant proteins provide a source of essential amino acids for many animals, which utilize the amino acids to synthesize their own cellular proteins. These animals in turn are a source of nitrogen for their predators.

Animals excrete their nitrogenous wastes largely as ammonia, urea, or uric acid. The latter two compounds and those arising from the decay of dead organisms are eventually transformed by decomposers to ammonia. Ammonia itself has a number of biological fates. It may be absorbed directly by the roots of higher plants and used for the synthesis of amino acids and proteins. It may also be oxidized to nitrate by the process called **nitrification,** principally as a result of the successive action of two groups of soil bacteria collectively called the **nitrifying bacteria.** The first group, *Nitrosomonas*, aerobically oxidizes ammonia to nitrite (NO_2), which is then further oxidized aerobically to nitrate by the second bacterial group, *Nitrobacter*. The production of nitrate from ammonia, and its utilization as a nitrogen source by green plants and numerous microorganisms, completes the nitrogen cycle (Fig. 21-13, p. 603).

The nitrogen cycle, however, is still a little more complex. Another group of soil bacteria, called **denitrifying bacteria,** converts nitrate and nitrite to molecular nitrogen (N_2), which is lost to the atmosphere. **Denitrification,** therefore, is a drain on the nitrogen cycle.

Nitrogen gas, we noted, is relatively inert. In biological fixation it combines with hydrogen to form ammonia, which is more reactive and can be used in a variety of biological

(a)

(b)

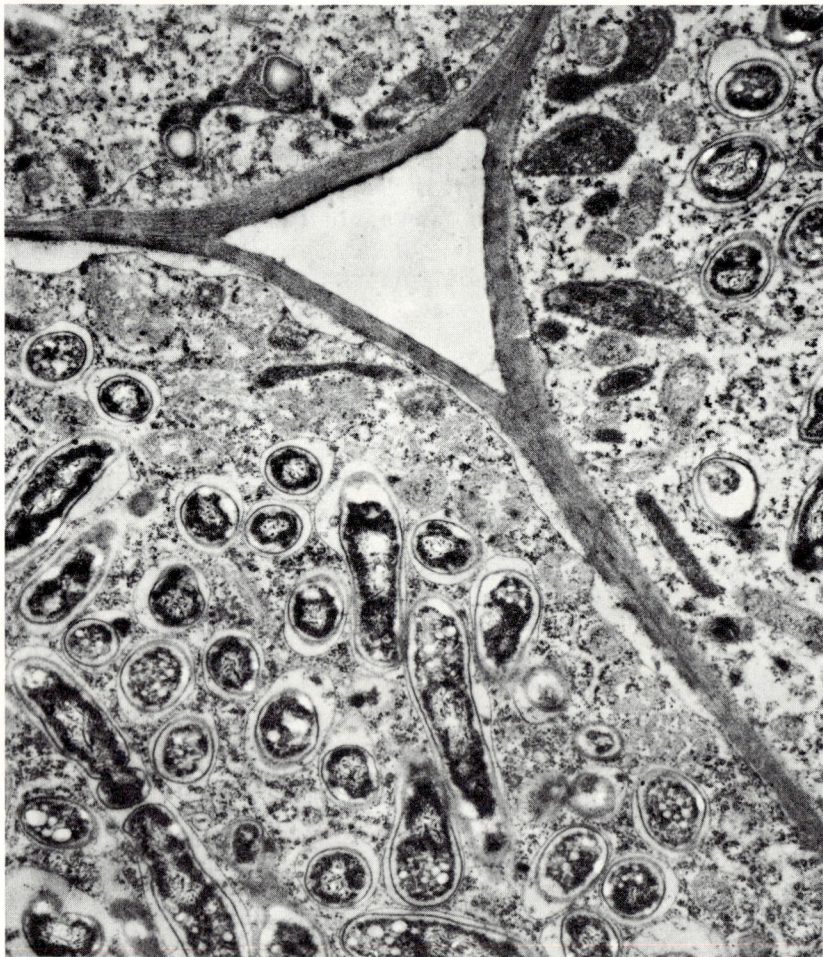

synthetic reactions. Combined into other compounds, it becomes so reactive as to be unstable. **Tri-nitrotoluol** (TNT), for example, is a powerful explosive used in conventional bombs.

It was the demand for explosives during World War I that provided the incentive for a German chemist, Fritz Haber, to invent an industrial process, called **catalytic fixation,** for fixing nitrogen. In this process nitrogen (from air) and hydrogen (usually from the methane of natural gas) are passed under pressure over hot nickel. The nickel acts as a catalyst for the combination of nitrogen and hy-

Figure 21-12   (a) The roots of a pea plant, like other legumes, contain nodules of nitrogen-fixing microorganisms. These specialized bacteria penetrate the root hairs and stimulate the cells to divide. Using nutrients provided by the root, the symbiotic bacteria convert the nitrogen of the air to ammonia or nitrates.
(b) Electron micrograph of a root nodule. Portions of three root-hair cells are shown, containing many nitrogen-fixing bacteria. Magnification 22,000×.

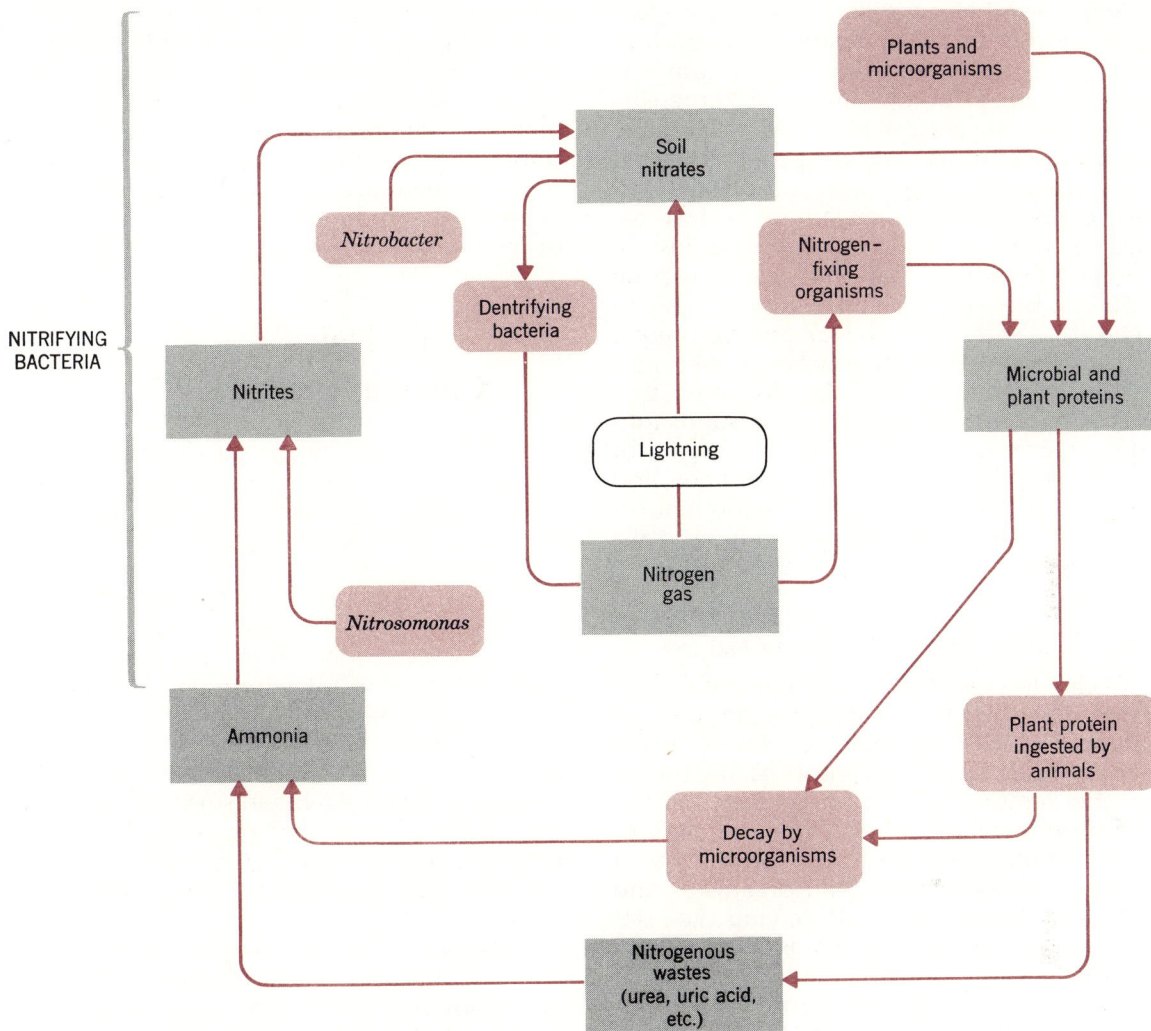

Figure 21-13 The nitrogen cycle.

drogen to form ammonia (NH_3). In the same process the ammonia can be combined with oxygen to form ammonium nitrate (NH_4NO_3), and with CO_2 to produce urea [$CO(NH_2)_2$]. Both of these compounds are used as fertilizers.

Of all man's recent interventions in the cycles of nature, the industrial fixation of nitrogen far exceeds all others in seriousness. Since 1950 the amount of nitrogen annually fixed for the production of fertilizer has in-

creased fivefold, until it now about equals the amount fixed by all biological ecosystems before the advent of agriculture. In 1968 the world's annual output of industrially fixed nitrogen amounted to about 30 million tons of nitrogen; by the year 2000 it is expected to exceed 100 million tons per year.[2]

[2] Delwiche, C. C., "The Nitrogen Cycle," *Scientific American* (September 1970), p. 137.

Before synthetic fertilizers were manufactured in large amounts, the nitrogen removed from the atmosphere by natural fixation was nicely balanced by the amount returned by denitrifying organisms that convert organic nitrates back to gaseous nitrogen; this was on the order of 30 million tons of nitrogen per year converted each way. Now, because we add an equal or greater amount of industrial nitrogen fixation to that cycle, we cannot be sure that the earth's denitrifying process can keep pace with fixation. Nor can we predict the consequences if an imbalance continues for an extended period. We do know that excessive runoff of nitrogen compounds into streams and rivers can result in blooms of algae that deplete the available oxygen and destroy fish and other oxygen-dependent organisms. The rapid **eutrophication** of Lake Erie is perhaps the best-known example (see p. 622).

Other Nutrient Element Cycles in Nature

Cycles for other nutrients comparable to those described for carbon, oxygen, and nitrogen also exist in nature. Cycles of ingestion, utilization, and excretion by organisms can be demonstrated for hydrogen and the various mineral elements. The availability of mineral nutrients is aided by the release of mineral elements from dead organisms and tissues by the action of **putrefying** (i.e., decomposing) microorganisms and from rocks by **weathering.** Among the most important of the mineral cycles are those for phosphorus, used in the production of DNA and the vital energy compound ATP; sulfur (Fig. 21-14), an essential component of proteins; and the trace elements that are required as coenzymes in so many biological reactions.

In undisturbed nature the element that is most critical for growth is usually phosphorus. It is much scarcer than carbon or nitrogen (Fig. 21-7), and is essentially insoluble in most natural (i.e., alkaline) waters. In a lake, for example, where the output of organic material may revolve around phosphorus at concentrations on the order of 50 mg per liter (or .05 ppm), if that concentration is doubled it

commonly results in a doubling of the "standing crop" of plankton and algae. Under these conditions the situation in a lake changes drastically (p. 623). If phosphorus is plentiful, nitrate may become the **limiting** (most critical) nutrient; in this case blooms of blue-green algae often foul the water because of the ability of these microorganisms to fix atmospheric nitrogen.

Ecological Communities

Every organism in any ecosystem lives in a particular place or region called its **habitat.** A habitat may be an area as large as an ocean or desert, or as small as a rose leaf or the intestine of a termite. Within its habitat and ecosystem, every organism also plays a variety of roles. It feeds on some organisms and serves as food for others. It extracts some nutrients from the environment and returns some to it. It is primarily either a producer or a consumer. The sum total of all of these roles for any organism is referred to as its ecological **niche.** An organism's habitat, then, is the place it lives; its niche includes all the physical, chemical, and biotic factors the organism needs to maintain itself and reproduce.

Organisms that live together in the same habitat usually have very different niches. A tidal pool on the seashore, for example, may contain starfish, sea anemones, mussels, and marine snails, as well as seaweed and smaller filamentous algae and phytoplankton. All share the same habitat (the sandy tidal pool), but the algae and phytoplankton are producers, the rest are consumers. Some feed mainly on the plant life (anemones, for example), whereas others are carnivorous or omnivorous (starfish). Thus an algal seaweed occupies a very different ecological niche from a starfish living next to it in the same habitat.

Every environment offers a large number of niches and habitats. Although two species

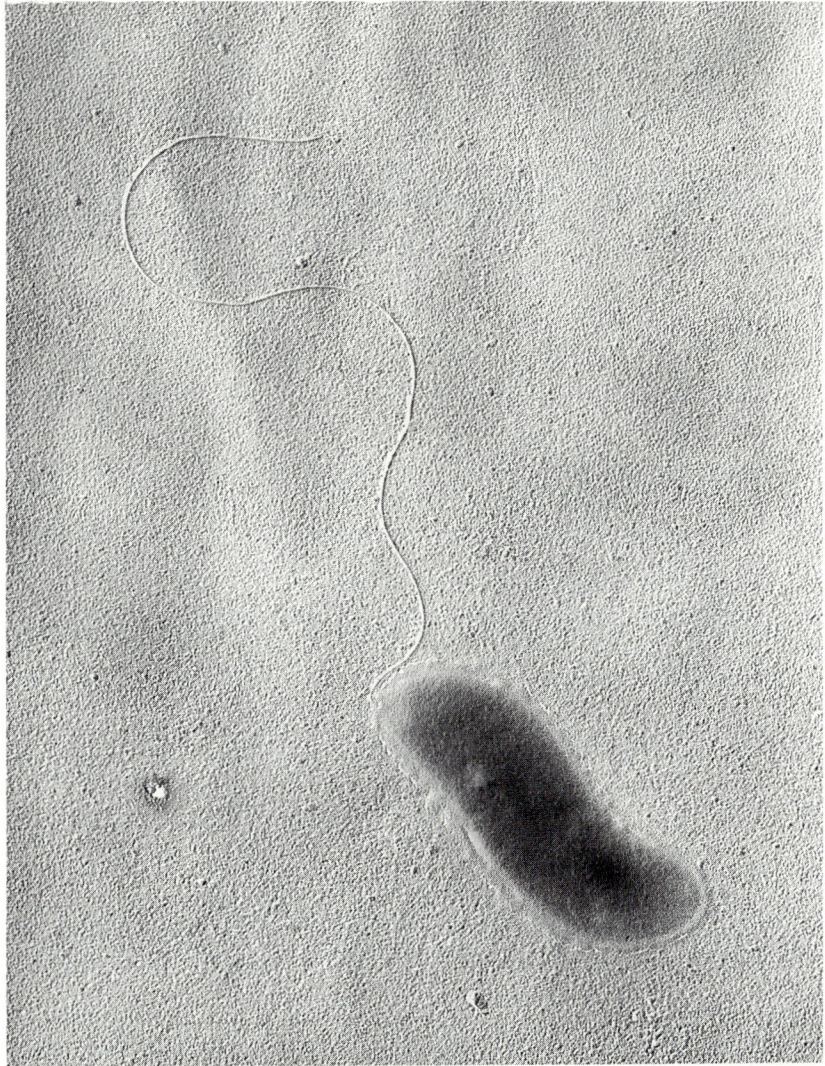

Figure 21-14   Electron micrograph of a sulfur-fixing bacterium that metabolizes the sulfates in sea water and releases sulfur as hydrogen sulfide (H_2S), in a form that can be used by other organisms for amino acid synthesis. The bacterium is shown here magnified about 22,000×; it is actually about 3.5 μ long.

can occupy the same habitat, they cannot occupy the same niche for very long. Occupying the same niche would mean competing on nearly every level of existence; such competition generally results in the elimination of one or the other of the species, as we saw in Chapter 20.

About three-fourths of the earth's surface is covered by ocean, in which cold arctic depths contrast with tropical coral reefs. On land we find forests, prairies, lakes, mountains; almost a third of the land mass is arid desert. In each of these regions life exists. Each has its food chains and energy flows and in each nutrients are cycled. Yet how very different each is from the others as a community of organisms and as an interrelated system of ecological niches.

The ocean must have been the first ecological habitat on earth; evidence is overwhelming that life first evolved there (Chapter 22). Most ocean life today, including representatives of all the major groups of living organisms, is found at the **seashore,** or **tidal zone** (Fig. 21-15), and in the shallow waters of the **continental shelf.** Some marine forms also inhabit the vast stretches of the open sea, to depths of about 600 ft. A few species have even adapted to the cold and darkness of the bottom regions.

The Ocean Food Chain

A quart of seawater dipped up from the middle of the ocean will contain more than a million organisms, mostly the microscopic floating plant cells that form phytoplankton (Fig. 21-4). Just as all land animals are dependent on the photosynthetic plants of field and forest, all the animals of the sea obtain their sustenance ultimately from this photosynthetic blanket of floating, or **pelagic,** plants. "Floating" is, however, a misleading term for plankton (which means, literally,

"drifting"). Most of the cells and tiny organisms that make up the plankton are slightly denser than seawater, and under quiet conditions would slowly sink to the bottom. In fact most ultimately do, forming a constant rain or organic matter. However, the upper layers never become depleted of plant cells—and thus of the capacity to generate food and oxygen—because of turbulence of the water. The plant cells slowly sink, but as they do they divide, and the population in the upper waters is continually replenished from below by turbulent upwelling water. This stirring action is also what prevents a given volume of water from being depleted of nutrients.

Drifting with the phytoplankton and feeding directly on them are the zoöplankton. The phytoplankton and zoöplankton together provide a rich broth on which many larger marine animals depend for their only diet. A wide variety of sea animals have evolved specialized gills or other fine-meshed structures that allow them to filter these organisms from the water; sponges, many mollusks and marine worms, some fish, and numerous other species, are such **filter feeders.** The largest of

Figure 21-15 Marine organisms either live on the bottom of the ocean (benthic zone), or swim or float near the surface (pelagic zone). The benthic zone includes the shoreline, or tidal zone, the littoral region, and the deep-sea bottom. Pelagic organisms live mainly within about 600 ft. of the surface, where sunlight penetrates.

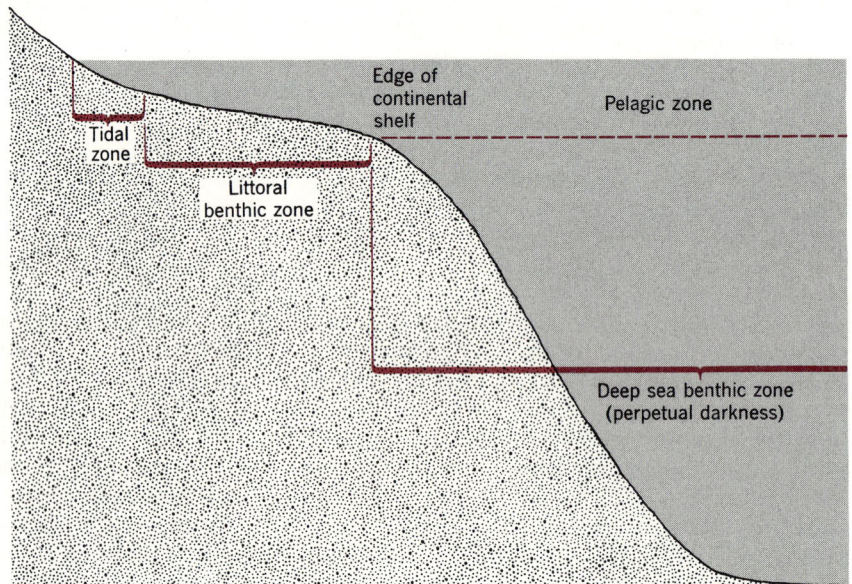

Tidal zone

Edge of continental shelf

Pelagic zone

Littoral benthic zone

Deep sea benthic zone (perpetual darkness)

Figure 21-16 The common mussel Mytilus edulis, *growing in a rocky tidal zone along the New England coast.*

all animals on earth, the blue whale—a mammal that may reach 90 ft in length and weigh 100 tons or more—lives entirely on plankton.

Most of the fish and other marine animals that are not plankton eaters are carnivores that eat smaller animals. There are few large herbivorous animals living in the oceans.

Habitats of the Sea

The seashore actually extends from 10 to 150 miles out into the ocean around all the continents (the continental shelf), forming an area of relatively shallow water (reaching a depth of about 500 ft). In this **littoral zone** (Fig. 21-15) life is much denser than in the open sea. Such creatures as sponges, corals, coelenterates abound, as do **benthic (bottom-dwelling)** predators such as mollusks, crustaceans, and starfish, and some bottom-dwell-

ing fish. Many small animals live in the grass and the filamentous algae that form seaweed. Snails, slugs, and worms crawl over the surface or burrow in the bottom.

The **tidal,** or **intertidal, zone** (the portion of the littoral zone that lies between the high- and low-tide boundaries) has its own local ecological structure. Periwinkles and related forms live at the high-tide level (Fig. 21-16); mussels, barnacles, and others inhabit a distinct area below the high-tide level; and seaweeds are found at the low-tide level. Crabs and other animals run in and out with the tides. The abundance of life is staggering (Fig. 21-17, p. 608); tens of thousands of individuals of a single species of mollusk may inhabit one square meter of this zone.

The strangest marine habitat, to our eyes, is the cold, still **ocean bottom,** or **benthic**

Figure 21-17 (above)   Tidal-zone mud flat, carpeted with sea snails.

(a)

zone. With modern submarine vessels called **bathyspheres,** specially designed to withstand the enormous pressures of the deep sea, marine biologists have begun to explore the deepest parts of the ocean, where the bottom extends 7 miles down (Fig. 21-18). Because sunlight never penetrates beyond about 2000 ft, there are no plants here; bottom-dwelling animals are all heterotrophs, feeding on particulate remains of organisms that sink from the surface waters, or carnivores, feeding on one another. Many species are **bioluminescent,** generating light similar to that of a firefly, with a special set of enzymes (Fig. 21-19, p. 610).

(b)

The Land

The major terrestrial biome types—prairie, desert, tundra, forest—are generally characterized by the kinds of plant that grow in them and by the physical traits that determine that vegetation (e.g., climate, altitude, length of day, amount of rain, type of soil).

A glance at a map of vegetation (Fig. 21-20, p. 610) shows that in a general way these land biomes are related to geography. North-south location (latitude) between the equator and the north and south poles controls temperature, day length, and length of growing season. Near the equator we find a zone of **tropical rain forests.** North of this zone, where the summer rainfall and temperatures diminish, **deciduous forests** are the major biotic type. Still further from the equator are **coniferous forests,** and encircling the polar ice cap is a band of **arctic tundra.**

Along with the effects of latitude and temperature, another controlling factor is that of moisture. Cold air holds less moisture, as water vapor, than warm air. At the equator the warm, moist air rises, cools, and produces heavy rainfalls. The cooler, drier air that remains moves north to about 30° latitude before beginning to fall. It is thus not accidental that most of the great desert areas are found between 20° and 40° latitude. Farther from the equator, where the air becomes progressively cooler, deserts and dry grasslands usually occur on the east side of mountain ranges. For example, from about 40° to 60° latitude on the west coast of North America, winds blowing in off the warm Pacific Ocean are saturated with moisture. When this air reaches the coastal mountain ranges it rises

Figure 21-18 Life in the ocean depths is represented by forms such as (a) (opposite), the angler fish which lives at depths of 2000 feet or more; (b) a polyp of the family Umellulidae. This 3-ft-tall organism, which looks like a plant, is actually an animal; it was recently photographed on the bottom of the South Atlantic at a depth of 15,900 ft.

Figure 21-19   Deep sea fish. Note the bioluminescent spots.

Figure 21-20   A vegetation map of North America, showing the major north-south zones. Note how these are modified by conditions of rainfall and climate, which are determined by altitude as well as by latitude.

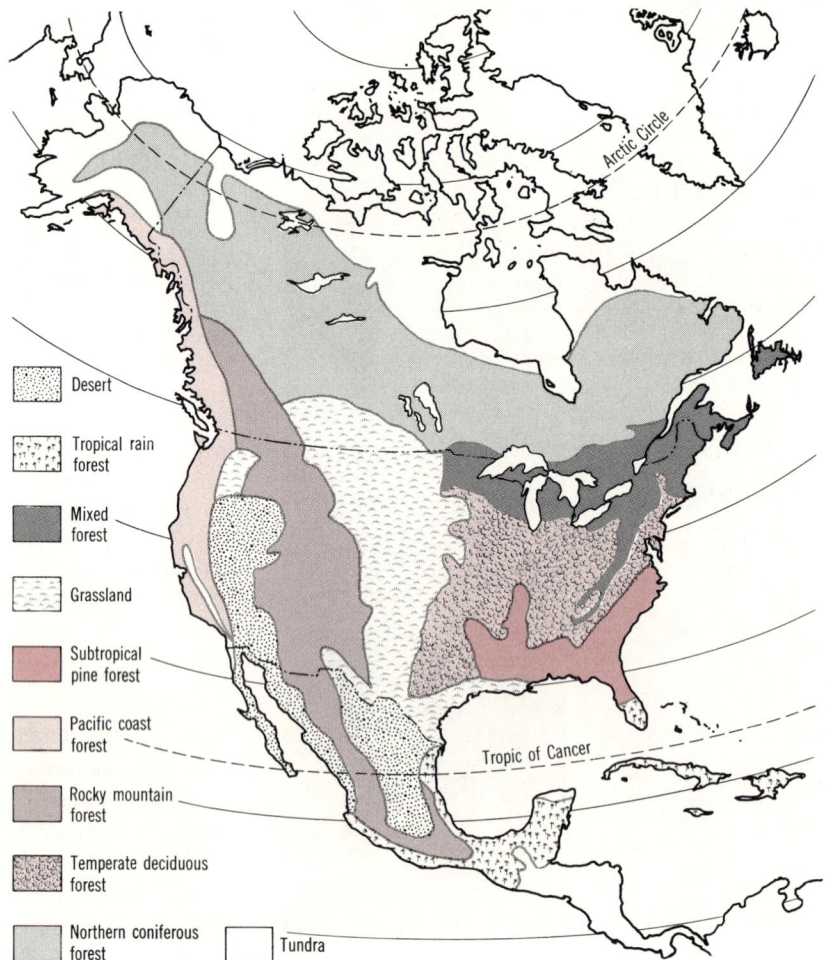

Desert

Tropical rain forest

Mixed forest

Grassland

Subtropical pine forest

Pacific coast forest

Rocky mountain forest

Temperate deciduous forest

Northern coniferous forest

Tundra

and cools, yielding heavy rainfalls on the western side of the mountains (Fig. 21-21). As much as 140 in. of rain per year fall on the Olympic rain forests of Washington and Oregon, and more falls as snow on the western slopes of the Cascade-Sierra Nevada range. This pattern of rainfall produces thick, lush forests, including giant sequoia stands. On the eastern slopes, in the **rain shadow,** deserts and plains begin.

Major Terrestrial Biome Types

Near the equator in low-lying areas are the **tropical rain forests.** Characterized by a warm, moist climate year round, with annual rainfalls of 100 in. or more, these forests nurture life in its most complex and luxuriant diversity; the number of species of plants and animals is astonishing. Perhaps 100 species of trees, for example, will be present per acre, in contrast to 10 or 15 species in a temperate deciduous forest. Growth is so heavy that the foliage forms a canopy, blocking out light from the moist ground. Little underbrush grows in the dimly lit soil. Leaves, flowers, and fruit that fall decompose quickly or are eaten by a profusion of small mammals and birds.

In the canopy itself brilliantly colored jungle birds such as parrots and toucans share a rich diet of flowers, fruits, and insects with a variety of monkeys, sloths, anteaters, and arboreal reptiles. Large herbivores like the elephant, deer, and tapir feed on leaves, twigs, and bark, and are eaten in turn by tigers, leopards, and other carnivores. For those of us who live in temperate zones, the numbers of species represented in a rain forest is hard to comprehend. In one study of a six-square mile area in the Panama Canal zone, for example, 20,000 different species of insects were found.

About 20° latitude north of the equator, in Florida and the middle southern states, for example, are the **warm-temperate pine forests.** Not so wet or lush as the rain forests, these regions are characterized by pines and live oak, magnolia, holly, cypress. Orchids and Spanish moss are common (Fig. 21-22, p. 612). A rich variety of birds, insects, reptiles, and small mammals are found in the pine

Figure 21-21 Moisture-laden on-shore winds traveling eastward from the Pacific Ocean encounter the high barrier of the coastal mountain ranges. As they rise, they cool, and their moisture condenses into rain or snow. Most of this moisture is deposited on the western slopes. Having thus lost its moisture, the dry air continuing eastward results in an arid area, or "rain shadow" east of the mountains, where, as a consequence, the desert begins.

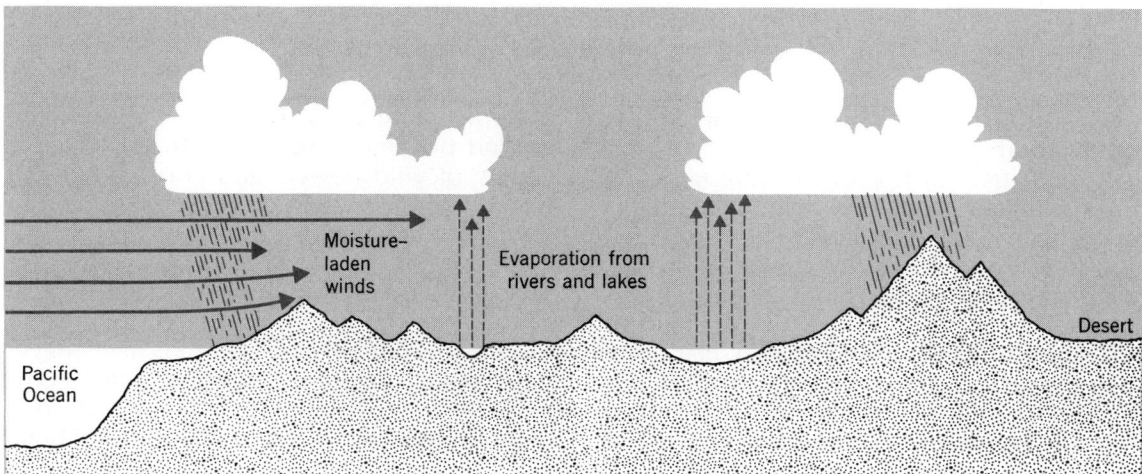

Figure 21-22 *Cypress trees anchor themselves in spongy soil, usually submerged beneath pond or swampland. Spanish moss hangs from the branches. Dense mats of water hyacinth are present.*

forests, though less than in the rain forests further south.

In dramatic contrast to the forests is the **desert biome.** In regions with annual rainfalls of less than 10 in., life depends on scattered cloudbursts and the ability of plants and animals to store water. Vegetation is sparse, characterized by cactus, sagebrush, creosote bush and other small shrubs (Fig. 21-23). Plants are widely spaced, an adaptation that allows each a maximum area from which to draw available water. In the California desert, for example, during the brief rainy season the ground becomes carpeted with an amazing variety of wild flowers and grasses, most of which complete their life cycle, from seed to seed, in a few weeks.

Animals of the desert—mainly reptiles, insects, and small rodents—all have special adaptations to the hot, dry life. The kangaroo rat and the gerbil, for example, need never drink water; eating moist desert plants provides all they require. Their excreta are dry; they rarely urinate; essentially all **metabolic** water—that produced in respiration—is conserved. In **absolute deserts,** like the central Sahara, where no rain ever falls, life is sparse indeed.

In the interiors of continents, where rainfall averages 10 to 30 in. per year, moisture is inadequate to support forests but greater than that of a desert. Here the open **grassland biome** is found: the prairies of the western United States or Australia, the savanna of Africa, the steppes of Russia and Siberia are examples. Trees and shrubs may occur as scattered individuals or along stream banks, but the major vegetation is a variety of species of grasses. These provide fodder for the major animals, mostly grazing or burrowing mammals—bison, antelope, zebras, rabbits, prairie dogs, gophers, wildebeests. The herds of grazing mammals support carnivores such as lions, tigers, and wolves. Smaller mammals provide food for hawks.

Generally at about the same latitudes as the prairies, (30°–50°) but in regions with greater rainfall, are found **temperate deciduous forests** (Fig. 21-24, p. 614). As the name implies, these forests exist in areas with moderate climates and distinct seasons. Deciduous trees—those that lose their leaves in the fall—such as maples, beech, oak, hickory, cottonwood, sycamore, and elm predominate. These are the forests that once covered most of eastern North America, western Europe, and eastern Asia, before the coming of cities and cultivated farmland.

During the relatively long growth season of spring and summer, foliage in these forests

Figure 21-23   The Sonoran desert of Arizona. Typical of the desert biome are the tall Saguaro cactus, prickly pear and sagebrush.

Figure 21-24 The temperate deciduous forest of the mid-Atlantic states. The dominant trees are oak, hickory, and maple.

becomes dense, and ample food in the form of leaves, bark, and fruits supports a variety of herbivorous animals: protozoa and worms in the moist soil; insects and a variety of small mammals such as the chipmunk, mole, squirrel, raccoon, and oppossum burrow or climb above and below the ground. At one time wolves, foxes, and mountain lions fed on these mammals, and on the deer that grazed at the edge of the forests. Few of these carnivores remain in the United States; many species are nearly extinct.

Farther from the equator the mean annual temperature falls and the forests gradually change character. Maple, elm and oak give way to needle-leaved evergreens of the **northern coniferous** forests (the great north woods of the outdoor adventurers), which stretch across the northern United States and Canada, northern Europe, and Siberia. Dominant trees are the coniferous forms: spruce, fir and pine (Fig. 21-25). Nonconifers such as aspen, poplar, and birch are also common.

In those areas where civilization has not yet encroached, moose and muledeer, black bear and grizzly are the characteristic large mammals. Smaller animals include porcupines, hares, and wolverines; and there is a variety of migratory birds. Wolves and deer are still common in some pine forest regions.

Between the northern edge of the coniferous forests and the polar ice cap, in a band about 60° north latitude, lies a zone of treeless **tundra.** Here the ground remains permanently frozen year-round **(permafrost)** from a few inches to several feet below the surface (Fig. 21-26). During the warmest summer months the temperature never rises above 10°C; the ground thaws only to a depth of a few inches. Some 5 million acres of this frozen, marshy plain stretch across the northern edge of North America, Europe and Siberia, inhabited by a few species of grassy sedge, moss, and lichens that are hardy enough to thrive in the cold, short growing season. Perhaps the most characteristic plant is the lichen known as reindeer moss, on which herds of caribou and reindeer graze (Fig. 21-27, p. 616). Other animals that have adapted to this inhospitable country are the arctic hare, fox, and lemming. Polar bear fish in the sea. The only common winter birds are the ptarmigan—a member of the grouse family—and the snowy owl.

The Freshwater Biome

Walking from field or forest toward a lake, we will note a progressive change in vegetation. A common sequence might be from terrestrial plants such as maples and elms to **semiterrestrial** (or **semiaquatic**) forms like willows and buttonbush, to various types of **aquatic** vegetation (reeds and sedges, cattails, floating water lilies, and completely submerged plants like elodea and the stonewort algae). Thus the land gradually merges with **fresh-water habitat.** Most large lakes are subdivided into zones similar to those of the ocean: the shallow water near the shore, or **littoral zone;** the surface waters away from the shore, or **limnetic zone;** and the deep water, or **profundal zone.**

614 *The Interactions of Organisms*

Figure 21-25 Northern coniferous forest of Alaska, showing dense stands of cedar and spruce.

Figure 21-26 Tundra permafrost near Point Barrow, Alaska. The polygons, seen here from a low-flying airplane, are 15 to 25 ft. across; they result from winter freezing when the cracks fill with ice and thawing in the brief summer.

Figure 21-27 Summer tundra with grazing caribou.

The littoral zone at the edge of a lake abounds with life. Here among the reeds and rushes, freshwater snails, crayfish, and insect larvae feed on the plants. Numerous small invertebrates such as worms and clams burrow in the rich organic bottom mud, and are prey to frogs, salamanders, and water turtles. Attracted by the rich diet, fish, too, are generally abundant at the lake margins, as are ducks, geese, and herons, otters, muskrats, and beavers.

Most lakes and ponds on earth are relatively young, geologically speaking. Ponds may range in life span from a few months to hundreds of years; larger lakes may be thousands of years old. Lake Baikal in Siberia is the oldest and deepest lake in the world; it was formed approximately 25 million years ago in the Mesozoic era (Chapter 22), when dinosaurs roamed the earth.

Ecological Succession

Each of the biomes described in the last section includes many communities. Each community is a living, dynamic system. A forest is not merely a stand of trees, or a pond a pool of water. We have tried to show that each is a complex, viable organization that has the constancy of a steady-state rather than a fixed structure. Like organisms, communities absorb energy, cycle materials, and produce feedback regulation of the interacting components. Like organisms, communities grow and attain a stable maturity.

Stability of Ecosystems

Food chains contribute to the stability of communities. By their very nature, food pyramids tend to maintain dynamically constant populations. For example, an increase in the number or production of an insect species may lead to an even faster increase in a parasite feeding on it, which kills a larger proportion of the host insect, and thus keeps the host population from growing out of control. Populations may be regulated by such processes of negative feedback involving other species, or environmental limits, or both. Thus internal population fluctuations in a natural community of organisms tend to be damped or minimized by the dynamic relationships that exist among living things in a food chain. It is a general ecological rule, in fact, that the greater the number of species in a community, and the greater the complexity of the food web, the more stable is the ecosystem.

How long a natural community of organisms persists depends mainly on the stability of the environment. If the physical environment—temperature, moisture, and local chemical factors—is maintained in its original condition from year to year, an ecosystem may change very slowly, at a rate measured in geological time. Until recently the oceans, large lakes, deserts, and prairies exhibited only long-term progressive changes, lasting hundreds of thousands or millions of years.

Nonetheless, most areas do tend to experience an orderly sequence of changes in biotic communities, with one community gradually replacing another, an ecological phenomenon known as **succession.** The initial community in an area, which is replaced in time by a sequence of succeeding communities, is called the **pioneer stage.**

The process of ecological succession, which is accompanied by physical, chemical, and biological environmental changes, ultimately culminates in the establishment of a relatively stable, mature community termed a **climax community.** The characteristics of community succession are so consistently orderly and regular that identification of any of the intermediate stages by a trained ecologist permits a fairly accurate prediction of future sequential changes. The presence of various plant species is one of the most useful criteria for recognizing a stage in a community succession.

One of the best examples of community succession is illustrated by the gradual transformation of a lake to a marsh, to culminate ultimately in a hardwood forest climax community. The pioneer community encompassed by a relatively young lake formed, say, by the flooding or damming of an area, is relatively simple (Fig. 21-28a); the plant and animal life is at first comparatively sparse. In due course more plants and animals begin to appear and thrive as they are introduced by the brooks and streams draining into the lake as well as by the activities of wind and animals.

Figure 21-28   *Ecological succession, as illustrated by the gradual transformation of the pioneer stage of a lake (a) to a climax hardwood forest (d).*

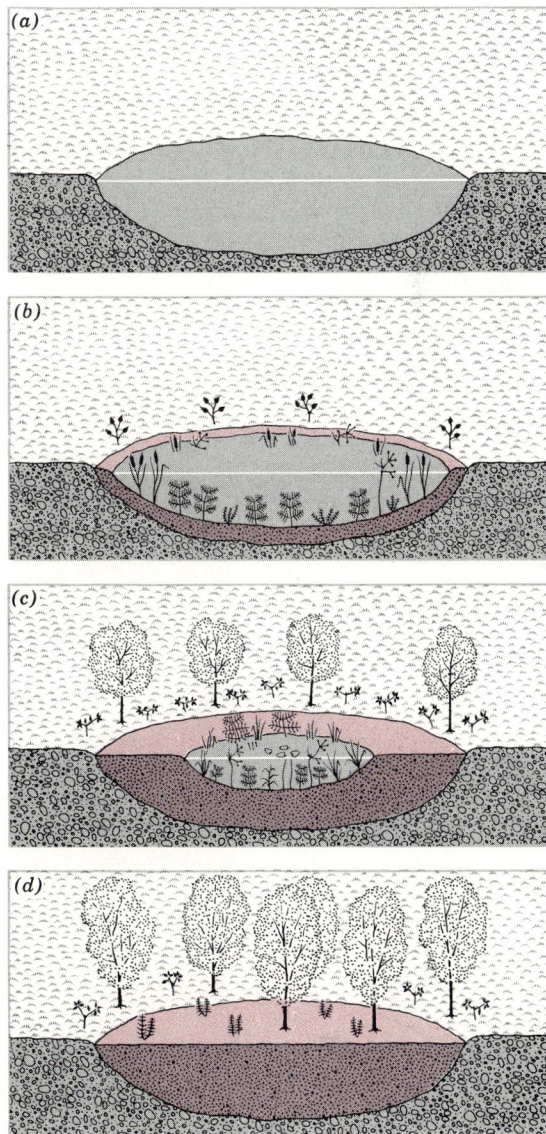

With the passage of time, sediment gradually accumulates on the lake bottom, the result of deposition of plant and animal remains and fine soil particles transported by brooks and streams. The number and kinds of plants and animals gradually increase, some becoming extinct and others becoming dominant. Older populations are gradually replaced by a sequence of new populations, and each new community helps modify its physical and chemical environment. As the vegetation at the edges of the water becomes thicker with grasses, water lilies, cattails, and so on, the dead remains of these plants, which we call **humus,** accumulate. Thus, in a matter of years, the lake edges become marshes and swamps, encroaching on the lake itself (Fig. 21-28b, 21-29a).

Proceeding from the lake edges toward dry land, we see a telescoped panorama of changes that will take place as the lake is progressively transformed to a smaller pond and finally to land. The swampy margins of the lake support the further growth of grasses, cattails, water lilies, and related plants. Further inland, these marshes are invaded by mosses, herbs, sedges, and shrubs. Willows take root, followed by evergreens such as tamaracks, spruce, and cedar. The climax community is reached with the establishment

Figure 21-29 Beach ponds on Presque Isle at Erie, Pa. (a) A newly formed pond less than a year old. (b) A similar pond about 50 years later.

(a)

(b)

of such hardwoods as maple and basswood trees (Fig. 21-28c; Fig. 21-29b). During this entire process of change, which takes place over the course of many years, the net effect is the gradual transformation of a water environment, with its attendant plant and animal life, through successive stages to yield a relatively stable climax community in the form of a hardwood forest. One of the most important factors contributing to this conversion of an aqueous to a terrestrial environment is the accumulation of plant remains (humus).

The main emphasis here is on the plant population because it often serves as the major controlling factor in community sustenance and change. Plants are also the best criteria for identifying a particular stage in the sequence of community succession. However, the succession of plant communities is usually accompanied by related changes in the animal populations. A young lake may at first contain a sparse, predominantly invertebrate population of zoöplankton and insects. Subsequent community succession usually proceeds to richer communities containing fish, amphibia, reptiles, birds, small land animals, and larger animals.

A climax community, although relatively stable compared to the communities that preceded it, results from an equilibrium of numerous factors in which any upheaval or significant change can lead to an imbalance. This imbalance will result in a replacement of the climax community by still another community or sequence of communities. Natural or artificial catastrophes such as fires or flood or invasion by destructive, disease-causing organisms, too, can easily shift the equilibrium.

Without a doubt one inexorable and important force in nature inevitably leading to the replacement of communities is biological evolution. It contributes significantly not only to the extinction of species but also to change and the appearance of new and different forms of life. Another and more potent factor in recent times has been the growth and activities of man.

Ecological Effects of Man

Human beings have roamed the earth for perhaps a million years, leading a precarious existence for most of that time. In the last 1000 years or so, we have achieved modest control of our immediate environment and some degree of security. But only in the incredibly short span of the last few decades, since we learned to use energy in concentrated form and developed the medical and agricultural knowledge that have resulted in our unchecked population growth, have our ecological innovations begun to work against us. Now our niche in the planetary ecosystem is, possibly, less secure than ever before. We are poisoning our world. In the words of Joseph Myler, "Man has managed to make his rivers rotten. He has transformed green pastures into deserts. He has clogged the air with chemicals which menace health and dust which is changing the climate. He is a menace to himself and other species."[3]

Until recently, the only materials that circulated through the biosphere were nutrients, the natural inorganic and organic components of protoplasm. Toxic elements such as arsenic, mercury, and lead lay buried deep in the earth, insulated from entry into life processes by their scarcity and essential insolubility. Each nutrient cycle remained in balance from one year to the next. Energy from the sun flowed slowly through every ecosystem, accumulating gradually in the packed humus of the forest and lake and the organic ooze of the ocean bottom. Oxygen consumed by animals was balanced by the carbon dioxide they produced, and both gases were returned to the ecosystem in like amounts through chlorophyll-bearing plants. Man, as a hunter-gatherer, had no more im-

[3]Myler, J. L., "The Dirty Animal—Man" in *Eco-Crisis*, C. E. Johnson (ed.), Wiley, New York, 1970, pp. 116–141.

Figure 21-30 "My God! There are traces of tuna fish in this shipment of mercury!" (Drawing by Dana Fradon; © 1971, The New Yorker Magazine, Inc.

pact on the continuing global cycles than did any other species.

Now all that is changed, perhaps irreversibly. Modern technology is deluging nature with tens of thousands of synthetic substances, many of which are very highly resistant to decay. Each year Americans junk 7 million cars, 100 million tires, 20 million tons of paper, 28 billion bottles, and 48 billion cans. Plastic sandwich bags, plastic-coated containers, and polystyrene cups dot the landscape. We dig vast holes in the earth to mine highly toxic metals such as lead, mercury, copper, and tin, and then discard large quantities of them as waste into our rivers, lakes, and atmosphere.

U. S. industries every year dump 165 million tons of solid waste and spew 172 million tons of smoke and fumes into the air. As a result of mining operations, vegetables in some parts of Montana contain 100 times the maximal concentrations of lead allowed in food for interstate shipment. Tuna and swordfish caught in the open ocean are declared unfit for human consumption because of excessive mercury in their tissues (Fig. 21-30). More than half of all the rainwater that falls on the United States is now used for industrial proc-

esses, mainly to carry off waste materials and waste heat. As a result of burning 5 billion tons of fossil carbon every year for energy production, we have increased the carbon dioxide content of the atmosphere appreciably. Industrial phosphates and nitrate compounds have largely replaced manure as fertilizers, and are now altering the balanced cycling of these elements in the biosphere.

The three major ecological problems man is creating involve alteration or contamination of Earth's atmosphere and its water, and the appalling accumulation of synthetic poisonous agents resulting mainly from the use of herbicides, pesticides, and detergents. Together, these developments have been termed man's **ecocrisis.**

The Atmosphere

Every year we burn fossil fuels at a greater rate than in the preceding year, while we remove millions of acres of forest and grassland, largely by paving, from the cycle of photosynthetic productivity. That is, we accelerate the combination of carbon with oxygen to produce carbon dioxide and reduce the rate at which oxygen in the atmosphere is regenerated. Although the best estimates indicate that we are not likely, for 10,000 years or more, to influence the oxygen concentration of air, the average CO_2 content has already increased—as we noted earlier—from 290 to 320 ppm (Fig. 21-11).

Most of the shortwave light radiation from the sun that strikes the earth is converted to a longwave heat energy, which is radiated back out to space. Both carbon dioxide and water vapor in the atmosphere are more transparent to light than to the longwave heat radiation. Therefore these substances act as a filter, allowing sunlight to get through and retaining the heat—the so-called **greenhouse** effect. In theory this should already have resulted in a general warming trend over the face of the planet. Even an increase of a few degrees in mean air temperature could have catastrophic effects; the polar icecaps hold enough water as ice to raise the sea level by 200 ft if they melted.

On the other hand, recent experiments in-dicate that when concentrations of carbon dioxide in the atmosphere are increased without permitting similar increases in particulate matter, plants grow more rapidly. Thus it is not inconceivable that continuing increases in atmospheric carbon dioxide could have beneficial effects in terms of net photosynthetic output of plants and increased food production if the atmosphere could be kept clean otherwise.

In any case, although the increase in CO_2 levels in the air is an indisputable fact, there has been no worldwide warming trend. This is generally attributed to other human activities, which have increased the density of particulate materials in the atmosphere; small particles reflect sunlight and reduce the amount of radiation that comes through the atmosphere.

American automobiles and industries, jet planes and fires, emit about 170 million tons of particulate matter as smoke and dust every year (Fig. 21-31). Worldwide emission of par-

Figure 21-31 Man's automobiles and industries, jet planes and fires emit many millions of tons per year of particulate matter as smoke and dust. The blanket of reflective matter filters out enough sunlight so that we would by now have experienced an appreciable fall in global temperature if this effect were not counteracted by the warming action of increased atmospheric carbon dioxide. This view of Denver, Colorado shows the effects of trapping such pollutants below a lid of warm air in a thermal inversion.

ticulates into the atmosphere is estimated at 800 to 1000 million tons per year; enough sunlight has apparently been filtered out to counteract the warming effect of the increased CO_2. Global temperature from 1940 to 1960, when a large fraction of the CO_2 increase occurred, actually fell slightly—by about .2°F. It is not possible to predict which of these two effects—warming from CO_2 or cooling by smog—will predominate in the future. It can, however, be stated with confidence that man's activities have now reached an order of magnitude that will inevitably influence the entire biosphere.

In addition to carbon dioxide and solid particles, major air pollutants (listed in order of their annual tonnage) are carbon monoxide, sulfur oxides, hydrocarbons, and nitrogen oxides. These also interact with the biosphere in various ways, not all well understood. **Carbon monoxide** (CO) appears to be almost entirely man-made; the only significant source known is incomplete combustion of fossil fuels, a common problem in internal combustion engines. Concentrations of CO rise dramatically in urban centers during peak traffic hours, reaching local levels of 30 to 50 ppm. CO has profound physiological effects; it has been implicated as a causal agent in coronary artery disease and emphysema. A maximum allowable concentration of 4 ppm in city air has been recommended. Although CO is emitted in large amounts, it does not seem to accumulate in the atmosphere; the mechanism of its removal is not known.

Sulfur occurs mainly as an impurity in fossil fuels, and in the form of **sulfur dioxide** is among the most troublesome of air pollutants. Sulfur dioxide reacts with moisture in the atmosphere to form **sulfuric acid** or **ammonium sulfate.** When washed from the air by rain, sulfur products increase the acidity of the rainfall. pH values as low as 4 have been found in the Netherlands and Sweden as a result of the extensive industrialization of western Europe. In Italy statues that have withstood weathering for hundreds of years have begun to disintegrate in the last decade as a result of the acid rainfall. Small lakes and rivers have begun to show increased acidity that could destroy their stability as ecosystems; many fishes, such as salmon, cannot survive at pH values below 5.5.

Only about 15 percent of the total **hydrocarbon** emission results from human activities. Hydrocarbon compounds, mainly as methane, are produced naturally from bacterial decomposition of organic matter. However, in combination with **nitrogen oxides** and ultraviolet rays in sunlight, hydrocarbons produce the photochemical smog that is responsible for the eye irritation and respiratory problems that plague Los Angeles, Philadelphia, New York, and most other major urban centers.

The biological effects of several of the products of these reactions, including ozone and other complex organic oxidants, can be severe. Ozone is highly detrimental to vegetation; it has been linked (along with other oxidants) to the spectacular rise of emphysema and lung cancer in recent years among urban dwellers.

Water Pollution

One of the major principles of ecology is that ecosystems, like organisms, age; that is, they change with time. They undergo succession. A lake ecosystem, as we have seen, proceeds inexorably to a terrestrial state. A young lake is a clear body of water, often with a stone or gravel bottom. There is little nutrient turnover, and the biotic community is simple, with short food chains. The lake is said to be **oligotrophic.** As lakes age they gain sediment and their nutrient content is enriched. This process of enrichment is called **eutrophication.** A eutrophic lake is turbid with plankton and algae, and has a muddy bottom; both the water and bottom are rich in nutrients and organic debris. The biotic community is complex, with many elements and involved food chains and webs, the members of which are a prelude to the eventual filling in and disappearance of the lake. This is the normal evolutionary fate of all lakes, but until recently the process took thousands or millions of years for lakes of moderate size.

Pollution with excessive nutrients, however, can accelerate eutrophication, with unpleasant and sometimes disastrous consequences. Young, oligotrophic lakes support the best food fish—trout, char, chub, whitefish. As eutrophication proceeds these are replaced by less valued species such as bass, perch, and pike, and still later by forms that are useless to man.

Eventually, if the eutrophication process is sufficiently rapid, there are periodic sudden increases in populations of some organisms, followed by massive die-offs. Beaches are littered with tons of foul-smelling algae or fish (Fig. 21-32).

In the American Great Lakes the natural process of eutrophication has been accelerated by man's misplacement of natural resources; that is, by pollution. This problem is most striking in Lake Erie, which is the smallest and shallowest. Lake Erie receives its main input of water from rivers that run past Detroit, Toledo, Cleveland, Erie, and Buffalo, and which drain several thousand square miles of agricultural land. Records show that over the last 100 years there have been significant increases in the amount of calcium, sodium, potassium, chlorides, phosphates, and sulfates, ranging from twice to five times the normal amounts in parts per million. Oxygen-demanding wastes have accumulated on the lake bottom, whereas normally critically scarce elements such as nitrogen and phosphorus have increased dramatically. The deeper parts of the lake are subject to severe oxygen depletion involving many hundreds of square miles. Near the surface, distinct signs of eutrophication exist.

Since 1930 bacterial counts in the shallow water at the ends of the lake have increased by three to four times the usual amount, and phytoplankton by 30 times. In recent summers three-quarters of the swimming beaches have been closed all season because the bacterial count exceeded safety levels.

The deep waters can no longer support commercially significant fish; the catch of species such as herring and blue pike dropped from 20 to 40 million lb per year before 1920 to a few hundred lb a year since 1965. Near the surface, where sunlight and CO_2 are available, blooms of photosynthetic microorganisms further deplete the dissolved oxygen and add to the turbidity and sediment. It has been estimated that if all pollution were discontinued *now*, it would take

Figure 21-32 Dead alewives on a Chicago lake-front beach, the result of artificially accelerated eutrophication.

about 20 years for 90 percent of the wastes to be cleared from Lake Erie. Lakes Superior and Michigan, which are larger, would require hundreds of years to cleanse themselves.

It is not for lack of ecological or technological expertise that water pollution abatement has made little progress. The barriers are mainly political and economic. A striking example of the sociopolitical problem involved in prevention of severe water pollution may be seen in sections of Wyoming, Utah, and Colorado where there are extensive outcroppings of Green River Shale in which new sources of petroleum are being developed. Because of the high organic content of these sedimentary rocks—up to 65 percent—it is expected that large quantities of oil can be extracted from them. Estimates run as high as 2 trillion barrels, enough for 500 years at present demand rates. To free oil from the shale, heat treatments of 500 to 550°C are required, which liberate enormous quantities of alkaline waste. Traditionally, shale processing plants dump such waste directly into nearby rivers. In Colorado the major available rivers, North Platte and Green, could be catastrophically polluted by such a procedure.

The shale deposits extend over 1600 square miles of public lands. Because the potential oil resources are valued at trillions of dollars, private interests are scrambling to get control of these lands from the federal government and stake out claims. Local governments are unable to cope with problems on this scale, and in fact often support these developments because of the potential increase in taxable income.

One of the most serious forms of water pollution results from contamination with heavy metals—lead, mercury, and cadmium—used by various industries. During this century in the United States about 75 million kg of mercury have been used by American technological processes: in the paper industry, in pesticides, for the production of chlor-alkali (an important industrial compound). World production now amounts to about 10,000 tons a year, about one-third of which is used in the United States. Recently it has become clear that compounds of mercury present a substantial hazard. Industrial wastes containing mercury have routinely been discarded in lakes and rivers. In the bottom mud mercury is converted by anaerobic microorganisms into methyl mercury ion (CH_3Hg^+), which is soluble in water. Moreover, it is concentrated by higher organisms, usually appearing in their body lipids. For example, fish absorb methyl mercury directly through their gills, and can concentrate it by as much as 3000 times (see below). Substantial mercury pollution in the Great Lakes became apparent in March 1970, when mercury concentrations as high as 1 to 5 ppm were found in pickerel and other fish taken from Lake Erie.

Physiological and cytological studies have revealed some of the behavior of methyl mercury. It tends to be associated with red blood cells and nervous tissue. It easily passes the placental barrier, becoming further concentrated in the fetus. It can cause chromosomal disorders, congenital defects of the young, and neurological damage. In Japan up until 1969, 137 persons had died or suffered serious brain damage as a result of eating fish and shellfish caught in mercury-contaminated areas. Among these were 19 congenitally defective babies born of mothers who had eaten these foods. In some cases the fish were found to contain mercury at concentrations of 5 to 20 ppm.

Even if all further mercury pollution were stopped immediately, it would be many years before natural processes could cleanse the waters. The major problems with mercury compounds, called **mercurials,** arise from two of their biological properties: their long residence time, or **half-life,** in the food chain, and their great toxicity even at very low concentrations.

After metallic mercury has been methylated by anaerobic bacteria in the lake bottoms, a process called **organic complexing** spreads the methyl mercury systematically throughout the aquatic environment. Algae eat the bacteria; fish eat the algae. As the fish grow, so does the methyl mercury in them, because mercury has a half-life in fish of about 200 days. That is, the amount eaten each day is only very slowly removed, so that half of it still remains 200 days later. However, the next day more is eaten—half of which remains 200 days—and the next day more, and so on, concentrating the poison thousands of times.

Humans who eat such fish also retain the mercury for an extended period; the half life of mercury in man is about 70 days. Thus, if fish form even an occasional part of the diet of humans they concentrate the poison still further. Of even greater import, however, is the recent finding that both marine and freshwater phytoplankton are sensitive to much lower concentrations of mercurial compounds. Levels as low as 10 parts per billion can reduce the amount of photosynthesis by these microorganisms by 50 to 80 percent. Recall that about half of the total biomass on earth is produced in the world's lakes and oceans, most by phytoplankton could seriously disrupt the entire earth's ecosystem.

An especially serious form of oceanic water pollution results from oil spills. The spill in the Santa Barbara channel in 1969, which involved about 10,000 tons, and the Torrey Canyon spill in 1967, involving 100,000 tons, produced intense local concentration of oil, which is toxic to many marine organisms. Besides these well-publicized events, there is an annual worldwide spillage that adds up to about a million tons from various oil operations, as well as natural oil seeps of unknown magnitude. Added to all of these is the dumping of waste motor oil, which is much more toxic than crude oil. In the United States alone about a million tons of motor oil is discarded every year.

Up to the present no worldwide effects of oil spills are detectable, other than the deaths of a few hundred thousand sea birds. However, because oil is degraded rather slowly in water, tarry mementos of such spills will continue to disrupt the ecology and scenery of the coasts of many nations for decades, at least. Moreover, when it is recognized that one gallon of oil is sufficient to cover four acres of water, and that even a thin film of oil retards the rate of oxygen uptake by the water beneath it, the potential danger of continued oil contamination becomes frighteningly apparent.

Biocides and Detergents

Pesticides

A **biocide** is any agent that kills living organisms. **Pesticides** have probably existed ever since the first cave dweller noticed that sea water killed plants. DDT was first synthesized in 1874; its insecticidal properties were discovered in 1939. However, it was mainly during World War II that scientists developed a diverse battery of **insecticides**—dieldrin, lindane, Malathion, and so on—for military uses, and **herbicides** (plant killers)—2,4-D, and 2,4,5-T—to destroy enemy crops. The war ended before most of these substances could be tested in the field, but the chemicals soon became available to farmers. It was found that, with weed and insect pests controlled, crop production burgeoned and yields per acre increased dramatically. In 1947 synthetic pesticide production in the United States was about 125 million lb; in 1969 it had soared to over 1 billion lb. The wholesale value of such products is now several billion dollars.

There are numerous short-term advantages to be gained from such agents. Before World War II, for example, malaria was the largest single infectious killer of man. Carried by the *Anopheles* mosquito, this disease could be prevented by ridding an area of that insect. In 1946 Ceylon initiated an antimalarial campaign with DDT, and a few months later the United States followed suit. In one year the disease was virtually eliminated in Ceylon. In America mortalities, which had ranged between 200,000 and 400,000 per year, mainly in the southern states, dropped to fewer than 100. In India 100,000,000 persons suffered from malaria every year before 1945; after an extensive DDT campaign the number of cases fell to fewer than 50,000 by 1966.

In 1955 the World Health Organization undertook to eradicate malaria altogether, and during the following decade spent al-

Figure 21-33 DDT being sprayed by helicopter in the Congo as part of a malaria-control program. Such programs led to a dramatic reduction in the disease, but also to the appearance of DDT-resistant strains of insects.

most $1 billion on the task (Fig. 21-33). Of the total world population of about 2.5 billion at the end of 1964, about 1.6 billion lived in areas that were or had been ridden with malaria. By that date 75 percent of these persons could enjoy a completely malaria-free environment or could observe eradication in progress. It has been estimated that the single largest factor in the spectacular increase in life expectancy since the early 1940s in the underdeveloped countries has been DDT.

There are two main problems, however, that make the use of such agents ecologically unsound and ultimately ineffective. One is that target organisms tend to develop **resistance** to any given pesticide. The other is that all such agents are **nonspecific;** that is, they kill or sicken species other than the target organism, often beneficial forms and sometimes man himself. By 1965, for example, of some 60 species of *Anopheles* that transmit malaria, 34 were found to be no longer affected by the usual doses of DDT. In Italy, Sweden, and the United States, houseflies,

fleas, lice, and mosquitos, which had initially been highly susceptible to DDT, developed resistances not only to that compound and a long list of related chlorinated hydrocarbons, but to other kinds of insecticide as well.

When insecticides are used for extended periods on food crops and near rivers and water supplies, these agents soon become widespread throughout the ecosystem. Some pesticides are subject to normal biological degradation by soil and water bacteria (i.e., are **biodegradable**), and are quickly destroyed. On the other hand, some, like DDT and dieldrin, are **nonbiodegradable.** A low, nonlethal dose of DDT absorbed by phytoplankton in a lake, for example, is absorbed in the fatty tissues of organisms higher up the food chain. Fish feeding on contaminated organisms concentrate the poison in much the same way we have described for mercury. During its lifetime a fish consumes much plankton; most of the fat-soluble DDT (and its metabolic product DDE, which is equally toxic) is retained in the fatty tissues of the

626   *The Interactions of Organisms*

fish. The story is repeated, with further concentration, when such fish are in turn eaten by predatory birds. In a recent study of DDT concentrations in organisms along the south shore of Long Island, where this insecticide has been used to control mosquitos for 20 years or more, carnivorous birds and mammals had concentrated the poison 1000 times over the organisms at the base of the food chain.

DDT and DDE are now ubiquitous features of the earth's environment. It is estimated that 1 billion lb of these substances are in the world ecosystem. They have been identified in animals everywhere, from polar bears in the Arctic to seals in the Antarctic. Mallard ducks in Colorado, prairie falcons from the southwest desert, and brown pelicans off the California coast carry DDE concentrations as high as 2500 ppm in their tissues. In fact the populations of many of these birds are being decimated because of reproductive failure and thinning of the eggshells caused by these poisons (Fig. 21-34).

The loss of species such as the peregrine falcon or golden eagle from the face of the earth as a result of DDT poisoning would be dramatic and tragic enough. But it is not only birds and wild animals that may be affected. Our own status is not comforting. In 1970 a typical value for DDT and its degradation products in human fat was 12 ppm. According to analyses by the U. S. Food and Drug Administration, residues of pesticide chemicals are found in about half of the thousands of food samples tested each year; about 3 percent of the samples contain levels in excess of accepted tolerances. During the period 1964 to 1966, for example, residues of chlorinated organic pesticides (DDT, dieldrin, lindane, etc.) were commonly found in all diet samples and all kinds of food: meat, fish, poultry, and dairy products; everything except beverages. There is little direct application of insecticides to these products; the poisons get there via the food chains. Although levels of these toxins were well below "acceptable" tolerances set by various agencies, there is no reason to think that intake will cease or that concentrations in the envi-

Figure 21-34   *This egg in the nest of a brown pelican off the California coast had such a thin shell that the weight of its parent's body crushed it. The concentration of DDT in these eggs was measured at 2500 ppm. None of the eggs of the entire 300-pair colony hatched.*

ronment will not continue to rise.

Although occasional fatalities from massive insecticide poisoning are reported—most often among children—there are as yet no known diseases in man that result from chronic low doses of these substances. Unfortunately, the same cannot be said of herbicides, which have been used just as widely. The weed-killers 2,4-D and 2,4,5-T have recently been shown to produce birth defects in laboratory animals, and have been implicated in a rise of such anomalies among children in Vietnam, where thousands of pounds of these substances have been used for military purposes. At least part of their toxicity derives from contaminants known as **dioxins,** which appear to be poisonous at very low levels and highly resistant to degradation. In a single decade production of these compounds rose from a few thousand dollars worth to a value of $1 billion in 1970. By 1975 they are expected to account for 60 percent of the pesticide sales in the United States.

Detergents

Two main classes of detergent have been used in the last 20 years or so. The first, known as **alkyl benzene sulfonate** (ABS) is nonbiodegradable and persists for long periods after sewage treatment. The group called **linear alkyl sulfonate** (LAS) is more readily decomposed by soil bacteria. Largely because of adverse public reaction to detergent foam in drinking water, lakes, and streams, the detergent industry has converted from ABS to LAS in a number of countries: England in 1962, West Germany in 1964, the United States in 1965. Although this conversion has dramatically reduced the amount of foaming and the concentration of detergent residues in lakes and rivers, it has revealed another, more important problem. Degradation of the detergent molecule releases large quantities of phosphates—perhaps 1 billion lb per year in the United States. Most heavy-duty detergents contain at least 50 percent phosphates; some of the "enzyme presoak" products are 80 percent phosphates (largely as "polyphosphates").

Phosphates are added to detergents for three main reasons. They combine with calcium and other dissolved minerals to soften water; they disperse and suspend dirt; and they maintain a desirable level of alkalinity in the water. But, as we have seen, they are also efficient stimulators of algae growth and of eutrophication.

It has been estimated that phosphate detergents in 1970 contributed 50 percent of the phosphate in municipal sewage in Canada and 70 percent in the United States. The average phosphate content in such sewage in America that year was 10 ppm, or 2000 times what might normally be present in an uncontaminated lake. Thus removal of phosphates from detergents would reduce sewage phosphate content by two- or threefold, but

would still leave substantial amounts from other sources.

Although polyphosphates promote eutrophication, they have a major advantage of being very low in toxicity to humans and other vertebrates. Recently, however, attention has been drawn to high concentrations of arsenic, as much as 70 ppm, in several common phosphate detergents. This relatively high concentration poses potential hazards from absorption through the skin following domestic use and from pollution of drinking water. As an alternative to phosphates in detergents, it has recently been suggested that nitrogen compounds be used. Some of these, such as nitrilotriacetic acid (NTA), are economical, effective, and highly biodegradable. However, nitrates too promote eutrophication, and some of their breakdown products seem to be carcinogenic.

The most obvious way to remove phosphates and other undesirable chemicals from waste water is by appropriate sewage treatment. Well-known processes of distillation, reverse osmosis, and chemical treatment can eliminate up to 95 percent of such contaminants of sewage. These processes are costly, however, and not now in wide use; most communities do not have such treatment facilities and it would take many years to build enough treatment plants to remove any substantial amount of pollutants from municipal sewage. The question, then, is not whether a polyphosphate or NTA is a better additive for detergents. The question is whether the use of these compounds or any synthetic chemicals should be permitted in ways that will inevitably result in the dumping of millions or billions of pounds into our waterways without adequate knowledge of the biological and ecological consequences.

The Message of Ecology

It is all too easy, after a consideration of man's disastrous ecological habits to date, to draw the simplistic conclusion merely that we must stop contaminating our atmosphere and our water with chemicals and poisons. It would seem that scientists and science-based technicians are unable to see the consequences of any given activity; that, no matter how well motivated, they inevitably produce undesired side-effects and wreak more havoc than they do good. It is feelings of this sort that have turned many of today's students away from science.

However, humanity now has itself in a dilemma. Continued pollution could lead to irreversible or at least unpredictable consequences on a worldwide scale. On the other hand, if we now suddenly withdraw the use of all herbicides and pesticides, for example, it is absolutely predictable that the populations of many nations (mainly the poorer ones, of course) will in a few years be ravaged by insect-borne diseases such as malaria, typhus, yellow fever, cholera, and smallpox, which have for years been under control. Food production would diminish drastically, and famines would soon follow. Clearly, the toll in human suffering makes this an unacceptable alternative.

Fortunately we have another choice, which is not to renounce science and the manipulation of the environment, but to learn more and to use our knowledge with greater wisdom. Work in recent years, for example, has been expanding in methods of pest control that do not require the use of broadly toxic chemicals. Insect-resistant varieties of plants and natural insect predators are being explored as possibilities for reducing the need for insecticides.

Many varieties of insects locate their food supplies and mating partners by chemical stimuli called pheromones (Chapter 19), and food attractants are probably responsible for the fact that certain insect species feed only on specific plant or animal hosts. If these attractants can be isolated and identified, they can be used to lead insects to capture and death without requiring broad distribution of poisonous materials. Recently a synthetic feeding lure **(methyl eugenol)** that attracts male fruit flies was used to eradicate the oriental fruit fly from the island of Rota in the Pacific. Similar results have been obtained with sex attractants. Male gypsy moths, for example, may be lured from as far as 2 miles away by a few milligrams of a synthetic attractant called **gyplure.** These compounds have the advantage of being enormously potent, extremely specific—they are totally ignored even by related species—and, so far as is known, completely nontoxic. As of 1970 some 20 insect attractants had been identified and isolated.

Another environmentally innocuous method of insect control, which has proved highly successful in eliminating the screw-worm fly in Texas and Florida, is the sterile-male technique. Flies are farmed in large numbers in the laboratory. Up to several million fly eggs are collected and reared on artificial food until pupation. The pupae are then sterilized by radiation and are permitted to develop into adults.

These millions of sterile flies are dispersed from low-flying aircraft over a wide area. The sterile females fail to reproduce; the irradiated males fertilize large numbers of non-treated females in the field with their sterile sperm. Once an egg has been fertilized it cannot again be penetrated by a second sperm. Thus if the first sperm is nonfunctional, the egg fails to develop.

The problems that arise from man's impact on his ecosystem will not be easy to solve. Man's hope, however, lies in his recognition of his niche as a natural and integral part of the ecosystem. He is a partner with other organisms in husbanding the earth's resources and managing his own cultural and biological waste. The future will depend on his ability to develop a body of knowledge regarding these matters, and applying that knowledge with wisdom. This demands from all of us a rethinking of our attitudes about man's place in nature, and the development—as Aldo Leopold put it—of a "new ethic for the land." Nature is no longer our adversary to be conquered, or servant to be exploited. We must learn that it is home; it is "spaceship earth."

Reading List

Brodine, V., "Episode 104," *Environment* (January/February 1971), pp. 2–27.

Broeker, W. W., "Man's Oxygen Reserves," *Science* (1970), **168,** p. 1538.

Brower, L. P., "Ecological Chemistry," *Scientific American* (February 1969), pp. 22–30.

Ehrenfeld, D. W., *Biological Conservation*. Holt, Rhinehart & Winston, New York, 1970.

Epstein, S. S., "A Family Likeness," *Environment* (July/August 1970), pp. 16–25.

Flanagan, D. (ed.), "The Biosphere," *Scientific American* (September 1970), pp. 44–208. Entire issue devoted to ecology, with eleven articles on the energy and nutrient cycles.

Flanagan, D. (ed.), "Energy and Power," *Scientific American* (September 1971), pp. 36–200.

Fritts, H. C., "Tree Rings and Climate," *Scientific American* (May 1972), pp. 92–100.

Goldwater, L. J., "Mercury in the Environment," *Scientific American* (May 1971), pp. 15–21.

Johnson, C. E. (ed.), *Eco-Crisis*. Wiley, New York, 1970.

Kormondy, E. J., *Concepts of Ecology*. Prentice-Hall, Englewood Cliffs, N. J., 1969.

McCaull, J., "The Black Tide," *Environment* (November 1969), pp. 2–16.

Novick, S. (ed.), "Project Survival," *Environment* (April, 1970), pp. 2–47. Entire issue devoted to articles on air and water pollution, pesticides, and radiation.

Peakall, D. B., "Pesticides and the Reproduction of Birds," *Scientific American* (April 1970), pp. 72–83.

Shea, K., "Infectious Cure," *Environment* (January/February 1971), pp. 43–45.

Shepard, P., and D. McKinley (eds.), *The Subversive Science*. Houghton Mifflin, Boston, 1969.

Watts, A., "The World is Your Body," in *The Book on the Taboo Against Who You Are*. Pantheon, New York, 1966.

"And what we now are witnessing is perhaps the most dramatic event in the slow evolution of life: the human brain scrutinizing itself and its origins. . . . We who are of nature are evolving to know nature."

A. G. LOEWY & P. SIEKEVITZ, 1969, CELL STRUCTURE AND FUNCTION

part nine | *Evolution and Classification*

chapter twenty-two | Biological Origins

"At the beginning all things were confused, infinite in number as well as in smallness, for only the infinitesimally small existed."

ANAXAGORAS, 5TH CENTURY B.C.,

Man is a curious animal. Humans will suffer hardships, go hungry, even willingly die to satisfy their curiosity. Other species also exhibit behavior that is not immediately directed toward satisfying a basic need such as hunger or reproduction; a monkey will work for the reward of being able to solve puzzles (p. 550). But only man philosophizes on his existence and wonders about his own origins. Even the most primitive cave dweller must have begun to contemplate the world around him. He looked up at the fixed and yet changing skies; he noted the cycle of seasons. He was aware of death; he observed life constantly being born around him. And he asked: "If I was born of parents, and they of parents before them, what was the beginning?"

Origin of the Universe

To the primitive mind, in order for things to exist they must have been created; this thought required that there must have been a creator. Thus most ancient concepts of the origin of the universe were based on supernatural phenomena and special acts of creation in which the world originated in very much the same state as we now find it (Fig. 22-1, p. 634). As long as this idea of an abrupt, special creation was prevalent, the world had to be viewed as arbitrary and illogical. Man was separate from the animals and unrelated to the rest of nature except in the most tenuous terms.

The concept of **evolution,** focused mainly through the monumental work of Charles Darwin in 1859, completely changed western

A collection of plant fossils from the Carboniferous Period.

Figure 22-1 Biblical account of the creation of the universe is here illustrated in a sixteenth-century bible printed in Venice. On the first day (di uno) God created heaven and earth. On the second day (di segundo) He separated the firmament (sky) from the waters. On the third day (di terzo) He made the dry land and plants. Only then did He make the sun, moon, and stars. On the fifth day He made the birds and the fishes. And, finally, on the sixth day He created the land animals and man. According to Genesis, rather than creating the animals and plants directly, God bade the earth and waters bring them forth. Thus there is no necessary conflict between this theological view and the modern ideas of organic evolution.

man's concept of the natural world. According to this view, life originated from the inanimate material of the universe as a result of slow natural changes occurring during the course of many millions of years. All living things are seen as descended from one or a few types of primitive organism that first appeared 2 or 3 billion years ago; these in turn arose a billion years or so before that from spontaneously formed molecular aggregates capable only of acting as templates for the formation of other aggregates like themselves. Thus the similarities among various animals and plants can be understood in rational terms, as signs of their common ancestry, just as brothers share a common likeness with their parents.

With the concept of evolution man becomes part of nature; he achieves relatedness to all living things; he represents not the product of an abrupt event, but the culmination of a grand and awesome natural process extending back to the dawn of the universe itself.

The Universe Expands

Perhaps the most startling and philosophically profound discovery made in science this century is that we live in a continually changing universe, a universe populated by billions of aggregations of stars, called **galaxies** (Fig. 22-2, p. 636), which are systematically receding from one another like spots of paint on an expanding rubber balloon. The astronomer Edwin Hubble, who made this discovery, calculated that if these galaxies have always moved with their present measurable directions and velocities, they must have started from a single point in space about 12 billion years ago. This calculation has led to a widely accepted **cosmological** hypothesis that the world began with an incredible explosion of a primordial "atom" containing all of the matter in the universe.

According to this **"big bang"** view, first examined in detail by physicist George Gamow in 1940, the universe consisted for the first few seconds of a gigantic primoridal fireball with a temperature of several billion degrees. As the fireball expanded, probably for the first 180,000 years or so, no structures such as galaxies or stars could have existed. All space would have been filled with radiation and hot gas composed of the nuclei of hydrogen and helium atoms and their accompanying electrons.

This continuously expanding gas mixture was the matter of the universe for an estimated 250 million years. As it expanded, it cooled rapidly, and condensed into isolated clouds or balls of gas. The huge separate masses of gas subsequently gave rise to large aggregations of stars by further condensation as they were drawn away from one another by the continuing expansion of the universe (Fig. 22-3, p. 637).

The high pressures caused by the rapid contraction of these huge fragments of gas resulted in exceedingly high temperatures, of the order of a few million degrees, in their more dense regions. These high temperatures were probably responsible for the final stage in the formation of stars, for when matter gets hot enough it releases nuclear energy in the form of heat and light. It is this energy that we see as starlight from distant galaxies and as light and heat from our own sun.

Gamow's theory has not met with unanimous acceptance. What was the origin of the dense primordial core? One explanation proposes that the universe experiences an unending cycle of expansion and contraction, extending itself to a certain maximal limit, then contracting to a dense core of subatomic material with successive expansion and contraction. A possible alternative to this **cyclic,** or **pulsating,** scheme is the proposition that the universe always existed as it does now. Galaxies are, however, expanding away from each other and new matter is continuously being created everywhere in the universe in between that which already exists. Thus new galaxies are constantly being formed as expansion proceeds.

Approximately a billion or so galaxies are now visible with our most advanced telescope, the great Mt. Palomar instrument in California. Its 200-in. mirror has permitted man to look at stars more than 40 sextillion miles away, or 7 billion light years[1] into space. This means that it takes 7 billion years for the light of these distant stars to reach us, traveling at a speed of 186,000 miles per second. Thus the most distant galaxies are seen by us as they were when the light left them 7 billion

[1] A light-year is the distance traveled by light in one year at the rate of 186,000 miles per second (or 670 million miles per hour) and is equal to almost 6 million million miles.

Figure 22-2   Spiral galaxy M101 in the constellation Ursa Major shows the loosely wound spiral arms that characterize this type of galaxy. Our own Milky Way galaxy is similar. M101 contains about 100 billion stars; it is about 100,000 light-years in diameter. This photograph was taken with the 200-in. Hale telescope at the Mt. Palomar Observatory.

years ago. We are looking not only far out into space, but also backward in time.

Our own galaxy, the Milky Way, is a collection of some 100 billion stars, with our sun—a rather average star among all the others—and its planets out toward the rim. Most galaxies are in a state of rotation around their axes, as indicated by their elliptical shapes and the spiral arms encircling condensed centers (Fig. 22-2). The stars (including our sun) themselves also rotate rapidly. They are made up almost completely of a gaseous mixture of hydrogen and helium. The gases are continually undergoing nuclear reactions (Chapter 2), liberating energy in much the same way as does a man-made hydrogen bomb.

The Solar System

Our solar system consists of the Sun, a medium-sized star with a diameter of about 850,000 miles, its nine planets, and a variety of smaller bodies ranging from the 31 planetary satellites (including our own moon) to smaller bodies and planetary dust. The Sun's mass is 330,000 times that of Earth, and its temperature ranges from 6000°C at its surface to an estimated 25 million°C at its core. The Sun is almost completely gaseous; the most prevalent gas is hydrogen, with traces of 65 other elements.

The planets rotate around their axes and revolve around the Sun in elliptical paths determined by the force of gravitation. The planet Mercury is closest to the sun, followed by Venus, Earth, Mars, Jupiter, Saturn, Uranus, Neptune, and Pluto, in that order; the mean diameter of the solar system is about 7 billion miles. The inner planets are close together; the outer ones are further apart.

The best current theory concerning the origin of the planets, known as the **dust-cloud hypothesis,** postulates that they formed from relatively smaller masses consisting of clouds of dust particles and gas torn away from the edges of the newly forming stars and held together by the mutual attraction of gravity. They proceeded to grow by the gradual accumulation of solid dust particles composed

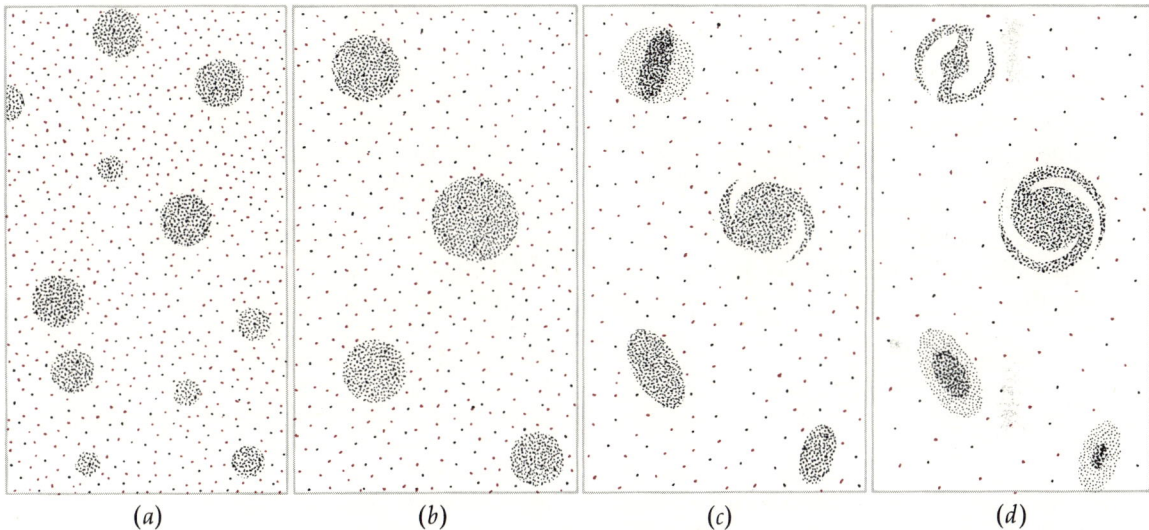

| (a) | (b) | (c) | (d) |

Figure 22-3 *Formation of galaxies by condensation is represented in this sequence of drawings. (a) For about the first 100,000 years after the "big bang," the temperature of the expanding fireball was so high that all matter (black stippling) was ionized (that is, dissociated into electrically charged particles) and background radiation, in the form of photons (colored stippling), was extremely dense. (b) When the fireball had cooled to about 3000°, electrons could be captured by protons to form hydrogen and helium atoms. Initial slight local differences in density acted as nuclei for condensation of matter, and radiation diminished. (c) Expansion and cooling of the fireball continued and matter was progressively concentrated by gravitational forces, first into "protogalaxies." (d) After about 500 million years, these further condensed into the first galaxies of the type seen today.*

637 *Biological Origins*

in large part of iron oxides, silicates, and water crystals. Growth occurred through collision and capture of smaller bodies by the larger ones to form still larger, rotating bodies called **protoplanets.** At some point in their evolution these protoplanets, which revolved around their stars in accordance with the laws of motion and gravitation, condensed to form planets. The heat generated by their contraction was probably sufficient to convert the newly formed planets to a molten state, but not sufficient to initiate nuclear reactions in their centers.

Strong supporting evidence for the dust-cloud theory is the present existence of gigantic clouds of microscopic dust and gas in interstellar space.

It is almost certain that many of the countless billions of stars in the various galaxies have their own systems of planets, probably similar to those in our solar system. As yet we have not been able to observe these planets directly, even with our most powerful telescopes. However, there are good reasons for estimating that as many as 100 million of these planets throughout the universe have environmental conditions that would permit the origin and existence of life as we know it on earth.

The Earth

The Age of the Earth

How do we know when evolutionary events happened on earth? How can we date events so remote in time? At one time it was usual to calculate a precise date for the beginning of the world from biblical narrations; it was estimated that creation must have taken place 4000 or 5000 years ago. Geological studies in the eighteenth and nineteenth centuries broke away from the theological restrictions, and suggested 10 million, 100 million, or even 1 billion years. In Darwin's time (100 years ago) the earth was estimated to be about 100 million years old. This may seem old enough by our human time scale, but Darwin was troubled by this estimate because it did not seem long enough for the

slow trial-and-error process of evolution he postulated. We now have much more accurate measures for dating the age of the earth, which indicate that it is at least 4.5 billion years old. This estimate derives mainly from the technique known as **radioactive dating.**

Recall from Chapter 2 that all of the atoms of any given element contain the same number of protons, but can have different numbers of neutrons and therefore different atomic weights. These are the isotopes of a given element. Carbon (C^{12}) normally has six neutrons and an atomic weight of 12. Isotope C^{14} has eight neutrons. Carbon^{14}, like many naturally occurring isotopes, is unstable in the sense that it tends to emit high-energy particles from its nucleus until it reaches a stable form; that is it is a **radioactive** isotope.

All radioactive elements disintegrate at a fixed rate, which is measured in terms of **half-life** (the time in which half of the atoms in a sample of the element lose their radioactivity and become stable). Half-lives vary widely: the radioactive isotope N^{13} has a half-life of 10 minutes; tritium (H^3) has a half-life of just over 12 years; the most common isotope of uranium (U^{238}) has a half-life of 4.5 billion years.

Many tests have been made to see if changes in temperature, pressure, or any other conditions affect the rate of isotopic decay; all results have been negative. That is, so far as can be determined, all radioactive elements break down at a constant rate that is unaffected by any environmental factors. As uranium emits nuclear particles, it undergoes a series of decays and is converted into lead. If we assume that uranium decay is the chief means by which lead originated, then by analyzing the relative concentrations of specific types of uranium and lead in a given sample of rock, we can compute the age of the rock. This idea can best be illustrated by comparing it with the breakdown of a ruined stone building. If we know that one stone falls from the ruins every year, we can count the fallen stones and determine the year that the structure began to fall to pieces.

The oldest rocks found so far have been in Europe and North America; these have been dated at about 3.5 billion years old. Meteorites, however, can generally be dated at about 4.5 billion years, and it is assumed that meteorites and the planets condensed at about the same time. Thus the present estimates of geologic time, like past ones, are based on certain assumptions and hypotheses. Although these rest on a great deal of well-verified experimental information from a variety of disciplines, they are nonetheless subject to modification with increasing knowledge.

Structure of the Earth

The earth, whose mean distance from the sun is 93 million miles, has a diameter of about 8000 miles and is surrounded by the envelope of its atmosphere. At some early point in the cooling process, the molten material at the earth's surface began to form a rocky **crust;** this change to the solid state is still not complete. The solidification process has so far yielded an outer shell, the earth's crust, which is about 20 to 25 miles thick under the continents and about 3 miles thick, or less, under the oceans. It is largely made of rock known as **basalt.** Protruding through this basaltic crust are isolated **continents,** chiefly made of a lighter rock called **granite.** Like icebergs in an ocean, the continents have more than 90 percent of their mass embedded in the basaltic material beneath the surface of the earth.

Formation of the earth's crust from molten material was undoubtedly accompanied by wrinkling, cracking, and shifting of the surface layers, phenomena that have never completely ceased, as indicated by present-day earthquakes. In earlier stages of the earth's history, the shifting was evinced by great cracks in the crust, upheavals, and folding of the land masses to form mountains.

The flow of molten lava from active volcanoes located in different parts of the world illustrates the present hot, liquid properties of matter underlying the relatively thin crust. This hot, semifluid molten rock, or **magma,** immediately below the earth's crust, is called the **mantle.** Denser than the crust, it has an upper and a lower layer, with depths of about 600 miles and 1200 miles, respectively. The upper mantle is probably the seat of most of the earth's great earthquakes and volcanoes. Below the mantle is the earth's two-layered **core,** believed to be made of molten iron and nickel (Fig. 22-4). It is estimated that the temperature of the earth's core is probably nearly the same as the temperature existing at the surface of the sun (about 6000°C).

The Oceans of the Earth

Until sufficient cooling had occurred, free water could not accumulate on the earth's surface. A good deal of the planet's water must have taken the form of huge layers of clouds mounting to tremendous altitudes, almost completely blocking out the light of the sun. The torrential downpours from the sky were immediately transformed to vapor and steam as they approached the hot earth. With further cooling of the earth, the rains

Figure 22-4 *The earth is composed of a molten core about 4400 miles in diameter, consisting mainly of iron and nickel, surrounded by an upper and lower mantle, in a semifluid state, about 1800 miles thick. Only the thin crust, ranging from 3 to 25 miles in thickness, is solid.*

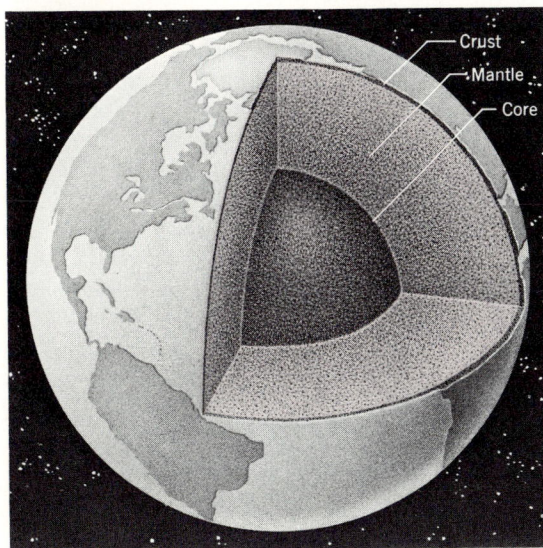

began to collect in basins and depressions of the earth's crust, to form oceans, lakes, rivers, and streams. With the passage of centuries and the accumulation of water on the planet's surface, the thick clouds grew correspondingly thinner, permitting more of the sun's light to reach the earth.

The primeval oceans must have been only slightly salty. Over time, salts were deposited, primarily by rivers and streams containing dissolved and suspended minerals washed out from the rocks of the earth's crust through the action of rain and other forms of precipitation. Salt concentrations in the oceans today amount to about 3 percent, or enough to cover an area the size of the United States with a layer of dry salt approximately 2 miles thick. Slowly but inexorably the stage was being set for the initial appearance of life, for it was undoubtedly in the primeval ocean, with its accumulating salts and its continuously reacting substances of ever-increasing complexity, that the first primitive living forms arose.

The Origin of Life

Ancient Beliefs

Throughout history men have believed that certain living things could arise suddenly and spontaneously from inanimate substances, a phenomenon they called **spontaneous generation.** In Babylonia and Egypt, in ancient Greece, in India, and in Europe during the Dark Ages and the Renaissance—and, in fact, until very recently—various forms of life were believed to originate directly from nonliving forms. It was thought that toads and snakes formed from the mud and silt of the Nile; fleas, beetles, and maggots from sweat; mice from refuse and damp earth; intestinal worms from dung and rotting meat; lice from decaying human bodies and excreta; and microorganisms from spoiling broth.

This idea was perfectly logical. It arose from common observations, and, in fact, was accepted by the most rational elements of society, the most prominent scientists and thinkers. William Harvey, the seventeenth-century physiologist and originator of the theory of blood circulation, Francis Bacon, the outstanding English spokesman of seventeenth-century materialism, and Descartes, the French philosopher of the same era, all considered the origin of living forms from inanimate material to be perfectly feasible and beyond dispute.

Our own religious traditions reconciled the principle of spontaneous generation with the will of a divine being. On the third day of creation, according to *Genesis*, God bade the earth and waters bring forth living creatures—first the plants, then fish, birds, land animals, and finally man. Thomas Aquinas, in his thirteenth-century classic, *Summa Theologica*, accepted spontaneous creation as a manifestation of animation by good or evil spiritual principles.

The acceptance of spontaneous generation was based essentially on preconceived and uncritical evaluations of nature. Observations of the origin of insects, rodents, microorganisms, and other living forms from nonliving substances were accepted readily, without careful examination and without adequate control of experimental conditions. Thus the famous seventeenth century Belgian physician Jean-Baptiste Van Helmont concluded, on the basis of a 21-day experiment utilizing a sweat-soaked shirt stored with wheat grains, that human sweat was the life-forming principle for creating mice from grain.

Step by step, however, in a great controversy that lasted 200 years, the belief was slowly whittled away by experimental evidence. In 1668 the Italian physician Francesco Redi demonstrated that meat placed under a protective covering, so that the eggs of flies could not be deposited on it, never developed maggots; maggots appeared only when the eggs fell on decaying meat. In the next century, after the invention of the microscope and discovery or microorganisms, John

Needham, an English Jesuit priest, described extensive experiments in which hermetically sealed vessels containing meat extracts soon swarmed with microorganisms despite previous exposure to elevated temperatures. Needham ascribed this phenomenon to the presence in each particle of organic matter of a special "vital force," which was responsible for the animation and therefore the appearance of living things.

Needham's views and results were experimentally challenged by Lazzaro Spallanzani in 1765. He found that after prolonged heating in hermetically sealed vessels vegetable broths and other organic substances never developed microorganisms. Spallanzani attributed Needham's results instead to inadequate heating, which failed to destroy completely the contaminating microorganisms in the vessels.

Needham countered that by too much boiling Spallanzani had "tortured" or destroyed the "vital force" in the broths as well as spoiled the small amount of air that remained in the vessels. Spallanzani responded with new experiments demonstrating that his heated broth could develop microorganisms when the flasks were subsequently opened to unsterilized air; but he was unable to prove that his boiling treatment had not altered the air within the vessel. The dispute remained unsettled, and in fact was regarded at the time as a victory for Needham.

The controversy was finally and decisively resolved by Louis Pasteur in 1862 through a series of rigorous and convincing experiments that are still accepted today as models of scientific acumen and experimental design. First, Pasteur demonstrated the presence of microorganisms in the air, a fact previously doubted by the proponents of spontaneous generation, by drawing air through a tube containing a gun-cotton plug and examining the trapped particles under the microscope. He then showed that by heating incoming air to high temperatures before allowing it to enter a flask of boiled broth, or by removing all accompanying dust particles, bacteria, molds, and microorganisms from incoming

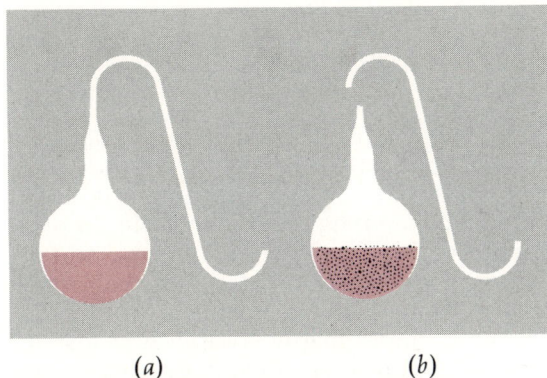

(a)   (b)

Figure 22-5   To disprove the possibility of spontaneous generation of microorganisms, Pasteur boiled nutrient broth in a flask with a long S-shaped neck. (a) While molecules of air could pass back and forth freely, heavier particles of dust, bacteria, and molds were trapped on the walls of the curved neck. In such flasks the broth remained uncontaminated. (b) When the neck was broken, growth of microorganisms soon became apparent.

air, the broth would remain uncontaminated and sterile indefinitely (Fig. 22-5). That is, if air and nutrient broth are free of organisms to start with, no organism ever appears spontaneously.

Today we often view this historic controversy as a triumph of reason over mysticism. In fact it is more nearly the opposite. The reasonable view was to believe in spontaneous generation; the only alternative at the time was to accept a supernatural act of creation. There was no third possibility. Only recently have ways been found to consider the spontaneous or natural origin of life again as a scientific problem, open to laboratory experimentation. What the experiments in the original controversy showed to be untenable is only the belief that living organisms arise spontaneously in a brief time interval under certain conditions. The question modern scientists ask is how life might have arisen spontaneously over hundreds of millions of years, under very different conditions.

The Modern Theory of Organic Evolution

Emergence of the Theory

The hypothesis of spontaneous generation oddly enough had served in an important theoretical role for two opposing schools of thought concerning the origin of life. To most scientists of the nineteenth century, no fundamental difference existed between living and nonliving matter. Organisms simply represented a more highly complicated and integrated arrangement of energy and matter, consisting of different kinds of inanimate materials collectively endowed with the characteristics of life by virtue of their organization. Many had favored this **mechanistic** view in contrast to the **vitalistic,** or religious, belief. The vitalists stated that it was beyond the human intellect to comprehend the life force and that an impenetrable barrier existed between living and inanimate matter. According to their outlook, the origin of life could only be explained on the basis of a special, mysterious "vital force," the result of a divine act of creation.

The great majority of scientists of that era represented the mechanistic approach; in view of Pasteur's conclusive work they were now left without any conceivable explanation for the origin of life. They were faced with a seemingly insoluble dilemma. Either life had been created by some supernatural or mystical force, a belief that they were unwilling to accept, or living things could arise spontaneously, a possibility that appeared to have been eliminated by Pasteur.

It was only toward the end of the nineteenth century that a small group of chemists and naturalists began to consider the possibility that organic materials might arise from simple carbon compounds in a manner analogous to the formation of complex salts from elements and minerals. This idea was given impetus when the German chemist Friedrich Wöhler synthesized an organic substance, urea, in the laboratory for the first time without the participation of living organisms; previously it had been generally accepted that such organic molecules could be formed only within living things.

Wöhler's work opened up the entirely new field of **organic chemistry;** since then several million different organic substances have been synthesized artificially. Then, in 1922, the Russian biochemist A. I. Oparin delivered before the Botonical Society of Moscow a paper dealing with the origin of life, in which he envisioned organic compounds as having formed first from simpler carbon-containing molecules such as CO_2 and methane (CH_4), and proposed that life evolved from these preexisting compounds. He viewed the origin of life as an evolutionary sequence from inorganic to organic compounds, to macromolecules, precells, and, finally, primitive cellular forms.

This concept of the **abiotic** formation of organic substances in the early prebiological history of the earth was also expressed by the British writer J. B. S. Haldane, who suggested in 1928 that before the origin of life organic compounds must have accumulated until the primitive oceans were like "hot dilute soup." It was he who recognized that the primitive atmosphere would have contained carbon dioxide, ammonia (NH_3), and water vapor, but no oxygen. Haldane claimed that such a mixture, exposed to ultraviolet light, would give rise to a variety of organic compounds.

Since the 1930s there has been evidence that simple organic compounds such as hydrocarbons are widespread in the universe—in interstellar dust and gas clouds—and on the larger planets of the solar system, supporting the idea of the origin of organic substances in nature without the mediation of living organisms. We know that chemical reactions for producing organic substances of increasing complexity are promoted by high temperatures and pressures, electrical discharges, and ultraviolet irradiation. In the laboratory it has been possible to produce some of the naturally occurring amino acids, the building blocks of proteins, by exposing a mixture of simple chemicals to such conditions. Experiments by the Nobel-prize-winning chemist Harold Urey, for example, showed that a number of amino acids could be randomly synthesized by passing electrical discharges through a gaseous mixture of ammonia, hydrogen gas, water vapor, and methane. Comparable experiments by other workers demonstrated the formation from simple substances even of purine and pyrimidine bases, which serve as structural units of the nucleic acids (Chapter 3).

Evolution of Complex Organic Materials

On the primeval earth there were many energy sources to drive such chemical reactions. The energy of the sun in the form of ultraviolet light and heat undoubtedly played a role in transforming simple organic compounds into more complex substances, as did electrical discharges in the atmosphere (e.g., lightning), cosmic radiation, and radioactive breakdown. It has been estimated that before the occurrence of living things the primeval ocean must have been at least 10 percent organic matter.

It is now well established by laboratory experiments that carbohydrates and lipids could have arisen spontaneously in the oceans by purely chance chemical reactions long before the appearance of the first form of life. The possibility, however, that proteins and nucleic acids could have been produced in a similar chance manner is considered less probable. It appears more likely that amino acids would spontaneously have formed polypeptides, and that nitrogenous bases would have linked together to build polynucleotides.

A recent summary of experiments in this field lists more than 100 organic compounds formed in reaction mixtures from simple, inorganic, carbon-containing molecules that were treated with heat, electrical discharges, or some form of radiation (ultraviolet, beta-rays, electrons, etc.). These compounds have ranged in complexity from those that are relatively simple, like urea, amino acids, three- to six-carbon sugars, purines, and pyrimidines, to much more complex molecules such as proteinoids up to 18 amino acids, ATP, polynucleotides several residues long, and even porphyrin compounds.

The inanimate forerunner of the first living forms, therefore must have been an organization of polypeptides, polynucleotides, carbohydrates, fats and other compounds, combined into some sort of an aggregate. Such molecular aggregates have been called **coacervates.** At the Institute of Molecular Evolution in Florida, Sidney W. Fox and his co-workers have found that hot aqueous solutions of artificially formed proteinoids separate spontaneously on cooling into a suspension of microscopic spheres, generally a few microns in diameter (Fig. 22-6, p. 644). Such **microspheres** absorb certain substances, such as ATP, from the medium, and they have been found to be mildly catalytic in a number of chemical reactions.

But we know that large-molecular-weight compounds are highly perishable. Even if complex organic compounds or microspheres were synthesized abiotically in the primordial oceans, why were they not destroyed before any substantial concentrations could accumulate? To answer this we must recognize that organic compounds on today's earth are perishable for two main reasons. First, they react slowly with molecular oxygen from the atmosphere and become oxidized spontaneously; second, they are decomposed or used for food by microorganisms. But in the **prebiotic** world (before life formed), the atmosphere was devoid of oxygen. Spontaneous oxidation did not occur and there were no organisms of any kind. Therefore neither oxidation nor decay would have destroyed the early organic materials. A third destructive agent—ultraviolet radiation from the sun— would have been much more significant in prebiotic times than today, because it is the oxygen content of the present atmosphere that filters out most ultraviolet; such radiation, however, is absorbed by water. Thus complex organic molecules that would have been destroyed quickly on land could accumulate over periods of millions or hundreds of millions of years in the primordial seas.

The complex nonliving systems, such as microspheres, that have been created in highly simplified form in the laoratory tend to remove and accumulate various substances from the surrounding medium. These substances may undergo a variety of chemi-

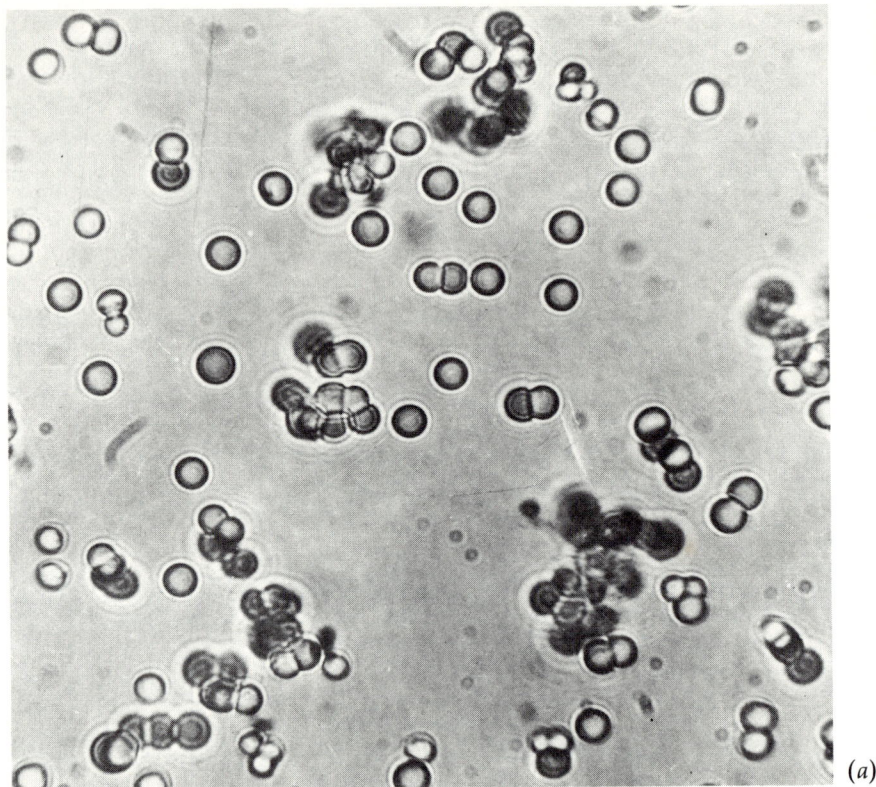

Figure 22-6 Microspheres produced from proteinoids, derived by heating mixtures of formaldehyde, ammonia, and water. (a) Microspheres of uniform size at low magnification. Magnification 1000×. (b) Electron micrograph of such a microsphere. Magnification 12,000×.

(a)

cal reactions, directed in part by the structural features of the microsphere, with some of the products of the reactions eventually released to the external environment. This type of inanimate system is therefore analogous to a living thing; both are essentially dynamic systems that undergo a continuous influx and outflow of matter and energy.

Undoubtedly, large numbers of different coacervate systems arose. Despite the protected conditions, most were probably too unstable to last long, but a few would have contained particularly favorable combinations of compounds (especially molecules with catalytic activity), and would thus promote reactions. These might tend to survive longer and therefore increase in internal complexity.

Perhaps the ability to catalyze certain reactions would cause a microsphere to absorb reaction substrates at a higher rate than other spheres. If such a product were retained in the aggregate, the aggregate would tend to increase in size, which in turn could lead to greater stability and greater complexity. Thus a primitive form of selection would take place whereby certain chemical properties would be favored because their existence in a coacervate would foster the stability and complexity of that aggregate. Just as organisms develop features that favor their ability to cope with their environment have an added advantage in survival as compared to other living things, so evolutionary selection can be applied to these inanimate complex multimolecular systems.

In this fashion systems must have evolved that had a capacity for self-renewal or self-preservation of parts by selected physical and chemical processes. We can only speculate

644   *Evolution and Classification*

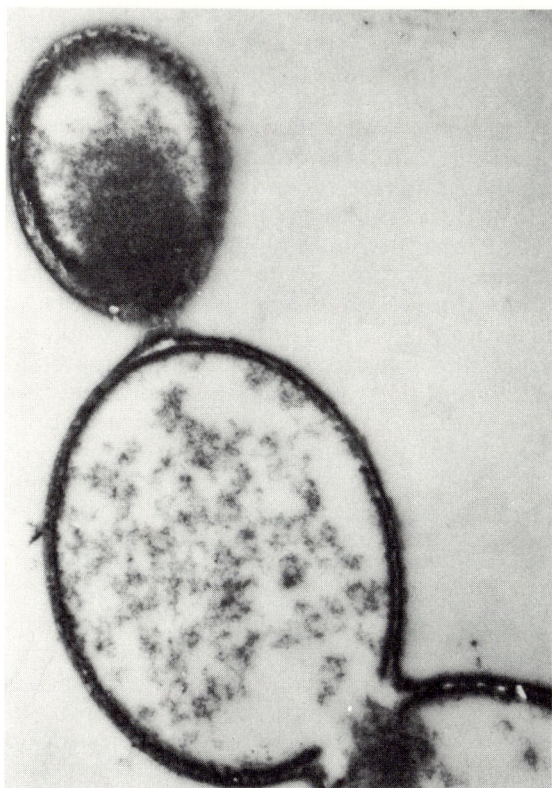

(b)

on the course of events that led from the complex inanimate systems to the first form of life on earth. The chance formation of catalysts and their selective evolution into protein-like catalysts of high specificity and activity undoubtedly played an important role in this transition. It has been suggested that the constant repetition of these coordinated and connected chains or cycles of reactions were somehow responsible for the appearance of the most unique characteristic of all: reproduction. How this occurred is not clear, but at that point the system would be regarded as alive. This is estimated to have occurred about 3 billion years ago.

Primitive Metabolic Reactions

The essential feature of all present-day metabolism is the step-by-step release of the energy stored in the chemical bonds of or-

ganic compounds and its utilization in primary life processes. This is accomplished in living cells by coordinated chains of many chemical reactions, each catalyzed by its own specific enzyme. Remarkably enough, all living organisms have essentially the same sequence of chemical reactions, catalyzed by similar enzymes, as the first stage in their energy metabolism: the anaerobic respiratory cycle (Chapter 4). The similarity of these enzymatic sequences suggests that all present-day living things share the remains of a primitive metabolism inherited from the early forms of life.

The first organisms must have been heterotrophs, dependent on the ready-made organic compounds that accumulated in the oceanic soup. With the later growth and multiplication of primitive life there ensued a greater consumption of organic materials from the environment. This must inevitably have led to greater competition among organisms for the limited supply of organic substances and to a natural selection in favor of the primitive organisms that were most efficient in utilizing the existing energy sources. Organisms that were predisposed to using simpler organic substances as an energy source (because they had developed a means of converting them to more complex substances) must also have had a distinct advantage in the gradually developing struggle for existence.

We now have good evidence to indicate when such primitive heterotrophic prokaryotes (cells without nuclei) first appeared on earth. In a geological formation in South Africa, called the Figtree Formation, fossilized, bacteria-like organisms have been found that are estimated by radioactive dating methods to be more than 3 billion years old. Similar samples of still older rocks contain carbon—a sign of organic compounds—but no such biological specimens.

Evolution of Photosynthesis

Some of the earliest coacervates must have included compounds that were pigments; that is, compounds that absorbed light energy in the visible range. Any such compound would offer a slight advantage to a primitive organism, if only by raising its temperature slightly as a result of the conversion of light to heat energy in the process of absorption.

If a pigment also had catalytic activity, it would be to the distinct advantage of the organism containing it, because all catalytic reactions are speeded up by elevated temperatures. The steps from such a primitive beginning to the evolution of chlorophyll and the process of photosynthesis are of course not known, but neither are they difficult to conceive. Clearly, as the concentration of organic materials decreased, the first organism capable of synthesizing an organic compound from dissolved carbon dioxide in the water—using energy from the sun to speed up the process—would have had very great chances of survival in competition with forms lacking this ability. Such an organism, presumably, was the primitive forerunner of a photosynthetic autotroph, which would no longer be dependent on a ready-made source of nutrients in the sea around it.

The evolution of photosynthetic organisms had another, even more profound effect on the further evolution of life. Such organisms produce not only organic compounds, but also molecular oxygen. With their competitive advantage and no natural checks on growth, such organisms would gradually, over millions of years, populate the oceans; as they did so, oxygen would begin to accumulate in the atmosphere. Fossilized cells very similar to present-day blue-green algae have recently been discovered by Elso Barghoorn in sedimentary rock from Ontario, Canada (Fig. 21-1). These are estimated to be more than 2 billion years old; thus such organisms are assumed to have formed between 2 and 3 billion years ago.

This agrees well with another geological indicator. Metallic iron and iron salts such as chlorides and sulfides are very soluble. Iron oxides, on the other hand, are insoluble; these could only have formed after molecular oxygen had appeared dissolved in the oceans, and would have resulted rather soon in the sedimentation of "red-beds" of iron ore. The oldest known beds of oxidized iron on earth are just over 2 billion years old, suggesting that it took only a few hundred million years from the time photosynthetic organisms evolved until oxidized iron began to precipitate out of the water.

The increasing concentration of free oxygen in the atmosphere as a result of photosynthesis altered the entire course of biological evolution. Molecular oxygen provided a means of obtaining energy from the organic products of aerobic metabolism that had theretofore been discarded as waste. Now heterotrophic organisms, which were able to utilize the energy stored in organic metabolic waste products by breaking them down further with the aid of oxygen, had a distinct advantage. They began to evolve more rapidly, and in time became a predominant group of living things on the planet; that is, they formed the primitive animals.

The Evolution of Cells and Metazoa

It apparently took almost a billion years from the appearance of the early photosynthetic prokaryotes until organisms recognizable as cells—with membranes and nuclei—were formed. Preston Cloud has found eukaryotic cells (containing true nuclei) with a fully developed mitotic mechanism and mitochondria, that are at least 1.2 billion years old (Fig. 22-7). If the first eukaryotes actually evolved no earlier than that, 600 to 800 million years was then required for the development of soft-bodied multicellular organisms; the first fossil animal skeletons were deposited around 600 million years ago.

Simple one-celled organisms probably gave rise to aggregates of cells and these in turn presumably evolved into multicellular forms. Some of the multicellular organisms developed highly specialized groups of cells with specialized functions such as reproduction, digestion, breathing, and excretion. Sponge-like and coral animals as well as algae evolved in the oceans, followed by advanced forms of ancient life resembling the jellyfish, worms, and hard-shelled creatures of the present.

The sterile, rocky continents continued to undergo cracking, shifting, mountain formation, and erosion; and the oceans in that ancient period continued to serve as the me-

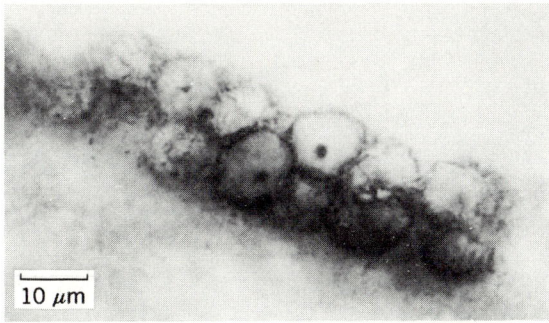

Figure 22-7 The oldest true cells that contained a nucleus and divided by mitosis were recently found in California. The fossilized cells shown here have an average diameter of about 14μ and are estimated to be 1.2 to 1.4 billion years old.

dium for the evolution of life. By 600 million years ago the ancestors of all the main groups of invertebrates, as well as the algae and ancient seaweed, had already developed in the seas. At about this time the first fossil records were entered in the rocks of the continents.

Terrestrial Organisms

It is believed that the first successful venture of organisms onto the land was accomplished approximately 350 million years ago by a hard-shelled invertebrate resembling the scorpion, a member of a group of organisms (the **arthropods**) that later gave rise to lobsters, crabs, and insects. The first **vertebrates** (animals with backbones) were beginning to make their appearance in the sea at about the same time, in the form of ancient fish now extinct. At this time, too, the first land plants, derived from the simple aquatic green plants, were slowly gaining a foothold; they spread and diversified, aiding in crumbling the rock and converting it into soil.

Some of the early fish evolved special structures such as internal gas bladders that functioned mainly in buoyancy but also served in the storage of air. These structures eventually evolved into air-breathing lungs, permitting the organisms bearing them to lead a terrestrial existence. Animals with fins were gradually replaced by organisms with legs. The first amphibians probably appeared about 300 million years ago, and during the next 75 million years or so they developed rapidly and became widespread.

Primitive insects, mosses, ferns, and seed plants also appeared on the land. New forms of life continued to evolve as environmental conditions changed. Other forms, old and new, which were placed at a disadvantage, declined and became extinct. At times gigantic catastrophes (e.g., floods, earthquakes, volcanic action, glacier movements, and rapid mountain formation) wiped out huge segments of the living population. An infinitesimally small portion of these ancient organisms, however, was preserved in the rocks and coal formations of the earth's crust, our only evidence of their past existence. The major coal beds of the world today represent some of the dense, widespread swamp forests of 250 to 300 million years ago.

Primitive reptiles probably developed from some ancient amphibian ancestor approximately 200 million years ago, and this was the period during which the early flowering plants also evolved. In the next 75 million years or so the reptiles became the predominant form of animal life on the continents. Flowering plants continued to diversify and spread. The first primitive mammal must have had its origin in this period, from some ancient reptile now extinct, as did the birds, approximately 30 million years later.

Some land animals, such as reptiles and mammals, returned to the sea. The descendants of huge reptiles, which 150 million years ago reentered the oceans, are now represented by the sea turtles and alligators. Contemporary ocean mammals (e.g., seals and whales) are descendants of mammals that returned to the ocean waters 50 million years ago.

Evolution of Man

Many of the higher plants and mammals developed during the last 60 million years. It is only during the last million years, however, that man, the most advanced and unique animal of them all, finally evolved.

Time, years ago	Events	
20–50 thousand	Modern man, modern plants and animals The Atomic Age	
1–60 million	Evolution of higher mammals and plants, even greater diversification and distribution of flowering plants	
50 million	Some mammals enter the oceans, continuing diversification and distribution of flowering plants	
100–150 million	First mammals and first birds, increasing diversification and distribution of flowering plants	
125–200 million	Rise and predominance of reptiles, development of extensive gymnosperm forests, rise of flowering plants	
250–300 million	First amphibians, insects, mosses, ferns; rise of dense swamp forests which later formed major coal beds of the world	
350 million	First succesful invasion of land by animals and plants, appearance of first vertebrates as ancient fish in the sea	
500 million	Ancestors of all major groups of invertebrates present in the oceans, also alga and ancient seaweed	
1 billion	Increasing population of unicellular and simpler multicellular organisms including many invertebrates in the seas, origin of photosynthesis $CO_2 + H_2O$	
2–4 billion	Formation of increasingly complex organic molecules in primeval seas, origin of life	
4–5 billion	Formation of the earth and other members of solar system	
10 billion	Origin of universe?	

648 *Evolution and Classification*

Man descended from a group of arboreal land mammals, animals that lived among the trees and moved principally by swinging from branch to branch with their hands. Acute stereoscopic vision and unusual manual dexterity must have been of distinct advantage in the survival of these tree-living forms. Later in their evolutionary history these animals descended to a terrestrial existence and eventually gave rise to a form with an exceptionally expanded brain. This animal was the forerunner of man.

Modern man evolved in his present form about 20,000 to 50,000 years ago, the culmination of a vast evolutionary process that began with the origin of the first primitive living forms some 3 billion years ago (Fig. 22-8). His two most distinctive biological attributes, a highly developed brain, which accounts for his amazing ability to reason, and his unusual manual dexterity, which is responsible for his ability to manipulate the environment about him, have made man the most successful of all terrestrial organisms. It has been a long and noble process. It would be a pity to end it now.

Reading List

Bernal, J. D., *The Origins of Life,* World, New York, 1968.

Bullard, Sir E., "The Origin of the Oceans," *Scientific American* (September 1969), pp. 66–75.

Denton, G. H., and S. C. Porter, "Neoglaciation," *Scientific American* (June 1970), pp. 100–111.

Iben, I., "Globular Cluster Stars," *Scientific American* (July 1970), pp. 26–39.

Keosian, J., *The Origin of Life.* Reinhold, New York, 1968.

Kurten, B., "Continental Drift and Evolution," *Scientific American* (March 1969), pp. 54–64.

Lawless, J. G., C. E. Folsome, and K. A. Kvenvolden, "Organic Matter in Meteorites," *Scientific American* (June 1972) pp. 38–53.

Margulis, L., "Symbiosis and Evolution," *Scientific American* (August 1971), pp. 48–61.

Newell, N. D., "The Evolution of Reefs," *Scientific American* (June 1972) pp. 54–69.

Rees, M. J., and J. Silk, "The Origin of Galaxies," *Scientific American* (June 1970), pp. 26–35.

Simpson, G. G., *The Meaning of Evolution* (rev. ed.). Yale University Press, New Haven, 1967.

Figure 22-8 (left)   A summary of the evolution of life on earth.

chapter twenty-three | *Mechanisms of Evolution*

Evolution as a Unifying Principle

There are certain broad concepts that draw together a large number of observations into a logical framework and make sense out of them, even though they appear to be unrelated. Gravitational attraction is such a concept; the equivalence of matter and energy is another. In biology the idea of genes as the carriers of heritable characteristics serves this purpose.

The concept that change, not immutability, is the rule of the universe is also a fundamental, unifying principle of nature. **Evolution,** a gradual and continuous orderly succession of changes, is the underlying theme that ties a vast welter of facts and information into a broad, cohesive, and unified picture of the universe—its past, its present, and to a certain extent, its future. Its effects are seen in every field of human knowledge and thought. Matter and energy are the basic components of the universe (Chapter 2), and by their very interchangeability and dynamic nature make change, or evolution, inevitable; biological evolution is simply one aspect of the over-all evolution of the cosmos. We conceive of evolution as having started from the nonbiological state with the origin of the expanding universe, the birth of galaxies and stars, and as the gradual change of inanimate matter by many steps over billions of years into living systems and thence to contemporary organisms (Chapter 22).

We have noted that the basic resemblance among all forms of life provides strong support for a concept of organic evolution, the

651

concept that all different kinds of plants and animals that exist today and have existed in the past are related, having arisen from pre-existing kinds by relatively slow changes.

There are perhaps 2 million different species of organism living today. If we could trace their ancestry far enough back, their origins would converge, like the branches of a tree, into fewer but larger branches that would ultimately fuse into a main trunk that represents the single primitive ancestral stock of all life.

Some organisms have arisen from a common ancestor rather recently as evolutionary time is measured. Modern birds and lizards probably arose from a common reptilian predecessor about 200 million years ago. Other classes, insects and crustaceans, for example, are obviously more distantly related, having shared a common ancestor in the deeper past of biological history (Fig. 23-1). But even the most seemingly unrelated organisms such as a man and a mold must have shared a common ancestor at some point in their evolution; in this example that predecessor would have lived some 2 to 3 billion years ago in the primeval seas.

We recognize this degree of evolutionary relatedness among organisms most clearly by the way we classify them. Thus we divide all living things into three different **kingdoms:** the bacteria and blue-green algae; the animals; and the plants. The relationships between these groups are real and fundamental. Recall that all three share common molecular structures and enzymatic sequences (Chapters 3, 4). Animals and plants also have similar cellular structures, and many of their basic life processes—such as mitotic division, growth and development, osmotic control, cytoplasmic movement, and so on— are regulated by essentially the same mechanisms.

In all outward appearances, however, and in their behavior and in the ways they relate to the environment, members of the three kingdoms are indeed very different. Within each kingdom the relationships are closer, the differences less striking. In each kingdom organisms are grouped, again according to evolutionary relatedness, into broad categories termed **phyla** (singular, **phylum**). Among the animals, for example, insects fall into the phylum Arthropoda, whereas mammals are in the phylum Chordata (the vertebrates). In each phylum more closely related forms are grouped into **classes;** each class is further subdivided into **families** and **genera** (singular, **genus**). Members of a genus are classified into different **species,** defined by the ability of its members to mate. Even among the members of a single species, however, systematic differences often require further subdivision into **subspecies** and **races.** In Chapter 24 we describe in greater detail how organisms are classified. We want here only to introduce the important terms of **taxonomy** (classification) and to emphasize the concept that taxonomy is based on evolutionary relationships.

Although we tend to think of evolution and natural selection as processes that occurred in the distant past, there is ample evidence that evolution has not stopped. The evolutionary process, both biological and nonbiological, continues, but is generally slow for higher animals and plants. For most organisms, under natural circumstances, evolutionary changes may require tens or even hundreds of generations to be detectable; thus organisms with generation times measured in years or decades appear not to have changed during the brief time men have recorded their observations. But in forms that multiply every few hours or days the evidence for evolution is obvious to all who care to look; its mechanisms are available as subjects for study in the laboratory. The strong tendency for evolutionary change, however, is balanced by **conservative forces** in nature, forces that tend to maintain the properties of organisms and their natural groupings constant from one generation to the next. How much evolutionary change actually takes place in a population of organisms under particular circumstances is determined by the state of balance that exists between these two antagonistic sets of forces.

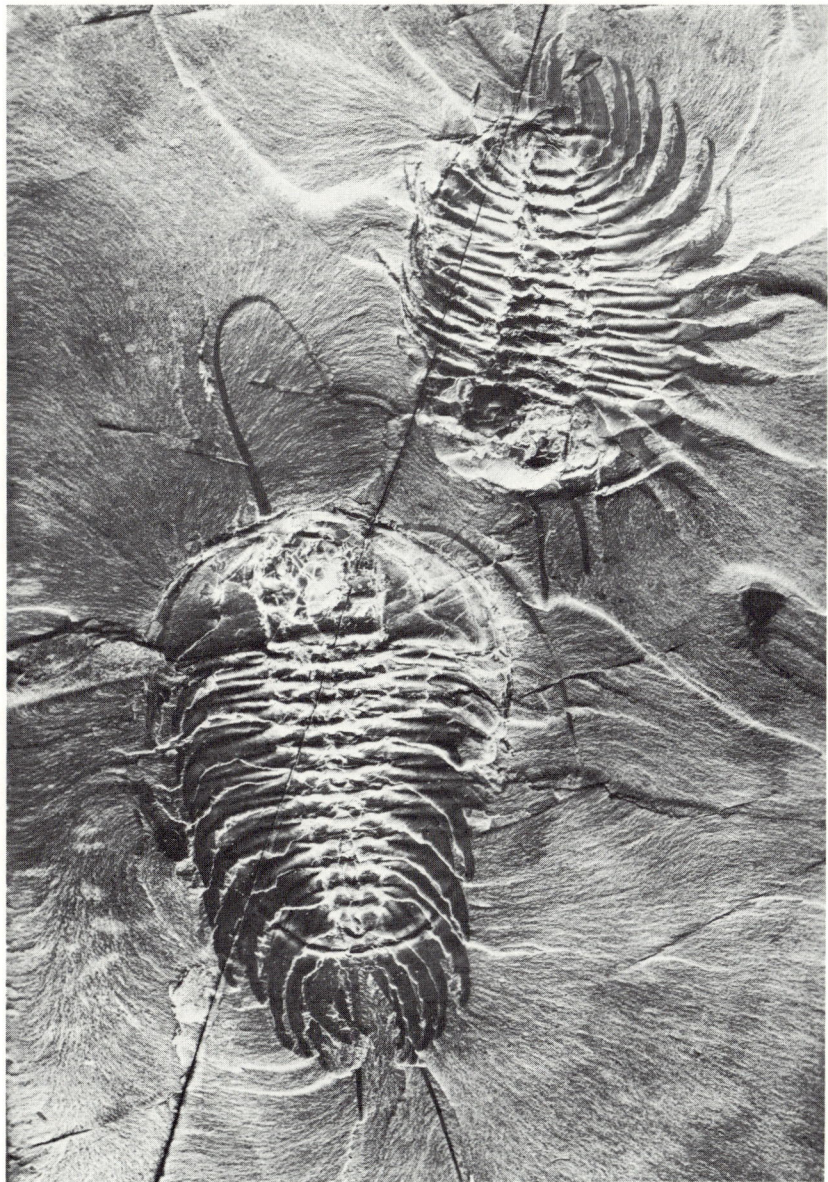

Figure 23-1 This trilobite is a fossil arthropod from the Cambrian Period of geologic time; it lived about 500 million years ago. Although already somewhat specialized, it was probably similar to the most primitive arthropod forms that evolved from earlier more generalized metazoa and was ancestral to modern crustaceans and insects.

The Evolution of Evolutionary Thought

Faint glimmerings of the idea of organic evolution have appeared at various times in the history of human thought, as far back as the ancient civilizations. The essence of the idea appears in the writings of the early Greek philosophers (600 B.C.). Aristotle, in the fourth century B.C., proposed a classification of living things that bordered on a concept of organic evolution. Nature, he suggested, "advances by small steps from inanimate things to animate . . . ."

Although it had become clear by the end of the seventeenth century that plants and animals fall into natural groups, or species, it was commonly accepted that they were unchangeable, and that each species was the result of an act of special creation. By the eighteenth century several biologists seriously began to question the concept that species were unchangeable. Studies with cultivated plants and domestic animals seemed to indicate the opposite.

The first great pioneering step toward the development of our modern theory of organic evolution was made by Jean Lamarck, who perceived that species and varieties were subject to change, and that the more complex organisms have in reality evolved from simpler ones. But Lamarck believed that evolution was caused by changes in traits acquired during the lifetime of an organism as a result of use or disuse; he contended that such acquired traits were somehow incorporated into the heredity of the individual and thus transmitted from generation to generation.

It was not until the middle of the nineteenth century that the English naturalist Charles Darwin laid the foundations for the modern concept of organic evolution. Not only did Darwin provide for the first time a mass of detailed and extensive evidence and reasonable explanations to show that biological evolution had taken place, but he also presented a mechanism, **natural selection,** to account for its operation.

During a five-year voyage around the world as naturalist on the British ship H. M. S. Beagle, Darwin had the opportunity to study a great many plants and animals. One of these important studies was an analysis of the birds of the Galapagos Islands (Fig. 23-2 and pp. 655–656). As a careful scientific observer he arrived at the conclusion that organisms were not immutable and that species did change, diverging into different species and different lines of descent in the course of time. When he had become convinced of this fact himself, he then sought an explanation or mechanism to account for the manner in which the evolutionary process operated.

By the 1850s he had formulated his theory

Figure 23-2   On the Galapagos Islands Darwin discovered "a most singular group of finches," 13 species, all clearly related, but each slightly different from the others and none similar to the finches of mainland North America. A few of these differences are shown. Seeing this diversity and gradation of structure in one small, related group of birds, Darwin wondered whether they might all have been derived from a single early species that had been "modified for different ends" by evolution. This was one of the observations that led him to conclude that species are not immutable.

of natural selection, which he published jointly with another English naturalist, Alfred Russell Wallace, in 1858. In the following year Darwin published his great volume entitled *On the Origin of Species by Means of Natural Selection, or the Preservation of Favored Races in the Struggle for Life* (usually called *Origin of Species*), which contained in detail his theory of evolution by natural selection and a vast array of evidence in support of it.

The Darwinian Theory |

When Darwin returned from his voyage on the Beagle, he happened on a copy of the writings of the English economist Thomas Malthus. In his *Essay on Population* Malthus contended that more children are born than reach maturity, and that such factors as limited food, war, and disease are important in the "struggle for existence" and in restricting the size of the human population.

Darwin realized that these principles could also apply to any other living species. He reasoned that under the competitive conditions in which all organisms live selection should work automatically, because inherited variations that favored an organism's survival and ability to produce fertile offspring would be maintained from generation to generation. Variations that were disadvantageous would be eliminated sooner or later, because those organisms would tend not to survive long enough to reproduce. Thus an automatic selection process was constantly in operation, tending to perpetuate any variations that conferred an advantage in terms of survival and production of fertile offspring.

As we have said, Darwin was not the first to suggest an evolutionary theory. In fact, exactly at the same time he was writing the *Origin of Species,* Wallace, working entirely independently and without any knowledge of Darwin's studies, arrived at the same theory of organic evolution and natural selection. Wallace's work was largely based on observations of animal and plant life in Indonesia. Darwin's formulation of the idea was, however, the first to be generally understandable and convincing. His collection and compilation of the evidence that species have changed and are changing was so extensive as to be virtually irrefutable. And the explanation he gave for these changes—natural selection—was so simple and logical as to be "obvious" after it was pointed out. Darwin's triumph was in seeing the relationship between natural selection and heritable changes in populations; with this insight he could not only demonstrate that evolutionary changes had occurred, but could also explain how they must have come about.

Let us summarize the essentials of the Darwinian theory of evolution with a series of quotations from the *Origin of Species*.

"There is no exception to the rule that every organic being naturally increases at so high a rate that, if not destroyed, the earth would soon be covered by the progeny of a single pair."

That is, the size of any population tends to increase exponentially when conditions permit. We discussed this idea at length in Chapter 20.

"In the case of every species, many different checks, acting at different periods of life, and during different seasons or years . . . concur in determining the average number or even the existence of the species."

By "checks" Darwin meant forces in the environment that decrease the net growth rate of a population, either by decreasing the rate of proliferation or increasing mortality.

"Hence, as more individuals are produced than can possibly survive, there must in every case be a struggle for existence . . ."

Note that "struggle for existence" in this context does not refer to combat among individuals, but to competition for those natural resources that permit some individuals to survive while others are eliminated.

"No one supposes that all the individuals of the same species are cast in the same actual mould. These individual differences are of the highest importance for us, for they are often inherited. . . . Any variation that is not inherited is unimportant to us."

Thus variation, in the form of individual differences, exists in every species or population; and differences that are heritable represent part of the mechanism of evolution.

"Can it then be thought improbable . . . that . . . variations useful to each being in the great complex battle for life, should occur? . . . Can we doubt (remembering that many more individuals are born than can possibly survive) that individuals having any advantage, however slight, over others, would have the best chance of surviving and procreating their kind . . . ?"

Darwin deduced that a selective elimination process was constantly taking place in nature, the organisms or varieties surviving being the ones that are more "fit." However, he did not define fitness merely as the ability to compete for food, space, or mates, or to escape predation. According to Darwin,

". . . I use this term in a large and metaphorical sense including dependence of one being on another, and . . . (which is more important) . . . success in leaving progeny."

Thus fitness is the capacity to survive long enough to reproduce offspring that can in turn survive long enough to repeat the process.

"Hence, I look at individual differences . . . as of the highest importance for us, as being the first steps towards such slight varieties as are barely thought worth recording. . . . And I look at varieties which are . . . more distinct and permanent, as steps . . . leading to subspecies, and then to species. The passage from one stage of difference to another may, in many cases, be . . . safely attributed to the cumulative action of natural selection. . . . A well-marked variety may therefore be called an incipient species."

Thus evolution represents a gradual change in the hereditary makeup of a segment of a population as it slowly becomes distinguishable from the original population. Evolution is seen as a two-part process involving the origin of variation and the enhancement of variation by natural selection.

Evolutionary Mechanisms

Population Genetics

In Chapters 8 and 10 we described genes as the fundamental units of biological heredity and noted their conservative nature. Genes tend to be stable elements, for the most part, capable of exact duplication in the process of reproduction. But the differences between various kinds of organism—between a man and a monkey, or a beaver and a beech tree—are reflections primarily of differences in their genomes. If all organisms are derived from an ancient single ancestral cell type, how did this divergence come to be? How did the sets of genes represented in various species change so radically?

A fundamental point in understanding how evolution works is in recognizing that genes are housed in individuals, and that neither an individual nor its genes evolves. Genes may be miscopied during reproduction to produce a **mutation,** but this occurs only during the formation of a new individual. Thus individuals do not evolve; **only populations evolve.**

Recall that a biological population is defined as all of the individuals of the same species in a given area at a particular time. In all diploid species, remember (pp. 203–212), each individual has pairs of homologous chromosomes with allelic pairs of genes controlling each hereditary trait. If both genes controlling a trait are the same, the individual is homozygous for that trait; if they are different, he is heterozygous. Moreover, within a species there may be 10 or 20 or 100 different genes for each trait represented among the population; an example is the variety of eye colors in *Drosophila* (p. 216). A **gene pool** is the total of all the genes for all of the different traits in a species. It is that gene pool that evolves, rather than any individual. At any time each gene has a given frequency in the pool. Evolution, in its simplest terms, is a change in gene frequency in the gene pool.

Genes as the Basis for Diversification

Genes in modified forms (i.e., mutations) and in different combinations are responsible for the inherited variations and diversities that are filtered, sifted, and tested in the course of natural selection to constitute evolutionary change. Most mutations are harmful; however, a mutation or genotype that is harmful in a particular environment might prove to be highly advantageous if the surroundings were changed. Under the new environmental circumstances a formerly unfavorable mutation might endow an organism with a better chance to survive and reproduce.

Sexual species have a tremendous potential reservoir for genetic variability available for recombination from the gene pool. Sexual reproduction as a means for creating new gene combinations may be responsible for the appearance of an almost infinite diversity of genotypes (and therefore phenotypes). It has in fact been established that if mutations were to cease the vast number of possible recombinations in most sexually reproducing populations, including man, would nevertheless result in a virtually unaltered and continuing rate of evolution for many generations to come.

In an asexually reproducing population mutation is the only source of inherited variation. By its very nature asexual reproduction does not permit the combination in an individual of favorable genes or mutations from the common gene pool of the species. For this reason asexual forms are less adaptable than sexual forms. The evolution of sexual from asexual means of reproduction obviously gave the sexual organisms decided advantages, for it endowed them with the increased possibilities for evolutionary adaptation.

The Basic Role of Natural Selection

The environment directs but does not cause evolutionary change. The environment does not produce inherited variations; it only determines which variations will survive and which will become extinct. This is accomplished by natural selection, the natural elimination of organisms with variations less suited to the environment and the perpetuation of those with more favorable ones. In other words, natural selection is the **isolation** procedure for sorting out the genotypes that better fit an organism for existence in a given environment.

There is little doubt that natural selection has brought about a vast increase in the diversity of living things and in the types of environment in which they can exist and propagate. This is borne out by the fossil record, which alone demonstrates the gradual transformation in the course of geologic time from the limited primordial life originally and totally confined to a watery existence to an almost infinite spectrum of land-inhabiting and airborne forms.

Molecular Basis of Evolution

The molecular basis of organic evolution resides primarily in DNA, for this is the key substance of genes and therefore the substance of heredity. The nucleic acids direct and control all life activities by directing and controlling protein synthesis (Chapter 10). The molecular basis of evolution, which rests primarily on the evolution of nucleic acids, therefore resides, by extension, in the evolution of the various proteins of the cell.

Molecular evolution is governed by the same principles that apply to the evolution of the whole organism: the production of a new or modified enzyme, for example, with new or different catalytic properties as a result of mutation, could under suitable environmental conditions confer on an organism an advantage in its existence. As an illustration, a new enzyme participating in melanin formation might result in a marked increase in skin pigmentation. Such a trait could tip the balance between survival and extinction by protecting against ultraviolet light irradiation damage in a region of intense and prolonged sunlight, say at the equator.

Conversely, the genetic modification of a catalytic protein could impose an unfavorable trait on an organism, which is more often the case. In humans the genetic disease galactosemia (p. 231) is a clear example of an inherited variation in one of the thousands of kinds of protein molecule in the cell; it puts the organism at a distinct disadvantage in terms of survival.

The Hardy-Weinberg Law

Most large, sexually reproducing populations exhibit a tendency toward genetic stability in the sense that inherited characteristics (and therefore genes) are maintained in more or less the same proportions from generation to generation. It seemed puzzling to the early geneticists that dominant characteristics did not continue to increase in the population and eventually replace the recessive ones, thus creating an increasingly "purer" population with time.

The explanation was independently discovered in 1908 by an English mathematician, G. H. Hardy, and a German geneticist, Wilhelm Weinberg. They proved by mathematical formulations that the gene frequencies or ratios of different genes in a total population remain constant in succeeding generations (a phenomenon now called the **Hardy-Weinberg Law**), provided all three of the following conditions are satisfied: (1) **mating is random:** (2) **mutations do not occur;** and (3) **the population is large.**

At first glance it might appear that the Hardy-Weinberg Law contradicts what we have said so far about evolution. It implies that any population should be in **genetic equilibrium** and that no evolution should take place. But we know that evolutionary change has occurred in large, sexually reproducing populations, and—as we shall see—is still occurring. The answer to the apparent contradiction is that the three conditions cited are rarely met.

For example, mating in most populations is not truly random in the statistical sense, for all genotypes do not contribute equally to the gene pool of the next generation. Some genotypes may be more fertile, others may be sterile, and others may fail to maintain the viability of individuals until the reproductive age, resulting in a statistically nonrandom mating. This inevitably results in a natural selection for those genotypes that are viable and better suited for reproduction, which leads then to a change of gene frequencies in successive generations and therefore to the occurrence of evolution.

The occurrence of mutations and their subsequent contribution to natural selection by favoring or handicapping the existence and reproduction of the organisms in which they occur can lead to a change in the frequencies or ratios of the genes in the population and therefore to evolutionary change.

Finally, small populations tend to experience greater percentages of fluctuations in gene frequencies purely on a chance basis. This leads to definite changes in the proportions of genes in the population, a phenomenon known as **genetic drift** (see p. 663).

The Patterns of Evolutionary Change

Evolution, then, involves small changes in gene frequencies in the gene pool of a population from one generation to the next. No single change leads to a totally new population, but no new generation is ever identical with its predecessor. To emphasize this gradual progressive change we speak of **sequential evolution,** which can be seen by paleontologists studying fossils, by geneticists in work with laboratory populations, and by field biologists investigating isolated natural populations (Fig. 23-3). Sequential evolution may produce random fluctuations over long periods of time, but show little cumulative change. It may also result in gradual shifts in gene combinations so that the descendant population is very different from its original ancestor.

Evolution can also result in the splitting of a population into two or more different species, with diverse body structures and habits, from a common ancestor. This phenomenon is known as **adaptive radiation,** or **divergent evolution.** For example, if some organisms of

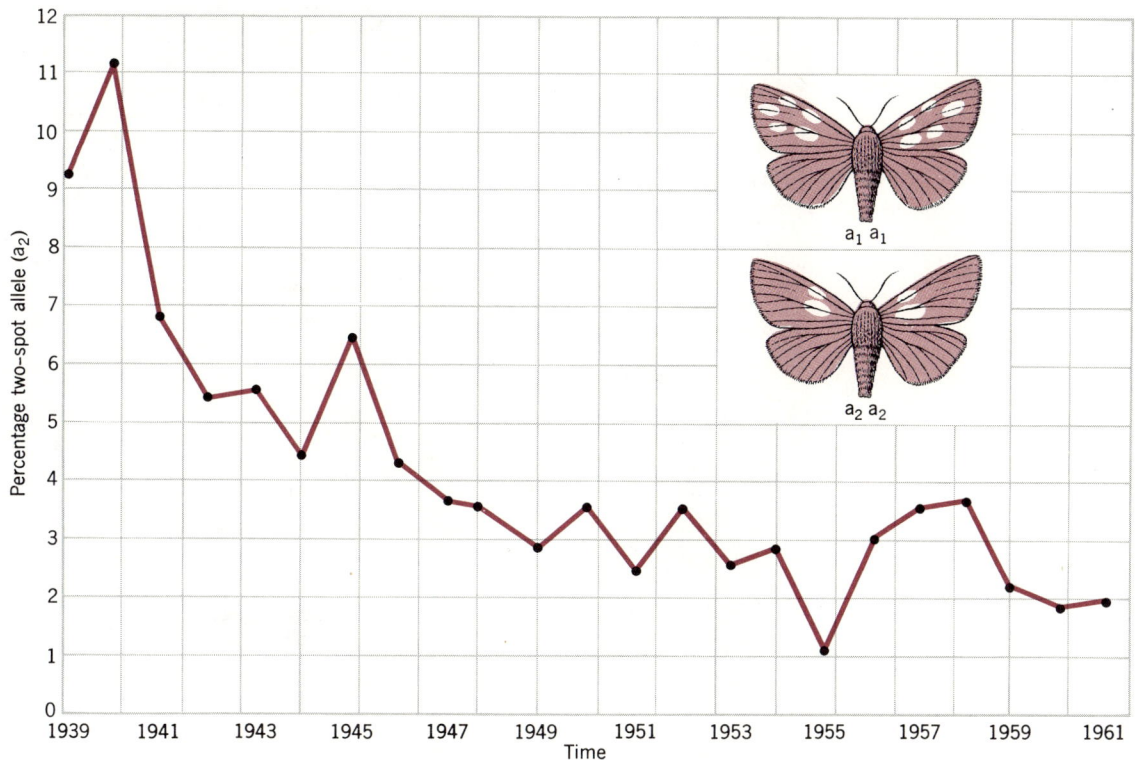

Figure 23-3 Sequential evolution; fluctuations in gene frequency in a population of scarlet tiger moths. It is believed that the original population had plain gray forewings. At some time a mutation that produced one or two white spots occurred (this allele is referred to as a_2). More recently, a second mutation (a_1) caused multiple spots in either the homozygous or heterozygous condition. Apparently conferring an adaptive advantage of camouflage, the a_1 gene has gradually spread throughout the population. The graph, from the results of a study by a group of British geneticists, shows a period of 23 years of that evolutionary change during which the mean percentage of moths showing the two-spot condition (a_2a_2) fell from about 10 percent to about 2 percent. No adults survive the winter in this species, so each year represents a new generation. Although there are much larger changes between some generations than others, no sign of genetic equilibrium exists.

a given population should migrate into a new environment, natural selection operating in the two different environments would favor different mutations; different gene combinations from those that were advantageous in the original environment would be selected. With the passage of time, therefore, we would expect differences to appear between the populations in separate environments, leading to the evolution of new species.

Many examples of adaptive radiation are known. One of these is seen in the many orders of mammals now living, most of which evolved, in the course of some 35 million

659   *Mechanisms of Evolution*

Figure 23-4 Convergent evolution occurs when members of different species adapt in similar ways to a given environment or mode of life. An example is the development of wings in (a) insects (the monarch butterfly); (b) reptiles (an ancient flying reptile called a pterosaur; (c) birds (a seagull); (d) mammals (a bat).

(a)

(b)

years, from just a few primitive ancestors.

Natural selection can also lead different species, living in similar kinds of environments or having similar ways of life, to evolve similar characteristics; this is called **convergent evolution** and results largely from the selective influence of a similar environment. Examples of convergent evolution are seen in the unusual similarity between the molluscan and vertebrate eye, the development of wings in at least four groups of animals (insects, flying reptiles—the so-called **pterosaurs**—birds, and bats (Fig. 23-4) and the adaptive formation in aquatic life of flipper-like appendages, as seen in the sea turtle (a reptile), penguin (a bird), and walrus (a mammal).

(c)

(d)

661 *Mechanisms of Evolution*

The Elemental Forces of Evolution

Because evolution is based on genetic variation and changes in gene frequency in populations, the ultimate source of evolutionary change is mutation. Mutation alone, however, results only in the introduction of variation into a population. The force primarily responsible for upsetting genetic equilibrium—that is, for causing the gene pool to disobey the Hardy-Weinberg law—is natural selection.

Natural selection brings about evolutionary change by favoring differential reproduction of genes and produces changes in gene frequency from one generation to the next. Furthermore, selection always encourages the genes that assure the highest level of **adaptive efficiency** between the population and its environment; it favors increased reproduction of gene combinations most efficient under the prevailing environmental circumstances. The idea commonly held of Darwinian selection as based on **differential mortality**—"survival of the fittest"—is only part of the evolutionary pressure applied by selection. The two-spot moths in Figure 23-3 did not just die off faster to allow the multispot allele to overrun the population. Selection may result from **differential viability** (if the multispot moths were naturally healthier); from **differential fertility** (if they laid more eggs or had a greater percentage of the eggs that hatched; or from **differential behavior** (e.g., if males preferred multispot females for mating). Examples of each of these mechanisms abound in nature, and are just as common as differential mortality.

A mutation, we have said, is a miscopy of the sequence of nucleotides that make up each chromosomal strand of DNA (Chapter 10). Evolution, therefore, is ultimately based on molecular processes, as are all other biological phenomena. The English geneticist John Maynard Smith has compared **molecular evolution** to a word game. The object of the game is to move from one word to another of the same length by changing one letter at a time, with the requirement that all the intermediate words are meaningful in the same language. Thus FISH can be converted into

BIRD in a series of single-letter changes, as follows: FISH, DISH, DASH, BASH, BASE, BARE, BARD, BIRD.

This is an analogue of evolution: each word represents a given gene locus capable of coding for a particular protein; the letters represent nucleotide triplets, coding for individual amino acids; the alteration of a single letter corresponds to the simplest evolutionary step, the substitution of one amino acid for another; and the requirement that each word be meaningful corresponds to the evolutionary requirement that each unit step must be from one functional protein or other structure to another. Normally, nonsense sequences or nonfunctional proteins would be selected against. This is a highly simplified model of the way one gene may change into another.

There are numerous examples known of mutations that involve only a single base pair in the DNA sequence; for instance, about 150 different kinds of mutant hemoglobin have been found in humans. Most of these differ from normal hemoglobin by only a single amino acid. Many such mutations are apparently deleterious; many are harmless. Presumably, many of the harmful ones are fairly recent in origin and will be selected against. This is not necessarily so, however, because of the **pleiotropic** effect of many genes—their tendency to influence more than one trait at a time.

The mutant hemoglobin responsible for sickle-cell anemia (Fig. 23-5 and Fig. 10-18), for example, differs from the normal by only one amino acid in the protein sequence: a glutamic acid is replaced by valine. In genetic code this represents a change in the nucleotide sequence from AUG to UUG. Homozygotes of this mutant suffer a severe anemia that is often fatal. Why, then, has natural selection not removed it from the population? The answer to that question became obvious when it was discovered that in the heterozygotic condition the mutant confers a substantial resistance to malaria on the individual carrying it. It becomes clear, therefore, why sickle-cell anemia has been found mainly in Africa, India, and other tropical regions where malaria is endemic.

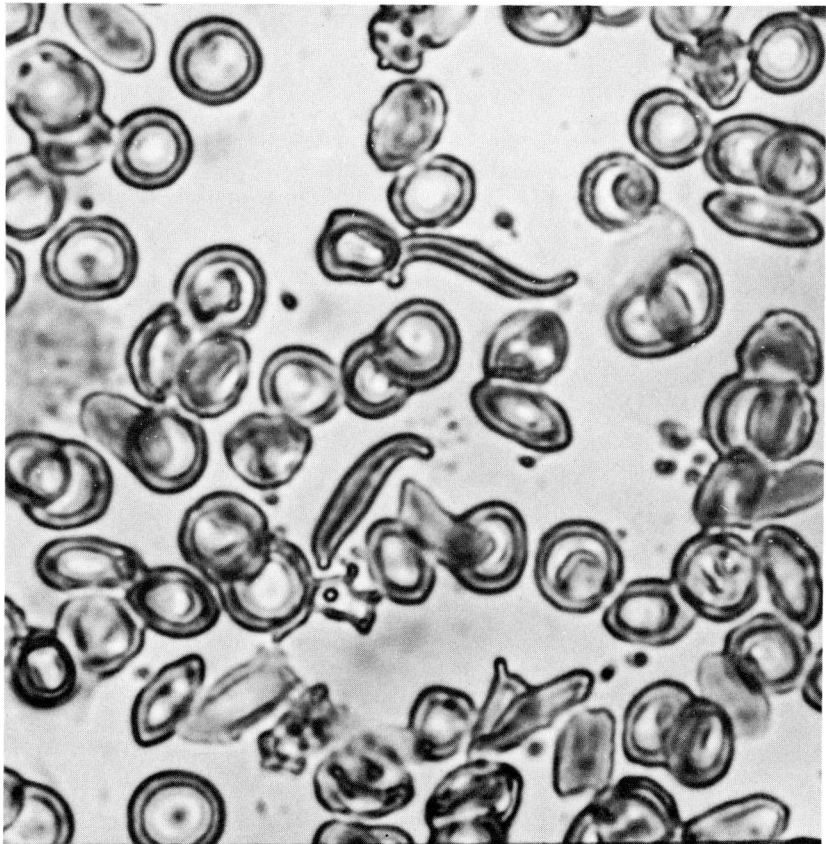

Figure 23-5   Red blood cells from a patient with the sickle-cell trait. Round cells are normal; the distorted, elongate cells contain sickle hemoglobin.

Genetic Drift

From what we have said so far it might appear that only mutations or gene recombinations that are selected for are retained in a population. We now know that this is not true. The number of amino acid substitutions among closely related proteins, for example, suggests a rate of molecular mutation far greater than that predicted by the known rates of evolution. That is, many changes in genes are not selected for or against; they seem to be **selectively neutral.** Such genes can spread throughout a population and become established, or **fixed,** in relatively few generations by purely random fluctuations in gene frequency, a process known as **genetic drift.**

To illustrate this point, suppose we establish a series of populations of eight fruit flies in the laboratory (Fig. 23-6, p. 664). Seven in each population are homozygous for some normal trait such as wing shape. Call these a_1a_1. One fly in each population is heterozygous for this allele, carrying a dominant mutant gene for an abnormal wing; its genotype is a_1a_2. Such a situation occurs in nature when a single new gene mutation has just arisen. In our laboratory, where a source of food is supplied and all other environmental conditions are maintained at an optimal level, this mutation is selectively neutral. In each population several hundred progeny are produced at each generation; however, let us take at random only eight newly hatched flies to represent each new generation.

At the first generation the heterozygote can mate only with a_1a_1 individuals, and will produce offspring of a_1a_1 or a_1a_2 genotype. Be-

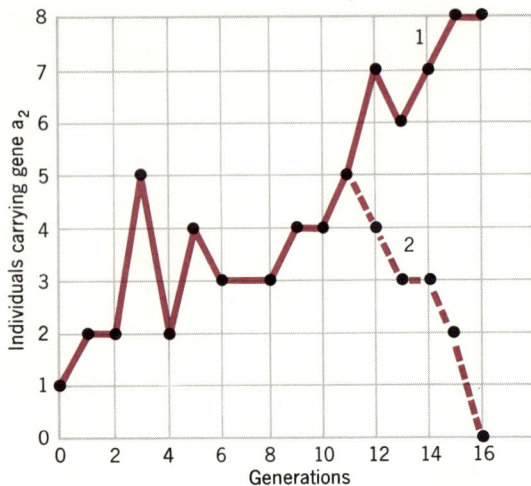

Figure 23-6   Genetic drift is illustrated by the introduction of a single mutation into a small population of fruit-flies (eight individuals), in which selection is completely random. As long as the population remains small, changes in gene frequency from one generation to the next will be erratic and random. If a large number of such populations are studied, in some the mutant gene will displace the wild-type allele and be fixed in the population (curve 1); in other populations, the mutant will be eliminated (curve 2).

cause the eight members of the second-generation population are selected at random from a large number of progeny, any one of nine combinations of wild-type (normal) or mutant flies may appear (8:0, 7:1, 6:2, 5:3, 4:4, 3:5, 2:6, 1:7, 0:8). There is a chance that all but one of the mutant flies may be lost, but any of the other combinations represents a change in gene frequency from the original population.

In one set of actual experiments 96 populations were studied in this fashion for 16 generations. At the end of the experiment only 26 of the populations still contained both genes a_1 and a_2. In 41 populations the wild-type gene had displaced the mutant, so that all of the individuals were homozygous a_1a_1 (as in curve 2, Fig. 23-6). In 29 populations

the mutant had become fixed; all of the individuals had the genotype a_2a_2 (curve 1, Fig. 23-6). The essential feature of genetic drift is that it can only take place in small populations; the smaller the population, the greater the random variations in gene frequencies from generation to generation.

Consider, for example, a very small population of only two mating pairs. If one individual contains a new mutation a_2, its genotype would be a_1a_2, whereas that of the other three individuals would be a_1a_1. Suppose that each pair produces 10 offspring, only two of which survive. On a purely chance basis, it is entirely possible that the surviving young of the mutant and its mate might both be a_1a_2. Thus, of the four members of the second generation, 50 percent would carry the mutant. By random combinations of gametes, it is again possible that the next generation would all be mutants, in which case in only two generations the new allele would have been fixed in the population. Of course, it is equally possible that the new allele a_2 would be eliminated from the population in one or two generations if the surviving offspring happened to be individuals of type a_1a_1. In contrast, in a large population of, say, 100,000 breeding couples, it would take hundreds or thousands of generations for a new mutation to spread to a substantial fraction of the individuals in the population (assuming relatively few surviving offspring per pair); the more likely eventuality, in the absence of positive selection for the mutant, would be its disappearance from the gene pool.

In natural circumstances, how can there be genes that are selectively neutral? Recent work demonstrating **gene amplification** suggests an answer to this question. In all eukaryotes that have been examined, genes are present in multiple copies, aligned end to end along the DNA strand. The nucleotide sequence of a particular gene may be repeated many times per genome. For example, the DNA that codes for ribosomal RNA in the South African clawed toad *Xenopus* is repeated 800 times per nucleus. It has been estimated that almost half of the DNA that comprises the calf genome consists of nucleotide sequences that are repeated 100,000 to 1 million times. These large numbers of copies permit a great deal of **latent variability** in the gene pool. A mutation in one copy of a gene would have to result in a gene product that was extremely favorable or highly toxic in order for it to be selected for or against. Otherwise the product would be diluted by the many normal representatives produced by the remaining copies of the sequence.

Formation of New Races and Species

What are the steps by which mutations and new combinations of genes in a population give rise to a new species? A variety of factors is involved. One of the most significant is **isolation.** We know that inherited variations occur at random and in all directions. For these new variations to become established as an evolutionary change, a certain degree of isolation of the population is necessary because free interbreeding tends to promote a relative homogeneity among the individuals of the population.

Isolation in nature may be obtained in several ways. If a population is split into smaller populations by natural physical barriers (e.g., by migration of several individuals and their subsequent isolation by mountains, glaciers, deserts, or bodies of water), **geographic isolation** is the result. Each of the geographically isolated populations originally possessed the same pool of genes, but each now continues to experience independent random mutations and gene combinations in all directions in the course of successive generations. It can be expected on the basis of chance that the gene pools, and therefore the inherited variations of the isolated populations, will gradually become different from one another. Thus chances are that the originally identical populations will in the course of time evolve in different directions. Moreover, the divergences in evolution will be appreciably influenced if the environmental conditions are different for each of the isolated populations.

When a species becomes separated into two **demes** (isolated breeding populations) by geographic barriers, we call the two populations **allopatric**—they occupy different ranges. A second pattern of isolation occurs in cases when related populations share a portion of their ecological range and become isolated from one another not by space but by the physiological expression of genetic differences. Such populations are **sympatric.** For example, the salamanders of California form a series of sympatric species, all of the genus *Ensatina* (Fig. 23-7, p. 666). *E. eschscholtzi* ranges from the southern tip of the state northward along the coast. *E. picta* is found throughout most of the northern third of California. The range of *E. oregonensis* overlaps both in the San Francisco Bay region. Thus the two populations are sympatric in central California. Even though these closely related newts breed at the same time and in the same streams, genetic exchange between them is prevented by differences in breeding behavior and egg deposition.

Races, or **subspecies,** are populations in the same species that have become different in one or a few of their genes as a result of isolation. In general, the differences between races are mostly quantitative rather than qualitative; that is, they usually differ in gene frequencies rather than in the presence or absence of genes. If the geographical barriers between two or more races of a species are removed, interbreeding will usually occur; they will soon share a common gene pool, and in time they thus become a single race or population again. If the barriers remain, the isolated populations may tend with the passage of time to become gradually and increasingly different, until a point is reached at which members of one population may be unable to breed with those of the others if they should come in contact. When this point is reached gene exchange between them cannot occur, and they are now considered separate species. Once a species is established its members breed only among themselves. This, of course, leads to the unity and continuity of the species.

In summary, then, the early phases of evolution lead first to differences that account for the appearance of several races or strains in a given species; these in turn may lead progressively, by further alterations in geno-

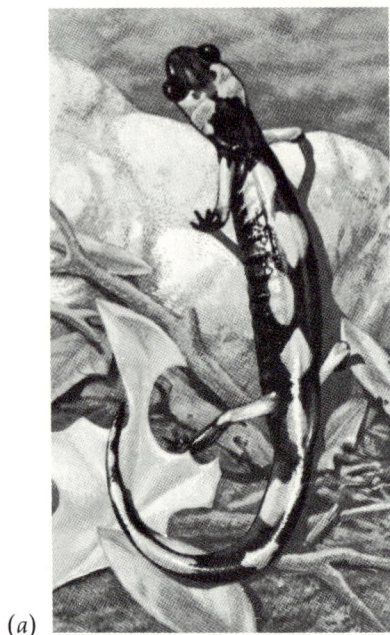

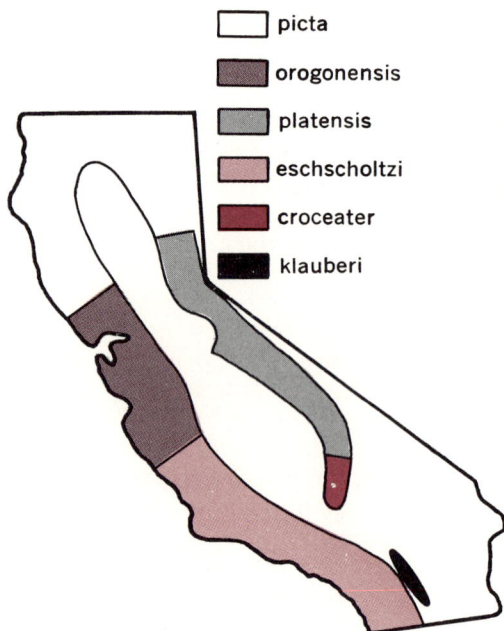

Legend:
- □ picta
- ▨ orogonensis
- ▦ platensis
- ▨ eschscholtzi
- ▦ croceater
- ■ klauberi

Figure 23-7 Salamanders of California, all of the genus Ensatina, form a series of sympatric species, four of which are pictured. (a) E. croceater, (b) E. platensis, E. eschscholtzi, (d) E. klauberi. Hybrids of all the species can be produced in the laboratory, but in nature they normally fail to interbreed, even when their ranges overlap, because of physiological and behavioral differences.

type, to the formation of new species, genera, families, and so on. From this examination of the mechanisms of the evolutionary process, it must already be obvious that the evidence in support of the concept of organic evolution is overwhelming. Since Darwin there has accrued an enormous body of information that can only be explained sensibly by the contention that evolution occurred in the past and is in operation today. Evidence independently arrived at in many different scientific disciplines (the study of fossils, called **paleontology;** comparative studies of the structure, function, and biochemistry of living animals and plants; studies of classification, or **taxonomy;** and modern molecular genetics) points in the same direction.

Evidence for Evolution

The Fossil Record

Fossils are the preserved remains or impressions of ancient plants and animals, usually in rocks (Fig. 23-8, p. 668). It is from fossil remains that paleontologists attempt to reconstruct a picture of the ancient extinct organisms as they actually existed and the environment in which they lived. When organisms die their bodies are usually decomposed by microorganisms. Occasionally when ancient plants or animals died they fell into sediments that later became a portion of the stratified rocks of the earth's crust, and thus were preserved in one form or another. Usually the skeletons of animals and woody parts of plants were infiltrated by water containing high concentrations of minerals such as silica or calcium, and the intercellular spaces of these hard tissues gradually became filled with insoluble mineral matter. Such fossils, collected and analyzed by paleontologists, have provided a record of ancient life and have furnished us with direct evidence that evolution has occurred by demonstrating the differences between organisms that lived millions of years ago and their descendants that exist today.

The earth's crust is made up in part of various layers or strata of **sedimentary rocks** that were formed over vast periods of time in the distant past by the slow settling out and compaction of sediments or deposits of sand, silt, mud, and volcanic ash from the seas, lakes, rivers, and air. The lowest rock layers are in general the first and oldest deposits; the succeeding upper strata reflect the sequence of their deposition very much like the bricks of a house; the top layer is the most recent. The exposed stratified rock walls of the Grand Canyon (Fig. 23-9, p. 669) are a good example of such a time sequence.

This sequence of the fossil-bearing rock strata in the earth's crust serves as another timetable of the ancient past in the study of primitive life. It has provided us with valuable information of the approximate ages of the fossils found in each stratum, and with important clues about the seccession and relationship of the changes in the evolution of organisms. The fossil record often shows a succession of animals and plants from the relatively simple aquatic types of older geologic times, in the deeper strata, to the progressively more complex and better adapted terrestrial forms of the more recent geologic eras, nearer the surface. It has also given us important clues about the relationships between extinct and living forms, about ancient climatic conditions, and about past distribution of land and seas.

Geologists have been able to relate the sequence of the formation of sedimentation of the rock strata of the earth to a record of geologic history often called the **geologic timetable** (see Table 23-1, pp. 670–671). The ages of the different sedimentary rock strata have been measured by several methods, including their order of arrangement (from the oldest strata at the bottom to the progressively more recent ones at the top) and measurements of the extent of decay of long-lived radioactive elements contained in them (that is, by radioactive dating).

The geologic timetable of the earth is based on the five major sedimentary rock strata that make up a significant portion of the earth's crust. Accordingly, the geologic history of the earth is divided into five major time intervals, called **eras,** which in turn are progressively

Figure 23-8 *A fossil plant and animal.* (a) *A giant fern that functioned 300 million years ago in the area that is now Illinois left the imprint of this frond in a layer of shale after it died.* (b) *A cephalopod lived about 150 million years ago in the ancient sea—at about the same time that dinosaurs ruled the land. It resembles a present-day chambered nautilus.*

subdivided into **periods** and **epochs.** The oldest geologic era is the **Archeozoic,** followed in order by the progressively younger periods as indicated in Table 23-1. We are now still in the Cenozoic. Biologists often refer to the oldest eras collectively as the **Precambrian Era** because of the virtual lack of fossils in rock strata older than about 600 million years.

At times there have been widespread and intense intervals of mountain building resulting from tremendous geologic disturbances whereby vast portions of the earth's crust were shifted and raised, causing enormous changes in distribution of land, oceans, and climates, and the death of many organisms. These are called geological **revolutions.** The revolutions that gave rise to the Appalachian Mountains marked the end of the Paleozoic

Era and the beginning of the Mesozoic Era. The more recent revolution that formed the Rockies, Andes, Alps, and Himalayas marks the transition between the end of the Mesozoic Era and the start of the Cenozoic.

Distribution of Fossils
in the Geologic Timetable

According to our latest estimates the earth originated at least 4.5 billion years ago. Fossils have been found in relative abundance in rock as old as 600 million years. Only a few fossils have been discovered in older rocks, for two principal reasons. First, the living forms of those eras were presumably very small and composed chiefly of soft tissues, features that are unfavorable for preservation as fossils. Second, the rock layers, under the enormous weight of all other overlying strata, have been greatly modified, or **metamorphosed,** by great pressures and high temperatures—conditions that are hardly conducive

Figure 23-9   The stratified walls of the Grand Canyon represent a geological timetable extending back almost a billion years into Precambrian times. The south wall of the canyon in some locations exposes a mile-thick section of stratified Paleozoic sediments.

table 23-1
The Geologic Timetable[a]

ERAS	PERIODS	EPOCHS	DATES[b]	AQUATIC LIFE	TERRESTRIAL LIFE
Cenozoic (Age of Mammals)	Quaternary	Recent	025		Man in the New World
		Pleistocene	1	--- Periodic glaciation ---	First men
	Tertiary	Pliocene	10		Hominids and pongids
		Miocene	25	All	Monkeys and ancestors of apes
		Oligocene	40	modern	Adaptive radiation of birds
		Eocene	60	groups	Modern mammals and herbaceous
		Paleocene	75	present	angiosperms
				--- Mountain building (e.g. Rockies, Andes) at end of period ---	
	Cretaceous		130	Modern bony fish	Extinction of dinosaurs, pterosaurs
				Extinction of ammonites, plesiosaurs, ichthyosaurs	Rise of woody angiosperms, snakes
				--- Inland seas, warm climate ---	
Mesozoic (Age of Reptiles)	Jurassic		165	Plesiosaurs, ichthyosaurs abundant	Dinosaurs dominant
				Ammonites again abundant	First lizards: Archeopteryx
				Skates, rays, and bony fish abundant	Insects abundant
					First angiosperms
				--- Warm climate, many deserts ---	
	Triassic		200	First plesiosaurs, ichthyosaurs	Adaptive radiation of reptiles (thecodonts, therapsids, turtles, crocodiles, first dinosaurs)
				Ammonites abundant at first	First mammals
				Rise of bony fish	
				--- Appalachian Mts. formed, periodic glaciation and arid climate ---	
	Permian		230	Extinction of trilobites, placoderms	Reptiles abundant (cotylosaurs, pelycosaurs)
					Cycads and conifers; ginkgos

Era	Period	Epoch	Age[b]	Marine life and climate	Terrestrial life
Paleozoic	Carboniferous (Age of Amphibians)	Pennsylvanian	250	— Warm, humid climate — Ammonites, bony fish	First reptiles Cool swamps
		Mississippian	280	— Warm, humid climate — Adaptive radiation of sharks	Forests of lycopsids, sphenopsids, and seed ferns. Amphibians abundant. Land snails
	Devonian (Age of Fish)		325	— Periodic aridity — Placoderms, cartilaginous and bony fish Ammonites, nautiloids	Ferns, lycopsids, and sphenopsids First gymnosperms and bryophytes First insects First amphibians
	Silurian		360	— Extensive inland seas — Adaptive radiation of ostracoderms; eurypterids	First land plants (psilopsids, lycopsids) Arachnids (scorpions)
	Ordovician		425	— Mild climate, inland seas — First vertebrates (ostracoderms) Nautiloids, Pilina, other mollusks Trilobites abundant	none
	Cambrian		525	— Mild climate, inland seas — Trilobites dominant First eurypterids, crustaceans Mollusks, echinoderms Sponges, cnidarians, annelids Tunicates	none
Pre-Cambrian	Proterozoic Archeozoic		3000	— Periodic glaciation — Fossils rare, but many protistan and invertebrate phyla probably present	none

[a]Adapted from Kimball, J. W., *Biology*. Addison-Wesley, Reading, Mass., 1968.
[b]Approximate dates, in millions of years, from beginning of indicated era, period, or epoch.

671

to the preservation of delicate organisms. The rocks where fossils first appear abundantly already include representatives of most of the modern sea animals (with the exception of the vertebrates); this must mean that life originated considerably earlier.

All of the earliest fossils found are of aquatic animals and plants, a most important line of evidence indicating that life originated in the sea (Fig. 23-10). In fact, fossils of the land-dwelling organisms only appear in rock strata deposited some 200 million years later (in the Silurian period). We now believe that the first life on our planet originated approximately 2 billion years ago in the primitive seas (Chapter 22).

By far the majority of the ancient living forms of the past—from 100 million years ago and earlier—has become extinct. The organisms of today are descendants of only a handful of the hundreds of thousands of species that once existed in the Mesozoic Era.

Descent of the Vertebrates

The main animal groups of the past can be represented as in Fig. 23-11, which indicates the "success" of each phylum during the various periods in terms of the number of species or genera that have been found as fossils. The Cambrian Period of the Paleozoic Era, which in effect is as far back as the usable fossil record extends, already contained most of the modern major invertebrate phyla living in the seas today. These included many of the marine protozoa, sponges, coelenterates, echinoderms, mollusks, annelids, and arthropods (Appendix 2).

The principal biological event of the Ordovician Period was the origin of the **verte-brates,** most likely from an ancestor similar to present-day tunicates (or sea squirts).

The oldest vertebrate fossils that have been discovered so far are those of the **jawless fish.** In the course of geologic time certain ancient jawless fish must have given rise to the now

Figure 23-10   Representation of life in the early seas 500–600 million years ago (during the Cambrian Period). The organisms include trilobites and other arthropods, a jellyfish (at left), a sea cucumber, annelid worm, various mollusks and sponges, and a variety of algal plant forms.

HISTORY OF ANIMALS

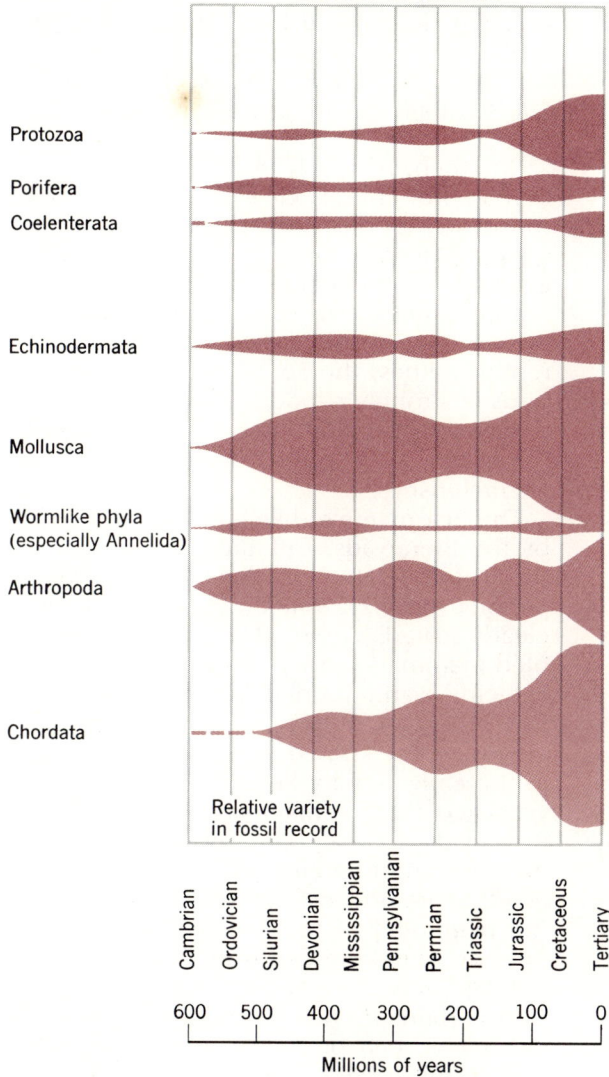

Protozoa
Porifera
Coelenterata
Echinodermata
Mollusca
Wormlike phyla (especially Annelida)
Arthropoda
Chordata

Relative variety in fossil record

Cambrian | Ordovician | Silurian | Devonian | Mississippian | Pennsylvanian | Permian | Triassic | Jurassic | Cretaceous | Tertiary

600   500   400   300   200   100   0

Millions of years

(a)

HISTORY OF THE VERTEBRATES

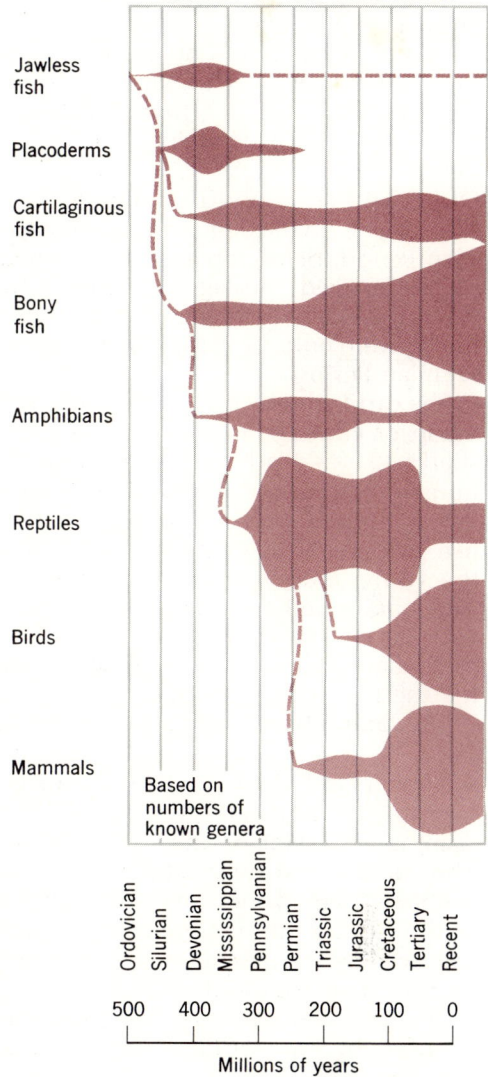

Jawless fish
Placoderms
Cartilaginous fish
Bony fish
Amphibians
Reptiles
Birds
Mammals

Based on numbers of known genera

Ordovician | Silurian | Devonian | Mississippian | Pennsylvanian | Permian | Triassic | Jurassic | Cretaceous | Tertiary | Recent

500   400   300   200   100   0

Millions of years

(b)

Figure 23-11   The history of the major animal phyla, represented by bands or pathways. The width of each band is proportional to the known variety (i.e., numbers of genera) in each of the geological periods since the Precambrian. Part (b) is an expansion in greater detail of the chordate (vertebrate) band in part (a).

673   *Mechanisms of Evolution*

extinct jawed fish, or **placoderms,** the only known extinct class of vertebrates. Certain ancestral descendants of the placoderms in turn probably evolved independently into the cartilaginous fish, or **chondrichthyans** and the bony fish, or **osteichthyans.** The placoderms became most abundant during the Devonian Period and then rapidly began to disappear with the beginning of the Mississippian Period, being replaced by the relatively newly evolved cartilaginous and bony fish. All the evidence indicates that the placoderms were entirely extinct by the end of the Permian Period.

The **cartilaginous fish** (those with skeletons made of cartilage instead of bone, such as sharks, skates, and rays) show few fluctuations in abundance during their long evolutionary history. By contrast, the ancient **bony fish** gave rise to several main lines, including the vastly abundant modern bony fish and the lobe-finned fish (Fig. 23-12), now also virtually extinct, from which the amphibians are believed to have evolved sometime during the Devonian Period.

The **amphibians** attained their peak in number, size, and diversity during the Pennsylvanian and Permian Periods. **Insects,** as terrestrial invertebrates, were also rapidly becoming abundant during this same interval. The sudden and near extinction of the amphibians, perhaps because of the appearance of the reptiles during the Jurassic Period, was followed by the evolution of new amphibian groups—the frogs and toads.

The amphibians were then subsequently eclipsed by the **reptiles,** which apparently arose as an offshoot of certain primitive amphibians during the later Mississippian or early Pennsylvanian Periods. The reptiles, the first truly terrestrial vertebrates, rapidly became abundant during the Permian Period, decreased somewhat during the subsequent Triassic Period, and then increased, achieving their greatest numbers during the Cretaceous Period. They adapted to an existence in the water, on the land, and in the air with correspondingly different body structures and ways of life. The most famous were, of course, the dinosaurs, some of which attained lengths of nearly 100 ft and weights of approximately 25 tons. The reptiles were probably the most successful of all the terrestrial animals living in the past. The Mesozoic Era is usually referred to as the Age of Reptiles.

But then, relatively suddenly, for reasons that are not entirely clear, many of the larger forms such as the dinosaurs disappeared and were replaced by birds and mammals, which evolved independently from two different reptilian lines. The evolutionary transition during the Jurassic Period of one of these reptilian lines, the so-called **thecodonts,** to birds is strongly supported by the fossil of a remarkable primitive bird, an **Archeopteryx** (Fig. 23-13, p. 676), with its numerous reptilian characteristics.

The first mammal-like reptiles, represented by the **therapsids,** had already arisen by the time of the later Permian or early Triassic periods some 250 million years ago. Though clearly scaled, like reptiles, therapsids resembled mammals in their skeletal structure and in being warmblooded. Whether they nourished their young via rudimentary mammary glands is not known. However, it was only in the early Tertiary Period, about 60 million years ago, that the **mammals** evolved into the numerous diverse lines representing every mammalian order alive today, including several that are now extinct. The time span from the beginning of the Tertiary Period to the present, called the Cenozoic Era, is known as the Age of Mammals.

In certain instances closely related fossils that have been preserved in a more or less continuous succession of rock strata have permitted us to trace in some detail the evolutionary progress of a species as it occurred over several million years. An outstanding example is the sequence of fossil horses found in the Cenozoic rock strata of western North America, which show a progressive modification of the limbs from a three-toed to a single-toed condition (Fig. 23-14, p. 677).

Like that of the horse family the detailed evolutionary histories of several other species of organisms (e.g., the elephant family and certain mollusks) have been equally well re-

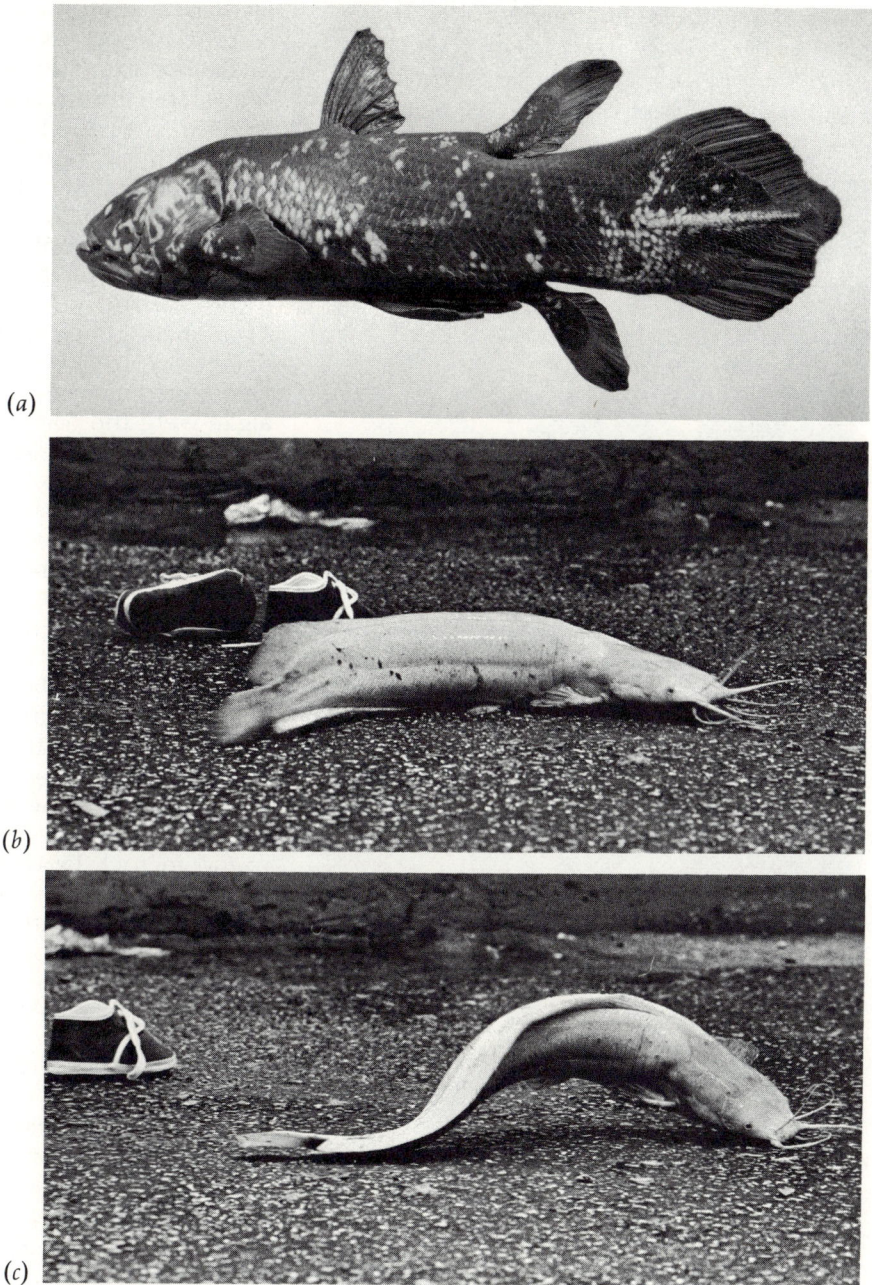

Figure 23-12 (a) A fossil lobe-finned lungfish (coelacanth), which was probably the type of animal that pulled itself out of the water some 400 million years ago and gave rise to the amphibians. (b) and (c) Modern-day "walking fish," a species of Asiatic catfish introduced into Florida in the 1960s, which is able to travel overland for considerable distances.

675 *Mechanisms of Evolution*

Figure 23-13 (a) Fossilized impression of the primitive reptile-like bird Archeopteryx. (b) A scientific restoration of Archeopteryx, depicting its actual appearance when killed some 200 million years ago.

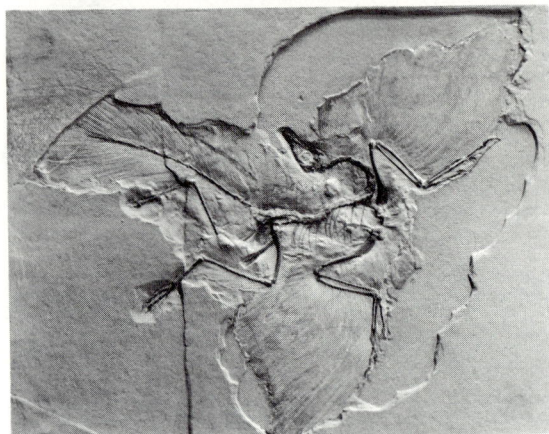

(a)

(b)

of these evolutionary branches extend back in time almost unchanged for millions of years. For example, the present-day horseshoe crab, Limulus (Fig. 23-15, p. 678), is often referred to as a living fossil. It appears to be very similar in structure to its remote ancestors of some 200 million years ago.

The Fossil Record of Plant Life

The history of the major groups of plants is summarized in Fig. 23-16 (p. 679). As is true of animals, the earliest reliable plant fossils date back to the Cambrian Period of the Paleozoic Era. They indicate that the **marine algae** were the earliest predominant form of plant life, including the blue-green algae, the green algae, and even certain brown algae resembling those of our present day.

Fossils of the first known land plants, the now extinct **Psilophytales** (Chapter 13) appear in the Silurian Period, approximately 50 million years before the appearance of the first land animals. These Psilophytales already contained vascular tissues (xylem and phloem), the vascular tissues much like of modern plants of the phylum Tracheophyta. For that reason they are presumed to be ancestors of most of the modern forms. The earliest **bryophyte** (moss) fossils are always found in rock layers formed later. Thus, although mosses are often considered to be more primitive than the vascular plants, they

constructed. In general, they illustrate clearly the basic rule that evolution has not proceeded at a steady rate in any single direction. Instead it has followed a complicated branching, or **radiating,** course leading to many diverse organisms. Man happens to be at the end of one of these branches, whereas other existing living plants, animals, and bacteria, are at the ends of other branches. Some

Figure 23-14 Evolution of the horse. The earliest known representative found in the fossil record was a cat-sized animal, Eohippus (now called Hyracotherium), which lived about 60 million years ago and had four padded toes on the front legs and three on the back. The main changes leading to the modern horse, Equus, include a substantial increase in size, elongation of the neck as an adaptation to grazing, and the dramatic change in foot structure to a single-toed hoof. As indicated by the many branching paths, descendants of the first horselike creatures radiated into many forms, including rhinoceros, donkey, and zebra.

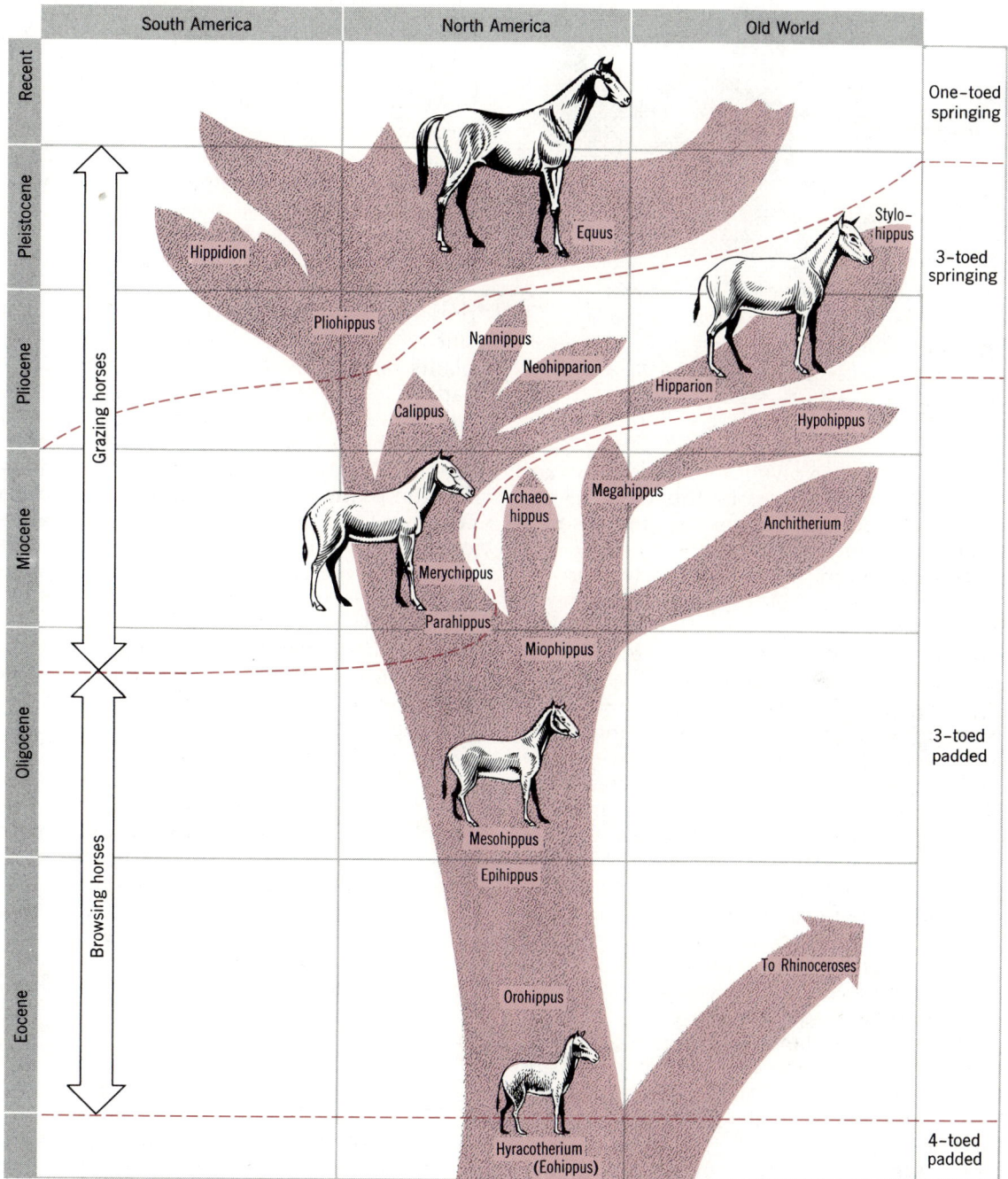

Figure 23-15   The horseshoe crab Limulus *is not a crab at all, but a relative of the spiders. In structure and appearance it is essentially unchanged from fossil ancestors that lived 200 million years ago.*

seem to have evolved independently and slightly more recently. In the ensuing Devonian Period the Psilophytales increased in abundance and distribution.

The Mississippian and Pennsylvanian Periods (the Carboniferous, or Coal Age) witnessed the development of dense, widespread, lowland or swamp forests made up of large, treelike **lycopods** (our modern "ground pine"), **horsetails,** the now extinct **seed ferns,** and other primitive **gymnosperms** (Chapter 13). The remains of these ancient swamp forests were gradually transformed through decomposition into the major coal beds of the world. With the closing of the Paleozoic Era, many ancient and dominant plants of the Coal Age declined into extinction and were replaced by extensive coniferous forests more closely resembling those of our present day. Thus in the course of a time span of some 350 million years, representing the entire Paleozoic Era, the early dominant plant life of our planet was progressively succeeded by land plants (the Psilophytales) and then the lycopods, horsetails, ferns, and seed ferns, culminating finally in extensive forests of gymnosperms.

The Mesozoic Era that followed was characterized by a wider distribution and diversi-fication of the gymnosperms, including the major living order Coniferales (or true conifers), and the extinction of the giant lycopods, giant horsetails, and seed ferns. It is in the early Mesozoic Period that we note the first fossil remains of **angiosperms,** or flowering plants, presumably evolved from an ancient group of seed ferns independently of the origin of the higher gymnosperms. The remarkably rapid rise, development, and distribution of the angiosperms resulted in their ascendancy as the dominant terrestrial vegetation (mostly woody plants) of the earth at the end of the Mesozoic Era, when the once-prevalent gymnosperms began to dwindle.

The Cenozoic Era, which extends to the present, has witnessed an even greater diversification and distribution of the flowering plants ranging over virtually all parts of the earth. The prevalent woody types of angiosperm began to give way to herbaceous plants (the flowering annuals). The Great Ice Age of 1.5 million years ago was responsible for the extinction of many warm-adapted woody plants; at the same time it provided cooler and drier environmental conditions that herbaceous plants could better withstand than woody ones. Herbaceous plants, as annuals, often grow sufficiently rapidly from seed in a short growing season to produce another generation of seeds before they are killed off by unfavorable weather; thus the species survives.

The fossil record of both plants and animals clearly demonstrates that evolution is a fact; that it has in general proceeded from simpler to progressively more complex groups; and that many species have become extinct over the course of time.

Evidence from Biochemistry and Molecular Genetics

DNA Hybridization

The most direct and elegant evidence that organisms are closely related through their evolutionary lines of descent comes from

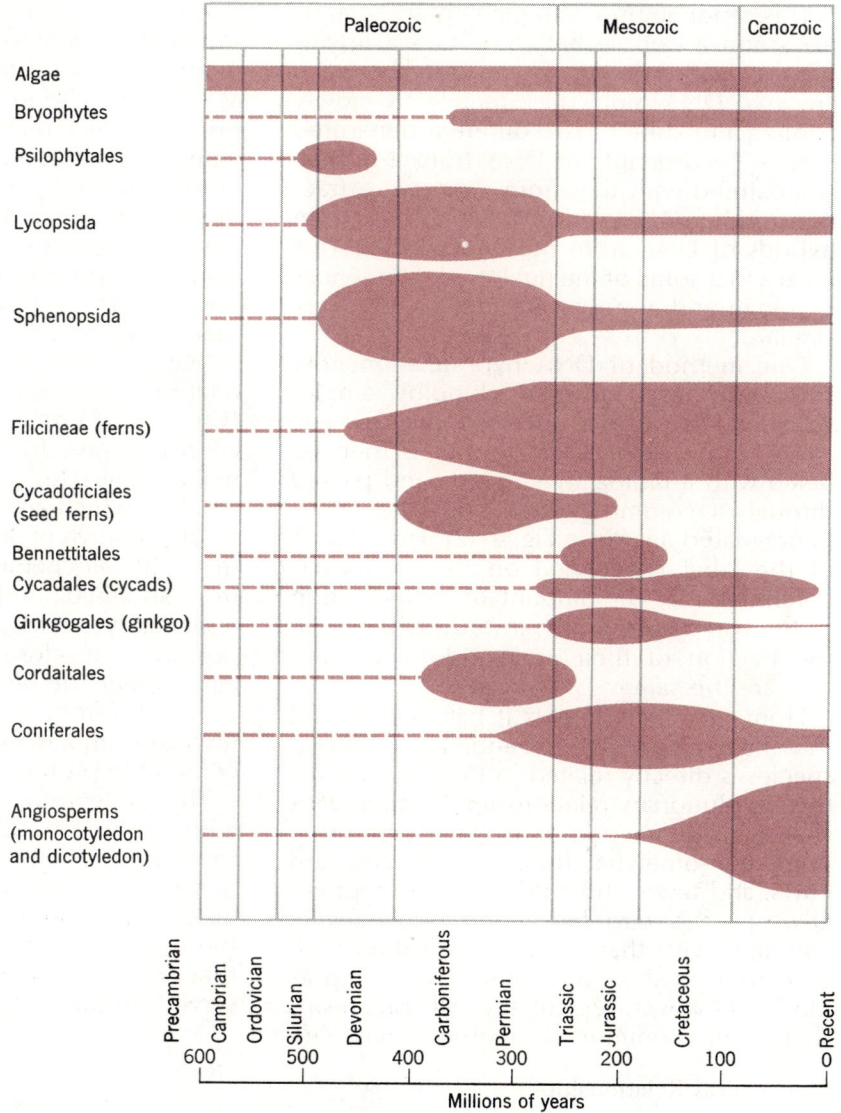

HISTORY OF THE PLANT KINGDOM

Figure 23-16   The history
of the major plant groups.
The width of each band is
proportional to the numbers
of known genera in each
of the geological periods,
analogous to the animal
history depicted in
Figure 23-11, p. 673.

recent demonstrations of different degrees of
similarity in the DNA of various species.
Chromosomal DNA can be extracted as a
long strand of double helix (Chapter 3) from
the cells of any organism. When these DNA
molecules are heated nearly to 100°C, the

hydrogen bonds binding the G...C and A...T
base pairs break and the two complementary
strands separate. This is referred to as **denaturation** of the DNA. If the heated solution
is slowly cooled, the DNA **renatures,** complementary strands rejoining to form double

helical molecules again (Chapter 4).

This tendency for complementary strands to reform a double helix provides a means for testing the degree of similarity between any two DNA molecules, merely by slowly cooling a mixture of two different denatured DNA's. For example, if DNA from a human is renatured with that from a mouse, a fraction (about 25 percent) of the strands form **hybrids** of DNA from the two species. This means that some of the nucleotide sequences in a man and a mouse are identical, or very similar.

One method of DNA hybridization now commonly used involves binding single-stranded DNA of one species to an agar gel. DNA from a second test species is then labeled with a radioactive isotope and passed through a column packed with the DNA-impregnated agar (see Fig. 4-12). How much of the label is retained on the column is determined by the amount of hybridization of the two DNAs, which in turn depends on the fraction of their nucleotide sequences that are the same.

From such experiments it has been found that the degree of hybridization between two species is directly related to the closeness of their evolutionary relationship. Human DNA shares a large fraction of its genes with other primates, somewhat fewer with horses and cows, and fewer still with birds and reptiles. DNA from a bird has more sequences in common with that from a lizard than with that from a fish or a mammal. Most importantly, however, essentially all organisms share some common nucleotide sequences.

Biochemical Relationships

Biochemical evidence of other kinds also points to similar evolutionary relationships. Recall that the cytochrome enzymes (Chapter 4) are found in essentially every organism, plant, animal, and protista. One of these, cytochrome c, is a polypeptide chain of just over 100 amino acids. In recent years the exact sequences of the amino acids in organisms as diverse as man, rabbit, penguin, rattlesnake, tuna, moth, bread mold *Neurospora*, and many others have been de-termined. Again, it is found that the more closely related the organisms, the greater the similarity in amino acid sequence. The sequence in man differs from that in monkey at only one site in the chain. Cytochrome c from a wheat plant, however, differs from man's in 35 of its amino acids; however, even in such distantly related species the majority of the amino acids are the same. In fact, among all species examined, 35 of the amino acids have proved to be invariant—including one section of 11 consecutive amino acids that are common to all organisms known.

Many more examples of such biochemical relationships could be cited. We have noted (Chapter 18) that the contractile proteins actin and myosin, similar to those in mammalian muscle, can be extracted from amebas, which are thought to be among the most primitive organisms in the evolutionary tree. The biochemical properties of myosin from an ameba and a rabbit are nonetheless remarkably similar; their amino acid compositions are closely related and they look alike under the electron microscope. Moreover, actin from an ameba can combine with myosin from a mammal to form a functional contractile protein.

These phenomena are virtually inexplicable without the concept of evolution. They make sense only when taken to mean that a large number of proteins evolved from common precursors, and that the configurations of these proteins (as well as the DNA sequences that code for them) have been largely conserved throughout evolution.

Evolution of Man

Man, like any other organism, is the product of a long evolution that inevitably extends from simpler and less complex forms. Modern man, or **Homo sapiens** ("wise man"), is presently classified in the family Hominidae, of the suborder Anthropoidea, of the order Primates, of the class Mammalia, of the subphylum Vertebrata, and of the phylum

Chordata. *Homo sapiens* is the only species in the family Hominidae.

The most closely related living group of animals is the so-called **anthropoid apes,** family Pongidae, which consists of chimpanzees, gorillas, gibbons, and orangutans (Fig. 23-17, p. 682). In contrast to man, who lives and walks erect on the ground, the anthropoid apes (except for the adult gorilla) are tree dwellers. An ape's maximal brain size is about 700 cubic centimeters, in comparison with a range of about 1200 to 1600 cm³ for modern man. However, the marked resemblances between living man and the anthropoid apes is seen in the similarities of brain and skull structure, absence of a tail, nearly identical reproductive features, including menstruation and embryonic development, and numerous other anatomical and physiological characteristics. According to recent biochemical evidence, including data on the blood types, the gorilla, chimpanzee, and man are more closely related to one another than to orangutans or gibbons.

The Races of Mankind

Present-day man, although he has spread to nearly all parts of the globe, is a single species, *Homo sapiens,* with several subspecies, or **races.** The races are constantly shifting and to a certain extent interbreeding with one another, thus experiencing an exchange of genes. Our ever-increasing mobility continues to reduce the chances of geographic isolation among human races and thus precludes the possibility of their evolution into new species.

There is no common agreement among anthropologists on the number of different races of mankind. Various classification systems have been proposed, ranging from only a few human races to as many as several hundred. One classification system has three major divisions: (1) White, or **Caucasoid,** consisting of light-skinned peoples, frequently with straight or wavy (but not woolly) hair and long noses; (2) **Negroid,** consisting of dark-skinned peoples with broad, flat noses, woolly hair, and long heads;

and (3) **Mongoloid,** consisting of yellowish-brown-skinned peoples with coarse black hair, a "slanted" eye caused by an overlapping skin fold that extends from the upper eyelid (the so-called **epicanthic fold**) and a flattened nose (Fig. 23-18, p. 683).

Origin of Man

Near-Human Ancestors

During the later Tertiary and early Quaternary Periods, especially in South Africa, the anthropoid apes underwent an almost explosive radiation into several different forms. One of these is believed to have given rise to the more immediate forebears of man. The discovery of fossil skulls possessing characteristics of both the ape family (Pongidae) and the family of man (Hominidae), with a brain capacity of some 600 cm³ (like that of the chimpanzee or gorilla) but with more manlike teeth, forehead, and posture, has led most authorities to consider these extinct ape-men, called **Australopithecus** as ancestors of man. They appear to be the earliest near-human fossils found so far. (The major stages in human evolution are summarized in Fig. 23-19, p. 684.) Current estimates are that they arose as far back as 2 million years in the past and existed up to as recently as the middle Pleistocene Epoch, approximately half a million years ago. They seem to have lived in caves and hunted animals.

According to some anthropologists, by the early Pleistocene (about 1.7 million years ago), two species of near-men had become established. (Other anthropologists classify these simply as two types of *Australopithecus.*) One, *Paranthropus* (or *Australopithecus*) *robustus,* was the larger of the two; he was mainly herbivorous and seems not to have used many tools. The other was *Australopithecus africanus,* now called *Homo transvaalensis.* This species was omnivorous, and used crude stone tools with jagged cutting edges to kill and skin small animals.

Once tool use was established in these near-humans, strong natural selection could operate toward greater levels of creativity in

Figure 23-17   *The anthropoid apes (family Pongidae), the living animals most closely related to man. (a) Chimpanzee; (b) gorilla; (c) gibbon; and (d) orangutan.*

(a)

(b)

(c)

Figure 23-18   The races of man, according to one common system of classification. (a) White, or Caucasoid race; Basque male from northern Spain; (b) Mongoloid race, female from Canton, South China; (c) Negroid race, male from Senegal, West Africa.

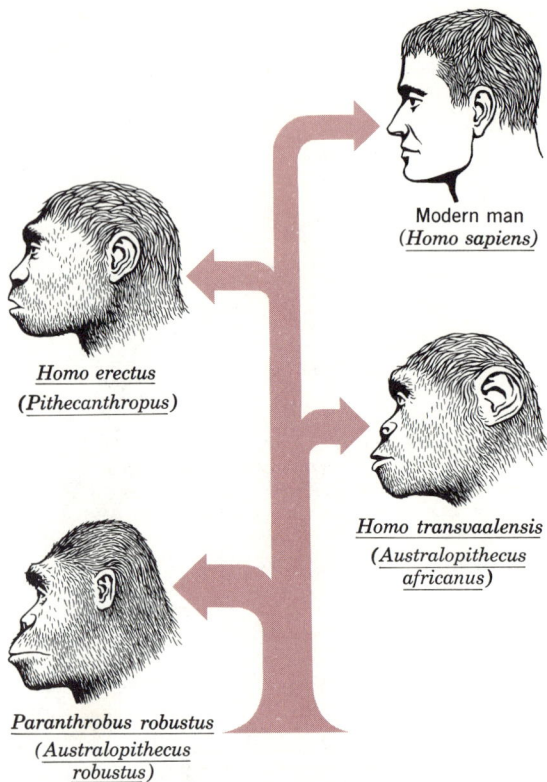

Modern man
(*Homo sapiens*)

Homo erectus
(*Pithecanthropus*)

Homo transvaalensis
(*Australopithecus
africanus*)

Paranthrobus robustus
(*Australopithecus
robustus*)

Figure 23-19 (*above*) *Significant stages in the evolution of Man.*

light on the more recent evolution of man himself. The oldest of these are the skulls and other bone fragments found first in Java in 1891, from what was originally called **Java Man** or ***Pithecanthropus*** (Fig. 23-20a), and in 1929 onward in caves near Peking, China, originally called **Peking Man** or ***Sinanthropus.*** Both date from the Pleistocene Epoch, 500,000 or more years ago, and are sufficiently similar to warrant their recent reclassification as simply different races of a single species with the name ***Homo erectus.***

The brain size in this early man ranged from about 750 to 1200 cm^3 (as compared to the average 1350 cm^3 of modern man), and he possessed thick skull bones, prominent ridges at the eyebrows, and protruding jaws (smaller than those of *Australopithecus*) and teeth. He walked erect, probably attained an average height of 5 ft, inhabited caves, hunted wild animals, used fire, rough stone tools, and weapons, and is thought to have been cannibalistic.

manipulation and design of tools and toward development of larger brains to expedite their control and invention. Probably correlated with tool manipulation and the increase in brain size was an increase in association centers (Chapter 17) in the brain. Very likely, even in the small brain of *Homo transvaalensis,* the amount of brain tissue devoted to sensory and motor activities was different from that in the apes; probably more tissue was already involved in association centers. The growth of association centers made possible an increasing ability to symbolize and generalize, and made speech possible.

Homo Erectus

Several types of early human fossils that still possess certain apelike features have been found and studied; these shed some

(*a*)

Neanderthal Man

Relatively numerous and more recent fossil remains have been found in many locations in Europe, western Asia, and North Africa of a race of man called **Neanderthal man** (Fig. 23-20b). This was a powerfully built race, averaging about 5 ft in height, with short legs and moderately long arms. The brain size was fully as large as that of modern man, although the structures of the skull and face, with thick bones, brow ridges, protruding jaw, and receding forehead and chin, were somewhat similar to but less extreme than those of Java Man and Peking Man. *H. neanderthalensis,* or *H. sapiens neanderthalensis,* inhabited Europe, western Asia, and North Africa, and probably became extinct some 50,000 to 75,000 years ago, being replaced by more contemporary forms. It is generally agreed, however, that Neanderthal man was a subspecies of *Homo sapiens.*

Cro-Magnon Man

The earliest known fossils of modern man—*Homo sapiens*—who were clearly like ourselves, were first found in a cave in Cro-

Figure 23-20 Early man. Reconstructed heads of: (a) (opposite) Homo erectus (Pithecanthropus); (b) Homo sapiens neanderthalensis (Neanderthal man); (c) Homo sapiens— Cro-Magnon man

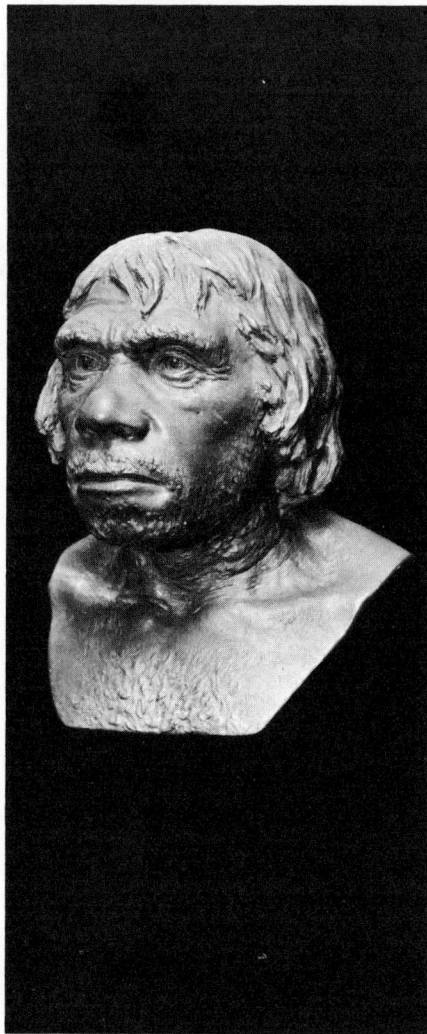

(b)

(c)

Magnon, France. They are representative of a race of modern man, now called **Cro-Magnon Man** (Fig. 23-20c), that dates back approximately 25 to 50 thousand years. They were tall and erect, attaining heights of 6 ft or more. They were also intelligent, possessing a brain size that was the same as that of present-day man, and were advanced culturally. Their caves in southern Europe are decorated with an artistic display of painting and carving of the men and animals of the time. It is possible that Cro-Magnon man originated in western Asia and migrated to southern Europe, replacing Neanderthal man in the process, in part by interbreeding.

Modern Man

The intermediate stages in the evolutionary progression of modern man from his apelike ancestors are not entirely clear. Fossil remains of other primitive men have been found in addition to the principal ones described. Together they indicate one indisputable fact: man did not evolve through an unbranched, straight-line sequence of evolutionary events.

Modern Europeans are probably descended from a mixture of different races of *Homo sapiens,* including certain Asiatic races and Cro-Magnon man. The first humans to reach North America are believed to have arrived from Asia during the late Pleistocene Epoch (via a then-existing narrow land connection at Bering Strait). With the passage of time, they moved southward, and by 25,000 B.C. were apparently well established in both North and South America.

The most significant developments in the evolution of man have been in those parts of the brain that involve intelligence, for it is man's intelligence that has been the most instrumental factor in his phenomenal success in controlling and dealing with his living and nonliving environment. It has endowed man with adaptive powers that permit him to live and thrive in virtually every naturally occurring environment on our planet, and, perhaps in the not-too-distant future, on other planets.

Man is the culmination in the animal kingdom, just as flowering plants are the culmination in the plant kingdom, of one of the most advanced and complex states in the incredibly long sequence of the evolutionary organization of energy and matter—a chain of events that started with the beginning of the expanding universe and evolved step by step from nonliving to living systems and thence to contemporary living organisms.

Reading List

Binford, S. R., and L. R. Binford, "Stone Tools and Human Behavior," *Scientific American* (April 1969), pp. 70–84.

Britten, R. J., and D. E. Kohne, "Repeated Segments of DNA," *Scientific American* (April 1970), pp. 24–31.

Cavalli-Sforza, L. L., "'Genetic Drift' in an Italian Population," *Scientific American* (August 1969), pp. 30–37.

Dayhoff, M. O., "Computer Analysis of Protein Evolution," *Scientific American* (July 1969), pp. 86–95.

deLumley, H., "Paleolithic Camp at Nice," *Scientific American* (May 1969), pp. 42–50.

Eckhardt, R. B., "Population Genetics and Human Origins," *Scientific American* (January 1972), pp. 94–102.

Hardy, Sir A., *The Living Stream: Evolution and Man.* World Publishing, Cleveland, 1968.

King, J. L., and T. H. Jukes, "Non-Darwinian Evolution," *Science* (1969), **164,** pp. 788–798.

Renfrew, C., "Carbon[14] and the Prehistory of Europe," *Scientific American* (October 1971), pp. 63–72.

Savage, J. M., *Evolution* (2nd ed.). Holt, Rhinehart & Winston, New York, 1969.

Shimer, J. A., *This Changing Earth: An Introduction to Geology.* Barnes & Noble, New York, 1968.

Sigurbjornsson, B., "Induced Mutations in Plants," *Scientific American* (January 1971), pp. 86–95.

Simpson, G. G., *The Meaning of Evolution* (rev. ed.). Yale University Press, New Haven, 1967.

chapter twenty-four | *The Basis of Classification*

"Understanding the world of organisms that surround us requires the ordering of the multitude of forms into some sort of rational system. Consequently systematics is as old as man's quest for knowledge, and the effort to classify and understand the variability of animals and plants has led to the development of all the other branches of biology. The information obtained by these more specialized approaches has in turn affected and profoundly modified the field of taxonomy, transforming it from a mere effort to classify into the scientific enterprise of discovering and understanding the reasons for the apparent order of nature.

OTTO T. SOLBRIG, 1966, *EVOLUTION AND SYSTEMATICS*

Artificial and Natural Systems

The earth today is inhabited by an immense number of living things, and no two individuals are identical. It is an important goal in biology that there be a systematic, orderly classification or cataloguing of this vast and changing quantity of biological material, just as a library requires that its many books be systematized and catalogued. The branch of biology that deals with classification is called **taxonomy,** or **systematics.**

Numerous and diverse systems of classification of living forms have been devised by man in the course of history. They have generally fallen into two main groups, artificial and natural. An **artificial** system of classification is based on arbitrary or artificial standards (e.g., an alphabetical listing); there is no recognition of relationships among the different kinds of organism in the sense of common or related descent. It therefore merely serves as a convenient filing system.

By contrast, modern taxonomists use a **natural** system of classification which is founded on the natural or evolutionary relationships among organisms; therefore, in addition to being convenient, it reflects as much as possible the probable evolution of these organisms. The basic premise is that the groups of organisms that possess the greatest number of characteristics in common are most closely related, and those that share the fewest features are more distantly related. An example of a natural classification system that attempts

689

to reflect these considerations is shown in Appendix 2.

There is ample evidence to indicate that one of the main trends in the sequence of evolution has been of the progressive type, from organisms relatively simple in structure to those of great complexity. Nevertheless, exceptions are known to exist for which the evolutionary trend has been retrogressive, from complexity of one or more features toward greater simplicity, or **reduction.** Obviously, then, extreme caution must be exercised in deciding whether a particular characteristic is simple because it is actually primitive or whether it has become simple by evolutionary reduction.

A second major hazard in interpreting evolutionary relationships is seen in the independent evolutionary acquisition of similar (although not necessarily identical) structures in organisms that are not closely related. This phenomenon is known as **convergent evolution** (Chapter 23). Such organisms may show resemblances that can easily be misinterpreted as evidences of relationship, although they have not risen from the same ancestral stock. Bees, birds, and bats all have functional wings but are not closely related.

The present status of natural classification systems is still unsettled. Many organisms can be readily and conveniently classified as either **animals** or **plants,** based primarily on the absence or presence of a cell wall and to a lesser extent on other characteristics. There are several groups (e.g., numerous microorganisms), however, whose relationships are not at all clear and therefore are open to diverse interpretations. Some taxonomists prefer to classify most unicellular organisms into a group called **Protista,** separate from either plants or animals; others file them in the plant or animal kingdoms according to selected characteristics.

The use of the electron microscope has provided a clearer picture of the cell structure of microorganisms and as a result it has become increasingly common in recent years to separate microorganisms into two groups depending on the nature of their cells. **Prokaryotic organisms,** lacking cell organelles and a nuclear membrane (e.g. bacteria and blue-green algae), are placed in a group called **Monera. Eukaryotic** microorganisms, with cells containing organelles such as mitochondria and chloroplasts and an organized nucleus with a nuclear membrane or envelope surrounding it, are placed in the **Protista,** or are divided between the plants and animals.

Approximately 1.3 million different kinds of living organism have been described. Many more undoubtedly exist. Taxonomy, like any other field of biology, is subject to constant change. We can expect that with the discovery of new information and new relationships classification systems will accordingly undergo revisions and new interpretations.

The Species as the Unit of Classification

Many billions of individual organisms exist on our planet, but we can immediately recognize that certain of these individual organisms closely resemble one another. We can proceed on the basis of their similarity to assemble these individuals into separate species. Nevertheless, any two individuals in a species are usually distinguishable from one another in at least several different ways.

The **species** is the basic unit employed in the classification of living forms. The species consists of a population of individuals or a group of populations closely related in ancestry in which effective exchange of genetic material (gene flow) can occur. Accordingly, its members display the same structural and functional characteristics and are capable of interbreeding with one another to produce fertile offspring.

Effective gene flow (i.e., the ability to mate and produce fertile offspring), is probably the most basic single criterion for deciding that individuals belong to the same species. It implies a close resemblance in genetic constitution and evolutionary history. The horse and the donkey, for example, may be mated to produce an offspring, the mule, which is nearly always infertile. For this reason the horse and donkey are often classified as different but closely related species. In nature interbreeding between members of different but closely related species is a rare occurrence. Species, therefore, exchange genes with other species rarely or not at all. In other words, different species are genetically separated, or **reproductively isolated,** from one another and are referred to as **genetically closed systems.**

We mentioned in Chapter 23 that a species is often divided into subunits such as races, varieties, or subspecies. Races consist of particular populations in a species that differ to a small extent in genetic composition. Several races may be included in a species. Such races are **genetically open** because they can and often do exchange genes by interbreeding, leading at times to a fusion of several races into a single population.

But the criterion of infertility for distinguishing one species from another is not entirely satisfactory, especially in those lower organisms in which reproduction occurs only by asexual means; the taxonomist must obviously rely on criteria other than ability to mate in determining classification. The same is true of classifying fossils. By necessity, these are assigned taxonomically on the basis of their morphological characteristics alone.

Nomenclature

The present-day system of nomenclature used in classifying living things was introduced some two centuries ago, principally by the Swedish physician and biologist Carolus Linnaeus, who is now called the father of taxonomy. He grouped similar organisms together into species, and combined similar groups of species, which he presumed to be more closely related to one another than to other species, into a higher unit of classification, the **genus.** He assigned to all known organisms two Latin names, one representing the genus and the other the species. For this reason his system is known as the **binomial system of nomenclature;** it is used today throughout the world by international agreement among biologists. All living human beings, for example, are members of the genus *Homo* and the species *sapiens* and are accordingly designated by the scientific name, *Homo sapiens.* (Note that the **generic name** always starts with a capital letter, whereas the **specific name** always starts with a lower-case letter; both are always italicized.) The obvious feline characteristics of such animals as the lion, puma, tiger, jaguar, and the common cat indicate a close relationship. All are sufficiently similar to belong to the same genus (*Felis*) but are sufficiently different to be classified in separate species. They have been assigned the scientific names *Felis leo, Felis concolor, Felis tigris, Felis onca,* and *Felis domestica,* respectively (Fig. 24-1, p. 692–693).

The generic name is a Latin noun and the specific name is an adjective. The specific name usually describes some characteristic of the organism that the original namer believed to be typical or unique. Occasionally the specific name is derived from the name of a person who first described the organism or is being honored for one reason or another.

The use of an internationally agreed-on scientific name for each kind of organism helps to avoid confusion and uncertainty in identifying and recording different kinds of organisms. The use of the common name of an organism, although convenient, has a distinct disadvantage. A given organism, especially if it is widely distributed, often has many common names that are strictly local in their use. The European white water lily, *Nymphaea odorata,* for example, is called by some 200 different common local names. Conversely, the name mahogany is applied to many kinds of tree.

Linnaeus assembled similar genera into larger groups called **orders,** which are further arranged into still larger related groups called **classes.** The system of units is comparable to that employed in our political organization of a nation: a town or city is the elementary unit, followed successively by progressively larger units such as county, state, and nation.

(a)

(b)

Figure 24-1 Some animals belonging to the same genus (Felis) but sufficiently different to be classified in separate species: (a) common cat; (b) tiger; (c); lion; (d) jaguar.

692 Evolution and Classification

(c)

(d)

693 *The Basis of Classification*

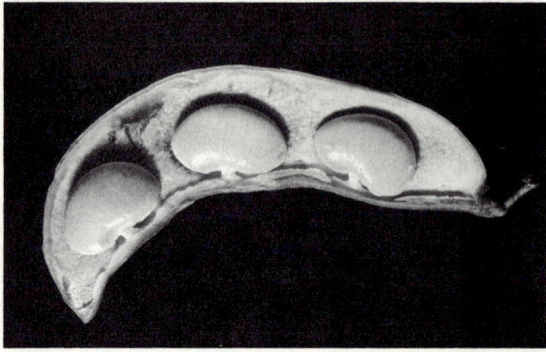

Figure 24-2   Lima beans in open pod, or legume, characteristic of the plant family Leguminosae.

Many of the names used for the higher classification units are often based on some outstanding characteristic of the group. For example, the term **Leguminosa** refers to the family of plants whose members bear their seeds in a pod called a **legume** (Fig. 24-2).

The original categories of Linnaeus have since been enlarged in view of the vast amount of new information and newly discovered organisms. The terms **family,** which falls between the genus and the order, and **phylum** (plural, phyla), which is the largest category of all, have subsequently been added. In addition, subdivisions such as **subphylum, subclass, suborder,** and **subfamily** have been included. The sequence of categories, in descending order of inclusiveness, is phylum, class, order, family, genus, species. The differences between individuals and groups become progressively less in the descending units.

On the basis of our modern interpretation of evolutionary relationships, a genus ideally designates a group of like or related species. Similarly, the family consists of a group of related genera, the order of related families, the class of related orders, and the phylum of related classes. In place of the term phylum the equivalent term **division** has been traditionally employed by botanists for the major groups within the plant kingdom. For simplicity, however, the term phylum is used in this book to designate the major groups of both plants and animals.

The modern classification of modern man according to this system is as follows:

Phylum—Chordata
  Subphylum—Vertebrata
    Class—Mammalia
      Subclass—Placentalia
        Order—Primates
          Suborder—Anthropoidea
            Family—Hominidae
              Genus—*Homo*
                Species—*sapiens*

The corresponding classification for the white oak tree is:

Division, or **Phylum**—Tracheophyta
  Subphylum—Pteropsida
    Class—Angiospermae
      Subclass—Dicotyledoneae
        Order—Fagales
          Family—Fagaceae
            Genus—*Quercus*
              Species—*alba*

Criteria for Classification

Anatomical and physiological characteristics are the most commonly used criteria in contemporary classification of living things, although biochemical features are becoming of increasing importance, especially for lower forms of life. Decisions as to which are the most important characteristics and what interpretations we can draw with regard to evolutionary relationships, and therefore classifications, can at times prove to be difficult and controversial. Needless to say, any valid judgement as to the relationship among several species of organisms is more reliable when based on several features rather than on a single characteristic alone.

Homology versus Analogy

Those structures of different organisms that correspond to one another in terms of inheritance and therefore in basic development and form are said to be **homologous.** Homologous parts in different organisms do not necessarily perform the same function. For example, the self-evident similarity in basic anatomy of the arm of man, the wing of a bird, and the foreleg of a frog (Fig. 24-3) strongly indicates that these structures, despite their differences in function, have a common inheritance and descent and are therefore homologous.

By contrast, structures of different organisms that possess a similar function but are basically different in origin and structure are said to be **analogous.** We used a classic example of analogy earlier (Fig. 23-4) in comparing the wings of a bird and a butterfly to illustrate convergent evolution. Although these structures are similar in function, they are entirely different in structure and in origin. The bird's feather-covered wings are supported by an internal skeletal framework; the wing of a butterfly essentially is made up of a stiffened membrane.

Obviously, homology, in contrast to analogy, indicates an evolutionary relationship by implying descent from the same ancestral line. It thus serves as an important basis in a natural classification of living things. The cataloguing of organisms on the basis of analogous structures would inevitably lead to an artificial system of classification and an incorrect interpretation of evolutionary relationships.

Criteria for Classification of Animals

Several specific criteria have been employed in establishing a natural system of classification for animals. The fact that different animals are superficially alike in appearance and share the same living conditions does not necessarily mean that they are closely related. For example, whales and fish live and swim in the sea and resemble one another in form. Closer examination soon reveals that whales have no gills but breathe by lungs, nourish their young on milk, and have all the essential structural and func-

Figure 24-3   Homologous bones in the forelimbs of several widely different vertebrates, illustrating basic similarity in form.

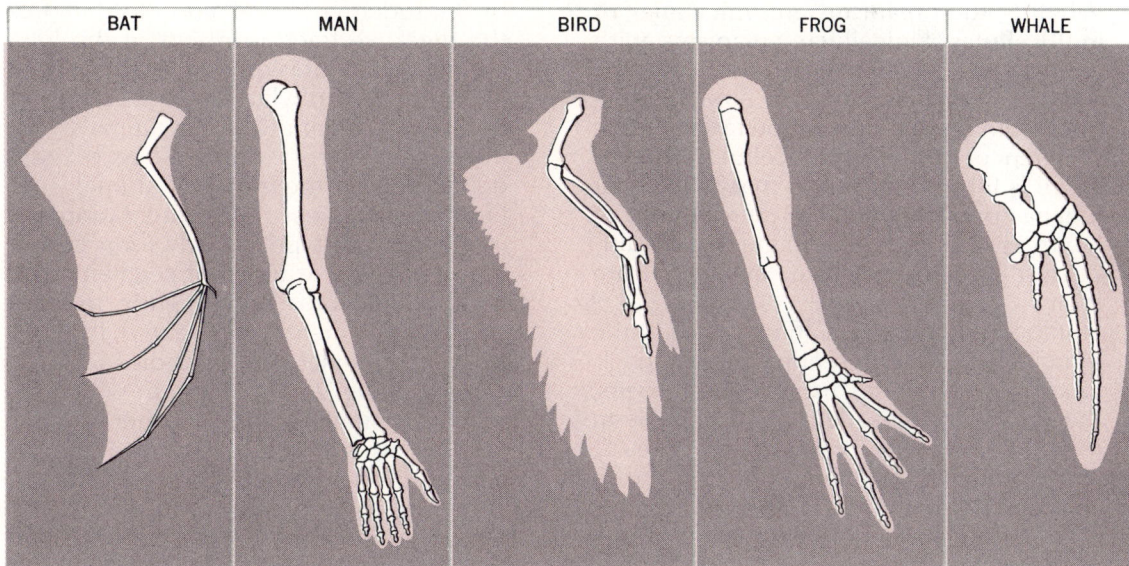

BAT	MAN	BIRD	FROG	WHALE

tional characteristics that unquestionably identify them as mammals. The indications are that at some time in the ancient past they evolved from a small group of terrestrial mammals that returned to a watery environment. Through the process of natural selection they have adapted to the way of life in water, as indicated by the presence of fins and the fishlike, streamlined shape of the body, distinctly advantageous modifications for survival in aquatic surroundings. But the basic body structures have not been significantly changed. The presence of mammary glands, the lungs, the typical four-chambered heart, and all the other characteristics (except an almost complete absence of hair) that mark them as true mammals are easily recognizable.

In large measure the major animal phyla have been distinguished from one another on the basis of several fundamental anatomical characteristics. Together these different anatomical features and their accompanying processes show a typical design of organization and function for each particular phylum; these designs fit into an over-all, cohesive evolutionary pattern. The more important of these distinguishing features are the following.

1. Whether the organisms are **unicellular or multicellular.** Unicellular organisms with animal cell characteristics are usually classified as a separate kingdom, subkingdom, or phylum (depending on the taxonomist's preference) from the multicellular animals. There is little or no doubt that unicellular organisms were among the first forms of life to appear on the earth and that multicellular organisms have evolved from them.
2. Number of **germ layers.** In nearly all of the principal phyla representing the multicellular animals, three distinct germ layers, the ectoderm, mesoderm, and endoderm (pp. 287-291) are present during early embryonic development. They give rise subsequently to all the tissues and organs of the fully formed individual. In some primitive animals only two germ layers,

ectoderm and endoderm, are present (Fig. 24-4).

3. Course and degree of development of certain **organ systems.** This applies particularly to the digestive, circulatory, and nervous systems, and to a lesser extent to the skeletal system. In several of the lower animal phyla digestion is accomplished in a central cavity (and in part in the cells lining the surface of this cavity), which connects to the outside by a single opening (Fig. 24-4). All other principal animal phyla possess a more complex digestive tube with an opening at each end (one for taking in food and the other for eliminating undigested residues), with varying degrees of specialization. Similarly, the circulatory and nervous systems display progressive degrees of structural and functional specialization, including muscular pumping vessels, or "hearts," and nerve cords, as well as anteriorly located masses of ganglia, or "brains." The skeletal system (including its origin in embryonic development, its orientation—whether external or internal—and its composition) varies among phyla and serves to a certain extent as a classification criterion.
4. Presence or absence of a **coelom.** A coelom, or true body cavity, surrounded by mesodermal tissue and distinct from the digestive tract, is absent in the lower animal phyla but is present in all the major higher animal phyla (see Fig. 12-15). The body cavity in multicellular animals varies in its characteristics and mode of origin depending on the group of animal.
5. Presence or absence of **segmentation.** Segmentation is a type of body form consisting of successive units, or segments each with a similar or only slightly modified structure. Segmentation occurs in three major animal phyla; it is most distinct in the Annelida (Fig. 24-5, p. 698), in which segments are similar to one another. Segmentation is also present but somewhat modified in the Arthropoda and the Chordata.
6. Type of **symmetry.** The proportionality or organization of most living things is such that they are symmetrical. That is, theoret-

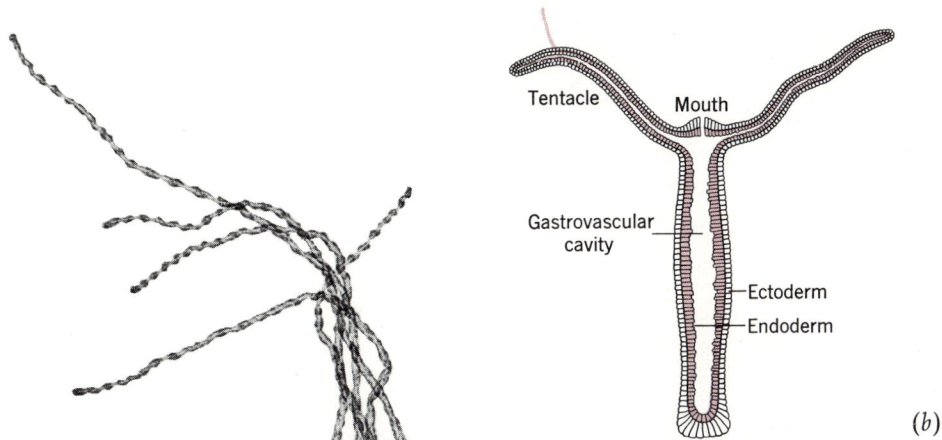

Figure 24-4 Hydra, a representative of the phylum Coelenterata. This tiny organism that lives in ponds is seldom more than 1 cm in length. (a) Intact organism; (b) Diagrammatic longitudinal section showing the two germ layers, ectoderm and endoderm, and a central cavity with a single opening (Courtesy Carolina Biological Supply Company); (c) cross-section.

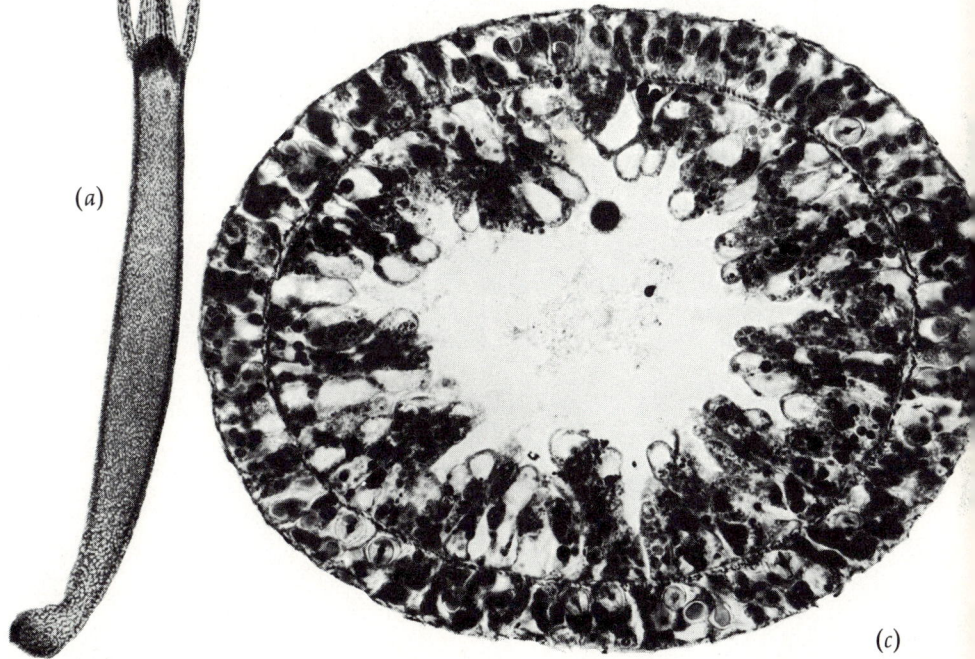

ically, they can be cut into two equal or equivalent halves, at least as far as external appearance is concerned. Most organisms display either radial or bilateral symmetry. In **radial symmetry** the body is so arranged that it possesses a hypothetical central axis, like the hub of a bicycle wheel, and the body parts are organized about this axis in a radiating pattern, like the spokes of the wheel. This form can be cut in any vertical plane through the axis to yield two equal or equivalent halves. The starfish is a classic example of radial symmetry in living things (Fig. 24-6a, p. 298).

In **bilateral symmetry** the presence of a distinguishable upper, or dorsal, surface and a lower, or ventral, surface as well as an anterior and posterior region restricts

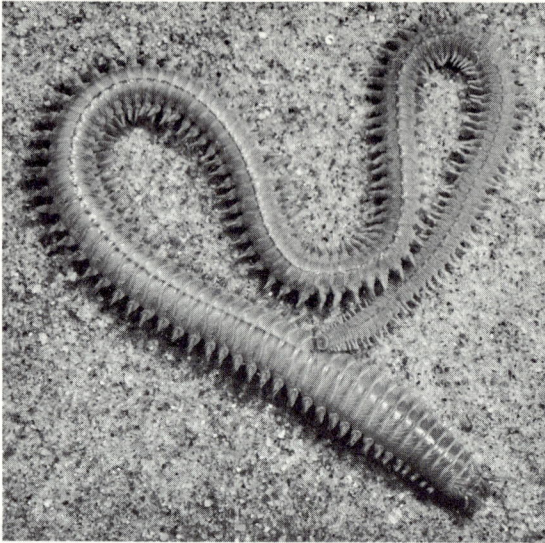

Figure 24-5   Nereis, a marine representative of the phylum Annelida. Notice that its body is made of many segments, each bearing bristled, paddle-like appendages used in locomotion.

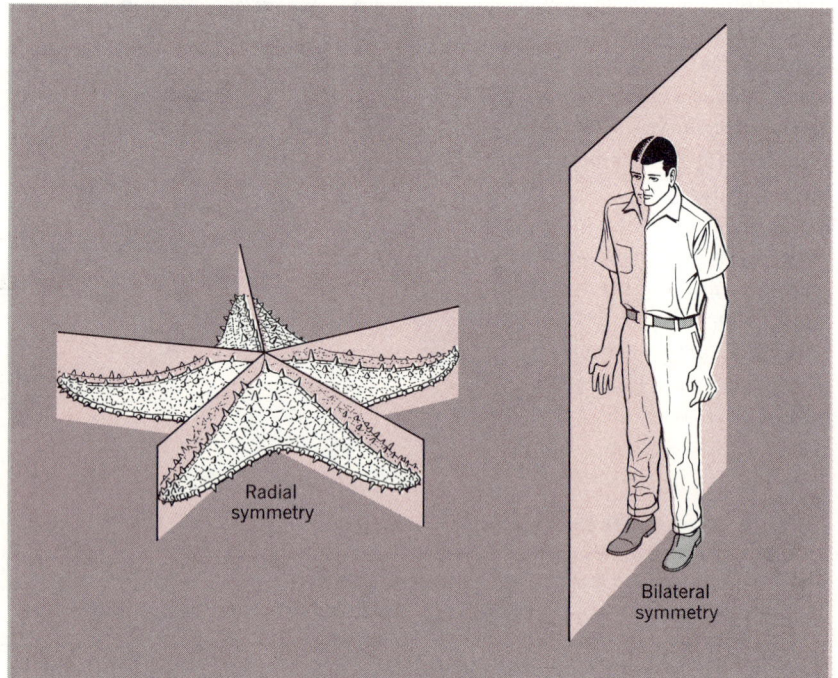

Radial symmetry

Bilateral symmetry

Figure 24-6   Diagram illustrating two principal types of symmetry in organisms: (a) radial symmetry; (b) bilateral symmetry.'

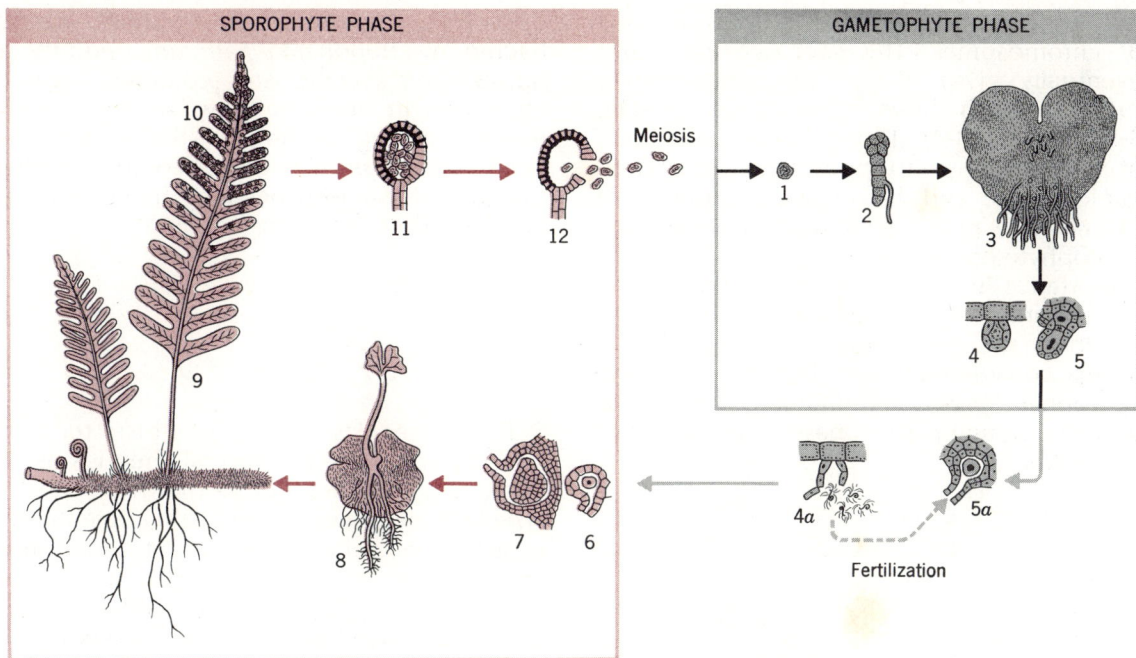

Figure 24-7 *The life cycle of a fern. The sporophyte, with its familiar fronds, or leaves, dominates the life cycle. The gametophyte, which is inconspicuous and short-lived, bears sex organs. (1) Spore resulting from meiosis. (2) Young gametophyte. (3) Mature gametophyte. (4) Male sex organ, liberating sperm (4a). (5) Egg cell. (6) Diploid zygote. (7) Embryo sporophyte. (8) Immature sporophyte still attached to gametophyte. (9) Mature sporophyte. (10) Sporangia on underside of sporophyte leaf. (11) Individual sporangium. (12) Sporangium releasing haploid spores formed by meiosis.*

the hypothetical division of the body into a single plane. The two equivalent halves are approximately mirror images of each other (Fig. 24-6b). In the human body this special plane is represented by a theoretical perpendicular midline, or midsagittal plane, of the body.

Criteria for Classification of Plants

Although specific criteria used in classifying plants are different from those used in the animal kingdom, the same principles apply. Every attempt is made to construct a classification system that is as natural as possible in reflecting evolutionary relationships. Within the plant kingdom itself the kinds of criteria used in classification differ; for example, with algae the types of chlorophyll and other pigments present in the cells are important, whereas with angiosperms taxonomists attach much importance to the characteristics of the flower.

Plants commonly show an alternation of generations (Fig. 24-7) between a gametophyte generation made up of individuals with a single set of chromosomes, which produces gametes, and a sporophyte generation of the same species, which arises with the fusion of two gametes and therefore contains two sets of chromosomes. The sporophyte in turn eventually gives rise to the gametophyte (p. 346) with the intervention of meiosis (Fig. 8-5 and Fig. 24-7). The nature of this cycle from gametophyte to sporophyte and back to gametophyte, and the relative importance or prominence of the gametophyte versus the sporophyte portion of the cycle, are of much importance in classifying plants. Among the lower forms of plant life either form may predominate, depending on the species or group considered. With higher plants the sporophyte always dominates the life cycle and the gametophyte phase is often barely discernible.

Classifying Prokaryotic Organisms

Prokaryotes (bacteria and blue green algae) are so fundamentally different from other forms of life in the nature of their cell structure that they are commonly placed in a separate kingdom, the Monera. Because of the relative simplicity of these organisms taxonomists can rely to only a very limited extent on structural characteristics, and even these are best observed with an electron microscope. Instead, their classification is based largely on physiological and biochemical characteristics. The bacterium *Ferrobacillus ferrooxidans,* for instance, obtains its needed energy by oxidizing iron rather than organic compounds, and obtains its carbon compound building blocks by using this energy to reduce CO_2. By contrast, the syphilis-causing bacterium, *Treponema pallidum,* requires a number of specific and complex organic chemicals in order to grow properly. With disease-causing pathogens the conditions necessary for growth are often so complex that growth can only occur in the tissues and cells of other organisms.

Future Problems in Taxonomy

The proper classification of viruses remains an enigma. There are several theories as to the origin of viruses, none of which can be conclusively documented at this time. Some experts consider them to be retrograde microorganisms that have gradually lost metabolic capability, or primitive cellular constituents that have been modified by evolution so that they can survive in the cells of other organisms. Others feel that their simplicity indicates that viruses represent the earliest forms of life. Many biologists do not consider viruses to be living organisms at all because, under normal conditions, they must reproduce in living cells. Until further information is available viruses must be given a very tentative assignment in a natural classification system.

The classification system in Appendix 2 is only one of a number of such systems that have been proposed by modern biologists. As the biological sciences continue to grow and more knowledge is added, present schemes will have to be modified appropriately.

Reading List

Allee, W. C., *Animal Aggregations.* University of Chicago Press, Chicago, 1931.

Dobzhansky, T., *Genetics and the Origin of Species.* Columbia University Press, New York, 1951.

Lawrence, G. H. M., *Taxonomy of Vascular Plants.* Macmillan, New York, 1951.

Mayr, E., *Animal Species and Evolution.* Harvard University Press, Cambridge, Mass., 1963.

Mayr, E., E. G. Lindsay, and R. L. Unsinger, *Methods and Principles of Systematic Zoology.* McGraw-Hill, New York, 1953.

Scagel, R. F., R. J. Bandoni, G. E. Rouse, W. B. Schofield, J. R. Stein, and T. M. C. Taylor, *An Evolutionary Survey of the Plant Kingdom.* Wadsworth, Belmont, Calif., 1965.

Solbrig, O. T., *Evolution and Systematics.* Macmillan, New York, 1966.

Stanier, R. Y., M. Doudorff, and E. A. Adelberg, *The Microbial World* (3rd ed.), Prentice-Hall, Englewood Cliffs, N.J., 1970.

Stebbins, G. L., *Variation and Evolution in Plants.* Columbia University Press, New York, 1950.

Exponential Notation

Throughout this book **exponential notation** is used to deal with very large and very small numbers. An **exponent** is a symbol written to the right and above a number or mathematical expression; it indicates the number of times that figure is to be multiplied by itself. An exponent is also called the **power** of a number. For numbers greater than one the exponent is positive and is equal to the number of zeros following the one.

$$10 = 10^1 = \text{ten}$$
$$100 = 10^2 = \text{one hundred}$$
$$1000 = 10^3 = \text{one thousand}$$
$$10{,}00 = 10^4 = \text{ten thousand}$$
$$1{,}000{,}000 = 10^6 = \text{one million}$$
$$1{,}000{,}000{,}000 = 10^9 = \text{one billion}$$

For numbers smaller than one the exponent is negative and is equal to the number of places the one is to the right of the decimal point.

$$.1 = 10^{-1} = 1 \text{ tenth}$$
$$.01 = 10^{-2} = 1 \text{ hundredth}$$
$$.001 = 10^{-3} = 1 \text{ thousandth}$$
$$.0001 = 10^{-4} = 1 \text{ ten-thousandth}$$
$$.000{,}001 = 10^{-6} = 1 \text{ millionth}$$
$$.000{,}000{,}001 = 10^{-9} = 1 \text{ billionth}$$

Any large or small number may be expressed as the product of a more convenient sized number and a power of 10. Thus 467 may be expressed as $4.67 \times 100 = 4.67 \times 10^2$. Similarly, $467,000,000 = 467 \times 10^6$, or 4.67×10^8.

In decimals smaller than one the decimal point is moved to the right, and the resulting number is multiplied by a negative power of 10 equal to the number of places the decimal point was moved: $.0467 = 4.67 \times 10^{-2}$ or 46.7×10^{-3}; $.0000467 = 4.67 \times 10^{-5}$ or 46.7×10^{-6}.

The Metric System of Weights and Measures

In this book we have generally used the metric system of measurement, instead of the more familiar British system of pounds, feet, and inches, because the metric system is used internationally in the sciences.

Length

Basic unit is the **meter (m),** which equals 39.37 in., or 3.28 ft. Common multiples and subdivisions are:

kilometer (km) $= 10^3$ m $= .62$ miles
decimeter (dm) $= 10^{-1}$ m $= 3.94$ in.
centimeter (cm) $= 10^{-2}$ m $= .39$ in.
millimeter (mm) $= 10^{-3}$ m $= .039$ in.
micrometer (μm) or micron (μ) $= 10^{-6}$ m $= 10^{-3}$ mm
nanometer (nm) $= 10^{-9}$ m $= 10^{-3}$ μ
angstrom (Å) $= 10^{-10}$ m $= 10^{-8}$ cm $= 10^{-1}$ nm

Mass

Basic unit is the **gram (g),** which equals .035 oz. At standard temperature and pressure, 1 g of water occupies a volume of 1 ml. Common multiples and subdivisions are:

kilogram (kg) $= 10^3$ gm $= 2.2$ lb
centigram (cg) $= 10^{-2}$ gm
milligram (mg) $= 10^{-3}$ gm
microgram (μg) $= 10^{-6}$ gm

Volume

Basic unit is the **liter (L),** which equals 1.06 qt. At standard temperature and pressure, a liter of pure water weighs 1 kg. Therefore 1 milliliter (1 ml $= 10^{-3}$L) weighs 1 gram, and has a volume of one cubic centimeter (cm^3 or cc); thus 1 ml $= 1$ cm^3, and the units are used interchangeably.

Temperature

Basic unit is the **Celsius degree, °C** (formerly known as **centigrade**). 0°C is the freezing point of water; 100°C is the boiling point of water. To convert from °C to **°F (Fahrenheit),** or vice versa: $°C = \frac{5}{9} (°F - 32)$; $°F = \frac{9}{5} °C + 32$.

Physicists measure temperature on the **Kelvin scale, °K.** 0°K $= -273°C = $ **absolute 0,** defined as the temperature at which molecular motion ceases.

Useful Equivalents	1 in. = 2.54 cm 1 ft = 30.48 cm 1 yd = .91 m 1 U. S. fluid oz = 29.57 ml 1 U. S. liquid qt. = .946 L

Common units of Energy

dyne (dyn) = force that produces an acceleration of 1 cm/sec^2 on a mass of 1 g

erg = work performed by 1 dyn of force moving an object 1 cm

Mechanical Energy

joule (J) = 10^7 erg

Heat Energy

cal (cal) = amount of heat required to raise the temperature of 1 g of water from 15.5°C to 16.5°C.

Electrical Energy

coulomb (C) = unit of electrical charge; the charge of 6.25 × 10^{18} electrons.

ampere (A) = unit of electric current; flow of one C/sec past a given point.

volt (V) = unit of electric potential energy; the work needed to cause a current of 1 A to flow against 1 **ohm (Ω)** of resistence.

Energy Conversions

Mechanical equivalent of heat:
   1 cal = 4.18 × 10^7 erg of work.
Electrical equivalent of heat:
   1 C passing between 1 V potential difference = 10^7 erg of work.
   1 **electron-volt (eV)** = 23,000 cal.
Chemical equivalent of electricity:
   1 **Farad** (1F = 96,500 coulombs) is that amount of charge capable of ionizing 1 mole of any compound.

Atomic Weights and Numbers[a]
Elements Essential for Protoplasm[b]

ELEMENT	SYMBOL	ATOMIC[c] NUMBER	ATOMIC[d] WEIGHT	PERCENTAGE OF HUMAN BODY WEIGHT
Actinium	Ac	89	(227)[e]	
Aluminum	Al	13	26.98	
Americium	Am	95	(243)	
Antimony	Sb	51	121.75	
Argon	Ar	18	39.948	
Arsenic	As	33	74.92	
Astatine	At	85	(210)	
Barium	Ba	56	137.34	
Berkelium	Bk	97	(247)	
Beryllium	Be	4	9.01	
Bismuth	Bi	83	208.98	
Boron	B	5	10.81	trace
Bromine	Br	35	79.90	
Cadmium	Cd	48	112.40	
Calcium	Ca	20	40.08	1.5
Californium	Cf	98	(249)	
Carbon	C	6	12.01	18.5
Cerium	Ce	58	140.12	
Cesium	Cs	55	132.90	
Chlorine	Cl	17	35.45	.2
Chromium	Cr	24	51.99	
Cobalt	Co	27	58.93	trace
Copper	Cu	29	63.55	trace
Curium	Cm	96	(247)	

ELEMENT	SYMBOL	ATOMIC NUMBER	ATOMIC WEIGHT	PERCENTAGE OF HUMAN BODY WEIGHT
Mercury	Hg	80	200.59	
Molybdenum	Mo	42	95.94	trace
Neodymium	Nd	60	144.24	
Neon	Ne	10	20.18	
Neptunium	Np	93	(237)	
Nickel	Ni	28	58.71	
Niobium	Nb	41	92.906	
Nitrogen	N	7	14.01	3.3
Nobelium	No	102	(254?)	
Osmium	Os	76	190.2	
Oxygen	O	8	15.99	65.0
Palladium	Pd	46	106.4	
Phosphorus	P	15	30.97	1.0
Platinum	Pt	78	195.09	
Plutonium	Pu	94	(242)	
Polonium	Po	84	(210)	
Potassium	K	19	39.10	.4
Praseodymium	Pr	59	140.90	
Promethium	Pm	61	(147)	
Protactinium	Pa	91	(231)	
Radium	Ra	88	(226)	
Radon	Rn	86	(222)	
Rhenium	Re	75	186.2	
Rhodium	Rh	45	102.90	

Element	Symbol	Atomic number	Atomic weight	%
Dysprosium	Dy	66	162.50	
Einsteinium	Es	99	(25)	
Erbium	Er	68	167.26	
Europium	Eu	63	151.96	
Fermium	Fm	100	(253)	
Fluorine	F	9	18.99	
Francium	Fr	87	(223)	
Gadolinium	Gd	64	157.25	
Gallium	Ga	31	69.72	
Germanium	Ge	32	72.59	
Gold	Au	79	196.96	
Hafnium	Hf	72	178.49	
Helium	He	2	4.00	
Holmium	Ho	67	164.93	
Hydrogen	H	1	1.01	9.5
Indium	In	49	114.82	
Iodine	I	53	126.90	trace
Iridium	Ir	77	192.2	
Iron	Fe	26	55.8	trace
Krypton	Kr	36	83.80	
Lanthanum	La	57	138.91	
Lawrencium	Lw	103	(257)	
Lead	Pb	82	207.19	
Lithium	Li	3	6.94	
Lutetium	Lu	71	174.97	
Magnesium	Mg	12	24.312	.1
Manganese	Mn	25	54.94	
Mendelevium	Md	101	(256)	trace
Rubidium	Rb	37	85.47	
Ruthenium	Ru	44	101.07	
Samarium	Sm	62	150.35	
Scandium	Sc	21	44.95	
Selenium	Se	34	78.96	trace
Silicon	Si	14	28.08	
Silver	Ag	47	107.86	
Sodium	Na	11	22.9	.2
Strontium	Sr	38	87.62	
Sulfur	S	16	32.06	.3
Tantalum	Ta	73	180.94	
Technetium	Tc	43	(97)	
Tellurium	Te	52	127.60	
Terbium	Tb	65	158.92	
Thallium	Tl	81	204.37	
Thorium	Th	90	232.04	
Thulium	Tm	69	168.93	
Tin	Sn	50	118.69	
Titanium	Ti	22	47.90	
Tungsten	W	74	183.85	
Uranium	U	92	238.03	
Vanadium	V	23	50.94	
Xenon	Xe	54	131.04	
Ytterbium	Yb	70	173.04	
Yttrium	Y	39	88.90	
Zinc	Zn	30	65.37	trace
Zirconium	Zr	40	91.22	

[a] Reference standard is carbon-12.

[b] Elements essential for protoplasm are printed in boldface.

[c] Atomic number equals the number of protons in an atom.

[d] Atomic weight is determined by the sum of the protons and neutrons that make up the nucleus of the atom.

[e] Atomic weights in parentheses are for the most stable or the best-known isotope for the element if it does not have a reasonably fixed isotopic composition.

Chemical Notation

Chemical compounds are named after their constituent atoms. By convention, the name of the positively charged atom appears first. Thus the compound formed when sodium and chloride atoms combine is termed sodium chloride. The shorthand notations, or **formulas,** for compounds indicate what atoms are present and in what ratio they occur, by a combination of the atomic symbols (as in the table above) and subscript numbers following those symbols. If no subscript follows an atomic symbol, it is understood to be 1. Thus the formula for sodium chloride is NaCl (not $Na_1 Cl_1$). Calcium atoms and chloride atoms combine in a ratio of 1 Ca to 2 Cl; they cannot combine in any other ratio. Therefore the formula for calcium chloride is

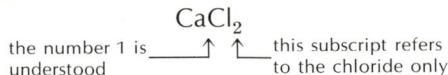

$$CaCl_2$$

the number 1 is ↑ ↑   this subscript refers
understood   to the chloride only

The smallest whole-number ratio is usually used for chemical formulas. The ratio of 1:2 in $CaCl_2$ or 2:1 in H_2O could also be expressed as Ca_2Cl_4 or $HO_{\frac{1}{2}}$ but this is not usually done except when balancing chemical equations.

A chemical **equation** is a shorthand statement about the way atoms or compounds react (or combine) with one another. An equation consists of the symbols or formulas of the reacting atoms and molecules; a set of numbers, called **coefficients,** that indicate the proportions of each atom or compound; one or more plus signs ($+$), which indicate the addition or mixing of one chemical with another; and at least one arrow ($-\longrightarrow$), which should be read as "yields" or "will produce." The chemical reaction $2H_2 + O_2 \longrightarrow 2H_2O$ can be read as "Two molecules of hydrogen (each consisting of two hydrogen atoms) will combine with one molecule of oxygen (which consists of two oxygen atoms) to yield two molecules of water." Note that the formula for the oxygen molecule (O_2) is not preceded by a coefficient; the coefficient 1 is understood. Thus the same number of each kind of atom is represented on each side of the arrow (4H and 2O), and the equation is said to be **balanced.**

appendix two Classification of Organisms

The following scheme represents an attempt to classify living organisms on the basis of phylogenetic (evolutionary) relationships, and is only one of many ways in which organisms can legitimately be classified. Because evidence of relationships among organisms is often sketchy, and disagreements are common as to the relative importance of various groups, a number of different schemes may be found in recent biological publications. Even within the foregoing text we have elected a pluralistic approach to taxonomy, employing the concept of the protista (p. 145), for example, whereas in this appendix we retain the more traditional distinction between unicellular plants such as the Euglenophytes and single-celled animal forms, or Protozoa. Although most major groups of organisms are included in this classification, some less familiar groups have been omitted.

Kingdom Monera *(prokaryotes)*	Cells lack a nuclear membrane, plastids, mitochondria, and other organelles. Cells are typically much smaller than eukaryotic cells and show little specialization. Both autotrophic (photosynthetic or chemosynthetic) and heterotrophic forms are known.
Phylum Cyanophyta (blue-green algae)	Organisms contain chlorophyll a and bluish pigments called phycobilins; these photosynthesize and produce oxygen. Motility is common, but the mechanisms are poorly understood. Flagellated cells are not found.
Phylum Schizophyta	A very diverse and ubiquitous group that is generally not photosynthetic; when there are photosynthetic forms, these do not produce oxygen and contain no phycobilins.

709

Class Schizomycetes (bacteria)

Individual cells may be in shape of spheres, rods, or spirals. Some forms are filamentous. Many are motile and contain flagella. Motion by gliding occurs in some forms. One order (Spirochaetales) moves by use of a contractile axial filament attached to each end of an elongate, spiral-shaped, flexible cell.

Class Rickettsiales (rickettsia)

Parasites of animals, usually more than $.1\mu$ in diameter.

Class Virales (viruses)

Parasites of plants, animals, and bacteria, usually less than $.1\mu$ in diameter.

Kingdom Plantae

Cells contain a nuclear membrane, mitochondria, other organelles, and a cell wall for at least part of their life cycle. Structural differentiation of cells is usually present. Most groups are photosynthetic.

Phylum Myxomycota (slime molds)

Heterotrophic ameboid organisms that develop cell walls during the formation of sporangia. Cell walls are absent at other times. Food is obtained by ingestion.

Class Myxomycetes (plasmodial slime molds)

Multinucleate unit that moves over surface as a mass of protoplasm (plasmodium) and eventually differentiates into multinucleate sporangia that form spores.

Class Acrasiomycetes (cellular slime molds)

Individual cellular amebas that eventually congregate to form a unit in which each cell retains its identity. The congregated cells differentiate and give rise to a sporangium and spores.

Phylum Eumycota (fungi)

Heterotrophic organisms that obtain nutrients primarily by absorption. Nuclei are found in a continuous, multinucleate, tubular, branching mycelium that forms cross-walls at certain stages in the life cycle. Many species are of economic importance as plant parasites, as agents in baking, brewing, cheese production, and antibiotic production. Some forms are edible.

Class Oömycetes

Mostly aquatic fungi that have flagellated, motile cells present at some stages of their life cycle. Their cell walls contain cellulose.

Class Zygomycetes

Terrestrial fungi that have no flagellated cells. The mycelium forms cross-walls only during the reproductive phase of the life cycle. Chitin is prevalent in the cell walls.

Class Ascomycetes

Both terrestrial and aquatic forms are known. The mycelium has perforated cross-walls. In sexual reproduction meiosis and the production of spores occur in a typical spore-case or ascus. Economically important members include the yeasts and *Penicillium*, the mold from which penicillin is derived.

Class Basidiomycetes

Terrestrial fungi with perforated cross-walls in the mycelium. Meiosis and the production of spores occurs in the basidium. Edible mushrooms are in this group.

Lichens

The classification of this group is complicated by the fact that individuals are actually a symbiotic combination of a fungus (usually an Ascomycetes) and a unicellular alga. They are ubiquitous on tree trunks and rocks. Lichens are very sensitive to air pollution and are rare in large cities.

Phylum Chrysophyta (diatoms and golden algae)

Photosynthetic organisms that contain chlorophyll and an accessory pigment fucoxanthin. Food-storage products are oils and the carbohydrate leucosin. Most members are unicellular or colonial. In diatoms the cell walls are impregnated with silicon.

Phylum Euglenophyta (euglena)

Unicellular or colonial photosynthetic organisms, with a flexible cell wall (pellicle) and a single apical flagellum.

Phylum Rhodophyta (red algae)

Primarily multicellular marine plants that contain reddish phycobilin pigments in addition to chlorophyll. The life cycles are often very complex, and no motile cell stages are known.

Phylum Phaeophyta (brown algae)

Multicellular marine plants that contain the brown pigment fucoxanthin in addition to chlorophyll. Organisms are often very large and may contain specialized conducting cells for distributing food materials away from actively photosynthesizing portions of the plant.

Phylum Chlorophyta (green algae)

Unicellular, colonial, or multicellular plants with pigments similar to those of higher plants (chlorophylls a and b and carotenoids). Starch is the primary food reserve. The degree of cell specialization is limited.

Phylum Bryophyta (mosses, liverworts, hornworts)

Small, multicellular plants commonly found in moist terrestrial habitats. The bryophytes contain chlorophyll a and b, carotenoids, and the reserve food, starch. Cell differentiation and specialization is common. The gametophyte generation dominates the life cycle. True roots, stems, and leaves are absent. Water is necessary for sperm to swim to egg over surface of plant body.

Phylum Tracheophyta (vascular plants)

Pigmentation and food reserves are the same as in bryophytes and green algae. The life cycle is dominated by the sporophyte, which typically shows a high level of differentiation into leaves, stems, and roots. Vascular tissue for the transport of water and organic materials is well developed.

SUBPHYLUM LYCOPSIDA (club mosses)

Leaves are typically small, with a single vein of vascular tissue, and are spirally arranged on the stem. Roots and stems show dichotomous branching pattern. Fossils of large, treelike, extinct forms have been found, but living forms are generally quite small.

Most forms are extinct and known only from the fossil record. One genus, *Equisetum*, survives and is quite widespread. The stems are jointed, leaves are small and arranged in whorls. They are sometimes referred to as "scouring rush" because of their rough, silica-impregnated stems.

This subgroup of the Tracheophyta has by far the largest number of living species. Members are wide spread geographically, existing in a variety of habitats. Leaves are generally large with numerous veins.

Class Filicineae (ferns)

Although tree ferns may be found in tropical regions, the typical fern of temperate climates consists of conspicuous leaves (fronds) arising from an underground stem (rhizome). This group, together with giant club mosses and horsetails, was once the earth's dominant vegetation.

Class Gymnospermae (gymnosperm)

The most ancient seed-bearing plants. Ovules, which mature into seeds, are borne "naked" on modified leaves (sporophylls), which are formed into cones. The order Coniferales contains the most familiar examples such as pines, spruce, fir, and cedar.

Class Angiospermae (flowering plants)

This group is characterized by a specialized reproductive organ, the flower, which is often showy and attracts insects that aid in pollination. Unlike the gymnosperms, the developing seeds, or ovules, are enclosed by a modified leaf, or carpel. The sexual phase of the life cycle is characterized by a division of the pollen nucleus into 2 sperm nuclei, one of which produces an embryo and the other a nutritive tissue called the edosperm.

Subclass Dicotyledonae (dicots)

The embryo has two cotyledons (seed leaves). Flower parts are usually present in fours or fives or multiples of these numbers. The leaves are usually net-veined, and the vascular bundles of the stem are arranged in a cylinder. A vascular cambium that produces secondary conductive tissue (xylem and phloem) is often present.

Subclass Monocotyledonae (monocots)

The embryo has a single cotyledon. Flower parts are usually present in threes and sixes. Leaves are usually parallel veined, and vascular bundles of the stem are scattered rather than cylindrical in arrangement. Secondary conductive tissue is formed only rarely.

Kingdom Animalia

Phylum Protozoa

Cells contain a nuclear membrane, mitochondria and other organelles, but no plastids and no cell wall. Nutrition is typically by ingestion. A high degree of cellular differentiation is present in most groups.

Unicellular or colonial animals.

Class Mastigophora (flagellates)

Cells are motile by means of one or more flagella; and generally are uninucleate. Cells show some differentiation of organelles.

Class Sarcodina

Organisms are motile by means of pseudopods, and are either uni- or multinucleate. Amebas are a familiar example.

Class Ciliphora

Cells are motile by means of cilia and contain a large macronucleus and a small micronucleus. Paramecia are a familiar example.

Class Sporozoa

Parasitic microorganisms with reproductive cycles that are often complicated. Some of these organisms may cause serious disease (e.g., malaria).

Phylum Porifera (sponges)

Aquatic animals (mostly marine) of loose structure and organization containing numerous body openings or pores. Sponges circulate water through these pores and a system of internal canals and filter out microscopic food particles.

Phylum Coelenterata (polyps, jellyfish, corals)

Organisms with a body formed from two germ tissues (endoderm and ectoderm) and a gastrovascular cavity with a single opening. They commonly have numerous tentacles with stinging cells used to immobilize prey. The group is primarily marine and carnivorous, catching worms, crustaceans and fish.

Phylum Platyhelminthes (flatworms)

The bodies of organisms in this group are flattened. They have no circulatory system, and the digestive, or alimentary, system lacks an anus. Parasites such as tapeworms and flukes are included. Most members are hermaphroditic (i.e., male and female organs are found on the same individual).

Phylum Aschelminthes

Wormlike animals with a cuticle. They often exhibit superficial segmentation. The female and male sex organs are usually on separate individuals. The number and arrangement of cells or nuclei are constant for individuals of a given species. A body cavity known as pseudocoel (false coelom) is present. Two of the more familiar classes are listed below.

Class Rotifera

Microscopic aquatic animals with an anterior wheel organ of cilia used in feeding and locomotion.

Class Nematoda (roundworms)

A large group of cylindrical, elongated worms, some of which are parasitic.

Phylum Mollusca (mollusks)

Soft-bodied animals with a bilateral body composed of a head, ventral foot, and a dorsal visceral hump that is covered by a mantle of muscular tissue that secretes the exoskeleton, or shell. Breathing takes place via gills. Male and female sex organs may be on separate individuals, or individuals may be hermaphroditic. The group is large and varied; only the better known classes are listed here.

Class Gastropoda (snails and related types)

A large class with marine, freshwater, and terrestrial species. The visceral hump is typically coiled. Individuals usually have a shell and a head with eyes and one or two pairs of tentacles.

Class Pelecypoda (clams, mussels, oysters, etc.)

Sedentary, mostly marine laterally compressed organisms with dorsally hinged valves. The head is rudimentary, without tentacles, the foot is tongue-shaped and used in burrowing.

Class Cephalopoda (squid, octopus, nautilus)

Marine organisms that have a head with tentacles, a well-developed nervous system, and a cartilaginous endoskeleton. Cephalopods are the most advanced and specialized class of mollusk.

Phylum Annelida

Metamerically segmented worms with organ systems segmentally arranged. A true body cavity (coelom) is present. Clamworms, earthworms, and leeches are included.

Phylum Arthropoda

Metamerically segmented animals with jointed legs and a chitinous exoskeleton, and characteristically having compound eyes. Paired limbs are present on some or all of the segments. More species have been identified in this phylum than in all other living forms—plants, animals, and Monera. Approximately 1 million species have been described, and it is estimated that 10 million may actually exist. Only some of the more important and better known subclassifications are described here.

SUBPHYLUM TRILOBITA (trilobites)

An extinct group of marine organisms with a body marked on the dorsal side into three lobes by two lengthwise furrows. Each segment contains paired identical ventral legs with the exception of the first two and the last. Compound eyes are present.

SUBPHYLUM CHELICERATA

The adult body of these organisms typically consists of an unsegmented cephalothorax and an abdomen that may or may not be segmented. The cephalothorax usually contains six pairs of appendages. The first pair are pincerlike and the last four pairs are walking legs. Jaws or antennae are never present.

Class Arachnida (spiders, ticks, mites, scorpions)

Arachnids are usually carnivorous, predatory and terrestrial. Eyes, if present, are not compound.

Class Xiphosura (Horseshoe crabs, Limulus)

SUBPHYLUM MANDIBULATA

Members of this subgroup are numerous and varied. The body may be comprised of a cephalothorax and abdomen; a head, thorax, and abdomen; or a head and trunk. The cephalothorax or head is not externally segmented but the more posterior body regions usually are. The head typically contains one or two pairs of antennae and a pair of mandibles (jaw parts). Abdominal appendages may or may not be present.

Class Crustacea (shrimps, lobsters, crayfish, crabs, water fleas, brine shrimps, barnacles)

The Crustacea are primarily aquatic, although terrestrial forms are known. The head and thorax are often covered with a protective carapace. Compound eyes and two pair of antennae are typically present.

Class Chilopoda (centipedes)

These terrestrial organisms are typically capable of rapid movement, and are carnivorous and predatory. The head contains a pair of antennae and a pair of mandibles. The trunk has 15 or more pairs of walking legs, with a single pair on each segment.

Class Diplopoda (millipedes)

These herbivorous, scavenging organisms are generally much slower moving than the centipedes. The first four trunk segments are single, the rest are double, and fused, with two pairs of legs generally arising from each fused segment pair.

Class Insecta

The insects are primarily, but not exclusively, terrestrial. In several orders more than 100,000 species have been described. The body is divided into three distinct regions, the head, thorax, and abdomen. The thorax typically contains three pairs of legs and two pairs of wings. Compound eyes and light-sensitive ocelli are present. The class is divided into many orders, several of which have 100,000 or more species. Some of the more prominent orders are listed here.

Order Orthoptera (grasshoppers, locusts, crickets, cockroaches)
Approximately 25,000 species.

Order Odonata (dragonflies, damsel flies)
Approximately 5000 species.

Order Anopleura (sucking lice, human body lice)
Approximately 500 species.

Order Hemiptera (bed bugs, stink bugs, water scorpions, water striders)
Approximately 40,000 species.

Order Homoptera (plant lice, scale insects, mealy bugs)
Approximately 30,000 species.

Order Lepidoptera (moths, butterflies)
Approximately 125,000 species.

Order Diptera (flies, gnats, mosquitos)
Approximately 100,000 species.

Order Coleoptera (beetles, weevils)
Approximately 300,000 species.

Order Hymenoptera (bees, ants, wasps, sawflies)
Approximately 110,000 species.

Phylum Echinodermata (starfishes, sea urchins, sand dollars, sea cucumbers)

This group of exclusively marine animal shows radial symmetry in the adult stage although the larvae are bilateral. The organisms have a spiny skin and a calcareous endoskeleton.

Phylum Chordata The chordates are characterized, in part, by a longitudinal semirigid structure called a notochord that serves as an internal skeleton in the embryo and sometimes in the adult. Pharyngeal gill slits and a dorsal hollow nerve cord are also present in immature stages and sometimes in the adult.

SUBPHYLUM
UROCHORDATA
(tunicates, sea squirts)
The body of these marine organisms is surrounded by a secreted envelope or tunic, and is unsegmented. Gill slits are used for breathing and filter feeding. The coelom is not well developed.

SUBPHYLUM
VERTEBRATA
This group consists of animals that in the adult stage possess a notochord and/or a vertebrate column of bone or cartilage. The coelom is well developed, the circulatory system is closed, and a two-, three-, or four-chambered heart is present. A cranium, or skull, encloses the brain.

Class Agnatha (lamprey, hagfish, etc.)

Jawless fish that retain a notochord throughout life. The internal skeleton is cartilaginous.

Class Chondrichthyes (sharks, rays, etc.)

These aquatic, primarily marine, animals possess a cartilaginous rather than a bony skeleton. The notochord becomes reduced, but does persist in the adult.

Class Osteichthyes (bony fish)

Aquatic animals usually covered with scales and with a skeleton mostly of bone. A lung or a swim bladder is present.

Class Amphibia (frogs, toads, salamanders, etc.)

Fresh-water and terrestrial animals with paired legs and skin without scales. The heart is three-chambered and breathing is via gills, lungs, skin, and the mouth cavity. The dividing fertilized egg typically produces a tadpole larval stage.

Class Reptilia (turtles, snakes, lizards, crocodiles)

Reptiles posses an epidermis with scales, five-toed limbs, and a four-chambered heart, although ventricular separation is usually incomplete. Eggs have extraembryonic protective membranes and shells.

Class Aves (birds)

The skin is covered with feathers that are sometimes brightly colored. The forelimbs are modified into wings. A four-chambered heart is present, and these animals are homeothermic.

Class Mammalia (mammals)

The young are nourished by milk from mammary glands of the adult female. The skin is covered with hair. Homeothermic.

Illustration Credits

Frontispiece: Jen and Des Bartlett/ Photo Researchers.

Part One Opener: H. Jelinek/Photo Researchers.

Figure 1-1 (a): Ward's Natural Science Establishment, Inc.

Figure 1-1 (b) and (d): R. H. Noailles.

Figure 1-1 (c), (e), and (f): Russ Kinne/Photo Researchers.

Figure 1-2: Russ Kinne/Photo Researchers.

Figure 1-3 (a) and (b): Courtesy T. Elliot Weier. From T. E. Weier, C. R. Stocking, and M. G. Barbour, *Botany* (4th ed.). Wiley, New York, 1970.

Part Two Opener: Courtesy George E. Palade, Rockefeller University.

Figure 2-2: From A. V. Crewe, R. B. Park, and J. Biggins, "Visibility of Single Atoms," *Science* (June 12, 1970), **168**, Fig. 4. © 1970 by the American Association for the Advancement of Science.

Figure 2-10: Redrawn after A. C. Wahl, "Chemistry by Computer," *Scientific American* (April 1970), p. 62.

Figure 3-11 (a): Courtesy John Rash, Department of Embryology, Carnegie Institution of Washington.

Figure 3-11 (b): Courtesy T. Elliot Weier. From T. E. Weier, C. R. Stocking, and M. G. Barbour, *Botany* (4th ed.). Wiley, New York, 1970.

Figure 3-19: Courtesy Henry S. Slayter, Children's Cancer Research Foundation, Inc., Boston.

Figure 3-20: Redrawn after B. Low and J. T. Edsall, *Currents in Biochemical Research*, Wiley-Interscience, New York, 1956.

Figure 3-22: Redrawn after J. C. Kendrew, "The Three-Dimensional Structure of a Protein Molecule," *Scientific American* (December 1961), p. 96.

Part Three Opener: Courtesy Perkin-Elmer Corporation.

Figure 5-1 (a): Courtesy Ivan Sorvall, Inc. Photo by Joseph Bonadonna.

Figure 5-2 (a): Courtesy Unitron Instrument Company.

Figure 5-2 (b): Courtesy Hitachi Perkin-Elmer.

717

Figure 5-3: Rare Book Division, The New York Public Library, Astor, Lenox, and Tilden Foundations.

Figure 5-4 (b): Courtesy R. Levi-Montalcini, Washington University, St. Louis.

Figure 5-5 (b to e): Robert L. DeHaan.

Figure 5-7 (a): Courtesy Walter Stoeckenius, University of California, San Francisco.

Figure 5-7 (b): Redrawn after A. G. Loewy and P. Siekevitz, *Cell Structure and Function*. Holt, Rinehart & Winston, New York, 1969, p. 472.

Figure 5-8 (a): Courtesy J. D. Robertson. From *Cellular Membranes in Development* (M. Locke, ed.). Academic, New York, 1964.

Figure 5-8 (b): Daniel Branton, *Proceedings of National Academy of Science* U. S. (1966), **552**, p. 1048.

Figure 5-9 (b): Courtesy J. D. Robertson. From *Cellular Membranes in Development* (M. Locke, ed.). Academic, New York, 1964.

Figure 5-9 (c): Daniel Branton, "Fracture Faces of Frozen Myelin," *Experimental Cell Research* (1967), **45**, pp. 703-707.

Figure 5-10 (a): Courtesy N. Scott McNutt and R. W. Weinstein, From *Journal of Cell Biology* (1970), **47**, p. 680.

Figure 5-11: Courtesy D. W. Fawcett, Harvard Medical School, Boston.

Figure 5-12: Courtesy J. G. Hirsch. From *Journal of Experimental Medicine* (1962), **116**, p. 827.

Figure 5-13: Courtesy Douglas Kelly. From Rockefeller Press, *Journal of Cell Biology* (1966), **28**, pp. 51-72.

Figure 5-14: Courtesy N. Scott McNutt and R. W. Weinstein. From *Journal of Cell Biology* (1970), **47**, p. 670.

Figure 5-15: Courtesy D. W. Fawcett, Harvard Medical School, Boston.

Figure 5-16: Courtesy D. Friend, University of California, San Francisco.

Figure 5-18 (a): Courtesy John Rash, Carnegie Institution of Washington.

Figure 5-19: Courtesy B. Chance, From *Science* (1963), **142**, p. 1176 (VII).

Figure 5-20: Courtesy M. M. K. Nass. From *Proceedings of the National Academy of Science* (1966), **56**, p. 1215.

Figure 5-21 (a): Courtesy Umberto Muscatello. From U. Muscatello, I. Pasquali-Ronchetti, and A. Barasa, *Journal of Ultrastructure Research* (1968) **23**, p. 44, Fig. 11.

Figure 5-22 (a): Courtesy Daniel Branton. From T. E. Weier, C. R. Stocking, and M. G. Barbour, *Botany* (4th ed.). Wiley, New York, 1970.

Figure 5-22 (b): Courtesy W. W. Franke and V. Scheer. From *Journal of Ultrastructure Research* (1970), **30**, p. 306, Fig. 20.

Figure 5-23: Courtesy Ernest J. DuPraw, *DNA and Chromosomes*. Holt, Rinehart & Winston, New York, 1970.

Figure 5-26 (a) and (c): Courtesy I. R. Gibbons, University of Hawaii, Honolulu.

Figure 5-26 (b): Courtesy Peter Satir, University of California, Berkeley.

Figure 5-28 (b): Courtesy Myron C. Ledbetter, Brookhaven National Laboratory.

Figure 5-29: Courtesy T. Elliot Weier. From T. E. Weier, C. R. Stocking, and M. G. Barbour, *Botany* (4th ed.). Wiley, New York, 1970.

Figure 5-30: Courtesy R. B. Park. From R. B. Park and J. Biggins, "Quantasome: Size and Composition," *Science* (May 22, 1964), **144**, Fig. 1. © 1964 by the American Association for the Advancement of Science.

Figure 5-31: Courtesy T. Bisalputra, From *Journal of Cell Biology* (1967), **33**, p. 511.

Figure 6-3 (a): Courtesy R. C. Williams, University of California, Berkeley.

Figure 6-3 (b): Courtesy R. W. G. Wyckoff, University of Arizona, Phoenix.

Figure 6-3 (c): Courtesy Louis W. Labaw, National Institutes of Health, Bethesda.

Figure 6-3 (d): Courtesy E. Boy de la Tour, University of Geneva.

Figure 6-4 (b): Courtesy A. K. Kleinschmidt, New York University Medical Center.

Figure 6-5 (a): Courtesy Judith F. M. Hoeniger, University of Toronto.

Figure 6-5 (b) and (c): Courtesy Stanley C. Holt, University of Massachusetts, Amherst.

Figure 6-8: Courtesy Richard C. Starr, Indiana University, Bloomington.

Figure 6-9: Redrawn after J. Brachet, *Biochemical Cytology*. Academic, New York, 1957, p. 303.

Figure 6-10 (a): Courtesy L. Feldman and E. Cutter. From T. E. Weier, C. R. Stocking, and M. G. Barbour, *Botany* (4th ed.). Wiley, New York, 1970.

Figure 6-11 (g) and (h): Courtesy D. W. Fawcett, Harvard Medical School, Boston.

Figure 6-13 (a): Courtesy M. Jakus, National Institutes of Health, Bethesda.

Figure 6-13 (b): Courtesy D. W. Fawcett, Harvard Medical School, Boston.

Figure 6-13 (c): Andreas Feininger/*Life* Magazine. © 1972 Time, Inc.

Figure 6-14: Courtesy D. W. Fawcett, Harvard Medical School, Boston.

Figure 6-15: Courtesy D. W. Fawcett, Harvard Medical School, Boston.

Figure 6-16 (b): Courtesy D. W. Fawcett, Harvard Medical School, Boston.

Figure 6-17: Robert L. DeHaan.

Figure 6-20: Courtesy Patricia N. Farnsworth. From Arthur K. Solomon, "The State of Water in Red Cells," *Scientific American* (February 1971), p. 89.

Figure 7-4: Redrawn after E. J. DuPraw, "Macromolecular Organization of Nuclei and Chromosomes," *Nature* (1965) **206**, p. 338.

Figure 7-6: Courtesy Elliot Robbins. From E. Robbins and R. Gonates, *Journal of Cell Biology* (1964), **21**, p. 429.

Figure 7-8 (a) and (b): Courtesy Y. Hiramoto, From *Journal of Cell Biology, Suppl.* (1965), **25**, p. 161.

Figure 7-9 (a) to (c): Courtesy Andrew S. Bajer, From *Chromosoma* (1968), **25**, p. 249 © Springer-Verlag, Berlin.

Figure 7-9 (d) and (e): Courtesy Peter K. Hepler. From P. K. Hepler and W. T. Jackson, *Journal of Cell Biology* (1968), **38**, p. 437.

Figure 7-11: Courtesy John Cairns, Cold Spring Harbor Laboratory.

Figure 7-12 (a) to (c): Courtesy D. M. Prescott. From D. M. Prescott and M. A. Bender, *Experimental Cell Research* (1963), **29**, pp. 430-442.

Figure 7-13: Redrawn after B. I.

Balinsky, *An Introduction to Embryology* (3rd ed.). Saunders, Philadelphia, 1970.

Part Four Opener: Courtesy Everett Anderson, Harvard Medical School, Boston.

Figure 8-3: Courtesy T. T. Puck and J. H. Tjio, Eleanor Roosevelt Institute for Cancer Research, Denver.

Figure 8-7 (a): Karl Gullers/Rapho Guillumette.

Figure 8-7 (b): Nancy Hays/Monkmeyer.

Figure 8-15: Data for linkage maps from C. B. Bridges and K. S. Brehme, "The Mutants of *Drosophila Melanogaster*," Carnegie Institution of Washington Publication 552, 1949.

Figure 8-16: Courtesy R. A. Boolootian.

Figure 8-17: Modified from BSCS, *Biological Science; Molecules to Man* (Blue Version). Houghton Mifflin, Boston, 1968.

Figure 9-1: Redrawn after Helena Curtis, *Biology*. Worth, New York, 1968, p. 369.

Figure 9-3: Modified from data prepared by the Eugenics Survey of Vermont.

Figure 9-4: Courtesy Victor A. McKusick, *Human Genetics* (2nd ed.). Prentice-Hall, Englewood Cliffs, N. J., 1969.

Figure 9-8 (a) and (b): Courtesy M. L. Barr, University of Western Ontario, London, Ont.

Figure 9-8 (c): From James German, "Studying Human Chromosomes Today," *American Scientist* (1970), **58**, p. 182. Courtesy *American Scientist.*

Figure 9-9: Courtesy Victor A. McKusick. From *Medical Genetics 1958-1960*, Mosby, St. Louis, 1961, and *Journal of Chronic Diseases* (July 1960).

Figure 9-10 (b): Courtesy Arthur D. Bloom and Shozo Iida, Department of Human Genetics, University of Michigan.

Figure 9-13: Data from L. Erlenmeyer-Kimling and L. F. Jarvik, *Science* (1964), **142**, pp. 1477-1479.

Figure 9-15 (a): Courtesy Victor A. McKusick. From V. A. McKusick, *Human Genetics* (2nd ed.). Prentice Hall, Englewood Cliffs, N. J., 1969; *Medical Genetics 1958-1960;* Mosby, St. Louis, 1961, and *Jour-*

nal of Chronic Diseases (July 1960).

Figure 9-15 (b): Courtesy Arthur D. Bloom and Shozo Iida, Department of Human Genetics, University of Michigan.

Figure 10-3: Courtesy Lee D. Simon, The Institute for Cancer Research, Philadelphia.

Figure 10-4: Courtesy Lee D. Simon, The Institute for Cancer Research, Philadelphia.

Figure 10-5: Courtesy Lee D. Simon, The Institute for Cancer Research, Philadelphia.

Figure 10-9: From James D. Watson, *The Double Helix*. Atheneum, New York, 1968. © 1968 by James D. Watson. By permission of the publishers.

Figure 10-10: From James D. Watson, *The Double Helix*. Atheneum, New York, 1968. © 1968 by James D. Watson. By permission of the publishers.

Figure 10-11: Courtesy Gunter F. Bahr, Armed Forces Institute of Pathology.

Figure 10-12 (a) and (b): Courtesy D. M. Prescott. From *Progressive Nucleic Acid Research* (1964), **III**, p. 35.

Figure 10-13: Courtesy O. L. Miller, Oak Ridge National Laboratory. From O. L. Miller et al., *Science* (July 24, 1970), **169**, pp. 392-395, Fig. 2. © 1970 by the American Association for the Advancement of Science.

Figure 10-14: Courtesy O. L. Miller, Oak Ridge National Laboratory. From O. L. Miller and B. R. Beatty, *Science* (May 23, 1969), **164**, pp. 955-957, Fig. 1. © 1969 by the American Association for the Advancement of Science.

Figure 10-15: Courtesy Peter C. Wensink and Donald D. Brown, Department of Embryology, Carnegie Institution of Washington.

Figure 10-16: Courtesy D. E. Wimber and Dale M. Steffensen. From *Science* (November 6, 1970), **170**, pp. 639-641, Fig. 2. © 1970 by the American Association for the Advancement of Science.

Part Five Opener: Courtesy Dr. Igor Dawid, Carnegie Institution of Washington, Baltimore.

Figure 11-1: Courtesy of the American Museum of Natural History.

Figure 11-2 (a) and (b): Courtesy

Carnegie Institution of Washington. From A. T. Hertig, J. Rock, and E. Adams, *American Journal of Anatomy* (1956), **98**, p. 435.

Figure 11-4: From C. A. Villee, W. F. Walker, and F. E. Smith, *General Zoology*. Saunders, Philadelphia, 1968.

Figure 11-7: From Richard J. Harrison and William Montagna, *Man.* © 1969 by Appleton-Century-Crofts, Educational Division, Meredith Corporation. By permission of Appleton-Century-Crofts.

Figure 11-9 (a): Redrawn after L. B. Arey, *Developmental Anatomy* (7th ed.). Saunders, Philadelphia, 1965.

Figure 11-9 (b): Courtesy D. W. Fawcett, Harvard Medical School, Boston.

Figure 11-13: After A. A. Maximow. From W. Bloom and D. W. Fawcett, *Textbook of Histology*. Saunders, Philadelphia, 1968.

Figure 11-15: Courtesy D. W. Fawcett, Harvard Medical School, Boston.

Figure 11-16: Courtesy D. W. Fawcett, Harvard Medical School, Boston.

Figure 11-17 (a) and (b): Courtesy Joseph Kennedy, Johns Hopkins University, Baltimore.

Figure 11-18 (a) to (c): Courtesy H. M. Seitz et al. From "Cleavage of Human Ova in Vitro," *Fertility and Sterility* (1971), **22**, p. 255. © 1971 by Williams & Wilkins, Baltimore.

Figure 11-20 (a) to (c): Courtesy Carnegie Institution of Washington.

Figure 11-21 (a) and (b): Courtesy Carnegie Institution of Washington. From A. T. Hertig, J. Rock, and E. Adams, *American Journal of Anatomy* (1956), 98, p. 435.

Figure 11-25 (a) to (c): Courtesy Department of Embryology, Carnegie Institution of Washington.

Figure 11-27: Wayne Miller/Magnum.

Figure 12-2: Redrawn after G. B. Moment, "Simultaneous Anterior and Posterior Regeneration and Other Growth Phenomena in Maldanid Polychaetes," *Journal of Experimental Zoology* (1951), **117**, p. 6.

Figure 12-3: Redrawn after C. W. Bodemer, *Modern Embryology*. Holt, Rinehart & Winston, 1968, p. 5.

Figure 12-4: Redrawn after L. B. Arey, *Developmental Anatomy* (7th ed.). Saunders, Philadelphia, 1965, p. 24.

Figure 12-5: Redrawn after L. B. Arey, *Developmental Anatomy* (7th ed.). Saunders, Philadelphia, 1965, p. 531.

Figure 12-6: Walter Dawn.

Figure 12-10 (a) to (c): Courtesy K. B. Raper, University of Wisconsin.

Figure 12-10 (d): Courtesy David Francis. From J. T. Bonner, *Cellular Slime Molds* (2nd ed.). Princeton University Press, Princeton, 1967.

Figure 12-10 (e) and (f): Courtesy J. T. Bonner, Princeton University.

Figure 12-12 (a) to (i): Courtesy A. C. Hildebrandt and V. Vasil, *Science* (November 12, 1965), **150**, pp. 889-892, Fig. 1. © 1965 by the American Association for the Advancement of Science.

Figure 12-13 (a) and (c): R. H. Noailles.

Figure 12-13 (b): Russ Kinne/Photo Researchers.

Figure 12-14 (a) to (p): Courtesy Irwin H. Krakoff. From D. A. Karnofsky and Eva B. Simmel, *Progressive Experimental Tumor Research* (1963), **33**, pp. 254-295. Karger, Basel/New York, 1963.

Figure 12-18: Redrawn after C. W. Bodemer, *Modern Embryology*. Holt, Rinehart & Winston, 1968, pp. 40 and 43.

Figure 12-19 (a) and (b): Courtesy I. R. Konigsberg, *Scientific American* (1964), **211**, p. 61.

Figure 12-19 (c) and (d): Courtesy I. R. Konigsberg, *Science* (1963), **140**, p. 1273. © 1963 by the American Association for the Advancement of Science.

Figure 12-21: Redrawn after B. I. Balinsky *An Introduction to Embryology* (3rd ed.). Saunders, Philadelphia, 1970.

Figure 12-22: Redrawn after H. Spemann, *Embryonic Development and Induction*. Yale University Press, New Haven, 1938, p. 27.

Figure 12-23: Redrawn after Helena Curtis, *Biology*. Worth, New York, 1968.

Figure 12-25: Redrawn after C. W. Bodemer, *Modern Embryology*. Holt, Rinehart & Winston, 1968, p. 41.

Figure 12-29 (a) to (d): Courtesy V. Hamburger and A. Hamilton. From *Journal of Morphology* (1951), **88**, pp. 49-92.

Figure 12-30: Redrawn after B. I. Balinsky, *An Introduction to Embryology* (3rd ed.). Saunders, Philadelphia, 1970.

Figure 12-33: Courtesy V. Hamburger and A. Hamilton. From *Journal of Morphology* (1951), **88**, pp. 49-92.

Part Six Opener: Walter Dawn.

Figure 13-2 (a): Hugh Spencer.

Figure 13-2 (b): Ross E. Hutchins.

Figure 13-23: Courtesy L. Rappaport, University of California, Davis.

Figure 13-24: Courtesy Agricultural Research Service, Plant Industry Station, U. S. Department of Agriculture.

Part Seven Opener: Courtesy Philadelphia Museum of Art. Gift of Charles Bregler.

Figure 14-2 (a) to (c): Courtesy O. L. Miller, Jr., and David M. Prescott, Biology Division, Oak Ridge National Laboratory.

Figure 14-6 (d): Courtesy D. W. Fawcett, Harvard Medical School, Boston.

Figure 14-9: Courtesy Picker International Corporation.

Figure 14-10: Courtesy UNICEF.

Figure 15-2 (b): Courtesy Keith Porter, University of Colorado, Boulder.

Figure 15-3 (a): Courtesy Dr. Norman M. Hodgkin, Physiological Laboratory, Cambridge, England.

Figure 15-3 (b): Courtesy D. W. Fawcett, Harvard Medical School, Boston.

Figure 15-3 (c): Courtesy Pfizer, Inc.

Figure 15-4 (b) to (e): Courtesy John Rash, Carnegie Institution of Washington.

Figure 15-5: Adapted from E. F. Adolph, "The Heart's Pacemaker," *Scientific American* (March 1967), p. 32.

Figure 15-14: Courtesy J. Rhodin. From *An Atlas of Ultrastructure*. Saunders, Philadelphia, 1963.

Figure 16-3: Courtesy R. W. Carlin.

Figure 16-4: Courtesy John H. Olive, The University of Akron.

Figure 16-5 (b): Courtesy J. R. McClintic. From J. E. Crouch and J. R. McClintic, *Human Anatomy and Physiology*. Wiley, New York, 1971.

Figure 16-9: Wide World Photos.

Figure 17-3 (a): Courtesy J. D. Robertson, From *Annals New York Academy of Science* (1961), p. 94.

Figure 17-4 (a): From B. Katz, *Nerve, Muscle and Synapse*. McGraw-Hill, New York, 1966, p. 12.

Figure 17-4 (b): Courtesy A. L. Hodgkin. From Hodgkin and Keynes, *Journal of Physiology* (1956), **131**, p. 592.

Figure 17-10: Redrawn after W. T. Keeton, *Biological Science*. Norton, New York, 1967, p. 416.

Figure 17-16: From T. A. Rogers, *Elementary Human Physiology*. Wiley, New York, 1961.

Figure 17-22: Courtesy Richard W. Young, UCLA.

Figure 18-4: From M. Brodel, *Johns Hopkins Hospital Bulletin* (1937), **61**, p. 295.

Figure 18-5: Based on a drawing by Sylvia Collard Keene, in W. Bloom and D. W. Fawcett, *A Textbook of Histology* (9th ed.). Saunders, Philadelphia, 1968, p. 273.

Figure 18-6: Courtesy H. E. Huxley, Medical Research Council, Cambridge, England.

Figure 18-7: Courtesy M. K. Reedy. From *American Zoologist* (1967), **7**, p. 469.

Figure 18-8: Redrawn after W. Bloom and D. W. Fawcett, *A Textbook of Histology* (8th ed.). Saunders, Philadelphia, 1968.

Figure 18-14: From R. J. Harrison and W. Montagna, *Man*. © 1969 by Appleton-Century-Crofts, Educational Division, Meredith Corporation. By permission of Appleton-Century-Crofts.

Part Eight Opener: Syd Greenberg/Photo Researchers.

Figure 19-2 (a) and (b): Ron Church.

Figure 19-3: Sdeuard C. Bisserot/Bruce Coleman.

Figure 19-4: Redrawn after Helena Curtis, *Biology*. Worth, New York, 1968.

Figure 19-5: Redrawn from W. G. Van der Kloot, *Behavior*. Holt-Rinehart & Winston, New York, 1968, p. 70.

Figure 19-6: Adapted from Noton and Stark, *Science* (January 1971), **171**, p. 309.

Figure 19-7: Courtesy Thomas Eisner,

Cornell University.

Figure 19-8: Courtesy Sol Mednick.

Figure 19-9 (a): Courtesy Thomas Schultze-Westrum.

Figure 19-9 (b): Leonard Lee Rue IV/Bruce Coleman.

Figure 19-9 (c): N. Meyers/Bruce Coleman.

Figure 19-9 (d): Jane Burton/Bruce Coleman.

Figure 19-13: From Jack Hailman, "How Instinct is Learned," *Scientific American* (December 1969), p. 99. © 1969 by Scientific American, Inc. All rights reserved.

Figure 19-14: Adapted from J. Hailman, "How Instinct is Learned," *Scientific American*, (1969), p. 102.

Figure 19-16: After M. Menaker, *Bioscience* (1969), **19**, p. 685, Fig. 6.

Figure 19-17: After M. Menaker, *Bioscience* (1969), **19**, p. 686, Fig. 7.

Figure 19-19: Redrawn after E. R. Kandel, *Scientific American* (July 1970), pp. 61 and 63.

Figure 19-20: The Bettman Archive.

Figure 19-21 (a) and (b): Nina Leen © Time-Life, Inc.

Figure 19-23 (a) to (d): Eric Hosking/Bruce Coleman.

Figure 19-24: Courtesy Yerkes Regional Primate Center.

Figure 19-25: Courtesy Ross. E. Hutchins.

Figure 19-26 (b): George Holton/Photo Researchers.

Figure 19-27: Ylla/Rapho Guillumette.

Figure 19-28: Courtesy George B. Schaller, New York Zoological Society.

Figure 20-1: Marc and Evelyne Bernheim/Rapho Guillumette.

Figure 20-2: Robert Hermes/National Audubon Society.

Figure 20-5: Wide World Photos.

Figure 20-7 (a) and (b): Courtesy Department of Lands, Queensland, Australia.

Figure 20-8: Joe Munroe/Photo Researchers.

Figure 20-11: Courtesy Bodeleran Library, Oxford, England.

Figure 20-12: Courtesy National Library of Medicine, Bethesda, Maryland.

Figure 20-13: Redrawn after E. J. Kormondy, *Concepts of Ecology*. Prentice-Hall, Englewood Cliffs, N. J., 1969, p. 172. Data from *Population Profile*, Population

Reference Bureau, Inc., Washington, D. C., March 1957.

Figure 20-14: Redrawn after L. B. Young, *Population in Perspective*. Oxford University Press, New York, 1968, p. 312. Data from Paul Woodring, *Saturday Review* (March 18, 1967).

Figure 20-15: Courtesy Bureau of the Census.

Figure 20-16: Baldev/PIX.

Figure 21-1: Courtesy E. S. Barghoorn. From G. E. Hutchinson, "The Biosphere," *Scientific American* (September 1970), p. 44.

Figure 21-2: Courtesy Graphische Samnlung Albertina, Vienna, Austria.

Figure 21-4 (a) and (b): Douglas P. Wilson.

Figure 21-7: Redrawn after Edward S. Deevey, Jr., "Mineral Cycles," *Scientific American* (September 1970), p. 150.

Figure 21-8: Courtesy Raytheon/Autometric. From *Scientific American* (September 1970), p. 98.

Figure 21-9: Aerofilms Ltd.

Figure 21-11: Redrawn after Burt Bolin, "The Carbon Cycle," *Scientific American* (September 1970), p. 128.

Figure 21-12 (a): Courtesy The Nitragin Company.

Figure 21-12 (b): Courtesy D. J. Goodchild. From *Scientific American* (September 1970), p. 144.

Figure 21-14: Courtesy Judith A. Murphy. From *Scientific American* (September 1970), p. 148.

Figure 21-16: Sdeuard C. Bisserot/Bruce Coleman.

Figure 21-17: Mary M. Thacher/Photo Researchers.

Figure 21-18 (a): Peter David/Photo Researchers.

Figure 21-18 (b): Courtesy U. S. Naval Oceanographic Office. From *Scientific American* (September 1970), p. 49.

Figure 21-19: Ron Church.

Figure 21-20: Redrawn after Helena Curtis, *Biology*. Worth, New York, 1968.

Figure 21-22: Courtesy David W. Ehrenfeld, Columbia University, New York.

Figure 21-23: Russ Kinne/Photo Researchers.

Figure 21-24: U. S. Forest Service.

Figure 21-25: U. S. Forest Service.

Figure 21-26: Courtesy William Campbell Steere, The New York Botanical Garden.

Figure 21-27: Steve and Dolores McCutcheon/Alaska Pictorial Service.

Figure 21-29 (a) and (b): Courtesy Edward J. Kormondy, The Evergreen State College, Olympia, Washington.

Figure 21-31: Courtesy Environmental Protection Agency. Photo by Charles E. Glover.

Figure 21-32: Reprinted by permission from the Chicago *Sun-Times*. Photo by Robert Langer.

Figure 21-33: Courtesy United Nations.

Figure 21-34: Courtesy Joseph R. Jehl, Jr. From *Scientific American* (April 1970), p. 72.

Part Nine Opener: John Gerard/Monkmeyer.

Figure 22-1: Courtesy Rare Book Division, The New York Public Library. Astor, Lenox, and Tilden Foundations.

Figure 22-2: Courtesy Hale Observatories.

Figure 22-3: Redrawn after M. J. Rees and J. Silk, "The Origin of Galaxies," *Scientific American* (June 1970), pp. 32-33.

Figure 22-6 (a) and (b): Courtesy S. W. Fox, Institute for Molecular and Cellular Evolution, University of Miami.

Figure 22-7: Courtesy Gerald R. Licari. From *Scientific American* (September 1970), p. 112.

Figure 23-1: Courtesy Smithsonian Institution.

Figure 23-3: Data from J. M. Savage, *Evolution* (2nd ed.). Holt, Rinehart & Winston, New York, 1969.

Figure 23-4 (a): Lynwood M. Chase/National Audubon Society.

Figure 23-4 (b): Courtesy American Museum of Natural History.

Figure 23-4 (c): Gordon S. Smith/National Audubon Society.

Figure 23-4 (d): Edgar Monch/Photo Researchers.

Figure 23-5: Walter Dawn/National Audubon Society.

Figure 23-7: From Gideon E. Nelson, Gerald G. Robinson, Richard A. Boolootion, *Fundamental Concepts of Biology* (2nd ed.). Wiley, New York, 1970.

Figure 23-8 (a): Courtesy Smithsonian Institution.

Figure 23-8 (b): Courtesy American Museum of Natural History.

Figure 23-9: U. S. Department of the Interior.

Figure 23-10: Courtesy American Museum of Natural History.

Figure 23-12 (a): Courtesy American Museum of Natural History.

Figure 23-12 (b) and (c): Courtesy Charles L. Trainer.

Figure 23-13 (a) and (b): Courtesy American Museum of Natural History.

Figure 23-14: Redrawn after E. M. Savage, *Evolution* (2nd ed.). Holt, Rinehart & Winston, New York, 1969.

Figure 23-15: Hugh Spencer.

Figure 23-17 (a) and (c): Ylla/Rapho Guillumette.

Figure 23-17 (b): Courtesy New York Zoological Society.

Figure 23-17 (d): Russ Kinne/Photo Researchers.

Figure 23-18 (a) to (c): Courtesy Field Museum of Natural History, Chicago. Malvina Hoffman, sculptress.

Figure 23-20 (a) to (c): Courtesy American Museum of Natural History.

Figure 24-1 (a): Ylla/Rapho Guillumette.

Figure 24-1 (b) and (d): R. Van Nostrand/National Audubon Society.

Figure 24-1 (c): A. W. Ambler/National Audubon Society.

Figure 24-2: Hugh Spencer.

Figure 24-5: Walter Dawn.

Glossary–Index

Abscissic acid, 368-369

Abscission: The falling of leaves, flowers, and fruits because of depletion of auxin. 368

Acetabularia, characteristics of, 152
  as a developing system, 305-307

Acetyl-CoA, in the Krebs cycle, 91
  as a link in metabolism, 84, 96-97
  in lipid metabolism, 95-96
  in protein and amino acid metabolism, 96

Acetylcholine, 474

Acid: A substance that liberates a proton (hydrogen ion). 39-40

Acromegaly, 453

Actin, 163, 504-505

Action potential, and heartbeat, 412
  and nerves, 465-470

Activation energy: The amount of energy, over the average energy of a given quantity of molecules, that is necessary for colliding molecules to undergo chemical reaction.
  and catalysts, 45
  and collision theory, 43-44
  and rates of reaction, 47

Active site: The part of an enzyme that is directly involved in its catalytic activity. 73-74

Active transport: The pumping of materials across a cell membrane with the aid of metabolic energy. 125
  and digestion, 388
  and nerve impulses, 469

Actomyosin, 504-505

Adaptation, 3

Adenosine triphosphate (ATP), bond energy of, 38-39
  as an energy source, 85
  formation of from ADP, 89
  and kinases, 75
  in muscle contraction, 505-511
  as an organophosphate, 58
  in photosynthesis, 363-365
  in protein synthesis, 103
  in respiration, 84
  role in trapping free energy, 47
  yield of in respiration, 93-94

Addison's disease, 450

Adenyl cyclase, 459

Adrenal gland, 448-451

Adrenal stress syndrome, 566

Adrenalin, 448-449

Adrenocorticotropic hormone (ACTH),

725

characteristic for each chromosome. 209-211

Chromoplast, 142

Chromosome: The gene-bearing nucleoprotein structure; becomes rod-shaped during mitosis and meiosis. 189-211
sex, 219-220

Chymotrypsin, 386

Cilia, 128-140, 149

Circadian rhythm: A biological rhythm that exhibits a period of approximately one day under constant conditions. 541-543
in bees, 530

Circulation, 403, 410

Circumcision, 278

Citric acid cycle (Krebs cycle), 91-92

Cleavage, in amphibians, 324-330
control of, 330-331
in echinoderms, 316-324
in humans, 283

Cleavage furrow, 174

Clitoris, 270

Clone: A group of cells all derived from a single, common ancestor and existing in the same environment. 313

Coacervate: An aggregate of molecules and macromolecules. 643-645

Codon: The site on the mRNA molecule, consisting of three bases, that is complementary to the anticodon on the tRNA molecule.
and mitosis, 174
and protein synthesis, 104-106, 356-359

Coelenterate, 463

Coelom: A true body cavity, distinct from the digestive tract, surrounded by mesodermal tissue.
and taxonomy, 696

Coenzyme: The organic component of a conjugated enzyme that is not composed of amino acids. 75

Coenzyme A, 84

Cohort: A limited subpopulation defined in terms of species, sex, and age. 570

Coitus, in humans, 268, 280-281, 555
in primates, 555

Colloidal solution: A system of particles of colloidal dimensions dispersed in a solvent medium. 51

Colon, 389

Colostrum, 297

Competition, among populations, 571-572

Compound: A substance composed of molecules that consist of two or more different kinds of atom. 28

Cones, 494

Conifer, 347-349

Conjugated protein: A specific complex between a protein and an organic molecule that does not consist of amino acids. 72

Conjugation, 305

Connective tissue, 159-162

Contraceptive, see Birth control

Cornea, 493

Corpus luteum, 272, 456

Cortex, in brain, 478
in plants, 352-355

Corticosterone, 450

Cortisol, 450-451

Cortisone, 450

Cotyledon, 349, 358

Covalent bond: A type of chemical bond that results from the sharing of electrons between two atoms. 32-39

Cowper's gland, 278

Cretinism, 445

Cristae, 133

Crossing-over: The exchange of genetic material between homologous chromosomes. 204-209

Cycad, 348

Cyclic adenosine monophosphate, 459

Cyclic electron transport pathway, 365

Cytochrome: A type of red-colored conjugated protein involved in electron transport, consisting of a heme and a specific globular protein.
and hemes, 75
and respiration, 92
sequence similarity and evolution, 680

Cytology: The study of cell structure.
and cellular dimensions, 114-115

Cytokinesis: The division of the cytoplasm following nuclear division. 169, 174-176

Cytoplasm: The material within a cell membrane, excluding the nucleus.
division of, 169-176
doubling of, 182-184
sol-gel properties of, 51
streaming of, 501
structure of, 120, 126, 142

Dalton: The unit of molecular weight. 32, 51

Dark phase, 95, 365

Darwin, Charles, 633-634, 655-656

Davson-Danielli model of cell membrane, 121

DDT, as a biocide, 625-627

Defecation: The elimination from the body of undigested material that has passed through the digestive tract. 431
reflex, 389

Dehydrogenase: An enzyme that catalyzes oxidation-reduction reactions. 74

Deme: A small, localized population of a single species. 567

Demography: The study of distribution and change in populations. 575-585

Dendrite, 164, 464

Denitrification, 601

Deoxyribonuclease: An enzyme that specifically hydrolyzes DNA. 102

Deoxyribonucleic acid (DNA): The macromolecule that contains the hereditary information for a cell's structure and function.
annealing of, 100
base sequence of, 78-80
biosynthesis of, 98-99
breakdown of, 102
cellular localization of synthesis, 108
chemical composition of, 78-80
in chloroplasts, 142
in chromatin, 172-174
function of, 8, 102
as the genetic material, 8, 239-246
hybridization of, 100, 678-680
in mitochondria, 134
polymerase, 98, 179
repair enzymes, 180
replication of, 177-184
sequence similarity of, and evolution, 678-680
structure of, 78-80, 178, 246-252
as a template, 252-261
transcription of, 99
in viruses, 148

Deoxyribose, 60

Dermis, 398

Desmosome: A specialized region of attachment between cell membranes of adjacent cells.
in epithelial tissue, 159

Development: The formation of a biologic unit or structure from a simpler unit as a result of a sequence of well-ordered changes.
and differentiation, 108
and heredity, 8
of an organism, 8

726

phosphates and, 604
process of, 622-623
Evolution: The gradual change in a
biological unit that results from
mutation and natural selection.
of cells, 646-647
of complex molecules, 4, 642-645
concept of, 633-634
convergent, 590
mutation and natural selection in, 9
evidence for, 667-672
of man, 647-649, 680-686
mechanisms of, 656-667
of organisms, in the present, 5
of photosynthesis, 645-646
theories of, 653-656
Excretion: The process by which an
organism rids itself of metabolic
waste. 431-438
Exergonic reaction: A chemical reaction that liberates energy. 46
Exponential growth: A period during which a population of cells
is doubling at a geometric
rate.
and cell division, 167-168
Eye, 492-498

Fecundity: The ability of a female
to bear young. 569
Fermentation, alcoholic: The process by which certain microorganisms produce ethyl alcohol as
the end product of anaerobic
respiration.
in yeast, 86-89
Ferredoxin, 364
Fertilization: The part of sexual
reproduction in which a sperm
and an egg fuse, resulting in the
mixing of the genetic information carried in the two cells.
discovery of, 196
and heredity, 191
phenomenon of, 202, 267, 281-283
in plants, 349, 358
Fetus: Embryo, 291-296
Fibrinogen, 406
Fibroblast: An undifferentiated cell
found in most tissues. 155
Fibrocyte: The major cell type of
fibrous connective tissue. 160
Filament, of myofibril, 505
First law of thermodynamics: The
principle that energy can be neither created nor destroyed. 22
Flagella, 138-140, 148
Florigen, 369
Flower, 355-358
Follicle-stimulating hormone (FSH),
and female reproductive system,

269-274, 443, 454, 456
and male reproductive system, 275,
443, 454
Food chain: The transfer of energy
from one organism to another—
by eating and being eaten—in a
series of trophic levels.
characteristics of, 590-596
and ecosystem stability, 616
Foreplay, 279-280
Fossil: The preserved remains or
impressions of an ancient plant
or animal, usually in rocks.
and evolution, 667-672
Founders: The original animals from
which a new population is formed.
567
Free energy: The useful energy
possessed by a chemical.
in biologically important molecules, 47-48
and chemical reactions, 46-47
Fructose, 59-60
Functional group: A reactive group
of atoms in an organic compound
that is chiefly responsible for
whatever reactivity the molecule
displays. 54
Fungi, 146

Galactosemia, 231
Gallbladder, 387
Gallstone, 387
Gametes: The cells of an organism
that are involved in sexual reproduction.
and heredity, 191
Gametogenesis, 199-204
Ganglion, 463, 483
Gap junction, see Nexal junction
Gastric juices, 382-383
Gastrulation, 320
Gel: A semisolid solution of a
solvent trapped in a loose
matrix consisting of the fibrous
solute, which has aggregated, or
polymerized. 51
Gene: The unit of DNA that is the
code for a polypeptide chain
having a definite structural or
functional role.
amplification, 664
constituting heredity, 8, 189-190
controlling development, 8
cytology of, 209-210
differential action and development, 314-316, 323-330
and evolution, 656-667
linkage of, 204-209
mapping of, 208-210
mutation of, 89, 656

and the one gene-one enzyme hypothesis, 102
pleiotropic effects, 662
pool, 657
Generation time: The time it takes
for a cell to double. 168
Generator potential: Depolarization
leading to a nerve impulse. 485
Genetic counseling, 234
Genetic drift, 658, 663-664
Genetic isolation, 567
Genome: The entire genetic constitution of a cell or individual.
196
Genotype: A cell's or organism's gene-
tic information that codes for a
specific trait or traits.
analysis of, 203-208
and heredity, 190
Geologic timetable: The record of
geologic history determined by
the sequence of the formation of
sedimentation of rock strata.
667
Geotropism, 367-368
Germ layer, as a criterion for classi-
fication, 696
formation of, 287-288
Germination, seed, 359
Gestation, see Pregnancy
Giant chromosome, 210
Gibberellin, 368
Gigantism, 453
Gill, 421-423
Glucagon, 446-448
Glucocorticoid hormone, 450, 452-
453
Glycogen, importance in man, 391
structure of, 61
Glycolysis, see Anaerobic respiration
Glycosidic bond, 60
Goiter, 445
Golgi apparatus: A system of flatten-
ed sacs and rounded vesicles in
the cytoplasm. 127
Gonadotropic hormone, 235-236
chorionic, 287
in female, 270
in male, 275
Gonium, 145
Graafian follicle (ovarian follicle), 269
Granum, 142
Growth: Increase in size and com-
plexity.
in cells, 167-168
differential, 303
hormone, 453-454
in living things, 3
of populations, 566-571
Gymnosperm, 346-349
fossil record of, 676-678

728

Habitat: The region normally occupied by a population or organism. 565, 604

Hair, 400

Haploid number (1n): The number of chromosome pairs in a cell; half the normal number of chromosomes in a somatic cell (*compare* Diploid number).
and heredity, 191
for some organisms, 197-199

Hardy-Weinberg law, 658

Hatch-Slack pathway, 367

Haversian canal, 161

Hearing, 491-492

Heart, formation of, 337-340
muscle, 163
structure and function of, 410-416

Heme: The iron containing prosthetic group of a number of conjugated proteins. 75

Hemoglobin: The red-colored conjugated protein that binds oxygen in the blood and consists of a heme and a specific globular protein.
and blood tissue, 164
and hemes, 75
mutant, 662
and oxygen-carrying capacity, 429

Heredity: The genetic material received from a parent that contains the information for developmental potential. 7-9

Heterochromatin: The chromatin that remains condensed during interphase.
Barr body, 219-226
characteristics of, 219

Heterotroph, 47, 95, 391, 591

Hierarchies, social, 557

High-energy phosphate bond: A covalent bond between an oxygen atom and a phosphorous atom; when broken, liberates a large amount of energy. 39

Hill reaction, 363

Histamine, 456

Histology: The study and use of techniques for cell and tissue preparation for examination in the microscope. 114

Homeostasis: The tendency for the internal environment to be maintained at a constant level when a change in the external environment occurs.
and living systems, 9, 441
and negative feedback, 9
among populations, 571-572

and the visceral nervous system, 482

Homologous chromosomes: Morphologically identical chromosomes that carry genes coding for the same characteristic.
and heredity, 191
and meiosis, 197-199

Homologous structures: Structures of different organisms that correspond to one another in basic developmental pattern and form. 694-695

Homosexual, 281

Homozygote: An individual or cell that has both alleles of a locus occupied by the same form of the gene.
distinguishing from heterozygote, 203-204
and segregation, 193

Hooke, Robert, 115

Hormones, action of, 459
caterpillar, 457
and courting behavior, 555
in *Drosophila*, 457
in man, 441-457
in plants, 367-371
in *Rhodnius*, 457
in roosters, 457
in tadpoles, 457

Human, evolution of, 647-649, 680-686
genetics of, 216-234
population growth of, 574-585

Huntington's chorea, 217

Hyaluronidase, 282

Hybrid: The progeny of a cross between two genetically unlike parents. 192, 193

Hybridization, molecular, 101, 255-256, 356

Hydrocarbon: An organic compound containing only carbon and hydrogen. 54

Hydrochloric acid, in the stomach, 382

Hydrogen bond: A weak chemical bond between a covalently bonded hydrogen atom and a covalently bonded electronegative atom.
in the alpha helix of proteins, 69
energy of, 36-37

Hydrogen ion: A proton, a hydrogen atom without its electon. 29

Hydrolase: An enzyme that catalyzes the splitting of molecules into smaller molecules by the introduction of water. 74

Hydrolysis: The chemical reaction of water with another molecule. 43

and the glycosidic bond, 60

Hydrophobicity: Insolubility in water.
of fatty acids, 62
and interactions in proteins, 72

Hydroxyl group, 54

Hymen, 270

Hyperthyroidism, 445

Hypocotyl, 349, 358

Hypothalamus, 477

Hypothesis: A tentative explanation put forth to account for an observed phenomenon. 11

Hypothyroidism, 445

Hysterectomy: Surgical removal of the uterus. 292

Immunity, 408

Immunoglobulin: The class of proteins that includes antibodies. 408
amino acid sequence of, 68

Immunology, 407-408

Incomplete dominance: A condition in which the heterozygous phenotype is intermediate between the homozygous parents. 214-215

Independent assortment: The independent inheritance of two traits that are on different chromosomes; results from the random assortment of chromosomes during meiosis. 194-196

Inert: Chemically nonreactive.
pertaining to carbon compounds, 53

Inguinal canal, 276

Inorganic compound: Substance that does not contain carbon-hydrogen bonds. 51-52
as component of protoplasm, 393-395

Instinct: A complex behavioral pattern that is genetically determined.
in birds, 535-536, 538-540
in fish, 536-538

Insulin: A hormone secreted by the pancreas and required for the proper utilization of sugar by the body. 446-448
amino acid sequence of, 68

Intelligence: The capacity to generalize.
and behavior, 552-553
inheritance of, 229-230

Intercalated disc, 163; 411

Interferon, 330

Interphase: The part of the cell cycle between the end of one division and the beginning of the M phase. 169

729

732

Rhodopsin, 497
Ribonuclease: An enzyme that specifically hydrolyzes RNA.
in nucleic acid breakdown, 102
Ribonucleic acid (RNA), amplification of, 316
biosynthesis of, 99-102
breakdown of, 102
cellular localization of synthesis of, 108
functions of, 100-101
hybridization of, 101
and memory, 552
polymerase, 100, 253
in protein synthesis, 252-261
structure of, 80
translation of, 99
types of, 80-81, 97, 101, 254
Ribose, 60
Ribosomal RNA, amount of, 80
amplification of, 316
biosynthesis of during embryogenesis, 328-329
function of, 101
Ribosome: An organelle made of ribonucleoprotein on which proteins are synthesized.
in protein synthesis, 97-98, 104-107
structure of, 127
RNA polymerase: An enzyme that catalyzes the formation of RNA from the ribonucleoside triphosphates, using RNA or DNA as a template. 100
Rods, 494
Root, cap, 355
hair, 355, 359
pressure, 361
structure, 354-355, 359

Salivary gland, 379
Salt, 39-40
Saponification: The formation of a soap by heating a fat in the presence of a strong alkali to form glycerol and its component fatty acids. 63
Sarcolemma: A sheath of connective tissue encasing a muscle fiber. 504-505
Sarcoplasm: The cytoplasm of a muscle fiber. 504-505
Sarcosome: Mitochondrion found in a muscle fiber. 504-505
Saturated bond, in fatty acids, 62
structure of, 53
Schleiden, Matthias, 115
Schwann, Theodor, 115
Schwann cell, 464
Science, methods of, 11-18, 522-

523
Scrotum, 276
Sea anemone, 524-525
Second law of thermodynamics: The principle that any process (such as a biochemical reaction) tends to lose useful energy in the form of entropy. 22-23
Secondary sexual characteristics, human female, 270-271
human male, 275-276
Secretion: The discharge from particular cells or tissues of substances that often are used in other parts of the organism. 431
Seed, 346-347, 358-359
Segmentation: A type of body form consisting of successive segments, each with a similar structure.
and taxonomy, 696
Segregation: The separation of two genes at the same locus or of two homologous chromosomes during meiosis.
and heredity, 190
law of, 192-194
Semen, 278-281
Seminiferous tubule, 275
Sense organ: A specialized tissue in contact with a nerve cell; sensitive to a specific stimulus or change in the environment. 485-499
Serotonin, 474
Sex chromosomes, 219-220
Sex hormones, 451, 455-456
Sex-linked inheritance: A condition in which a trait is controlled by a gene carried on a sex chromosome. 226-229
Sexual behavior, 279-281
Short-lived plant, 369-370
Sickle-cell anemia, 215
Sieve tube cell, 352, 360
Sigma factor: A protein required for the initiation of RNA biosynthesis.
in transcription, 100
Sino-atrial node, 163, 412
Sino-ventricular conduction system, 163
Skeletal muscle, 162-163, 502-503
Skeleton: The hard, supportive connective tissues around or within which an organism is built. 511-515
Skin, 397-400
Skinner, B. F., 522, 548-549
Slime mold, as a model for differentiation, 308-312

Smell receptor, 485-487
Smooth muscle, 163, 502
Sol: A liquid solution of colloid-sized particles in a solvent. 40, 50-51
Solar system, origin of, 637-638
Solute: A substance that is dissolved in a liquid.
size of in true solution, 51
in water, 50
Solution: A mixture of a solute in a solvent.
colloidal solutions, 51
true solutions, 51
of water, 50
Solvent: A substance in which a solute is dissolved.
water as, 50
Somatic cells: The cells of an organism that are not involved in sexual reproduction.
and heredity, 191
Sound reception, and behavior, 530-532
Species: A population of similar individuals closely related in ancestry in which effective exchange of genetic material can occur.
formation of, 665
interspecific interactions, 570-571
intraspecific interactions, 571-574
as a unit of classification, 690-691
Sperm: Male gamete.
and heredity, 191
number of, 281
and spermatogenesis, 199-200
structure of, 278
Spermatogenesis, 199-200, 276
Spinal cord, 475
Spleen, 420
Sporangium, 345-347
Spore, plant, 346-347
Stamen, 349, 356
Starch, as a product of photosynthesis, 363
structure of, 61
Stationary phase: A period of no net change in the number of cells in a population.
and cell division, 168
Stem, plant, 350-351
Steroid, 54, 63
Stimulus: Any qualitative or quantitative change in the environment that can be detected by an organism. 462, 470-471
Stomach, 381-383
Stomata, 355-356
functions of, 359-361
Strain: descendant line.

734

heredity, 190
Strobilus, 348
Strontium 90, and bone damage, 515
Succession, ecological, 616-619
Sucrose, 60
Symmetry, 696-699
Sympathetic nervous system, 483
Sympathin, 474
Synapse: A specialized junction at which the terminal branches of an axon come together with the terminal branches of the dendrites of another neuron. 164, 470
Synapsis: The pairing of homologous chromosomes during meiosis. 199
Syncytium: A multinucleated cell formed by the fusion of many single cells and the breakdown of their individual cell membranes.
in fungi, 146
in muscle, 162
Syndrome: A group of symptoms or disorders that usually occur together. 222-224
Systole, 413-414, 419

Taste receptor, 485-486
Taxonomy: A system of classification of organisms, based on like characteristics, into groups such as kingdoms, phyla, classes, families, etc. 652, 689-700, 709-716
Telophase, 174 (*see* Meiosis, Mitosis)
Temperature receptor, 485-488
Tendon: A tough, fibrous cord that connects a muscle to a bone. 161, 504
Territoriality: The behavior of an animal to defend an area from intrusion by another animal of the same species. 531, 552-555, 557-559, 563
Test-cross, 203-204
Testis: Male gonad. 268, 276
Testosterone, 275, 456
Tetanus: A sustained contraction of a muscle fiber. 510
Tetrad: A unit of chromosomes present during meiosis; consists of paired homologous chromosomes, each of which consists of 2 chromatids. 199
Tetraploid number (4n): Twice the normal number of chromosome pairs. 198
Thalamus, 477

Thallophyta, 345
Thermodynamics laws of, *see* First law of, thermodynamics Second law of thermodynamics
Thrombin, 406
Thyroid gland, 443-445
Thyroid-stimulating hormone (thyrotropic hormone), 445, 451-452
Thyroxin, 444-445, 457
Tight junction: A cell-to-cell contact characterized by the fusion of the unit membranes in epithelial tissue. 159
Tissue: An organized group of cells that are similar in structure and function.
animal, 156-165
characteristics of, 155
plant, 156
Tongue, 379
Tooth, 379
Touch receptor, 485-488
Trachea, insect, 422
man, 425
Tracheid, 351-352
Tracheophyta, 346
Tranquilizer, 474
Transaminase: An enzyme that catalyzes the transfer of an amino group. 74-75
Transcription: The process by which the base sequence of a unit of DNA is used as the template to form a complementary base sequence of a unit of RNA. 99
Transduction, 242-243
Transfer RNA (tRNA), in protein synthesis, 103-107
structure of, 80
Transferase: An enzyme that catalyzes the transfer of a chemical group from one substrate to another. 74-75
Transformation, 244-246
Translocation, by the phloem, 352, 360
Transpiration, process of, 360-361
and stomata, 359
Tricuspid valve, 411
Triglyceride fat, roles of, 95-96
structure of, 62
Trophoblast, 285-286
Tropism: The oriented growth or movement of organisms toward or away from a specific environmental agent. 367-368
Trypsin, 387
Turner's syndrome, 222
Twins, 202

Ulcer, 382
Ultracentrifuge: A laboratory instrument used to separate components of a colloidal solution or suspension. 51
Umbilical cord, 287, 291
Unit membrane, 121-124
Universe, origin of, 635-636
Unsaturated bond, in fatty acids, 63
structure of, 53
Urea, 393, 431, 436-438
Urinary system, 432-438
Urine, 393, 431-432, 437
Uterus, 270, 295-296

Vacuole, 126-127
Vagina, 270
Vas deferens, 278
Vascular tissue, in plants, 346, 350-352
Vasopressin, 454-455
Vein, 418-419
Vena cava, 411
Vertebrata, origin of, 672-674
Vessel, in plants, 352
Vestibule: The common opening chamber of the vagina and the urethra. 270
Villi, 383
Virchow, Rudolph, 115
Virus, 147
Visceral nervous system, 482-485
Visual perception, in bees, 527
patterns of, 528-530
in pigeons, 527-528
Vitamin, as a coenzyme, 75, 393
role in metabolism, 393
types of, 393
Volvox, 146, 151
von Frisch, Karl, 527-530

Water, amount of in living things, 50
and the biosphere, 397
in photosynthesis, 94, 363-365
pollution of, 622-625
as a solvent, 50-51
Wax, in the cell wall, 141
chemistry of, 63
White blood cell, 406, 408
Wild-type population: A population as it normally occurs in nature. 190

X-ray analysis, in determining protein structure, 72
Xylem, 351-355, 359-361

Yolk: Stored food in the egg cytoplasm.
and oögenesis, 201

sac, 287

Zero growth condition: A population in stationary phase, during which death rate equals reproductive rate. 369, 384-385

Zona pellucida, 282
Zona radiata, 282
Zoology: The study of animals. 5-6
Zwitterion: A molecule that possesses both a negative and a positive charge.
  amino acids, 58
Zygote: Fertilized egg.
  in *Acetabularia*, 154
  and heredity, 191